Data Often Used

Earth		
	Mean radius	6.37×10^6 m
	Mass	5.98×10^{24} kg
	Mean distance to sun	1.49×10^{11} m
Moon		
	Mean radius	1.74×10^6 m
	Mass	7.36×10^{22} kg
	Mean distance to earth	3.84×10^8 m
Sun		
	Mean radius	6.96×10^8 m
	Mass	1.99×10^{30} kg

"Standard" acceleration due to gravity	9.80665 m/s^2
Standard atmospheric pressure	1.013×10^5 Pa
Density of air (0 °C and 1 atm)	1.293 kg/m^3
Density of water (0 °C − 20 °C)	1000 kg/m^3
Specific heat of water	4186 J/kg.K
Speed of sound (0 °C)	331.5 m/s
(20 °C)	343.4 m/s

PREFIXES FOR POWERS OF TEN

Power	Prefix	Abbreviation	Power	Prefix	Abbreviation
10^{-18}	atto	a	10^1	deka	da
10^{-15}	femto	f	10^2	hecto	h
10^{-12}	pico	p	10^3	kilo	k
10^{-9}	nano	n	10^6	mega	M
10^{-6}	micro	μ	10^9	giga	G
10^{-3}	milli	m	10^{12}	tera	T
10^{-2}	centi	c	10^{15}	peta	P
10^{-1}	deci	d	10^{18}	exa	E

MATHEMATICAL SYMBOLS

$\propto$	is proportional to
$>$ ($<$)	is greater (less) than
$\geq$ ($\leq$)	is greater (less) than or equal to
$>>$ ($<<$)	is much greater (less) than
$\approx$	is approximately equal to
Δx	the change in x
$\sum\limits_{i=1}^{N} x_i$	$x_1 + x_2 + x_3 + \cdots + x_N$
$|x|$	the magnitude, or absolute value, of x
$\Delta x \rightarrow 0$	Δx approaches zero
$n!$	$n(n-1)(n-2) \ldots$

UNIVERSITY
PHYSICS

UNIVERSITY
PHYSICS

HARRIS BENSON
Vanier College

John Wiley & Sons, Inc.

New York   Chichester   Brisbane   Toronto   Singapore

COVER: Photograph of heavy ion accelerator by Achim Zschau, GSI, Darmstadt

To my family, Frances, Coleman and Emily, and to my mentors, D. L. Mills and A. A. Maradudin.

ACQUISITIONS EDITOR   Cliff Mills
DEVELOPMENTAL EDITOR   John Haber
MANAGING EDITOR   Joan Kalkut
PRODUCTION MANAGER   Joe Ford
DESIGN SUPERVISOR   Ann Renzi
MANUFACTURING MANAGER   Lorraine Fumoso
COPY EDITOR   Virginia Dunn
PHOTO RESEARCHER   John Schultz
PHOTO RESEARCH MANAGER   Stella Kupferberg
ILLUSTRATION SUPERVISOR   John Balbalis
ILLUSTRATION COORDINATOR   Sigmund Malinowski

Recognizing the importance of preserving what has been written, it is a policy of John Wiley & Sons, Inc. to have books of enduring value published in the United States printed on acid-free paper, and we exert our best efforts to that end.

Library of Congress Cataloging in Publication Data:
Benson, Harris
    University physics / Harris Benson.
        p.   cm.
    Includes index.
    ISBN 0-471-60528-X (cloth)
    1. Physics.   I. Title.
QC21.2.B459   1991
530—dc20   90-21696
                                                                    CIP
Printed in the United States of America
1 2 3 4 5 6 7·8 9 10

Preface

This text is intended for a calculus-based introductory physics course for science and engineering students. It contains sufficient material for three semesters and, with appropriate omissions, it is easily adapted to a two-semester sequence. Students should be familiar with algebra and trigonometry (a review is included in Appendix B). Ideally, they should have completed one semester of calculus; however, calculus may also be taken concurrently with this course. Derivatives are used sparingly in the early chapters in mechanics, and integrals appear for the first time in Chapter 7 (Work and Energy). The book is based on the SI system of units; the British system is rarely mentioned.

Some of the features that are intended to enhance the appeal and effectiveness of the text are outlined below.

Accuracy

My primary objective has been to present concepts and principles clearly and correctly. I hope that the physics is free of even subtle misconceptions. In some optional sections I try to deal correctly with topics, such as the work–energy theorem for systems, that are inadequately treated in other texts. Attention is given to details such as the subtle question of signs in the application of Coulomb's law, in Faraday's law, or in Kirchhoff's loop rule for ac circuits. The concepts of emf and potential difference are clearly distinguished. The acceleration due to gravity and the gravitational field strength are given different symbols (although I do not dwell on this distinction).

Writing Style

I have tried to write in a simple, clear, and concise manner. This approach applies as much to the language as to the presentation of the mathematics and the notation used. The examples emphasize the important, or conceptually difficult, steps. Although the coverage is fairly complete, this textbook is significantly shorter than many that have appeared in recent years.

Pedagogy

I focus on central issues and highlight as few equations as possible (in the second color). Special cases, such as the "range formula" in projectile motion, are often discussed in an example and do not appear in the chapter summary. Also, I prefer not to present multiple versions of the same equation. For example, the intensity variation in the double-slit interference pattern appears only in terms of the phase difference (ϕ), and not in terms of the angular position (θ) or the vertical coordinate on the screen (y).

Questions, Exercises, and Problems

There is a large selection of questions, exercises, and problems. The questions are limited to those that students should find useful in improving their understanding. I avoid questions with no clear-cut answers. The exercises are keyed to the sections, whereas the problems are not. As an aid to students and instructors, both the exercises and the problems are graded according to two levels of difficulty: I and II. The answers to the odd-numbered exercises and problems are given in the text. A solutions manual containing brief solutions for all the exercises and problems is available only to instructors, not to students.

Content and Structure

The text includes almost all the traditional topics in classical physics. The last six chapters cover selected topics in modern physics. Basic material appears in a single-column format, whereas optional sections are in a two-column format. The overall structure is conventional. The discussions of the dot and cross products fit neatly into Chapter 2 but are easily postponed until they are needed. Chapters 15 to 17 on oscillations and waves may be combined with optics for a unified treatment of waves. The dynamics and energy aspects of satellite motion are discussed in Chapters 6 and 8, respectively. These topics may be delayed until Chapter 13 for a unified treatment of gravitation, but, as it stands, the whole of Chapter 13 may be omitted.

Pedagogical Aids

Major Points

These are placed at the beginning of each chapter and serve as a brief overview of the important concepts, laws, principles, and phenomena in that chapter.

In-Chapter Exercises

These are straightforward exercises that test a student's grasp of the preceding material. They encourage the active involvement of the student. Brief solutions, not just the answers, are provided at the end of each chapter.

Problem-solving Guides

Problem solving is arguably the most important aspect of a physics course. Throughout the text, but especially in the earlier chapters, there are problem-solving guides that outline a step-by-step approach for the analysis of problems.

The *chapter summaries* contain only the most important equations and a brief reiteration of the central principles and concepts. *Marginal notes* highlight important items throughout the text.

Worked Examples

A basic student complaint about physics texts is either that they do not have enough examples or that the examples do not adequately prepare them for the end-of-chapter problems. I address this issue by including a large number of nontrivial examples. Furthermore, their range of difficulty often encompasses the level of the tougher, multiple-step *problems,* rather than just that of the *exercises*. I occasionally address the frustrations likely to be encountered, such as false starts, extraneous roots, irrelevant data, difficulties in notation, and so on.

History

A distinctive feature of this text is the inclusion of historical material, which is included both as a pedagogical aid and as a source of enrichment. Historical material is used in the following ways:

1. To show how an idea, such as the conservation of energy, or a theory, such as relativity or quantum mechanics, emerged and developed.
2. To portray physics more realistically as a human endeavor.

3. To present some items for their intrinsic interest (e.g., in occasional anecdotes).

Short historical accounts are blended into the text; they serve to illuminate the subject and to provide insight into the concepts. Longer case histories are placed in distinct *Historical Notes* (set in a two-column format with a different typeface). Some accounts deal with the development of topics such as the concept of inertia, Newton's law of gravitation, or electromagnetic induction. Others outline the elegant reasoning used, for example, by Huygens in his study of collisions, or by Einstein in his first derivation of $E = mc^2$. None of the historical material is tested in any way.

Students often have incorrect preconceived notions about the physical world. Consequently, a lucid presentation of some physical concept or theory may not, by itself, be sufficient to erase ideas that fit well into their worldviews. It is possible to directly address student misconceptions—for example, regarding inertia or heat—by outlining the historical development of the concept.

Introductory physics can easily seem to be a litany of conclusions—the end products of the labors of brilliant minds. Not only is this intimidating, but it also falsely portrays the subject as being established, rather than as a developing body of knowledge. Historical material can address this issue and also serve to reassure students that even great minds are confused before concepts get sorted out. Indeed, some of their own (incorrect) ideas were shared by great thinkers, such as Aristotle and Galileo.

The historical material has been kept fairly simple, and so proper credit has not been given to every scientist involved. Also, the discussions do not travel up all the original blind alleys. The history should be accurate, informative, and interesting, but it is not meant to be complete. This is a text with an historical flavor, not a book on the history of physics.

Special Topics

The *Special Topic* sections deal with interesting phenomena that relate directly to the material covered in the text. Some deal with the physics of everyday phenomena such as the tides, the rainbow, cat twists, atmospheric electricity, and the earth's magnetism. Others discuss up-to-date topics such as holography, superconductivity, magnetic levitation, scanning tunneling microscopy, and fusion power. Chapter 44, Elementary Particles, is written as a Special Topic; it has no examples or end-of-chapter material.

Color

Wherever it is appropriate, the clarity and effectiveness of the art has been improved by the use of multicolor diagrams. The presentation has been made more vivid and attractive by the inclusion of a large number of full-color photographs.

Reviewers

The comments and suggestions that I received from the reviewers listed below were invaluable in improving the manuscript. They showed great sensitivity to the needs of students and I am truly grateful to them for their help and advice.

Richard J. Anderson	University of Arkansas
Robert P. Bauman	University of Alabama
Richard A. Bartels	Trinity University
James W. Calvert	Emporia State University
Roger W. Clapp, Jr.	University of South Florida
Albert C. Claus	Loyola University of Chicago
James T. Cushing	University of Notre Dame
Philip C. Eastman	University of Waterloo
James C. Eckert	Harvey Mudd College
Melvin Eisner	University of Houston
Lewis A. Ford	Texas A & M University
J. D. Gavenda	University of Texas, Austin
James B. Gerhart	University of Washington
J. L. Heilbron	University of California, Berkeley
Robert H. Hankla	Georgia State University
Michael J. Hones	Villanova University
William H. Ingham	James Madison University
Alvin W. Jenkins	North Carolina State University
Edwin R. Jones, Jr.	University of South Carolina
Roger Judge	University of California, San Diego
Sanford Kern	Colorado State University
Jesusa Kinderman	University of California, Los Angeles
Paul D. Kleiber	University of Iowa
John V. Lockhead	University of Massachusetts
Stefan Machlup	Case Western Reserve University
Howard C. McAllister	University of Hawaii, Manoa
Douglas M. McKay	University of Kansas
John K. McIver	University of New Mexico
Howard M. Miles	Washington State University
George K. Miner	University of Dayton
David Olsen	Metropolitan State College
Martin G. Olsson	University of Wisconsin, Madison
William C. Parkinson	University of Michigan, Ann Arbor
John R. Sabin	University of Florida
James H. Smith	University of Illinois, Urbana
Clifford E. Swartz	S.U.N.Y., Stony Brook
Charles M. Waddell	University of Southern California
Robert P. Wolf	Harvey Mudd College
Edward J. Zimmerman	University of Nebraska.

Phil Eastman kindly agreed to check the wording of all the exercises and problems and to solve them independently. I am grateful for his prolonged efforts and useful comments. I thank Sid Freudenstein for his input to the solutions.

Consulting Editors

I was fortunate that two individuals agreed to act as consulting editors. Stephen G. Brush, a noted science historian, offered many suggestions regarding the historical material. I was able to address only some of the issues he raised. Kenneth W. Ford, an accomplished physicist and author himself, offered valuable advice on the pedagogy and the physics. I am grateful for his interest in this project and the encouragement he provided.

Acknowledgments

The development of such a project requires the contributions of many people. I am especially grateful to Heidi Udell and Logan Campbell who had enough confidence in me as an author to get this project started. Cathy Faduska strongly supported me while she was physics editor and subsequently guided the promotion of the book. I am greatly indebted to Cliff Mills for his unswerving confidence in my efforts. His kindness and sensitivity to an author's point of view made many tasks much easier than they might have been. I appreciate the input of John Haber, the developmental editor, who worked hard to improve the text. His many detailed comments were incisive and insightful.

I am thankful to Lorraine Burke and Ed Burke who did the page layout and coordinated many aspects of the production process. The black-and-white photographs were acquired by Mary Schoenthaler, while the photo research for the color photographs was expertly handled by John Schultz. Stella Kupferberg supervised the quality of the photographic reproduction. The art program was supervised by John Balbalis. I appreciate his suggestions for improving several figures. I also wish to acknowledge the help or contributions provided by Ann Renzi, Virginia Dunn, Cathy Donovan, Deborah Herbert, Joe Ford, Joan Kalkut, and Barbara Heaney.

I am grateful to my colleagues for their support. In particular, I must thank Luong Nguyen who encouraged my efforts from the beginning. He, along with David Stephen and Paul Antaki, provided me with much useful reference material. I appreciate several useful discussions with Michael Cowan and Jack Burnett.

Finally, I owe much to my wife, Frances, and to my children, Coleman and Emily. Without their patience, love, and tolerance over many years, this book could not have been completed. In future, my time with them will not be so measured.

I hope that students find this text makes their study of physics interesting and enjoyable. Any comments or corrections from students or faculty would be most welcome.

HARRIS BENSON
Vanier College,
821 Ste. Croix Blvd.,
Montreal, H4L 3X9

Contents

UNIVERSITY
PHYSICS

CHAPTER 1

Introduction

The study of bodies such as the Horsehead Nebula requires an understanding of the principles of physics.

Major Points

1. The meaning of the terms **model, principle,** and **theory.**
2. The **SI** system of **base units.** The conversion of units.
3. The use of **significant figures** to indicate the precision of data.
4. The use of **dimensional analysis** to check equations and to obtain relationships between physical quantities.
5. Reference frames. The **Cartesian** and **polar** coordinate systems.

1.1 WHAT IS PHYSICS?

Children have an insatiable curiosity about everything around them. Sights, sounds, and smells are a constant source of wonder and amazement. They are eager to learn about nature by looking at plants, birds, and insects, and by trying all sorts of experiments with straws, bottles, pebbles, water, paint, balls, and of course food and mud. They also love to take apart a watch or a mechanical toy to see what is inside and how it works. A scientist is a person who retains some of this childlike sense of curiosity and wonder about nature.

A scientist tries to make sense of how nature operates and to discover some underlying order in the vast array of natural phenomena. This can be done at various levels, each of which reveals a different layer of reality. Social science deals with the behavior of groups, psychology with individuals, biology with the structure and function of organisms, chemistry with the combinations of atoms.

Physics deals with the behavior and composition of **matter** and its **interactions** at the most fundamental level. It is concerned with the nature of *physical* reality, that is, only with things that can be measured by instruments. Its domain stretches from inside the tiny nucleus of an atom to the vast expanses of the universe. Geology, chemistry, engineering, and astronomy all require an understanding of the principles of physics. Physics also finds many applications in biology, physiology, and medicine.

Between 1600 and 1900, three broad areas were developed in what is called **classical physics:**

Classical physics

1. **Classical Mechanics:** The study of the motion of particles and fluids.
2. **Thermodynamics:** The study of temperature, heat transfer, and the properties of aggregations of many particles.
3. **Electromagnetism:** Electricity, magnetism, electromagnetic waves, and optics.

These three areas encompass virtually all the physical phenomena with which we are familiar. However, by 1905 it became apparent that classical ideas failed to explain several phenomena. Three important theories in **modern physics** are:

Modern physics

4. **Special Relativity:** A theory of the behavior of particles moving at high speeds. It led to a radical revision of our ideas of space, time, and energy.

5. **Quantum Mechanics:** A theory of the submicroscopic world of the atom. It also required a profound upheaval in our vision of how nature operates.

6. **General Relativity:** A theory that relates the force of gravity to the geometrical properties of space.

The goal of physicists is to explain physical phenomena in the simplest and most economical terms. For example, we want to discover the "ultimate" building blocks of matter. According to our present state of knowledge, ordinary matter is constructed from atoms, the atoms from nuclei and electrons, the nuclei from neutrons and protons, the neutrons and protons from quarks. Indeed all elementary particles (of which there are hundreds) can be constructed from just *two* basic types of particle: quarks and leptons.

As another example of the drive for economy, consider the apparently wide variety of forces we encounter in nature: forces exerted by ropes, springs, fluids, electric charges, magnets, the earth and the sun, chemical forces, nuclear forces, and so on. Despite this great variety, physicists can explain all physical phenomena in terms of just four basic interactions. Their ranges and relative strengths are summarized in Table 1.1.

TABLE 1.1 THE BASIC INTERACTIONS

Interaction	Relative Strength	Range
Strong	1	10^{-15} m
Electromagnetic	10^{-2}	Infinite
Weak	10^{-6}	10^{-17} m
Gravitational	10^{-38}	Infinite

The **gravitational** interaction produces an attractive force between all particles. It is responsible for our weight, causes apples to fall, and holds the planets in their orbits around the sun. The **electromagnetic** interaction between electric charges is manifested in chemical reactions, light, radio and TV signals, X rays, friction, and all the other forces we experience every day. It also governs the transmission of signals along nerve fibers. The **strong** interaction between quarks and most other subnuclear particles holds particles within the nucleus. The **weak** interaction between quarks and leptons is associated with radioactivity. In 1983 it was confirmed that the electromagnetic and weak interactions are different manifestations of a more basic **electroweak** interaction. Progress has also been made in attempts to combine the strong and electroweak interactions in a single **grand unified theory.** Clearly, the dream of physicists is to discover a *single* fundamental interaction from which all forces can be derived.

The basic interactions

1.2   CONCEPTS, MODELS, AND THEORIES

In physics we deal with concepts, laws, principles, models, and theories. Let us briefly consider the meaning of each of these terms.

A **concept** is an idea or a physical quantity that is used to analyze natural phenomena. For example, the abstract idea of *space* is a concept and so is the measurable physical quantity, *length*. In physics we use concepts such as mass, length, time, acceleration, force, energy, temperature, and electric charge. One can define a physical quantity by the procedure used to measure it. For example, temperature can be defined as the reading on a ''standard'' thermometer, or electric charge by the force that electrified bodies exert on each other. Our intuitive understanding of such *operational definitions* is often enhanced by their relation to human perception. For example, the concept of temperature is based on the sensations of hot and cold, a force is a push or a pull, and so on. However, some concepts, such as energy, are difficult to define precisely in words. A concept such as electric charge is completely mysterious. One can measure charge and say what it *does*, but one cannot say what it *is*.

Laws and Principles

A physicist tries to establish mathematical relationships, called **laws,** between physical quantities through experimentation or theoretical analysis. Mathematics is the natural language of physics because it allows us to state the relationships concisely. Once a mathematical statement has been made, it can be manipulated according to the rules of mathematics. If the initial equations in an analysis are correct, then mathematical logic can lead to new insights and new laws.

A law is a mathematical relationship

Whereas a law may be restricted to a limited area of physics, a **principle** is a very general statement about how nature operates. It spans the whole subject and is part of its foundation. Consider a boat floating steadily down a river. The principle of relativity states that the laws of physics deduced by people on the boat must be the same as those discovered by people on land. It makes no reference to specific laws, but does force us to ensure that the laws we formulate do not violate this principle. As we will see, this seemingly innocuous statement has profound consequences. Sometimes the terms law and principle are used interchangeably. For example, we often refer to the law of conservation of energy, when really it should be the principle of conservation of energy. Such subtle differences in terminology are unimportant.

A principle encompasses many areas

Models

A **model** is a convenient analog or representation of a physical system. The phenomena occurring in the system are analyzed *as if* the system were designed according to the model. A model may merely replace the real thing to simplify the analysis. For example, in some problems one might treat the earth and the moon as if they were point particles. A model is often a mental picture of the structure or workings of a system. Here are some examples: Light has been modeled as a stream of discrete particles and also as a continuous wave; heat and electric charge have been treated like fluids; matter was considered to be composed of tiny indivisible atoms long before there was evidence for these entities; more recently the atom itself was pictured as a tiny planetary system. Even great theoretical physicists have used mechanical models as a way of relating abstract ideas to something concrete and familiar. Once the theory is complete, the model may be revealed to others, or quietly forgotten.

This orrery is a mechanical model of part of our solar system.

There are also purely mathematical models whose mathematical properties reflect those of the real system. In some cases we begin to suspect that there is more to the model than pure mathematics. That is, perhaps the mathematical entities represent actual physical quantities. Quarks, for example, were first proposed as part of a mathematical model of elementary particles. There is now so much supportive evidence that we regard them as "real." But the reality of quarks cannot be guaranteed since we cannot actually look inside a nucleus. It is a quark *model* that successfully accounts for a range of phenomena.

A model can be useful as a stepping stone even if it is incomplete or later shown to be incorrect. For example, in the model of the hydrogen atom proposed in 1913 by Niels Bohr, an electron orbits a proton, just like a planet orbits the sun. Although we now know that this is an unrealistic picture, Bohr used it to explain features of the optical spectrum of hydrogen and other atoms. It was later refined by the introduction of new concepts and used to explain the basis of the periodic table. Its shortcomings gradually became apparent and it was superseded by quantum mechanics around 1925. Although the models that considered heat and electric charge to be fluids are incorrect, they nonetheless led scientists to establish important results. Unfortunately, concrete models are not always available. The theory of quantum mechanics accounts for the strange behavior of atoms and subatomic particles, but nothing in our everyday experience comes even close to mimicking an atomic system.

A model can be useful even if it is incomplete or incorrect

Theories

A **theory** uses a combination of principles, a model, and initial assumptions (called *postulates*) to deduce specific consequences or laws. By organizing data from different areas, or by tying together concepts mathematically, a theory reveals an underlying unity in diverse phenomena. For example, Newton's theory of gravitation explained why an apple falls to the earth, the motion of the planets about the sun, why the tides occur, and even the shape of the earth. His theory showed that the same laws of physics apply to objects on earth as to the celestial bodies.

A physical theory must make precise numerical predictions, and its validity rests ultimately on the experimental verification of these predictions. A theory is considered to be plausible and accepted only if it has passed every experimental test. Yet even if no disagreement has ever been found, one cannot be sure that a theory has been proved "absolutely" correct. For over 200 years, classical mechanics was perfectly adequate for dealing with the motion of particles. Then, in 1905, the special theory of relativity showed that it is not correct when particles move at very high speeds. Newton's law of gravitation explains the motion of the planets perfectly well. However, the general theory of relativity is a more profound explanation of gravitation. Therefore, one should keep in mind that theories are always tentative. Within their limits, classical mechanics and Newton's law of gravitation are still extremely useful. In fact, they are precise enough for us to use them to send a probe to another planet.

Theories are always tentative

Contrary to common belief, theories do not follow inexorably from experimental observations. Consider the following observation: A rolling ball comes to a stop. The Greek philosopher Aristotle (ca. 340 B.C.) noted that since the ball eventually stops, it must need something to keep it going. The Italian physicist Galileo (ca. A.D. 1600) was struck by the fact that the ball keeps going for so long. He believed that if one could eliminate friction, it would go on forever. The same "fact" is interpreted in ways that are diametrically opposed. Both views are justifiable, but the second one marks the beginning of physics. Consider another example: In the geocentric model of the universe, the sun, the stars, and the planets revolve around a stationary earth. In the heliocentric model, the earth and the other planets orbit the sun. The heliocentric model, which we now accept, was not inferred directly from the astronomical data, because the earth and the other planets certainly do not *appear* to go around the sun.

Thus, although experiments serve to stimulate the creation of new theories and also to test them, the "facts" alone do not lead to theory. The formulation of a theory requires a creative mind that can see beyond the facts to make intuitive leaps and inspired guesses. Although science is a rational way of looking at the world, the creation of theories is not a rational procedure. This is the only way that one can transcend the confines of existing knowledge. It may involve an unexpected flash of insight whose origin even the scientist cannot explain. The formulation of physical theories is often guided by such esthetic notions as beauty, simplicity, and mathematical elegance. If two theories have the same range and predictive power, the simpler, or more elegant one, is usually preferred.

Strictly speaking, a theory can only *describe* natural phenomena, not explain them. But when a theory begins with a small number of assumptions and then accounts for a wide range of phenomena, it is natural to say that it has explained them. And indeed it has, but only in terms of the basic postulates and concepts. Suppose one begins with Coulomb's law for the force between two charges and derives some results that are confirmed experimentally. One could still ask why charges attract or what charge is. One can explain only how the charges interact, not why. A theory *accounts* for phenomena in terms of ultimately inexplicable quantities such as mass and charge.

The formulation of a theory requires both observation and imagination

1.3 UNITS

The value of any physical quantity must be expressed in terms of some standard or **unit**. For example, we might specify the distance between two posts in meters or in feet. Such units are necessary for us to compare measurements and also to distinguish between different physical quantities. All physical quantities can be expressed in terms of three fundamental quantities: mass, length, and time. In the **Système International (SI)** the *base* units for mass, length, and time are the **kilogram** (kg), the **meter** (m), and the **second** (s). It is convenient to define additional base units: the **kelvin** (K) for temperature, the **ampere** (A) for electric current, and the **candela** (cd) for luminous intensity. A base unit must have a precise and reproducible standard. For the moment we consider only mass, length, and time.

SI base units

Mass

The SI unit of mass, the kilogram (kg), was originally defined as the mass of one liter of water at 4 °C. Practical difficulties in obtaining pure water and the fact that this definition involved another quantity, namely temperature, led to its replacement. The SI unit of mass (1 kg) is now defined to be the mass of a platinum–iridium cylinder kept in the International Bureau of Weights and Measures in Sèvres, France (see Fig. 1.1). With this standard one can measure mass to a precision of 1 in 10^8. At the atomic level it is convenient to have a *secondary* unit of mass called the **unified atomic mass unit** (u). The mass of an atom of carbon-12 is defined to be exactly 12 u. The relation between these units is $1 \text{ u} = 1.66 \times 10^{-27}$ kg.

Time

The SI unit of time is the second (s). This was originally defined as 1/84,600 of a mean solar day. (The interval between the times at which the sun reaches the highest point in the sky on successive days is called a solar day. Because of

FIGURE 1.1 The standard kilogram is a platinum–iridium cylinder.

FIGURE 1.2 A cesium atomic clock at the National Bureau of Standards.

The value of the speed of light is defined

seasonal variations and random fluctuations, the mean value over a year is taken.) Because the rate of rotation of the earth has been gradually decreasing, the mean solar day was chosen to be the value in 1900. This is hardly a reproducible standard! In 1967, the second was redefined in terms of certain radiation emitted by atoms of cesium-133. Specifically, in one second there are 9,162,631,770 vibrations in the radiation. The cesium atomic clock, shown in Fig. 1.2, is so stable that it is accurate to within 1 s in 30,000 years. Secondary units of time include the hour, the day, the year, and the century.

Length

The SI unit of length is the meter (m). The meter was originally defined (in the eighteenth century) to be one ten-millionth (10^{-7}) of the distance from the equator to the North Pole. In this century, but prior to 1960, it was defined as the distance between two fine scratches on a platinum–iridium bar stored under controlled conditions in Sèvres, France. The use of the standard bar had two drawbacks. First, although copies of the bar are available in major industrialized countries (see Fig. 1.3), it is preferable to have a standard that can be produced in any well-equipped laboratory. Second, the width of the scratches became a limiting factor. Thus, in 1960 the standard meter was measured as precisely as possible in terms of the number of wavelengths of the orange light emitted by krypton-86. The meter was then *defined* as 1,650,763.73 wavelengths of this light. As techniques improved (through the development of lasers), the precision with which the krypton wavelength could be specified itself became a limitation. In 1983 the meter was redefined as the distance traveled by light in a vacuum in 1/299,792,458 second. This length standard, which depends on the definition of the second, effectively defines the speed of light in vacuum to be *exactly* 299,792,458 m/s. The speed of light has become a primary standard, and any improvement in measuring either the meter or the second is automatically reflected in the other.

In the British system, still used in the United States, the base units are the pound (lb) for force, the foot (ft) for length, and the second for time. Virtually all scientific data are now expressed in SI units.

Derived Units

The units of physical quantities other than mass, length, and time are combinations of the base units and are called *derived* units. For example, the unit of speed is m/s, for acceleration it is m/s^2, and for density (mass per unit volume) it is kg/m^3. Sometimes a derived unit is given a special name to honor someone. For example, Newton's second law relates the acceleration a of a body of mass m to the force F acting on it: $F = ma$. The unit of force is kg·m/s^2. This combination is called the newton (N).

Conversion of Units

It is often necessary to convert the unit of a physical quantity. Suppose we wish to convert miles per hour (mi/h) to meters per second (m/s) given that 1 mi = 1.6 km. The ratio (1.6 km)/(1 mi), which has the value one, is called a *conversion factor*. By proper use of such ratios, one can eliminate the unwanted unit and obtain a new one. For example,

$$5.0 \, \frac{\text{mi}}{\text{h}} = \left(\frac{5.0 \text{ mi}}{1 \text{ h}}\right)\left(\frac{1.6 \text{ km}}{1 \text{ mi}}\right)\left(\frac{10^3 \text{ m}}{1 \text{ km}}\right)\left(\frac{1 \text{ h}}{3600 \text{ s}}\right) = 2.2 \, \frac{\text{m}}{\text{s}}$$

When you substitute into any equation, do not mix SI units and British units.

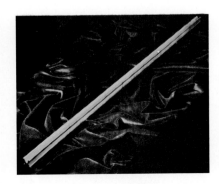

FIGURE 1.3 Prior to 1960 the meter was defined as the distance between two scratches on a platinum–iridium bar.

1.4 POWER OF TEN NOTATION AND SIGNIFICANT FIGURES

Suppose you were asked to compare the size of an atom specified as 0.000,000,000,2 m with that of a nucleus given as 0.000,000,000,000,005 m. In this form, these numbers are difficult to handle. Very large or very small numbers should be expressed in *power of ten notation*. The size of the atom is 2×10^{-10} m and that of the nucleus is 5×10^{-15} m; thus the ratio of the sizes is

Power of ten notation

$$\frac{2 \times 10^{-10} \text{ m}}{5 \times 10^{-15} \text{ m}} = \frac{2}{5} \times 10^{5} = 4 \times 10^{4}$$

It is often convenient to denote the power of ten by a prefix to the unit. For example *kilo* means thousand, so 2.36 kN = 2.36×10^{3} N; *milli* means one thousandth, so 6.4 ms = 6.4×10^{-3} s. Other prefixes are listed inside the front cover.

Numerical values obtained from measurement always have some uncertainty. For example, the result of a measurement may be 15.6 m with an uncertainty of 2%. Since 2% of 15.6 is approximately 0.3, the result is 15.6 ± 0.3 m. The true value is likely to lie between 15.3 m and 15.9 m. Instead of an explicit statement of uncertainty, the precision of a result is often indicated by the number of digits retained. We say that 15.6 m has three **significant figures,** with the understanding that the last figure (6) may not be certain. The result 15.624 has five significant figures, with the 4 uncertain. Zeros that serve only to indicate the power of ten are not counted, but those at the end are. For example 0.002560 has four significant figures. The number of significant figures in 12,000 is not clear, whereas 12,000.0 definitely has six significant figures. Power of ten notation is useful in such cases. Thus, 1.2×10^{4} has two significant figures, whereas 1.200×10^{4} has four significant figures.

Significant figures indicate the precision of data

To ensure that the results of computations are not stated to unwarranted precision, the following simple rule of thumb should be used: In products and divisions, the number of significant figures in the final result should equal that of the factor with the least number of significant figures. Thus, for example,

$$\frac{36.479 \times 2.6}{14.85} = (6.387) = 6.4$$

Although extra figures may be retained in the intermediate steps, we state the final answer only to the two significant figures in 2.6. In additions and subtractions, only the least number of decimal places should be retained. Thus, 17.524 + 2.4 − 3.56 = (16.364) = 16.4. Unless otherwise indicated in the text, you may assume that all given values are precise enough for the final answer to be stated to three significant figures. Thus 5 m may be taken to mean 5.00 m.

EXERCISE 1. Express the following in power of ten notation: (a) 1500×400; (b) 24,000/(0.006).

EXERCISE 2. Evaluate the following: (a) the volume V of a circular cylinder of radius $r = 1.26$ cm and height $\ell = 7.3$ cm, where $V = \pi r^{2}\ell$; (b) the sum $0.056 \times 10^{2} + 11.8 \times 10^{-1}$.

1.5 ORDER OF MAGNITUDE

We often hear people say there are "trillions of stars in the universe" or perhaps a "mountain weighs a billion tons." The words "trillion" and "billion" are really substitutes for "many, many, . . . ," without any intuitive feel for the reasonableness of the statements. Although such numbers are beyond our imaginations, it is often possible to arrive at a rough estimate of the size of some quantity.

To do this, a scientist thinks in terms of **orders of magnitude.** This means that he or she wants to "guesstimate" the size of something only to within a factor of 10. Obtaining an order of magnitude estimate for some complex phenomenon often involves insight and experience concerning what is important and what is irrelevant. It is ironic that in this "exact" science of physics, a physicist is often respected most for the ability to give quick order of magnitude estimates, that is, to be quite inexact. This ability allows him or her to cut through all the verbiage of some presentation, and to decide with a "back of the envelope" calculation whether a theory is reasonable.

To obtain an order of magnitude estimate, the input data need have just one significant figure. For example,

An order of magnitude estimate

$$\frac{193.7 \times 39.64}{8.71} \approx \frac{(2 \times 10^2)(4 \times 10^1)}{9} \approx 1 \times 10^3$$

For certain purposes this would be close enough to the correct answer, which is about 881.

A scientist or engineer may wish to make a measurement of some physical quantity or to build an instrument. By making an order of magnitude calculation based on the sensitivity of the instruments, the properties of the materials available, the size of the phenomenon itself, and so on, one can judge the feasibility of the project. Here is an example.

EXAMPLE 1.1: An engineer is designing a pacemaker for cardiac patients. For a 20-year-old woman, how many times should the device have to beat for her to have a normal life expectancy?

Solution: We require several estimates.

(a) If she lives to 75 years, the device must last at least 60 years.

(b) How many times per second must the device beat? Our normal pulse is about 76 beats per minute; so let's say 1 beat per second.

(c) How many seconds in a year?

$$(365 \text{ d/y})(24 \text{ h/d})(3600 \text{ s/h}) \approx (400 \text{ d/y})(20 \text{ h/d})(4000 \text{ s/h})$$
$$\approx 3 \times 10^7 \text{ s/y}$$

(Do the exact calculation and compare.) The total number of beats is

$$(1 \text{ beat/s})(60 \text{ y})(3 \times 10^7 \text{ s/y}) \approx 2 \times 10^9 \text{ beats}$$

It would be wise to include a safety factor of, say, 2. Therefore, the cardiac pacemaker should operate for 4×10^9 beats before breaking down.

You should cultivate the habit of knowing the order of magnitude value of frequently encountered physical quantities, such as the size of an atom or a nucleus, the mass and charge of an electron, the speed of light, the mass and radius of the earth, the distance to the sun, and so on. This will help you develop your insight, and will also prevent ludicrous answers. Quite often, after a calculation that involves a small mistake, a student will state that the deflection of an electron in a TV tube is 10^{+12} m. A moment's reflection would reveal that this is greater than the earth–sun distance!

1.6   DIMENSIONAL ANALYSIS

Each derived unit in mechanics can be reduced to factors of the base units mass, length, and time. If one ignores the unit system, that is, whether it is SI or British, then the factors are called **dimensions.** When referring to the dimensions of a quantity x, we place it in square brackets: $[x]$. For example, an area A is the product of two lengths so its dimensions are $[A] = L^2$. The dimensions of speed are $[v] = LT^{-1}$, of force $[F] = MLT^{-2}$, and so on.

An equation such as $A = B + C$ has meaning only if the dimensions of all the three quantities are identical. It makes no sense to add a distance to a speed. The equation must be *dimensionally consistent*. Consider the equation $s = \frac{1}{2}at^2$, in which s is the distance moved in time t by a particle that starts from rest with acceleration a. We have $[s] = L$, whereas $[at^2] = (LT^{-2})(T^2) = L$. Both sides have the dimension L, so the equation is dimensionally consistent.

An equation must be dimensionally consistent

When an algebraic expression has been derived, a check for dimensional consistency should always be performed. This does not guarantee that the equation is correct, but at least it will eliminate any equation that is not dimensionally consistent. Dimensional analysis can be used to obtain the functional form of relations, as the following example illustrates.

EXERCISE 3. If P and Q have different dimensions, which of the following operations are possible: (a) $P + Q$; (b) PQ; (c) $P - \sqrt{Q}$; (d) $1 - P/Q$?

EXAMPLE 1.2: The period P of a simple pendulum is the time for one complete swing. How does P depend on the mass m of the bob, the length ℓ of the string, and the acceleration due to gravity g?

Solution: We begin by expressing the period P in terms of the other quantities as follows:

$$P = k\, m^x\, \ell^y\, g^z$$

where k is a constant and x, y, and z are to be determined. Next we insert the dimensions of each quantity:

$$T = M^x L^y L^z T^{-2z}$$
$$= M^x L^{y+z} T^{-2z}$$

and equate the powers of each dimension on either side of the equation. Thus,

$$
\begin{aligned}
\text{T:} \quad & 1 = -2z \\
\text{M:} \quad & 0 = x \\
\text{L:} \quad & 0 = y + z
\end{aligned}
$$

These equations are easily solved and yield $x = 0$, $z = -\frac{1}{2}$, and $y = +\frac{1}{2}$. Thus,

$$P = k \sqrt{\dfrac{\ell}{g}}$$

This analysis will not yield the value of k, but we have found, perhaps surprisingly, that the period does not depend on the mass. A derivation in terms of the forces acting on the bob shows that $k = 2\pi$.

The results of such dimensional analysis depend on insight into what the important parameters are. It would seem, at least at first sight, that the angle through which the pendulum swings should also be included. Since an angle is the ratio of two lengths, it is a dimensionless quantity, so its effect would not show up anyway. A careful derivation shows that the period does depend to some extent on the angle of swing, but the above expression is quite adequate for small angles.

1.7 REFERENCE FRAMES AND COORDINATE SYSTEMS

The position of a body has meaning only in relation to a **frame of reference,** which is something physical, such as a tabletop, a room, a ship, or the earth itself. The position is specified with respect to a **coordinate system** that consists of a set of axes, each of which specifies a direction in space. In the **Cartesian** coordinate system, the axes are labeled x, y, and z. They are mutually perpendicular and intersect at the origin. In two dimensions, a point P may be located by its Cartesian coordinates (x, y), as shown in Fig. 1.4. A scale is marked on each axis. Starting at O, x and y are the number of units (+ or −) one must move in the direction of each axis to reach P.

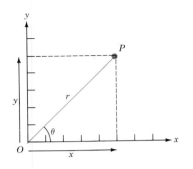

FIGURE 1.4 The *Cartesian* coordinates of point *P* are (*x*, *y*). The *polar* coordinates are (*r*, *θ*).

In **plane polar** coordinates, the length of the line *OP* and the angle *θ* with respect to the reference direction (+*x* axis) are given. These two types of coordinates are related as follows:

$$x = r \cos \theta \tag{1.1}$$
$$y = r \sin \theta \tag{1.2}$$

where

$$r = \sqrt{x^2 + y^2} \tag{1.3}$$

and

$$\tan \theta = \frac{y}{x} \tag{1.4}$$

Note that *θ* is measured counterclockwise from the +*x* axis. We will not need to use any other type of coordinate system in this book.

SUMMARY

The value of a physical quantity is expressed in terms of a **unit** of measurement. The SI **base unit** of mass is the kilogram (kg), the unit of length is the meter (m), and the unit of time is the second (s). Other physical quantities may be expressed in terms of *derived units*, which are combinations of the base units. To convert from one derived unit to another, for example, km/h to m/s, we use *conversion factors* whose value is one, for example, (3600 s/1 h) or (1 mi/1.609 km).

Numerical values obtained from experiment always have some uncertainty. The uncertainty may be stated explicitly as a percentage or a range, as in 17 ± 2 m. Or, it may be implicit in the number of **significant figures** retained. In multiplication and division, the number of significant figures in the answer is that of the factor with the least number of significant figures. In addition and subtraction, the number of decimal places in the answer is that of the value with the least number of decimal places.

In an **order of magnitude** calculation, one is concerned with estimating the value of some quantity to within a factor of ten. The data that are used to obtain the estimate need have only one significant figure.

A derived unit may be reduced to factors of the base units. The **dimensions** of a physical quantity indicate the powers to which each base unit is raised. Any equation must be dimensionally consistent.

ANSWERS TO IN-CHAPTER EXERCISES

1. (a) 6×10^5; (b) 4×10^6
2. (a) $V = (3.142)(1.26)^2(7.3) = (36.4) = 36$ cm³ (two significant figures); (b) $5.6 + 1.18 = (6.78) = 6.8$ (one decimal place).
3. (a) No; (b) Yes; (c) Yes, but only if *P* and $\sqrt{Q}$ have the same dimensions; (d) No, *P*/*Q* must be dimensionless.

QUESTIONS

1. What features are desirable in the choice of a standard of measurement?
2. What happens if someone steals the standard kilogram in Sèvres, France? Is this a safe standard? Can you suggest an alternative mass standard?
3. The United States is among only about half a dozen nations that have not switched to the metric system. Are there any advantages to the British system based on pounds, feet, and seconds?
4. (a) What are the drawbacks in using a pendulum as a time

standard? (b) List some natural phenomena that would be better choices as time standards.

5. What are the problems in using a bar as a standard of length?

6. Atomic clocks indicate that the length of a day fluctuates. How do we know that it is not the rate of the clocks that is fluctuating?

7. What is your height in meters?

8. It would be simpler to define the speed of light to be exactly 3×10^8 m/s instead of 2.99792458×10^8 m/s. Why is this not done?

9. Mass has been defined as the "quantity of matter" in a body. Could you use this to set up a base unit? If so, how?

10. What is the difference between a reference frame and a coordinate system? Give examples of each.

EXERCISES

1.3 Units

1. (I) Express the U.S. speed limit of 55 miles per hour in (a) ft/s; (b) m/s.

2. (I) A furlong is 220 yards and a fortnight is 14 days. A person walks at 5 mph. Express this in furlongs per fortnight.

3. (I) The density of water is about 1 g/cm³. What is this in SI units?

4. (I) How many seconds are there in a year, which is 365.25 days?

5. (I) (a) The distance traveled by light in a year is called a *light-year*. Given that the speed of light is 3.00×10^8 m/s, express the light-year in kilometers. (b) The average distance between the earth and the sun is called an *astronomical unit* (AU) and its value is about 1.5×10^{11} m. What is the speed of light in AU/h?

6. (I) (a) Express the mass of a proton, 1.6726×10^{-27} kg, in terms of the unified mass unit (u). (b) The mass of a neutron is 1.00867 u. What is this in kilograms?

7. (I) A knot is a nautical unit of speed: 1 knot = 1.15 mph. What is a knot in m/s?

8. (I) Given that 1 in. = 2.54 cm exactly, express the speed of light, 3.00×10^8 m/s, in (a) ft/ns; (b) mi/s.

9. (I) A small car has a 2.2-L engine. Convert this to cubic inches.

10. (II) The fuel consumption of cars is specified in Canada in terms of liters per 100 km. Convert 30 miles per gallon to this unit. Note that 1 gallon (U.S.) = 3.79 L.

1.4 Power of Ten Notation, Significant Figures

11. (I) Specify the number of significant figures in each of the following values: (a) 23.001 s; (b) 0.500×10^2 m; (c) 0.002030 kg; (d) 2700 kg/s.

12. (I) Express the following values without prefixes to the units: (a) 6.5 ns; (b) 12.8 μm; (c) 20,000 MW; (d) 0.3 mA; (e) 1.5 pA.

13. (I) Given that $\pi = 3.14159$, find: (a) the area of a circle of radius 4.20 m; (b) the surface area of a sphere of radius 0.46 m; (c) the volume of a sphere of radius 2.318 m.

14. (I) Express the following numbers in power of ten notation: (a) 1.002/4.0; (b) $(8.00 \times 10^6)^{-1/3}$; (c) 0.00076300.

15. (I) Evaluate $[(3.00 \times 10^{12})(1.20 \times 10^{-20})/(4.00 \times 10^{-1})]^{-1/2}$.

16. (I) Evaluate (a) $1.075 \times 10^2 - 6.37 \times 10 + 4.18$; (b) $402.1 + 1.073$.

17. (I) Convert the following to scientific notation: (a) the distance to the sun, 149,500,000,000 m; (b) the wavelength of yellow sodium light, 0.000,000,5893 m; (c) the radius of an atom 0.000,000,000,2 m; (d) the radius of a nucleus 0.000,000,000,000,004 m.

18. (I) Evaluate (a) $15.827 - (2.30 \times 10^{-4})/(1.70 \times 10^{-3})$; (b) $88.894/11.0 + 2.222 \times 8.00$.

19. (I) One inch is defined to be 2.54 cm exactly. Convert (a) 100.00 yd to meters; (b) one acre (4840 yd²) to hectares (10^4 m).

20. (I) Express the precision of the following results by using just the appropriate number of significant figures: (a) 6237 ± 42 m; (b) 27.34 ± 0.09 s; (c) 600 ± 0.003 kg.

21. (II) If the radius of a sphere is 10 ± 0.2 cm, what is the percentage uncertainty in (a) its radius; (b) its surface area; and (c) its volume? (d) Do you perceive a pattern? If so, what is it? (*Hint:* For (b) and (c) first find the minimum and maximum possible values.)

22. (II) The dimensions of a board are measured to be 17.6 ± 0.2 cm by 13.8 ± 0.1 cm. What is its area?

1.5 Order of Magnitude

23. (I) (a) What is the surface area of the earth? (b) What is the volume of the earth? (c) How many times larger is the volume of the sun compared to that of the earth?

24. (I) How many hairs does a normal person have on the head?

25. (I) A watch is advertised as being 99% accurate. Would you buy it?

26. (I) How fast is a person at the equator moving relative to the North Pole?

27. (I) How many frames are there in a 2-h feature film? (What information do you need?)

28. (I) Use a meter stick to estimate the thickness of a sheet of paper in this book.

29. (I) In an average lifetime: (a) How many kilometers does a

person living in a city walk? (b) How many kilograms of food are consumed by each person?

30. (I) How much more light does the 200-in.-diameter Mount Palomar telescope collect compared to the pupil of your eye?

31. (I) How many liters of water would be needed to raise the level of Lake Superior by 1 cm?

32. (I) How many grains of uncooked rice are there in one cup?

33. (I) What is the volume of your body? How could you roughly check your estimate?

1.6 Dimensional Analysis

34. (I) According to Newton's second law of motion, the force F acting on a particle is related to its mass m and acceleration a according to $F = ma$. According to Newton's law of gravitation there is an attractive force between particles given by $F = Gm_1m_2/r^2$, where r is the distance between them. What are the dimensions of G?

35. (I) Check the following equations for dimensional consistency where v is speed (m/s), a is acceleration (m/s^2), and x is position (m): (a) $x = v^2/(2a)$; (b) $x = \frac{1}{2}at$; (c) $t = (2x/a)^{1/2}$.

36. (I) The speed of a particle varies in time according to $v = At - Bt^3$. What are the dimensions of A and B?

37. (I) Convert the following polar coordinates (r, θ) to Cartesian coordinates: (a) (3.50 m, 40°); (b) (1.80 m, 230°); (c) (2.20 m, 145°); (d) (2.60 m, 320°).

38. (I) Convert the following Cartesian coordinates to plane polar coordinates: (a) (3 m, 4 m); (b) (−2 m, 3 m); (c) (2.5 m, −1.5 m); (d) (−2 m, −1 m).

39. (II) The argument of a trigonometric function must be a dimensionless quantity. If the speed v of a particle of mass m as a function of time t is given by $v = \omega A \sin[(k/m)^{1/2} t]$, find the dimensions of ω and k, given that A is a length.

PROBLEMS

1. (I) What thickness of rubber is worn off a car tire in each revolution? Given that the size of an atom is about 10^{-10} m, how many atoms does this correspond to?

2. (I) A particle moving in a circle of radius r at constant speed v undergoes an acceleration a. Use dimensional analysis to express the acceleration in terms of v and r.

3. (I) A block of mass m vibrates at the end of a spring. The spring is characterized by a quantity k called the spring constant that is measured in N/m. Express the period P of the vibration in terms of m and k. (1 N = 1 kg·m/s^2)

4. (I) Express the position x reached by a particle in terms of its acceleration a and the elapsed time t in the form $x = k\, a^m t^n$. Find m and n through dimensional analysis.

HISTORICAL NOTE: The Geocentric Theory Versus The Heliocentric Theory

The origins of physics as we know it today can be traced to the confrontation between two views of the earth's place in the universe. In the *geocentric* view, the earth is at the center of the universe, with the sun, the planets, and the stars revolving around it. In the *heliocentric* view, the earth and the planets orbit the sun. The ingenious arguments produced by advocates on both sides of this great debate served to sharpen our understanding of nature and its mechanisms.

The Greek philosopher Plato (ca. 400 B.C.) advocated the geocentric view. He believed that the celestial bodies (the stars and planets) were "perfect." To him, this meant that their natural motion had to be uniform (steady) motion in a circle. However, planets sometimes appear to reverse their motion temporarily. Clearly, a single uniform circular motion cannot explain such *retrograde* motion. Plato also believed that our senses do not perceive the "real" world, and so truth should be attained through reasoning alone. He posed the following question: What *combinations* of uniform circular motions are needed to reproduce the observed planetary paths? Or, as he put it, "to save the appearances."

Aristotle, a student of Plato, suggested a system in which each planet was assigned a number of spheres concentric with the earth. The spheres (a total of 55) rotated about axes oriented in various directions. The combinations of different axes and rates of rotation could produce quite complicated motions. He still could not explain why the brightness of some planets varies, but he dismissed this as a minor discrepancy. (As we will see, one person's trivial detail can lead to glory for another, not so cavalier.) Aristotle disagreed with Plato in a fundamental way. He felt that it is through observation of nature, rather than pure reasoning, that one gains knowledge of the world. Accordingly, he embarked on gathering an impressive collection of information in all fields. Although his ideas on motion and astronomy have proven to be false, many of his

contributions to science (especially biology), politics, ethics, and law have stood the test of time.

To explain the variations in apparent brightness, speed, and size of some planets, Hipparchus (ca. 150 B.C.) invented a new system of uniform circular motions that were not all concentric with the earth. The path of a planet was composed of a *deferent* on which was superimposed an *epicycle,* as shown in Fig. 1.5a. Different rates of rotation could produce various paths such as those in Fig. 1.5b. To improve agreement with observation, Ptolemy (ca. A.D. 130) added other refinements. For example, he shifted the center of the deferent from the earth to another point called the *eccentric*. The Ptolemaic system was used by astronomers for many centuries.

Earlier, Aristarchus of Samos (ca. 310 B.C.) had proposed a heliocentric theory that correctly placed the earth as the third planet from the sun. The proposed orbital motion of the earth around the sun should cause the apparent positions of the stars to change during the course of the orbit. But this so-called *stellar parallax* was not seen. (It requires careful measurements with a modern telescope.) Because of the apparent absence of stellar parallax and any sensation of the earth's motion, the idea lay dormant for 1800 years before it sparked the imagination of Nicholas Copernicus.

THE COPERNICAN REVOLUTION

Copernicus left his native Poland in 1496 to bask in the light of the Italian Renaissance. He studied law and astronomy in Padua and Florence. He found the Ptolemaic system of deferents, epicycles, and so on, "not pleasing to the mind." He felt that the heliocentric system of Aristarchus was basically simpler in conception. It was clear to Copernicus that what appears to be the sun's circular motion around the earth could be

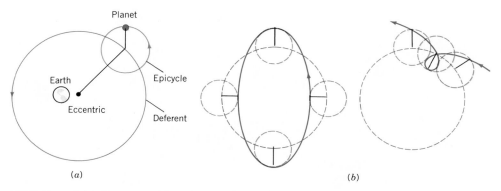

FIGURE 1.5 (a) In the geocentric system each planet was assumed to move in a circular *epicycle* superimposed on a *déferent* whose center was at the *eccentric*. (b) Various paths could be produced by suitable choices for the rates of revolution.

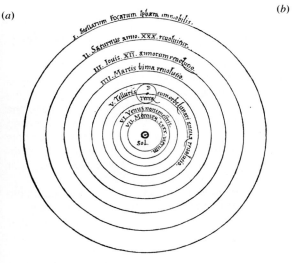

FIGURE 1.6 (a) Nicholas Copernicus (1473–1543). (b) In the heliocentric system advocated by Copernicus, the planets orbit the sun in circular paths. However, to improve accuracy Copernicus had to employ the devices of the epicycle and eccentric.

explained instead by a daily rotation of the earth. Still adhering to Plato's rule of uniform circular motion, he developed a heliocentric theory that was published in 1543, just before he died (see Fig. 1.6).

Copernicus easily explained retrograde motion and the apparent variation in the brightness and sizes of the planets. However, to improve agreement with observations, he was forced to use epicycles and eccentrics. It was also realized that if the Copernican model were correct, Venus should display phases: Its appearance should vary from a crescent to a full circle, as we see in our moon. These phases, shown in Fig. 1.7, could not be observed without telescopes. In its final form the Copernican system was neither simpler, nor more precise, than the Ptolemaic system. The neat explanation of retrograde motion was balanced by the apparent absence of stellar paral-

lax and the phases of Venus. Nonetheless, Copernicus had demonstrated that the heliocentric system could also be used to predict the positions of planets.

It is not surprising that people did not rush to embrace the theory. The claim that the earth moves did violence to common sense. People did not feel a strong wind as the earth rotates under the atmosphere. Copernicus responded that the earth drags along its atmosphere. He correctly explained the absence of stellar parallax by stating that the stars were simply too far away for observers to detect it, but this was not a convincing argument at the time. Another important objection to the idea that the earth moves was based on the fact that an arrow shot vertically up falls back to the firing point. At the time, people expected that it should be left behind. This objection Copernicus could not tackle. Different aspects of Coperni-

FIGURE 1.7 The phases of Venus. There was no explanation for this phenomenon in a geocentric system.

FIGURE 1.8 Johannes Kepler (1571–1630).

FIGURE 1.9 Galileo Galilei (1564–1642).

cus' theory were taken up by the German astronomer Johannes Kepler (Fig. 1.8) and the Italian physicist Galileo Galilei (Fig. 1.9).

The Danish astronomer Tycho Brahe spent twenty years making very precise measurements (without a telescope) of the positions of stars and the planets. (This was done partly to enable him to cast more accurate horoscopes, which was part of his duties.) After his death, his assistant Johannes Kepler acquired the data. Kepler, who believed in the heliocentric system, wanted to determine the exact orbit of Mars given data taken from a moving earth, whose own orbit was unknown. After six years of calculations he found a circle that looked quite good. However, there were tiny discrepancies of 8 minutes of arc: Mars was either inside or outside the perfect circle by this amount. A less dedicated person would have brushed this aside as so much experimental error. But Kepler had faith in Brahe's data, which were accurate to within 4 minutes of arc, which is close to the limit of resolution of the human eye. So Kepler discarded those six years of work. (His writings do not spare the reader any of the agony.) He came to suspect that the idea of uniform circular motion was false. After another two years of work, in 1609 he published two laws which stated: (i) The planets move in elliptical, not circular, orbits; and (ii) their speeds are not constant. (These laws are more fully discussed later.) Kepler had replaced the 48 intertwined circles of Copernicus by just seven beautiful and unadorned ellipses (one for each planet then known). As we will see, these laws (plus one more) were extremely important for the development of mechanics.

Galileo Galilei, a contemporary of Kepler, also advocated the Copernican system. Although he gained moral support from Kepler, he was skeptical of the latter's interest in mysticism and numerology. Hence, he was also suspicious of Kepler's elliptical orbits. In 1609 word reached him that two lenses could be combined to produce a magnified image. He soon devised a telescope and immediately turned it skyward. Among other things, he discovered moons orbiting Jupiter—something not envisaged either by Aristotle or by Copernicus. When viewed through the telescope, the stars remained as tiny spots of light. This observation lent support to Copernicus' assertion that they are very far away. The most damaging evidence against geocentricity came when Galileo actually observed the phases of Venus. If Venus orbited the earth there would be no reason for these phases to appear, or for the apparent size of this planet to change.

Since the heliocentric theory conflicted with the Christian belief that humans were at the center of the universe, Galileo's brilliant defense of it made the Vatican uneasy. In 1618 it passed an injunction that forbade Galileo to defend the Copernican system. In 1623, Barbarini, a friend of Galileo, became Pope Urban VIII and agreed to let Galileo teach the new system—but only as a hypothesis. However, in 1632 Galileo published the *Dialogues on the Two Chief World Systems,* which left little doubt as to his true opinions. The Vatican summoned him before the Holy Inquisition and forced him to recant his belief in the Copernican system as a true description of the world. Because of his age (70 years) and his eminence, he was sentenced to a relatively mild house arrest. It was then that Galileo did his most important work: He overthrew Aristotle's ideas on motion and thereby laid the foundation for mechanics.

CHAPTER 2

Vectors

Major Points

1. The distinction between **scalars** and **vectors.**
2. (a) The rectangular **components** of a vector. Unit vector notation.
   (b) The addition of vectors by using components.
3. (a) The **scalar product** of two vectors (not needed until Chapter 7).
   (b) The **vector product** of two vectors (not needed until Chapter 12).

Sailors are always concerned with the magnitude and the direction of the wind velocity.

Some of the physical quantities mentioned in Chapter 1 are specified simply by a number and a unit. For example, a mass may be stated as 4 kg, a temperature as 15 °C, and a time interval as 25 min. These quantities are called **scalars;** they have magnitude but no direction. Other physical quantities, called **vectors,** are specified by both a magnitude and a direction. For example, a wind velocity might be given as 20 km/h from the southeast. Almost all the quantities we will encounter in this book are classified as either scalar or vector. An important property of vectors is that they do not combine according to the rules of ordinary algebra. In this chapter we use the vector quantity **displacement,** which is defined as a change in position, to present the rules of vector addition and vector products.

Vector notation is a powerful tool. First, it enables us to express many physical laws quite neatly and concisely. The number of equations or mathematical operations one has to keep track of can be drastically reduced. In fact, vector analysis was developed in the nineteenth century by the physicists J. W. Gibbs in the United States and O. Heaviside in England, precisely because they felt that vector notation was an efficient way to represent many relationships between physical quantities. A second advantage is that when an equation is expressed in vector notation, it retains its form even if the coordinate system is changed. This property reflects the important fact that the laws of physics do not depend on the choice of coordinate system.

2.1 SCALARS AND VECTORS

Figure 2.1 depicts the journey of a car from position P_1 to position P_2 along some route. The length of the path taken is the **distance** traveled, for example, 420 km. Distance is an example of a scalar.

A *scalar* is a quantity that is completely specified by a number and its unit. It has a magnitude but no direction. Scalars obey the rules of ordinary algebra.

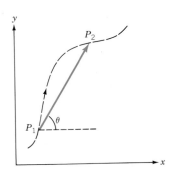

FIGURE 2.1 When a particle moves along the dashed path, the displacement from P_1 to P_2 is represented by the arrow.

If the car is driven another 105 km, the total distance traveled is simply 420 km + 105 km = 525 km. We will come across other scalars such as speed, mass, temperature, energy, and electric charge. Some scalars, such as charge and energy, can be negative, but their signs have nothing to do with direction in space: The value of a scalar does not depend on the orientation of the coordinate axes.

The *change in position* of the car is called a **displacement.** The displacement depends only on the coordinates of the initial and final positions, not on the path taken. It has a magnitude given by the length of the straight line joining P_1 to P_2, say, 360 km, and a direction given by the angle θ, for example, 30° north of east. In general, the distance traveled between two points is not equal to the magnitude of the displacement between the same two points. Displacement is an example of a vector.

Displacement

> A *vector* is a quantity that is specified by both a magnitude and a direction in space. Vectors obey the laws of vector algebra.

A vector quantity is usually printed in boldface, **A**, and is typed or written with an arrow on top, $\vec{A}$. The magnitude of a vector is a positive scalar and is written as either $|\mathbf{A}|$ or A. We will come across many vector quantities, such as force, torque, momentum, and electric field strength.*

A vector may be represented geometrically as a *directed line segment* whose length is proportional to the magnitude of the vector. It is drawn as an arrow at the appropriate angle to a reference line, with the arrowhead placed either at one end or in the middle, as shown in Fig. 2.2. To specify a vector in two dimensions requires two numbers, such as its magnitude and one angle. A three-dimensional vector requires three numbers, such as its magnitude and two angles. (A different choice is discussed in Section 2.3.) In drawing a vector, its tail may be placed anywhere relative to the coordinate axes. However, in a physical situation, the location of a vector quantity, such as the point of application of a force, may be important.

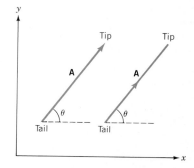

FIGURE 2.2 The geometrical representation of a vector. The tail of a vector may be placed at any point relative to the coordinate system.

A vector changes when either its magnitude or its direction changes. Of course, both may change. The vector equality **A** = **B** implies that both the magnitudes and the directions of the vectors are the same: That is, $A = B$ and $\theta_A = \theta_B$, as shown in Fig. 2.3. Note also that the tails of the vectors are at different locations.

It was pointed out in Chapter 1 that when we add, subtract, or equate physical quantities, they must have the same unit. An equally important condition is that they must have the same character: All the terms on both sides of the equation must be either scalar or vector. For example, the equation $A = \mathbf{B}$ and the sum $A +$ **B** are meaningless.

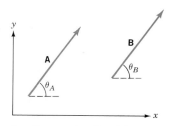

A = **B** means $A = B$ and $\theta_A = \theta_B$

FIGURE 2.3 The vector equality **A** = **B** means that $A = B$ and $\theta_A = \theta_B$.

A vector may be multiplied by a pure number or by a scalar. Multiplication by a pure number merely changes the magnitude of the vector, as shown in Fig. 2.4. If the number is negative, the direction is reversed. The negative of a vector **A**, written as the vector $-\mathbf{A}$, has the same magnitude as **A**, but is in the opposite direction. When a vector is multiplied by a scalar, the new vector also becomes a different physical quantity. For example, when velocity, a vector, is multiplied by time, a scalar, we obtain a displacement. When velocity is multiplied by mass, a scalar, we obtain another vector called linear momentum. Products of two vectors are discussed later.

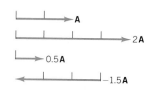

FIGURE 2.4 When a vector is multiplied by a scalar its magnitude and/or its direction changes.

* Scalars and vectors are actually mathematical entities that obey certain rules (see Problem 2.4). A physical quantity may *behave* like a scalar, or like a vector. For simplicity and convenience it is referred to simply as a scalar, or a vector.

EXERCISE 1. (a) Given that **A**, **B**, and **C** are vectors, is the equation $A\mathbf{B} = \mathbf{C}$ mathematically acceptable? (b) Under what condition would it be physically acceptable?

2.2 VECTOR ADDITION

To illustrate the special nature of vector addition, we consider the sum of two displacements. Starting from point O in Fig. 2.5, a woman walks 4 m east to point Q, and then 3 m north to point P. We have labeled the first displacement **A** and the second one **B**. The net effect of these two displacements is equivalent to a single displacement **C** from O to P, which is called the **vector sum** or **resultant** of the vectors **A** and **B**. As an equation this is expressed as

$$\mathbf{C} = \mathbf{A} + \mathbf{B}$$

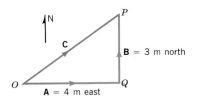

FIGURE 2.5 The net effect of two displacements **A** and **B** is equivalent to a single displacement **C** called the *sum* or *resultant* of **A** and **B**.

This is not an ordinary algebraic equation, as can easily be demonstrated with Fig. 2.5. We know that $A = 4$ m and $B = 3$ m. Since **C** forms the hypotenuse of the right-angled triangle OPQ, we have $C = (3^2 + 4^2)^{1/2} = 5$ m. Clearly $C \neq A + B$. In general, the magnitude of the sum of two vectors is not equal to the sum of their magnitudes:

$$|\mathbf{A} + \mathbf{B}| \neq A + B$$

(What is the exception to this rule?) All vectors add in the way just demonstrated for displacements.

Suppose we wish to add the vectors **A** and **B** by a graphical method. In the *tail-to-tip method*, shown in Fig. 2.6a, we first draw **A** to scale and then place the tail of **B** at the tip of **A**. Then we connect the tail of **A** to the tip of **B** to obtain the sum **A** + **B**. By comparing Figs. 2.6a and b we see that vectors may be added in any order; that is, vector addition is *commutative:*

$$\mathbf{A} + \mathbf{B} = \mathbf{B} + \mathbf{A}$$

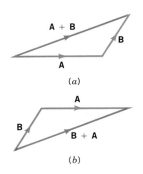

FIGURE 2.6 To add the vectors **A** and **B** by the tail-to-tip method, place the tail of the second vector at the tip of the first. The sum is drawn from the tail of the first to the tip of the second. Comparison of (a) and (b) shows that the vectors may be added in any order.

The extension of the tail-to-tip method to several vectors is straightforward. We could add the vectors a pair at a time. In Fig. 2.7a, **B** is first added to **A** and then **C** is added to their sum. This yields (**A** + **B**) + **C**. Instead, we could first add **C** to **B** and then add **A** to their sum to obtain (**B** + **C**) + **A**, as in Fig. 2.7b. The resultant does not depend on how the vectors are grouped; that is, vector addition is *associative:*

$$(\mathbf{A} + \mathbf{B}) + \mathbf{C} = \mathbf{A} + (\mathbf{B} + \mathbf{C})$$

Figure 2.7 also shows that it is not necessary to add the vectors a pair at a time. It is easier simply to repeat the tail-to-tip method: The tail of each vector is placed at

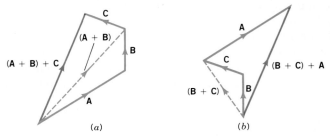

FIGURE 2.7 To add several vectors, the tail of each vector is placed at the tip of the preceding one. The resultant is drawn from the tail of the first vector to the tip of the last.

the tip of the preceding one. The resultant is drawn from the tail of the first vector to the tip of the last.

Vector Subtraction

It is sometimes necessary to find the difference of two vectors. In Fig. 2.8a, the difference $\mathbf{A} - \mathbf{B}$ is treated as a special case of addition, that is,

$$\mathbf{A} - \mathbf{B} = \mathbf{A} + (-\mathbf{B})$$

A different approach is based on the equation

$$\mathbf{A} = \mathbf{B} + (\mathbf{A} - \mathbf{B})$$

In this case we draw $\mathbf{A}$ and $\mathbf{B}$ starting from the same point as in Fig. 2.8b. The difference $\mathbf{A} - \mathbf{B}$ is drawn from the tip of $\mathbf{B}$ to the tip of $\mathbf{A}$. Notice that $\mathbf{A}$ is the resultant vector in this diagram.

EXERCISE 2. Under what conditions would we obtain the following results for $|\mathbf{A} + \mathbf{B}|$: (a) $A + B$; (b) $A - B$; (c) $B - A$; (d) $(A^2 + B^2)^{1/2}$?

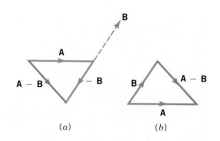

FIGURE 2.8 Two methods of finding the difference of two vectors. In (a), the subtraction is treated as a special case of addition. In (b), $\mathbf{A}$ is the resultant.

2.3   COMPONENTS AND UNIT VECTORS

The graphical addition of vectors is cumbersome and limited in precision. Furthermore, it would be impossible in three dimensions. For these reasons we now present an analytic method of vector addition that involves only the usual rules of algebra. Figure 2.9 shows a two-dimensional vector $\mathbf{A}$ that lies in the xy plane. From its ends we drop lines perpendicular to the x and y axes. The projections A_x and A_y onto the x and y axes are called the rectangular **components** of $\mathbf{A}$. Instead of specifying the vector by its magnitude and direction (A, θ), we may specify it by its components (A_x, A_y). These two representations of the vector are easily related by using Fig. 2.9. Since $\cos \theta = A_x/A$ and $\sin \theta = A_y/A$, we see that

$$A_x = A \cos \theta \tag{2.1a}$$

$$A_y = A \sin \theta \tag{2.1b}$$

One can also express the magnitude and direction in terms of the components:

$$A = \sqrt{A_x^2 + A_y^2} \tag{2.2}$$

$$\tan \theta = \frac{A_y}{A_x} \tag{2.3}$$

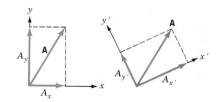

FIGURE 2.9 The rectangular components of vector $\mathbf{A}$ are A_x and A_y.

Note that whereas the components of a vector depend on the orientation of the coordinate axes, as is shown in Fig. 2.10, the magnitude of the vector does not. In any given problem, we deal either with the vector itself or with its components. To avoid taking both the vector and its components into account, one or the other, might be drawn with dashed lines as in Fig. 2.11.

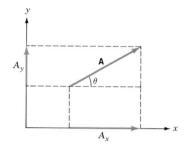

FIGURE 2.10 The components of a vector change when the coordinate axes change. However, the magnitude stays the same.

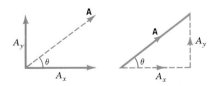

FIGURE 2.11 A vector and its components may be distinguished by the use of solid and dashed lines.

Two ways of finding direction in space: a compass and the stars.

The correct signs of the components may be found in two ways. First, they are given directly by Eq. 2.1 *provided* θ is always measured from the $+x$ axis toward the $+y$ axis (counterclockwise in Fig. 2.9). Second, we may note that a component is positive if it points in the direction of the positive axis and negative if it points in the opposite direction. Both ways are indicated in Fig. 2.12. Clearly, they must both lead to the same numerical value, but the second approach is usually easier.

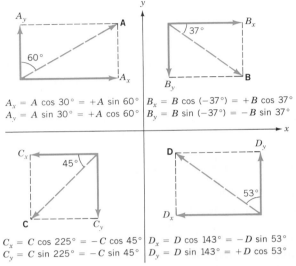

$$A_x = A \cos 30° = +A \sin 60° \quad B_x = B \cos (-37°) = +B \cos 37°$$
$$A_y = A \sin 30° = +A \cos 60° \quad B_y = B \sin (-37°) = -B \sin 37°$$

$$C_x = C \cos 225° = -C \cos 45° \quad D_x = D \cos 143° = -D \sin 53°$$
$$C_y = C \sin 225° = -C \sin 45° \quad D_y = D \sin 143° = +D \cos 53°$$

FIGURE 2.12 The components of a vector may be found (i) by using Eq. 2.1 with the angle always measured from the $+x$ axis, or (ii) by using the given angles and determining the sign of a component by its direction relative to the positive axis.

The benefit of using components is immediately apparent when we consider vector addition. From Fig. 2.13 it is easy to see how the components of the resultant $\mathbf{R} = \mathbf{A} + \mathbf{B}$ can be found from the components of $\mathbf{A}$ and $\mathbf{B}$. The vector equation

$$\mathbf{R} = \mathbf{A} + \mathbf{B}$$

implies that

$$R_x = A_x + B_x \tag{2.4a}$$
$$R_y = A_y + B_y \tag{2.4b}$$

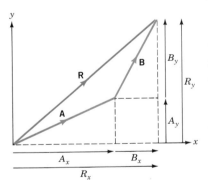

FIGURE 2.13 The components of the resultant $\mathbf{R} = \mathbf{A} + \mathbf{B}$ are $R_x = A_x + B_x$ and $R_y = A_y + B_y$.

The *geometric* addition of the vectors has been replaced by the simpler *algebraic* addition of their components. If they are needed, the magnitude and direction of $\mathbf{R}$ may be found from Eqs. 2.2 and 2.3:

$$R = \sqrt{R_x^2 + R_y^2} \tag{2.5}$$
$$\tan \theta = \frac{R_y}{R_x} = \frac{A_y + B_y}{A_x + B_x} \tag{2.6}$$

Equation 2.6 gives us the angle to the $+x$ axis. Since there are two possible angles for a given $\tan \theta$, the proper choice is made by looking at the signs of R_x and R_y, as is shown in the example on the following page.

EXAMPLE 2.1: A man walks 5 m at 37° north of east and then 10 m at 60° west of north. What is the magnitude and direction of his net displacement?

Solution: The x and y axes in Fig. 2.14 point east and north, respectively. We label the first displacement **A**, the second one **B**, and the resultant **R**. The vector diagram for the sum **R** = **A** + **B** is shown in Fig. 2.14. The components of **R** are

$$R_x = A_x + B_x = 5 \cos 37° - 10 \sin 60° = -4.66 \text{ m}$$
$$R_y = A_y + B_y = 5 \sin 37° + 10 \cos 60° = +8.00 \text{ m}$$

The magnitude is

$$R = \sqrt{R_x^2 + R_y^2} = 9.26 \text{ m}$$

and the direction is given by

$$\tan \theta = \frac{R_y}{R_x} = \frac{+8.00}{-4.66} = -1.72$$

The angle could be either 120° or −60°. Since R_x is negative and R_y is positive, the vector lies in the second quadrant and so θ = 120° to the +x axis. The net displacement is 9.26 m at 30° west of north.

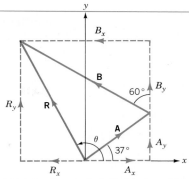

FIGURE 2.14 The direction of the resultant **R** is found from $\tan \theta = R_y/R_x$. This leads to two values for the angle θ relative to the +x axis. The choice is made by considering the signs of R_x and R_y.

EXERCISE 3. You are given that vector **A** is 5 m at 37° N of E and vector **B** is 10 m at 53° W of N. Use the components to find the magnitude and direction of (a) **A** + **B**, and (b) **B** − **A**.

Unit Vectors

To simplify the manipulation of vectors, it is convenient to introduce the **unit vectors i, j**, and **k**, that lie along the x, y, and z axes, respectively. A unit vector is a dimensionless quantity that serves only to specify a direction in space. It has unit length; that is, $|\mathbf{i}| = |\mathbf{j}| = |\mathbf{k}| = 1$. In Fig. 2.15, the unit vectors are drawn at the origin only for convenience.

In general, any vector may be expressed as the sum of three vectors parallel to each of the coordinate axes, as shown in Fig. 2.16:

$$\mathbf{A} = A_x \mathbf{i} + A_y \mathbf{j} + A_z \mathbf{k} \qquad (2.7)$$

For example, $A_x \mathbf{i}$ is a vector along the x axis whose magnitude and direction are given by the component A_x. The magnitude of **A** is given by the extension of the Pythagorean theorem to three dimensions:

$$A = \sqrt{A_x^2 + A_y^2 + A_z^2} \qquad (2.8)$$

In unit vector notation, the equality **A** = **B** is written

$$A_x \mathbf{i} + A_y \mathbf{j} + A_z \mathbf{k} = B_x \mathbf{i} + B_y \mathbf{j} + B_z \mathbf{k}$$

Since **i, j**, and **k** are mutually perpendicular, this equation is satisfied only if

$$A_x = B_x \qquad A_y = B_y \qquad A_z = B_z$$

Thus, if two vectors are equal, then each of their components must also be equal. From this result, we see that in three dimensions, the vector equation **R** = **A** + **B** is equivalent to three equations in terms of the components:

If **R** = **A** + **B**, then
$$R_x = A_x + B_x$$
$$R_y = A_y + B_y$$
$$R_z = A_z + B_z$$

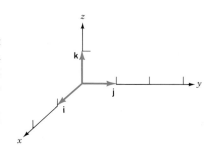

FIGURE 2.15 The unit vectors **i, j**, and **k** are directed along the x, y, and z axes, respectively.

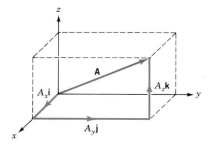

FIGURE 2.16 A vector **A** may be expressed as the sum of three vectors parallel to each of the axes: **A** = $A_x \mathbf{i}$ + $A_y \mathbf{j}$ + $A_z \mathbf{k}$.

The difference between these two vectors would be

$$\mathbf{S} = \mathbf{A} - \mathbf{B} = (A_x - B_x)\mathbf{i} + (A_y - B_y)\mathbf{j} + (A_z - B_z)\mathbf{k}$$

EXAMPLE 2.2: A girl walks 3 m east and then 4 m south. What is her net displacement?

Solution: We choose the x and y axes to point east and north, respectively. The first displacement is $\mathbf{A} = 3\mathbf{i}$ m and the second is $\mathbf{B} = -4\mathbf{j}$ m, as shown in Fig. 2.17. The resultant is

$$\mathbf{R} = \mathbf{A} + \mathbf{B} = 3\mathbf{i} - 4\mathbf{j} \text{ m}$$

The vector $\mathbf{R}$ is completely specified by its components. However, if its magnitude and direction are required, we can use

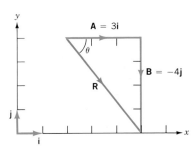

FIGURE 2.17 The resultant displacement is $\mathbf{R} = 3\mathbf{i} - 4\mathbf{j}$ m.

Eqs. 2.5 and 2.6 to find $R = (3^2 + 4^2)^{1/2} = 5$ m and $\tan \theta = -\frac{4}{3}$. Therefore $\theta = -53°$ or $143°$. Since R_x is positive and R_y is negative, we choose $\theta = -53°$.

EXAMPLE 2.3: Given the vectors $\mathbf{A} = 2\mathbf{i} - 3\mathbf{j} + 6\mathbf{k}$ m and $\mathbf{B} = \mathbf{i} + 2\mathbf{j} - 3\mathbf{k}$ m, find: (a) $A + B$; (b) $|\mathbf{A} + \mathbf{B}|$; (c) $2\mathbf{A} - 3\mathbf{B}$.

Solution: (a) Notice that what is asked for is the sum of the magnitudes, which are found from Eq. 2.8:

$$A = \sqrt{2^2 + 3^2 + 6^2} = 7.00 \text{ m}$$
$$B = \sqrt{1^2 + 2^2 + 3^2} = 3.74 \text{ m}$$

Hence, $A + B = 10.7$ m.
(b) The vector sum is

$$\mathbf{A} + \mathbf{B} = (2 + 1)\mathbf{i} + (-3 + 2)\mathbf{j} + (6 - 3)\mathbf{k} = 3\mathbf{i} - \mathbf{j} + 3\mathbf{k} \text{ m}$$

The magnitude of the sum is $|\mathbf{A} + \mathbf{B}| = (3^2 + 1^2 + 3^2)^{1/2} = 4.36$ m. Clearly, the magnitude of the sum is not equal to the sum of the magnitudes found in part (a).
(c) $2\mathbf{A} - 3\mathbf{B} = 2(2\mathbf{i} - 3\mathbf{j} + 5\mathbf{k}) - 3(\mathbf{i} + 2\mathbf{j} - 3\mathbf{k}) = \mathbf{i} - 12\mathbf{j} + 19\mathbf{k}$ m.

EXERCISE 4. Given the vector $\mathbf{R} = 3\mathbf{i} + 4\mathbf{j}$ m, find (a) R; (b) the unit vector $\hat{\mathbf{R}}$ in the direction of $\mathbf{R}$.

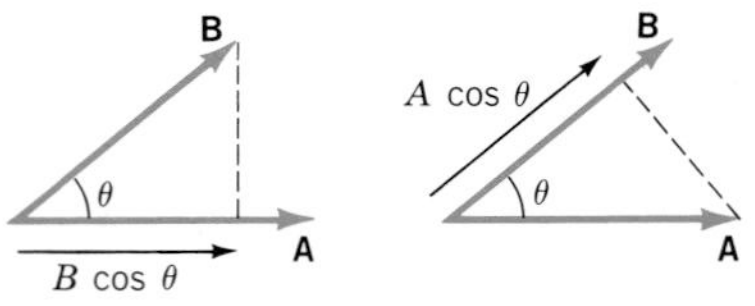

FIGURE 2.18 The scalar product $\mathbf{A} \cdot \mathbf{B} = AB \cos \theta$ of two vectors is the product of one vector and the component of the second vector along the direction of the first.

2.4   THE SCALAR (DOT) PRODUCT*

In the analysis of physical situations, one often encounters an expression that involves the multiplication of the magnitudes of two vectors and a trigonometric function. This has led to the definition of two types of products for vectors. The scalar product is discussed in this section, and the vector product in the next section.

The **scalar** or **dot** product of two vectors $\mathbf{A}$ and $\mathbf{B}$ (see Fig. 2.18) is defined as

$$\mathbf{A} \cdot \mathbf{B} = AB \cos \theta \tag{2.9}$$

where A and B are the magnitudes of the vectors and θ is the angle between them. The product is *defined* to be a scalar and therefore does not depend on the choice of coordinate axes. If Eq. 2.9 is expressed as $A(B \cos \theta)$ or $B(A \cos \theta)$, we see that a scalar product is the product of the first vector and the component of the second vector in the direction of the first. The dot product arises naturally in many cases. For example (as we will see in Chapter 7), the work W done by a constant force $\mathbf{F}$ when its point of application undergoes a displacement

---

* This section is not needed until Chapter 7.

s is $W = \mathbf{F} \cdot \mathbf{s}$. Note that the order of the vectors does not matter; that is,

$$\mathbf{B} \cdot \mathbf{A} = \mathbf{A} \cdot \mathbf{B}$$

The scalar products of the unit vectors **i**, **j**, and **k**, are

$$\mathbf{i} \cdot \mathbf{i} = \mathbf{j} \cdot \mathbf{j} = \mathbf{k} \cdot \mathbf{k} = 1$$
$$\mathbf{i} \cdot \mathbf{j} = \mathbf{i} \cdot \mathbf{k} = \mathbf{j} \cdot \mathbf{k} = 0$$

(2.10)

One can show that the scalar product is distributive, that is,

$$\mathbf{A} \cdot (\mathbf{B} + \mathbf{C}) = \mathbf{A} \cdot \mathbf{B} + \mathbf{A} \cdot \mathbf{C}$$

This allows us to express the scalar product solely in terms of the components of the two vectors. Thus, using Eq. 2.10,

$$\mathbf{A} \cdot \mathbf{B} = (A_x\mathbf{i} + A_y\mathbf{j} + A_z\mathbf{k}) \cdot (B_x\mathbf{i} + B_y\mathbf{j} + B_z\mathbf{k})$$

becomes

$$\mathbf{A} \cdot \mathbf{B} = A_xB_x + A_yB_y + A_zB_z$$

(2.11)

Notice that if $\mathbf{B} = \mathbf{A}$, then $\mathbf{A} \cdot \mathbf{A} = A^2 = A_x^2 + A_y^2 + A_z^2$, a result we already know.

EXAMPLE 2.4: Find the scalar product of $\mathbf{A} = 8\mathbf{i} + 2\mathbf{j} - 3\mathbf{k}$ and $\mathbf{B} = 3\mathbf{i} - 6\mathbf{j} + 4\mathbf{k}$.

Solution: From Eq. 2.11,

$$\mathbf{A} \cdot \mathbf{B} = A_xB_x + A_yB_y + A_zB_z = (8 \times 3 - 2 \times 6 - 3 \times 4) = 0$$

Since neither A nor B is zero, this means that $\cos \theta = 0$. That is, $\theta = 90°$.

EXAMPLE 2.5: Find the angle between $\mathbf{A} = 2\mathbf{i} + \mathbf{j} + 2\mathbf{k}$ and $\mathbf{B} = +4\mathbf{i} - 3\mathbf{j}$.

Solution: From Eq. 2.9,

$$\cos \theta = \frac{\mathbf{A} \cdot \mathbf{B}}{AB}$$
$$= \frac{2 \times 4 - 1 \times 3 + 0}{3 \times 5} = \frac{1}{3}$$

Thus, $\theta = \cos^{-1}(1/3) = 70.5°$. The geometrical approach to finding the angle between two lines in arbitrary directions is considerably more difficult.

EXAMPLE 2.6: Derive the law of cosines using the scalar product.

Solution: Consider the difference $\mathbf{C} = \mathbf{A} - \mathbf{B}$ of the two vectors shown in Fig. 2.19. We take the dot product $\mathbf{C} \cdot \mathbf{C}$:

$$\mathbf{C} \cdot \mathbf{C} = (\mathbf{A} - \mathbf{B}) \cdot (\mathbf{A} - \mathbf{B}) = A^2 + B^2 - 2\mathbf{A} \cdot \mathbf{B}$$

Thus,

$$C^2 = A^2 + B^2 - 2AB \cos \theta$$

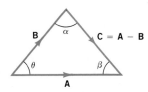

FIGURE 2.19 In drawing the vector $\mathbf{C} = \mathbf{A} - \mathbf{B}$ the method shown in Fig. 2.8b has been used.

EXERCISE 5. Given two vectors **A** and **B** of nonzero magnitude, what is the angle between them if $\mathbf{A} \cdot \mathbf{B}$ has the value $-AB/2$?

2.5 THE VECTOR (CROSS) PRODUCT*

The **vector** or **cross** product of two vectors **A** and **B** is defined as

$$\mathbf{A} \times \mathbf{B} = AB \sin \theta \, \hat{\mathbf{n}}$$

(2.12)   Vector or cross product

The product is *defined* to be a vector of magnitude $AB \sin \theta$ that points in the

---

* This section is not needed until Chapter 12.

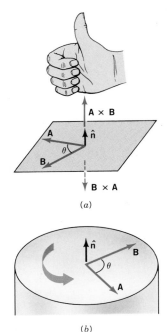

(a)

(b)

FIGURE 2.20 The direction of the vector product $\mathbf{A} \times \mathbf{B} = AB \sin \theta \, \hat{\mathbf{n}}$ may be found in two ways. (a) According to the *right-hand rule* curve your fingers and stick out your thumb as if you were hitch-hiking. Note that the fingers are directed from the first vector to the second. The thumb indicates the direction of $\hat{\mathbf{n}}$. (b) To use the *bottlecap rule* imagine that the vectors lie on a bottlecap. As the cap is turned from the first to the second vector, it either opens or closes the bottle. The motion of the cap along its axis indicates the direction of $\hat{\mathbf{n}}$.

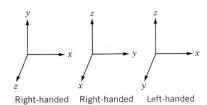

Right-handed   Right-handed   Left-handed

FIGURE 2.21 Two right-handed coordinate systems and one left-handed coordinate system.

direction of the unit vector $\hat{\mathbf{n}}$ normal (perpendicular) to the plane of **A** and **B**. The angle θ is the smaller angle between the vectors. The definition of the cross product reflects the behavior of many physical quantities, such as torque, angular momentum, and the magnetic force on a moving charged particle.

The direction of $\hat{\mathbf{n}}$ is still ambiguous. This ambiguity is removed by using a convention called the **right-hand rule.** Curl the fingers of your right hand and stick out your thumb as if you were hitch-hiking as in Fig. 2.20a. The sense of rotation of the fingers should be from the first vector **A** to the second vector **B**—through the *smaller* angle between them. The thumb indicates the direction of $\hat{\mathbf{n}}$. A related method is the *bottlecap rule.* Imagine that both vectors lie on a bottlecap, as in Fig. 2.20b. As you turn the cap, with the tips of your fingers, from the first vector to the second, you are either opening or closing the bottle. The direction of motion of the cap along its axis specifies $\hat{\mathbf{n}}$.

Because of the right-hand rule, the order of the vectors in the cross product is important. The vector product is *noncommutative:*

$$\mathbf{B} \times \mathbf{A} = -\mathbf{A} \times \mathbf{B}$$

These products have the same magnitude but they point in opposite directions. One can show that the vector product is *distributive:*

$$\mathbf{A} \times (\mathbf{B} + \mathbf{C}) = \mathbf{A} \times \mathbf{B} + \mathbf{A} \times \mathbf{C}$$

In order for the definition of the vector product, including the right-hand rule, to be applied to the unit vectors **i**, **j**, and **k**, the coordinate system must be *right-handed.* As you imagine rotating the x axis toward the y axis, the right-hand rule should give you the z axis. Figure 2.21 shows one left-handed and two right-handed coordinate systems. For a right-handed system, the vector products are

$$\mathbf{i} \times \mathbf{i} = 0 \qquad \mathbf{j} \times \mathbf{j} = 0 \qquad \mathbf{k} \times \mathbf{k} = 0$$
$$\mathbf{i} \times \mathbf{j} = \mathbf{k} \qquad \mathbf{j} \times \mathbf{i} = -\mathbf{k} \qquad \mathbf{k} \times \mathbf{i} = \mathbf{j}$$
$$\mathbf{i} \times \mathbf{k} = -\mathbf{j} \qquad \mathbf{j} \times \mathbf{k} = \mathbf{i} \qquad \mathbf{k} \times \mathbf{j} = -\mathbf{i}$$

The general expression for the vector product $\mathbf{C} = \mathbf{A} \times \mathbf{B}$ in terms of the components is somewhat complicated:

$$\mathbf{C} = (A_x\mathbf{i} + A_y\mathbf{j} + A_z\mathbf{k}) \times (B_x\mathbf{i} + B_y\mathbf{j} + B_z\mathbf{k})$$
$$= (A_xB_y\mathbf{k} - A_xB_z\mathbf{j}) + (-A_yB_x\mathbf{k} + A_yB_z\mathbf{i}) + (A_zB_x\mathbf{j} - A_zB_y\mathbf{i})$$

After collecting terms, we find

$$C_x\mathbf{i} + C_y\mathbf{j} + C_z\mathbf{k}$$
$$= (A_yB_z - A_zB_y)\mathbf{i} + (A_zB_x - A_xB_z)\mathbf{j} + (A_xB_y - A_yB_x)\mathbf{k}$$

Equating each component yields

$$C_x = A_yB_z - A_zB_y$$
$$C_y = A_zB_x - A_xB_z$$
$$C_z = A_xB_y - A_yB_x$$

If you happen to be familiar with determinants, note that the cross product is neatly expressed as

$$\mathbf{A} \times \mathbf{B} = \begin{vmatrix} \mathbf{i} & \mathbf{j} & \mathbf{k} \\ A_x & A_y & A_z \\ B_x & B_y & B_z \end{vmatrix}$$

EXAMPLE 2.7: Find the vector product of $\mathbf{A} = 3\mathbf{i} - 2\mathbf{j} + \mathbf{k}$ and $\mathbf{B} = \mathbf{i} + 4\mathbf{j} - 2\mathbf{k}$.

Solution: The vector product is

$$\mathbf{A} \times \mathbf{B} = (3\mathbf{i} - 2\mathbf{j} + \mathbf{k}) \times (\mathbf{i} + 4\mathbf{j} - 2\mathbf{k})$$
$$= (12\mathbf{k} + 6\mathbf{j}) + (2\mathbf{k} + 4\mathbf{i}) + (\mathbf{j} - 4\mathbf{i})$$
$$= 7\mathbf{j} + 14\mathbf{k}$$

EXAMPLE 2.8: Derive the law of sines using the cross product.

Solution: Consider the difference of two vectors: $\mathbf{C} = \mathbf{A} - \mathbf{B}$, as shown in Fig. 2.19. Take the cross product of each side with $\mathbf{C}$:

$$\mathbf{C} \times \mathbf{C} = \mathbf{A} \times \mathbf{C} - \mathbf{B} \times \mathbf{C}$$
$$0 = AC \sin \beta \, \hat{\mathbf{n}} - BC \sin \alpha \, \hat{\mathbf{n}}$$

Thus,

$$\frac{\sin \alpha}{A} = \frac{\sin \beta}{B}$$

Another cross product supplies the third term, $\sin \theta / C$, in the law of sines. Notice that we used the same unit vector $\hat{\mathbf{n}}$ for each product. Why is this correct?

SUMMARY

A **scalar** is specified by a number and its unit. It has magnitude but no direction. It obeys the rules of ordinary algebra. A **vector** has magnitude and direction. It obeys the laws of vector algebra.

In the *tail-to-tip* method of vector addition, the tail of each vector is placed at the tip of the preceding one. The resultant is drawn from the tail of the first vector to the tip of the last vector, as in Fig. 2.22.

In two dimensions, the **rectangular components** of the vector $\mathbf{A}$ of Fig. 2.23 are

$$A_x = A \cos \theta \qquad A_y = A \sin \theta$$

The magnitude and direction of the vector may be related to its components:

$$A = \sqrt{A_x^2 + A_y^2} \qquad \tan \theta = \frac{A_y}{A_x}$$

The angle θ found from $\tan \theta = A_y/A_x$ is measured from the $+x$ axis. In three dimensions a vector may be expressed in **unit vector** notation:

$$\mathbf{A} = A_x\mathbf{i} + A_y\mathbf{j} + A_z\mathbf{k}$$

Its magnitude is

$$A = \sqrt{A_x^2 + A_y^2 + A_z^2}$$

The components depend on the choice of coordinate axes, but the magnitude of the vector does not.

The vector equation $\mathbf{R} = \mathbf{A} + \mathbf{B}$ is equivalent to three equations:

$$R_x = A_x + B_x \qquad R_y = A_y + B_y \qquad R_z = A_z + B_z$$

The **scalar (dot) product** of two vectors is

$$\mathbf{A} \cdot \mathbf{B} = AB \cos \theta = A_xB_x + A_yB_y + A_zB_z$$

The **vector (cross) product** of two vectors is

$$\mathbf{A} \times \mathbf{B} = AB \sin \theta \, \hat{\mathbf{n}}$$

where the direction of $\hat{\mathbf{n}}$ is given by the right-hand rule.

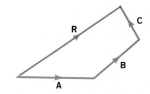

FIGURE 2.22 The resultant $\mathbf{R}$ of several vectors may be found by the tail-to-tip method.

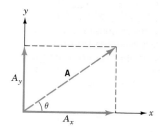

FIGURE 2.23 The x and y components of $\mathbf{A}$ are A_x and A_y.

ANSWERS TO IN-CHAPTER EXERCISES

1. (a) Yes. (b) The units must be the same on both sides.
2. (a) Parallel; (b) antiparallel with $A > B$; (c) antiparallel with $B > A$; (d) perpendicular.
3. The x axis points east and the y axis points north.

$$A_x = A \cos 37° = 4 \text{ m} \qquad A_y = A \sin 37° = 3 \text{ m}$$

$$B_x = -B \sin 53° = -8 \text{ m} \qquad B_y = B \cos 53° = 6 \text{ m}$$

(a) $\mathbf{A} + \mathbf{B} = -4\mathbf{i} + 9\mathbf{j}$ m; (b) $\mathbf{B} - \mathbf{A} = -12\mathbf{i} + 3\mathbf{j}$ m.

4. (a) $R = \sqrt{R_x^2 + R_y^2} = 5$ m; (b) $\hat{\mathbf{R}} = \mathbf{R}/R = 0.6\mathbf{i} + 0.8\mathbf{j}$.
5. We are given that $\cos \theta = -0.5$; thus $\theta = 120°$.

QUESTIONS

1. Classify the following quantities as scalar, vector, or neither: (a) Cartesian coordinates (x, y); (b) the rectangular components of a vector; (c) time; (d) temperature; (e) volume; (f) an electron; (g) light.
2. True or false: (a) The components of a vector depend on the choice of coordinate system. (b) The magnitude of a vector depends on the choice of coordinate system. (c) The magnitude of a vector cannot be less than that of a nonzero component.
3. Is it possible to have $|\mathbf{A} + \mathbf{B}| = |\mathbf{A} - \mathbf{B}|$? If so, illustrate with a diagram.
4. True or false: If $\mathbf{A} - \mathbf{B} = \mathbf{C} - \mathbf{D}$, then $\mathbf{A} = \mathbf{C}$ and $\mathbf{B} = \mathbf{D}$.
5. Which of the following are valid statements? (a) $A = 5$ m; (b) $B = -6$ km; (c) $\mathbf{A} + \mathbf{B} = \mathbf{C} + \mathbf{D}$; (d) $|\mathbf{A} + \mathbf{B}| = C$; (e) $AB = \mathbf{C}$.
6. What can you conclude if the component of $\mathbf{A}$ along $\mathbf{B}$ is (a) 0; (b) $-A$; (c) $A/2$?
7. If the magnitudes of two vectors are equal, that is, $A = B$, is it possible that (a) $|\mathbf{A} + \mathbf{B}| = A$, or (b) $|\mathbf{A} - \mathbf{B}| = A$? If so, illustrate with a diagram.

8. Draw a vector diagram to illustrate each of the following results for $|\mathbf{A} + \mathbf{B}|$: (a) 0; (b) $A + B$; (c) $A - B$; (d) $B - A$; (e) $(A^2 + B^2)^{1/2}$.
9. Draw a vector diagram to illustrate each of the following results for $|\mathbf{A} - \mathbf{B}|$: (a) 0; (b) $A - B$; (c) $A + B$; (d) $B - A$; (e) $(A^2 + B^2)^{1/2}$.
10. What is the unit vector in the direction of $(\mathbf{i} + \mathbf{j} + \mathbf{k})$?
11. Can the magnitude of the difference of two vectors be greater than the magnitude of their sum? If so, illustrate with a diagram.
12. Does a unit vector have a unit?
13. What can you conclude about vectors $\mathbf{B}$ and $\mathbf{C}$ given that (a) $\mathbf{A} \cdot \mathbf{B} = \mathbf{A} \cdot \mathbf{C}$; (b) $\mathbf{A} \times \mathbf{B} = \mathbf{A} \times \mathbf{C}$?
14. What can you conclude about the components of vectors $\mathbf{A}$ and $\mathbf{B}$ given that (a) $\mathbf{A} = \mathbf{B}$, (b) $A = B$?
15. True or false: If $\mathbf{A} \cdot \mathbf{A} = \mathbf{B} \cdot \mathbf{B}$, then $\mathbf{A} = \pm\mathbf{B}$.
16. Given two vectors $\mathbf{A}$ and $\mathbf{B}$ of nonzero magnitude, illustrate with a diagram how the following results for $\mathbf{A} \cdot \mathbf{B}$ would arise: (a) AB; (b) $-AB$; (c) 0; (d) $AB/2$; (e) $-AB/2$.

EXERCISES

2.2 Vector Addition

1. (I) The magnitudes of the vectors $\mathbf{A}$ and $\mathbf{B}$ shown in Fig. 2.24 are $A = 3$ m and $B = 2$ m. Find graphically: (a) $\mathbf{A} + \mathbf{B}$; (b) $\mathbf{A} - \mathbf{B}$.

2. (I) The magnitudes of the vectors $\mathbf{C}$ and $\mathbf{D}$ shown in Fig. 2.25 are $C = 4$ m and $D = 2.5$ m. Find graphically: (a) $\mathbf{C} + \mathbf{D}$; (b) $\mathbf{C} - \mathbf{D}$.

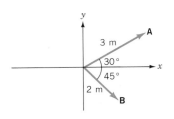

FIGURE 2.24 Exercise 1.

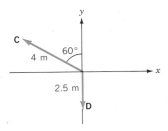

FIGURE 2.25 Exercise 2.

3. (I) For the three vectors shown in Fig. 2.26, take $A = 1.5$ m, $B = 2$ m, and $C = 1$ m. Find graphically: (a) $\mathbf{A} + \mathbf{B} + \mathbf{C}$; (b) $\mathbf{A} - \mathbf{B} - \mathbf{C}$.

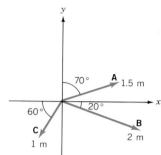

FIGURE 2.26 Exercises 3 and 4.

4. (I) Consider the three vectors shown in Fig. 2.26. Find the vector $\mathbf{D}$ which when added to $\mathbf{A} + \mathbf{B} - \mathbf{C}$, produces a null vector.

5. (I) Three vectors have equal magnitudes of 10 m. Draw a vector diagram to illustrate how the magnitude of their resultant can be: (a) 0; (b) 10 m; (c) 20 m; (d) 30 m.

6. (II) Two vectors have equal magnitudes of 2 m. Find graphically the angle between them if the magnitude of their resultant is (a) 3 m; (b) 1 m. In each case use the law of cosines to confirm your answer.

7. (I) The resultant of two vectors $\mathbf{A}$ and $\mathbf{B}$ is 40 m due north. If $\mathbf{A}$ is 30 m in the direction 30° S of W, find $\mathbf{B}$ graphically.

2.3 Components and Unit Vectors

In the following problems take the x axis to point east, the y axis to point north, and, if needed, the z axis upward. An unknown vector should be expressed in terms of its components: $\mathbf{V} = V_x\mathbf{i} + V_y\mathbf{j}$. The magnitude and direction can be found from the components.

8. (I) A person undergoes a displacement of 4 m in the direction 40° W of N followed by a displacement of 3 m at 20° S of W. Find the magnitude and direction of the resultant displacement.

9. (I) Three vectors are specified as follows: $\mathbf{A}$ is 5 m at 45° N of E, $\mathbf{B}$ is 7 m at 60° E of S, and $\mathbf{C}$ is 4 m at 30° W of S. Find the magnitude and direction of their sum.

10. (I) A person walks 5 m south and then 12 m west. What is the net displacement?

11. (I) An insect walks 50 cm in a straight line along a wall. If its horizontal displacement is 25 cm, what is its vertical displacement?

12. (II) An airplane is flown in the direction 30° W of N. If the magnitude of the westerly component of the displacement is 100 km, how far north does it travel?

13. (I) Four vectors, each of magnitude 2 m, are shown in Fig. 2.27. (a) Express each in unit vector notation. (b) Express their sum in unit vector notation. (c) What is the magnitude and direction of their sum?

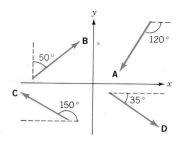

FIGURE 2.27 Exercise 13.

14. (I) Each of the vectors in Fig. 2.28 has a magnitude of 4 m. (a) Express each in unit vector notation. (b) Express their sum in unit vector notation. (c) What is the magnitude and direction of their sum?

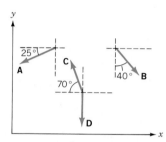

FIGURE 2.28 Exercise 14.

15. (I) Given two vectors, $\mathbf{A} = 2\mathbf{i} - 3\mathbf{j} + \mathbf{k}$ m and $\mathbf{B} = -\mathbf{i} + 2\mathbf{j} - \mathbf{k}$ m, find: (a) $\mathbf{R} = \mathbf{A} + \mathbf{B}$; (b) R; (c) $\hat{\mathbf{R}}$.

16. (I) Given two vectors, $\mathbf{C} = 4\mathbf{i} + \mathbf{j} - 3\mathbf{k}$ m and $\mathbf{D} = 2\mathbf{i} - 3\mathbf{j} + 5\mathbf{k}$ m, find: (a) $\mathbf{S} = \mathbf{C} - \mathbf{D}$; (b) S; (c) $\hat{\mathbf{S}}$.

17. (II) The vector $\mathbf{A}$ has a magnitude of 6 m and vector $\mathbf{B}$ has a magnitude of 4 m. What is the angle between them if the magnitude of their resultant is (a) the maximum possible; (b) the minimum possible; (c) 3 m; and (d) 8 m. Do each part graphically and by components. (Let $\mathbf{A}$ lie along the x axis.)

18. (I) The resultant $\mathbf{R}$ of two displacements is 10 m at 37° W of N. If the second displacement was 6 m at 53° N of E, what was the first?

19. (I) In a yacht race the boats sail around three buoys as shown in Fig. 2.29. What is the displacement from the last buoy to the starting point? Express your answer (a) in unit vector notation, and (b) as a magnitude and direction.

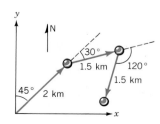

FIGURE 2.29 Exercise 19.

20. (II) The displacement **A** is 6 m east. Find displacement **B** if the magnitude of **A** − **B** is half that of **A** and points in the direction 30° N of E.

21. (II) A ship sails from a point at a distance of 4 km and a bearing of 40° N of E relative to a lighthouse to a point 6 km away at 60° N of W. (a) What is its displacement? (b) What is the least distance between them?

22. (I) A submarine sails 40 km north and then 30 km west. What third displacement would produce a net displacement of 20 km at 30° S of W?

23. (II) In Fig. 2.30, **A** and **B** are position vectors. Show that the position vector **C** of the midpoint of the line joining the tips of these vectors is **C** = (**A** + **B**)/2.

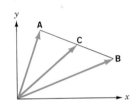

FIGURE 2.30 Exercise 23.

24. (II) Given two vectors **A** = 3**i** − 2**j** and **B** = −**i** + 5**j**, find the vector **C** in the *xy* plane such that |**C**| = |**A** + **B**|, but whose direction is perpendicular to **A** + **B**.

25. (I) Given two vectors, **A** = 2**i** + **j** − **k** and **B** = −3**i** + 2**j** + **k**, find the unit vector in the direction of **S** = 2**B** − 3**A**.

26. (I) Given two vectors, **A** = 5**i** + 2**j**, and **B** = −2**i** − 3**j**, find: (a) *A* + *B*; (b) |**A** + **B**|; (c) |**A** − **B**|; (d) *A* − *B*.

27. (I) If vector **A** = 6**i** − 2**j** + 3**k** m, find: (a) a vector in the same direction as **A** with magnitude 2*A*; (b) a unit vector in the direction of **A**; (c) a vector opposite to **A** with a magnitude of 4 m.

28. (I) Given two vectors, **A** = 2**i** − 3**j** + **k** and **B** = −4**i** + **j** − 5**k**, find a third vector, **C**, such that **A** − 2**B** + **C**/3 = 0.

29. (II) Show that if the sum of three vectors is zero, they must all lie in the same plane. Does the same restriction apply to the zero sum of four vectors?

30. (I) The vectors **A** and **B** have the following components: $A_x = 2$ m, $A_y = -3.5$ m, $B_x = -1.5$ m, and $B_y = -2.5$ m. Find the magnitude and direction of **C** = 3**A** − 2**B**.

31. (I) Find the components of the following vectors: (a) **P** of length 5 m directed at 150° counterclockwise from the +*x* axis; (b) **Q** of length 3.6 m directed at 120° clockwise from the +*y* axis.

32. (I) A body moves from a position with coordinates (3 m, 2 m) to (−4 m, 4 m). Find its displacement (a) in unit vector notation, and (b) in terms of its magnitude and direction.

33. (I) Given that vector **A** is 5 m at 37° N of E, find vector **B** such that their sum is directed along the negative *x* axis and has a magnitude of 3 m.

34. (II) The hour hand of a clock is 6 cm long. Take 12 noon to

lie along the *y* axis and 3 p.m. to lie along the *x* axis. Find the displacement (in unit vector notation) of the tip between each of the following times: (a) 1 p.m. to 4 p.m.; (b) 2 p.m. to 9:30 p.m.

35. (II) Figure 2.31 shows the directions of three vectors whose magnitudes in arbitrary units are *W* = 20, *F* = 10, and *T* = 30. The *x* and *y* axes are tilted as shown. Find: (a) the components of the vectors; (b) their sum in unit vector notation.

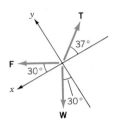

FIGURE 2.31 Exercise 35.

36. (I) The instructions to a treasure hunt state: Start at the oak tree and walk 5 m in a straight line at 30° W of N. Turn to your right through 45° and walk 4 m. Dig a hole 2 m deep. Where is the treasure relative to the base of the tree?

37. (II) Given the vector **A** = 2**i** + 3**j** m, find the vector **B** of length 5 m that is perpendicular to **A** and lies in the following planes: (a) the *xz* plane; (b) the *xy* plane.

38. (I) A helicopter rises 100 m from its pad and travels a horizontal distance of 200 m at 25° S of W. What is its displacement?

2.4 Dot Product

39. (I) What is the angle between the vectors **A** = **i** − 2**j** and **B** = 2**i** + 3**j**?

40. (I) Given the vectors **A** = −2**i** + **j** − 3**k** and **B** = 5**i** + 2**j** − **k**, find: (a) **A** · **B**, and (b) (**A** + **B**) · (**A** − **B**).

41. (I) The dot product of two vectors, whose magnitudes are 3 m and 5 m, is −4 m². What is the angle between them?

42. (I) The components of two vectors are $A_x = 2.4$, $A_y = -1.2$, $A_z = 4.0$, and $B_x = -3.6$, $B_y = 1.8$, $B_z = -2.6$. Find the angle between them.

43. (I) The vectors **A** and **B** are in the *xy* plane where **A** is 3.2 m at 45° to the +*x* axis, and **B** is 2.4 m at 290° to +*x* axis. Find **A** · **B**.

44. (II) The vectors **A** and **B** in Fig. 2.32 define two sides of a parallelogram. (a) Express the diagonals in terms of **A** and **B**. (b) Show that the diagonals are perpendicular if *A* = *B*.

FIGURE 2.32 Exercises 44 and 51.

45. (II) Show that the angles α, β and γ between a vector **A** and the x, y, and z axes, respectively, are given by

$$\cos \alpha = \frac{\mathbf{A} \cdot \mathbf{i}}{A} \qquad \cos \beta = \frac{\mathbf{A} \cdot \mathbf{j}}{A} \qquad \cos \gamma = \frac{\mathbf{A} \cdot \mathbf{k}}{A}$$

If $\mathbf{A} = 3\mathbf{i} + 2\mathbf{j} + \mathbf{k}$, find the angle between **A** and each axis.

46. (II) Given the three vectors $\mathbf{A} = \mathbf{i} - 4\mathbf{j}$, $\mathbf{B} = 3\mathbf{i}$, and $\mathbf{C} = -2\mathbf{j}$ evaluate the following expressions *if* they are allowed mathematically: (a) $\mathbf{C} \cdot (\mathbf{A} + \mathbf{B})$; (b) $\mathbf{C} \cdot (\mathbf{A} \cdot \mathbf{B})$; (c) $C + \mathbf{A} \cdot \mathbf{B}$; (d) $C(\mathbf{A} \cdot \mathbf{B})$, (e) $\mathbf{C}(\mathbf{A} \cdot \mathbf{B})$.

47. (II) What is the component of the vector $\mathbf{A} = \mathbf{i} - 2\mathbf{j} + \mathbf{k}$ m in the direction of the vector $\mathbf{B} = -3\mathbf{i} + 4\mathbf{k}$ m?

2.5 Cross Product

48. (I) Given two vectors, $\mathbf{A} = \mathbf{i} + 2\mathbf{j} - 4\mathbf{k}$ and $\mathbf{B} = 3\mathbf{i} - \mathbf{j} + 5\mathbf{k}$, find $\mathbf{A} \times \mathbf{B}$.

49. (I) (a) Show, for arbitrary vectors **A** and **B**, that

$$\mathbf{A} \cdot (\mathbf{A} \times \mathbf{B}) = 0$$

(b) How could you have arrived at this conclusion without any computation?

50. (I) Vectors **A** and **B** are in the xy plane with $A = 3.6$ m at 25° counterclockwise from the $+x$ axis, and $B = 4.4$ m at 160° counterclockwise from the $+x$ axis. Find $\mathbf{A} \times \mathbf{B}$.

51. (II) Show that the area of a parallelogram, as in Fig. 2.32, is $|\mathbf{A} \times \mathbf{B}|$.

52. (II) Given three vectors, $\mathbf{A} = 2\mathbf{i} - 5\mathbf{j}$, $\mathbf{B} = 4\mathbf{j}$, and $\mathbf{C} = 3\mathbf{i}$, evaluate the following expressions *if* they are mathematically allowed:
(a) $C(\mathbf{A} \times \mathbf{B})$; (b) $\mathbf{C} \cdot (\mathbf{A} \times \mathbf{B})$; (c) $\mathbf{C} \times (\mathbf{A} \cdot \mathbf{B})$; (d) $\mathbf{C} \times (\mathbf{A} \times \mathbf{B})$; (e) $\mathbf{C} + \mathbf{A} \times \mathbf{B}$.

53. (II) Vector **A** has a magnitude of 4 m and lies in the xy plane directed at 45° counterclockwise from the $+x$ axis, whereas **B** has a magnitude of 3 m and lies in the yz plane directed at 30° clockwise from the $+z$ axis; see Fig. 2.33. Find $\mathbf{A} \times \mathbf{B}$.

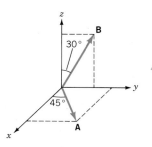

FIGURE 2.33 Exercise 53.

54. (II) Find a vector of length 5 m that is perpendicular to both vectors $\mathbf{A} = 3\mathbf{i} - 2\mathbf{j} + 4\mathbf{k}$ m and $\mathbf{B} = 4\mathbf{i} - 3\mathbf{j} - \mathbf{k}$ m.

PROBLEMS

1. (I) Find a vector of length 5 m in the xy plane that is perpendicular to $\mathbf{A} = 3\mathbf{i} + 6\mathbf{j} - 2\mathbf{k}$ m. (*Hint:* Consider the dot product.)

2. (I) The vector **A** has a magnitude of 2 m and is directed toward the northeast. Find the vector **B** such that $|\mathbf{A} + \mathbf{B}| = 2|\mathbf{A}|$ given the following: (a) **B** has the maximum possible magnitude; (b) **B** has the minimum possible magnitude; (c) **B** points northwest; (d) $\mathbf{A} + \mathbf{B}$ points south.

3. (I) The magnitudes of vectors **A** and **B** are equal and the angle between them is θ. Show that
(a) $|\mathbf{A} + \mathbf{B}| = 2A \cos(\theta/2)$, and (b) $|\mathbf{A} - \mathbf{B}| = 2A \sin(\theta/2)$.

4. (II) A rectangular coordinate system with axes x' and y' is rotated by angle θ from axes x and y as shown in Fig. 2.34.

(a) What are the components of the position vector **r** in the two coordinate systems? (b) Use part (a) to show that the coordinates of a point P in the two systems are related by

$$x' = x \cos \theta + y \sin \theta \qquad y' = -x \sin \theta + y \cos \theta$$

(*Hint:* You will need to expand $\cos(\phi - \theta)$.) These equations show how the coordinates (x, y) transform under the rotation of the axes. A vector is defined as a quantity whose components transform in this way.

5. (II) The edges of a cube of side L lie along the x, y, and z axes, respectively. A face diagonal that lies along a face, and a body diagonal that goes through the cube are shown in Fig. 2.35. Find the angle between: (a) the body diagonal

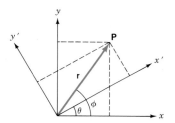

FIGURE 2.34 Problem 4.

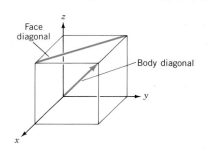

FIGURE 2.35 Problem 5.

shown and the z axis; (b) two face diagonals on adjacent faces; (c) a face diagonal and a body diagonal that share a corner. (Assign vectors to each line and express each in unit vector notation.)

6. (I) Personnel at an airport control tower track a UFO. At 11:02 a.m. it is located at a horizontal distance of 2 km in the direction 30° N of E at an altitude of 1200 m. At 11:15 a.m. the location is 1 km at 45° S of E at an altitude of 800 m; see Fig. 2.36. What was the displacement of the UFO?

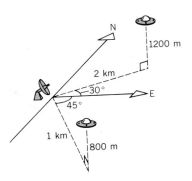

FIGURE 2.36 Problem 6.

7. (II) Show that the volume of the parallelopiped in Fig. 2.37, whose edges are defined by the vectors **A**, **B**, and **C**, is given by $\mathbf{A} \cdot (\mathbf{B} \times \mathbf{C})$.

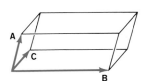

FIGURE 2.37 Problem 7.

8. (II) Show that $\mathbf{A} \times (\mathbf{B} \times \mathbf{C}) = \mathbf{B}(\mathbf{A} \cdot \mathbf{C}) - \mathbf{C}(\mathbf{A} \cdot \mathbf{B})$. (Without loss in generality you can assume that **B** lies along one axis.)

9. (I) Show that the polar unit vectors $\hat{\mathbf{r}}$ and $\hat{\boldsymbol{\theta}}$ in Fig. 2.38 are related to the Cartesian unit vectors according to

$$\hat{\mathbf{r}} = \cos\theta\,\mathbf{i} + \sin\theta\,\mathbf{j} \qquad \hat{\boldsymbol{\theta}} = -\sin\theta\,\mathbf{i} + \cos\theta\,\mathbf{j}$$

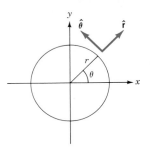

FIGURE 2.38 Problem 9.

10. (II) The position vector of a particle is $\mathbf{r} = x\mathbf{i} + y\mathbf{j} + z\mathbf{k}$. The angles between this vector and the x, y, and z axes respectively are α, β, and γ. Show that

$$\cos^2\alpha + \cos^2\beta + \cos^2\gamma = 1$$

11. (II) A three-dimensional vector **A** has a length of 10 m and makes the angles 65° and 40° with the $+x$ and $+z$ axes, respectively. Find the magnitudes of its Cartesian components.

CHAPTER 3

One-Dimensional Kinematics

Major Points

1. (a) The definitions of **speed, velocity,** and **acceleration.**
   (b) The distinction between **average** and **instantaneous** values of velocity and acceleration.
2. **Graphical analysis** of motion:
   (a) The use of tangents to obtain instantaneous values.
   (b) The use of areas to obtain displacements and changes in velocity.
3. The **equations of kinematics** for constant acceleration.
4. Vertical **free-fall.**

3.1 PARTICLE KINEMATICS

Many objects that physicists study, from atoms to galaxies, are in motion. Motion may be orderly or random, steady or intermittent, or even a confusing mixture of these. We cannot hope to gain insight into how nature works unless we can clearly define and then measure motion. The subject of **kinematics** is concerned with the description of how a body moves in space and time. In **translational** motion, depicted in Fig. 3.1a, all parts of a body undergo the same change in position. In **rotational** motion, shown in Fig. 3.1b, the body changes its orientation in space. In **vibrational** motion, as in Fig. 3.1c, the shape or size of the body changes rhythmically. This chapter is restricted to translational motion along a line; that is, *one-dimensional kinematics*.

The translational motion of an object can be completely described by the movement of any single point on it. Consequently, the object may be treated as a **particle.** In everyday language, the word "particle" implies something tiny, almost invisible. In physics, however, a particle is an idealized model of a real object that allows us to locate it at a single point in space. The particle model is a useful simplification when the size, shape, and internal structure of the system under study are of no interest. For example, when we study the orbital motion of the earth about the sun, the earth may be considered to be a particle. However, if we wish to study earthquakes, tides, or hurricanes, the earth's internal structure and rotational motion become significant and it would not be adequate to treat the earth as a particle.

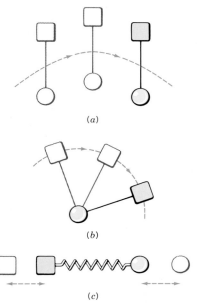

FIGURE 3.1 The motion of a body may involve (a) translation, (b) rotation, (c) vibration, or a combination of these.

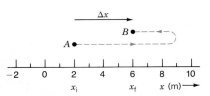

FIGURE 3.2 When a particle moves from A to B via any path its displacement is $\Delta x = x_f - x_i$.

3.2   DISPLACEMENT AND VELOCITY

Suppose we wish to study the translational motion of a car along a straight road. We may treat the car as a particle and use the road as a frame of reference. For one-dimensional motion, we need just one axis to locate the particle. We take our x axis to lie along the road (see Fig. 3.2). In Chapter 2 we defined displacement as a change in position and noted that it is a vector. When a particle moves from an initial coordinate x_i to a final coordinate x_f, its **displacement** is written as

$$\Delta x = x_f - x_i \tag{3.1}$$

The Greek letter Δ (capital delta) is usually used to signify the change in the variable that follows it. Note that Δx always means the final value minus the initial value, not the greater minus the lesser. The sign of Δx indicates its direction relative to the $+x$ axis. (In general, Δx is the x *component* of a displacement. For one-dimensional motion we will usually omit the phrase "x component of . . .") In Fig. 3.2 the particle starts at point A ($x_i = 2$ m) and stops at point B ($x_f = 6$ m), having turned around at $x = 9$ m. Its displacement, $\Delta x = 6 - 2 = +4$ m, depends only on the initial and final positions, not on the details of the journey. The **distance** traveled—that is, the length of the actual path—is a positive scalar. Its value for the journey in Fig. 3.2 is 10 m. In general, the distance traveled between two points is not equal to the magnitude of the displacement between the same two points.

Distance

A basic question regarding the motion of a particle is how fast it moves. The **average speed** for a finite time interval is defined as

Average speed

$$\text{Average speed} = \frac{\text{Distance traveled}}{\text{Time interval}} \tag{3.2}$$

Since it is defined in terms of distance, average speed is also a positive scalar. (We will not use a symbol for average speed.) In contrast, the **average velocity** for a finite time interval is defined as

$$\text{Average velocity} = \frac{\text{Displacement}}{\text{Time interval}}$$

Average velocity is a vector in the direction of the displacement. The average velocity depends only on the net displacement and the time interval; the details of the journey in between are of no consequence. Since we are concerned only with motion along the x axis, the x component of the average velocity is

Average velocity

$$v_{av} = \frac{\Delta x}{\Delta t} \tag{3.3}$$

where $\Delta t = t_f - t_i$. The SI unit of both speed and velocity is m/s. The sign of v_{av} is the same as that of the displacement. Positive v_{av} means that the net motion is in the direction of the $+x$ axis. (From now on, the phrase "x component of . . ." will be omitted for simplicity.) For the motion depicted in Fig. 3.2, suppose it took 4 s to go from A to B. The average speed would be (10 m)/(4 s) = 2.5 m/s, whereas the average velocity would be $v_{av} = (+4$ m$)/(4$ s$) = +1$ m/s.

EXAMPLE 3.1: A bird flies east at 10 m/s for 100 m. It then turns around and flies at 20 m/s for 15 s. Find: (a) its average speed; (b) its average velocity.

Solution: Let us take the x axis to point east. A sketch of the path is shown in Fig. 3.3. To find the required quantities, we need the total time interval. The first part of the journey took

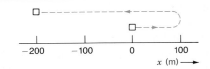

FIGURE 3.3 The displacement of the bird is -200 m.

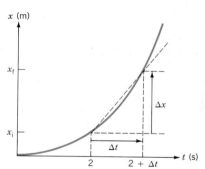

FIGURE 3.9 To find the instantaneous velocity at 2 s by the limiting process one must evaluate the ratio $\Delta x/\Delta t$ for smaller and smaller values of Δt. The limiting value as $\Delta t \to 0$ is the instantaneous velocity.

Ct^2 where C is a constant. From t to $t + \Delta t$, the displacement is

$$\Delta x = C(t + \Delta t)^2 - Ct^2$$
$$= 2Ct(\Delta t) + C(\Delta t)^2$$

Dividing by Δt gives

$$\frac{\Delta x}{\Delta t} = 2Ct + C(\Delta t)$$

As Δt approaches zero, the second term vanishes. Consequently,

$$\frac{dx}{dt} = \lim_{\Delta t \to 0} \frac{\Delta x}{\Delta t} = 2Ct$$

With $C = 3$ and $t = 2$ s, we find $v = dx/dt = 12$ m/s, as in part (a). Now, however, the above expression allows us to find the instantaneous velocity at any time. According to the rules of calculus, the derivative of the frequently encountered power function

$$x = Ct^n$$

where C and n are any constants, is

$$\frac{dx}{dt} = nCt^{n-1}$$

The derivatives of other functions are listed in Appendix C.

EXERCISE 2. The position of a particle is given by $x = 40 - 5t - 5t^2$ m. (a) What is its average velocity between 1 and 2 s? (b) Find its instantaneous velocity at $t = 2$ s by calculating the derivative dx/dt.

3.4   ACCELERATION

It is commonly understood that acceleration is associated with changes in speed. Although this is true, it is only part of the concept of acceleration. Suppose that during some time interval, a ball reverses its direction of motion but its initial and final speeds are equal. The quantity "change in speed" would not reflect the fact that the ball was struck by a bat. In physics we must broaden the concept of acceleration to include such cases. We say that a body accelerates when its *velocity* changes in magnitude or in direction, or both. The **average acceleration** for a finite time interval is defined as

$$\text{Average acceleration} = \frac{\text{Change in velocity}}{\text{Time interval}}$$

Average acceleration is a vector quantity whose direction is the same as that of the change in velocity. For one-dimensional motion, the average acceleration is

$$a_{\text{av}} = \frac{\Delta v}{\Delta t} \tag{3.6}$$

The SI unit of acceleration is m/s². If a car goes from rest to 90 km/h in 15 s, $a =$ (90 km/h)/(15 s) = 6 km·h⁻¹/s. This means that on average the velocity increases by 6 km/h in each 1-s interval. On a graph of v versus t, such as Fig. 3.10, the average acceleration is found from the slope of the line joining the initial and final points. The sign of a_{av} is determined by the sign of Δv.

By analogy with Eqs. 3.4 and 3.5, the **instantaneous acceleration** is defined as the derivative of v with respect to t:

$$a = \frac{dv}{dt} \tag{3.7}$$

Graphically, the instantaneous acceleration at a particular instant is found from the slope of the tangent to the v versus t graph at that instant, as shown in

Average acceleration

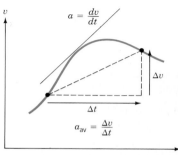

FIGURE 3.10 On a velocity–time graph, the slope of the line joining two points is the average acceleration for that time interval. The instantaneous acceleration at a given time is the slope of the tangent at that time.

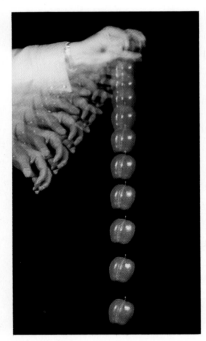

Fig. 3.10. Positive acceleration points in the direction of the $+x$ axis, while negative acceleration points in the opposite direction. *Do not assume that a negative acceleration means a deceleration.* The word "deceleration" means only a slowing down; it tells us nothing about direction. When v and a have the same sign, the body speeds up; when they have opposite signs, the body slows down.

If the positions of a particle undergoing constant acceleration are recorded at equal time intervals, they appear as in Fig. 3.11. Recall that for constant velocity the displacement between dots is fixed (see Fig. 3.5a). For constant acceleration, it is the *change* in the displacement from one time interval to the next that is fixed. The tape in Fig. 3.11 shows that in successive time intervals the displacement Δx increases by 4 m.

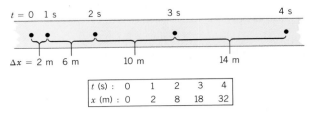

t (s) :	0	1	2	3	4
x (m) :	0	2	8	18	32

FIGURE 3.11 A record of the positions of a particle undergoing constant acceleration. In successive 1-s time intervals the displacement increases by a constant amount, which in this case is 4 m.

Accelerated motion in the vertical direction.

Note that a sudden ($\Delta t = 0$) change in velocity is not physically reasonable: A finite change in velocity requires a finite time interval. It is more realistic to round off the corners in the v versus t graphs, as in Fig. 3.12a. When the change in velocity is very rapid, the a versus t graph looks like the spike in Fig. 3.12b.

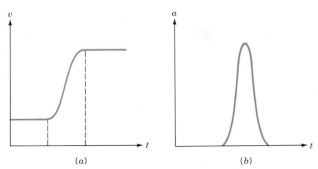

(a)   (b)

FIGURE 3.12 (a) The velocity of a particle cannot change instantaneously; it must occur over a time interval. (b) When Δt is very short, the acceleration–time graph has a sharp peak.

EXAMPLE 3.4: At $t = 0$ a car is moving east at 10 m/s. Find its average acceleration between $t = 0$ and each of the following times at which it has the given velocities: (a) $t = 2$ s, 15 m/s east; (b) $t = 5$ s, 5 m/s east; (c) $t = 10$ s, 10 m/s west; (d) $t = 20$ s, 20 m/s west.

Solution: We take east as the $+x$ direction. In each case we have to find

$$a_{av} = \frac{v_f - v_i}{\Delta t}$$

(a) $a_{av} = (15 - 10)/2 = +2.5$ m/s²   (east; $|v|$ changes)

(b) $a_{av} = (5 - 10)/5 = -1$ m/s²   (west; $|v|$ changes)

(c) $a_{av} = (-10 - 10)/10 = -2$ m/s²   (west; direction of v changes)

(d) $a_{av} = (-20 - 10)/20 = -1.5$ m/s²   (west, $|v|$ and direction change)

The sign of a_{av} is determined by the direction of Δv relative to the chosen $+x$ axis. It is not determined solely by whether the speed has increased or decreased.

EXERCISE 3. With the data of Example 3.4, find a_{av} between (a) 2 and 10 s; (b) 10 and 20 s.

3.5   THE USE OF AREAS

The previous sections have shown how to obtain the velocity from a position–time graph and the acceleration from a velocity–time graph. We now discuss the inverse problems of determining x from a graph of v versus t and v from a graph of a versus t. For motion at constant velocity, the v versus t graph is a horizontal line, as shown in Fig. 3.13a. Since $v = \Delta x/\Delta t$, the displacement Δx in a time interval Δt is given by $\Delta x = v\Delta t$. This is just the area of the shaded rectangle of height v and width Δt. Notice that the unit of this area is (m/s)(s) = m.

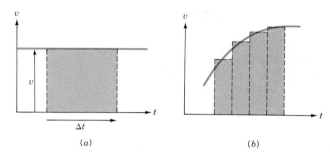

(a)   (b)

FIGURE 3.13 (a) When a particle travels at constant velocity v, the displacement in time interval Δt is $\Delta x = v\Delta t$. This equals the rectangular area under the v versus t graph for this time interval. (b) When the velocity is not constant, the actual area may be approximated by the sum of rectangular areas.

The area under any curve may be approximated by the sum of the areas of rectangles of appropriate heights, as in Fig. 3.13b. The approximation improves as the number of rectangles is increased. In general, *the displacement Δx in some time interval is given by the area under the v versus t graph for that interval.* Since a negative velocity leads to a negative displacement, areas below the time axis are negative.

In an analogous way, the equation $\Delta v = a\Delta t$ leads to the following conclusion: For a given time interval, the area under the a versus t graph gives the *change* in velocity Δv during that interval.

Consider now the case of a body whose velocity increases with constant acceleration, as shown in Fig. 3.14. The initial velocity is v_i and the final velocity is v_f after a time interval Δt. The area of the shaded trapezoid is the sum of the area of the rectangle and a triangle: $(v_i\Delta t) + \frac{1}{2}(v_f - v_i)\Delta t = \frac{1}{2}(v_i + v_f)\Delta t$. This is the same as the area of a rectangle of height $(v_i + v_f)/2$ and width Δt. From Eq. 3.3, the displacement is

$$\Delta x = v_{av}\Delta t = \tfrac{1}{2}(v_i + v_f)\Delta t \tag{3.8}$$

Equation 3.8 shows that for the special case of constant acceleration, we may write $v_{av} = (v_i + v_f)/2$.*

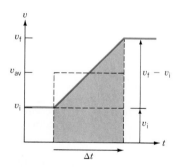

FIGURE 3.14 During the interval Δt, the velocity of the particle increases linearly with time. The area under the trapezoid is equal to the area under the rectangle of base Δt and height $v_{av} = (v_i + v_f)/2$.

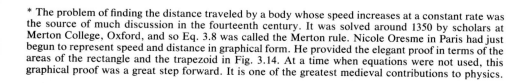

* The problem of finding the distance traveled by a body whose speed increases at a constant rate was the source of much discussion in the fourteenth century. It was solved around 1350 by scholars at Merton College, Oxford, and so Eq. 3.8 was called the Merton rule. Nicole Oresme in Paris had just begun to represent speed and distance in graphical form. He provided the elegant proof in terms of the areas of the rectangle and the trapezoid in Fig. 3.14. At a time when equations were not used, this graphical proof was a great step forward. It is one of the greatest medieval contributions to physics.

EXAMPLE 3.5: At $t = 0$ a particle is at rest at the origin. Its acceleration is 2 m/s² for the first 3 s and −2 m/s² for the next 3 s. Plot the x versus t and v versus t graphs.

Solution: We are given that $x = 0$ and $v = 0$ at $t = 0$. The a versus t graph is plotted in Fig. 3.15a. The area between the function and the time axis is also indicated for each 1-s interval. This is the *change* in velocity for the corresponding interval. In the first second $\Delta v = +2$ m/s, and since $v = 0$ at $t = 0$, we find that $v = 2$ m/s at $t = 1$ s. Between 1 and 2 s, $\Delta v = +2$ m/s; thus, $v = 4$ m/s at $t = 2$ s, and so on.

The v versus t graph is plotted in Fig. 3.15b. The area under this function for each 1-s interval is again indicated and represents the displacement Δx for that interval. Starting at $x = 0$ at $t = 0$, we add each displacement (with the proper sign) to the previous value of x to obtain the next value of x. In this way we can plot the x versus t graph shown in Fig. 3.15c. Remember that the slope of the tangent to the x versus t graph at any instant should give the velocity at that instant.

You can easily verify that the values of x and t for the first three seconds are related by the equation $x = t^2$, which is the equation of a parabola. In general, a *linear v versus t graph* leads to a *parabolic x versus t graph*.

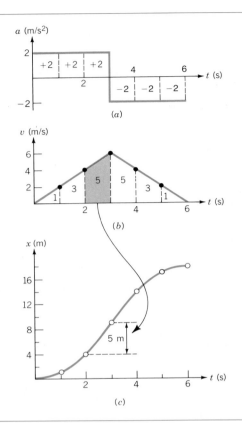

FIGURE 3.15 (a) The areas indicated under the a versus t graph are the *changes* in velocity in each 1-s interval. (b) The areas under the v versus t graph are the displacements for each 1-s interval. (c) The x versus t graph obtained from the v versus t graph.

3.6 THE EQUATIONS OF KINEMATICS FOR CONSTANT ACCELERATION

The use of graphs to deal with motion can become tedious. It is more efficient to relate the quantities position, velocity, acceleration, and time with equations. For the special case of constant acceleration, these equations are easily derived. We will use the notation indicated in the table in the margin. "Initial" and "final" refer to the values at the beginning and at the end of any time interval you choose to examine. In a given problem, the final value for one part may be the initial value for a later part. To keep the notation simple, we take the initial values of position x_0 and velocity v_0 to be at $t = 0$. The final values, x and v, occur at a later time t.

When the acceleration is constant, its average and instantaneous values are identical, so we may write $a = (v_f - v_i)/(t_f - t_i)$. When we set $t_i = 0$ and $t_f = t$, and use the new notation, we obtain $a = (v - v_0)/t$, or

$$v = v_0 + at \tag{3.9}$$

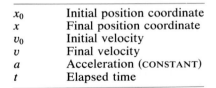

x_0	Initial position coordinate
x	Final position coordinate
v_0	Initial velocity
v	Final velocity
a	Acceleration (CONSTANT)
t	Elapsed time

On a v versus t graph, this is the equation of a straight line whose slope is a (see Fig. 3.16). With t as the elapsed time, the displacement $\Delta x = x - x_0$ is given by Eq. 3.8:

$$x = x_0 + \tfrac{1}{2}(v_0 + v)t \tag{3.10}$$

If Eq. 3.9 for v is used in Eq. 3.10, we find

$$x = x_0 + v_0t + \tfrac{1}{2}at^2 \tag{3.11}$$

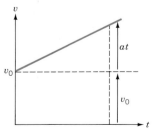

FIGURE 3.16 The v versus t graph for constant (positive) acceleration.

This is the equation of a parabola, as shown in Fig. 3.17. We can eliminate time from Eq. 3.11 by using $t = (v - v_0)/a$ from Eq. 3.9. After some rearrangement (go through the algebra yourself) we obtain

$$v^2 = v_0^2 + 2a(x - x_0) \qquad (3.12)$$

Equations 3.9–3.12 are the **equations of kinematics** for *acceleration that is constant both in magnitude and in direction along the x axis.* Of the five quantities x, v_0, v, a, and t, we need to be given at least three to solve any problem. The value of x_0 depends on where the origin is placed. We usually try to have $x_0 = 0$ to simplify the problem. Keep in mind that the symbols v, v_0, and a represent the x components of the associated vectors $\mathbf{v}$, $\mathbf{v_0}$, and $\mathbf{a}$. Their signs are determined by their directions relative to the chosen coordinate axes.

The steps outlined below should help you approach the solution of kinematics problems in a systematic way. Although they cannot encompass every situation you may encounter, they should give you a good start on most problems. As you gain insight and self-confidence, you will likely develop your own shortcuts.

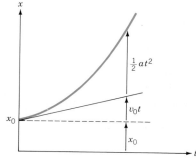

FIGURE 3.17 The x versus t graph for constant (positive) acceleration is a parabola. The slope of the tangent at $t = 0$ is the initial velocity v_0.

Problem-Solving Guide for Kinematics

1. Make a simple *sketch* of the situation described.

2. Set up a *coordinate system* and clearly indicate the origin.

3. (a) List the *given* quantities. Assign each a value, including its sign.
   (b) List the *unknown* quantities. Identify the one(s) you are solving for.

4. Find the equation that has the quantity you need as the *only* unknown. (This is not always possible; see Example 3.7.)

5. It is often helpful to obtain a rough *graphical solution*. Do this either before setting up the algebra or afterward as a check on it.

EXAMPLE 3.6: A car accelerates with constant acceleration from rest to 30 m/s in 10 s. It then continues at constant velocity. Find: (a) its acceleration; (b) how far it tavels while speeding up; (c) the distance it covers while its velocity changes from 10 m/s to 20 m/s.

Solution: A sketch and coordinate system are shown in Fig. 3.18a. Note that $x_0 = 0$.
(a) GIVEN: $v_0 = 0$; $v = 30$ m/s; $t = 10$ s.
UNKNOWN: $a = ?$; $x = ?$
From Eq. 3.9 we have

$$a = \frac{v - v_0}{t} = +3 \text{ m/s}^2$$

(b) GIVEN: $v_0 = 0$; $v = 30$ m/s; $t = 10$ s; $a = 3$ m/s².
UNKNOWN: $x = ?$
The position coordinate x appears as the only unknown in Eqs. 3.10–3.12. If we had not found the acceleration in (a), we would have to use Eq. 3.10:

$$x = x_0 + \tfrac{1}{2}(v_0 + v)t$$
$$= 0 + \tfrac{1}{2}(0 + 30)(10) = 150 \text{ m}$$

(c) GIVEN: $v_0 = 10$ m/s; $v = 20$ m/s; $a = 3$ m/s².
UNKNOWN: $x_0 = ?$; $x = ?$; $t = ?$
If we maintain the origin as in Fig. 3.18a, we have to find x_0 for

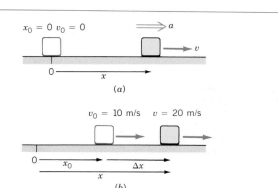

FIGURE 3.18 (a) A sketch that includes the coordinate system, with the origin clearly indicated. The acceleration is drawn with a double-stemmed arrow. Both a and v are positive. (b) The initial values, x_0 and v_0, for the second part of the problem are not the same as those for the first part.

this part of the trip. However, we need only the difference $\Delta x = x - x_0$, which can be found from Eqs. 3.10–3.12. The first two equations also contain the unknown t, so we use Eq. 3.12:

$$v^2 = v_0^2 + 2a\Delta x$$
$$20^2 = 10^2 + 2(3)\Delta x$$
$$\Delta x = 50 \text{ m}$$

EXAMPLE 3.7: A particle is at $x = 5$ m at $t = 2$ s and has a velocity $v = 10$ m/s. Its acceleration is constant at -4 m/s^2. Find the initial position at $t = 0$.

Solution: GIVEN: $x = 5$ m; $v = 10$ m/s; $a = -4$ m/s^2; $t = 2$ s; UNKNOWN: $x_0 = ?$; $v_0 = ?$ See Fig. 3.19.

In this case none of the equations of kinematics yields x_0 immediately. The quantity x_0 appears in three equations, but always with the other unknown, v_0. We have to find v_0 first. From Eq. 3.9,

$$v = v_0 + at$$
$$10 = v_0 + (-4)(2)$$

Thus, $v_0 = 18$ m/s. Any of the other equations will give x_0. From Eq. 3.10,

$$x = x_0 + \tfrac{1}{2}(v_0 + v)t$$
$$5 = x_0 + \tfrac{1}{2}(18 + 10)(2)$$

Thus, $x_0 = -23$ m.

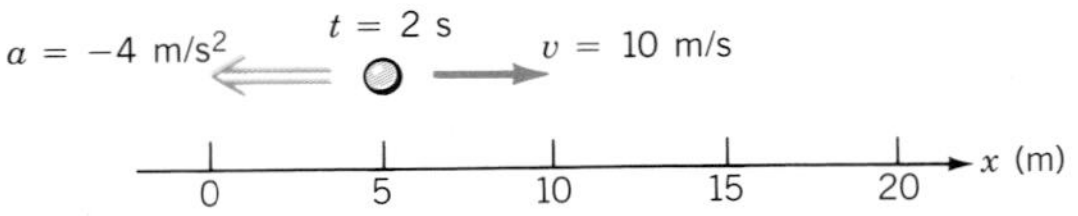

$a = -4$ m/s^2 ⟸ $t = 2$ s $v = 10$ m/s ⟶

x (m)
0   5   10   15   20

FIGURE 3.19 The position, velocity, and acceleration of the particle at 2 s.

EXERCISE 4. Is the equation $x = x_0 + vt - \tfrac{1}{2}at^2$ correct? Could it be used in Example 3.7 to find x_0?

EXAMPLE 3.8: A speeder moves at a constant 15 m/s in a school zone. A police car starts from rest just as the speeder passes it. The police car accelerates at 2 m/s^2 until it reaches its maximum velocity of 20 m/s. Where and when does the speeder get caught?

Solution: When two particles are involved in the same problem, we use simple subscripts to distinguish the variables, as shown in Fig. 3.20a. The motion of the police car has two phases: one at constant acceleration and one at constant velocity. In such problems, it is convenient to use Δt instead of t in the equations. The police may or may not catch the speeder during the acceleration phase. This has to be checked. We set the origin at the police lookout, which means $x_{0S} = x_{0P} = 0$.

Acceleration Phase. Let us say this takes a time interval Δt_1.
GIVEN: $v_S = 15$ m/s; $a_P = 2$ m/s^2; $v_{0P} = 0$; $v_P = 20$ m/s.
UNKNOWN: $x_S = ?$; $x_P = ?$; $\Delta t_1 = ?$
From $v = v_0 + at$, we have $20 = 0 + (2)\Delta t_1$, thus $\Delta t_1 = 10$ s. At this time, the positions are given by $x = x_0 + v_0 t + \tfrac{1}{2}at^2$:

$$x_S = (15)(10) = 150 \text{ m} \qquad x_P = \tfrac{1}{2}(2)(10)^2 = 100 \text{ m}$$

The speeder is still ahead.

Constant Velocity Phase. Let us say this takes a time interval Δt_2.
GIVEN: $x_{0S} = 150$ m; $x_{0P} = 100$ m; $v_S = 15$ m/s; $v_P = 20$ m/s; $a_S = a_P = 0$.
UNKNOWN: $x_S = ?$; $x_P = ?$; $\Delta t_2 = ?$
The cars meet when they have the same position, that is, $x_S = x_P$. However, we cannot find *where* until we find *when*. From Eq. 3.11,

$$x_S = 150 + 15\Delta t_2 \qquad x_P = 100 + 20\Delta t_2$$

On setting $x_S = x_P$ we find $\Delta t_2 = 10$ s. Substituting into either equation gives $x = 300$ m. The speeder is caught at 300 m after a period of 20 s. The graphical solution is depicted in Fig. 3.20b.

EXERCISE 5. Repeat Example 3.8 assuming that the police car continues to accelerate at 2 m/s^2.

EXAMPLE 3.9: Two cars approach each other on a straight road. Car A moves at 16 m/s and car B moves at 8 m/s. When they are 45 m apart, both drivers apply their brakes. Car A slows down at 2 m/s^2, while car B slows down at 4 m/s^2. Where and when do they collide?

Solution: Figure 3.21a is a simple sketch of the situation. We choose the origin at A's initial position and point the positive axis in the direction of its velocity.
GIVEN: $x_{0A} = 0$; $v_{0A} = 16$ m/s; $a_A = -2$ m/s^2; $x_{0B} = 45$ m; $v_{0B} =$

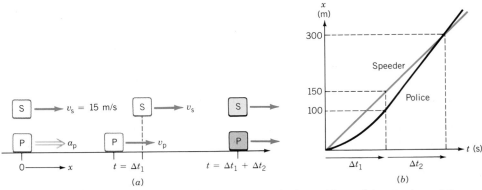

FIGURE 3.20 (*a*) Initially, the sketch can include only the positions of the speeder and the police officer at $t = 0$. Their positions after a time interval Δt_1, when the police car reaches its maximum velocity, can be indicated only after the calculation. (*b*) In the graphical solution, the acceleration phase of the police car is represented by a parabolic curve.

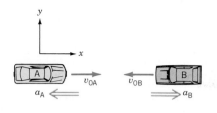

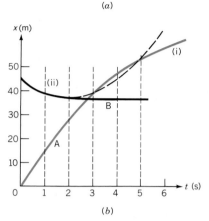

FIGURE 3.21 (a) The origin is at the original position of car A. Note that although the accelerations have opposite signs, both cars are slowing down. (b) The x versus t plots are parabolas until each car stops. The dashed parabola would be meaningful only if the acceleration of car B had remained constant.

-8 m/s; $a_B = 4$ m/s^2.
UNKNOWN: $x_A = ?; v_A = ?; x_B = ?; v_B = ?; t = ?$
Carefully check the signs. The cars meet when $x_A = x_B$, so we set up general expressions for these quantities using $x = x_0 + v_0t + \frac{1}{2}at^2$:

$$x_A = 16t - t^2 \qquad \text{(i)}$$
$$x_B = 45 - 8t + 2t^2 \qquad \text{(ii)}$$

When we set $x_A = x_B$, we find $3t^2 - 24t + 45 = 3(t - 5)(t - 3) = 0$. We seem to have two possible times for the collision: $t = 3$ s and $t = 5$ s.
STOP! Try to find the flaw in the above argument.

Let us look at the velocities to see what has happened. At $t = 3$ s,

$$v_A = v_{0A} + a_At = 16 + (-2)(3) = 10 \text{ m/s}$$
$$v_B = v_{0B} + a_Bt = -8 + (4)(3) = +4 \text{ m/s}$$

Does this give you a hint of the difficulty? v_A shows nothing unusual; A has slowed down. But look at the sign of v_B. We seem to have found that when the brakes were applied the car reversed its velocity! You can easily verify that B stops at 2 s and then stays at rest. This means that (ii) is not valid after 2 s. Similarly, (i) is not valid after 8 s. Of course, we could have checked this at the outset, but that's hindsight.

This is a good point to look at the graphical solution. Figure 3.21b shows that parabola (i) stops at 8 s, whereas parabola (ii) stops at 2 s. From then on the graphs are horizontal lines. The solutions 3 and 5 s are the intersections of the two parabolas. These would have been physically acceptable if the accelerations had remained constant in both magnitude and direction.

The graph helps us find the proper solution. We must find the intersection of the horizontal line for B and the parabola for A. Let us find where B stops. At $t = 2$ s, (ii) gives

$$x_B = 45 - 8(2) + 2(2)^2 = 37 \text{ m}$$

B stays at this position until it is hit by A. The condition $x_A = x_B$ becomes

$$16t - t^2 = 37$$

Thus, $t = 2.8$ s; 13.2 s. We reject 13.2 s since there can be only one collision. The collision occurs at 2.8 s and 37 m.

This example illustrates some of the complications that can arise in an apparently straightforward problem. You should not place blind faith in the logic of mathematics to lead you to physically acceptable solutions. By whatever means you have at your disposal, you should check the reasonableness of any answer. For example, having found two positive times, you could evaluate the corresponding positions and velocities to check for consistency or validity.

EXERCISE 6. Repeat Example 3.9 but assume that both cars speed up.

HISTORICAL NOTE: Falling Bodies

The Greek philosopher Aristotle (ca. 340 B.C.) considered all terrestrial objects to be composed of four elements, each with its natural place, one above the other. They were, in vertical order starting from the bottom: earth, water, air, and fire. If this order were disturbed, each element would "seek its natural place." The natural place of a particular object depended on the relative proportions of the elements it contained. The scheme is not unreasonable. Flames rise in air, air bubbles rise in water, and stones, which were thought to consist mainly of earth, fall through air and water. Aristotle believed that after a brief period, during which its speed increases, a body falls

at a constant speed proportional to its weight. Thus, one body with twice the weight of another would take half as long to fall from a given height. In a liquid, a large stone does reach a greater constant speed than a small one, but Aristotle extended this observation too readily to bodies falling through the air.

Around A.D. 500, the scholar John Philoponous challenged this view:

For if you let fall from the same height two weights of which one is many times as heavy as the other, you will see that the ratio of the times required for the motion does not depend on the ratio of the weights, but that the difference in time is a very small one.

A thousand years later, in 1586, the Dutch mathematician Simon Stevin dropped two lead balls, one ten times the weight of the other, and noted that their impacts with a board on the ground produced "a single sensation of sound."

While he was in Pisa (ca. 1590), Galileo is reputed to have dropped a large iron ball and a small iron ball from the leaning tower and showed that they landed at the same time. Even if this did occur, it was merely a repetition of Stevin's work and is not the source of Galileo's fame. At that time Galileo believed that most of the fall of a body occurs at a constant speed—not far from Aristotle's view. In fact, he also stated that balls of the same size but made of different materials (lead and wood) would move at different speeds.

Over the next few years, Galileo tried to explain vertical motion, but made little progress. He finally abandoned his search for the *cause* of motion, and focused instead on obtaining a "true description" of the motion of falling bodies. He had come to realize that bodies do not fall at constant speed and wanted to determine *how* the speed varies. Does the speed increase in proportion to the distance fallen or to the time taken? Galileo's greatness, and one justification for calling him the father of physics, lies in his replacement of idle speculation by experimental test.

Since objects fell too fast for direct measurements, he had the brilliant idea of "diluting" gravity by having balls roll down a smooth incline instead. Galileo had convinced himself that the speed of a body depends only on the *vertical* distance through which it falls, not on the actual path it takes. (The evidence is discussed in Section 4.1.) He was unable to measure speed directly, so he had to find the relationship between distance and time instead.* The data showed that for times that were in the ratios $1:2:3:4$, the total distances traveled, starting from rest, were in the ratios $1:4:9:16$ (see Fig. 3.22). Clearly, $x \propto t^2$. Five years later, in 1609, by working backward from this result, he was able to infer that the speed increases in direct proportion to the time taken, and not to the distance traveled. It took this great mind nearly two decades to find out that acceleration defined as $a = \Delta v/\Delta t$, rather than $a = \Delta v/\Delta x$, is constant for a falling body (in the absence of air resistance).

* Galileo's measurement of time is a story in itself. See S. Drake, *Am. J. Phys.* 54: 302 (1986).

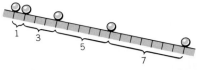

FIGURE 3.22 Galileo discovered that in successive equal time intervals the distances traveled by a ball rolling down an incline were in the ratios $1:3:5:7$, and so on. He inferred that the total distance x increased as the square of the total elapsed time t; that is, $x \propto t^2$.

In the seventeenth century, Robert Boyle learned how to produce a crude vacuum and was able to present a famous demonstration involving the fall of a coin and a feather depicted in Fig. 3.23. The astronaut David Scott repeated the experiment on the moon in 1971. When he released a hammer and a feather at the same time, millions of viewers witnessed that they landed simultaneously.

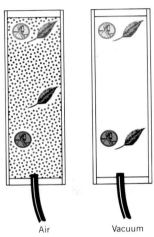

Air   Vacuum

FIGURE 3.23 (a) In the presence of air resistance a feather lags behind a coin dropped at the same time. (b) When the air is evacuated from the chamber, the coin and the feather fall with the same acceleration.

3.7 VERTICAL FREE-FALL

Free-fall

Motion that occurs solely under the influence of gravity is called **free-fall.** This term applies as much to satellites orbiting the earth as to bodies moving vertically up or down. Even today, with stopwatches that can measure to 0.01 s, it is not easy to determine how the speed of a falling body varies. Thus, it is hardly surprising that the nature of this motion was unclear for centuries (see the essay on Falling Bodies). In the early seventeenth century, Galileo established an important fact:

In the absence of air resistance, all falling bodies have the same acceleration due to gravity, regardless of their sizes or shapes.

The value of the acceleration due to gravity depends on both latitude and altitude. It is approximately 9.8 m/s² near the surface of the earth. In the presence of air resistance, the acceleration of a falling body is only approximately constant. The acceleration diminishes with time and can even become zero (see Section 3.8). For low speeds and short time intervals one can neglect this complication, and assume that bodies are in free-fall with a constant acceleration. In this case, we may apply the equations of kinematics.

Because we will later use the x axis for horizontal motion, we use the y axis for vertical motion. If the y axis points upward, as in Fig. 3.24, the acceleration due to gravity is $\mathbf{a} = -g\mathbf{j}$, where $g = 9.8$ m/s² (a positive scalar). With $a_y = -g$, the equations of kinematics now read

FIGURE 3.24 If the y axis is chosen to point upward, the acceleration of a particle in free-fall is $a_y = -g$, where $g = 9.8$ m/s² is the magnitude of the acceleration due to gravity.

$$v = v_0 - gt \tag{3.13}$$
$$y = y_0 + \tfrac{1}{2}(v_0 + v)t \tag{3.14}$$
$$y = y_0 + v_0 t - \tfrac{1}{2}gt^2 \tag{3.15}$$
$$v^2 = v_0^2 - 2g(y - y_0) \tag{3.16}$$

Free-fall equations of kinematics

The quantities v_0 and v are the y *components* of the vectors $\mathbf{v}_0$ and $\mathbf{v}$. Their signs are determined by their directions relative to the chosen $+y$ axis. Note in particular that the sign of the acceleration does *not* depend on whether the body is going up or coming down. It is important to clearly indicate the origin so that a value may be assigned to the initial vertical position coordinate y_0.

EXAMPLE 3.10: A ball thrown up from the ground reaches a maximum height of 20 m. Find: (a) its initial velocity; (b) the time taken to reach the highest point; (c) its velocity just before hitting the ground; (d) its displacement between 0.5 and 2.5 s; (e) the time at which it is 15 m above the ground.

Solution: A coordinate system is shown in Fig. 3.25. Note that at the highest point the ball is instantaneously at rest, that is, $v = 0$.

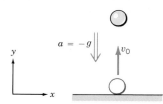

FIGURE 3.25 With the indicated coordinate axes, the free-fall acceleration is negative, irrespective of whether the particle is moving upward or downward.

(a) GIVEN: $y_0 = 0$; $y = 20$ m; $v = 0$; $a = -9.8$ m/s².
UNKNOWN: $v_0 = ?$; $t = ?$
Both Eq. 3.13 and Eq. 3.15 involve the other unknown, namely t. So from Eq. 3.16,

$$0 = v_0^2 + 2(-9.8)(20 - 0)$$

Thus, $v_0^2 = 392$ m²/s², and $v_0 = \pm 19.8$ m/s. Since "up" is positive, $v_0 = +19.8$ m/s.
(b) GIVEN: All the information in (a) and $v_0 = 19.8$ m/s.
UNKNOWN: $t = ?$

Either Eq. 3.13 or Eq. 3.15 would lead to t. Since Eq. 3.15 involves a quadratic it is quicker to use Eq. 3.14:

$$0 = 19.8 + (-9.8)t$$

Thus, $t = 2.02$ s.
(c) GIVEN: $y_0 = 0$; $y = 0$; $v_0 = 19.8$ m/s.
UNKNOWN: $v = ?$; $t = ?$
From Eq. 3.16 we have

$$v^2 = v_0^2 + 2(-9.8)(0 - 0)$$

Thus, $v = v_0$; $-v_0$. At the single point $y = 0$, the velocity has two values: $+v_0$ initially and $-v_0$ when it lands. (Use another approach to find the velocity on landing.)
(d) To find the displacement, $\Delta y = y_2 - y_1$, we need Eq. 3.15:

$$y_1 = 19.8(0.5) - 4.9(0.5)^2 = 8.68 \text{ m}$$
$$y_2 = 19.8(2.5) - 4.9(2.5)^2 = 18.9 \text{ m}$$

Hence, $\Delta y = +10.2$ m.
(e) GIVEN: $y = 15$ m; $y_0 = 0$; $v_0 = 19.8$ m/s.
UNKNOWN: $t = ?$; $v = ?$
Since Eqs. 3.13 and 3.16 also contain the other unknown, we use Eq. 3.15:

$$15 = 0 + 19.8t - 4.9t^2$$

The solutions of the quadratic equation are

$$t = \frac{19.8 \pm \sqrt{19.8^2 - 4 \times 4.9 \times 15}}{9.8} = 1.01 \text{ s}, 3.03 \text{ s}$$

Both solutions are acceptable. At $t = 1.01$ s the ball is on its way up, and at $t = 3.03$ s, it is on the way down.

EXERCISE 7. With the data of Example 3.10, find (a) the average velocity, and (b) the average speed, between 0.5 and 2.5 s.

EXAMPLE 3.11: A ball is thrown upward with an initial velocity of 12 m/s from a rooftop 40 m high. Find: (a) its velocity on hitting the ground; (b) the time of flight; (c) the maximum height; (d) the time to return to roof level; (e) the time it is 15 m below the rooftop.

Solution: Figure 3.26 shows the origin at ground level so that all positions are positive.
GIVEN: $y_0 = 40$ m; $v_0 = +12$ m/s; $a = -9.8$ m/s^2.
(a) When the ball lands, its final position coordinate is $y = 0$. The final velocity v appears as the only unknown in Eq. 3.16:

$$v^2 = 12^2 + 2(-9.8)(0 - 40) = 928 \text{ m}^2/\text{s}^2$$

For this part $v = -30.5$ m/s is appropriate. (Why?)
(b) Since v is now known, we can use Eq. 3.13:

$$-30.5 = 12 - 9.8t$$

which gives $t = 4.34$ s. If v were unknown we could use Eq. 3.15:

$$0 = 40 + 12t - 4.9t^2$$

which yields $t = 4.34$ s, -1.89 s. Since the motion starts at $t = 0$, we reject the negative solution. Note that it is not necessary to break this question into smaller parts, first to find the time to reach the highest point and then to obtain the time from the highest point to the ground.
(c) At the maximum height $v = 0$, so from Eq. 3.16,

$$0 = 12^2 + 2(-9.8)(y - 40)$$

Thus, $y = 47.3$ m. For the time taken we use Eq. 3.13:

$$0 = 12 - 9.8t$$

Thus, $t = 1.22$ s.
(d) At the roof level, the final position is $y = 40$ m. From Eq. 3.15,

$$40 = 40 + 12t - 4.9t^2$$

Therefore, $t = 0$ s, 2.45 s. Of course, we pick $t = 2.45$ s. This is just double the time needed to reach the maximum height.

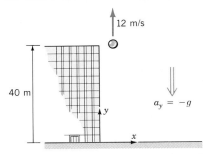

FIGURE 3.26 The position of the origin must always be clearly indicated. In the present case it is at the base of the building.

(e) Again from Eq. 3.15, with $y = 25$ m,

$$25 = 40 + 12t - 4.9t^2$$

We find $t = -0.91$ s; 3.36 s. Since the ball was thrown at $t = 0$, we pick $t = 3.36$ s.

EXAMPLE 3.12: Two balls are thrown toward each other: ball A at 16.0 m/s upward from the ground, ball B at 9.00 m/s downward from a roof 30.0 m high, one second later. (a) Where and when do they meet? (b) What are their velocities on impact?

Solution: Recall that in this kind of problem we have to find *when* before *where*. We need to write the general expression for the position coordinates. The coordinate system is shown in Fig. 3.27.

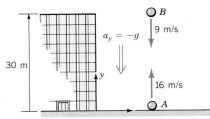

FIGURE 3.27 The motions of both balls are referred to the same coordinate system. They have the same (negative) acceleration.

(a) GIVEN: $y_{0A} = 0$; $v_{0A} = +16$ m/s; $y_{0B} = 30$ m; $v_{0B} = -9$ m/s; $a = -9.8$ m/s^2. If A has been in motion for time t, then B has been in motion for time $(t - 1)$. From Eq. 3.15,

$$y_A = 16t - 4.9t^2$$
$$y_B = 30 - 9(t - 1) - 4.9(t - 1)^2$$

They meet when $y_A = y_B$. This condition immediately leads to $t = 2.24$ s. Substituting into either y_A or y_B gives $y = 11.3$ m.
(b) Since $t = 2.24$ s, we have $v_A = 16.0 + (-9.8)(2.24) = -5.95$ m/s and $v_B = -9.00 + (-9.8)(1.24) = -21.2$ m/s. Notice that A is already moving downward when it collides with B.

EXERCISE 8. In Example 3.12, where is B when A is at its maximum height?

EXTRANEOUS ROOTS

Some of the above calculations illustrate the constant danger inherent in applying mathematics to physical situations. We often obtain more than one solution that is mathematically correct. We then have to choose the results that are relevant and reject those that are physically unacceptable. For example, if we find $v^2 = 64$ m^2/s^2, then mathematically $v = \pm 8$ m/s. If the question merely asks for the velocity at $y = 25$ m without further restriction, then both solutions are valid, and both should be stated. Otherwise, we have to choose the correct sign, based on the direction we know the velocity must have.

In part (e) of Example 3.11 we encountered a similar problem concerning time. We rejected the negative time because we knew that the problem started at $t = 0$. However, negative values are not completely meaningless. They are often solutions to another problem that is somehow related to the given problem. Suppose in Example 3.11 that the ball were thrown from the ground such that its velocity happened to be 12 m/s at $y = 40$ m. Part (e) tells us that $y = 25$ m at a time 0.91 s before the ball reaches roof level. There is another interpretation. If the ball were thrown downward at 12 m/s, then it would take 0.91 s to drop 15 m.

3.8   TERMINAL SPEED

Thus far in our discussion of falling bodies, the effects of air resistance have been ignored. An object dropped from a great enough height does not accelerate indefinitely. Figure 3.28a shows how the speed varies with time. The object ultimately reaches a **terminal speed**, v_T, and then continues to fall at this constant rate. If this were not so, hailstones would be lethal! The value of v_T depends on the weight and shape of the falling object and the density of the air. For snowflakes or feathers, the terminal speed is quite low.

Therefore, there is some truth in Aristotle's assertion that a body falls at a constant speed that depends on its weight. He missed the fact that it takes far longer to reach the terminal speed in air than it does in a liquid. With greater insight, Galileo realized that air resistance was merely a complication and not a fundamental aspect of motion under the influence of gravity.

Skydivers can control the amount of resistance they encounter by orienting their bodies in different ways. In a vertical posture, it takes about 15 s to reach the terminal speed of about 300 km/h (83 m/s). In a spread-eagle posture, shown in Fig. 3.28b, v_T is about 200 km/h (55 m/s). (During a practice drop in April 1987 in Phoenix, Arizona, skydiver Debbie Williams collided with a fellow skydiver. She lost consciousness and tumbled toward the ground at about 240 km/h. The instructor, Gregory Robertson, who was following his students, sped up to about 290 km/h and managed to release her chute a few seconds before landing!) At great altitudes, over 25 km, the lower density of the atmosphere results in much higher terminal speeds. There are remarkable stories of people who have fallen from great heights and survived the landing. They usually landed into something soft, such as a large haystack or a snow-covered slope. The record belongs to a Yugoslav airline hostess who fell from a DC9 aircraft that exploded at 10,610 m (33,330 ft)!

Consider a ball dropped from the height of the tower of Pisa, 54.6 m. For a sphere, v_T depends on the product of the density of the object and its radius (see Section 6.4). If there were no air resistance, the time to land would be 3.34 s and the ball would strike the ground at 32.7 m/s. These values would apply well to an iron ball (30 kg, radius 0.1 m, $v_T = 180$ m/s) and to a wooden ball (3.4 kg, radius 0.1 m, $v_T = 60$ m/s). However, a 0.3-kg ball of radius 0.1 m would barely reach its terminal speed of 18.4 m/s and its time to land would be about 4.3 s, a second longer. For a Ping-Pong ball (2.4 g, radius 1.9 cm) the terminal speed is only about 8.7 m/s. (See M.S. Greenwood et al., *Phys. Teacher*, 24(3): 153(1986).)

In the *Two New Sciences* (1638), Galileo claimed that if a lead ball and a wooden (ebony) ball were dropped from a height of about 100 m, the lead ball would land ahead of the wooden ball by only 4 in. In a modern experiment, lead and ebony balls with the same radius of 10 cm were dropped from 100 m. (See C. G. Adler and B. L. Coulter, *Am. J. Phys.* 46: 199 (1978).) It was found that the ebony ball was about 7 m above the ground when the lead ball landed. When balls of the same material, but of different radius, are dropped the difference is much less. For a 100-lb iron ball and a 1-lb iron ball, the difference would be about 1.3 m (not 2 in. as Galileo claimed).

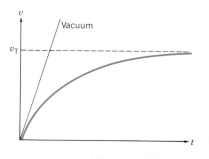

(a)

(b)

FIGURE 3.28 (a) In the presence of air resistence the speed of a falling object reaches a terminal speed v_T. (b) In the spread-eagle posture the terminal speed for a person is about 200 km/h.

SPECIAL TOPIC: Physiological Effects of Acceleration

In the early days of mechanized travel, people feared the effects of moving at high speed. In fact, physiologically, we are sensitive not to speed but to acceleration. When an elevator accelerates upward we feel heavier. Even a mild downward acceleration can make us queasy. Of course, many amusement park rides capitalize on the thrill of acceleration.

In the course of their jobs, jet pilots, astronauts, and parachutists are subjected to large accelerations. The firing of manned rockets and ejection seats and the opening of parachutes need to be carefully controlled to avoid injury. Hence, the physiological effects of acceleration and the associated limits of human tolerance are matters of active research. Accelerations are produced by rocket sleds (see Fig. 3.29a), catapults, centrifuges, and even by falling from a high tower into some material of known stiffness.

The response of the human body to acceleration de-pends on its magnitude, its duration, and its direction. The magnitude is measured in g's where $g = 9.8$ m/s^2. The direction is specified as either longitudinal (a_L) for the head-to-foot axis or transverse (a_T) for the front-to-back axis (see Fig. 3.29b). By convention, when you stand upright $a_L = 1g$. (Although you are stationary, the physiological effect is the same as if you were being accelerated at 9.8 m/s^2 in a rocket in outer space.) If you hang upside down, $a_L = -1g$.

As a first approximation we consider the body to be a semisolid object (bone and muscles) in which fluid (the blood) can flow. For short accelerations, under 0.2 s, the stress limits of the bones, vertebrae, and internal organs are important. For sustained accelerations, over 0.2 s, the movement of large volumes of blood causes either engorgement or depletion in various parts of the anatomy. Naturally, this is most serious for positive longitudinal acceleration. The drop in blood pressure in the head quickly

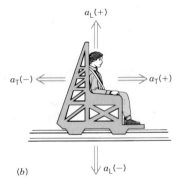

(b)

FIGURE 3.29 (a) Colonel John Stapp being subjected to a large acceleration in a rocket sled. (b) The acceleration of the human body is specified as either longitudinal (along the head-to-toe axis) or transverse (along the front-to-back axis).

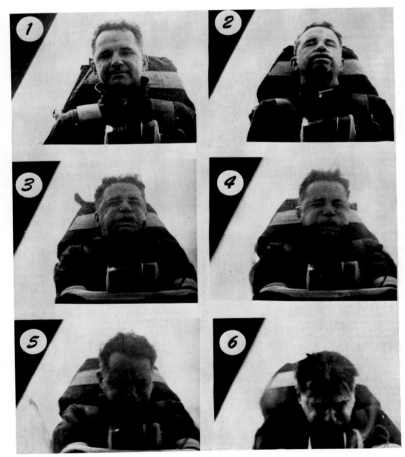

(a)

leads to loss in vision and unconsciousness.

Table 3.1 lists various sources of acceleration and their severity. Humans can tolerate 45g for 0.1 s without injury. The limit drops to 10g for a period of 1 s. An acceleration of 100g for 0.1 s leads to severe, perhaps fatal, injury. In an auto or plane crash, a seat belt and a tight shoulder harness give the passenger the same acceleration as the vehicle, perhaps about 100g for 0.03 s. If it were not restrained, the body would hurtle through the air and suffer a much larger acceleration (500g) on hitting an obstacle, such as a windshield.

TABLE 3.1

Source of Acceleration	a (g)	Duration (s)
Elevators	0.2	3
Car (panic stop)	1	3
Parachute landing	2–6	0.2–0.3
Catapult	5	0.1
Parachute opening	8–30	0.2–0.4
Ejection seat	15–20	0.2
Fall into firefighter's net	20	0.1
Auto or plane crash (possibly nonfatal)	20–100	0.02–0.1
Rocket sled	45	0.2–0.4
Free-fall landing (survived)	150	0.02
Auto or plane crash (fatal)	150–1000	0.01–0.001

Table 3.2 lists some responses to sustained acceleration. Combat pilots can experience accelerations of 6g in tight maneuvers. To raise the threshold for blackout, they are trained to tense their muscles and to grunt. This helps by about 0.5g. So-called "anti-g suits," which consist of bandages or inflatable cuffs to prevent the pooling of blood in the legs and abdomen, help by about 1.5g. The tolerance limit is about 7g for 15 s.

Although we have concentrated our attention on acceleration, safety engineers also examine the rate of change of acceleration, that is, da/dt. This quantity, called jerk, is thought to be even more important in its effects.

TABLE 3.2

Acceleration (g)	Effect on Human Body
	Positive longitudinal (a_L)
2.5	Difficult to raise oneself
3–4	Impossible to raise oneself, dimming of vision after 3 s
6	Blackout in 5 s, followed by unconsciousness without training or anti-g suit
	Negative longitudinal (a_L)
−1	Unpleasant facial congestion
−2 to −3	Severe facial congestion, throbbing headache, blurring of vision
−5	Rarely tolerated
	Positive transverse (a_T)
2–3	Increased abdominal pressure, difficulty in focusing
4–6	Difficulty in breathing, chest pain
6–12	Severe breathing difficulty and chest pain. Arms and legs immobile at 8g

[Ref: Bioastronautics Data Book, J. F. Parker Jr. and V. R. West, NASA SP-3006, 2nd edition (1973).]

SUMMARY

The **displacement** of a particle is its change in position,

$$\Delta x = x_\mathrm{f} - x_\mathrm{i}$$

The **average velocity** for a time interval Δt is

$$v_\mathrm{av} = \frac{\Delta x}{\Delta t}$$

It is the slope of the line joining the initial and final points on an x versus t graph. The **instantaneous velocity**

$$v = \frac{dx}{dt}$$

is the rate of change of position with respect to time. It is given by the slope of the tangent to the x versus t graph.

The **average acceleration** for a time interval Δt is

$$a_\mathrm{av} = \frac{\Delta v}{\Delta t}$$

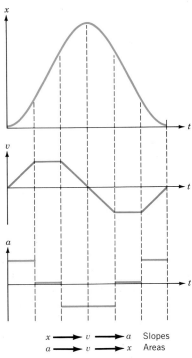

FIGURE 3.30 The relation between the x versus t, v versus t, and a versus t graphs for different constant accelerations.

It is the slope of the line joining the initial and final points on a v versus t graph. The **instantaneous acceleration**

$$a = \frac{dv}{dt}$$

is the rate of change of velocity with respect to time. It is given by the slope of the tangent to the v versus t graph.

In a given time interval, the area under the v versus t graph is the displacement. The area under the a versus t graph is the change in velocity in that time interval. The graphical relations between x, v, and a are summarized in Fig. 3.30. The **equations of kinematics for constant acceleration** are

$$v = v_0 + at$$
$$x = x_0 + \tfrac{1}{2}(v_0 + v)t$$
$$x = x_0 + v_0 t + \tfrac{1}{2}at^2$$
$$v^2 = v_0^2 + 2a(x - x_0)$$

where v_0, v, and a are the components of the vectors $\mathbf{v}_0$, $\mathbf{v}$, and $\mathbf{a}$. Their signs are determined by their directions relative to the chosen positive axis.

In the absence of air resistance, all bodies fall with the same acceleration due to gravity, regardless of their sizes or shapes. If the y axis is chosen to point upward, then $\mathbf{a} = -g\mathbf{j}$ where $g = 9.8$ m/s^2 is a positive scalar. In the presence of air resistance, the acceleration is not constant. It decreases as the body speeds up and may even drop to zero at the terminal speed.

ANSWERS TO IN-CHAPTER EXERCISES

1. (a) Yes, for a straight-line journey, no turnaround; (b) no.

2. (a) $x(1) = 30$ m and $x(2) = 10$ m. Thus, $v_{av} = \Delta x/\Delta t = -20$ m/s. (b) $v = dx/dt = -5 - 10t = -25$ m/s.

3. (a) $(-10 - 15)/8 = -3.13$ m/s^2; (b) $(-20 + 10)/10 = -1$ m/s^2.

4. The equation is correct. Substitute $v_0 = v - at$ into Eq. 3.11. For Example 3.7 use $x_0 = x - vt + \tfrac{1}{2}at^2 = 23$ m.

5. $x_S = 15t$ and $x_P = t^2$. Setting $x_S = x_P$ yields $t = 15$ s and $x = 225$ m.

6. With the same axes, $x_A = 16t + t^2$ and $x_B = 45 - 8t - 2t^2$. Setting $x_A = x_B$ yields 1.57 s and $x = 27.5$ m.

7. (a) From part (d), $v_{av} = \Delta y/\Delta t = (10.2 \text{ m})/(2 \text{ s}) = 5.1$ m/s. (b) The total distance traveled is (20 m $-$ 8.68 m) + (20 m $-$ 18.9 m) = 12.4 m. The average speed is (12.4 m)/(2 s) = 6.2 m/s.

8. To find the time at which ball A reaches its maximum height, use $v = v_0 + at$ to find $t = 1.63$ s. At this time, the position of B is $y = 30 - 9(0.63) - 4.9(0.63)^2 = 22.4$ m.

QUESTIONS

1. Describe a physical situation, for example, with a ball or a car, for each of the following cases. State whether the object is speeding up or slowing down. (a) $a = 0$, $v \neq 0$; (b) $v = 0$, $a \neq 0$; (c) $v < 0$, $a > 0$; (d) $v < 0$, $a < 0$.

2. Can a body have (a) zero instantaneous velocity and yet be accelerating; (b) zero average speed but nonzero average velocity; (c) negative acceleration and yet be speeding up?

3. (a) Can the magnitude of the average velocity equal the average speed? If so, under what condition? (b) Can the direction of the velocity reverse if the acceleration is constant?

4. A journey consists of two segments, each at constant velocity. Under what condition would the average velocity for the whole trip equal the average of the two velocities?

5. An object is dropped from a high tower. Make a sketch of how the speed varies with the distance fallen when air resistance is (a) ignored, or (b) taken into account. (c) Can an object speed up if its acceleration is decreasing?

6. Make a rough sketch of the v versus t graph for one foot of a person walking at constant velocity v_0. What is the average velocity of the foot?

7. True or false: (a) Positive slope on the x versus t graph implies motion away from the origin. (b) Negative slope on the v versus t graph means that the body is slowing down.

8. The Greek philosopher Zeno of Elea (495–435 B.C.) posed the following paradox. Achilles can run ten times as fast as a tortoise and so gives it a head start of 1 km. When Achilles covers the kilometer, the tortoise has moved ahead by $\frac{1}{10}$ km. When Achilles covers this $\frac{1}{10}$ km, the tortoise is ahead by $\frac{1}{100}$ km. This process is repeated ad infinitum: The tortoise is always some fraction of a kilometer ahead, which means that Achilles does not win the race. How would you respond?

9. An object thrown vertically up is instantaneously at rest at the highest point. What is its acceleration at this point?

10. Ball A is thrown up, whereas ball B is thrown down with the same initial speed, from the same rooftop. Compare the speeds at which they land—in the absence of air resistance.

11. If air resistance is taken into account for a body fired vertically up, would the time to rise be greater or less than the time to fall? (*Hint:* What can you deduce about the initial and final velocities?)

12. A ball is thrown up and then falls back to the ground. Which of the graphs in Fig. 3.31 best represents the variation of its velocity with time?

13. A body is fired upward with speed v_0. It takes time T to reach its maximum height H. True or false: (a) It reaches $H/2$ in $T/2$. (b) It has speed $v_0/2$ at $H/2$. (c) It has speed $v_0/2$ at $T/2$. (d) It has speed v_0 at $2T$.

14. A simple way to measure your reaction time is to have someone release a meter stick between your fingers. How does this work? (This method gives an optimistic estimate since you are tensed and ready for the event.)

15. Two balls are dropped one after the other from a tall tower. As a function of time does the distance between them increase, decrease, or stay constant?

16. The x versus t graph of Fig. 3.32 depicts the journeys of three bodies, A, B, and C. (a) At 1 s, which has the greatest

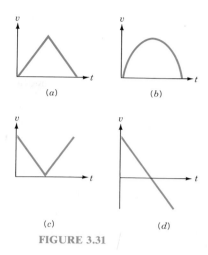

(a)   (b)

(c)   (d)

FIGURE 3.31

velocity? (b) At 2 s, which has traveled the farthest? (c) When A meets C, is B moving faster or slower than A? (d) Is there any time at which the velocity of A is equal to that of B? If so, what is it?

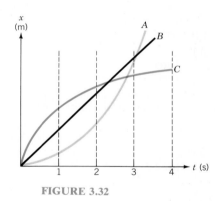

FIGURE 3.32

EXERCISES

3.2 and 3.3 Displacement and Velocity

1. (I) In September 1988 in Seoul, Carl Lewis set a new world record by running 100 m in 9.92 s. What was his average speed?

2. (I) It takes a sprinter 4 min to run a mile (1.6 km) and a long-distance runner 2.25 h to run a marathon (42 km). (a) Find the average speeds. (b) How long would the marathon take if it were run at the pace of the mile?

3. (I) A road journey takes 4 h 30 min at 80 km/h, including a half hour break for lunch. How much time would be saved by traveling at 100 km/h instead?

4. (I) A cyclist moves at 12 m/s for 1 min and at 16 m/s for 2 min. Find the average velocity if the second part of the

motion is (a) in the same direction as the first, and (b) in the opposite direction.

5. (I) In 1979, Bryan Allen pedalled the human-powered airplane *Gossamer Albatross* from Folkestone, England, to Cap Gris-Nez, France. He covered a straight-line distance of 38.5 km in 2 h 49 min. What was his average speed?

6. (I) From the x versus t graph of Fig. 3.33, find the average velocity between the following times: (a) 0 and 2 s, and (b) 1 and 3 s.

7. (I) From the x versus t graph of Fig. 3.34, find the average velocity for each of the following intervals: (a) 0 to 2 s; (b) 1 to 3 s, (c) 2 to 4 s, (d) 4 to 6 s.

8. (II) In her qualifying lap for the British Grand Prix in July 1986, Andrea de Cesaris drove around the 4.2-km track in 1

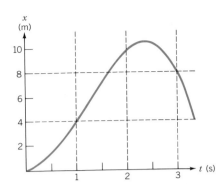

FIGURE 3.33 Exercises 6 and 11

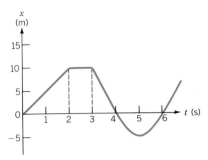

FIGURE 3.34 Exercises 7 and 12

min 13 s. The 75-lap race was won by Nigel Mansell at an average speed of 208 km/h. If she could have maintained her qualifying speed throughout the race, by what distance would she have won or lost?

9. (I) Consider a 500-km car race on a 10-km track. Car A finishes the race in 4 h and is 1.5 laps ahead of B at this time. What is B's time for the race?

10. (I) On its maiden voyage in July 1952, the liner *United States* won the coveted Blue Ribbon for the fastest crossing of the Atlantic from New York to Cornwall, U.K. The trip took 3 d 10 h 40 min at an average speed of 34.5 knots (65.5 km/h). This was 10 h 2 min less than the 14-year-old record held by the *Queen Mary*. What was the average speed of the *Queen Mary*?

11. (I) From the x versus t graph of Fig. 3.33 estimate the instantaneous velocity at the following times: (a) 1 s; (b) 2 s; (c) 3 s.

12. (I) From the x versus t graph of Fig. 3.34 estimate the instantaneous velocity at the following times: (a) 1 s; (b) 2.5 s; (c) 3.5 s; (d) 4.5 s; (e) 5 s;

3.4 Acceleration

13. (I) A bird flies north at 20 m/s for 15 s. It rests for 5 s and then flies south at 25 m/s for 10 s. For the whole trip, find: (a) the average speed; (b) the average velocity; (c) the average acceleration.

14. (I) Find the average acceleration in each of the following cases: (a) a DC10 jumbo jet starts from rest and reaches its takeoff speed of 360 km/h in 50 s; (b) a naval Corsair jet approaches an aircraft carrier at 180 km/h and is brought to rest in 4 s by a net; (c) a rocket sled hits 1440 km/h in 2 s.

15. (I) A baseball is thrown at 30 m/s. The batter hits it and gives it a velocity of 40 m/s in the opposite direction. If the ball and bat are in contact for 0.04 s, what is the average acceleration of the ball during this interval?

16. (I) At $t = 3$ s a particle is at $x = 7$ m and has velocity $v = 4$ m/s. At $t = 7$ s, it is at $x = -5$ m and has velocity $v = -2$ m/s. Find: (a) its average velocity; (b) its average acceleration.

17. (I) In the 1988 Summer Nationals of the NHRA, drag racer

Shirley Muldowney completed a $\frac{1}{4}$ mile (0.4 km) from a standing start in 5.40 s. At the end of her run, the speed was 256 mph (410 km/h). Find: (a) the average velocity; (b) the average acceleration.

18. (II) A driver in the Le Mans rally has to maintain an average speed of 75 km/h for a 300-km run. He travels at 100 km/h for the first 180 km and then rests for 12 min. What average speed does he need for the rest of the journey?

19. (I) The following data were recorded for a 1986 Toyota Camry. Find the acceleration for each run assuming that it is constant: (a) 0 to 50 km/h in 4.9 s; (b) 0 to 100 km/h in 14.8 s; (c) 70 km/h to 105 km/h in 9.2 s; (d) braked to a stop from 100 km/h in 4 s.

20. (II) The position of a particle is given by $x = 5 + 7t - 2t^2$ m. (a) Evaluate x at 1-s intervals between 0 and 5 s and then plot x versus t. (b) What is the average velocity between 1 and 2 s? (c) Find the instantaneous velocity at 3 s by drawing a tangent to the curve. (d) Use the limiting process of Eq. 3.4 to find the instantaneous velocity at 3 s.

21. (II) The position of a particle is given by $x = 5 \sin(120t)$ m, where t is measured in seconds. (The argument of the sine function is in degrees.) (a) Evaluate x at 0.1-s intervals from $t = 0.1$ to 0.8 s and make a plot of x versus t. (b) What is the average velocity between 0.5 and 0.6 s? (c) Find the instantaneous velocity at 0.5 s by drawing a tangent to the curve. (d) Use the limiting process of Eq. 3.4 to find the instantaneous velocity at 0.5 s.

22. (II) The position of a particle is given by $x = 10e^{-0.2t}$ m. Plot x versus t between 0 and 10 s. Find: (a) the average velocity between 2 and 3 s; (b) the instantaneous velocity at 2 s from the tangent to the curve; (c) the instantaneous velocity at 2 s by using the limiting process of Eq. 3.4.

23. (II) The position of a particle is given by $x = 4 - 5t + 3t^2$ m. (a) What is its instantaneous velocity and acceleration at $t = 3$ s? (b) At what time is the particle at rest?

24. (I) Use the v versus t graph of Fig. 3.35 to estimate (a) the average acceleration for the first 5.0 s, and (b) the instantaneous acceleration at 2.0 s. (c) Is there any time at which the particle reverses the direction of its motion? If so, what is it?

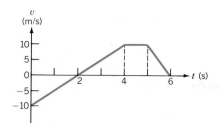

FIGURE 3.35 Exercises 24, 29 and 31

25. (I) From the v versus t graph of Fig. 3.36 find: (a) the time(s) at which the particle is at rest; (b) at what time, if any, the particle reverses the direction of its motion; (c) the average acceleration between 1 and 4 s; (d) the instantaneous acceleration at 3 s.

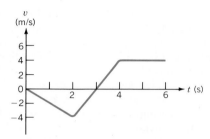

FIGURE 3.36 Exercises 25, 30 and 32

26. (I) In the x versus t graph of Fig. 3.37, is there any time, or time interval, for which the following hold? (a) $v = 0$, $a = 0$; (b) $v = 0$, $a \neq 0$; (c) $v \neq 0$, $a = 0$; (d) $v > 0$, $a > 0$; (e) $v > 0$, $a < 0$; (f) $v < 0$, $a < 0$; (g) $v < 0$, $a > 0$.

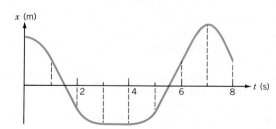

FIGURE 3.37 Exercise 26

3.5 Use of Areas

27. (I) From the v versus t graph of Fig. 3.38, estimate: (a) the displacement between 2 and 3 s; (b) the average velocity for the first 3 s.

28. (I) Use the v versus t graph of Fig. 3.39 to estimate the average velocity between 1 and 4 s.

29. (I) Use the v versus t graph of Fig. 3.35 to estimate: (a) the average velocity for the first 6 s; (b) the average speed for the first 6 s.

30. (I) From the v versus t graph of Fig. 3.36 estimate: (a) the average velocity for the first 5 s; (b) the average speed for the first 5 s.

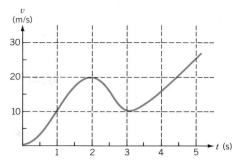

FIGURE 3.38 Exercise 27

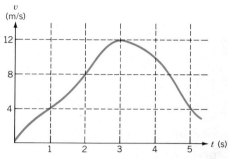

FIGURE 3.39 Exercise 28

31. (II) From the v versus t graph of Fig. 3.35, plot the following graphs: (a) a versus t; (b) x versus t. Assume $x = 0$ at $t = 0$. (c) What is the average acceleration for the first 6 s? (d) What is the instantaneous acceleration at 2 s?

32. (II) From the v versus t graph of Fig. 3.36, plot (a) the a versus t graph, and (b) the x versus t graph. Take $x = 0$ at $t = 0$. (c) What is the average acceleration from 1 s to 4 s? (d) What is the instantaneous acceleration at 3 s?

33. (II) The positions of a particle were recorded as follows:

x (m): 0.0 0.6 1.8 3.5 6.5 9.6 11.1 12.0 12.5 12.8 13.0
t (s): 0 1 2 3 4 5 6 7 8 9 10

Use these data to plot the x versus t, v versus t, and a versus t graphs.

34. (II) A bus starts from rest at the origin and accelerates at 2 m/s² for 3 s. It moves at constant velocity for 2 s and then has an acceleration of −3 m/s² for 2 s. Plot the v versus t, and x versus t graphs. Take $x = 0$ and $v = 0$ at $t = 0$.

35. (II) The following data for a Corvette were reported in the Nov. 1986 issue of *Car and Driver*.

v (mph): 30 40 50 60 70 80 90 100 110 120
t (s): 1.9 2.6 3.7 4.7 6.3 7.6 9.8 11.6 14.0 19.2

(a) Plot v versus t. (b) What is the average acceleration (in mph/s) between the first two readings? (c) If the value found in part (b) were maintained, how long would it take to reach 120 mph? (d) Estimate how long it takes to complete $\frac{1}{4}$ mi starting from rest and the speed at this time.

36. (II) Figure 3.40 shows the v versus t graphs for cars A and B. At $t = 0$ both are at $x = 0$. Estimate: (a) where and when they meet again; and (b) their velocities when they meet.

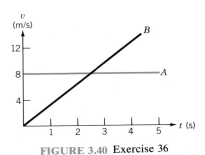

FIGURE 3.40 Exercise 36

37. (II) A v versus t graph from the Nov. 1986 issue of *Road and Track* for an Alfa Romeo is shown in Fig. 3.41. (a) Estimate the average acceleration in mph/s in each of the first three gears. (b) Suppose the acceleration were constant from rest until the shift from third to fourth gear. What would be the distance covered? (c) Estimate the actual distance covered up until the shift from third to fourth.

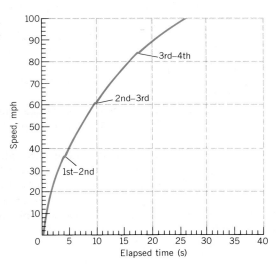

FIGURE 3.41 Exercise 37

3.6 Equations of Kinematics

In the following exercises assume that the acceleration is constant.

38. (I) A bullet emerges from the 60-cm barrel of a Winchester rifle at 900 m/s. Find: (a) its acceleration; (b) the time of travel through the barrel.

39. (I) A particle is 5 m east of the origin and moving west at 2 m/s. Five seconds later it is 11 m east of the origin. What was its acceleration?

40. (I) A Ford Taurus is initially moving at 112 km/h. Find its acceleration and the time taken to stop given that: (a) it

brakes to a stop in 64 m; (b) it crashes head-on into a barrier and crumples by 1 m.

41. (I) A Ferrari Testarossa can accelerate at from zero to 60 mph (96 km/h) in 5.0 s. Compare the times taken and the distances covered when its speed changes as follows: (a) 10 to 20 m/s; (b) 20 to 30 m/s. Is it reasonable to assume that the acceleration is constant?

42. (I) If an object could accelerate continuously at 10 m/s², how far would it travel and how long would it take to reach: (a) the speed of sound, 330 m/s; (b) the escape speed of a rocket from the earth, 11.2 km/s; (c) 3×10^7 m/s, which is 10% of the speed of light? Assume it starts from rest.

43. (I) The length of a train is 44 m. Its front is 100 m from a pole. It accelerates from rest at 0.5 m/s². (a) How long does it take to go past the pole? (b) At what speeds do its front and rear pass the pole?

44. (II) A car moving at 60 km/h meets a train of length 1 km moving at 40 km/h along a track parallel to the road. What distance does the car travel in passing the length of the train given that they travel (a) in the same direction, or (b) in opposite directions?

45. (II) The driver of a truck moving at 30 m/s suddenly notices a moose 70 m straight ahead. If the driver's reaction time is 0.5 s and the maximum deceleration is 8 m/s², can he avoid hitting the moose without steering to one side?

46. (I) A bus slows with constant acceleration from 24 m/s to 16 m/s and moves 50 m in the process. (a) How much further does it travel before coming to a stop? (b) How long does it take to stop from 24 m/s?

47. (II) A cyclist is initially moving at 12 m/s. He covers 32 m in the next 4 s. Find: (a) his acceleration; (b) his speed after 4 s.

48. (II) A car has an initial velocity of 20 m/s and an acceleration of -5 m/s². Find its average velocity in the time interval during which its displacement is 30 m from the initial position.

49. (II) At $t = 3$ s, the position of a particle is $x = 2$ m and its velocity is $v = 4$ m/s. At $t = 7$ s, $v = -12$ m/s. Find: (a) the position and the velocity at $t = 0$; (b) the average speed from 3 s to 7 s, and (c) the average velocity from 3 s to 7 s.

3.7 Vertical Free-fall

50. (I) Water emerges vertically from a hose at ground level and rises to a height of 3.2 m. (a) At what speed does it emerge from the nozzle? (b) How long is a given drop in the air?

51. (I) A stone thrown vertically up from the ground rises to a height of 25 m. How high would it rise on the moon if given the same initial speed? The acceleration due to gravity on the moon is one-sixth that on earth.

52. (I) A stone released at the top of the shaft of a well hits the water 1.5 s later. (a) How deep is the well? (b) At what speed does the stone hit the water?

53. (I) The maximum height from which a person can safely jump is 2.45 m. What is the maximum allowable landing speed for a parachutist?

54. (I) (a) In an exceptional vertical jump, a person can raise the torso 50 cm above its normal height. (a) What initial velocity does this entail? (b) A dolphin can rise 6.0 m above the water. What is its initial vertical velocity?

55. (I) A parachutist lands at 7 m/s. What is her acceleration after making contact with the ground given that she lands (a) by flexing her knees and rolling to a stop in 0.6 m, or (b) stiffly, in 0.1 m.

56. (I) An arrow fired vertically up lands 8 s later. Find: (a) its maximum height, and (b) its initial speed.

57. (I) A toy rocket is rising at a constant speed of 20 m/s. When it is 24 m above the ground, a bolt comes loose. (a) How long does the bolt take to land? (b) What is its maximum height? (c) At what speed does it hit the ground?

58. (I) A baseball is thrown up at 30 m/s from the ground. Find: (a) its velocity at a height of 25 m; (b) when its speed is 15 m/s; (c) when its height is 40 m.

59. (I) A stone is thrown vertically up at 20 m/s from ground level. Find the times at which (a) it is at half its maximum height; (b) it has half its maximum speed.

60. (I) A ball thrown up from the ground reaches 30 m in 2 s. Find the next time it is at the same height.

61. (II) A juggler has three oranges that rise 1.8 m above his hands. He takes 0.3 s to transfer an orange from one hand to the other. When one orange is at its maximum height, where are the other two? Assume they are equally spaced in time.

62. (II) A tennis ball is dropped from a height of 5 m and rebounds to a height of 3.2 m. If it is in contact with the floor for 0.036 s, what is its average acceleration during this period?

63. (I) An object is thrown up from the top of a building 50 m high. It rises to a maximum height of 20 m above the roof. (a) When does it land on the ground? (b) At what velocity does it land? (c) When is it 20 m below the roof?

64. (I) A ball thrown up from a rooftop of height 40 m lands on the ground in 4 s. (a) What is its maximum height above the ground? (b) What is its velocity 15 m below the rooftop?

65. (I) From the data sent back by the *Voyager* spacecraft in 1979, engineer Linda Morabito discovered the first nonter-

restrial volcanic activity on Io, a moon of Jupiter. The plume rose about 280 km above the surface. Although the acceleration due to gravity varies with height, take it to be constant at 1.5 m/s^2. Estimate the time taken to reach the maximum height. See Fig. 3.42.

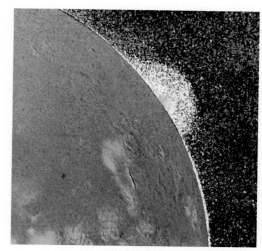

FIGURE 3.42 Exercise 65

66. (I) A ball thrown down from a balcony lands in 0.8 s at a speed of 13 m/s. Find: (a) the initial velocity; (b) the height from which it was thrown; (c) the time to land if it were thrown up from the balcony with the same initial speed.

67. (II) Find the maximum height and the time of flight of a body fired vertically up from the ground given that it loses 60% of its initial speed in rising 4.2 m.

68. (II) An object is fired vertically up from the ground. Find its maximum height and time of flight given that it rises to 50% of its maximum height in 2 s.

69. (II) An object is fired vertically up from the ground. Find its maximum height and the time of flight given that it loses 30% of its initial speed in 1.8 s and is moving upward.

70. (II) When a body is fired vertically up from the ground it reaches 75% of its maximum height with a speed of 30 m/s. Find its maximum height and the time of flight.

PROBLEMS

1. (I) The fastest speed ever recorded for a human was for Robert Hayes who ran at 12.5 m/s for part of a race. A Porsche 911 Turbo can reach 60 mph (96 km/h) in 4.6 s. Suppose Hayes can maintain his top speed and that the car starts just as he passes it. (a) Where and when do they meet? (b) What are their velocities at this point? (c) Sketch the *x* versus *t* graph.

2. (I) A truck starts from rest and accelerates at 1 m/s^2. Ten seconds later, a car accelerates from rest at the same point with an acceleration of 2 m/s^2. (a) Where and when does the car catch the truck? (b) What are their velocities when they meet?

3. (II) A car and a truck are initially moving in the same direction at 20 m/s with the truck 38 m ahead. The car accelerates at a constant 2 m/s^2, passes the truck, and returns to the right lane when it is 11 m ahead of the truck. How far has the truck moved during this time?

4. (II) Two ships are 10 km apart and moving toward each other at 6 km/h and 4 km/h, respectively. A bird flies continuously back and forth between them at 20 km/h. What is the total distance traveled by the bird when the ships meet?

5. (I) Train A has a length of 1 km and travels at 50 m/s. Train B has a length of 0.5 km and starts just as the rear of train A passes the front of train B. Train B has an acceleration of 3

m/s^2 and a maximum speed of 60 m/s. (a) When does B pass A; that is, when does the rear of B pass the front of A? (b) How far has A traveled in this time?

6. (I) The driver of a car moving at 30 m/s suddenly sees a truck that is moving in the same direction at 10 m/s and is 60 m ahead. The maximum deceleration of the car is 5 m/s^2. (a) Will a collision occur if the driver's reaction time is zero? If so, when? (b) If the car driver's reaction time of 0.5 s is included, what is the minimum deceleration required to avoid a collision?

7. (I) Ball A is thrown vertically up at 5 m/s from a rooftop of height 100 m. Ball B is thrown down from the same point 2 s later at 20 m/s. (a) Where and when do they meet? (b) What are their velocities when they meet?

8. (I) A bus undergoing constant deceleration covers successive distances of 300 m in 10 and 15 s, respectively. (a) How much farther does it travel before stopping? (b) How much longer does it take to stop?

9. (I) A car moving with constant acceleration passes two poles 100 m apart with speeds of 15 and 25 m/s. (a) What is its speed at the next pole, 100 m down the road? (b) How long does it take to travel from the second to the third pole?

10. (II) A late riser runs at a constant 4.5 m/s to catch a bus. She reaches the bus stop 2 s after the bus starts from rest with an acceleration of 1 m/s^2. (a) Where and when does the person catch the bus? (b) How much longer could she have slept in and still made it to work on time? (*Hint:* Sketch the x versus t graph.)

11. (I) Cyclist A is moving at 20 m/s whereas cyclist B is moving at 12 m/s in the same direction and is initially ahead of A. When they are abreast, they both start to accelerate. Twelve seconds later, B overtakes A when B's speed is 36 m/s. What is A's speed at this point?

12. (II) A bus with a maximum speed of 20 m/s takes 21 s to travel 270 m from stop to stop. Its acceleration is twice as great as its deceleration. Find: (a) the acceleration; (b) the distance traveled at maximum speed.(*Hint:* Draw the v versus t graph.)

13. (II) A car starts from rest and accelerates uniformly for 200 m. It moves at constant speed for 160 m and then decelerates to rest in 50 m. The whole trip takes 33 s. How long did it move at constant speed? (*Hint:* Draw the v versus t graph.)

14. (II) Car A moves along a straight road at 16 m/s while car B moves in the opposite direction at 8 m/s. When they are 48 m apart, both drivers apply their brakes. Car A slows at 2.4 m/s^2 while car B slows at 4 m/s^2. Is there a collision? If so, where and when?

15. (II) A cheetah can hit 105 km/h in 2 s and maintain this speed for 15 s. After this time it must rest. An antelope can reach 90 km/h in 2 s and sustain this for a long time. Suppose they are initially separated by 100 m and the antelope reacts in 0.5 s. (a) Can the cheetah catch the antelope? (b) If not, how close does it get? Assume both start from rest.

16. (I) A ball is thrown up from a rooftop with an initial speed of 15 m/s. Two seconds later, another ball is dropped from the same point. (a) Assuming that neither has landed, where and when do they meet? (b) What are their speeds when they meet?

17. (I) Ball A is thrown up from the ground at 25 m/s and ball B is thrown down at 15 m/s 1 s later from a roof of height 95 m. (a) Where and when do they meet? (b) What are their velocities when they meet?

18. (I) A rocket rises vertically from the ground with an acceleration of 4 m/s^2. Its fuel burns out in 8 s. (a) What is its maximum altitude? (b) What is its time of flight?

19. (I) An elevator of height 3 m moves upward at 2 m/s. A ball is dropped from the roof. (a) When does it hit the floor? (b) What total distance did it travel relative to the ground?

20. (I) An open elevator moves upward at 7 m/s. When its floor is 25 m above ground, a ball is thrown up at 20 m/s relative to the ground. (a) What is the maximum height of the ball above the ground? (b) How long is it in free-fall?

21. (I) A flower pot falls off a balcony. It takes 0.1 s to pass a window of height 1.25 m. From what height above the bottom of the window did it fall?

22. (II) A body covers 64% of the total distance fallen in the last second. From what height did it fall?

23. (II) A climber can estimate the height of a cliff by dropping a stone and noting the time at which he hears the impact on the ground. Suppose this time is 2.5 s. Find the height of the cliff under the following conditions: (a) by assuming the speed of sound is large enough to be ignored; (b) by taking the speed of sound to be 330 m/s.

CHAPTER 4

Inertia and Two-Dimensional Motion

Multi-colored trajectories in a fireworks display.

Major Points

1. (a) The concept of **inertia.**
   (b) **Newton's first law.**
2. In **projectile motion** the horizontal and vertical motions are independent.
3. In **uniform circular motion** there is a (radially inward) **centripetal** acceleration.
4. Newton's first law is valid only in an **inertial reference frame.**
5. **Relative velocity:** The velocity of one body relative to another.
6. (a) The **Galilean transformation** of coordinates.
   (b) The **Galilean principle of relativity:** The laws of mechanics are the same in all inertial frames.

There appears to be a natural tendency of moving bodies to come to rest if left alone. For example, a rolling ball eventually slows down and comes to a stop. Objects seem to require external effort just to maintain their motion: A book moves along a table only as long as it is pushed. It seems that the "natural" state of bodies is to be at rest. But wait a moment. When the brakes are applied on an icy road, a car continues to keep going. So what *is* the natural tendency of bodies? Do they slow down unless pushed, or do they continue to move once there is motion? It is no exaggeration to say that the birth of physics depended on this crucial choice. Casual observation of everyday phenomena tends to support the incorrect views of Aristotle, who believed that the natural state of bodies was to be at rest. It was not until 2000 years later that Galileo disentangled himself from the web of ancient misconceptions.

According to Galileo's "principle of inertia" a body on a frictionless horizontal surface would continue to move indefinitely at constant speed. The term **inertia** is used to describe the tendency of a body to resist any change in its velocity. Galileo's clarification of the concept of inertia paved the way for an understanding of the kinematics of projectile motion near the surface of the earth and of circular motion. (See the historical note on p. 69.)

4.1 NEWTON'S FIRST LAW

Based on the work of Galileo and the French philosopher René Descartes, Isaac Newton published his first law of motion in 1687. According to **Newton's first law:**

Newton's first law

> Every body continues in its state of rest or of uniform motion in a straight line unless it is compelled to change that state by forces impressed upon it.

This law involves a property of bodies called **inertia:**

Inertia

> The inertia of a body is its tendency to resist any change in its state of motion.

In other words, objects at rest tend to stay at rest, and if moving, they tend to keep moving at constant velocity. They display the same resistance to slowing down as to speeding up. Later we will connect the concept of inertia with the concept of mass, which is a measure of the inertia of a body.

The first law implies that a *change* in velocity, and therefore an acceleration, is produced by "forces." It does so without defining what force is. At this stage we have to use our intuitive understanding of force as either a push or a pull. The first law does not imply anything about the functional relationship between force and acceleration. Nonetheless, it does tell us when *no net force* acts on an object. It does not distinguish between cases for which there are no external forces at all, and those for which the forces balance to produce a zero resultant. For example, in Fig. 4.1 a block is being pulled along a rough floor. It experiences two horizontal forces—the tension in the rope and the frictional force due to the floor. If the magnitudes of these two forces are equal, the block will move at constant velocity.

Consider a stone attached to a rope being swung in a circle at constant speed, as shown in Fig. 4.2*a*. At any point in its circular path, the instantaneous velocity of the stone is directed along the tangent. Because of its inertia, the stone tends to continue moving in this direction. However, the inward pull of the rope draws it away from its natural inertial path. When the rope is released, the stone has no force acting on it. Thus, it will obey the first law and continue to travel along the tangent line at constant velocity. (Try it.) Figure 4.2*b* shows sparks flying off tangentially from a grinding wheel.

Let us leave the last word to Sir Arthur Eddington* whose version of the first law was "Every body continues in its state of rest or of uniform motion in a

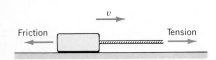

Friction ← Tension

FIGURE 4.1 If the two horizontal forces acting on the block are balanced, which means that there is no net force, the block will move at constant velocity.

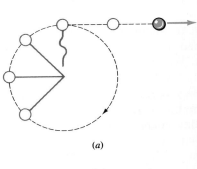

(a)

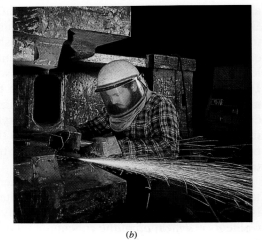

(b)

FIGURE 4.2 (*a*) A particle moves in a circular path at the end of a rope. When the rope is released, the particle moves along the tangent to the circle. (*b*) Sparks from a grinding wheel fly off tangentially.

* *The Nature of the Physical World,* Ann Arbor Books, University of Michigan Press, Ann Arbor, 1967.

straight line; except insofar as it doesn't!'' What he meant by this is that one cannot "prove" the first law since one can never guarantee that *no force whatso-ever,* such as some unknown influence from outer space, acts on the particle. To dismiss the first law because of this is to deny the very core of this subject. In physics we have to idealize. Our given world is not ideal, and so we have to transcend this handicap to formulate laws that are precise—even beyond our ability to test them experimentally. Ultimately, Newton's first law is a statement of our belief in how nature operates. If we find some evidence that the law is slightly faulty, rather than abandon it, we prefer to search for the mysterious agent that is causing the deviations.

4.2 TWO-DIMENSIONAL MOTION

The quantities velocity and acceleration were introduced in Chapter 3 for one-dimensional motion. We must generalize them to two and three dimensions and emphasize their vector nature. In three dimensions the **position** vector $\mathbf{r}$ of a particle whose coordinates are (x, y, z) is

$$\mathbf{r} = x\mathbf{i} + y\mathbf{j} + z\mathbf{k} \tag{4.1}$$

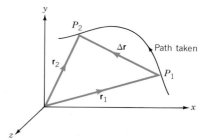

FIGURE 4.3 When a particle moves along the curved path from P_1 to P_2, its displacement is $\Delta\mathbf{r} = \mathbf{r}_2 - \mathbf{r}_1$.

The components of $\mathbf{r}$ are the Cartesian coordinates. If the particle moves from P_1 at position $\mathbf{r}_1$ to P_2 at position $\mathbf{r}_2$, as shown in Fig. 4.3, its **displacement,** that is, the change in position, is

$$\Delta\mathbf{r} = \mathbf{r}_2 - \mathbf{r}_1 = \Delta x\mathbf{i} + \Delta y\mathbf{j} + \Delta z\mathbf{k} \tag{4.2}$$

The following should help you to draw the direction of $\Delta\mathbf{r}$ correctly: $\Delta\mathbf{r}$ is the vector that must be added to the initial position $\mathbf{r}_1$ to give the final position $\mathbf{r}_2$, that is, $\mathbf{r}_2 = \mathbf{r}_1 + \Delta\mathbf{r}$. As in Chapter 3, the **average velocity** is defined as the ratio of the displacement over the time interval:

$$\mathbf{v}_{av} = \frac{\Delta\mathbf{r}}{\Delta t} \tag{4.3}$$

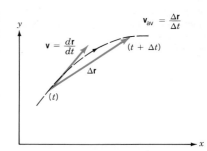

FIGURE 4.4 The dashed line represents the path of a particle in the xy plane. The instantaneous velocity $\mathbf{v}$ is directed along the tangent to the path, but its magnitude is *not* the slope of that line.

The average velocity $\mathbf{v}_{av}$ is in the direction of $\Delta\mathbf{r}$, which is a chord across the path of the particle shown in Fig. 4.4. As Δt gets smaller, the chord approaches the tangent line and $\Delta\mathbf{r}$ becomes parallel to the path. The **instantaneous velocity** is

$$\mathbf{v} = \lim_{\Delta t \to 0} \frac{\Delta\mathbf{r}}{\Delta t}$$

or

$$\mathbf{v} = \frac{d\mathbf{r}}{dt} = v_x\mathbf{i} + v_y\mathbf{j} + v_z\mathbf{k} \tag{4.4}$$

where $v_x = dx/dt$, $v_y = dy/dt$, and $v_z = dz/dt$. The direction of $\mathbf{v}$ is along the tangent to the path. Notice, however, that the diagram is not a position–time graph. Consequently, the magnitude of $\mathbf{v}$, the instantaneous speed, is not given by the slope of the tangent.

The **instantaneous acceleration** is the rate of change of the velocity with respect to time:

$$\mathbf{a} = \frac{d\mathbf{v}}{dt} = a_x\mathbf{i} + a_y\mathbf{j} + a_z\mathbf{k} \tag{4.5}$$

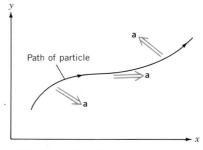

FIGURE 4.5 Possible directions for the acceleration of a particle that travels along a curved path with varying speed.

where $a_x = dv_x/dt$, $a_y = dv_y/dt$, and $a_z = dv_z/dt$. In general one cannot determine $\mathbf{a}$ directly from the path of the particle. We need to know how each component of the velocity varies as a function of space and time. Figure 4.5 illustrates how the acceleration might vary (the speed is not constant).

Constant Acceleration

When a body moves with constant acceleration in two or three dimensions, Eqs. 3.9–3.12 can be written in vector notation:

$$\mathbf{v} = \mathbf{v}_0 + \mathbf{a}t \tag{4.6}$$

$$\mathbf{r} = \mathbf{r}_0 + \tfrac{1}{2}(\mathbf{v}_0 + \mathbf{v})t \tag{4.7}$$

$$\mathbf{r} = \mathbf{r}_0 + \mathbf{v}_0 t + \tfrac{1}{2}\mathbf{a}t^2 \tag{4.8}$$

For two-dimensional motion in the xy plane, the x and y components of these equations are

$$v_x = v_{0x} + a_x t \qquad\qquad v_y = v_{0y} + a_y t$$
$$x = x_0 + \tfrac{1}{2}(v_{0x} + v_x)t \qquad y = y_0 + \tfrac{1}{2}(v_{0y} + v_y)t$$
$$x = x_0 + v_{0x}t + \tfrac{1}{2}a_x t^2 \qquad y = y_0 + v_{0y}t + \tfrac{1}{2}a_y t^2$$
$$v_x^2 = v_{0x}^2 + 2a_x(x - x_0) \qquad v_y^2 = v_{0y}^2 + 2a_y(y - y_0)$$

The fourth equation is included because it is often useful. We next deal with two-dimensional projectile motion near the surface of the earth.

4.3   PROJECTILE MOTION

As late as the sixteenth century, the path of a projectile was drawn as in Fig. 4.6. It was believed that when a cannonball was fired, it was given an "impressed force" that produced "violent" motion in a straight line. Because of air resistance, there followed a region of mixed motion ("violent" plus "natural" motion vertically down). Finally, the "natural" motion vertically down prevailed. This description was partly the result of the inability of medieval scholars to combine two forces that were not parallel: The "force of the cannon" along the line of firing, and the pull of gravity downward. In any case, the problem itself was based on the false premise that the cannon has some role in determining the motion of the ball *after* it has been fired. In fact, if we ignore air resistance, once the ball has been fired it is subject *only* to the force of gravity. Initially, Galileo also believed that the motion of a projectile was governed by an "impressed force" that gradually diminishes. (Indeed, it is still a common belief that the force used to throw a ball up somehow stays with it. The "force of the hand" is supposed to be gradually overcome by the force of gravity, which ultimately causes the ball to fall. This misconception has a long history!) It was only after Galileo had developed his principle of inertia that he could tackle the problem of projectile motion properly.

Galileo suggested the following experiment to illustrate the essential feature of projectile motion. Suppose that a ball is dropped from the top of the mast of a ship moving at constant velocity, as shown in Fig. 4.7a. Where would it land? Galileo pointed out that at the moment the ball is released, it has the same horizontal motion as the ship. Ignoring the small effect of air resistance, it will *maintain* this horizontal component of velocity, even as it accelerates in the vertical direction. As a result, the ball would land at the base of the mast. Figure 4.7b shows the same phenomenon for bombs being released from an aircraft.

Galileo had arrived at the crucial insight that a projectile near the surface of the earth has *two independent motions:* A horizontal motion at constant speed and a vertical motion subject to the acceleration due to gravity.

In order to deal with any problem in projectile motion, one has to choose a coordinate system and also to clearly specify the origin. If the x axis is horizontal and the y axis points vertically upward, then

$$a_x = 0; \quad a_y = -g$$

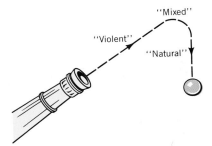

FIGURE 4.6 Prior to the sixteenth century the path of a projectile was assumed to consist of an initial "violent" motion in a straight line followed by a region of "mixed motion" and finally "natural motion" vertically down.

The acceleration of a projectile

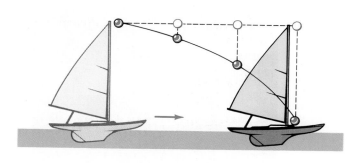

(a)

(b)

FIGURE 4.7 (a) When a ball is released from the top of the mast of a boat moving at constant velocity, it will land at the base (ignoring air resistance). (b) Bombs released by an aircraft maintain the horizontal component of their velocity.

One can usually pick the origin such that the initial horizontal (x) coordinate is zero, that is, $x_0 = 0$. The equations of kinematics for projectiles near the earth's surface take the form:

$$x = v_{0x}t \tag{4.9}$$
$$v_y = v_{0y} - gt \tag{4.10}$$
$$y = y_0 + v_{0y}t - \tfrac{1}{2}gt^2 \tag{4.11}$$
$$v_y^2 = v_{0y}^2 - 2g(y - y_0) \tag{4.12}$$

Equations of kinematics for projectile motion

It is essential that you maintain the x and y subscripts in these equations. Notice in particular that it is the vertical *component, v_{0y},* of the initial velocity, rather than v_0 itself, that appears in the last three equations.

In Chapter 3 it was pointed out that because of air resistance, a falling body may reach a terminal speed and stop accelerating. Therefore, the predictions of the above equations are valid only when the speed of a projectile is much less than its terminal speed. They do not really apply even to such commonplace projectiles as baseballs or golfballs, let alone arrows, bullets, or ballistic missiles. Spinning Frisbees even seem to defy gravity temporarily. Computations based on these equations are accurate only when the effects of the air can be neglected and the acceleration due to gravity is constant both in magnitude and in direction. They may be applied to slower projectiles, such as a shotput. Nonetheless, in other cases they do provide a good first approximation to a complete, and usually far more complex, solution.

EXAMPLE 4.1: A ball is projected horizontally at 15 m/s from a cliff of height 20 m. Find: (a) its time of flight; (b) its horizontal range R. (The horizontal displacement from the point of firing.)

Solution: In Fig. 4.8a we have chosen the origin to be at the base of the cliff. Note that the path (solid line) is a plot of y versus x, not y versus t. The figure shows that in equal time intervals the horizontal displacements are equal, that is, $v_x = v_{0x} = $ constant. At any instant, the y coordinate and the vertical component of the velocity are the same as that of a ball merely dropped at the same time (see Fig. 4.8b).

GIVEN: $x_0 = 0$; $y_0 = 20$ m; $v_{0x} = 15$ m/s; and $v_{0y} = 0$. The coordinates at a later time are given by Eqs. 4.9 and 4.11:

$$x = 15t \tag{i}$$
$$y = 20 - 4.9t^2 \tag{ii}$$

(a) When the ball lands, its vertical coordinate is zero, that is, $y = 0$. From (ii) the time of flight is given by

$$0 = 20 - 4.9t^2$$

Thus, $t = -2.02$ s or 2.02 s. We reject the negative solution

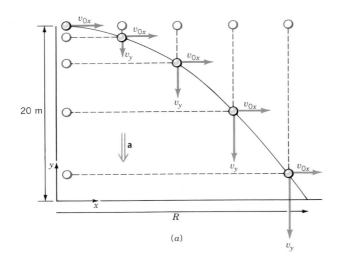

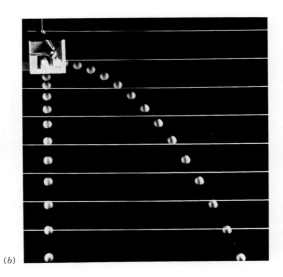

FIGURE 4.8 (a) The horizontal motion of a projectile is at constant velocity while the vertical motion occurs at constant acceleration (provided air resistance is negligible). (b) The vertical component of the motion of a ball projected horizontally is the same as that of a ball that is simply dropped.

since we assume that the ball is thrown at $t = 0$. The time of flight does not depend on the value of the horizontal component of the initial velocity. A ball merely dropped from the same height would also land in 2.02 s.

(b) To find the horizontal range, we use the time of flight in (i). Thus,

$$R = v_{0x}t = 30.3 \text{ m}$$

EXAMPLE 4.2: A projectile is fired from the ground with an initial velocity v_0 at an angle θ above the horizontal. Find: (a)

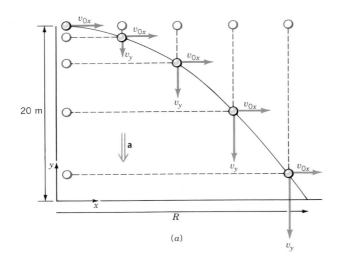

$v_{0x} = v_0 \cos \theta$
$v_{0y} = v_0 \sin \theta$

(a)

FIGURE 4.9 In the absence of air resistance the path of a projectile is parabolic. The path is symmetrical about the highest point only if the particle lands at the level from which it was fired.

the time of flight; (b) the horizontal range R; (c) the shape of the path.

Solution: The coordinate system and the components of the velocity at various points are illustrated in Fig. 4.9. From Eqs. 4.9 and 4.11 the coordinates at time t are

$$x = (v_0 \cos \theta)t \qquad \text{(i)}$$
$$y = (v_0 \sin \theta)t - \tfrac{1}{2}gt^2 \qquad \text{(ii)}$$

(a) To find the time of flight, we note that $y = 0$ at this time. From (ii) we find $t = 0$ (which is the first time that $y = 0$) and

$$t = \frac{2v_0 \sin \theta}{g} \qquad \text{(iii)}$$

which is the next time that $y = 0$.
(b) To find the horizontal range we substitute (iii) into (i): $R = (v_0 \cos \theta)(2v_0 \sin \theta)/g$. By using $\sin 2\theta = 2\sin \theta \cos \theta$, we obtain

$$R = \frac{v_0^2 \sin 2\theta}{g} \qquad \text{(iv)}$$

Note that Eq. (iv) is valid *only* when the projectile returns to the initial vertical level, that is, $\Delta y = 0$. For a given initial speed v_0, the range is a maximum when $\sin 2\theta = 1$, that is, when $\theta = 45°$. In general, for given values of R and v_0, there are two possible values for θ. For example, if $v_0 = 20$ m/s and $R = 30$ m, then $\sin 2\theta = Rg/v_0^2 = 0.735$. Thus $\theta = 23.7°$ or $66.3°$. Notice that $\theta = 45° \pm \alpha$ where $\alpha = 21.3°$ (see Fig. 4.10).
(c) To find the shape of the path, we need to express y in terms

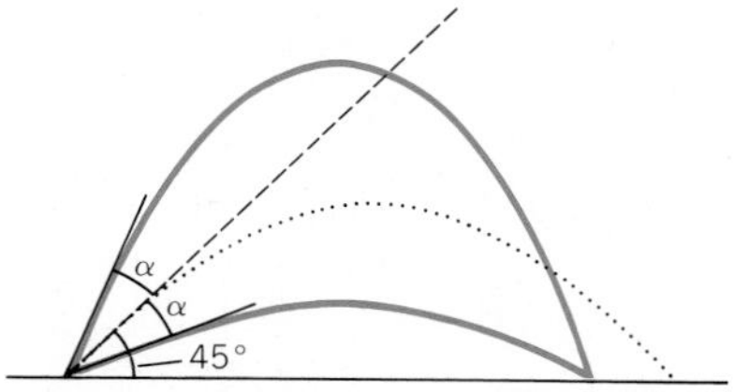

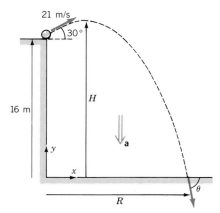

FIGURE 4.10 The maximum range for a projectile that lands at the initial level occurs for an angle of projection of 45°. Galileo proved that the horizontal ranges for angles of projection 45° − α and 45° + α are equal.

of x. To do this we take $t = x/(v_0 \cos \theta)$ from (i) and substitute it into (ii). This leads to

$$y = (\tan \theta)x - \frac{g}{2(v_0 \cos \theta)^2} x^2 \qquad \text{(v)}$$

This is of the form $y = Ax + Bx^2$, which is the equation of a parabola. Galileo was the first to prove that in the absence of air resistance the path of a projectile is parabolic. He was also the first to show that for initial angles 45° ± α, the ranges are equal.

EXAMPLE 4.3: A ball is thrown at 21 m/s at 30° above the horizontal from the top of a roof 16 m high (see Fig. 4.11). Find: (a) the time of flight; (b) the horizontal range; (c) the maximum height; (d) the angle at which the ball hits the ground; (e) the velocity when it is 2 m above the roof.

Solution: The origin and coordinate system are shown in Fig. 4.11. GIVEN: $x_0 = 0$; $y_0 = 16$ m; $v_0 = 21$ m/s and $\theta = 30°$. Thus, $v_{0x} = v_0 \cos \theta = 18.2$ m/s and $v_{0y} = v_0 \sin \theta = 10.5$ m/s. From Eqs. 4.9 and 4.11 we have

$$x = 18.2t \qquad \text{(i)}$$
$$y = 16 + 10.5t - 4.9t^2 \qquad \text{(ii)}$$

FIGURE 4.11 When the landing point is above or below the initial point the path is not symmetrical about the highest point. The angle of impact is given by $\tan \theta = v_y/v_x$.

(a) The flight ends when $y = 0$. So from (ii),

$$t = \frac{10.5 \pm \sqrt{10.5^2 + 64 \times 4.9}}{9.8} = 3.17 \text{ s}, \ -1.03 \text{ s}.$$

Therefore, the time of flight is 3.17 s.
(b) Using $t = 3.17$ s in (i), the range $R = (18.2 \text{ m/s})(3.17 \text{ s}) = 57.7$ m
(c) At the maximum height H we know that $v_y = 0$. From Eq. 4.12 we have

$$0 = (10.5)^2 - 2(9.8)(H - 16)$$

Thus, $H = 21.6$ m. Note that the maximum height is not reached at half the time of flight because the path is not symmetrical about the highest point, as it was in Example 4.2.
(d) For the angle of impact we need to find the direction of the velocity vector. The horizontal component remains unchanged, that is, $v_x = v_{0x} = 18.2$ m/s. The vertical component is $v_y = v_{0y} - gt = 10.5 - (9.8)(3.17) = -20.6$ m/s. The angle to the x axis is given by

$$\tan \theta = \frac{v_y}{v_x}$$
$$= \frac{-20.6}{18.2} = -1.13$$

Thus, $\theta = 48.5°$ below the horizontal.
(e) Since Eq. 4.10 has another unknown (namely t), we use Eq. 4.12 with $y = 18$ m to find v_y:

$$v_y^2 = 10.5^2 - 2(9.8)(18 - 16)$$

Thus, $v_y = 8.43$ m/s or -8.43 m/s. Since the question does not rule out either case, both are acceptable. Thus, $\mathbf{v}_1 = 18.2\mathbf{i} + 8.4\mathbf{j}$ m/s and $\mathbf{v}_2 = 18.2\mathbf{i} - 8.4\mathbf{j}$ m/s.

EXERCISE 1. In part (a) of Example 4.3 what significance, if any, does the solution $t = -1.03$ s have?

EXAMPLE 4.4: The archerfish in Fig. 4.12a shoots a drop of water directly at a beetle. At the same instant the beetle starts to fall. Show that the beetle will be hit provided the trajectory of the drop intersects the line of the beetle's fall.

Solution: Let us say that the beetle is at a height H above the fish and a horizontal distance L away. The beetle (B) and the drop (D) will meet when their (x, y) coordinates are identical; see Fig. 4.12b. So when $x_B = x_D = L$, we need to show that y_B is the same as y_D. Since $H = L \tan \theta$, the vertical coordinate of the beetle is given by

$$y_B = L \tan \theta - \tfrac{1}{2}gt^2$$

The vertical coordinate of the drop is given by

$$y_D = (v_0 \sin \theta)t - \tfrac{1}{2}gt^2$$

The horizontal position of the drop is $x_D = (v_0 \cos \theta)t$. Thus,

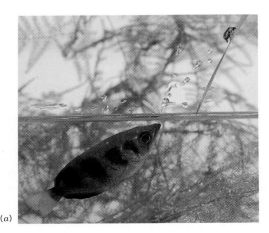

(a)

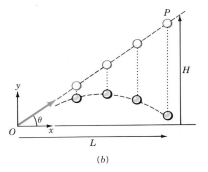

(b)

FIGURE 4.12 (a) An archerfish shoots a drop of water directly at a beetle. If the beetle falls at the instant the water is projected, it will be hit. (b) The vertical displacements of the beetle and the drop from the initial line of fire are always equal.

when $x_D = L$, we have $t = L/(v_0 \cos \theta)$. When this t is substituted into the first term of y_D, the expressions for y_D and y_B look identical. The drop will hit the beetle, provided v_0 is large enough for the parabolic path of the drop to intersect the vertical line of fall of the beetle.

The explanation of this curiosity is that the beetle and the drop both lie on the line OP in the beginning. Their vertical displacements from this line ($-\frac{1}{2}gt^2$) are the same at all times.

Therefore, when $x_D = x_B$, the vertical displacements of the drop and the beetle from point P are also the same; in other words, $y_D = y_B$.

EXERCISE 2. A ball is thrown from the roof of a building of height 45 m with a speed v_0 at angle θ below the horizontal. It lands 2 s later at a point 30 m from the base of the building. Find v_0 and θ.

4.4 UNIFORM CIRCULAR MOTION

In Fig. 4.13a, a car initially moves eastward at velocity $\mathbf{v}_1$, then it turns south and has velocity $\mathbf{v}_2$. If the speed is constant, that is, $v_1 = v_2 = v$, the change in velocity arises solely because the direction has changed. The diagram shows that $\Delta \mathbf{v} = \mathbf{v}_2 - \mathbf{v}_1$ points toward the inside of the curve. Therefore, the average acceleration $\mathbf{a}_{av} = \Delta \mathbf{v}/\Delta t$ also points in the same direction. If the turn is broken into two stages, we obtain two accelerations, as shown in Fig. 4.13b. As the number of stages increases, the accelerations appear as in Fig. 4.13c. When the line segments merge into an arc of a circle, the instantaneous acceleration is directed radially inward, toward the center. It is called the **centripetal** (center-seeking) acceleration.

Figure 4.14 shows a particle moving at constant speed v in a circle of radius r.

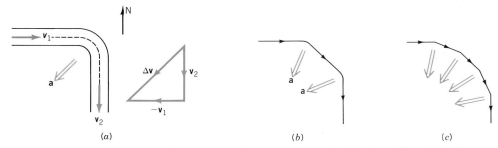

(a)   (b)   (c)

FIGURE 4.13 (a) A particle traveling at constant speed changes the direction of its velocity from east to south. Its average acceleration is directed as shown. In (b) the turn is accomplished in two steps, whereas in (c) it takes four steps. Each time the direction of $\mathbf{v}$ changes the direction of the acceleration is along the bisector of the angle between the initial and final directions of $\mathbf{v}$. When the line segments merge into an arc of a circle, the acceleration is directed toward the center of the circle.

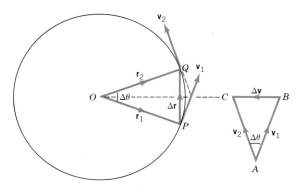

FIGURE 4.14 A particle moves at constant speed v in a circle of radius r. In any time interval Δt, the directions of the position vector $\mathbf{r}$ and of the velocity $\mathbf{v}$ change by the same angle $\Delta\theta$. The displacement $\Delta\mathbf{r}$ in Δt is perpendicular to the change in velocity $\Delta\mathbf{v}$. Triangles OPQ and ABC are both isosceles triangles with the same angles.

Suppose that in a short time interval Δt its position vector rotates through the angle $\Delta\theta$ and the particle's displacement, $\Delta\mathbf{r} = \mathbf{r}_2 - \mathbf{r}_1$, is vertical. Since $\mathbf{v}$ is always perpendicular to $\mathbf{r}$, these two vectors change their directions by the same angle in any time interval. In the vector diagram for the equation $\mathbf{v}_2 = \mathbf{v}_1 + \Delta\mathbf{v}$, we know that $v_2 = v_1 = v$. The direction of $\Delta\mathbf{v}$ is horizontal and radially inward—along the bisector of the angle $\Delta\theta$ drawn within the circle. The triangles OPQ and ABC are both isosceles triangles with the same angles. (Why?) Thus,

$$\frac{|\Delta\mathbf{r}|}{r} = \frac{|\Delta\mathbf{v}|}{v}$$

from which we obtain $|\Delta\mathbf{v}| = (v/r)|\Delta\mathbf{r}|$. Since $|\Delta\mathbf{r}| \approx v\Delta t$, we see that $|\Delta\mathbf{v}|/\Delta t \approx v^2/r$. From the definition $a = \lim_{\Delta t \to 0}(|\Delta\mathbf{v}|/\Delta t)$, we find that the centripetal acceleration is

$$a_{\mathrm{r}} = \frac{v^2}{r} \qquad (4.13) \qquad \text{Centripetal acceleration}$$

The subscript r indicates that the acceleration is radial. As a vector equation we would write

$$\mathbf{a}_{\mathrm{r}} = -\frac{v^2}{r}\,\hat{\mathbf{r}}$$

where $\hat{\mathbf{r}}$ is the radial unit vector shown in Fig. 4.15. Note that although the magnitude of $\hat{\mathbf{r}}$ is constant ($=1$), its direction changes in time. (Check the dimensions of Eq. 4.13.)

The *period T* is the time it takes to complete one revolution, a distance of $2\pi r$, so the speed is $v = (2\pi r)/T$. Equation 4.13 may be written as

$$a_{\mathrm{r}} = \frac{4\pi^2 r}{T^2} \qquad (4.14)$$

A derivation of the centripetal acceleration that employs calculus is suggested in Problem 4.12.

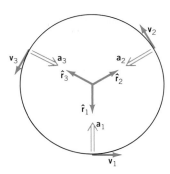

FIGURE 4.15 When a particle moves in uniform circular motion its centripetal acceleration is constant in magnitude but not in direction.

EXAMPLE 4.5: A pilot guides a plane in a horizontal circular turn with a centripetal acceleration of $5g$. If the speed of the plane is Mach 2 (twice the speed of sound, which is 340 m/s), what is the radius of the turn?

Solution: If we take $g = 9.8$ m/s^2, we are given $a_{\mathrm{r}} = 5g = 49$ m/s^2. The speed is $v = 680$ m/s. From Eq. 4.13,

$$r = \frac{v^2}{a_{\mathrm{r}}} = 9.44 \text{ km}$$

EXAMPLE 4.6: The moon orbits the earth with a period of 27.3 d at a distance of 3.84×10^8 m from the center of the earth. Find its centripetal acceleration.

Solution: First we must convert the period to seconds: $T = 2.36 \times 10^6$ s. Then from Eq. 4.14 we find,

$$a = \frac{4\pi^2 r}{T^2} = 2.72 \times 10^{-3} \text{ m/s}^2$$

EXAMPLE 4.7: Estimate the period of a low-altitude reconnaissance satellite. Ignore the effects of air resistance.

Solution: The radius of the orbit is essentially that of the earth, that is, $r \approx R_E = 6.4 \times 10^6$ m. Since the satellite is in free-fall near the surface of the earth, its acceleration is $a_r \approx g = 9.8$ m/s^2. Since $a_r = v^2/r$, we have

$$v^2 = gR_E$$

The speed is $v = [(9.8 \text{ m/s}^2)(6.4 \times 10^6 \text{ m})]^{1/2} = 7.9$ km/s. The period is given by

$$T = \frac{2\pi r}{v} = 2\pi \sqrt{\frac{R_E}{g}}$$
$$= 84 \text{ min}$$

For an actual low-altitude satellite the period is about 90 min.

EXERCISE 3. In uniform circular motion, the magnitude of the acceleration is constant. Can we use the equations of kinematics for constant acceleration? Explain why or why not.

EXERCISE 4. A particle moves at constant speed in a circle of radius 2 cm. It makes 5 revolutions per second. What is its centripetal acceleration?

4.5   INERTIAL REFERENCE FRAMES

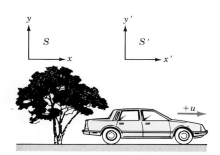

FIGURE 4.16 The road and the car serve as reference frames S and S', respectively.

Inertial reference frame

The position or velocity of a particle has meaning only in relation to other bodies. When told the speed of a car, we all assume that it was measured relative to the road. The road is an example of a **reference frame**. A reference frame is something physical, such as a road, a train, a tabletop, or even the earth itself. In Fig. 4.16 the road forms the frame S with coordinate axes x and y. A car moving at constant velocity $+u$ along the x axis is a frame S' with axes x' and y'. A ball at rest in S' moves with velocity $+u$ relative to S. A tree at rest in S, moves at velocity $-u$ relative to S'. Although observers in each frame would describe the motion of the ball or the tree differently, they would agree that both objects have no acceleration. Since either a state of rest or a state of motion at constant velocity implies that no net force acts on a body, the first law is obeyed in both S and S'.

A reference frame in which Newton's first law is valid is called an **inertial reference frame**. In fact, the first law is used to define such a frame:

In an inertial reference frame, a body subject to no net force will either stay at rest or move at constant velocity.

Any frame moving at constant velocity relative to a known inertial frame is also an inertial frame. If the acceleration of a particle is zero in one inertial frame, it is zero in all inertial frames.

If the car of Fig. 4.16 were to accelerate, it would not qualify as an inertial frame. Suppose a ball is on the frictionless floor of the car as the brakes are applied. Since there is no net force on the ball, an observer on the ground will see the ball continue to move at the velocity of the car just before the brakes were applied. However, relative to observers in the car, the ball accelerates in the forward direction, even though there is no net force on it. Objects in *noninertial* reference frames do not obey Newton's first law. (Noninertial frames are discussed in more detail in Section 6.5.)

Because of the daily rotation of the earth, a frame based on its surface is continually changing the direction of its velocity. Hence, it is not really an inertial frame. Even if the earth were not rotating, it would still be in orbit around the sun, which itself has an acceleration relative to other stars. Our innocent little definition of an inertial frame has led us into quicksand. We have developed Newton's first law and the concept of an inertial frame without ever having direct experience with one.

All is not lost, however. The acceleration of an object at the surface of the earth, and the earth's acceleration in its orbit around the sun, are quite small. For experiments conducted over a small area, such as a laboratory, a frame based on the surface of the earth is a perfectly adequate approximation to an inertial frame. For interplanetary travel, a frame based on the sun or nearby stars serves us very well. To the question of whether there is a "real" inertial frame, there is no clear-cut answer. Newton proposed that a frame based on the "fixed stars" could serve as the standard inertial frame. This is as good as we are ever likely to need.

4.6 RELATIVE VELOCITY

The motion of any body has to be described relative to some frame of reference, such as the ground. Sometimes it is necessary to examine the motion of one body relative to another body that is also moving relative to the ground. For one-dimensional motion, it is easy to determine the velocity of one body relative to another. Consider car A which is moving north at 35 m/s and is ahead of car B, which is moving north at 30 m/s. These velocities are relative to the ground. Relative to observers in car B, car A is moving north at 5 m/s, whereas relative to car A, car B is moving south at 5 m/s. We now consider how to determine relative motion in two dimensions.

In Fig. 4.17 the position of a particle P with respect to frame A is $\mathbf{r}_{PA}$. The position of P with respect to frame B is $\mathbf{r}_{PB}$. Finally, the position of frame B with respect to frame A is $\mathbf{r}_{BA}$. The vector triangle shows that

$$\mathbf{r}_{PA} = \mathbf{r}_{PB} + \mathbf{r}_{BA} \tag{4.15}$$

Now suppose both the particle P and the frame B are moving with respect to frame A. Since $\mathbf{v} = d\mathbf{r}/dt$, Eq. 4.15 leads to

$$\mathbf{v}_{PA} = \mathbf{v}_{PB} + \mathbf{v}_{BA} \tag{4.16}$$

$$\begin{array}{ccc} \text{Vel. of P} & = & \text{Vel. of P} + \text{Vel. of B} \\ \text{rel. to A} & & \text{rel. to B} \quad \text{rel. to A} \end{array}$$

Notice that Eq. 4.16 is a vector *sum*—we worry about signs only when we take components. The order of the subscripts should be carefully noted. In general,

$$\mathbf{v}_{AB} = -\mathbf{v}_{BA} \tag{4.17}$$

(Can you prove this?)

Let us consider a concrete example. In Fig. 4.18 a man walks across a train with velocity $\mathbf{v}_{MT}$, relative to the train. The train itself moves relative to the ground with velocity $\mathbf{v}_{TG}$. The ground forms a stationary frame (x_G, y_G), while the train forms a moving frame (x_T, y_T). We wish to find the velocity of the man relative to the ground, that is, $\mathbf{v}_{MG}$. From Eq. 4.16 we have

$$\mathbf{v}_{MG} = \mathbf{v}_{MT} + \mathbf{v}_{TG}$$

An observer on the train will see the man walk along the y_T axis. An observer on the ground will see him move in a direction given by $\tan\theta = v_{MT}/v_{TG}$ to the x_G axis.

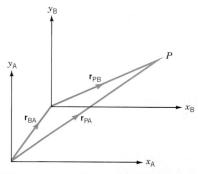

FIGURE 4.17 The position of point P relative to frame B is $\mathbf{r}_{PB}$ and relative to frame A it is $\mathbf{r}_{PA}$. They are related by the equation $\mathbf{r}_{PA} = \mathbf{r}_{PB} + \mathbf{r}_{BA}$, where $\mathbf{r}_{BA}$ is the position of B relative to A.

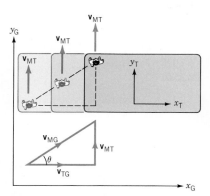

FIGURE 4.18 The diagram shows the positions of a man (M) at three times as he walks across a train (T) with velocity $\mathbf{v}_{MT}$ relative to the train. His velocity relative to the ground (G) is $\mathbf{v}_{MG} = \mathbf{v}_{MT} + \mathbf{v}_{TG}$, where $\mathbf{v}_{TG}$ is the velocity of the train relative to the ground.

EXAMPLE 4.8: A motor boat can travel at 10 m/s relative to the water. It starts at one bank of a river that is 100 m wide and flows eastward at 5 m/s. If the boat is pointed directly across, find: (a) its velocity relative to the bank; (b) how far downstream it travels.

Solution: (a) The train has been replaced by the river (R) and the man by the boat (B). We will refer to the bank as the ground (G). What does one mean by the "velocity relative to the water"? Imagine, if you will, a set of beer barrels tied together and floating down the river (Fig. 4.19). This forms the reference frame of the river (x_R, y_R). The boat will acquire the river's velocity as it is carried along with the current. We are given $\mathbf{v}_{RG}$ = 5 m/s due east and $\mathbf{v}_{BR}$ = 10 m/s due north. We need to find $\mathbf{v}_{BG}$. From Eq. 4.16, we have

$$\mathbf{v}_{BG} = \mathbf{v}_{BR} + \mathbf{v}_{RG}$$

The vector triangle is shown in Fig. 4.19. In terms of the magnitudes,

$$v_{BG} = \sqrt{10^2 + 5^2} = 11.2 \text{ m/s}$$

The direction is given by

$$\tan \theta = \frac{5}{10} = 0.5$$

Thus, θ = 26.5° E of N.

(b) First, we need to find the time taken to cross. We note that the component of the boat's velocity perpendicular to the bank is 10 m/s. Since the river is 100 m wide, the time taken is 10 s. In this time the boat drifts a distance (5 m/s)(10 s) = 50 m.

EXAMPLE 4.9: The captain of the boat in Example 4.8 real-izes his mistake. (a) In which direction must he point the boat to get directly across? (b) How long does this take?

Solution: (a) Clearly, the boat must be pointed upstream. The question gives us the direction of the velocity of the boat relative to the ground: We know $\mathbf{v}_{BG}$ is directly across. We also know that $\mathbf{v}_{RG}$ is 5 m/s due east and that v_{BR} = 10 m/s. From Eq. 4.16,

$$\mathbf{v}_{BG} = \mathbf{v}_{BR} + \mathbf{v}_{RG}$$

The vector triangle is shown in Fig. 4.20. In terms of the magnitudes

$$v_{BG} = \sqrt{10^2 - 5^2} = 8.7 \text{ m/s}$$

The direction is given by

$$\sin \theta = \frac{5}{10} = 0.5$$

Thus, θ = 30° west of north.

(b) For the time to cross we need only the component of $\mathbf{v}_{BG}$ perpendicular to the bank. This is just v_{BG}. Thus, the time is (100 m)/(8.7 m/s) = 11.5 s.

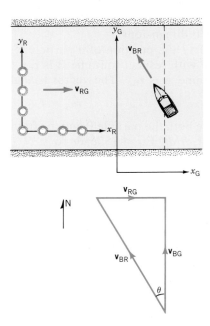

FIGURE 4.20 To get directly across the river, the boat must be pointed upstream.

EXAMPLE 4.10: The pilot of an aircraft has to get to a point 320 km due north in 1 h. Ground control reports that there is a crosswind of 80 km/h toward 37° S of W. What is the required heading of the plane?

Solution: Our two frames are the air (A) and the ground (G). We assume that the plane (P) has acquired the full velocity of the wind. We are given $\mathbf{v}_{PG}$ = 320 km/h north and $\mathbf{v}_{AG}$ = 80 km/h at 37° S of W. We need to find $\mathbf{v}_{PA}$, the velocity of the plane

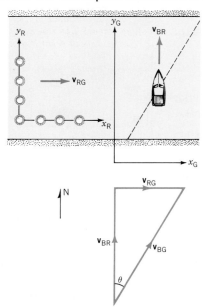

FIGURE 4.19 When a boat is pointed directly across a river it is carried downstream. The connected circles represent beer barrels floating down the river. They help us to visualize "the frame of the river."

relative to the air. From Eqs. 4.16 and 4.17, we have

$$\mathbf{v}_{PA} = \mathbf{v}_{PG} + \mathbf{v}_{GA}$$
$$= \mathbf{v}_{PG} - \mathbf{v}_{AG} \qquad \text{(i)}$$

To prevent ourselves from drowning in subscripts, let us change the notation: $\mathbf{v}_{PA} = \mathbf{P}$ (plane); $\mathbf{v}_{AG} = \mathbf{W}$ (wind); and $\mathbf{v}_{PG} = \mathbf{R}$ (resultant). Equation (i) now reads

$$\mathbf{P} = \mathbf{R} - \mathbf{W} \qquad \text{(ii)}$$

The vector triangle is drawn in Fig. 4.21. The components of (ii) are

$$P_x = R_x - W_x = 0 - (-80 \cos 37°) = +64 \text{ km/h}$$
$$P_y = R_y - W_y = 320 - (-80 \sin 37°) = +368 \text{ km/h}$$

Thus, $\mathbf{P} = 64\mathbf{i} + 368\mathbf{j}$ km/h. The correct heading is given by

$$\alpha = \tan^{-1}\left(\frac{P_x}{P_y}\right) = \tan^{-1}(0.174) = 9.9° \text{ E of N}$$

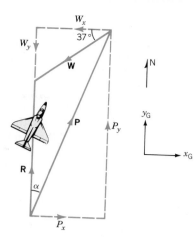

FIGURE 4.21 For the plane to travel due north in the presence of a wind with velocity **W**, the required heading is $\alpha°$ E of N.

4.7   THE GALILEAN TRANSFORMATION

We now consider how the values of the position, velocity, and acceleration of a particle differ in two inertial frames that move relative to each other at constant velocity. Figure 4.22 shows two such frames, S and S'. Frame S' moves at constant velocity $\mathbf{u}$ relative to frame S. We assume that the origins O and O' coincided at $t = 0$. The positions of a point P relative to the two frames are related by

$$\mathbf{r}' = \mathbf{r} - \mathbf{u}t \qquad (4.18)$$

This vector equation is equivalent to three equations in terms of the components. An often encountered special case occurs when the x and x' axes coincide and frame S' moves at constant velocity $+u\mathbf{i}$ along the x axis of S, as shown in Fig. 4.23. The y and z coordinates of P are the same for both coordinate systems. The figure shows that $x' = x - ut$. Therefore, for this special case,

$$x' = x - ut \qquad y' = y \qquad z' = z \qquad t' = t \qquad (4.19)$$

Equation 4.19 relates the coordinates of a particle in two inertial frames moving relative to each other at constant velocity. It is called the **Galilean transformation** of coordinates.*

The velocity of the particle may be found by taking the derivative of Eq. 4.18 with respect to time:

$$\mathbf{v}' = \mathbf{v} - \mathbf{u} \qquad (4.20)$$

(We saw this equation in the form $\mathbf{v}_{PA} = \mathbf{v}_{PB} + \mathbf{v}_{BA}$ in Eq. 4.16.) Equation 4.20 indicates that the velocity assigned to the particle will be different in the two frames. However, observers in both frames would agree that it moves at constant velocity and is, therefore, not subject to a net force. Note that an observer in

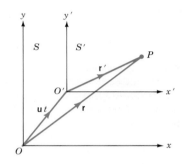

FIGURE 4.22 Frame S' moves at velocity $\mathbf{u}$ relative to frame S. The position vectors of point P relative to the two frames are related by $\mathbf{r}' = \mathbf{r} - \mathbf{u}t$.

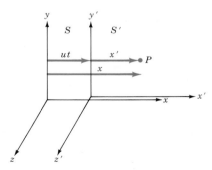

FIGURE 4.23 When frame S' moves at velocity $u\mathbf{i}$ along the x axis of frame S, the coordinates of point P are related by the Galilean transformation: $x' = x - ut$, $y' = y$, $z' = z$, $t' = t$.

---

* We have made the "obvious" assumption that observers in both frames record the same elapsed time, that is, $t = t'$. Once their clocks are synchronized, they remain that way indefinitely. Therefore, in Newtonian mechanics, there is a single *universal* or *absolute* time for all frames. Experimental confirmation of the special theory of relativity has shown this assumption to be false. But it is an excellent approximation for everyday phenomena and when speeds are much lower than the speed of light.

frame S can measure the velocity of frame S' relative to S, but cannot claim that either frame is "really" at rest or "really" moving.

The time derivative of Eq. 4.20 yields the accelerations (note $d\mathbf{u}/dt = 0$):

$$\mathbf{a} = \mathbf{a}' \qquad (4.21)$$

Observers in all inertial frames would assign the same acceleration to the particle. To see the meaning of Eq. 4.21, consider a simple experiment. In Fig. 4.24a, a ball is thrown up vertically from a railcar (frame S'). An observer in this frame will see the ball move only along the y' axis with the acceleration due to gravity. Now suppose the railcar moves at constant velocity $+u$ along the x axis of the ground frame S, as in Fig. 4.24b. In this frame, the ball always has the fixed horizontal velocity $+u$. The path described in S is a combination of the fixed horizontal velocity and the vertical accelerated motion. The path is parabolic. Although the horizontal velocity of the ball is different in S and S', its acceleration is the same in both. If someone on the ground were to throw a ball vertically up, an observer in S' would see a parabolic path—this time traced backward since the ball's horizontal component of velocity would be $-u$ relative to S'.

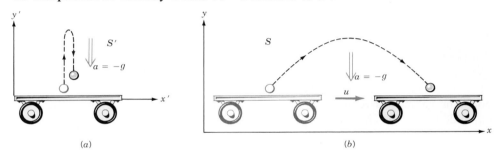

FIGURE 4.24 (a) The path of a ball thrown vertically upward relative to a platform moving at velocity u relative to the ground. (b) The path of the ball as observed by someone on the ground is a parabola. The acceleration of the ball is the same in both cases.

In both frames, either path, vertical or parabolic, is associated with the same acceleration. This simple experiment would not allow us to claim that one frame is fixed, while the other is moving. In fact no experiments allow us to distinguish between inertial frames. This is the essence of the **Galilean principle of relativity** which states:

The laws of mechanics have the same form in all inertial reference frames.

The laws formulated as the result of a series of experiments on a boat moving at constant velocity would be identical to those formulated on land.

4.8 NONUNIFORM CIRCULAR MOTION

Consider a particle moving along a curved path, as shown in Fig. 4.25. In general, both the magnitude and the direction of the velocity may vary along its path. The radial acceleration associated with changes in the direction of the velocity is

$$a_\mathrm{r} = \frac{v^2}{r}$$

where r is the radius of curvature of the path at the given point. This (centripetal) acceleration is directed toward the center of curvature. When the speed varies, there is also an acceleration along the tangent to the path:

$$a_\mathrm{t} = \frac{dv}{dt}$$

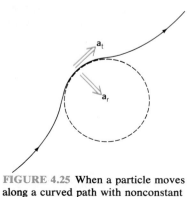

FIGURE 4.25 When a particle moves along a curved path with nonconstant speed, its acceleration has radial and tangential components.

This tangential acceleration is in the direction of the velocity if the speed is increasing, and opposite to **v** if the speed is decreasing. The resultant acceleration is the vector sum of these two components:

$$\mathbf{a} = \mathbf{a}_r + \mathbf{a}_t \qquad (4.22)$$

Since $\mathbf{a}_r$ and $\mathbf{a}_t$ are always perpendicular, the magnitude of the resultant acceleration is $a = (a_r^2 + a_t^2)^{1/2}$.

In the special case of motion in a circle, it is sometimes convenient to use the unit vectors $\hat{\boldsymbol{\theta}}$ and $\hat{\mathbf{r}}$ shown in Fig. 4.26, where $\hat{\mathbf{r}}$ is directed radially outward from the center and $\hat{\boldsymbol{\theta}}$ is in the direction of increasing θ. Although the magnitudes of these unit vectors are constant ($=1$), their directions change in time. The net acceleration would be expressed as

$$\mathbf{a} = \mathbf{a}_r + \mathbf{a}_t = -\frac{v^2}{r}\hat{\mathbf{r}} + \frac{dv}{dt}\hat{\boldsymbol{\theta}} \qquad (4.23)$$

In uniform circular motion $dv/dt = 0$, so the acceleration has only the radial term.

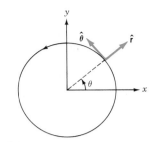

FIGURE 4.26 The directions of the radial and tangential unit vectors $\hat{\mathbf{r}}$ and $\hat{\boldsymbol{\theta}}$ change as the particle moves along the circle.

HISTORICAL NOTE: The Development of the Concept of Inertia

Central to Aristotle's philosophy was the dictum: All that moves is moved by something else. He felt that a "motive force" was required just to maintain motion. This seems quite reasonable: A cart does not move by itself; it has to be pulled by a horse. "Natural" motion occurred when an object moved vertically up or down toward its "natural place" (see Chapter 3). Any motion away from the natural place was "violent" and needed a cause to explain it. Horizontal motion was also classified as being violent and had to be explained in terms of something in contact with the object—for example, a hand pushing a book. The only bodies that did not need a cause for their motion were the celestial bodies—the planets and the stars. Their natural motion was assumed to occur at constant speed in perfectly circular paths.

The prolonged horizontal motion of an arrow posed a problem for Aristotle. Why does it go so far? What keeps it moving for so long? He had painted himself into a corner by requiring some agent to be in contact with the object. So he devised the following bizarre scheme. As the arrow moves forward, it disturbs the air ahead of it. To prevent the formation of a vacuum at the rear, the air rushes to the tail of the arrow, as shown in Fig. 4.27. The resulting turbulence propels the arrow forward. Ultimately, the resistance of the air causes the arrow to slow down and fall to the ground.

Around A.D. 500, John Philoponous challenged the absurdity of this explanation. How could the air both propel and resist the arrow at the same time? In his opinion, "Such a view is quite incredible, and borders on the fantastic." (To be fair, Aristotle also had doubts about it.) Philoponous suggested that an "impressed force," which originated in the bow, kept an arrow moving; the air played no part as the cause of motion. He assumed that any impressed force gradually dies out, even in a void. The possibility that motion could continue indefinitely did not occur to him.

Eight centuries later, Jean Buridan (ca. 1350) proposed that an arrow is given an "impetus" by the bow. Unlike im-

pressed force, this impetus was not expected to die out. The impetus it received from the bow would continue to move an arrow indefinitely—if it were not expended in overcoming the resistance of the air. The size of the impetus was some function of the weight and speed of the arrow. Buridan's view marks an important shift from *external* agents propelling the arrow, to some acquired *internal* property or state.

It was Galileo's defense of the Copernican system that triggered the major breakthrough. People had raised an apparently reasonable objection to Copernicus' assertion that the earth rotates. They pointed out that if this were true, then an arrow fired vertically up should land some distance to the west and not at the same point, as it is seen to do. Their reasoning had its roots in Aristotle: The arrow has nothing to keep it moving horizontally, so it should be left behind. Starting in 1592, Galileo tried to explain vertical motion by using the ideas of impressed force and impetus, but made little progress. But he did perform an important experiment with a simple pendulum. As the bob swings back and forth, it rises to the same vertical level on either side of its swing. Galileo placed a peg to obstruct the swing, as shown in Fig. 4.28, and discovered that the bob *still* rose to the same vertical level—pro-

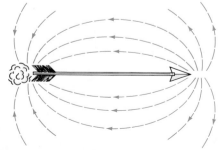

FIGURE 4.27 Aristotle suggested that the air in front of an arrow rushes to its rear to prevent the formation of a vacuum. The resultant turbulence was supposed to propel the arrow forward.

vided the peg was not set too low. He concluded that the speed of the bob at the lowest point depends only on the *vertical* distance through which it falls, not on the actual path. The pendulum result led him to extend his inclined plane experiments in the hope of learning something new. He let the balls roll up a second incline, as in Fig. 4.29, and discovered that they rose to the same height from which they were released, after making a small allowance for friction. The angle of the second incline did not matter.

Galileo's approach in dealing with this result made him the first true physicist. He wondered what happens when the second incline becomes horizontal. The answer came in the form of a "thought" experiment in which he did two very important things. First, he imagined an infinite horizontal plane, and second, he made the simplifying assumption of ignoring the ever-present friction. He knew he was on the trail of something important and was not going to let friction confuse the issue. And so, he conceived of an *ideal*—a frictionless horizontal plane. Aristotle would not have done this. He would have insisted (correctly!) that friction is always present in the real world.

Galileo believed that a ball slows down while moving up an incline because it is moving away from its natural position, which is as close as possible to the center of the earth. This part of his thinking is not much beyond that of Aristotle. Anyway, he reasoned that on a horizontal plane, the ball would be moving neither away from, nor toward, the center of the earth. Hence it should neither speed up nor slow down. His principle of inertia was included in the *Dialogues Concerning Two New Sciences* (1638) (paraphrased):

> *A body will continue to move with constant speed on a frictionless infinite horizontal plane.*

Galileo considered a large flat surface to be a good approximation to an "infinite horizontal plane." On such a surface a body would move at constant speed along a straight line. However, he used the word horizontal to mean equidistant from the center of the earth, as shown in Fig. 4.30. This is really a pity, for when his principle was applied on a large scale, the straight line became a circular path. Furthermore, he thought that the circular motion of the planets was "natural" and therefore required no further explanation. He had not fully broken the shackles of Aristotelian thought. Nonetheless, he was the

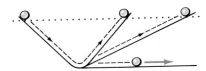

FIGURE 4.29 When Galileo let balls roll down one incline and then up a second incline they rose to the level from which they were released. The slope of the second incline did not matter.

first to realize that an *external influence or "force" is needed only to change velocity, not simply to maintain it.*

Galileo explained why an arrow fired vertically lands back at the firing point as follows. In addition to its vertical motion, the arrow also has the "horizontal" rotational motion of the surface of the earth. Since there is no horizontal force acting on the arrow, it will *maintain* this component of its motion. You can easily verify this statement. While traveling in a car, throw an object vertically up. If the car is moving at constant velocity, the object will land back in your hand. This can happen only if the object maintains the same horizontal velocity as the car.

The French philosopher and mathematician René Descartes made two contributions to this topic. First, he extended the idea of inertia to *all* bodies, including the celestial ones. Second, he pointed out that any circular motion is constrained motion. For example, one must pull inward to keep a stone attached to a rope moving in a circle. He stated:

> *A body free of external influence will move in a straight line at constant speed.*

Another scholar, Pierre Gassendi, had also changed Galileo's "horizontal plane" into a "straight line." It seems almost inconceivable that this minor playing with words was so crucial to the development of physics. Yet it defines clearly for the first time what is "natural" motion and distinguishes it from motion that needs explanation. The vertical fall of an object does not occur at constant speed, so it must be caused by some external influence—a point missed even by Galileo. By applying the law of inertia to all bodies, Descartes erased the centuries-old division that had always been made between terrestrial and celestial phenomena. The realization that the circular motion of the moon is not "natural" motion, and therefore requires an explanation, led to the most profound physical insight in 2000 years—Newton's theory of gravitation. Descartes did not proceed beyond his correct statement of the law of inertia, and so credit goes to Newton, who used it to build a foundation for mechanics.

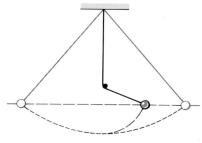

FIGURE 4.28 Galileo discovered that even when the swing of a pendulum was interupted by a peg, the bob rose to the same maximum height (provided the peg was not set too low).

FIGURE 4.30 When Galileo's conception of inertia was applied on a large scale, it implied that a ball on a frictionless surface would "naturally" go around the earth. Thus, motion in a circle was "natural motion"—which is not correct.

SPECIAL TOPIC: Real Projectiles

It was pointed out earlier that the equations of kinematics for constant acceleration must be applied with caution even to commonplace projectiles. Projectile motion plays a central role in ballistics and many sports activities. We now consider some of the complications that can occur.

SHOTPUT

One of the few sports events in which the effects of the air are small is the shotput. In Example 4.2 we found that maximum range is produced with an initial angle of 45° *only* when the projectile returns to its initial level. The shot, however, leaves the hand at a height h above the ground with an initial speed v_0 at an angle θ to the horizontal, and lands on the ground. At what angle should the shot be projected to obtain maximum range? (Why can we dismiss $\theta > 45°$?) When the initial angle is less than 45°, the shot returns to its original height at a lesser horizontal range, but the horizontal component of its velocity is greater than it would be at 45°, as shown in Fig. 4.31. This may be sufficient to carry it farther. The angle θ_m for maximum range is given by*

$$\tan \theta_m = \frac{1}{\sqrt{1 + 2gh/v_0^2}}$$

This result shows that θ_m depends on both v_0 and h. If $h = 0$, then $\tan \theta_m = 1$ and $\theta_m = 45°$, in agreement with our earlier analysis. Typical values for a world-class shotputter are $h = 2$ m and $v_0 = 14$ m/s. The above equation yields $\theta_m \approx 42.5°$. The horizontal range predicted is about 23 m. (What can you conclude about θ_m for amateur shotputters?)

AIR RESISTANCE

A body moving through a fluid, such as air, experiences a "drag force" that is directed opposite to its velocity. In

* See J. S. Thomsen, *Am. J. Phys.*, 52:881 (1984).

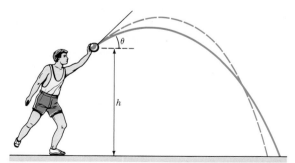

Dashed path is for 45°

FIGURE 4.31 When a projectile lands below its initial level, the maximum horizontal range is attained at an angle of projection less than 45°.

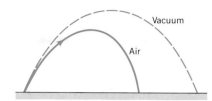

FIGURE 4.32 In the presence of air resistance the path of a projectile is not parabolic. The range and the maximum height are reduced and the angle of impact is greater than the angle of projection.

many instances the drag force is proportional to the square of the speed (see Chapter 6). For a 0.145-kg baseball, this results in a terminal speed of about 40 m/s. At 35 m/s, which is often achieved by a pitcher, the drag force is about two-thirds of the weight of the ball. Figure 4.32 shows the effect of air drag on the path of a projectile by comparing it with the idealized parabolic path. Both the maximum height and the horizontal range are reduced. The maximum height is achieved at an earlier time, and the direction of motion on landing is more nearly vertical. The medieval sketches of the paths of cannonballs were at least correct in depicting the paths as unsymmetrical.

Since the horizontal component of velocity decreases gradually and the vertical component drops to zero, the effect of drag is greater on the range than on the maximum height. The initial angle for maximum range is less than 45° and decreases as the initial speed increases. For example, if the initial speed is 35 m/s, the optimum angle is 44°, for a range of 112 m.

ARTILLERY SHELLS AND BALLISTIC MISSILES

Even when air resistance is ignored, parabolic trajectories occur only when the direction of the acceleration due to gravity is fixed. When the initial velocity is large, as it is for an artillery shell or ballistic missile, one cannot ignore the curvature of the earth. The acceleration is always directed toward the center of the earth and so it changes direction along the path of the shell or missile. In the absence of air resistance the actual path is elliptical (see Chapter 13), as shown in Fig. 4.33a.

A further complication arises because the earth is rotating. Suppose a projectile is fired from the North Pole toward a target along a line of longitude (Fig. 4.33b). During the flight of the shell or missile, the rotation of the earth causes the target to move from its initial position relative to the point of firing. Consequently, the projectile will miss if it is aimed directly at the moving target. The missile must be launched in a direction toward the east to compensate for this effect (see Chapter 6). The calculation of realistic trajectories for long-range artillery shells, which included the effect of the air and the earth, was one of the first jobs assigned to ENIAC, the first modern computer.

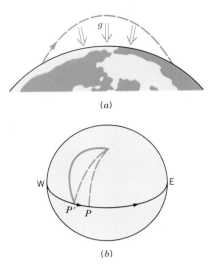

FIGURE 4.33 (a) In the case of long-range projectiles, such as artillery shells, the change in the direction of the acceleration due to gravity must be taken into account. The path is elliptical, not parabolic. (b) A missile is fired along a meridian line toward the point P on the equator. Because of the rotation of the earth, it will land at point P' to the west of P.

EFFECTS OF SPIN

When the game of golf was invented, players soon discovered that rough balls traveled much farther than smooth ones. In 1896, the physicist P. G. Tait realized that the range of a rough ball was much greater than theory would predict, even if air resistance were ignored. He correctly deduced that the ball is given a backspin that results in its being given a lift force.

A professional golfer can give the ball an initial speed of 60 m/s (135 mph) and a spin rate of up to 120 rev/s (see Fig. 4.34a). Although the spinning ball interacts with the air in a complex way, one can obtain a qualitative understanding of why lift occurs. When a ball has backspin, its bottom part has a greater speed than its upper part. As a result the upward force due to the impact of air molecules at the bottom (F_b) is greater than the downward force at the top (F_t), as shown in Fig. 4.34b. The component of the net force that is perpendicular to the velocity is called lift, whereas that in the direction opposite to the velocity is called drag. Both lift and drag increase with the speed of the ball, but lift is more sensitive to backspin. The dimples on the ball give it a controlled amount of roughness.

When the ball is struck, it is possible for the lift force to be greater than the weight of the ball. Consequently, the path may actually curve upward slightly. In any case, it is relatively straight for over half the trajectory. Figure 4.35 shows the effect of backspin on the trajectory by comparing it to the path predicted without spin. The maximum range for a spinning golfball does not occur at a projection angle of 45°, because the net force has a large backward component. The optimum angle is about 20°, although in practice

(a)

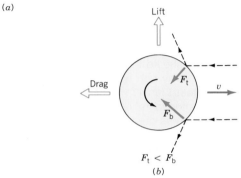

(b)

FIGURE 4.34 (a) When a golf ball is struck it acquires a "backspin." (b) As a golf ball with a backspin moves through the air the impact of the air molecules produces a greater force on the bottom half (F_b) than on the top half (F_t). This results in a lift force perpendicular to the direction of motion.

10° is the usual initial angle for a long-range drive. At the 10° angle and an initial speed of 60 m/s, the simple theory predicts a horizontal range of 130 m. In fact, the range is closer to 180 m.

Spin also plays a role in other sports, such as baseball, tennis, and cricket. The fuzz on a tennis ball and the ridges on a baseball or cricket ball are needed to produce effects due to backspin, topspin, or spin about a vertical axis. A baseball can be pitched with a spin of about 20 rev/s. Although the effect of spin on the range is not great, the subtle deflections that occur between the pitcher and the batter are part of the lure of the game. In tennis and cricket the situation is further complicated for the receiving player because a spinning ball bounces in an unpredictable way.

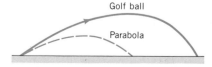

FIGURE 4.35 For the same initial velocity, the horizontal range and maximum height of a spinning golf ball are *greater* than that of an ordinary projectile in the absence of air resistance.

SUMMARY

Newton's first law states that if the net force on a body is zero, it will either stay at rest, or, if moving, it will travel in a straight line at constant speed.

The **inertia** of a body is its tendency to resist any *change* in its velocity. In an **inertial reference frame** a body obeys Newton's first law.

The horizontal and vertical motions of projectiles are independent:

$$a_x = 0 \qquad v_x = v_{0x}$$
$$a_y = -g \qquad v_y = v_{0y} - gt$$

A particle moving at constant speed v in a circle of radius r, experiences a **centripetal** (inward) acceleration:

$$a = \frac{v^2}{r}$$

The velocity $\mathbf{v}_{AB}$ of a body A relative to a body B is given by

$$\mathbf{v}_{AB} = \mathbf{v}_{AC} + \mathbf{v}_{CB}$$

where C is a third body (reference frame) such as the ground.

The coordinates of a point (x', y') in frame S' that is moving at constant velocity v along the x axis of frame S are related to its coordinates (x, y) in frame S according to the **Galilean transformation**

$$x' = x - vt \qquad y' = y \qquad z' = z \qquad t' = t$$

According to the **Galilean principle of relativity** the laws of mechanics are the same in all inertial frames.

ANSWERS TO IN-CHAPTER EXERCISES

1. Let's try throwing at 30° *below* the horizontal:

$$y = 16 - 10.5t - 4.9t^2 = 0$$

leads to $t = 1.03$ s and -3.17 s.

2. With the origin at the base:

$$x = (v_0 \cos \theta)t; \qquad y = 45 - (v_0 \sin \theta)t - 4.9t^2$$

Inserting the given values we find

$$15 = v_0 \cos \theta; \qquad 12.7 = v_0 \sin \theta$$

Square these equations, add them, and use the identity $\sin^2 \theta + \cos^2 \theta = 1$. We find $v_0 = 19.7$ m/s and $\theta = 40.4°$.

3. No. The acceleration is not constant in direction.

4. The period is $\frac{1}{5}$ s $= 0.2$ s; thus, $v = 2\pi r/T = 0.628$ m/s. The centripetal acceleration is $a = v^2/r = 19.7$ m/s².

QUESTIONS

1. A hot splinter is dislodged from a spinning grinding wheel. Which of the paths drawn in Fig. 4.36 best represents its subsequent motion given that the axis of rotation is (a) vertical, or (b) horizontal?

2. Can a particle move at constant speed and yet be accelerating? If so, give an example.

3. If you pull slowly on a tablecloth, the dishes crash to the floor. If you pull fast enough, they stay put (see Fig. 4.37). Explain.

4. The driver of a large freight train may back up a bit before starting on a journey. Why would he do this?

5. True or false: When a particle moves in uniform circular motion its acceleration is constant.

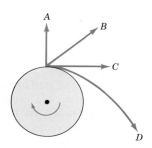

FIGURE 4.36 Question 1

FIGURE 4.37 Question 3

6. A child in a car that moves at constant speed throws a ball vertically upward relative to herself. Where does the ball land given that the car is traveling (a) in a straight line; (b) around a curve?

7. How could you estimate the speed of raindrops?

8. True or false: Motion in a circle *always* involves acceleration.

9. David twirls a stone attached to a rope in a horizontal circle of radius *r*. Goliath is standing at a distance *D* due east of David (see Fig. 4.38). At what point should the rope be released?

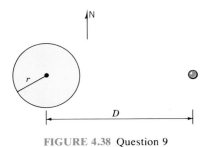

FIGURE 4.38 Question 9

10. What are the reasons why the path of a high-speed projectile is not a parabola?

11. Is a satellite in free-fall? If so, why doesn't it land?

12. Would you expect the effects of air resistance to be greater on the horizontal or on the vertical motion of a projectile fired at 45° to the horizontal? Why?

EXERCISES

4.3 Two-Dimensional Motion

1. (I) The position of a particle as a function of time is given by $\mathbf{r} = (3t^2 - 2t)\mathbf{i} - t^3\mathbf{j}$ m. Find: (a) its velocity at $t = 2$ s; (b) its acceleration at 4 s; (c) its average acceleration between $t = 1$ s and $t = 3$ s.

2. (I) (a) The position of a particle changes from $\mathbf{r}_1 = 3\mathbf{i} + 2\mathbf{j} - \mathbf{k}$ m to $\mathbf{r}_2 = 4\mathbf{i} - \mathbf{j} + 3\mathbf{k}$ m in 2 s. What is its average velocity? (b) A particle has an acceleration of $\mathbf{a} = -7\mathbf{i} + 2\mathbf{j}$ m/s² for a period of 5 s. After this time the velocity is $\mathbf{v}_2 = 5\mathbf{i} - 2\mathbf{k}$ m/s. What was the initial velocity?

3. (I) At $t = 0$ a particle at the origin has a velocity of 15 m/s at 37° above the horizontal x axis. At $t = 5$ s it is at $x = 20$ m and $y = 35$ m and its velocity is 30 m/s at 53° above the horizontal. Find: (a) its average velocity; (b) its average acceleration.

4. (I) A car travels east at 20 m/s for 10 s and then north at 10 m/s for 15 s. Find: (a) the distance covered; (b) the displacement; (c) the average velocity; (d) the average speed; (e) the average acceleration.

5. (I) A child on a merry-go-round starts on the $+y$ axis at $t = 0$ and moves toward the $+x$ axis in a circle of circumference 8 m at a constant speed of 1 m/s. Find: (a) the displacement from $t = 0$ s to $t = 1$ s; (b) the average velocity from $t = 0$ s to $t = 3$ s; (c) the average acceleration between 1 and 3 s.

6. (II) On May 20–21, 1927, Charles Lindberg flew the *Spirit of St. Louis* nonstop from New York to Paris, a distance of 5810 km, in 33.5h. What was the magnitude of his average velocity? (Radian measure is needed.)

4.4 Projectile Motion

7. (I) A rock is thrown with an initial velocity of 25 m/s at 53° above the horizontal from the top of a cliff 100 m high. Find: (a) the time of flight; (b) the maximum height; (c) the horizontal range; (d) the velocity on hitting the ground.

8. (I) Amorosa, who is on a balcony 40 m above the ground, throws the key to her heart at 37° below the horizontal to Pedro on the ground. He catches the key two seconds after it was thrown. (a) How far from the base of the building was he standing? (b) At what angle was the key moving when he caught it?

9. (I) A ball is thrown at 14.1 m/s at 45° above the horizontal. Someone located 30 m away along line of the path starts to run just as the ball is thrown. How fast, and in which direction, must the person run to catch the ball at the level from which it was thrown?

10. (II) Emanuel Zacchini, a human cannonball, was projected 53 m, measured horizontally to the initial level, during a Ringling Brothers Circus performance in 1940. His muzzle velocity was 87 km/h. What was the firing angle?

11. (II) A baseball is pitched horizontally at 30 m/s to the catcher 18.3 m away. (a) How far below its initial level is it caught? (b) At what distance above the initial height should the ball be aimed to reach the catcher at the original level?

12. (II) A dartgun is pointed horizontally at a target 3 m away. (a) If the dart strikes 5 cm too low, what is the muzzle velocity? (b) At what angle to the horizontal should the gun be pointed?

13. (I) If a baseball player can throw a ball at 45° to a point 100 m away horizontally to the initial vertical level, how high could he throw it vertically upward?

14. **(I) A plane dives at 37° to the horizontal and releases a package at an altitude of 200 m. If the load is in the air for 4 s, find: (a) the speed of the plane; (b) the horizontal distance traveled by the package after it is released.**

15. (I) (a) An athlete can broad jump 8.5 m. Assume his waistline returns to the initial level. If the takeoff is at 30° to the horizontal, what is the initial speed? (b) An impala can leap 12 m horizontally. If it jumps at 50° to the horizontal, find its initial speed.

16. (I) Electrons are moving initially in the horizontal direction at 3×10^6 m/s. Through what horizontal distance would they have to travel for the vertical fall to be 0.1 mm? Assume they move in a vacuum.

17. (II) A basketball is thrown at 45° above the horizontal. The hoop is located 4 m away horizontally at a height of 0.8 m above the point of release. What is the required initial speed?

18. (I) A daredevil motorcyclist plans to jump across a gorge of width 60 m. He takes off on a 15° ramp. What minimum speed does he require if he lands at the initial level?

19. (I) A baseball is hit at a height of 1 m with an initial velocity of 27 m/s at 32° above the horizontal. A field player is located 50 m from the home plate along the line of flight. (a) If he remains stationary, at what height does the ball pass him? (b) If he runs, is there a chance he could catch it? Assume his reaction time is 0.5 s and that he would catch it at the initial level.

20. (I) A ball rolls off a 1-m-high table and lands 1.6 m along the floor. Find (a) its initial speed and (b) the time of flight.

21. (II) A rocket launched at 70° to the horizontal has a constant net acceleration of 8 m/s² along this direction for 6.5 s and then is in free-fall. Find: (a) the maximum height; (b) the horizontal range.

22. (I) A projectile fired from the ground has a velocity $\mathbf{v} = 24\mathbf{i} - 8\mathbf{j}$ m/s at a height of 9.8 m. Find: (a) the initial velocity; (b) the maximum height.

23. (II) An Olympic athlete completes a long jump of 8.3 m with an initial speed of 9.7 m/s. (a) At what angle did he take off? (b) What was the maximum increase in height of his waistline? (c) How long was he in the air? Assume that on landing his waistline is at the takeoff level.

24. (I) A ball is thrown from ground level. Three seconds later it is moving horizontally at 15 m/s. Find: (a) the horizontal range; (b) the angle of impact.

25. (I) A boy on a 10-m-high balcony throws a ball at 20 m/s directly at a target on the ground 40 m from the base of the building. By what horizontal distance does the ball miss?

26. (I) A body is launched from the ground and 3 s later its velocity is $\mathbf{v} = 20\mathbf{i} - 4\mathbf{j}$ m/s. Find: (a) the horizontal range; (b) the maximum height.

27. (I) Electrons are initially traveling at 2.4×10^6 m/s in the horizontal direction. They enter a region between two charged plates of length 2 cm and experience an acceleration of 4×10^{14} m/s² vertically upward. Find: (a) the vertical position as they leave the region between the plates; (b) the angle at which they emerge from between the plates.

28. (II) Water from a firehose emerges at 18 m/s. At what two angles can the hose be pointed so that the water hits a spot 30 m away at the same level as the spout?

29. (II) A javelin thrown from a height of 2 m at 30° above the horizontal lands on the ground 42 m away horizontally. Find: (a) the initial speed; (b) the time of flight; (c) the maximum height.

30. (II) A stone thrown at 20 m/s at angle θ below the horizontal from a cliff of height H lands 70 m from the base 4 s later. Find θ and H.

31. (II) A medieval catapult, as in Fig. 4.39, could project a 75-kg stone at 50 m/s at 30° above the horizontal. Suppose the target is a fortress wall of height 12 m at a horizontal distance of 200 m. (a) Would the stone hit the wall? (b) If so, at what height, and (c) at what angle?

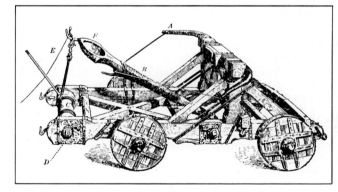

FIGURE 4.39 Exercise 31

32. (II) (a) A projectile is fired from ground level with an initial velocity v_0 at angle θ above the horizontal. Show that the maximum height H is given by

$$H = \frac{(v_0 \sin\theta)^2}{2g}$$

(b) A kangaroo is reputed to have jumped over a 9-ft (2.75-m) fence. If its torso rose 2 m above the initial level and it jumped at 30° to the horizontal, what was the initial speed?

33. (II) A shotputter projects the shot at 42° to the horizontal from a height of 2.1 m. It lands 17 m away horizontally. Next, he gives it the same initial speed but changes the angle to 40°. What effect does this have on the horizontal range?

34. (II) A tennis ball is served at a height of 2.4 m at 30 m/s in the horizontal direction. The net is 0.9 m high at a horizontal distance of 12 m. Does the ball clear the net? If so, by how much? If not, where does it strike the net?

35. (II) A baseball is hit at a height of 1 m with an initial velocity v_0 at 35° above the horizontal. It just clears a barrier of height 29 m at a horizontal distance of 64 m. Find (a) v_0; (b) the time to reach the barrier; (c) the velocity at the barrier.

36. (II) A stone is thrown at 25 m/s at 50° above the horizontal. At what time(s) is its velocity at 30° to the horizontal?

4.5 Uniform Circular Motion

37. (I) Find the centripetal acceleration for each of the following: (a) a point on the earth's equator (ignore its orbital motion); (b) the earth in its orbit around the sun; (c) the sun's orbit around the center of the Milky Way (period = 2×10^8 y; radius = 3×10^{20} m).

38. (I) (a) The electron in a hydrogen atom has a speed of 2.2×10^6 m/s and orbits the proton at a distance of 5.3×10^{-11} m. What is its centripetal acceleration? (b) A neutron star of radius 20 km is found to rotate once per second. What is the centripetal acceleration of a point on its equator?

39. (I) Compute the acceleration for the following in terms of g's where $g = 9.8$ m/s²: (a) a car moving at 100 km/h round a curve of radius 50 m; (b) a jet flying at 1500 km/h and making a turn of radius 5 km; (c) a stone being twirled every 0.5 s at the end of a rope of length 1 m; (d) a speck of dust at the rim of a 12-in. (30-cm) LP turning at $33\frac{1}{3}$ rpm; (e) a molecule in a centrifuge rotating at 30,000 rpm at a radius of 15 cm.

40. (I) A geosynchronous satellite is one that appears to be fixed in the sky, a condition useful for telecommunications. If such a satellite orbits the earth at an altitude of 35,800 km above the earth's surface, what is its centripetal acceleration?

41. (I) A doughnut-shaped space station has an outer rim of radius 1 km. With what period should it rotate for a person at the rim to experience an acceleration of $g/5$?

42. (I) Suppose the earth's rotation speeded up such that the centripetal acceleration at the equator equaled g. What would be the length of a "day"?

43. (I) In a conical pendulum, a bob is suspended at the end of a string and describes a horizontal circle at a constant speed of 1.21 m/s (see Fig. 4.40). If the length of the string is 1.2 m and it makes an angle of 20° with the vertical, find the acceleration of the bob.

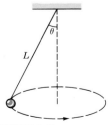

FIGURE 4.40

44. (I) The cornering ability of a car is specified in terms of the lateral acceleration it can sustain while going round a curve of radius 45.7 m. A 1986 Corvette was rated at 0.84g. At what speed did it travel?

45. (I) A particle travels in a circular path of circumference 8 m and makes 5 revolutions per second. What is its centripetal acceleration?

46. (I) A stone moves in a circle of radius 60 cm and has a centripetal acceleration of 90 m/s². How long does it take to make 8 revolutions.

47. (I) The French TGV (*train à grande vitesse*), shown in Fig. 4.41, travels at a speed of 300 km/h. If passengers are not to be subject to more than 0.05g, what is the minimum radius of the track?

FIGURE 4.41 Exercise 47

4.7 Relative Velocity

48. (I) (a) Object A has a velocity of $2\mathbf{i} + \mathbf{j}$ m/s while object B has a velocity of $-\mathbf{i} + 5\mathbf{j}$ m/s. What is the velocity of B relative to A? (b) Object A is moving east at 3 m/s while object B is moving north at 4 m/s. What is the velocity of A relative to B?

49. (I) A ship is sailing west at 5 km/h relative to land. To people on the ship, a balloon appears to move away horizontally at 10 km/h at 37° S of E. What is the wind velocity relative to land?

50. (I) A boat can move at 4 m/s relative to still water. It has to cross a river flowing east at 3 m/s. The river is 100 m wide. (a) If the boat is pointed straight across, how long does the crossing take? (b) In what direction should the boat be pointed to get straight across? How long does this take?

51. (I) Two sailors decide to have an unusual race. Each has a powerboat that travels at 10 m/s in still water. However, the river, of width 100 m, flows at 5 m/s. Sailor A will go to a point directly across the starting point and return. Sailor B will travel 100 m downstream and then return. Who wins?

52. (I) Rain falls vertically at a constant 10 m/s. A tube is mounted on a railcar moving horizontally at 20 m/s. At what angle should the tube be tilted so that the water does not touch the sides?

53. (II) A plane can fly at 200 km/h in still air and has to reach a destination 600 km NE. A 50-km/h wind blows from the west. (a) What is the required heading of the plane? (b) How long does the journey take? (*Hint*: Use the law of sines.)

54. (II) A pilot has to reach a town 400 km due north in 1 h. A crosswind of 80 km/h is blowing toward 37° N of E. (a) In which direction should the plane be headed? (b) What is the plane's airspeed?

55. (II) A boat can move at 5 m/s relative to still water. It has to cross a river of width 100 m that flows east at 4 m/s. The boat has to get to a point 50 m downstream. (a) In which direction should it be pointed? (b) How long does the trip take? (*Hint*: Use the law of sines.)

56. (II) A boat is traveling east at 12 m/s. A flag flaps at 53° N of W at the bow. Another flag on shore flaps due north. Find the speed of the wind as measured (a) on land, and (b) on the boat.

57. (II) An airplane compass indicates a heading of 30° N of W and the airspeed is measured to be 180 km/h. Ground control informs the pilot that there is a 40 km/h wind blowing toward the NE. What is the plane's velocity relative to the ground?

58. (II) Two go-carts, A and B, travel at 10 and 15 m/s, respectively, around a circular path of circumference 100 m. They start together due north of the center and travel clockwise. (a) Find the velocity of B relative to A after 8 s. (b) What is the first time at which the velocity of B relative to A has its greatest magnitude?

59. (II) Two ships approach each other with constant velocities that are not parallel. Show that if the bearing of one ship relative to another is constant, they will collide.

60. (II) A fish swimming east at 3 km/h is spotted by a penguin that is 50 m due south. If the penguin is capable of moving at 30 km/h under water, how quickly could it have a meal?

4.9 Nonuniform Circular Motion

61. (I) A car goes around a curve of radius 40 m. When its velocity points north its speed is changing at 2 m/s² and its total acceleration is at 30° N of W. (a) Is the car speeding up or slowing down? (b) What is its speed at this instant?

62. (II) The speed of a car moving along a circular track of radius 50 m is increased uniformly in time. When the car is east of the center, its total acceleration is 10 m/s² at 37° W of N. (a) Find its radial and tangential accelerations. (b) How long does it take to return to the same point?

63. (I) A particle travels in a circle of radius 4 m. Its speed increases at 2 m/s² and its centripetal acceleration is 6 m/s². Find: (a) the magnitude of its total linear acceleration; (b) its speed.

PROBLEMS

1. (I) Two balls are thrown from the top of a cliff with equal initial speeds. One starts at θ above the horizontal while the other starts at θ below. Show that the difference in their ranges is $v_0^2 \sin 2\theta/g$.

2. (I) A ball is thrown with speed v_0 at an angle θ above the horizontal from the top of a building of height $2R$. It lands at a distance R from the base. Show that

$$R = \frac{2v_0^2 \cos^2 \theta (2 + \tan \theta)}{g}$$

3. (I) A plane is flown horizontally at a speed V at an altitude H. The pilot has to drop supplies to a liferaft. (a) At what angle to the vertical must she sight the target at the time of dropping the load? (b) At this instant, what is the horizontal distance to the liferaft?

4. (I) A football player kicks a ball on the ground. The horizontal bar of the goalpost is 3 m high and at a horizontal distance of 10 m. If the initial speed of the ball is 15 m/s, what range of initial angles is available to him?

5. (I) A projectile lands at the level from which it was fired. (a) For what initial angle is the horizontal range equal to the maximum height? (b) For what initial angle is the range equal to half the maximum height?

6. (II) A projectile is fired up an incline with an initial speed v_0 at an angle θ to the horizontal, as shown in Fig. 4.42. (a) If the angle of the incline is α to the horizontal, show that the range along the incline is

$$R = \frac{2v_0^2 \sin(\theta - \alpha) \cos \theta}{g \cos^2 \alpha}$$

(b) Recast this expression for R by using a trigonometric identity for $\sin A \cos B$. For what value of θ is R a maximum? (c) Show that the maximum value of R is

$$R = \frac{v_0^2}{g(1 + \sin \alpha)}$$

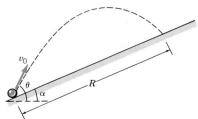

FIGURE 4.42 Problem 6

7. (II) Show that if the projectile in Problem 6 is fired down the incline, the maximum range along the incline is

$$R = \frac{v_0^2}{g(1 - \sin \alpha)}$$

8. (II) For a given horizontal range, less than the maximum, there are two possible projection angles, θ_1 and θ_2. If $\theta_1 < \theta_2$, show that the ratio of the times of flight is

$$\frac{T_1}{T_2} = \tan \theta_1$$

(*Hint*: See Example 4.2. Obtain two expressions for T_1/T_2, equate them, and then solve a quadratic equation in $(T_1/T_2)^2$.)

9. (I) A particle is projected at angle α below 45° and another at α above 45°. They land at the level from which they were fired. Show that the difference in their times of flight is

$$T_2 - T_1 = \frac{2\sqrt{2}\, v_0 \sin \alpha}{g}$$

10. (I) A stone is thrown from a clifftop of height H. Show that the speed at which it lands is independent of the angle of projection.

11. (I) A cannon on the ground can fire a ball at 50 m/s at 53° above the horizontal. The target is 60 m above the ground. At what horizontal distance from the target should the cannon be located?

12. (II) The position of a particle as a function of time t is given by

$$\mathbf{r} = A \cos(\omega t)\mathbf{i} + A \sin(\omega t)\mathbf{j}$$

where A and ω are constants. (a) What is the shape of the path? (Let $A = 1$ m and $\omega = 0.1\pi$ rad/s.) (b) Find the velocity and the acceleration. (c) Calculate the speed. (d) Show that

$$\mathbf{a} = -\omega^2 \mathbf{r} = -\frac{v^2}{r}\hat{\mathbf{r}}$$

13. (II) Ship A sails east at 3 m/s while ship B is 100 km to the northeast of A and heading south at 4 m/s. (a) What is the velocity of B relative to A? (b) If the velocities are maintained, what is their distance of closest approach? Use a reference frame in which A is at rest. (c) If A has a radio with a 20-km range, how long can the ships communicate?

14. (II) A skier takes off at 108 km/h at 10° above the horizontal. How far along the 20° slope does she land? See Fig. 4.43.

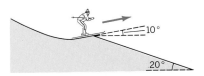

FIGURE 4.43 Problem 14

15. (I) A soccer player kicks a ball at 15 m/s at 20° above the horizontal to another member of the team. At what speed should the second player run in order to reach the ball just as it lands? The initial separation of the players is 25 m.

16. (II) A basketball player throws the ball with initial velocity v_0 at θ above the horizontal to the hoop which is located a horizontal distance L and at a height h above the point of release, as shown in Fig. 4.44. (a) Show that the initial speed required is given by

$$v_0^2 = \frac{gL}{2 \cos^2 \theta(\tan \theta - h/L)}$$

(b) Show that the angle α to the horizontal at which it reaches the hoop is given by $\tan \alpha = 2h/L - \tan \theta$.

FIGURE 4.44 Problem 16

17. (II) A car goes out of control and slides off a steep embankment of height h at θ to the horizontal (see Fig. 4.45). It lands in a ditch at a distance R from the base. Show that the speed at which the car left the slope was

$$v_0 = \frac{R}{\cos \theta} \sqrt{\frac{g}{2(h - R \tan \theta)}}$$

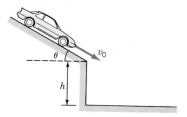

FIGURE 4.45 Problem 17

18. (II) A projectile is fired with initial speed v_0 at angle θ above the horizontal from a point at height h, as in Fig. 4.46. Show

that the horizontal range is

$$R = \frac{v_0^2 \sin 2\theta}{2g} \left(1 + \sqrt{1 + \frac{2gh}{v_0^2 \sin^2 \theta}} \right)$$

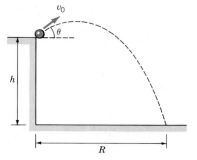

FIGURE 4.46

19. (I) At a certain instant two cars are each 10 km from the intersection of two roads that are perpendicular. Car A is moving east at 30 km/h while car B moves north at 50 km/h, both toward the intersection. Find: (a) the velocity of B relative to A; (b) their distance of closest approach. Use a reference frame in which A is at rest. (c) Where are A and B when they are closest?

20. (I) A ship sails east at 3 m/s. Initially it is 2 km at 40° W of N relative to a tug that is capable of 6 m/s. (a) In which direction should the tug head to rendezvous in the shortest time? (b) How long does this take? (*Hint*: See Exercise 4.59. Use the law of sines.)

21. (I) Prove Galileo's prediction, depicted in Fig. 4.14, that for initial angles of projection, $\theta_1 = 45° - \alpha$ and $\theta_2 = 45° + \alpha$, the horizontal ranges are the same.

CHAPTER 5

Particle Dynamics I

A space-shuttle lift-off.

Major Points

1. The concepts of **force** and **mass.**
2. **Newton's second law** relates the acceleration of a particle to the net force acting on it.
3. The distinctions between **mass, weight,** and **apparent weight.**
4. **Newton's third law.** The identification of action–reaction pairs.
5. The use of a **free-body diagram** to analyze the forces acting on a body.

This chapter marks an important transition. We progress from the descriptions of kinematics (*how* bodies move) to the explanations of **dynamics** (*why* bodies move). Dynamics is a branch of mechanics in which the concept of force is used to account for the accelerated motion of bodies.

In everyday language, the term ''force'' is used quite broadly. For example, the expressions ''to apply force'' and ''to apply pressure'' are used interchangeably. An explosion may be said to involve tremendous force, or power, or energy, as if all these terms mean the same thing. Even among scientists, this vagueness existed for a long time. Galileo associated the term ''force'' only with the action of machines, such as levers and pulleys. He considered free-fall to be ''natural motion,'' so he did not classify weight (the pull of gravity) as a force. According to his contemporary, Descartes, the ''forces of machines,'' the ''forces of impact'' in collisions, the ''centrifugal force'' in circular motion, and the ''force of gravity'' were all distinct physical quantities. He and others referred to inertia as ''the force of a moving body'' (to continue its motion). In this case, force was something *possessed* by a body, rather than something that *acts* on an object. In other fields one would come across ''vital forces'' in living things, or the catch-all ''forces of nature.'' The word ''force'' covered quite a range!

In his second law of motion, published in 1687, Isaac Newton (Fig. 5.1) presented a clear definition of what he meant by ''motive force.'' He then demonstrated its usefulness by applying it brilliantly to a variety of problems, including the motion of the planets about the sun. Nonetheless the word ''force'' was pregnant with so many other meanings that even in scientific circles it took nearly 200 years before Newton's version was universally accepted. This is not to imply that the other meanings were wrong, but simply that a choice had to be made. Confusion regarding the exact meaning of scientific terms invariably impedes the

FIGURE 5.1 Sir Isaac Newton (1642–1727).

understanding and development of any subject. The present scientific meaning of force is derived from, but not identical to, Newton's definition.

Newtonian dynamics was the cornerstone for the development of physics for nearly two centuries. However, in this century its limitations have become apparent. The theory of special relativity, published in 1905, showed that classical mechanics is not correct when applied to a particle whose speed is an appreciable fraction of the speed of light (3×10^8 m/s). Furthermore, classical mechanics completely fails to account for the behavior of atoms. The theory of quantum mechanics, developed in the 1920s, has proven successful in this domain. Despite these restrictions, classical mechanics applies to a wide range of problems and so it still merits careful study.

Newton's second law of motion introduces two concepts, force and mass, in the same equation. It appears that one unknown has been defined in terms of another. It is difficult to avoid this logical flaw in presenting the laws of dynamics. Indeed, it was part of Newton's genius that he deftly skirted the issue, and got on with the job. We will first try to distinguish clearly between these two concepts and then present the second law as the result of a simple experiment. Newton's own approach is discussed in Chapter 9.

5.1 FORCE AND MASS

In physics, we build on our intuitive understanding of **force** as either a push or a pull. Although forces are not visible, we do see and experience their effects. Forces cause bodies to deform: They expand springs, compress balloons, and bend beams. Also, as Newton's first law implies, a net (unbalanced) force will cause a body to change its velocity. These two effects are often seen together. For example, a ball struck by a bat is both deformed and accelerated.

Although every force is a manifestation of one of the fundamental interactions mentioned in Section 1.1, it is sometimes convenient to refer to a force as being either a **contact** force or an **action at a distance.** Contact forces arise when one body is in physical contact with another. Examples include the forces exerted by ropes or springs, the forces involved in collisions, the force of friction between two surfaces, and the force exerted by a fluid on its container. Contact forces are the result of the electromagnetic interaction between the surface atoms of the two bodies. An action at a distance is manifested when two bodies, such as the earth and the sun, interact without any material medium between them. Magnets and electric charges can also interact in a vacuum.

In order to measure and compare forces, we may use the expansion of a "standard" spring A, shown in Fig. 5.2a. We might define a force of one unit as the action that extends the spring by one unit of length. To measure larger forces, spring A must be calibrated. We take springs B and C and apply the same force (one unit) to each. By the simple method depicted in Fig. 5.2b, a force of two units is applied to A, and so a "2" can be marked on the scale. This procedure is easily repeated for larger forces. (Do springs B and C have to be identical to A?)

The above definition of force in terms of the static deformation of a spring seems reasonable. It lets us flex our muscles and feel the force, and it is not intermingled with any other new concept. However, it is also impractical. It would be difficult to produce identical "standard" springs for use in different laboratories and to maintain them under closely specified conditions. More importantly, one cannot attach a spring to a tiny atom to measure the force acting on it. Before we abandon this definition, we can use it to establish an important result.

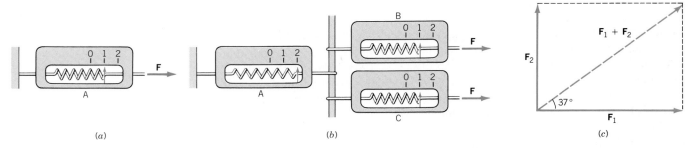

FIGURE 5.2 (*a*) Force may be defined in terms of the static extension of a "standard" spring A. (*b*) A procedure for calibrating the standard spring. (*c*) A force of 3 units directed toward the east and a force of 4 units directed north act on an object. Experiment shows that the same effect is produced by a force of 5 unit directed at 37° N of E. This confirms the vector nature of force.

Force is clearly a quantity that has both magnitude and direction. But before one can call it a vector, one must verify that it obeys the law of vector addition. Figure 5.2*c* shows a force F_1 of 4 units directed due east and a force F_2 of 3 units directed due north. Experiment shows that the combined effect of these two forces is the same as that of a force of 5 units in the direction 37° N of E. This is simply the magnitude and direction of the vector sum of these two forces, which confirms the vector nature of force.

Mass

Newton defined mass as the "quantity of matter" in a body. This expresses an intuitive feeling, but is not useful. How, for example, are we to compare quantities of matter? Newton's first law provides the clue to a better definition:

> The mass of a body is a measure of its inertia, that is, its resistance to change in velocity.

The mass of a body is a scalar quantity that tells us how difficult it is to change the magnitude, or direction, of its velocity. Mass is an intrinsic property of a body, independent of its location.

To compare the masses of two pucks, we place them on an airtable—which supports them on a cushion of air. If they either stay at rest or move at constant velocity, we conclude that they are subject to no net force. (The downward pull of gravity is balanced by the upward push due to the air.) The pucks are then placed on either side of a spring that is held in a compressed state by a thread, as shown in Fig. 5.3*a*. When the thread is cut, the pucks are pushed apart and undergo changes in velocity $\Delta\mathbf{v}_A$ and $\Delta\mathbf{v}_B$, shown in Fig. 5.3*b*. We discover that for any time interval, the ratio $|\Delta\mathbf{v}_A|/|\Delta\mathbf{v}_B|$ is a *fixed* number for the two given pucks. The nature of the interaction does not matter. They can interact via the spring, via magnets attached to them, or directly with each other as in a collision. Since mass is a measure of resistance to change in velocity, we define the ratio of the masses to be

$$\frac{m_A}{m_B} = \frac{|\Delta\mathbf{v}_B|}{|\Delta\mathbf{v}_A|} \tag{5.1}$$

Once the standard kilogram has been chosen, we may in principle determine the mass of any other body. (Are the subscripts in Eq. 5.1 in their proper positions?)

Mass

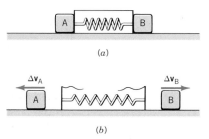

FIGURE 5.3 (*a*) Two pucks are placed on either side of compressed spring. (*b*) When the thread is cut, the ratio of the changes in velocity of the pucks may be used to define the ratio of their masses.

Determining the mass

5.2 NEWTON'S SECOND LAW

From Newton's first law we know that a net force acting on a body will produce an acceleration. To determine how the acceleration depends on the force and the mass of the body, these quantities must be varied one at a time. When the mass is held fixed and the force is varied, we find that $a \propto F$. When the force is kept constant and the mass is varied, we find $a \propto 1/m$. These results, depicted in Fig. 5.4, tell us that $a \propto F/m$. We conclude that $F = kma$, where k is a constant of proportionality. In SI units, a force of one newton (N) applied to mass of 1 kg produces an acceleration of 1 m/s², so $k = 1$:

$$1 \text{ N} = 1 \text{ kg·m/s}^2$$

The weight of an orange is about 1.5 N.

When several forces act on a particle, we must take their vector sum, so **Newton's second law** of motion is written as

$$\Sigma \mathbf{F} = m\mathbf{a} \qquad (5.2)$$

The net force, $\Sigma\mathbf{F}$, acting on a particle of mass m produces an acceleration $\mathbf{a} = \Sigma\mathbf{F}/m$ in the direction of the net force.

Equation 5.2 is equivalent to three equations in terms of the components:

$$\Sigma F_x = ma_x \qquad \Sigma F_y = ma_y \qquad \Sigma F_z = ma_z \qquad (5.3)$$

Notice in Fig. 5.5 that the direction of motion of a particle (given by its velocity) does not in general coincide with the direction of the force acting on it. It is the *rate of change* of the velocity that is related to the force. Note also that the acceleration must be measured with respect to an inertial reference frame, that is, one in which Newton's first law is valid. We may now abandon the definition of force in terms of the spring's extension and use Newton's second law to indicate the meaning of force. In Chapter 4 we saw that the acceleration of a body is the same in all inertial reference frames. Since $\Sigma\mathbf{F} = m\mathbf{a}$, Newton's second law of motion has the same form in all inertial frames.

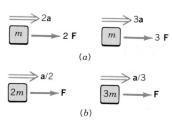

FIGURE 5.4 (*a*) For a fixed mass, the acceleration is proportional to the force; that is, $a \propto F$. (*b*) For a fixed force, the acceleration is inversely proportional to the mass; that is, $a \propto 1/m$.

Newton's second law

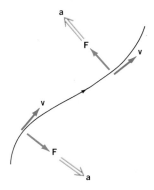

FIGURE 5.5 In general, the direction of motion of a particle (given by the velocity) does not coincide with the direction of its acceleration.

EXAMPLE 5.1: The total thrust of the engines on a Boeing 747 is 8.8×10^5 N. The maximum lift-off mass of this aircraft is 3.0×10^5 kg. (a) What is the maximum acceleration possible during the takeoff run? (b) Starting from rest, how fast would the plane be moving after 10 s? Ignore the retarding forces exerted by the air and the ground.

Solution: (a) Since we assume that the thrust is the only horizontal force acting, the acceleration is

$$a = \frac{F}{m} = \frac{8.8 \times 10^5 \text{ N}}{3.0 \times 10^5 \text{ kg}}$$
$$= 2.9 \text{ m/s}^2$$

(b) The velocity is given by $v = v_0 + at = 0 + (2.9 \text{ m/s}^2)(10 \text{ s}) = 29$ m/s. This is about 104 km/h.

EXAMPLE 5.2: A 1200-kg car is stalled on an icy patch of road. Two ropes attached to it are used to exert forces $F_1 = 800$ N at 35° N of E and $F_2 = 600$ N at 25° S of E. What is the acceleration of the car? Treat it as a particle and assume friction is negligible.

Solution: To apply Newton's second law, one must first choose a coordinate system. This is indicated in Fig. 5.6*a*. Next, it is necessary to resolve the forces along the axes and to write the component form of the law, Eq. 5.3. It is best to redraw the forces (dashed lines) and their components (solid lines) on a separate diagram as is done in Fig. 5.6*b*. This is called a *free-body diagram* and will be discussed further in Section 5.5.

The vector form of Newton's second law is

$$\Sigma\mathbf{F} = \mathbf{F}_1 + \mathbf{F}_2 = m\mathbf{a}$$

The components of this equation are

$$\Sigma F_x = F_1 \cos \theta_1 + F_2 \cos \theta_2 = ma_x \qquad (i)$$
$$\Sigma F_y = F_1 \sin \theta_1 - F_2 \sin \theta_2 = ma_y \qquad (ii)$$

On substituting the given values we find

$$a_x = \frac{800 \text{ N} \times 0.819 + 600 \text{ N} \times 0.906}{1200 \text{ kg}} = 1.00 \text{ m/s}^2$$

$$a_y = \frac{800 \text{ N} \times 0.574 - 600 \text{ N} \times 0.423}{1200 \text{ kg}} = 0.17 \text{ m/s}^2$$

The resultant acceleration is $\mathbf{a} = 1.00\mathbf{i} + 0.17\mathbf{j}$ m/s^2.

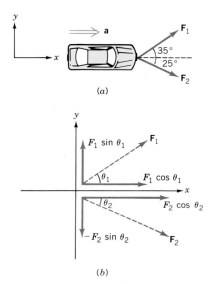

(a)

(b)

FIGURE 5.6 (a) The forces acting on the car and the coordinate system are indicated. (b) The forces (dashed lines) are resolved into their components (solid lines).

EXAMPLE 5.3: An electron of mass 9.1×10^{-31} kg has an initial velocity $\mathbf{v}_0 = 10^6\mathbf{i}$ m/s. It enters a region in which it experiences a force $\mathbf{F} = 8 \times 10^{-17}\mathbf{j}$ N for a period of 10^{-8} s. What is its velocity as it emerges from the region?

Solution: Since the force is constant, the acceleration is also constant and the final velocity can be found from the equation $\mathbf{v} = \mathbf{v}_0 + \mathbf{a}t$, where $\mathbf{a} = \mathbf{F}/m$. There is no acceleration in the x direction, so $v_x = v_{0x} = 10^6$ m/s. In the y direction $a_y = F_y/m$; therefore,

$$v_y = v_{0y} + \frac{F_y}{m} t$$

$$= 0 + \frac{8 \times 10^{-17} \text{ N}}{9.1 \times 10^{-31} \text{ kg}} \times 10^{-8} \text{ s}$$

$$= 8.8 \times 10^5 \text{ m/s}$$

The final velocity is $\mathbf{v} = 10^6\mathbf{i} + 8.8 \times 10^5\mathbf{j}$ m/s. The angle at which it emerges relative to the x axis is found from

$$\tan \theta = \frac{v_y}{v_x} = 0.88$$

which yields $\theta = 41.3°$. Figure 5.7 shows a sketch of the path of the electron from a time before it enters the region to some time after it has emerged. (Identify the shape of each segment of the path.)

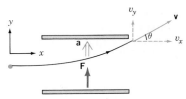

FIGURE 5.7 A electron experiencing constant acceleration as it passes between two charged plates. The angle at which it emerges from the plates is given by $\tan \theta = v_y/v_x$.

5.3 WEIGHT

As part of an attempt to explain the motion of the planets around the sun, in 1687 Newton put forward a **law of universal gravitation.** It states that between any two point particles with masses m and M, separated by a distance r as in Fig. 5.8a, there is an attractive force whose magnitude is given by

Newton's law of gravitation

$$F = \frac{GmM}{r^2} \qquad (5.4)$$

where $G = 6.67 \times 10^{-11}$ N·m^2/kg^2.

Weight

The **weight** of an object is the net gravitational force acting on it.

It turns out (see Chapter 13) that for the special case of a body whose mass distribution is spherically symmetric (a spherical shell or a solid sphere), Eq. 5.4 may be used with r taken as the distance to the center. Thus, if the earth is taken to be a uniform sphere of mass M_E and radius R_E, as in Fig. 5.8*b*, the weight of an object of mass m at its surface is

$$W = \frac{GmM_E}{R_E^2} \tag{5.5}$$

This equation is usually written in the form

$$\mathbf{W} = m\mathbf{g} \tag{5.6}$$

where $g = GM_E/R_E^2$, measured in N/kg, is the magnitude of the gravitational *force per unit mass* on an object at the surface and is called the **gravitational field strength** at the surface. Since $W = mg$ has the same form as $F = ma$ and the unit N/kg reduces to m/s², g is often called the "acceleration due to gravity"—for which we have been using the symbol g. There are two pitfalls in doing this. First, it might create the impression that (true) weight depends on the acceleration of the object. In fact, as shown in Fig. 5.9*a*, the weight of an apple is the same whether it is in free-fall or at rest on a table. Second, because of the earth's rotation, the measured value of the acceleration due to gravity is not the same as it would be if the earth were not spinning. It is only in an inertial frame that the measured acceleration due to gravity (g) is exactly equal to the gravitational force per unit mass (g). In most situations one can ignore the numerical difference between these quantities. Keep in mind however, that they are different concepts.

Mass versus Weight

The concepts of mass (a measure of the inertia of a body) and weight (the net gravitational force on a body) are often confused. The commercial and industrial use of the kilogram as a unit of weight further complicates the issue. The first point to note is that mass is a scalar measured in kilograms, whereas weight is a vector measured in newtons. Second, whereas mass is an intrinsic property that *does not change with position,* the weight of a body depends on the local value of g, which

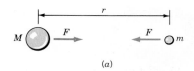
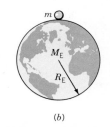

FIGURE 5.8 (*a*) The magnitude of the gravitational force exerted by two point particles on each other is given by $F = GmM/r^2$. (*b*) If the earth is considered to be a uniform sphere, one can use Newton's law of gravitation by assuming all the mass is concentrated at the center.

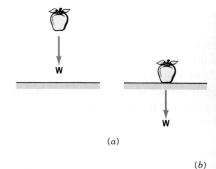

FIGURE 5.9 (*a*) The weight of an apple is the same whether it is in free-fall or at rest on the ground. (*b*) When astronaut Bill Fisher pulled *Leasat 3* into the payload bay of the space shuttle, he had to contend with its mass not its weight.

Although the mass of Edwin Aldrin, Jr., had not changed, his weight on the moon was roughly one-sixth his weight on earth.

varies because of the nonspherical shape of the earth, oil and mineral deposits, altitude, and so on. If a body were taken to the moon, its weight would be only one-sixth of its value on earth. In outer space, far from any stars or planets, it would be essentially weightless. However, if we were to kick the body in its weightless state, it would still hurt our toes. Its mass, that is, its resistance to change in velocity, would not have changed (see Fig. 5.9b).

Strange as it may seem, it is very difficult to measure weight directly. The common equal arm balance only compares an unknown weight with a standard weight. The spring or bathroom scale does not measure (true) weight as we have defined it. Instead it measures the *apparent* weight (see Section 5.6), a quantity that depends on the acceleration of the body. The scale would read true weight only in an inertial frame. But, because of its rotation and orbital motion, the earth is not an inertial frame. For this reason, some physicists actually define weight in terms of the scale reading. In practice the corrections needed to obtain the true weight from the scale reading are of the order of 0.1% and can usually be ignored. (Corrections for the buoyancy of air are often made in chemical laboratories.)

EXERCISE 1. True or false: To start moving a book along a frictionless table, you must first overcome its weight.

EXERCISE 2. The label on a bottle specifies its contents as 1 lb (454 g). Comment on this.

FIGURE 5.10 According to Newton's third law, the force exerted on A by B is equal and opposite to the force exerted on B by A: $\mathbf{F}_{AB} = -\mathbf{F}_{BA}$

Newton's third law

5.4   NEWTON'S THIRD LAW

When you push against a wall, or pull at it, you experience a force in the opposite direction. The harder you push (or pull), the harder the wall pushes (or pulls) back. In fact, the force exerted by the wall on you is exactly equal in magnitude and opposite in direction to the force exerted by you on the wall. This is an example of **Newton's third law,** the law of action and reaction. Consider the two bodies shown in Fig. 5.10. With the notation $\mathbf{F}_{AB}$ for the force exerted on A by B, the third law may be written as

$$\mathbf{F}_{AB} = -\mathbf{F}_{BA} \tag{5.7}$$

The force exerted on A by B is equal and opposite to the force exerted on B by A.

Notice that a force is exerted *by* one body *on* another; one should not refer to the force *of* a body. The third law says that there is no such thing as an isolated force; forces always occur in pairs.

The popular expression "To every action there is an equal and opposite reaction" does not emphasize that the forces act on *different* bodies. It is a bit startling to realize that an apple pulls the earth just as hard as the earth pulls the apple, as shown in Fig. 5.11. Because of its obvious motion when dropped, it seems that only the apple is affected by the force of gravity. However, since the equal and opposite forces act on different bodies, the consequences of the forces may be vastly different. (When someone steps on your toes, it really does not help to know that your toes exerted an equal and opposite force on the other person's shoe.) Since $a = F/m$ and $\Delta y = \frac{1}{2}at^2$, the ratio of the displacement of the apple to that of the earth is $\Delta y_A/\Delta y_E = m_E/m_A \approx 10^{24}$.

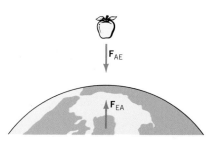

FIGURE 5.11 The apple pulls the earth with the same force that the earth pulls the apple.

In Fig. 5.12a, an apple is at rest on a table. The net force on it is zero: Its weight **W** is balanced by the "contact" force **N** exerted by the table normal (perpendicular) to its plane. Therefore, $\mathbf{N} + \mathbf{W} = 0$ and $\mathbf{N} = -\mathbf{W}$. These forces are equal and opposite, but **N** and **W** are *not* an action–reaction pair, because they act on the *same* body. The magnitudes of the forces, N and W, happen to be equal only because the apple has no acceleration, which is a consequence of Newton's *second* law. If the table were in an elevator accelerating up or down, N would no longer equal W (see Section 5.6). Furthermore, they are completely different kinds of force: $\mathbf{W} = \mathbf{F}_{AE}$ is the gravitational force exerted on the apple by the earth, whereas $\mathbf{N} = \mathbf{F}_{AT}$ is an electromagnetic force exerted on the apple by the surface atoms of the table. Figure 5.12*b* identifies the proper action–reaction pairs.

The third law is valid only in inertial frames, although the interacting particles themselves may be accelerating. This is equivalent to saying that it deals only with real forces associated with the basic interactions discussed in Chapter 1. When the brakes are applied on a bus, a passenger might claim that a "force" throws him forward. There is no source or agent responsible for this force and hence neither is there a reaction to it, as would be required by the third law.

5.5 APPLICATIONS OF NEWTON'S LAWS

To analyze the stability of a structure or the operation of machinery, one must be able to isolate its various components and to identify the forces acting on them. Although we deal mostly with simple, idealized situations, the following problem-solving guide should help you develop the techniques needed to analyze more complex situations.

Problem-Solving Guide for Dynamics

1. Draw a large and neat *diagram* to represent the physical situation. Identify the *system,* one or more bodies, whose dynamics you are studying.

2. Draw all the *external* forces acting *on* each body. If you can imagine yourself in its place, ask what forces are exerted on you by your environment. If you cannot identify the source or agent responsible for a force, such as the earth, a table, or a rope, do not include it.

3. Select an *inertial* reference frame. Each particle may have its own coordinate axes. Usually, the dynamics is most clear if one axis coincides with the direction of the acceleration, given or assumed.

4. Draw a **free-body diagram** for each mass: The particle is put at the origin and all the forces acting on it are resolved along the axes.

   NOTE: The quantity ma is *not* to be included in the free-body diagram. It is equal to the *resultant* of the forces acting on the particle.

5. Use the free-body diagram to write the second law in *component* form:

$$\Sigma F_x = ma_x; \qquad \Sigma F_y = ma_y$$

Solve these equations for the unknowns.

6. Are your results *reasonable?* Here are some points worth checking:
   (a) Does a negative sign have an obvious interpretation or does it indicate an earlier mistake or false assumption?
   (b) If you have an algebraic expression, check its dimensions. Try extreme values of the variables that appear to see if they lead to results you already know.

FIGURE 5.12 (*a*) The "equal and opposite" forces on an apple at rest on a table are *not* an action–reaction pair. (*b*) The action–reaction pairs for an apple on a tabletop.

Can you estimate the acceleration of these riders on the Magnum roller coaster at Sandusky, Ohio?

FIGURE 5.13 The tension in a rope is the force exerted by one segment, such as B, on an adjacent segment, such as A or C. If the rope is massless, then $T_1 = T_2$. That is, *the tension is the same at all points along a massless rope*.

$\Sigma \mathbf{F} = 0$ for a particle in equilibrium

Before we proceed, we must clarify the meaning of the term "tension." The **tension** in a rope is the pulling force exerted by one section of the rope on an adjacent section or on an object attached to its end. If the rope were cut and a spring scale inserted, its reading would indicate the tension at that point. In Fig. 5.13 we imagine a rope is cut at two points. If the rope has mass and is accelerating, or hanging vertically, the tensions T_1 and T_2 acting on the middle segment will be different. Unless otherwise indicated, we will assume that ropes and strings are massless, which means that the *tension has the same value at all points*.

We begin the applications of Newton's second law with a special case. When the sum of the forces acting on a particle is zero, it is said to be in **translational equilibrium.** If the particle is at rest, it is in **static equilibrium,** whereas if it is moving at constant velocity, it is in **dynamic equilibrium.** In either case, $\Sigma\mathbf{F} = 0$.

EXAMPLE 5.4: A picture frame of weight 20 N is suspended by two ropes as shown in Fig. 5.14a. Find the tensions in the ropes given that $\theta_1 = 30°$ and $\theta_2 = 45°$.

Solution: We identify the picture frame as our system. In addition to its weight $\mathbf{W}$ (an action at a distance), the frame is subject to the tensions $\mathbf{T}_1$ and $\mathbf{T}_2$ (contact forces). Using the indicated coordinate axes, the components of the forces are drawn in the free-body diagram, Fig. 5.14b. This is also called an *isolation diagram* because the system is extracted from its surroundings in order that we might concentrate solely on its dynamics. The vector form of Newton's second law is

$$\Sigma\mathbf{F} = \mathbf{T}_1 + \mathbf{T}_2 + \mathbf{W} = 0$$

The components of this equation are

$$\Sigma F_x = T_1 \cos 30° - T_2 \cos 45° = 0 \qquad \text{(i)}$$
$$\Sigma F_y = T_1 \sin 30° + T_2 \sin 45° - W = 0 \qquad \text{(ii)}$$

From (i) we have $T_2 = T_1 \cos 30°/\cos 45° = 1.22\ T_1$. Substituting this into (ii) we find $T_1(0.5 + 1.22 \times 0.707) - 20 = 0$. Finally, $T_1 = 14.7$ N and $T_2 = 17.9$ N.

EXAMPLE 5.5: A skier of mass 60 kg slides down an icy (frictionless) slope which is inclined at 20° to the horizontal. Find her acceleration and the force exerted on her by the slope. Use the following coordinate systems: (a) the x axis is horizontal; (b) the x axis points down along the incline.

Solution: A sketch is shown in Fig. 5.15a. The skier is subject only to her weight $\mathbf{W}$ and the normal force $\mathbf{N}$ perpendicular to the incline. N is an example of a *constraint force:* It forces the particle to travel along a certain path. The skier's acceleration is determined by the net force:

$$\Sigma\mathbf{F} = \mathbf{N} + \mathbf{W} = m\mathbf{a} \qquad \text{(i)}$$

Since there is no motion perpendicular to the slope, the acceleration must point down along the incline.

(a) When the x axis is horizontal, the acceleration has components along both the x and the y axes: $a_x = a \cos\theta$ and $a_y = -a \sin\theta$. The free-body diagram is drawn in Fig. 5.15b. The components of (i) are

$$\Sigma F_x = N \sin\theta + 0 = ma \cos\theta \qquad \text{(ii)}$$
$$\Sigma F_y = N \cos\theta - W = -ma \sin\theta \qquad \text{(iii)}$$

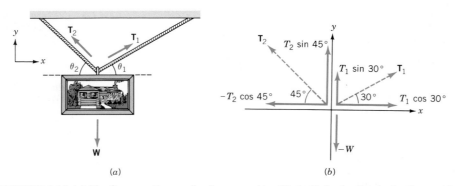

(a)   (b)

FIGURE 5.14 (a) The forces acting on the frame are identified. (b) In the free-body diagram the tails of the forces (dashed lines) are placed at the origin and resolved into the components (solid lines) along the chosen axes.

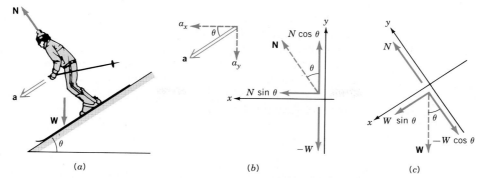

(a)   (b)   (c)

FIGURE 5.15 (a) The forces acting on a skier on a frictionless slope. (b) A poor choice of coordinate system since the acceleration has components along both axes. (c) In this tilted coordinate system the acceleration is directed along the +x axis and thus has only one (positive) component.

The acceleration appears in both equations and is multiplied by a negative factor. Although these two equations can be solved for the unknowns N and a, they are needlessly complicated. Let us try the other coordinate system.

(b) When the +x axis points down the incline, the acceleration appears in only one component equation: $a_x = a$ and $a_y = 0$. (The condition $a_y = 0$ is called a constraint equation.) The free-body diagram is shown in Fig. 5.15c. The components of (i) are

$$\Sigma F_x = 0 + mg \sin \theta = ma \qquad \text{(iv)}$$
$$\Sigma F_y = N - mg \cos \theta = 0 \qquad \text{(v)}$$

From (iv), we see immediately that $a = g \sin \theta = (9.8) \sin 20° = 3.3$ m/s². From (v), we find $N = mg \cos \theta = 550$ N. These axes are clearly more suitable for this problem. This example illustrates that a careful choice of coordinate axes can save labor and reduce errors.

EXAMPLE 5.6: A sled of mass 8 kg is on a frictionless slope inclined at 35° to the horizontal. It is pulled by a rope whose tension is 40 N and which makes an angle of 20° with the slope, as shown in Fig. 5.16a. Find the acceleration of the sled and the normal force due to the incline.

Solution: The forces on the sled are shown in Fig. 5.16a. We assume the acceleration is up the incline and so we take this direction to be the +x axis. The second law in vector form is

$$\Sigma \mathbf{F} = \mathbf{T} + \mathbf{N} + \mathbf{W} = m\mathbf{a} \qquad \text{(i)}$$

The components of the forces are shown in the free-body diagram of Fig. 5.16b. The component equations are

$$\Sigma F_x = T \cos \alpha - W \sin \theta = ma \qquad \text{(ii)}$$
$$\Sigma F_y = T \sin \alpha + N - W \cos \theta = 0 \qquad \text{(iii)}$$

When the values $\theta = 35°$, $\alpha = 20°$, $W = (8$ kg$)(9.8$ N/kg$) = 78.4$ N, and $T = 40$ N are inserted, (i) leads to $a = -0.92$ m/s² and from (ii) we find $N = 50.5$ N. The assumption regarding the direction of the acceleration was incorrect. The component of the tension up the incline is less than the component of the

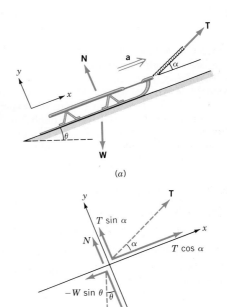

(a)

(b)

FIGURE 5.16 (a) The forces acting on a sled. The acceleration is assumed to be up the incline. (b) The free-body diagram.

weight down the incline. The net force, and, therefore, the acceleration, are down the incline.

EXERCISE 3. Does the negative sign of the acceleration in Example 5.6 tell us whether the sled is moving up or down the incline? If so, what is the direction?

EXAMPLE 5.7: Two railcars, A and B, with masses $m_A = 1.2 \times 10^4$ kg and $m_B = 8 \times 10^3$ kg can roll freely on a horizontal track; see Fig. 5.17a. A locomotive of mass 10^5 kg exerts a force

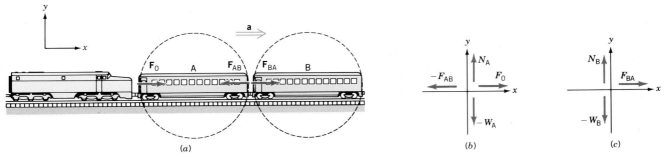

FIGURE 5.17 (*a*) When two bodies are in contact it can help to draw a bubble around each. Only forces that cross the boundary are included in the corresponding free-body diagram. For simplicity the vertical forces have not been drawn. (*b*) The free-body diagram for A. (*c*) The free-body diagram for B.

F_0 on A that produces an acceleration of 2 m/s². Find: (a) F_0; (b) the force exerted on A by B.

Solution: When two or more bodies are involved in a problem, it can help to draw a bubble around the chosen system. Only those forces that cross the boundary are to be considered. For simplicity only the horizontal forces experienced by the railcars are shown in Fig. 5.17*a*. Notice that B does *not* experience the force F_0; that is, $\mathbf{F}_0$ is not "transmitted" through A to B. Railcar B does experience the contact force $\mathbf{F}_{BA}$ exerted on it by A. We use the free-body diagrams (which must include *all* the forces) of Fig. 5.17*b* and Fig. 5.17*c* to write the *x* component of the second law:

Railcar A $\sum F_x = F_0 - F_{AB} = m_A a$ (i)
Railcar B $\sum F_x = F_{BA} = m_B a$ (ii)

From (ii) we find $F_{BA} = (8 \times 10^3 \text{ kg})(2 \text{ m/s}^2) = 1.6 \times 10^4$ N. Since $F_{AB} = F_{BA}$, (i) becomes

$$F_0 - 1.6 \times 10^4 = (1.2 \times 10^4 \text{ kg})(2 \text{ m/s}^2)$$

from which we find $F_0 = 4.0 \times 10^4$ N.

EXERCISE 4. What is the horizontal force exerted on the locomotive by the track?

EXAMPLE 5.8: An amusing but instructive illustration of the third law arises in the so-called "cart–horse" paradox. The horse, knowledgeable in the ways of the world, protests that the harder he pulls forward, the harder the cart will pull backward. Hence, there is no point in wasting his effort. Explain why the cart does move forward.

Solution: The horizontal forces on the cart and the horse are shown in Fig. 5.18. (The vertical forces are omitted because they are not relevant.) The horse is correct in asserting that $F_{CH} = F_{HC}$. However, this pair of forces is internal to the cart–horse system and so does not affect the acceleration of the system as a whole. The external forces are due to the road. The horse pushes backward on the road with F_{RH} and the road pushes the horse forward with F_{HR}. Similarly, the cart will ex-

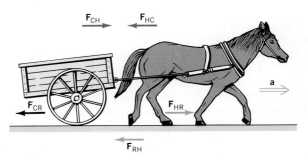

FIGURE 5.18 The forces exerted by the cart and by the horse on each other are internal to the cart–horse system. They cannot affect the motion of the system as a whole. The system moves forward because the force exerted on the horse by the road, F_{HR}, is greater than the force exerted on the cart by the road, F_{CR}.

perience a resistive force F_{CR}. The second law applied to the cart–horse system is

$$F_{HR} - F_{CR} = (m_H + m_C)a$$

Since F_{HR} is greater than F_{CR}, the system accelerates forward.

EXAMPLE 5.9: Three blocks with masses $m_1 = 3$ kg, $m_2 = 2$ kg, and $m_3 = 1$ kg are connected by two ropes, one of which hangs over a light, frictionless pulley as shown in Fig. 5.19*a*. Find the acceleration of the blocks and the tension in the ropes. Take $\theta = 25°$.

Solution: First, the forces acting on each block are shown in the sketch. Next, we assume that m_1 accelerates downward and choose the axes for each block accordingly. In problems involving two particles it is important to retain proper subscripts on variables. However, there is no point in introducing subscripts when they serve no purpose. In the present example $a_{1y} = a_{2x} = a_{3x} = a$. (This is a constraint equation.) Note also that there is only one value of the tension for each (massless) rope. Only the most complicated free-body diagram, that for m_2, is drawn in Fig. 5.19*b*. (You should draw the other two before proceeding.)

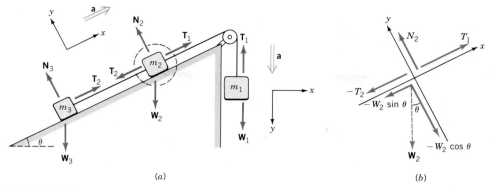

(a)

(b)

FIGURE 5.19 (a) Different coordinate axes are chosen for the vertical block and the blocks on the incline. (b) The free-body diagram for m_2.

The component equations are

Block 1 $\sum F_y = W_1 - T_1 = m_1 a$ (i)

Block 2 $\sum F_x = T_1 - T_2 - W_2 \sin \theta = m_2 a$ (ii)

Block 3 $\sum F_x = T_2 - W_3 \sin \theta = m_3 a$ (iii)

On adding these equations we obtain

$$W_1 - (W_2 + W_3) \sin \theta = (m_1 + m_2 + m_3)a$$ (iv)

Although this equation was not derived by a direct application of the second law (it is not a component of $\Sigma\mathbf{F} = m\mathbf{a}$), it does have a simple physical interpretation. The expression on the left is the net external force along the direction in which the system of the three blocks is free to move, that is, along the rope. The sign (direction) of the acceleration is determined by the relative magnitudes of the two terms. Notice that the internal forces between the blocks, that is, the tensions, do not appear. On inserting the given values into (iv) we find $a = 2.83$ m/s². Using this in (i) and (iii) yields $T_1 = 20.9$ N and $T_2 = 7.0$ N.

EXERCISE 5. Two blocks with masses $m_1 = 3$ kg and $m_2 = 5$ kg are connected by a rope that hangs over a pulley. (See Fig. 5.43.) Find the acceleration of the masses and the tension in the rope.

EXAMPLE 5.10*: A block of mass m is placed on a wedge of mass M that is on a horizontal table. All surfaces are frictionless. Find the acceleration of the wedge.

Solution: We know that the block will slide down the incline, but the contact force on the wedge due to the block will cause the wedge to accelerate horizontally. The direction of the block's acceleration relative to the table is not known. The choice of an appropriate inertial coordinate system is not obvious.

We may choose an inertial coordinate system with axes along and perpendicular to the incline and resolve the acceleration along these axes. (See Problem 8.) We may also introduce the acceleration of the block relative to the *incline* and combine this with the acceleration of the wedge to obtain the block's acceleration relative to the *table*. We illustrate this approach.

In Fig. 5.20a, $\mathbf{A}$ is the acceleration of the wedge relative to the table (an inertial frame), whereas $\mathbf{a}'$ is the block's accelera-

* This example is considerably more difficult than those preceding. It may be omitted since only a few of the end-of-chapter problems are as complex.

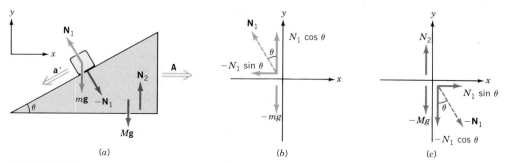

(a)

(b)

(c)

FIGURE 5.20 (a) The acceleration of the wedge relative to the table is A, while a' is the acceleration of the block relative to the wedge. The coordinate system is in the inertial frame of the table. (b) The free-body diagram for the block. (c) The free-body diagram for the wedge.

tion relative to the wedge. An observer moving with the wedge will see the block accelerate downward at θ below the horizontal. An observer on the table will assign horizontal and vertical components to the block's acceleration given by $a_x = A - a' \cos \theta$ and $a_y = -a' \sin \theta$. From the free-body diagrams of Figs. 5.20b and c we find

Wedge $\quad \Sigma F_x = N_1 \sin \theta = MA$ $\qquad\qquad$ (i)

Block $\quad \Sigma F_x = -N_1 \sin \theta = m(A - a' \cos \theta)$ $\quad$ (ii)

$\qquad\qquad \Sigma F_y = +N_1 \cos \theta - mg = -ma' \sin \theta$ $\quad$ (iii)

Adding (i) and (ii) eliminates N_1 and leads to

$$(m + M)A = ma' \cos \theta \qquad\qquad \text{(iv)}$$

Multiply (ii) by $\cos \theta$ and (iii) by $\sin \theta$ and add, to eliminate N_1 again:

$$mg \sin \theta = ma' - mA \cos \theta \qquad\qquad \text{(v)}$$

By eliminating a' from (iv) and (v) we find (go through the steps yourself)

$$A = \frac{mg \sin \theta \cos \theta}{M + m \sin^2 \theta}$$

Are the dimensions of this equation correct? What happens to A when θ is either close to zero or close to 90°? What if we had found M, rather than m, in the numerator?

5.6   APPARENT WEIGHT

The primary indication we have as to how much we weigh is the force exerted on us by a supporting surface, such as a chair or a floor. In an elevator moving at constant velocity, we seem to have our normal weight. However, when the elevator accelerates upward we seem to weigh more, and when it accelerates downward we seem to weigh less. Our *true* weight ($\mathbf{W} = m\mathbf{g}$) does not depend on the acceleration of the elevator, but our **apparent weight** does.

> The magnitude of the apparent weight of a body is the magnitude of the resultant force exerted on it by a supporting surface.*

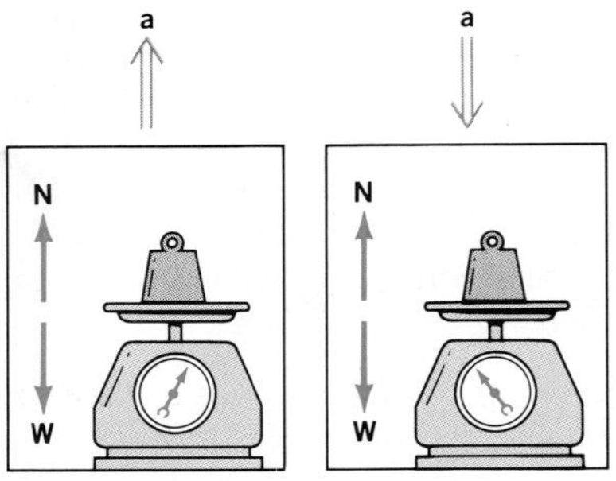

When an object is placed on a scale, the reading is the magnitude of the normal force between the object and the pan. This normal force is a measure of the apparent weight.

Figure 5.21 illustrates the forces acting on the object and the two possible directions of the acceleration. In either case the vector form of the second law is $\mathbf{N} + m\mathbf{g} = m\mathbf{a}$. When the velocity is constant, $N = mg$: The apparent weight is equal to the true weight. If the acceleration is upward, as in Fig. 5.21a, we have $N - mg = ma$ and so $N = m(g + a)$: The apparent weight is *greater* than the normal weight. When the acceleration is downward, as in Fig. 5.21b, we have $mg - N = ma$ and so $N = m(g - a)$: The apparent weight is *less* than the normal weight.

If the supporting cables were to break, the elevator would be in free-fall, which means $a = g$ and so $N = 0$. The object would be *apparently weightless*. In this condition, the object would "float" in the elevator. The condition of apparent weightlessness occurs when an object is in free-fall. If you jump up from the ground, you are apparently weightless while you are in the air.

FIGURE 5.21 (a) When the elevator accelerates upward the apparent weight is greater than the weight; that is, $N > mg$. (b) When the elevator accelerates downward, $N < mg$.

EXERCISE 6. A 60-kg person is in an elevator that is moving upward at 4 m/s and slowing down at 2 m/s². (a) What is the person's apparent weight? (b) According to Newton's third law, what is the magnitude of the "reaction" to the person's weight? On what body is this force exerted?

---

* The term apparent weight is used differently when we deal with the buoyancy of fluids. See Chapter 14.

not valid

SUMMARY

The **inertia** of a body is its tendency to resist any change in its velocity. The **mass** of a body is a measure of its inertia. Mass is an intrinsic property, independent of location.

A **force** is an action that produces a deformation and/or an acceleration. A net (unbalanced) force acting on a body always causes an acceleration. **Newton's second law** relates the acceleration of a particle to the net force acting on it. It is valid only in an inertial frame.

$$\Sigma \mathbf{F} = m\mathbf{a}$$

In terms of components,

$$\Sigma F_x = ma_x \qquad \Sigma F_y = ma_y \qquad \Sigma F_z = ma_z$$

The **weight** of a body

$$\mathbf{W} = m\mathbf{g}$$

is the net gravitational force acting on it.

According to **Newton's third law**

$$\mathbf{F}_{AB} = -\mathbf{F}_{BA}$$

The force $\mathbf{F}_{AB}$ exerted on body A by body B is equal in magnitude and opposite in direction to $\mathbf{F}_{BA}$, the force exerted on B by A. The third law implies that forces always occur in action–reaction pairs.

The **apparent weight** of a body is the force exerted on it by a supporting surface. A body in free-fall is *apparently* weightless, but its (true) weight is not zero.

ANSWERS TO IN-CHAPTER EXERCISES

1. False. The force you exert is *perpendicular* to the weight. You have to contend with the inertia (mass) of the object and any friction that is present.

2. The pound is the British unit of *force* whereas the kilogram is the SI unit of *mass*.

3. The sign of the acceleration does *not* tell us whether the block is moving up or down the incline.

4. The force exerted on the locomotive by the track produces the acceleration of the locomotive and the railcars. Thus, $F = ma = (1.2 \times 10^5 \text{ kg})(2 \text{ m/s}^2) = 2.4 \times 10^5$ N.

5. m_2 accelerates downward, so its y axis is downward, whereas m_1 accelerates upward, so its y axis is upward.

$$\Sigma F_y = T - m_1 g = m_1 a \qquad \Sigma F_y = m_2 g - T = m_2 a$$

On adding these equations we find

$$a = \frac{(m_2 g - m_1 g)}{(m_1 + m_2)} = 2.45 \text{ m/s}^2.$$

Substituting into either equation yields $T = 36.8$ N. Notice that $T > m_1 g$ and $T < m_2 g$.

6. (a) The acceleration is downward, so the y axis is also downward. The forces on the person are as in Fig. 5.21b:

$$\Sigma F_y = mg - N = ma$$

So her *apparent weight* is $N = m(g - a) = 468$ N.
(b) The magnitude of the reaction to her *weight* (mg) is 588 N. According to the third law $\mathbf{F}_{PE} = -\mathbf{F}_{EP}$, where P is the person and E is the earth. The reaction force acts *on the earth*.

QUESTIONS

1. Answer the following true/false questions with a brief explanation:
(a) A nonzero net force acting on a body always changes its speed.

(b) A person in a free-falling elevator is weightless.

(c) While pushing a crate along the floor, you must exert a greater force on it than it does on you.

2. Would it be easy to play "catch" with a large bowling ball in outer space since the ball would have no weight?

3. A block is hung from a support by string A, as in Fig. 5.22. String B is attached to the underside of the block. Explain the following: If B is pulled with slowly increasing tension, then A breaks; but if B is pulled with a jerk, then B breaks.

FIGURE 5.22 Question 2

4. A ball thrown vertically upward momentarily comes to rest at its highest point. Is it in equilibrium at this instant? Explain why or why not.

5. A small car and a large truck collide. At any instant, which experiences (a) the greater force? (b) the greater acceleration?

6. A truck with several birds in its closed cargo compartment is a little too heavy for a bridge. The driver makes a lot of noise to keep the birds flying. Does he make it across? What if the birds were in a wire cage instead?

7. Without moving your feet off the floor, can you exert a force on it that differs from your weight? Is there a simple way you can check your response?

8. Consider the statement, "In a tug-of-war, the two teams exert equal forces on each other." Do you accept this? Explain why one team wins.

9. (a) On a calm day a sailor decides to use a large fan to blow wind onto the sail, as in Fig. 5.23. Does this scheme work? (b) Is there another way to use the fan to move the boat?

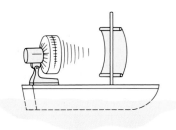

FIGURE 5.23 Question 9

10. You and a friend are having difficulty in pulling out a car that is stuck in mud. She suggests tying the rope to a tree and pulling perpendicular to the rope, as in Fig. 5.24. Is this more effective? Explain why or why not.

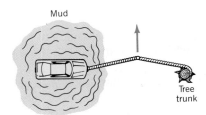

FIGURE 5.24 Question 10

11. Figure 5.25 shows two blocks with masses m_1 and m_2 connected by a massless string that hangs over a frictionless pulley. The horizontal surface is frictionless and the masses are initially held at rest and then are released. (a) Will the system start to move if $m_2 < m_1$? (b) If $m_2 > m_1$, how does the magnitude of the tension in the string compare with the weight of m_2?

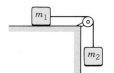

FIGURE 5.25 Question 11

12. Someone asserts that an object falling through the air at its terminal speed is in equilibrium. Explain why you agree or disagree with this statement.

13. Cars with the engine in the rear tend to spin around if they enter into a skid. Why is this?

14. Why is the traction of a front-wheel-drive car intrinsically better than that of a rear-wheel-drive car? Is the braking also better?

15. What is the reading on the spring scale for each of the situations depicted in Fig. 5.26? Each of the blocks has a mass of 5 kg.

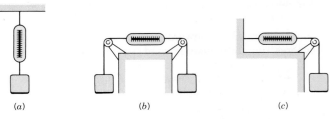

FIGURE 5.26 Question 15

16. The action of any particle has an equal and opposite reaction. Sometimes these can be considered as canceling each other. When is this approach valid?

EXERCISES

5.2 and 5.3 Newton's Second Law; Weight

1. (I) A 7-kg block is suspended with two ropes, as shown in Fig. 5.27. Find the tension in each rope.

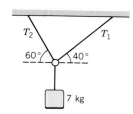

FIGURE 5.27 Exercise 1

2. (I) A 3-kg block is suspended with two ropes, one of which is horizontal, as in Fig. 5.28. Find the tension in each rope.

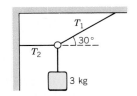

FIGURE 5.28 Exercise 2

3. (I) A 2-kg block is suspended by a single rope. A horizontal force holds the rope at 37° to the vertical. Find: (a) the force and (b) the tension in the rope.

4. (I) A 70-kg tightrope walker stands at the midpoint of a rope whose extended length is 100 m. If the rope sags by 1.5 m, find the tension in it.

5. (I) A 2-kg particle is subject to two forces that produce a resultant acceleration $\mathbf{a} = 4\mathbf{i} - 3\mathbf{j}$ m/s². If $\mathbf{F}_1 = -\mathbf{i} + 2\mathbf{j} + 3\mathbf{k}$ N, find $\mathbf{F}_2$.

6. (I) Two forces $\mathbf{F}_1 = \mathbf{i} + 2\mathbf{j}$ N and $\mathbf{F}_2$ which is 4 N directed at 37°, measured from the x axis toward the y axis, act on a 200-g particle. What is its acceleration?

7. (I) A particle of mass 1.5 kg has an initial velocity of $2\mathbf{i} + 3\mathbf{j}$ m/s. It is acted on by a force $4\mathbf{i} - \mathbf{j}$ N for 2 s. What is its final velocity?

8. (I) A 10-g bullet moving at 400 m/s is brought to a stop in 3 cm in a block of wood. Find the force on the bullet, assuming that it is constant. Compare this force with the weight of a 60-kg person.

9. (I) An electron of mass 9.11×10^{-31} kg is accelerated by a constant force along the direction of its motion from 2×10^6 m/s to 8×10^6 m/s within 2 cm. (a) What is the magnitude of the force? (b) How long did it act?

10. (I) A 20-kg child starts from rest and travels 3 m down a slide that is inclined at 35° to the horizontal. If her speed at the bottom is 1 m/s, what was the frictional force along the slide?

11. (I) The 50-kg torso of a sprinter starts at rest and reaches 6 m/s within 80 cm, both measured horizontally. Estimate (a) the acceleration, and (b) the horizontal force exerted on the torso at the hip joint.

12. (I) What is the constant force required to accelerate a 1225-kg Saab 900S in each of the following cases: (a) It starts from rest and reaches 96 km/h in 10 s; (b) it is braked to a stop from 112 km/h in 64 m. In each case, what is the origin of the force?

13. (I) Find the constant force acting on a 12,500-kg F4 Phantom jet in the following situations: (a) It is catapulted from rest to 250 km/h in 2.2 s; (b) it is stopped from 180 km/h by a net within 40 m.

14. (I) A length of rope has a mass of 30 g. It is used to accelerate a 200-g object vertically upward at 4 m/s². What is the tension at the midpoint of the rope?

15. (I) A person lowers his 50-kg torso by 15 cm and jumps vertically. If the torso rises 40 cm above normal, find the force (assumed to be constant) exerted on the torso at the hip joint.

16. (I) A block of mass 1 kg is on a frictionless 37° incline and is subject to a horizontal force of 5 N, as in Fig. 5.29. (a) What is its acceleration? (b) If it is initially moving up the incline at 4 m/s, what is its displacement along the incline in 2 s?

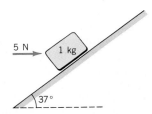

FIGURE 5.29 Exercise 16

17. (I) A 30-kg child on roller skates starts up a 10° incline at 15 km/h. Assuming she does not propel herself, how far up the incline does she travel before stopping? Ignore frictional losses.

18. (I) A man pushes a 20-kg lawn mower with a force of 80 N directed along the handle, which is inclined at 30° to the horizontal, as in Fig. 5.30. (a) If he moves at constant velocity, what is the retarding force due to the ground? (b) What force along the handle would produce an acceleration of 1 m/s² given the same retarding force?

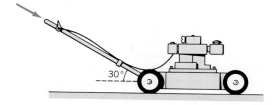

FIGURE 5.30 Exercise 18

19. (I) A 70-kg sprinter starts from rest and covers 6 m in 1 s. What constant horizontal force would produce the same result?

20. (I) A 2-kg object is moving east at 10 m/s. It experiences a constant force of 20 N which is directed south and lasts for 4 s. What is the final velocity?

21. (I) A car is moving at 25 m/s. The 75-kg driver is held firmly in the seat by a belt. Find the force on the driver (assumed to be constant) if the car crashes: (a) by crumpling through a distance of 75 cm; (b) rigidly within 25 cm.

22. (I) A 150-g baseball moving at 30 m/s is hit by a bat. Its final velocity is 40 m/s in the opposite direction. Given that the ball and bat were in contact for 5 ms, what is the force on the ball, assuming that it is constant?

23. (I) A 12,500-kg jet has engines with a total thrust of 1.6×10^5 N. (a) If the takeoff speed is 220 km/h, what is the minimum length of runway needed? (b) How long would it take the plane to rise vertically by 1 km assuming its initial vertical velocity is zero?

24. (I) A Saturn V rocket has a mass of 2.7×10^6 kg and a thrust of 3.3×10^7 N. What is its initial vertical acceleration?

25. (I) A Polaris missile has a mass of 1.4×10^4 kg and its engine has a thrust of 2×10^5 N. If its engines fire in the vertical direction of 1 min starting from rest, to what vertical height would it rise in the absence of air resistance?

26. (I) A child drops from a ledge 1.0 m above the floor. Estimate the force on her 40-kg torso when she lands: (a) by bending her knees and stopping her torso in 30 cm; (b) stiffly and her torso is stopped within 4 cm.

27. (I) Find the force (assumed to be constant) on a 0.15-kg baseball given that: (a) It is thrown at 25 m/s by a hand that moves through 2 m; (b) it is hit by a bat that reverses the direction of motion and changes the speed to 35 m/s in 5 ms; (c) it is caught by an outfielder who moves his mitt through 15 cm after the ball has slowed to 20 m/s.

28. (II) Estimate the force on a 7.25-kg shotput if it is thrown at 45° and lands 16 m away horizontally. Assume the hand moved through 1.5 m and that the shot lands at the height at which it left the hand.

29. (II) A hot-air balloon of mass M descends with an acceleration a ($<g$). How much ballast should be thrown out for it to accelerate upward with the same acceleration? Assume that the buoyant force is unchanged.

30. (I) A light rope can withstand a maximum tension of 600 N. With what minimum acceleration should a 75-kg person slide down?

5.4 Newton's Third Law

31. (I) The two blocks shown in Fig. 5.31 have masses $m_A = 2$ kg and $m_B = 3$ kg. They are in contact and slide over a frictionless horizontal surface. A force of 20 N acts on B as shown. Find: (a) the acceleration; (b) the force on B due to A; (c) the net force on B; (d) the force on B due to A if the blocks are interchanged.

FIGURE 5.31 Exercise 31

32. (I) A 0.7-kg toy bulldozer (B) pushes a 0.2-kg car (C) that rolls freely on the ground (G) with an acceleration of 0.5 m/s². Find the magnitude and direction of the horizontal component of each of the following forces: (a) F_{BG}; (b) F_{CG}; (c) F_{BC}. (F_{AB} means the force exerted on A by B.)

33. (I) A 60-kg parachutist and her 7-kg parachute fall at a constant 6 m/s. Find: (a) the force on the woman due to the chute; (b) the force on the chute due to the air. (Ignore the force on the woman due to the air.)

34. (I) A train has 10 railcars, each of mass 4×10^4 kg. The engine has a mass of 2.2×10^5 kg and it pulls the first railcar with a force of 8×10^5 N. Find: (a) the tension in the coupling between the first and second cars; (b) the tension in the coupling between the last two cars; (c) the horizontal force exerted by the track on the engine. Assume that the railcars roll freely.

5.5 Applications of Newton's Laws

35. (I) A boy of mass 25 kg and a girl of mass 20 kg are connected by a light rope, as shown in Fig. 5.32. They move horizontally on a frictionless skating rink. The boy is pulled by a horizontal force of 200 N. Find their acceleration and the tension in the rope.

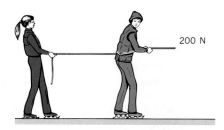
FIGURE 5.32 Exercise 35

36. (II) Two blocks of masses $m_1 = 5$ kg and $m_2 = 6$ kg are on either side of the wedge in Fig. 5.33. Find their acceleration and the tension in the rope. Ignore friction and the pulley.

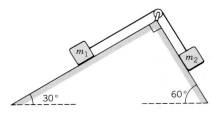
FIGURE 5.33 Exercise 36

37. (II) Two blocks with masses $m_A = 0.2$ kg and $m_B = 0.3$ kg hang one under the other, as shown in Fig. 5.34. Find the tensions in the (massless) ropes in the following situations: (a) The blocks are at rest; (b) they move upward at 5 m/s; (c) they accelerate upward at 2 m/s²; (d) they accelerate downward at 2 m/s². (e) If the maximum allowable tension is 10 N, what is the maximum possible upward acceleration?

FIGURE 5.34 Exercise 37

38. (II) Three blocks with masses $4M$, $2M$, and $8M$ are connected as shown in Fig. 5.35. The tensions in the ropes are T_1 and T_2. In terms of M, g, and θ, obtain expressions for: (a) the acceleration; (b) $T_1 - T_2$. Ignore friction. (c) Evaluate your expressions for $M = 1$ kg and $\theta = 45°$.

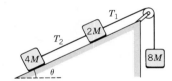

FIGURE 5.35 Exercise 38

39. (II) A 9-kg block is held by a pulley system shown in Fig. 5.36. What force must the person exert in the following cases: (a) to hold the block at rest; (b) to lower it at 2 m/s; (c) to raise it with an acceleration of 0.5 m/s²?

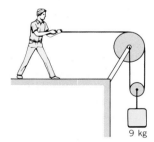

FIGURE 5.36 Exercise 39

40. (II) A 10-kg monkey holds a rope that slides over a pulley and is connected to a 12-kg bunch of bananas. Can the monkey climb the rope in such a way so as to raise the bananas off the ground? If so, explain how.

41. (II) A 5-kg block has a rope of mass 2 kg attached to its underside and a 3-kg block is suspended from the other end of the rope (see Fig. 5.37). The whole system is accelerated upward at 2 m/s² by an external force F_0. (a) What is F_0? (b) What is the net force on the rope? (c) What is the tension at the middle of the rope?

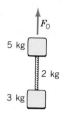

FIGURE 5.37 Exercise 41

42. (I) Two blocks are connected by a massless rope as shown in Fig. 5.38. The horizontal surface is frictionless. If $m_1 = 2$ kg, for what value of m_2 will (a) the acceleration of the system be 4 m/s², or (b) the tension in the rope be 8 N?

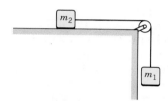

FIGURE 5.38 Exercise 42

43. (I) Two blocks with masses $m_1 = 3$ kg and $m_2 = 5$ kg are connected by a light rope and slide on a frictionless surface as in Fig. 5.39. A force $F_0 = 10$ N acts on m_2 at 20° to the horizontal. Find the acceleration of the system and the tension in the rope.

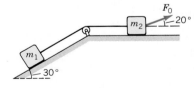

FIGURE 5.39 Exercise 43

5.6 Apparent Weight

44. (I) A 70-kg person stands on a scale in an elevator that accelerates upward at 2 m/s². (a) What is the scale reading? (b) What is the person's weight? (c) According to the third law, what is the reaction to the person's weight? On what body does it act?

45. (I) A person of mass 75 kg stands on a scale in an elevator. What can you infer about the motion if the scale reads: (a) 735 N? (b) 600 N? (c) 900 N?

46. (I) A 3-kg block is suspended from the ceiling of an elevator. Find the tension in the rope given the following: (a) The elevator is moving up at 5 m/s and slowing down at 4 m/s^2. (b) The elevator is moving down at 3 m/s and speeding up at 2 m/s^2.

47. (II) A rocket has a net acceleration of 2.4 m/s^2 directed at 60° above the horizontal. What is the apparent weight of an 80-kg astronaut near the earth's surface.

48. (I) A 12,500-kg F4 Phantom jet has engines with a total thrust of 160 kN. It is catapulted from the deck of a carrier and reaches its takeoff speed of 80 m/s in 2.1 s. (a) What (constant) force does the steam catapult exert? (b) What is the apparent weight of a 70-kg pilot?

49. (I) The rocket engine of the 4800-kg ascent stage of the Lunar Module used in the Apollo program had a thrust of 15.5 kN. What was the apparent weight of a 70-kg astronaut on takeoff from the moon? The weight of an object on the surface of the moon is one-sixth that on earth.

PROBLEMS

1. (I) A painter of mass M = 75 kg stands on a platform of mass m = 15 kg. He pulls on a rope that passes around a pulley, as in Fig. 5.40. Find the tension in the rope given that (a) he is at rest, or (b) he accelerates upward at 0.4 m/s^2. (c) If the maximum tension the rope can withstand is 700 N, what happens when he ties the rope to a hook on the wall?

acting at the axis of the pulley accelerates the system upward. Find: (a) the acceleration of each mass; (b) the tension in the string.

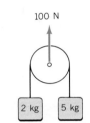

FIGURE 5.42 Problem 3

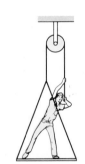

FIGURE 5.40 Problem 1

2. (I) Find the tension in the rope and the acceleration of each of the blocks in Fig. 5.41. Ignore the pulleys and friction. Take m_1 = m_2 = 1 kg. (*Hint: m_1 and m_2 have different accelerations.*)

4. (I) Two blocks of masses 3 and 5 kg hang over a pulley, as shown in Fig. 5.43. The 5-kg block is initially held 4 m above the floor and then released. What is the maximum height reached by the 3-kg block?

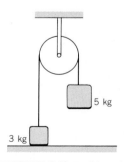

FIGURE 5.43 Problem 4

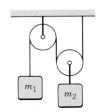

FIGURE 5.41 Problem 2

3. (II) Two blocks of masses m_1 = 2 kg and m_2 = 5 kg hang over a massless pulley as in Fig. 5.42. A force F_0 = 100 N

5. (II) Here is a problem first solved by Galileo. A bead slides on a frictionless rod whose ends lie on a circle, as in Fig. 5.44. Show that if the bead starts at the top, the time required to slide down any chord is independent of the particular chord chosen.

FIGURE 5.44 Problem 5

6. (II) In Fig. 5.45, a block of mass m is on a wedge of mass M. All surfaces are frictionless. Suppose the wedge is accelerated toward the right by an external force at 5 m/s². What is the acceleration of the block relative to the incline? Take $\alpha = 37°$. (Choose the axes carefully.)

FIGURE 5.45 Problems 6 and 8

7. (II) Atwood's machine was an apparatus that allowed a direct verification of Newton's second law. It could also be used to measure g. Two equal blocks of mass M hang on either side of a pulley; see Fig. 5.46. A small square rider of mass m is placed on one block. When the block is released, it accelerates for a distance H till the rider is caught by a ring that allows the block to pass. From then on the system moves at a constant speed which is measured by timing the fall through a distance D. Show that

$$g = \frac{(2M + m)D^2}{2mHt^2}$$

where t is the time moved at constant speed.

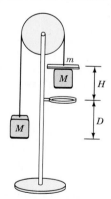

FIGURE 5.46 Problem 7

8. (II) A block of mass m is placed on a wedge of mass M, as in Fig. 5.45. All surfaces are frictionless. (a) Find the horizontal and vertical components of the acceleration of the block and of the wedge. (b) Show that the normal force between the block and the wedge is

$$\frac{Mmg \cos \theta}{M + m \sin^2 \theta}$$

9. (II) A flexible chain of length L slides off the edge of a frictionless table, as in Fig. 5.47. Initially a length y_0 hangs over the edge. (a) Find the acceleration of the chain as a function of y. (b) Show that the velocity as the chain becomes completely vertical is

$$v = \sqrt{g(L - y_0^2/L)}$$

(This problem requires integration.)

FIGURE 5.47 Problem 9

10. (II) Three masses are hung over two pulleys, as shown in Fig. 5.48. The pulleys are massless and frictionless. Assume that $m_1 > (m_2 + m_3)$ and $m_2 > m_3$. (a) Show that the tension in the rope supporting the fixed pulley is

$$T = \frac{16m_1m_2m_3g}{m_1(m_2 + m_3) + 4m_2m_3}$$

(b) Identify the mass that has the following acceleration:

$$a = \frac{[m_1(m_2 + m_3) - 4m_2m_3]g}{m_1(m_2 + m_3) + 4m_2m_3}$$

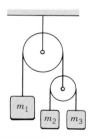

FIGURE 5.48 Problem 10

CHAPTER 6

Particle Dynamics II

A skateboarder experiences circular motion inside a pipe.

Major Points

1. (a) The distinction between **static friction** and **kinetic friction**.
   (b) The **coefficients of friction.**
2. The dynamics of **circular motion.**
3. Satellite **orbits.** Kepler's third law relates the radius and the period of an orbit.

6.1 FRICTION

Friction plays dual roles in our lives. It impedes the motion of objects, causes abrasion and wear, and converts other forms of energy into heat. On the other hand, without it we could not walk, drive cars, climb ropes, or use nails. Friction is a contact force that opposes the relative motion of two bodies. It is a complex phenomenon for which there is no fundamental theory. However, the basic facts concerning friction between dry, unlubricated surfaces are fairly simple and were established long ago. In 1508, Leonardo da Vinci discovered two features:

(i) The force of friction is proportional to the load.

(ii) The force of friction is independent of the area of contact.

The laws of friction   By "load" we mean the force pressing the two surfaces together. Da Vinci never published his results but they were independently discovered by a French scientist, Amontons, in 1699. Amontons also found a third feature:

(iii) The force of friction is independent of the speed.

These three facts are sometimes called Amontons laws.

    The fact that the force of friction is proportional to the load seems reasonable. Its independence of the area of contact is surprising. To understand this we must distinguish between the apparent area of contact and the actual area of contact. As Fig. 6.1 shows, even highly polished surfaces are not smooth at the microscopic level. The irregularities can vary from 10^{-5} to 10^{-4} mm. The two surfaces meet

1 μ

10 μ

FIGURE 6.1 Even a polished surface is uneven on a microscopic scale.

only at the hilltops (called asperities), which means that the actual area of contact may be less than one ten-thousandth of the apparent area of contact. The weight of the block in Fig. 6.2 is distributed over fewer points of contact in position B as compared with position A. However, the hilltops are more flattened in position B, so the load is spread over the same real area of contact. Therefore, friction does increase with the real area of contact, but it is independent of the apparent area of contact.

It is natural to assume that friction is caused by roughness of the surfaces in contact. In 1785 Charles Coulomb suggested that the surface irregularities mesh with each other, like the bristles of a brush, and therefore work is constantly needed to lift the moving surface over the bumps. For visibly rough surfaces, this is partly true. (A coarse surface has bumps about 0.01 mm high, whereas for a fine surface they would be about 1% of this). As two rough surfaces are sanded, the friction between them does drop. However, as they are polished further, something strange happens; the friction starts to increase. It is quite incorrect to consider a ''smooth'' surface to be frictionless. If this were so, the polished wheels of a locomotive could not develop traction on shiny steel rails! Clearly roughness is an inadequate explanation. A modern theory of friction is discussed in the Special Topic at the end of this chapter.

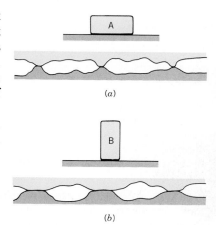

FIGURE 6.2 The real area of contact between two surfaces is much less than the apparent area of contact. In position A there are more points of contact than in position B but the actual area of contact is the same.

Static and Kinetic Friction

It is common experience that it takes a certain minimum force to start an object sliding over a surface. Once the sliding has started, the force needed to keep it moving at constant velocity is lower. In 1748 Leonhard Euler distinguished between static friction and kinetic friction. In Fig. 6.3 a force is applied to a block on a horizontal surface. The force of **static friction** opposes the tendency of the block to move relative to the surface. If the block does not move, the force of static friction f_s must be exactly equal to the applied force F_{app}. As the applied force is increased, f_s also increases and stays equal to F_{app}, as shown in Fig. 6.4, but only until a critical value, $f_{s(max)}$, is reached. For a larger applied force, the block starts to slide and is then subject to **kinetic friction**. As sliding commences, the frictional force rapidly falls at low speeds. At higher speeds, the force of kinetic friction f_k either stays constant or decreases gradually as the speed increases. In many cases the friction at low speeds is characterized by a combination of static and kinetic friction. This produces a jerky ''stick–slip'' motion that can often be heard in creaking doors or planks, tire squeal, and squeaky wheels.

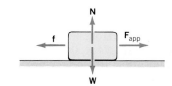

FIGURE 6.3 The force of friction **f** opposes the relative motion of two surfaces in contact.

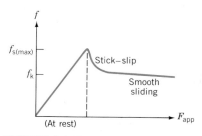

FIGURE 6.4 The variation of the force of friction with the applied force. When the block is at rest the force of static friction f_s balances the applied force F_{app} until it reaches a maximum value. When the block moves, it is subject to the force of kinetic friction.

The fact that the frictional force is proportional to the load allows us to define two coefficients of friction. The force of kinetic friction f_k may be related to the normal force between the sliding surfaces by the equation

$$f_k = \mu_k N \qquad (6.1)$$

where μ_k, the **coefficient of kinetic friction,** is a dimensionless number. Notice that Eq. 6.1 is not a vector equation. (Why?) As was mentioned above, the force of static friction does not have a fixed value. However, its *maximum* value is simply related to the normal force:

$$f_{s(max)} = \mu_s N \quad \text{or} \quad f_s \leq \mu_s N \qquad (6.2)$$

Force of static friction

where μ_s is the **coefficient of static friction.** Make a careful note of the difference in form between Eqs. 6.1 and 6.2. In general, $\mu_s > \mu_k$, but there are exceptions.

The above relations are not exactly true. The coefficients are not really constants for any pair of surfaces; they depend on roughness, cleanliness (grease or oxide films), temperature, humidity, and so on. In some cases it is found that $f \propto N^{0.9}$, instead of $f \propto N$. Kinetic friction depends on speed, and static friction can depend on the length of time the two surfaces have been in contact. When a lubricant is present, friction depends in a complicated way on both the load and the speed.

EXAMPLE 6.1: A 5-kg block is on a horizontal surface for which $\mu_s = 0.2$ and $\mu_k = 0.1$. It is pulled by a 10-N force directed at 55° above the horizontal, as shown in Fig. 6.5a. Find the force of friction on the block given that: (a) it is at rest; (b) it is moving.

Solution: (a) When the block is at rest, the direction of the force of static friction is unambiguously specified: f_s is opposite to the horizontal component of the applied force. The forces on the block are shown in Fig. 6.5a. The normal force is less than the weight:

$$N = mg - F \sin 55°$$
$$= 5 \text{ kg} \times 9.8 \text{ N/kg} - 10 \text{ N} \times 0.819 = 40.8 \text{ N}$$

Thus, the maximum value possible for the force of static friction is

$$f_{s(max)} = \mu_s N = 8.16 \text{ N}$$

Since the horizontal component of the applied force is only $10 \cos 55° = 5.74$ N, the maximum value of f_s is not needed. The force of static friction required to keep the block at rest is just 5.74 N, directed to the left.
(b) Since the block is moving, the magnitude of the frictional force is $f_k = \mu_k N = 4.08$ N. If the block moves to the right, f_k is toward the left, as in Fig. 6.5b. If the block moves to the left, f_k is to the right, as in Fig. 6.5c.

EXERCISE 1. Find the acceleration of the block in Example 6.1 given that it is moving (a) to the right, and (b) to the left.

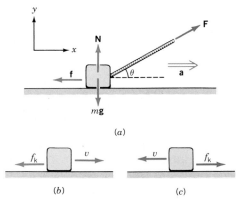

(a)

(b)   (c)

FIGURE 6.5 (a) The forces acting on a block. In (b) and (c) the force of kinetic friction on the block is opposite to its velocity relative to the surface on which it slides.

EXAMPLE 6.2: Two blocks with masses $m_1 = 7$ kg and $m_2 = 4$ kg are connected with a rope and move on two surfaces of a right-angled wedge as shown in Fig. 6.6. Given that $\theta_1 = 37°$,

$\theta_2 = 53°$, $\mu_s = 0.2$, and $\mu_k = 0.1$, find the acceleration of the blocks.

Solution: This is a poorly worded problem. There is no unique answer because the initial conditions have not been specified. The direction of the frictional force depends on the direction of motion of the blocks. Let us consider two possibilities: (a) The blocks are initially at rest; (b) m_1 is moving down the slope.
(a) The forces acting on the blocks, *excluding* friction, are shown in Fig. 6.6. Before we can assign a direction to the force of friction we have to find out which way the system tends to move. That is, we need to find the net force along the direction of the rope (see Example 5.7). The components of the weights along the inclines are $W_1 \sin 37° = 42$ N and $W_2 \sin 53° = 32$ N. Thus, in the absence of friction m_1 would move down and m_2 would move up. The next question is whether this net force (42 N − 32 N = 10 N) is large enough to overcome the maximum force of static friction on both blocks:

$$f_{s(max)} = \mu_s(N_1 + N_2)$$
$$= \mu_s(W_1 \cos 37° + W_2 \cos 53°) = 16 \text{ N}$$

The system does not even start to move.
(b) Although friction plays a role in determining the acceleration, it is the direction of the *velocity* of a block relative to the surface, *not* its acceleration, that determines the direction of the friction force. We are given that m_1 is moving down and so the directions of the forces of kinetic friction on both blocks are clear. From the second law we have

$$W_1 \sin \theta_1 - T - \mu_k(W_1 \cos \theta_1) = m_1 a$$
$$T - W_2 \sin \theta_2 - \mu_k(W_2 \cos \theta_2) = m_2 a$$

where we have assumed that the direction of the acceleration is the same as that of the initial motion. When numbers are inserted we find $a = +0.18$ m/s². The positive sign means that our assumption was correct.

If you obtain a negative acceleration in a problem that includes friction, make sure your assumptions about the direction of the frictional force were correct.

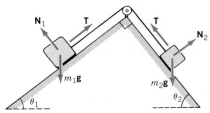

FIGURE 6.6 In the presence of friction one cannot determine the direction of the acceleration unless the system starts at rest or the direction of motion is specified.

EXAMPLE 6.3: A block of mass $m_1 = 2$ kg is placed on a block of mass $m_2 = 4$ kg. The lower block is on a frictionless horizontal surface and is subject to a force $F_0 = 30$ N as shown in Fig. 6.7a. Find the minimum value of the coefficient of friction such that m_1 does not slide on m_2.

Solution: Let us first consider the more general situation in which the upper block does slide relative to the lower one. Their accelerations toward the right will be different. Because of its inertia, m_1 will tend to slip backward relative to m_2. Hence, the force of friction on m_1 is in the *forward* direction, which is the same as the direction of motion. This also follows from the fact that friction is the only horizontal force acting on m_1, so it must be responsible for m_1's acceleration to the right. From Newton's third law, the frictional force exerted by m_1 on m_2 will be to the left. In the free-body diagrams for each block, shown in Figs. 6.7b and c, N_1 is the magnitude of the normal force between the blocks and N_2 is the normal force on m_2 due

to the horizontal (frictionless) surface. The horizontal components of Newton's second law applied to each block are

Block 1 $\qquad\qquad f = m_1 a_1$ (i)

Block 2 $\qquad\qquad F_0 - f = m_2 a_2$ (ii)

If m_1 does slide on m_2, f will be the force of kinetic friction. However, we are told that m_1 does not slide relative to m_2, so f is the force of static friction and the accelerations are the same, $a_1 = a_2 = a$. On adding (i) and (ii) we find

$$F_0 = (m_1 + m_2)a$$

which gives $a = (30\text{ N})/(6\text{ kg}) = 5$ m/s². Since $f_{s(max)} = \mu_s N_1 = \mu_s(m_1 g)$, from (i) we have

$$\mu_s(m_1 g) = m_1 a$$

Therefore, $\mu_s = a/g \approx 0.5$. (Why is this the minimum value?)

EXERCISE 2. If $F_0 = 40$ N and $\mu_k = 0.2$, what is the acceleration of each block?

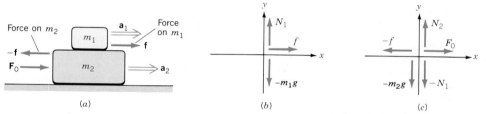

(a)   (b)   (c)

FIGURE 6.7 (a) When the lower block accelerates, the force of friction on the upper block must be in the same direction as *its* acceleration. (b) The free-body diagram for m_1. (c) The free-body diagram for m_2.

6.2 DYNAMICS OF CIRCULAR MOTION

An important example of accelerated motion is that of a particle moving at a constant speed v in a circular path of radius r, as shown in Fig. 6.8. From Eq. 4.18, the centripetal acceleration of a particle in such uniform circular motion is

$$\mathbf{a} = -\frac{v^2}{r}\,\hat{\mathbf{r}}$$

This acceleration is constant in magnitude but not in direction. It is centripetal—always directed radially inward, toward the center. From the second law, $\mathbf{F} = m\mathbf{a}$, we know that the particle must be subject to a **centripetal force** of magnitude

$$F = \frac{mv^2}{r} \qquad (6.3)$$

The term "centripetal" merely indicates the direction of the force; it tells us nothing about its nature or origin. The centripetal force may be a single force due to a rope, a spring, the force of gravity, friction, and so forth, or it may be the resultant of several such forces. It is *not* a new force to be added to the free-body diagram.

The discussion of uniform circular motion is often complicated by the mistaken introduction of a *centrifugal* (center-fleeing) force. Suppose you twirl a

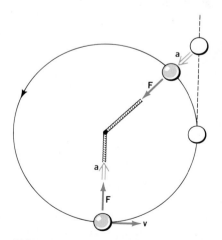

FIGURE 6.8 A particle moving with uniform circular motion is constantly being accelerated from its inertial path (dotted line). The centripetal force (toward the center) is provided by a rope.

stone in a circle at the end of a string. Your hand will experience an outward force. This leads some to believe that the stone is trying to move radially outward. They regard uniform circular motion as an *equilibrium* situation in which the inward pull balances an outward centrifugal force. This is *false!* The outward force on the hand and the inward pull on the stone are equal and opposite forces but they act on *different* bodies. At any instant, the velocity of the stone is along the tangent to the circle. The action of a centripetal force is to pull (or push) a particle from its natural inertial path along the tangent, and to make it travel in a circular path. When the string is released, the stone is no longer subject to a force, and so it moves at constant velocity along the tangent line.

The foregoing discussion was based on a description of events as seen by observers in an *inertial* reference frame—the ground, for example. The centrifugal force is an invention of observers in a rotating (accelerated) reference frame, such as a car going round a curve. Motion in *noninertial* frames is discussed in Section 6.5.

A motocyclist travels at high speed around a banked curve.

EXAMPLE 6.4: A small coin is placed at the rim of a turntable of radius 15 cm which rotates at 30 rev/min. Find the minimum coefficient of friction for the coin to stay on.

Solution: In such problems it is important to make a sketch from the proper perspective. A view from the top would not show the weight or the normal force on the coin. A side view, as in Fig. 6.9, shows the three forces acting on the coin. The acceleration is toward the center so we choose the $+x$ axis in this direction. The necessary centripetal force is provided by the friction force **f**. This is static friction since the coin does not slip on the table. The vector form of the second law is

$$\Sigma \mathbf{F} = \mathbf{N} + \mathbf{f} + \mathbf{W} = m\mathbf{a}$$

Its components are

$$\Sigma F_x = f = \frac{mv^2}{r}$$
$$\Sigma F_y = N - mg = 0$$

Since $f = \mu N = \mu(mg)$, we find

$$\mu mg = \frac{mv^2}{r}$$

or

$$\mu = \frac{v^2}{rg}$$

This is the minimum coefficient of static friction for the coin not to slip. Notice that the mass of the coin does not appear. In one revolution a point on the rim travels a distance $2\pi r$; hence, the speed of the coin is $v = (30 \text{ rev/min})(2\pi r \text{ m/rev})/(60 \text{ s/min}) = \pi r$ m/s. Using this in the above expression, we find $\mu = \pi^2 r/g = 0.15$.

EXAMPLE 6.5: In a carnival ride called the rotor, people stand on a ledge inside a large cylinder that rotates about a vertical axis. When it reaches a high enough rotational speed, the ledge drops away. Find the minimum coefficient of friction for the people not to slide down. Take the radius to be 2 m and the period to be 2 s.

Solution: A sketch that includes the forces acting *on* a person is shown in Fig. 6.10. (We are not interested in the outward force exerted on the cylinder by the person.) Notice that compared

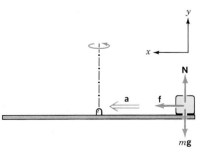

FIGURE 6.9 A coin on a rotating turntable. Friction is the only horizontal force. It must act *inward* to provide the centripetal force.

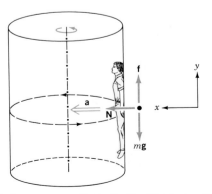

FIGURE 6.10 A person on a "rotor" ride. The force of friction balances the weight. The centripetal force is provided by the normal force **N**.

with Example 6.4, the roles of the normal force N and the force of friction f are interchanged. In this case N provides the centripetal force, whereas f prevents vertical sliding. The vector form of the second law is

$$\Sigma \mathbf{F} = \mathbf{N} + \mathbf{f} + \mathbf{W} = m\mathbf{a}$$

Its components are

$$\Sigma F_x = N = \frac{mv^2}{r}$$
$$\Sigma F_y = f - mg = 0$$

Since $f = \mu N = \mu(mv^2/r)$, we find

$$\mu \frac{mv^2}{r} = mg$$

or

$$\mu = \frac{rg}{v^2}$$

The period T is the time for one revolution; therefore, $v = 2\pi r/T = \pi r$. Using this in the above expression, we find $\mu = g/(\pi^2 r) \approx 0.5$.

EXAMPLE 6.6: A 50-kg woman is on a ferris wheel of radius 9 m that rotates in a vertical circle at 6 rev/min. What is the magnitude of her apparent weight when she is halfway up?

Solution: The woman is subject to three forces, as shown in Fig. 6.11. The vertical force N_1 due to the chair merely balances her weight since she is not accelerating in the vertical direction. The horizontal normal force N_2, due to the back of the chair, provides the centripetal acceleration. The vector form of the second law is

$$\Sigma \mathbf{F} = \mathbf{N}_1 + \mathbf{N}_2 + m\mathbf{g} = m\mathbf{a}$$

Its components are

$$\Sigma F_x = N_2 = \frac{mv^2}{r}$$
$$\Sigma F_y = N_1 - mg = 0$$

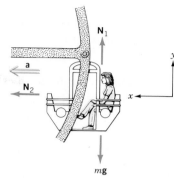

FIGURE 6.11 A woman on ferris wheel moves in a vertical circle. Her apparent weight is the resultant of the normal forces $\mathbf{N}_1$ and $\mathbf{N}_2$ exerted on her by the seat and the backrest.

We see that $N_1 = mg = 50$ kg $\times$ 9.8 N/kg $= 490$ N. Since she makes 6 rev/min, her speed is $v = (6$ rev/min$)(2\pi \times 9$ m/rev$)/(60$ s/min$) = 1.8\pi$ m/s. Therefore, $N_2 = (50$ kg$)(1.8\pi$ m/s$)^2/(9$ m$) = 178$ N. **The magnitude of her apparent weight is the magnitude of the resultant force exerted on her by the chair; that is,**

$$N = \sqrt{N_1^2 + N_2^2}$$
$$= \sqrt{490^2 + 178^2} = 521 \text{ N}$$

EXAMPLE 6.7: On fast highways or indoor bicycle tracks, the outside edge of curves is raised. Such *banking* prevents the vehicle from slipping sideways if there is not enough friction to provide the centripetal force. A car of mass 1000 kg rounds a circular curve of radius 10 m that is banked at 37° to the horizontal. The road is slippery, so the coefficient of static friction is only 0.1. Find the maximum safe speed at which the car can travel.

Solution: Figure 6.12a shows a top view of the curve. We want the car to travel in a horizontal circle—which means its acceleration is also horizontal. This determines the $+x$ axis, as shown in Fig. 6.12b. Since it is helping to prevent sliding up the in-

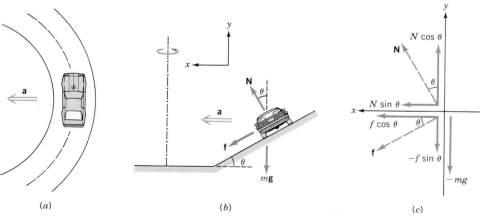

(a)   (b)   (c)

FIGURE 6.12 (*a*) A car going around a banked curve. (*b*) The forces acting the car. If the car does not slip sideways the acceleration is horizontal. (*c*) The free-body diagram for the car. Note the orientation of the axes.

cline, toward the outer edge of the circle, the frictional force must point down the incline. (Compare Fig. 6.12b with Fig. 6.9.) The vector form of the second law is

$$\sum \mathbf{F} = \mathbf{N} + \mathbf{f} + \mathbf{W} = m\mathbf{a}$$

From the free-body diagram in Fig. 6.12c, we see that its components are

$$\sum F_x = N \sin \theta + f \cos \theta = \frac{mv^2}{r} \qquad \text{(i)}$$

$$\sum F_y = N \cos \theta - f \sin \theta - mg = 0 \qquad \text{(ii)}$$

With $f = \mu N = 0.1N$, $\cos 37° = 0.8$, and $\sin 37° = 0.6$, (i) and (ii) become

$$0.6N + 0.08N = 10^2 v_{max}^2 \qquad \text{(iii)}$$

$$0.8N - 0.06N - 10^4 = 0 \qquad \text{(iv)}$$

From (iv), $N = 10^4/(0.74) = 1.35 \times 10^4$ N. Using this in (iii) gives $v_{max} = 9.6$ m/s.

EXERCISE 3. Is it possible for a car to go around a banked curve even if there is no friction?

EXAMPLE 6.8: A stone attached to the end of a rope moves in a vertical circle solely under the influence of gravity and the tension in the rope. Find the tension in the rope at the following points: (a) at the lowest point; (b) at the highest point; (c) when the rope is at angle θ to the vertical.

Solution: Since the rope is flexible, it cannot exert any force perpendicular to its length (unlike a rod, which can). The net acceleration is not constant in magnitude or in direction. The stone is subject to two forces, so the vector form of the second law is

$$\sum \mathbf{F} = \mathbf{T} + m\mathbf{g} = m\mathbf{a} \qquad \text{(i)}$$

(a) At the lowest point, shown in Fig. 6.13a, the acceleration is vertically up. If v_B is the speed at the bottom, then

$$\sum F_y = T - mg = \frac{mv_B^2}{r} \qquad \text{(ii)}$$

Thus, $T = mv_B^2/r + mg$. The tension supports the weight and also provides the centripetal force.

(b) At the highest point, shown in Fig. 6.13b, both T and mg contribute to the centripetal acceleration, which is vertically down:

$$\sum F_y = T + mg = \frac{mv_T^2}{r} \qquad \text{(iii)}$$

where v_T is the speed at the top. We find $T = mv_T^2/r - mg$. If it happens that $v_T^2 = rg$, then $T = 0$. This would mean that the weight alone is sufficient to provide the centripetal force.

(c) When the rope is at an angle θ to the vertical as in Fig. 6.13c, the stone has an acceleration tangential to its path because the force of gravity has a component in this direction. The radial and tangential components of (i) are

$$\sum F_x = mg \sin \theta = ma_t \qquad \text{(iv)}$$

$$\sum F_y = T - mg \cos \theta = \frac{mv^2}{r} \qquad \text{(v)}$$

Thus, $a_t = g \sin \theta$ and $T = mv^2/r + mg \cos \theta$. (ii) and (iii) are special cases of (v) for $\theta = 0°$ and $\theta = 180°$. (This is a check on (v).)

EXERCISE 4. It is possible to twirl a bucket containing water in a vertical circle without the water spilling out. (a) What are the forces acting on the water at the highest point? (b) Why doesn't the water spill out?

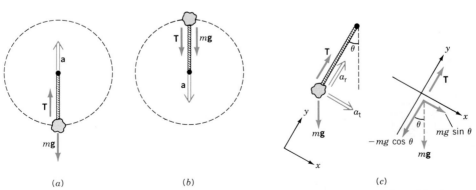

(a)   (b)   (c)

FIGURE 6.13 A stone traveling in a vertical circle. The forces and the acceleration when the stone is (a) at the bottom, (b) at the top, and (c) at an arbitrary position.

6.3   SATELLITE ORBITS

An important example of circular motion occurs in the orbits of satellites. The centripetal force is provided by the force of gravity. It was Newton who first described the connection between short-range trajectories and orbital motion. He imagined a cannon placed atop a high mountain firing balls with initial velocities tangential to the surface of the earth, as in Fig. 6.14. If it is given a low initial speed, the ball travels in an approximately parabolic path (ignoring air resistance). At higher initial speeds, it travels farther before it lands. The direction of the acceleration due to gravity (radially inward) varies along the trajectory. The general shape of the path is elliptical. If the initial speed is high enough, the ball travels right around the earth (back to the starting point). It would then be in orbit. Although the ball is always in free-fall from its straight-line inertial path, the curvature of the earth matches the curvature of the orbit. Consequently, the path of the satellite never intersects the surface, that is, it never lands. The following discussion is restricted to circular orbits, which are a special case of elliptical orbits.

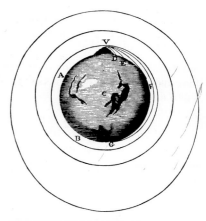

FIGURE 6.14 Newton's illustration showing the trajectories of balls fired with different initial speeds. At low initial speeds the ball will land on earth. At a high enough initial speed the ball travels completely around the earth; that is, it goes into orbit.

We assume that the mass of the central body, for example, the sun if we are looking at the oribts of planets, is very much greater than that of the orbiting body. This allows us to treat the central body as being fixed. We ignore any retarding forces due, for example, to the atmosphere of the earth. Figure 6.15 shows a particle of mass m in stable circular orbit around a stationary body of mass M. In Newton's second law, $F = ma$, the inward force of gravity provides the centripetal acceleration. Equation 6.3 takes the form

$$\frac{GmM}{r^2} = \frac{mv^2}{r} \tag{6.4}$$

The orbital speed is, therefore,

$$v_{\text{orb}} = \sqrt{\frac{GM}{r}} \tag{6.5}$$

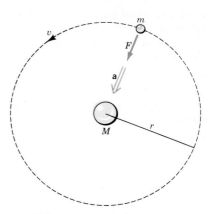

FIGURE 6.15 A satellite of mass m in orbit around a central body of mass M. The force of gravity provides the centripetal force.

Notice that the mass of the satellite does not appear and that the orbital speed decreases as the radius of the orbit increases. The period of the orbit is $T = 2\pi r/v_{\text{orb}}$, so

$$T = \frac{2\pi}{\sqrt{GM}}\sqrt{r^3}$$

or

$$T^2 = \frac{4\pi^2}{GM}r^3 = \kappa r^3 \tag{6.6}$$

This is called **Kepler's third law** after the German astronomer Johannes Kepler, who first discovered this relationship in 1619. It states that the square of the period of the orbit is proportional to the cube of the radius of the orbit. Notice that the constant κ involves only the mass M of the *central* body. Consequently, the orbits of all the planets have the same constant $\kappa_S = (4\pi^2/GM_S)$, where M_S is the mass of the sun. For our moon and other earth satellites, $\kappa_E = (4\pi^2/GM_E)$, where M_E is the mass of the earth. Kepler's third law is useful in determining the mass of the central body, such as a planet, if it has a satellite, since measurement of T and r yields M.

Kepler's third law

EXAMPLE 6.9: For reconnaissance, prospecting, or surveying, earth satellites are sometimes launched into orbits only about 150 km above the earth's surface. Find the period. Take $\kappa_E = 9.9 \times 10^{-14}$ s²/m².

Solution: Since the radius of the earth is about 6400 km, we may take this to be the radius of the orbit, that is, $r \approx R_E$. From Eq. 6.6,

$$T \approx \sqrt{\kappa_E R_E^3}$$
$$= \sqrt{(9.9 \times 10^{-14} \text{ s}^2/\text{m}^2)(6.4 \times 10^6 \text{ m})^3}$$
$$= 5.1 \times 10^3 \text{ s}$$

The period is approximately 85 min.

EXAMPLE 6.10: For telecommunication we need a satellite that appears fixed in the sky. Find the radius of its orbit.

Solution: The satellite must be in an equatorial orbit and have a period $T = 1$ d $= 8.64 \times 10^4$ s. From Eq. 6.6,

$$r = \left(\frac{T^2}{\kappa_E}\right)^{1/3}$$

Using $\kappa_E = 9.9 \times 10^{-14}$ s²/m², we find $r = 4.2 \times 10^7$ m.

This astronaut is apparently weightless, but his true weight is not zero.

"Weightlessness" in Orbit

Astronauts in stable orbit float freely in their cabin and are usually referred to as being "weightless." A spacecraft in orbit is in free-fall just as much as an elevator whose supporting cable has broken, and so the situation is similar to that of a person in a freely falling elevator. When the acceleration of the person and the enclosure are equal, the person is *apparently* weightless, even though the (true) weight is not zero. (In fact, the weight provides the centripetal force!)

In the same vein, one often hears of astronauts orbiting in the space shuttle having to "lift" some object. The astronaut does not have to contend with the weight of the object since it is also in free-fall. He or she has to deal with the mass or inertia of the object (see Fig. 5.9b).

6.4 MOTION IN RESISTIVE MEDIA (Optional)

When an object moves through a fluid, either liquid or gaseous, it experiences a resistance to its motion. This *drag force* depends on the size, shape, and orientation of the body and the density of the fluid. It also depends on the velocity of the body relative to the fluid. You can easily verify that the force depends on orientation and velocity by putting a hand out of a moving car. A qualitative discussion of the effects of air resistance on falling bodies was presented in Section 3.8.

Resistance Proportional to v

When the velocity of a body relative to a fluid is low, the fluid streams past it in a smooth and continuous way. In such **laminar** flow, a thin "boundary layer" of fluid forms around the body. The resistance experienced by the body is due to friction that arises as one layer of fluid flows over another. Under such conditions, the drag force is proportional to the speed:

$$F_D = \gamma v \tag{6.7}$$

where γ is a constant that depends on the dimensions of the body (for a sphere γ is proportional to the radius) and the ease with which the fluid flows, which is specified by a property called the *viscosity*. This viscous drag force occurs for dust particles or fog droplets falling through air or for small particles

falling in a thick liquid. Newton's second law applied to a body falling vertically through the fluid (Fig. 6.16) is*

$$mg - \gamma v = m\frac{dv}{dt} \tag{6.8}$$

FIGURE 6.16 When a body falls through a fluid it experiences a drag force $\mathbf{F}_D$. As a result, its acceleration is not constant.

We see that the acceleration of the body decreases as v increases. Figure 6.17 shows how the speed varies with time. When $v = 0$, dv/dt is the free-fall acceleration. During the initial part of the fall, the acceleration is large, but less than the free-fall value. The acceleration drops to zero at the **terminal speed** v_T, which from Eq. 6.8 is given by

$$v_T = \frac{mg}{\gamma} \tag{6.9}$$

* We ignore a force due to the buoyancy of the fluid.

Once the body reaches its terminal speed, it continues to fall at this constant rate.

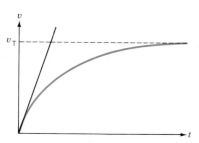

FIGURE 6.17 In falling through a resistive fluid an object reaches a terminal speed v_T.

Resistance Proportional to v^2

For large bodies moving at ordinary speeds, such as a falling stone or a car moving through the air, the flow of fluid past a body becomes **turbulent.** The motion of the molecules becomes erratic and a *turbulent wake* forms behind the body. The drag is caused by a difference in fluid pressure between the front and the rear of the body. The drag force is usually expressed in the form

$$F_D = \tfrac{1}{2} C_D \rho A v^2 = k v^2 \qquad (6.10)$$

where ρ is the density of the fluid, A is the area of the body "projected" onto a plane perpendicular to the motion, and C_D, called the **drag coefficient,** is a dimensionless number that depends on the shape and orientation of the body. It also depends on the velocity and the roughness of the surface. Typical values of C_D are indicated in Table 6.1.

TABLE 6.1[a] DRAG COEFFICIENTS

Smooth Sphere	Baseball	Cylinder	Disk	Person	Airfoil	Car
0.5	0.3	1.2	1.1	0.9	0.01	0.4

[a] The cylinder moves perpendicular to its long axis, the plane of the disk is perpendicular to the motion, and the person is in a spread-eagle position.

Newton's second law applied to a body falling through the air is (again neglecting buoyancy)

$$mg - k v^2 = m \frac{dv}{dt} \qquad (6.11)$$

The acceleration again decreases as v increases. Its dependence on time is similar to that of the linear drag force shown in Fig. 6.17. In this case the terminal speed is

$$v_T = \sqrt{\frac{mg}{k}} \qquad (6.12)$$

Table 6.2 lists terminal speeds in air for a few bodies. (For a smooth sphere of radius r falling in air, $k = 0.87 r^2$ in SI units.)

TABLE 6.2 TERMINAL SPEEDS

Body	v_T (m/s)
Skydiver	
Vertical	85
Spread eagle	55
Parachutist	6.5
Ping-Pong ball	7
Baseball	40
Golfball	30
Iron ball (2-cm radius)	80
Stone (1-cm radius)	30
Raindrop	10

6.5 NONINERTIAL FRAMES (Optional)

Our development of dynamics has been based on inertial frames, in which the first law is valid. There are, however, many examples of accelerated **noninertial frames** in which the first law is not valid. The earth's daily rotation and its orbital motion about the sun involve acceleration; thus, the earth is not really an inertial frame. Neither is a car rounding a curve, or a bus coming to a stop. We now examine motion as measured in such noninertial frames.

Consider a ball on the frictionless floor of a box that is accelerating and therefore forms a noninertial frame (see Fig. 6.18). In the inertial frame S of the laboratory, the box has an acceleration $\mathbf{a}$. If the ball is initially at rest with respect to the S frame, it will stay at rest since it experiences no net force. However, with respect to the noninertial frame S' of the box, the ball has an acceleration $\mathbf{a}' = -\mathbf{a}$. Therefore, in the S' frame, an observer who believes in Newton's second law, will say that the ball experiences a force, of unknown origin, given by

$$\mathbf{F}' = m\mathbf{a}' = -m\mathbf{a} \qquad (6.13)$$

An observer in the noninertial frame must invent a fictitious "inertial force" to explain the acceleration of the body.

This fictitious force is real enough to "throw" you forward when a bus suddenly stops. It is ficticious in the sense that it has no physical origin; that is, it is not caused by one of the basic interactions in nature. This "action" does not have the "reaction" required by the third law.

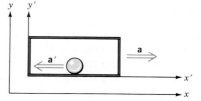

FIGURE 6.18 When the box accelerates at $\mathbf{a}$ relative to an inertial frame, the ball accelerates at $\mathbf{a}'$ ($= -\mathbf{a}$) relative to the box.

Figure 6.19 shows a pendulum suspended from the roof of a truck that has a constant acceleration **a** relative to the inertial frame of the road. An inertial observer would say that the bob has acceleration **a**, and so the second law is $\mathbf{T} + m\mathbf{g} = m\mathbf{a}$ (see Fig. 6.19a). An observer in the noninertial frame of the truck (Fig. 6.19b) sees the bob in static equilibrium and explains the deflection of the rope in terms of the inertial force **F'**. This noninertial observer's version of the second law is $\mathbf{T} + m\mathbf{g} + \mathbf{F'} = 0$.

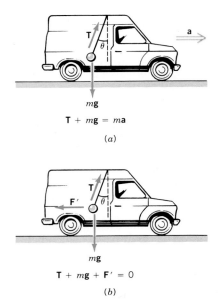

$$\mathbf{T} + m\mathbf{g} = m\mathbf{a}$$

(a)

$$\mathbf{T} + m\mathbf{g} + \mathbf{F'} = 0$$

(b)

FIGURE 6.19 (a) In an inertial frame the suspended bob has an acceleration **a** caused by the horizontal component of the tension **T**. (b) In the noninertial frame of the truck, the bob is in static equilibrium, and is also subject to the fictitious force **F'**.

The Centrifugal Force

Figure 6.20 shows a particle in uniform circular motion at the end of a string. In an inertial frame (Fig. 6.20a) the particle has an inward (centripetal) acceleration, so the second law is $\mathbf{T} = m\mathbf{a}$. In the rotating noninertial frame (Fig. 6.20b) in which the particle is at rest, the second law is $\mathbf{T} + \mathbf{F'} = 0$: There is an outward inertial force called the **centrifugal force.** The observer in the rotating frame can measure the tension in the string and therefore has to invent the centrifugal force to explain why the particle is at rest.

The centrifugal force seems "real" as we go around a curve in a car; we seem to be thrown toward the outer edge of the road. To understand the origin of this phenomenon, consider a ball that is initially held at rest with respect to a box that moves in a circle at constant speed, as in Fig. 6.21. The ball is released when the box is at position A. In our inertial frame the ball obeys the first law and travels in a straight line tangent to the path. An observer in the noninertial frame of the box sees

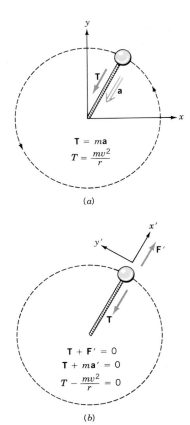

$$\mathbf{T} = m\mathbf{a}$$
$$T = \frac{mv^2}{r}$$

(a)

$$\mathbf{T} + \mathbf{F'} = 0$$
$$\mathbf{T} + m\mathbf{a'} = 0$$
$$T - \frac{mv^2}{r} = 0$$

(b)

FIGURE 6.20 (a) In an inertial frame a particle moves in a circle and has a centripetal acceleration produced by the tension **T**. (b) In the noninertial frame moving with the particle, the particle is in equilibrium. It is subject to the fictitious centrifugal force **F'**.

the ball move toward the outer face and explains this acceleration by the centrifugal force: $\mathbf{F'} = m\mathbf{a'} = -m\mathbf{a} = +(mv^2/r)\hat{\mathbf{r}}$.

Before we proceed, we must introduce the concept of an-

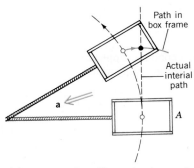

FIGURE 6.21 A box moves in uniform circular motion. A particle released at the center travels in a straight line relative to an inertial frame. Relative to the noninertial frame of the box, the particle moves toward the outside of the curve.

gular velocity. Consider a particle moving at a constant speed v in a circle of radius r, Fig. 6.22. In a time interval Δt, the angle $\Delta\theta$ (in radians) through which the radial line rotates is related to the arc Δs according to $\Delta\theta = \Delta s/r$. The rate at which the radial line rotates is called the angular velocity $\omega = \Delta\theta/\Delta t$ (rad/s). Using the expression for $\Delta\theta$ we have $\omega = (\Delta s/\Delta t)(1/r) = v/r$ and $v = \omega r$.

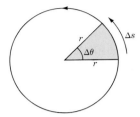

FIGURE 6.22 In radian measure $\Delta\theta = \Delta s/r$. The angular velocity is defined to be $\omega = \Delta\theta/\Delta t$.

The Coriolis Force

The centrifugal force does not depend on the velocity of a particle relative to a rotating noninertial frame. G. G. Coriolis investigated an additional inertial force that appears if the particle has a velocity relative to a rotating frame.

Figure 6.23a shows a platform (frame S') of radius R rotating with respect to an inertial frame S. At $t = 0$, a person at the center O throws a ball at speed v' toward a person P whose

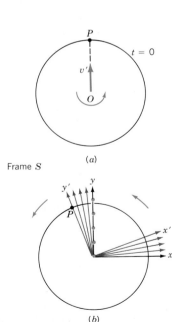

FIGURE 6.23 (a) A ball is thrown with speed v' from the origin of a rotating platform. (b) In the inertial frame (x, y) of the ground, the ball moves along the y axis.

position at this instant is also shown. We assume that the (x, y) axes coincide with the (x', y') axes at this time. In the inertial frame S, the ball moves in a straight line, as in Fig. 6.23b. By the time t when the ball reaches the rim, P has moved a distance $(\omega R)t$. In the rotating frame S', Fig. 6.24, as the ball moves toward the rim, it develops an increasing tangential velocity, so its path is curved. If we assume that the ball's speed is large enough ($v' \gg \omega R$) that the platform rotates only through a small angle, we may treat the arc along the rim as a straight line. When we equate $s = \frac{1}{2}a't^2$ to $s = (\omega R)t$, we find $\frac{1}{2}a't = \omega R$. Since $t = R/v'$, we have

$$a' = 2\omega v' \qquad (6.14)$$

This is called the **Coriolis acceleration.** Notice that the ball is deflected to the right of its direction of motion. It can be shown that $\mathbf{a}'$ is always perpendicular to $\mathbf{v}'$.

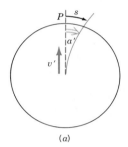

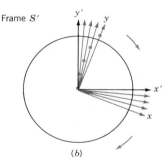

FIGURE 6.24 (a) In the noninertial frame (x', y') of the rotating platform, the path of the ball is curved. This is explained in terms of the fictitious Coriolis force. (b) As the platform rotates in a counterclockwise sense, the Coriolis acceleration a' deflects it toward the right.

Suppose the person P at the rim were to throw a ball toward the origin O, as in Fig. 6.25. At the moment it is released, it has a tangential velocity ωR with respect to the origin. Thus, it will again be deflected toward the right and will miss the origin. We conclude that in a frame that is rotating *counterclockwise* a particle is deflected to the *right* of its direction of motion. In practical terms, if a person were to try to walk from the center of the platform and aim for a post fixed in the ground, she would feel a push toward her right. If the platform were

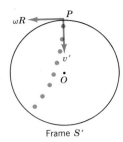

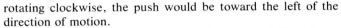

FIGURE 6.25 A person at point P on a rotating platform throws a ball toward the origin O. In this noninertial frame it will appear to travel along a curve path.

rotating clockwise, the push would be toward the left of the direction of motion.

We can obtain the expressions for the Coriolis and the centrifugal accelerations in the following way. Suppose a person walks along the rim of the platform in Fig. 6.26 at speed v' relative to the rim. His path is also circular in the ground frame, but his speed is $v = v' + \omega R$. In the inertial frame, his acceleration ($a = v^2/R$) is

$$a = \frac{v'^2}{R} + 2\omega v' + \omega^2 R$$

In the rotating frame $a' = v'^2/R$; thus,

$$a' = a - 2\omega v' - \omega^2 R \qquad (6.15)$$

The second term on the right side is the Coriolis acceleration and the third is the centrifugal acceleration. In this case, the Coriolis acceleration is in the same direction as the centrifugal acceleration—radially outward.

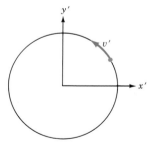

FIGURE 6.26 A person walks along the rim at speed v' relative to a rotating platform. The speed relative to the ground is $v = v' + \omega R$.

Foucault Pendulum

A young French scientist, Leon Foucault, realized that the work of Coriolis had an important physical implication. Since the earth forms a rotating frame, a particle moving horizontally in the Northern Hemisphere will experience an acceleration to the right *independent of the direction of its velocity* (see Fig. 6.27). The path of the bob of a simple pendulum in the inertial

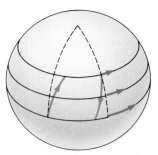

FIGURE 6.27 In the Northern Hemisphere, a particle will be deflected to the right of its direction of motion by the Coriolis force.

frame of the stars lies in a fixed plane, since the string does not exert any force perpendicular to the swing. However, at the start of each swing, the velocity of the bob relative to the earth is not the same as that of the (expected) end point in the swing. Thus, the bob misses the "target" no matter which way it swings. The deflection of the bob during one half of the swing is *not* canceled when the direction of motion is reversed. This means that the tiny Coriolis effect would accumulate with each complete swing and could be used to settle the question of whether the earth really rotates—a question that had bothered scientists and philosophers for centuries. As the earth rotates, the plane of swing changes its orientation relative to the earth (see Fig. 6.28). At the North Pole the plane makes one complete revolution per day. At a latitude ϕ, the time for a complete revolution is $(24 \text{ h})/(\sin \phi)$ and so the plane of the swing will rotate less than 360° in one day. In 1851, Foucault suspended a 28-kg bob from a dome of height 70 m and thereby settled the question of the reality of the earth's rotation.

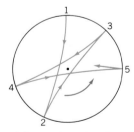

FIGURE 6.28 As a pendulum swings the plane of its motion rotates relative to the earth (which is a rotating reference frame).

Projectiles

The Coriolis acceleration affects the trajectories of artillery shells and ballistic missiles, even though its magnitude is very small. For the earth $\omega = 7 \times 10^{-5}$ rad/s. If $v' = 1000$ m/s, then $a' = 2\omega v' = 0.14$ m/s². As we pointed out in the special topic on real projectiles, if the shell or missile is aimed directly at the target, it will miss. When a body is dropped from a high tower, its initial tangential velocity is greater than that of the point on the ground vertically below. As a result it will fall toward the east of the expected landing point.

Weather

The Coriolis effect also plays a role in our weather. Figure 6.29 shows a low-pressure area in the Northern Hemisphere. Far from this area, the air starts to flow from the surrounding regions of higher pressure toward the center of the low-pressure area. The direction of flow is perpendicular to the isobars (lines of equal pressure). However, the Coriolis force, which is perpendicular to the pressure force in this region, deflects the air to the right. Near the center of the "low," the air flows around, rather than toward, the center. Close to the center, the Coriolis force is nearly opposite to the pressure force. The air reaches a state in which it moves almost tangentially to the isobars. The resulting counterclockwise circulation can be mild, as in a large weather system, or intense, as in a hurricane or tornado. Perhaps the most dramatic illustration of the Coriolis effect is in the famous "red spot" on Jupiter, shown in Fig. 6.30.

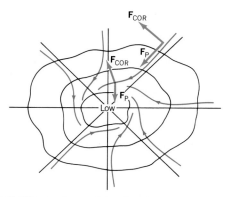

FIGURE 6.29 The resultant of the Coriolis force and the force due to the atmospheric pressure gradient causes the air to flow around a low-pressure region.

FIGURE 6.30 The red spot on Jupiter.

SPECIAL TOPIC: Friction Phenomena

It was pointed out earlier that surface roughness is not an adequate explanation for friction. It is estimated that roughness accounts for less than 10% of the total friction. It was only in the 1950s that F. P. Bowden and D. Tabor* showed that *surface adhesion* is the primary factor that determines the frictional force. As Fig. 6.31 shows, junctions are formed between metals by "cold welding." Friction arises when these junctions are sheared. During sliding, these bonds are continually made and broken. Often the break occurs within the bulk of the softer metal, and so tiny fragments are ripped off and can be detected by radioisotope techniques. The adhesion theory is given support by the fact that two very clean surfaces, preferably in a vacuum, can bond together without external pressure.

Another theory involves *electrostatic forces*. It is believed that as the surfaces slide over each other, charges are transferred from one to the other. Electrical attraction between the oppositely charged surfaces then produces the frictional drag. We know that objects are often charged by friction, for example, by rubbing a glass rod with fur. This charge separation undoubtedly occurs for many pairs of substances—when one is a nonconductor. However, it cannot occur for two clean surfaces of the same metal. So, at best, electrostatic attraction accounts for only some cases. In moving over granular snow or sand, there is obviously a "ploughing" contribution to the friction.

STICK–SLIP

The phenomenon of stick–slip (named by F. P. Bowden in 1939) occurs when the relative speed of the sliding surfaces is low and when one of the objects can vibrate. Imag-

* F. P. Bowden and D. Tabor, *Friction and Lubrication of Solids.* Clarendon Press, Oxford, 1950.

ine a flexible rod dragged along a surface (Fig. 6.32). As it is pulled, it bends because of the static friction at its base. At some point it will break away and move forward rapidly, only to be caught again. It goes through cycles of static (stick) and kinetic (slip) friction. The period of the cycle depends on the elastic properties of the object and the average speed of the motion. Often these vibrations are audible. For example, a fingernail scraped across a blackboard is an effective way of getting the attention of a class. The relationship between the pitch of the sound and the speed of the motion is easily heard by rubbing a wet finger on a bathtub. The pitch of the squeal rises with the speed. However, at higher speeds, it stops. This indicates that the motion has become purely sliding. Lest you get a bad impression of stick–slip, we should point out that all bowed instruments depend on it.

SNOW AND ICE

The friction associated with snow and ice is of great interest to skiers, skaters, and dogsledders. The friction due to snow depends on several factors: load, temperature, wax on the skis, wood versus metal skis, whether the snow is "wet" or "dry," and so on. The low value of the kinetic coefficient is due to a film of water (about 0.01 cm thick) that is produced by frictional heating between the ski and the snow or between the skate and the ice. (It is true that the melting temperature of ice is lowered when pressure is applied, but this plays a secondary role in skating.) At very low temperatures, this film is not formed, and so the friction is quite large. Because of the "ploughing" contribution, friction in wet snow is greater than in dry snow.

Wooden skis experience lower friction than metal ones because of their poor heat conduction. The metal carries away the heat generated and prevents the formation of the film. Waxes and tar are used to prevent the water from soaking the wood, which would lead to freezing of the film.

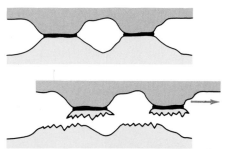

FIGURE 6.31 (a) The points of contact between two surfaces are "cold-welded" together. (b) The force of friction arises when the bonds are broken. Often, part of the softer material is removed.

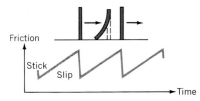

FIGURE 6.32 At low speeds friction is characterized by a mixture of static and kinetic friction and is referred to as "stick–slip." The vibration of one of the bodies in contact is often audible as a squeak.

The friction of ice on ice depends on both temperature and sliding speed. At low temperature ($-40\ °C$), the static coefficient is 0.45 and drops as the temperature rises. The kinetic coefficient varies from 0.03 to 0.2 depending on speed and temperature. Because of the film that is formed, the kinetic friction is not directly proportional to the load.

ROLLING FRICTION

A ball or wheel is subject to **rolling friction,** which is usually very much smaller than kinetic friction. Before we discuss the mechanism of rolling friction we should distinguish between a wheel that is rolling freely and a "driven" wheel.

When a bicycle is pedalled, the bottom of the rear wheel is pushed backward against the road. This is the force F_{RW} on the road due to the wheel, as in Fig. 6.33. The reaction to this, F_{WR}, is in the forward direction and causes the bike to move forward. If there is no relative motion between the bottom of the wheel and the road, the maximum force that the road can exert on the wheel is the static frictional force $f_{s(max)}$. If the wheel is driven harder, it slips. If the brakes are applied but the wheels are allowed to roll, the maximum force the road can exert is again $f_{s(max)}$, directed backward. If all the wheels are locked, they slide and experience the lower kinetic friction.

Rolling friction arises from several causes. Adhesion must play some part. As a wheel rolls, it has to peel itself off the surface. Sometimes it also slips and so there is some kinetic friction associated with this. Another contribution comes from plastic deformation. The surface on which the wheel rolls is indented under the wheel and so the wheel is

in a tiny crater. This is most pronounced when a hard sphere rolls on a softer surface, such as rubber or billiard cloth, in which case the surface ahead of the wheel bunches up, as in Fig. 6.34a. The force exerted by the bump has a rearward component. In the case of a rubber tire on concrete, it is the tire that deforms, as in Fig. 6.34b. The part of the tire that is ahead of the center "collides" with the road, whereas the portion that is behind the tire is lifting off. In effect the net force on the tire can be taken to act ahead of the center and tilted toward the rear, as shown in the figure.

A major contribution to rolling friction comes from elastic effects. As the surfaces compress under increasing load, the force versus deformation curve follows path 1 in Fig. 6.35. As the load is removed, path 2 is traced. The material does not return to its original state and retains some deformation. This effect is called *elastic hysteresis*. The area within the hysteresis loop represents energy that is temporarily stored within the wheel and ultimately lost as heat. This explains why lubricants have little effect on rolling friction. In the case of a tire on concrete, rolling friction amounts to about 2% of the load. Thus, in the case of a 10^3-kg car, the rolling resistance is about 200 N. Maintaining proper tire pressure and using radial tires rather than bias ply tires reduces rolling resistance.

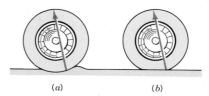

(a)   *(b)*

FIGURE 6.34 A rolling wheel that is not driven. (a) A hard wheel rolls on a soft surface. (b) A soft wheel rolls on a hard surface. In either the case, the net force on the wheel acts ahead of the center of the wheel in the direction shown.

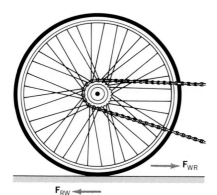

FIGURE 6.33 On a wheel driven by a chain, the force of friction on the wheel exerted by the road, **F**$_{WR}$, is in the forward direction.

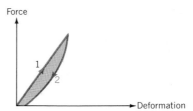

FIGURE 6.35 The force versus deformation curve for a material such as rubber. As the force is increased from zero, curve 1 is traced. As the force is removed, curve 2 is traced. The effect is called elastic hysteresis. The area within the loop represents lost energy.

SUMMARY

Friction is a contact force that opposes the relative motion of two surfaces. When a block is at rest on a surface and a force is applied tangential to the surface, the force of static friction automatically adjusts its value to equal the applied force, but only up to a critical value given by

$$f_{s(max)} = \mu_s N$$

where μ_s is the **coefficient of static friction** and N is the normal force—which is not necessarily equal to the weight of the block. When the block slides over the surface, the force of kinetic friction is

$$f_k = \mu_k N$$

where μ_k is the **coefficient of kinetic friction.**

A particle moving in a circular path of radius r at constant speed v is subject to a net radially inward force called the **centripetal force:**

$$F = \frac{mv^2}{r}$$

The source of the centripetal force may be, for example, the tension in a rope, friction, the force of gravity, or a combination of these. The centripetal force is *not* a new force to be included in the free-body diagram.

An observer in a noninertial (accelerated) frame must introduce fictitious **inertial forces** in order to apply Newton's second law. An inertial force does not obey Newton's third law, since it is not exerted by any physical object. When a particle is at rest in a rotating frame, it appears to be subject to a radially outward **centrifugal force.** If it is moving relative to the rotating frame, it also experiences a **Coriolis force.**

ANSWERS TO IN-CHAPTER EXERCISES

1. (a) $\Sigma F_x = F \cos \theta - f_k = ma$, so $a = 0.33$ m/s². The block is speeding up. (b) $\Sigma F_x = F \cos \theta + f_k = ma$, so $a = 1.96$ m/s². The block is slowing down.

2. The force of kinetic friction is $f_k = \mu_k m_1 g = 3.92$ N. Using this in (i) and (ii) we find $a_1 = 1.96$ m/s² and $a_2 = 9.02$ m/s².

3. Yes, see Exercise 31.

4. (a) The weight mg and the normal force N due to the bottom of the bucket. Both forces, and the acceleration, are directed downward. (b) If the bucket were released at the highest point it would be in free-fall and would travel in a parabolic path. Instead, it has a centripetal acceleration greater than the free-fall acceleration. The bucket is "falling" faster than it would under the influence of gravity alone.

QUESTIONS

1. An astronaut in circular orbit around the earth feels weightless. Why is this?

2. A coin is at the rim of a rotating turntable. When the rate of rotation reaches a certain value, why does the coin fly off?

3. Can astronauts in stable circular orbit around the earth be "weighed"? Can they be "massed"?

4. Is Newton's third law valid in a noninertial frame? Give an example to support your answer.

5. A motorcyclist claims to get better acceleration by lifting the front wheel off the ground. Does this make sense?

6. Does it help a car's stability to accelerate gently while rounding a banked curve? If so, explain how.

7. When a plane moves in a vertical circle, at what point is the danger of pilot blackout greatest?

8. What is the magnitude and direction of the average frictional force on a person who walks at constant velocity?

9. Why doesn't a satellite in stable orbit fall back to earth? Ignore air friction.

10. Given the period and the radius of the earth's orbit and the radius of the moon's orbit, could one find the period of the moon's orbit? No other information is available.

11. Can an astronaut in stable orbit determine that the spacecraft is under the influence of gravity if there are no windows? If so, how?

12. Suggest locations at which the (true) weight of an astronaut is essentially zero.

13. The force on the moon due to the sun is greater than the force due to the earth. So why doesn't the moon leave the earth?

14. When the brakes on a bus are suddenly applied, the passengers are thrown forward. How would you explain this?

EXERCISES

6.1 Friction

1. (I) A 90-g hockey puck with an initial speed of 10 m/s slows to 8 m/s in 12 m. Find: (a) the frictional force on it; (b) the coefficient of friction.

2. (I) A standard car has two-wheel drive. Assume that the weight is equally distributed over all four wheels. If the coefficient of static friction is 0.8, find the maximum acceleration possible when the car (a) starts from rest, or (b) comes to a stop.

3. (I) Two blocks are connected as shown in Fig. 6.36 and move at constant velocity (the 5 kg-block moves down the incline). Find: (a) the coefficient of kinetic friction assuming it is the same for both blocks; (b) the tension in the rope.

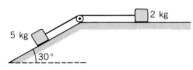

FIGURE 6.36 Exercise 3

4. (II) A 1-kg block is at rest on an incline for which $\mu_k = 0.6$ and $\mu_s = 0.8$; see Fig. 6.37. (a) Does the block move if it is released? (b) If a horizontal force $F_0 = 40$ N acts on it, what is its acceleration?

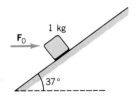

FIGURE 6.37 Exercise 4

5. (I) An 80-kg person pushes a 20-kg crate over a rough surface. Take $\mu_s = 0.8$ for the person and $\mu_k = 0.4$ for the crate. (a) What is the maximum possible acceleration of the crate? (b) For the condition in part (a) find the force exerted by the crate on the person.

6. (I) A 5-kg block is subject to a horizontal force of 30 N, as in Fig. 6.38. It lies on a surface for which $\mu_k = 0.5$ and $\mu_s = 0.7$. (a) If the block is at rest, what is the frictional force on it? What is the acceleration of the block if it is moving (b) to the left, or (c) to the right?

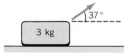

FIGURE 6.38 Exercise 6

7. (I) A 3-kg block is acted on by a 25-N force that acts 37° below the horizontal, as shown in Fig. 6.39. Take $\mu_k = 0.2$ and $\mu_s = 0.5$. (a) Does the block move if it is initially at rest? (b) If it moves to the right, what is its acceleration?

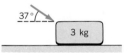

FIGURE 6.39 Exercise 7

8. (I) Repeat Exercise 7, with the same force acting at 37° above the horizontal, as shown in Fig. 6.40.

FIGURE 6.40 Exercise 8

9. (I) A 2.5-kg block is on a 53° incline for which $\mu_k = 0.25$ and $\mu_s = 0.5$. Find its acceleration given that: (a) it is initially at rest; (b) it is moving up the slope; (c) it is moving down the slope.

10. (I) A crate lies on the flat bed of a truck that decelerates at 6 m/s². What is the required minimum coefficient of friction for it not to slide?

11. (II) The minimum stopping distance for a car from an initial 100 km/h is 60 m on level ground. What is the stopping distance when it moves (a) down a 10° incline; (b) up a 10° incline? Assume that the initial speed and the surface are unchanged.

12. A downhill skier starts from rest down a 40° slope for which $\mu_k = 0.1$. (a) How long does it take to reach 80 km/h? (b) What is the distance covered? Ignore air resistance.

13. (I) A skier approaches a 10° incline at 80 km/h. If $\mu_k = 0.1$ and air resistance is neglected, how far up the slope does she reach? Assume that the poles are not used.

14. (II) Approximately 60% of the weight of any car is carried by the front wheels. Assume all other factors for a front-wheel-drive car and a rear-wheel-drive car are the same. (a) Compare the minimum times needed to accelerate from rest to a given speed. (b) Compare the minimum distances required to stop from a given speed. Express each answer as a ratio.

15. (I) Three skaters, A, B, and C, shown in Figure 6.41 have masses $m_A = 30$ kg, $m_B = 50$ kg, and $m_C = 20$ kg. They are pulled along a horizontal surface for which $\mu_k = 0.1$ by grabbing a horizontal rope. The tension in the rope ahead of A is 200 N. Find: (a) the acceleration; (b) the tension T_1 in the section of rope between A and B; (c) the tension T_2 in the section of rope between B and C.

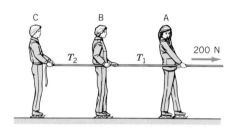

FIGURE 6.41 Exercise 15

16. (I) A block is released at the top of a 20° incline. Determine the coefficient of kinetic friction given that it slides 2.4 m in 3 s.

17. (II) A block of mass M is subject to the force F shown in Fig. 6.42. The friction coefficients are μ_k and μ_s. (a) What minimum value of F will start the block moving from rest? (b) Show that if $\mu_s = \cot \theta$, the block cannot be moved for any value of F. (c) To what situation does the condition $\mu_k = \cot \theta$ apply?

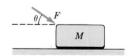

FIGURE 6.42 Exercise 17

18. (I) A 5-kg block is on a 37° incline for which $\mu_k = 0.1$. It is acted on by a horizontal force of 25 N, as in Fig. 6.43. (a) What is the acceleration of the block if it is moving up the incline? (b) If its initial speed is 6 m/s up the incline, how far does it travel in 2 s?

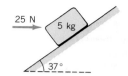

FIGURE 6.43 Exercise 18

19. (II) Block A of mass $m_A = 2$ kg is on block B of mass $m_B = 5$ kg; see Fig. 6.44. The lower block is on a frictionless surface while $\mu_s = 0.25$ between the two blocks. (a) If they are moving at constant velocity, what is the frictional force between A and B? (b) What is the maximum horizontal force that can be applied to B without A slipping?

FIGURE 6.44 Exercise 19

20. (II) Block A of mass $m_A = 2$ kg is on the front face of cart B of mass $m_B = 3$ kg (see Fig. 6.45). A force of 60 N acts on B. What is the minimum coefficient of friction needed for A not to slide down?

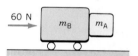

FIGURE 6.45 Exercise 20

21. (II) Two blocks with equal masses $m_1 = m_2 = 5$ kg are connected via a pulley, as shown in Fig. 6.46. Take $\mu_k = 0.25$. Find the acceleration given that (a) m_1 is moving down, (b) m_1 is moving up. (c) If $m_2 = 6$ kg, for what values of m_1 will the pair move at constant speed?

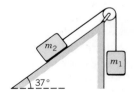

FIGURE 6.46 Exercise 21

22. (II) In Fig. 6.47, when $m_1 = 3$ kg, its acceleration is 0.6 m/s². When $m_1 = 4$ kg, its acceleration is 1.6 m/s². Find m_2 and the frictional force acting on it. (Note that m_1 and m_2 do not have the same acceleration.)

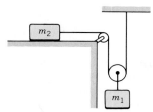

FIGURE 6.47 Exercise 22

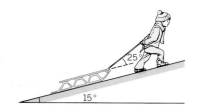

FIGURE 6.50 Exercise 27

23. (II) The horizontal force F in Fig. 6.48 accelerates the 4-kg block at 1 m/s^2 to the left. Assume $\mu_k = 0.5$. Given that the 5 kg block moves up the slope find (a) the value of F, and (b) the tension in the rope. (c) If $F = 10$ N and the 5-kg block moves down the slope, what is the acceleration?

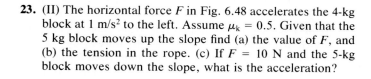

FIGURE 6.48 Exercise 23

24. (II) A block of mass $m = 2$ kg is attached to a wall by a rope and is placed on a block of mass $M = 6$ kg, as in Fig. 6.49. When a force of 24 N is applied to the lower block it accelerates at 3 m/s^2. Given that all surfaces are the same, find the coefficient of kinetic friction.

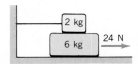

FIGURE 6.49 Exercise 24

25. (I) A 580-g mass is suspended from the roof of a truck. When the truck accelerates horizontally, the tension in the rope is 6 N. Find the acceleration.

26. (I) A car travels at 108 km/h on a road with a static frictional coefficient μ_s. What is the minimum stopping distance if (a) $\mu_s = 0.9$ (dry road), or (b) $\mu_s = 0.3$ (wet road). Why is the friction static rather than kinetic?

27. (II) A child pulls a 3.6-kg sled at 25° to a slope that is at 15° to the horizontal, as in Fig. 6.50. The sled moves at constant velocity when the tension is 16 N. What is the acceleration of the sled if the rope is released?

6.2 Dynamics of Circular Motion

28. (I) A pail of water is moved in a vertical circle of radius

80 cm. What is the minimum speed required at the top for the water not to spill out of the inverted pail?

29. (I) A car travels along a hilly road as in Fig. 6.51. The top of a hill has a radius of curvature of 20 m. (a) What is the maximum speed possible if the car is not to lose contact with the road? (b) At this speed, what would be the apparent weight of a 75-kg passenger when the car is at the bottom of a valley of radius of curvature 20 m?

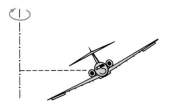

FIGURE 6.51 Exercise 29

30. (I) A circular off ramp has a radius of 60 m and a posted speed limit of 60 km/h. If the road is horizontal, what is the minimum coefficient of friction required?

31. (II) A car travels at speed v around a frictionless curve of radius r that is banked at an angle θ to the horizontal. Show that the proper angle of banking is given by $\tan \theta = v^2/rg$.

32. (II) A 70-kg pilot flies an airplane at 400 km/h in a turn of radius 2 km, as shown in Fig. 6.52. (a) At what angle to the horizontal are the wings? (b) What is the pilot's apparent weight? The aerodynamic lift force is perpendicular to the wings.

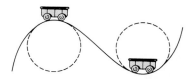

FIGURE 6.52 Exercises 32 and 38

33. (II) A small block is placed inside a cylinder of radius $R = 40$ cm that rotates with a period of 2 s about a horizontal axis as in Fig. 6.53. Show that the maximum angle θ reached by the block before it starts to slip is given by

$$g \sin \theta = \mu_s \left(g \cos \theta + \frac{v^2}{R} \right)$$

where $\mu_s = 0.75$ is the coefficient of static friction and v is the speed of the block. (b) Determine θ. (*Hint:* Use $\sin^2 \theta = 1 - \cos^2 \theta$ and solve a quadratic equation in $\cos \theta$.)

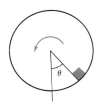

FIGURE 6.53 Exercise 33

34. (II) A daredevil bicyclist moves at 7 m/s in a horizontal circle in a cylindrical "well of death" of radius 4 m (see Fig. 6.54). What is the minimum coefficient of friction required?

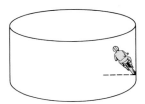

FIGURE 6.54 Exercise 34

35. (I) A car turns through a flat circular curve of radius 40 m at 15 m/s. What is the magnitude of the apparent weight of the 70-kg driver?

36. (II) A 30-kg child is on a "rotor" ride at a carnival (see Fig. 6.10). The radius of the rotor is 3 m and it completes 0.4 revolutions in one second. The coefficient of static friction is 0.6. What is the magnitude of the child's apparent weight?

37. (I) The electron in a hydrogen atom orbits a stationary proton at a distance of 5.3×10^{-11} m at a speed of 2.2×10^6 m/s. Find: (a) the period; (b) the force on the electron.

38. (II) A pilot flies a plane at 800 km/h in a horizontal circle (see Fig. 6.52). The magnitude of the apparent weight of the pilot is 40% greater than his normal weight. What is the radius of the turn?

39. (I) What would be the length of a day if the people at the equator were apparently weightless?

40. (I) A 0.2-kg mass moves in a vertical circle at the end of a string of length 30 cm. What is the tension in the string at the following points: (a) the top of the circle where the speed is 2.00 m/s; (b) the bottom of the circle where the speed is 3.96 m/s; (c) halfway up, where the speed is 3.14 m/s?

41. (I) A 60-kg woman is on a Ferris wheel that takes her in a vertical circle of radius 20 m at constant speed. (a) At what speed would she feel weightless at the top? (b) At this speed, what is her apparent weight at the bottom?

42. (I) A button is at the rim of a turntable of radius 15 cm rotating at 45 rpm. What is the minimum coefficient of friction needed for it to stay on?

43. (II) A car is moving at speed v perpendicular to a wall. The static coefficient of friction is μ_s. What is the minimum distance from the wall within which the driver must act if (a) she brakes in a straight line, or (b) turns in a circle? (c) Is there a strategy that could minimize damage?

44. (I) A doughnut-shaped space station has a diameter of 2 km. What should be the period of its rotation for the astronauts to experience an apparent weight 20% of that on earth?

45. (I) In a vertical loop-the-loop ride at an amusement park, shown in Fig. 6.55, the radius of curvature at the highest point is 6.5 m. (a) What is the minimum speed for the train not to lose contact with the track at this point? (b) If the actual speed is 9.5 m/s, what is the apparent weight of a 40-kg child at the highest point?

FIGURE 6.55 Exercise 45

6.3 Satellite Orbits; Kepler's Third Law

46. (I) (a) Use the moon's period of 27.32 d and the radius of its orbit, 3.84×10^5 km, to find the mass of the earth. (b) Use the earth–sun distance and the period of the earth to find the mass of the sun.

47. (I) The period and radius of the orbit of our moon are 27.3 d and 3.84×10^5 km. The corresponding values for a moon of Jupiter are 3.5 d and 6.7×10^5 km. Compute the ratio of the masses of the earth and Jupiter, M_J/M_E.

48. (I) The sun, whose mass is 2×10^{30} kg, orbits around the center of the Milky Way at a distance of 2.4×10^{20} m. Its period is 2.5×10^8 y. (a) Estimate the mass of the galaxy within the sun's orbit. What assumption must you make? (b) If all the stars are comparable to our sun, how many are there within the orbit?

49. (II) A neutron star rotates with a period of 1 s. If its radius is 20 km, what must its mass be for it to keep an object at the equator bound to the surface?

50. (I) (a) Show that the speed of a satellite in a circular orbit close to the earth of radius R_E is $(gR_E)^{1/2}$, where g is the gravitational force per unit mass at the surface; see Section 5.3. (b) Estimate the period of a low-altitude spy satellite.

51. (I) The moon Io of Jupiter is in a circular orbit of radius 4.22×10^5 km with a period of 1.77 d. (a) The period of another moon, Europa, is 3.55 d. What is the radius of its orbit? (b) What is the mass of Jupiter?

52. (I) What would be the orbital speed of a tiny insect in "low-altitude" orbit around a 2000-kg elephant? Treat the pachyderm as a uniform sphere of radius 0.8 m.

53. (I) During an Apollo mission, the radius of the orbit of the 10^4-kg Lunar Command module was 1.8×10^6 m. (a) What was the period? (b) What force did the moon exert on the craft?

54. (II) Planet A has twice the radius of planet B but only half the density. Compare the periods of low-altitude satellites of these planets.

55. (I) In November 1984 the space shuttle *Discovery* was placed in a circular orbit at an altitude of 315 km in order to catch up with the disabled *Westar* 6 satellite in circular orbit at an altitude of 360 km. Suppose that these two objects were initially on opposite sides of the earth. How many orbits would the satellite have made for the shuttle to be beneath the satellite (that is, along a radial line and closer to the earth)? Take $R_E = 6370$ km.

6.4 Motion in Resistive Media

56. (I) A 20-kg package falls out of a helicopter and reaches a terminal speed of 30 m/s. What is the drag force when the speed is (a) v_T, and (b) $0.5v_T$? Assume $F_D \propto v^2$.

57. (I) A 5-g ball bearing falling in oil reaches a terminal speed of 2 cm/s. Find the drag force at 1 cm/s. Assume $F_D \propto v$.

58. (I) When the resistive force on a body of mass m is proportional to the speed v, the speed as a function of time is given by $v = v_T(1 - e^{-\gamma t/m})$, where v_T is the terminal speed and γ is a constant. Find the acceleration as a function of time.

6.5 Noninertial Frames

59. (I) A pendulum suspended from the roof of a truck is deflected by 8° from the vertical. What is the acceleration of the truck?

60. (I) A pendulum of length 80 cm with a bob of mass 0.4 kg is suspended from the roof of a truck that accelerates at 2.6 m/s². Find: (a) the horizontal deflection of the bob; (b) the tension in the rope.

61. (I) A car goes around a flat curve of radius 40 m at constant speed. The level of water in a vertical glass of diameter 3 cm rises from the horizontal position by 0.5 cm on one side. What is the speed of the car?

62. (II) A person is at the rim of a carousel of radius 4.5 m that rotates at 0.8 rad/s. He throws a ball toward the center at 30 m/s relative to the carousel. By what distance does it miss?

PROBLEMS

1. (I) In Fig. 6.56 a block of mass m rests on a table for which the coefficient of static friction is μ_s. (a) Show that the minimum value of the force F needed to start the block moving occurs at an angle given by $\tan \theta = \mu_s$. (b) Show further that this minimum force is $mg \sin \theta$.

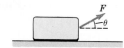

FIGURE 6.56 Problem 1

2. (I) A block of mass $m = 2$ kg is placed on a block of mass $M = 4$ kg as shown in Fig. 6.57. The coefficient of kinetic friction for all surfaces is $\mu_k = 0.2$. Ignore the pulley and the rope. For what value of the horizontal force F_0 will the blocks (a) move at constant speed, (b) accelerate at 2 m/s²?

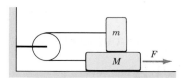

FIGURE 6.57 Problem 2

3. (I) (a) A 3000-kg truck coasts (in neutral) down a 5° slope at constant velocity. What force is needed to move it uphill at the same speed? (b) A car can climb a maximum slope of 10° at constant velocity. What is its maximum acceleration on a level road? Assume that the road and air resistances are unchanged.

4. (I) A car travels around a circular banked curve of radius 40 m at 35° to the horizontal. If the coefficient of static friction is 0.4, what are the minimum and maximum safe speeds?

5. (I) A toy car of mass m can travel at a fixed speed. It moves in a circle on a horizontal table. The centripetal force is provided by a string attached to a block of mass M that hangs as shown in Fig. 6.58. The coefficient of static friction is μ. Show that the ratio of the maximum radius to the minimum radius possible is

$$\frac{M + \mu m}{M - \mu m}$$

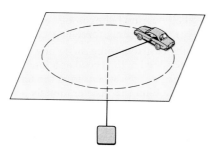

FIGURE 6.58 Problem 5

6. (I) A simple pendulum consists of a 2-kg mass suspended at the end of a rope of length 4 m. When the rope is at 20° to the vertical, the speed of the bob is 3 m/s. Find: (a) the radial and tangential components of the acceleration; (b) the tension.

7. (I) A block of mass 0.4 kg is attached to a vertical rotating spindle by two strings of equal length, as in Fig. 6.59. The period of rotation is 1.2 s. Determine the tensions in the strings.

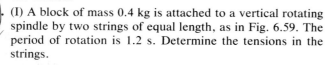

FIGURE 6.59 Problem 7

8. (I) A car rounds a circular curve of radius 80 m banked at 15° to the horizontal at 108 km/h. (a) In which direction (up or down the slope) is the lateral force of friction? (b) What is the coefficient of friction required for the car not to skid?

9. (I) Two blocks of masses 3 and 5 kg are attached by a string and slide down a 30° incline; see Fig. 6.60. The coefficient of kinetic friction for the 3-kg block is 0.4 whereas for the

5-kg block it is 0.3. Find the acceleration of the blocks and the tension in the string.

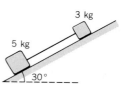

FIGURE 6.60 Problem 9

10. (I) In a conical pendulum, shown in Fig. 6.61, a bob moves in a horizontal circle. Show that the period is

$$T = 2\pi \sqrt{\frac{L \cos \theta}{g}}$$

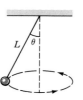

FIGURE 6.61 Problem 10

11. (I) A block rests on an incline that is at an angle θ to the horizontal, as in Fig. 6.62. The coefficients of friction are μ_s and μ_k. (a) Show that the block starts to slide when $\mu_s = \tan \theta_s$, where θ_s is the maximum value of θ. (b) Show that if θ is slightly greater than θ_s, the time taken to slide a distance d along the incline is

$$t = \sqrt{\frac{2d}{g \cos \theta_s(\mu_s - \mu_k)}}$$

(c) Could one find μ_k in terms of θ? If so, how?

FIGURE 6.62 Problem 11

12. (II) In Fig. 6.63 a block of mass $m = 0.5$ kg is on a wedge of mass $M = 2$ kg. The wedge is subject to a horizontal force F_0 and slides on a frictionless surface. The coefficient of static friction between block and wedge is 0.6. Find the range of values of F_0 for which the block does not slide on the incline. Take $\theta = 40°$.

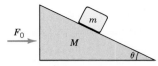

FIGURE 6.63 Problem 12

13. (II) A car travels in a horizontal circle around a curve of radius r banked at an angle θ to the horizontal. If μ is the coefficient of static friction, show that the maximum speed possible without sliding sideways is

$$v_{max} = \sqrt{\frac{rg(\mu + \tan\theta)}{1 - \mu\tan\theta}}$$

Assume that $\mu\tan\theta < 1$.

14. (II) Use Eq. 6.8 to show that when the drag force is proportional to speed, that is, $F_D = \gamma v$, the speed of a falling body of mass m increases according to

$$v = v_T(1 - e^{-\gamma t/m})$$

where v_T is the terminal speed ($= mg/\gamma$).

CHAPTER 7

Work and Energy

A horse doing work as it pulls a cart.

Major Points

---

1. (a) The definition of **work** done by a constant force.
   (b) Work as the area under the *F* versus *x* graph.
2. **Kinetic energy:** The energy an object has by virtue of its motion.
3. The **work–energy theorem** relates the work done by the net force on a particle to the change in its kinetic energy.
4. The definition of **power.**

In principle, all problems in classical mechanics can be solved by using Newton's laws. Given the initial positions and velocities of the particles in a system and all the forces acting on them, one can predict the future development of the system. In practice, we often have little information on the forces involved in a given situation. Even if we know how a force varies with position or time, a direct application of Newton's laws can be cumbersome. In this chapter we introduce a different approach to the solution of problems in mechanics that is based on the concepts of **work** and **energy.**

In everyday language we speak of mental work, physical work, medical work, scientific work, and so on. In physics and engineering we are concerned with *mechanical* work. The muscular exertion involved in lifting an object clearly depends both on its weight and the height through which it is lifted. This idea formed the basis for comparing the capabilities of steam engines—which were first used to pump water from mines. It was agreed that the product "weight of water × height lifted," would be a useful measure of what an engine accomplished for a given amount of coal. This developed into our modern definition of work.

Energy is sometimes defined as the capacity to do work. Although this is a logically acceptable definition, it has limited usefulness. Energy can be transferred on a microscopic scale from one system to another without the performance of measurable work. In many systems the stored energy cannot be readily or completely converted into work. Consequently, one cannot always measure the capacity to do work. All one can say is that energy is not a substance; it is an abstract concept whose importance arises from the fact that it is a conserved quantity: Although energy can change form, it can be neither created nor destroyed. This is called the **principle of the conservation of energy** and is discussed in the next chapter.

7.1 WORK DONE BY A CONSTANT FORCE

The **work** W done by a constant force $\mathbf{F}$ when its point of application undergoes a displacement $\mathbf{s}$ is defined to be

$$W = Fs \cos \theta \qquad (7.1a)$$

where θ is the angle between $\mathbf{F}$ and $\mathbf{s}$ as indicated in Fig. 7.1. Only the component of $\mathbf{F}$ along $\mathbf{s}$, that is, $F \cos \theta$, contributes to the work done. Strictly speaking, the work is done by the *source* or *agent* that applies the force. Work is a scalar quantity and its SI unit is the joule (J). From Eq. 7.1a we see that

$$1 \text{ J} = 1 \text{ N} \cdot \text{m}$$

In British units, the unit of work is the foot pound, where $1 \text{ J} = 0.738 \text{ ft} \cdot \text{lb}$. Equation 7.1a may be expressed in terms of the dot product defined in Eq. 2.9:

$$W = \mathbf{F} \cdot \mathbf{s} \qquad (7.1b)$$

In terms of rectangular components, the two vectors are $\mathbf{F} = F_x\mathbf{i} + F_y\mathbf{j} + F_z\mathbf{k}$ and $\mathbf{s} = \Delta x\mathbf{i} + \Delta y\mathbf{j} + \Delta z\mathbf{k}$; hence, Eq. 7.1b may be written as (see Eq. 2.11)

$$W = F_x\Delta x + F_y\Delta y + F_z\Delta z \qquad (7.1c)$$

The work done by a given force on a body depends only on the force, the displacement, and the angle between them. It does not depend on the velocity or the acceleration of the body, or on the presence of other forces. Since work is a scalar, its value also does not depend on the orientation of the coordinate axes. However, since the magnitude of a displacement in a given time interval depends on the velocity of the frame of reference used to measure the displacement, the calculated work also depends on the reference frame.

From Eq. 7.1a we see that when the force and the displacement are perpendicular, no work is done. Thus, when a block slides along a plane, as in Fig. 7.2a, the normal force $\mathbf{N}$ does no work. Similarly, a centripetal force, such as the tension $\mathbf{F}$ in a rope (Fig. 7.2b), does no work because it is always perpendicular to the motion. Work was initially required to give the object its speed, but none is needed to maintain it. This is a situation in which a force accelerates a particle, yet does no work.

Although the definition of work seems reasonable, it does conflict with our everyday notions. Figure 7.3a shows a ball being held stationary in a hand. The force F exerted by the hand is equal to the weight of the ball. When we just hold the ball, the pain and the heat generated in our muscles lead us to claim that

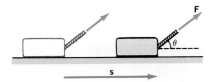

FIGURE 7.1 The work W done by the force $\mathbf{F}$ when its point of application undergoes a displacement $\mathbf{s}$ is $W = \mathbf{F} \cdot \mathbf{s} = Fs \cos \theta$.

Work done by a constant force

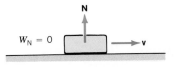

(a)

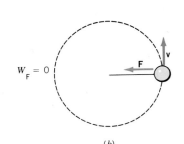

(b)

FIGURE 7.2 When the force and the displacement are perpendicular, as is the case for (a) the normal force and (b) the centripetal force, no work is done.

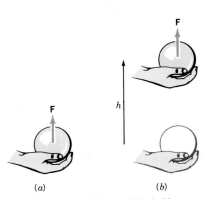

(a)   (b)

FIGURE 7.3 (a) No work is done by the hand if the ball is held at rest. (b) Work is done by the hand when it lifts the ball.

"work" is being done. However, one must be careful to distinguish between mechanical work, as we have defined it, and such physiological work. By definition, mechanical work is done only when the point of application of the force moves through a distance. If the hand does not move, as in Fig. 7.3a, it does no work. Actually, when a muscle is under tension, the individual muscle fibers contract repeatedly and so they do perform microscopic work. In the present case, there is no macroscopic displacement, and therefore no "useful" mechanical work is done. As far as a physicist or engineer is concerned, the hand can be replaced by a passive support, such as a table. If the ball were lifted through a height h, as in Fig. 7.3b, the work done by the hand on the ball would be $W = Fh$.

When several forces act on a body one may calculate the work done by each force individually. The net work done on the body is the algebraic sum of the individual contributions:

$$W_{NET} = \mathbf{F}_1 \cdot \mathbf{s}_1 + \mathbf{F}_2 \cdot \mathbf{s}_2 + \cdots + \mathbf{F}_N \cdot \mathbf{s}_N$$
$$= W_1 + W_2 + \cdots + W_N$$

If rotation or vibration are involved, or if the body is not rigid, the particles undergo different displacements. We assume for the moment that all the particles have the same displacement. That is, the body as a whole undergoes pure translation and may be treated as a particle. The expression for the work done on the body then takes the form

(Translation only) $\qquad W_{NET} = \mathbf{F}_{NET} \cdot \mathbf{s}$ $\qquad$ (7.2)

where $\mathbf{F}_{NET} = \Sigma\mathbf{F}_i$ is the *resultant* of all the external forces acting on the body.

A weightlifter does work to lift weights but not to hold them at rest.

Negative Work and Work Done by Friction

Figure 7.4 shows a block B being pushed by a hand A. From Newton's third law we know that the forces exerted by these objects on each other are equal and opposite, $\mathbf{F}_{AB} = -\mathbf{F}_{BA}$. Since the point of application of each force undergoes the same displacement, we infer that when A does positive work on B, B does an equal amount of negative work on A:

$$W_{By\ A\ on\ B} = -W_{By\ B\ on\ A}$$ (7.3)

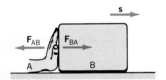

FIGURE 7.4 The hand does positive work on the block. The block does an equal amount of negative work on the hand.

Consider the work done by the force of kinetic friction on a block that slides over a rough tabletop, as in Fig. 7.5a. The frictional force f_k on the block and the displacement are in opposite directions. From Eq. 7.1a, the work done by the force of friction on the block is

Work done by the force of friction

$$W_f = f_k s \cos 180° = -f_k s$$ (7.4)

Since the work done by friction on the block is negative, it means that positive work was done by the block on, or against, friction. This work done by the block is needed to break the bonds that form between the two surfaces.

There is a common misconception that the force of kinetic (sliding) friction always does negative work. Consider the case of block A placed on block B, on which a force F_0 acts as shown in Fig. 7.5b. We saw in Example 6.3 that as A slides backward relative to B, the force of friction on A is in the forward direction. The displacement of A relative to the floor is also in the forward direction. Clearly, the work done by kinetic friction on A is *positive*.

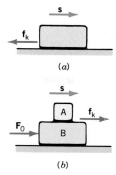

(a)

(b)

FIGURE 7.5 (a) When a block slides on a stationary surface, the work done by the force of kinetic friction is negative. (b) Block A slides backward relative to block B, which is being accelerated by a force $\mathbf{F}_0$. The displacement of A relative to the table is in the forward direction. The work done by kinetic friction on block A is positive.

Work Done by Gravity

Figure 7.6 shows a block that undergoes a displacement along an incline. Let us find the work done by the force of gravity on the block. To use Eq. 7.1c, we need to choose coordinate axes to specify the components of the two vectors—although the value of the work will not depend on the orientation of the axes. With the choice of axes indicated, $m\mathbf{g} = -mg\mathbf{j}$. In terms of rectangular components, a finite displacement is always expressed in the form

$$\mathbf{s} = \Delta x\mathbf{i} + \Delta y\mathbf{j} + \Delta z\mathbf{k}$$

Thus, the work done by gravity on the block is (with $\Delta z = 0$)

$$W_g = m\mathbf{g} \cdot \mathbf{s} = (-mg\mathbf{j}) \cdot (\Delta x\mathbf{i} + \Delta y\mathbf{j})$$

Since $\Delta y = y_f - y_i$, we have

$$W_g = -mg(y_f - y_i) \qquad (7.5)$$

In Fig. 7.6, $\Delta y = y_f - y_i = -h$, so $W_g = +mgh$. (Use Eq. 7.1a to obtain this result.) If the displacement were in the opposite direction, Eq. 7.5 would still have the same form; one would simply substitute the appropriate values for y_f and y_i.

From Eq. 7.5 we infer an important result: *The work done by the force of gravity depends only on the initial and final vertical coordinates, not on the path taken.* Furthermore, if $y_f = y_i$, then $W_g = 0$. That is, *the work done by gravity is zero for any path that returns to the initial point.* These facts will assume great importance in the next chapter.

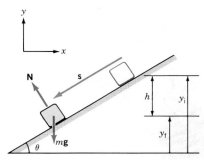

FIGURE 7.6 The work done by gravity is $W_g = -mg(y_f - y_i) = +mgh$.

Work done by the force of gravity (assumed to be constant)

EXAMPLE 7.1: A skier of mass $m = 40$ kg is given a displacement of 20 m along a slope inclined at $\theta = 15°$ to the horizontal. The tension in the towrope is $T = 250$ N and acts at an angle $\alpha = 30°$ to the incline (see Fig. 7.7). Given that $\mu_k = 0.1$, find the work done by each force and the net work on the skier.

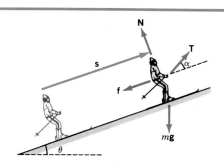

FIGURE 7.7 The tension in the rope does positive work on the skier; the forces of friction and gravity do negative work.

Solution: The four forces acting on the skier are shown in the figure. From the condition $\Sigma F_y = 0$, we find

$$N = mg \cos \theta - T \sin \alpha = 379 \text{ N} - 125 \text{ N} = 254 \text{ N}$$

so that $f = \mu N = 25.4$ N. From Eq. 7.1a, the work done by each force is

$$W_T = \mathbf{T} \cdot \mathbf{s} = Ts \cos 30° = +4330 \text{ J}$$
$$W_f = \mathbf{f} \cdot \mathbf{s} = -fs = -508 \text{ J}$$

$$W_N = \mathbf{N} \cdot \mathbf{s} = 0$$
$$W_g = m\mathbf{g} \cdot \mathbf{s} = -mgs \sin 15° = -2030 \text{ J}$$

Note that the vertical displacement is $\Delta y = +s \sin 15°$. Another way to determine W_g is to note that the component of mg along the incline is $mg \sin 15°$ and that it is opposite to the displacement. The net work is

$$W_{NET} = W_T + W_f + W_N + W_g = +1.79 \text{ kJ}$$

7.2   WORK BY A VARIABLE FORCE IN ONE DIMENSION

In many situations the force acting on a particle varies both in magnitude and in direction. In this section we consider cases in which the force and the displacement lie along the same line, say, the x axis. (The general case is dealt with in Section 7.5.) We begin by showing that work may be calculated from the area under the F versus x graph.

Let us first consider the work done by a constant force $\mathbf{F} = F_x \mathbf{i}$ on a body whose displacement $\mathbf{s} = \Delta x \mathbf{i}$; that is,

$$W = \mathbf{F} \cdot \mathbf{s} = F_x \Delta x$$

The work is positive if F_x and Δx are in the same direction, and it is negative if they are in opposite directions. We may represent this work by the area under the graph of F_x versus x. In Fig. 7.8 we have taken $F_x = F_0$. The work done in a displacement from x_A to x_B is $F_0(x_B - x_A)$, which is the shaded area in the graph. A displacement from x_B to x_A would mean that the area has a negative sign.

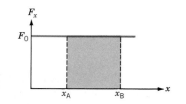

FIGURE 7.8 The work done by the constant force F_0 in a displacement from x_A to x_B is $W = F_0 (x_B - x_A)$. This is the area of the shaded rectangle

EXERCISE 1.   What is the work done by the force $F_x = -5$ N if the displacement is from $x = 7$ m to $x = 3$ m?

Work Done by a Spring

The relationship between the force F_{sp} exerted by an ideal spring and its extension, or compression, was discovered in the seventeenth century by Robert Hooke and is therefore called Hooke's law:

Hooke's law

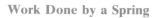

$$F_{sp} = -kx \tag{7.6}$$

where k, measured in N/m, is called the *spring constant*. The quantity x is the displacement of the free end of the spring from its equilibrium position—the point at which the spring has its natural length (see Fig. 7.9). The negative sign signifies that the force always opposes the extension ($x > 0$) or the compression ($x < 0$) of the spring. In other words, the force always tends to restore the system to its equilibrium position.

The work done by the spring in a displacement from x_i to x_f is the shaded area in Fig. 7.10, which is the difference in the areas of two triangles. Thus, $W_{i \to f} = \frac{1}{2}F_f x_f - \frac{1}{2}F_i x_i$. The work done by the spring when the displacement of the free end from its equilibrium position changes from x_i to x_f is

Work done by a spring

$$W_{sp} = -\tfrac{1}{2}k(x_f^2 - x_i^2) \tag{7.7}$$

This work is positive when the force and the displacement are in the same direction. Note that sometimes Hooke's law is expressed in terms of the external force, $F_{EXT} = +kx$, needed to produce a given x. For a given displacement, the work W_{EXT} done by the external agent (e.g., you) on the spring would have a sign opposite to that of W_{sp}.

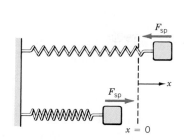

FIGURE 7.9 The force exerted by an ideal spring is given by Hooke's law: $F_{sp} = -kx$, where x is the extension or compression of the spring.

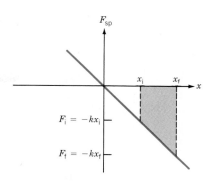

FIGURE 7.10 The work done by the spring when the displacement of its free end changes from x_i to x_f is the area of the trapezoid: $W_{sp} = -\frac{1}{2}k(x_f^2 - x_i^2)$.

As we found with the force of gravity, Eq. 7.7 shows that the *work done by the force due to an ideal spring depends only on the initial and final points.* Also, if $x_f = x_i$, then $W_{sp} = 0$. That is, *the net work done by the ideal spring is zero for any path that returns to the initial point.*

EXAMPLE 7.2: A block of mass 100 g is attached to the end of a spring whose spring constant is $k = 40$ N/m. The block slides on a horizontal surface for which $\mu_k = 0.1$. The spring is extended by 5 cm and then released. (a) Find the work done by the spring up to the point at which it is compressed by 3 cm. (b) Find the net work done on the block up to this point.

Solution: From Eq. 7.7,

$$W_{sp} = -\frac{1}{2}k(x_f^2 - x_i^2)$$
$$= -(20 \text{ N/m})(9 \times 10^{-4} \text{ m}^2 - 25 \times 10^{-4} \text{ m}^2) = +0.032 \text{ J}$$

(b) The work done by the force of friction, $f_k = \mu_k mg = 0.098$ N, is

$$W_f = -f_k s$$
$$= -(0.098 \text{ N})(0.08 \text{ m}) = -0.0078 \text{ J}$$

The net work done on the block is
$$W_{NET} = W_{sp} + W_f$$
$$= +0.024 \text{ J}.$$

Work as an Integral of the F_x versus x Graph

When the force is an arbitrary function of position, we need the techniques of calculus to evaluate the work done by it. Figure 7.11a shows F_x as some function of the position x. We begin by replacing the actual variation of the force by a series of small steps. The area under each segment of the curve is approximately equal to the area of a rectangle. The height of the rectangle is a constant value of force, and its width is a small displacement Δx. Thus, the nth step involves an amount of work $\Delta W_n = F_n \Delta x_n$. The total work done is given approximately by the sum of the areas of the rectangles:

$$W \approx \Sigma F_n \Delta x_n$$

As the size of the steps is reduced, the tops of the rectangles more closely trace the actual curve shown in Fig. 7.11b. In the limit $\Delta x \to 0$, which is equivalent to letting the number of steps tend to infinity, the discrete sum is replaced by a continuous integral:

$$\lim_{\Delta x_n \to 0} (\Sigma F_n \Delta x_n) = \int F_x \, dx$$

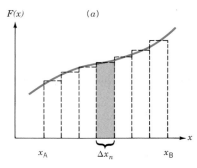

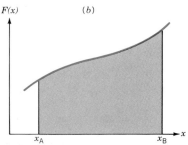

FIGURE 7.11 (a) The work done by a nonconstant force is approximately equal to sum of the areas of the rectangles. (b) The area under the curve is given by the integral $W = \int F_x \, dx$.

The net work done on the javelin is equal to the change in its kinetic energy.

Thus, the work done by a force F_x from an initial point A to final point B is

$$W_{A \to B} = \int_{x_A}^{x_B} F_x \, dx \tag{7.8}$$

The sign of this work depends on F_x *and* the limits of the integral. Interchanging the limits of the integral is equivalent to reversing the direction of the displacement. Thus,

$$W_{B \to A} = -W_{A \to B} \tag{7.9}$$

Let us use Eq. 7.8 to find the work done by the force of a spring, $F_{sp} = -kx$, when the displacement from equilibrium changes from x_i to x_f:

$$W_{sp} = \int_{x_i}^{x_f} (-kx) \, dx = -\tfrac{1}{2}k(x_f^2 - x_i^2)$$

This is the same as Eq. 7.7.

7.3   WORK–ENERGY THEOREM IN ONE DIMENSION

We now consider what physical quantity changes when net work is done on a particle. For the moment we restrict the discussion to one-dimensional translational motion and a constant force. If a constant force F acts through a displacement Δx, it does work $W = F\Delta x = ma\Delta x$ on the particle. Since the acceleration is constant, we may use Eq. 3.12, $v_f^2 = v_i^2 + 2a\Delta x$, to replace $a\Delta x$ in the expression for W:

$$W = \tfrac{1}{2}mv_f^2 - \tfrac{1}{2}mv_i^2 \tag{7.10}$$

The quantity

$$K = \tfrac{1}{2}mv^2 \tag{7.11}$$

is a scalar and is called the **kinetic energy** of the particle. Kinetic energy is energy that a particle possesses by virtue of its motion. Equation 7.10 may be written in the form

$$W_{NET} = \Delta K \tag{7.12}$$

where W_{NET} is the work done by the *resultant* of all the forces acting on the particle. Equation 7.12 is called the **work–energy theorem.** It states that the net work done on a particle is equal to the resulting change in its (translational) kinetic energy. From Eq. 7.12 we see that the kinetic energy of an object is a measure of the amount of work needed to increase its speed from zero to a given value. Equivalently, the translational kinetic energy of an object is the work it can do on its surroundings in coming to rest. This fits the idea of energy as the capacity to do work. Note that since the displacement and the velocity of a particle depend on the frame of reference, the numerical values of the work and the kinetic energy also depend on the frame.

Since Eq. 7.12 is based on Newton's second law, it has the same information content, but the change in the motion of a particle is described in different terms. Whereas force and acceleration are vectors, work and kinetic energy are scalars, which makes them easier to deal with. Although Eq. 7.12 has been derived for a constant force acting in one dimension, it is valid for a variable force acting in three dimensions (see Problem 12).

EXAMPLE 7.3: A block of mass $m = 4$ kg is dragged 2 m along a horizontal surface by a force $F = 30$ N acting at 53° to the horizontal (Fig. 7.12). The initial speed is 3 m/s and $\mu_k = 1/8$. Find: (a) the change in kinetic energy of the block; (b) its final speed.

Solution: (a) The forces acting on the block are shown in Fig. 7.12. Clearly, $W_N = 0$ and $W_g = 0$, whereas $W_F = Fs \cos \theta$ and $W_f = -\mu_k Ns$. In this case the condition $\Sigma F_y = 0$ leads to $N = mg - F \sin \theta$. The work–energy theorem,

$$\Delta K = W_{NET} = W_F + W_f$$

is therefore

$$\Delta K = Fs \cos \theta - \mu_k(mg - F \sin \theta)s$$
$$= (30 \text{ N})(2 \text{ m})(0.6) - \tfrac{1}{8}(40 \text{ N} - 24 \text{ N})(2 \text{ m}) = 32 \text{ J}$$

(b) From (a), $\Delta K = \tfrac{1}{2}mv_f^2 - \tfrac{1}{2}mv_i^2 = 32$ J. We are given $v_i = 3$ m/s; therefore, $v_f = 5$ m/s.

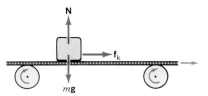

FIGURE 7.12 The change in the kinetic energy of the block is given by the net work done on it.

EXERCISE 2. If the force in Example 7.3 is removed once the block reaches the speed in part (b), how far does it move before it stops?

EXAMPLE 7.4 A block of mass $m = 2$ kg is attached to a spring whose spring constant is $k = 8$ N/m (see Fig. 7.13). The block slides on an incline for which $\mu_k = \tfrac{1}{8}$ and $\theta = 37°$. If the block starts at rest with the spring unextended, what is its speed when it has slid a distance $d = 0.5$ m down the incline?

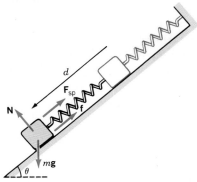

FIGURE 7.13 The work done by gravity is positive; the work done by the spring and by friction are negative.

Solution: This problem can be solved by using Newton's second law. However, the force exerted by the spring varies with

position and, therefore, so does the acceleration. We avoid this difficulty by using the work–energy theorem. (Unless you are solving a problem in static equilibrium, it is a good idea to think of the energy approach rather than dynamics when you see a spring in a problem.) The work done by the force due to the spring was found in Eq. 7.7. In this example $x_i = 0$ and $x_f = +d$. The work done by each of the forces on the block is

$$W_g = m\mathbf{g} \cdot \mathbf{s} = +mgd \sin \theta$$
$$W_f = \mathbf{f} \cdot \mathbf{s} = -\mu_k(mg \cos \theta)d$$
$$W_{sp} = -\tfrac{1}{2}kd^2$$

Of course, $W_N = 0$. The work–energy theorem, with $\Delta K = \tfrac{1}{2}mv^2 - 0$, tells us

$$mgd \sin \theta - \mu_k(mg \cos \theta)d - \tfrac{1}{2}kd^2 = \tfrac{1}{2}mv^2 \qquad \text{(i)}$$

On inserting the given values into (i), we find $v = 2$ m/s.

EXERCISE 3. What is the maximum extension of the spring in Example 7.4?

EXAMPLE 7.5: A crate of mass m is dropped onto a conveyor belt that moves at a constant speed v (Fig. 7.14). The coefficient of kinetic friction is μ_k. (a) What is the work done by friction? (b) How far does the crate move before reaching its final speed? (c) When the crate reaches its final speed, how far has the belt moved?

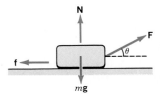

FIGURE 7.14 A block is dropped onto a conveyor belt. The force of kinetic friction and the displacement of the block are in the same direction. Hence, the work done by the force of kinetic friction on the block is positive.

Solution: When the crate is first placed on the belt there will be slipping between the two. But the force of friction on the crate and its displacement are in the *same* direction. Consequently, the work done by kinetic friction is positive. Since the final speed of the crate is v,

$$W_f = \Delta K = +\tfrac{1}{2}mv^2 \qquad \text{(i)}$$

(b) The force of friction is $f = \mu_k N = \mu_k mg$ and $W_f = +fd$. Thus, from (i)

$$+\mu_k mgd = +\tfrac{1}{2}mv^2 \qquad \text{(ii)}$$

from which we find $d = v^2/(2\mu_k g)$.

(c) If the crate takes a time t to reach speed v, then $v = at$ where a is the crate's acceleration. In this time it will move $d = \tfrac{1}{2}at^2 = \tfrac{1}{2}vt$. Since the belt's speed is fixed, in time t it moves a distance $vt = 2d$. The belt moves *twice* as far as the crate while the crate is accelerating.

7.4   POWER

As the pace of the industrial revolution quickened, it was no longer sufficient to know how much work an engine could do; it became necessary to specify how fast it could work. **Mechanical power** is defined as the rate at which work is done. If an amount of work ΔW is done in a time interval Δt, then the average power is defined to be

Average power
$$P_{av} = \frac{\Delta W}{\Delta t} \tag{7.13}$$

The SI unit of power is J/s which is given the name watt (W) in honor of James Watt; thus, 1 W = 1 J/s. The instantaneous mechanical power is the limiting value of P_{av} as $\Delta t \rightarrow 0$; that is

Instantaneous power
$$P = \frac{dW}{dt} \tag{7.14}$$

The work done by a force **F** on a object that has an infinitesimal displacement $d\mathbf{s}$ is $dW = \mathbf{F} \cdot d\mathbf{s}$. Since the velocity of the object is $\mathbf{v} = d\mathbf{s}/dt$, the instantaneous mechanical power may be written as $P = dW/dt = \mathbf{F} \cdot d\mathbf{s}/dt$ or

$$P = \mathbf{F} \cdot \mathbf{v} \tag{7.15}$$

Another unit of power is the *horsepower* (hp), where 1 hp = 746 W. It was originally defined by Watt as the power output of a horse over an extended period, and its value was taken to be 550 ft lb/s. Watt used this unit to rate his steam engines.

Since work and energy are closely related, a more general definition of power is the rate of energy transfer from one body to another, or the rate at which energy is transformed from one form to another:

$$P = \frac{dE}{dt} \tag{7.16}$$

where E stands for any form of energy. This form is often required because energy transfer—for example, in the form of heat—need not involve macroscopic mechanical work.

EXAMPLE 7.6: A pump raises water from a well of depth 20 m at a rate of 10 kg/s and discharges it at 6 m/s. What is the power of the motor?

Solution: The pump does work to lift the water and also to change its kinetic energy. The total work on a mass m would be

$$W = mgh + \tfrac{1}{2}mv^2$$

Since h and v are constant, the instantaneous power is

$$P = \frac{dW}{dt} = \frac{dm}{dt}\left(gh + \frac{v^2}{2}\right)$$
$$= (10 \text{ kg/s})(200 \text{ m}^2/\text{s}^2 + 18 \text{ m}^2/\text{s}^2) = 2180 \text{ W}$$

EXAMPLE 7.7: A 10^3-kg car requires 12 hp to cruise at a steady 80 km/h on a level road. What would be the power required to move up to a 10° incline at the same speed? Assume that the total frictional force due to the road and air resistance is fixed.

Solution: At constant speed, the driven wheels (actually the *road!*) provide a forward force F that just balances the rolling friction f_R of the tires and the drag force f_D due to the air. That is, $f = f_R + f_D$. Hence, the power delivered to the wheels of the car is $P = fv$. Converting to SI units we have $v = 80$ km/h = $(80 \times 10^3 \text{ m})/(3600 \text{ s}) = 22.2$ m/s, and 12 hp = $12 \times 746 = 8.95 \times 10^3$ W. Thus,

$$f = \frac{P}{v} = \frac{8.95 \times 10^3 \text{ W}}{22.2 \text{ m/s}} = 403 \text{ N}$$

To go up the incline at constant speed, the car requires a force $F = f + mg \sin 10°$ (see Fig. 7.15). The power required would be

$$P = (f + mg \sin 10°)v$$
$$= [403 \text{ N} + (10^3 \text{ kg})(9.8 \text{ N/kg})(0.174)] (22.2 \text{ m/s})$$
$$= 46.6 \times 10^3 \text{ W} = 62.7 \text{ hp}$$

The power required to climb an incline at a given constant speed is considerably greater than on a horizontal surface.

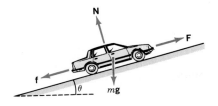

FIGURE 7.15 The force $F = mg \sin \theta + f$ exerted on the driven wheel of the car by the road keeps the speed v of the car constant. The power required of the engine is $P = Fv$.

EXAMPLE 7.8: A rocket of mass 2×10^5 kg starts from rest at ground level and accelerates vertically upward at 4 m/s². What is the instantaneous power of the engines when the speed is 50 m/s? Ignore air resistance and changes in the mass of the rocket.

Solution: The rocket is subject to two forces, the thrust **T** up-

ward and its weight $m\mathbf{g}$. Since the acceleration is upward we have $(T - mg) = ma$. From Eq. 7.15 the instantaneous power is

$$P = Tv = m(g + a)v$$
$$= 1.4 \times 10^8 \text{ W}$$

This is approximately 190,000 hp, which is the power of a large ship, such as the *Queen Mary*.

We may approach this problem slightly differently. When the rocket is at a height y and has speed v, the engines have done a total amount of work given by $W = mgy + \frac{1}{2}mv^2$. The instantaneous power is

$$P = \frac{dW}{dt} = mg\frac{dy}{dt} + mv\frac{dv}{dt}$$
$$= m(g + a)v$$

which is the same expression as before. (In fact, the mass of the rocket is not constant because fuel is being ejected and often one or two stages are discarded.)

TABLE OF ENERGIES (J)

Supernova explosion	10^{44}
Annual solar output	1.2×10^{34}
Rotational energy of earth	2×10^{29}
Initial fossil fuels (ca. 1800)	2×10^{23}
Annual global use	2.7×10^{20}
Annual loss tidal friction	10^{20}
Annual U.S. consumption	5×10^{19}
One megaton explosion	4.2×10^{15}
Annual output of power station	10^{16}
Lightning flash	5×10^{9}
One (U.S.) gallon of gasoline	1.3×10^{8}
Daily intake of a person	1.3×10^{7}
Proton in supercollider	3×10^{-7}
Fission of uranium nucleus	3×10^{-11}
Proton in nucleus	10^{-13}
Hydrogen bond	6×10^{-18}
Minimum detectable by eye	3×10^{-18}
Electron in atom	10^{-18}
Kinetic energy of molecule at room temperature	10^{-21}

TABLE OF POWERS (W)

Quasar (3C273)	4×10^{40}
Solar output	4×10^{26}
Solar power incident on earth	1.74×10^{17}
U.S. consumption	2.7×10^{12}
Photosynthesis (global)	4×10^{12}
Hydroelectric (potential)	3×10^{12}
Tidal friction loss	3×10^{12}
Tidal power in use	7×10^{10}
Geothermal power in use	1.5×10^{9}
Grand Coulee Dam	6.5×10^{9}
Boeing 747	1.4×10^{8}
Aircraft carrier	1.2×10^{8}
Locomotive	3×10^{6}
Powerful laser	10^{6}
Automobile	10^{5}
Radio transmitter	5×10^{4}
Human output (short duration)	2×10^{3}
Person (resting)	75

7.5   WORK AND ENERGY IN THREE DIMENSIONS

When the magnitude and direction of a force vary in three dimensions, it can be expressed as a function of the position vector $\mathbf{F}(\mathbf{r})$, or in terms of the coordinates $\mathbf{F}(x, y, z)$. The work done by such a force in an infinitesimal displacement $d\mathbf{s}$ is

$$dW = \mathbf{F} \cdot d\mathbf{s} \tag{7.17}$$

Following the discussion in Section 7.3, the total work done in going from point A to point B in Fig. 7.16 is

$$W_{A \to B} = \int_A^B \mathbf{F} \cdot d\mathbf{s} = \int_A^B F \cos \theta \, ds \tag{7.18}$$

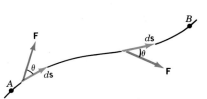

FIGURE 7.16 A particle moves along a curved path subject to a nonconstant force. The work done by the force in a displacement $d\mathbf{s}$ is $dW = \mathbf{F} \cdot d\mathbf{s}$.

In terms of rectangular components, $\mathbf{F} = F_x\mathbf{i} + F_y\mathbf{j} + F_z\mathbf{k}$ and $d\mathbf{s} = dx\mathbf{i} + dy\mathbf{j} + dz\mathbf{k}$; therefore,

$$W_{A \to B} = \int_{x_A}^{x_B} F_x \, dx + \int_{y_A}^{y_B} F_y \, dy + \int_{z_A}^{z_B} F_z \, dz \qquad (7.19)$$

Each integral in Eq. 7.19 depends on the actual path taken between A and B, and so it is called a *path* or *line* integral. The sign of the integral is determined both by the signs of the components and by the limits on the integral—which specify the direction along the path. For example, in Fig. 7.17 the force of gravity has only a y component. Since $\mathbf{F} = F_y\mathbf{j} = -mg\mathbf{j}$, we have $\mathbf{F} \cdot d\mathbf{s} = F_y \, dy = -mg \, dy$. Therefore,

$$W_{A \to B} = \int_A^B m\mathbf{g} \cdot d\mathbf{s} = -\int_{y_A}^{y_B} mg \, dy$$
$$= -mg(y_B - y_A)$$

which agrees with Eq. 7.6. In Fig. 7.17*a* the work is positive, whereas in Fig. 7.17*b* it is negative.

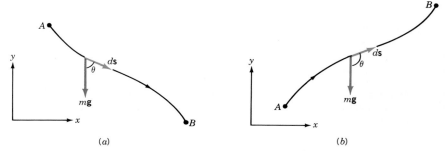

FIGURE 7.17 Since the force has just one component, the work done in a displacement $d\mathbf{s}$ is $dW = \mathbf{F} \cdot d\mathbf{s} = F_y \, dy$ where, in this case, $F_y = -mg$. In (*a*) the work done is positive; in (*b*) it is negative.

EXAMPLE 7.9: A horizontal force F very slowly lifts the bob of a simple pendulum from a vertical position to a point at which the string makes an angle θ_0 to the vertical. The magnitude of the force is varied so that the bob is essentially in equilibrium at all times. What is the work done by the force on the bob?

Solution: Figure 7.18*a* is a sketch of the system and shows the forces acting on the bob. Since the acceleration is zero, both the vertical and horizontal components of the forces balance:

$$\Sigma F_x = F - T \sin \theta = 0$$
$$\Sigma F_y = T \cos \theta - mg = 0$$

Eliminating T we find

$$F = mg \tan \theta \qquad (i)$$

This is how the force must vary as a function of angle in order for the bob to be in equilibrium. The work done by $\mathbf{F}$ in a

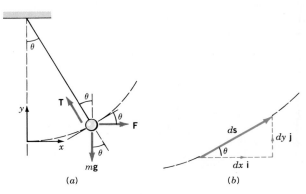

FIGURE 7.18 (*a*) To move the bob at constant speed the force must vary with the angle θ. (*b*) An infinitesimal displacement has vertical and horizontal components.

differential displacement ds along the circular arc is ($F = F_x$ and $F_y = F_z = 0$)

$$dW = \mathbf{F} \cdot d\mathbf{s} = F_x\, dx$$
$$= mg \tan \theta\, dx \qquad \text{(ii)}$$

From Fig. 7.18b we see that $dy/dx = \tan \theta$; thus, $dy = \tan \theta\, dx$. Equation (ii) becomes $dW = mg\, dy$; therefore, the total work done from $y = 0$ to $y = y_0$ is

$$W = \int_0^{y_0} mg\, dy = mgy_0$$
$$= mgL(1 - \cos \theta_0)$$

An interesting feature of this result is that it is the *vertical* displacement $y_0 = L(1 - \cos \theta_0)$ that appears (see Exercise 4 below).

EXERCISE 4. In Example 7.9 what is the work done by the force of gravity for the same displacement?

EXERCISE 5. Repeat Example 7.9 given that the force is always tangential to the circular arc. What is the work done by the force from $\theta = 0$ to $\theta = \theta_0$? (*Hint*: A displacement along the arc is $L\, d\theta$.)

SUMMARY

The **work** done by a constant force $\mathbf{F} = F_x\mathbf{i} + F_y\mathbf{i} + F_z\mathbf{k}$ when its point of application undergoes a displacement $\mathbf{s} = \Delta x\mathbf{i} + \Delta y\mathbf{j} + \Delta z\mathbf{k}$ is

$$W = \mathbf{F} \cdot \mathbf{s} = Fs \cos \theta$$
$$= F_x \Delta x + F_y \Delta y + F_z \Delta z$$

In one dimension, the work done by a variable force is

$$W_{A \to B} = \int_A^B F_x\, dx$$

The direction of the displacement is specified by the limits on the integral.

The work done by the (constant) force of gravity $m\mathbf{g} = -mg\mathbf{j}$ is given by

$$W_g = -mg(y_f - y_i)$$

This work depends only on the y coordinates of the initial and final points.

The work done by the force exerted by a spring, $F_{sp} = -kx$, is

$$W_{sp} = -\tfrac{1}{2}k(x_f^2 - x_i^2)$$

This work also depends only on the initial and final points.

The **work–energy theorem** relates the work done by the net force on a particle to the resulting change in its kinetic energy:

$$W_{NET} = \mathbf{F}_{NET} \cdot \mathbf{s} = \Delta K$$

where $K = \tfrac{1}{2}mv^2$.

The **instantaneous mechanical power** is the rate at which work is being done:

$$P = \frac{dW}{dt} = \mathbf{F} \cdot \mathbf{v}$$

where $\mathbf{F}$ is the force exerted by some agent on the body, whose velocity is $\mathbf{v}$. Power is also expressed as the rate at which one form of energy is transformed into another.

In three dimensions, the work done by a variable force is

$$W_{A \to B} = \int_A^B \mathbf{F} \cdot d\mathbf{s} = \int_A^B F \cos \theta\, ds$$

where $d\mathbf{s} = dx\mathbf{i} + dy\mathbf{j} + dz\mathbf{k}$ is an infinitesimal displacement.

ANSWERS TO IN-CHAPTER EXERCISES

1. $W = F_x \Delta x = (-5 \text{ N})(3 \text{ m} - 7 \text{ m}) = +20$ J. An area below the x axis is not necessarily negative.

2. When F is removed, the only horizontal force is $f = \mu_k N$ and $\Delta K = 0 - \frac{1}{2}mv_i^2$ since $v_f = 0$. The work–energy theorem becomes

$$-fs = 0 - \tfrac{1}{2}mv_i^2$$

In this part $N = mg$, so that $f = \mu_k mg$. Using this and $v_i = 5$ m/s, we find $s = v_i^2/(2\mu_k g) = 10.2$ m.

3. When the spring has its maximum extension, the speed of the block is zero. From (i) we find $d = 2.45$ m.

4. $W_g = -mg\Delta y = -mgL(1 - \cos\theta_0)$, which is the negative of the work done by the force. The net work on the block is zero.

5. The force must equal the component of the weight along the arc, that is, $F = mg\sin\theta$. For a displacement $ds = L\,d\theta$, the work done by F is $dW = F(L\,d\theta) = mgL\sin\theta\,d\theta$. The total work done by F on the block is

$$W = mgL \int_0^{\theta_0} \sin\theta\,d\theta = mgL(1 - \cos\theta_0)$$

QUESTIONS

1. True/false: If the kinetic energy of a body is fixed, the net force on it is zero. Explain your response.

2. Do devices such as levers, pulleys, and gears, which allow us to use smaller forces, also save us work? Explain.

3. You lift a box from the floor and place it on a table. Does the amount of work you do on the box depend on how fast you lift?

4. An incline is a sort of machine because it allows us to use a smaller force to accomplish some task. Does this also mean that we do less work? Does the presence or absence of friction affect your answer?

5. Compare the work done by a person who rises at constant velocity from the ground to the first floor (a) by using the stairs, and (b) by climbing a rope. Does the amount of energy he expends depend on the method?

6. If the speed of a car increases by 50%, by what factor does this increase the stopping distance?

7. True/false: The area below the x axis on an F versus x graph represents negative work.

8. Does the work done by a force depend on the frame of reference? Can the work be positive in one frame and negative in another?

9. True/false: The landing speed of a projectile fired from a cliff is independent of the angle of projection.

10. Can work be done (a) by the force of static friction, or (b) by a centripetal force?

11. If the net work done on a particle is zero, what can you conclude about the following: (a) its acceleration; (b) its velocity?

12. Give an example in which the work done by the force of kinetic friction is (a) negative, and (b) positive.

EXERCISES

7.1 Work by a Constant Force

1. (I) A 2-kg block is pulled 3 m along a frictionless horizontal plane by a 10-N force that acts at 37° above the horizontal. What is the work done by the force on the block?

2. (I) A 0.3-kg particle moves from an initial position $\mathbf{r}_1 = 2\mathbf{i} - \mathbf{j} + 3\mathbf{k}$ m to a final position $\mathbf{r}_2 = 4\mathbf{i} - 3\mathbf{j} - \mathbf{k}$ m while a force $\mathbf{F} = 2\mathbf{i} - 3\mathbf{j} + \mathbf{k}$ N acts on it. What is the work done by the force on the particle?

3. (I) A lawn mower is pushed at constant speed by a horizontal force of 40 N. The blade is 50 cm in diameter. What is the work done by the person on the mower in mowing a 10 m × 20 m lawn?

4. (I) A person pushes a 10-kg crate 3 m up a 30° incline with an 80-N force parallel to the incline. The frictional force is 22 N. Find the work done (a) by the person, (b) by gravity, and (c) by friction.

5. (I) A tool being sharpened is held against a grinding wheel of radius 4 cm with a force of 20 N directed radially inward, as shown in Fig. 7.19. If the coefficient of kinetic friction is 0.4, how much work is done by the wheel on the tool in 12 revolutions?

FIGURE 7.19 Exercise 5.

6. (II) A 1.8-kg block is moved at constant speed over a surface for which $\mu_k = 0.25$. The displacement is 2 m. It is pulled by a force directed at 45° to the horizontal as shown

in Fig. 7.20. Find the work done on the block by: (a) the force F; (b) friction; (c) gravity.

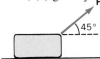

FIGURE 7.20 Exercise 6.

7. (II) Repeat Exercise 6 with the force pushing the block at 45° below the horizontal, as shown in Fig. 7.21.

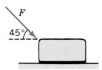

FIGURE 7.21 Exercise 7.

8. (I) A 1.5-kg block is moved at constant speed in a vertical plane from position 1 to position 3 via several routes shown in Fig. 7.22. Compute the work done by gravity on the block for each segment indicated, where W_{ab} means work done from a to b. (a) W_{13}; (b) $W_{12} + W_{23}$; (c) $W_{14} + W_{43}$; (d) $W_{14} + W_{45} + W_{53}$.

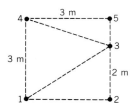

FIGURE 7.22 Exercise 8.

9. (I) A 60-kg skier slides 200 m down a 25° slope. What is the work done on the skier (a) by gravity, and (b) by the force of friction which is 20 N?

10. (II) What is the work needed to lift 15 kg of water from a well 12 m deep. Assume the water has a constant upward acceleration of 0.7 m/s².

11. (II) An inclined conveyor belt lowers a block of mass $M = 20$ kg at a constant speed $v = 3$ m/s, as shown in Fig. 7.23. (a) What is the work done by the motor on the block when the block moves 2 m? (b) What would be the work done by the motor to raise the block through the same distance at constant speed?

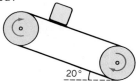

FIGURE 7.23 Exercise 11.

12. (II) A pulley of radius $r = 3$ cm has a handle that moves in a circle of radius $R = 10$ cm as shown in Fig. 7.24. If a bucket of mass 5 kg is raised at a constant 2 m/s, how much work must you do in 8 s?

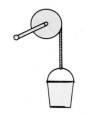

FIGURE 7.24 Exercise 12.

7.2 Work by a Variable Force in One Dimension

13. (I) A force varies with position as shown in Fig. 7.25. Find the work done by it from (a) $x = -4$ to $+4$ m, (b) $x = 0$ to -2 m.

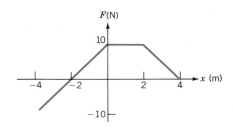

FIGURE 7.25 Exercise 13.

14. (I) The variation of a force with position is depicted in Fig. 7.26. Find the work done from (a) $x = -A$ to $x = 0$, (b) $x = +A$ to $x = 0$.

15. (I) The variation of a force with position is depicted in Fig. 7.27. Find the work from (a) $x = 0$ to $x = -A$, (b) $x = +A$ to $x = 0$.

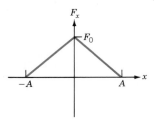

FIGURE 7.26 Exercise 14.

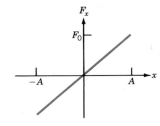

FIGURE 7.27 Exercise 15.

16. (I) A force varies with position as shown in Fig. 7.28. Find the work done from (a) $x = A$ to $x = 0$, (b) $x = -A$ to $x = 0$.

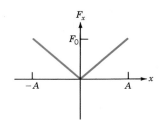

FIGURE 7.28 Exercise 16.

17. (I) The spring constant of a spring is 40 N/m. Find the work needed (a) to extend it from 0 to 0.1 m; (b) to change its compression from 0.1 m to 0.2 m.

18. (II) When two 75-kg persons sit in a 1000-kg car, it is lowered by 2 cm. Estimate the spring constant of the suspension. What simplifying assumptions do you need to make?

19. (II) Two springs have stiffness constants such that $k_1 > k_2$. Compare the work done on them if they (a) are given the same extension, or (b) are subject to the same force.

7.3 Work–Energy Theorem

20. (I) The velocity of a 2-kg particle changes from $2\mathbf{i} - 3\mathbf{j}$ m/s to $-5\mathbf{i} + 2\mathbf{j}$ m/s. What is the change in its kinetic energy?

21. (I) Find the kinetic energy of the earth due to its orbital motion. Express your answer in megatons. (A ''ton'' of energy is released by a metric ton (10^3 kg) of TNT and is equal to 4.2×10^9 J.)

22. (I) Compute the kinetic energy for each of the cases below. Through what distance would a 800-N force have to act to stop each object? (a) A 150-g baseball moving at 40 m/s; (b) a 13-g bullet from a Remington rifle moving at 635 m/s; (c) a 1500-kg Corvette moving at 250 km/h; (d) a 1.8×10^5-kg Concorde airliner moving at 2240 km/h.

23. (I) A 200-g ball thrown vertically up with an initial speed of 20 m/s reaches a maximum height of 18 m. Find: (a) the change in its kinetic energy; (b) the work done by gravity. (c) Are the two quantities just calculated equal? Explain why or why not.

24. (I) Compute the kinetic energies for each of the following. What force would be required to stop each object in 1 km? (a) The 8×10^7-kg carrier *Nimitz* moving at 55 km/h; (b) a 3.4×10^5-kg Boeing 747 moving at 1000 km/h; (c) the 270-kg *Pioneer 10* spacecraft moving at 51,800 km/h.

25. (I) Compute the energy needed to move a 10^3-kg car against a constant retarding force of 250 N under the following conditions: (a) at a constant speed of 20 m/s for 10 s; (b) with constant acceleration from rest to 20 m/s in 10 s; (c) with constant acceleration from 20 m/s to 40 m/s in 10 s.

26. (I) The maximum speed for a sprinter, 12.5 m/s, was attained briefly by Bob Hayes. The record for a downhill skier, Franz Weber, is 201 km/h. Assume they have the same mass of 70 kg. (a) What are the kinetic energies? (b) Suppose Weber was subject to a net drag force of 20 N.

How far along a 45° slope would he need to travel to attain his speed?

27. (I) An engine can pull a train with a force of 3×10^4 N. It pulls 20 railcars each of mass 2×10^4 kg. The railcars are subject to rolling friction equal to 0.5% of their weight. (a) What is the work done by the engine on the train (not including the engine) when it starts from rest and moves 1 km? (b) What is the speed at 1 km?

28. (I) A crane lifts a 200-kg bucket of cement 6 m vertically with an acceleration of 0.2 m/s². Find: (a) the work done by the crane on the bucket; (b) the work done by gravity on the bucket; (c) the change in kinetic energy of the bucket.

29. (II) How much work is needed to push a 1100-kg Chrysler K car with a constant force from rest to 2.5 m/s within a distance of 30 m if the total resistive force is 200 N?

30. (I) A 12-g bullet emerges at 850 m/s from a Winchester rifle with a 60-cm barrel. Use energy considerations to estimate the force that acted on it.

31. (I) A 10-g bullet moving at 400 m/s is stopped in 2.5 cm by a block held firmly in place. What is the force on the bullet (assumed constant)? Compare this with your weight.

32. (I) The aircraft carrier *Dwight D. Eisenhower* has a mass of 9.3×10^7 kg and can cruise at 55 km/h. (a) What is the work required to stop it? (b) What constant force could stop it in 5 km? (c) If only half the work required to stop it is done, what is the final speed?

33. (II) A horizontal force is used to push a 6-kg sled 4 m up a 30° incline at constant speed. If $\mu_k = 0.2$, find the work done on the sled (a) by the force, (b) by gravity, and (c) by friction. (d) What is the net work done on the sled?

34. (I) A 2-kg block slides down a rough 30° incline. In traveling 50 cm its speed changes from 1 m/s to 2 m/s. Find the work done (a) by gravity, and (b) by friction. (c) What is the coefficient of kinetic friction?

35. (I) A 2-kg block is projected at 3 m/s up a 15° incline for which $\mu_k = 0.2$. Use the work–energy theorem to find the distance it travels before coming to rest.

36. (I) A spring gun with $k = 45$ N/m is compressed by 10 cm. What is the exit speed of a 2-g projectile?

37. (II) A person applies a horizontal force of 200 N to a 40-kg crate that moves 2 m along a 5° incline. Find the work done on the crate (a) by the person, (b) by friction if $\mu_k = 0.25$, and (c) by gravity. (d) Find the final speed if the crate starts from rest.

38. (II) Use energy methods to show that the minimum stopping distance for a car moving at speed v is $v^2/(2\mu_s g)$, where μ_s is the static coefficient of friction. Evaluate the expression for $v = 30$ m/s and $\mu_s = 0.8$.

39. (II) The force experienced by a 0.25-kg particle is depicted in Fig. 7.29. The particle approaches the origin from the right at a speed of 20 m/s. (a) Find the work done by the force as the particle moves from $x = 6$ m to $x = 0$. (b) What is the kinetic energy of the particle at the origin?

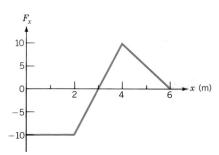

FIGURE 7.29 Exercise 39.

40. (II) A block of mass $m = 0.5$ kg is held against a spring which is compressed by $A = 20$ cm as in Fig. 7.30. Take the friction coefficient to be $\mu_k = 0.4$ and the spring constant to be $k = 80$ N/m. When the system is released, find: (a) the work done by the spring on the block until the point the block leaves it; (b) the work done by friction until the point the block leaves the spring; (c) the speed at which the block leaves the spring; and (d) how far the block travels from its initial position before coming to a stop.

FIGURE 7.30 Exercise 40.

41. (II) A force $F = 24$ N acts at 60° above the horizontal on a 3-kg block that is attached to a spring whose stiffness constant is $k = 20$ N/m (see Fig. 7.31). Assume $\mu_k = 0.1$. The system starts at rest with the spring unextended. If the block moves 40 cm, find the work done (a) by F, (b) by friction, (c) by the spring. (d) What is the final speed of the block?

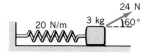

FIGURE 7.31 Exercise 41.

7.4 Power

42. (I) (a) A Chevrolet Caprice wagon requires 20 hp delivered to the wheels to maintain a speed of 80 km/h. (a) What is the retarding force on the car? (b) Where does it originate?

43. (I) A 2000-kg elevator is attached to a 1800-kg counterweight. What power must the motor supply to raise the elevator at 0.4 m/s?

44. (I) A grandfather clock has a 1.8-kg cylinder that compen-

sates for frictional losses. If the cylinder falls 6 cm per day, what is the frictional power loss in the mechanism?

45. (1) A winch drags a 200-kg crate along a 15° incline at 0.5 m/s (see Fig. 7.32). The coefficient of kinetic friction is $\mu_k = 0.2$. What is the power required as the crate moves up the slope?

FIGURE 7.32 Exercise 45.

46. (I) In 1970 the rocket-powered car *Blue Flame* reached a record 1002 km/h. The thrust of its engine was 13,000 lb (58 kN). What was its maximum horsepower?

47. (I) The F5.E fighter has 10,000 lb (44.5 kN) of thrust and can fly at Mach 1.6. The newer F5.G has 16,000 lb (71 kN) of thrust and can fly at Mach 2.1. If the speed of sound is 330 m/s, find the peak horsepower of each aircraft.

48. (I) What is the average power output (in hp) of a weight-lifter who (a) raises 250 kg through a distance of 2.1 m in 3 s, or (b) benchpresses 60 kg through a vertical distance of 40 cm, ten times in 1 min.

49. (I) A 60-kg skydiver is falling at the terminal speed of 55 m/s. How much power (in hp) is dissipated due to air resistance?

50. (I) If electrical energy costs 7 cents per kWh, how much does it cost to run a 0.25-hp motor for 2 h?

51. (I) Vigorous exercise requires a metabolic rate (release of stored chemical energy) of 600 kcal/h. How long would it take to lose 0.1 kg if the metabolism of 1 g of fat releases 9 kcal?

52. (I) A 40-hp motorboat moves at a constant 30 km/h. What would be the tension in the rope if it were being towed at the same speed by another boat?

53. (I) A champion cyclist can sustain an output of 0.5 hp for 10 min. How far could he travel at constant speed if the net retarding force is 18.5 N?

54. (I) (a) The United States, with a population of 2.2×10^8 people, consumes 5×10^{19} J per year. What is the per capita consumption in watts? (b) The sun's radiation provides the earth with 1000 W/m². Assuming solar energy can be converted to electrical energy with a 20% efficiency, how much area is needed to serve the energy needs of each U.S. citizen?

55. (II) A locust can propel its body (≈ 3 g) from rest to 3.4 m/s in 4 cm. Estimate the average power output of its legs.

56. (II) A 1050-kg car requires 12 hp delivered at the wheels to cruise at a steady 80 km/h. (a) What is the total resistive force on it? (b) What horsepower does it require to climb a 10° incline at the same speed?

57. (II) A 7.25-kg shotput and a 0.8-kg javelin are thrown at 45° and land at the same level at which they are thrown. The shotput lands 20 m away and the javelin 90 m away. The shot is thrown through a distance of 1.5 m while the javelin is moved through 2.2 m. Compute the power required of each athlete. Assume both projectiles start at rest in the thrower's hand and have constant acceleration.

58. (II) On June 12, 1979, Bryan Allen pedalled the human-powered *Gossamer Albatross* 38.5 km across the English Channel in 2 h 49 min. His average mechanical power output was 0.33 hp. Leg muscles convert the chemical energy released by the metabolism of fat to mechanical work with an efficiency of about 22%. The metabolism of 1 g of fat releases 9 kcal (3.76×10^4 J). How much fat did he lose? (It had been estimated earlier that it would have been impossible for him to have continued beyond 2 h 55 min!)

59. (II) The heat of combustion of gasoline is 3.4×10^7 J/L. The fuel economy of a car is rated at 12 km/L at 100 km/h. If the mechanical output at this speed is 25 hp, what is the efficiency (power output/power input) of the motor?

60. (I) Two horses pull a barge along a canal at a steady 6 km/h, as shown in Fig. 7.33. The tension in each rope is 450 N and each is at 30° to the direction of motion. What is the horsepower provided by the horses?

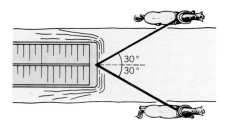

FIGURE 7.33 Exercise 60.

61. (I) One ton of TNT releases 4.6×10^9 J. If the energy of a 1-megaton explosion could be harnessed with 10% efficiency, for how long would it supply the needs of the United States—which requires 5×10^{19} J per year?

PROBLEMS

1. (I) A 2-kg block travels on a tabletop in a circle of radius 1.2 m. In one revolution its speed drops from 8 to 6 m/s. How many more revolutions does it make before stopping?

2. (II) A 0.3-kg block is attached to a spring ($k = 12$ N/m) and slides on a horizontal surface for which $\mu_k = 0.18$. The block is released with the spring initially extended by 20 cm. Use the work–energy theorem to find (a) the speed when the spring returns to its natural length, and (b) where the block first comes to rest momentarily.

3. (II) If the power P supplied to an object of mass m is contant, show that the distance traveled in a straight line is $(8Pt^3/9m)^{1/2}$. The object starts at rest.

4. (I) The external force needed to produce an extension x in a spring is given by $F(x) = 16x + 0.5x^3$ N. What is the external work needed to extend it from $x = 1$ m to $x = 2$ m?

5. (I) In Fig. 7.34 a 2-kg block on an incline is attached to a spring whose stiffness constant is $k = 20$ N/m. The coefficient of kinetic friction is $\mu_k = \frac{1}{6}$. The block starts at rest with the spring unextended. After it slides 40 cm, find: (a) W_{sp}; (b) W_f; (c) W_g; (d) the speed of the block. (e) What is the maximum extension of the spring?

6. (II) The horsepower required at the wheels of a car to maintain various speeds are as follows: 5 hp (3.73 kW) at 13.5 m/s and 13 hp (9.70 kW) at 22.2 m/s. Assume that the drag force on the car moving at speed v can be expressed in the form $F_D = a + bv^2$, where a represents the rolling resistance of the tires and bv^2 represents the air resistance. (a)

FIGURE 7.34 Problem 5.

What are the constants a and b? (b) What is the "road horsepower" required at the wheels at 30 m/s?

7. (I) A person lifts a 25-kg crate hanging over a pulley by walking horizontally, as shown in Fig. 7.35. As the person walks 2 m, the angle of the rope to the horizontal changes from 45° to 30°. How much work does the person do if the crate rises at constant speed?

8. (I) A spring of natural length ℓ_0 and whose stiffness constant is k is fixed at one end. The other end is attached to a ring of mass m that can slide on a frictionless rod as shown in Fig. 7.36. If the initial length of the spring is ℓ, show that the speed of the ring when the spring returns to its natural length ℓ_0 and is perpendicular to the rod is $\sqrt{k/m}\,(\ell - \ell_0)$.

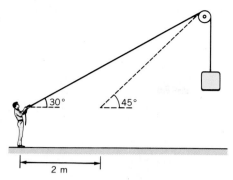

FIGURE 7.35 Problem 7.

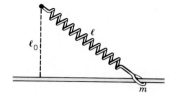

FIGURE 7.36 Problem 8.

9. (II) The retarding force on a 1100-kg car is given by $f = 200 + 0.8v^2$ N, where v is in m/s. (a) What is the power required at the wheels at 20 m/s on a horizontal road? (b) What is the "road power" needed to travel up a 5° incline at 0.5 m/s² when $v = 20$ m/s? (c) Given that only 15% of the energy supplied by the combustion of fuel is delivered to the wheels, what is the "fuel input power" (in hp) required for part (b)?

10. (II) A 1200-kg car starts to climb a 10° incline at 30 m/s. After 500 m its speed is 20 m/s. What was the average power (in hp) delivered to the wheels? Assume that the retarding force due to the air and the road is constant at 500 N.

11. (II) Prove the work–energy theorem for motion in three dimensions with a nonconstant force. (*Hint:* First show that the work done by the force **F** in a differential displacement $d\mathbf{s}$ may be expressed in the form $dW = \mathbf{F} \cdot d\mathbf{s} = m d\mathbf{v} \cdot \mathbf{v}$, where **v** is the velocity of the particle. Then show that $d\mathbf{v} \cdot \mathbf{v} = \frac{1}{2}d(v^2)$ and perform a simple integration.)

SPECIAL TOPIC: Energy and the Automobile

The family car has changed our life-styles by giving us great mobility. Unfortunately it is not a very efficient device. Even under ideal conditions, only about 15% of the energy available from the combustion of gasoline is used to propel the vehicle. In stop-and-go driving the percentage drops even lower. No wonder the approximately 100 million cars in the United States are responsible for about half the total energy consumption in all forms of transportation.

The chart below shows how the chemical energy in a given quantity of gasoline is expended during the federal EPA driving cycle, which involves a mixture of city and highway driving. All items are expressed as percentages of the fuel input power (FIP). Indicated HP is the mechanical power developed within the cylinders. After combustion, 62% of FIP is lost to the exhaust and cooling systems. After mainly frictional losses within the engine, the mechanical output power of the engine is 25% of FIP. This is called brake horsepower (BHP). In the process of transmitting the BHP to the wheels, 3% is lost due to fluid friction in the transmission and rear axle. In city driving, a further 7.5% is wasted in braking and idling. The accessories require

about 2.5%. We are left with just 12% of FIP of "road power" at the wheels. This road power is available to accelerate the car and to overcome the frictional losses due to rolling resistance of the tires and aerodynamic drag. (At about 70 km/h, these two losses are equal.)

The rolling resistance of the tires f_r is roughly constant. (In fact, it decreases slightly as the velocity increases because of aerodynamic lift forces on the car.) The power loss is therefore $P_r = f_r v$. In Chapter 5 we noted that the drag force due to a fluid may be expressed as

$$F_D = \tfrac{1}{2} C_D \rho A v^2$$

where ρ is the density of the fluid, A is the projected area normal to the velocity v, and C_D is the drag coefficient. The total power loss is the sum of the contributions of rolling resistance and aerodynmc drag:

$$P = P_r + P_a = f_r v + k v^3$$

Note the strong dependence of the second term, $P_a \propto v^3$, on speed. At 70 mph (31.3 m/s), twice as much power is re-

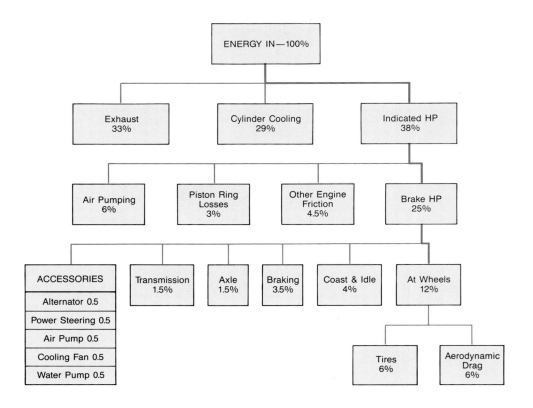

quired to overcome air resistance as at 55 mph (24.6 m/s). At 80 mph it would be three times as much. If we compare a car ($C_D \approx 0.5$, $A \approx 2.3$ m²) and a truck ($C_D \approx 1.4$, $A \approx 4.5$ m²) both moving at 30 m/s, we find that the air drag losses are 20 kW for the car and 115 kW for the truck. (20 kW would satisfy the electrical power requirement of an average home.)

These large losses can be reduced in several ways. Obviously, we can drive at lower speeds. Also, the shape of the vehicle can be streamlined to reduce C_D. A modern sedan can have $C_D = 0.35$. Trucks nowadays employ a "spoiler" above the cab to reduce C_D. The frontal area can be reduced by building smaller cars and trucks. Even closing windows can decrease air drag by a few percent. The losses due to the tires can be minimized by maintaining higher pressures to reduce tire flexing and by using radial tires instead of bias ply tires.

Usually we measure fuel economy in miles per gallon or in metric units, km/L. Of equal interest is fuel consumption, which is stated in L/km. The energy required to move at constant velocity is given by $\Delta E = \int P\, dt = P\Delta t$ since P is a constant. The distance traveled in Δt is $\Delta x = v\Delta t$. Thus, the fuel consumption is

$$\frac{\Delta E}{\Delta x} = f_r + kv^2 \quad \text{(J/m)}$$

The energy content (released during combustion) of 1 L of gasoline is about 3×10^7 J, so we may convert this expression to L/km.

EXAMPLE: A car requires 12 BHP to cruise at a constant 80 km/h. Take $C_D = 0.4$ and $A = 1.5$ m². The density of air is 1.29 kg/m³. Find: (a) the fuel consumption in L/km and the fuel economy in mpg; (b) the power used to overcome air resistance.

Solution: (a) First we convert the speed to useful units: $v = 22 \times 10^{-3}$ km/s. Second, 12 BHP = 9 kW. From the chart we see that BHP is 25% of fuel input power, and, hence, the fuel input power 36 kW.

$$\text{FIP} = (36 \times 10^3 \text{ J/s})[1 \text{ L}/(3 \times 10^7 \text{ J})] = 12 \times 10^{-4} \text{ L/s}$$

The rate of fuel consumption at the given speed is

$$\frac{\Delta E}{\Delta x} = (12 \times 10^{-4} \text{ L/s})(1 \text{ s}/22 \times 10^{-3} \text{ km}) = 0.055 \text{ L/km}$$

which is equivalent to 18 km/L. Next we use 1 mi = 1.6 km and 1 gal (U.S.) = 3.78 L to find the fuel economy to be 42.5 mpg. Note that 0.055 L/km equals 1.7×10^6 J/km. In contrast, a bicyclist would require only about 7000 J of food energy to cover 1 km, which is a factor of 250 less!

(b) $P_a = \frac{1}{2}C_D\rho Av^3 = \frac{1}{2}(0.4)(1.29)(1.5)(22)^3 = 5$ kW = 6.6 hp

References

1. O. Pinkus and D. F. Wilcock, *Strategy for Energy Conservation through Tribology.* Am. Soc. Mech. Eng., 1977.
2. G. Waring, *Phys. Teacher* 18: 94 (1980).

CHAPTER 8

Conservation of Energy

A waterfall illustrates the conversion of potential energy to kinetic energy.

Major Points

1. The distinction between **conservative** and **nonconservative** forces.
2. (a) **Potential energy:** Energy shared by two interacting particles by virtue of their positions.
   (b) The relation between potential energy and conservative forces.
3. (a) Gravitational potential energy near the earth's surface.
   (b) The potential energy of a spring.
4. (a) The **conservation of mechanical energy.**
   (b) Modification of the conservation law in the presence of nonconservative forces.
5. Gravitational potential energy and Newton's law of gravitation.

FIGURE 8.1 Galileo discovered that when the swing of a pendulum was interrupted by a peg, the bob still rose to its initial level. In addition to leading him to clarify the concept of inertia, this finding was an important step in the formulation of the conservation of energy.

In the last chapter we introduced the concept of kinetic energy, which is energy that a system possesses by virtue of its motion. In this chapter we introduce **potential energy,** which is energy that a system possesses by virtue of the positions of its interacting particles. The work–energy theorem then leads to the principle of the **conservation of mechanical energy:** Under certain conditions, the sum of the kinetic and potential energies in a system stays constant. This principle provides us with a powerful alternative to Newton's laws in the solution of problems in mechanics.

The idea that the motion of an object is governed by something that does not change had its origin in the pendulum experiments of Galileo. It was clear that the bob of a freely swinging pendulum rises to the same height on either side of its swing. When Galileo placed a peg in the path of the string to interrupt its motion, as shown in Fig. 8.1, the bob still returned to the same height on either side. He concluded that the speed that a body acquires as it falls depends only on the *vertical* drop, and not on the actual path it travels (in the absence of friction). The Dutch scientist Christiaan Huygens later showed that the speed v reached as a result of a fall from a height h is given by $v^2 \propto h$.

One can interpret these results by saying that a falling body acquires something that can "carry" it back to its original level. The mathematician Gottfried Leibnitz (see Fig. 8.2) called this something "force" and believed that it kept all natural processes running. (His "force" is what we now call energy.) Since the bob always rises to the same level, he took this as evidence that "force" is never destroyed. In 1667 Huygens had found that the total value of the quantity mv^2 does not change when there is a collision between two hard balls. Leibnitz called it *vis viva* or "living force" and used the relation $v^2 \propto h$ to state: "The force of a moving body is proportional to the square of its speed or to the height to which it would rise against gravity."

Both Huygens and Leibnitz focused on the conservation of *vis viva*. They did not care that it could temporarily vanish during a collision, or when a pendulum bob reaches its maximum height, because it reappeared later. It took many decades for other workers to develop the idea that energy could be associated with the deformation of the balls during a collision, with the compression or expansion of a spring, or with the position of a body relative to the ground. This form of energy is now called potential energy.

FIGURE 8.2 Gottfried W. Leibnitz (1646–1716).

8.1   POTENTIAL ENERGY

In order to raise an apple from the ground to some height, one must either lift it by hand or throw it with enough initial kinetic energy. When it reaches its new height, what happens to the lost kinetic energy or the work done by the hand? We know from our study of free-fall that when an apple is thrown up, it returns to its starting point with a speed equal to the initial value. The initial kinetic energy, or the work done by the hand, is somehow stored and is later fully recovered in the form of kinetic energy. The apple must have something at the new height that it does not have at ground level. By virtue of its position it has **potential energy:** *Potential energy is energy associated with the relative positions of two or more interacting particles.*

In Fig. 8.3*a*, the two interacting bodies are the apple and the earth. The work done by an external agent (for example, you) to lift the apple is stored as gravitational potential energy. The potential energy belongs to the *system* "apple + earth." Nonetheless, we still tend to refer to the "potential energy of the apple," as if it had sole possession. This is because we see the apple move when it is released, whereas the earth appears motionless. However, from Newton's third law, we know that the earth must also move toward the apple—albeit very slightly. A spring, as in Fig. 8.3*b*, is another system that can store potential energy. External work used to expand or compress a spring is stored as elastic potential energy—which is actually electrical potential energy shared by the atoms.

Potential energy fits well the idea of energy as the capacity to do work. For example, the gravitational potential energy of an object raised off the ground can be used to compress or expand a spring, to lift another weight, or to drive a stake into the ground. As a coil spring unwinds, or a straight spring returns to its natural length, the stored potential energy can be used to do work. For example, if a block is attached to the spring, the elastic potential energy can be converted to kinetic energy of the block.

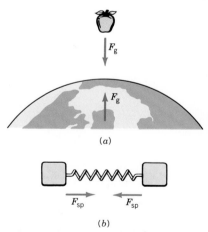

FIGURE 8.3 Potential energy is shared by two interacting bodies. In (*a*) the energy belongs to the apple–earth system. In (*b*) the energy is stored in the spring.

Potential Energy and External Work

As we have just seen, work done by an external agent can change the potential energy of a system. We may use this fact to present an initial, tentative definition of potential energy. If the external force on an apple (say) exactly balances the gravitational force, the speed and kinetic energy of the apple does not change. In

Potential energy defined in terms of work done by an external agent

this case, *all* the external work must appear as a change in the potential energy U:

$$(v \text{ constant}) \qquad W_{\text{EXT}} = +\Delta U = U_f - U_i \qquad (8.1)$$

Positive external work leads to an increase in the potential energy of a system.

Equation 8.1 shows that only *changes* in potential energy are significant. This gives us the freedom to assign $U = 0$ to whatever configuration is convenient. (By configuration we mean the arrangement of the particles in a system, or equivalently, the shape of a deformable body.) When a particle moves near the surface of the earth, we set the zero of gravitational potential energy, $U_g = 0$, at any convenient horizontal level, such as the ground or a tabletop. In the case of a spring, it is customary to choose the zero of the spring's potential energy, $U_{\text{sp}} = 0$, at $x = 0$, where there is neither expansion nor compression. If the initial configuration has zero potential energy, then $W_{\text{EXT}} = U_f - 0 = U_f$:

> The potential energy of a system is the external work needed to bring the particles from the $U = 0$ configuration to the given positions at constant speed.*

This statement (and Eq. 8.1) is helpful in introducing the concept of potential energy, but it suffers from two drawbacks. First, it refers to an unspecified external agent. Second, the speed of each particle must be kept constant by the application of *two* equal and opposite forces—one external, one internal. It is preferable to refer, instead, only to the internal forces between the particles in the system. We must first examine more fully the condition under which potential energy may be defined.

The climber has done work to increase his potential energy.

8.2 CONSERVATIVE FORCES

In the last chapter we found that the work done on a body by the force of gravity, $W_g = -mg(y_f - y_i)$, or the work done by the force due to a spring, $W_{\text{sp}} = -\frac{1}{2}k(x_f^2 - x_i^2)$, depends only on the initial and final positions, not on the path taken. In contrast, the work done by the force of friction—say, on a block that slides on a rough floor—depends on the length of the path, not just the end points. The force of gravity and the force exerted by an ideal spring are called **conservative forces,** whereas the force of friction is a **nonconservative force.**

The expressions for W_g and W_{sp} also show that if the final point coincides with the initial point, then $W_g = 0$ and $W_{\text{sp}} = 0$. That is, the work done in a round-trip journey is zero. For example, consider a block that is projected up a frictionless incline and then returns to its starting point, as in Fig. 8.4. The work done by gravity on the block on the way up is $W_g = -mgh$, and on the way down it is $W_g = +mgh$. The net work for the round-trip is $W_g = 0$. If the incline is rough, the work done by the force of friction on the way up is $W_f = -fd$, and on the way down it is $W_f = -fd$. The net work for the round-trip is $W_f = -2fd$.

Let us state these ideas formally. When a particle moves under the influence of a conservative force from A to B, as in Fig. 8.5, the work done on the particle by the conservative force is the same for path 1 and for path 2:

$$W_{A \to B}^{(1)} = W_{A \to B}^{(2)} \qquad (8.2)$$

The work done by a conservative force is independent of the path taken.

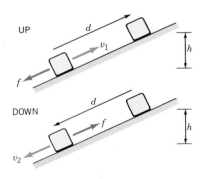

UP : $W_g = -mgh$ $\quad W_f = -fd$
DOWN : $W_g = +mgh$ $\quad W_f = -fd$

FIGURE 8.4 A block slides up and then down a rough incline. In the round trip, the work done by gravity is zero whereas the work done by friction is negative.

* Strictly speaking the speed should be close to zero. For example, if the object moves very fast, one cannot neglect the effects of the air.

If we reverse the direction of travel along path 2 in Fig. 8.5, the force is unchanged but each of the infinitesimal displacements is in the opposite direction. Therefore, the sign of the work will change: $W_{A \to B}^{(2)} = -W_{B \to A}^{(2)}$. Equation 8.2 may therefore be rewritten as

$$W_{A \to B}^{(1)} + W_{B \to A}^{(2)} = 0 \qquad (8.3)$$

The work done by a conservative force around *any* closed path is zero.

In order for the work done by a conservative force not to depend on the path, the force must be a function *only of position,* not of velocity or time. The magnetic force on a moving charge and fluid resistance are velocity dependent, which means they are nonconservative. The force exerted by a hand can vary in time; hence, it is also nonconservative.

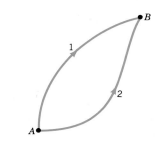

FIGURE 8.5 The work done by a conservative force from point A to point B is the same for *any* two paths such as 1 and 2.

8.3 POTENTIAL ENERGY AND CONSERVATIVE FORCES

As we will soon see, potential energy can be defined only for a conservative force. Since $\Delta K = 0$ in the definition of potential energy in Eq. 8.1, the net work done on the particle—by the external force (W_{EXT}) and by the internal conservative force (W_c)—is zero. That is, $W_{EXT} + W_c = 0$, or, equivalently, $W_c = -W_{EXT}$. Therefore, one can define the change in potential energy in terms of the work done by the conservative force:

$$W_c = -\Delta U = -(U_f - U_i) \qquad (8.4)$$

Potential energy defined in terms of work done by the associated conservative force

Equation 8.4 is preferable to Eq. 8.1 because there is no reference to any external agent. Furthermore, the speed of the particles need not be constant since the conservative force may be the only force present. The negative sign indicates that positive work by a conservative force leads to a decrease in the associated potential energy. Conservative forces tend to minimize the potential energy within any system: If allowed to, an apple falls to the ground and a spring returns to its natural length.

In three dimensions, a conservative force may vary both in magnitude and in direction. From Eq. 8.4, the infinitesimal change in potential energy dU associated with an infinitesimal displacement $d\mathbf{s}$ is

$$dU = -dW_c = -\mathbf{F}_c \cdot d\mathbf{s} \qquad (8.5)$$

The change in potential energy as a particle moves from point A to point B is equal to the negative of the work done by the associated conservative force:

$$U_B - U_A = -\int_A^B \mathbf{F}_c \cdot d\mathbf{s} \qquad (8.6)$$

Potential energy can be defined only for a conservative force, because only the work done by such a force does not depend on the path taken. From Eq. 8.6 we see that starting with potential energy U_A at point A, we obtain a unique value U_B at point B, because W_c *has the same value for all paths.* When a block slides along a rough floor, the work done by the force of friction on the block depends on the length of the path taken from point A to point B. There is no unique value for the work done, so one cannot assign unique values for potential energy at each point.

When the forces within a system are conservative, external work done on the system is stored as potential energy and is fully recoverable. When external work is done in the presence of friction, some of it is not fully recoverable because it is used to increase the vibrational energy of the atoms of the sliding surfaces. (This is manifested as a rise in temperature.) Similarly, when a real spring is extended beyond its elastic limit, it becomes permanently deformed and does not return to its original length when the external force is removed. Some of the external work done to stretch the spring is stored as "internal" energy and cannot be fully recovered. The force exerted by a nonideal spring is nonconservative.

Because friction is a common example of a nonconservative force, the term "nonconservative" is often taken to mean "dissipative," which implies that the force leads to a permanent loss in kinetic energy. This is incorrect. For example, the nonconservative magnetic force on a moving charged particle does not change the kinetic energy of the particle moving in a circular path. The nonconservative force exerted by a hand can either increase or decrease the kinetic energy of a particle in a round-trip journey. The distinction between conservative and nonconservative forces is best stated as follows:

Distinction between conservative and nonconservative forces

> A conservative force may be associated with a scalar potential energy function, whereas a nonconservative force cannot.

The meaning of the phrase "associated with" is clarified in Section 8.7.

8.4 POTENTIAL ENERGY FUNCTIONS

Potential energy is clearly a function of position. We now find this function for the (approximately constant) gravitational force near the surface of the earth and for the force exerted by an ideal spring. (The potential energy associated with the more general inverse square law of gravitation is derived in Section 8.9.)

Gravitational Potential Energy (Near the Earth's Surface)

From Eq. 7.5, the work done by gravity on a particle of mass m whose vertical coordinate changes from y_f to y_i is

$$W_g = -mg(y_f - y_i)$$

From Eq. 8.4 we have $W_g = -\Delta U_g = -(U_f - U_i)$. We deduce that the gravitational potential energy near the surface of the earth is given by

Gravitational potential energy near the earth's surface

$$U_g = mgy \qquad (8.7)$$

Implicit in this expression is the choice $U_g = 0$ at $y = 0$.

Spring Potential Energy

The work done by the force due to a spring when the displacement of the free end changes from x_i to x_f is given by Eq. 7.6:

$$W_{sp} = -\tfrac{1}{2}k(x_f^2 - x_i^2)$$

From $W_{sp} = -\Delta U_{sp} = -(U_f - U_i)$, we deduce that the potential energy of the spring is

$$U_{sp} = \tfrac{1}{2}kx^2 \qquad (8.8)$$

Note that $U_{sp} = 0$ at $x = 0$. Since Hooke's law is also valid for compressions, the same potential energy function, shown in Fig. 8.6, applies for negative x (so long as the coils do not touch).

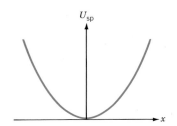

FIGURE 8.6 The potential energy of an ideal spring is a parabolic function of the displacement x from equilibrium.

EXAMPLE 8.1: How much energy does a 70-kg man need to climb a flight of stairs 30 m high at contant speed?

Solution: The man will expend his internal (chemical) energy in the process of increasing his gravitational potential energy. From Eq. 8.7,

$$\Delta U_g = mg\Delta y = 2.1 \times 10^4 \text{ J}$$

The "energy content" of foods is specified in Calories (Cal) where 1 Cal = 4180 J. Thus, the man needs about 5 Cal, which is the energy content of just half a cookie. In fact, he will generate heat and dissipate about 20 Cal of stored chemical energy. To reduce one's weight, one must exercise *and* consume fewer food calories.

EXAMPLE 8.2: How much work is needed to change the extension of a spring from $\frac{1}{3}$ m to $\frac{1}{2}$ m? Take the spring constant to be $k = 12$ N/m.

Solution: The work done is stored as potential energy. From Eq. 8.8,

$$\Delta U_{sp} = \tfrac{1}{2}k(x_f^2 - x_i^2) = 0.833 \text{ J}$$

Note that ΔU_{sp} is *not* $\frac{1}{2}k(x_f - x_i)^2$. The result would be the same had we started with a compression instead of an extension.

8.5   CONSERVATION OF MECHANICAL ENERGY

Consider a particle that is subject only to conservative forces. We may combine the work–energy theorem, $W_{NET} = \Delta K$, and the definition of potential energy, $W_c = -\Delta U$. Since $W_{NET} = W_c$, we have $\Delta K = -\Delta U$, or

$$\Delta K + \Delta U = 0 \qquad\qquad (8.9a)$$

Since $\Delta K = K_f - K_i$ and $\Delta U = U_f - U_i$, this may be written as

$$K_f + U_f = K_i + U_i \qquad\qquad (8.9b)$$

Equation 8.9b says that although the kinetic energy and the potential energy change individually, their sum has the same value at any position. The **mechanical energy** E is defined as

$$E = K + U \qquad\qquad (8.10)$$

In terms of this function, Eqs. 8.9a and 8.9b take the forms

$$\Delta E = 0 \qquad\qquad E_f = E_i \qquad\qquad (8.11) \qquad \text{Conservation of mechanical energy}$$

Equations 8.9 and 8.11 express the **principle of the conservation of mechanical energy.**

   In order for us to apply the conservation of mechanical energy there should be *no net work done by any external force or any internal nonconservative force.* Each potential energy function, such as U_g or U_{sp}, takes into account the work done by the (internal) conservative force. A second condition is that the energies must be measured in the same inertial frame. This is necessary because velocity, and hence the kinetic energy, depend on the frame of reference.

   The conservation of mechanical energy provides us with an approach to problems that is often simpler than a direct application of Newton's laws. It offers several advantages. First, whereas force is a vector, work and energy are scalars, which makes them easier to deal with. Second, only the initial and final states of a system are considered. This is useful since we are often not concerned with how a system evolves in time. Third, the concept of energy is useful even when Newton's second law is inadequate. For example, in modern physics and chemistry one can measure energies in atoms and molecules, but not the forces involved.

Gravitation

Let us see how the motion of an object in free-fall may be described in terms of the

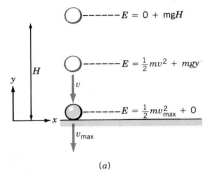

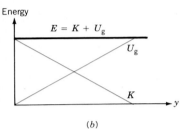

FIGURE 8.7 (a) As an object falls from height H, its potential energy is converted to kinetic energy. At height H, the energy is $E = mgH$. Just as it lands, $E = \frac{1}{2}mv_{max}^2$. (b) The potential energy and the kinetic energy vary linearly with vertical height y. The mechanical energy $E = K + U = \frac{1}{2}mv^2 + mgy$ stays constant.

conservation of mechanical energy. Since $U_g = mgy$, the mechanical energy is

$$E = \tfrac{1}{2}mv^2 + mgy$$

and the conservation law takes the form

$$\tfrac{1}{2}mv_f^2 + mgy_f = \tfrac{1}{2}mv_i^2 + mgy_i \qquad (8.12)$$

Suppose that the particle starts from rest at a height H above the ground, as in Fig. 8.7a. At this point it has no kinetic energy but has potential energy $U = mgH$. As it falls, its height decreases and its speed increases. That is, it loses potential energy and gains kinetic energy, but the sum, $E = K + U$, remains constant. Just before it lands, the object has its maximum kinetic energy, $K = \frac{1}{2}mv_{max}^2$, and its potential energy is zero. The variation of kinetic energy and potential energy with y is shown in Fig. 8.7b. In summary,

$$\begin{aligned}E &= \tfrac{1}{2}mv^2 + mgy \\ &= \tfrac{1}{2}mv_{max}^2 = mgH \qquad (8.13)\end{aligned}$$

When the object lands it experiences a force due to the ground, and so its mechanical energy is no longer conserved.

Equation 8.13 is a general result that does not depend on the path taken between the initial and final points. For example, in Fig. 8.8 a skier moves along an icy (frictionless) slope. Since the normal force N does no work (why?), the conservation of mechanical energy may be applied to find the speed at any point along the slope.

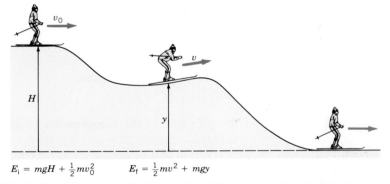

$$E_i = mgH + \tfrac{1}{2}mv_0^2 \qquad E_f = \tfrac{1}{2}mv^2 + mgy$$

FIGURE 8.8 A skier travels along frictionless slope. The conservation of mechanical energy may be used to find the speed at any point.

Springs

In Fig. 8.9a, a block of mass m on a frictionless surface, is attached to a spring that is initially extended to $x = A$ and then released. The force exerted by the spring will accelerate the block toward the equilibrium position $x = 0$. At an arbitrary point, the energy of the block–spring system is

$$E = \tfrac{1}{2}mv^2 + \tfrac{1}{2}kx^2$$

and the conservation of mechanical energy states

$$\tfrac{1}{2}mv_f^2 + \tfrac{1}{2}kx_f^2 = \tfrac{1}{2}mv_i^2 + \tfrac{1}{2}kx_i^2$$

At its initial point there is no kinetic energy and the system has its maximum potential energy: $E = K + U = 0 + \frac{1}{2}kA^2$. As the block speeds up and moves

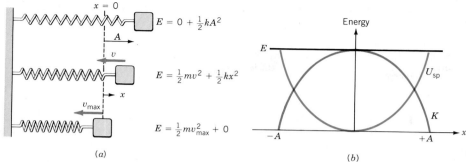

FIGURE 8.9 (a) A block connected to a spring. At the maximum extension $x = A$, the energy is $E = \frac{1}{2}kA^2$. When $x = 0$, the energy is $E = \frac{1}{2}mv_{max}^2$. (b) The variation of K and U with x. The energy of the block–spring system, $E = \frac{1}{2}mv^2 + \frac{1}{2}kx^2$, is constant.

toward the origin, the gain in kinetic energy exactly balances the loss in potential energy so that their sum stays constant. At $x = 0$, the block has its maximum kinetic energy and zero potential energy, or $E = K + U = \frac{1}{2}mv_{max}^2 + 0$. In summary,

$$E = \frac{1}{2}mv^2 + \frac{1}{2}kx^2$$
$$= \frac{1}{2}mv_{max}^2 = \frac{1}{2}kA^2 \qquad (8.14)$$

Assuming that the coils of the spring do not touch, the block continues its motion past $x = 0$. Its kinetic energy decreases as its potential energy increases, until finally at $x = -A$ the energy is again purely potential: $E = K + U = 0 + \frac{1}{2}kA^2$. The variation of the kinetic and potential energies is shown in Fig. 8.9b. Since the force due to the spring is always directed toward the origin, the block oscillates back and forth. The *turning points* are at $x = \pm A$, where $K = 0$ and $E = U_{max}$.

Can you identify the various kinds of energy involved in a pole vault?

Problem-solving Guide for Energy Conservation

A few points need to be kept in mind when applying the conservation of mechanical energy.

1. In general more than one particle will contribute to the kinetic energy, and there may be more than one type of potential energy:

$$E = K + U_g + U_{sp} \qquad (8.15)$$

2. Decide which form of the conservation law, Eq. 8.9a or 8.9b, you will use.
   (a) $K_f + U_f = K_i + U_i$: This form requires you to specify the $U = 0$ reference position. Always set $U_{sp} = 0$ at $x = 0$. The $U_g = 0$ level is chosen at any convenient level such as the ground or a tabletop. We often set $U_g = 0$ at the lowest point in the motion so that all other values are positive. If both U_g and U_{sp} are involved, their reference levels need not coincide.
   (b) $\Delta K + \Delta U = 0$: In this form, there is no need to specify a reference level for the potential energy since only *changes* are needed. Care must be taken with signs.

EXAMPLE 8.3: A ball is projected from the top of a cliff of height H with an initial speed v_0 at some angle above the horizontal, as shown in Fig. 8.10. Discuss its motion in terms of (a) the equations of kinematics, and (b) the conservation of mechanical energy.

Solution: (a) We use the equation of kinematics, Eq. 3.12, with $a_y = -g$ and $\Delta y = -H$. The vertical component of the velocity as the ball lands is given by

$$v_y^2 = v_{0y}^2 + 2gH \qquad (i)$$

(b) With $U_g = 0$ at the base of the cliff and v as the final speed, the initial and final energies are

$$E_i = \tfrac{1}{2}mv_0^2 + mgH; \qquad E_f = \tfrac{1}{2}mv^2 + 0$$

On setting $E_f = E_i$, we find

$$v^2 = v_0^2 + 2gH \qquad (ii)$$

Equation (i) looks superficially like Eq. (ii). However, there is a subtle, yet crucial, difference between them. In (i), v_y is only the *vertical component* of the final velocity, whereas in (ii), v is the final *speed*. The two equations are equivalent only in the special case of one-dimensional (vertical) motion.

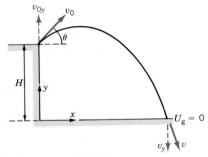

FIGURE 8.10 A projectile fired at speed v_0 lands at speed v.

EXERCISE 1. Show that kinematics can be used to derive Eq. (ii) in Example 8.3.

EXAMPLE 8.4: Two blocks with masses $m_1 = 3$ kg and $m_2 = 5$ kg are connected by a light string that slides over two frictionless pegs as in Fig. 8.11. Initially m_2 is held 5 m off the floor while m_1 is on the floor. The system is then released. At what speed does m_2 hit the floor?

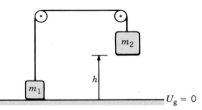

FIGURE 8.11 As m_2 falls, it loses potential energy and gains kinetic energy, whereas m_1 gains both potential and kinetic energy.

Solution: It is convenient to set $U_g = 0$ at the floor. Initially only m_2 has potential energy. As it falls, it loses potential energy and gains kinetic energy. At the same time, m_1 gains potential energy and kinetic energy. Just before m_2 lands, it has only kinetic energy. Although we are asked only for the final speed v of m_2, the kinetic energy of the system includes contributions

from *both* particles. The conservation of mechanical energy tells us

$$K_f + U_f = K_i + U_i$$
$$\tfrac{1}{2}(m_1 + m_2)v^2 + m_1gh = 0 + m_2gh$$
$$v^2 = \frac{2(m_2 - m_1)gh}{m_1 + m_2}$$

(Check the dimensions of this equation.) We find $v = 4.95$ m/s.

EXERCISE 2. For the data in Example 8.4, use $\Delta K + \Delta U = 0$ to find the speed of m_1 when it is at a height of 1.8 m.

EXAMPLE 8.5: A block of mass $m = 0.8$ kg is attached to a spring whose constant is $k = 20$ N/m. It slides on a frictionless surface. The spring is extended by 12 cm and then released. (a) Find the maximum speed of the block. (b) Find its velocity when the spring is compressed by 8 cm. (c) At what points are the kinetic energy and potential energy equal? (d) At what points is the speed half its maximum value?

Solution: (a) The maximum speed occurs where all the initial potential energy has been converted to kinetic energy, that is, at $x = 0$. The initial and final energies are

$$E_i = \tfrac{1}{2}kA^2; \qquad E_f = \tfrac{1}{2}mv_{max}^2$$

On equating $E_f = E_i$, we obtain $v_{max} = (k/m)^{1/2}A = \pm 0.6$ m/s. We were asked for speed so the sign does not matter.
(b) The initial energy is still $E_i = \tfrac{1}{2}kA^2$, while $E_f = \tfrac{1}{2}mv^2 + \tfrac{1}{2}kx^2$. The conservation law yields

$$0 + \tfrac{1}{2}kA^2 = \tfrac{1}{2}mv^2 + \tfrac{1}{2}kx^2 \qquad (i)$$

Thus,

$$v = \sqrt{\frac{k(A^2 - x^2)}{m}}$$

With $A = 0.12$ m and $x = -0.08$ m, we find $v = \pm 0.45$ m/s. The two signs indicate that at a given value of x, the velocity may be in either direction. Notice that the same two values of velocity would occur at an extension $x = +0.08$ m.
(c) If the potential energy and kinetic energy are equal, then each must be half the total energy; that is, $K = U = E/2$. Since we are looking for a value of x, we use $U = \tfrac{1}{2}kx^2 = \tfrac{1}{2}(\tfrac{1}{2}kA^2)$. Thus, $x = \pm A/\sqrt{2} = \pm 0.085$ m.
(d) We need to find out where $v = v_{max}/2 = \pm 0.3$ m/s. From the conservation equation (i), we find

$$x = \sqrt{\frac{(kA^2 - mv^2)}{k}} = \pm 0.1 \text{ m}$$

A given value of the speed occurs at two positions located symmetrically on either side of the origin.

EXERCISE 3. Continuing Example 8.5: (a) What is the speed of the block at $x = -A/2$? (b) What is its speed when the kinetic energy is equal to the potential energy?

EXAMPLE 8.6: Two blocks with masses $m_1 = 2$ kg and $m_2 = 3$ kg hang on either side of a pulley as shown in Fig. 8.12. Block m_1 is on an incline ($\theta = 30°$) and is attached to a spring whose constant is 40 N/m. The system is released from rest with the spring at its natural length. Find: (a) the maximum extension of the spring; (b) the speed of m_2 when the extension is 0.5 m. Ignore friction and the pulley.

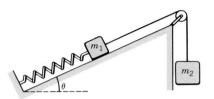

FIGURE 8.12 As m_2 falls, m_1 rises and the spring is extended.

Solution: To use $E_f = E_i$ we would need to assign the initial heights of the blocks arbitrary values h_1 and h_2. The corresponding potential energies, m_1gh_1 and m_2gh_2, would appear in both E_i and E_f and hence would cancel. We avoid this process by using the form $\Delta K + \Delta U = 0$ instead, since it does not require a $U = 0$ reference level.

(a) At the maximum extension D, the blocks come to rest, and thus $\Delta K = 0$. Next, we must find the changes in U_g and U_{sp}. When m_2 falls by D, the spring extends by D and m_1 rises by $D \sin \theta$. Therefore,

$$\Delta K + \Delta U_g + \Delta U_{sp} = 0$$
$$0 + (-m_2gD + m_1gD \sin \theta) + \tfrac{1}{2}kD^2 = 0$$

(Note carefully the signs.) Thus,

$$D = \frac{2g}{k}(m_2 - m_1 \sin \theta) = 0.98 \text{ m}$$

(b) In this case the change in kinetic energy is $\Delta K = \tfrac{1}{2}(m_1 + m_2)v^2$. The change in potential energy has the same form as in part (a), but with D replaced by $d = 0.5$ m:

$$\tfrac{1}{2}(m_1 + m_2)v^2 + (-m_2gd + m_1gd \sin \theta) + \tfrac{1}{2}kd^2 = 0$$

Inserting the given values we find $v = 1.39$ m/s.

EXERCISE 4. At what value of extension is the speed 1.0 m/s?

EXAMPLE 8.7: The bob of a simple pendulum of length $L = 2$ m has a mass $m = 2$ kg and a speed $v = 1.2$ m/s when the string is at 35° to the vertical. Find the tension in the string at: (a) the lowest point in its swing; (b) the highest point.

Solution: This problem requires us to use dynamics *and* the conservation of mechanical energy. The forces on the bob are shown in Fig. 8.13. Newton's second law in vector form is

$$\mathbf{T} + m\mathbf{g} = m\mathbf{a}$$

where $\mathbf{a}$ has both radial and tangential components. The equation for the tangential component, $\Sigma F_x = mg \sin \theta = ma_t$, does not concern us here. Since the bob is moving in a circular path of radius L, the equation for the radial component is

$$\Sigma F_y = T - mg \cos \theta = \frac{mv^2}{L} \tag{i}$$

To find the tension we need the speed, which can be found from the conservation law. We set $U_g = 0$ at the lowest point and note that the height is $y = L - L \cos \theta$. The mechanical energy is

$$E = \tfrac{1}{2}mv^2 + mgL(1 - \cos \theta) \tag{ii}$$
$$= \tfrac{1}{2}(2 \text{ kg})(1.2 \text{ m/s})^2 + (2 \text{ kg})(9.8 \text{ N/kg})(2 \text{ m})(1 - 0.82) = 8.5 \text{ J}$$

(a) At the lowest point $\theta = 0$; hence, (ii) becomes

$$E = \tfrac{1}{2}mv_{max}^2 + 0$$

Since $E = 8.5$ J we find $v_{max} = \pm 2.9$ m/s. Now that we have the speed we can find the tension at $\theta = 0$. From (i), $T - mg = mv_{max}^2/L$, from which we find $T = 19.6 + 8.5 = 28.1$ N.

(b) At the highest point $v = 0$; hence, (ii) becomes

$$E = 0 + mgL(1 - \cos \theta_{max})$$

Using $E = 8.5$ J leads to $\cos \theta_{max} = 0.783$. Since $v = 0$, $T = mg \cos \theta_{max} = 15.3$ N.

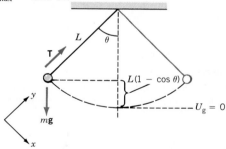

FIGURE 8.13 The potential energy of the pendulum bob is determined by the vertical position $L - L \cos \theta$.

EXERCISE 5. What is the tension in the string at $\theta = 15°$?

8.6 MECHANICAL ENERGY AND NONCONSERVATIVE FORCES

The conservation of mechanical energy may be applied to a system only when there is no work done by any internal or external nonconservative force. The work done by the internal conservative forces is taken into account in the poten-

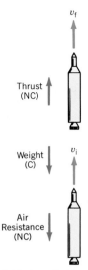

FIGURE 8.14 A rocket is subject to both conservative and nonconservative forces.

tial energy term ($W_c = -\Delta U$). In general, however, the change in kinetic energy of a particle depends on *all* the forces acting on it ($W_{\text{NET}} = \Delta K$). Therefore, when nonconservative forces are present, the work–energy theorem reads

$$W_{\text{NET}} = W_c + W_{\text{nc}} = \Delta K$$

Since $W_c = -\Delta U$, the above equation may be written as $W_{\text{nc}} = \Delta K + \Delta U$ or

$$\Delta E = E_f - E_i = W_{\text{nc}} \tag{8.16}$$

Equation 8.16 is the modification of the conservation of mechanical energy when work is done by nonconservative forces. It expresses the fact that work done by a nonconservative force changes the mechanical energy of a system. Consider a rocket fired vertically upward, as in Fig. 8.14. In addition to its weight, which is a conservative force, it experiences two nonconservative forces: air resistance and the thrust due to the engines. The thrust tends to increase the mechanical energy, whereas the air resistance tends to decrease it.

EXAMPLE 8.8: A block of mass m is attached to a spring and moves on a rough incline as in Fig. 8.15. Initially, the block is at rest with the spring unextended. A force F acting at an angle α to the incline pulls the block. Write the modified form of the work–energy theorem.

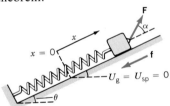

FIGURE 8.15 The levels at which U_g and U_{sp} are zero are the same in this example. The work done by the nonconservative force **F** leads to an increase in the energy of the system.

Solution: As usual, $U_{\text{sp}} = 0$ at $x = 0$. Since the motion begins at $x = 0$ it is convenient to set $U_g = 0$ at this point also. The mechanical energy has three terms:

$$E = K + U_g + U_{\text{sp}} = \tfrac{1}{2}mv^2 + mgh + \tfrac{1}{2}kx^2 \tag{i}$$

Note that h is a vertical displacement, whereas x is along the incline. In fact, $h = x \sin \theta$. We are given $K_i = 0$ and we have chosen both initial U's to be zero; therefore, $E_i = 0$. The force F and the force of friction f are nonconservative. The work done by them is

$$W_{\text{nc}} = Fx \cos \alpha - fx \tag{ii}$$

The equation we seek is $E_f = 0 + W_{\text{nc}}$, where E_f is the expression in (i).

EXAMPLE 8.9: A block of mass $m = 0.2$ kg is held against, but not attached to a spring ($k = 50$ N/m) which is compressed by 20 cm, as in Fig. 8.16. When released, the block slides 50 cm

up the rough incline before coming to rest. Find: (a) the force of friction; (b) the speed of the block just as it leaves the spring.

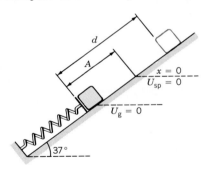

FIGURE 8.16 In this example U_g and U_{sp} are chosen to be zero at different points.

Solution: We could set $U_g = 0$ at $x = 0$, but if we use the lowest point instead, all subsequent values are positive. Again the total energy has three terms $E = K + U_g + U_{\text{sp}}$.
(a) We set $A = 0.2$ m and $d = 0.5$ m. Both K_i and K_f are zero, so $E_i = \tfrac{1}{2}kA^2$ and $E_f = mgd \sin \theta$. From Eq. 8.16,

$$E_f - E_i = W_{\text{nc}}$$
$$mgd \sin \theta - \tfrac{1}{2}kA^2 = -fd$$

Thus, $f = 0.82$ N.
(b) The initial energy E_i is the same as above, but the final value at $x = 0$ is $E_f = \tfrac{1}{2}mv^2 + mgA \sin \theta$. From Eq. 8.16,

$$\tfrac{1}{2}mv^2 + mgA \sin \theta - \tfrac{1}{2}kA^2 = -fA$$

From this we find $v = 2.45$ m/s.

EXERCISE 6. When the block slides down again, what is the maximum compression A of the spring? (The block moves down a distance $A + 0.3$ m along the incline.)

8.7   CONSERVATIVE FORCES AND POTENTIAL ENERGY FUNCTIONS

We now consider how we can find a conservative force if we are given the associated potential energy function. According to Eq. 8.5, an infinitesimal change in potential energy dU is related to the work done by the conservative force $\mathbf{F}_c$ in an infinitesimal displacement $d\mathbf{s}$ as follows:

$$dU = -\mathbf{F}_c \cdot d\mathbf{s} \tag{8.5}$$

For simplicity we confine ourselves to potential energy functions that involve just one coordinate, for example, $U(x)$ or $U(r)$. In one dimension the above equation reduces to $dU = -F_x\, dx$; thus,

$$F_x = -\frac{dU}{dx} \tag{8.17}$$

In general, a particular component (such as x, y, z, or r) of a conservative force is given by the negative derivative of the potential energy function along that axis. The negative sign indicates that the force points in the direction of decreasing potential energy. Let us see how Eq. 8.17 works for potential energy functions we have already seen:

$$U_g = mgy; \qquad F_y = -\frac{dU}{dy} = -mg$$

$$U_{sp} = \tfrac{1}{2}kx^2; \qquad F_x = -\frac{dU}{dx} = -kx$$

In both cases we obtain the correct force function. Equation 8.17 is useful as a means of defining a conservative force:

A conservative force can be derived from a scalar potential energy function.

Consider the hypothetical potential energy function $U(r)$ depicted in Fig. 8.17. We assume that one particle is fixed at the origin and that r is the radial coordinate of the other particle. (If U were gravitational potential energy, the curve would actually be the profile of the surface on which a ball rolls.) The radial component of the associated conservative force is the negative of the slope of the potential energy function; that is,

$$F_r = -\frac{dU}{dr} \tag{8.18}$$

One can obtain a qualitative idea of the force experienced by the second particle when it is placed at various points. We examine the slope, dU/dr, of the curve, and note that the sign of F_r is opposite to that of dU/dr. If $F_r > 0$, the force is directed toward positive r, which means repulsion, whereas $F_r < 0$ means attraction.

$(r > r_2)$: $F_r > 0$. The particle is weakly repelled.

$(r = r_2)$: $F_r = 0$. At the maximum point of the potential energy function, the particle would be in *unstable* equilibrium. If the particle were slightly displaced either to the left or to the right, it would tend to move away from this point.

$(r_0 < r < r_2)$: $F_r < 0$. The force is attractive, being strongest at r_1 where the slope is greatest.

$(r = r_0)$: $F_r = 0$. At the minimum point of the potential energy function, the particle

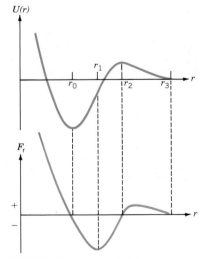

FIGURE 8.17 From the given potential energy function $U(r)$ one can find the radial component of the force from $F_r = -dU(r)/dr$, which is the negative of the slope of the $U(r)$ curve. A positive force means repulsion, and a negative force means attraction.

would be in *stable* equilibrium. If slightly displaced in either direction, it would tend to return to this point.

($r < r_0$): $F_r > 0$. The particles repel each other. The repulsive force becomes stronger as r is reduced (since the slope of $U(r)$ gets steeper).

Although it has not been depicted in the figure, when the potential energy function is constant over a region, the force on the particle is zero. If it is slightly displaced, it stays in the new position. This is called *neutral* equilibrium.

8.8 ENERGY DIAGRAMS

If we are given a potential energy diagram for a particle, we can deduce several aspects of its motion. The potential energy function of Fig. 8.18 is characterized by a **potential well** of depth $-U_0$. We assume one particle is fixed at the origin and consider the behavior of a second particle when it has different values for its total mechanical energy $E = K + U$. The particle is in a **bound state** when $E < 0$, and it is unbound when $E > 0$ (note that $U = 0$ at $r = \infty$). We assume that for a given state E is a constant, so it is drawn as a horizontal line on the diagram. This is called an **energy level.**

The minimum of the potential energy is $-U_0$ and occurs at r_0. Typical values for a diatomic molecule are $U_0 = 5 \times 10^{-20}$ J and $r_0 = 3 \times 10^{-10}$ m. If the particle had no kinetic energy, this would be a point of stable equilibrium. The shape of the potential well is approximately parabolic in the region close to the point r_0; it is similar to that of an ideal spring (Fig. 8.6). At points farther away, the shape is not symmetrical about r_0.

The kinetic energy (and speed) of the particle at any point is determined by the difference $K = E - U$. This is the vertical distance between the potential energy curve and the energy level. For any value of E, the kinetic energy (and speed) is a maximum at r_0, where $U(r)$ has its minimum value. When the particle is in a bound state, its motion is restricted to lie between two *turning points*. At these points, the total energy is equal to the potential energy, and, therefore, $K = E - U = 0$. (E cannot be less than U because this would imply that K is negative.) At a turning point the particle comes to rest momentarily and then reverses the direction of its motion. The speed increases as it approaches r_0, at which point it is a maximum, and then decreases to zero at the other turning point. The back and forth motion repeats itself indefinitely.

When the particle is in a bound state with energy E_1, the turning points are at r_1 and r_1'. When it has a higher energy E_2, the turning points are farther apart, at r_2 and r_2'. It is clear that the motion of the particle is not symmetrical about the equilibrium point r_0. The average value of its position is greater than r_0 and increases as the energy E increases.

When the particle has energy $E_3 > 0$, it is unbound and there is only one turning point at r_3. As it approaches from large values of r, its speed increases until r_0 and then drops to zero at r_3. After it turns around, it does not return again. Even when the particle has a positive energy E, it can become a bound particle if it loses energy in a collision with another particle or by the emission of radiation. This process may be repeated and its energy may then drop to a lower level.

The **binding energy** of a particle in a bound state is the minimum energy that must be supplied by an external agent to make it an unbound particle. For the potential energy function of Fig. 8.18, the binding energy of the nth state would be $|E_n|$. For an electron in an atom, the binding energy of the lowest state is called the ionization energy. In the case of a molecule, it is called the dissociation energy.

Potential well

Bound state

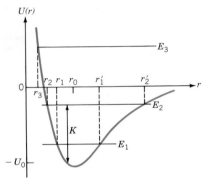

FIGURE 8.18 A potential energy function $U(r)$. The horizontal lines represent different energy levels E. The points of intersection of U and E are the turning points in the motion.

8.9 GRAVITATIONAL POTENTIAL ENERGY, ESCAPE SPEED*

The expression $U_g = mgh$ for gravitational potential energy is valid only near the surface of the earth, where one can assume that the force of gravity is constant. We now find the proper potential energy function when the variation in the force of gravity with distance from the earth is taken into account. From Eq. 8.6 we know that the change in potential energy between two points is given by the negative of the work done by the conservative force:

$$U_B - U_A = -\int_A^B \mathbf{F}_c \cdot d\mathbf{s} \tag{8.6}$$

The force of gravity is an example of a **central force,** which means that it is always directed along the line joining the two particles. The force is also *spherically symmetric* since it depends only on the radial coordinate r. If one particle is at the origin, the force on the other has the form $\mathbf{F} = F_r\hat{\mathbf{r}}$ where the sign of F_r may be positive or negative. The force varies in both magnitude and direction, but it is always along the radial axis. Since there is only a radial component, the work done by the conservative force in an infinitesimal displacement $d\mathbf{s}$ is $dW_c = \mathbf{F}_c \cdot d\mathbf{s} = F_r \, dr$. The total work done by the conservative force from r_A to r_B is

$$W_c = \int_A^B F_r \, dr$$

According to Newton's law of gravitation, Eq. 5.4, we know that $F_r = -GmM/r^2$, so the work done by this force from point A to point B is

$$W_g = \int_{r_A}^{r_B} (-GmM)\frac{dr}{r^2} = +\frac{GmM}{r_B} - \frac{GmM}{r_A}$$

Clearly, the work done depends only on the initial and final points, as is characteristic of a conservative force. Using this result in Eq. 8.6 we find the gravitational potential energy of two point particles separated by a distance r to be

$$U(r) = -\frac{GmM}{r} \tag{8.19}$$

Gravitational potential of point particles

Implicit in this definition is our choice of the zero level of potential energy, $U = 0$ at $r = \infty$. The negative sign means that work must be done *by* an external agent on the particles to increase the separation between them: The potential energy becomes less negative as r increases. Equation 8.19 is also valid when the interacting objects are not point particles but are spheres with uniform mass distribution. In such cases, r is the distance to the center.

* This section may be delayed until Chapter 13.

EXAMPLE 8.10: What is the connection between Eq. 8.19 and the equation $U_g = mgh$?

Solution: In earlier sections we used the expression $U_g = mgh$ for the potential of objects close to the earth's surface (Fig. 8.19a). We assumed that the force of gravity is constant. When Newton's law of gravitation is taken into account, the potential

energy associated with an apple of mass m at the surface of the earth (mass M_E, radius R_E), as in Fig. 8.19b, is

$$U(R_E) = -\frac{GmM_E}{R_E}$$

while at a height h above the surface it is

$$U(R_E + h) = -\frac{GmM_E}{R_E + h}$$

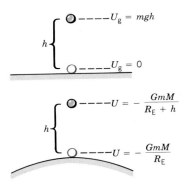

FIGURE 8.19 (a) When the force of gravity is taken to be constant, the potential energy at height h is $U_g = mgh$, with the choice $U_g = 0$ at the ground. (b) At large distances, the proper expression is $U = -GmM/(R_E + h)$, with the choice $U = 0$ at $r = \infty$.

The change in potential energy is

$$\Delta U = U(R_E + h) - U(R_E) = \frac{GmM_E h}{R_E(R_E + h)}$$

When $h \ll R_E$, we have $(R_E + h) \approx R_E$, so the above expression becomes

$$\Delta U = \frac{GmM_E}{R_E^2} h = mgh$$

where we have used $g = GM_E/R_E^2$ from Eq. 5.6. Figure 8.20 illustrates that the equation $U_g = mgh$ represents the *change* in the gravitational potential energy between two points close to the surface of the earth. When you deal with the motion of rockets and satellites you *must* use the expression $U(r) = -GmM/r$ since it takes into account the fact that the force of gravity varies with distance from the earth.

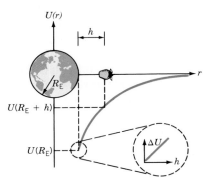

FIGURE 8.20 For small distances from the surface, the *change* in potential energy varies linearly with height: $\Delta U = mgh$. As the apple, or a rocket, moves away from the earth, it climbs a "potential well."

Mechanical Energy

We may obtain a simple expression for the mechanical energy of a two-particle system if we assume that one mass is very much larger than the other. If $M \gg m$, then the change in kinetic energy of M is negligible (as is the case when an apple falls to the earth). The total mechanical energy of the system is the sum of its potential energy and the kinetic energy of the smaller mass:

$$E = \tfrac{1}{2}mv^2 - \frac{GmM}{r} \tag{8.20}$$

From Eq. 6.5 we know that the orbital speed of a satellite in stable circular orbit is $v_{orb} = (GM/r)^{1/2}$. Therefore, the kinetic energy of the satellite is

$$K = \tfrac{1}{2}mv_{orb}^2 = +\frac{GmM}{2r}$$

Using Eq. 8.20 we see that the mechanical energy of the satellite is

Mechanical energy in a circular orbit   (Orbit) $$E = K + U = -\frac{GmM}{2r} \tag{8.21}$$

The negative sign of the energy means that the satellite is in a bound state. The quantity $|E|$ is the *binding energy* of the particle (see Section 8.8), which is the minimum energy that must be supplied by an external agent to make it an unbound particle—whose mechanical energy is greater than zero.

EXAMPLE 8.11: A rocket with a payload of mass m is at rest at the surface of the earth. Calculate the work needed to raise the payload to the following states: (a) at rest at a maximum altitude equal to R_E; (b) in circular orbit at an altitude R_E.

Solution: In both cases the initial energy of the payload is purely potential energy:

$$E_1 = K_1 + U_1 = 0 - \frac{GmM}{R_E}$$

(a) At the maximum altitude the distance to the center of the earth is $r = 2R_E$.

$$E_2 = K_2 + U_2 = 0 - \frac{GmM}{2R_E}$$

The work needed is $E_2 - E_1 = +GmM/2R_E$.
(b) From Eq. 8.21 the energy in orbit is

$$E_3 = K_3 + U_3 = -\frac{GmM}{4R_E}$$

The work needed is $E_3 - E_1 = +3GmM/4R_E$. Naturally, it takes more work to put the satellite into orbit than merely to raise it to the same altitude.

Escape Speed

A particle at rest at the surface of the earth, or in stable orbit around it, is in a bound state. We wish to find the minimum initial speed it must have to escape from the earth, never to return. From Fig. 8.20 we see that as a rocket moves away from the earth, it climbs a potential energy well. To make it an unbound particle, it must be given enough initial kinetic energy for it to reach the point of maximum potential energy with zero (or greater) speed. In the case of gravitation, the maximum value of the potential energy is zero at the point $r = \infty$. Therefore, a particle is bound if its mechanical energy is negative and it is unbound if $E = K + U > 0$.

When the rocket is fired with the minimum **escape speed** v_{esc} it will reach $r = \infty$ with zero speed; that is, $E_f = 0$. Its initial energy at the earth's surface is

$$E_i = \tfrac{1}{2}mv^2 - \frac{GmM_E}{R_E}$$

On setting $E_f = E_i$, we find

$$v_{esc} = \sqrt{\frac{2GM_E}{R_E}} \qquad (8.22) \quad \text{Escape speed}$$

Notice that the escape speed does not depend on the mass of the rocket. For a particle at the earth's surface, $v_{esc} = 11.2$ km/s relative to the center of the earth. It does not depend on the direction of the launching. (Why?)

EXAMPLE 8.12: A rocket is fired vertically with half the escape speed. What is its maximum altitude in terms of the radius of the earth R_E? Ignore the earth's rotation.

Solution: Such a problem is easily solved using the conservation of energy. However, you must take care *not* to use $U_g = mgh$ since this equation is valid only when g may be taken to be a constant. The initial energy is

$$E_i = \tfrac{1}{2}m\left(\frac{v_{esc}}{2}\right)^2 - \frac{GmM_E}{R_E}$$

$$= \frac{GmM_E}{4R_E} - \frac{GmM_E}{R_E} = -\frac{3GmM_E}{4R_E}$$

where we have used Eq. 8.22 for v_{esc}. At the maximum altitude, the kinetic energy is zero:

$$E_f = 0 - \frac{GmM}{R_E + h}$$

On setting $E_f = E_i$, we find $h = R_E/3$.

8.10 GENERALIZED CONSERVATION OF ENERGY

Thus far we have taken a rather limited view of energy. We have considered only macroscopic (bulk) kinetic and potential energies. In effect all objects were treated as particles without structure. However, consider a block that slides to a stop on a rough surface. What happens to its kinetic energy? The work done by the force of friction on the block transforms the bulk kinetic energy of the block into *internal* kinetic and potential energies associated with the *random* motion of the atoms. This added **thermal energy** is manifested as a rise in temperature of the block. After a while, the block cools down. The extra thermal energy is removed by **heat,** which is a transfer of energy associated with the temperature difference between the warm block and the cooler surroundings.

Thermal energy and heat

The concept of energy is used in such diverse fields as chemistry, biology, engineering, and physiology. This is because it appears in many forms, such as electromagnetic, acoustic, gravitational, thermal, nuclear, chemical, and mechanical. Furthermore, it can transform from one kind to another, and so it plays a central role in countless physical and biological processes. In a flashlight, chemical energy in the battery is converted into electrical energy in the wires. This in turn produces heat and light. In the process of photosynthesis, plants convert the sun's radiant energy into chemical energy in the cells. When the plants are eaten by animals, this chemical energy is converted to heat and mechanical energy.

Around 1845, several scientists independently concluded that all natural processes are subject to an important constraint called the **principle of the conservation of energy:**

Conservation of energy

Energy can change its form, but it can neither be created nor destroyed.

The formulation of this principle required the identification of heat as a form of energy. The fact that the internal energy of a system can be changed by the performance of work or by the transfer of heat is the essence of the **first law of thermodynamics** (Chapter 19).

In general, one can regard the relation between work and energy as follows: Work is a mode of energy transfer from one body to another in which a force moves through a displacement. If the work done by object A on object B is positive, energy is transferred from A to B. If it is negative, energy is transferred from B to A. Work is also a measure of the transformation of one form of energy into another. For example, the work done by our muscles in pulling a block represents a transformation of chemical energy into kinetic energy. Energy transfer without the performance of macroscopic work occurs in a solar cell when light energy is converted directly into electrical energy, or in a lightbulb, when thermal energy is converted directly into light energy.

The connection between work and energy

SUMMARY

The work done by a **conservative force** is independent of the path taken between two points A and B; that is,

$$W_{A \to B}^{(1)} = W_{A \to B}^{(2)}$$

Equivalently, the work done by a conservative force in *any* closed loop is zero.

Potential energy is energy associated with the relative positions of a system of interacting particles. It belongs to the *system*. Potential energy can be defined only for a conservative force. The change in potential energy between two points is related to the work done by the associated conservative force:

$$\Delta U = -W_c$$

The negative sign means that positive W_c leads to a decrease in potential energy. Since only changes in potential energy are significant, the $U = 0$ position may be chosen at any convenient point.

Gravitational potential energy $\quad U_g = mgy \quad$ (g constant)

Spring potential energy $\quad\quad U_{sp} = \frac{1}{2}kx^2$

Mechanical energy $\quad\quad\quad\quad E = K + U$

According to the principle of **conservation of mechanical energy,**

$$K_i + U_i = K_f + U_f \quad \text{or} \quad \Delta E = \Delta K + \Delta U = 0$$

provided no work is done by nonconservative forces. In the presence of work done by nonconservative forces, the conservation law is modified:

$$\Delta E = E_f - E_i = W_{nc}$$

A component of a conservative force may be obtained from the derivative of the associated potential energy. For example,

$$F_x = -\frac{dU}{dx}$$

A nonconservative force cannot be derived from a potential energy function.

When the variation in the force of gravity is taken into account, the **gravitational potential energy** of two particles of masses m and M is

$$U = -\frac{GmM}{r}$$

ANSWERS TO IN-CHAPTER EXERCISES

1. Note that $v^2 = v_x^2 + v_y^2$, where $v_x = v_{0x}$ and v_y is given by (i).

2. The kinetic energy of both blocks increases, so $\Delta K = \frac{1}{2}(m_1 + m_2)v^2 = 4v^2$. The potential energy of m_1 increases, while that of m_2 decreases, so $\Delta U = +m_1gD - m_2gD$, where $D = 1.8$ m. The equation $\Delta K + \Delta U = 0$ leads to $v = 2.97$ m/s.

3. (a) From (i): $\frac{1}{2}kA^2 = \frac{1}{2}mv^2 + \frac{1}{2}k(-A/2)^2$; so $v^2 = 3kA^2/4m$ and $v = 0.52$ m/s. (b) When $K = U = E/2$, we have $\frac{1}{2}mv^2 = \frac{1}{4}kA^2$; so $v^2 = kA^2/2m$ and $v = 0.42$ m/s.

4. Equation (i) becomes $20d^2 - 19.6d + 2.5 = 0$ from which we find $d = 0.15$ m and $d = 0.83$ m.

5. From (ii): $8.5 = v^2 + (39.2)(1 - 0.966)$; thus, $v = 2.68$ m/s. From (i), $T = mg \cos \theta + mv^2/L = 18.9 + 7.2 = 26.1$ N.

6. The equation $E_f - E_i = W_{nc}$ leads to $\frac{1}{2}kA^2 - mgD \sin \theta = -fD$, where $D = A + 0.3$ m. The solution of $25A^2 = 0.36(A + 0.3)$ is $A = 7.3$ cm. Notice that we were not concerned with kinetic energy at all.

QUESTIONS

1. True or false: (a) Conservation of mechanical energy does not apply to nonconservative forces. (b) The net force acting on a ball being lifted at constant velocity is zero; hence, its energy is conserved.

2. How can one reconcile energy production from oil, gas, and coal with the principle of conservation of energy?

3. A body is dropped from rest. Sketch the variation of the kinetic energy and potential energy as functions of (a) height, and (b) time.

4. We use chemical energy in walking up stairs and thereby gain potential energy. Do we regain this energy as we walk down the stairs? If not, where does it go?

5. The drag force due to fluid resistance may be written $f = -av - bv^2$. Is this a conservative force?

6. True or false: If a body is dropped from height H, then at $H/4$: (a) v is $v_{max}/2$; (b) the kinetic energy is $K_{max}/4$; (c) $K/U = 3/4$.

7. When you descend at constant speed in an elevator, what happens to your lost potential energy?

8. Is it possible to drop an object and have it bounce to a higher level? If so, explain how.

9. Name three forms of energy available on earth whose origin is not ultimately traceable to the sun.

10. Does the escape speed for a rocket depend on the angle of launch?

11. True or false: If a body is launched with twice the escape speed, its speed at infinity is v_{esc}.

12. Does the maximum altitude attained by a rocket depend on the angle of launch?

13. How would Fig. 8.6 be modified if the coils of a compressed spring were to touch?

14. Can the rotation of the earth help in the launching of a satellite? If so, explain how.

EXERCISES

8.5 Conservation of Mechanical Energy

1. (I) Two blocks of masses $m_1 = 5$ kg and $m_2 = 2$ kg hang on either side of a frictionless cylinder as in Fig. 8.21. If the system starts at rest, what is the speed of m_1 after it has fallen 40 cm?

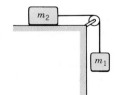

FIGURE 8.21 Exercise 1.

2. (I) Two blocks with masses $m_1 = 0.5$ kg and $m_2 = 1.5$ kg are connected by a rope, as shown in Fig. 8.22. The horizontal surface is frictionless and the pulley is massless. If the blocks start from rest, what is the speed of m_1 after it has fallen by 60 cm?

FIGURE 8.22 Exercise 2.

3. (I) Two blocks with masses $m_1 = 4$ kg and $m_2 = 5$ kg are connected by a light rope and slide on a frictionless wedge as shown in Fig. 8.23. Given that it starts at rest, what is the speed of m_2 after it has moved 40 cm along the incline?

4. (I) The metabolism of 1 g of fat releases 9 kcal (3.76×10^4 J). A 70-kg person walks to the top of the 433-m Sears tower. Assuming a 15% efficiency in converting the released chemical energy to mechanical energy, how much mass could the person lose?

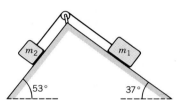

FIGURE 8.23 Exercise 3.

5. (I) A simple pendulum has a length of 75 cm and a bob of mass 0.6 kg. When the string is at 30° to the vertical, the bob has a speed of 2 m/s. (a) What is the maximum speed of the bob? (b) What is the maximum angle to the vertical?

6. (I) A pendulum bob of mass 0.7 kg is suspended by a string of length 1.6 m. The bob is released from rest when the string is at 30° to the vertical. The swing is interrupted by a peg 1 m vertically below the support as shown in Fig. 8.24. What is the maximum angle to the vertical made by the string after it hits the peg?

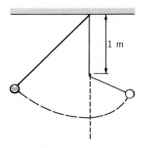

FIGURE 8.24 Exercise 6.

7. (I) A block of mass $m = 0.25$ kg can slide over a frictionless horizontal surface. It is attached to a spring whose stiffness constant is $k = 10$ N/m. The block is pulled 40 cm and let go. (a) What is its maximum speed? (b) What is its speed

when the extension is 20 cm? (c) At what position is the kinetic energy equal to the potential energy?

8. (I) A 2-kg particle is subject to a single conservative force directed along the x axis. The work done by the conservative force on the particle when its position changes from $x = -1$ m to $x = 3$ m is $+60$ J. Find: (a) the change in kinetic energy; (b) the change in potential energy; (c) the final speed if the initial speed is 4 m/s.

9. (II) A 500-g block is dropped from a height of 60 cm above the top of a vertical spring whose stiffness constant is $k = 120$ N/m (see Fig. 8.25). Find the maximum compression. (You will need to solve a quadratic equation.)

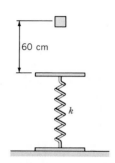

FIGURE 8.25 Exercise 9.

10. (II) A simple pendulum has a length of 1.25 m and a bob of mass 0.5 kg. It is released when the rope is horizontal. If the rope has a breaking tension of 6 N, at what angle to the vertical does the break occur?

11. (II) A 2-kg block slides on a frictionless horizontal surface and is connected on one side to a spring ($k = 40$ N/m) as shown in Fig. 8.26. The other side is connected to a 4-kg block that hangs vertically. The system starts from rest with the spring unextended. (a) What is the maximum extension of the spring? (b) What is the speed of the 4-kg block when the extension is 50 cm?

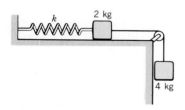

FIGURE 8.26 Exercise 11.

12. (II) A 50-g body compresses a vertical spring by 10 cm. It is pushed down a further 20 cm and released. (a) Relative to this position, what is the maximum height reached by the body if it is not connected to the spring? (b) What is the maximum extension of the spring if the body is glued to the spring? (You will need to solve a quadratic equation.)

13. (II) Two blocks with masses $m_1 = 5$ kg and $m_2 = 3$ kg are connected by a light thread that passes over frictionless pegs, as in Fig. 8.27. The smaller mass is attached to a spring ($k = 32$ N/m). If the system starts at rest with the spring unextended, find: (a) the maximum displacement of the larger mass; (b) the speed of the larger mass after it has fallen 1 m.

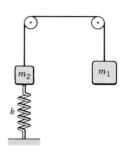

FIGURE 8.27 Exercise 13.

14. (II) A 100-g block starts from rest and slides 4 m down a frictionless 30° incline. Its motion is halted by a spring ($k = 5$ N/m) as shown in Fig. 8.28. (a) What is the speed of the block just as it reaches the spring? (b) Find the maximum compression of the spring. (You will need to solve a quadratic equation.)

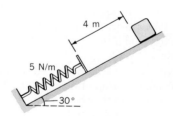

FIGURE 8.28 Exercises 14 and 34.

15. (I) A 3.2-kg trolley initially moving at 5 m/s at a height of 4 m encounters a hill of height 5 m, as shown in Fig. 8.29. At a later point there is a horizontal spring ($k = 120$ N/m) at a height of 2 m. (a) Does the trolley reach the spring? (b) If so, what is the maximum compression? Ignore frictional losses and the rotational energy of the wheels.

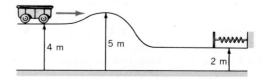

FIGURE 8.29 Exercise 15.

16. (II) A simple pendulum has a length of 1.2 m and a bob of mass 0.8 kg. It is released at 90° to the vertical. What is the speed of the bob and the tension in the string when (a) the string is vertical, and (b) when the string is at 37° to the vertical?

17. (I) A body fired vertically up reaches a maximum height H. Where is the kinetic energy 75% of the potential energy? Take $U_g = 0$ at $y = 0$.

18. (I) A ball is fired vertically up with an initial speed of 40 m/s. (a) At what point is $K = U$? (b) Where is $K = U/2$?

19. (I) A carriage on a roller coaster has a mass of 600 kg including passengers. Its speed is 12 m/s at point A, a height of 30 m (see Fig. 8.30). Find the speed: (a) at point B; (b) at point C. Neglect frictional losses.

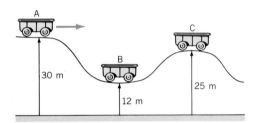

FIGURE 8.30 Exercise 19.

20. (I) A block of mass $m = 32$ g is attached to a vertical spring with stiffness constant $k = 2.8$ N/m. (a) What is the equilibrium extension of the spring if the block is held in the hand and slowly lowered to its final position? (b) What is the maximum extension if the block is released at the point the spring has its natural length?

21. (I) A projectile is fired at 25 m/s in a direction 60° above the horizontal from a rooftop of height 40 m. Use energy considerations to find: (a) the speed with which it lands on the ground; (b) the height at which its speed is 15 m/s.

22. (I) When a high jumper clears 2 m, he raises his torso by 1 m. He takes off at 75° to the horizontal. (a) What is the minimum speed with which he must take off from the ground? (b) What is his speed at the highest point?

23. (I) About 5×10^6 kg of water fall over Niagara Falls each second. Its height is about 50 m. (a) How much potential energy does the water lose per second? (b) Assuming a 95% efficiency in converting mechanical to electrical energy, how many 100-W light bulbs could this light?

24. (I) A pump has to raise water from a depth of 50 m and eject it at 10 m/s. If the flow rate is 2 kg/s, what horsepower is needed?

25. (I) A rocket of mass 5×10^4 kg acquires a speed of 5000 km/h in 1 min. In so doing it rises to an altitude of 25 km. What is its average horsepower? Ignore frictional losses and the change in mass due to the ejected gases. Assume that the force of gravity is constant.

26. (II) Coal is dropped with essentially zero speed at a rate of 8 kg/s onto a conveyor belt that moves at 1 m/s and is inclined at 10° to the horizontal, as shown in Fig. 8.31. The coal is carried a distance of 40 m and then discharged. What is the mechanical power delivered to the coal?

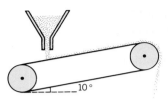

FIGURE 8.31 Exercise 26.

27. (II) A ski lift carries people at 1 m/s for a distance of 0.5 km up a 20° slope. The chairs are 5 m apart and each carries a single skier. If all the chairs are occupied, what power is required? Assume that the average mass of a skier is 70 kg.

28. (I) A 70-kg pole vaulter converts his kinetic energy to potential energy with the help of the pole. Given that his initial speed is 9 m/s and that he goes over the bar at 0.5 m/s, estimate the change in height of his waistline.

8.6 Mechanical Energy and Nonconservative Forces

29. (I) A 75-kg parachutist is attached to an 8-kg parachute. She jumps from a plane flying at 140 km/h at an altitude of 1 km and immediately opens the parachute. If she lands vertically at 7 m/s, find the work done by the parachute on the air?

30. (I) A 20-kg child starts from rest and reaches 4 m/s at the bottom of a 20° slide of length 4 m. What is the coefficient of kinetic friction?

31. (I) A 2-kg particle is projected with an initial speed of 4 m/s along a surface for which $\mu_k = 0.6$. Find the distance it travels given that: (a) the surface is horizontal; (b) the particle moves up a 30° incline; (c) the particle moves down a 30° incline.

32. (II) A 1-kg particle at a height of 4 m has a speed of 2 m/s down a 53° incline; see Fig. 8.32. It slides on a horizontal section of length 3 m at ground level and then up a 37° incline. All surfaces have $\mu_k = 0.4$. How far along the 37° incline does the particle first come to rest?

FIGURE 8.32 Exercise 32.

33. (I) A 60-g tennis ball thrown vertically up at 24 m/s rises to a maximum height of 26 m. What was the work done by resistive forces?

34. (II) For the situation depicted in Fig. 8.28 assume the following values: $m = 2$ kg, $k = 60$ N/m, $\theta = 37°$ and $\mu_k = 0.5$.

Given that the block starts at rest, find the maximum compression of the spring. (You will need to solve a quadratic equation.)

8.7 and 8.8 Conservative Forces; Energy Diagrams

35. (II) Find the potential energy function $U(x)$ for the force $F_x = Cx^3$. Take $U = 0$ at $x = 0$.

36. (II) Find the potential energy function for the force $F_x = b/x^2$. Take $U = 0$ at $x = \infty$.

37. (II) Given the force $F_x = ax/(b^2 + x^2)^{3/2}$, find the associated potential energy function. Take $U = 0$ at $x = \infty$.

38. (I) (a) Given the potential energy function $U(x, y) = A/(x^2 + y^2)^{1/2}$ find the force $\mathbf{F}(x, y)$. (b) From the potential energy function $U(r) = Ae^{-Br}/r$, find the force F_r.

39. (I) A constant force $\mathbf{F} = 2\mathbf{i} - 5\mathbf{j}$ N acts on a particle that changes its position from $3\mathbf{i} + 5\mathbf{j}$ m to $-2\mathbf{i} + 11\mathbf{j}$ m. What is the change in potential energy associated with this displacement?

40. (II) Use the potential energy function $U(x)$ shown in Fig. 8.33 to sketch the corresponding F_x versus x graph.

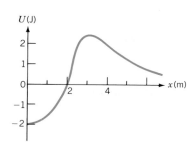

FIGURE 8.33 Exercise 40.

41. (II) Use the F_x versus x graph in Fig. 8.34 to sketch the corresponding $U(x)$ versus x graph. Take $U(0) = 0$.

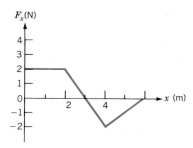

FIGURE 8.34 Exercise 41.

42. (II) Which of the following forces acting along the x axis is conservative? (a) $F_x = -kx + bx^2$; (b) $F_x = Ae^{-bx}$; (c) $F_x = cx^3$. (d) Is any force that is a function of only one coordinate conservative?

43. (I) A particle is subject to a constant frictional force of 10 N directed opposite to its motion. Compute the work done along each of the following paths in Fig. 8.35: (a) W_{OAB} and W_{OCB}. Do your values satisfy the criterion of Eq. 8.2? Why or why not? (b) $W_{OAB} + W_{BCO}$. Does this satisfy the criterion of Eq. 8.3?

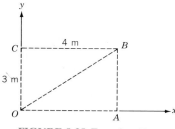

FIGURE 8.35 Exercise 43.

44. (I) Repeat Exercise 43 for a constant force $\mathbf{F} = 2\mathbf{i} - 3\mathbf{j}$ N.

45. (II) Consider the potential energy function in Fig. 8.36. What are the turning points if the energy of the particle is (a) E_1 or (b) E_2? (c) What is the kinetic energy of the particle at position 2×10^{-10} m when its total energy is E_1?

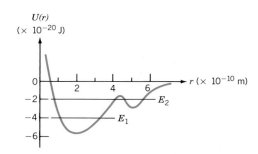

FIGURE 8.36 Exercise 45.

8.9 Gravitational Potential Energy, Escape Speed

46. (II) A rocket is fired vertically from the earth with speed v_{esc}/N, where N is some number greater than one. Show that the maximum altitude is

$$h = \frac{R_E}{N^2 - 1}$$

where R_E is the radius of the earth. Ignore the rotation of the earth and air resistance.

47. (I) A rocket fired vertically reaches a maximum altitude of $4R_E$, where R_E is the radius of the earth. What was its initial speed? Ignore the motion of the earth and air resistance.

48. (I) Find the escape speed from: (a) the moon; (b) Mars (mass = 6.42×10^{23} kg, radius = 3.37×10^3 km); (c) Jupiter (mass = 1.90×10^{27} kg, radius = 6.99×10^4 km).

49. (I) What is the escape speed for a tiny insect stranded on a

2000-kg elephant in outer space? Treat the elephant as a uniform sphere of radius 1 m.

50. (II) To what radius would the earth have to collapse such that the work needed to remove an object of mass m from the surface would equal mc^2, where c is the speed of light?

51. (II) A projectile is fired vertically upward from the surface of the earth, of radius R_E, with an initial speed v_0. Show that its maximum altitude H is

$$H = \frac{v_0^2}{2g_0 - v_0^2/R_E}$$

where $g_0 = GM/R_E^2$ is the gravitational force per unit mass at the surface. Ignore the rotation of the earth.

52. (II) A satellite is in stable circular orbit with a speed v_{orb}. Show that to escape from orbit it requires a speed of $\sqrt{2}v_{orb}$.

53. (II) (a) What is the minimum (net) energy needed to take a 1-kg object from the surface of the earth to the surface of the moon. Take $g_M = 0.16g_E$, where $g = GM/R^2$ is the gravitational force per unit mass at the surface. (b) Show that this is roughly twice the work needed to put the object into an orbit close to the earth's surface.

54. (II) An object is released from an altitude h above the earth of radius R_E. Show that it lands at a speed given by

$$v^2 = \frac{2g_0R_Eh}{(R_E + h)}$$

where $g_0 = GM/R_E^2$ is the gravitational force per unit mass at the surface. Ignore the earth's rotation and air resistance.

55. (II) What is the energy required to place a 2100-kg communications satellite into a geostationary orbit?

PROBLEMS

1. (I) The potential energy associated with two particles has the form $U(x) = C/(a^2 + x^2)^{1/2}$ where C and a are constants. (a) Find the force F_x. (b) Where is F_x a maximum?

2. (I) The potential energy shared between an infinite line of charge and a charged particle has the form $U(r) = C \ln(r/a)$ where r is the distance from the line and C and a are constants. What is the force F_r on the charged particle?

3. (II) The potential energy shared by two atoms separated by a distance r in a diatomic molecule is given by the Lennard-Jones function (U_0 and r_0 are constants):

$$U(r) = U_0 \left[\left(\frac{r_0}{r}\right)^{12} - 2 \left(\frac{r_0}{r}\right)^6 \right]$$

(a) Where is $U(r) = 0$? (b) Show that the minimum potential energy is $-U_0$ and that it occurs at r_0. (c) Where is $F_r = 0$? (d) Sketch $U(r)$.

4. (II) The Yukawa potential energy function for the interactions of neutrons and protons in a nucleus is

$$U(r) = -(r_0/r) U_0 e^{(-r/r_0)}$$

(a) What is the force function F_r? (b) Compute the value of the force at $r = r_0$, and $r = 3r_0$. Take $U_0 = 5 \times 10^{-12}$ J and $r_0 = 1.5 \times 10^{-15}$ m.

5. (II) A ball of mass m moves in a vertical circle at the end of a string. Show that the tension at the bottom is greater than that at the top by $6mg$.

6. (I) A pendulum bob of length L has its motion interrupted by a peg vertically beneath the support at a distance $3L/4$ from the support; see Fig. 8.37. (a) If the bob is released from a horizontal position, what is the tension in the rope at the top of its motion after it has hit the peg? (b) Show that the angle to the vertical at which the pendulum should be released so that the tension at the top of the circle just vanishes is given by $\cos \theta = 3/8$.

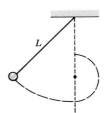

FIGURE 8.37 Problems 6 and 7.

7. (I) A pendulum of length L has its motion interrupted by a peg vertically beneath the support at a distance y below the support; see Fig. 8.37. The bob is released when the string is horizontal. Show that for the bob to swing in a complete circle, the minimum value of y is $3L/5$.

8. (I) A block slides on a frictionless surface starting at height H, as in Fig. 8.38. It encounters a hill of radius r. What minimum value of H will allow the block to just skim over the peak without touching it?

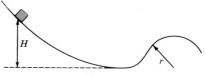

FIGURE 8.38 Problem 8.

9. (I) A Prony brake, shown in Fig. 8.39, is a device that measures the horsepower of engines. The engine rotates a shaft of radius R at N rev/sec. The rotation is opposed by a belt which is attached to two tension measuring devices, such as springs. (a) What is the force of friction acting on the engine? (b) Obtain an expression for the power developed at the given rotation rate.

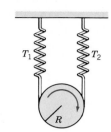

FIGURE 8.39 Problem 9.

10. In designing an amusement park ride, an engineer considers a particle of mass m that is released at height H and then slides on a frictionless surface that becomes a vertical circle of radius R, as shown in Fig. 8.40. (a) What is the minimum value of H for the particle not to lose contact at the highest point of the circle? (b) If it is released at twice this minimum height, what is the force exerted by the track on the particle at the highest point?

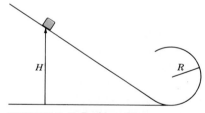

FIGURE 8.40 Problem 10.

11. (II) An Eskimo child slides on an icy (frictionless) hemispherical igloo of radius R, as in Fig. 8.41. She starts with a negligible speed at the top. (a) At what angle to the vertical does she lose contact with the surface? (b) If there were friction, would contact be lost at a higher or a lower point?

FIGURE 8.41 Problem 11.

12. (II) Show that any force of the form $\mathbf{F}(r) = F(r)\,\hat{\boldsymbol{\theta}}$, where $\hat{\boldsymbol{\theta}}$ is a unit vector perpendicular to the radius vector, is not a conservative force. In plane polar coordinates an infinitesimal displacement is $d\mathbf{s} = dr\hat{\mathbf{r}} + d\theta\hat{\boldsymbol{\theta}}$.

13. (II) A force varies as $\mathbf{F}(x, y) = xy^2\mathbf{i}$. Perform the integration $\int F_x\,dx + \int F_y\,dy$ from O to B in Fig. 8.42, along the following paths: (a) OA then AB; (b) OC then CB. Does this function satisfy the criterion for being conservative?

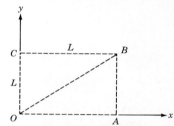

FIGURE 8.42 Problems 13 and 14.

14. (II) Repeat Problem 13 with $\mathbf{F}(x, y) = 2y^2\mathbf{i} + 3x\mathbf{j}$.

15. (I) A vertical spring extends 40 cm when a 1-kg block is suspended from it. It is pulled down a further 10 cm and released. (a) Write an expression for the total potential energy $U(x)$, where x is the extension of the spring. Set $U = 0$ at $x = 0$. (b) What is its speed when it returns to the equilibrium position? (c) What is its maximum height above the equilibrium point? (d) Where would the potential energy of the system be zero? (e) Sketch $U(x)$.

16. (I) *Echo I* was a 100-ft-diameter balloon satellite of mass 75 kg that was placed in an approximately circular orbit 150 km above the earth's surface. (a) What was its orbital speed? (b) What was its mechanical energy? (c) It lost 0.75% of its mechanical energy in the first year. What was the power loss? (d) Estimate the frictional drag force exerted on it by the atmosphere.

CHAPTER 9

Linear Momentum

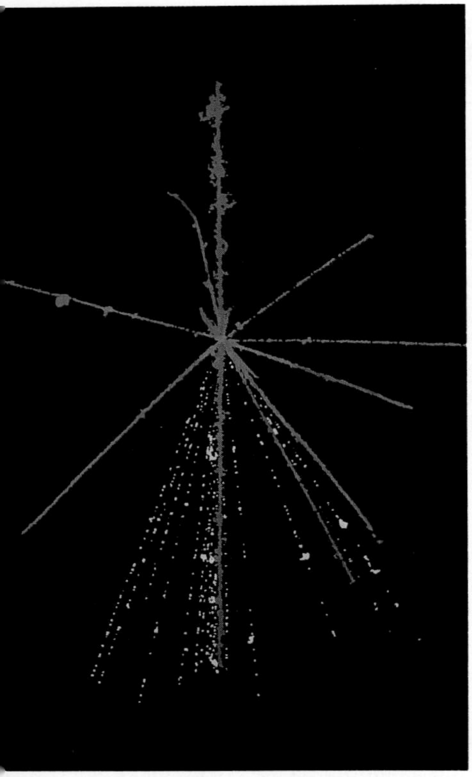

A false-color rendition of tracks in an emulsion records the collision of an incoming particle (red track) with a nucleus. The collision produces a spray of other particles.

Major Points

1. The **conservation of linear momentum.**
2. (a) In an **inelastic** collision, only momentum is conserved.
   (b) In an **elastic** collision, both momentum and kinetic energy are conserved.
3. (a) The definition of **impulse.**
   (b) The use of impulse to find the average force on a body.

9.1 LINEAR MOMENTUM

By the middle of the seventeenth century it was clear that a body free of external influence moves at constant velocity. So what "external influence" causes a change in its velocity? René Descartes (Fig. 9.1) suggested that it was the impact of another body, which itself suffers a change in velocity. Furthermore, these changes were not arbitrary. Descartes was a "mechanical" philosopher: He believed that God had created the world like a perfect and never-changing clockwork mechanism that would always have a fixed amount of "matter" and "motion." For example, in a collision between two particles, the individual speeds would change, but the total "quantity of motion," which he defined as mass $\times$ speed, would remain constant.

Let us consider the collision of two putty balls of equal mass m and speed u that initially travel in opposite directions. The initial quantity of motion is $2mu$. After they collide and come to rest, the final quantity of motion is zero. This quantity of motion is clearly not conserved. Descartes also presented several rules of impact, most of which were incorrect. He asserted, for example, that when a small body strikes a large one, it rebounds with the same speed, leaving the large body unmoved. The assertion is approximately true when a Ping-Pong ball collides with a bowling ball, but it is not *exactly* correct. Clearly, more work was required to formulate the correct "laws of impact."

Although Descartes' analysis of specific examples was weak, he had sown the seed of an extremely important idea in physics: Despite all the complexity of some event or process within an isolated system, there is some physical quantity that does not change. This is called a principle of conservation.

In 1668, the newly formed Royal Society in London requested any scientists who had worked on the problem of collisions to communicate their results. It was soon realized that if the definition of quantity of motion were changed from the

FIGURE 9.1 René Descartes (1596–1650).

scalar mass × speed, to the vector mass × velocity, its net value would be zero both before and after the collision of two identical putty balls. The product of the mass m of a particle and its velocity $\mathbf{v}$ is called its **linear momentum:**

$$\mathbf{p} = m\mathbf{v} \qquad (9.1) \qquad \text{Linear momentum}$$

In 1669, when the results were presented, they were not clear-cut. John Wallis announced that when two isolated bodies collide and stick together, the total linear momentum is conserved. Thus, if $\mathbf{p}_1$ and $\mathbf{p}_2$ are the momenta of the bodies, then

$$\mathbf{p}_1 + \mathbf{p}_2 = \text{constant} \quad \text{or} \quad \Delta\mathbf{p}_1 + \Delta\mathbf{p}_2 = 0 \qquad (9.2)$$

However, the Dutch scientist Christiaan Huygens and Christopher Wren, architect of St. Paul's Cathedral, independently concluded that the quantity mv^2 (twice the kinetic energy) is conserved in collisions between hard balls. Later, Newton conducted careful experiments on collisions between all kinds of substances, such as glass, wood, steel, and putty, and found that the vector $m\mathbf{v}$ was *always* conserved, but that the scalar quantity mv^2 was conserved only in the *special* case of collisions between hard spheres. These experimentally derived conservation laws replaced Descartes' metaphysical notion of a clockwork world. Newton used the result stated in Eq. 9.2 to formulate his second and third laws.

In Newton's own version of his second law, he stated that the "motive force" exerted on a particle is equal to the change in its linear momentum, $\Delta\mathbf{p}$.* However, this definition does not fully correspond to our intuitive understanding of the term "force." We know, for example, that a small force acting for a long time (a person pushing a car) can produce the same change in momentum as a large force acting for a short time (a towtruck pulling the car). In 1752, the mathematician L. Euler modified Newton's definition to include the aspect of time explicitly. A modern statement of Newton's second law is

$$\mathbf{F} = \frac{d\mathbf{p}}{dt} \qquad (9.3)$$

The net force acting on a particle is equal to the time rate of change of its linear momentum.

If the mass of the body is constant, then $\mathbf{F} = d(m\mathbf{v})/dt = m\,d\mathbf{v}/dt = m\mathbf{a}$. Thus, Eq. 9.3 is consistent with Eq. 5.2.

The changes in momentum in Eq. 9.2 are produced by the forces acting on each body. With the notation $\mathbf{F}_{12}$ = force on 1 due to 2, and the second law in the form $\Delta\mathbf{p} = \mathbf{F}\Delta t$, we see that

$$\Delta\mathbf{p}_1 = \mathbf{F}_{12}\Delta t; \qquad\qquad \Delta\mathbf{p}_2 = \mathbf{F}_{21}\Delta t$$

From Eq. 9.2, we can state that $\Delta(\mathbf{p}_1 + \mathbf{p}_2) = 0$ for any Δt; therefore, $(\mathbf{F}_{12} + \mathbf{F}_{21})\Delta t = 0$, or $\mathbf{F}_{12} = -\mathbf{F}_{21}$, which of course is the third law.

9.2   CONSERVATION OF LINEAR MOMENTUM

Figure 9.2 shows a collision between two particles of masses m_1 and m_2 with initial velocities $\mathbf{u}_1$ and $\mathbf{u}_2$ and final velocities $\mathbf{v}_1$ and $\mathbf{v}_2$. When they interact, they may physically touch, as would two billiard balls, or merely repel each other, as would

---

* $\Delta\mathbf{p}$ is now called *impulse*. See Section 9.4.

two electrical charges of the same sign. The initial and final velocities are related according to the principle of the **conservation of linear momentum:**

$$m_1\mathbf{u}_1 + m_2\mathbf{u}_2 = m_1\mathbf{v}_1 + m_2\mathbf{v}_2 \qquad (9.4)$$

Since this is a *vector* equation, *each* component of momentum is independently conserved:

$$m_1u_{1x} + m_2u_{2x} = m_1v_{1x} + m_2v_{2x}$$
$$m_1u_{1y} + m_2u_{2y} = m_1v_{1y} + m_2v_{2y}$$
$$m_1u_{1z} + m_2u_{2z} = m_1v_{1z} + m_2v_{2z} \qquad (9.5)$$

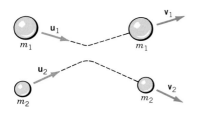

FIGURE 9.2 A collision between two particles. The initial velocities are $\mathbf{u}_1$ and $\mathbf{u}_2$; the final velocities are $\mathbf{v}_1$ and $\mathbf{v}_2$.

In order for us to apply the conservation of linear momentum, there must be no net external force acting on the system. To see this, suppose that external forces $\mathbf{F}_{1e}$ and $\mathbf{F}_{2e}$ also act on m_1 and m_2, respectively. From Eq. 9.3, we have $(\mathbf{F}_{1e} + \mathbf{F}_{12}) = d\mathbf{p}_1/dt$ for m_1, and $(\mathbf{F}_{2e} + \mathbf{F}_{21}) = d\mathbf{p}_2/dt$ for m_2. When we add these equations and use Newton's third law in the form $\mathbf{F}_{12} + \mathbf{F}_{21} = 0$, we find

$$(\mathbf{F}_{1e} + \mathbf{F}_{2e}) = \frac{d(\mathbf{p}_1 + \mathbf{p}_2)}{dt}$$

This result can be extended to a system with any number of particles. The internal forces between the particles in the system cancel in pairs and Newton's second law takes the form

Newton's second law for a system of particles

$$\mathbf{F}_{\text{EXT}} = \frac{d\mathbf{P}}{dt} \qquad (9.6)$$

where $\mathbf{F}_{\text{EXT}} = \Sigma\mathbf{F}_{ie}$ is the net external force on the system and $\mathbf{P} = \Sigma\mathbf{p}_i$ is the total linear momentum of the particles.* From Eq. 9.6 we infer the condition for the applicability of the conservation of linear momentum:

$$\text{If } \mathbf{F}_{\text{EXT}} = 0, \quad \text{then} \quad \mathbf{P} = \Sigma\mathbf{p}_i = \text{constant}$$

If the net external force on a system is zero, the total linear momentum is constant.

If a net external force acts in the x and z directions, say, but not in the y direction, then the y component of linear momentum is still conserved.

The principle of the conservation of linear momentum is remarkably simple and general. It is valid for all types of interaction and may be applied to such diverse phenomena as collisions, explosions, radioactive decay, nuclear reactions, and the emission and absorption of light. It also helps us analyze such commonplace phenomena as the kick of a rifle and rocket propulsion.

The conservation of linear momentum may be applied as a first approximation even in situations in which the net external force is not zero. This is allowable if the internal forces, such as those in an explosion or in a collision, are much greater than the external force, for example, the force of gravity. If the duration of the event is short, the external force does not act long enough to significantly alter the total momentum of the system. The momenta of the particles just after the event are determined primarily by the internal forces.

---

* To convince yourself that internal forces cannot affect the momentum of a system as a whole, try a little experiment: Bend over and pull hard on your shoelaces. If you lift yourself off the floor, you have a whole new career ahead of you.

Types of Collision

Before we begin to apply the conservation of linear momentum, we first clarify what is meant by a collision and distinguish between two types of collision. The term "collision" is usually applied to a brief, strong interaction between two bodies. The duration of the interaction is short enough to allow us to limit our discussion to times both before and after the event. However, the duration of a collision depends on the time scale of our interest. A collision between elementary particles may last 10^{-23} s, whereas a collision between galaxies takes millions of years. Collisions involving objects such as balls and cars last from about 10^{-3} s to 1 s.

Collisions may be either **elastic** or **inelastic.** Linear momentum is conserved in both cases. A *perfectly elastic* collision is defined as one in which the total kinetic energy of the particles is also conserved:

(Elastic) $$\tfrac{1}{2}m_1u_1^2 + \tfrac{1}{2}m_2u_2^2 = \tfrac{1}{2}m_1v_1^2 + \tfrac{1}{2}m_2v_2^2 \qquad (9.7)$$ Elastic collision

The word "perfectly" is often omitted. Notice that this is a scalar equation. During an elastic collision the kinetic energy of the particles is wholly or partly stored as potential energy and then completely recovered as kinetic energy. Collisions between hard steel balls come close to being elastic. In atomic and nuclear systems, elastic collisions are quite common.

In an *inelastic* collision, the total kinetic energy of the particles changes. Some of the kinetic energy is stored as potential energy associated with a change in internal structure or state, and is not immediately recovered. Some of the kinetic energy may be used to raise the system (e.g., an atom) to a state with higher energy. Or, it may be converted into thermal energy of vibrating atoms and molecules or into light, sound, or some other form of energy. (The *total* energy, which includes *all* forms, is always conserved.) In a *completely* inelastic collision, the two bodies couple or stick together. You may come across the term superelastic collision. This refers to the possibility that the total kinetic energy actually increases as a result of the collision. This may occur because a compressed spring, or an explosive charge, is triggered and releases stored energy.   Inelastic collision

EXERCISE 1. (a) You can easily demonstrate that a collision between two wooden balls is not perfectly elastic. How? (b) How could you determine whether the collision between a "superball" and a floor is elastic?

EXERCISE 2. Express the kinetic energy of a particle in terms of its mass m and the magnitude of its linear momentum p.

Problem-Solving Guide for Conservation of Linear Momentum

---

1. (a) Make a sketch that includes the directions of all velocities both before and after the event.
   (b) Set up coordinate axes.
2. (a) Momentum is a *vector*. You must write the conservation law for each *component*.
   (b) Kinetic energy is a *scalar*. It is conserved only in elastic collisions.
3. The signs in the component equations should be consistent with the directions of the velocities drawn in your sketch. An unknown velocity is written as $\mathbf{v} = v_x\mathbf{i} + v_y\mathbf{j}$. The signs of v_x and v_y will come out of the solution.

EXAMPLE 9.1: A 2000-kg Cadillac limousine moving east at 10 m/s collides with a 1000-kg Honda Prelude moving west at 26 m/s. The collision is completely inelastic and takes place on an icy (frictionless) patch of road. (a) Find their common velocity after the collision. (b) What is the fractional loss in kinetic energy?

Solution: The sketch and axes are drawn in Fig. 9.3, where the label "1" refers to the Cadillac. We assume the unknown common velocity $\mathbf{V}$ is in the $+x$ direction, that is, $\mathbf{V} = +V\mathbf{i}$. The vector and component forms of the conservation law are

$\Sigma\mathbf{p}$: $m_1\mathbf{u}_1 + m_2\mathbf{u}_2 = (m_1 + m_2)\mathbf{V}$
Σp_x: $m_1 u_1 - m_2 u_2 = (m_1 + m_2)V$

With the given values we find $V = -2$ m/s, which means $\mathbf{V} = -2\mathbf{i}$ m/s.
(b) The initial and final kinetic energies are

$$K_i = \tfrac{1}{2}m_1 u_1^2 + \tfrac{1}{2}m_2 u_2^2 = 4.38 \times 10^5 \text{ J}$$
$$K_f = \tfrac{1}{2}(m_1 + m_2)V^2 = 6000 \text{ J}$$

The fractional *change* in kinetic energy is

$$\frac{\Delta K}{K_i} = \frac{K_f - K_i}{K_i} = -0.99$$

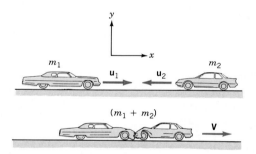

FIGURE 9.3 A completely inelastic collision. The unknown final velocity $\mathbf{V}$ is assumed to be in the $+x$ direction.

EXAMPLE 9.2: A 3.24-kg Winchester Super X rifle, initially at rest, fires a 11.7-g bullet with a muzzle speed of 800 m/s. (a) What is the recoil velocity of the rifle? (b) What is the ratio of the kinetic energies of the bullet and the rifle?

Solution: The kick of a rifle when it is fired is a consequence of the conservation of linear momentum. Strictly speaking, the rifle and the bullet are not isolated since the rifle butt rests against a shoulder and the person is likely to be on firm ground. However, one can estimate what would happen if the rifle were held lightly and away from the shoulder. (A dangerous procedure!) The initial momentum of the system is zero, so from Fig. 9.4 we have

$\Sigma\mathbf{p}$: $0 = m_1\mathbf{v}_1 + m_2\mathbf{v}_2$
Σp_x: $0 = -m_1 v_1 + m_2 v_2$

Using the given values we find

$$v_2 = \frac{m_1}{m_2}v_1 = \frac{11.7 \times 10^{-3} \text{ kg}}{3.24 \text{ kg}}(800 \text{ m/s})$$

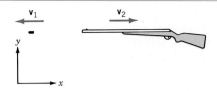

FIGURE 9.4 When a rifle at rest fires a bullet, its recoil momentum is equal and opposite to the momentum of the bullet.

Therefore, $\mathbf{v}_2 = 2.89\mathbf{i}$ m/s.
(b) It is convenient to express the kinetic energy in terms of linear momentum. Since $p = mv$ and $K = \tfrac{1}{2}mv^2$, we have $K = p^2/2m$. Since $p_1 = m_1 v_1$, and $p_2 = m_2 v_2$, and $p_1 = p_2$, the ratio of the kinetic energies is

$$\frac{K_2}{K_1} = \frac{m_1}{m_2}$$

This is an interesting result. It shows that even though the two bodies have equal and opposite momenta, the kinetic energy each carries away is inversely proportional to its mass: $K \propto 1/m$. In the present case, $K_2/K_1 = (11.7 \times 10^{-3} \text{ kg})/(3.24 \text{ kg}) = 3.8 \times 10^{-3}$. The kinetic energy of the rifle is only 0.38% of the kinetic energy of the bullet.

Example 9.2 covers the essential physics of rocket propulsion. A rocket is like a machine gun that fires bullets in rapid succession. The "bullets" are actually gas molecules ejected at high speed relative to the rocket. Thus, the principle governing rocket propulsion is the conservation of linear momentum. The internal forces between the rocket and the gas particles cannot change the total momentum of the rocket–gas system. However, from Newton's third law we know that as the rocket pushes the gas particles backward, their reaction pushes the rocket forward.

EXAMPLE 9.3: A puck of mass $m_1 = 3$ kg has an initial velocity of 10 m/s at 20° S of E. A second puck of mass $m_2 = 5$ kg has a velocity of 5 m/s at 40° W of N. They collide and stick together. Find their common velocity after the collision.

Solution: A sketch with the axes is shown in Fig. 9.5. A most serious error in this kind of problem would be to treat momen-

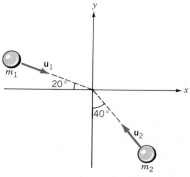

FIGURE 9.5 In a two-dimensional collision, each component of momentum is conserved.

HISTORICAL NOTE:
Robert Goddard
and Early Rocketry

In the early 1920s, the American physicist and space pioneer Robert H. Goddard (1882–1945) of Clark University was developing his ideas on rocket propulsion (see Fig. 9.6a). In a 1919 article, he suggested that a rocket could travel in the vacuum of space, even as far as the moon. An editorial in the *New York Times* on January 13, 1920, scoffed: "That Professor Goddard, with his 'chair' in Clark College and the countenancing of the Smithsonian Institution, does not know the relation of action to reaction, and the need to have something better than a vacuum against which to react—to say that would be absurd. Of course, he only seems to lack the knowledge ladled out daily in high schools." The popular press had a field day, lampooning him as the "Moon man." To counter such arguments, Goddard attached a .22-caliber pistol to a spindle free to rotate inside a bell jar from which the air had been evacuated (see

Fig. 9.6b). When a blank cartridge was fired, the gun recoiled in the direction opposite to that of the escaping gases. The analogy to the proposed rocket was obvious.

On March 16, 1926, he successfully launched the first liquid-fueled rocket (using liquid oxygen and gasoline). It fired for all of 2.5 s and had an average speed of 96 km/h. It rose to a height of 12.5 m and landed 56 m away in a cabbage patch. On July 17, 1969, when Neil Armstrong, Edwin Aldrin, and Michael Collins took off for the first moon landing, the *Times* published the following tongue-in-cheek retraction: "Further investigation and experimentation have confirmed the findings of Isaac Newton in the 17th century and it is now definitely established that a rocket can function in a vacuum as well as in an atmosphere. The *Times* regrets the error."

FIGURE 9.6 (a) Robert H. Goddard (1882–1945) with his first rocket. (b) By firing a blank cartridge from a pistol in an evacuated jar, Goddard showed that a rocket would work in empty space.

tum as a scalar. In two dimensions there are two independent equations in the components.

$\Sigma \mathbf{p}$:
$$m_1\mathbf{u}_1 + m_2\mathbf{u}_2 = (m_1 + m_2)\mathbf{V}$$

Σp_x:
$$m_1 u_1 \cos 20° - m_2 u_2 \sin 40° = (m_1 + m_2)V_x$$

Σp_y:
$$-m_1 u_1 \sin 20° + m_2 u_2 \cos 40° = (m_1 + m_2)V_y$$

On inserting the given values we find $\mathbf{V} = 1.52\mathbf{i} + 1.11\mathbf{j}$ m/s. Notice that we represented the components of the unknown velocity as (V_x, V_y) rather than $(V \cos \theta, V \sin \theta)$. This reduced

the number of unknowns in each component equation from two to one and saved us the trouble of solving the equations simultaneously. V and θ are easily found from the rectangular components. (This technique offers no great advantage in elastic collisions.)

EXAMPLE 9.4: According to an accident report, a Chevrolet Chevette of mass $m_1 = 950$ kg was moving east and a Ford LTD of mass $m_2 = 1350$ kg was moving north as shown in Fig. 9.7.

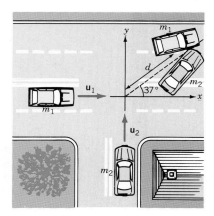

FIGURE 9.7 A completely inelastic collision between two cars. It is assumed that during the impact the force exerted by the cars on each other is much greater than the force due to the road. This allows the application of the conservation of linear momentum.

Although the brakes were applied, the cars collided and joined together. The skid marks after the collision were straight and 6 m long, in a direction 37° north of east. The coefficient of kinetic friction was estimated to be 0.6. Was either car exceeding the 15-m/s speed limit?

Solution: We have to work backward from the given information to find the speeds of the cars just before the collision. First, we use the skid marks and the friction information to find the common speed of the cars just after they collided. From Newton's second law we have $f_k = ma$, where $f_k = \mu_k N$. Thus,

$$\mu_k(m_1 + m_2)g = (m_1 + m_2)a$$

from which we find $a = \mu_k g$. From $v^2 = v_0^2 + 2a\Delta x$, we have

$$0 = v_0^2 + 2(-\mu_k g)d$$

where v_0 is the initial speed in the direction of the skid and d is the length of the skid. The negative sign is needed because v_0 and a are in opposite directions. We infer that

$$v_0 = (2\mu_k g d)^{1/2} = 8.5 \text{ m/s} \tag{i}$$

This is the common speed just after the collision.

Next we apply the conservation of linear momentum to the collision. This assumes that the (internal) forces between the cars during the collision are much larger than the (external) frictional force due to the road. The vector and component forms of the conservation of linear momentum are

$\Sigma \mathbf{p}$: $m_1\mathbf{u}_1 + m_2\mathbf{u}_2 = (m_1 + m_2)\mathbf{v}_0$

Σp_x: $m_1 u_1 + 0 = (m_1 + m_2)v_0 \cos\theta \tag{ii}$

Σp_y: $0 + m_2 u_2 = (m_1 + m_2)v_0 \sin\theta \tag{iii}$

Using (i) in (ii) and (iii) we find $u_1 = 16.4$ m/s and $u_2 = 8.7$ m/s. Clearly, the Chevette was speeding. Since the brakes were ap-

plied for some time before the collision, this car was traveling even faster than the above calculation indicates.

EXAMPLE 9.5: In 1742, Benjamin Robins devised a simple, yet ingenious, device, called a *ballistic pendulum,* for measuring the speed of a bullet. Suppose that a bullet, of mass $m = 10$ g and speed u, is fired into a block of mass $M = 2$ kg suspended as in Fig. 9.8. The bullet embeds in the block and raises it by a height $H = 5$ cm. (a) How can one determine u from H? (b) What is the thermal energy generated?

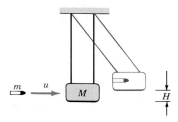

FIGURE 9.8 A ballistic pendulum. The height to which the block rises may be used to determine the speed of the bullet. This requires the application of two conservation laws.

Solution: If the collision occurs within a very short time interval, the ropes will remain essentially vertical as the bullet comes to rest. Thus, there will be no external horizontal force. We may apply the conservation of linear momentum along this direction:

$$mu = (m + M)V \tag{i}$$

where V is the common velocity after the collision. Since the collision is completely inelastic, part of the bullet's initial kinetic energy, $\frac{1}{2}mu^2$, is converted into thermal energy. Only the kinetic energy of the bullet–block system that remains *after* the collision is available to raise the system by height H. From conservation of mechanical energy we have

$$\tfrac{1}{2}(m + M)V^2 = (m + M)gH \tag{ii}$$

Hence, $V = (2gH)^{1/2}$. Substituting this into (i), we find

$$u = \frac{(m + M)\sqrt{2gH}}{m}$$

With the given values, we find $u \approx 200$ m/s.
(b) The kinetic energies before and after the collision are

$$K_i = \tfrac{1}{2}mu^2 = 200 \text{ J}$$
$$K_f = \tfrac{1}{2}(m + M)V^2 = 1 \text{ J}$$

The change in kinetic energy as a result of the collision is -199 J. Virtually all the bullet's kinetic energy is converted into thermal energy.

EXERCISE 3. Estimate the force of friction on the bullet assuming that it stops in 4 cm.

9.3   ELASTIC COLLISIONS IN ONE DIMENSION

Figure 9.9 depicts a one-dimensional elastic collision between two balls. All the velocities are drawn in the same direction to simplify the analysis. The quantities u and v are the components; that is, $\mathbf{u} = u\mathbf{i}$ and $\mathbf{v} = v\mathbf{i}$. Of course, in practice we would need $u_1 > u_2$, or $u_2 < 0$, for a collision to occur. The x component of the linear momentum and the kinetic energy are both conserved. We rewrite the component form of Eq. 9.4 and Eq. 9.7 as

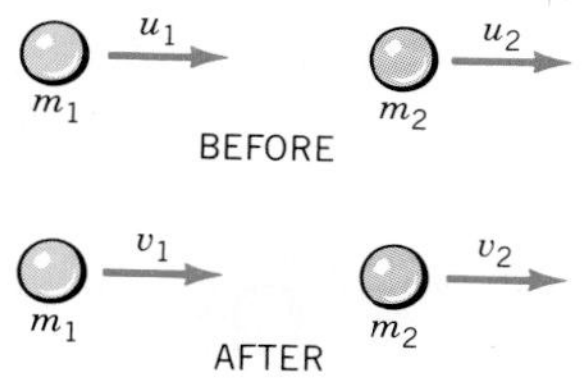

Σp_x: $m_1(u_1 - v_1) = m_2(v_2 - u_2)$ (9.8)

ΣK: $m_1(u_1^2 - v_1^2) = m_2(v_2^2 - u_2^2)$ (9.9a)

Next we use the identity $a^2 - b^2 = (a - b)(a + b)$ to rewrite Eq. 9.9a in the form

$$m_1(u_1 - v_1)(u_1 + v_1) = m_2(v_2 - u_2)(v_2 + u_2)   (9.9b)$$

When Eq. 9.9b is divided by Eq. 9.8, we find $(u_1 + v_1) = (u_2 + v_2)$, which is equivalent to

(Elastic) $(v_2 - v_1) = -(u_2 - u_1)$ (9.10)

In a one-dimensional elastic collision, the relative velocity of the particles is unchanged in magnitude but is reversed in direction.

To appreciate the physical meaning of Eq. 9.10 you might imagine yourself moving with one of the particles. According to this equation, the other particle will approach and later recede from you, with exactly the same speed.

Note that while Eq. 9.10 will not by itself provide a complete solution to most problems, it always serves as a check on any solution. In fact, when dealing with elastic collisions in one dimension, it is easier to use the conservation of momentum and the derived result Eq. 9.10, instead of the conservation of kinetic energy. This approach avoids the squared terms.

EXAMPLE 9.6: A block of mass $m_1 = 2$ kg and initial velocity $\mathbf{u}_1 = 4\mathbf{i}$ m/s makes a one-dimensional elastic collision with a block of mass $m_2 = 3$ kg moving at $\mathbf{u}_2 = 2\mathbf{i}$ m/s. Find their final velocities.

Solution: The x component of the initial momentum is $(2 \text{ kg} \times 4 \text{ m/s}) + (3 \text{ kg} \times 2 \text{ m/s}) = 14 \text{ kg} \cdot \text{m/s}$. The component form of the conservation of linear momentum states

$$2v_1 + 3v_2 = 14   \text{(i)}$$

The initial velocity of m_2 relative to m_1 is

$$(u_2 - u_1) = -2 \text{ m/s}.$$

From Eq. 9.10, we have

$$v_2 - v_1 = 2   \text{(ii)}$$

Substituting $v_2 = v_1 + 2$ from (ii) into (i), we first find $v_1 = 1.6$ m/s, and then $v_2 = 3.6$ m/s.

EXERCISE 4. Confirm that kinetic energy is conserved in Example 9.6.

Let us consider two special cases of collisions: First, when the particles have equal mass, and second, when one of them, say m_2, is initially at rest.

(i) Equal Masses: $m_1 = m_2 = m$

Equation 9.4 takes the form $u_1 + u_2 = v_1 + v_2$, and from Eq. 9.10 we have $u_1 - u_2 = -v_1 + v_2$. These are easily solved:

$$v_1 = u_2 \qquad v_2 = u_1$$

FIGURE 9.9 An elastic collision in one dimension. To simplify the analysis, all the velocities have been drawn in the same direction.

The velocities are *exchanged,* as shown in Fig. 9.10a. For example, if m_2 is initially at rest ($u_2 = 0$), then $v_1 = 0$ and $v_2 = u_1$. That is, m_1 comes to a stop and m_2 moves off with the initial velocity of m_1. This is often seen in billiards.

(ii) **Unequal masses $m_1 \neq m_2$. Target at Rest:** $u_2 = 0$

Equation 9.4 becomes $m_1 u_1 = m_1 v_1 + m_2 v_2$, and Eq. 9.10 is $u_1 = -v_1 + v_2$. One can use these two equations to express the final velocities in terms of u_1 (see Exercise 5 below):

$$v_1 = \frac{(m_1 - m_2)u_1}{m_1 + m_2} \tag{9.11}$$

$$v_2 = \frac{2m_1 u_1}{m_1 + m_2} \tag{9.12}$$

(a) When $m_1 \gg m_2$, we may ignore the mass of m_2 in comparison with m_1. This leads to $v_1 \approx u_1$ and $v_2 \approx 2u_1$, which means that m_1 maintains its initial velocity u_1 but it imparts *double* this value to m_2 (see Fig. 9.10b). A common example is a golf club hitting a golf ball.

(b) When $m_1 \ll m_2$ we may ignore m_1 in comparison with m_2. We then find that $v_1 \approx -u_1$ and $v_2 \approx 0$. Thus, m_1 reverses its velocity, leaving m_2 essentially unmoved as in Fig. 9.10c. A Ping-Pong ball colliding with a bowling ball illustrates this case.

EXERCISE 5. Derive the expressions in Eqs. 9.11 and 9.12.

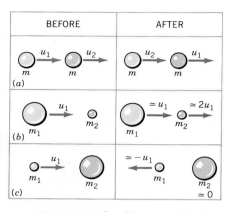

BEFORE	AFTER

FIGURE 9.10 Special cases of elastic collisions in one dimension. (*a*) When the masses are equal, the particles exchange velocities. (*b*) With $m_1 \gg m_2$ and $u_2 = 0$, the final velocity of m_2 is about double the initial velocity of m_1. (*c*) With $m_1 \ll m_2$ and $u_2 = 0$, m_1 bounces back with its initial speed.

EXAMPLE 9.7: A particle of mass m_1 makes a one-dimensional elastic collision with a particle of mass m_2 initially at rest. What fraction of the initial kinetic energy is transferred to the target particle?

Solution: The initial and final kinetic energies of m_1 are $K_{1i} = \frac{1}{2}m_1 u_1^2$ and $K_{1f} = \frac{1}{2}m_1 v_1^2$. Since $u_2 = 0$, we can use Eq. 9.11 to find the fraction f_1 of the initial kinetic energy retained by m_1:

$$f_1 = \frac{K_{1f}}{K_{1i}} = \frac{v_1^2}{u_1^2}$$

$$= \frac{(m_1 - m_2)^2}{(m_1 + m_2)^2}$$

We see immediately that $K_{1f} = 0$ when $m_1 = m_2$, which means that all of the initial kinetic energy of m_1 is transferred to m_2.

The result that the maximum transfer of kinetic energy occurs when the particles have the same mass has important applications in the control of nuclear reactors. After the fission (splitting) of the nucleus of ^{235}U into smaller fragments, neutrons are emitted with speeds of about 10^7 m/s. However, slow neutrons (10^3 m/s) are more likely to be captured and to initiate fission in other nuclei, thereby creating a chain reaction. The above discussion shows that particles whose mass is close to that of the neutrons will be most effective in slowing them.

The most obvious candidate for a "moderator" is water, since it has a large supply of protons. Unfortunately, the neutrons and protons tend to combine to form deuterons: $n + p \rightarrow D$. For this reason many reactors use "heavy water" (D_2O). The neutrons are effectively slowed down without being captured. Another frequently used moderator is carbon. Graphite rods are not as effective as heavy water but offer the advantage of operation at higher temperatures with low probability of neutron capture. The effectiveness of any moderator is less than that implied by the above equation because not all collisions are one dimensional.

EXERCISE 6. Obtain an expression for the fraction f_2 of energy transferred from m_1 to m_2. (*Hint:* Use f_1 found above.) Evaluate f_2 for (a) $m_2 = 0.5m_1$, (b) $m_2 = m_1$, (c) $m_2 = 2m_1$.

9.4 IMPULSE

The **impulse I** experienced by a particle is defined as the change in its linear momentum:

Impulse $$\mathbf{I} = \Delta \mathbf{p} = \mathbf{p}_f - \mathbf{p}_i \tag{9.13}$$

Impulse is a vector quantity with the same unit as linear momentum (kg · m/s). Its direction is that of the *change* in momentum. We may relate impulse to the net force acting on the particle with Newton's second law in the form $\mathbf{F} = d\mathbf{p}/dt$. Since $\Delta\mathbf{p} = \int d\mathbf{p} = \int \mathbf{F}\,dt$, we have

$$\mathbf{I} = \int_{t_i}^{t_f} \mathbf{F}\,dt \qquad (9\text{-}14)$$

This equation is valid for any time interval $\Delta t = t_f - t_i$, but it is most often applied to so-called impulsive forces. Such forces start to act at a specific time t_i, rise in value in some unknown fashion, and stop abruptly at t_f. Such a variation is depicted in Fig. 9.11, although in practice there may be more than one peak. Impulsive forces act for a very short time interval and are very large compared to other forces that may be acting. For example, while a tennis ball is being struck by a racket, as in Fig. 9.12, the fact that it is also subject to gravity and air resistance is not significant. The change in momentum of the ball is determined almost exclusively by the impulsive force due to the racket. Thus, even though $\mathbf{F}$ in Eq. 9.14 should be the net force on the ball, we make little error by considering just the force exerted by the racket. This is sometimes called the impulse approximation.

In Fig. 9.11 we may interpret the impulse as the area under the curve. We usually have little information on how an impulsive force varies in time. Thus, it is convenient to define the *average* force acting on the particle by (see Eq. 9.3)

$$\mathbf{I} = \Delta\mathbf{p} = \mathbf{F}_{av}\Delta t \qquad (9.15)$$

This is nothing but Newton's second law in disguise. In effect we replace the true variation of the force by a constant value that has the same area in the given time interval. This is the rectangle in Fig. 9.11.

Figure 9.13 illustrates another point: A given change in momentum may be produced by a large force acting for a short time or by a small force acting for a long time. If you have to stop something, such as a ball coming at you, it is better to take as long as possible to do this: Instead of holding your arms stiffly, you should let them "give" as the ball makes contact. The same applies to a fall: You can reduce the chance of injury by prolonging the fall by flexing the knees or by rolling.

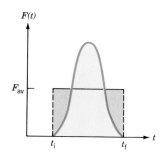

FIGURE 9.11 The area under the F versus t curve indicates the impulse experienced by the particle. The average force is defined such that the area under the rectangle is equal to the area under the true function.

FIGURE 9.12 A tennis ball is subject to an impulsive force from the racket.

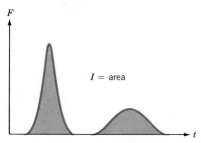

FIGURE 9.13 The impulse due to a large force acting for a short time can be produced by a weaker force acting for a longer time.

EXAMPLE 9.8: A 150-g ball is thrown at 30 m/s. It is struck by a bat which gives it a velocity of 40 m/s in the opposite direction. If the time of contact is 10^{-2} s, what is the average force on the ball?

Solution: If we choose the original direction as the $+x$ axis, then

$$\Delta\mathbf{p} = m\mathbf{v}_f - m\mathbf{v}_i = m(-40\mathbf{i} - 30\mathbf{i})$$

The average force is

$$\mathbf{F}_{av} = \frac{\Delta\mathbf{p}}{\Delta t}$$

$$= \frac{-0.15 \text{ kg} \times 70\mathbf{i} \text{ m/s}}{10^{-2} \text{ s}} = -1050\mathbf{i} \text{ N}$$

Notice that this is much larger than the weight (1.5 N) of the ball.

9.5   COMPARISON OF LINEAR MOMENTUM WITH KINETIC ENERGY

Both linear momentum and kinetic energy are functions of mass and velocity and so one might wonder why two such functions are needed. Let us highlight some of the differences between momentum and kinetic energy.

(i) Both mv and mv^2 first arose as conserved quantities in the study of collisions in the late seventeenth century. However, the conservation of linear momentum is a *generally* valid law, whereas conservation of kinetic energy is true only in the *special* case of elastic collisions.

(ii) Momentum is a vector, whereas kinetic energy is a scalar. This distinction is forced upon us. If momentum is treated as a scalar or we try to take components of kinetic energy, the analysis of collisions and other phenomena makes no sense.

(iii) Both linear momentum and kinetic energy are related to the "effort" or "exertion" needed to change the velocity of a particle. The change in momentum is the *impulse* on the body $\Delta p = F\Delta t$, whereas the change in kinetic energy is the *work* done on it $\Delta K = F\Delta x$. We infer that

$$F = \frac{\Delta p}{\Delta t} \qquad\qquad F = \frac{\Delta K}{\Delta x}$$

Force is either the rate of change of momentum with respect to *time*, or the rate of change of kinetic energy with respect to *position*. If the force is not constant, these two expressions lead to different values of the "average" force. One would be an average over time, while the other would be an average over space.

EXERCISE 7. (a) Can one change the linear momentum of a particle without changing its kinetic energy? (b) Can one change the kinetic energy of a particle without changing its linear momentum? If so, give examples.

9.6   ELASTIC COLLISIONS IN TWO DIMENSIONS (Optional)

We now present an example of an elastic collision between two particles that is not head-on. We consider a situation that arises frequently in nuclear and high-energy physics in which one particle is initially at rest ($u_2 = 0$), as in Fig. 9.14. We need to apply the conservation of momentum only in two dimensions.*

After the collision the particles move off at angles θ_1 and θ_2 to the original direction of $\mathbf{u}_1$. The conservation of two components of momentum and the conservation of kinetic energy yield

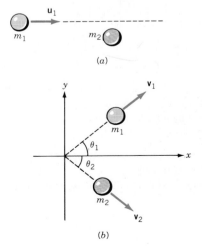

(a)

(b)

FIGURE 9.14 An elastic collision in two dimensions with m_2 initially at rest.

Σp_x: $m_1 u_1 = m_1 v_1 \cos \theta_1 + m_2 v_2 \cos \theta_2$

Σp_y: $0 = m_1 v_1 \sin \theta_1 - m_2 v_2 \sin \theta_2$

ΣK: $\frac{1}{2}m_1 u_1^2 = \frac{1}{2}m_1 v_1^2 + \frac{1}{2}m_2 v_2^2$

There are four unknowns in these three equations: v_1, v_2, θ_1, and θ_2. Often, either θ_1 or θ_2 is measured.

* Suppose the initial momentum is along the x axis. The lines of motion of the particles after the collision define a plane which we call the xy plane.

EXAMPLE 9.9: A proton moving at speed $u_1 = 5$ km/s makes an elastic collision with another proton initially at rest. Given that $\theta_1 = 37°$, find v_1, v_2, and θ_2.

Solution: A sketch and our axes are shown in Fig. 9.14. With the given values, the equations written above become

$$5 = 0.8v_1 + v_2 \cos \theta_2 \tag{i}$$
$$0 = 0.6v_1 - v_2 \sin \theta_2 \tag{ii}$$
$$25 = v_1^2 + v_2^2 \tag{iii}$$

We isolate the trigonometric functions in (i) and (ii): $v_2 \cos \theta_2 = 5 - 0.8v_1$ and $v_2 \sin \theta_2 = 0.6 v_1$. When we square both sides of these equations and add we find

$$v_2^2 = 25 - 8v_1 + v_1^2 \tag{iv}$$

Using this in (iii) leads to $v_1 = 4$ m/s and from this we find $v_2 = 3$ m/s and $\theta_2 = 53°$. Since $\theta_1 + \theta_2 = 90°$, the final velocities are perpendicular. Figure 9.15 illustrates two such collisions between particles of equal mass.

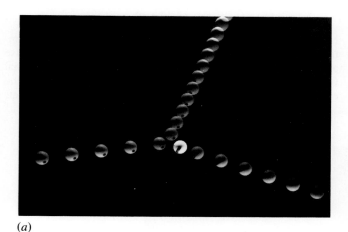

(a)

(b)

FIGURE 9.15 In an elastic collision between two particles of equal mass, one initially at rest, the final velocities are perpendicular. (a) A collision between two pucks. (b) A collision between two nuclear particles.

EXERCISE 8. Use the expression $K = p^2/2m$ and the vector form of the conservation of linear momentum to prove that in an elastic collision between two equal masses, one initially at rest, the final velocities are perpendicular.

9.7 ROCKET PROPULSION (Optional)

The essential physics of rocket propulsion was covered in Example 9.3 concerning the recoil of a rifle. We now consider specifically how the speed of the rocket varies as the burnt fuel is expelled. When dealing with rockets we are usually given the velocity of the exhaust gases relative to the rocket; that is, $\mathbf{v}_{ex} = \mathbf{v}_{GR}$. If the rocket is moving at velocity $\mathbf{v}_{RE}$ relative to the earth (say), then the velocity of the gases relative to the earth is $\mathbf{v}_{GE} = \mathbf{v}_{GR} + \mathbf{v}_{RE}$.

Consider a rocket of mass M with fuel of mass Δm. Their common velocity is $\mathbf{v}$ relative to some inertial frame; see Fig. 9.16a. When the rocket engines are fired, the gases are expelled backward with an exhaust velocity $\mathbf{v}_{ex} = -v_{ex}\mathbf{i}$ *relative to the rocket*. This is a fixed quantity determined by the design of the engine and the type of fuel. If the rocket's velocity changes to $\mathbf{v} + \Delta\mathbf{v}$ relative to the inertial frame, the velocity of the exhaust gas relative to this frame will be $\mathbf{v}_{gas} = \mathbf{v}_{ex} + \mathbf{v} + \Delta\mathbf{v} = (-v_{ex} + v + \Delta v)\mathbf{i}$, as shown in Fig. 9.16b. The component form of the law of conservation of momentum becomes

$$\Sigma p_x: \quad (M + \Delta m)v = M(v + \Delta v) + \Delta m(-v_{ex} + v + \Delta v)$$

After some cancellation we find

$$0 = M\Delta v + \Delta m(-v_{ex} + \Delta v)$$

If both Δv and Δm are small quantites relative to v and M, respectively, their product, $\Delta v \Delta m$, is negligible in comparison with the other terms, and we are left with

$$\Delta v = v_{ex} \frac{\Delta m}{M} \tag{i}$$

FIGURE 9.16 (a) A rocket of mass M with fuel Δm moving at velocity $\mathbf{v}$ relative to an inertial frame. (b) After the fuel is expelled at velocity $\mathbf{v}_{ex}$ relative to the rocket, the velocity of the rocket is $\mathbf{v} + \Delta\mathbf{v}$ relative to the inertial frame.

To proceed further, we must reduce the number of variables. Since an increase in the mass of the expelled gases corresponds exactly to the loss in mass of the rocket system, we have $\Delta m = -\Delta M$. In the limit as $\Delta M \to 0$, (i) becomes $dv = -v_{ex}dM/M$. On integrating both sides

$$\int_{v_i}^{v_f} dv = -\int_{M_i}^{M_f} v_{ex} \frac{dM}{M}$$

we find

$$v_f - v_i = v_{ex} \ln \frac{M_i}{M_f} \qquad (9.16)$$

We see that the change in the velocity of the rocket is directly proportional to the exhaust speed. Since the change in velocity depends on the ratio of M_i = mass of rocket + fuel to M_f = mass of rocket alone, the mass of the fuel should be as large as possible. Through the use of multiple stages the mass of the rocket is reduced as each empty fuel tank is discarded. When the fuel is expelled as a continuous stream, as it is here, the final velocity is independent of the rate at which mass is expelled.*

It is interesting to note that a rocket gains more kinetic energy from a given quantity of fuel when the rocket is traveling at high speed than when it is moving at low speed. The combustion of the fuel releases a certain amount of energy that is shared between the rocket and the exhaust gases. When the rocket is moving slowly, the gases travel at a speed close to v_{ex} relative to an inertial frame and take a large fraction of the energy available. When the rocket is traveling at high speed, the gases move at lower speed relative to the inertial frame, and the rocket gains a larger fraction of the energy. When the speed of the rocket reaches v_{ex}, the expelled gases are brought to rest relative to the inertial frame. At this point, the rocket takes *all* the energy.

* This is not so if the mass is expelled in discrete units. See Problem 2.

EXAMPLE 9.10: The mass of a rocket without fuel is $M_R = 10^4$ kg and the exhaust speed of the gases is $v_{ex} = 10^3$ m/s relative to the rocket. If the rocket starts at rest with respect to the earth, what mass of fuel M_F is needed to reach the exhaust speed?

Solution: If we set $v_i = 0$ and drop the subscript on v_f for simplicity, Eq. 9.18 becomes

$$\frac{v}{v_{ex}} = \ln \frac{M_i}{M_f} \qquad (i)$$

It is convenient to use the relation between the exponential and the natural logarithm: By definition, if $e^x = A$ then $x = \ln A$. Thus, (i) may be transformed to

$$M_f = M_i \exp\left(-\frac{v}{v_{ex}}\right)$$

When the rocket reaches the exhaust speed, $v = v_{ex}$; therefore, $M_f = M_i e^{-1}$, or $M_i = eM_f = 2.718\, M_f$. Since $M_i = M_R + M_F$ and $M_f = M_R$, we find that $M_R + M_F = 2.718\, M_R$, or

$$M_f = 1.718\, M_R = 1.7 \times 10^4 \text{ kg}$$

Note that it was not really necessary to convert to the exponential; we did so merely to illustrate a technique that we will use quite often. Notice that for a larger amount of fuel, the final velocity of the rocket can exceed v_{ex}. When this happens, the exhaust gases actually move in the same direction as the rocket.

Harold Edgerton at the Massachusetts Institute of Technology suggested this as a way to make applesauce. The exposure time was 0.33 μs.

HISTORICAL NOTE: Collisions and Galilean Relativity

It is interesting to examine the reasoning used by Huygens (Fig. 9.17) to investigate elastic collisions. His approach was based on the Galilean principle of relativity which says that the laws of mechanics should be the same in all inertial frames. He considered how two observers, one on land and one on a boat moving at velocity W relative to land, would describe the same event. Figure 9.18 shows one of his illustrations from *On the Motion of Bodies in Percussion* (1703).

FIGURE 9.17 Christiaan Huygens (1629–1695).

FIGURE 9.18 An illustration from Huygens' *On the Motion of Bodies in Percussion* (1703).

Figures 9.19 and 9.20 show the velocities before and after the collision as measured in the boat frame and in the land frame. We use lowercase letters (u, v) for velocities relative to the boat, and uppercase letters (U, V) for velocities relative to land. Note that $U = u + W$ and $V = v + W$.

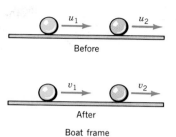

FIGURE 9.19 The velocities of two spheres before and after a collision as viewed in the boat frame.

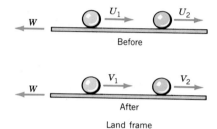

FIGURE 9.20 The velocities of two spheres before and after a collision as viewed in the land frame.

Huygens defined an elastic collision as one in which two spheres with equal masses and opposite initial velocities would merely reverse their directions of motion. He then used the following axiom for unequal masses: If the velocity of one mass merely reverses its direction, the same is true for the other mass. With the choice of $W = -(u_2 + v_2)/2$, the velocities relative to land are

$$U_1 = u_1 - \frac{u_2 + v_2}{2} \qquad U_2 = u_2 - \frac{u_2 + v_2}{2}$$

$$V_1 = v_1 - \frac{u_2 + v_2}{2} \qquad V_2 = v_2 - \frac{u_2 + v_2}{2}$$

We see that $V_2 = -U_2$, which means that m_2 has merely reversed its velocity. From the axiom, it follows that $V_1 = -U_1$, which leads to Eq. 9.10:

$$(V_2 - V_1) = -(U_2 - U_1)$$

From this result Huygens deduced that mv^2 is conserved in elastic collisions. He did not explicitly state that linear momentum is conserved in elastic collisions, but his notes show that he knew that it is. He was uncertain about whether linear momentum was conserved in inelastic collisions. Fortunately, the work of Wallis, and later experiments by Newton, nicely complemented Huygens' elegant theoretical analysis.

SUMMARY

The **linear momentum** of a particle of mass m and velocity $\mathbf{v}$ is defined as

$$\mathbf{p} = m\mathbf{v}$$

Newton's second law applied to a system of particles is

$$\mathbf{F}_{\text{EXT}} = \frac{d\mathbf{P}}{dt}$$

where $\mathbf{F}_{\text{EXT}} = \Sigma\mathbf{F}_i$ is the net external force acting on the system and $\mathbf{P} = \Sigma\mathbf{p}_i$ is the total momentum of the particles. From this we infer the principle of **conservation of linear momentum:**

$$\text{If } \mathbf{F}_{\text{EXT}} = 0, \quad \text{then} \quad \mathbf{P} = \Sigma\mathbf{p}_i = \text{constant}$$

In particular, for an isolated two-particle system, where $\mathbf{u}$ and $\mathbf{v}$ are the initial and final velocities:

$$m_1\mathbf{u}_1 + m_2\mathbf{u}_2 = m_1\mathbf{v}_1 + m_2\mathbf{v}_2$$

This is a vector equation and so each *component* is independently conserved.

In an elastic collision, kinetic energy is also conserved:

(Elastic) $$\tfrac{1}{2}m_1 u_1^2 + \tfrac{1}{2}m_1 u_2^2 = \tfrac{1}{2}m_1 v_1^2 + \tfrac{1}{2}m_2 v_2^2$$

In a one-dimensional elastic collision the relative velocity of the particles is unchanged in magnitude but is reversed in direction:

$$v_2 - v_1 = -(u_2 - u_1)$$

The **impulse** experienced by an object is its change in linear momentum:

$$\mathbf{I} = \Delta\mathbf{p} = \mathbf{F}_{\text{av}}\Delta t$$

This equation may be used to find the "average force" acting on a particle.

ANSWERS TO IN-CHAPTER EXERCISES

1. (a) The fact that a sound is produced means that some kinetic energy has been lost. (b) Release it above the floor and see if it returns to the initial height.

2. Since $p = mv$ and $K = \tfrac{1}{2}mv^2$, we have $K = p^2/2m$.

3. Since $W_f = -fd = -199$ J and $d = 0.04$ m, we find $f = 5 \times 10^3$ N. (Compare this with your weight.)

4. Both the initial and final kinetic energies are 22 J.

5. Substituting $v_2 = u_1 + v_1$ into $m_1 u_1 = m_1 v_1 + m_2 v_2$ we find $(m_1 - m_2)u_1 = (m_1 + m_2)v_1$. This leads to Eq. 9.11. Substituting $v_1 = v_2 - u_1$ into the conservation of momentum leads to Eq. 9.12. Note that although we have assumed that m_2 is initially at rest ($u_2 = 0$), these equations can be used to

handle the general case in which u_2 is not zero. It is necessary to transform the initial velocities to a frame moving with m_2, then to calculate the final velocities in this frame, and finally to transform these back to the original frame.

6. The fraction of energy transferred to m_2 is simply $f_2 = 1 - f_1 = 4m_1m_2/(m_1 + m_2)^2$. (a) 8/9; (b) 1; (c) 8/9.

7. (a) Yes, as in uniform circular motion. (b) No.

8. Say the initial momentum is $\mathbf{p}_0$ and the final momenta are $\mathbf{p}_1$ and $\mathbf{p}_2$; thus, $\mathbf{p}_0 = \mathbf{p}_1 + \mathbf{p}_2$. The conservation of kinetic energy takes the form $p_0^2 = p_1^2 + p_2^2$. These equations are satisfied only if $\mathbf{p}_1$ is perpendicular to $\mathbf{p}_2$.

QUESTIONS

1. When a ball strikes the floor and rebounds, it appears that momentum is not conserved. Is this true? Explain why or why not.

2. A sailor in a small boat has a fan and a sail available to him. Is there some way the fan can be used to propel the boat? If so, give details.

3. Suppose you are stranded on a frozen frictionless lake. Is there anything you can do to reach shore? If so, what?

4. A car and a truck collide completely inelastically. Which experiences the greater change in: (a) velocity; (b) momentum; (c) kinetic energy?

5. (a) If a particle's momentum is doubled, how does its kinetic energy change? (b) If its kinetic energy is doubled, how does its momentum change?

6. Consider two bodies with unequal masses, $m_1 > m_2$. (a) If their kinetic energies are equal, which has the larger momentum? (b) If their momenta are the same, compare their kinetic energies.

7. Suppose that ball A collides elastically with ball B which is initially at rest. How would you choose the mass of B compared to that of A such that B recoils with the greatest: (a) kinetic energy; (b) momentum; (c) speed?

8. A car collides head-on with a truck. Is the damage to the car greater if they have (a) the same kinetic energy, or (b) the same linear momentum?

9. An unfortunate individual is at the center of a railway car that is at the edge of a cliff, as in Fig. 9.21. Which way should the person walk to minimize the danger?

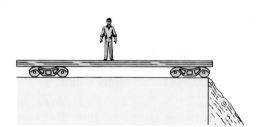

FIGURE 9.21 Question 9.

10. (a) Can the speed of a rocket ever exceed the exhaust speed? (b) Can the speed of a jet ever exceed the exhaust speed?

11. Why is it advisable to press the butt of a rifle against the shoulder?

12. Someone states that in a fall a drunk is less likely to be injured than a sober person. Is this plausible? If so, explain why.

13. (a) With the same muscular effort, why is a steel-headed hammer more effective in driving a nail than one with a rubber head. (b) Why is standing on the nail ineffective?

14. A salesperson claims that an airbag absorbs the force of a collision. Explain why you agree or disagree with this statement.

15. A ball is thrown at a block of wood. In which case will the ball exert the greater impulse: (a) when it sticks, or (b) when it rebounds with the same speed?

16. Explain why a balloon flies off when air escapes from it.

17. A flower pot falls off a balcony. Can one apply the conservation of linear momentum to its fall? Explain your answer.

18. Cannons on ships are mounted on bases attached to springs. Exactly what purpose do the springs serve?

19. Is all the kinetic energy lost in a completely inelastic collision?

20. Since $v^2 = v_x^2 + v_y^2 + v_z^2$, why don't we say $\frac{1}{2}mv_x^2$ is the x component of the kinetic energy $\frac{1}{2}mv^2$?

21. The explosion of 1 kg of TNT releases 4.1×10^6 J of energy. Express the loss in kinetic energy in Example 9.1 in terms of kilograms of TNT.

22. A car moving at speed v is brought to a halt by the constant frictional force due to the road. How do the stopping *distance* and the stopping *time* depend on the initial speed?

EXERCISES

9.1 and 9.2 Conservation of Linear Momentum

1. (I) A fast athlete, of mass 70 kg, can run at 10 m/s. At what speed would the following have the same momentum: (a) a 20-g bullet; (b) a 1500-kg car?

2. (I) A 10,000-kg truck moves at 30 m/s. At what speed would a 1200-kg car have the same value of (a) linear momentum, (b) kinetic energy?

3. (I) Consider a 20-g bullet (B) and a 60-kg runner (R). (a) If they have the same momentum, what is the ratio of their kinetic energies, K_B/K_R? (b) If they have the same kinetic energy, what is the ratio of their momenta, p_B/p_R?

4. (I) An object at rest explodes into three pieces of equal mass. One moves east at 20 m/s, while a second moves northwest at 15 m/s. What is the velocity of the third piece?

5. (I) A 10-kg object with a velocity $6\mathbf{i}$ m/s explodes into two equal fragments. One flies off with a velocity $2\mathbf{i} - \mathbf{j}$ m/s. What is the velocity of the other fragment?

6. (I) A 6-kg bomb moving at 5 m/s in the direction 37° S of E explodes into three pieces. A 3-kg piece moves off at 2 m/s at 53° N of E, whereas a 2-kg piece moves west at 3 m/s. What is the velocity of the third fragment? Assume all motion occurs in a horizontal plane.

7. (I) A ball of mass $m_1 = 3$ kg moving south at 6 m/s collides with a ball of mass $m_2 = 2$ kg initially at rest. The incoming ball is deflected in the direction 60°S of W and the target ball moves off at 25°E of S. What are the final speeds?

8. (II) A 500-g ball of putty moving horizontally at 6 m/s collides with and sticks to a block lying on a frictionless horizontal surface. If 25% of the kinetic energy is lost, what is the mass of the block?

9. (I) A 200-g ball of putty falls vertically into a 2.5-kg cart that is rolling freely at 2 m/s on a horizontal surface. What is the final speed of the cart?

10. (II) A particle of mass $m_1 = 2$ kg moving at speed u_1 makes a one-dimensional completely inelastic collision with a particle of mass $m_2 = 3$ kg, initially at rest. If 60 J of kinetic energy are lost, find u_1.

11. (I) A neutron at rest decays into a proton, an electron, and a neutrino. If the proton's momentum is 3×10^{-24} kg · m/s in the direction 37° N of E and the electron's momentum is 4×10^{-24} kg · m/s in the direction 53° S of W, what is the momentum of the neutrino?

12. (I) Particle A has a mass of 1.2 kg and an initial velocity of $-\mathbf{i} + 3\mathbf{j}$ m/s, whereas particle B has a mass of 1.8 kg and an initial velocity of $3\mathbf{i} + 4\mathbf{j}$ m/s. After they collide, A's velocity is $2\mathbf{i} + 1.5\mathbf{j}$ m/s. (a) What is B's final velocity? (b) What is the total change in kinetic energy?

13. (I) An 80-kg hunter with a 4-kg rifle is on a frozen frictionless lake. The rifle fires a 15-g bullet at 600 m/s relative to the ice. (a) What is the recoil speed of the rifle assuming it is held loosely, away from his shoulder? (b) After the rifle strikes the shoulder, what is the hunter's velocity? Assume the collision is completely inelastic. (c) What would be the hunter's speed if the rifle were held firmly against the shoulder?

14. (I) A particle of mass $m_1 = 1$ kg moving along the x axis collides with a particle of mass $m_2 = 2$ kg initially at rest. The incoming particle is deflected in the direction 30° above the x axis, whereas the target particle moves off at 10 m/s at 45° below the x axis. What are the initial and final speeds of the 1-kg particle?

15. (I) A railcar of mass 2×10^4 kg moving at 6 m/s collides with, and couples to, a railcar of mass 4×10^4 kg at rest. (a) What fraction of the initial kinetic energy is lost? (b) If the roles of the railcars are reversed, what is the fractional loss in kinetic energy?

16. (II) An object of mass 1 kg makes a completely inelastic collision with an object of unknown mass at rest. If 60% of the kinetic energy is lost, what is the unknown mass?

17. (II) A 1000-kg Subaru GL at rest at a stoplight is struck from the rear by a 1400-kg Pontiac 6000. They couple together and leave skid marks 4 m long. The coefficient of kinetic friction is 0.6. (a) What was their common speed just after the collision? (b) What was the speed of the Pontiac just prior to the collision?

18. (II) A 1500-kg Buick Electra moving at 20 m/s makes a completely inelastic one-dimensional collision with a stationary 1000-kg Ford Escort. If the coefficient of kinetic friction is $\mu_k = 0.5$, estimate the distance they move after the collision. Assume the wheels are locked.

19. (I) A 90-kg quarterback runs due north at 8 m/s to tackle a 110-kg player running east at 7.5 m/s. If the completely inelastic collision occurs while their feet are briefly off the

ground, find: (a) their common velocity just after the collision; (b) the loss in kinetic energy.

20. (I) An ancient ship of mass 1.5×10^6 kg carried twenty cannons on each side. They could fire 8-kg cannonballs at 400 m/s. If all the cannons on one side were fired simultaneously, what would be the recoil speed of the ship? Ignore the resistance of the water.

21. (I) A nucleus of radioactive radium (^{226}Ra), initially at rest, decays into a radon nucleus (^{222}Rn) and an α particle (a ^{4}He nucleus). If the kinetic energy of the α particle is 6.72×10^{-13} J, what is (a) the recoil speed of the radon nucleus, and (b) its kinetic energy? The superscripts indicate, roughly, the mass of each nucleus in unified mass units (u), where 1 u $= 1.66 \times 10^{-27}$ kg.

22. (I) A meteor of mass 5×10^8 kg moving at 10 km/s relative to the earth, strikes the earth in a completely inelastic collision. (a) What is the recoil speed of the earth, relative to the frame in which it was initially at rest? (b) What is the loss in kinetic energy measured in megatons? (One ton of TNT releases 4.2×10^9 J.)

23. (I) On July 25, 1956, the liner *Andrea Doria*, of mass 4.1×10^7 kg, was heading west at 40 km/h. It collided off Nantucket Island with the *Stockholm*, of mass 1.7×10^7 kg, sailing at 30 km/h in the direction 20° east of north. The bow of the *Stockholm* temporarily lodged in the side of the *Andrea Doria*; that is, the collision was completely inelastic. (a) Find their common velocity just after the collision. (b) What was the loss in kinetic energy due to the collision? (The *Andrea Doria* subsequently sank.)

24. (I) Jack, of mass 75 kg, and Jill, of mass 60 kg, are at rest on a frictionless frozen lake. Find their final speeds if Jack throws a 0.5-kg ball to Jill at 24 m/s relative to the ice and she catches it. Assume the ball moves horizontally.

25. (I) A railcar of mass 2×10^4 kg moving at $3\mathbf{i}$ m/s collides with, and couples to, another railcar of mass 3×10^4 kg. What is the common final velocity and the loss in kinetic energy if the second railcar was initially moving at (a) $2\mathbf{i}$ m/s, (b) $-3.5\mathbf{i}$ m/s?

26. (I) A particle of mass $m_1 = 2$ kg moves south at 4 m/s and a particle of mass $m_2 = 3$ kg moves east at 5 m/s. After the collision m_1 moves at 3 m/s in the direction 30° S of E. (a) What is the final velocity of m_2? (b) Was the collision elastic?

27. (II) A 15-g bullet strikes and embeds in a 2-kg block suspended at the end of a 1.2-m string. After the collision the string rises to a maximum angle of 20° to the vertical. Find: (a) the speed of the bullet; (b) the percentage loss in kinetic energy due to the collision.

28. (II) A projective of mass $m = 200$ g strikes a stationary block of mass $M = 1.3$ kg from below with speed $u = 30$ m/s (see Fig. 9.22.) The projective embeds in the block. (a) To what height does the block rise? (b) What is the loss in kinetic energy due to the collision?

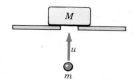

FIGURE 9.22 Exercise 28.

29. (II) A 10-g bullet moving at 400 m/s strikes a ballistic pendulum of mass 2.5 kg. The bullet emerges with a speed of 100 m/s. (a) To what height does the pendulum bob rise? (b) How much work was done by the bullet in passing through the block?

30. (II) A projectile of mass 0.25 kg moving at 24 m/s collides with and sticks to a 1.75-kg block that is connected to a spring for which $k = 40$ N/m, as in Fig. 9.23. The block is initially on a frictionless part of a horizontal surface but starts to slide on a rough section immediately after the collision. If the maximum compression of the spring is 0.5 m, what is the force of friction on the block?

FIGURE 9.23 Exercise 38.

9.3 Elastic Collisions in One Dimension

31. (I) An α particle of mass 4 u moving at 1.5×10^7 m/s collides elastically with a gold nucleus of mass 197 u initially at rest. Find the recoil velocity of the nucleus given that the α particle returns along its original path.

32. (I) The head of a golf club moving at 160 km/h collides elastically with a 46-g golf ball initially at rest. Find the final velocities of both for the following clubheads: (a) 7 oz (mass 46 g); (b) 14 oz (mass 92 g). (Even though the club is held in the hand, the clubhead behaves as if it were free. This has been proved by allowing the head to pivot freely at the end.)

33. (II) A pendulum bob of mass m is released from a height H above the lowest point. It collides at the lowest point with another pendulum of the same length but with a bob of mass $2m$ initially at rest. Find the heights to which the bobs rise given that the collision is (a) completely inelastic, (b) perfectly elastic.

34. (I) A particle of mass m_1 and with speed u makes a one-dimensional elastic collision with a particle of mass m_2 initially at rest. Find their final velocities given that: (a) $m_1 = 3m_2$; (b) $m_2 = 3m_1$.

35. (II) A particle of mass m_1 makes a head-on elastic collision with a particle of mass m_2 at rest. If the initial kinetic energy is K_0, and the final kinetic energy of m_2 is K_2, evaluate the ratio K_2/K_0 for a collision between a neutron of mass 1 u

and each of the following: (a) a deuteron of mass 2 u; (b) a carbon nucleus of mass 12 u; (c) a lead nucleus of mass 208 u.

36. (I) A particle of mass m_1 and with initial velocity $+ u$ makes a head-on elastic collision with a particle of mass m_2 at rest. Find m_2/m_1 given that the final velocity of m_1 is: (a) $-u/3$; (b) $+u/2$.

37. (II) In 1932, James Chadwick fired neutrons of unknown speed and mass into different substances. He found that protons (of mass 1 u) were given a speed 7.5 times that given to nitrogen nuclei (of mass 14 u). If the collisions were elastic and head on, what can you deduce about the mass of the neutron?

38. (II) A particle of mass $m_1 = 2$ kg moving at velocity $u\mathbf{i}$ makes a one-dimensional elastic collision with a particle of mass m_2 at rest. Find m_2 given the following: (a) It moves off with velocity $0.5u\mathbf{i}$. (b) It has one-third the initial kinetic energy of m_1.

9.4 Impulse

39. (I) The head of a golf club strikes a 46-g golf ball at rest. If the collision lasts 0.5 ms and the ball is given a speed of 220 km/h, estimate the average force on the ball.

40. (I) A 1-kg ball falls vertically and hits the floor at 20 m/s and rebounds vertically with a speed of 15 m/s. If the ball was in contact with the floor for 0.1 s, what was the average force on it?

41. (I) A 10-g bullet traveling at 400 m/s strikes a wooden block and emerges at 100 m/s. It was in the block for 0.01 s. What was the force on the block?

42. (I) A 0.15-kg baseball moving at $30\mathbf{i}$ m/s is struck by a bat in a collision that lasts 1 ms. What is the average force on the ball if (a) it merely reverses its velocity, or (b) it moves off at $40\mathbf{j}$ m/s.

43. (I) A hammer with a head of mass 0.5 kg strikes a nail at 4 m/s and is brought to rest. If the collision lasts 10^{-3} s, what is the average force on the nail? Compare this force with your weight.

44. (I) Car A, with a mass of 2×10^3 kg and a speed of 15 m/s, collides head-on with car B of mass 10^3 kg initially at rest. They stick together after the collision. (a) What is their common speed after the collision? (b) If the collision lasts 0.2 s, what is the magnitude of the average force on each car? (c) Estimate the force exerted by the seatbelt on a 70-kg passenger in each car. (Assume that the passenger is held firmly by the seatbelt.)

45. (I) Water from a hose emerges at 10 m/s horizontally and strikes a wall. It then dribbles down. The flow rate is 1.5 kg/s. What is the average force on the wall? Is this estimate likely to be too large or too small?

46. (II) Steel pellets moving at 12 m/s strike a plate set at 45° to their line of motion. The pellets are thereby deflected by 90° without a change in speed. If the flow rate is 0.5 kg/s, what is the average force on the plate?

47. (I) A machine gun fires 15-g bullets at 450 m/s at a rate of 600 rounds/min. What is the average force on the support?

48. (II) A 0.4-kg projectile is fired at 20 m/s at 37° above the horizontal. It lands at the initial level at 16 m/s at 53° below the horizontal. (a) What impulse did it experience? (b) What was the source of the impulse? (c) Can you find the average force on the projectile?

49. (II) A ball of mass 200 g falls from 4 m and rebounds to 3 m. If it is in contact with the ground for 10 ms, find the average force on it. Assume the motion is vertical.

50. (I) During a serve, a 60-g tennis ball, initially at rest, is given a speed of 30 m/s in 0.04 s. Find the average force on the ball.

51. (I) A 60-g tennis ball strikes the ground at 25 m/s at 40° to the horizontal. It bounces off at 20 m/s at 30° to the horizontal; see Fig. 9.24. (a) Find the impulse exerted on the ball. (b) If the collision lasted 5 ms, find the average force exerted on the ball by the court.

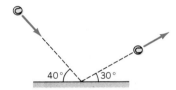

FIGURE 9.24 Exercise 51.

52. (I) From the F versus t curve shown in Fig. 9.25, find: (a) the impulse; (b) the average force.

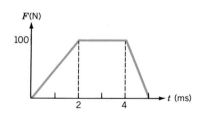

FIGURE 9.25 Exercise 52.

53. (II) The F versus t curve for someone jumping vertically is shown in Fig. 9.26. The force exerted by the floor increases from the person's normal weight of 650 N to 1350 N in 0.3 s, stays constant for 0.1 s and then drops to zero as the feet lose contact with the floor. (a) Find the impulse exerted by the floor. (b) What is the impulse due to gravity in 0.4 s? (c) At what speed did the feet leave the floor? (d) What was the maximum height attained? (e) Show that the peak power generated was 4.78 hp.

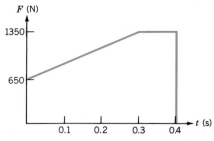

FIGURE 9.26 Exercise 53.

54. (II) Figure 9.27 shows the F versus t curve for the force exerted by the hip joints on the 50-kg torso of a sprinter as he starts to run. (a) What is the impulse exerted on the torso? (b) Estimate the sprinter's change in speed? Assume that the force and the motion are horizontal.

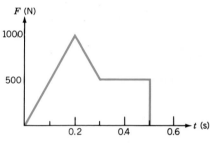

FIGURE 9.27 Exercise 54.

55. (I) A 1200-kg car moving east at 16 m/s turns south and 8 s later is moving at 12 m/s. (a) What was the impulse on the car? (b) What was the average force on the car?

9.6 Elastic Collisions in Two Dimensions

56. (II) A particle of mass $m_1 = 2$ kg moving at 8i m/s collides elastically with a particle of mass $m_2 = 1$ kg initially at rest. The incoming particle moves off at 30° above the $+x$ axis. Find (a) the final speed of m_1; (b) the velocity of m_2.

57. (II) A deuteron of mass M moving at speed u collides elastically with an α particle of mass $2M$ initially at rest. The deuteron is scattered through 90° from the initial direction of motion. Find: (a) the angle at which the α particle moves off; (b) the final speeds in terms of u; (c) the percentage of the deuteron's kinetic energy transferred to the α particle.

58. (II) A particle moving at 20 m/s collides elastically with an identical particle at rest. The incoming particle moves off at 30° to the original direction. What is the final speed of each particle?

59. (II) A billiard ball moving at 3 m/s collides elastically with an identical ball at rest. The final speed of the first ball is 2 m/s. At what angles to the original direction do the balls move off?

9.7 Rocket Propulsion

60. (I) What fraction of the initial mass of a rocket should be fuel for the final speed of the rocket to equal (a) the exhaust speed; (b) 2.5 times the exhaust speed? Ignore gravity and assume the rocket is initially at rest.

61. (I) The speed of a rocket in space increases from 2×10^3 m/s to 4×10^3 m/s while its mass decreases by 60%. What is the exhaust speed of the gases?

PROBLEMS

1. (I) A tennis racket (held firmly in hand) moving at speed u_1 hits a ball moving at speed u_0. Show that the maximum possible speed of the ball after it is hit is $u_0 + 2u_1$.

2. (II) A 60-kg girl can throw a 0.5-kg ball horizontally at 6 m/s *relative to herself*. Assume that she carries two such balls and is initially at rest on a frozen frictionless lake. Find her final velocity given that: (a) she throws both balls simultaneously; (b) she throws them one after the other in the same direction. (c) Use Eq. 9.16 to find her final velocity if she carried a device to continuously discharge 1 kg of fluid with the same relative velocity.

3. (I) Two particles with masses m_1 and m_2 travel toward each other with speeds u_1 and u_2. They collide and stick together. Show that the loss in kinetic energy is

$$\frac{m_1 m_2 (u_1 + u_2)^2}{2(m_1 + m_2)}$$

4. (II) A block of mass m is released on a wedge of mass M at a height h above the floor as in Fig. 9.28. All surfaces are frictionless. Show that the speed of the wedge when the block hits the floor is given by

$$\sqrt{\frac{2m^2 gh \cos^2 \theta}{(M + m)(M + m \sin^2 \theta)}}$$

(*Hint:* If u is the speed of the block relative to the wedge, what are the components of its velocity relative to the floor? See Example 5.10.)

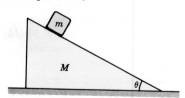

FIGURE 9.28 Problem 4.

5. (II) Figure 9.29 shows a cannon that is mounted at an angle α to the horizontal on a flatcar initially at rest. The mass of the car and cannon is M. A cannonball of mass m is fired at speed V relative to the cannon. (a) Show that the recoil speed of the flatcar is

$$\frac{mV \cos \alpha}{M + m}$$

(b) Show that the angle θ to the horizontal at which the ball emerges from the cannon is given by

$$\tan \theta = \frac{(M + m)}{M} \tan \alpha$$

FIGURE 9.29 Problem 5.

6. (I) A pendulum bob of mass 500 g is suspended by a string of length 1 m and released when the string is horizontal. It collides elastically with a block of mass M on a frictionless horizontal surface, as in Fig. 9.30. To what height does the bob rise given that: (a) $M = 2.5$ kg; (b) $M = 200$ g?

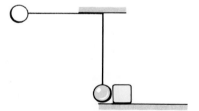

FIGURE 9.30 Problem 6.

7. (I) A particle of mass m_1 moving at velocity $\mathbf{u}_1 = u_1 \mathbf{i}$ makes a one-dimensional elastic collision with a particle of mass m_2 moving at velocity $\mathbf{u}_2 = u_2 \mathbf{i}$. The final velocities are $\mathbf{v}_1 = v_1 \mathbf{i}$ and $\mathbf{v}_2 = v_2 \mathbf{i}$. Show that

$$v_1 = \frac{(m_1 - m_2)u_1 + 2m_2 u_2}{m_1 + m_2}$$

$$v_2 = \frac{2m_1u_1 + (m_2 - m_1)u_2}{m_1 + m_2}$$

8. (II) A 10-kg bomb moving east at 20 m/s explodes into three pieces. The explosion provides an additional 10^4 J of energy. The 5-kg piece moves off at 20 m/s in the direction 37° N of E, and the 3-kg piece moves off due south. What is the velocity of the 2-kg piece? Assume all motion occurs in a horizontal plane.

9. (I) Two pendulum bobs of masses m_1 and m_2 (= $2m_1$) collide elastically at the lowest point in their motion, when the centers are at the same level, as in Fig. 9.31. If both are released from height H above the lowest point, to what heights do they rise the first time? (See Problem 7.)

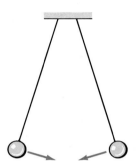

FIGURE 9.31 Problem 9.

10. (II) Figure 9.32 shows a cone of mass 200 g suspended by a vertical jet of water from a garden hose. The flow rate is 0.7 kg/s. The water rises 4 m if unimpeded. What is the height of the cone? Assume the water emerges horizontally from holes at the top of the cone.

FIGURE 9.32 Problem 10.

11. (II) A one-dimensional inelastic collision may be characterized by a *coefficient of restitution* e that relates the relative velocities before and after the collision: $(v_1 - v_2) = -e(u_1 - u_2)$. Show that the final velocities are

$$v_1 = \frac{(m_1 - em_2)u_1 + m_2(1 + e)u_2}{(m_1 + m_2)}$$

$$v_2 = \frac{m_1(1 + e)u_1 + (m_2 - em_1)u_2}{(m_1 + m_2)}$$

12. (II) Use the expressions given in Problem 11 to show that the loss in kinetic energy in an inelastic collision is

$$\frac{1}{2}\frac{m_1m_2}{m_1 + m_2}(u_1 - u_2)^2(1 - e^2)$$

where u_1 and u_2 are the initial velocities and e is the coefficient of restitution.

13. (I) Two particles of equal mass $4M$ are initially at rest. A particle of mass M moving at speed u collides elastically with one of the larger balls (see Fig. 9.33). How many collisions occur?

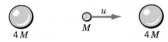

FIGURE 9.33 Problem 13.

14. (I) A ball of mass $m_1 = 3M$ moving at speed u collides elastically with a ball of mass $m_2 = 2M$ at rest. Another ball of mass $m_3 = 3M$ lies along the same line as shown in Fig. 9.34. After two impacts, what are the velocities?

FIGURE 9.34 Problem 14.

15. (I) A particle of mass m makes a one-dimensional elastic collision with another particle at rest. The first particle rebounds with 25% of its initial kinetic energy. What is the mass of the other particle?

16. (II) A bomb of mass $3M$ moving with velocity $10\mathbf{i}$ m/s explodes into two parts with masses M and $2M$. The explosion supplies 100 J of kinetic energy. Find the final speeds given that M moves in the $-x$ direction and $2M$ moves in the $+x$ direction.

17. (I) A particle of mass $m_1 = 2$ kg moving at $8\mathbf{i}$ m/s collides with a particle of mass $m_2 = 6$ kg at rest. The first particle rebounds with velocity $-1.5\mathbf{i}$ m/s. Find: (a) the final velocity of m_2; (b) the coefficient of restitution. (See Problem 11.)

18. (II) An ideal spring with stiffness constant $k = 400$ N/m is attached to a stationary block of mass 4 kg as in Fig. 9.35. A 2-kg block approaches at 8 m/s. (a) What is the maximum compression of the spring? (b) What are the final velocities of the two blocks? The motion occurs on a horizontal frictionless surface.

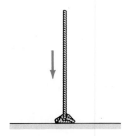

FIGURE 9.35 Problem 18.

19. (II) A vertical chain has a length L and a mass M. It is released with the bottom just touching a table; see Fig. 9.36. (a) Find the force on the table as a function of the distance fallen by the top end. (b) Show that the maximum force is $3Mg$.

FIGURE 9.36 Problem 19.

20. (I) A block of mass $m_1 = 5$ kg moving at 8 m/s collides with and sticks to a block of mass $m_2 = 3$ kg at rest on a horizon-

tal frictionless surface; see Fig. 9.37. How far along the rough incline ($\mu_k = \frac{1}{4}$) do the combined blocks slide before stopping?

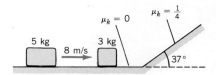

FIGURE 9.37 Problem 20.

21. (II) A particle of mass M_1 collides elastically with a particle of mass M_2 ($<M_1$) initially at rest. Show that the maximum angle θ_1 to the original direction of motion at which M_1 can move off is given by

$$\sin \theta_{1(max)} = \frac{M_2}{M_1}$$

(*Hint:* Obtain a quadratic equation in v_1. What is the condition for a real solution?)

CHAPTER 10

Systems of Particles

During a grand jeté, a ballet dancer appears briefly to "float in air". However, the center of mass still follows a parabolic path.

Major Points

1. The location of the **center of mass** (CM) of a system.
2. **Newton's first law for a system:** The velocity of the CM of an isolated system stays constant.
3. **Newton's second law for a system** relates the external force to the acceleration of the center of mass.
4. The kinetic energy of a system may be separated into energy of the CM motion and energy of motion relative to the CM.

10.1   CENTER OF MASS

Until now we have dealt mainly with single particles. The particle model was adequate since we were concerned only with translational motion. When the motion of a body involves rotation and vibration, we must treat it as a *system* of particles. A system is any well-defined assembly of particles which may, or may not, interact or be connected to each other. Despite the complex motions of which a system is capable, there is a single point, the **center of mass** (CM), whose translational motion is characteristic of the system as a whole.

The existence of this special point can be demonstrated as follows. In Fig. 10.1 two masses m_1 and m_2 are connected by a rod of negligible mass. If a force F is applied to the rod at an arbitrary point, as in Figs. 10.1a and b, the system rotates. When the force is applied at the CM however, there is only translation, as shown in Fig. 10.1c. In this sense, the system behaves as if all its mass were concentrated at the CM. The relation between the distances ℓ_1 and ℓ_2 of the

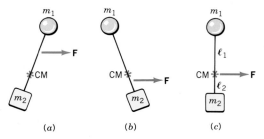

(a)   (b)   (c)

FIGURE 10.1 (a) When a net force is applied to a system of an arbitrary point, it rotates and translates. (b) When the force is applied at the CM, there is only translation.

particles from the CM is found to be $\ell_2/\ell_1 = m_1/m_2$, or

$$m_1\ell_1 = m_2\ell_2$$

One can express the position of the CM in terms of the coordinate system shown in Fig. 10.2. We see that $\ell_1 = x_{CM} - x_1$ and $\ell_2 = x_2 - x_{CM}$. When these are used in the above equation, we find

$$m_1(x_{CM} - x_1) = m_2(x_2 - x_{CM})$$

which leads to

$$x_{CM} = \frac{m_1x_1 + m_2x_2}{m_1 + m_2}$$

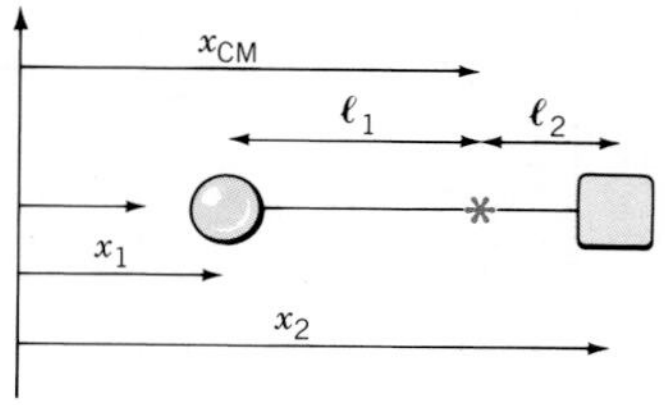

The CM is a sort of average position of the particles. One could not simply write $x_{av} = (x_1 + x_2)/2$ because the position of the larger mass should count more heavily. The position x_{CM} is a *weighted* average in which each coordinate is (mathematically) weighted by the mass located at that point. The same logic extends to any number of particles and to three dimensions. For N particles the (vector) position of the CM is

$$\mathbf{r}_{CM} = \frac{m_1\mathbf{r}_1 + m_2\mathbf{r}_2 + \cdots + m_N\mathbf{r}_N}{m_1 + m_2 + \ldots + m_N}$$

$$\mathbf{r}_{CM} = \frac{\Sigma\, m_i\mathbf{r}_i}{M} \qquad (10.1) \qquad \text{Position of the center of mass (CM)}$$

where $M = \Sigma m_i$ is the total mass of the system. The components of Eq. 10.1 are

$$x_{CM} = \frac{\Sigma\, m_ix_i}{M}; \qquad y_{CM} = \frac{\Sigma\, m_iy_i}{M}; \qquad z_{CM} = \frac{\Sigma\, m_iz_i}{M} \qquad (10.2)$$

The position of the CM of a symmetric body can often be found by inspection. For example, if we imagine the rectangular plate in Fig. 10.3 divided into thin strips parallel to the y axis, then for every strip at $+x$ there is a corresponding strip at $-x$. Consequently, the sum Σm_ix_i vanishes and $x_{CM} = 0$. By the same logic $y_{CM} = 0$. The x and y axes are examples of *axes of symmetry.** A circular cylinder, shown in Fig. 10.4, has an axis of symmetry along its central axis. It also appears unchanged when "reflected" in the *plane of symmetry* shown. The center of mass of a symmetric body always lies on an axis or a plane of symmetry. The centers of mass of a uniform sphere, a uniform circular disk, or a uniform rod are at their geometric centers.

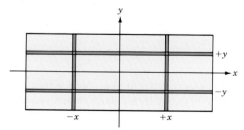

FIGURE 10.3 A thin rectangular plate. For every strip at $+x$ there is a corresponding strip at $-x$. Consequently, the center of mass must be at $x = 0$. The x and y axes are axes of symmetry.

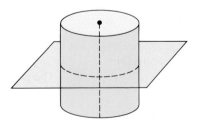

FIGURE 10.4 The center of mass of a cylinder lies at the intersection of the central axis of symmetry and the plane of symmetry.

* The body appears unchanged after a rotation about an axis of symmetry through an angle of $360°/n$, where $n = 2$, or 3, or 5 (or any integer multiple of these).

FIGURE 10.2 The position of the center of mass may be expressed in terms of the positions of the particles in the system.

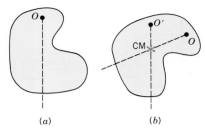

FIGURE 10.5 Locating the center of mass of a planar body.

The CM of a planar body can be determined experimentally as follows. When the body is pivoted freely about an arbitrary point, it behaves as if it were a simple pendulum with its mass concentrated at the CM. The body will therefore rotate unless the CM lies vertically below the pivot, as shown in Fig. 10.5a. In this position we draw a vertical line through the pivot. Next we suspend the body about another point, let it come to rest, and draw another vertical line. The intersection of the two lines, as indicated in Fig. 10.5b, locates the CM. For a nonplanar body, one would use three points of suspension that do not lie in one plane.

EXERCISE 1. The gravitational potential energy of the ith particle in a system is $U_i = m_i g y_i$. Express the potential energy of the whole system of particles in terms of y_{CM}, the position of the CM.

EXAMPLE 10.1: Find the CM of the four point masses shown in Fig. 10.6.

Solution: The total mass is $M = 12$ kg. From Eq. 10.2, we have

$$x_{CM} = \frac{(2 \text{ kg})(3 \text{ m}) + (4 \text{ kg})(3 \text{ m}) + (5 \text{ kg})(-4 \text{ m}) + (1 \text{ kg})(-3 \text{ m})}{12 \text{ kg}}$$

$$= -\frac{5}{12} \text{ m}$$

$$y_{CM} = \frac{(2 \text{ kg})(-1 \text{ m}) + (4 \text{ kg})(3 \text{ m}) + (5 \text{ kg})(4 \text{ m}) + (1 \text{ kg})(-2 \text{ m})}{12 \text{ kg}}$$

$$= \frac{28}{12} \text{ m}$$

The position of the CM is $r_{CM} = -0.42i + 2.3j$ m

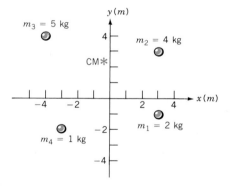

FIGURE 10.6 The position of the center of mass of the four particles is indicated by the star.

EXAMPLE 10.2: A thin rod of length $3L$ is bent at right angles at a distance L from one end (see Fig. 10.7). Locate the CM with respect to the corner. Take $L = 1.2$ m.

Solution: The CM of each arm is at its midpoint. The CM of the two arms can be found by treating each arm as a point particle at its CM. From the diagram we see that $x_1 = L/2$, $y_1 = 0$ and

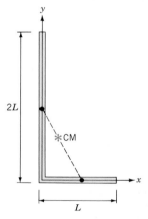

FIGURE 10.7 In determining the center of mass of the object, the two uniform arms may be replaced by point particles at their midpoints.

$x_2 = 0$, $y_2 = L$. If we take $m_1 = m$, then $m_2 = 2m$. From Eq. 10.2 we have

$$x_{CM} = \frac{m_1 x_1 + m_2 x_2}{M} = \frac{L}{6}$$

$$y_{CM} = \frac{m_1 y_1 + m_2 y_2}{M} = \frac{2L}{3}$$

The position of the CM is $r_{CM} = 0.2i + 0.8j$ m. Notice that the CM does not lie within the body itself. Similarly, when you are standing straight, your CM is within your body; when you bend over to touch your toes, your CM is outside your body.

EXERCISE 2. A rod of the same type as in Example 10.2 is used to complete the triangle in Fig. 10.7. Where is the center of mass of the triangle? (What is the mass of the rod and where is its CM?)

10.2   CENTER OF MASS OF CONTINUOUS BODIES

We now consider situations in which symmetry considerations alone are not enough to determine the position of the CM of a continuous body. The approach is to first divide the body into suitably chosen infinitesimal elements. The choice is usually determined by the symmetry of the body. For the element of mass dm in Fig. 10.8 the quantity $m_i\mathbf{r}_i$ in Eq. 10.1 is replaced by $\mathbf{r}\,dm$, and the discrete sum over particles, $\Sigma m_i\mathbf{r}_i/M$, becomes an integral over the body:

$$\mathbf{r}_{CM} = \frac{1}{M}\int \mathbf{r}\,dm \qquad (10.3)$$

The components of this equation are

$$x_{CM} = \frac{1}{M}\int x\,dm; \qquad y_{CM} = \frac{1}{M}\int y\,dm; \qquad z_{CM} = \frac{1}{M}\int z\,dm$$

To evaluate these integrals, one must express the variable m in terms of the spatial coordinates x, y, z, or r, as illustrated in the examples below.

Consider the thin rod of mass M and length L in Fig. 10.9a. The infinitesimal element in this case is a slice of length dx. The rod has to be thin enough to ensure that all the particles of the element are at the same distance from the origin. If the *volume* mass density (mass per unit volume) of the rod is ρ (kg/m³), the mass of the element of volume dV is $dm = \rho\,dV = \rho A\,dx$. If we define $\lambda = \rho A$, we have $dm = \lambda\,dx$. The quantity $\lambda = M/L$ is called the *linear* mass density (mass per unit length) and is measured in kg/m.

For a disk or cylinder, the appropriate element is a ring of width dr and area dA, which extends through the body of the solid, as indicated in Fig. 10.9b. Its mass is $dm = \rho\,dV = \rho h\,dA$. If we define $\sigma = \rho h$, we have $dm = \sigma\,dA$. The quantity $\sigma = M/A$ is called the *areal* mass density (mass per unit area) and is measured in kg/m². Note that A is the cross-sectional area in a plane of symmetry.

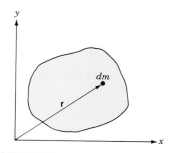

FIGURE 10.8 To find the center of mass of a continuous body one must integrate the contributions of each mass element dm.

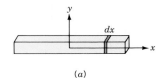

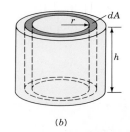

FIGURE 10.9 (a) The mass of the element of length dx is $dm = (M/L)dx$. (b) The mass of the element of area dA and height h is $dm = (M/A)dA$, where $dA = 2\pi r\,dr$.

EXAMPLE 10.3: Find the CM of a semicircular rod of radius R and linear density λ kg/m as shown in Fig. 10.10.

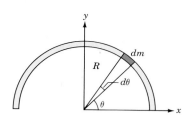

FIGURE 10.10 The element is an arc of length $R\,d\theta$.

Solution: From the symmetry of the body we see at once that the CM must lie along the y axis, so $x_{CM} = 0$. In this case it is convenient to express the mass element in terms of the angle θ, measured in radians. The element, which subtends an angle $d\theta$ at the origin, has a length $R\,d\theta$ and a mass $dm = \lambda R\,d\theta$. Its y coordinate is $y = R\sin\theta$. Therefore, $y_{CM} = \int y\,dm/M$ takes the

form

$$y_{CM} = \frac{1}{M}\int_0^\pi \lambda R^2 \sin\theta\,d\theta = \frac{\lambda R^2}{M}\Big[-\cos\theta\Big]_0^\pi$$
$$= \frac{2\lambda R^2}{M}$$

The total mass of the ring is $M = \pi R\lambda$; therefore, $y_{CM} = 2R/\pi$.

EXERCISE 3.   A thin rod is bent to form a quarter of a circle. Where is its CM relative to the center of the circle?

EXAMPLE 10.4: Find the CM of a uniform solid cone of height h and semiangle α, as in Fig. 10.11.

Solution: We place the apex of the cone at the origin. It is clear that the CM will lie along the y axis. We divide the cone into disks of radius x and thickness dy. The volume of such a disk is $dV = \pi x^2\,dy = \pi(y\tan\alpha)^2\,dy$. The mass of the disk is $dm = \rho\,dV$. Let us first find the total mass of the cone.

$$M = \int dm = \pi\rho\tan^2\alpha \int_0^h y^2\,dy = \pi\rho\tan^2\alpha\,\frac{h^3}{3} \qquad (i)$$

The position of the CM is given by $y_{CM} = \int y \, dm / M$:

$$y_{CM} = \frac{1}{M} \pi \rho \tan^2\alpha \int_0^h y^3 \, dy = \frac{1}{M} \pi \rho \tan^2\alpha \frac{h^4}{4} \qquad \text{(ii)}$$

Using (i) in (ii) we find $y_{CM} = 3h/4$.

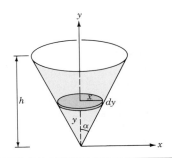

FIGURE 10.11 In finding the center of mass of a cone, it is divided into disks as shown.

10.3 MOTION OF THE CENTER OF MASS

The instantaneous velocity of a particle is $\mathbf{v} = d\mathbf{r}/dt$. Hence, if we take the time derivative of Eq. 10.1 we find

Velocity of the center of mass

$$\mathbf{v}_{CM} = \frac{\sum m_i \mathbf{v}_i}{M} \qquad (10.4)$$

This may be written in a form that highlights the importance of the CM:

Total linear momentum of a system of particles

$$\mathbf{P} = M\mathbf{v}_{CM} = m_1\mathbf{v}_1 + m_2\mathbf{v}_2 + \cdots + m_N\mathbf{v}_N \qquad (10.5)$$

The total momentum $\mathbf{P} = \Sigma\mathbf{p}_i$ of a system of particles is equivalent to that of a single (imaginary) particle of mass $M = \Sigma m_i$ moving at the velocity of the center of mass $\mathbf{v}_{CM}$.

This result provides us with an enormous simplification: We may deal with the *translational* motion of extended objects or systems of particles, as if they were point particles with all the mass concentrated at the CM.

If we take the derivative of Eq. 10.5 we find $M\mathbf{a}_{CM} = \Sigma m_i\mathbf{a}_i = \Sigma\mathbf{F}_i$, where $\mathbf{F}_i$ is the net force on the ith particle. In Section 9.2 we saw that in evaluating the sum $\Sigma\mathbf{F}_i$ over all the particles, the internal forces between them cancel in pairs leaving just the net external force; that is, $\Sigma\mathbf{F}_i = \mathbf{F}_{EXT}$. We conclude that **Newton's second law** for a system of particles is

$$\mathbf{F}_{EXT} = M\mathbf{a}_{CM} \qquad (10.6)$$

The CM accelerates as if it were a point particle of mass $M = \Sigma m_i$ and the net external force were applied at this point.

If we had started with the second law in the form $\mathbf{F} = d\mathbf{p}/dt$, we would have found

$$\mathbf{F}_{EXT} = \frac{d\mathbf{P}}{dt} \qquad (10.7)$$

The rate of change of the total momentum of a system is equal to the net external force.

Equations 10.6 and 10.7 allow us to apply the second law to a system of particles in a very simple way—provided we are interested only in the translational motion of the CM. For a complete description of the motion of the system, we would have to apply the second law to each individual particle, which can be a formidable task. Equation 10.6 is, in effect, what we used in earlier chapters.

Figure 10.12 shows an acrobat somersaulting. According to Eq. 10.6 the acceleration of her CM is the acceleration due to gravity. Nothing she does during the stunt can alter the parabolic shape of the trajectory of her CM (ignoring air

FIGURE 10.12 The center of mass of the acrobat moves along a parabolic path.

resistance). As another example, consider an artillery shell that explodes at some point on its trajectory, as shown in Fig. 10.13. The fragments experience large internal forces due to the explosive, and each follows its own new parabolic path. However, since these forces are internal to the system, the motion of the CM is not affected: It continues along its original path—at least until one fragment lands and experiences a new force due to the ground.

We may deduce **Newton's first law** as it applies to a system of particles either from the conservation of linear momentum or from the second law. Using either Eq. 10.6 or 10.7, we may state

$$\text{If } \mathbf{F}_{EXT} = 0, \quad \text{then} \quad \mathbf{v}_{CM} = \text{constant}$$

If the net external force on a system of particles is zero, the velocity of the center of mass remains constant.

Figure 10.14 shows a spinning wrench moving over a frictionless surface. Although the motion of any given particle is quite complex, the velocity of the CM stays constant.

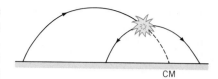

FIGURE 10.13 When a projectile explodes, the center of mass of the fragments moves along the original trajectory.

Newton's first law for a system of particles

FIGURE 10.14 Newton's first law applied to a system says that the velocity of the center of mass of an isolated system is constant.

EXAMPLE 10.5: A man of mass $m_1 = 60$ kg is at the rear of a stationary boat of mass $m_2 = 40$ kg and length 3 m, which can move freely on the water; see Fig. 10.15a. The front of the boat is 2 m from the dock. What happens when the man walks to the front? Treat the boat as a uniform object.

Solution: The initial position of the CM is indicated by a star in Fig. 10.15a. Since there is no net external force, x_{CM} is fixed. We treat the boat as a point particle at its center. In terms of the initial positions,

$$x_{CM} = \frac{m_1 x_1 + m_2 x_2}{M}$$

$$= \frac{60 \text{ kg} \times 5 \text{ m} + 40 \text{ kg} \times 3.5 \text{ m}}{100 \text{ kg}} = 4.4 \text{ m} \qquad \text{(i)}$$

After the man walks to the front, let us say the front is at a distance d from the dock as in Fig. 10.15b. In terms of the new positions,

$$x_{CM} = \frac{m_1 d + m_2(d + 1.5)}{100} \qquad \text{(ii)}$$

On equating (i) and (ii) we find $d = 3.8$ m. The negative displacement of the man is accompanied by a positive displacement of the boat such that the CM stays fixed. The total momentum of the system stays zero at all times.

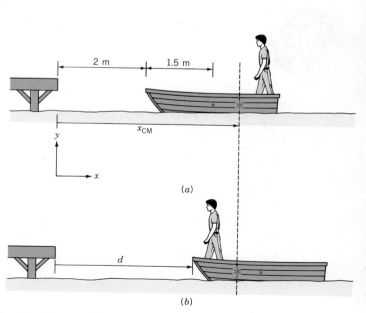

FIGURE 10.15 (a) A person at the stern of a boat. (b) When the person walks to the bow, the position of the center of mass of the system remains unchanged.

EXAMPLE 10.6: Two balls with masses $m_1 = 3$ kg and $m_2 = 5$ kg have initial velocities $v_1 = v_2 = 5$ m/s in the directions shown in Fig. 10.16. They collide at the origin. (a) Find the velocity of the CM 3 s before the collision. (b) Find the position of the CM 2 s after the collision.

Solution: (a) The given time is of no consequence since $\mathbf{v}_{CM}$ is fixed for all times. From Eq. 10.4 in component form

$$v_{CMx} = \frac{m_1 v_{1x} + m_2 v_{2x}}{M}$$

$$= \frac{(3\text{ kg})(-5\cos 37°\text{ m/s}) + (5\text{ kg})(0\text{ m/s})}{8\text{ kg}} = -1.5\text{ m/s}$$

$$v_{CMy} = \frac{m_1 v_{1y} + m_2 v_{2y}}{M}$$

$$= \frac{(3\text{ kg})(-5\sin 37°\text{ m/s}) + (5\text{ kg})(5\text{ m/s})}{8\text{ kg}} = +2\text{ m/s}$$

Thus,

$$\mathbf{v}_{CM} = -1.5\mathbf{i} + 2\mathbf{j}\text{ m/s}$$

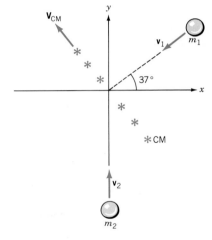

FIGURE 10.16 A collision between two particles. The velocity of the center of mass stays constant.

(b) Since the collision occurs at the origin, the position of the CM 2 s later is

$$\mathbf{r}_{CM} = \mathbf{v}_{CM}t = -3\mathbf{i} + 4\mathbf{j}\text{ m}$$

EXAMPLE 10.7: A 75-kg man sits at the rear end of a platform of mass 25 kg and length 4 m, which moves initially at $4\mathbf{i}$ m/s over a frictionless surface. At $t = 0$, he walks at 2 m/s relative to the platform and then sits down at the front end. During the period he is walking, find the displacements of: (a) the platform; (b) the man; (c) the center of mass.

Solution: Initially the man, the platform, and the CM have the same velocity, $4\mathbf{i}$ m/s, as in Fig. 10.17a. When he starts to walk forward, his increase in momentum must be compensated for by a decrease in the platform's momentum. Let us say the velocity of the platform relative to the ground while he is walking is $\mathbf{v}_{PG} = v_p\mathbf{i}$, as in Fig. 10.17b. The man's velocity relative to the ground is then $\mathbf{v}_{MG} = \mathbf{v}_{MP} + \mathbf{v}_{PG} = (2 + v_p)\mathbf{i}$. From the conservation of momentum we have

$$\Sigma p_x: \qquad 100 \times 4 = 75(2 + v_p) + 25v_p$$

Thus, the velocity of the platform is $v_p = 2.5$ m/s, and the velocity of the man is $v_m = 4.5$ m/s. It takes the man (4 m)/(2 m/s) = 2 s to walk from the rear to the front. In this time the displacement of the platform is $\Delta x_p = v_p\Delta t = 5$ m and that of the man is $\Delta x_m = v_m\Delta t = 9$ m. The velocity of the CM is always 4 m/s; thus, $\Delta x_{CM} = v_{CM}\Delta t = 8$ m.

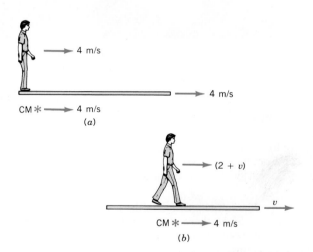

FIGURE 10.17 (a) A person at the rear of a platform moving at 4 m/s. (b) The person walks at 2 m/s relative to the platform. The velocity of the center of mass does not change.

EXERCISE 4. Repeat Example 10.7 with the same values but with the man walking from the front to the rear.

10.4 KINETIC ENERGY OF A SYSTEM OF PARTICLES

We now show that the kinetic energy of a system of particles can in general be divided into two terms: the kinetic energy of the CM and kinetic energy relative to the CM. In Fig. 10.18, the position of the ith particle with respect to the fixed origin O is $\mathbf{r}_i = \mathbf{r}_{CM} + \mathbf{r}'_i$, where $\mathbf{r}_{CM}$ is the position of the CM and $\mathbf{r}'_i$ is the position of the particle with respect to the CM. Taking the time derivative of this equation,

HISTORICAL NOTE: Mass–Energy Equivalence, $E = mc^2$

The belief that the motion of the CM of an isolated system cannot be affected by any internal process led to a profound discovery in 1905. Einstein imagined a closed and isolated box with a light bulb at one end and a detector at the other. It was known from classical electromagnetic theory that light carries linear momentum. Consequently, when the bulb emits a flash of light toward the detector, the box will recoil just as a rifle does. When the flash is received by the detector, the box will experience an equal and opposite impulse, and so the whole system will again come to rest—but at a new position. It appeared that x_{CM} for this isolated system would not be fixed.

Einstein was faced with the unpleasant prospect that the conservation of linear momentum could be violated in this simple thought experiment. Instead of abandoning the conservation law, he noted that while the box has been displaced in one direction, there has been a transfer of energy in the opposite direction from the bulb to the detector. From Example 10.5 we see that for the CM to stay fixed there has to be a transfer of mass from the bulb to the detector. Einstein concluded that the transfer of energy must be equivalent to a transfer of mass. He then went on to derive the equation $E = mc^2$ that relates the mass of a particle to its total energy (see Section 39.7). His profound faith in the conservation of linear momentum, and astonishing insight, led to the most famous result of the theory of special relativity.

---

we find $\mathbf{v}_i = \mathbf{v}_{CM} + \mathbf{v}_i'$. The kinetic energy of the ith particle with respect to O is, therefore, $K_i = \frac{1}{2}m_i(\mathbf{v}_i \cdot \mathbf{v}_i) = \frac{1}{2}m_i(v_{CM}^2 + v_i'^2 + 2\mathbf{v}_{CM} \cdot \mathbf{v}_i')$. The total kinetic energy of the system, $K = \Sigma K_i$, is

$$K = \tfrac{1}{2}(\Sigma m_i)v_{CM}^2 + \Sigma \tfrac{1}{2}m_i v_i'^2 + \mathbf{v}_{CM} \cdot (\Sigma m_i \mathbf{v}_i')$$

The last term, $\Sigma\, m_i\mathbf{v}_i'$, is the total momentum of the system relative to the CM, which from Eq. 10.4 is $M\mathbf{v}_{CM}'$. But the velocity of the CM relative to the CM is obviously zero, so the last term vanishes. We are left with

$$K = K_{CM} + K_{rel} \tag{10.8}$$

where

$K_{CM} = \frac{1}{2}Mv_{CM}^2$ Kinetic energy of the CM relative to the fixed origin O

$K_{rel} = \Sigma\, \frac{1}{2}m_i v_i'^2$ Kinetic energy of the particles relative to the CM

The term K_{rel} may involve translation, rotation, or vibration relative to the CM.

The division in Eq. 10.8 is useful in the analysis of collisions between two particles. In an isolated system $\mathbf{v}_{CM}$, and, therefore, K_{CM}, must have the same value before and after the collision. This means that only K_{rel} is available to take part in reactions or to induce transitions between energy levels.

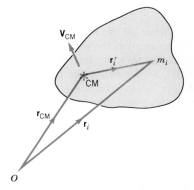

FIGURE 10.18 The position of the ith particle relative to the origin O is $\mathbf{r}_i$ and relative to the center of mass it is $\mathbf{r}_i'$.

EXAMPLE 10.8: A particle of mass $m_1 = 4$ kg moves at $5\mathbf{i}$ m/s, while $m_2 = 2$ kg moves at $2\mathbf{i}$ m/s, Fig. 10.19a. Find K_{CM} and K_{rel}.

Solution: First we evaluate the velocity of the CM using Eq. 10.4:

$$v_{CM} = \frac{(4 \text{ kg})(5 \text{ kg}) + (2 \text{ kg})(2 \text{ m/s})}{6 \text{ kg}} = 4 \text{ m/s}$$

As Fig. 10.19b shows, the velocities relative to the CM are

$$v_1' = v_1 - v_{CM} = +1 \text{ m/s}$$
$$v_2' = v_2 - v_{CM} = -2 \text{ m/s}$$

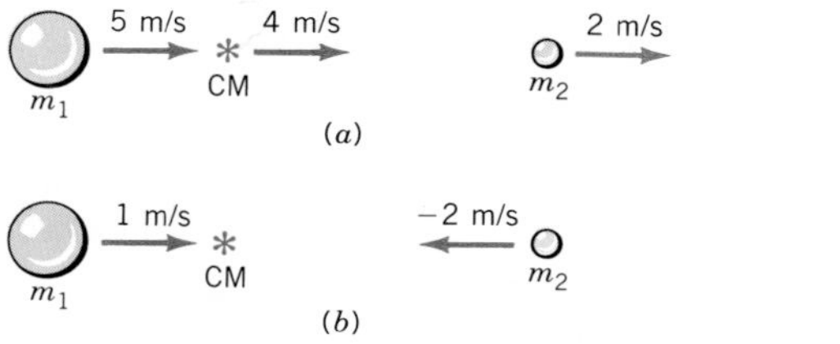

FIGURE 10.19 The velocities of two particles and of their center of mass (a) relative to the laboratory frame, and (b) relative to the frame in which the center of mass is at rest.

The two terms for the total kinetic energy are

$$K_{CM} = \tfrac{1}{2}(m_1 + m_2)v_{CM}^2 = 48 \text{ J}$$
$$K_{rel} = \tfrac{1}{2}m_1v_1'^2 + \tfrac{1}{2}m_2v_2'^2 = 6 \text{ J}$$

EXERCISE 5. For a given amount of available energy, in which case would you learn more about the structure of two bodies: When the larger body strikes the smaller body at rest or vice versa?

10.5 WORK–ENERGY THEOREM FOR A SYSTEM OF PARTICLES (Optional)

In Section 7.3 we derived the work–energy theorem, $W_{NET} = \Delta K$, for a single particle. We now generalize it to systems of particles, for which $K = K_{CM} + K_{rel}$. From Newton's third law we know that the internal forces cancel in pairs, that is, $\Sigma F_{INT} = 0$. However, if the particles can move relative to each other, the net work done by these internal forces on the particles is not necessarily zero; that is, $\Sigma W_{INT} \neq 0$. Consider, for example, a stationary, isolated system of two equal blocks held against a compressed spring. When the spring is released, the work done by the internal force of the spring changes the kinetic energy relative to the CM, while the CM itself stays fixed. The work–energy theorem for a system must include both external and internal work:

$$W_{EXT} + W_{INT} = \Delta K_{CM} + \Delta K_{rel} \qquad (10.9)$$

Since all the basic interactions are conservative, one can always express the work done by the internal forces in terms of a change in some internal potential energy function: $W_{INT} = -\Delta U_{INT}$. We define the **internal energy** of a system as $E_{INT} = K_{rel} + U_{INT}$, where K_{rel} is the internal kinetic energy which includes translation and rotation relative to the CM. Equation 10.9 may then be rewritten as

$$W_{EXT} = \Delta K_{CM} + \Delta E_{INT} \qquad (10.10)$$

This equation says that external work on a system can change the translational kinetic energy of the CM and whatever forms of internal energy the system possesses. The internal energy includes elastic potential energy, gravitational potential energy, electromagnetic energy, chemical energy stored in chemical bonds, nuclear energy within the nucleus, thermal energy associated with the random motion of the particles within the system, and so on.

The CM Equation

From Fig. 10.18 we know that the position of the ith particle is $r_i = r_{CM} + r_i'$. If F_i is the net external force on the ith particle, the work it does is $W_i = \int F_i \cdot dr_i = \int F_i \cdot dr_{CM} + \int F_i \cdot dr_i'$. The total external work, $W_{EXT} = \Sigma W_i$, is therefore the sum of two terms:

$$W_{EXT} = W_{CM} + W_{rel} \qquad (10.11)$$

where

$$W_{CM} = \int F_{EXT} \cdot dr_{CM} \qquad \text{External work associated with the displacement of the CM}$$

$$W_{rel} = \Sigma \int F_i \cdot dr_i' \qquad \text{External work associated with displacements relative to the CM}$$

One can relate W_{CM} to ΔK_{CM} through Newton's second law as it applies to the motion of the CM: $F_{EXT} = Ma_{CM}$. We omit the details of the integration. The result, as expected, is simply

(CM equation) $\qquad W_{CM} = \Delta K_{CM} \qquad (10.12)$

Before we discuss specific cases, let us recall the meaning of $F_{EXT} = Ma_{CM}$. According to this equation, the CM accelerates *as if F_{EXT} were applied at the CM*. It does not matter whether or not F_{EXT} is *really* applied at the CM. Furthermore, this equation deals only with the translational kinetic energy of the CM and ignores all internal energies. For this reason, Eq. 10.12 is only *part* of the work–energy theorem: It has exactly the same information content as $F_{EXT} = Ma_{CM}$, from which it was derived. W_{CM} is the part of W_{EXT} that changes the translational kinetic energy of the CM. Equation 10.12 is actually what we used in Chapter 7 where we treated all bodies as particles—with no possibility of internal energy. The CM equation has interesting implications when $W_{EXT} = 0$ but $\Delta K_{CM} \neq 0$.

EXAMPLE 10.9: Figure 10.20 shows a skater on a frictionless surface as she pushes off a wall. Use the work–energy theorem to discuss her motion.

FIGURE 10.20 When a skater pushes off a wall, the external force due to the wall does no work.

Solution: The net external force on her is due to the wall. But the point of application of this F_{EXT} has zero displacement, so $W_{EXT} = 0$. Her kinetic energy increases, yet the force that accelerates her CM does no work! From the point of view of dynamics, there can be no change in the motion of the CM without an external force. The skater is a nonisolated system in which the internal forces (due to muscles) bring into play external forces (due to the wall). To explain ΔK_{CM} from the point of view of energy, we set $W_{EXT} = 0$ in Eq. 10.10 and find

$$\Delta K_{CM} = -\Delta E_{INT}$$

A decrease in the (internal) chemical energy in her muscles accounts for the increase in her K_{CM}. The chemical energy must also pay for the internal kinetic energy of her arms relative to her CM as she pushes off, and some heat production.

10.6   WORK DONE BY FRICTION (Optional)

A block moving with an initial velocity v_{CM} on a rough surface, as in Fig. 10.21a, eventually comes to a stop. The CM equation, $\mathbf{F}_{EXT} \cdot \mathbf{s}_{CM} = \Delta K_{CM}$, states

$$-f_k s_{CM} = -\tfrac{1}{2} m v_{CM}^2$$

One might say that the change in the translational kinetic energy of the block is equal to the "work done by friction." Yet there is something obviously missing in this treatment. We know that both the block and the surface become warmer. Although the above equation is correct, we can gain insight by applying Eq. 10.10.

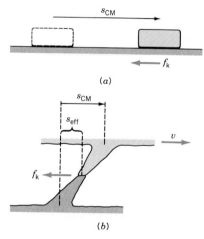

(a)

(b)

FIGURE 10.21 (a) When a block slows down because of friction, $\Delta K_{CM} = -f_k s_{CM}$. (b) The work done by friction is $-f_k s_{eff}$, where s_{eff} is an unknown displacement less than s_{CM}.

Block: $\Delta K_{CM} + \Delta E_{INT} = W_{EXT}$

where W_{EXT} is the external work done by friction on the block. But we know that $\Delta K_{CM} = -f_k s_{CM}$, which means the work done by friction is NOT $(-f_k s_{CM})$! The fact that the block and the surface become warmer means that their internal (thermal) energy has increased, that is, $\Delta E_{INT} > 0$. We conclude that the magnitude of W_{EXT}, the work done by friction, must be somewhat *less* than $f_k s_{CM}$. Since f_k is a given quantity, it means that s_{CM} is *not* the displacement to use in calculating the work done by friction; the appropriate displacement is less. To understand this peculiar situation consider the welds that form between the bottom of the block and the surface as shown in Fig. 10.21b. As the block moves, the displacement of the weld, the point at which the force of friction acts, is less than that of the block.

Thus, the work done by friction is $-f_k s_{eff}$ where s_{eff} is some effective displacement whose magnitude we do not know.*

Consider the system that includes both the block (B) and the plane (P). There is no external work and the CM of the plane does not move. So,

(Block + Plane) $\Delta K_{CM} + (\Delta E_{INT})_B + (\Delta E_{INT})_P = 0$

Thus, the term $f_k s_{CM} (= -\Delta K_{CM})$ represents the gain in internal energy (thermal energy) of the block and the plane, which is manifested as a rise in temperature.

Heat

After a while, both the block and the plane cool down. The extra internal energy produced by the sliding leaves the system as **heat.** Heat is a transfer of energy that results from a difference in temperature between two bodies. When the possibility of heat transfer is taken into account, Eq. 10.10 takes the form

$$\Delta K_{CM} + \Delta E_{INT} = W_{EXT} + Q \qquad (10.13)$$

where Q is the heat transfer into, or out of, the system. We take Q as positive when it enters a system. For historical reasons, Eq. 10.13 is called the first law of thermodynamics and will be discussed in detail in Chapter 19. Now let us consider the block and plane after the block has come to a stop. Clearly, $\Delta K_{CM} = 0$ and $W_{EXT} = 0$, so Eq. 10.13 yields

$$\Delta E_{INT} = Q$$

The process of heat transfer stops when the system and its surroundings, such as the air, reach the same temperature.

EXAMPLE 10.10: Apply the foregoing analysis to the motion of a car.

Solution: Consider a car that accelerates from rest. It accelerates because the external force of friction due to the road acts on the "driven" front (or rear) tires, as drawn in Fig. 10.22a. If a rolling tire does not slip, the part in contact with the road is instantaneously at rest. Since there is no relative displacement between the bottom of the tire and the road, the force of static friction does no work; that is, $W_{EXT} = 0$. Nonetheless we may use Eq. 10.12 to find the change in the CM kinetic energy and the distance required to reach a certain speed. Since $\mathbf{F}_{EXT} \cdot \mathbf{s}_{CM} = f s_{CM}$, where f is the net frictional force, we have

(CM equation) $f s_{CM} = \tfrac{1}{2} M v_{CM}^2$

Note that $f s_{CM}$ is *not* the work done by friction on the car. In fact, there is *no external work* done by friction on the car; that is, $W_{EXT} = 0$. Now let us see what the first law of thermodynamics (FLT) has to offer.

---

* This subtle point was first discussed by B. A. Sherwood and W. H. Bernard, *Am. J. Phys.* 52: 1001 (1984).

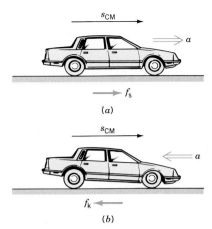

FIGURE 10.22 (a) When a car accelerates the net frictional force is in the forward direction. (b) When a car is braked with all wheels locked, the force of kinetic friction is backward on all wheels.

(FLT) $$\Delta K_{CM} + \Delta E_{INT} = 0 + Q$$

where ΔE_{INT} consists of a loss in chemical energy in the fuel and an increase in the internal kinetic energy of the moving parts of the engine, transmission, and wheels. Since various parts of the car, including the tires, get hotter than the surrounding air, Q leaves the car. The loss in chemical energy pays for all the other terms.

When the car is braked to a halt with all wheels locked, as in Fig. 10.22b, the friction is kinetic. The CM equation and the FLT yield:

(CM equation) $$-f_k s_{CM} = -\tfrac{1}{2}Mv_{CM}^2$$
(FLT) $$\Delta K_{CM} + \Delta E_{INT} = W_{EXT} + Q$$

where ΔK_{CM} is given by the CM equation. Both the work done by friction (W_{EXT}) and Q are negative. In this case the CM kinetic energy is dissipated as heat at the tire–road interface. Heat Q leaves the car from the warm tires, hot engine, and so forth.

10.7   SYSTEMS OF VARIABLE MASS (Optional)

Until now we have dealt with the dynamics only of systems whose mass is constant. We now wish to discuss the dynamics of a system, such as a rocket, whose mass varies. One may try to apply Eq. 10.7 to such a case. With $\mathbf{P} = M\mathbf{v}$,

$$\mathbf{F}_{EXT} = \frac{d\mathbf{P}}{dt} = M\frac{d\mathbf{v}}{dt} + \mathbf{v}\frac{dM}{dt} \qquad (10.14)$$

However, this equation is correct in only a few very special cases. This approach is *not* correct when mass actually enters or leaves the system.

Let us examine the motion of a body of mass M, moving with velocity $\mathbf{v}$ (Fig. 10.23a). Another small body of mass ΔM approaches with velocity $\mathbf{u}$ along the same line. We assume

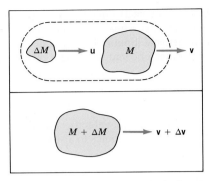

FIGURE 10.23 The system consists of a principal part of mass M and a body of mass ΔM. They undergo a completely inelastic collision.

$\mathbf{u} > \mathbf{v}$ and that after the collision the two bodies stick together and move at velocity $\mathbf{v} + \Delta\mathbf{v}$, as in Fig. 10.23b. By defining our system to include both bodies we can use $\mathbf{F}_{EXT} = d\mathbf{P}/dt$, where $\mathbf{P}$ is the total momentum of the system of *constant* mass, $M + \Delta M$. The change in momentum of the system in time Δt is

$$\Delta\mathbf{P} = (M + \Delta M)(\mathbf{v} + \Delta\mathbf{v}) - M\mathbf{v} - \Delta M\mathbf{u}$$
$$= M\Delta\mathbf{v} - (\mathbf{u} - \mathbf{v})\Delta M$$

Now $(\mathbf{u} - \mathbf{v}) = \mathbf{v}_{rel}$ is the velocity of ΔM relative to M before the collision. We divide both sides of this equation by Δt and take the limit as $\Delta t \to 0$:

$$\mathbf{F}_{EXT} = \frac{d\mathbf{P}}{dt} = M\frac{d\mathbf{v}}{dt} - \mathbf{v}_{rel}\frac{dM}{dt} \qquad (10.15)$$

Notice how Eq. 10.15 differs from Eq. 10.14: The sign of the second term is negative and $\mathbf{v}$ is replaced by $\mathbf{v}_{rel}$. The reason for this discrepancy is that the expression $\mathbf{P} = M\mathbf{v}$ for the total momentum is valid only if we take $\mathbf{v}$ to be $\mathbf{v}_{CM}$. In Eq. 10.15, $\mathbf{v}$ refers to the velocity of just the "principal" part of the system. It is convenient to rewrite it in the form

$$M\frac{d\mathbf{v}}{dt} = \mathbf{F}_{EXT} + \mathbf{v}_{rel}\frac{dM}{dt} \qquad (10.16)$$

The acceleration $d\mathbf{v}/dt$ of the principal part, whose instantaneous mass is M, is determined (i) by $\mathbf{F}_{EXT}$, the net external force on the *whole* system (of constant mass $M + \Delta M$), and (ii) by $\mathbf{v}_{rel}\, dM/dt$, which is the rate at which momentum is being transferred into, or out of, the principal part.

Rocket Thrust

In the case of a rocket, $\mathbf{F}_{EXT}$ would be the weight plus the force due to air resistance and $\mathbf{v}_{rel}$ is what we called $\mathbf{v}_{ex}$ in Section 9.7. The second term in Eq. 10.16 is called the **thrust**:

$$\text{Thrust} = \mathbf{v}_{ex}\frac{dM}{dt} \qquad (10.17)$$

Since the exhaust gases are expelled backward, $\mathbf{v}_{ex} < 0$. Since the mass of the rocket is decreasing ($dM/dt < 0$), the thrust is in the positive direction. The thrust is the reaction force due to the gases as they are expelled from the engines. From Eq. 10.17, we see that the thrust is directly proportional to the exhaust velocity and to the rate at which the gases are expelled.

EXAMPLE 10.11: The mass of the Saturn V rocket is 2.8×10^6 kg at launch time. Of this, 2×10^6 kg is fuel. The exhaust speed is 2500 m/s and the fuel is ejected at the rate of 1.4×10^4 kg/s. (a) Find the thrust of the rocket. (b) What is its initial acceleration at launch time? Ignore air resistance.

Solution: (a) The magnitude of the thrust is given by

$$\text{Thrust} = v_{\text{ex}} \frac{dM}{dt} = 3.5 \times 10^7 \text{ N}$$

(b) From Eq. 10.16 the acceleration is given by dividing both sides by M:

$$\frac{dv}{dt} = -g + \frac{1}{M} v_{\text{ex}} \frac{dM}{dt} = -9.8 + 12.5 = +2.7 \text{ m/s}^2$$

EXAMPLE 10.12: A hopper releases grain at a rate dm/dt onto a conveyor belt that moves at a constant speed v. What is the power of the motor driving the belt? See Fig. 10.24.

Solution: The system is some arbitrary length of belt whose mass we can call M. The mass of the system increases at the

FIGURE 10.24 Grain falls from a hopper onto a conveyor belt that is kept moving at constant speed.

same rate that the grain falls, so $dM/dt = dm/dt$. Since the grain falls vertically, $\mathbf{u} = 0$ and $\mathbf{v}_{\text{rel}} = \mathbf{u} - \mathbf{v} = -\mathbf{v}$. Since the speed is constant, $dv/dt = 0$. Thus, from Eq. 10.16,

$$0 = \mathbf{F}_{\text{EXT}} - \mathbf{v} \frac{dm}{dt}$$

where $\mathbf{F}_{\text{EXT}}$ is the force needed to maintain constant speed because the mass is increasing. The power required ($P = \mathbf{F} \cdot \mathbf{v}$) is

$$P = v^2 \frac{dm}{dt}$$

It is interesting to compare this with the rate at which the kinetic energy of the grain increases:

$$\frac{dK}{dt} = \frac{d}{dt}\left(\frac{1}{2} mv^2\right) = \frac{1}{2} v^2 \frac{dm}{dt}$$

This is only half the power input. The other half is dissipated as heat when the grain lands on the belt and slips relative to it.

SUMMARY

The coordinates of the **center of mass** of a system of discrete particles are given by

$$x_{\text{CM}} = \frac{\Sigma m_i x_i}{M}; \qquad y_{\text{CM}} = \frac{\Sigma m_i y_i}{M}; \qquad z_{\text{CM}} = \frac{\Sigma m_i z_i}{M}$$

where $M = \Sigma m_i$ is the total mass. The velocity of the CM is given by

$$\mathbf{v}_{\text{CM}} = \frac{\Sigma m_i \mathbf{v}_i}{M}$$

The **total linear momentum** of the system is

$$\mathbf{P} = \Sigma m_i \mathbf{v}_i = M\mathbf{v}_{\text{CM}}$$

It is the same as that of an imaginary particle of mass M moving at $\mathbf{v}_{\text{CM}}$.

According to **Newton's second law for a system,** the net external force acting on a system is equal to the rate of change of its total linear momentum:

$$\mathbf{F}_{\text{EXT}} = \frac{d\mathbf{P}}{dt} = M\mathbf{a}_{\text{CM}}$$

The CM accelerates as if all the mass were concentrated at the CM and as if $\mathbf{F}_{\text{EXT}}$ acted at this point.

Newton's first law for a system of particles is equivalent to the conservation of linear momentum:

$$\text{If } \mathbf{F}_{\text{EXT}} = 0, \quad \text{then} \quad \mathbf{P} = M\mathbf{v}_{\text{CM}} = \text{constant}$$

The kinetic energy of a system of particles may be divided into two terms

$$K = K_{CM} + K_{rel}$$

where $K_{CM} = \frac{1}{2}Mv_{CM}^2$ is the kinetic energy of the CM motion and K_{rel} is the kinetic energy of the particles relative to the CM.

ANSWERS TO IN-CHAPTER EXERCISES

1. $U = \Sigma m_i g y_i = g\Sigma m_i y_i = Mgy_{CM}$, where M is the total mass.

2. Since the length of the rod is $\sqrt{5}L$, its mass is $m_3 = \sqrt{5}m$. Its CM is at its midpoint: $x_3 = L/2$ and $y_3 = L$. Thus,

$$x_{CM} = \frac{m(L/2) + 0 + \sqrt{5}m(L/2)}{3m + \sqrt{5}m} = 0.31L$$

$$y_{CM} = \frac{0 + 2m(L) + \sqrt{5}m(L)}{3m + \sqrt{5}m} = 0.81L$$

3. Change the limits on the integral: $\theta = \pi/4$ to $3\pi/4$. So $x_{CM} = 0$ and

$$y_{CM} = \frac{\lambda R^2}{M}\left[-\cos\theta\right]_{\pi/4}^{3\pi/4} = \frac{2\sqrt{2}R}{\pi}$$

where we have used $M = \lambda\pi R/2$.

4. The velocity of the man relative to the ground is $(-2 + v_p)$.

$$\Sigma p_x: 400 = 75(-2 + v_p) + 25v_p$$

Thus, $v_p = 5.5$ m/s and $v_m = 3.5$ m/s, so $\Delta x_p = 11$ m, $\Delta x_m = 7$ m, and $\Delta x_{CM} = 8$ m.

5. When the smaller body strikes the larger body. In this case a smaller fraction of the available energy is taken up by the center of mass motion and a larger fraction is available as energy relative to the CM.

QUESTIONS

1. Explain the relation between Newton's first law and the conservation of linear momentum.

2. A towel slides off a rack. How does its CM move?

3. Is it possible for the center of mass of a high jumper or pole vaulter to pass under the bar while the torso passes over it? If so, how?

4. Stand with your heels touching a wall. Try to bend over to touch your toes. Explain what happens.

5. A rowboat is stranded without oars on a lake. Can the sailor move the boat through actions confined within the boat? If so, explain what might be done.

6. When you stand or lie down, your center of mass is approximately at your belly button. Where is it when you bend to touch your toes?

7. (a) Is it possible to have a system with nonzero kinetic energy but zero total linear momentum? If so, give an example. (b) How about nonzero linear momentum but no kinetic energy?

8. Can the speed of a rocket exceed the exhaust speed?

9. True/false: The velocity of the CM of a system stays constant only if the external force on each particle is zero.

10. The velocity of the CM of an isolated system must stay constant. So how does a rocket accelerate in free space?

11. Consider an arbitrary triangular plate. Divide it into thin strips parallel to one side. What can you say about the location of the CM of the triangle?

12. Only a net external force can change the velocity of the CM of a system. So what purpose does the engine of a car serve?

13. A person is on a railcar that rolls freely. There is a vertical board at one end. Can he propel the railcar by bouncing balls off the board? If so, explain how.

14. Two systems with total masses M_1 and M_2 have their centers of mass located at $\mathbf{R}_1$ and $\mathbf{R}_2$, respectively. Where is the CM of the two systems?

EXERCISES

10.1 Center of Mass

1. (I) Locate the CM of the following molecules: (a) The HCl molecule has a dumbbell shape, as in Fig. 10.25a, with the atoms separated by 1.3×10^{-10} m. The mass of the H atom is 1 u and that of the Cl atom is 35 u. (b) The H_2O molecule has the shape shown in Fig. 10.25b. The mass of the O atom is 16 u. The H and O atoms are separated by 10^{-10} m and the bond angle is 105°.

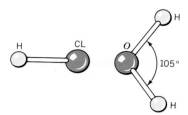

FIGURE 10.25 Exercise 1.

2. (I) The masses and positions of three particles in the xy plane are as follows: 2 kg at $(-2$ m, 3 m); 3 kg at $(-3$ m, $+4$ m); and 5 kg at $(3$ m, -1 m). What is the position of the CM?

3. (I) A fishing rod consists of three uniform 80-cm lengths with masses 10g, 20g, and 30 g. Locate the CM with respect to the end of the 30 g section.

4. (II) A uniform disk of radius R has a circular hole of radius $R/2$ removed as shown in Fig. 10.26a. Locate the CM with respect to the center of the original disk. (*Hint:* Treat the hole as an object of negative mass.)

5. (II) A square of side $2R$ has a circular hole of radius $R/2$ removed. Relative to the center of the square the center of the hole is located at $(R/2, R/2)$ as shown in Fig. 10.26b. Locate the CM with respect to the center of the square. (*Hint:* Treat the hole as an object of negative mass.)

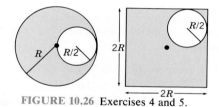

FIGURE 10.26 Exercises 4 and 5.

6. (II) A uniform sphere of radius R has a spherical hole of radius r removed. The center of the hole is at a distance d from the center of the original sphere (see Fig. 10.27). Locate the CM relative to the center of the sphere. (*Hint:* Treat the hole as an object of negative mass.)

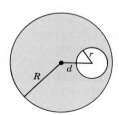

FIGURE 10.27 Exercise 6.

7. (I) Figure 10.28 shows a uniform plate in the shape of an equilateral triangle of side L attached to a square of side L. Locate the CM with respect to the lower left corner of the square. (The CM of a uniform triangular plate is located at one-third the distance from the midpoint of a side to the opposite corner.)

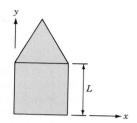

FIGURE 10.28 Exercise 7.

8. (I) (a) The shape in Fig. 10.29a is cut from a uniform sheet of material. Each square of side 2 cm has a mass of 10 g. Locate the CM with respect to the origin. (b) Repeat (a) for the shape in Fig. 10.29b.

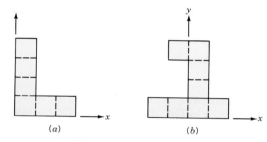

FIGURE 10.29 Exercise 8.

9. (I) Where is the CM of the earth–moon system relative to the center of the earth?

10. (I) Two spheres of radii R and $2R$, made from the same material, are placed in contact. Locate the CM relative to the center of the larger sphere.

11. (I) Jack, of mass 75 kg, is at $x = 0$, whereas Jill, of mass 60 kg, is at $x = 5$ m on a frictionless frozen lake. When they pull on a rope, Jill moves 1.5 m. (a) How far apart are they at this time? (b) Where do they finally meet?

12. (I) A mechanic balances a 20-kg wheel by placing a small lead clip on the wheel rim 18 cm from its center. If the CM of the wheel is 0.3 mm from the center, how much mass should be put on the rim?

13. (II) (a) A uniform thin rod is bent into the shape shown in Fig. 10.30a. Locate the CM with respect to the axes shown. (b) Repeat (a) for the shape in Fig. 10.30b.

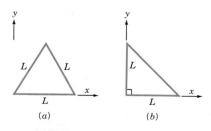

FIGURE 10.30 Exercise 13.

10.3 Motion of the Center of Mass

14. (I) A block of mass $m_1 = 2$ kg has velocity $\mathbf{u}_1 = 5\mathbf{i} - 3\mathbf{j} + 4\mathbf{k}$ m/s and another block of mass $m_2 = 6$ kg has velocity $\mathbf{u}_2 = -3\mathbf{i} + 2\mathbf{j} - \mathbf{k}$ m/s. (a) What is the velocity of the CM? (b) What is the total momentum of the system of two blocks?

15. (II) Two blocks of masses m and $2m$ are held against a massless compressed spring within a box of mass $3m$ and length $4L$ whose center is at $x = 0$ (see Fig. 10.31). All surfaces are frictionless. After the blocks are released they are each at a distance L from the ends of the box when they lose contact with the spring. Show that the position of the center of the box shifts by $L/6$ after both blocks collide with and stick to it.

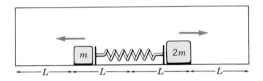

FIGURE 10.31 Exercise 15.

16. (I) The instantaneous positions and velocities of three particles are shown in Fig. 10.32. Find: (a) the position of the CM; (b) the velocity of the CM; (c) the position of the CM 3 s later if there is no external force acting.

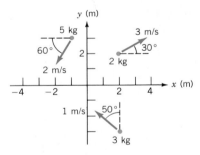

FIGURE 10.32 Exercise 16.

17. (II) A block of mass $m_1 = 1$ kg is at the origin at $t = 0$ and moving at velocity $2\mathbf{i}$ m/s. It is subject to a force $\mathbf{F}_1 = 10\mathbf{j}$ N. Another block of mass $m_2 = 2$ kg is at $x = 10$ m at $t = 0$ and moves at $4\mathbf{j}$ m/s and is subject to a force $\mathbf{F}_2 = 8\mathbf{i}$ N. (a) Find the position and velocity of the CM at $t = 0$. (b) What are the individual accelerations? (c) Use part (b) to find the acceleration of the CM. (d) Use Newton's second law for a system, $\mathbf{F}_{EXT} = M\mathbf{a}_{CM}$, to verify your answer to part (c). (e) Where is the CM at $t = 2$ s?

18. (I) A particle of mass $m_1 = 2$ kg has a position $\mathbf{r}_1 = 2\mathbf{i} + 3\mathbf{j}$ m and a velocity $\mathbf{v}_1 = -\mathbf{i} + 5\mathbf{j}$ m/s, whereas a particle of mass $m_2 = 5$ kg has a position $\mathbf{r}_2 = -5\mathbf{i} + \mathbf{j}$ m and a velocity $\mathbf{v}_2 = 3\mathbf{i} - 4\mathbf{j}$ m/s. Find (a) $\mathbf{r}_{CM}$; (b) $\mathbf{v}_{CM}$; (c) the total linear mo-

mentum; (d) the position of the CM 2 s later if there is no external force.

19. (I) Jack, of mass 75 kg, and Jill, of mass 60 kg, are 10 m apart on a frictionless frozen lake. When they pull on a rope, she moves at 0.3 m/s. (a) How fast does he move? (b) Where do they meet relative to Jack's initial position?

20. (II) The front of a uniform 4-m raft of mass 25 kg is 6 m from the pier. The 60-kg sailor is initially at the rear. (a) Locate the CM of the system with respect to the pier. (b) Where is she with respect to the pier when she walks to the front? Ignore the resistance of the water.

21. (II) A 6-kg object is projected from the top of a cliff 100 m high with an initial velocity of 50 m/s at 53° above the horizontal. At some point in its path it explodes into two pieces. The 4-kg part hits the ground 200 m from the base of the cliff. Assuming both parts land simultaneously, where does the other piece land? Take $g \approx 10$ m/s².

22. (I) A 60-kg girl is sliding east at 2 m/s on a light platform over a frozen frictionless lake. She throws a 1-kg ball such that it travels northward at 5 m/s relative to the ice. (a) What is her new velocity? (b) What is the displacement of the CM of the girl-ball system 4 s after the ball is thrown?

23. (I) A 1000-kg Honda Accord heading east at 15 m/s makes a completely inelastic collision with a 1800-kg Jaguar XJ-6 moving north at 10 m/s on an icy road. (a) What is the velocity of the CM before the collision? (b) Where is the CM 3 s after the collision?

24. (I) A particle of mass $m_1 = 5$ kg moves at $3\mathbf{i} + 2\mathbf{j}$ m/s and a particle of mass $m_2 = 2$ kg moves at $-4\mathbf{i} - 3\mathbf{j}$ m/s. They collide at the origin. Find the position of the CM 3 s before the collision.

10.4 Kinetic Energy of a System of Particles

25. (II) A particle of mass $m_1 = 0.8$ kg moves at $3\mathbf{i}$ m/s and a particle of mass $m_2 = 1.2$ kg moves at $-5\mathbf{i}$ m/s. Find: (a) $\mathbf{v}_{CM}$; (b) the velocities of the particles relative to the CM; (c) the total kinetic energy; (d) the kinetic energy of the CM motion; (e) the kinetic energy relative to the CM.

26. (II) Find the kinetic energy of the CM motion and the kinetic energy relative to the CM in the following cases: (a) a 5-kg block moves at 12 m/s toward a 1-kg block at rest; (b) a 1-kg block moves at 12 m/s toward a 5-kg block at rest.

27. (II) A 10-kg bomb moving at $7\mathbf{i}$ m/s explodes into two parts that continue to move along the x axis. The 4-kg part has velocity $10\mathbf{i}$ m/s. Find: (a) the velocity of the other part; (b) the initial total kinetic energy; (c) the final total kinetic energy; (d) the kinetic energy of the CM motion before the explosion; (e) the kinetic energy of the CM motion after the explosion; (f) the initial kinetic energy relative to the CM; (g) the final kinetic energy relative to the CM.

28. (II) An atom of mass 20 u requires 8×10^{-19} J of excitation energy for a certain reaction that can be induced by a collision with a proton of mass 1 u. Calculate the minimum initial kinetic energies required in the following circumstances: (a) the atom is at rest and the proton is fired at it;

(b) the proton is at rest and the atom is fired at it. (*Hint:* Find the kinetic energy relative to the CM.)

29. (II) A particle of mass $m_1 = 4$ kg moving at $6\mathbf{i}$ m/s collides perfectly elastically with a particle of mass $m_2 = 2$ kg moving at $3\mathbf{i}$ m/s. Find: (a) $\mathbf{v}_{CM}$; (b) the velocities of the particles relative to the CM before the collision; (c) the velocities relative to the CM after the collision.

30. (II) A particle of mass $m_1 = 2$ kg moving at $6\mathbf{i}$ m/s collides elastically with a particle of mass $m_2 = 3$ kg moving at $-5\mathbf{i}$ m/s. Find: (a) $\mathbf{v}_{CM}$; (b) the initial velocities relative to the CM; (c) the final velocities relative to the CM.

31. (II) A particle of mass $m_1 = 5$ kg moving at $4\mathbf{j}$ m/s makes a one-dimensional elastic collision with a particle of mass $m_2 = 3$ kg at rest. Find the kinetic energy (a) relative to the CM, and (b) of the CM motion.

10.7 Variable Mass Systems

32. (I) The *Saturn V* rocket has a mass of 2.5×10^6 kg. It expels gas at the rate of 1.5×10^4 kg/s with an exhaust speed of 2600 m/s relative to the rocket. (a) Find the thrust of the rocket. (b) What is its initial acceleration at the launch pad?

33. (II) A rocket plus its fuel has an initial mass of 135,000 kg and a thrust of 1.8×10^6 N. The exhaust speed is 2 km/s relative to the rocket which is initially at rest on earth. What is the speed of the rocket 10 s after it is launched vertically? Ignore the earth's rotation and air resistance. (See Section 9.7.)

34. (II) Grain from a hopper fills a 10^4-kg railcar at 100 kg/s. What is the acceleration of the railcar after 1 min given that it is pulled by a constant force of 10^3 N and that it moves at 10 cm/s?

PROBLEMS

1. (I) Use integration to locate the CM of the triangular plate of base b and height h shown in Fig. 10.33. The plate has a uniform areal mass density σ kg/m^2.

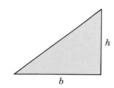

FIGURE 10.33 Problem 1.

2. (I) Find the CM of the flat semicircular plate of radius R shown in Fig. 10.34. (*Hint:* Use the result in Example 10.3.)

FIGURE 10.34 Problem 2.

3. (I) Locate the CM of a thin wire frame that consists of a quarter of a circle and two radial lines of length R, as shown in Fig. 10.35.

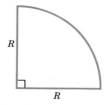

FIGURE 10.35 Problem 3.

4. (I) A thin rod of length L varies in its composition in such a manner that its linear mass density is $\lambda(x) = a + bx$, where x is the distance from one end. Locate the CM relative to $x = 0$.

5. (I) Locate the CM of a hollow cone of height h given that its base is a circle of radius a.

6. (I) What is the change in the earth–sun distance in one-half lunar orbit (between the times the earth, moon, and sun lie along a straight line)?

7. (I) A 75-kg man is at one end of a uniform 25-kg platform of length 4 m, initially at rest. He can walk at 2 m/s relative to the platform, which can roll freely. (a) Where is the CM of the system initially relative to the man? (b) What is the speed of the platform when the man walks to the other end? (c) What is the distance moved by the platform when he reaches the other end?

8. (I) Two boys, each of mass 50 kg, are at the ends of a uniform 4-m platform of mass 25 kg that is moving at 2 m/s. The boy at the rear rolls a 5-kg ball toward the front at 4 m/s relative to himself. (a) What is the speed of the platform while the ball is rolling? (b) How far does the platform move by the time the ball is caught? (c) How far has the CM of the whole system moved in this time?

9. (I) A rocket has a mass of 50,000 kg which includes 45,000 kg of fuel that is expelled at 100 kg/s at 2000 m/s relative to the rocket. (a) What is the thrust of the engine? (b) If the engines fire for 30 s, and the initial velocity is 1 km/s, find the final velocity. The rocket is in free space.

10. (I) A 1200-kg spacecraft is moving in free space at 3×10^4 m/s. Its engine expels the burned gases at 2 km/s relative to the engine. At what rate should the fuel burn to produce an acceleration of 2 m/s^2?

11. (I) A box of side L has three faces removed (see Fig. 10.36). The three remaining edges lie along the x, y, and z axes, respectively. Locate the CM.

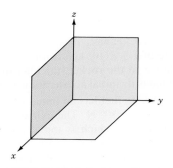

FIGURE 10.36 Problem 11.

12. (II) Figure 10.37 shows a spring within a tube that has a latch to hold the spring in a compressed position. The total mass is M. A particle of mass m and speed u collides with the endplate of the spring. If the energy of the spring when compressed is E. Show that the minimum energy required of the particle to make an inelastic collision is $\left(\dfrac{M+m}{M}\right) E$.

FIGURE 10.37 Problem 12.

CHAPTER 11

Rotation of a Rigid Body about a Fixed Axis

Many rides at an amusement park based on rotation about a fixed axis.

Major Points

1. The equations of **rotational kinematics.**
2. The definition of **moment of inertia,** which is a measure of resistance to angular acceleration.
3. The **rotational kinetic energy** of a rigid body.
4. The definition of **torque.** The concept of a *lever arm.*
5. **Newton's second law for rotation** of a rigid body about a fixed axis relates the net torque to the angular acceleration.

In previous chapters we were concerned only with translational motion. We now broaden our interest to include the rotation of a **rigid body** about a fixed axis of rotation. A rigid body is defined as an object that has fixed size and shape. In other words, the relative positions of its constituent particles remain constant. Although a perfectly rigid body does not exist, it is a useful idealization. By "fixed axis" we mean that the axis must be fixed relative to the body and fixed in direction relative to an inertial frame. When the axis of rotation is also fixed in position, for example, by an axle, as in Fig. 11.1a, the body undergoes pure rotational motion: All particles of the body move in circular paths centered on the axis of rotation. If the axis is not fixed in position, as is the case for a cylinder rolling down an incline (Fig. 11.1b), then it passes through the center of mass. The discussion of general rotation, in which both the position and the direction of the axis change, is quite complex. The only such case we will discuss is that of the gyroscope (spinning top) in the next chapter.

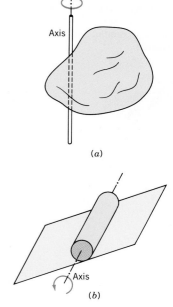

FIGURE 11.1 (a) The axis of rotation is fixed in position and in direction. (b) The axis is fixed only in direction.

11.1 ROTATIONAL KINEMATICS

Figure 11.2 shows a body rotating about a fixed axis at O. In a given time interval all the particles on line OA move to corresponding positions on OB. Although the particles of the body have different linear displacements, they all have the same angular displacement. From the definition of a radian (arc length/radius) we know

$$\theta = s/r \qquad (11.1)$$

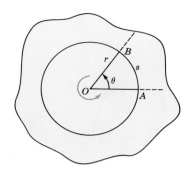

FIGURE 11.2 When a body rotates about an axis fixed at O, each particle travels in a circular path. In radian measure $\theta = s/r$. The angular velocity is defined to be $\omega = d\theta/dt$.

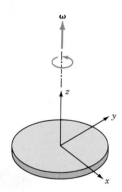

FIGURE 11.3 The direction of the (vector) angular velocity $\boldsymbol{\omega}$ is given by the right-hand rule. In this chapter only the sense of rotation will be considered.

The **average angular velocity** of the body for a finite time interval is defined as

$$\omega_{\text{av}} = \frac{\Delta\theta}{\Delta t} = \frac{\theta_f - \theta_i}{t_f - t_i} \tag{11.2}$$

The unit of angular velocity is rad/s. The **instantaneous angular velocity,** ω, is defined as

$$\omega = \lim_{\Delta t \to 0} \frac{\Delta\theta}{\Delta t} = \frac{d\theta}{dt} \tag{11.3}$$

The angular velocity is the rate of change of the angular position θ with respect to time. We will see later that it is a vector quantity.

Figure 11.3 shows a disk rotating about an axis perpendicular to its flat surface. The only unique direction that characterizes the rotational motion lies along the axis. We associate the (vector) angular velocity with the direction of the axis. To be specific, we adopt a right-hand rule such that when the fingers of the right hand curve in the sense of the rotation, the thumb points in the direction of $\boldsymbol{\omega}$, as shown in Fig. 11.3. In situations that do not involve the full vector nature of angular velocity, it is often convenient to refer to the sense of rotation instead of the direction of the vector. On this basis a counterclockwise rotation, seen by looking down the z axis in Fig. 11.3, may be taken as positive, and a clockwise rotation as negative.

The *period T* is the time for one revolution and the *frequency f* is the number of revolutions per second (rev/s). The relation between period and frequency is $f = 1/T$. If the angular velocity is constant, the instantaneous and average values are equal. In one revolution the body rotates through 2π rad, and so from Eq. 11.2 we have

$$\omega = \frac{2\pi}{T} = 2\pi f \tag{11.4}$$

Note that a frequency $f = 1$ rev/s corresponds to $\omega = 2\pi$ rad/s. We may relate the linear speed of a particle $v = ds/dt$ to the angular velocity ω by using Eq. 11.1. Since $\omega = d\theta/dt = (ds/dt)(1/r)$, we have

$$v = \omega r \tag{11.5}$$

Although all particles have the same angular velocity, their speeds increase linearly with distance from the axis of rotation.

When the angular velocity changes, the **average angular acceleration** is defined as

Angular acceleration

$$\alpha_{\text{av}} = \frac{\Delta\omega}{\Delta t}$$

and the **instantaneous angular acceleration** as

$$\alpha = \frac{d\omega}{dt} \tag{11.6}$$

Angular acceleration is a vector measured in rad/s^2. For rotation about a fixed axis, all the particles of a rigid body have the same angular velocity and angular acceleration. When ω increases, α is in the same sense (direction) as ω. Since virtually all our discussions in this chapter will involve fixed axes, we will not be concerned with the true vector nature of $\boldsymbol{\omega}$ or $\boldsymbol{\alpha}$.

When the angular acceleration is constant, we can find the change in angular velocity from Eq. 11.6. We rewrite it in the form $d\omega = \alpha\,dt$ and integrate:

$$\int_{\omega_0}^{\omega} d\omega = \int_0^t \alpha \, dt$$

to find

$$\omega - \omega_0 = \alpha t \tag{11.7}$$

Next we use this result in Eq. 11.3, $d\theta = \omega \, dt = (\omega_0 + \alpha t)dt$, and again integrate

$$\int_{\theta_0}^{\theta} d\theta = \int_0^t \omega \, dt$$

to find

Equations of rotational kinematics

$$\theta - \theta_0 = \omega_0 t + \tfrac{1}{2}\alpha t^2 \tag{11.8}$$

Finally, we can eliminate t by substituting $t = (\omega - \omega_0)/\alpha$ from Eq. 11.7 into Eq. 11.8. After some algebra, we find

$$\omega^2 = \omega_0^2 + 2\alpha(\theta - \theta_0) \tag{11.9}$$

These **equations for rotational kinematics** for constant angular acceleration are identical in form to the equations of linear kinematics. Both sets of equations are displayed in Table 11.1 to emphasize the analogies between the various quantities.

TABLE 11.1 EQUATIONS OF KINEMATICS

$v = v_0 + at$	$\omega = \omega_0 + \alpha t$
$x = x_0 + v_0 t + \tfrac{1}{2}at^2$	$\theta = \theta_0 + \omega_0 t + \tfrac{1}{2}\alpha t^2$
$v^2 = v_0^2 + 2a(x - x_0)$	$\omega^2 = \omega_0^2 + 2\alpha(\theta - \theta_0)$

A particle moving in a circular path at speed v has a centripetal acceleration $a_r = v^2/r$. In terms of ω (see Eq. 11.5),

$$a_r = \frac{v^2}{r} = \omega^2 r \tag{11.10}$$

If there is angular acceleration, the linear speed v of each particle also changes. We can find the tangential (linear) acceleration $a_t = dv/dt$ by taking the time derivative of Eq. 11.5:

$$a_t = \alpha r \tag{11.11}$$

The net *linear acceleration* is $\mathbf{a} = \mathbf{a}_r + \mathbf{a}_t$. As Fig. 11.4 shows, the two contributions are perpendicular, so the magnitude of the linear acceleration is

$$a = \sqrt{a_r^2 + a_t^2}$$

Note that the terms "linear speed" and "linear acceleration" do not necessarily mean that the particle is traveling in a straight line.

FIGURE 11.4 When a body undergoes angular acceleration, the linear acceleration of each particle has radial and tangential components.

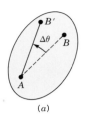 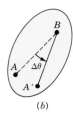

(a)　　　　(b)

FIGURE 11.5 The angular velocity of a body rotating about an axis fixed in direction is the same about any point on the body.

What do you notice about the front wheel?

For future reference let us establish the following: *The angular velocity of a rotating body is the same relative to any point on it.* Consider two points A and B on a rotating body, as in Fig. 11.5. In a given time interval, relative to A, the line AB rotates counterclockwise through the angle $\Delta\theta$ to AB' (Fig. 11.5a). Relative to B, the line BA turns counterclockwise through $\Delta\theta$ to BA' (Fig. 11.5b). The angular displacement, and hence also the angular velocity, is the same about A and B, or about any other point on the body.

Rolling

A common example of rotation is that of a ball or wheel rolling on a surface. Figure 11.6 shows a wheel of radius R rolling without slipping. In one revolution it covers a distance equal to its circumference and takes a time equal to one period T. Thus, the speed of the center is $v_c = (2\pi R)/T = \omega R$, where ω is the angular velocity of the wheel. From Eq. 11.5 this is equal to the tangential speed v_t of a point on the rim relative to the center:

$$v_c = v_t = \omega R \tag{11.12}$$

Rolling is a combination of translation of the center and rotation about the center. The velocity of any point on the rim is the vector sum $\mathbf{v} = \mathbf{v}_c + \mathbf{v}_t$. At the top of the wheel these two velocities are in the same direction, so $v = 2\omega R$, as shown in Fig. 11.7a. At the bottom, they are in opposite directions, so $v = 0$. Since the wheel does not slip, the point of contact with the surface is instantaneously at rest and the wheel momentarily rotates about this point: The particles appear to describe circular paths with angular velocity ω about P as the center, as indicated in Fig. 11.7b. The increase in speed with distance from the point of contact is easily seen in the spokes of a bicycle wheel: The ends of the spokes near the road are fairly distinct, whereas those at the top are blurred.

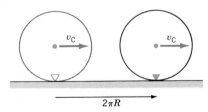

FIGURE 11.6 When a wheel of radius R rolls without slipping at angular velocity ω, the speed of its center is $v_c = \omega R$.

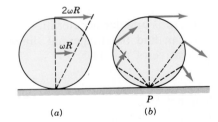

(a)　　　　(b)

FIGURE 11.7 When a wheel rolls without slipping the point of contact with the floor is instantaneously at rest and acts as an instantaneous center of rotation.

EXAMPLE 11.1: A flywheel of radius 20 cm starts from rest, and has a constant angular acceleration of 60 rad/s². Find: (a) the magnitude of the net linear acceleration of a point on the rim after 0.15 s; (b) the number of revolutions completed in 0.25 s.

Solution: (a) The tangential acceleration is constant and given by

$$a_t = \alpha r = (60 \text{ rad/s}^2)(0.2 \text{ m}) = 12 \text{ m/s}^2$$

In order to calculate the radial acceleration we first need to find the angular velocity at the given time. From Eq. 11.7 we have

$$\omega = \omega_0 + \alpha t = 0 + (60 \text{ rad/s}^2)(0.15 \text{ s}) = 9 \text{ rad/s}$$

Then using Eq. 11.10, we have

$$a_r = \omega^2 r = (81 \text{ rad}^2/\text{s}^2)(0.2 \text{ m}) = 16.2 \text{ m/s}^2$$

The magnitude of the net linear acceleration is

$$a = \sqrt{a_r^2 + a_t^2} = 20.2 \text{ m/s}^2$$

(b) From Eq. 11.9,

$$\theta = \tfrac{1}{2}\alpha t^2 = \tfrac{1}{2}(60 \text{ rad/s}^2)(0.25 \text{ s})^2 = 1.88 \text{ rad}$$

This corresponds to $(1.88 \text{ rad})(1 \text{ rev}/2\pi \text{ rad}) = 0.3 \text{ rev}.$

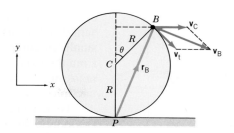

FIGURE 11.8 The dot product of the position and velocity vectors of point B on the rim of a wheel is zero, that is, $\mathbf{r}_B \cdot \mathbf{v}_B = 0$.

EXAMPLE 11.2: Show that the velocity of a point on the rim of a wheel that rolls without slipping is perpendicular to the line joining it to the point of contact.

Solution: From Fig. 11.8 we see that the position vector of point B with respect to the point of contact P is

$$\mathbf{r}_B = R \sin \theta \mathbf{i} + (R + R \cos \theta)\mathbf{j}$$

Its velocity is

$$\mathbf{v}_B = \mathbf{v}_c + \mathbf{v}_t = v\mathbf{i} + (v \cos \theta \mathbf{i} - v \sin \theta \mathbf{j})$$

The dot product $\mathbf{r}_B \cdot \mathbf{v}_B = 0$. (Check this.) Since neither r_B nor v_B is zero, we conclude that these vectors are perpendicular. This confirms that the point of contact acts as an instantaneous center of rotation.

11.2 ROTATIONAL KINETIC ENERGY AND MOMENT OF INERTIA

Figure 11.9 shows a rigid body of arbitrary shape rotating about an axis fixed both in position and in direction. The body consists of point particles of mass m_i at distances r_i from the axis. Note that the r_i are the *perpendicular* distances to the axis, not the distances from an origin. The kinetic energy of the ith particle is $K_i = \tfrac{1}{2}m_i v_i^2$. Since all the particles have the same angular velocity $v_i = \omega r_i$, we have $K_i = \tfrac{1}{2}m_i r_i^2 \omega^2$. The total kinetic energy

$$K = \Sigma K_i = \tfrac{1}{2}\Sigma \, m_i r_i^2 \omega^2$$

may be written in the form

$$K = \tfrac{1}{2}I\omega^2 \tag{11.13}$$

where

$$I = \Sigma m_i r_i^2 \tag{11.14}$$

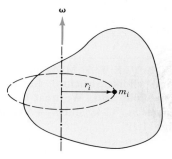

FIGURE 11.9 A rigid body rotating about a fixed axis. The kinetic energy of the ith particle is $K_i = \tfrac{1}{2}m_i v_i^2 = \tfrac{1}{2}m_i r_i^2 \omega^2$.

Moment of inertia

I is called the **moment of inertia** of the body about the given axis. For rotation about a fixed axis, it may be taken to be a scalar. The value of I depends on the location of the axis, that is, on how the mass of the body is distributed relative to the axis. Thus, a body does not possess a unique moment of inertia; different axes through the body are associated with different moments of inertia.

When $K = \tfrac{1}{2}I\omega^2$ is compared with $K = \tfrac{1}{2}mv^2$, we see that moment of inertia is analogous to mass. That is, I plays the same role in rotational motion that m plays in translational motion.

The moment of inertia of a body is a measure of its rotational inertia, that is, its resistance to change in its angular velocity.

Figure 11.10 shows a cylinder, a disk, and a thin ring, all of the same mass. In

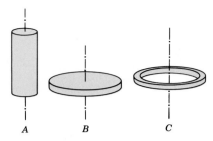

FIGURE 11.10 A cylinder, a disk, and a ring with the same mass. The moments of inertia about the central axis depend on how the mass is distributed relative to the axis: $I_C > I_B > I_A$.

each case the axis of rotation is indicated. The radii of the disk and the ring are the same. In the case of the cylinder, all its particles are close to the axis, so the r_i are small. In comparison, the particles of the disk are spread out to larger distances from the axis. Without any calculation we may conclude that $I_A < I_B$. Whereas the mass of the disk is uniformly spread out, all the particles of the ring are at the maximum possible distance from the axis. Consequently, $I_B < I_C$.

To get a feel for the concept of moment of inertia, hold a hammer at the end of the handle with one hand. Using just your wrist, swing it back and forth as shown in Fig. 11.11a. You will experience the resistance of the hammer to the angular acceleration needed to produce the given motion. The moment of inertia is large since the majority of particles (at the head) are far from the axis (your wrist). You will find it easier to repeat the motion when you hold the head of the hammer, as in Fig. 11.11b. In this case the moment of inertia is small since the majority of particles are close to the axis.

FIGURE 11.11 The moment of inertia of a hammer about an axis through the end of the handle, as in (a), is larger than the moment of inertia about an axis through the head, as in (b).

EXAMPLE 11.3: Many molecules have a simple diatomic, dumbbell-like structure. Let us find the moments of inertia about four axes. We treat the bodies as point particles with masses $m_1 = 3$ kg and $m_2 = 5$ kg. Take $d_1 = 1$ m and $d_2 = 2$ m in Fig. 11.12.

EXAMPLE 11.4: Four point masses lie at the corners of a rectangle with sides of length 3 m and 4 m, as shown in Fig. 11.13. Find the moment of inertia about each of the diagonals. Take $M = 1$ kg.

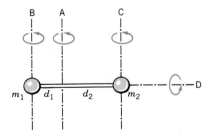

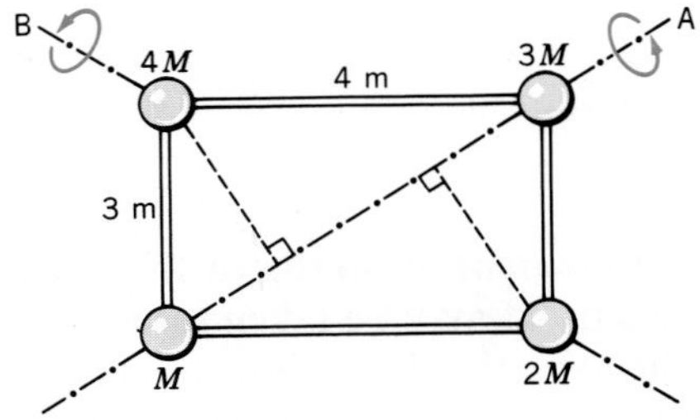

FIGURE 11.12 If the spheres are treated as point particles, the moment of inertia about axis D is zero.

FIGURE 11.13 In finding the moment of inertia one must use the perpendicular distance of each mass from the axis.

Solution:

Axis A: $I_A = m_1 d_1^2 + m_2 d_2^2 = (3\text{ kg})(1\text{ m})^2 + (5\text{ kg})(2\text{ m})^2$
$= 23$ kg·m^2

Axis B: $I_B = m_1(0) + m_2(d_1 + d_2)^2 = 45$ kg·m^2

Axis C: $I_C = m_1(d_1 + d_2)^2 + m_2(0) = 27$ kg·m^2

Axis D: $I_D = 0$

I_D is zero because we treated the masses as point particles.

Solution: For each mass we need its perpendicular distance from the axis. For each axis, two masses do not contribute to the moment of inertia. The other two are at the same distance $3 \sin 53° = 2.4$ m.

$$I_A = (4\text{ kg})(2.4\text{ m})^2 + (2\text{ kg})(2.4\text{ m})^2 = 34.6 \text{ kg·m}^2$$
$$I_B = (1\text{ kg})(2.4\text{ m})^2 + (3\text{ kg})(2.4\text{ m})^2 = 23.0 \text{ kg·m}^2$$

Figure 11.14 shows a rigid body rotating with angular velocity ω about an axis that is located at a perpendicular distance h from the CM. From Eq. 10.8 we know that the kinetic energy of the body has two terms, $K = K_{CM} + K_{rel}$. The first term, K_{CM}, is the kinetic energy associated with the motion of the CM, and K_{rel} is the kinetic energy of motion relative to the CM. In Section 11.1 it was shown that the angular velocity of a rotating body is the same about any point on it. In particular, although the body is rotating about O, the angular velocity about the CM is also ω. Therefore, the total kinetic energy of a body rotating about an axis fixed in direction is

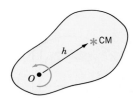

FIGURE 11.14 The moment of inertia I of a body rotating about an axis at O is related to the moment of inertia I_{CM} about a parallel axis passing through the center of mass, $I = I_{CM} + Mh^2$, where h is the distance between the parallel axes.

$$K = K_{CM} + K_{rel}$$

or

$$K = \tfrac{1}{2}Mv^2_{CM} + \tfrac{1}{2}I_{CM}\omega^2 \tag{11.15}$$

From Eq. 11.5, $v_{CM} = \omega h$, and so $K_{CM} = \tfrac{1}{2}M(\omega h)^2$. The total kinetic energy becomes $K = \tfrac{1}{2}(I_{CM} + Mh^2)\omega^2$. This may be expressed as $K = \tfrac{1}{2}I\omega^2$ where

$$I = I_{CM} + Mh^2 \tag{11.16}$$ Parallel axis theorem

This relationship, called the **parallel axis theorem,** relates the moment of inertia I about any axis to the moment of inertia I_{CM} about a parallel axis through the CM. This result is very helpful since it is often easier to evaluate the moment of inertia about the CM.

11.3   MOMENTS OF INERTIA OF CONTINUOUS BODIES

When the distribution of mass of a system of particles is continuous, the discrete sum $I = \Sigma m_i r_i^2$ is replaced by an integral. We have to sum the contributions of infinitesimal mass elements dm shown in Fig. 11.15, each of which contributes $dI = r^2\,dm$ to the moment of inertia. The mass element should be chosen such that all the particles on it are at the same perpendicular distance from the axis. The moment of inertia of the whole body takes the form

$$I = \int r^2\,dm \tag{11.17}$$

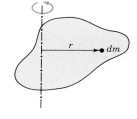

FIGURE 11.15 The mass element dm contributes $dI = r^2 dm$ to the moment of inertia of the system, where r is the perpendicular distance to the axis.

Keep in mind that here the quantity r is the perpendicular distance to an axis, not the distance to an origin. To evaluate this integral, we must express m in terms of r. The following examples illustrate how this is done.

EXAMPLE 11.5: Find the moment of inertia of a *thin rod* of mass M and length L about an axis at one end and perpendicular to the rod, as shown in Fig. 11.16.

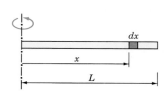

FIGURE 11.16 The rod rotates about an axis through one end.

Solution: The mass of an element of length dx is $dm = \lambda\,dx$, where $\lambda = M/L$ is the linear mass density. The moment of

inertia of the mass element that is at a distance x from the axis is $dI = r^2\,dm = x^2(\lambda\,dx)$. For the entire rod the moment of inertia is

$$I = \int_0^L \lambda x^2\,dx = \frac{\lambda L^3}{3}$$

Since $M = \lambda L$,

(Rod) $$I_{END} = \frac{ML^2}{3} \tag{11.18}$$

EXERCISE 1. What is the moment of inertia of a thin rod about an axis through its CM and perpendicular to the rod?

EXAMPLE 11.6 Find the moment of inertia of a *circular disk or solid cylinder* of radius R about the following axes: (a) through the center and perpendicular to the flat surface; (b) at the rim and perpendicular to the flat surface.

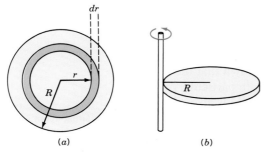

FIGURE 11.17 (a) A disk or cylinder rotates about the central axis perpendicular to the flat surfaces. (b) A disk or cylinder rotates about an axis at its rim.

Solution: (a) Figure 11.17a shows that the appropriate mass element is a circular ring of radius r and width dr. Its area is $dA = 2\pi r\, dr$ and its mass is $dm = \sigma\, dA$, where $\sigma = M/A$ is the areal mass density. The moment of inertia of this element is

$$dI = r^2\, dm = 2\pi\sigma r^3\, dr$$

For the whole body,

$$I = 2\pi\sigma \int_0^R r^3 dr = \tfrac{1}{2}\pi\sigma R^4$$

The mass of the whole disk or cylinder is $M = \sigma A = \sigma\pi R^2$, and so

(Disk or solid cylinder) $I_{CM} = \tfrac{1}{2}MR^2$ (11.19)

(b) The evaluation of the moment of inertia about an axis at the rim, as in Fig. 11.17b, by integration is difficult. The parallel axis theorem, Eq. 11.16, with $h = R$, provides the answer with remarkable simplicity:

$$I_{rim} = I_{CM} + Mh^2 = \tfrac{1}{2}MR^2 + MR^2 = \tfrac{3}{2}MR^2$$

EXAMPLE 11.7:
Find the moment of inertia of a uniform *solid sphere* of mass M and radius R about a diameter.

Solution: The sphere may be divided into disks perpendicular to the given axis, as in Fig. 11.18. The disk at a distance x from the center of the sphere has a radius $r = (R^2 - x^2)^{1/2}$ and a thickness dx. If $\rho = M/V$ is the volume mass density (mass per unit volume), the mass of this elemental disk is $dm = \rho\, dV = \rho\pi r^2\, dx$, or

$$dm = \rho\pi(R^2 - x^2)dx$$

From Eq. 11.19 the moment of inertia of this elemental disk is

$$dI = \tfrac{1}{2}dm\, r^2 = \tfrac{1}{2}\rho\pi(R^2 - x^2)^2\, dx$$

The total moment of inertia is

$$I = \tfrac{1}{2}\rho\pi \int_0^R (R^4 - 2R^2x^2 + x^4)dx$$

$$= \tfrac{1}{2}\rho\pi\left[R^4x - \tfrac{2}{3}R^2x^3 + \tfrac{1}{5}x^5\right]_0^R$$

$$= \tfrac{8}{15}\rho\pi R^5$$

The total mass of the sphere is $M = \rho(\tfrac{4}{3}\pi R^3)$, so the moment of inertia may be written as

(Solid sphere) $I = \tfrac{2}{5}MR^2$ (11.20a)

This is the moment of inertia of a solid sphere about an axis through its center. In Problem 6 you are asked to show that the moment of inertia of a thin spherical shell, of mass M and radius R, about an axis through its center is

(Spherical Shell) $I = \tfrac{2}{3}MR^2$ (11.20b)

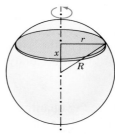

FIGURE 11.18 A uniform sphere rotating about a diameter.

11.4   CONSERVATION OF MECHANICAL ENERGY INCLUDING ROTATION

The principle of conservation of mechanical energy may be applied to problems in rigid body rotation. From Eq. 11.15, we know that the total kinetic energy of a body that rotates about an axis fixed in direction is

$$K = \tfrac{1}{2}Mv_{CM}^2 + \tfrac{1}{2}I_{CM}\omega^2$$

For a body that rolls without slipping, Eq. 11.12 shows that $v_{CM} = \omega R$.

EXAMPLE 11.8: A solid sphere and a disk are released from the same point on an incline, as shown in Fig. 11.19. Given that they roll without slipping, which has the greater speed at the bottom? Ignore dissipative effects.

Solution: If we set the gravitational potential energy $U_g = 0$ at the bottom of the incline, the initial energy is purely potential, and the final energy is purely kinetic:

$$E_i = MgH$$

$$E_f = \tfrac{1}{2}Mv_{CM}^2 + \tfrac{1}{2}I_{CM}\omega^2$$

When we set $E_f = E_i$ and use $v_{CM} = \omega R$, we find

$$v_{CM}^2 = \frac{2MgH}{M + I_{CM}/R^2}$$

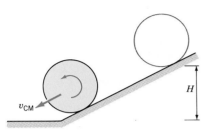

FIGURE 11.19 The rolling sphere has both translational and rotational kinetic energy.

Since $I_{sphere} = \tfrac{2}{5}MR^2$ and $I_{disk} = \tfrac{1}{2}MR^2$, we find $v_{sphere} = \sqrt{10gH/7}$ and $v_{disk} = \sqrt{4gH/3}$. Notice that the mass M and radius R do not appear in the final expressions for the speed. Since $\tfrac{10}{7} > \tfrac{4}{3}$, a sphere is faster than a disk. The object with the smaller moment of inertia is faster; a lower fraction of its energy is rotational energy.

EXAMPLE 11.9: A block of mass $m = 4$ kg is attached to a spring ($k = 32$ N/m) by a rope that hangs over a pulley of mass $M = 8$ kg; see Fig. 11.20. If the system starts from rest with the spring unextended, find the speed of the block after it falls 1 m. Treat the pulley as a disk, so $I = \tfrac{1}{2}MR^2$.

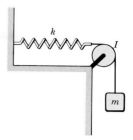

FIGURE 11.20 When the hanging weight falls, the rotational kinetic energy of the pulley must be included.

Solution: Since the rim of the pulley moves at the same speed as the block, the speed of the block and the angular velocity of the pulley are related by $v = \omega R$. When the block falls by a distance x, its potential energy decreases ($\Delta U_g = -mgx$); the potential energy of the spring increases ($\Delta U_{sp} = +\tfrac{1}{2}kx^2$), and both the block and the pulley gain kinetic energy ($\Delta K = \tfrac{1}{2}mv^2 + \tfrac{1}{2}I\omega^2$). From the conservation of mechanical energy, $\Delta K + \Delta U = 0$, we have

$$\tfrac{1}{2}mv^2 + \tfrac{1}{2}I\left(\frac{v}{R}\right)^2 + \tfrac{1}{2}kx^2 - mgx = 0$$

$$\tfrac{1}{2}\left(m + \frac{M}{2}\right)v^2 + \tfrac{1}{2}kx^2 - mgx = 0$$

Notice that R was not needed. Using the given values we find $v = 2.4$ m/s.

11.5   TORQUE

We now turn our attention to rotational dynamics. When Newton's second law is applied to the rotational motion of a body, the analysis is greatly simplified by the introduction of a quantity called **torque.** As we will see, torque is the rotational analog of force: Force causes linear acceleration; torque causes angular acceleration.

In order to lift a stone by using a lever, as shown in Fig. 11.21, a force is required. The effectiveness of the force depends both on its direction and where it is applied relative to the pivot point. The "turning ability" of a force about an axis or pivot is called its *torque*. Figure 11.22 shows two unequal weights hanging on either side of a balanced rod. The forces act at distances r_1 and r_2 from the pivot. Archimedes (ca. 250 B.C.) saw that the condition for the rod to be balanced is $r_2/r_1 = F_1/F_2$, or

$$r_1F_1 = r_2F_2$$

The product rF is the torque of the force F about the pivot. Clearly, the torque increases linearly with distance from the pivot.

Leonardo da Vinci extended this concept of torque to cases in which the force does not act perpendicular to the lever. In Fig. 11.23 a force is used to turn a rod

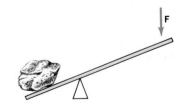

FIGURE 11.21 The effectiveness of the force F in lifting the stone depends on its direction and its point of application.

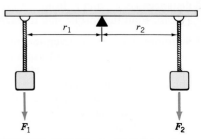

FIGURE 11.22 The rod is balanced when $r_1F_1 = r_2F_2$.

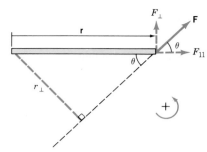

FIGURE 11.23 The magnitude of the torque exerted by the force F is $\tau = rF \sin \theta$ where θ is the angle between the vectors $\mathbf{r}$ and $\mathbf{F}$. One can also use the expressions $\tau = r_\perp F$, where $r_\perp$ is the lever arm, or $\tau = rF_\perp$, where $F_\perp$ is the component of $\mathbf{F}$ perpendicular to $\mathbf{r}$.

Lever arm

pivoted at one end. The component of the force parallel to the rod merely pulls at the pivot, so only the perpendicular component contributes to the turning effect. Thus, the torque is $rF_\perp$. Da Vinci pointed out that one could also consider the force as acting at the effective distance $r_\perp$, called the *lever arm*. The lever arm is the perpendicular distance from the origin (pivot or axis) to the line of action of the force—which is obtained by extending its arrow either forward or backward. The torque is also $r_\perp F$. From the figure we see that $r_\perp = r \sin \theta$ and $F_\perp = F \sin \theta$, so the two expressions, $r_\perp F$ and $rF_\perp$, are equivalent:

$$\tau = r_\perp F = rF_\perp \tag{11.21}$$

Torque

The torque of a force F that acts at a distance r from the origin is defined to be

$$\tau = rF \sin \theta \tag{11.22}$$

where θ is the angle between the vectors $\mathbf{r}$ and $\mathbf{F}$. The SI unit of torque is N·m. Although this has the same dimensions as energy, these two concepts are unrelated. We will see in Chapter 12 that whereas energy is a scalar, torque is a vector.

Since this chapter is restricted to the special case of a rigid body rotating about a fixed axis, we may postpone having to deal with the full vector nature of torque. Instead, we specify its sense rather than its proper direction. We may adopt the convention that a torque tending to produce a counterclockwise rotation is positive, as indicated by the circular arc and plus sign in Fig. 11.23. In any case, the choice must be consistent with the convention chosen for angular velocity.

EXAMPLE 11.10: Three forces, F_1, F_2, and F_3, act on a rod at distances r_1, r_2, and r_3 from the pivoted end, as shown in Fig. 11.24a. Find the torque due to each force about the pivot.

Solution: The sign convention for torque is indicated in Fig. 11.24a. The sign of each torque is determined by considering which way the rod would rotate if the given force were the only one acting. We may apply the expression $\tau = rF \sin \theta$ directly to Fig. 11.24a. Notice, however, that the given angles are not necessarily the angle θ between $\mathbf{r}$ and $\mathbf{F}$ required in Eq. 11.22.

Figure 11.24b illustrates the use of $\tau = r_\perp F$ for F_1 and $\tau = rF_\perp$ for F_2 and F_3. In any problem, you may use whatever combination of the expressions for τ you find convenient.

$$\tau_1 = -r_1 F_1 \sin (90° + \theta) = -r_1 F_1 \cos \theta$$
$$\tau_2 = +r_2 F_2 \sin (180° - \alpha) = +r_2 F_2 \sin \alpha$$
$$\tau_3 = +r_3 F_3 \sin (90° - \phi) = +r_3 F_3 \cos \phi$$

EXERCISE 2. A horizontal force F_4 acts toward the right at a distance r_4 from the pivot. What is its torque?

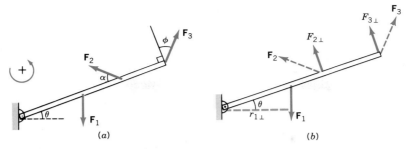

FIGURE 11.24 (a) Three forces, F_1, F_2, and F_3, acting on a rod at distances r_1, r_2, and r_3, respectively. (b) The lever arm $r_\perp$ has been used to find τ_1, whereas the components of $\mathbf{F}_2$ and $\mathbf{F}_3$ perpendicular to the rod have been used to find τ_2 and τ_3.

11.6 ROTATIONAL DYNAMICS OF A RIGID BODY (FIXED AXIS)

Figure 11.25 shows a rigid body rotating about a fixed axis, where $\mathbf{F}_i$ is the net external force on the ith particle of mass m_i. Any component of $\mathbf{F}_i$ parallel to the axis is counteracted by the reaction of the supports. For the same reason any radial component is also balanced. Only the component F_{it}, tangential to the circular path, will accelerate the particle. We may relate the linear acceleration of the particle to the angular acceleration of the body with Eq. 11.11, $a_t = \alpha r$. Thus, the second law, $\mathbf{F}_i = m\mathbf{a}_i$, becomes

$$F_{it} = m_i a_{it} = m_i r_i \alpha$$

The torque on the particle about the axis is

$$\tau_i = r_i F_{it} = m_i r_i^2 \alpha$$

When we add the torques on all the particles, we find

(Rigid body, fixed axis) $\tau = I\alpha$ (11.23)

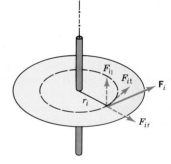

FIGURE 11.25 The ith particle on a rigid body rotating about a fixed axis is acted on by a force $\mathbf{F}_i$. The radial component and the component parallel to the axis are counteracted by the supports of the axle. Only the component F_{it} tangential to the motion can produce angular acceleration.

where $\tau = \Sigma \tau_i$ is the net external torque on the body and $I = \Sigma m_i r_i^2$ is the moment of inertia about the given axis.

Equation 11.23 has the same form as $F = Ma$. Thus, torque is the rotational analog of force: Torque causes angular acceleration and force causes linear acceleration. Notice, however, that Eq. 11.23 is *not* a vector equation. We assert without proof that it is valid in two situations:

(i) The axis is fixed in position and direction.

(ii) The axis passes through the CM and is fixed in direction only. The equation, $\tau_{CM} = I_{CM}\alpha_{CM}$, is valid even if the CM is accelerating.

EXAMPLE 11.11: A flywheel of mass $M = 2$ kg and radius $R = 40$ cm rotates freely at 600 rpm. Its moment of inertia is $\frac{1}{2}MR^2$. A brake applies a force $F = 10$ N radially inward at the edge as shown in Fig. 11.26. If the coefficient of friction is $\mu_k = 0.5$, how many revolutions does the wheel make before coming to rest?

Solution: We choose the initial sense of the angular velocity as positive. The force of friction is $f = \mu_k F$ and its (counterclockwise) torque is $\tau = -fR$. From $\tau = I\alpha$, we have

$$-(\mu_k F)R = (\tfrac{1}{2}MR^2)\alpha$$

$$\alpha = -\frac{2\mu_k F}{MR} = -12.5 \text{ rad/s}^2$$

To find the number of revolutions, we need the angular displacement, which can be found from

$$\omega^2 = \omega_0^2 + 2\alpha\Delta\theta$$

Since 600 rev is equivalent to $600 \times 2\pi$ rad, we see that $\omega_0 =$

20π rad/s, and we have

$$0 = (20\pi \text{ rad/s})^2 + 2(-12.5 \text{ rad/s}^2)\Delta\theta$$

Thus, $\Delta\theta = 16\pi^2$ rad. The number of revolutions is $(16\pi^2 \text{ rad})(1 \text{ rev}/2\pi \text{ rad}) = 8\pi$ revolutions.

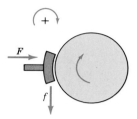

FIGURE 11.26 A wheel is slowed down by the application of force **F**. With the chosen positive sense, the frictional torque is negative.

EXAMPLE 11.12: A disk-shaped pulley has mass $M = 4$ kg and radius $R = 0.5$ m. It rotates freely on a horizontal axis, as in Fig. 11.27. A block of mass $m = 2$ kg hangs by a string that is tightly wrapped around the pulley. (a) What is the angular velocity of the pulley 3 s after the block is released? (b) Find the speed of the block after it has fallen 1.6 m. Assume the system starts at rest.

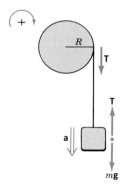

FIGURE 11.27 The acceleration of the block is determined by $F = ma$. The angular acceleration of the pulley is determined by $\tau = I\alpha$.

Solution: This problem may be tackled by using either dynamics or energy conservation. We illustrate the former. The coordinate axes and the sign convention for the torque and angular velocity are indicated in Fig. 11.27. Since the string is tangential to the pulley, the torque on it due to the tension is $\tau = TR$. The two forms of Newton's second law for the block and the pulley yield

Block	$(F = ma)$	$mg - T = ma$	(i)
Pulley	$(\tau = I\alpha)$	$TR = (\frac{1}{2}MR^2)\alpha$	(ii)

where we have used $I = \frac{1}{2}MR^2$ for a disk. Since the block and the rim of the pulley have the same speed (the string does not slip), we have $v = \omega R$ and $a = \alpha R$. Using this in (ii) we find

$$T = \tfrac{1}{2}Ma \qquad \text{(iii)}$$

Adding (i) and (iii) leads to

$$a = \frac{mg}{m + M/2}$$
$$= 5 \text{ m/s}^2 \qquad \text{(iv)}$$

where we have used $g \approx 10$ N/kg.
(a) To find ω after 3 s, we use Eq. 11.7

$$\omega = \omega_0 + \alpha t = 0 + \left(\frac{a}{R}\right)t = 30 \text{ rad/s}$$

(b) To find the speed of the block we use

$$v^2 = v_0^2 + 2a\Delta y = 0 + 2(5 \text{ m/s}^2)(1.6 \text{ m})$$

Thus, $v = 4$ m/s.

EXERCISE 3. Repeat part (b) of Example 11.12 by using the conservation of energy. Use $g \approx 10$ N/kg.

EXAMPLE 11.13: Figure 11.28 shows a sphere of mass M and radius R that rolls without slipping down an incline. Its moment of inertia about a central axis is $\frac{2}{5}MR^2$. (a) Find the linear acceleration of the CM. (b) What is the minimum coefficient of friction required for the sphere to roll without slipping?

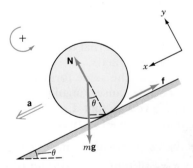

FIGURE 11.28 A sphere rolls down an incline. Since the sphere is not "driven," the force of friction is directed up the incline.

Solution: Since the sphere is not "driven" by a chain or an axle the force of friction must be directed backward, up the slope. If there is no slipping, the point of contact is instantaneously at rest and so the friction is static. (We ignore the effects of rolling friction. This is equivalent to assuming that there are no energy losses.)

(a) We apply the second law in the form $F_{\text{EXT}} = Ma_{\text{CM}}$ to the translational motion of the CM and $\tau_{\text{CM}} = I_{\text{CM}}\alpha_{\text{CM}}$ to the rotational motion about the CM. The coordinate axes and the sign convention for torque are indicated in the figure. The subscript CM is dropped for simplicity.

(ΣF_x) $Mg \sin \theta - f = Ma$ (i)

$(\Sigma \tau)$ $fR = I\alpha$ (ii)

Since the sphere rolls without slipping, the speed of the center is $v = \omega R$, which means $a = \alpha R$. Using this and $I = \frac{2}{5}MR^2$ in (ii) leads to

$$f = \tfrac{2}{5}Ma \qquad \text{(iii)}$$

When (i) and (iii) are added, we find

$$a = \tfrac{5}{7}g \sin \theta \qquad \text{(iv)}$$

(b) Substituting (iv) into (iii) yields

$$f = \tfrac{2}{7}Mg \sin \theta \qquad \text{(v)}$$

We use (v) to find the minimum coefficient of friction required for the sphere to roll without slipping. By definition, $f = \mu N$ where $N = Mg \cos \theta$. Combining this with (v) we have $\mu = \frac{2}{7} \tan \theta$. If the coefficient of static friction is less than this value, the sphere will slip as it rolls down the incline.

EXERCISE 4. Use (iv) to find the speed of the sphere after it falls a vertical distance H, starting from rest.

EXAMPLE 11.14: A car goes around an unbanked curve of radius r at speed v. Find the critical speed at which it tends to overturn.

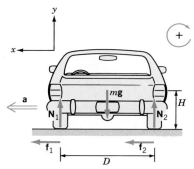

FIGURE 11.29 Forces acting on a car that is going around a curve.

Solution: Figure 11.29 shows the forces on the car, the coordinate axes, and the sign convention for torque. In general, the normal forces and the frictional forces will not be the same at the inner and outer wheels. We assume the CM is at height H and the wheels are a distance D apart. Since we want all four wheels to stay on the road, the angular acceleration of the car about its CM should be zero. That is, the car should be in rotational equilibrium. We apply the two forms of the second law to the uniform circular motion and to the rotation about the CM:

(ΣF_x) $f_1 + f_2 = \dfrac{mv^2}{r}$ (i)

(ΣF_y) $N_1 + N_2 - mg = 0$ (ii)

$(\Sigma \tau)$ $(f_1 + f_2)H + (N_1 - N_2)\dfrac{D}{2} = 0$ (iii)

We substitute (i) into (iii) and $N_2 = mg - N_1$ from (ii) into (iii) to find

$$N_1 = m \left(\frac{g}{2} - \frac{v^2 H}{rD} \right) \qquad \text{(iv)}$$

When $N_1 = 0$, the inner wheel has lost contact with the road. From (iv) we see that this will occur when

$$v_{max}^2 = \frac{grD}{2H}$$

For speeds greater than this, the car will tend to overturn. The equation for v_{max} shows that stability improves as the width D increases and as the CM height H decreases. With some typical values, such as $r = 50$ m, $D = 1.5$ m, and $H = 0.5$ m, we find $v_{max} \approx 27.5$ m/s or 100 km/h.

EXAMPLE 11.15: A uniform rod of length L and mass M is pivoted freely at one end. (a) What is the angular acceleration of the rod when it is at angle θ to the vertical? (b) What is the tangential linear acceleration of the free end when the rod is horizontal? The moment of inertia of a rod about one end is $\frac{1}{3}ML^2$.

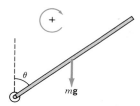

FIGURE 11.30 The angular acceleration of the rod is produced by the torque due to its weight.

Solution: Figure 11.30 shows the rod at an angle θ to the vertical. If we take torques about the pivot we need not be concerned with the force due to the pivot. The torque due to the weight is $mgL/2 \sin \theta$, so the second law for the rotational motion is

$$\frac{mgL}{2} \sin \theta = \frac{ML^2}{3} \alpha$$

Thus,

$$\alpha = \frac{3g \sin \theta}{2L}$$

(b) When the rod is horizontal $\theta = \pi/2$ and $\alpha = 3g/2L$. From Eq. 11.11 the tangential linear acceleration is

$$a_t = \alpha L = \frac{3g}{2}$$

This is *greater* than the acceleration of an object in free-fall!

EXERCISE 5. How could you demonstrate that the tip of the pivoted rod has an acceleration greater than that in free-fall?

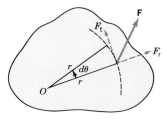

FIGURE 11.31 A force **F** acts on a rigid body and exerts a torque $\tau = rF_t$, where F_t is the component tangential to the motion. The work done by the torque in an infinitesimal rotation through angle $d\theta$ is $dW = \tau\,d\theta$.

11.7   WORK AND POWER

Figure 11.31 shows a body rotating about a fixed axis at O, normal to the page. It is acted on by an external force **F** that has a radial component F_r and a tangential component F_t. In an infinitesimal time interval dt, the point of application of the force moves in circular arc of length $r\,d\theta$. The work done by the tangential component is

$$dW = (F_t)(r\,d\theta) = \tau\,d\theta$$

since $\tau = F_t r$ in this case. From the definition of instantaneous power, $P = dW/dt$, applied to the above equation we find

$$P = \tau\omega \tag{11.24}$$

which is analogous to $P = Fv$.

In order to derive the work–energy theorem for rotational motion, we first express torque in a convenient form. Using the chain rule we have

$$\tau = I\frac{d\omega}{dt} = I\frac{d\omega}{d\theta}\frac{d\theta}{dt} = I\frac{d\omega}{d\theta}\omega$$

Work–energy theorem for rotational motion

We next use this result in $dW = \tau\,d\theta = I\omega\,d\omega$ and integrate to find

$$W = \tfrac{1}{2}I\omega_f^2 - \tfrac{1}{2}I\omega_i^2 \tag{11.25}$$

The work done by a torque on a rigid body rotating about a fixed axis leads to a change in its rotational kinetic energy.

EXAMPLE 11.16: A motor rotates a pulley of radius 25 cm at 20 rpm. A rope around the pulley lifts a 50-kg block, as shown in Fig. 11.32. What is the power output of the motor?

Solution: The tension in the rope is equal to the weight since there is no acceleration; thus, $T = 500$ N. The torque required of the motor is simply $\tau = TR = (500)(0.25) = 125$ N·m. For the power to be in watts, the angular velocity must be in rad/s. In this case 20 rpm is equivalent to $2\pi/3$ rad/s. The power required is

$$P = \tau\omega = (125 \text{ N} \cdot \text{m})\left(\frac{2\pi}{3}\text{ rad/s}\right) = 260 \text{ W}$$

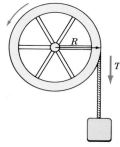

FIGURE 11.32 The motor must supply a torque TR to keep the block rising at constant speed.

11.8   DYNAMICS OF ROLLING FRICTION
(Optional)

Figure 11.33 shows a wheel of mass M and radius R that is rolling without slipping on a horizontal surface. The forces on the wheel are usually drawn as shown, where f is the force of rolling friction that causes the wheel to roll. According to $\mathbf{F}_{EXT} = M\mathbf{a}_{CM}$ we would write $f = Ma$, where a is directed opposite to the velocity; that is, f is the retarding force that slows down the wheel. From the rotational form of the second law, $\tau_{CM} = I_{CM}\alpha_{CM}$, we would write $fR = I\alpha$. But this torque tends to *increase* the angular velocity and therefore to speed up the wheel! This description is seriously flawed.

In fact, as we pointed out in Chapter 6, the wheel and the surface are not perfectly rigid and therefore are deformed. The net force on the wheel due to the surface effectively acts at a point *ahead* of the center and is directed backward. Its vertical

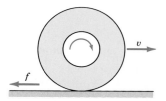

FIGURE 11.33 The force of friction on a rolling (undriven) wheel is backward.

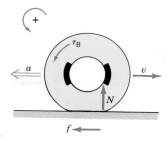

FIGURE 11.35 The brakes are applied to a wheel. The torque due to the brakes is in the same sense as that due to **N** and opposite to the torque due to friction.

and horizontal components are N and f, as drawn in Fig. 11.34, where N does not pass through the center. The torque due to f tends to speed up the wheel, but it is counteracted by a larger torque τ_N due to N (with an unknown lever arm). With the axes and sign convention shown, the two forms of Newton's second law are

$$f = Ma \tag{i}$$
$$\tau_N - fR = I\alpha \tag{ii}$$

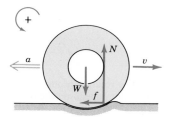

FIGURE 11.34 A rolling wheel is slowed down by the torque due to the vertical force **N**, which effectively acts at a point ahead of the center.

If there is no slipping then $v = \omega R$ and so $a = \alpha R$. Using this and $I = \frac{1}{2}MR^2$ in (ii), one finds $\tau_N = \frac{3}{2}fR$. That is, the torque due to N is greater than the torque due to f. From the point of view of dynamics, rolling friction is due to the combined effect of f and N. As was also pointed out in Chapter 6, rolling friction is associated with the energy losses that accompany the deformation of the wheel.

Now suppose that another torque acts on the wheel. This happens when the wheel is driven, or when the brakes are applied. In either case, the torque due to the additional force is added to those of the forces f and N. However, the magnitude and/or direction of the net frictional force changes.

When the brakes are applied, as in Fig. 11.35, their torque, τ_B, is in the same sense as τ_N and considerably greater. In order to slow down the translational motion of a bicycle (say), the frictional force must be directed rearward and is much larger than for a freely rolling wheel. If the wheel keeps rolling, then from the second law we have

$$f = Ma; \qquad \tau_B + \tau_N - fR = I\alpha$$

where M is the mass of the bicycle and I is the moment of inertia of the wheel. If the wheel keeps rolling, the point of contact with the road is instantaneously at rest; hence, the force

on the wheel is one of static friction. The maximum value of f is the maximum force of static friction $f_{s(max)}$. If the brakes are applied too hard, the wheel locks and then slides over the road. In this case, f is the force of kinetic friction. Maximum braking action is obtained when the wheel is just on the verge of slipping, but in practice this condition is difficult to achieve.

When the wheel is driven, as in Fig. 11.36, the torque due to the axle τ_A is in the opposite sense to that of τ_N. Since the bottom of the wheel is pushing backward against the road, the force of friction is in the forward direction. From the two forms of the second law,

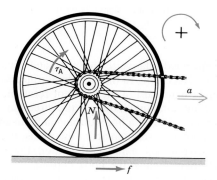

FIGURE 11.36 In a "driven" wheel, the torque due to the axle is opposite to the torques due to friction and the force **N**.

$$f = Ma; \qquad \tau_A - \tau_N - fR = I\alpha$$

Again, if the wheel does not slip, the maximum value of f is $f_{s(max)}$. If the wheel is driven harder, the bottom of the wheel slips and the friction becomes kinetic.

11.9 VECTOR NATURE OF ANGULAR VELOCITY (Optional)

To show that angular velocity, $\omega = d\theta/dt$, is a vector, we will consider infinitesimal angular displacements, $d\theta$. An angular displacement is a quantity that has both magnitude and direction. The magnitude is measured in radians (say), and its direc-

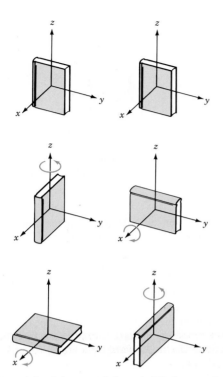

FIGURE 11.37 A book is rotated through 90° about each of the x and z axes. The final orientation depends on the order in which the rotations are performed.

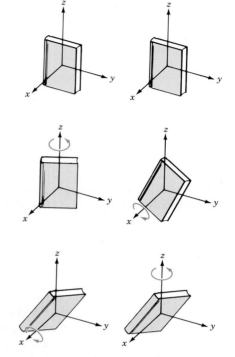

FIGURE 11.38 The book is rotated through infinitesimal angles. The final orientation does not depend on the order of the rotations.

tion may be specified relative to the x, y, and z axes with the right-hand rule. According to the right-hand rule, when the fingers of the right hand curl in the sense of rotation, the thumb (held out as in hitch-hiking) indicates the direction.

Consider a rotation of a book through a finite angle of $\pi/2$. The angle θ_1 involves a rotation about the z axis; θ_2 is about the x axis. Starting from a given initial orientation, Fig. 11.37 shows that the final orientation of the book depends on whether the rotation through θ_1, or θ_2, is performed first. One can express this as

$$\theta_1 + \theta_2 \neq \theta_2 + \theta_1$$

That is, finite angular displacements do not obey the commutative property of vector addition and therefore are not vectors.

Now consider small rotations about the same axes. Figure 11.38 suggests that in the limit of infinitesimal rotations, the final orientation of the book does not depend on the order in which the rotations occur. Thus,

$$d\boldsymbol{\theta}_1 + d\boldsymbol{\theta}_2 = d\boldsymbol{\theta}_2 + d\boldsymbol{\theta}_1$$

Infinitesimal rotations *do* obey the laws of vector addition and therefore are vector quantities. Since the angular velocity is defined as $\boldsymbol{\omega} = d\boldsymbol{\theta}/dt$, where dt is a scalar quantity, $\boldsymbol{\omega}$ is also a vector.

SUMMARY

The **equations of rotational kinematics** for constant angular acceleration are

$$\omega = \omega_0 + \alpha t$$
$$\theta = \theta_0 + \omega_0 t + \tfrac{1}{2}\alpha t^2$$
$$\omega^2 = \omega_0^2 + 2\alpha(\theta - \theta_0)$$

The **moment of inertia** of a system of discrete particles about a given axis is

$$I = \Sigma m_i r_i^2$$

where r_i is the *perpendicular distance to the axis*. The moment of inertia is a measure of the object's rotational inertia; its resistance to change in angular velocity.

The **kinetic energy of a rigid body** rotating about a *fixed axis* is

$$K = \tfrac{1}{2}I\omega^2$$

When a body rolls without slipping, the linear speed of its center is related to the angular velocity about the center by $v_c = \omega R$.

The total kinetic energy of a rigid body is the sum of its translational and rotational kinetic energies:

$$K = \tfrac{1}{2}Mv_{CM}^2 + \tfrac{1}{2}I_{CM}\omega^2$$

where I_{CM} is the moment of inertia about an axis through the CM.

The **torque** exerted by a force F that acts at a distance r from the origin is

$$\tau = rF \sin \theta$$
$$= rF_\perp = r_\perp F$$

where θ is the angle between the vectors $\mathbf{r}$ and $\mathbf{F}$. The quantity $F_\perp$ is the component of the force perpendicular to the vector $\mathbf{r}$. The quantity $r_\perp$, called the *lever arm*, is the perpendicular distance from the origin to the line of action of the force.

When a torque τ acts on a rigid body that is free to rotate about a *fixed axis*, the angular acceleration is determined by **Newton's second law for rotation:**

(Fixed axis) $$\tau = I\alpha$$

where I is the moment of inertia about the given axis.

When a torque τ acts on a body rotating at angular velocity ω the power delivered is

$$P = \tau\omega$$

Table 11.2 displays the similarities in the forms of the equations for translational and rotational motion.

TABLE 11.2

Translation	Rotation
$K = \tfrac{1}{2}mv^2$	$K = \tfrac{1}{2}I\omega^2$
$F = ma$	$\tau = I\alpha$
$P = Fv$	$P = \tau\omega$

ANSWERS TO IN-CHAPTER EXERCISES

1. We merely change the limits on the integral:

$$I = \int_{-L/2}^{L/2} \lambda x^2 \, dx = \frac{\lambda L^3}{12} = \frac{ML^2}{12}$$

since $M = \lambda L$.

2. The angle between F_4 and the rod is θ; thus, $F_{4\perp} = F_4 \sin \theta$ and

$$\tau_4 = -r_4 F_{4\perp} = -r_4 F_4 \sin \theta$$

3. The initial and final energies are

$$E_i = mgH$$

$$E_f = \tfrac{1}{2}mv^2 + \tfrac{1}{2}I\omega^2 = \left(\frac{m}{2} + \frac{M}{4}\right)v^2$$

Equating these we find $v = 4$ m/s.

4. Using $v^2 = v_0^2 + 2a\Delta x$ with $\Delta x = H/\sin \theta$, we find $v = \sqrt{10gH/7}$. This is the same as the result in Example 11.5.

5. If a coin is placed at the end, it will lose contact with the rod.

QUESTIONS

1. True or false: A quick way of computing the moment of inertia of a body is to consider its mass as being concentrated at the center of mass.

2. (a) How can one tell the difference between a hard-boiled egg and a raw egg by trying to spin them? (b) After they are spinning, explain what happens if each is stopped for an instant and then released.

3. A straight uniform rod tumbles from a vertical position on a frictionless surface. What is the path of the CM?

4. The book in Fig. 11.39 has the same shape as this text. About which axis is the moment of inertia (a) the largest; (b) the smallest?

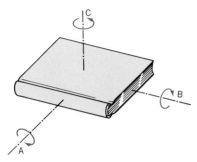

FIGURE 11.39 Question 4.

5. Two identical cans of concentrated orange juice are released at the top of an incline. One is frozen and the other has defrosted. Which reaches the bottom first?

6. The spokes of wagon wheels sometimes appear to rotate backward on film or on TV. Why is this?

7. The Volvo in Fig. 11.40 is going round the curve at high speed. Is there anything unusual about the picture? If so, what is happening?

FIGURE 11.40 Question 7.

8. Are the equations of rotational kinematics valid if θ is measured in degrees? If not, can they be modified?

9. Snow tires have a slightly greater diameter than summer tires. Is the reading of the speedometer affected?

10. The linear motion of the pistons in an automobile engine is converted to rotational motion of the crankshaft. There is a large flywheel in the linkage between the crankshaft and the transmission. What purpose does it serve?

11. About what axis is the moment of inertia of a person (a) the greatest? (b) the least? Are your answers subject to conditions?

12. The angular velocity of a bicycle wheel points north while its angular acceleration points south. (a) In which direction is the bicycle moving? (b) Is it speeding up or slowing down?

13. A spool of thread is pulled in opposite directions by equal and opposite forces as shown in Fig. 11.41. Does the spool move? If so, in which direction? (The surface is rough.)

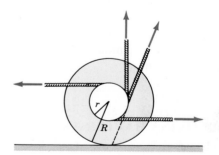

FIGURE 11.41 Question 13.

14. A spool of thread is on a rough surface (see Fig. 11.42). The axle has radius r while the rim has radius R. Discuss the motion of the spool for the various directions in which the thread is pulled. (a) For what angle is there sliding but no rolling? (b) What is the condition for the spool to wind up?

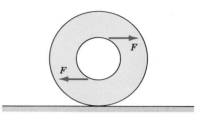

FIGURE 11.42 Question 14.

15. A uniform rod is freely pivoted at one end. Consider its motion as it is released from a horizontal position. (a) Is there kinetic energy associated with motion of the center of mass? (b) Is there kinetic energy relative to the center of mass? (c) Can one use the formula $\frac{1}{2}I\omega^2$ for the total kinetic energy? Explain why or why not.

16. What does the parallel axis theorem tell us about the minimum moment of inertia of a body?

EXERCISES

11.1 Rotational Kinematics

1. (I) A 12-in. (30-cm) diameter turntable starts from rest and takes 2 s to reach its final rotation rate of $33\frac{1}{3}$ rpm. Find: (a) the angular acceleration; (b) the number of revolutions completed in 5 s; (c) the time needed to complete 2 revolutions; (d) the radial and tangential accelerations of a point on the rim at 1 s. (e) Repeat (d) at 3 s.

2. (I) A 12-in. (30-cm) diameter LP rotates at 45 rpm and plays for 20 min. Find: (a) the speed of a point at the beginning when $r = 14.5$ cm; (b) the speed of a point at the end when $r = 6.2$ cm; (c) the radial velocity of the needle; (d) the average width of the groove.

3. (I) The earth spins in the same sense as it orbits around the sun. Find: (a) the earth's spin angular velocity about its internal axis; (b) its orbital angular velocity about the sun; (c) the linear speed of points closest to and farthest from the sun, measured relative to the sun. (Assume that the two axes of rotation are parallel.)

4. (I) Find the linear speed of the following points relative to the earth's center: (a) Quito, Ecuador (roughly) at the equator; (b) New York City, at latitude 41°. Take the radius of the earth to be 6370 km.

5. (I) The angular position of a line on a disk of radius $r = 6$ cm is given by $\theta = 10 - 5t + 4t^2$ rad. Find: (a) the average angular speed between 1 and 3 s; (b) the linear speed of a point on the rim at 2 s; (c) the radial and tangential accelerations of a point on the rim at 2 s.

6. (I) A particle moves in a circle of radius r with angular velocity ω and angular acceleration α. Show that the total linear acceleration is $a = r(\omega^4 + \alpha^2)^{1/2}$.

7. (I) At $t = 0$ a flywheel is rotating at 50 rpm. A motor gives it a constant acceleration of 0.5 rad/s² until it reaches 100 rpm. The motor is then disconnected. How many revolutions are completed at $t = 20$ s?

8. (I) A car with tires of radius 30 cm starts at rest and reaches 108 km/h in 10 s. (a) What is the angular acceleration of the wheels? (b) How many revolutions do they make? (There is no slipping.) (c) What is the radial acceleration relative to the center of a point on the rim when the speed is 108 km/h?

9. (I) A bullet travels through the 60-cm barrel of a Winchester rifle at 850 m/s. A protrusion within the barrel follows a spiral path. It cuts a groove in the bullet and forces it to rotate. The "rifling" involves one turn in 25 cm. What is the rate of spin of the emerging bullet in rpm?

10. (I) A clock has a second hand of length 8 cm. Find (a) the angular speed, and (b) the linear speed of the tip.

11. (I) Magnetic tape moves at a constant 4.8 cm/s in a cassette. What is the angular speed of the takeup spool when it is (a) empty, with radius $r = 0.75$ cm, and (b) full, with $r = 1.75$ cm? (c) What is the average angular acceleration with a 86.4 m tape?

12. (I) The angular position of a rotating object is given by $\theta = 2t - 5t^2 + 2t^4$ rad. Find: (a) the angular acceleration at 1 s; (b) the average angular acceleration between 1 and 2 s; (c) the average angular speed between 1 and 2 s.

13. (I) The wheel of a car has a radius of 20 cm. It initially rotates at 120 rpm. In the next minute it makes 90 revolutions. (a) What is the angular acceleration? (b) How much further does the car travel before coming to rest? There is no slipping.

14. (I) A car has wheels of radius 25 cm. The rate of rotation drops from 1250 rpm to 500 rpm in 25 s. How much further does the car move before stopping? There is no slipping and the acceleration is constant.

15. (I) A car with tires of radius 25 cm comes to a stop from 100 km/h in 50 m without any slipping of the tires. Find: (a) the angular acceleration of the wheels; (b) the number of revolutions made while coming to rest.

16. (II) A car with tires of radius 25 cm accelerates from rest to 30 m/s in 10 s. When the car's speed is 2 m/s find the linear acceleration of the top of the wheel relative to (a) the center of the wheel, (b) the road.

17. (I) A wheel makes 40 revolutions in 5 s and rotates at 100 rpm at the end of this period. What was the angular acceleration—assumed to be constant?

18. (I) A circular saw of diameter 18 cm starts from rest and reaches 5300 rpm in 1.5 s. (a) What is its angular acceleration? (b) What are the radial and tangential accelerations of a point on the rim at 1 s?

19. (I) Find the magnitude of the linear acceleration of an astronaut in (a) a centrifuge of radius 15 m (Fig. 11.43a) that has an angular velocity of 1.2 rad/s and an angular acceleration of 0.8 rad/s²; (b) in a doughnut-shaped space station (Fig. 11.43b) of radius 1.0 km that rotates at a constant 0.5 rpm.

FIGURE 11.43 Exercise 19.

20. (I) A wheel starts from rest and accelerates uniformly. As the rate of rotation changes from 20 to 50 rpm the wheel makes 40 revolutions. Find: (a) the angular acceleration; (b) the number of revolutions completed by the time the wheel reaches 20 rpm.

21. (I) The platter in a belt drive turntable is driven by a belt that wraps around a hub, of radius 3 cm, below the platter and around the shaft of the motor. If the platter rotates at $33\frac{1}{3}$ rpm and the motor rotates at 60 rpm, what is the required radius of the shaft?

11.2 and 11.3 Rotational Kinetic Energy; Moment of Inertia

22. (I) Four particles in the xy plane have the following masses and coordinates: 1 kg at (3 m, 1 m); 2 kg at (−2 m, 2 m); 3 kg at (1 m, −1 m); and 4 kg at (−2 m, −1 m). Find the total moment of inertia about (a) the x axis, (b) the y axis, and (c) the z axis.

23. (I) Four particles with equal masses M are connected by rods whose mass can be neglected. Find the moment of inertia, about the indicated axis, of the following: (a) the cross in Fig. 11.44a, (b) the square in Fig. 11.44b.

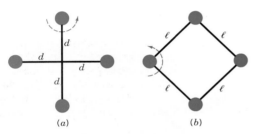

FIGURE 11.44 Exercise 23.

24. (I) What is the moment of inertia of a thin ring of mass M and radius R about an axis through its center and perpendicular to the plane of the ring?

25. (I) Two particles with masses 2 and 5 kg are connected by a light rod of length 2 m. Find the moment of inertia of the system about an axis perpendicular to the rod and passing through (a) the midpoint and (b) the center of mass.

26. (II) The wheel shown in Fig. 11.45 has a central hub of radius 2 m and a mass of 2 kg. Each of the four spokes is 4 m long and has a mass of 1 kg. The outer thin ring has a radius of 6 m and a mass of 2 kg. Find the moment of inertia about an axis through the center perpendicular to the plane of the wheel. Treat the hub as a disk.

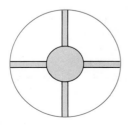

FIGURE 11.45 Exercise 26.

27. (II) Find the moments of inertia of the following shapes, made from the same thin wire, about a central axis perpendicular to the plane of the figure: (a) a circle of diameter $2a$; (b) a square of side $2a$; (c) an equilateral triangle of side $2a$. The mass per unit length of the wire is λ.

28. (II) Repeat Exercise 27 with the same dimensions but now take the masses to be identical. (Wires of different densities could be used.)

29. (II) Two solid spheres of mass m and radius R are stuck to the ends of a thin rod of mass m and length $3R$. Find the moment of inertia of the system about the axis at the midpoint of the rod and perpendicular to it, as shown in Fig. 11.46.

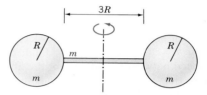

FIGURE 11.46 Exercise 29.

30. (I) In a water molecule the distance between the oxygen and hydrogen atoms is 9×10^{-11} m and the masses of the atoms are $m_O = 16m_H$, where $m_H = 1.67 \times 10^{-27}$ kg. The angle between the two H–O bonds is 105° (see Fig. 11.47). Find the moment of inertia of the molecule about (a) an axis along the H–O bond, and (b) an axis through the O atom parallel to the line joining the two H atoms.

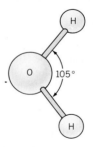

FIGURE 11.47 Exercise 30.

31. (II) A can is a hollow cylinder of radius R and height h. Its ends are sealed and it has no seams (see Fig. 11.48). The can is made from sheet metal of areal mass density σ kg/m². What is its moment of inertia about the central axis of symmetry?

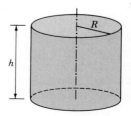

FIGURE 11.48 Exercise 31.

32. (II) A skater raises a leg at angle α to the vertical. By treating it as a straight rod of mass M and length L, as in Fig. 11.49, find its moment of inertia about the vertical axis passing through the top end.

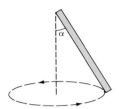

FIGURE 11.49 Exercise 32.

33. (II) A uniform rod of mass M and length L is pivoted about an axis perpendicular to the rod at a distance $L/4$ from one end. Find the moment of inertia by (a) integration; (b) the parallel axis theorem. (See In-Chapter Exercise 1.)

34. (II) Consider a planar body in the xy plane, as shown in Fig. 11.50. (a) Find the moment of inertia of a particle of mass m at (x, y) about the x, y, and z axes, respectively. (b) Show that the moments of inertia of the body about each of the axes are related by

$$I_z = I_x + I_y$$

This is called the *perpendicular axis theorem.*

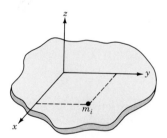

FIGURE 11.50 Exercise 34.

35. (I) Use the perpendicular axis theorem (see Exercise 34) to find the moment of inertia of a thin ring of radius R and mass M about a diameter.

36. (II). Consider a uniform thin rod of mass M and length L. Show by integration that the moment of inertia about an axis at a distance h ($>L/2$) from the center and perpendicular to the rod is

$$I = M \left(\frac{L^2}{12} + h^2 \right)$$

11.4 Conservation of Energy

37. (I) A uniform rod of length L and mass M is pivoted freely about a horizontal axis at one end. It falls from a vertical position. What is the speed of the tip when the rod is horizontal?

38. (I) A large part of the energy supplied by gasoline to a car or bus is wasted during braking. Flywheels can store rotational energy at such times, which can later be used to drive the vehicle. Figure 11.51 shows a flywheel-driven car. Suppose a car is equipped with a 15-kg cylinder of radius 0.2 m that spins at 14,000 rpm. What is the kinetic energy of the flywheel?

FIGURE 11.51 Exercise 38.

39. (II) Figure 11.52 shows a block of mass 4 kg suspended by a rope that passes over a pulley of mass 2 kg and radius 5 cm. The rope is connected to a spring whose stiffness constant is 80 N/m. (a) If the block is released from rest, what is the maximum extension of the spring? (b) What is the speed of the block after it has fallen 20 cm? Treat the pulley as a disk.

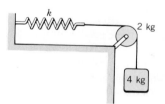

FIGURE 11.52 Exercise 39.

40. (I) What is the kinetic energy of the earth associated with its daily rotation? Since the earth is not a uniform sphere, its moment of inertia is $I = 0.33MR^2$.

41. (I) A solid sphere and a disk of the same mass and radius roll up an incline. Find the ratio the heights, h_S/h_D, to which they rise if, at the bottom, they have the same (a) kinetic energy, (b) speed.

42. (I) The propellor on an aircraft engine has four vanes, each of length 1.0 m and mass 10 kg. The engine rotates at 3000 rpm while the plane flies at a constant 200 km/h. What is the total kinetic energy of the propellor? Treat the vanes as uniform rods.

43. (II) A thin disk of mass M and radius R can rotate freely about a pivot on its rim, as in Fig. 11.53. It is released when its center is at the level of the pivot. What is the speed of the lowest point of the disk when the center is vertically below the pivot?

FIGURE 11.53 Exercise 43.

44. (II) The pendulum of a grandfather clock consists of a uniform rod of mass 1.2 kg and length 60 cm. At its end is a disk of radius 5 cm and mass 0.4 kg (see Fig. 11.54). It is released when the rod is 30° to the vertical. (a) What is the moment of inertia about the axis at the top of the rod? (b) What is the speed of the lowest point when the rod is vertical?

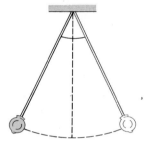

FIGURE 11.54 Exercise 44.

45. (II). A car has four tires each of mass 25 kg and radius 30 cm. The mass of the car without tires is 10^3 kg. (a) What is the total kinetic energy of the car and tires if the speed is 30 m/s? (b) How far along a 10° incline would the car roll (in neutral) before coming to rest if the initial speed were 30 m/s? Ignore frictional losses. Take $I = 1/2 MR^2$.

11.6 and 11.7 Torque; Rotational Dynamics

46. (I) For each of the forces depicted in Fig. 11.55 find the torque about the pivot. Take $F_1 = 10$ N, $F_2 = 15$ N, $F_3 = 8$ N, and $L = 8$ m.

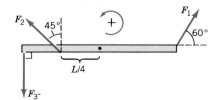

FIGURE 11.55 Exercise 46.

47. (I) A wheel of a locomotive is driven by a rod that is connected 50 cm from the center of the wheel. The rod is linked to a piston in a steam chamber, Fig. 11.56. The compressed steam applies at force of 10^5 N to the piston. What is the torque on the wheel when the rod is in the position shown?

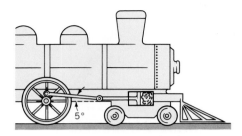

FIGURE 11.56 Exercise 47.

48. (I) A cyclist exerts a vertical force of 120 N on a pedal rod of length 20 cm that is inclined at angle θ to the horizontal, as in Fig. 11.57. Find the torque about the axle for the following values of θ: (a) 0; (b) 30°; (c) 45°; (d) 60°.

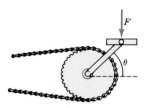

FIGURE 11.57 Exercise 48.

49. (I) A wheel, whose moment of inertia is 0.03 kg·m², is accelerated from rest to 20 rad/s in 5 s. When the external torque is removed, the wheel stops in 1 min. Find: (a) the frictional torque; (b) the external torque.

50. (I) A torque of 40 N·m acts on a wheel of moment of inertia 10 kg·m² for 5 s and then is removed. (a) What is the angular acceleration of the wheel? (b) How many revolutions does it make in 10 s if it starts at rest?

51. (I) A grindstone of radius 10 cm and moment of inertia 0.2 kg·m² rotates at 200 rpm. A tool is pressed against the rim with force of 50 N along a radial direction. The coefficient of kinetic friction is 0.6. (a) What is the power needed to keep the stone rotating at a fixed rate? (b) If the drive is removed, how long would it take for the stone to stop? (c) How many revolutions does it make while slowing down?

52. (II) A flywheel initially rotating at 1200 rpm stops in 4 min when only friction acts. If an additional torque of 300 N·m is applied, it stops in 1 min. (a) What is the moment of inertia of the wheel? (b) What is the frictional torque?

53. (I) A wheel starts from rest and rotates through 150 rad in 5 s. The net torque due to the motor and friction is constant

at 48 N·m. When the motor is switched off, the wheel stops in 12 s. Find the torque due to (a) friction, and (b) the motor.

54. (I) A block of mass $m = 2$ kg hangs vertically from a frictionless pulley of mass $M = 4$ kg and radius $R = 15$ cm. Find: (a) the acceleration of the block; (b) the tension in the rope; (c) the speed of the block after it has fallen 40 cm—assuming it started at rest. Treat the pulley as a disk.

55. (II) A block of mass $m = 2$ kg can slide down a frictionless 53° incline but it is connected to a pulley of mass $M = 4$ kg and radius $R = 0.5$ m, as shown in Fig. 11.58. The pulley can be treated as a disk. Find: (a) the angular acceleration of the pulley; (b) the speed of the block after it has slid 1 m, starting from rest.

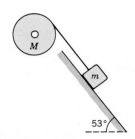

FIGURE 11.58 Exercise 55.

56. (II) A solid cylinder of mass M and radius R unwinds on a vertical string (see Fig. 11.59). (a) Use the energy approach to show that the speed of the spool after it falls a distance h starting from rest is $\sqrt{4gh/3}$. (b) Use the result of part (a) to find the linear acceleration of the CM. (c) Use dynamics to find the linear acceleration of the spool. (d) What is the tension? (e) With what force should the string be pulled to have the spool spin but not fall? What is its angular acceleration in this case?

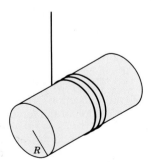

FIGURE 11.59 Exercise 56.

57. (II) The pulley in Fig. 11.60 consists of two disks of different diameters attached to the same shaft. The rope connected to the block of mass $m_1 = 1$ kg passes over a smooth peg, while the block of mass $m_2 = 3$ kg hangs vertically from one disk. The moment of inertia of the pulley is 0.2 kg·m²; $r_1 = 5$ cm and $r_2 = 10$ cm. Find the tensions in the ropes and the accelerations of the blocks.

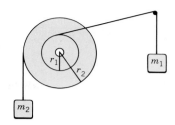

FIGURE 11.60 Exercise 57.

58. (II) A lawn roller is a solid cylinder of mass M and radius R. As shown in Fig. 11.61, it is pulled at its center by a horizontal force F and rolls without slipping on a horizontal surface. Find: (a) the acceleration of the cylinder; (b) the force of friction acting on it.

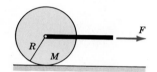

FIGURE 11.61 Exercise 58.

59. (II) A solid cylinder is released on an incline and rolls without slipping. (a) Find the acceleration of its center of mass. (b) What is the minimum coefficient of friction needed to prevent slipping?

60. (II) A meter stick of mass 40 g is pivoted at the 35-cm mark. What is its angular acceleration when it is inclined at 20° to the horizontal? Treat the stick as a thin rod.

11.8 Work and Power

61. (II) A turntable has a mass of 2 kg and a radius of 15 cm. When the motor is switched off it takes 20 s to stop from an initial $33\frac{1}{3}$ rpm. What is the power needed to maintain $33\frac{1}{3}$ rpm? Treat the turntable as a solid disk.

62. (I) A wheel whose moment of inertia is 45 kg·m² is to be accelerated from 20 to 100 rpm in 10 s. What is the average power needed?

63. (I) A 15-kg bucket of water is being raised from a well at a constant 20 cm/s. The rope wraps around a spindle of radius 3 cm. The spindle is turned by a handle of length 40 cm (see Fig. 11.62). (a) What is the power required to raise the bucket? (b) If the applied force is always perpendicular to the handle, what is the force required?

64. (I) The liner *Normandie* (82,800 gross tons) was possibly the largest passenger ship ever built. Its four engines developed 60,000 hp at 225 rpm. What torque did each engine provide?

65. (I) In November 1984, astronaut Joe Allen of the shuttle *Discovery* attached a device to the disabled *Palaba B* satellite that was spinning at 2 rpm (see Fig. 11.63). The satellite was a solid cylinder of mass 7000 kg and had a radius of 80

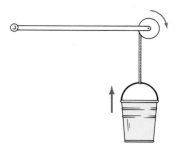

FIGURE 11.62 Exercise 63.

FIGURE 11.63 Exercise 65.

cm. He fired a 20-N jet tangential to the cylindrical surface to halt the rotation. (a) What was the initial kinetic energy of the satellite? (b) How long did it take to stop the spinning?

66. (I) In August 1985, astronauts William Fisher and James Van Hoften of the shuttle *Discovery* made repairs to the *Leasat 3* satellite (see Fig. 11.64). In order to stabilize it, prior to the firing of its rockets, Fisher pushed a handle at the rim to set it rotating and noted that it returned in 30 s. Treat the 7.6×10^3-kg satellite as a solid cylinder of diameter 4.3 m and assume the force was exerted tangentially. (a) How much work did Fisher do to rotate the structure? (b) If his hands moved through 1.2 m, estimate the force he applied.

FIGURE 11.64 Exercise 66.

PROBLEMS

1. (I) A disk of radius B has a concentric hole of radius A drilled through its center. The mass of the doughnut is M. Find its moment of inertia about the axis through the center of the disk perpendicular to its plane.

2. (II) A marble of radius r rolls without slipping down an incline and then up along a vertical circular track of radius R, as shown in Fig. 11.65. What is the minimum height H from which the ball must start so that it barely stays in contact at the top of the circle? Assume $r \ll H$ and $r \ll R$.

FIGURE 11.65 Problem 2.

3. (II) A uniform rod of length L is held vertically on a frictionless floor. A very slight nudge causes it to fall. (a) What is

its angular velocity on landing? (b) Find the speeds of its ends when it lands.

4. (II) Consider a very tall tower of height h at the equator, (see Fig. 11.66). The top will have a greater linear speed than the bottom. If the angular velocity of the earth is ω, show that a ball dropped from the top of the tower will not land at the base. It will be deflected by a distance from the base approximately given by $\omega h(2h/g)^{1/2}$. In which direction (east or west) is the deflection? (This expression is not exact because the direction of the force of gravity changes along the path.)

FIGURE 11.66 Problem 4.

5. (I) A uniform sphere of radius a has a concentric spherical cavity of radius b. Find the moment of inertia about a diameter. The mass of the object is M.

6. (I) (a) Show that the moment of inertia of a thin spherical shell of mass M and radius R about a diameter is $\frac{2}{3}MR^2$. (*Hint:* Break the shell into rings and use the angle to the axis as the variable.) (b) Use part (a) to find the moment of inertia of a solid sphere of mass M and radius R.

7. (I) A uniform disk of radius R has a hole of radius a drilled through it, as in Fig. 11.67. The center of the hole is at a distance b from the center of the original disk. What is the moment of inertia of the disk about an axis through the center perpendicular to its plane?

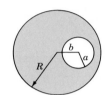

FIGURE 11.67 Problem 7.

8. (I) The angular acceleration of an object is given by $\alpha = 12t - 3t^2$ rad/s². The angular speed is 10 rad/s at $t = 1$ s and the angular displacement is 5 rad at 2 s. Write expressions for the angular speed and displacement as functions of time.

9. (I) Find the moment of inertia of a hollow cone of mass M, height h and apex angle 2α about the central axis of symmetry. (*Hint:* Use the distance from the apex along the surface as the variable.)

10. (I) At a given temperature the molecules in a gas have a range of speeds which can be measured as follows. Suppose the molecules leave an oven at speed v. They pass through two disks which have slots in them. The slots are offset from each other by θ degrees, as shown in Fig. 11.68. The disks rotate at N rpm and are separated by a distance D. Find an expression for the value of v at which the molecules pass through both slits.

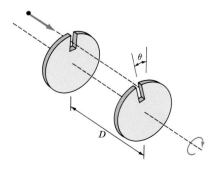

FIGURE 11.68 Problem 10.

11. (II) A person assumes the posture shown in Fig. 11.69. Estimate the moment of inertia about the central vertical axis. Each part of the body is approximated by a simple shape, the masses and dimensions of which are as follows:

Head: 5 kg   A solid sphere of radius 8 cm
Arm: 4.5 kg A rod of length 70 cm
Leg: 12 kg   A cylinder of radius 6 cm
Torso: 32 kg   A rectangular cylinder of horizontal dimensions $a \times b$. Use $I = M(a^2 + b^2)/12$ with $a = 30$ cm and $b = 15$ cm

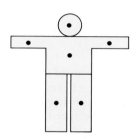

FIGURE 11.69 Problem 11.

You will have to use the parallel axis theorem in some cases.

12. (I) A solid cylinder rolls without slipping down a plane inclined at angle θ to the horizontal. Find: (a) the acceleration of the CM; (b) the minimum coefficient of friction required to prevent slipping.

13. (II) Use the perpendicular axis theorem (Exercise 34) and what you know about the moments of inertia of a thin rod, to find the moment of inertia of a thin rectangular plate with sides of lengths a and b about a central axis perpendicular to its plane, as shown in Fig. 11.70. (The final result is not restricted to a thin plate. Why not?)

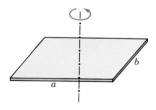

FIGURE 11.70 Problem 13.

14. (I) The density of a sphere of radius R varies according to $\rho = \rho_0(1 - r/2R)$, where ρ_0 is the density at the center. (a) What is its total mass M? (b) Show that its moment of inertia about a central axis is $0.37 \, MR^2$. (See Problem 6.)

15. (I) Figure 11.71 shows a disk with eight evenly spaced marks around its perimeter. It makes N rotations per second and is illuminated by a strobe light that flashes f_0 times per second. (a) What is the minimum value of N such that the disk appears stationary? (b) What is the apparent motion of the disk if N is 5% higher than the value found in (a)? (c) What is the apparent motion of the disk if N is 12.5% lower than the value found in (a)?

FIGURE 11.71 Problem 15.

16. (I) A bicycle has pedal rods of length 16 cm connected to a sprocketed disk of radius 10 cm. The bicycle wheels are 70 cm in diameter and the chain runs over a gear of radius 4 cm (see Fig. 11.72). The speed is constant. (a) How many revolutions does the bicycle wheel make for each complete

revolution of the pedals? If a force of 100 N is applied perpendicular to the pedal rod, what is (b) the torque on the sprocketed disk, (c) the tension in the upper portion of the chain and (d) the torque on the rear bicycle wheel? If the pedals are rotating at two revolutions per second, find (e) the power delivered by the cyclist, (f) the speed of the bicycle, and (g) the net force of friction due to the road.

FIGURE 11.72 Problem 16.

CHAPTER 12

Angular Momentum and Statics

Major Points

1. The definition of **torque** as a vector.
2. (a) The **angular momentum** of a *particle*.
   (b) The angular momentum of a *rigid body* about a fixed axis.
3. **Rotational dynamics.**
4. The **conservation of angular momentum.**
5. The conditions for **static equilibrium.**

In the last chapter we discussed the rotational dynamics of a rigid body rotating about a fixed axis. This allowed us to specify just the sense, clockwise or counterclockwise, of a torque. In this chapter torque is defined in a more general way that properly expresses its vector nature. Just as force is related to linear momentum, we will see that torque is related to a quantity called **angular momentum.** The importance of angular momentum lies in the fact that it is a conserved quantity. We will also consider the conditions under which a rigid body at rest stays in translational and rotational equilibrium.

This stunning feat is discussed in terms of the conservation of angular momentum on p. 250.

12.1 THE TORQUE VECTOR

In Eq. 11.22 the magnitude of a torque was defined as $\tau = rF \sin \theta$, where θ is the angle between the vectors $\mathbf{r}$ and $\mathbf{F}$. Torque is in fact a vector quantity, and so its direction must be specified relative to a coordinate system. We recall the definition of the vector product (Section 2.5), $\mathbf{A} \times \mathbf{B} = AB \sin \theta \, \hat{\mathbf{n}}$, and note that its magnitude has the same form as that of τ. The definition of torque as a vector quantity is

$$\boldsymbol{\tau} = \mathbf{r} \times \mathbf{F} = rF \sin \theta \, \hat{\mathbf{n}} \tag{12.1}$$

where $\hat{\mathbf{n}}$ is a unit vector normal to the plane of $\mathbf{r}$ and $\mathbf{F}$. Its direction is given by the right-hand rule, as shown in Fig. 12.1. Since the position vector $\mathbf{r}$ is measured relative to an origin O, the torque is also measured relative to this *point*. (In contrast, the r that appears in the moment of inertia formula is the perpendicular distance to an *axis* of rotation. In many instances the origin is chosen to lie on the axis.)

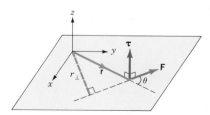

FIGURE 12.1 The torque exerted by the force $\mathbf{F}$ is $\boldsymbol{\tau} = \mathbf{r} \times \mathbf{F}$. Its magnitude is $\tau = rF \sin \theta$. The perpendicular distance $r_\perp = r \sin \theta$, from the origin to the line of action of the force is the lever arm or moment arm.

EXERCISE 1. A force $\mathbf{F} = 2\mathbf{i} + 3\mathbf{j}$ N acts at a point $\mathbf{r} = -\mathbf{i} + 5\mathbf{k}$ m. Calculate the torque about the origin.

12.2 ANGULAR MOMENTUM

Consider a single particle that has linear momentum $\mathbf{p}$ and is located at position $\mathbf{r}$ relative to an origin O, as in Fig. 12.2. Its **angular momentum** ℓ is defined as

Single particle $\qquad\qquad \ell = \mathbf{r} \times \mathbf{p}$ $\qquad\qquad$ (12.2)

Angular momentum is measured with respect to a *point*, the origin of the position vector $\mathbf{r}$. It is often convenient to express the magnitude of ℓ in terms of the "moment arm," $r_\perp = r \sin \theta$, which is the perpendicular distance from the origin to the line of motion of the particle. Thus,

Single particle $\qquad\qquad \ell = rp \sin \theta = r_\perp p$ $\qquad\qquad$ (12.3)

The SI unit of angular momentum is kg·m²/s. Let us apply these definitions to motion along a straight line and to motion in a circle.

Motion along a Straight Line

From Eq. 12.3 it follows that a particle moving in a straight line has angular momentum about any origin that does not lie directly on the line of motion. The "angular" aspect of straight line motion is associated with the rotation of the position vector about its origin. In Fig. 12.3 the particle moves with constant velocity. The angular momentum at both points A and B is directed out of the page. Its magnitude at these points is $\ell_A = r_A p \sin \theta_A$ and $\ell_B = r_B p \sin \theta_B$. The diagram shows that $r_A \sin \theta_A = r_B \sin \theta_B = r_\perp$, which is the moment arm. Since $r_\perp$ and p are constant, the angular momentum is also constant.

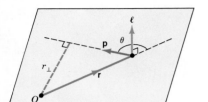

FIGURE 12.2 A single particle is located at position $\mathbf{r}$ and has linear momentum $\mathbf{p}$. Its angular momentum about the origin O is $\ell = \mathbf{r} \times \mathbf{p}$. The magnitude of the angular momentum is $\ell = rp \sin \theta$. An equivalent expression is $\ell = r_\perp p$, where $r_\perp$, the moment arm, is the perpendicular distance from the origin to the line of motion.

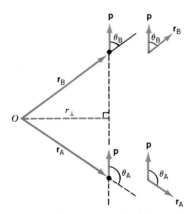

FIGURE 12.3 A particle moving in a straight line has angular momentum about any point not on its line of motion.

EXAMPLE 12.1: What is the angular momentum of a particle of mass $m = 2$ kg that is located 15 m from the origin in the direction 37° S of W and has a velocity $v = 10$ m/s in the direction 30° W of N? See Fig. 12.4a.

Solution: In Fig. 12.4a, the x axis points east. We know $r =$

15 m and $p = mv = 20$ kg·m/s. The angle between $\mathbf{r}$ and $\mathbf{p}$ is $(180° - 23°) = 157°$ (see Fig. 12.4a). Thus,

$$\ell = rp \sin \theta = (15)(20) \sin 157° = 120 \text{ kg·m}^2/\text{s}$$

We could also have used the moment arm $r_\perp = 15 \sin 23° =$

5.9 m and $\ell = r_\perp p$. From the right-hand rule we see that ℓ is into the page, along $-z$. (Imagine the vectors placed tail-to-tail.)

In unit vector notation, $\mathbf{r} = -15 \cos 37°\, \mathbf{i} - 15 \sin 37°\, \mathbf{j}$ m and $\mathbf{p} = 20 \sin 30°\, \mathbf{i} + 20 \cos 30°\, \mathbf{j}$ kg·m/s. Therefore,

$$\ell = (-12\mathbf{i} - 9\mathbf{j}) \times (10\mathbf{i} + 10\sqrt{3}\mathbf{j}) = -120\mathbf{k} \text{ kg·m}^2\text{/s}$$

EXERCISE 2. What is the angular momentum of a particle of mass $m = 0.5$ kg that is located at $(0, 2$ m$)$ and has a velocity $v = 10$ m/s in the direction 37° N of E? See Fig. 12.4b.

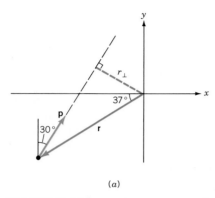

(a)

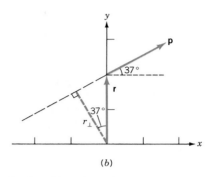

(b)

FIGURE 12.4 The angular momentum of each particle may be found by using unit vector notation or by finding the magnitude from $r_\perp p$ and the direction from the right-hand rule.

Motion in a Circle

In Fig. 12.5 a particle moves in a circle of radius R at constant speed v and angular velocity ω. We choose the origin to be at the center. The angle between $\mathbf{r}$ and $\mathbf{p}$ is always 90°, so the magnitude of the angular momentum is $\ell = rp \sin 90° = Rp$. Since $v = \omega R$, we have

$$\ell = mvR = mR^2\omega \tag{12.4}$$

For this special case of a single particle with the origin at the center, the vectors ℓ and ω are parallel. (Use the right-hand rule to confirm the direction of ℓ.) One can easily show that this is not true in general.

In Fig. 12.6 the origin is not at the center of the circle. The vector ω is still perpendicular to the plane of the circle, whose radius is R. In this case the position vector $\mathbf{r}$ is *not* along a radial line, so the product $\ell = \mathbf{r} \times \mathbf{p}$ is not perpendicular to the plane of the circle, which means that ℓ and ω are not parallel. However, the z component of ℓ does lie along ω. Since $\mathbf{r}$ is perpendicular to $\mathbf{p}$, the magnitude of the angular momentum is $\ell = rp \sin 90° = mvr$, and its z component is $\ell_z = \ell \sin \phi = (mvr)(R/r) = mR^2\omega$:

$$\ell_z = mR^2\omega \tag{12.5}$$

Note that both Eqs. 12.4 and 12.5 involve the radius of the circular path, but Eq. 12.5 refers only to the *component* of the angular momentum along the direction of ω.

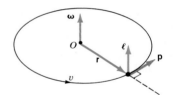

FIGURE 12.5 A particle moves in a circle. Since the origin is chosen to be at the center, the vectors ℓ and ω are parallel.

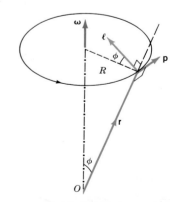

FIGURE 12.6 A particle moves in a circle. Since the origin does not lie at the center, the vectors ℓ and ω are not parallel.

SYSTEM OF PARTICLES

The total angular momentum $\mathbf{L}$ of a system of particles relative to a given origin is the sum of the angular momenta of the particles. From Eq. 12.2,

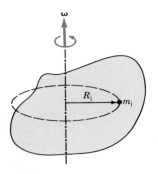

FIGURE 12.7 A rigid body rotating about a fixed axis (the z axis). The z component of the angular momentum of the ith particle is $\ell_{iz} = m_i R_i^2 \omega$.

(System of particles)
$$\mathbf{L} = \Sigma \boldsymbol{\ell}_i = \Sigma \mathbf{r}_i \times \mathbf{p}_i \qquad (12.6)$$

This sum takes a simple form in the special case of a rigid body rotating about a fixed axis. The origin can be chosen at any point on the axis of rotation, which we take to be the z axis (see Fig. 12.7). From Eq. 12.5 the *z component* of the angular momentum of the *i*th particle is $\ell_{iz} = m_i R_i^2 \omega$, where R_i is the radius of its circular path. Thus, the z component of the total angular momentum of a rigid body rotating about a fixed axis is

$$L_z = \Sigma \ell_{iz} = \Sigma m_i R_i^2 \omega$$

Since the moment of inertia is $I = \Sigma m_i R_i^2$, we have

(Rigid body, fixed axis) $\qquad L_z = I\omega \qquad (12.7)$

The subscript z is often dropped. Equation 12.7 is not a vector equation; it represents only the *component* of the angular momentum along the angular velocity. Note that $L = I\omega$ has the same form as $p = mv$: the product of an inertial property and a speed. Thus, angular momentum is the rotational analog of linear momentum.

EXAMPLE 12.2: A disk of mass M and radius R is rotating at angular velocity ω about an axis perpendicular to its plane at a distance $R/2$ from the center, as shown in Fig. 12.8. What is its angular momentum? The moment of inertia of a disk about the central axis is $\frac{1}{2}MR^2$.

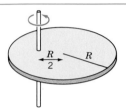

FIGURE 12.8 The axis of rotation is at a distance $R/2$ from the center of the disk.

Solution: The moment of inertia of the disk about the given axis may be found from the parallel axis theorem, Eq. 11.16: $I = I_{CM} + Mh^2$, where h is the distance between the given axis and a parallel axis through the center of mass. In the present example $h = R/2$. Therefore, the moment of inertia is

$$I = \tfrac{1}{2}MR^2 + M\left(\frac{R}{2}\right)^2 = \tfrac{3}{4}MR^2$$

The angular momentum is

$$L = I\omega = \tfrac{3}{4}MR^2\omega$$

12.3   ROTATIONAL DYNAMICS

We have seen that torque and angular momentum are the rotational analogs of force and linear momentum. Since $\mathbf{F} = d\mathbf{p}/dt$ one could make a good guess at the relationship between $\boldsymbol{\tau}$ and $\boldsymbol{\ell}$. Naturally we feel more secure in deriving it. The angular momentum of a single particle is $\boldsymbol{\ell} = \mathbf{r} \times \mathbf{p}$, and therefore its rate of change with time is*

$$\frac{d\boldsymbol{\ell}}{dt} = \mathbf{r} \times \frac{d\mathbf{p}}{dt} + \frac{d\mathbf{r}}{dt} \times \mathbf{p}$$
$$= \mathbf{r} \times \mathbf{F} + m\mathbf{v} \times \mathbf{v}$$

where we have used $\mathbf{v} = d\mathbf{r}/dt$ and $\mathbf{p} = m\mathbf{v}$. The first term is the torque, whereas the second term vanishes by definition of the vector product. We are left with

---

* Note that $d(\mathbf{A} \times \mathbf{B})/dt = \mathbf{A} \times d\mathbf{B}/dt + d\mathbf{A}/dt \times \mathbf{B}$.

(Single particle) $$\tau = \frac{d\ell}{dt}$$ (12.8)

The torque acting on a particle is equal to the time rate of change of its angular momentum. Note that both vectors τ and ℓ must be measured with respect to the *same* origin in an inertial reference frame. This equation is the rotational analog of $\mathbf{F} = d\mathbf{p}/dt$.

EXAMPLE 12.3: Show that Eq. 12.8 can be applied to the motion of a projectile.

Solution: In Fig. 12.9 we take the initial point as the origin. At a later time, $\mathbf{r} = x\mathbf{i} + y\mathbf{j}$. Since the force on the particle is $\mathbf{F} =$

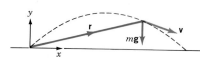

FIGURE 12.9 The change in the angular momentum of a projectile is produced by the torque exerted by the force of gravity.

$-mg\mathbf{j}$, the gravitational torque on it is

$$\tau = (x\mathbf{i} + y\mathbf{j}) \times (-mg\mathbf{j}) = -mgx\mathbf{k}$$

The rate of change of the angular momentum $\ell = \mathbf{r} \times \mathbf{p}$ is $d\ell/dt = \mathbf{r} \times d\mathbf{p}/dt = m\mathbf{r} \times d\mathbf{v}/dt$. But the acceleration is $d\mathbf{v}/dt = -g\mathbf{j}$. Thus,

$$\frac{d\ell}{dt} = m\mathbf{r} \times \frac{d\mathbf{v}}{dt}$$
$$= m(x\mathbf{i} + y\mathbf{j}) \times (-g\mathbf{j})$$
$$= -mgx\mathbf{k}$$

We see that $\tau = d\ell/dt$, which means that Eq. 12.8 is applicable—even though it is not the most transparent way to handle the problem!

SYSTEM OF PARTICLES

When we dealt with the dynamics of systems of particles, we saw that only external forces had to be considered; the internal forces between the particles cancel in pairs. A similar cancellation occurs with internal torques (see Fig. 12.10), which means that the rotational motion of a system is determined only by external torques. If the *i*th particle experiences an external force $\mathbf{F}_i$ and has linear momentum $\mathbf{p}_i$, then the torque on it is $\tau_i = \mathbf{r}_i \times \mathbf{F}_i$ and its angular momentum is $\ell_i = \mathbf{r}_i \times \mathbf{p}_i$. Since Eq. 12.8 applies to each particle, for the whole system we have

(System of particles) $$\tau_{\text{EXT}} = \frac{d\mathbf{L}}{dt}$$ (12.9)

where $\tau_{\text{EXT}} = \Sigma\tau_i$ is the net external torque on the system and $\mathbf{L} = \Sigma\ell_i$ is the total angular momentum of the particles. Equation 12.9 is the general equation for the rotational motion of any system and is the rotational analog of $\mathbf{F}_{\text{EXT}} = d\mathbf{P}/dt$. It is valid only when both the torque and the angular momentum are measured (i) with respect to the *same origin in an inertial frame,* or (ii) relative to the *center of mass* of the system—even if this point is accelerating.

For the special case of a rigid body rotating about a fixed axis we know from Eq. 12.7 that the *component* of $\mathbf{L}$ along the axis is $L = I\omega$. Since I is constant, $dL/dt = I\, d\omega/dt = I\alpha$, so Eq. 12.9 takes the (scalar) form

(Rigid body, fixed axis) $$\tau = I\alpha$$ (12.10)

which we used in Chapter 11.

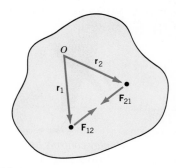

FIGURE 12.10 Two particles in a body exert equal and opposite forces ($\mathbf{F}_{12} = -\mathbf{F}_{21}$) on each other. The net torque due to these two forces relative to the origin O is $\tau_1 + \tau_2 = \mathbf{r}_1 \times \mathbf{F}_{12} + \mathbf{r}_2 \times \mathbf{F}_{21} = (\mathbf{r}_1 - \mathbf{r}_2) \times \mathbf{F}_{12}$. Since $(\mathbf{r}_1 - \mathbf{r}_2)$ and $\mathbf{F}_{12}$ are antiparallel, the net torque is zero. All the internal torques cancel in pairs.

EXAMPLE 12.4: Two blocks with masses $m_1 = 3$ kg and $m_2 = 1$ kg are connected by a rope that passes over a pulley of radius $R = 0.2$ m and mass $M = 4$ kg; see Fig. 12.11. The moment of inertia of the pulley about its center is $I = \frac{1}{2}MR^2$. Use Eq. 12.9 to find the linear acceleration of the blocks. There is no friction. Assume that the CM of the block of mass m_2 is at a distance R above the center of the pulley.

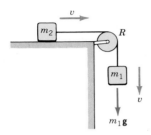

FIGURE 12.11 The torque due to the weight of m_1 produces the change in angular momentum of the system.

Solution: This problem could be tackled by applying $F = ma$ to the blocks and $\tau = I\alpha$ to the pulley. Instead, this example illustrates the use of angular momentum. From Eq. 12.3 we know that the angular momentum of a particle moving in a straight line is $\ell = r_\perp p$. If we take the origin at the center of the pulley, the angular momenta of the blocks are m_1vR and m_2vR. The angular momentum of the pulley is given by Eq. 12.7, $L = I\omega$. Therefore, the total angular momentum is

$$L = m_1vR + m_2vR + I\omega \qquad (i)$$

If the rope does not slip, then $v = \omega R$. The net external torque about the center of the pulley is due to the weight of m_1:

$$\tau_{EXT} = r_\perp F = R(m_1g) \qquad (ii)$$

In order to apply Eq. 12.9, we must find the time derivative of L and equate it to τ_{EXT}. Since $a = dv/dt$ and $d\omega/dt = \alpha = a/R$ we find

$$Rm_1g = (m_1 + m_2)Ra + I\frac{a}{R}$$

Using $I = \frac{1}{2}MR^2$ leads to an expression independent of R:

$$a = \frac{m_1g}{m_1 + m_2 + M/2} = 4.9 \text{ m/s}^2$$

Elizabeth Manley controls her angular speed by varying her moment of inertia.

12.4   CONSERVATION OF ANGULAR MOMENTUM

From Eq. 12.9 we infer the following:

If $\tau_{EXT} = 0$,   then   **L** = constant

> If the net external torque on a system is zero, the total angular momentum is constant in magnitude and direction.

This is the principle of the **conservation of angular momentum.** We begin by applying it to the special case of rigid bodies rotating about fixed axes, in which case the angular momentum is $L = I\omega$. The condition $L_f = L_i$ then takes the form

(Rigid body) $\qquad I_f\omega_f = I_i\omega_i \qquad (12.11)$

If a body is able to redistribute its mass and thereby change its moment of inertia, its angular velocity will also change. Although the body changes shape, we are justified in using a rigid body formula if, at any given time, all the particles have the same angular velocity.

EXAMPLE 12.5: A disk of moment of inertia 4 kg·m² is spinning freely at 3 rad/s. A second disk of moment of inertia 2 kg·m² slides down a spindle and they rotate together. (a) What is the angular velocity of the combination? (b) What is the change in kinetic energy of the system? See Fig. 12.12.

Solution: Since there are no external torques acting, we may apply the conservation of angular momentum:

$$I_f\omega_f = I_i\omega_i$$
$$(6 \text{ kg·m}^2)\omega_f = (4 \text{ kg·m}^2)(3 \text{ rad/s})$$

Thus, $\omega_f = 2$ rad/s.

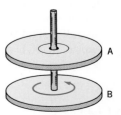

FIGURE 12.12 Disk A, initially not rotating, slips down a spindle onto disk B that is initially rotating freely.

(b) The kinetic energies before and after the collision are

$$K_i = \tfrac{1}{2}I_i\omega_i^2 = 18 \text{ J}; \qquad K_f = \tfrac{1}{2}I_f\omega_f^2 = 12 \text{ J}$$

The change is $\Delta K = K_f - K_i = -6$ J. In order for the two disks to spin together at the same rate, there had to be friction between them. The lost kinetic energy is converted to thermal energy.

EXAMPLE 12.6: A man is on a platform that rotates at 0.5 rev/s. With arms outstretched he holds two 4-kg blocks at a distance of 1 m from the axis of rotation—which passes through him; see Fig. 12.13. He then reduces the distance of the blocks from the axis to 0.5 m. Assume that the moment of inertia of the system "man + platform" is constant at 4 kg·m². (a) What is the new angular velocity? (b) What is the change in kinetic energy?

FIGURE 12.13 A person stands on a rotating platform and holds two blocks on outstretched arms. When the arms are brought in, the angular velocity increases.

Solution: (a) The moment of inertia of the system includes the contributions of the man, the platform, and the blocks.

$$I_i = 4 + 2mr_i^2 = 12 \text{ kg·m}^2$$
$$I_f = 4 + 2mr_f^2 = 6 \text{ kg·m}^2$$

The initial frequency of 0.5 rev/s is equivalent to $\omega_i = \pi$ rad/s. On setting

$$I_f\omega_f = I_i\omega_i$$

we find $\omega_f = 2\pi$ rad/s.
(b) The initial and final kinetic energies are

$$K_i = \tfrac{1}{2}I_i\omega_i^2 = 6\pi^2 \text{ J}; \qquad K_f = \tfrac{1}{2}I_f\omega_f^2 = 12\pi^2 \text{ J}$$

Thus, $\Delta K = K_f - K_i = +6\pi^2 \approx 60$ J. This increase in kinetic energy comes from the work done by the man in pulling in the blocks. Notice that even though there is no net external work done on the system, we cannot apply the conservation of mechanical energy because of the nonconservative work done by internal forces.

Skaters and ballet dancers make use of the conservation of angular momentum to perform their spins. They start with their arms outstretched and one foot as far as possible from the axis of rotation in order to maximize the moment of inertia. When the arms and legs are brought as close as possible to the axis, the moment of inertia decreases. Consequently, the angular velocity must increase in order for the angular momentum to stay constant.

EXAMPLE 12.7: A man stands on a stationary platform with a spinning bicycle wheel in his hands, as in Fig. 12.14. The moment of inertia of the man plus platform is $I_M = 4$ kg·m², and for the bicycle wheel it is $I_B = 1$ kg·m². The angular velocity of the wheel is 10 rad/s counterclockwise as viewed from above. Explain what occurs when the man turns the wheel upside down. The system is isolated in the sense that there are no external torques acting.

FIGURE 12.14 A man holds a spinning bicycle wheel. (a) The initial angular momentum of the wheel is directed upward. (b) When the wheel is flipped over, its angular momentum is directed downward. The conservation of angular momentum explains why the man must develop an angular momentum directed upward.

Solution: In this problem we must explicitly include the fact that angular velocity and angular momentum are vectors. The clockwise or counterclockwise specification is not sufficient. In both cases, the direction is given by the right-hand rule. Let us pick the $+z$ axis to point vertically up. The initial angular momentum is that of the wheel alone: $L_i = +I_B\omega_B$ (up) as shown in Fig. 12.14a. When the wheel is flipped over, its angular momentum is in the opposite direction: $-I_B\omega_B$ (down), as in Fig. 12.14b. Since these two terms are unequal (they have opposite signs), there has to be another contribution to the angular momentum for it to be conserved. When the man exerts a torque on the wheel to flip it over, he experiences the wheel's reaction to this. As a result, when the angular momentum of the wheel is reversed, the man and platform acquire an angular momentum in the original $+z$ direction—they turn counterclockwise. The final angular momentum is $L_f = -I_B\omega_B + I_M\omega_M$. From the conservation law,

$$-I_B\omega_B + I_M\omega_M = +I_B\omega_B$$

The man acquires an angular velocity $\omega_M = 2I_B\omega_B/I_M = +5$ rad/s. The positive sign means that ω_M is along the $+z$ axis.

EXAMPLE 12.8: According to *Kepler's second law* of planetary motion, the line joining the sun to a planet sweeps out equal areas in equal time intervals. Show that this is a consequence of the conservation of angular momentum. The path of the planet is an ellipse.

Solution: The gravitational force exerted by the sun on a planet is a *central force*—it acts along the line joining the two bodies, as shown in Fig. 12.15. The torque on the planet is $\boldsymbol{\tau} = \mathbf{r} \times \mathbf{F} =$

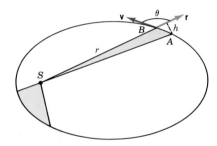

FIGURE 12.15 A planet moves in its elliptical path. Kepler's law of "equal areas in equal times" is a consequence of the conservation of angular momentum.

$rF \sin 180° \, \hat{\mathbf{n}} = 0$, which means that its angular momentum is constant. The constancy in direction implies that the plane of the orbit does not change. The constancy in magnitude leads to the "law of areas." In a time interval Δt, the planet moves from A to B by a distance $AB = v\Delta t$. The height of the triangle SAB is $h = AB \sin (180 - \theta)° = v\Delta t \sin \theta$. Thus, its area is

$$\Delta A = \tfrac{1}{2}rh = \tfrac{1}{2}rv\Delta t \sin \theta$$

so,

$$\frac{\Delta A}{\Delta t} = \tfrac{1}{2}rv \sin \theta \qquad \text{(i)}$$

The angular momentum of the planet is

$$\ell = rp \sin \theta = mrv \sin \theta \qquad \text{(ii)}$$

Combining (i) and (ii) we see that

$$\frac{\Delta A}{\Delta t} = \frac{\ell}{2m} = \text{constant}$$

The rate at which the radial line sweeps out area is constant. Or, equivalently, the radial line sweeps out equal areas in equal time intervals.

EXAMPLE 12.9: A man of mass $m = 80$ kg runs at a speed $u = 4$ m/s along the tangent to a disk-shaped platform of mass $M = 160$ kg and radius $R = 2$ m. The platform is initially at rest but can rotate freely about an axis through its center. Take $I = \tfrac{1}{2}MR^2$. (a) Find the angular velocity of the platform after the man jumps on. (b) He then walks to the center. Find the new angular velocity. Treat the man as a point particle.

Solution: Notice first that one cannot apply the conservation of linear momentum because the axle exerts an external force on the system "man + platform." (The man is in the air just before he lands on the platform.) However, since the axle does not exert any torque, we may use the conservaton of angular momentum. (Can we apply the conservation of kinetic energy for the collision between the man and the platform?)

(a) We choose the origin at the center of the platform shown in Fig. 12.16. When the man runs in a straight line, his initial angular momentum about this origin is $L = r_\perp p$, where in this case $r_\perp = R$, so

$$L_i = muR$$

After he jumps on, one must take into account his contribution mR^2 to the moment of inertia. The final angular momentum, $L = I\omega$, is

$$L_f = (\tfrac{1}{2}MR^2 + mR^2)\omega_1$$

When we set $L_f = L_i$ and use the given values, we find

$$\omega_1 = \frac{mu}{(M/2 + m)R} = 1 \text{ rad/s}$$

(b) When the man reaches the center, his contribution to the moment of inertia is zero. The "final" angular momentum of part (a) is the "initial" value for (b):

$$L_i = (\tfrac{1}{2}MR^2 + mR^2)\omega_1 = 640 \text{ kg·m}^2/\text{s}$$
$$L_f = (MR^2)\omega_2 = 320 \, \omega_2$$

The conservation condition $L_f = L_i$ yields $\omega_2 = 2$ rad/s.

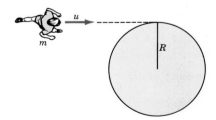

FIGURE 12.16 A man runs along a tangent to a circular platform and jumps on. The initial angular momentum relative to O is that of the man alone.

EXERCISE 3. What is the angular velocity when the man is midway between the rim and the center?

12.5 CONDITIONS FOR STATIC EQUILIBRIUM

The subject of **statics** is concerned with the forces and torques that act on bodies at rest. It is important that bridges or buildings retain their structures even when subjected to unusual stresses due to earthquakes or high winds. Knowledge of the forces acting at different points allows one to choose the appropriate construction materials. A bridge or aircraft designer also needs to know how the structure vibrates, but we will assume the body is rigid. (To minimize damage from an earthquake it is important that the structure of a building *not* be rigid.)

In Chapter 5 we saw that the condition for a point particle to be in equilibrium is that the sum of the forces acting on it be zero. We now examine the conditions under which a rigid body remains at rest. Clearly, the acceleration of its center of mass and its angular acceleration must be zero. When $a = 0$, the body is in **translational equilibrium;** when $\alpha = 0$ it is in **rotational equilibrium.** The special case in which the body is at rest is referred to as **static equilibrium.**

In order for a rigid body at rest to be in translational equilibrium, the net external force must be zero:

$$\Sigma \mathbf{F} = 0 \qquad (12.12)$$

In general, this is equivalent to three equations in terms of the components. However, if the forces acting on the object are confined to a plane, say, the xy plane, then we require

$$\Sigma F_x = 0; \qquad \Sigma F_y = 0 \qquad (12.13)$$

This condition is not sufficient for equilibrium, as Fig. 12.17 shows. Even when two equal and opposite forces act on an object, it will rotate unless the forces act along the same line. For it to be in rotational equilibrium, the net external torque must be zero:

$$\Sigma \tau = 0 \qquad (12.14)$$

In the special case of a rigid body that can rotate only about a fixed axis, which we take to be the z axis, the vector condition reduces to

$$\Sigma \tau_z = 0 \qquad (12.15)$$

The sign of this component of torque is determined by its direction relative to the $+z$ axis. However, for our purposes, it is convenient to employ a sign convention in which either the clockwise or the counterclockwise sense is taken to be positive. Since the body is in static equilibrium, its angular acceleration is zero about any point. For this reason, the torque may be evaluated about *any* convenient point.

12.6 CENTER OF GRAVITY

Consider a light rod that has two unequal weights suspended as shown in Fig. 12.18. The system can be held in both translational and rotational equilibrium by a single force applied at an appropriately chosen pivot point.

The **center of gravity** (CG) of a body is the point about which the net gravitational torque is zero.

From Archimedes' "law of levers" applied to two weights w_1 and w_2 hung on a pivoted rod, we know that this point is located by the condition $w_1 \ell_1 = w_2 \ell_2$,

Chinese acrobats maintaining static equilibrium.

FIGURE 12.17 Two equal and opposite forces acting on a body. In (*a*) the body is not in equilibrium, whereas in (*b*) it is in equilibrium.

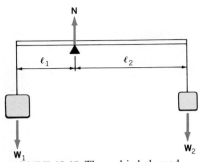

FIGURE 12.18 The rod is balanced when $w_1\ell = w_2\ell$. The single force at the center of gravity maintains the system in equilibrium.

where ℓ_1 and ℓ_2 are measured from the CG, which is at the pivot. Just as the center of mass is the point at which the *mass* of a system appears to be concentrated, the total *weight* of a system may be taken to act at the center of gravity. About any other point, the net gravitational torque on all the particles is the same as that of the total weight acting at the CG. The sum of the torques about the origin is

$$\Sigma\tau_i = w_1x_1 + w_2x_2 + \cdot\cdot\cdot + w_Nx_N$$

where $w_i = m_ig_i$—we leave open the possibility that g is not the same at all points. The torque due to the total weight acting at the CG is

$$(\Sigma w_i)x_{CG}$$

where x_{CG} is the position of the CG. Equating the above expressions we have

$$x_{CG} = \frac{\Sigma w_ix_i}{\Sigma w_i} \qquad (12.16)$$

If we assume that the g_i are the same, as is the case for objects close to the earth's surface, this expression reduces to

$$x_{CG} = \frac{\Sigma m_ix_i}{M} \qquad (12.17)$$

which we recognize as the position of the center of mass (CM). The CM of a rigid body is fixed in relation to the particles, but the CG will shift if the values of g vary from point to point. For practical purposes we take the CG and CM to be at the same position. For a uniform symmetrical body, the CG lies at the geometric center (see Section 10.1).

We have just seen that the translational and rotational equilibrium of a rigid body subject to gravity may be determined as if the total weight acts at the CG. Thus, the CG of a planar body may be located by pivoting it first about one axis, as shown in Fig. 12.19a. The torque due to the weight acting at the CG will cause the body to rotate until the CG lies vertically below the pivot point. A vertical line is drawn on the body. The body is then suspended from a different axis, and another such line is drawn. The CG is located at the intersection of the two lines (see Fig. 12.19b). For a nonplanar body one would require three points of suspension not all lying in a plane.

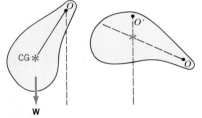

FIGURE 12.19 Determining the center of gravity of a planar body by suspending it at two different points.

EXERCISE 4. How does an object behave when it is suspended with the axis through the CG?

Problem-solving Guide for Statics

---

1. Choose the body whose equilibrium you are considering. Identify all external forces acting on it. You can make a good guess about the direction of the force of friction by imagining what would happen if friction were not present.

2. Choose a coordinate system and draw a free-body diagram that indicates the components of the forces.

3. Choose a convenient axis about which to evaluate torques and indicate the positive sense by a circular arrow: ↻. Note that if the point of application of a force passes through the axis, its torque is zero. This is one way of reducing the number of unknowns.

4. Write the conditions for equilibrium:

$$\Sigma F_x = 0; \qquad \Sigma F_y = 0; \qquad \Sigma\tau = 0$$

You may choose to apply the condition $\Sigma\tau = 0$ more than once about different axes. The number of unknowns must equal the number of independent equations.

EXAMPLE 12.10: A uniform rod of weight $W_1 = 35$ N is supported at its ends as shown in Fig. 12.20. A block of weight $W_2 = 10$ N is placed one-quarter of the distance from one end. What are the forces exerted by the supports?

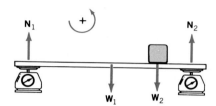

FIGURE 12.20 The net torque about *any* point on the rod is zero.

Solution: Our system is the rod. The forces acting on it are shown in the figure, which also indicates the coordinate system and the sign convention for torque. The first condition of equilibrium is

$$\Sigma F_y = N_1 + N_2 - W_1 - W_2 = 0 \qquad \text{(i)}$$

Taking torques about the center of the rod, we have

$$\Sigma\tau = -\frac{N_1 d}{2} - \frac{W_2 d}{4} + \frac{N_2 d}{2} = 0$$

or

$$-N_1 + N_2 - \frac{W_2}{2} = 0 \qquad \text{(ii)}$$

On adding (i) and (ii) we find

$$2N_2 - W_1 - \frac{3W_2}{2} = 0$$

Thus, $N_2 = 25$ N. Using this in either (i) or (ii) yields $N_1 = 20$ N.

EXERCISE 5. It is more straightforward to use $\Sigma\tau = 0$ about each end of the rod in Example 12.11 to find N_1 and N_2. Do so.

EXAMPLE 12.11: Figure 12.21 shows a diagram of an arm and the biceps muscle. The muscle is attached at a point about 4 cm from the socket which acts as a pivot point. If a weight of 50 N is held in the hand, what is the tension in the muscle? Assume that the forearm is a horizontal uniform rod of weight 15 N and length $L = 30$ cm. The force exerted by the muscle acts at 10° to the vertical.

Solution: Figure 12.21 shows the forces acting on the forearm, where **H** and **V** are the horizontal and vertical components of the force exerted by the pivot on the forearm. By taking torques about the pivot we avoid the force exerted by the pivot. Thus,

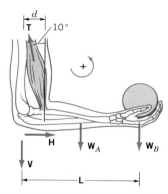

FIGURE 12.21 Since the muscle is close to the pivot point, it must exert a force much larger than the weight of the ball.

$$\Sigma\tau = (T \cos\theta)d - \frac{W_A L}{2} - W_B L = 0$$

$$(T \cos 10°)(0.04 \text{ m}) - (15 \text{ N})(0.15 \text{ m}) - (50 \text{ N})(0.30 \text{ m}) = 0$$

Thus, $T = 438$ N, which is considerably greater than the weight W_B.

EXERCISE 6. Determine the forces H and V in Example 12.11.

EXAMPLE 12.12: A ladder of length L and weight W rests on a rough floor and against a frictionless wall, as shown in Fig. 12.22. The coefficient of static friction at the floor is $\mu_s = 0.6$. (a) Find the maximum angle θ to the wall such that the ladder does not slip, (b) the force exerted by the wall at this θ.

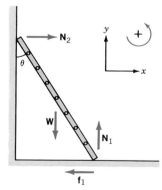

FIGURE 12.22 The ladder is prevented from slipping by the frictional force at the floor.

Solution: The coordinate system and sign convention for

torques are shown in Fig. 12.22. The forces exerted by the floor and the wall are drawn in terms of the normal component and the force of friction. Since the bottom of the ladder tends to slip toward the right, the direction of the force of friction is as shown.

The first condition of equilibrium yields

$$\Sigma F_x = N_2 - f_1 = 0 \tag{i}$$
$$\Sigma F_y = N_1 - W = 0 \tag{ii}$$

Applying the condition $\Sigma \tau = 0$ about axes, first through the top of the ladder and then through the bottom, yields

$$\Sigma \tau = -WL/2 \sin\theta - f_1 L \cos\theta + N_1 L \sin\theta = 0 \tag{iii}$$
$$\Sigma \tau = +WL/2 \sin\theta - N_2 L \cos\theta = 0 \tag{iv}$$

Just prior to slipping the force of static friction will have its maximum value $f_1 = \mu_s N_1$. From (ii) we see that $W = N_1$. Next,

we substitute for f_1 and W in (iii). After canceling the common factors N_1 and L, we find

$$\sin\theta - 2\mu_s \cos\theta = 0$$

This leads to the condition

$$\tan\theta = 2\mu_s$$

Since $\mu_s = 0.6$, we have $\tan\theta = 1.2$, which yields $\theta = 50.2°$. This is the maximum angle because we assumed that the frictional force had its maximum value.

(b) The force exerted by the wall is

$$N_2 = f_1 = \mu_s N_1$$

Since $N_1 = W$, $N_2 = 0.6\,W$. We could also have used (iv).

12.7 DYNAMIC BALANCE (Optional)

In general, when a body is supported by an axle, its weight distribution may produce a static imbalance. Static imbalance in a car tire is corrected by placing small weights at the rim while the tire is supported on a pivot. When a tire rotates, there may also be a "dynamic imbalance." A tire that is not both statically and dynamically balanced leads to a vibration at the steering wheel. Let us see how dynamic balance is reached.

Figure 12.23a shows a dumbbell consisting of two equal masses connected by a massless rod that rotates about a fixed axis (the z axis) at an angle to the rod. Each particle rotates in the xy plane. At the instant depicted in Fig. 12.23a, the rod is in the yz plane with m_1 moving into the page and m_2 out of the page. With the origin at the middle of the rod, the vectors $\ell_1 = \mathbf{r}_1 \times \mathbf{p}_1$

and $\ell_2 = \mathbf{r}_2 \times \mathbf{p}_2$ are in the same direction perpendicular to the rod; thus, the total angular momentum $\mathbf{L} = \ell_1 + \ell_2$ is also perpendicular to the rod. (See Fig. 12.6 for a different perspective.) As the dumbbell rotates, the vector $\mathbf{L}$ traces a cone about the z axis, along which $\boldsymbol{\omega}$ is directed.

The magnitude of $\mathbf{L}$ is constant but its direction is changing. From Eq. 12.9 we infer that there must be an external torque acting on the system. This torque is provided by an axle or bearings. Since this effect occurs only when the rod is rotating, this condition is called dynamic imbalance.

If the rod is bent as shown in Fig. 12.23b and rotates about the axis indicated, then $\mathbf{L}$ is parallel to $\boldsymbol{\omega}$, which means that the system is dynamically balanced. However, this arrangement is not *statically* balanced. A system that is both statically and dynamically balanced is shown in Fig. 12.23c.

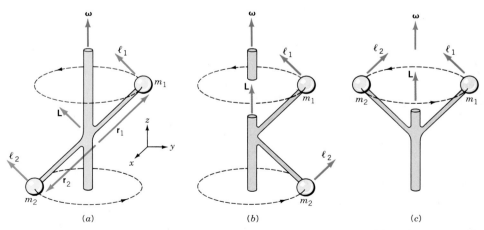

FIGURE 12.23 (a) The tip of the angular momentum vector traces a cone. The axle must provide the torque necessary to change the direction of $\mathbf{L}$. The system is not dynamically balanced. (b) With this configuration, $\mathbf{L}$ and $\boldsymbol{\omega}$ are parallel but the system is not statically balanced. (c) A configuration that is both statically and dynamically balanced.

12.8   SPIN AND ORBITAL ANGULAR MOMENTUM (Optional)

The earth has two distinct rotational motions. As Fig. 12.24 shows, it orbits the sun and it spins about an internal axis through its center of mass (CM). In general it is possible to split the angular momentum of a system into two terms representing these motions—the orbital angular momentum and the spin angular momentum.

> The **orbital angular momentum** L_O is the angular momentum of the CM motion about an origin O in an inertial frame.

> The **spin angular momentum** L_{CM} is the angular momentum relative to the CM.

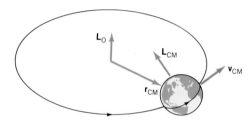

FIGURE 12.24 The angular momentum of a body can be separated into *orbital* angular momentum (L_O) of the center of mass relative to an external origin, and *spin* angular momentum (L_{CM}) relative to the center of mass.

The orbital term treats the system as a point particle at the CM, whereas the spin term is the sum of the angular momenta of the particles relative to the CM. The total angular momentum relative to an origin O in an inertial frame is the sum:

$$L = L_O + L_{CM}$$

Suppose a particle (such as the earth) moves with speed v_{CM} in a circular orbit of radius r_{CM}. The orbital term is $L_O = r_{CM} \times p_{CM}$, whereas the spin term is $I_{CM}\omega_{CM}$, where I_{CM} is the moment of inertia about the axis through the CM. Note that v_{CM} here is not related to ω_{CM}. The total angular momentum is

$$L = L_O + L_{CM} = r_{CM} \times p_{CM} + I_{CM}\omega_{CM}$$

The term "orbital" is a bit misleading since it is not restricted to orbital motion. A rigid body in motion has orbital angular momentum about any axis that does not pass through the CM. For example, in Fig. 12.25 a rigid body is rotating about an axis O at a distance h from the CM. Its angular momentum is $L = I\omega$. From the parallel axis theorem, $I = I_{CM} + Mh^2$; therefore, $L = Mh^2\omega + I_{CM}\omega = L_O + L_{CM}$.

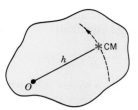

FIGURE 12.25 A disk rotates about an axis at a distance h from its center. It has both spin and orbital angular momentum.

EXAMPLE 12.13: A dumbbell consists of two equal masses connected by a light rod of length $2a$. It is cemented to a turntable that rotates at angular velocity ω. The dumbbell lies along a radius with its midpoint at a distance R from the center of the turntable. Find the total angular momentum of the dumbbell.

Solution: We see from Fig. 12.26 that each particle moves in a circle—one of radius $R + a$ and one of $R - a$. Since $\ell = mr^2\omega$, we have

$$
\begin{aligned}
L &= m(R + a)^2\omega + m(R - a)^2\omega \\
&= 2mR^2\omega + 2ma^2\omega \\
&= L_O + L_{CM}
\end{aligned}
$$

The first term is the angular momentum of a particle of mass $2m$ at the CM. The second term is the angular momentum of two particles, each at a distance a from the CM, about the CM. In this example, the orbital and spin angular velocities happen to be equal. In general, they will be different.

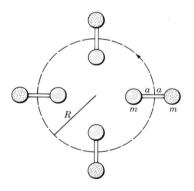

FIGURE 12.26 The dumbbell has both orbital and spin angular momentum.

ROTATIONAL DYNAMICS

We assert without proof that the rotational dynamics of a system may be uncoupled into orbital and spin parts. The net external torque is the sum of two terms:

$$\tau = \tau_O + \tau_{CM} \tag{12.18}$$

where τ_O is due to the net external force applied at the CM and is measured with respect to an origin O in an inertial frame, and τ_{CM} is the torque due to the external forces about the CM—

even if this point is accelerating. The orbital and spin motions then obey the separate equations:

$$\tau_O = \frac{d\mathbf{L}_O}{dt}; \qquad \tau_{CM} = \frac{d\mathbf{L}_{CM}}{dt} \qquad (12.19)$$

This uncoupling allows us to deal with whichever aspect of the motion is of interest in a given problem. An important implication of this split is that when either torque is zero, the corresponding angular momentum is independently conserved.

In Fig. 12.27 a spinning dumbbell is thrown as a projectile from the origin O. Its CM follows a parabolic path. The orbital angular momentum about O changes because of the force of gravity. However, the gravitational torque about the CM is zero and so the spin angular momentum is constant. This result is not restricted to symmetrical shapes. Divers employ the conservation of spin angular momentum to increase or decrease their rates of rotation. The rifling of the bore of a gun is used to impart a spin to the bullet or shell. The conservation of the spin angular momentum stabilizes the orientation of the projectile. A navigation aid called inertial guidance is also based on the gyrocompass, which is a rapidly spinning flywheel supported on frictionless (magnetic) bearings. Because of the conservation of spin angular momentum, the gyrocompass maintains its orientation in space (relative to the fixed stars). The motions of a craft (ship or plane) relative to the axis of rotation are recorded and fed to a computer that calculates its exact location.

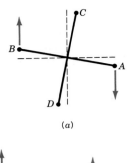

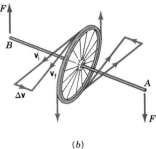

FIGURE 12.28 (a) Vertical forces applied at A and B cause the points C and D to move horizontally. (b) In a short time interval, the vertical forces applied at A and B cause the velocity of the top of the wheel to change in the horizontal direction.

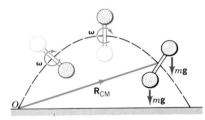

FIGURE 12.27 Although the orbital angular momentum of the dumbbell changes, its spin angular momentum is constant.

12.9   GYROSCOPIC MOTION (Optional)

A spinning top has the uncanny ability to resist falling down even when its axis is inclined to the vertical. Instead, the axis of the top traces the surface of an imaginary cone in a motion called **precession.** This seems to defy both gravity and our sense of what is reasonable. The rotational form of the second law, $\tau = d\mathbf{L}/dt$, leads to an adequate explanation of this strange phenomenon. However, one can gain some insight into the problem by using just the concepts of force and linear velocity.

Figure 12.28a shows two rigidly connected perpendicular rods. When equal and opposite *vertical* forces are applied to the horizontal rod at points A and B, the endpoints C and D of the vertical rod undergo *horizontal* displacements in opposite directions. Now consider a bicycle wheel spinning about a horizontal

axis, as shown in Fig. 12.28b. Suppose vertical forces are applied for a short time at the points A and B of the axle. The change in velocity of any small section will be in the direction of the force acting *on it*. The initial veloc··v of the top of the wheel is in the negative y direction. But the ιυrce on this section, and therefore the change in its velocity, is in the +x direction. The final velocity of the top section will remain essentially horizontal but it will point in a new direction. The change in velocity of the bottom will be in the opposite direction. The sections on opposite sides of the horizontal diameter continue to move vertically. Thus, the *vertical* forces at A and B cause the plane of rotation of the spinning wheel to change. Now let us examine this from the point of view of rotational dynamics.

Figure 12.29a shows a spinning wheel with one end of its axle supported on a frictionless support. We assume that the magnitude of the spin angular momentum L_s stays constant. Initially the axle is held horizontal, along the x axis. When the axle is released, the wheel experiences two forces: its weight and the reaction of the support. The net torque about the support is

$$\tau = \mathbf{r} \times \mathbf{F} = r\,\mathbf{i} \times (-mg\mathbf{k}) = mgr\,\mathbf{j}$$

Since $\tau = d\mathbf{L}/dt$, the change in angular momentum is

$$\Delta\mathbf{L} = \tau\Delta t = mgr\Delta t\,\mathbf{j}$$

The change $\Delta\mathbf{L}$ is associated with rotation about the y axis—which corresponds to a falling motion. If the wheel were not

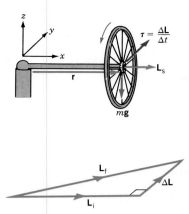

FIGURE 12.29 (a) The direction of the gravitational torque is horizontal. The change in angular momentum ∠**L** must be in the same direction. The spin angular momentum **I**ₛ also happens to be horizontal. (b) Since the initial (spin) angular momentum **L**ᵢ and the change in angular momentum Δ**L** are horizontal, the final angular momentum **L**f = **L**ᵢ + Δ**L** is also horizontal.

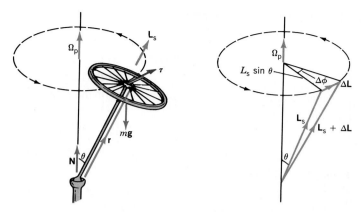

FIGURE 12.30 The spin angular momentum is at an angle θ to the vertical. The gravitational torque τ and the associated change in angular momentum ΔL are horizontal.

spinning, this angular momentum would increase until the wheel falls off the support. However, if the wheel is spinning fast, it has a large (spin) angular momentum directed along the x axis. The magnitude of the new (total) angular momentum, $L_f = L_i + \Delta L$, is essentially $L_i (= L_s)$, but its direction is slightly displaced in the horizontal plane, as shown in Fig. 12.29b. Since the action of the torque is continuous, the axle will continue to precess in a horizontal plane. (What holds the wheel up?)

We now obtain an expression for the precessional angular velocity Ω_p when the axle is at some arbitrary angle θ to the vertical, as shown in Fig. 12.30. The torque exerted by the force of gravity about the point of support is $\tau = mgr \sin \theta$, so the change in angular momentum is

$$\Delta L = \tau \Delta t = mgr \sin \theta \, \Delta t$$

From the diagram we see that $\Delta L = L_s \sin \theta \, \Delta \phi$ (from the definition of radian measure). Equating these two expressions we find $\Delta \phi = (mgr/L_s)\Delta t$, so the precessional angular velocity is

$$\Omega_p = \frac{\Delta \phi}{\Delta t} = \frac{mgr}{L_s} \qquad (12.20)$$

where $L_s = I\omega_s$ is the spin angular momentum. Equation 12.20 may be expressed in terms of the cross product:

$$\boldsymbol{\tau} = \boldsymbol{\Omega}_p \times \mathbf{L}_s \qquad (12.21)$$

There is an analogy between precessional motion and uniform circular motion. In uniform circular motion a force changes the direction of the linear momentum of a particle without changing its magnitude. In precessional motion, the torque

changes the direction of the spin angular momentum without changing its magnitude. The relation between the force and linear momentum in uniform circular motion may also be expressed in terms of the cross product. Since $F = mv^2/r = m\omega v = \omega p$, we see from Fig. 12.31 that

$$\mathbf{F} = \boldsymbol{\omega} \times \mathbf{p} \qquad (12.22)$$

which is analogous to Eq. 12.21.

It is worth pointing out that the tip of the axle has to precess below the original level for two reasons. First, since there are no torques in the z direction, L_z must remain zero at all times. The precessional motion (L_O) has a component along the z axis, which means that the component of $\mathbf{L}_s$ along this axis must decrease to compensate. Second, the energy of the system must be conserved. The rotational kinetic energy associated with the precession comes at the expense of gravitational potential energy of the wheel.

The above analysis is valid only when $\omega_s \gg \Omega_p$. In general, the motion of the top or wheel is complicated. In addition to the precession, the tip of the axle also moves up and down in a motion called *nutation*. The sense (clockwise or counterclockwise) of the precession may even be reversed temporarily.

FIGURE 12.31 A particle travels in uniform circular motion. The centripetal force **F** and the linear momentum **p** are related according to $\mathbf{F} = \boldsymbol{\omega} \times \mathbf{p}$, where $\boldsymbol{\omega}$ is the angular velocity.

SPECIAL TOPIC: Twists and Somersaults

Divers, acrobats, and ballet dancers perform many graceful rotational movements. The problem of reorienting the human body is also of particular interest to an astronaut who should not always have to rely on a gas gun. All these movements have their explanation in terms of the physical concepts developed in this chapter. Yet divers perform certain stunts that do seem—at first glance—to violate the conservation of angular momentum. The most spectacular and puzzling maneuver is executed by a cat. If it is held upside down and released from a height of about 1 m, it is able to right itself and land on its paws. Such finite rotations must be analyzed carefully to show that they do not violate any physical principle.

Moments of Inertia

The rotation of the human body can be related to three mutually perpendicular axes which pass through the CM as shown in Fig. 12.32. Rotation about the transverse y axis is called a *somersault*, about the longitudinal z axis it is a *twist*, and about the medial x axis it is a *pinwheel*.

The moments of inertia about these axes depend on the positions of the arms and legs relative to the torso. In general, I_z is much smaller than I_x or I_y. For the "pretwist layout" position depicted in Fig. 12.32a, the values are $I_z = 3.4$ kg·m²; $I_x = 19.2$ kg·m², and $I_y = 16.4$ kg·m². If the arms are brought to the sides, then I_z drops to 1.1 kg·m². In the "tuck" position of Fig. 12.32b, the values are $I_z = 2$ kg·m² and $I_x = I_y = 3.7$ kg·m².

Twists and Somersaults on Takeoff

The most obvious way a diver can acquire a twist or somersault motion is by using the board. Figure 12.33 shows two ways a diver can acquire angular momentum about the transverse axis, that is, a somersault. In Fig. 12.33a she merely leans forward as she jumps. The normal force N due to the board produces a torque about her CM. In Fig. 12.33b, she runs toward the end of the board, but lands with both feet for the takeoff. At takeoff the torso is erect but the body is moving forward while the feet are at rest. The frictional force due to the board produces the needed torque.

Twists may be obtained in two ways. If the feet apply unequal forces on the board, the reaction forces exert a torque tending to rotate the body about the longitudinal axis. A second method involves swinging the arms in the direction of the required twist while both feet are still on the board. After takeoff, the rotational motion of the upper body is transferred to the whole body. (You can easily do this while jumping up vertically from the floor.)

Contrary Rotations

Figure 12.34a shows a person in free-fall with no spin angular momentum. Suppose he lifts his legs to touch his feet. We know that the internal torques within a system cannot change its angular momentum. In this case the axis of rotation is at the hips. The counterclockwise torque needed to

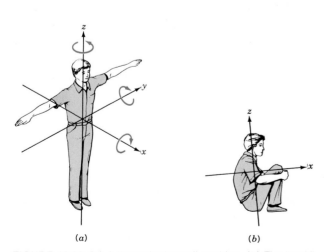

(a)   (b)

FIGURE 12.32 (a) A "pretwist layout" position. (b) The "tuck" position. The moments of inertia about each of the axes are different.

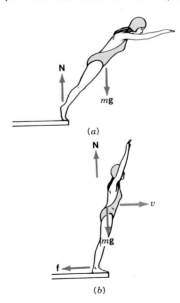

(a)

(b)

FIGURE 12.33 Two ways of obtaining angular momentum for a somersault. In (a) the diver merely leans over the edge. In (b) she runs along the board and lands with both feet at the edge.

lift the legs is accompanied by an equal and opposite (clockwise) torque on the upper body—which causes it to tilt forward. (The motion of the CM will not be affected.) Since the moment of inertia of the upper body about the hips is about three times that of the legs, the angular displacements are in the inverse ratio, as depicted in Fig. 12.34b. The spin angular momentum is zero at all times.

We will now see how one can produce a *finite* rotation even though the total angular momentum is always zero. Consider a person standing on a platform with frictionless bearings, as in Fig. 12.35a. (Nowadays it could be an astronaut floating in the space shuttle.) First the person twists at the waist. The upper body and the arms rotate clockwise as seen from the top (Fig. 12.35b). In order to conserve the spin angular momentum, the lower body and legs will rotate in the opposite (counterclockwise) sense. Next, the arms are brought to the sides—a "flapping" motion that involves no net angular momentum. Finally, the upper body is rotated counterclockwise. This is accompanied by a clockwise rotation of the lower body and legs (Fig. 12.35c). But, and this is the important point, the moment of inertia of the upper body was considerably reduced when the arms were brought down. Therefore, the lower body does not have to turn through as large an angle to keep the angular momentum zero. The body will have performed a finite twist with zero angular momentum at all times. The principle is that when the moment of inertia of one part changes, two "contrary" rotations do not cancel.

Somersaults with Zero Angular Momentum

Consider again a person in free-fall. The body is held rigid and the arms are brought rapidly forward in a "windmill" motion as shown in Fig. 12.36. As the arms are brought

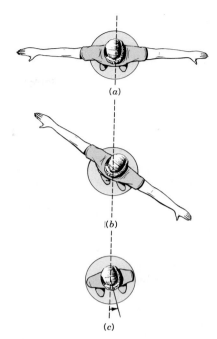

FIGURE 12.35 (a) A person stands on a frictionless platform with arms outstretched. (b) When the arms and upper body are rotated clockwise, the lower body and feet rotate counterclockwise. (c) The arms are brought down to reduce the moment of inertia of the upper body. When the twist at the waist is removed, the whole body has rotated through a finite angle.

down, the body will rotate in the opposite sense. The axis of rotation is at the shoulders. Since the moment of inertia of the arms is much less than that of the body, the body's rotation is slow. The somersault will continue as long as the arms perform the windmill motion. In practice the arms cannot move behind the body as much as they can in front and so a complete butterfly stroke produces only a 20° rotation of the body. We use this windmill technique instinctively when we are about to slip. If our feet start to slip foward, a

FIGURE 12.34 (a) A person in free-fall. (b) When the legs are brought to the horizontal position the upper torso rotates in the opposite sense.

FIGURE 12.36 The body can be kept in constant rotation by moving the arms in a "windmill" motion.

rapid windmill motion opposite to that in Fig. 12.36 will produce a countertorque on the body which helps to restore balance.

Instead of a windmill motion, a conical motion of the arms with the apex at the shoulder will also work. In the tuck position with 2.5-kg weights in the hands, an astronaut can produce a 50° somersault per cycle of arm rotation.

Twists with Zero Angular Momentum

Again we start with a person in free-fall with zero spin angular momentum. We pointed out that a conical motion of the arms can initiate a continuous somersaulting motion. This is quite slow because of the large moment of inertia of the torso. Figure 12.37 shows how a supple athlete could produce a continuous twisting motion with zero angular momentum at all times. The body must be bent at the hips or waist. The legs are made to trace the surface of a cone. (This will not be quite symmetrical because we cannot bend the spine backward to any great extent.) The moment of inertia of the upper body is kept small. As the legs rotate, the upper body will twist in the opposite sense. A diver can easily complete a 180° twist during a dive—which may include somersaulting motion. 360° twists are also possible.

Cat Twists

A cat is able to produce a 180° twist in about $\frac{1}{8}$ s without the benefit of a course in mechanics or diving techniques! It does so by varying the moments of inertia of its front and rear sections at the appropriate times. Figure 12.38 shows that the cat is initially upside down. We divide the fall into four stages:

1. The front paws are brought in to minimize the moment of inertia of the front section I_F. The rear paws and tail are kept perpendicular to the lower body to maximize the

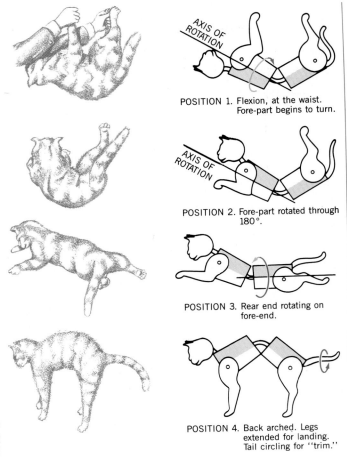

POSITION 1. Flexion, at the waist. Fore-part begins to turn.

POSITION 2. Fore-part rotated through 180°.

POSITION 3. Rear end rotating on fore-end.

POSITION 4. Back arched. Legs extended for landing. Tail circling for "trim."

FIGURE 12.38 Four stages in the fall of the cat.

moment of inertia of the rear I_R. Note that the body is bent and so there are separate axes of rotation for the front and rear sections.

2. The front end has twisted almost 180°; the face and upper body face downward. The rear end has rotated in the opposite sense. But because of its large moment of inertia, the angle is less than 10°.

3. The front paws are brought out in preparation for the landing. This action very nicely increases the moment of inertia of the front I_F. At the same time I_R is minimized by stretching the hind paws parallel to the body. The rear end is twisted, causing some counterrotation in the front.

4. The tail may be used to correct the final landing posture. But cats without tails can also perform the stunt.

The cat performs this stunning feat by a combination of the motions described in Fig. 12.35 and Fig. 12.37. (Reference: C. Frohlich, *Am. J. Phys.* 47: 583 (1979).)

FIGURE 12.37 By tracing a cone with the legs, the upper torso rotates in the opposite sense.

SUMMARY

The **torque** exerted by a force **F** acting at a position **r** is

$$\tau = \mathbf{r} \times \mathbf{F}$$

Its magnitude is $\tau = rF \sin \theta = r_\perp F = rF_\perp$.

The **angular momentum** of a single particle at position **r** with linear momentum **p** is

$$\ell = \mathbf{r} \times \mathbf{p}$$

Its magnitude is $\ell = rp \sin \theta = r_\perp p$.

The rotational analog of **Newton's second law**, $\mathbf{F}_{EXT} = d\mathbf{P}/dt$, is

$$\tau_{EXT} = \frac{d\mathbf{L}}{dt}$$

where $\tau_{EXT} = \Sigma \tau_i$ is the net external torque on the system and $\mathbf{L} = \Sigma \ell_i$ is the total angular momentum of the system. Both the torque and the angular momentum are measured relative to the origin of **r**. The equation

(Fixed axis) $\tau = I\alpha$

is a special case of the above equation for a rigid body rotating about a fixed axis.

According to the principle of **conservation of angular momentum,**

If $\tau_{EXT} = 0$,   then $\mathbf{L}$ = constant

The conditions for **static equilibrium** of a rigid body pivoted about an axis are

$$\Sigma \tau = 0; \qquad \Sigma F_x = 0; \qquad \Sigma F_y = 0$$

The **center of gravity** of a body is the point about which the net gravitational torque is zero.

ANSWERS TO IN-CHAPTER EXERCISES

1. $\tau = \mathbf{r} \times \mathbf{F} = (-\mathbf{i} + 5\mathbf{k}) \times (2\mathbf{i} + 3\mathbf{j}) = -15\mathbf{i} + 10\mathbf{j} - 3\mathbf{k}$ N·m.

2. In Fig. 12.4b the $+x$ axis points east. We are given $r = 2$ m, and $p = mv = 5$ kg·m/s. The angle between **r** and **p** is $53°$. Thus,

$$\ell = rp \sin \theta = (2)(5) \sin 53° = 8 \text{ kg·m}^2/\text{s}$$

We could have used the moment arm $r_\perp = 2 \sin 53°$ m and $\ell = r_\perp p$. The right-hand rule gives the direction of ℓ as being into the page, along the $-z$ direction.

In unit vector notation $\mathbf{r} = 2\mathbf{j}$ m and $\mathbf{p} = 5 \cos 37° \mathbf{i} + 5 \sin 37° \mathbf{j} = 4\mathbf{i} + 3\mathbf{j}$ kg·m/s. Thus,

$$\ell = (2\mathbf{j}) \times (4\mathbf{i} + 3\mathbf{j}) = -8\mathbf{k} \text{ kg·m}^2/\text{s}$$

3. The initial angular momentum is $L_i = 640$ kg·m²/s. The final angular momentum is

$$L_f = \left[\tfrac{1}{2} MR^2 + m \left(\frac{R}{2} \right)^2 \right] \omega_2 = 400\omega_2$$

Thus, $\omega_2 = 1.6$ rad/s.

4. There is no torque about the CG so the object is in neutral equilibrium—it remains in whatever position it is placed.

5. About the left end: $-W_1(d/2) - W_2(3d/4) + N_2 = 0$, so $N_2 = 25$ N.

About the right end: $+W_1(d/2) + W_2(d/4) - N_1 d = 0$, so $N_1 = 20$ N.

6. $\Sigma F_x = H - T \sin 10° = 0$; thus, $H = 76$ N.

$\Sigma F_y = T \cos 20° - V - W_A - W_B = 0$; thus, $V = 366$ N.

QUESTIONS

1. It is easier to balance a baseball bat with your finger at the thin end. Try it and explain why.

2. Why does spreading out both arms help one to balance on a tightrope? Why is holding a long pole even better?

3. Torque and energy have the same dimensions. Why do we distinguish between them?

4. (a) Consider an old rear-wheel-drive car with a worn out suspension. Why does the stationary car tilt to one side when the engine is revved in neutral? (b) Why does the front end of any car tilt upward when making a fast start from rest? See Fig. 12.39.

FIGURE 12.39 Question 4.

5. What can happen if only the front brakes are applied on a bicycle? Why does it happen?

6. If you were given two spheres of the same mass and radius that appear identical, could you determine whether either is solid or a shell?

7. A puck is moving in a circle on a horizontal frictionless plane. When the string that holds it is cut, what happens to the angular momentum of the puck?

8. (a) Why does a helicopter usually have a spinning blade at its tail? (b) Why do some helicopters have two rotors? See Fig. 12.40.

FIGURE 12.40 Question 8.

9. Figure 12.41 shows a monkey trying to get a bunch of bananas that hangs on the other side of a pulley. The monkey and the bananas have the same mass. Use angular momentum to discuss what happens when the monkey climbs the rope.

FIGURE 12.41 Question 9.

10. A billiard ball is struck at its center. Does it initially slide or roll?

11. Does the rotation of the propellors on a single-engine airplane have any effect when the pilot tries to turn in a horizontal circle?

12. Why does a rolling bicycle have greater stability than one at rest?

13. The earth and the moon rotate about their common center of mass. Are their angular momenta the same about this point?

14. The barrel of a gun is rifled to give the emerging bullet a spin. What purpose does this serve?

15. A cyclist going around a curve at high speed tilts toward the center of the circle, whereas a car tilts toward the outside. Explain the difference.

16. Explain qualitatively why a spinning gyroscope precesses instead of falling over.

17. What is the difference between static equilibrium and dynamic equilibrium?

18. What is the difference between the concepts of center of mass and of center of gravity? Given an example in which these two points are coincident and one in which they are different.

19. Can the center of gravity of a body lie outside the body? If so, how could its position be located experimentally?

20. When a person walks with a heavy suitcase, the body tilts. To which side does it tilt? Why?

21. How does the position of the center of gravity and the width of the "track" (the distance between the two front tires) affect the stability of a car?

22. Is a ladder more likely to slip when the person is near the bottom or near the top? Why?

23. The toy shown in Fig. 12.42 is a figure balanced on a rod. It recovers its vertical position if tilted to one side. Why does this occur?

24. The claw of a hammer is used to remove nails from wood. Explain the principle of its operation.

25. With the help of a diagram explain how a bottle cap opener helps in prying off a bottle cap.

26. Support a uniform meter stick with two fingers at different distances from the center. Slowly try to reduce the separation between the fingers. Explain what happens.

FIGURE 12.42 Question 23.

EXERCISES

12.2 Angular Momentum

1. (I) (a) Find the angular momentum about the origin of each of the particles shown in Fig. 12.43. (b) What is the magnitude and direction of the total angular momentum?

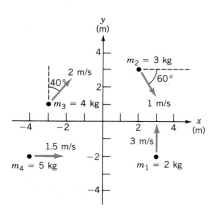

FIGURE 12.43 Exercise 1.

2. (I) A particle of mass 4 kg has coordinates $x = 2$ m, $y = -3$ m and velocity components $v_x = 5$ m/s and $v_y = 7$ m/s. Find its angular momentum about the origin.

3. (I) Two particles of equal mass have the same speed and travel in opposite directions along two parallel lines. Show that the total angular momentum is independent of the choice of origin.

4. (I) A rigid body rotates with constant angular velocity about a fixed axis. Show that its kinetic energy K and angular momentum L are related according to $K = L^2/2I$, where I is the moment of inertia.

5. (I) It was pointed out in the text that the relationship between L and ω is not in general a vector one. However, for a symmetric distribution of mass about the axis of rotation,

it is correct to write $\mathbf{L} = I\boldsymbol{\omega}$. Show this by adding a second particle to Fig. 12.6.

6. (I) A particle moves in a circle of radius r with angular velocity ω and angular acceleration α. (a) Express the radial acceleration $\mathbf{a}_r$ in terms of $\mathbf{v}$ and $\boldsymbol{\omega}$. (b) Express the tangential acceleration $\mathbf{a}_t$ in terms of $\boldsymbol{\alpha}$ and $\mathbf{r}$.

7. (II) A particle is executing uniform circular motion with speed v and angular velocity ω. The origin is on the axis of rotation but not at the center of the circle. Find the vector relationships that express: (a) $\mathbf{v}$ in terms of $\mathbf{r}$ and $\boldsymbol{\omega}$; (b) the radial acceleration $\mathbf{a}_r$ in terms of $\mathbf{v}$ and $\boldsymbol{\omega}$.

8. (II) In the Bohr theory of the hydrogen atom, an electron of mass m and charge $-e$ orbits a stationary proton whose charge is $+e$. The centripetal force is $F = ke^2/r^2$, where k is a constant and r is the radius of the orbit. In addition the angular momentum is quantized; it is restricted to only discrete values given by $mvr = nh/2\pi$, where h is called Planck's constant. Use the above to show that the nth allowed radius is

$$r_n = \frac{(nh/2\pi)^2}{mke^2}$$

12.3 Rotational Dynamics

9. (I) Two blocks with masses $m_1 = 3$ kg and $m_2 = 5$ kg are connected by a string that passes over a pulley of radius $R = 8$ cm and mass $M = 4$ kg (Fig. 12.44). Ignore friction and treat the pulley as a disk. Use the center of the pulley as the origin. (a) What is the net torque on the system? (b) What is the angular momentum of the system when the blocks have speed v? (c) Find the acceleration of the blocks by applying the equation $\tau = dL/dt$.

10. (I) A block of mass $m = 2$ kg hangs vertically from the rim of a pulley of mass $M = 4$ kg and radius $R = 20$ cm (see Fig. 12.45). Taking the center of the pulley as origin, use the equation $\tau = dL/dt$ to find the linear acceleration of the block. Ignore friction and treat the pulley as a disk.

11. (II) A particle of mass M moves in the xy plane. Its coordi-

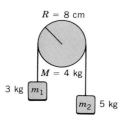

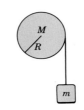

FIGURE 12.44 Exercise 9.   **FIGURE 12.45** Exercise 10.

nates as a function of time are given by $x(t) = At^3$; $y(t) = Bt^2 - Ct$, where A, B, and C are constants. (a) Find its angular momentum about the origin. (b) What force acts on it?

12. (II) A cyclist is turning to the left as she moves at constant speed v in a circular path of radius r. Assume that the rear wheel of radius R and moment of inertia I stays vertical. (a) What is the direction of the angular momentum? (b) Draw the angular momentum vectors at two times separated by a short interval. What is the rate of change of the angular momentum? (c) Show that $dL/dt = 2KR/r$, where K is the rotational kinetic energy of the wheel.

12.4 Conservation of Angular Momentum

NOTE: The moment of inertia of a uniform disk of mass M and radius R about an axis through its center and perpendicular to the plane of the disk is $\frac{1}{2}MR^2$.

13. (I) A turntable with a moment of inertia of 0.012 kg·m^2 rotates freely at 2 rad/s. A circular disk of mass 200 g and diameter 30 cm, and initially not rotating, slips down a spindle and lands on the turntable. (a) Find the new angular speed. (b) What is the change in kinetic energy?

14. (II) Three point particles with masses $3m$, m, and $2m$ are connected by rods of length d as shown in Fig. 12.46. The system rotates at angular velocity ω about one end. What is the angular momentum of the system (a) if the rods are massless? (b) if each rod has a mass M? (See In-Chapter Exercise 1 of Chapter 11 and Eq. 11.16.)

FIGURE 12.46 Exercise 14.

15. (I) A circular disk of mass 0.2 kg and radius 15 cm, initially not rotating, slips down a thin spindle onto a turntable (disk) of mass 1.6 kg and the same radius, rotating freely at 4 rad/s. (a) Find the angular velocity of the combination. (b) Is kinetic energy conserved? If not, what is the change? (c) If the motor is switched on after the disk has landed, what is the constant torque needed to regain the original speed in 2 s?

16. (I) A turntable of mass 2 kg and radius 15 cm rotates freely at 2 rad/s. A ball of putty of mass 500 g lands 10 cm from the center. (a) What is the new angular speed? (b) What is the change in kinetic energy? Treat the turntable as a disk.

17. (I) A 60-kg person runs with a velocity of 5 m/s along a tangent to a stationary circular platform, of radius 3 m and mass 100 kg, and jumps on (see Fig. 12.47). The platform (a disk) can rotate about a vertical axis. Find: (a) the angular velocity after the person jumps on; (b) the loss in mechanical energy.

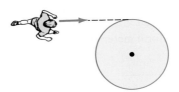

FIGURE 12.47 Exercise 17.

18. (II) Two beads, each of mass M, can slide on a frictionless rod of mass M and length L. The beads are initially at $L/4$ from the center about which the rod rotates freely in a horizontal plane at 20 rad/s. Find the angular speed (a) when the beads reach the ends of the rod, and (b) when they fly off the ends. (c) Why do the beads move at all? (See In-Chapter Exercise 1 of Chapter 11.)

19. (II) A child of mass 40 kg stands beside a circular platform of mass 80 kg and radius 2 m spinning at 2 rad/s. Treat the platform as a disk. The child steps onto the rim. (a) Does the angular speed change? If so, what is the new value? (b) She then walks to the center. Does the angular velocity change? If so, what is the new value? (c) Does the kinetic energy change when she walks from the rim to the center? If so, by how much?

20. (II) A girl of mass 60 kg is at the rim of a 100-kg platform of radius 2 m initially at rest. She starts walking at a speed of 2 m/s relative to the ground in a clockwise sense as seen from above. (a) What is the angular velocity of the platform? (b) She decides to walk to the center and stay there. What is the angular velocity of the platform? Treat the platform as a disk.

21. (II) An 80-kg man stands at the rim of a 100-kg carousel of radius 2 m initially at rest. He starts to walk along the rim at 1 m/s relative to the carousel. What is the angular speed of the carousel? Treat the carousel as a disk.

22. (II) (a) A rigid body is rotating freely with a period T. Show that if its moment of inertia changes by an infinitesimal amount dI, then $dI/I = dT/T$. (b) The Saudi Arabian government once considered a plan to tow icebergs from the Antarctic to aid their water supply. Estimate how the length of a year would be changed if 10^{12} kg of ice were moved from the pole to the equator without melting? The moment of inertia of the earth is $0.33MR^2$.

23. (II) Figure 12.48 shows a thin ring of mass $M = 1$ kg and radius $R = 0.4$ m spinning about a vertical diameter. (Take

$I = \frac{1}{2}MR^2$). A small bead of mass $m = 0.2$ kg can slide without friction along the ring. When the bead is at the top of the ring, the angular velocity is 5 rad/s. (a) What is the angular velocity when the bead slips halfway to the horizontal (when $\theta = 45°$)?

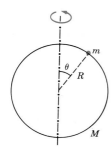

FIGURE 12.48 Exercise 23.

24. (II) Two skaters, each of mass 60 kg, approach each other along parallel lines at 2 m/s. One skater holds one end of a massless 2-m pole horizontal and perpendicular to the direction of motion. The other skater grabs the other end of the pole as they meet. (a) What is the angular velocity of their subsequent motion? (b) If they pull until their separation is halved, what is the new angular velocity? (c) Is there any change in kinetic energy between (a) and (b)? If so, find it and explain its origin. Treat the skaters as point particles.

25. (II) A particle of mass m travels in a horizontal circle on a frictionless table. The centripetal force is provided by a string, attached to two bodies of equal mass, which passes through a hole in the table (see Fig. 12.49). Show that if one of the hanging masses is removed, the radius of the motion changes by a factor of 1.26.

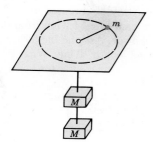

FIGURE 12.49 Exercise 25.

12.5 Static Equilibrium

26. (I) A person has to exert equal and opposite forces of magnitude 20 N perpendicular to the arms of a nutcracker to crack a nut as shown in Fig. 12.50. What is the force exerted on each side of the nut?

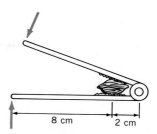

FIGURE 12.50 Exercise 26.

27. (I) Figure 12.51 shows a 7-g phono cartridge attached to the end of a uniform tonearm of length 26 cm and mass 12 g. The tonearm is pivoted 20 cm from one end. (a) At what point should the center of a 60-g cylindrical counterweight be placed for the arm to be balanced? Assume that the CG of the cartridge is at the end of the arm. (b) How far should the counterweight be moved for the needle at the end of the arm to exert a force of 0.02 N on an LP? Do you need to assume the arm stays horizontal?

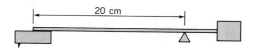

FIGURE 12.51 Exercise 27.

28. (I) A uniform plank of mass 3 kg and length 4 m is pivoted freely at its center. A 2-kg block is placed 50 cm from the pivot and a 2.4-kg block is placed 1.5 m from the pivot on the other side. Where should a 1.5-kg block be placed for the system to be in equilibrium? Does the plank have to be horizontal?

29. (I) A uniform plank of mass 5 kg and length 3.6 m is supported by vertical ropes at its ends as in Fig. 12.52. A 60-kg painter is located 0.5 m to the left of the center and an 8-kg bucket is 1 m to the right of the center. Find the tensions T_1 and T_2 in the ropes.

FIGURE 12.52 Exercise 29.

30. (I) The hooks on a 3-kg sign for a tavern are 72 cm apart and located at equal distances from the center of the sign (see Fig. 12.53). The right hook is 20 cm from the right end of a horizontal rod of mass 2 kg and length 1.2 m that pivots freely. Find: (a) the tension in the chain; (b) the horizontal and vertical forces exerted by the pivot.

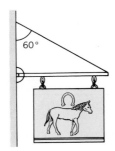

FIGURE 12.53 Exercise 30.

31. (I) In Fig. 12.54 a 20-kg boom of length 4 m is supported by a cable that has a breaking tension of 1000 N. The cable is perpendicular to the boom and is attached 3 m from the pivot. Find: (a) the maximum load that can be suspended from the end of the boom; (b) the horizontal and vertical forces exerted by the pivot in this case.

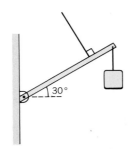

FIGURE 12.54 Exercise 31.

32. (II) A 3-kg plank of length $L = 2$ m is pivoted freely at one end and has a rope attached at the other end, as in Fig. 12.55. The rope is attached to the wall at a distance $h = 0.8$ m above the pivot and the rod is at an angle $\theta = 20°$ below the horizontal. Find: (a) the tension in the rope; (b) the horizontal and vertical forces exerted by the pivot.

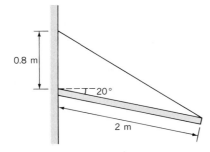

FIGURE 12.55 Exercise 32.

33. (I) A uniform board of length 2.4 m and mass 4 kg is pivoted at one end and has a rope attached at the other end, as shown in Fig. 12.56. A bucket of mass 2 kg is suspended 40 cm from the rope. Find: (a) the tension in the rope; (b) the horizontal and vertical forces exerted by the pivot.

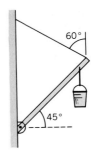

FIGURE 12.56 Exercise 33.

34. (I) A person holds a 5-kg mass in the hand while the forearm is at 30° below the horizontal. The biceps are connected 4 cm from the pivot point and act at 5° to the vertical (see Fig. 12.57). Take the forearm to be a uniform rod of mass 2 kg and length 30 cm. What is the tension in the muscle?

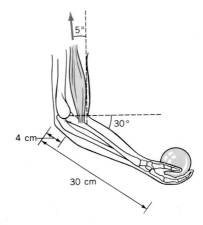

FIGURE 12.57 Exercise 34.

35. (I) Figure 12.58 shows a person pulling down on a rope with a force of 50 N with the forearm 30° above the horizontal. The triceps are connected 1.2 cm from the pivot point and exert a vertical force. Assume the forearm is a uniform rod of mass 2 kg and length 30 cm. What is the tension in the muscle?

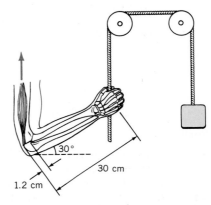

FIGURE 12.58 Exercise 35.

36. (I) A 60-kg person stands on tiptoes with the weight equally distributed between the feet (see Fig. 12.59). The muscle that causes the rotation of the foot is connected 4 cm from the ankle joint, and exerts a vertical force. What is the tension in the muscle?

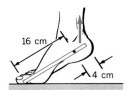

FIGURE 12.59 Exercise 36.

37. (I) The winch of a crane holds a 200-kg bucket of concrete, Fig. 12.60. The boom is at 60° to the horizontal. (a) What is the tension in the cable? (b) What are the horizontal and vertical forces added by the load to the base of the boom?

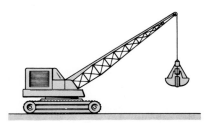

FIGURE 12.60 Exercise 37.

38. (II) Consider the pulley system shown in Fig. 12.61. What is the force required at the free end of the chain to hold the weight? Treat all the ropes as being vertical and the pulleys as massless and frictionless.

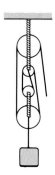

FIGURE 12.61 Exercise 38.

39. (I) A person bends over to pick up a 100-N package. The weight of the torso is 450 N. The back is supported by a muscle that attaches to the spinal column and makes a 12° angle with it, as in Fig. 12.62. The torso pivots about the base of the spine. By treating the upper body as a rod, find:

(a) the tension in the muscle; (b) the magnitude of the total force exerted at the base of the spine.

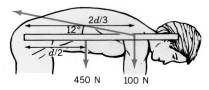

FIGURE 12.62 Exercise 39.

40. (I) A 60-kg diver stands at the end of a 3 m board of negligible mass that is attached to two supports 50 cm apart (see Fig. 12.63). What is the magnitude and direction of the force exerted by each support? Neglect the flexing of the board.

FIGURE 12.63 Exercise 40.

41. (II) A uniformly packed crate of weight $W = 200$ N and of dimensions $a = 0.4$ m and $b = 1.0$ m is placed on a dolly. What horizontal force applied at P is needed to hold the system in rotational (but not translational) equilibrium in the position shown in Fig. 12.64? Take $h = 1.1$ m and $\theta = 30°$.

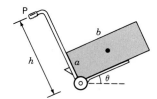

FIGURE 12.64 Exercise 41.

42. (I) A uniform meter stick of mass 20 g is used to suspend three objects as follows: 20 g at 15 cm, 40 g at 40 cm, and 50 g at 85 cm. Where could you place your finger to balance the stick in the horizontal position?

43. (I) A person of height 1.6 m lies on a light plank supported

FIGURE 12.65 Exercise 42.

at the head and the feet as shown in Fig. 12.65. The scales read 350 N at the head and 300 N at the feet. Where is the center of gravity?

PROBLEMS

1. (II) A particle of mass $m = 0.5$ kg moving at speed $u = 4$ m/s strikes a dumbbell consisting of two blocks of equal mass $M = 1$ kg separated by a massless rod of length 2 m (see Fig. 12.66). The dumbbell and the particle are free to slide on a horizontal surface. Find: (a) the speed of the center of mass of the system after the particle sticks to one of the blocks; (b) the angular velocity of the system about the center of mass.

FIGURE 12.66 Problem 1.

2. (II) Consider a baseball bat to be a uniform rod of length L (see Fig. 12.67). Show that the distance d from the center at which the bat should strike a ball to minimize the impact on the hands (assumed to be at the very end) is $d = L/6$. This point is called the *center of percussion* or the "sweet spot." Assume that the motion of the ball is perpendicular to the bat. (*Hint:* Apply the conservation of both linear and angular momenta. What is the condition for the held end to be at rest?)

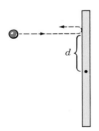

FIGURE 12.67 Problem 2.

3. (II) Figure 12.68 shows a particle of mass m moving in a circle with the centripetal force provided by a rope that passes through a hole in the table. The initial angular momentum is L_0. The force is changed in such a way that the radius of the motion decreases from r_1 to r_2. (a) How does

the force vary as a function of r? (b) Calculate the work done by the force in changing the radius. (c) What is the change in kinetic energy of the particle? (d) Does the work–energy theorem apply?

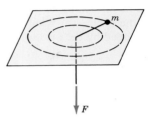

FIGURE 12.68 Problem 3.

4. (I) A dumbbell consists of two equal particles of mass m separated by a distance $2a$. It is initially attached along a radial line to a disk that is rotating at ω rad/s about a vertical axis (see Fig. 12.69). The center of the dumbbell is at a distance R from the center of the disk. What is the kinetic energy of the dumbbell after it is suddenly freed. Assume all the motion occurs on a horizontal frictionless plane.

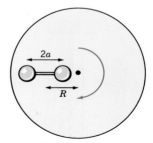

FIGURE 12.69 Problem 4.

5. (II) A cylinder of mass M and radius R is rotating at angular velocity ω_0 when it is placed vertically on a horizontal surface for which the coefficient of kinetic friction is μ_k. (a) Write the linear and the rotational forms of the second law. (b) Show that it takes a time $\omega_0 R/3\mu_k g$ for the cylinder to start rolling without slipping. (c) How far does it travel before it rolls without slipping?

6. (II) Consider the angular momentum of the system described in Problem 12.5. Choose the origin at the initial point of contact. (a) What is the initial total angular momentum about the origin? (b) What is the torque about the origin? (c) What is the final angular momentum when the cylinder rolls without slipping? (d) Show that when pure rolling starts the speed of the center is $\omega_0 R/3$.

7. (I) A solid sphere moves at speed v_0 without rolling. It encounters a rough surface whose coefficient of friction is μ. (a) Show that pure rolling commences when the speed of the center of mass is $5v_0/7$. (b) How far does the sphere travel before pure rolling starts? Take $I_{CM} = \frac{2}{5}MR^2$

8. (II) Figure 12.70 shows a spool of mass M with an axle of radius r and endplates of radius R. The moment of inertia about an axis through the center is I. The thread is pulled with a force as shown. If there is no slipping, find: (a) the acceleration of the center of mass; (b) the force of friction. (c) Discuss what happens to these quantities when r is less than, equal to, and greater than I/MR.

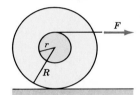

FIGURE 12.70 Problem 8.

9. (I) An α particle (a He nucleus) of mass m approaches a fixed nucleus at speed u_0 along a line displaced by a distance b from that for a head-on collision, as in Fig. 12.71. The distance b is called the impact parameter. The force between the nucleus and the α particle is central (it acts along the line joining them). The potential energy has the form $U(r) = C/r$ where C is a constant. (a) Is angular momentum conserved in this collision? If so, why? (b) Find the distance of closest approach r_{min} in terms of the given quantities. (You will encounter a quadratic equation.)

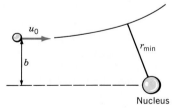

FIGURE 12.71 Problem 9.

10. (II) A spool of mass M and radius R has an axle of radius r around which a string is wrapped; see Fig. 12.72. The moment of inertia about an axis through the center is $\frac{1}{2}MR^2$. Show that the maximum value of the tension such that the spool rolls without slipping is

$$T = \frac{3\mu MgR}{R + 2r}$$

where μ is the coefficient of friction.

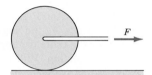

FIGURE 12.72 Problem 10.

11. (I) A uniform thin rod of mass M and length L is freely pivoted at one end and held horizontal by a finger at the other end. At the instant the finger is removed what is the force on the pivot? (See Example 11.5.)

12. (I) In Fig. 12.73 a lawn roller of mass M and radius R is being pulled by a horizontal force at its center. Show that the minimum coefficient of friction needed to prevent slipping is $F/3Mg$.

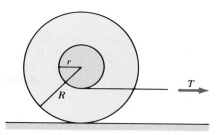

FIGURE 12.73 Problem 12.

13. (II) The wheelbase (distance between the front and rear axles) of a 1200-kg car is 3.0 m. On a horizontal surface the front tires carry 60% of the weight. The CG is 0.8 m above the road. What are the normal forces on the front and rear wheels when the car is at rest on a 20° incline? The car is facing up the incline. (Try Exercise 41 first.)

14. (I) An aluminum stepladder, of negligible mass, has legs of length 4 m. They pivot freely at the top and are connected at the midpoint by a cord, as shown in Fig. 12.74. The floor is frictionless. Find the tension in the cord and the vertical forces at the base of the legs in the following cases: (a) a 70-kg woman sits at the top; (b) she stands 1 m along the ladder from the top.

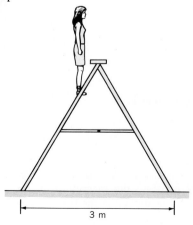

FIGURE 12.74 Problem 14.

15. (II) Four identical blocks of length L are stacked as shown in Fig. 12.75. (a) Find the maximum value of the distance d_1 such that block A does not tip over B. (b) Find the maximum value of d_2 such that A and B do not tip over C. (c) Find the maximum value of d_3 such that A, B, and C do not tip over D.

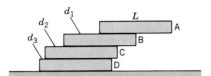

FIGURE 12.75 Problem 15.

16. (I) A crate with a square base of side $b = 70$ cm and height h 1.2 m is placed on an incline, as in Fig. 12.76, whose coefficient of kinetic friction is μ_k. As the angle θ is increased, find: (a) the condition for the crate to topple; (b) the condition for the crate to start sliding. Will the crate topple or slide given: (c) $\mu_k = 0.2$; or (d) $\mu_k = 0.7$?

17. (I) A horizontal force $F = 140$ N is applied to a 25-kg crate of width $b = 50$ cm and height $h = 1.1$ m that is on a horizontal surface for which $\mu_k = 0.4$ (see Fig. 12.77). (a) What is the maximum distance above the floor that F can be applied without tipping the crate? (b) If F is applied at the level of the center of gravity of the crate, what is the effective point of application of the resultant normal force due to the floor?

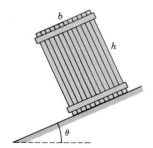

FIGURE 12.76 Problem 16.

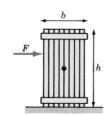

FIGURE 12.77 Problem 17.

18. (I) Figure 12.78 shows a cylinder of mass $M = 10$ kg and radius $R = 0.4$ m that has to be raised over a step of height $h = 0.02$ m. Find the magnitude of the horizontal force required if it is applied at the following points: (a) F_1 at the axle; or (b) F_2 at the top.

19. (I) In Fig. 12.79 a 3-m ladder of mass 10 kg rests against a

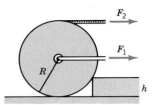

FIGURE 12.78 Problem 18.

frictionless wall at an angle $\theta = 70°$ to the horizontal. Find the vertical and horizontal forces exerted by the wall and the floor when a 50-kg person stands vertically at a point 1 m along the ladder from the bottom.

FIGURE 12.79 Problems 19 and 20.

20. (I) A ladder of mass M and length 3 m rests against a frictionless wall at an angle $\theta = 50°$ to the horizontal as in Fig. 12.79. The coefficient of static friction at the floor is 0.6. What is the maximum distance along the ladder a person of mass $12M$ can climb before the ladder starts to slip?

21. (I) A ladder of length L and negligible mass rests against a frictionless wall at an angle θ to the horizontal. The coefficient of static friction is 0.5 at the floor. A person of mass M stands at a point $2L/3$ from the bottom. What is the minimum value of θ for the ladder to be in equilibrium?

22. (I) A uniform 10-kg door with dimensions 75 cm × 200 cm is pivoted at hinges 25 cm from the top and from the bottom (see Fig. 12.80). Find the horizontal force exerted by each hinge.

FIGURE 12.80 Problem 22.

CHAPTER 13

Gravitation

The motion of the stars in a spiral galaxy is governed by the gravitational force.

Major Points

1. **Newton's law of gravitation.**
2. The **principle of superposition.**
3. The distinction between **inertial mass** and **gravitational mass.**
4. (a) The concept of a **field.**
   (b) The distinction between **gravitational field strength** and the acceleration due to gravity.
5. **Kepler's laws** of planetary motion.
6. Continuous distributions of mass, the **point mass theorem.**

Until the seventeenth century, it was believed that celestial bodies, such as the moon and the planets, were governed by laws different from those on earth. Even Galileo thought that the circular motion of the moon was "natural" and therefore required no further explanation. Newton, who extended the concept of inertia to *all* bodies, realized that the moon is accelerating and is therefore subject to a centripetal force. According to legend, Newton thought of his theory of gravitation in 1665 when his imagination was sparked by the fall of an apple. He guessed that the force that keeps the moon in its orbit has the *same* origin as the force that causes the apple to fall (Fig. 13.1). He recalled some thirty years later:

> *And in the same year . . . I began to think of gravity extending to the orb of the Moon . . . I deduced that the forces that keep the planets in their orbs must be reciprocally as the squares of their distances from the centers about which they revolve, and thereby compared the force required to keep the Moon in her orb with the force of gravity at the surface of the Earth, and found them to answer pretty nearly.*

Newton used the period of the moon's orbit (27.3 days) to calculate its centripetal acceleration: $a_M = \frac{1}{360}$ m/s^2. The acceleration of an apple at the earth's surface is $g \approx 10$ m/s^2, so the ratio of these accelerations is $a_M/g = \frac{1}{3600}$. Next, he assumed that the gravitational force between two bodies varies as the inverse square of the distance between them, that is, $F \propto 1/r^2$—an idea that had been around since about 1640. Thus, the ratio of the force on the moon to that on the apple at the surface of the earth should be R_E^2/r_M^2, where R_E is the radius of the earth and r_m is the distance to the moon. He knew that $r_M \approx 60\, R_E$, which means

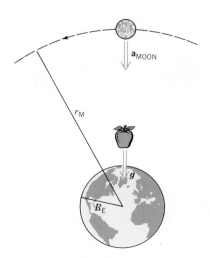

FIGURE 13.1 Newton realized (i) that the moon is accelerating and (ii) that the force of gravity is the cause.

that this ratio is $\frac{1}{3600}$—the same as the ratio of the accelerations. This is what he found to "answer pretty nearly."

Newton was hesitant about publishing the work for several reasons. First, he had considered only circular orbits although Kepler had shown them to be elliptical. Second, the value of the radius of the earth was not accurately known. Third, he treated the earth as if all its mass were concentrated at its center but he could not justify this at the time. (Its proof required great ingenuity and was not completed until 1684.) The full theory of gravitation was published in 1687 in the *Principia*.*

13.1 NEWTON'S LAW OF GRAVITATION

Consider two point particles with masses m_1 and m_2 separated by a distance r. Newton postulated that there is an attractive (gravitational) force between the particles. Its form can be deduced as follows. From Newton's second law, the force F_{12} acting on m_1 must be proportional to m_1: $F_{12} \propto m_1$. Similarly, $F_{21} \propto m_2$. From Newton's third law we know that $F_{12} = F_{21}$. Therefore, one can conclude that the force of interaction between the particles has the form $F \propto m_1 m_2$. We combine this with the (correct) guess that $F \propto 1/r^2$, to obtain **Newton's law of universal gravitation** in modern notation:

$$\mathbf{F}_{12} = - \frac{Gm_1 m_2}{r^2} \hat{\mathbf{r}}_{21} \tag{13.1}$$

where the gravitational constant $G = 6.67 \times 10^{-11}$ N·m²/kg². (Newton did not have a value for G.) Notice in Fig. 13.2a that the unit vector $\hat{\mathbf{r}}_{21}$ points from m_2 toward m_1. The negative sign in Eq. 13.1 therefore indicates that the force is always attractive. Equation 13.1 is often written without subscripts in the form

$$\mathbf{F} = - \frac{Gm_1 m_2}{r^2} \hat{\mathbf{r}}$$

where one must remember that $\hat{\mathbf{r}}$ has its origin at the *source* of the force. In many situations it is convenient to work with just the magnitude

$$F = \frac{Gm_1 m_2}{r^2} \tag{13.2}$$

It is important to realize that Newton's law of gravitation is stated only for *point* particles. For two arbitrary bodies, as in Fig. 13.2b, there is no unique value for the separation r. To compute the force between them requires integral calculus. However, for the special case of a uniform spherical mass distribution, r may be taken as the distance to the center. (See Example 13.5.) Also, when the separation between two objects is very much larger than their sizes, they may be approximated as point masses and so Eq. 13.2 may be used.

Experiment shows that when several particles interact, the force between a given pair is independent of the other particles present. The net force $\mathbf{F}_1$ on the point mass m_1 due to the other particles in Fig. 13.3 is found by first calculating its interaction with each of the other particles one at a time. This is the **principle of superposition:**

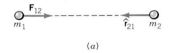
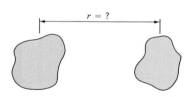

FIGURE 13.2 (a) The gravitational force $\mathbf{F}_{12}$ exerted on m_1 by m_2. The origin of the unit vector $\hat{\mathbf{r}}_{21}$ is at the source of the force, m_2. (b) The equation $F = GmM/r^2$ cannot be used for arbitrary bodies.

Principle of superposition

$$\mathbf{F}_1 = \mathbf{F}_{12} + \mathbf{F}_{13} + \cdots + \mathbf{F}_{1N}$$

The net force on m_1 is the vector sum of the pairwise interactions.

* The background to the publication of this work is discussed in the Historical Note on p. 271.

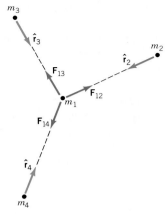

FIGURE 13.3 The net force on m_1 is $\mathbf{F}_1 = \mathbf{F}_{12} + \mathbf{F}_{13} + \mathbf{F}_{14}$.

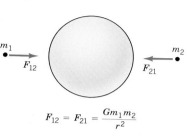

$$F_{12} = F_{21} = \frac{Gm_1 m_2}{r^2}$$

FIGURE 13.4 The gravitational force between m_1 and m_2 is not affected by the presence of a body placed between them.

If a large body of mass M were placed between two point masses, as in Fig. 13.4, the force between these point masses would still be given by Eq. 13.2. However, the net force on each particle would change.

EXAMPLE 13.1: Three point particles with masses $m_1 = 4$ kg, $m_2 = 2$ kg, and $m_3 = 3$ kg are at the corners of an equilateral triangle of side $L = 2$ m, as in Fig. 13.5. Find the net force on m_2.

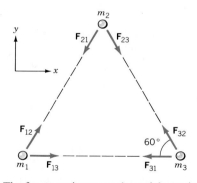

FIGURE 13.5 The forces acting on each particle are indicated.

Solution: When applying the superposition principle, it is useful to employ a systematic procedure.

1. Set up a convenient coordinate system.

2. Indicate the directions of the forces acting *on* the particle under consideration.

3. Calculate the (scalar) magnitudes of the forces.

4. Find the net force by using the component method.

The first two steps have already been taken in Fig. 13.5. The magnitudes are

$$F_{21} = Gm_2m_1/L^2 = 1.33 \times 10^{-10} \text{ N}$$
$$F_{23} = Gm_2m_3/L^2 = 1.01 \times 10^{-10} \text{ N}$$

The net force on m_2 is

$$\mathbf{F}_2 = \mathbf{F}_{21} + \mathbf{F}_{23}$$

Its components are

$$F_{2x} = -F_{21} \cos 60° + F_{23} \cos 60° = -1.6 \times 10^{-11} \text{ N}$$
$$F_{2y} = -F_{21} \sin 60° - F_{23} \sin 60° = -2.03 \times 10^{-10} \text{ N}$$

Thus,

$$\mathbf{F}_2 = -(1.6\mathbf{i} + 20.3\mathbf{j}) \times 10^{-11} \text{ N}$$

EXERCISE 1. Find $\mathbf{F}_3$.

13.2   GRAVITATIONAL AND INERTIAL MASS

Our use of the word "mass" has been a bit cavalier. It has appeared in two entirely different contexts: Newton's second law of motion and Newton's law of gravitation. The second law of motion should be written as

$$F = m_I a$$

Inertial mass

where m_I is the **inertial mass** of the body. It is a measure of the body's resistance to being accelerated. When Newton's second law is applied, the nature or origin of the force is not relevant. In contrast, the law of gravitation should be written as

$$F = \frac{Gm_G M_G}{r^2}$$

Gravitational mass

where m_G and M_G are the **gravitational masses** of the particles. "Gravitational charge" would be an appropriate name for this mysterious property that causes all particles to attract each other. When this law is used to calculate the gravitational force, the motion of the particles is not relevant. If the gravitational force is expressed as $F = m_G g$ and substituted into the second law, we see that

$$a = \frac{m_G}{m_I} g$$

The free-fall acceleration of a body thus depends on the ratio m_G/m_I. Within the limits of their techniques, Galileo and Stevin found that *all* bodies have the same free-fall acceleration. Therefore, this ratio must be the same for all bodies. It is convenient to choose the constant of proportionality to be one. (This affects the value of G to be used in the law of gravitation.)

Newton tried to distinguish between these two types of mass. If m_I and m_G are kept distinct, the period of a simple pendulum of length L has the form $T = 2\pi\sqrt{m_I L/m_G g}$. By using bobs of different materials he tried to find a discrepancy from the usual formula, which is $T = 2\pi\sqrt{L/g}$, but found none. He concluded that m_I and m_G are equal to about 1 part in 10^3. The latest experiments show that m_I and m_G are equivalent to within 1 part in 10^{12}.

This is a strange and tantalizing coincidence! To see the connection between the inertia of a body and gravitation, consider an astronaut in a rocket, as shown in Fig. 13.6. When the rocket is on the earth, as in Fig. 13.6a, the astronaut measures the weight of an apple and its free-fall acceleration (9.8 m/s²). Now suppose the rocket is far enough from any other body to be in gravity-free space (Fig. 13.6b) and is given an acceleration of 9.8 m/s² relative to an inertial reference frame. When the astronaut places the apple on a scale, it will register a reading

FIGURE 13.6 When a ball is released, it accelerates toward the floor of the cabin. The astronaut cannot tell whether the acceleration relative to the rocket was (a) caused by the force of gravity, or (b) a consequence of the acceleration of the rocket in gravity-free space.

equal to that found on earth. When he releases the apple, it will move toward the rear of the rocket with an acceleration of 9.8 m/s². In the accelerated frame of the rocket, the astronaut will describe these results in terms of the *inertial* force $F' = ma'$ (Eq. 6.13). He cannot tell whether he is on earth or accelerating in gravity-free space. According to the **principle of equivalence:**

> No experiment can distinguish the effects of a gravitational force from that of an inertial force in an accelerated frame.

Principle of equivalence

This idea was the inspiration for Einstein's general theory of relativity.

13.3 THE GRAVITATIONAL FIELD STRENGTH

Newton's law of gravitation involves some philosophical difficulties. It implies that two particles can interact directly with each other through free space. Newton was not happy with such "action at a distance." In a letter to Richard Bentley (who later helped with the publication of the second edition of the *Principia*) he wrote:

> *That gravity should be innate, inherent and essential to matter, so that one body may act upon another at a distance through a vacuum, without the mediation of anything else, by and through which their action and force may be conveyed from one to another, is to me so great an absurdity that I believe that no man who has in philosophical matters a competent faculty of thinking can ever fall into.*

The same problem of interaction without actual contact occurs with electric charges and magnets. This is where the matter stood till the 1830s when Michael Faraday developed the concept of a **field.** *

Weather charts, such as Fig. 13.7, usually display how pressure, wind velocity, and temperature vary over some region. Each point in space is assigned a particular value for each of these quantities. *The distribution of values over a region of space is called a field.* Pressure and temperature form *scalar* fields, whereas velocity and force give rise to *vector* fields.

Suppose we measure the force exerted by a stationary particle of mass M on a particle of mass m as it is moved to different points. Each point in the region around M is associated with a single (vector) force as in Fig. 13.8. Since the magnitudes in this force field apply only to the mass m, it is convenient to consider the *force per unit mass* $\mathbf{F}/m$. Using Eq. 13.1 we have

$$\mathbf{g} = \frac{\mathbf{F}}{m} = - \frac{GM}{r^2} \hat{\mathbf{r}} \qquad (13.3)$$

where the origin of $\hat{\mathbf{r}}$ is at M. The quantity $\mathbf{g}$, measured in N/kg, is called the **gravitational field strength** at position r with respect to M. The field still looks like Fig. 13.8, but the lengths of the vectors are changed by the factor $1/m$. Once the field strength is known, the gravitational force on any particle of mass m, that is, its weight, is given by

$$\mathbf{W} = m\mathbf{g}$$

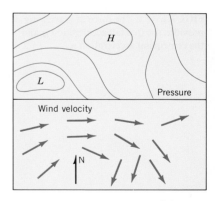

FIGURE 13.7 (*a*) A pressure field. The lines, called isobars, join points of equal pressure. They help us visualize the pattern of the field. (*b*) A velocity field. Each arrow indicates the magnitude and direction of the velocity at one point.

* This idea is briefly discussed here. We will return to it in Chapter 23.

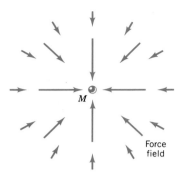

FIGURE 13.8 A force field. The arrows represent the forces exerted by a particle of mass M on a particle of mass m placed at various points.

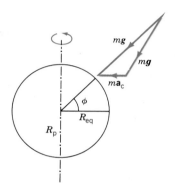

FIGURE 13.9 The force of gravity $m\boldsymbol{g}$ provides the free-fall acceleration $\boldsymbol{g}$ and the centripetal acceleration $\mathbf{a}_c$.

The magnitude of the field strength at the surface of the earth is

$$g(R_E) = \frac{GM_E}{R_E^2}$$

To see how the field concept resolves the problem of action at a distance, notice that M, but not m, appears in Eq. 13.3. We say that the "source" particle of mass M creates a gravitational field in the space around it, much like a candle creates a "light field" around it. When the particle of mass m approaches, it interacts with the preexisting field and experiences a force $m\mathbf{g}$, where $\mathbf{g}$ is the field strength at a particular point. From Newton's third law, M responds to the field created by m.

Acceleration due to Gravity

Although the unit N/kg reduces to m/s², the gravitational field strength $\boldsymbol{g}$ is a concept different from the acceleration due to gravity $\boldsymbol{g}$. To highlight the distinction consider a particle of mass m at some latitude ϕ on the earth, as in Fig. 13.9. We assume that the earth is a uniform sphere rotating about its north–south axis. The gravitational force $m\mathbf{g}$ is directed to the center and serves two functions: It causes the particle to fall with acceleration $\boldsymbol{g}$ and it produces the centripetal acceleration $\mathbf{a}_c$. From the second law $\mathbf{F} = m\mathbf{a}$, we have

$$m\mathbf{g} = m(\boldsymbol{g} + \mathbf{a}_c)$$

The vectors $\mathbf{g}$ and $\boldsymbol{g}$ are parallel only at the poles and at the equator. Thus, in general, a plumb line, which aligns itself with $\boldsymbol{g}$, will be at a small angle to the true vertical that passes through the center of the earth. At the pole $a_c = 0$ and so $g_p = g_p$. At the equator $a_c = v_e^2/R = 3.4$ cm/s²; thus, $g_e = g_e + 3.4$ cm/s². Combining these two results we find $(g_p - g_e) = (g_p - g_e) + 3.4$ cm/s². Measurements, however, show that $(g_p - g_e) = 5.2$ cm/s².

The difference of 1.8 cm/s² is only partly explained by the fact that the earth is not a perfect sphere. If we use $g = GM/R^2$ with $R_p = 6357$ km and $R_e = 6378$ km, we find $(g_p - g_e) = 6.5$ cm/s², which is far more than we need. The problem lies in our use of an equation that is valid only for a *uniform spherical* distribution. A more advanced calculation shows that the nonsphericity of the earth contributes about 0.5 cm/s². The remaining discrepancy is due to the variation in the earth's density along any radial line. The main reason that $g_p > g_e$ is that at the pole a particle is closer to the earth's dense inner core.

13.4 KEPLER'S LAWS OF PLANETARY MOTION

In 1543, Copernicus had changed the description of the motion of the planets from a geocentric view to a heliocentric one. Between 1601 and 1619, Kepler discovered three laws of planetary motion that further strengthened the idea that the earth orbits the sun rather than vice versa. They were found through a laborious analysis of data left by the astronomer Tycho Brahe. The first two laws were published in *Astronomia Nova* (1609).

Kepler's first law

LAW 1. The planets move around the sun in elliptical orbits with the sun at one focus.

An ellipse, drawn in Fig. 13.10, is a curve along which the sum of the distances to two fixed points F and F', called the *foci*, is constant. It may be traced by tying a thread to the foci and keeping it taut with a pencil. A circle is a special case in

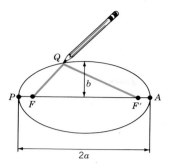

FIGURE 13.10 An ellipse may be traced by pinning the ends of a string to the foci F and F'. A pencil (shown at Q) moves over a sheet while the string is held taut. The length of the major axis is $2a$; the length of the minor axis is $2b$.

After the first landing on the moon in July 1969, the Apollo II Lunar Module rises to rendezvous with the Command Module which had stayed in lunar orbit.

which the foci coincide at the center. The short dimension is called the *minor axis* and has length $2b$; the long dimension is called the *major axis* and has length $2a$. The closest point, P, in the orbit to the sun is called the *perihelion;* the farthest point, A, is called the *aphelion.*

LAW 2. The line joining the sun to a planet sweeps out equal areas in equal times.

Suppose that in a given time interval a planet moves from A to B in Fig. 13.11, and from C to D during another time interval. According to the second law the areas SAB and SCD are equal. The speed of the planet, therefore, must vary during its orbit. It is greatest at the perihelion, and least at the aphelion. In Example 12.8 this law was shown to be a consequence of the conservation of angular momentum. It took another decade for Kepler to discover a mathematical relationship between the orbits of the various planets. The third law was published in *Harmonium Mundi* (1619).

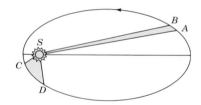

FIGURE 13.11 In equal times, the areas SAB and SCD swept out by the radial line to a planet are equal.

LAW 3. The square of the period of a planet is proportional to the cube of its mean distance from the sun.

The "mean distance" turns out to be the semimajor axis, a. As an equation,

$$T^2 = \kappa a^3 \tag{13.4}$$

Kepler's third law.

where κ is a constant that applies to all the planets (see p. 107).

Energy in an Elliptical Orbit

In the course of an orbit both the mechanical energy and the angular momentum are conserved. We can use Kepler's laws to obtain expressions for the speeds at perihelion and aphelion (see Fig. 13.12). At the perihelion and aphelion **r** is at $90°$ to **p** so the conservation of angular momentum tells us $mr_A v_A = mr_P v_P$, or

$$r_A v_A = r_P v_P \tag{13.5}$$

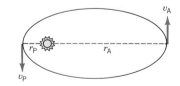

FIGURE 13.12 A planet in orbit around the sun. From the conservation of angular momentum we find $r_A v_A = r_P v_P$.

Recall from Eq. 8.19 that gravitational potential energy is $U = -GmM/r$. The conservation of mechanical energy, $\frac{1}{2}mv_A^2 - GmM/r_A = \frac{1}{2}mv_P^2 - GmM/r_P$, may be written in the form

$$2GM \left(\frac{1}{r_P} - \frac{1}{r_A} \right) = v_P^2 - v_A^2 \tag{13.6}$$

Equations 13.5 and 13.6 can be solved for either v_P or v_A in terms of r_P and r_A. If we substitute $v_P = r_A v_A / r_P$ in Eq. 13.6 and use the fact that $r_A + r_P = 2a$, we find (see Exercise 31):

$$v_A^2 = \frac{GM}{a} \frac{r_P}{r_A} \tag{13.7a}$$

$$v_P^2 = \frac{GM}{a} \frac{r_A}{r_P} \tag{13.7b}$$

If we use either the aphelion or the perihelion we find the mechanical energy, $E = K + U$, is given by (see Exercise 31):

Mechanical energy in an elliptical orbit

$$E = -\frac{GmM}{2a} \tag{13.8}$$

We see that the mechanical energy is determined by the length of the major axis.

Bound and Unbound Trajectories

Suppose that a cannonball is fired from the peak of a very tall tower with speed v (see Fig. 13.13). Let us consider the shapes of the paths for various values of v.

If the ball is located at a distance r from the center of the earth, the speed that it requires to escape from that point is found by setting the energy equal to zero: $E = \frac{1}{2}mv^2 - GmM/r$, and so $v_{esc} = \sqrt{2GM/r}$. Note also that for a circular orbit of radius r the orbital speed is $v_c = \sqrt{GM/r}$.

(a) When $v < v_c$, the orbit is elliptical, with the peak as the *apogee* (the farthest point from earth). If v is too small, the projectile will strike the earth. When $v = v_c$, the orbit is circular. When $v_{esc} > v > v_c$, the orbit is again an ellipse, but now the peak is the *perigee* (the closest point to earth).

(b) When $v = v_{esc} = \sqrt{2} \, v_c$, the path is parabolic and is not a closed orbit. The object is not bound.

(c) When $v > v_{esc}$, the path is a hyperbola and not closed. The object is not bound.

These comments are summarized in Table 13.1.

TABLE 13.1

$v < v_{esc}$	$E < 0$ (ellipses, circle)
$v = v_{esc}$	$E = 0$ (parabola)
$v > v_{esc}$	$E > 0$ (hyperbola)

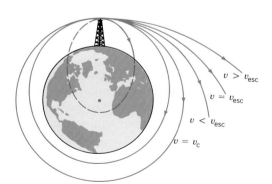

FIGURE 13.13 A redrawing of one of Newton's diagrams (Fig. 6.14). A projectile is fired with initial speed v from the top of a high tower. If $v < v_{esc}$, the trajectory is a closed elliptical orbit (which might intersect the surface). If $v = v_{esc}$, the trajectory is parabolic. If $v > v_{esc}$, the trajectory is hyperbolic.

13.5 CONTINUOUS DISTRIBUTIONS OF MASS

In order to calculate magnitude of the gravitational field strength g due to an extended object, it must be divided into infinitesimal mass elements, as in Fig. 13.14, each of which contributes

$$dg = \frac{G\,dm}{r^2} \tag{13.9}$$

The total field strength at a given point is the integral of this over the object. To perform the integration, it is necessary to express the mass element dm in terms of r—as we did in finding the CM (Section 10.2) or the moment of inertia (Section 11.4) of an extended body.

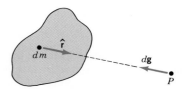

FIGURE 13.14 The contribution of the mass element dm to the field strength at point P is $d\mathbf{g}$.

EXAMPLE 13.2: Find the field strength at a point along the axis of a *thin rod* of length L and mass M, at a distance d from one end, see Fig. 13.15.

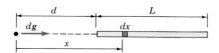

FIGURE 13.15 In finding the field due to a thin rod we first consider the contribution of an element of length dx.

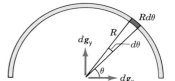

FIGURE 13.16 A semicircular ring. The x component of the resultant field at the center is zero by symmetry.

Solution: First we need to find the field due to an element of length dx. The rod must be thin if we are to assume that all points of the element are at the same distance from the field point. The mass of the element is $dm = (M/L)dx$, so its contribution to the field is

$$dg = G\,\frac{M}{L}\frac{dx}{x^2}$$

The total field strength is

$$g = \frac{GM}{L}\int_d^{L+d}\frac{dx}{x^2} = \frac{GM}{L}\left[\frac{1}{d} - \frac{1}{L+d}\right]$$
$$= \frac{GM}{d(L+d)}$$

Notice that when $d \gg L$, we find $g \to GM/d^2$, the result for a point particle.

EXAMPLE 13.3: Find the field strength at the center of a thin *semicircular ring* of radius R and mass M, as shown in Fig. 13.16. The linear mass density is λ kg/m.

Solution: We choose an element that is an arc of length $ds = R\,d\theta$. Its mass is $dm = \lambda R\,d\theta$. For every element at $+x$ there is an equivalent element at $-x$ whose contribution to the field at the center is in the opposite direction. From this symmetry we

immediately see that the field strength has no x component at the center. For the y component,

$$dg_y = dg\,\sin\theta = \frac{G\,dm\,\sin\theta}{R^2}$$

The total field strength is

$$g = \frac{G\lambda}{R}\int_0^{\pi}\sin\theta\,d\theta$$
$$= \frac{2G\lambda}{R}$$

EXAMPLE 13.4: The *point mass theorem* derived by Newton states: A *uniform spherical mass distribution attracts an external point particle as if all its mass were concentrated at its center.* Prove this theorem by finding the force exerted by a uniform spherical shell, of radius R and surface mass density σ kg/m², on a point particle of mass m.

Solution: Since the forces exerted by the various particles on the sphere have different magnitudes and directions, this result is not at all obvious. Indeed, it gave Newton a great deal of trouble. For every point particle above the x axis in Fig. 13.17, there is an equivalent particle below the axis. From this symmetry we see that the net force must lie along the line joining the point mass to the center of the sphere.

The shell can be divided into rings of width $R\,d\theta$. Notice that all points on such a mass element are at the same distance from the field point. The mass of an element is

$$dM = \sigma(2\pi R \sin\theta)(R\, d\theta)$$

We could continue by finding the component along the central line of the force exerted by this element. It is easier, however, to first calculate the potential energy of the point mass m and the element. From Eq. 8.19 we have

$$dU = -\frac{Gm\, dM}{s}$$

This involves two variables, θ and s. We eliminate θ by using the law of cosines:

$$s^2 = R^2 + r^2 - 2Rr\cos\theta$$

from which we obtain

$$s\, ds = Rr\sin\theta\, d\theta$$

Using this in the expression for dM we see that

$$\frac{dM}{s} = \frac{2\pi\sigma Rs\, ds}{r}$$

The variable s ranges from $s = r - R$ (at $\theta = 0$) to $s = r + R$ (at $\theta = \pi$) and so the total potential energy is

$$
\begin{aligned}
U &= -\frac{Gm2\pi\sigma R}{r}\int_{r-R}^{r+R} ds \\
&= -\frac{Gm2\pi\sigma R}{r}[(r + R) - (r - R)] \\
&= -\frac{GmM}{r}
\end{aligned}
$$

where we have used the fact that the surface area of the shell is $4\pi R^2$, and so its total mass is $M = \sigma(4\pi R^2)$. The force on the point mass is found from Eq. 8.18:

$$F_r = -\frac{dU}{dr} = -\frac{GmM}{r^2}$$

The only distance that appears is r, the distance to the center, which proves the point mass theorem. A solid sphere with a spherically symmetric distribution of mass may be treated as a collection of shells, so the same result applies.

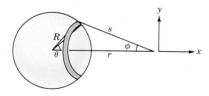

FIGURE 13.17 In finding the field due to a uniform spherical shell, it is divided into thin rings. It is easier to first find the scalar potential energy and then to use $F_r = -dU/dr$.

EXERCISE 2. By changing the limits of integration, from $s = R - r$ to $s = R + r$, show that the potential energy inside the spherical shell has the constant value $U = -GmM/R$. What is the force exerted on a point particle inside the shell?

EXAMPLE 13.5: How does the field strength vary inside a *uniform solid sphere* of density ρ kg/m³ and radius R?

Solution: At a distance r from the center, all those shells with radii greater than r will not contribute to the field strength (see Exercise 2 above). Only the sphere of radius r "below" a particle produces a net force. Its mass is $M(r) = 4\pi\rho r^3/3$, so from Eq. 13.3 we find the magnitude of the field strength is given by

$$g(r) = \frac{GM(r)}{r^2} = Cr$$

where C is a constant. The field strength increases linearly with distance from the center as indicated in Fig. 13.18.

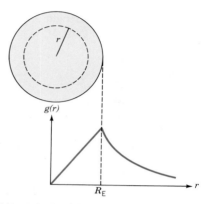

FIGURE 13.18 Inside a uniform sphere the field varies linearly with distance from the center.

Actually, the field strength is larger at the bottom of a mine than at the surface. The reason is that the earth is not a uniform sphere. Whereas the average density is 2.5 g/cm³ at the surface, it rises to about 15 g/cm³ at the core. Thus, although the shells above the bottom of a mine do not contribute to the field at that point, we would be closer to the dense core. The predicted variation is shown in Fig. 13.19.

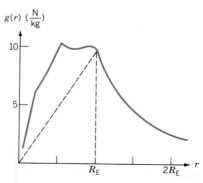

FIGURE 13.19 The predicted variation of the field inside the earth, whose density is not constant.

HISTORICAL NOTE: Background to the *Principia*

There is considerable doubt that Newton made the calculations discussed in the introduction in 1665. It seems he did not connect the apple to the moon until 1675 and that he held other views on gravitation until 1679. In any case, he could not have developed the full theory of gravitation without first clarifying his second and third laws (sometime after 1669).

In 1673, Huygens published the equation for centripetal acceleration: $a = 4\pi^2 r/T^2$. By 1679, Robert Hooke, Edmund Halley, and Christopher Wren had combined this result with Kepler's third law, $T^2 = \kappa r^3$, for circular orbits (Eq. 6.6), to deduce that $a \propto 1/r^2$. They independently proposed that the gravitational force varies as $F \propto 1/r^2$.

In 1679, Robert Hooke became secretary of the Royal Society in London. Newton had kept himself aloof from the society for many years, partly because of the criticism he had suffered at the hands of Hooke and others regarding his ideas on the nature of light. Hooke wrote to him to renew contact with the society. In November 1679, despite his dislike of Hooke, Newton replied to the conciliatory gesture. He threw in a few thoughts on the old problem of proving that the earth rotates. Recall that people had long felt that an object falling from a high tower would be left behind and so it would fall toward the west. Galileo had argued that it should land at the base, but Newton was more sophisticated. Because of the greater tangential speed of the top of the tower compared to the surface of the earth, the object should fall *ahead* of the tower—toward the east. Newton included a sketch of the path (Fig. 13.20). He continued it below the surface of the earth with the assumption that it met no resistance. It was a sort of spiral that ended at the earth's center. But Newton had made a mistake. Hooke read the letter to the society and corrected the error in public. He pointed out that the spiral was the consequence of assuming a constant force of gravity. If instead it varied as $1/r^2$ as everyone believed, the path would be an ellipse.

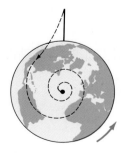

FIGURE 13.20 Newton's drawing of the eastward deflection of a body dropped from a high tower. The path drawn below the surface was incorrect.

Newton was furious at having been publicly humiliated—to make matters worse, by his old adversary. He had thrown in the spiral only as an afterthought, for he was more concerned with the easterly direction of the fall. Anyway, he replied with some corrections to Hooke's statement. In a second letter, in January 1680, Hooke raised some issues on whether it could be proved that an inverse square law leads to elliptical orbits. He pointed out the difficulty in justifying that the earth and the moon could be treated as point masses. He also noted that circular motion is the result of an attraction *toward* the center—most scientists had thought in terms of "centrifugal tendencies." Newton did not reply.

Five years later, in 1684, Christopher Wren, architect of St. Paul's cathedral in London, offered a prize for a proof that the inverse square law leads to elliptical orbits. Hooke claimed that he had the solution but would not divulge it, "in order that others may appreciate the difficulty involved." Wren was unconvinced. In the summer of the same year, the young astronomer Edmund Halley visited Newton at Cambridge. He asked, "What would be the path of a planet subject to an inverse square force?" "Why, an ellipse," replied Newton. Halley was overjoyed and asked to see the proof. Well, this had been mislaid. What the rest of the scientific community was searching for, Newton claimed to have lost! He promised to redo the calculations and send them on. Thus began eighteen months of intense and creative work that culminated in the publication of the *Principia* in 1687—at Halley's expense.

After the *Principia* was published, Hooke claimed to be the true discoverer of the law of gravitation, while Newton had merely filled in the details. Newton was contemptuous:

Now is this not very fine? Mathematicians that find out, settle and do all the business, must content themselves with being nothing but dry calculators and drudges, and another that does nothing but pretend and grasp at all things, must carry away all the invention.

Hooke did not appreciate the difference between speculation and proof. His hypotheses showed brilliant insight, but lacked foundation. In the second edition of the *Principia*, Newton acknowledged that Hooke, Halley, and Wren had independently shown that the inverse square law follows from Kepler's third law.

The combination of Kepler's laws of celestial bodies with Newton's terrestrial mechanics into one law of universal gravitation is called the *Newtonian synthesis*. Yet, while most of the scientific community was bedazzled by Newton's theory, Huygens and Leibniz still felt that the concept of attraction through empty space (action at a distance) was an "occult" quality that explained nothing and reflected medieval thinking.

I have not been able to discover the cause of the properties of gravity from phenomena, and I feign no hypotheses. . . . To us it is enough that gravity does really exist, and acts according to the laws which we have explained, and abundantly serves to account for all the motions of the celestial bodies, and of our sea.

Newton did explain a wide range of phenomena:

1. Kepler's laws and the motions of satellites
2. The variation of *g* with height
3. The nonspherical shape of the earth
4. The precession of the earth's spin axis
5. The tides

The most stunning success of the theory came much later in the prediction of hitherto unknown planets.

The *Principia* is perhaps the most important work in the history of science because its approach served as a model for the subsequent development of science. It demonstrated the power of a theory to explain a wide range of phenomena. It was also an inspiration to nonscientists: If mere mortals could comprehend the motion of the heavenly bodies, no earthly problem was beyond our capabilities. It is rarely read nowadays because of the complexity of its geometrical proofs.

Discovery of Neptune and Pluto

The development of the telescope in the seventeenth and eighteenth centuries provided astronomical data far more precise than those available to Kepler. (Actually, it is fortunate that Kepler's data were not better. The details might only have confused the basic correctness of his ideas.) Each planet interacts not only with the sun but also with the other planets. This leads to irregularities in the planetary orbits. Since the planets are continually changing their relative positions, the calculations of these perturbing effects was a major task for astronomers. Newton's law of gravitation accounted satisfactorily for the detailed motion of several planets.

In 1781, the amateur astronomer William Herschel discovered the planet Uranus while making a systematic study of the sky. Many decades later it was found that even when the effects of all the known planets were taken into account, there still remained discrepancies of the order of 2′ of arc in the expected position of Uranus. This led J. J. Leverrier and J. E. Adams to independently predict the existence of another planet. The Greenwich observatory more or less ignored the suggestion from the young and unknown Adams. Leverrier, who met with the same indifference in France, then wrote to the Berlin observatory. On September 23, 1846, Dr. Galle received the letter and, that very night, found Neptune within 1° of where Leverrier had told him to look!

Over the years, it became clear that Neptune's orbit also contained unexplained deviations. P. Lowell used these to calculate the position of yet another planet. His own search from 1906 to 1915 was unsuccessful. After his death, the search was resumed and in 1930 Clyde Tombaugh discovered Pluto. Recently a moon has been discovered orbiting this planet.

Newton provided the first correct explanation of the tides. High tide and low tide at the Bay of Fundy, where the record variation in the water level is 16.3 m.

SPECIAL TOPIC: The Tides

The daily ebb and flow of the tides is caused by the gravitational forces exerted by the moon and the sun. Overall, the effect of the moon is about two and a half times that of the sun, and so in the following discussion we will refer mainly to the moon. Twice each lunar month (one revolution of the moon around the earth) the earth, moon, and sun are roughly along the same line. The "spring" tides at these times are particularly strong. When the lines from the earth to the moon and from the earth to the sun are perpendicular, weak "neap" tides result.

Earth's Center of Rest

Let us treat the earth as a uniform sphere covered by a layer of water. Let us see what would happen if the earth's center were fixed in space. The ocean closer to the moon would experience a larger gravitational pull than that on the far side of the earth. Since water can flow, this variation in the gravitational force across the dimensions of the earth would cause the ocean to heap up "under" the moon, as in Fig. 13.21. If we include the earth's daily rotation, there would be one huge tide each day that would submerge many of today's coastal areas. In fact, there are two tides per day and they do not have such catastrophic effects. The model with the earth's center at rest is clearly inadequate.

Earth and Moon Revolve About Their CM

The earth does not have infinite mass and so it does not remain fixed as the moon orbits around it. (We are not interested in the earth's orbit around the sun.) Both the earth and the moon revolve around their common CM (see Fig. 13.22). In fact, both are in free-fall. The CM is about 4500 km from the center of the earth. If we ignore the daily rotation, each point on the earth describes a circle of radius 4500 km each lunar month (27.3 d). (To visualize its motion, take a circular pad and move it in a small circle as you would to polish a surface. Notice that each point moves in a circle with its *own* center.)

The force exerted by the moon on a given particle at the surface of the earth is directed along the line joining the moon and particle. The centripetal force the particle requires for its circular path is provided by the component of the gravitational force $\mathbf{F}_G$ due to the moon, along the earth–moon line. The difference between $\mathbf{F}_G$ and the centripetal force $\mathbf{F}_C$ is the **tide generating force $\mathbf{f} = \mathbf{F}_G - \mathbf{F}_C$**. The tide generating force is drawn with a red arrow in Fig. 13.23. The directions of $\mathbf{f}$ at the equatorial points are opposite because $\mathbf{F}_G$ varies as $1/r^2$. The earth–moon distance is about $60R_E$, where R_E is the radius of the earth. Thus, on the side closest to the moon $F_G \propto 1/(59R_E)^2$, whereas on the far side $F_G \propto 1/(61R_E)^2$. In general it is the *rate of change* in F_G across the dimensions of a body that determines the magnitude of the tide generating force. The gravitational force due to the sun is about 175 times stronger, but it does not change as much across the diameter of the earth. It turns out that the magnitudes of $\mathbf{f}_1$ and $\mathbf{f}_2$ are roughly equal and so the pattern in the figure above is symmetrical about the north–south line.

A tidal force creates a tension in the body that tends to tear it apart. In the present case, the components of $\mathbf{f}$ along the surface will cause water to move toward the points E_1 and E_2 on the earth–moon line as shown in Fig. 13.24. To a first approximation the tidal bulges stay along the earth–

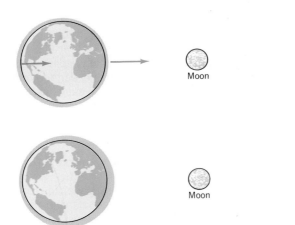

FIGURE 13.21 If the earth's center were at rest, there would be one huge tide each day.

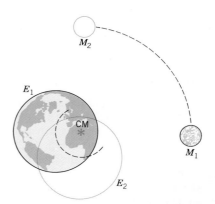

FIGURE 13.22 The dashed paths indicate that both the earth and the moon both revolve around their common center of mass.

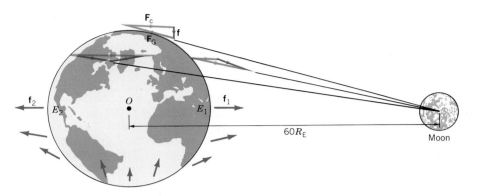

FIGURE 13.23 The tide-generating force **f** (drawn as red arrow) is the difference between the gravitational force $\mathbf{F}_G$ and the centripetal force $\mathbf{F}_c$ required for the circular motion about the center of mass.

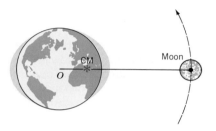

FIGURE 13.24 The tide-generating force causes the oceans to bulge along the earth–moon line.

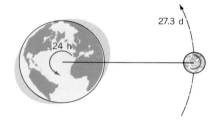

FIGURE 13.25 Because of the earth's rotation and tidal friction, the tidal bulges do not lie along the earth–moon line.

moon line—which rotates with the period of the moon's orbit, 27.3 d. As the earth spins in 24 h, a given point will encounter two high tides and two low tides. Since the moon rises 50 min later each day, the tides are actually 12 h 25 min apart.

The variation in level of the water is about 50 cm in the middle of an ocean. Depending on coastal terrain, this can be magnified. The famous Bay of Fundy tides rise and fall through 15 m! Since the earth is not really a rigid body, there are also earth tides. The land can heave up and down by about 20 cm. This causes minute changes in the acceleration due to gravity (2.5×10^{-6} m/s²) which have been detected.

Tidal Friction

Friction between the water and the solid earth carries the tidal bulge in the direction of the earth's rotation (eastward). As a result, the bulge is directed at a constant angle ahead of the earth–moon line, as shown in Fig. 13.25. Thus, at any given point on earth, high tide occurs after the moon passes "overhead." As the earth spins under the bulges, it experiences a braking torque that tends to slow its angular velocity and to lengthen the day. Evidence from growth rings on fossil corals shows that a day was 22 h long in the Devonian period, 370 million years ago. This means that the length of a day is increasing at the rate of 2 ms per century. It is hard

to measure this long-term trend because it is masked by much larger fluctuations associated with the earth's fluid core.

A similar braking torque due to the earth on the moon has caused the moon's spin angular velocity to equal its orbital angular velocity. In this condition, called "tidal lock," there is no further braking action. This is why the moon presents the same face to the earth.

History of the Earth–Moon System

The slowing down of the earth's rate of spin has interesting implications for the evolution of the earth–moon system. The total angular momentum of the system has two spin and two orbital (about the CM of the earth–moon system) contributions:

$$\mathbf{L} = \mathbf{L}_{oE} + \mathbf{L}_{sE} + \mathbf{L}_{oM} + \mathbf{L}_{sM}$$

The spin angular momentum of the moon, $\mathbf{L}_{sM}$, and the orbital angular momentum of the earth, $\mathbf{L}_{oE}$, are small compared to the other terms. In any case, the spin angular velocity of the moon will be kept equal to the orbital angular velocity by tidal lock. If we treat the system as isolated, its angular momentum is conserved. Consequently, as $\mathbf{L}_{sE}$ decreases because of tidal friction, the orbital angular momentum of the moon $\mathbf{L}_{oM}$ must increase. Since $L \propto r^{1/2}$ (Exercise 21), the earth–moon distance must increase.

From Newton's second law, the equation of motion of the moon is $m\omega^2 r = GmM/r^2$; that is,

$$r^3 = GM\omega^{-2}$$

$$\frac{dr}{dt} = -\frac{2r}{3\omega}\frac{d\omega}{dt}$$

Recent measurements indicate that $d\omega/dt = -10^{-23}$ rad/s² and so $dr/dt = 10^{-7}$ cm/s or about 3 cm/year. Modern laser ranging techniques can measure the distance to a reflector on the moon to an astonishing precision of 0.4 m. But this is still not good enough to detect the effect under discussion.

If we assume this rate of recession has been constant since the origin of the system about 4.5×10^9 y ago, the moon would have been much closer to the earth. The moon will continue to recede until it reaches about 150% of its present distance. At that time, 10^{10} y from now, the earth will be in tidal lock and the length of one day will equal one lunar month. At that time this will be 46 of our present days.

The development of the above scenario is complicated by the presence of the sun. In addition to causing ocean tides, the sun also generates "thermal tides" in our atmosphere. The net effect of these is to make both the earth and the moon spin faster. We may have reached a point at which the 24-h day represents a balance between these two opposing tendencies.

SUMMARY

Newton's law of universal gravitation states that there is an attractive force between any two point particles of masses m and M separated by a distance r:

$$\mathbf{F} = -\frac{GmM}{r^2}\hat{\mathbf{r}}$$

According to the *point mass theorem,* the above equation may also be applied to uniform spherical distributions of mass.

The **gravitational field strength** is the force per unit mass (N/kg) experienced by a particle:

$$\mathbf{g} = \frac{\mathbf{F}}{m} = -\frac{GM}{r^2}\hat{\mathbf{r}}$$

In general, value of the field strength, $\mathbf{g}$ (N/kg), is not the same as the acceleration due to gravity, g (m/s²).

Kepler's three laws of planetary motion state:

1. The planets move in elliptical paths with the sun at one focus.

2. The line joining the sun to a planet sweeps out equal areas in equal times.

3. The square of the period is proportional to the cube of the semimajor axis. That is, $T^2 = \kappa a^3$, where κ has the same value for all planets.

The mechanical energy of a body in an elliptical orbit is given by

$$E = -\frac{GmM}{2a}$$

ANSWERS TO IN-CHAPTER EXERCISES

1. The magnitudes of the forces are

$$F_{32} = Gm_3 m_2/L^2 = 1 \times 10^{-10} \text{ N}$$
$$F_{31} = Gm_3 m_1/L^2 = 2 \times 10^{-10} \text{ N}$$

The components of $\mathbf{F}_3$ are

$$F_{3x} = -F_{31} - F_{32}\cos 60° = -2 \times 10^{-10} \text{ N}$$
$$F_{3y} = 0 + F_{32}\sin 60° = 8.66 \times 10^{-11} \text{ N}$$

2. The potential energy within the shell is

$$U = -\frac{2Gm\pi\sigma R}{r}\int_{R-r}^{R+r} ds$$

$$= -\left(\frac{2Gm\pi\sigma R}{r}\right)[(R+r) - (R-r)]$$

$$= -\frac{GmM}{R}$$

The force on a particle at *any* point inside a uniform spherical shell is zero. This is a consequence of the fact that the gravitational force varies as the inverse square of the distance.

QUESTIONS

1. The force exerted by the sun on the oceans is much greater than that due to the moon. Why is the moon primarily responsible for the ocean tides?

2. A sidereal day (23 h 56 m 4 s) is the time for the earth to make one rotation relative to the distant stars. A solar day (24 h) is the interval between the times the sun is at the highest point in the sky. Why are these different?

3. The earth is closer to the sun in December than it is in June. Why is New York colder in December than in June?

4. What method(s) could we use to determine the mass of the moon?

5. For a fixed price, would you prefer to buy a pound of gold at the pole or at the equator? Does it matter whether a spring scale or an equal arm balance is used?

6. Suppose the gravitational force law varied as r^{-3} instead of r^{-2}. Which of Kepler's laws would still be valid?

7. A satellite is in stable circular orbit. What is the effect of firing its rockets for a brief time in the following directions: (a) forward; (b) backward; (c) radially inward; (d) radially outward. Answer qualitatively.

8. A ship sails along the equator. Would the reading on a spring balance depend on whether it is headed east or west? If so, explain why.

9. Tidal friction slows the earth's rotation. How does this affect the moon's motion?

10. What are the reasons for believing that the moon's orbit about the earth is circular? Sketch the path of the moon relative to a reference frame based on the sun.

11. At new moon and full moon, the earth, the moon, and the sun lie approximately along a line. Why are the "spring" tides that occur at these times particularly high?

12. The frictional drag exerted on a satellite by the atmosphere results in an *increase* in speed of the satellite. How is this possible?

13. Suppose it was proposed that the gravitational force between two particles is given by $F = G(m_1 + m_2)/r^2$. Would this conflict with either Newton's second or third law?

EXERCISES

13.1 Newton's Law of Gravitation

1. (II) Two point particles, each of mass 100 kg, are initially at rest 1 m apart in outer space. (a) What is their initial acceleration? (b) What are their speeds when their separation is 0.5 m?

2. (I) Estimate the magnitude of the force exerted on the moon (a) by the earth, and (b) by the sun.

3. (I) Estimate the magnitude of the force exerted on a 70-kg person on earth (a) by the moon, and (b) by the sun.

4. (I) At what point between the earth and the moon does the net force on a spacecraft due to these two bodies vanish? Assume that the mass of the earth is 81 times that of the moon.

5. (I) In Fig. 13.26 a particle of mass $M_1 = 20$ kg is at the origin while a particle of mass $M_2 = 80$ kg is at (0, 1 m). Find the force on a third particle of mass $M_3 = 10$ kg at (2 m, 0).

6. (I) A particle of mass $4M$ is at the origin while a particle of mass $9M$ is at $x = 1$ m. Where would the net force on a third particle be zero?

7. (I) Figure 13.27 shows four point particles with masses M, $2M$, $3M$, and $4M$ placed at the corners of a square of side L. Find the net force on (a) $2M$, and (b) $3M$.

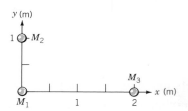

FIGURE 13.26 Exercise 5.

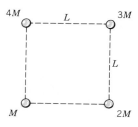

FIGURE 13.27 Exercise 7.

8. (I) In Fig. 13.28 three point particles with masses $2M$, $3M$,

and $5M$ lie at the corners of an equilateral triangle of side L. Find the force on (a) $3M$, and (b) $5M$.

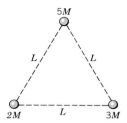

FIGURE 13.28 Exercise 8.

13.3 Gravitational Field Strength

9. (I) The equatorial radius of the earth is $R_e = 6378$ km and the polar radius is $R_p = 6357$ km. Estimate the difference in the gravitational field strengths at the pole and at the equator? Assume that the earth has a uniform mass distribution.

10. (II) Figure 13.29 shows a solid sphere of mass M and radius R with a sphere of radius $R/2$ removed. The center of the hole is $R/2$ from the center of the original sphere. What is the force exerted on a point mass m at a distance d from the center of the original sphere?

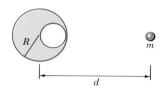

FIGURE 13.29 Exercise 10.

11. (II) The period T of a simple pendulum of length L is given by $T = 2\pi \sqrt{L/g}$. It is adjusted to keep correct time at sea level where $g = 9.810$ N/kg. When it is then taken to a mountain top it loses 1 min per day. (a) What is g at this point? (b) What is the height of the mountain?

12. (II) At what altitude h above the surface of the earth does the gravitational field strength g fall to the value g_0/N, where N is some number and g_0 is the field strength at the surface? Express your answer in terms of R_E, the radius of the earth.

13. (I) The period T of a simple pendulum of length L is given by $T = 2\pi\sqrt{L/g}$, where g is the field strength. A simple pendulum has a period of 2 s on earth. What would its period be on the moon?

14. (II) A dumbbell consisting of two particles of equal mass m separated by a massless rod of length $2a$ lies along a radial line of a solid sphere of mass M, as shown in Fig. 13.30. The center of the dumbbell is at a distance r from the center of the sphere. Show that if $r \gg a$ the difference in the

forces exerted on the two particles is $\Delta F = 4GmMa/r^3$. (The resulting tension in the rod connecting the particles is a *tidal force*.)

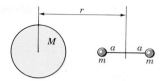

FIGURE 13.30 Exercise 14.

15. (I) Make a rough estimate of the gravitational force between the following bodies by treating them as point particles: (a) two aircraft carriers of mass 8×10^7 kg separated by 1 km; (b) two stars of mass 10^{30} kg separated by 1 light-year; (c) two people of mass 65 kg separated by 1 m.

16. (I) Calculate the gravitational field strength at the surface of the following bodies (assumed to be uniform spheres): (a) Neptune (1.03×10^{26} kg, radius 2.43×10^7 m); (b) Jupiter (1.9×10^{27} kg, radius 7.14×10^7 m); (c) a neutron star of mass 10^{30} kg and radius 20 km.

17. (II) (a) How does the gravitational field strength g at the surface of a planet depend on its radius R and its density ρ? How would g change in the following cases: (b) the mass is kept fixed and the radius halved; (c) the density is halved and the radius doubled; (d) the density is kept fixed but the volume is doubled?

18. (II) The variation in g was first demonstrated in an expedition to Cayenne (French Guiana) led by Jean Richer in 1672. He found that a simple pendulum lost 2.5 min per day compared to its rate in Paris. The period T of a simple pendulum of length L is given by $T = 2\pi \sqrt{L/g}$. (a) Show that $dT/T = -dg/2g$. (b) Find the field strength at Cayenne given that $g = 9.807$ N/kg at Paris. (c) How could Richer have determined that the period of the pendulum had changed?

19. (II) A ship sails along the equator at speed v relative to the surface. A particle of mass m is suspended on a spring scale. Show that when the ship makes a U-turn the reading changes by $4m\omega v$, where ω is the angular velocity of the earth.

13.4 Kepler's Laws

20. (I) Assume that a ball of mass 1 kg is launched into a circular orbit close to the earth's spherical surface. Ignore air resistance. Find: (a) its speed; (b) its period; (c) the energy required to launch it into orbit; (d) its total mechanical energy in orbit.

21. (I) A satellite is in stable circular orbit of radius r. How do the following quantities depend on r: (a) the speed; (b) the period; (c) the linear momentum; (d) the kinetic energy; (e) the angular momentum?

22. (I) A 10^4-kg lunar orbiter is in circular orbit 100 km above the moon's surface. The period is 118 min. What is the density of the moon (assumed to be a uniform sphere)?

23. (I) In its elliptical orbit, the speed of the earth at perihelion is $v_P = 3.03 \times 10^4$ m/s. If the distances to the sun at perihelion and aphelion are $r_P = 1.47 \times 10^{11}$ m and $r_A = 1.52 \times 10^{11}$ m, find v_A.

24. (I) The orbit of Halley's comet has a perihelion of 8.8×10^{10} m and the speed at this point is 5.5×10^4 m/s. Find its speed at aphelion given that $r_A = 5.3 \times 10^{12}$ m.

25. (II) The first Russian satellite, *Sputnik I*, of mass 83.5 kg, was launched on October 4, 1957, and placed in an orbit for which the distances from the center of earth at perigee and apogee were $r_P = 6610$ km and $r_A = 7330$ km. Find: (a) the mechanical energy of the satellite; (b) its period; (c) the speed at perigee.

26. (II) The first U.S. satellite, the 14-kg *Explorer I*, launched in March 1958, was placed in an orbit for which the distances from the center of the earth at perigee and apogee were $r_P = 6650$ km and $r_A = 9920$ km. Find: (a) the mechanical energy of the satellite; (b) the period; (c) the speed at perigee.

27. (II) Three stars of equal mass m rotate in a circular path of radius r about their center of mass, as shown in Fig. 13.31. They are equidistant from each other. Show that the angular velocity of the motion is given by

$$\omega^2 = \frac{Gm}{\sqrt{3}r^3}$$

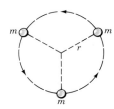

FIGURE 13.31 Exercise 27.

28. (II) On the *Apollo 17* mission, astronaut Eugene Cernan orbited the moon in the command module in an orbit for which the minimum and maximum altitudes were 100 km and 125 km. Find (a): the highest and lowest speeds in the orbit; (b) the period.

29. (II) The 7.6×10^4-kg *Leasat 3* satellite was placed in an orbit around the earth for which the minimum and maximum altitudes were 315 and 450 km. Find: (a) the period; (b) the mechanical energy; (c) the speeds at perigee and apogee.

30. (II) The 75-kg *Echo I* satellite was a 100-ft-diameter Mylar balloon. It was placed in an orbit around the earth for which the minimum and maximum altitudes above sea level were 1480 and 1603 km. Find: (a) its period; (b) its mechanical energy; (c) the speeds at perigee and apogee.

31. (II) Use Eqs. (13.5) and (13.6) to derive Eq. 13.7 and Eq. 13.8.

PROBLEMS

1. (I) Show that the period of a low-altitude satellite depends on the density of a planet but not on its radius.

2. (I) Show that the change in the gravitational field strength between the surface of the earth and a height h ($\ll R_E$) is

$$\Delta g \approx -2g_0 h/R_E$$

where g_0 is the value at the surface and R_E is the earth's radius. Evaluate for this change for the following: (a) the top of the Sears Tower of height 433 m; (b) the top of Mt. Everest of height 8850 m; (c) a satellite at an altitude of 100 km.

3. (II) Figure 13.32 shows two point particles of equal mass M at $y = +a$ and $y = -a$. A particle of mass m is at $(x, 0)$. (a) What is the potential energy $U(x)$ of the system? (b) From $U(x)$ find the force F_x on the particle of mass m. (c) For what value of x is the force a maximum? (d) Make a sketch of F_x versus x.

4. (II) Consider a particle of mass m inside a solid sphere of mass M and radius R; see Fig. 13.33. (a) How does the force F_r on the particle vary as a function of the distance r from the center? (b) Use Eq. 8.6 in the form

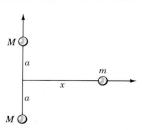

FIGURE 13.32 Problem 3.

$$U(r) - U(R) = -\int_R^r F_r \, dr$$

to show that the potential energy is given by

$$U(r) = \frac{GmM}{2R}\left(\frac{r^2}{R^2} - 3\right)$$

5. (II) A rocket is fired from the earth with 85% of the escape speed. Find its maximum distance from the earth's center if it is fired: (a) vertically, (b) horizontally. Ignore friction and the earth's rotation. (*Hint:* You need two conservation laws for (b). You will encounter a quadratic equation.)

6. (II) A rocket is fired at 60° to the local vertical with an initial speed $v_0 = \sqrt{GM/R}$, where M is the mass of the earth and

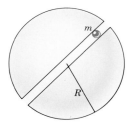

FIGURE 13.33 Problem 4.

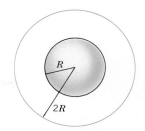

FIGURE 13.35 Problem 11.

R is its radius. Show that its maximum distance from the earth's center is $3R/2$. (You need two conservation laws.)

7. (I) Obtain an expression for the energy needed to send an object initially at rest on the earth's surface (a) vertically to a maximum height H, (b) into orbit at the same height. (c) For what value of H, in terms of R_E, the radius of the earth, would the value for part (b) be twice that in (a). Ignore the earth's rotation.

8. (II) Two planets have masses $M_1 = 3 \times 10^{24}$ kg and $M_2 = 5 \times 10^{24}$ kg and radii $R_1 = 2 \times 10^6$ m and $R_2 = 4 \times 10^6$ m respectively. They are initially at rest with their centers a distance $d = 10^8$ m apart. They start to move subject only to their mutual attraction. What are their speeds just as they touch? (*Hint:* You need two conservation laws.)

9. (I) Calculate the gravitational force on a particle of mass m at a point on the axis of a circular ring of mass M and radius R at a distance b from the center, as in Fig. 13.34? What is the form of your expression when $b \gg R$?

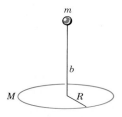

FIGURE 13.34 Problem 9.

10. (II) A space capsule is in a circular orbit of radius r_C about the earth. A short burst from a rocket in the backward direction increases the kinetic energy by 13% and causes the capsule to enter an elliptical orbit with the point of firing as the perigee; that is $r_P = r_C$. Show that the distance to the center of the earth at the apogee is $r_A = 1.3r_C$.

11. (I) Figure 13.35 shows a uniform solid sphere of mass M and radius R at the center of a thin spherical shell of radius $2R$ and mass M. What is the force on a point particle at the following distances from the center of the spheres: (a) $0.5R$, (b) $1.5R$, and (c) $2.5R$?

12. (II) (a) Show that the total energy of a planet of mass m in an elliptcal orbit about the sun of mass M can be expressed in the form

$$E = \tfrac{1}{2}m\left(v_r^2 + \frac{L^2}{(mr)^2}\right) - \frac{GmM}{r}$$

where L is the angular momentum about one focus and v_r is the radial component of its velocity. (b) How could this expression be used to find the perihelion and aphelion distances for given values of E and L?

13. (II) A solid sphere of radius R has a density that varies as $\rho = \rho_0(1 - r/2R)$ where r is the distance from the center. Determine the variation of the field strength with r within the sphere ($r < R$).

14. (I) A person can rise to a height H in a vertical jump on earth. What is the radius of the largest asteroid from which he could escape? Assume its density is that of the earth.

15. (I) A binary star system consists of two stars of masses m_1 and m_2 that are in circular orbits of radii r_1 and r_2 about their center of mass (see Fig. 13.36). (a) Write Newton's second law for each star. (b) Show that Kepler's third law takes the form

$$T^2 = \frac{4\pi^2}{G(M_1 + M_2)}(r_1 + r_2)^3$$

(c) Could observations lead to any information regarding the masses?

16. (II) A geosynchronous satellite is incorrectly positioned. What is its angular velocity relative to earth if it is 1 km too far from the earth? (*Hint:* Show that $d\omega/\omega = -1.5\,dr/r$.)

17. (II) Show that the deflection θ of a plumb line from the true vertical (radial line) at latitude λ is given by (see Fig. 13.37.)

$$\sin\theta = \frac{\omega^2 R \sin 2\lambda}{2g}$$

provided $\theta \ll \lambda$.

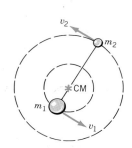
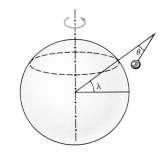

FIGURE 13.36 Problem 15.   FIGURE 13.37 Problem 17.

CHAPTER 14

Solids and Fluids

A vortex produced by the passage of a wingtip through smoke.

Major Points

1. The elastic properties of a material are characterized by **Young's modulus, the shear modulus,** and the **bulk modulus.**
2. **Pascal's principle** applies to an enclosed fluid subject to an external pressure.
3. **Archimedes' principle** relates the buoyant force on a body immersed in a fluid to the weight of the fluid displaced by the body.
4. The **equation of continuity** in fluid flow is a statement of the conservation of mass.
5. **Bernoulli's equation** applies to the *laminar* flow of an *ideal, incompressible* fluid.

Matter is usually classified into one of three states or phases: *solid, liquid,* or *gas.* Because they flow easily, both liquids and gases are called *fluids.* A solid has a fixed shape which it tends to retain, whereas fluids have no fixed shape. A liquid sinks to the bottom of its container, and a gas expands to fill the available volume. The atoms in a solid vibrate about fixed equilibrium positions, whereas the atoms or molecules in a liquid move about relatively freely and collide frequently with each other. The atoms in a solid or a liquid are quite closely packed, which makes it difficult to reduce their volume; they are almost incompressible. On the average, the atoms or molecules in a gas are far apart, typically about ten atomic diameters at room temperature and pressure. They collide much less frequently than those in a liquid. Gases in general are compressible.

The distinction between solid and liquid is not always clear-cut. For example, how should one classify asphalt or cold molasses? Glass, which is solid, and even brittle, can flow over a long period of time. Rock material under the earth's mantle is subject to high pressure and temperature and flows slowly over millennia. In this chapter we will discuss the elastic properties of solids and the behavior of fluids at rest and in motion.

14.1 DENSITY

At some time in the third century B.C., King Heiron of Syracuse supplied a certain weight of gold to a goldsmith to make a crown. When the task was completed, the king was uneasy. He asked Archimedes to find a way of determining whether or not the gold had been mixed with silver. On entering a bathtub one day, Archimedes noticed that the level of water would rise or fall depending on how deeply he immersed himself. He was immediately struck by the connection to his problem. According to legend, he shouted "Eureka!" (I have found it) and ran home naked. Archimedes had realized that even though the crown had a complicated shape, he could measure its volume by the volume of water it displaced. This could then be compared to the volume of water displaced by an equal weight of pure gold. Archimedes had discovered a use for the concept of **density.** The average density ρ of an object of mass m and volume V is defined as

$$\rho = \frac{m}{V} \tag{14.1}$$

If the density varies from point to point, one must use the definition

Definition of density

$$\rho = \frac{dm}{dV}$$

where dm is an infinitesimal mass element and dV is the infinitesimal volume element that it occupies. The SI unit of density is kg/m^3. Sometimes densities are stated in the cgs unit g/cm^3, where $1 \ g/cm^3 = 10^3 \ kg/m^3$. Table 14.1 shows the densities of a few substances. The density of a material depends on the pressure and temperature, although the variation is much larger for a gas than for a solid or a liquid. In addition to determining the purity of gold crowns, measurements of density are used to determine the condition of the electrolyte in an automobile battery or the antifreeze solution in the radiator. It is also one step in blood or urine analysis.

TABLE 14.1 DENSITIES AT 0°C AND 1 atm (kg/m^3)

Air	1.29	Pine	0.43×10^3
H	0.09	Al	2.70×10^3
He	0.18	Fe	7.86×10^3
O	1.43	Ag	10.5×10^3
Hg	13.6×10^3	Pb	11.3×10^3
Au	19.3×10^3	Pt	21.4×10^3
Cu	8.9×10^3	Ethyl Alcohol	0.8×10^3
Seawater	1025	Blood	1.05×10^3

The **specific gravity** of a substance is the ratio of its density to that of water at 4 °C, which is 1000 kg/m^3. Specific gravity is a dimensionless quantity numerically equal to the density quoted in g/cm^3. For example, the specific gravity of mercury is 13.6, and the specific gravity of water at 100 °C is 0.998.

14.2 ELASTIC MODULI

A force applied to an object can change its shape. In general, the response of a material to a given type of deforming force is characterized by an **elastic modulus,** which is defined as

$$\text{Elastic modulus} = \frac{\text{Stress}}{\text{Strain}} \tag{14.2}$$

The precise definition of **stress** depends on the particular situation being considered, but in general it is a force per unit area. The **strain** indicates some fractional change in a dimension or volume. The unit of stress is N/m², whereas strain is a dimensionless number. We will discuss three elastic moduli: Young's modulus for solids, the shear modulus for solids, and the bulk modulus for solids and fluids. In some instances the elastic moduli of a solid depend on direction in the material. For example, wood has different properties along or across the grain. We ignore such complications and assume that the materials are *isotropic;* that is, their properties are the same in all directions.

Young's Modulus

Young's modulus is a measure of the resistance of a solid to a change in its length when a force is applied perpendicular to a face. Consider a rod with an unstressed length L_0 and cross-sectional area A, as in Fig. 14.1a. When it is subject to equal and opposite forces F_n along its axis and perpendicular to the end faces, as in Fig. 14.1b, its length changes by ΔL. These forces tend to stretch the rod. The *tensile stress* on the rod is defined as

$$\text{Tensile stress} = \frac{F_n}{A} \tag{14.3}$$

Forces acting in the opposite direction, as in Fig. 14.1c, would produce a *compressive stress*. The resulting *strain* is defined as the dimensionless ratio

$$\text{Strain} = \frac{\Delta L}{L_0} \tag{14.4}$$

Young's modulus Y for the material of the rod is defined as the ratio

Young's modulus

$$\text{Young's modulus} = \frac{\text{Tensile stress}}{\text{Tensile strain}}$$

$$Y = \frac{F_n/A}{\Delta L/L_0} \tag{14.5}$$

By rewriting Eq. 14.5 in the form $F = YA(\Delta L/L_0)$, we see that if Y is a constant then the force required to produce a given strain is proportional to the strain and to the cross-sectional area of the rod.

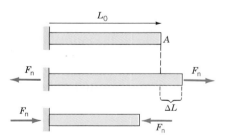

FIGURE 14.1 A force applied normal to the end face of a rod causes a change in length.

Figure 14.2 shows the relationship between tensile stress and strain for a typical metal. Below the *proportional* point, which typically corresponds to a strain of 0.01, stress is directly proportional to strain, which means Y is a constant. In this region the material obeys Hooke's law. Provided the strain is below the *yield* point, the material returns to its original shape and size when the force is removed. Beyond the yield point, the material retains a permanent deformation

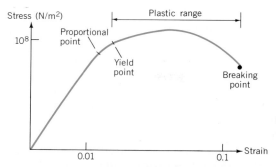

FIGURE 14.2 The stress–strain relationship for a metal. Beyond the yield point the material stretches with a small increase in the stress. The curve goes down toward the breaking point because the calculation of the stress is based on the original cross-sectional area, whereas the sample usually "necks down"; that is, its cross-sectional area decreases.

after the stress is removed. For stresses beyond the yield point, the material exhibits *plastic flow,* which means that it continues to elongate for little increase in the stress. The material fractures at a strain of perhaps 0.1.

EXERCISE 1. A copper wire has a length of 1.5 m and a radius of 0.5 mm. What is the change in its length when it is subject to a tension of 2000 N? Take $Y = 1.4 \times 10^{11}$ N/m².

Shear Modulus

The shear modulus of a solid indicates its resistance to a *shearing force,* which is a force applied tangentially to a surface, as shown in Fig. 14.3. (Since the bottom of the solid is assumed to be at rest, there is an equal and opposite force on the lower surface.) The top surface is displaced by Δx relative to the bottom surface. The shear stress is defined as

$$\text{Shear stress} = \frac{\text{Tangential force}}{\text{Area}} = \frac{F_t}{A}$$

where A is the area of the surface. The shear strain is defined as

$$\text{Shear strain} = \frac{\Delta x}{h}$$

where h is the separation between the top and bottom surfaces. The **shear modulus** S is defined as

$$\text{Shear modulus} = \frac{\text{Shear stress}}{\text{Shear strain}}$$

$$S = \frac{F_t/A}{\Delta x/h} \tag{14.6}$$

An ideal fluid cannot sustain a shear stress. Although a real fluid cannot sustain a permanent shearing force, there are tangential forces between adjacent layers in relative motion. This produces an internal friction called *viscosity.*

Bulk Modulus

The bulk modulus of a solid or a fluid indicates its resistance to a change in volume. Consider a cube of some material, solid or fluid, as shown in Fig. 14.4. We assume that all faces experience the same force F_n normal to each face. (One

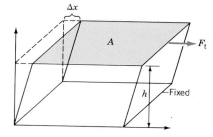

FIGURE 14.3 A shearing force F_t applied tangentially to one face causes the body to deform as shown. Such a deformation is easily produced in a textbook.

Shear modulus

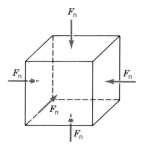

FIGURE 14.4 A cube of some material is subject to equal forces normal to each face. This condition may be achieved by immersing the body in a fluid.

way to accomplish this is to immerse the body in a fluid—as long as the change in pressure over the vertical height of the cube is negligible.) The **pressure** on the cube is defined as the normal force per unit area

$$P = \frac{F_n}{A}$$

The SI unit of pressure is N/m² and is given the name pascal (Pa). Pressure is a scalar because on any infinitesimal volume, it acts in all directions; it has no unique direction.

When the pressure on a body is increased, its volume decreases. The change in pressure ΔP is called the *volume stress* and the fractional change in volume $\Delta V/V$ is called the *volume strain*. The **bulk modulus** B of the material is defined as

Bulk modulus

$$\text{Bulk modulus} = \frac{\text{Volume stress}}{\text{Volume strain}}$$

$$B = -\frac{\Delta P}{\Delta V/V} \tag{14.7}$$

The negative sign is included to make B a positive number since an increase in pressure ($\Delta P > 0$) leads to a decrease in volume ($\Delta V < 0$). The inverse of B is called the **compressibility**, $k = 1/B$. Table 14.2 shows that the bulk moduli of liquids are comparable to those of solids. From this we can infer that the atoms in a liquid are almost as close to each other as they are in a solid. The bulk moduli of gases are quite low; they are easily compressed. The bulk moduli of solids and liquids are approximately constant for small changes in pressure. The bulk modulus of a gas depends on pressure (see Section 19.8).

TABLE 14.2 ELASTIC MODULI ($\times 10^9$ N/m²)

	Y	S	B
Cast iron	100	40	90
Steel	200	80	140
Aluminum	70	25	70
Concrete	20		
Pine	7.6		
Water			2.1
Mercury			2.6

14.3   PRESSURE IN FLUIDS

We now consider several features of the pressure exerted by a fluid at rest. Figure 14.5 shows the forces on a "particle" of a fluid at rest in a container. The particle is a tiny volume element that contains many molecules, yet it is small in comparison with the total volume of fluid. An ideal (nonviscous) fluid cannot exert a shearing force and so it exerts only a force normal to any given area. If the fluid is at rest, each particle is in equilibrium. Consequently, the pressure on a tiny volume element exerted by the surrounding fluid *is the same in all directions*. If this were not the case, it would contradict our assumption of equilibrium. (The effect of gravity is negligible for such a tiny volume.)

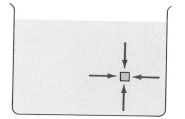

FIGURE 14.5 The pressure on an element of fluid in equilibrium is the same in all directions. The pressure may be measured by the contraction of a spring attached to a piston within a cyclinder.

The air around us exerts a pressure of about 10^5 N/m² on us. This fact is easily demonstrated by a tube, about a meter long, that is filled with mercury and then inverted into a bowl of mercury, as in Fig. 14.6. The column of liquid is supported by the pressure of the air on the open surface. It was first pointed out by E.

Torricelli that this pressure is due to the weight of the atmosphere above. In 1645, Otto von Guericke presented a dramatic demonstration of the forces due to atmospheric pressure. He placed together two hemispheres and evacuated the space within them. As Fig. 14.7 shows, two teams of eight horses could not pull the hemispheres apart.

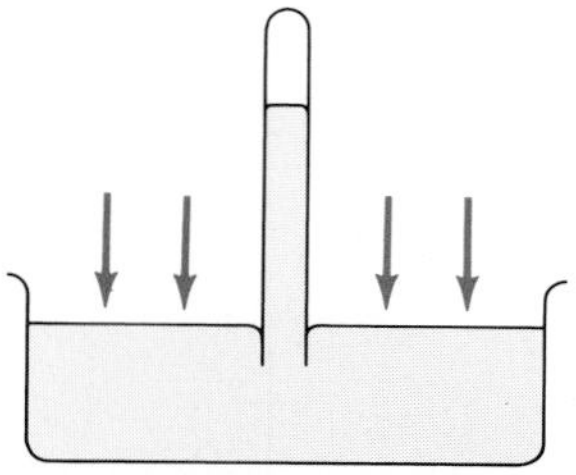

FIGURE 14.6 The fact that the atmosphere exerts a pressure may be demonstrated by inverting a tube of liquid, such as mercury, into a bowl of the same liquid. A liquid column is supported by the pressure at its base. At normal atmospheric pressure, a column of mercury is 76 cm high.

FIGURE 14.7 In 1645 Von Guericke placed two hemispheres together and evacuated the space inside. He then demonstrated that two teams of horses could not pull the hemispheres apart.

Variation of Pressure with Depth in a Liquid

An important feature of the pressure in a liquid may be inferred from the apparatus illustrated in Fig. 14.8. Several vessels of different shapes and cross sections are connected at the bottom by a tube. When liquid is poured into the vessels, it reaches the same level in all the vessels. The pressure at each top surface is that of the atmosphere. It is clear that the pressure in a liquid increases with depth because, as we go lower, each successive volume element has to support a greater body of liquid above it. The demonstration of Fig. 14.8 shows that the pressure is a function *only* of depth and does not depend on the shape of the container. For example, if the pressure at A were greater than that at B, the liquid would flow from A to B.

A B

FIGURE 14.8 Liquid fills several interconnected vessels to the same level. This demonstrates that pressure depends only on depth.

If we ignore the slight increase in density of ordinary liquids at ordinary depths, we can easily determine how the pressure increases with depth. Figure 14.9 shows a column of liquid of length h and cross-sectional area A. The weight of the column, $W = mg = (\rho Ah)g$, is supported by the net pressure force $(P - P_0)A$, where P_0 is the atmospheric pressure at the top surface and P is the pressure at its

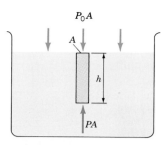

FIGURE 14.9 The weight of a column of liquid is balanced by the net pressure force.

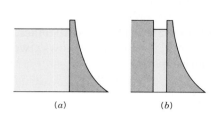

FIGURE 14.10 The design of a dam is determined only by the depth of the water—which may be in a large reservoir or just a narrow channel.

Variation of pressure with depth

base. From the condition $\Sigma F_y = 0$ we find

$$P = P_0 + \rho g h \tag{14.8}$$

We see that the pressure in a liquid increases linearly with depth and that all points at a given depth are at the same pressure.

It is natural to think of a dam as "holding back" the huge reservoir behind it, as in Fig. 14.10a. What is surprising is that the dam would have to be constructed in just the same manner if it had to contain a small body of water of the same depth, as in Fig. 14.10b. Around 1650, Blaise Pascal demonstrated this fact by inserting a thin, long tube into a closed barrel. As he filled the tube from the top, the pressure at the base increased until the barrel finally burst.

Pascal's Principle

Pascal's principle

Based on the fact that the static pressure in a fluid depends *only* on depth, Pascal stated the following principle in 1653:

An external pressure applied to a fluid in an enclosed container is transmitted undiminished to all parts of the fluid and the walls of the container.

This principle has many practical applications. For example, in a hydraulic jack or lift, shown in Fig. 14.11, the pressure due to a small force F_1 applied to a piston of area A_1 is transmitted to the larger piston of area A_2. The pressure at the two pistons is the same: $P = F_1/A_1 = F_2/A_2$. Consequently, the force on the larger piston is $F_2 = (F_1/A_1)A_2$, or

$$\frac{F_2}{F_1} = \frac{A_2}{A_1}$$

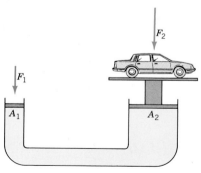

FIGURE 14.11 A hydraulic lift is based on the fact that the pressure due to a small force applied at a narrow tube is equal to the pressure caused by a large force applied at a wide tube.

Thus, a small force F_1 acting on a small area A_1 results in a larger force F_2 acting on a larger area A_2. Automobiles use a "hydraulic" fluid to activate the brakes. Pedal pressure is transmitted to a master cylinder and from there to the calipers or wheel cylinders. Controls in aircraft also use hydraulic lines. In an automobile garage, compressed air is often used to lift cars.

Measurement of Pressure

A simple way to measure pressure is with a *manometer*, shown in Fig. 14.12. One side of the U tube, which is filled with a suitable liquid such as mercury, is open to the atmosphere while the other side is connected to the fluid whose pressure is to be measured. The pressure is calculated from the difference in levels of the mercury. The absolute pressure P is the sum of the atmospheric pressure P_0 and the gauge pressure ρgh. Many pressure measuring devices record *gauge pressure,* which is the excess over atmospheric pressure. Thus, a gauge pressure of 200 kPa implies that the absolute pressure is about 300 kPa.

The inverted tube of mercury shown in Fig. 14.6 is a simple *barometer,* an instrument that measures atmospheric pressure. It was devised by E. Torricelli, a student of Galileo. It is sensitive to temperature changes and is affected by the presence of mercury vapor in the region above the column.

Although the SI unit of pressure is N/m^2 or Pa, one often encounters other units such as the *atmosphere* (atm):

$$1 \text{ atm} = 1.013 \times 10^5 \text{ Pa} = 101.3 \text{ kPa}$$

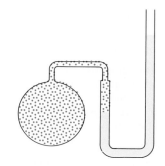

FIGURE 14.12 A simple manometer. The pressure in the vessel may be calculated from the difference in levels of the liquid in the U-tube.

Pressure is also specified in terms of the length of the column of mercury it can support. From Eq. 14.8 the pressure due to a column of mercury 1 mm high is

$$P = \rho gh = (13.6 \times 10^3 \text{ kg/m}^3)(9.8 \text{ m/s}^2)(10^{-3} \text{ m}) = 133 \text{ Pa}$$

A pressure of 1 atm is equivalent to a reading of 760 mm Hg.

EXERCISE 2. How high would water rise in a closed inverted column, as in Fig. 14.6, at an atmospheric pressure of 1 atm?

14.4   ARCHIMEDES' PRINCIPLE

An object that is partially or wholly immersed in a liquid seems to weigh less than it does in air. From this we conclude that fluids exert a buoyant force. This buoyant force causes wood to float on water and helium-filled balloons to rise in air.

Figure 14.13 shows a body immersed in a liquid. In general it will not be in equilibrium. An immersed body rises if its weight is less than the buoyant force and it sinks if its weight is greater than the buoyant force. The magnitude of the buoyant force is easily found by imagining that a body in equilibrium is removed and its place is taken by the liquid. We might think of this body of liquid as being separated by a thin membrane from the rest of the liquid. In addition to its weight, the ''new'' liquid experiences pressure forces from all directions. The sum of these forces, due to the surrounding liquid, must support the weight of the new liquid. Thus, the buoyant force is equal to the weight of the liquid within the volume of the membrane. When the solid body is immersed in the liquid, it displaces the same volume of liquid. Since the forces exerted by the surrounding liquid are unchanged, we arrive at **Archimedes' principle:**

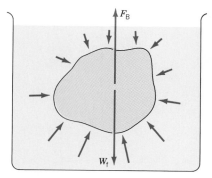

FIGURE 14.13 According to Archimedes' principle, the buoyant force F_B acting on a body is equal to the weight of the fluid displaced.

$$\text{Buoyant force} = \text{Weight of fluid displaced}$$
$$F_B = \rho_f V g \qquad (14.9)$$

where V is the volume of the body and ρ_f is the density of the fluid. Note that the buoyant force arises because the pressure in the fluid is not uniform; it increases with depth.

An object floats on water if it can displace a volume of water whose weight is greater than that of the object. If the density of the material is less than that of the liquid, it will float even if the material is a uniform solid, such as a block of wood. If the density of the material is greater than that of water, such as iron, the object can be made to float provided it is not a uniform solid. An iron-hulled ship is a common example.

A dirigible over an Amazon forest.

EXERCISE 3. How is it arranged for submarines to rise or to sink?

EXAMPLE 14.1: An iceberg with a density of 920 kg/m³ floats on an ocean of density 1025 kg/m³. What fraction of its volume is submerged?

Solution: Suppose the volume of the iceberg is V_i and that of the submerged portion is V_s. The weight of the water displaced is $\rho_f V_s g$, and hence this is also the buoyant force. The weight of the iceberg is $\rho_i V_i g$ where ρ_i is the density of the iceberg. Equating the weight of the iceberg to the buoyant force, we see that

$$\rho_f V_s g = \rho_i V_i g$$

or

$$\frac{V_s}{V_i} = \frac{\rho_i}{\rho_f}$$

The fraction of the volume of the iceberg submerged equals the ratio of the densities. With the given numbers we find that V_s/V_i is approximately 90%.

EXAMPLE 14.2: When a 3-kg crown is immersed in water, it has an apparent weight of 26 N. What is the density of the crown?

Solution: The apparent weight, $W' = 26$ N, is the difference between the real weight, $W = mg = (3 \text{ kg})(9.8 \text{ m/s}^2) = 29.4$ N,

and the buoyant force F_B:

$$W' = W - F_B$$

Thus, the buoyant force is

$$F_B = W - W' \tag{i}$$

If V is the volume of the object and ρ is its density, then $W = \rho g V$ and $F_B = \rho_f g V$, where ρ_f is the fluid density. Hence,

$$\frac{W}{F_B} = \frac{\rho}{\rho_f} \tag{ii}$$

Combining (i) and (ii) we find

$$\rho = \frac{\rho_f W}{W - W'}$$

$$= \frac{(10^3 \text{ kg/m}^3)(29.4 \text{ N})}{3.4 \text{ N}} = 8.6 \times 10^3 \text{ kg/m}^3$$

The metal may be mostly copper.

EXERCISE 4. A spherical balloon filled with helium at 1 atm just lifts a 2-kg load (which includes the mass of the balloon). What is its radius?

EXERCISE 5. A 5-kg ball of density $\rho_b = 6$ g/cm³ is completely submerged in water. What is the tension in a string attached to the ball?

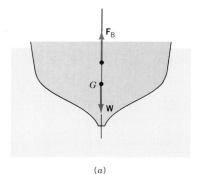

(a)

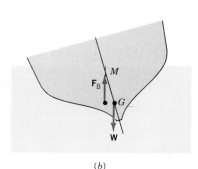

(b)

FIGURE 14.14 (a) The buoyant force acts at the center of gravity of the displaced fluid. (b) When the boat tilts, the line of action of the buoyant force intersects the axis of the boat at the metacenter M. In a stable boat, M is above the center of gravity of the boat.

The Stability of Boats

The stability of a boat depends on the effective point of application of the buoyant force. The weight of the boat acts at its center of gravity. The buoyant force acts at the center of gravity of the *displaced* liquid. (Why?) This is called the *center of buoyancy*. Under equilibrium conditions the center of gravity G and the center of buoyancy B lie along the vertical axis of the boat, as in Fig. 14.14a.

When the boat tilts to one side, the center of buoyancy shifts relative to the center of gravity, as shown in Fig. 14.14b. The two forces act along different vertical lines. As a result, the buoyant force exerts a torque about the center of gravity. The line of action of the buoyant force crosses the axis of the boat at the point M, called the *metacenter*. If G is below M, the torque will tend to return the boat to its equilibrium position. If M is below G, the boat will be unstable. (Draw a diagram to show what happens when M is below G.)

14.5 THE EQUATION OF CONTINUITY

In order to describe the motion of a fluid, in principle one might apply Newton's laws to a "particle" (a small volume element of fluid) and follow its progress in time. This is a difficult approach. Instead, we consider the properties of the fluid, such as velocity and pressure, at fixed points in space.

The motion of a fluid may be either **laminar** or **turbulent**. Laminar flow may be represented by **streamlines**, which can be made visible by injecting smoke or a colored dye into the fluid, as in Fig. 14.15a. The velocity of a particle at a given point is along the tangent to a streamline. For steady flow, each particle that passes through a given point follows the same streamline. Therefore, streamlines never cross. Because of this last property it is convenient to introduce the concept of a tube of flow. The fluid flows through a tube of flow as if it were confined to a

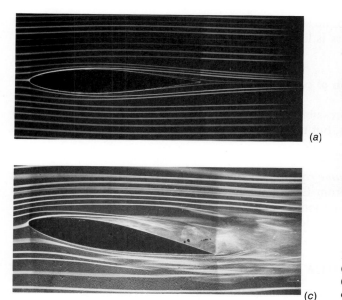

(a)

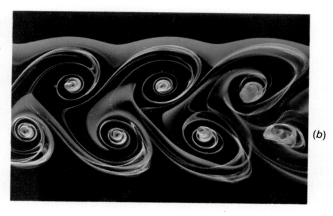

(b)

(c)

FIGURE 14.15 (*a*) Streamlines in a fluid flowing past an airfoil. (*b*) When the airfoil is tilted, the flow behind it becomes turbulent. (*c*) Eddies or vortices formed in a fluid as it flows past a cylindrical object.

real tube (whose cross-sectional area may change). When the fluid velocity is large or when the fluid encounters most obstacles, the flow becomes turbulent, as in Fig. 14.15*b*. Under certain conditions eddies, which are like little whirlpools, form behind the object (Fig. 14.15*c*). Turbulent flow involves loss of mechanical energy.

In order to simplify the discussion we make several assumptions:

1. **The fluid is nonviscous:** There is no dissipation of energy due to internal friction between adjacent layers in the liquid.

2. **The flow is steady:** The velocity and pressure at each point are constant in time.

3. **The flow is irrotational:** A tiny paddle wheel placed in the liquid will not rotate. In rotational flow, for example, in eddies, the fluid has net angular momentum about a given point.

In general, the velocity of a particle will not be constant along a streamline. The density and the cross-sectional area of a tube of flow will also change. Consider two sections of a tube of flow, as shown in Fig. 14.16. The mass of fluid contained in a small cylinder of length $\Delta \ell_1$ and area A_1 is $\Delta m_1 = \rho_1 A_1 \Delta \ell_1$. Since fluid does not leave the tube of flow, this mass will later pass through a cylinder of length $\Delta \ell_2$ and area A_2. The mass in this cylinder is $\Delta m_2 = \rho_2 A_2 \Delta \ell_2$. The lengths

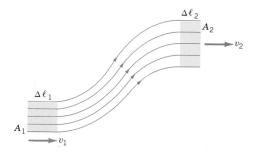

FIGURE 14.16 A "tube of flow." The fluid contained in the lower cylinder of length $\Delta \ell_1$ is later contained in the upper cylinder of length $\Delta \ell_2$.

$\Delta\ell_1$ and $\Delta\ell_2$ are related to the speeds at the respective locations: $\Delta\ell_1 = v_1\Delta t$ and $\Delta\ell_2 = v_2\Delta t$. Since no mass is lost or gained, $\Delta m_1 = \Delta m_2$, and

(Steady flow) $$\rho_1 A_1 v_1 = \rho_2 A_2 v_2 \tag{14.10}$$

This is called the **equation of continuity.** It is a statement of the conservation of mass.

If the fluid is incompressible, its density remains unchanged. This is a good approximation for liquids, but not for gases. If $\rho_1 = \rho_2$, then Eq. 14.10 becomes

(Incompressible) $$A_1 v_1 = A_2 v_2 \tag{14.11}$$

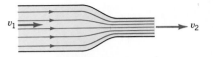

FIGURE 14.17 A fluid flowing through a pipe whose cross section changes. Notice that the streamlines are closer together in the narrower section. This indicates that the fluid is moving faster.

The product Av is the *volume rate of flow* (m³/s). Figure 14.17 shows a pipe whose cross section narrows. From Eq. 14.11 we conclude that the speed of a fluid is greatest where the cross-sectional area is the least. Notice that the streamlines are closer together where the speed is higher.

14.6   BERNOULLI'S EQUATION

An important theorem concerning fluid flow may be derived when the fluid is incompressible and nonviscous and the flow is steady and laminar. Figure 14.18 shows a tube of flow that varies in height and cross-sectional area. We focus our attention on the motion of the shaded region. This is our "system." The lower cylindrical element of fluid of length $\Delta\ell_1$ and area A_1 is at height y_1, and moves at speed v_1. After some time, the leading section of our system fills the upper cylinder of fluid of length $\Delta\ell_2$ and area A_2 at height y_2, and is then moving at speed v_2. A pressure force F_1 acts on the lower cylinder due to fluid to its left (not shown), and a pressure force F_2 acts on the upper cylinder in the opposite direction. The net work done on the system by F_1 and F_2 is

$$\begin{aligned} W &= F_1\Delta\ell_1 - F_2\Delta\ell_2 \\ &= P_1 A_1 \Delta\ell_1 - P_2 A_2 \Delta\ell_2 \\ &= (P_1 - P_2)\Delta V \end{aligned}$$

where we have used the relations $F = PA$ and $\Delta V = A_1\Delta\ell_1 = A_2\Delta\ell_2$. The net effect of the motion of the system is to raise the height of the lower cylinder of

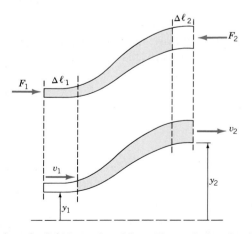

FIGURE 14.18 The motion of a fluid in a tube of flow. The work done by the pressure forces equals the change in energy of the shaded volume of fluid.

mass Δm and to change its speed. The changes in the potential and kinetic energies are

$$\Delta U = \Delta m\, g(y_2 - y_1)$$
$$\Delta K = \tfrac{1}{2}\Delta m\, (v_2^2 - v_1^2)$$

These changes are brought about by the net work done on the system, $W = \Delta U + \Delta K$:

$$(P_1 - P_2)\Delta V = \Delta m\, g(y_2 - y_1) + \tfrac{1}{2}\Delta m\, (v_2^2 - v_1^2)$$

Since the density is $\rho = \Delta m/\Delta V$, we have

$$P_1 + \rho g y_1 + \tfrac{1}{2}\rho v_1^2 = P_2 + \rho g y_2 + \tfrac{1}{2}\rho v_2^2$$

Since the points 1 and 2 can be chosen arbitrarily, we can express this result as **Bernoulli's equation:**

$$P + \rho g y + \tfrac{1}{2}\rho v^2 = \text{constant} \qquad (14.12) \quad \text{Bernoulli's equation}$$

Daniel Bernoulli derived this equation in 1738. It applies to all points *along a streamline in a nonviscous, incompressible fluid.* It is a disguised form of the work–energy theorem. Notice that when $v = 0$, we obtain Eq. 14.8.

EXAMPLE 14.3: Water emerges from a hole at the bottom of a large tank, as shown in Fig. 14.19. If the depth of water is h, what is the speed at which the water emerges?

Solution: If the tank is large and the hole small, the speed of the particles at the top surface will be essentially zero. The pressure at the top surface and at the hole is atmospheric. Thus, Bernoulli's equation reduces to

$$\rho g h = \tfrac{1}{2}\rho v_2^2$$

or

$$v_2 = \sqrt{2gh}$$

The speed of the emerging fluid is the same as that of a particle

that falls freely through the same vertical distance. This rather surprising result is called *Torricelli's theorem.*

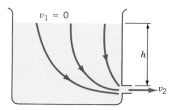

FIGURE 14.19 Water emerges from a hole in a tank. Its speed is the same as if it had fallen through the height h.

Figure 14.20 shows a tube whose cross section changes from A_1 to A_2. Since the heights of the sections are the same, Bernoulli's equation takes the form

$$P_1 + \tfrac{1}{2}\rho v_1^2 = P_2 + \tfrac{1}{2}\rho v_2^2$$

We see that where the pressure is high, the speed is low, and vice versa. This lowering of the pressure where the speed is greater is called the **Bernoulli effect.**

The Bernoulli effect is the basis of the *Venturi meter,* which is a device used to measure the speed of flow. From the equation of continuity we know that $A_1v_1 = A_2v_2$. On substituting $v_1 = v_2(A_2/A_1)$, we find

$$v_2^2 = \frac{2A_1^2(P_1 - P_2)}{\rho(A_1^2 - A_2^2)}$$

Since v_2^2 must be positive, it is necessary that $P_1 > P_2$. That is, the pressure is lower in the narrow section. The streamlines are closer together where the speed is higher and the pressure is lower (see Fig. 14.17).

Perfume atomizers and automobile carburetors employ the Bernoulli effect. In an atomizer, shown in Fig. 14.21*a*, liquid in a container is at atmospheric pressure. When air is blown through tube A connected to a rubber bulb, the

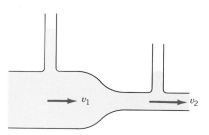

FIGURE 14.20 A fluid flowing through a tube whose cross section decreases. The pressure in the narrower tube, where the fluid is moving faster, is lower.

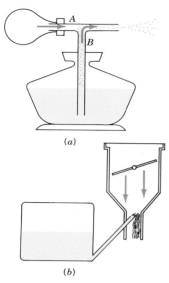

FIGURE 14.21 (a) A perfume atomizer. When air is blown through tube A past the top of tube B, the pressure at the top of B is lower than normal. Liquid in the container is forced into B, mixes with the air in A, and emerges as a fine mist through a nozzle at the end of A. (b) The flow of fuel to an engine is controlled by a throttle valve in a carburetor. The pressure of the air is lowered as it flows through the constriction and allows gasoline to emerge from a small pipe connected to the fuel bowl.

pressure at the top of tube B decreases. This drop in pressure causes the liquid to rise in tube B and mix with the fast-moving air. A fine mist emerges from the nozzle. In a carburetor, sketched in Fig. 14.21b, gasoline in the fuel bowl can flow through a tube connected to a narrow section of the throat. Air is sucked into the carburetor because of the partial vacuum created by the motion of the pistons. Since the air speeds up at the constriction, its pressure is lower than that of the fuel in the bowl and so gasoline is drawn into the engine. The flow of air is controlled by the throttle valve, which is connected to the accelerator pedal.

SUMMARY

The response of a material to an external stress is characterized by an **elastic modulus** defined as

$$\text{Elastic modulus} = \frac{\text{Stress}}{\text{Strain}}$$

where the stress is a measure of the force per unit area on a surface, and the strain, a dimensionless quantity, is a measure of the deformation produced. A force F_n applied normal to both end faces of a rod of length L produces a change in length ΔL. **Young's modulus** Y is defined as

$$Y = \frac{F_n/A}{\Delta L/L}$$

When a force F_t is applied tangential to a surface of area A, it produces a relative displacement Δx between two surfaces separated by h. The **shear modulus** S is defined as

$$S = \frac{F_t/A}{\Delta x/h}$$

When a force F is applied normal to a surface of area A, the **pressure** exerted on the surface is

$$P = \frac{F}{A}$$

When a pressure is applied to a body its volume changes. The **bulk modulus** of the material is defined as

$$B = -\frac{\Delta P}{\Delta V/V}$$

The pressure at a depth h in a liquid of density ρ in a container exposed to the atmosphere, whose pressure is P_0, is

$$P = P_0 + \rho gh$$

According to **Pascal's principle,** an external pressure applied to a confined fluid is transmitted undiminished to all parts of the fluid.

Archimedes' principle states that a body wholly or partially immersed in a fluid experiences a buoyant force given by

$$\text{Buoyant force} = \text{Weight of fluid displaced}$$

In the steady, laminar flow of an incompressible fluid through a pipe whose cross-sectional area changes from A_1 to A_2, the volume flow rate is constant:

$$A_1 v_1 = A_2 v_2$$

This is a special case of the equation of continuity (Eq. 14.10).

Bernoulli's equation for the laminar flow of an incompressible fluid of density ρ is

$$P + \tfrac{1}{2}\rho v^2 + \rho g y = \text{constant}$$

This equation is an expression of the work–energy theorem.

ANSWERS TO IN-CHAPTER EXERCISES

1. The cross-sectional area of the wire is $A = \pi r^2 = 7.84 \times 10^{-7}$ m^2. From Eq. 14.5,

$$\Delta L = \frac{F L_0}{A Y}$$

$$= \frac{(2 \times 10^3 \text{ N})(1.5 \text{ m})}{(7.84 \times 10^{-7} \text{ m}^2)(1.4 \times 10^{11} \text{ N/m}^2)}$$

$$= 2.73 \text{ cm}$$

2. The pressure at the base of the water column, $\rho g h$, is 1 atm: 1.01×10^5 N/m^2 = $(10^3$ kg/m$^3)(9.8$ m/s$^2)(h)$; thus, $h = 10.3$ m.

3. A submarine is made to sink when water is allowed into ballast tanks. When the water is pumped out, the submarine rises.

4. The buoyant force is $F_B = \rho_a g V$ where ρ_a is the density of the air and V is the volume of the balloon. This force must equal the weight of the load and the helium in the balloon: $F_B = mg + \rho_{He} g V$. Thus,

$$V = \frac{m}{\rho_a - \rho_{He}} = \frac{2 \text{ kg}}{1.11 \text{ kg/m}^3} = 1.8 \text{ m}^3$$

Since $V = 4\pi r^3/3$, we find $r = 0.75$ m.

5. The volume of the ball is $V_b = m/\rho_b = 8.33 \times 10^{-4}$ m^3. The tension is equal to the apparent weight; that is,

$$T = W - F_B = mg - \rho_w g V_b$$
$$= 49 \text{ N} - 8.16 \text{ N} = 40.8 \text{ N}$$

QUESTIONS

1. Given the densities of two liquids could you conclude anything with respect to the following: (a) the relative masses of the molecules; (b) the number of molecules per unit volume; (c) the number of molecules in 1 kg?

2. Why might airplane passengers be advised to remove ink from fountain pens?

3. The three vessels in Fig. 14.22 have bases with the same area and are filled to the same level with a liquid. (a) Compare the forces exerted by the base of each vessel on the liquid. (b) Compare the forces exerted by the base on the table. (c) If your answers to (a) and (b) are different, explain why. (This is called the "hydrostatic paradox.")

FIGURE 14.22 Question 3.

4. A ball floats on water in a jar. If the top of the jar is sealed and the air pressure increased, can the ball be made to sink? Assume that both the ball and the liquid are incompressible.

5. (a) Does the buoyant force exerted by a liquid change with depth? (b) Would the buoyant force on an object in a given liquid be the same on the moon as on earth? (c) Is Archimedes' principle valid if the fluid is in a container that has a vertical acceleration?

6. Reinforced concrete has steel rods embedded in it. The rods are held under tension while the concrete cures and the tension is removed after the concrete has set. In what way is the concrete "stronger"?

7. How is it possible for a person to lie on a bed of nails?

8. When a person's blood pressure is measured, why is the inflatable cuff attached to an arm at the level of the heart rather than, say, at an ankle?

9. Why do our bodies not collapse under the enormous pressure to which the atmosphere subjects us?

10. (a) Does the buoyant effect of the air act on an object lying on the ground? (b) Is a diver's buoyancy affected by the amount of air in the lungs?

11. Why does a suction cup stick to a smooth surface?

12. An ice cube floats in a glass of water filled to the brim. What happens to the level of the water as the ice melts?

13. A beach ball can be held in stable equilibrium in the jet of air from a vacuum cleaner, even if the tube is tilted from the vertical; see Fig. 14.23a. How is this possible?

14. A Ping-Pong ball can be held in the stream of air directed

downward to an inverted funnel, as in Fig. 14.23*b*. Explain how this is possible.

FIGURE 14.23 Questions 13 and 14.

15. Grain silos have reinforcing circular steel bands; see Fig. 14.24. Why are the bands not uniformly distributed?

FIGURE 14.24 Question 15.

16. Explain why a liquid rises when you suck on a straw.

17. Explain why, in laminar flow, the vertical stream from a faucet gets narrower (see Fig. 14.36).

18. (a) Hold two thin strips of paper vertically, as in Fig. 14.25, and blow between them. Explain what happens. (b) Hold a single strip of paper below your lower lip and blow. Explain what you see.

19. A roof may be detached when the wind of a hurricane flows over it. Would it help to have larger vents around the perimeter?

20. Why does the fabric top of a convertible bulge outward when the car moves at high speed?

21. Compare the weights of 1 kg of lead and 1 kg of feathers as measured on a spring scale. How would you know that you have exactly 1 kg?

FIGURE 14.25 Question 18.

22. What is the maximum depth at which a suction pump, Fig. 14.26, can operate? Could two or more pumps be used in tandem? If so, how?

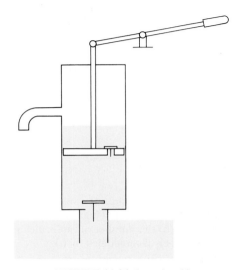

FIGURE 14.26 Question 22.

23. A rubber raft floats in a swimming pool. A rock is taken from the raft and dropped into the water. Does the water level of the pool change? If so, how?

24. Why does a hot air balloon rise? How does the pressure of the air in the balloon compare with that of the surrounding cool air?

25. Is there a limit to the height to which a helium-filled balloon can rise?

26. Why is a tall chimney more effective than a short one in producing a draft for a fireplace?

27. Compared to the values attained in a vacuum, the air reduces the horizontal range of a baseball but increases that of a golf ball. Explain why.

28. Suppose that a little water is boiled in a tin can and the can is sealed. What happens when the can cools down?

29. How would you determine the density of an object that (a) sinks in water; (b) floats in water?

EXERCISES

14.1 Density

1. (II) The antifreeze in a radiator consists of 70% ethylene glycol of density 0.8 g/cm^3 and 30% water. Find the density of the mixture if the percentages refer to: (a) volume; (b) mass. (Ignore the fact that the volume of the mixture is somewhat less than the sum of the original volumes.)

2. (I) A bottle has a mass of 25 g when empty and 125 g when filled with water. When filled with another liquid, the total mass is 140 g. What is the density of the liquid?

3. (I) A nucleus has a mass of 3×10^{-26} kg and a radius of 2×10^{-14} m. (a) What is its density? (b) What would be the radius of the earth if it had the density of nuclear matter?

14.2 Elastic Moduli

4. (I) A rod of length 2.5 m and cross-sectional area 0.3 cm^2 stretches by 0.1 cm when a tension of 800 N is applied. What is its Young's modulus?

5. (I) A circular steel wire of length 1.8 m must not stretch more than 1.5 mm when a load of 400 N is applied. What is the minimum diameter required?

6. (I) A sample of a liquid has an initial volume of 1.5 L. The volume is reduced by 0.2 mL when the pressure increases by 140 kPa. What is the bulk modulus of the liquid?

7. (I) An 800-kg elevator hangs by a steel cable for which the allowable stress is 1.2×10^8 N/m^2. What is the minimum diameter required if the elevator accelerates upward at 1.5 m/s^2?

8. (I) The pressure at the bottom of the Marianas Trench in the Pacific Ocean is about 1.08×10^8 Pa. What is the fractional change in volume if a given mass of water is moved from the surface to this depth?

9. (I) A human bone has a Young's modulus of about 10^{10} N/m^2. It fractures when the compressive strain exceeds 1%. What is the maximum load that can be sustained by a bone of cross-sectional area 3 cm^2?

10. (II) A steel bolt of diameter 1.2 cm is used to join two plates, as shown in Fig. 14.27. What equal and opposite forces applied to the plates will cause the bolt to undergo shearing failure? The ultimate shear strength is 3.5×10^8 N/m^2.

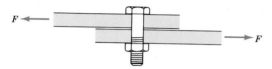

FIGURE 14.27 Exercise 10.

11. (II) A steel bolt connects two parts of a machine, as shown in Fig. 14.28. The shank of the bolt has a diameter of 1.2 cm and a head of height $h = 0.8$ cm. What is the maximum tensile load that the bolt head can withstand if its ultimate shear strength is 3.5×10^8 N/m^2?

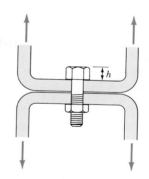

FIGURE 14.28 Exercise 11.

14.3 Pressure in Fluids

12. (II) (a) Estimate the mass of the earth's atmosphere given that the pressure at the surface is 101 kPa. Assume that the force of gravity does not vary with height. (b) If the density of air is 1.29 kg/m^3, what is the "effective" height of the atmosphere?

13. (I) The pressure at the center of a tornado is 0.4 atm. If the tornado suddenly passes over a house, what is the net force on a windowpane whose dimensions are 1.2 m $\times$ 1.4 m? Assume that the house is airtight and that the pressure inside is 1 atm.

14. (I) The cabin pressure in an aircraft is 90 kPa, whereas the external atmospheric pressure is 70 kPa. What is the force on a window of dimensions 15 cm $\times$ 20 cm?

15. (I) The piston in a hypodermic syringe has a radius of 0.5 cm, and the needle has a hole of radius 0.15 mm. What force must be applied to the plunger in order to inject fluid into a vein in which the blood pressure is 20 mm Hg?

16. (I) What is the absolute pressure in water at the following depths: (a) 3 m in a swimming pool; (b) 100 m in a lake; (c) 10.9 km in the Marianas Trench in the Pacific Ocean?

17. (I) A U tube of inner radius 0.4 cm contains 60 mL of mercury. When 25 mL of water is added to one arm, what is the difference in the levels of the liquid–air interfaces?

18. (I) Suppose that a phonograph stylus is a rod of radius 2×10^{-5} m with a flat end face. The tonearm is adjusted to apply a force of 15 mN to a flat surface. What height of a column of water would produce the same pressure?

19. (I) Given that the density of blood is 1.05 g/cm^3, what is the difference in hydrostatic pressure between the head and the feet of a standing person whose height is 1.8 m? Assume that the veins and arteries can be treated as ordinary tubes. (In fact, the distribution of blood in the circulatory system is controlled by several physiological mechanisms.)

20. (I) As a parachutist falls, the eardrums "pop" each time the pressure within the inner ear is made equal to the external pressure. Suppose this did not happen. What would be the

force on the eardrum of area 0.5 cm² as a result of a change in altitude of 1000 m? Assume that the density of air is constant at 1.29/g/m³.

21. (I) The densities of two liquids can be compared with the apparatus shown in Fig. 14.29. The liquids rise when the tubes are partially evacuated. One liquid is water which rises to a height $h_1 = 4.8$ cm. What is the density of the other liquid if $h_2 = 4.4$ cm?

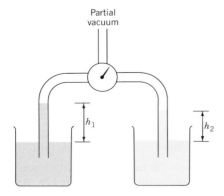

FIGURE 14.29 Exercise 21.

22. (II) What is the absolute pressure as a function of depth in a liquid that is in a container which has a vertical acceleration a upward?

23. (I) What is the minimum gauge pressure required at the base of a building of height 200 m for water to reach a closed faucet at the top of the building at a gauge pressure of 500 kPa?

24. (I) By inhaling, a person can create a gauge pressure of −60 mm Hg. To what height could the person raise water in a straw?

25. (I) The manometer shown in Fig. 14.30 contains oil of density 850 kg/m³. What is the absolute pressure of the gas in the bulb? The atmospheric pressure is 101 kPa.

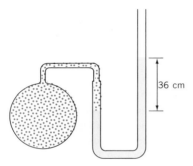

FIGURE 14.30 Exercise 25.

26. (I) In an intravenous infusion of fluid, a container is held at a height h above the arm. If the blood pressure is 20 mm Hg

and the density of the fluid is 1025 kg/m³, what is the minimum value of h for the fluid to enter the vein?

27. (II) The gauge pressure in the tires of a car is 200 kPa. The area of each tire in contact with the road is 120 cm². What is the mass of the car?

28. (II) In Otto von Guericke's famous experiment, Fig. 14.7, the hemispheres had a radius of about 30 cm. Assuming that the pressure inside the sphere was 0.1 atm, what was the force required to pull the two hemispheres apart? (The net force exerted by the air on a hemisphere is the same as that on a flat disk of the same radius.)

14.4 Archimedes' Principle

29. (II) A glass tube of radius 0.8 cm floats vertically in water, as shown in Fig. 14.31. What mass of lead pellets would cause the tube to sink a further 3 cm?

FIGURE 14.31 Exercise 29.

30. (I) A raft with dimensions 3 m × 3 m × 0.16 m is made of wood of density 600 kg/m³. What uniformly distributed load (in kg) would cause it to be 80% submerged in water?

31. (II) A sphere floats in water with 60% of its volume submerged. It floats in oil with 70% of its volume submerged. What is the density of the oil?

32. (I) A 2-kg block of copper, of density 9000 kg/m³, has an apparent weight of 17 N when completely submerged in a liquid. What is the density of the liquid?

33. (I) A 400-g cubic block of wood floats with 40% of its volume submerged. How much weight should be placed on the wood to cause it to just become fully submerged?

34. (II) The weight of an object is 12 N in air. It has an apparent weight of 8 N when totally immersed in water. Find its density and volume. Ignore the buoyant effect of the air.

35. (II) A body of weight 30 N requires a vertical force of 10 N to just submerge it in water. What is its density?

36. (I) A block of wood of density 600 kg/m³ has a length of 40 cm, a width of 30 cm, and a height of 20 cm (the vertical dimension). To what depth does it sink in oil of density 950 kg/m³?

37. (II) An object of mass 0.5 kg sinks in oil of density 800 kg/m³. It has an apparent weight of 4.2 N when fully immersed in the oil. What is its density?

38. (I) An oil freighter has an approximately rectangular horizontal cross section of 15 m × 200 m. It sinks by an additional 5 m in seawater when it is loaded. What is the mass of the load? The density of seawater is 1025 kg/m³.

39. (II) A 60-kg person floats vertically in a pool with just her head, of volume 2.5 L, exposed. What is her (average) density?

40. (I) An iceberg of density 920 kg/m³ floats in seawater of density 1025 kg/m³ with an exposed volume of 10⁶ m³. What is its total mass?

41. (I) A barge has a draft (submerged depth) of 3 m at sea. Assume that it has a rectangular horizontal cross-section. By how much does this change when it reaches a fresh-water lake? The density of seawater is 1025 kg/m³.

42. (II) A hydrometer is used to measure the density of liquids. It consists of a bulb weighted with lead pellets and a long stem; see Fig. 14.32. The stem has a graduated scale from which density may be read off. The bulb has volume of 4 mL and the stem has a 5 mm diameter. The hydrometer has a mass of 5 g. If the bulb sinks 1.5 cm lower when it is moved from water to another fluid, what is the density of the fluid?

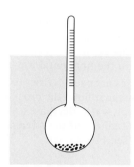

FIGURE 14.32 Exercise 42.

43. (I) In accurate measurements of mass using a spring scale, corrections have to be made for the buoyancy of the air. If the mass of a copper bar is measured by the spring scale to

be 200.00 g, what is its real mass? The density of copper is 9 g/cm³.

44. (I) A blimp in the form of a cylinder of radius 5 m and length 40 m is filled with helium at a pressure of 1 atm. The density of the helium is 0.18 kg/m³. What maximum load (including its own mass) can the blimp lift?

14.5 Equation of Continuity

45. (I) Water flows at 1.2 m/s through a hose of diameter 1.59 cm. How long does it take to fill a cylindrical pool of radius 2 m to a height of 1.25 m?

46. (I) A duct of square cross section (0.5 m × 0.5 m) is used to change the air in a room of dimensions 4 m × 3 m × 3 m every 20 min. What is the required speed of the airflow through the duct?

47. (I) Water flows at 2.4 m/s through a garden hose of diameter 1.59 cm and emerges from a nozzle of radius 0.64 cm. If the nozzle is directed vertically upward, to what height would the water rise?

14.6 Bernoulli Equation

48. (II) A fountain sprays water through the top end of a vertical uniform pipe of radius 0.6 cm and length 2 m to a height of 10 m. What is the gauge pressure at the pump connected to the lower end of the pipe?

49. (II) A 40-m/s wind blows past a roof of dimensions 10 m × 15 m. Assuming that the air under the roof is at rest, what is the net force on the roof?

50. (II) The wing of an airplane has an area of 80 m². Air flows over the top at 200 m/s and under the wing at 180 m/s. What is the net force on the wing due to the Bernoulli effect?

51. (II) Water enters a basement inlet pipe of radius 1.5 cm at 40 cm/s. It flows through a pipe of radius 0.5 cm at a height of 35 m at a gauge pressure of 0.2 atm. (a) What is the speed of the water at the higher point? (b) What is the gauge pressure at the basement?

52. (II) The diameter of a horizontal pipe through which water flows gradually decreases to one-half its original value. The initial speed and absolute pressure are 2.4 m/s and 160 kPa. Find the final speed and absolute pressure.

53. (II) Water flows at 2.4 m/s through a 1.59-cm diameter garden hose and emerges at atmospheric pressure from a nozzle of radius 0.64 cm. What is the gauge pressure at the faucet?

PROBLEMS

1. (II) A dam has a height H and a width W (see Fig. 14.33). Assuming that the water level reaches the top, show that the net pressure force exerted on the dam is

$$F = \frac{\rho g W H^2}{2}$$

Evaluate your expression for $H = 60$ m and $W = 200$ m. (*Hint:* Consider first the force on a horizontal strip.)

2. (II) Assuming that the water level reaches the top of the dam in Problem 1, calculate the torque experienced by the dam about a point at its base. Show that the torque would

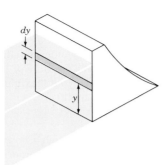

FIGURE 14.33 Problem 1.

be the same if the total force exerted on the dam has an "effective" point of application $H/3$ above the base.

3. (I) Show that the increase in density of a liquid as a function of depth h is

$$\Delta\rho = \frac{\rho^2 gh}{B}$$

where B is the bulk modulus. Estimate the density of water at the bottom of the Marianas Trench at depth of 10.9 km given $B = 2.1 \times 10^9$ N/m². (*Hint:* Show that $dV/V = -d\rho/\rho$.)

4. (II) A liquid of density ρ is in a bucket that spins with angular velocity ω as shown in Fig. 14.34. Show that the pressure at a radial distance r from the axis is

$$P = P_a + \frac{\rho\omega^2 r^2}{2}$$

where P_a is the pressure at the axis at the same level below the bottom of the curved surface. (*Hint:* Write Newton's second law for an elemental ring.)

FIGURE 14.34 Problem 4.

5. (I) A cylindrical can of length L and radius R is made of sheet metal of density ρ_s and thickness t ($<<R$). It is floated in a liquid of density $\rho\ell$ with the end faces vertical. Show that the fraction f of the volume submerged is

$$f = \frac{2t(L + R)\rho_s}{RL\rho_\ell}$$

6. (I) The training of astronauts includes simulated weightlessness in a large pool; see Fig. 14.35. An astronaut of mass 90 kg (including spacesuit) can float vertically with just the helmet, of volume 3 L, exposed. How much weight needs to be added to the spacesuit to give neutral buoyancy (i.e., an average density equal to that of the water)?

FIGURE 14.35 Problem 6.

7. (I) A 3-kg block of wood in the form of a cube floats with 60% of its volume submerged in water. What is the work required to submerge it completely? Assume a vertical force is applied to the horizontal top face.

8. (II) Water emerges at speed v_0 from the opening of a faucet of radius R. In laminar flow, the cross-sectional area of the vertical stream of water decreases as it falls. Obtain an equation for the radius r of the stream as a function of the vertical drop y (see Fig. 14.36).

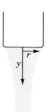

FIGURE 14.36 Problem 8.

9. (I) Water drains through an opening of area A_1 in a container of cross-sectional area A_2; see Fig. 14.37. If the motion of the water surface in the container is not ignored, show that the speed at which the water emerges is given by

$$v^2 = \frac{2gh}{1 - A_1^2/A_2^2}$$

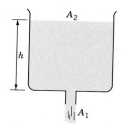

FIGURE 14.37 Problem 9.

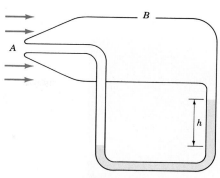

FIGURE 14.38 Problem 10.

10. (II) A Pitot tube, shown in Fig. 14.38, is used to measure the speed of an airplane relative to the air, or of a ship relative to water. Fluid entering at inlet A is brought to rest, whereas fluid flows past the opening at B. The difference in pressure is measured by the manometer which contains a liquid of density ρ_m. Show that the speed of the fluid of density ρ_f moving past B is given by

$$v = \sqrt{\frac{2gh\rho_m}{\rho_f}}$$

11. (I) Water emerges from a small opening at a height h from the bottom of a large container, as in Fig. 14.39, which is filled to a constant depth H. (a) Show that the distance R from the base at which water hits the ground is given by $R = 2\sqrt{h(H - h)}$.
(b) At what other height would a similar opening lead to the same point of impact?

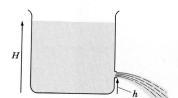

FIGURE 14.39 Problem 11.

CHAPTER 15

Oscillations

A torsion pendulum designed to detect the possible existence of a "fifth force." (See *Physics Today,* July 1988, p 21.)

Major Points

1. In a **simple harmonic oscillation** the amplitude is constant and the period is independent of the amplitude.
2. **Simple harmonic motion** occurs when the restoring force is proportional to the displacement from equilibrium and in the opposite direction.
3. The variation of the **kinetic** and **potential energies** with time and with position.
4. The behavior of **simple pendulums, physical pendulums,** and **torsional pendulums.**

Any motion or event that repeats itself at regular intervals is said to be **periodic.** In some periodic motions a body moves back and forth along a given path between two extreme positions. Examples include the vibration of a guitar string or a speaker cone, the swinging of a pendulum, the motion of a piston in an engine and the vibrations of the atoms in a solid. Such periodic motions are examples of **oscillation.** In general, an oscillation is a periodic fluctuation in the value of a physical quantity above and below some central or equilibrium value.

Definition of an oscillation

In mechanical oscillations, such as those cited above, the body undergoes linear or angular displacement. Nonmechanical oscillations involve the variation of quantities such as voltage or charge in electrical circuits, or the electric and magnetic fields in radio and TV signals. In this chapter we discuss only mechanical oscillations, but the techniques we develop are applicable to other kinds of oscillatory behavior.

The first quantitative observations of oscillations were probably made by Galileo. The chandeliers in the cathedral at Pisa had to be pulled toward a balcony for them to be lit. After they were released, they would swing back and forth for some time. On one occasion, Galileo used his pulse to time the swings and was surprised to learn that even as the oscillations diminished in size, the time for each complete swing did not change. This property of isochronism (*iso* = same, *chronos* = time) was the basis of early pendulum clocks.

We begin by discussing examples of **simple harmonic oscillation** which is an oscillation that occurs without loss of energy. If friction, or some other mechanism, causes the energy to decrease, the oscillations are said to be **damped.** Finally, we discuss the response of a system to an external driving force that varies sinusoidally in time. Such **forced oscillations** exhibit the phenomenon of **resonance** when the frequency of the driving force is close to the natural frequency of oscillation of the system.

15.1   SIMPLE HARMONIC OSCILLATION

A convenient system for studying oscillations is a block attached to a spring. To see how the displacement from equilibrium, x, varies with time, we can record the motion on a strip of paper that moves at a constant speed, as shown in Fig. 15.1. We find that a sinusoidal pattern is traced out. In the absence of friction, the block oscillates between the exreme values $x = +A$ and $x = -A$, where A is called the **amplitude** of the oscillation. As depicted in Fig. 15.1, the displacement from equilibrium is given by

$$x(t) = A \sin \omega t$$

where ω, measured in rad/s, is called the **angular frequency,** rather than the angular velocity, since it does not refer to the rotation of something physical. One cycle corresponds to 2π radians and is completed in one period, T. Therefore, $2\pi = \omega T$, or

$$\omega = \frac{2\pi}{T} = 2\pi f \tag{15.1}$$

Angular frequency

where $f = 1/T$, called the **frequency,** is measured in s^{-1}, or hertz (Hz). In Fig. 15.1 the block was at $x = 0$ at $t = 0$. In general this will not be the case, so we write

$$x(t) = A \sin (\omega t + \phi) \tag{15.2}$$

Simple harmonic oscillation

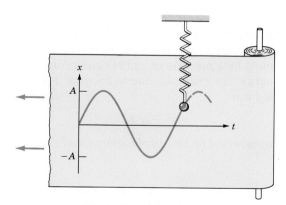

FIGURE 15.1 An oscillating block leaves a sinusoidal trace on a moving strip of paper.

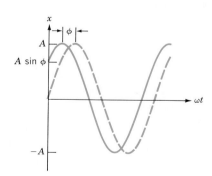

FIGURE 15.2 The function $x = A \sin(\omega t + \phi)$ (solid curve) is shifted by ϕ to the left relative to $x = A \sin \omega t$ (dashed curve). *Note that the horizontal axis is* ωt *(measured in radians).* The position at $t = 0$ is $x = A \sin \phi$.

The argument $\omega t + \phi$ is called the **phase,** while ϕ is called the **phase constant** (or in some contexts, the phase angle). Both the phase and the phase constant are measured in radians. The specific values of A and ϕ in a given problem are determined by the values of x and the velocity, $v = dx/dt$, at some specific time, such as $t = 0$.

According to Eq. 15.2, we see that $x = A \sin \phi$ at $t = 0$, and that $x = 0$ when $\sin(\omega t + \phi) = 0$. That is, $x = 0$ when $\omega t = -\phi$ or $t = -\phi/\omega$. As Fig. 15.2 shows, this means that when ϕ is positive, the curve shifts to the left in comparison with $x = A \sin \omega t$.

Any system in which the variation in time of a physical quantity is given by Eq. 15.2 is called a **simple harmonic oscillator.** In the case of oscillations in electrical circuits the displacement x might be replaced by charge or voltage. In the case of light and radio waves, x is replaced by electric and magnetic fields. A simple harmonic oscillator has the following characteristics:

Properties of a simple harmonic oscillator

1. The amplitude A is constant. (The oscillation is *simple.*)
2. The frequency and period are independent of the amplitude: Large oscillations have the same period as small ones. (The property of *isochronism.*)
3. The time dependence of the fluctuating quantity can be expressed in terms of a sinusoidal function of a single frequency. (The oscillation is *harmonic.*)

The first and second derivatives of Eq. 15.2 are

$$\frac{dx}{dt} = \omega A \cos(\omega t + \phi) \tag{15.3}$$

$$\frac{d^2x}{dt^2} = -\omega^2 A \sin(\omega t + \phi) \tag{15.4}$$

As Fig. 15.3 shows, the extreme values of the velocity, $v = \pm\omega A$, occur when $x = 0$, whereas the extreme values of the acceleration, $a = \pm\omega^2 A$, occur when $x = \mp A$.

Comparing Eq. 15.4 with Eq. 15.2 we see that

$$\frac{d^2x}{dt^2} + \omega^2 x = 0 \tag{15.5}$$

This form of *differential equation* characterizes all types of simple harmonic oscillation—mechanical or nonmechanical. The techniques we develop to deal with this equation may be applied to all examples of simple harmonic oscillation. Equation 15.2 is a *solution* of this differential equation.

The term **simple harmonic motion** (SHM) is applied to mechanical examples of simple harmonic oscillation. For SHM to occur, three conditions must be satisfied. First, there must be a position of stable equilibrium. Second, there must be no dissipation of energy, for example, due to friction. Third, as we can see by writing Eq. 15.5 in the form

$$a = -\omega^2 x$$

the *acceleration is proportional to the displacement and opposite in direction.*

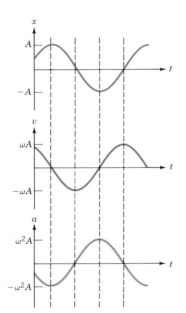

FIGURE 15.3 The variation in time of the position, the velocity, and the acceleration in a simple harmonic motion. Note that $a = -\omega^2 x$.

EXAMPLE 15.1: The position of a particle moving along the x axis is given by

$$x = 0.08 \sin(12t + 0.3) \text{ m}$$

where t is in seconds. (a) What are the amplitude and period of the motion? (b) Determine the position, velocity, and acceleration at $t = 0.6$ s.

Solution: (a) On comparing the given equation with the form in Eq. 15.2 we see that the amplitude is $A = 0.08$ m and the angular frequency is $\omega = 12$ rad/s. Thus, the period is $T = 2\pi/\omega = 0.524$ s.

(b) The velocity and acceleration at any time are given by

$$v = \frac{dx}{dt} = 0.96 \cos(12t + 0.3) \text{ m/s}$$

$$a = \frac{dv}{dt} = -11.5 \sin(12t + 0.3) \text{ m/s}^2$$

At $t = 0.6$ s, the phase of the motion is $(12 \times 0.6 + 0.3) = 7.5$ rad. When this is used in the above expressions, we find $x = 0.075$ m, $v = 0.333$ m/s, and $a = -10.8$ m/s^2.

EXERCISE 1. What is the acceleration when the position is $x = -0.05$ m?

15.2   THE BLOCK–SPRING SYSTEM

We begin by considering the dynamics of a block that is oscillating at the end of a massless spring, as in Fig. 15.4. We assume that the net force acting on the block is that exerted by the spring, which is given by Hooke's law:

$$F_{sp} = -kx$$

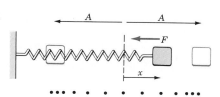

where x is the displacement from the equilibrium position. When x is positive, F_{sp} is negative, the force is directed to the left. When x is negative, F_{sp} is positive, the force is directed to the right. Thus, the force always tends to restore the block to its equilibrium position $x = 0$. Newton's second law ($F = ma$) applied to the block is $-kx = ma$, which means

FIGURE 15.4 A block oscillating at the end of a spring. The restoring force is proportional to the displacement from equilibrium.

$$a = -\frac{k}{m} x$$

The acceleration is directly proportional to the displacement, but is in the opposite direction, as is required for SHM. Since $a = d^2x/dt^2$, we have

$$\frac{d^2x}{dt^2} + \frac{k}{m} x = 0 \qquad (15.6)$$

This differential equation is merely another way of writing Newton's second law. When Eq. 15.6 is compared with Eq. 15.5, we see that the block–spring system executes simple harmonic motion with an angular frequency

$$\omega = \sqrt{\frac{k}{m}} \qquad (15.7)$$

or a period

$$T = \frac{2\pi}{\omega} = 2\pi \sqrt{\frac{m}{k}} \qquad (15.8) \qquad \text{Period of a block–spring system}$$

As is required for SHM, the period is independent of the amplitude. For a given spring constant, the period increases with the mass of the block: A more massive block oscillates more slowly. For a given block, the period decreases as k increases: A stiffer spring produces quicker oscillations.

EXAMPLE 15.2: A 2-kg block is attached to a spring for which $k = 200$ N/m. It is held at an extension of 5 cm and then released at $t = 0$. Find: (a) the displacement as a function of time; (b) the velocity when $x = +A/2$; (c) the acceleration when $x = +A/2$.

Solution: (a) We need to find A, ω, and ϕ in Eq. 15.2. The amplitude is the maximum extension; that is, $A = 0.05$ m. From Eq. 15.7 the angular frequency is

$$\omega = \sqrt{\frac{k}{m}} = 10 \text{ rad/s}$$

To find ϕ we note that at $t = 0$ we are given $x = +A$ and $v = 0$. Thus, from Eq. 15.2 and Eq. 15.3,

$$A = A \sin(0 + \phi)$$
$$0 = 10A \cos(0 + \phi)$$

Since $\sin \phi = 1$ and $\cos \phi = 0$, it follows that $\phi = \pi/2$ rad. Thus,

$$x = 0.05 \sin\left(10t + \frac{\pi}{2}\right) \text{ m} \qquad (i)$$

(b) In order to find the velocity we have to find when $x = A/2$. Equation (i) yields $\frac{1}{2} = \sin(10t + \pi/2)$, from which we infer that $(10t + \pi/2) = \pi/6$ or $5\pi/6$. (We need just the phase, not the time.) The velocity is given by

$$v = \frac{dx}{dt} = 0.5 \cos\left(10t + \frac{\pi}{2}\right)$$

$$= 0.5 \cos \frac{\pi}{6} \quad \text{or} \quad 0.5 \cos \frac{5\pi}{6}$$

$$= +0.43 \text{ m/s} \quad \text{or} \quad -0.43 \text{ m/s}$$

At a given position, there are two velocities of equal magnitude but opposite directions.

(c) The acceleration at $x = A/2$ may be found from Eq. 15.5:

$$a = -\frac{k}{m} x = -\omega^2 x$$
$$= -(10 \text{ rad/s})^2 (0.05/2 \text{ m}) = -2.5 \text{ m/s}^2$$

Figure 15.5 depicts values of a and v at intervals of $T/4$.

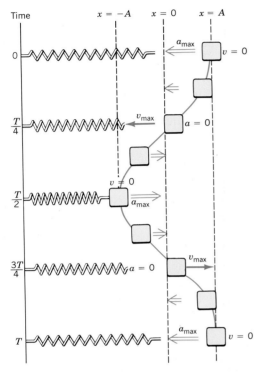

FIGURE 15.5 The acceleration and velocity of a block oscillating at the end of a spring at intervals of $T/4$.

EXERCISE 2. (a) Use part (b) above directly to find a when $x = A/2$. (b) What is the force on the block at $t = \pi/15$ s?

EXAMPLE 15.3: In a block–spring system, $m = 0.2$ kg and $k = 5$ N/m. At $t = \pi/10$ s, the spring has a compression of 6 cm and the block has a velocity $v = -40$ cm/s. (a) What is the displacement as a function of time? (b) What is the first time (>0) at which the velocity is positive and 60% of its maximum value?

Solution: (a) We need to find ω, A, and ϕ in Eq. 15.2. From Eq. 15.7, the angular velocity is

$$\omega = \sqrt{\frac{k}{m}} = \sqrt{\frac{5 \text{ N/m}}{0.2 \text{ kg}}} = 5 \text{ rad/s}$$

Using the given information in Eq. 15.2 and Eq. 15.3 we find

$$-0.06 = A \sin\left(\frac{5\pi}{10} + \phi\right) \qquad \text{(i)}$$

$$-0.08 = A \cos\left(\frac{5\pi}{10} + \phi\right) \qquad \text{(ii)}$$

When we square both equations and add them we find $A = 0.10$ m. (Note that $\cos^2\theta + \sin^2\theta = 1$.) The ratio of the equations can be used to find ϕ:

$$\tan\left(\frac{\pi}{2} + \phi\right) = \frac{3}{4} \qquad \text{(iii)}$$

which means $(\pi/2 + \phi) = \arctan \frac{3}{4}$. (One could also substitute $A = 0.1$ m into either (i) or (ii)) There are two possibilities, $(\pi/2 + \phi) = 37\pi/180$ rad or $215\pi/180$ rad. Since both the sine and the cosine in (i) and (ii) are negative, the angle is in the third quadrant, so we pick $(\pi/2 + \phi) = 215\pi/180$ radians. Thus, $\phi = 127\pi/180 = 2.2$ rad. The displacement as a function of time is given by

$$x = 0.1 \sin(5t + 2.2) \text{ m} \qquad \text{(iv)}$$

This function is plotted in Fig. 15.6. Notice that the horizontal axis is plotted in terms of ωt (in radians), not t.
(b) The derivative of (iv) is

$$v = 0.5 \cos(5t + 2.2) \text{ m/s}$$

We are given that v is 60% of the maximum value; thus,

$$0.6 = \cos(5t + 2.2)$$

which means $5t + 2.2 = \cos^{-1} 0.6$. Thus, $5t + 2.2 = 53\pi/180$ rad or $307\pi/180$ rad. The first possibility leads to $t < 0$, which is unacceptable. The other leads to $5t = (5.4 - 2.2)$ rad, which yields $t = 0.64$ s.

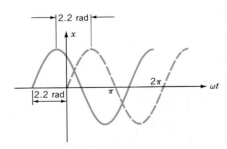

FIGURE 15.6 The function $x = A \sin(\omega t + 2.2 \text{ rad})$. Note that the horizontal axis is ωt, not t.

EXAMPLE 15.4: Show that a block hanging from a vertical spring, as in Fig. 15.7, executes simple harmonic motion.

Solution: The block is subject to two forces: the upward force exerted by the spring and the downward force due to gravity. If x_0 is the equilibrium extension at which these two forces balance, then

$$mg = kx_0$$

For any extension x, the net force on the block is

$$F = mg - kx = -k(x - x_0)$$
$$= -kx'$$

where $x' = x - x_0$ is the displacement *from the equilibrium position*. Since the restoring force is linearly proportional to the displacement from equilibrium, the motion will be simple harmonic.

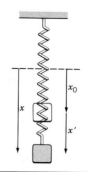

FIGURE 15.7 A block oscillating on a vertical spring executes simple harmonic motion about the equilibrium position.

15.3   ENERGY IN SIMPLE HARMONIC MOTION

The force exerted by an ideal spring is conservative, which means that, in the absence of friction, the energy of the block–spring system is constant. Thus, we may examine the motion of the block from the viewpoint of energy conservation. We can use Eq. 15.2 to express the potential energy as

$$U = \tfrac{1}{2}kx^2 = \tfrac{1}{2}kA^2 \sin^2(\omega t + \phi) \qquad (15.9)$$

Using Eq. 15.3, the kinetic energy is

$$K = \tfrac{1}{2}mv^2 = \tfrac{1}{2}m\omega^2 A^2 \cos^2(\omega t + \phi) \qquad (15.10)$$

Since $\omega^2 = k/m$ and $\cos^2\theta + \sin^2\theta = 1$, the total mechanical energy, $E = K + U$, is

$$E = \tfrac{1}{2}mv^2 + \tfrac{1}{2}kx^2 = \tfrac{1}{2}kA^2 \qquad (15.11)$$

Energy of a block-spring system

The total energy of any simple harmonic oscillator is constant and proportional to the square of the amplitude. The variation of K and U as functions of x is shown in Fig. 15.8. When $x = \pm A$, the kinetic energy is zero and the total energy is equal to the maximum potential energy, $E = U_{max} = \tfrac{1}{2}kA^2$. These are the turning points of the SHM. At $x = 0$, $U = 0$ and the energy is purely kinetic; that is, $E = K_{max} = \tfrac{1}{2}m(\omega A)^2$. Figure 15.9 shows how K and U vary in time, assuming that $\phi = 0$.

In Fig. 15.8 we see that the block is in a "potential well" created by the spring (see Section 8.8). *All SHM is characterized by a parabolic potential well.* That is, the potential energy is proportional to the square of the displacement. If a well is not parabolic, the simple harmonic approximation is often used as a first step to a complete solution. This is particularly important in dealing with interatomic potentials in molecules and crystals.

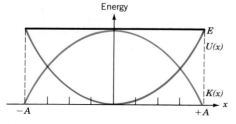

FIGURE 15.8 The variation of the kinetic energy, potential energy, and total energy as a function of position.

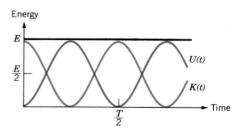

FIGURE 15.9 The variation of the kinetic energy, potential energy, and total energy as a function of time.

EXAMPLE 15.5: In Example 15.2 the displacement of a 2-kg particle attached to a spring for which k = 200 N/m was given by

$$x = 0.05 \sin\left(10t + \frac{\pi}{2}\right) \text{ m}$$

(a) Find K, U, and E at $t = \pi/15$ s. (b) What is the speed at $x = A/2$?

Solution: (a) The total energy is simply the maximum potential energy. Since $A = 0.05$ m, we have

$$E = \tfrac{1}{2}kA^2 = \tfrac{1}{2}(200 \text{ N/m})(0.05 \text{ m})^2 = 0.25 \text{ J}$$

The potential and kinetic energies are

$$U = \tfrac{1}{2}kx^2 = \tfrac{1}{2}(200 \text{ N/m})\left[0.05 \sin\left(\frac{2\pi}{3} + \frac{\pi}{2}\right) \text{ m}\right]^2$$

$$= \tfrac{1}{16}\text{J}$$

$$K = \tfrac{1}{2}mv^2 = \tfrac{1}{2}(1 \text{ kg})\left[0.5 \cos\left(\frac{2\pi}{3} + \frac{\pi}{2}\right) \text{ m/s}\right]^2$$

$$= \tfrac{3}{16}\text{J}$$

Clearly $E = K + U$, as it must.
(b) Substituting $x = A/2$ into Eq. 15.11 we have

$$\tfrac{1}{2}mv^2 + \tfrac{1}{2}k\left(\frac{A}{2}\right)^2 = \tfrac{1}{2}kA^2$$

Therefore,

$$v^2 = \frac{3kA^2}{4m} = \frac{3(200 \text{ N/m})(0.05 \text{ m})^2}{4 \times 2 \text{ kg}} = 0.188 \text{ m}^2/\text{s}^2$$

from which we find $v = 0.43$ m/s.

EXERCISE 3. For what value(s) of x is $K = U$? Express your answer in terms of A and compare it with Fig. 15.8.

EXERCISE 4. Use energy considerations to find A in part (a) of Example 15.3.

EXAMPLE 15.6: Show that the differential equation for simple harmonic motion can be obtained from the expression for the energy of the system.

Solution: The total energy of a simple harmonic oscillator is constant both in time and in space. Thus, from Eq. 15.11 we have

$$\frac{dE}{dt} = mv\frac{dv}{dt} + kx\frac{dx}{dt} = 0$$

By eliminating the common factor $v = dx/dt$, we obtain

$$m\frac{dv}{dt} + kx = 0$$

Since $dv/dt = d^2x/dt^2$, this equation is equivalent to Eq. 15.5.

EXERCISE 5. Show that the condition $dE/dx = 0$ also leads to Eq. 15.5. (You will need the chain rule.)

15.4 PENDULUMS

The Simple Pendulum

A simple pendulum is an idealized system in which a point mass is suspended at the end of a massless string. Figure 15.10 shows a simple pendulum length of L and a "pendulum bob" of mass m. The distance along the arc from the lowest point is $s = L\theta$, where θ is the angle (in radians) to the vertical. The net force on the bob along the tangent is the component of the weight along the tangent. Newton's second law applied along this direction is

$$-mg \sin \theta = m\frac{d^2s}{dt^2}$$

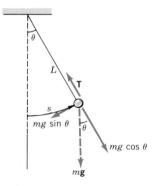

FIGURE 15.10 A simple pendulum. The only force along the tangent is the component of the weight: $mg \sin \theta$. For small angles, the restoring force is proportional to the displacement, so the motion is simple harmonic.

The negative sign arises from the way s is defined. Its physical meaning is that the component of the weight acts as a restoring force. Clearly this equation does not have the form associated with SHM. However, for small angles $\sin \theta \approx \theta$, where θ is in radians. Since $s = L\theta$, we substitute $d^2s/dt^2 = Ld^2\theta/dt^2$ and $\sin \theta \approx \theta$ into the above equation to find

$$\frac{d^2\theta}{dt^2} + \frac{g}{L}\theta = 0 \tag{15.12}$$

Comparing this with Eq. 15.5 for SHM we see that within the small-angle approximation, a simple pendulum executes simple harmonic motion with an angular frequency

(Simple pendulum)
$$\omega = \sqrt{\frac{g}{L}}$$

and a period

$$T = 2\pi \sqrt{\frac{L}{g}}$$ (15.13)   Period of a simple pendulum

The period does not depend on either the mass or the amplitude (Galileo's discovery). The solution to Eq. 15.12 is based on Eq. 15.2:

$$\theta = \theta_0 \sin(\omega t + \phi)$$ (15.14)

where θ_0 is the angular amplitude. Note that θ is a physical angular displacement, whereas ϕ is a mathematical phase constant that depends on the initial conditions. Furthermore, the angular frequency ω should not be confused with the (physical) instantaneous angular velocity $d\theta/dt$.

The Physical Pendulum

In Fig. 15.11 an extended body is pivoted freely about an axis that does not pass through its center of mass. Such an arrangement forms a **physical pendulum** that executes simple harmonic motion for small angular displacements. A practical example of a physical pendulum is an arm or a leg. If d is the distance from the pivot to the center of mass, the restoring torque is $-mgd \sin \theta$ (toward decreasing θ). The rotational form of Newton's second law, $\tau = I\alpha$, is

$$-mgd \sin \theta = I \frac{d^2\theta}{dt^2}$$

where I is the moment of inertia about the given axis. If we make the small-angle approximation, $\sin \theta \approx \theta$, then

$$\frac{d^2\theta}{dt^2} + \frac{mgd}{I} \theta = 0$$ (15.15)

which is the equation for simple harmonic oscillation. Comparison with Eq. 15.5 shows that

(Physical pendulum)
$$\omega = \sqrt{\frac{mgd}{I}}$$ (15.16)

and

$$T = 2\pi \sqrt{\frac{I}{mgd}}$$ (15.17)   Period of a physical pendulum

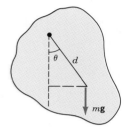

FIGURE 15.11 A physical pendulum pivoted about a point other than its center of mass.

If the location of the center of mass and d are known, then a measurement of the period allows us to determine the moment of inertia of the body.

EXAMPLE 15.7: The angular displacement of a simple pendulum is given by

$$\theta = 0.1\pi \sin\left(2\pi t + \frac{\pi}{6}\right) \text{ rad}$$

The mass of the bob is 0.4 kg. When $\theta = 0.05\pi$ find: (a) the length of the simple pendulum; (b) the velocity of the bob at $t = 0.25$ s.

Solution: (a) We are given $\theta_0 = 0.1\pi$ rad, $\phi = \pi/6$ rad, and $\omega = 2\pi$ rad/s. Since $\omega^2 = (g/L)$, we have

$$L = \frac{g}{\omega^2} = \frac{9.8 \text{ m/s}^2}{(2 \times 3.14 \text{ rad/s})^2} = 0.25 \text{ m}$$

(b) Since $s = L\theta$, the velocity of the bob, $v = ds/dt$, is

$$v = L \frac{d\theta}{dt}$$

$$= (0.25 \text{ m})(0.1\pi)(2\pi) \cos\left(\frac{\pi}{4} + \frac{\pi}{6}\right)$$

$$= 0.064 \text{ m/s}$$

EXAMPLE 15.8: A uniform rod of mass m and length L is freely pivoted at one end. (a) What is the period of its oscillation? (b) What is the length of a simple pendulum with the same period?

Solution: (a) The moment of inertia of a rod about one end is $I = \frac{1}{3}mL^2$ (Eq. 11.18). The center of mass of a uniform rod is at its center, so $d = L/2$ in Eq. 15.17. The period is

$$T = 2\pi \sqrt{\frac{mL^2/3}{mgL/2}} = 2\pi \sqrt{\frac{2L}{3g}}$$

(b) Comparing Eq. 15.17 with $T = 2\pi\sqrt{L/g}$ for a simple pendulum, we see that the period of a physical pendulum is the same as that of an "equivalent" simple pendulum of length

$$L_{eq} = \frac{I}{md}$$

For the uniform rod

$$L_{eq} = \frac{mL^2/3}{(mL/2)} = \frac{2L}{3}$$

If the angular amplitude of a pendulum is large, the small-angle approximation $\sin \theta \approx \theta$ ceases to be valid. As a result, the oscillations are no longer simple harmonic. In fact, the period increases as the angular amplitude increases (see Problem 11). In practice, frictional losses cause the amplitude, and therefore the period, of a pendulum to diminish in time. In a grandfather clock a falling weight drives a mechanism that compensates for such energy losses. By keeping the amplitude constant, it also ensures more accurate timekeeping.

Torsional Pendulum

Consider a body, such as a disk or a rod, suspended at the end of a wire, as shown in Fig. 15.12. When the end of the wire is twisted by an angle θ, the restoring torque τ obeys Hooke's law: $\tau = -\kappa\theta$, where κ is called the *torsional constant*. If the wire is twisted and released, the oscillating system is called a **torsional pendulum.** The rotational form of Newton's second law, $\tau = I\alpha$, is

$$-\kappa\theta = I \frac{d^2\theta}{dt^2}$$

which may be written in the form

$$\frac{d^2\theta}{dt^2} + \frac{\kappa}{I} \theta = 0$$

This is the equation of a simple harmonic oscillator with an angular frequency

(Torsional pendulum) $$\omega = \sqrt{\frac{\kappa}{I}} \qquad (15.18)$$

and a period

Period of a torsional pendulum $$T = 2\pi \sqrt{\frac{I}{\kappa}} \qquad (15.19)$$

Notice that we did not use the small-angle approximation. As long as the elastic limit of the system is not exceeded, it will execute simple harmonic motion. The balance wheel connected to the hairspring in a wristwatch is also a torsional pendulum.

FIGURE 15.12 A torsional pendulum. The restoring torque due to a twisted wire or fiber is proportional to the angle of twist. Thus, the motion is simple harmonic.

15.5   DAMPED OSCILLATIONS (Optional)

Thus far we have ignored the inevitable energy losses that occur in real situations. Such losses may arise from external fluid resistance or from "internal friction" within a system. The energy, and consequently the amplitude, of such a *damped* oscillator decrease in time. To formulate the equation for damped oscillations, we consider the situation depicted in Fig. 15.13 which shows a block immersed in a liquid. When the velocity is low, the damping is due to a resistive force **f** that is proportional to the velocity (see Section 6.4):

$$\mathbf{f} = -\gamma \mathbf{v} \qquad (15.20)$$

where γ, measured in kg/s, is the *damping constant*. If we ignore the buoyancy of the fluid, Newton's second law applied to the block is

$$F = -kx - \gamma \frac{dx}{dt} = m \frac{d^2x}{dt^2}$$

where x is the displacement from equilibrium. (The prime used in Example 15.4 is dropped for simplicity.) This equation may be written in the form

$$m \frac{d^2x}{dt^2} + \gamma \frac{dx}{dt} + kx = 0 \qquad (15.21)$$

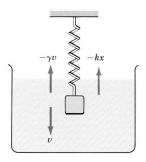

FIGURE 15.13 The oscillations of a block are damped when it is immersed in a fluid. In a real system, energy losses within the spring itself also lead to damping.

This form of differential equation arises in other mechanical or nonmechanical damped oscillations. Experience tells us that the mass will oscillate with ever-decreasing amplitude. As you can verify by substitution, the solution to Eq. 15.21 is

$$x = A_0 e^{-\gamma t/2m} \cos(\omega' t + \phi) \qquad (15.22)$$

The **damped angular frequency** ω' is given by

$$\omega' = \sqrt{\omega_0^2 - \left(\frac{\gamma}{2m}\right)^2} \qquad (15.23)$$

The damped angular frequency ω' is less than the **natural angular frequency,** $\omega_0 = \sqrt{k/m}$.

For ω' to be real, the condition $\gamma/2m < \omega_0$, or equivalently

$\gamma < 2m\omega_0$, must be satisfied. When ω' is real, the oscillations are **underdamped,** as illustrated in Fig. 15.14. The amplitude decays according to

(Underdamped) $\qquad A(t) = A_0 e^{-\gamma t/2m} \qquad (15.24)$

The damped period is $T' = 2\pi/\omega'$.

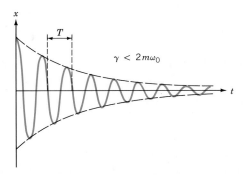

FIGURE 15.14 In an underdamped oscillation the system oscillates with an exponentially decaying amplitude.

When the damping is so large that $\gamma > 2m\omega_0$, ω' is an imaginary number. In this case there is no oscillation and the system moves slowly back to its equilibrium position, as shown in Fig. 15.15. Hinged doors that close automatically and cueing devices on tonearms of turntables are **overdamped.**

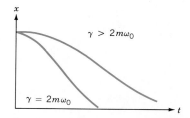

FIGURE 15.15 In critical damping ($\gamma = 2m\omega_0$) the system approaches the equilibrium position most rapidly. In an overdamped system ($\gamma > 2m\omega_0$), the system approaches equilibrium slowly.

When $\gamma = 2m\omega_0$, we have $\omega' = 0$ and again there is no oscillation. This condition of **critical damping** leads to the shortest time for the system to return to equilibrium and is also shown in Fig. 15.15. Critical damping is used in the movements of electrical meters to damp the oscillations of the needle. The suspension system of a car is adjusted to have somewhat less than critical damping. When a fender is pressed down and released, the car executes perhaps one and a half oscillations before coming to rest.

EXAMPLE 15.9: A 0.5-kg block is attached to a spring ($k = 12.5$ N/m). The damped frequency is 0.2% lower than the natural frequency. (a) What is the damping constant? (b) How does the amplitude vary in time? (c) What is the critical damping constant?

Solution: (a) The natural angular frequency is $\omega_0 = \sqrt{k/m} = 5$ rad/s. The damped angular frequency is $\omega' = 0.998\,\omega_0 = 4.99$ rad/s. From Eq. 15.23,

$$\gamma^2 = 4m^2(\omega_0^2 - \omega'^2)$$

This yields $\gamma = 0.316$ kg/s.
(b) From Eq. 15.24

$$A(t) = A_0 e^{-0.316t}$$

(c) The critical damping constant is

$$\gamma = 2m\omega_0 = 5 \text{ kg/s}$$

This is considerably larger than that found in part (a).

15.6 FORCED OSCILLATIONS (Optional)

The loss in energy of a damped oscillator may be compensated for by work done by an external agent. For example, a child on a swing can be kept in motion by appropriately timed pushes (Fig. 15.16). In many instances, the external driving force varies sinusoidally at some angular frequency ω_e. We assume $F(t) = F_0 \cos \omega_e t$. Newton's second law applied to such a **forced** or **driven** oscillator yields

$$m \frac{d^2x}{dt^2} + \gamma \frac{dx}{dt} + kx = F_0 \cos \omega_e t \qquad (15.25)$$

FIGURE 15.16 A child can be kept swinging by appropriately timed pushes.

When the force is first applied, the motion is complex. However, the system ultimately settles into a *steady-state* oscillation. At this stage the energy dissipated by the damping is exactly balanced by the external input. The steady-state solution to Eq. 15.25 is

$$x = A \cos(\omega_e t + \delta) \qquad (15.26)$$

where δ is the phase angle between the displacement x and the external force F. Notice that the amplitude is constant in time and that ω_e is the angular frequency of the *external driving force*. When Eq. 15.26 is substituted into Eq. 15.25, we are led to conclude (the details are omitted)

$$A(\omega_e) = \frac{F_0/m}{\sqrt{(\omega_0^2 - \omega_e^2)^2 + (\gamma\omega_e/m)^2}} \qquad (15.27)$$

Each driving frequency is characterized by its own amplitude, as shown in Fig. 15.17. At $\omega_e = 0$, the amplitude is merely the static extension $F_0/m\omega_e = F_0/k$. As the external angular frequency ω_e is increased, the amplitude rises until it reaches a maximum at ω_{max}, which is somewhat below ω_0. At higher frequencies, the amplitude again decreases. Such a response is called **resonance** and ω_{max} is called the **resonance angular frequency.** When γ is small, the resonance curve is narrow and the peak occurs close to the **natural angular frequency** ω_0. For large γ, the resonance is broad and the peak is shifted to lower frequencies. The value of γ may be so large that there is no resonance. At the resonance frequency the external force and the velocity of the particle are in phase. As a result, the power transfer ($P = \mathbf{F} \cdot \mathbf{v}$) to the oscillator has its maximum value. At frequencies above or below the resonance value, the force and velocity are not in phase, so the power transfer is lower.

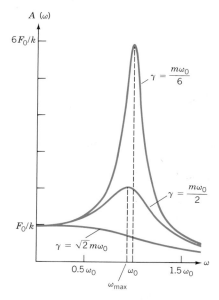

FIGURE 15.17 The amplitude of a driven oscillator displays a resonance as the angular frequency of the external agent is varied. For large damping, the peak occurs below the natural angular frequency ω_0 and resonance is broad.

Even large-scale structures, such as towers, bridges, and airplanes, can oscillate. If the frequency of the driving mechanism is close to the natural frequency, the object can literally be shaken to pieces. A dramatic example of resonance is the collapse of the Tacoma Narrows Bridge in Washington. The wind flowing past the structure set the bridge into one of its natural modes of vibration, as in Fig. 15.18a. After a couple of hours the amplitude became so large that the center span disintegrated (Fig. 15.18b). In later chapters we will discuss resonance in electrical circuits, which is vital to the transmission and reception of radio and TV signals. Resonance also plays a role in atomic and nuclear processes.

(a)   (b)

FIGURE 15.18 (a) In July 1940, high winds set the Tacoma Narrows Bridge into oscillation. (b) After a couple of hours, the center span collapsed.

SUMMARY

In a **simple harmonic oscillation** the amplitude A is constant and the period T is independent of the amplitude. The variation of the physical quantity is given by

$$x = A \sin(\omega t + \phi)$$

where ω is the angular frequency. The phase constant ϕ is determined by the values of x and dx/dt at a given time, such as $t = 0$. For a mechanical system to execute simple harmonic motion, the force or torque restoring the system to equilibrium must obey Hooke's law. The total energy of a simple harmonic oscillator is constant in time.

All simple harmonic oscillators obey a differential equation of the form

$$\frac{d^2x}{dt^2} + \omega^2 x = 0$$

In mechanical examples, this equation is a disguised form of Newton's second law. The angular frequency and period of the oscillation of a block of mass m attached to a spring whose constant is k is given by

$$\omega = \sqrt{\frac{k}{m}}$$

$$T = \frac{2\pi}{\omega} = 2\pi \sqrt{\frac{m}{k}}$$

The **energy** of the block–spring system is

$$E = \tfrac{1}{2}mv^2 + \tfrac{1}{2}kx^2 = \tfrac{1}{2}mv_{\max}^2 = \tfrac{1}{2}kA^2$$

The energy of any simple harmonic oscillator is proportional to the square of the amplitude.

Within the small-angle approximation, the angular frequency and period of a **simple pendulum** of length L are

$$\omega = \sqrt{\frac{g}{L}}; \qquad\qquad T = 2\pi\sqrt{\frac{L}{g}}$$

The angular frequency of a **physical pendulum** of mass m and moment of inertia I is

$$\omega = \sqrt{\frac{mgd}{I}}$$

where d is the distance from the pivot to the center of mass. The angular frequency of a **torsional pendulum** of moment of inertia I is

$$\omega = \sqrt{\frac{\kappa}{I}}$$

where κ is the torsional constant.

In the presence of friction, or some other mechanism that causes the energy to decrease, the oscillations are **damped.** That is, the amplitude decreases in time. It is possible to compensate for the energy loss by applying an external force that varies sinusoidally in time. As the frequency of the driving force is changed, the amplitude reaches a maximum at a certain **resonance frequency** that is close to the **natural frequency** of oscillation.

ANSWERS TO IN-CHAPTER EXERCISES

1. From Eq. 15.5 we know $a = -\omega^2 x = -(12 \text{ rad/s})^2(-0.05 \text{ m}) = 7.2 \text{ m/s}^2$.

2. (a) The acceleration is $a = dv/dt = -5\sin(10t + \pi/2) = -2.5$ m/s^2 at *both* times.
   (b) $F_{sp} = -kx = -(200 \text{ N/m})(0.05 \text{ m})\sin(10\pi/15 + \pi/2) = +5$ N.

3. Since $E = K + U$ and $K = U$, we have $U = E/2$. Thus, $\frac{1}{2}kx^2 =$ $\frac{1}{4}kA^2$, which leads to $x = \pm A/\sqrt{2} \approx 0.7A$.

4. From $E = \frac{1}{2}mv^2 + \frac{1}{2}kx^2 = \frac{1}{2}kA^2$, we have
   $$(0.2 \text{ kg})(0.4 \text{ m/s})^2 + (5 \text{ N/m})(0.06 \text{ m})^2 = (5 \text{ N/m})A^2$$
   which yields $A = 0.1$ m.

5. Set $dE/dx = mv\,dv/dx + kx = 0$ and use the chain rule $dv/dx = (dv/dt)(dt/dx)$.

QUESTIONS

1. Does either of the following execute simple harmonic motion: (a) an arm or leg allowed to swing freely; (b) a tonearm tracing a warped record?

2. If the amplitude of a simple harmonic oscillator is doubled, how are the following quantities affected: (a) frequency; (b) phase constant; (c) maximum speed; (d) maximum acceleration; (e) total energy?

3. A block–spring system undergoes simple harmonic motion at frequency f. How many times per cycle do the following conditions occur: (a) the speed is a maximum; (b) the acceleration is zero; (c) the kinetic energy equals 50% of the potential energy; (d) the potential energy equals the total energy?

4. A simple pendulum is suspended from the roof of an elevator. How is its period affected when the elevator's acceleration is (a) upward? (b) downward?

5. A block oscillates at the end of a vertical spring suspended from the roof of an elevator. How is the period affected when the elevator accelerates (a) upward? (b) downward?

6. A particle executes simple harmonic motion with a period T. It takes a time $T/4$ to travel from $x = -A$ to $x = 0$. Is the time to travel from $x = -A/2$ to $x = A/2$ (a) less, (b) the same, or (c) greater?

7. An open cart oscillates on a frictionless horizontal surface at the end of spring. How are the total energy and period affected if a block with the same mass is dropped vertically

into the cart (a) when $x = A$, or (b) when $x = 0$?

8. Two suspended balls undergo repeated elastic collisions at the lowest point in their swings, as shown in Fig. 15.19. Is this motion simple harmonic?

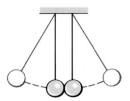

FIGURE 15.19   Question 8.

9. If you were given a stopwatch and a ruler, how could you roughly estimate the mass of an arm, or a leg?

10. A particle executes simple harmonic motion in one dimension with amplitude A and period T. What is the average speed (a) for one quarter cycle between $x = 0$ and $x = \pm A$, and (b) for one complete oscillation?

11. Even in the absence of air resistance, a mass oscillating at the end of spring eventually stops. Why does this occur?

12. Use qualitative reasoning to show that a simple pendulum cannot execute true simple harmonic motion. (*Hint:* Consider the restoring force at a large angular displacement from the vertical.)

13. Why are marching soldiers ordered to break step when they cross a small bridge?

14. The displacement of a particle is given by $x = A \cos \omega t$. What is the phase constant in terms of the standard form $x = A \sin(\omega t + \phi)$ used in the text?

15. A block oscillates at the end of a spring. The spring is cut in half and the block is attached to one of the smaller springs. Is the new period longer or shorter? Explain your answer qualitatively.

16. Simple harmonic motion occurs when the potential energy is quadratic in the displacement from equilibrium. Does a particle sliding without friction on the inside of a bowl with a parabolic shape execute simple harmonic motion?

17. A simple pendulum is suspended from the roof of a truck. How is the period affected when the truck accelerates horizontally?

18. Discuss qualitatively the effect of the mass of a real spring on the period of a block–spring system.

19. Figure 15.20 shows a method by which the mass of an astronaut in stable orbit can be determined. What is the procedure?

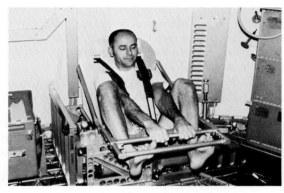

FIGURE 15.20   Question 19.

20. A ball rolls down one incline and up another as shown in Fig. 15.21. Ignore frictional losses. (a) Is the motion periodic? (b) Is there a point of stable equilibrium? (c) Is the motion simple harmonic?

FIGURE 15.21   Question 20.

EXERCISES

15.1 and 15.2 Simple Harmonic Motion; The Block–Spring System

1. (I) The displacement from equilibrium of a particle is given by $x = A \cos(\omega t - \pi/3)$. Which, if any, of the following are equivalent expressions:

(a) $x = A \cos(\omega t + \pi/3)$; (c) $x = A \sin(\omega t + \pi/6)$;
(b) $x = A \cos(\omega t + 5\pi/3)$; (d) $x = A \sin(\omega t - 5\pi/6)$?

2. (I) The displacement of a particle is given by $x =$ 0.03 $\sin(20\pi t + \pi/4)$ m. What are the first times (>0) at which the following quantities have their maximum (>0) values: (a) the displacement; (b) the velocity; (c) the acceleration?

3. (II) When two adults of total mass 150 kg enter a car of mass 1450 kg, the car is lowered by 1 cm. (a) What is the spring constant of the suspension system? (b) What is the period of oscillation when the loaded car hits a bump?

4. (I) The displacement of a block attached to a spring is given by $x = 0.2 \sin(12t + 0.2)$ m. Find: (a) the acceleration when $x = 0.08$ m; (b) the earliest time (>0) at which $x = +0.1$ m with $v < 0$.

5. (I) The condition $|v| = 0.5v_{max}$, where v_{max} is the maximum speed, occurs four times in each cycle in the oscillation of a block-spring system. Determine the first four times (>0) given that the displacement from equilibrium is $x = 0.35 \cos(3.6t - 0.5)$ m.

6. (I) A block attached to a spring is displaced from equilibrium to the position $x = +A$ and released. The period is T. At what positions and times during the first complete cycle do the following conditions occur: (a) $|v| = 0.5v_{max}$, where v_{max} is the maximum speed; (b) $|a| = 0.5a_{max}$, where a_{max} is the maximum magnitude of the acceleration? Specify your answer in terms of A and T.

7. (II) A block of mass $m = 0.5$ kg is attached to a horizontal spring whose spring constant is $k = 50$ N/m. At $t = 0.1$ s, the displacement $x = -0.2$ m and the velocity $v = +0.5$ m/s. Assume $x(t) = A \sin(\omega t + \phi)$. (a) Find the amplitude and the phase constant. (b) Write the equation for $x(t)$. (c) When does the condition $x = 0.2$ m and $v = -0.5$ m/s occur for the first time?

8. (II) In a block–spring system $m = 0.25$ kg and $k = 4$ N/m. At $t = 0.15$ s, the velocity is $v = -0.174$ m/s and the acceleration $a = +0.877$ m/s^2. Write an expression for the displacement as a function of time, $x(t)$.

9. (II) A block of mass 0.5 kg attached to a vertical spring extends it by 0.16 m. It is pulled down a further 0.08 m and released. (a) Write the equation for the displacement from equilibrium, $x(t)$. (b) Find the speed and acceleration when the spring extension is 0.1 m.

10. (II) With a block of mass m, the frequency of a block–spring system is 1.2 Hz. When 50 g is added, the frequency drops to 0.9 Hz. Find m and the spring constant.

11. (I) A block of mass 30 g oscillates with an amplitude of 12 cm at the end of a horizontal spring whose constant is 1.4 N/m. What are the velocity and acceleration when the displacement from equilibrium is (a) -4 cm; (b) 8 cm?

12. (II) Find the period for each of the spring combinations shown in Fig. 15.22. Assume each block slides on a frictionless horizontal surface.

13. (II) A particle moves at constant speed in a circle. Show that the projection of the tip of the position vector of the particle onto a diameter exhibits simple harmonic oscillation.

15.3 Energy in SHM

14. (II) The displacement of a 50-g block attached to a horizontal spring ($k = 32$ N/m) is given by $x = A \cos \omega t$, where $A = 20$ cm. Find: (a) the kinetic and potential energies at $t = 0.2T$, where T is the period; (b) the kinetic and potential energies at $x = A/2$; (c) the times at which the kinetic and potential energies are equal.

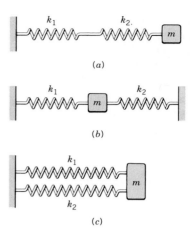

FIGURE 15.22 Exercise 12.

15. (II) The displacement of a block of mass 80 g attached to a spring whose spring constant is 60 N/m is given by $x = A \sin \omega t$, where $A = 12$ cm. In the first complete cycle find the values of x and t at which the kinetic energy is equal to one-half the potential energy.

16. (I) An atom of mass 10^{-26} kg executes simple harmonic oscillation about its equilibrium position in a crystal. The frequency is 10^{12} Hz and the amplitude is 0.05 nm. Find: (a) its total energy; (b) its maximum speed; (c) its maximum acceleration; (d) the effective spring constant.

17. (II) The displacement of a block attached to a horizontal spring whose spring constant is 12 N/m is given by $x = 0.2 \cos(4t - 0.8)$ m. Find: (a) the mass of the block; (b) the total energy; (c) the earliest time ($t > 0$) when the kinetic energy equals one-half the potential energy; (d) the acceleration at $t = 0.1$ s.

18. (I) A cart of mass m is attached to a horizontal spring and oscillates with amplitude A. When $x = A$, a block of mass $m/2$ is dropped vertically into the cart. How are the following quantities affected: (a) the amplitude; (b) the total energy; (c) the period; (d) the phase constant?

19. (II) A 50-g block is attached to a vertical spring whose stiffness constant is 4 N/m. The block is released at the position where the spring is unextended. (a) What is the maximum extension of the spring? (b) How long does it take the block to reach the lowest point?

20. (I) A 60-g block attached to a horizontal spring is held at 8 cm from its equilibrium position and released at $t = 0$. Its period is 0.9 s. Find: (a) the displacement x at 1.2 s; (b) the velocity when $x = -5$ cm; (c) the acceleration when $x = -5$ cm; (d) the total energy.

21. (I) Show that for any given value of the displacement x of a block attached to a spring, the velocity is given by

$$v = \pm\omega\sqrt{A^2 - x^2}$$

where ω is the angular frequency and A is the amplitude.

15.4 Pendulums

22. (II) A simple pendulum consists of a bob of mass 40 g and a string of length 80 cm. At $t = 0$, the angular displacement is $\theta = 0.15$ rad and the velocity is $v = 60$ cm/s. Find: (a) the angular amplitude and phase constant; (b) the total energy; (c) the maximum height above the equilibrium position.

23. (II) Determine the period of a meter stick when it is pivoted about a horizontal axis at (a) one end, and (b) at the 60-cm mark. The moment of inertia of a uniform rod of mass M and length L about an axis through the center and perpendicular to the rod is $I_{CM} = ML^2/12$. (You will need the parallel axis theorem, Eq. 11.16.)

24. (II) Determine the period of a uniform disk of mass M and radius R pivoted about a horizontal axis at the rim. The moment of inertia is $I = 3MR^2/2$.

25. (I) A wire has a torsional constant $\kappa = 2$ N·m/rad. A disk of radius $R = 5$ cm and mass $M = 100$ g is suspended at its center (see Fig. 15.23). What is the frequency of torsional oscillations? The moment of inertia of the disk is $I = \frac{1}{2}MR^2$.

FIGURE 15.23 Exercise 25.

26. (I) A rod of length $L = 50$ cm and mass $M = 100$ g is suspended at its midpoint from a wire whose torsional constant is 2.5 N·m/rad, as shown in Fig. 15.24. What is the period of torsional oscillations? The moment of inertia of the rod is $I = ML^2/12$.

FIGURE 15.24 Exercise 26.

27. (I) A simple pendulum of length 0.4 m is released when it makes an angle of 20° with the vertical. Find: (a) its period; (b) the speed at the lowest point. (c) If the mass of the bob is 50 g, what is the total energy?

28. (II) A rod suspended at its center oscillates as a torsional pendulum with a period of 0.3 s. The moment of inertia is $I = 0.5$ kg·m². The period changes to 0.4 s when an object is attached to the rod. What is the moment of inertia of the object?

29. (I) A rod suspended at its midpoint oscillates as a torsional pendulum with a period of 0.9 s. If another rod with twice the mass but half the length were used, what would be the period? Take $I = ML^2/12$.

30. (I) (a) What is the length of a simple pendulum whose period is 2.0 s? (b) If the pendulum were taken to the moon's surface, where the weight of the bob is $\frac{1}{6}$ that on earth, what would be its period?

31. (I) The 20-g bob of a simple pendulum of length 0.8 m is released at 30° to the vertical. Find: (a) the period; (b) the angular displacement as a function of time, $\theta(t)$; (c) the total energy; (d) the speed of the bob at $\theta = 15°$.

32. (I) A simple pendulum oscillates with an amplitude of 20° and a period of 2 s. How long does it take to travel directly from −10° to +10°?

PROBLEMS

1. (I) A block of mass 0.5 kg moving on a horizontal frictionless surface at 2.0 m/s collides with and sticks to a massless pan at the end of a horizontal spring whose stiffness constant is 32 N/m, as shown in Fig. 15.25. Obtain an expression for $x(t)$, the displacement from the equilibrium position as a function of time.

FIGURE 15.25 Problem 1.

2. (I) A coin rests on the top of a piston that executes simple harmonic motion vertically with an amplitude of 10 cm. At what minimum frequency does the coin lose contact with the piston?

3. (II) A block of mass m is attached to a vertical spring via a string that hangs over a pulley ($I = \frac{1}{2}MR^2$) of mass M and radius R, as shown in Fig. 15.26. The string does not slip. Show that the angular frequency of oscillations is given by $\omega^2 = 2k/(M + 2m)$. (*Hint:* Use the fact that the total energy is constant in time. See Example 15.6.)

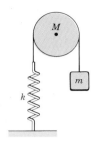

FIGURE 15.26 Problem 3.

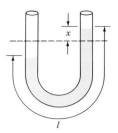

FIGURE 15.29 Problem 6.

4. (II) A block of mass $m = 1$ kg is placed on top of another block of mass $M = 5$ kg that is attached to a horizontal spring ($k = 20$ N/m), as shown in Fig. 15.27. The coefficient of static friction between the blocks is μ, whereas the lower block slides on a frictionless horizontal surface. The amplitude of oscillation is $A = 0.4$ m. What is the minimum value of μ such that the upper block does not slip relative to the lower block?

FIGURE 15.27 Problem 4.

5. (I) A small particle slides inside a frictionless spherical bowl of radius R, as shown in Fig. 15.28. (a) Show that the motion is simple harmonic for small displacements from the lowest point. (b) What is the period?

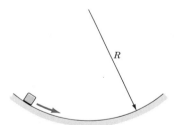

FIGURE 15.28 Problem 5.

6. (II) Water fills a length ℓ of a U tube, as in Fig. 15.29. The water is slightly displaced and then allowed to move freely. (a) Show that the liquid executes simple harmonic motion. (b) What is the period?

7. (II) Show that the angular frequency ω_{max} at which the amplitude of a driven and damped harmonic oscillator reaches a maximum is given by

$$\omega_{max} = \sqrt{\omega_0^2 - \frac{\gamma^2}{2m^2}}$$

(*Hint:* Take the derivative of Eq. 15.27.)

8. (II) A block of density ρ_B has a horizontal cross-sectional area A and a vertical height h. It floats in a fluid of density ρ_f. The block is pushed down and released. Show that it executes SHM with angular frequency

$$\omega = \sqrt{\frac{\rho_f g}{\rho_B h}}$$

9. (II) Figure 15.30 shows a block of mass M on a frictionless surface attached to a horizontal spring of mass m. (a) Show that when the speed of the block is v, the kinetic energy of the spring is $\frac{1}{6}mv^2$. (b) What is the period of oscillation? (*Hint:* First consider the kinetic energy of an element of length dx. Assume that its speed is proportional to the distance from the fixed end. All parts of the spring are in phase. For part (b) use the fact that the energy is constant.)

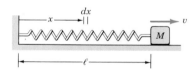

FIGURE 15.30 Problem 9.

10. (I) (a) What are the dimensions of the torsional constant κ in the equation $\tau = -\kappa\theta$? (b) Start with the assumption that the period of a torsional pendulum is a function only of the moment of inertia I and κ. Express the period in the form $T = I^x \kappa^y$ and use dimensional analysis to determine x and y. (See Example 1.5.)

11. (I) When the angular amplitude θ_0 of a simple pendulum or a physical pendulum is not small, the first few terms in the formula for the period are

$$T = T_0 \left(1 + \frac{1}{4} \sin^2 \frac{\theta_0}{2} + \frac{9}{64} \sin^4 \frac{\theta_0}{2} + \cdots \right)$$

where T_0 is the period for the simple harmonic motion. Assume that $T_0 = 1$ s. Use the above equation to calculate the period for the following values of θ_0: (a) 15°; (b) 30°; (c) 45°; and (d) 60°.

12. (I) (a) Write an expression for the mechanical energy E of a system in which a block is attached to a vertical spring (as in Fig. 15.7). Choose the position at which the extension is zero as the zero for the gravitational and spring potential energies U_g and U_{sp}. (b) Use the condition $dE/dt = 0$ to show that oscillations of the system are simple harmonic.

13. (II) Figure 15.31 shows a tunnel in a uniform planet of mass M and radius R. At a distance r from the center, the gravitational attraction is due only to the sphere of radius r (see Example 13.5). Thus,

$$F = \frac{GmM(r)}{r^2} = \frac{mgr}{R}$$

where $M(r) = Mr^3/R^3$ and $g = GM/R^2$. Show that Newton's

second law for the motion along the tunnel leads to the differential equation for simple harmonic motion:

$$\frac{d^2x}{dt^2} + \frac{g}{R}x = 0$$

Estimate the period of the oscillation for the earth.

14. (I) A uniform rod of mass M and length L is pivoted about a vertical axis at one end and attached to a horizontal spring whose constant is k (see Fig. 15.32). Show that for small angular displacements from the equilibrium position (indicated by the dashed line) the oscillations are simple harmonic. What is the period? The moment of inertia of the rod is $I = ML^2/3$.

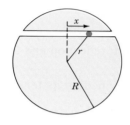

FIGURE 15.31 Problem 13.

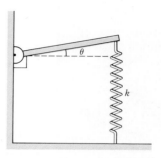

FIGURE 15.32 Problem 14.

CHAPTER 16

Mechanical Waves

World champion figure skater Katarina Witt makes a wave in her skirt.

Major Points

1. **Wave characteristics;** linear superposition; reflection and transmission of pulses.
2. The **speed** of a pulse on a string is determined by its tension and its mass density.
3. (a) **Traveling harmonic waves.**
   (b) The distinction between particle velocity and wave velocity.
4. (a) **Standing waves** are produced by the superposition of two harmonic waves of equal amplitude and period traveling in opposite directions.
   (b) **Resonant standing waves** in strings.
5. **Energy transport** by a wave on a string.
6. The **wave equation** is a differential equation satisfied by traveling waves.

In previous chapters we studied the motion of particles, rigid bodies, and fluids. We now begin the study of wave motion. A **wave** is a disturbance that travels, or *propagates,* without the transport of matter. For example, when a gust of wind blows a wave across a field of corn, the stalks are not carried with the wave; they bend momentarily and then swing back to their original positions. The ripples on a pond (Fig. 16.1), the sounds we hear, visible light, and radio signals are just a few examples of waves. Sound, light, and radio waves provide us with an effective means of transmitting and receiving energy and information. We learn about the structure of molecules by the light they emit or absorb; the passage of earthquake waves tells us about the core of the earth; the analysis of light from a star yields information on its motion and its chemical composition. Among other applications, X rays are used in diagnosis and treatment, high-frequency (ultrasonic) sound waves are used to monitor fetal development and in burglar alarm systems, and microwaves are used in telecommunications and in cooking.

Mechanical waves, such as water waves or sound waves, travel within, or on the surface of, a material with elastic properties: There must be some mechanism that tends to restore the medium to its normal or equilibrium state. In contrast, **electromagnetic waves,** such as light and TV signals, are *nonmechanical,* and can propagate through a vacuum. Electromagnetic waves are discussed in Chapter 34. In this century it has been discovered that elementary particles, such as the electron and proton, can also display wavelike behavior. These **matter waves** are

FIGURE 16.1 Ripples produced by small spheres dropped into water.

discussed in Chapter 41. In this chapter we confine our attention to waves on strings, and in the next chapter we consider sound waves.

16.1   WAVE CHARACTERISTICS

Figure 16.2a shows a string fixed at one end. If the free end is displaced momentarily, a disturbance will propagate along the string. This momentary disturbance from the equilibrium state is called a *wave pulse*. After the pulse passes, each segment of the string returns to its equilibrium position. When a piston in a tube is suddenly pushed to the right, as in Fig. 16.2b, the air in front of it is compressed, causing a local increase in density and pressure above the normal value. The collisions between the air molecules transmit this *compression* along the tube. If the piston is suddenly pulled to the left, a *rarefaction,* a region of lower than normal density and pressure, propagates along the tube.

In a *transverse* wave, shown in Fig. 16.2a, the displacement of the particles is perpendicular to the direction of travel of the wave. In a *longitudinal* wave, shown in Fig. 16.2b, the displacement of the particles is along the direction of the wave velocity. A solid can sustain both kinds of wave, as may be easily demonstrated with a Slinky (see Fig. 16.3). A fluid has no well-defined form or structure to maintain and offers far more resistance to compression than to a shearing force (see Fig. 14.3). Consequently, only longitudinal waves can propagate through a gas or within the body of an ideal (nonviscous) liquid. However, transverse waves can exist on the surface of a liquid. In the case of ripples on a pond, the force restoring the system to equilibrium is the surface tension of the water, whereas for ocean waves, it is the force of gravity. In an ocean wave, the water molecules actually move in circular or elliptical paths (Fig. 16.4) which involve both transverse and longitudinal displacements.

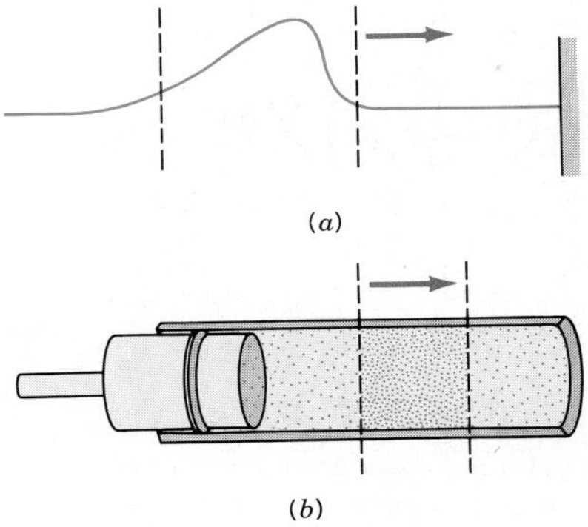

Motion of the Medium

If you watch a twig or a leaf on a pond as a ripple goes by, a surprising thing happens. The object does not travel with the wave; it merely bobs up and down in the same spot. Figure 16.5a shows how a small segment of a string moves as a transverse pulse passes. As the leading edge of the pulse reaches it, the segment moves perpendicular to the string's equilibrium position. Its displacement reaches a maximum as the peak of the pulse passes and then returns to its equilibrium position after the pulse has moved on. As indicated in Fig. 16.5b, the particles on

(a)

(b)

FIGURE 16.2 (a) A transverse pulse on a string. (b) A longitudinal pulse in a column of air.

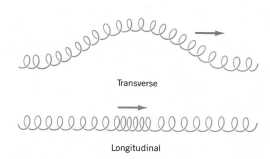

Transverse

Longitudinal

FIGURE 16.3 Both transverse and longitudinal pulses are propagated in a Slinky.

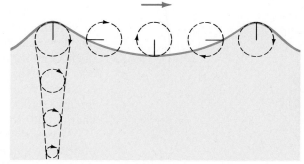

FIGURE 16.4 In a large ocean wave, the water particles move in circular (or elliptical) paths as the wave propagates.

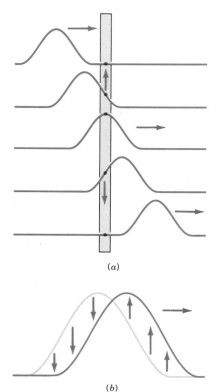

(a)

(b)

FIGURE 16.5 (a) As a transverse pulse passes, a given element of a string undergoes only transverse displacements. The medium (the rope) is not carried along with the wave. (b) On the leading edge of a pulse the particles are moving upward, whereas on the trailing edge the particles are moving downward.

the leading edge are moving upward while those on the trailing edge are moving downward.

We see that the particles of the medium are not carried along with the wave: They undergo small displacements about an equilibrium position, whereas the wave itself can travel a great distance. As we will see later, sound waves involve tiny longitudinal oscillations of air molecules. When the molecules in a given region move through a large distance, this creates a wind, not a sound wave. Hence, a wave could be defined as follows:

A wave is any disturbance from a normal or equilibrium condition that propagates without the transport of matter.

The disturbance created by a wave is represented by a **wave function.** For a string, the wave function is a (vector) displacement; whereas for sound waves it is a (scalar) pressure or density fluctuation. In the case of light or radio waves, the wave function is either an electric or a magnetic field vector.

Since the particles of the medium do not travel with the wave, what does? Consider again the leaf on the pond as a ripple passes by. As the leaf bobs up and down both its kinetic energy and gravitational potential energy change. On a larger scale, an ocean wave can lift a large ship or wreak havoc on a coastal town. Our eyes respond to light waves and our ears to sound waves because of the energy they supply. Thus, in general: *A wave transports both energy and momentum.* We will discuss momentum transport only for the case of electromagnetic waves (Chapter 34).

16.2   SUPERPOSITION OF WAVES

When two or more waves overlap in a given region, they are said to be *superposed*. The resultant wave function is given by the **principle of linear superposition:** The total wave function y_T at any point is the linear sum of the individual wave functions y; that is,

$$y_T = y_1 + y_2 + y_3 + \cdots + y_N$$

Depending on the nature of the wave, this may be an algebraic sum or a vector sum. Figure 16.6 shows what happens when two pulses are sent along a rope toward each other.

The superposition of two or more waves in a given region may give rise to **interference.** When the wave functions have the same sign, as in Fig. 16.6a, the interference is *constructive* and the resultant displacement is larger than that of either wave. When they have opposite signs, as in Fig. 16.6b, the interference is *destructive* and the resultant displacement is smaller than that of either wave. One might expect that when a crest (positive displacement) and a trough (negative displacement) of identical shape overlap they would cancel and that would be the end of it. However, as Fig. 16.6b shows, after they have been superposed, they continue on unchanged in shape. Although waves can interfere, they do not *interact* with each other, as particles would. Each pulse has a certain amount of energy and it would be unreasonable for this to vanish.

Waves that have a scalar wave function (sound waves are an example) always interfere when they are superposed. In the case of waves with vector wave functions, such as those on a string, only the component of one wave function along the direction of the other can give rise to interference. For example, if a string lies

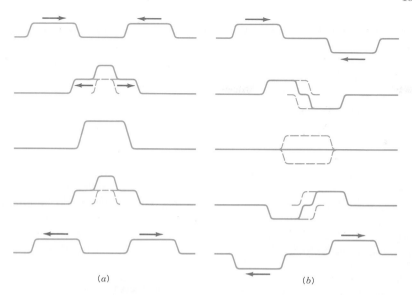

FIGURE 16.6 When two pulses (dashed) are superposed, the resultant displacement is the sum of the individual displacements.

along the x axis, then a pulse with displacements along the y axis would not interfere with a pulse with displacements along the z axis.

Linear superposition is valid for mechanical waves only if the amplitude of the oscillation does not exceed the elastic (Hooke's law) limit of the medium. Linear superposition does not apply to the huge ocean breakers used by surfers or to the shock waves produced by supersonic aircraft. In all our examples we assume that linear superposition is valid.

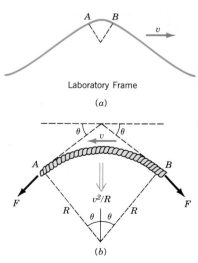

Surfing in Hawaii.

16.3   SPEED OF A PULSE ON A STRING

The speed at which a pulse propagates depends on the properties of the medium. A simple derivation of the wave velocity of a pulse on a string was devised by P. G. Tait in 1883. Our idealized string is uniform and perfectly flexible. We also assume that the pulse height is so small that the tension in the string is not changed by it. In our stationary *laboratory* reference frame (Fig. 16.7a), the pulse moves to the right at speed v. A derivation in this frame is given in Section 16.11. For the moment it is easier to use the frame that moves with the pulse. In this *pulse* frame, Fig. 16.7b, the pulse is stationary while the string moves to the left at speed v.

Any small segment, such as AB, may be treated as a circular arc of some radius R. If θ is the angle between the vertical and the radial line to B, then the length of AB is $R(2\theta)$. If μ is the linear mass density (in kg/m), then the mass of segment AB is $m = 2\mu R\theta$. The tension F in the string must provide the centripetal force needed for the circular motion. The angles between the tangents at A and B and the horizontal are also equal to θ. (Why?) The horizontal components of the forces cancel, and therefore the net force acting on the segment is $2F \sin \theta$, directed vertically down. From Newton's second law we have

$$2F \sin \theta = \frac{mv^2}{R}$$

FIGURE 16.7 (a) In the laboratory frame a pulse on a rope moves toward the right. (b) In a frame moving with the pulse, the rope moves toward the left. The centripetal force is provided by the tension in the rope.

Since the pulse height is small, we use the small-angle approximation, $\sin \theta \approx \theta$, and $m = 2\mu R\theta$ to find

$$2F\theta = 2\mu R\theta \frac{v^2}{R}$$

which leads to

Speed of a pulse on a string

$$v = \sqrt{\frac{F}{\mu}} \qquad (16.1)$$

Note that the velocity is measured with respect to the medium.

Equation 16.1 is valid for a pulse of any shape, provided it has small amplitude. This form for the wave velocity of a mechanical wave will be encountered again. The tension F tells us how strongly the string tries to restore itself to its equilibrium position. The mass density is a measure of the string's inertia. In general, the wave velocity of a mechanical wave propagating through a medium has the form

$$v = \sqrt{\frac{\text{Restoring force factor}}{\text{Inertia factor}}} \qquad (16.2)$$

We will see later that in the case of sound waves the restoring force factor is an elastic constant and the inertia factor is the mass density.

EXAMPLE 16.1: One end of a string is fixed. It hangs over a pulley and has a block of mass 2.00 kg attached to the other end, as in Fig. 16.8. The horizontal part has a length of 1.60 m and a mass of 20.0 g. What is the speed of a transverse pulse on the string?

Solution: The tension is simply the weight of the block; that is, $F = 19.6$ N. The linear mass density is $(2.00 \times 10^{-2}$ kg$)/(1.60$ m$) = 1.25 \times 10^{-2}$ kg/m. From Eq. 16.1, the wave speed is

$$v = \sqrt{\frac{F}{\mu}} = \sqrt{\frac{19.6 \text{ N}}{1.25 \times 10^{-2} \text{ kg/m}}} = 39.6 \text{ m/s}$$

EXERCISE 1. Assume that the speed of a pulse on a string is a function only of the tension and the linear mass density. Set $v =$

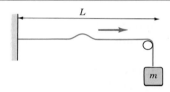

FIGURE 16.8 The tension in a string is produced by a hanging weight.

$F^x \mu^y$ and use dimensional analysis to determine x and y. (See Example 1.5.)

EXERCISE 2. Two strings are made of the same material. String 1 has twice the diameter of string 2, but is under half the tension. Find v_2/v_1.

16.4   REFLECTION AND TRANSMISSION

When a pulse traveling along a string reaches the end, it is reflected. If the end is *fixed*, as in Fig. 16.9a, the pulse returns inverted. This is because as the leading edge reaches the wall, the string pulls up on the connection. According to Newton's third law, the wall pulls down with equal force. The string, being much lighter, moves downward to form the inverted pulse. One can study the motion of a *free* end by letting a ring at the end of the string slide on a rod, as in Fig. 16.9b. In this case, the string has no vertical constraint; consequently, the reflected pulse is not inverted.

The details of the reflection process may be reconstructed by superposing the actual pulse and an imaginary pulse that approaches from the right, as in Fig. 16.10. The net displacement at any point is given by the principle of superposition. Notice that at one instant, the displacement of the free end is double the pulse height.

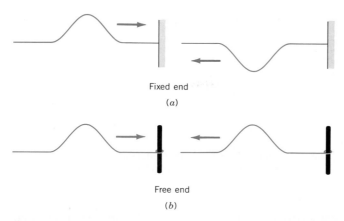

FIGURE 16.9 (a) When a pulse on a rope is reflected at a fixed end, it returns inverted. (b) At a free end, the reflected pulse is not inverted.

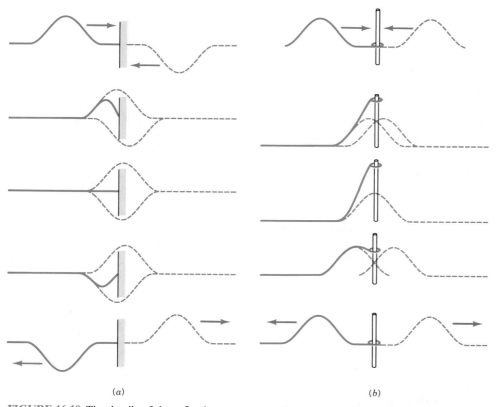

(a)   (b)

FIGURE 16.10 The details of the reflection process may be constructed by superposing an imaginary pulse and the real pulse.

Between the extreme cases of a fixed end or a free end, a pulse may encounter the boundary between a light string and a heavy string. This results in partial reflection and partial transmission. Since the tensions are the same, the relative magnitudes of the wave velocities are determined by the mass densities. In Fig. 16.11a, the pulse approaches from the light string. The heavy string behaves somewhat like a wall but it can move, and so part of the original pulse is transmit-

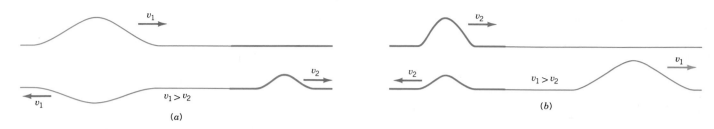

FIGURE 16.11 When a pulse encounters the boundary between two different ropes, it is partly reflected and partly transmitted. (*a*) If the second rope is heavier, the reflected pulse is inverted. (*b*) If the second rope is lighter, the reflected pulse is not inverted.

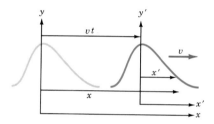

FIGURE 16.12 A pulse traveling at speed v relative to the xy frame. In the $x'y'$ frame of the pulse, it is at rest and the shape is described by $f(x')$. In the xy frame, the pulse is described by $f(x - vt)$.

ted to the heavy string. In Fig. 16.11*b*, the pulse approaches from the heavy string. The light string offers little resistance and now approximates a free end. Consequently, the reflected pulse is not inverted.

EXERCISE 3. What can you say about the relative lengths of the reflected and transmitted pulses in Fig. 16.11?

16.5 TRAVELING WAVES

How can one mathematically express the fact that a pulse is moving? Let us look at the pulse from two different reference frames. In Fig. 16.12, (x, y) are the coordinates of a point in our stationary frame, whereas (x', y') refer to a frame moving with the pulse. We assume the origins coincided at $t = 0$. In the moving frame the pulse is at rest, so at any time the vertical displacement y' at position x' is given by some function $f(x')$ that describes the shape of the pulse:

$$y'(x') = f(x')$$

In the stationary frame, the pulse has the same shape but is moving at a velocity v, which means that the displacement y is a function of both x and t, so we write $y(x, t)$. The coordinates of any given feature of the pulse as measured in the two frames are related by the Galilean transformation (Section 4.8): $x' = x - vt$ and $y' = y$. Using $f(x') = f(x - vt)$ and $y' = y$, the above equation becomes

$$y(x, t) = f(x - vt) \tag{16.3}$$

Equation 16.3 describes a pulse moving in the $+x$ direction. Any given phase (feature) of the pulse, for example, its peak, has a fixed value of x', which means

$$x - vt = \text{constant}$$

The quantity $x - vt$ is called the phase of the wave function. Taking the derivative of this with respect to time we find

$$\frac{dx}{dt} = v$$

where v is the phase velocity, although we often call it the **wave velocity.** It is the velocity at which a particular phase of the disturbance travels through space. A pulse moving in the $-x$ direction is represented by

$$y(x, t) = f(x + vt) \tag{16.4}$$

In order for the function to represent a wave traveling at speed v, the three quantities x, v, and t *must* appear in the combinations $(x + vt)$ or $(x - vt)$. Thus, $(x - vt)^2$ is acceptable, but $(x^2 - v^2t^2)$ is not.

EXAMPLE 16.2: At $t = 0$ a pulse is described by

$$y(x) = \frac{A}{(B + x^2)}$$

What is the function that describes it at an arbitrary time given that it moves in the $+x$ direction at 3 m/s? Sketch the pulse at $t = 0$, 1 s, and 2 s.

Solution: From Eq. 16.3 we have

$$y(x, t) = \frac{A}{B + (x - 3t)^2}$$

At $t = 0$, the displacement is $y(x, 0) = A/(B + x^2)$. At $t = 1$ s, we have $y(x, 1) = A/[B + (x - 3)^2]$; at $t = 2$ s, the displacement is

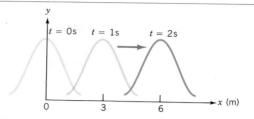

FIGURE 16.13 The positions of a pulse at three different times.

$y(x, 2) = A/[B + (x - 6)^2]$. The pulse is sketched in Fig. 16.13 at these three times.

16.6   TRAVELING HARMONIC WAVES

Thus far we have discussed various properties of single pulses. We now go on to discuss continuous waves. A simple way to produce continuous waves is to attach a string to a vibrating rod as in Fig. 16.14a. If the source of the waves (the rod) is a simple harmonic oscillator, the function $f(x \pm vt)$ is sinusoidal and it represents a **traveling harmonic wave.** When such a wave passes through a given region, the particles of the medium (the string) execute simple harmonic motion. The study of harmonic waves is particularly important because a disturbance of any shape may be formed by adding together suitable harmonic components of different frequencies and amplitudes (see Section 17.7).

Figure 16.14b shows the displacement $y(t)$ of a particle as a function of time as a wave passes by. The period T corresponds to a change of 2π radians in the phase. Thus, an arbitrary time t corresponds to a phase change of $2\pi(t/T)$. If $y = 0$ and $dy/dt > 0$ at $t = 0$, the phase constant is zero, and we may write

(Fixed x)
$$y(t) = A \sin \omega t$$

where

$$\omega = \frac{2\pi}{T}$$

is the **angular frequency** measured in rad/s.

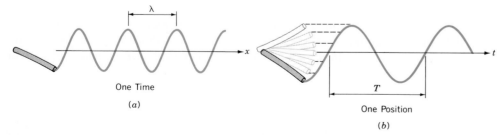

One Time

(a)

T

One Position

(b)

FIGURE 16.14 (a) A vibrating rod generates a wave in a string. The curve shows the transverse displacement of the string at one time. (b) The end of a string is attached to a rod vibrating in simple harmonic motion. The curve shows the transverse displacement at one position as a function of time.

A snapshot of a harmonic wave at a given instant (Fig. 16.14a) shows that the variation in the displacement as a function of position, $y(x)$, is also sinusoidal. The distance between two successive points with the same phase (for example, two crests) is called the **wavelength,** λ. Since λ corresponds to a change of 2π in the phase, the phase at any position x is $(2\pi)(x/\lambda)$. If $y = 0$ and $dy/dx > 0$ at $x = 0$, we may write

(Fixed t)
$$y(x) = A \sin kx$$

where

$$k = \frac{2\pi}{\lambda}$$

is the **wave number,** whose SI unit is rad/m. (Do not confuse this k with the spring constant.) The above discussion shows that a harmonic wave exhibits two periodicities: one in time, the other in space.

Let us focus our attention on a particular feature—for example, a crest—of a traveling harmonic wave (Fig. 16.15). In one period T the wave advances by one wavelength λ. Thus, the wave velocity is $v = \lambda/T$. Or, since the frequency is $f = 1/T = \omega/2\pi$, we have for the speed of a harmonic wave

$$v = f\lambda = \frac{\omega}{k} \tag{16.5}$$

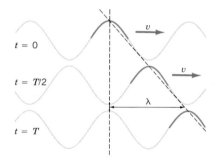

FIGURE 16.15 In one period a wave travel a distance of one wavelength; thus, its speed is $v = \lambda/T = f\lambda$.

Note that the frequency of a wave is determined by the *source,* whereas the wave velocity is determined by the properties of the *medium.* In the case of mechanical waves, the wave velocity is measured with respect to the medium. The wave velocity also depends on the mode of propagation. For example, transverse and longitudinal acoustic waves within a solid have different velocities.

The wave function that represents a traveling harmonic wave is obtained by combining the equation for a harmonic wave, $y(x) = A \sin kx$, and Eq. 16.3, $y(x, t) = f(x - vt)$, for a traveling wave:

$$y(x, t) = A \sin[k(x - vt)]$$
$$= A \sin(kx - \omega t) \tag{16.6}$$

This represents a harmonic wave moving in the $+x$ direction. A wave moving in the $-x$ direction is represented by

$$y(x, t) = A \sin(kx + \omega t) \tag{16.7}$$

In both of these equations $y = 0$ at $x = 0$ and $t = 0$. In general, this will not be the case, so we must incorporate a phase constant, ϕ:

$$y(x, t) = A \sin(kx - \omega t + \phi) \tag{16.8}$$

It is important to distinguish between the wave velocity v and the velocity of a particle of the medium $\partial y/\partial t$. Figure 16.16 illustrates the difference for a transverse wave. From Eq. 16.8 the velocity and acceleration of a particle at a given position are given by

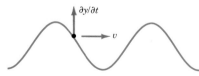

FIGURE 16.16 The velocity of a particle on a string, given by $\partial y/\partial t$, is perpendicular to the wave velocity v.

$$\frac{\partial y}{\partial t} = -\omega A \cos(kx - \omega t + \phi) \tag{16.9}$$

$$\frac{\partial^2 y}{\partial t^2} = -\omega^2 A \sin(kx - \omega t + \phi) \tag{16.10}$$

The partial derivative is needed because $y(x, t)$ is a function of two variables and we are keeping x fixed. The maximum speed of the particle is ωA, and its maximum acceleration is $\omega^2 A$.

EXAMPLE 16.3: The equation of a wave is

$$y(x, t) = 0.05 \sin\left[\frac{\pi}{2}(10x - 40t) - \frac{\pi}{4}\right] \text{ m}$$

Find: (a) the wavelength, the frequency, and the wave velocity; (b) the particle velocity and acceleration at $x = 0.5$ m and $t = 0.05$ s.

Solution: (a) The equation may be rewritten as

$$y(x, t) = 0.05 \sin\left(5\pi x - 20\pi t - \frac{\pi}{4}\right) \text{ m}$$

Comparing this with Eq. 16.8 we see that the wave number is $k = 2\pi/\lambda = 5\pi$ rad/m and so $\lambda = 0.4$ m. The angular frequency is $\omega = 2\pi f = 20\pi$ rad/s; thus, $f = 10$ Hz. The wave velocity is $v = f\lambda = \omega/k = 4$ m/s in the $+x$ direction.

(b) The particle velocity and acceleration are

$$\frac{\partial y}{\partial t} = -(20\pi)(0.05)\cos\left(\frac{5\pi}{2} - \pi - \frac{\pi}{4}\right)$$
$$= 2.22 \text{ m/s}$$
$$\frac{\partial^2 y}{\partial t^2} = -(20\pi)^2(0.05)\sin\left(\frac{5\pi}{2} - \pi - \frac{\pi}{4}\right)$$
$$= 140 \text{ m/s}^2$$

EXERCISE 4. Express Eq. 16.8 in terms of λ and T.

EXERCISE 5. Consider the wave function

$$y(x, t) = 0.02 \sin\left(\frac{x}{0.05} + \frac{t}{0.01}\right) \text{ m}$$

Find: (a) the wave velocity, and (b) the particle velocity at $x = 0.2$ m and $t = 0.3$ s.

16.7   STANDING WAVES

We now consider what happens when two harmonic waves of equal frequency and amplitude travel through a medium in opposite directions. The two component waves, $y_1 = A \sin(kx - \omega t)$ and $y_2 = A \sin(kx + \omega t)$, are drawn as orange and blue lines in Fig. 16.17. Their sum is

$$y(x, t) = A \sin(kx - \omega t) + A \sin(kx + \omega t)$$

By using the identity $\sin A + \sin B = 2 \sin[(A + B)/2] \cos[(A - B)/2]$, we obtain

$$y(x, t) = 2A \cos(\omega t) \sin(kx) \qquad (16.11)$$

Equation 16.11 represents a stationary harmonic wave, $y_T = A(t) \sin kx$, whose amplitude, $A(t) = 2A \cos \omega t$, varies in time. Such a **standing wave,** drawn as a green line in Fig. 16.17, does not travel. At the **nodes** the medium is permanently at rest, whereas at the **antinodes** the amplitude is double that of either wave. With the exception of the nodes, each point on the string executes simple harmonic motion with an amplitude that varies along the string. The nodes occur at points where $\sin kx = 0$, that is, where $kx = 0, \pi, 2\pi$, etc. The antinodes occur at points where $\sin kx = \pm 1$, that is, where $kx = \pi/2, 3\pi/2, 5\pi/2$, etc. In either case, the separation between nodes, or between antinodes, is $\lambda/2$.

16.8   RESONANT STANDING WAVES ON A STRING

In an unbounded continuous medium, there is no restriction on the frequencies or wavelengths of the standing waves. However, if the waves are confined in space—for example, when a string is tied at both ends—standing waves can be set up only for a discrete set of frequencies or wavelengths.

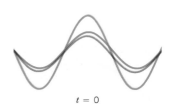
$t = 0$

$t = T/8$

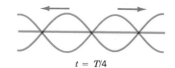
$t = T/4$

$t = 3T/8$

$t = T/2$

$t = 5T/8$

FIGURE 16.17 Two waves (orange and blue curves) of equal amplitude and frequency traveling in opposite directions produce a standing wave (green curve). The points of permanent zero displacement are called *nodes;* the points of maximum displacement are called *antinodes,*

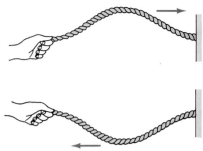

FIGURE 16.18 (*a*) A hand generates a crest in a rope fixed to a wall. (*b*) The inverted pulse, a trough, returns and will be reflected as a crest. If the hand starts to generate a second crest just as the trough reaches it, the reflected crest and the second crest reinforce each other.

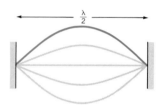

FIGURE 16.19 The fundamental mode of a string fixed at both ends. The distance between the ends is one half-wavelength.

Figure 16.18 shows a string fixed at one end to a wall, while the other end is held in a hand. Suppose that the hand generates a crest, as in Fig. 16.18*a*. At the fixed end the crest is inverted and returns as a trough, as shown in Fig. 16.18*b*. The returning trough reflects at the hand as a crest. If the hand starts to generate a second crest just as the leading edge of the trough arrives at the hand, the second crest will reinforce the pulse undergoing reflection. Now suppose the hand vibrates. If the time taken by a pulse for a round trip is an integer multiple of the period of the hand, the system will be in resonance. Continuous waves traveling in opposite directions along the string then set up a standing wave. The difference between this standing wave and the general discussion of the previous section is that the string has a finite length and its ends are fixed. (The amplitude of the hand is much less than that of the standing waves, so it is essentially a fixed point.) These boundary conditions impose constraints on the frequencies or wavelengths. They are called **resonant standing waves.**

If the string is perfectly flexible and the shape of the pulse is sinusoidal, the first resonance condition occurs when the distance between the fixed ends is one half-wavelength, as shown in Fig. 16.19. Using $L = \lambda/2$ and $v = f\lambda$, the **fundamental** frequency, or **first harmonic,** is $f_1 = v/2L$. This is the lowest frequency at which the string can vibrate. One full wavelength between the ends, as in Fig. 16.20, corresponds to the second harmonic with $\lambda = L$ and frequency $f_2 = v/L = 2f_1$. Figure 16.20 also shows how the displacements and velocities of different segments vary during one half-period. A resonant standing wave can exist only when the length is an integral number of half-wavelengths. The wavelength and frequency of the *n*th harmonic are given by

$$\lambda_n = \frac{2L}{n} \tag{16.12}$$

$$f_n = \frac{nv}{2L} \tag{16.13}$$

Figure 16.21 illustrates the first three harmonics of a rubber tube fixed at both ends. Each resonant standing wave pattern is called a **normal mode** of oscillation. When a string is plucked or bowed, we hear the fundamental and varying amounts of the higher harmonics. The number and relative strengths of the higher harmonics determine the tonal quality of a musical note. The different harmonic structures allow us to distinguish between two instruments playing the same fundamental note. Curiously, it is the "impurities" in a note that actually make the sound pleasing.

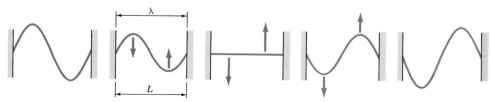

FIGURE 16.20 The second harmonic of a string fixed at both ends. The distance between the ends is one wavelength. This mode has double the frequency of the fundamental.

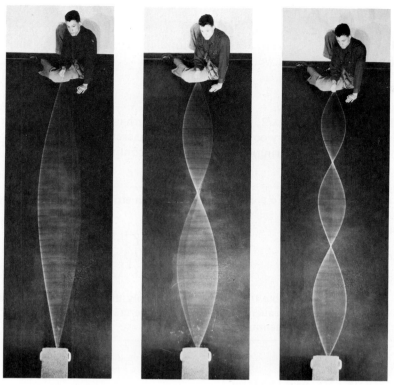

FIGURE 16.21 The first three modes of a rubber tube fixed at both ends.

A simple way to set up resonant standing waves is shown in Fig. 16.22. One prong of a tuning fork is attached to one end of the string. The string hangs over a pulley and a weight determines the tension in it. From Eq. 16.1, $v = \sqrt{F/\mu}$, so Eq. 16.13 for the resonant (normal mode) frequencies becomes

$$f_n = \frac{n}{2L} \sqrt{\frac{F}{\mu}} \qquad (16.14)$$

Notice that the frequency depends on the tension—a feature used in the tuning of stringed instruments.

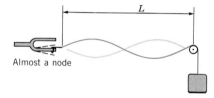

FIGURE 16.22 A string attached to a prong of a tuning fork may be set into resonance when the tension and length are appropriately chosen (to satisfy Eq. 16.14).

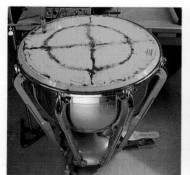

Some normal modes of vibrations of a drumhead. The black powder collects along lines of zero displacement.

Equation 16.11 leads to the discrete number of frequencies of Eq. 16.13 when the *boundary conditions* are applied. In the example of the string fixed at both ends, the displacement is zero at the ends for all modes; that is, $y = 0$ at $x = 0$ and $x = L$ for all times. Thus, we must have

$$\sin kL = 0$$

This can be true only if $kL = n\pi$ or equivalently $\lambda = 2L/n$, which is simply Eq. 16.12. Expressed as $L = n\lambda/2$, we see that the length is an integral number of half-wavelengths. The boundary conditions lead to a discrete set of allowed vibrational modes in a system (the string) that is otherwise continuous.

EXAMPLE 16.4: In the equal-tempered musical scale the ratio of the frequencies of the notes A and D is $f_A/f_D = 3/2$. Determine the ratio of the tensions in two piano strings, F_A/F_D, given that the ratio of the lengths is $L_A/L_D = 4/5$. The strings are made of the same wire and vibrate in their fundamental modes.

Solution: From Eq. 16.14 we obtain an expression for the tension

$$F = \left(\frac{2Lf}{n}\right)^2 \mu$$

We are given $n = 1$ for both strings and $\mu_A = \mu_D$. Therefore,

$$\frac{F_A}{F_D} = \left(\frac{L_A}{L_D}\right)^2 \left(\frac{f_A}{f_D}\right)^2$$
$$= \left(\frac{4}{5}\right)^2 \left(\frac{3}{2}\right)^2 = 1.44$$

Note that we were given only the *ratio* of the lengths and the frequencies. It would not be correct to assign values to these quantities.

16.9 THE WAVE EQUATION

Traveling waves satisfy a differential equation called the **linear wave equation.** In the case of mechanical waves, the wave equation arises when Newton's second law is applied to the motion of an element of the medium through which a wave propagates. One can obtain the form of this equation by taking the partial derivatives of the wave function for a traveling harmonic wave (Eq. 16.8)

$$y = y_0 \sin(kx - \omega t + \phi)$$

with respect to t and x (check these):

$$\frac{\partial^2 y}{\partial x^2} = -k^2 y_0 \sin(kx - \omega t + \phi)$$

$$\frac{\partial^2 y}{\partial t^2} = -\omega^2 y_0 \sin(kx - \omega t + \phi)$$

By comparing these derivatives, we see that

The wave equation

$$\frac{\partial^2 y}{\partial x^2} = \frac{1}{v^2}\frac{\partial^2 y}{\partial t^2} \tag{16.15}$$

where $v = \omega/k$ is the wave velocity. Although we will not prove it here, this **linear wave equation** is satisfied by any traveling wave of the form $y = f(x \pm vt)$. Equation 16.15 is a *linear* differential equation since the derivatives are raised only to the first power; that is, there are no terms such as $(dy/dx)^2$. This means that if y_1 and y_2 are separate solutions, then any linear combination, such as $ay_1 + by_2$, where a and b are constants, is also a solution. When Eq. 16.15 is satisfied, the principle of linear superposition is valid.

16.10 ENERGY TRANSPORT ON A STRING

As a wave propagates along a string, it transports energy. Figure 16.23 shows an element of a string under the influence of a wave. If μ is the linear mass density, the kinetic energy of the element of mass $\mu\,dx$ is

$$dK = \tfrac{1}{2}(\mu dx)\left(\frac{\partial y}{\partial t}\right)^2$$

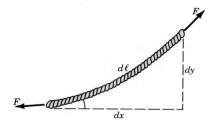

FIGURE 16.23 An element of a string under the action of a wave.

The potential energy of the element is equal to the work done to stretch it from dx to $d\ell$. Assuming the tension stays constant, $dU = F(d\ell - dx)$. We assume that the disturbance is small. This means that the slope, $\partial y/\partial x$, is small. By the Pythagorean theorem,

$$d\ell = \sqrt{dx^2 + dy^2} = dx\,\sqrt{1 + \left(\frac{\partial y}{\partial x}\right)^2}$$

$$\approx dx\left[1 + \frac{1}{2}\left(\frac{\partial y}{\partial x}\right)^2 + \cdots\right]$$

where, in the second line, we have used the binomial expansion $(1 + z)^n \approx 1 + nz$ for $z \ll 1$ (see Appendix   ). Thus,

$$dU \approx \tfrac{1}{2}Fdx\left(\frac{\partial y}{\partial x}\right)^2$$

The potential energy of the stretched element is related to its *slope* and not directly to its displacement.

The total mechanical energy of the element, $dE = dK + dU$, then becomes

$$dE = \frac{1}{2}\left[\mu\left(\frac{\partial y}{\partial t}\right)^2 + F\left(\frac{\partial y}{\partial x}\right)^2\right]dx$$

For a harmonic wave, $y(x, t) = y_0 \sin(kx - \omega t)$, so

$$dE = \tfrac{1}{2}[\mu(\omega y_0)^2 + F(ky_0)^2]\cos^2(kx - \omega t)dx$$

Since $\omega = vk$, and for a string $v = \sqrt{F/\mu}$, the two terms in the square bracket are equal. That is, the *instantaneous* values of the kinetic and potential energies of any element are equal, so

$$dE = \mu(\omega y_0)^2 \cos^2(kx - \omega t)dx$$

The quantity dE/dx is called the *linear energy density* (measured in J/m). As Fig. 16.24 shows, this is a maximum where $y = 0$ and a minimum where $y = y_0$. At any point, such as $x = 0$, the average value of $\cos^2 \omega t$ over one period is 1/2; thus,

$$dE_{av} = \tfrac{1}{2}\mu(\omega y_0)^2 dx$$

The **average power** transmitted by the wave is $P_{av} = (dE/dt)_{av}$; therefore,

$$P_{av} = \tfrac{1}{2}\mu(\omega y_0)^2 v \qquad (16.16)$$

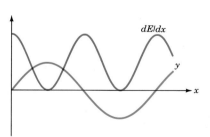

FIGURE 16.24 The linear energy density dE/dx (J/m) (red curve) associated with a wave (blue curve) traveling on a string. Note that the energy density is a maximum where the displacement is zero.

where $v = dx/dt$ is the wave velocity. The power is proportional to the square of the frequency and the square of the amplitude. This power must be supplied by the source of the waves.

EXAMPLE 16.5: A rod vibrating at 12 Hz generates harmonic waves with an amplitude of 1.5 mm in a string of linear mass density 2 g/m. If the tension in the string is 15 N, what is the average power supplied by the source?

Solution: The wave velocity is

$$v = \sqrt{\frac{F}{\mu}} = \sqrt{\frac{15 \text{ N}}{(2 \times 10^{-3} \text{ kg/m})}} = 86.6 \text{ m/s}$$

The angular frequency is $\omega = 2\pi f = 75.4$ rad/s. Thus, from Eq. 16.16, the average power is

$$P_{av} = \tfrac{1}{2}\mu(\omega y_0)^2 v$$

$$= \tfrac{1}{2}(2 \times 10^{-3} \text{ kg/m})(75.4 \text{ rad/s})^2(1.5 \times 10^{-3} \text{ m})^2(86.6 \text{ m/s})$$

$$= 1.1 \text{ mW}$$

16.11   VELOCITY OF WAVES ON A STRING

Figure 16.25 shows an element of a string under the influence of a disturbance. We assume that the string is perfectly flexible and that the displacement is purely transverse. We also assume that the displacement is so small that the tension is unchanged. If μ is the linear mass density, then the mass of the element of length Δx is $\mu \Delta x$. Newton's second law applied to the motion in the y direction is

$$F[\sin(\theta + \Delta\theta) - \sin\theta] = \mu\Delta x \frac{\partial^2 y}{\partial t^2} \qquad (16.17)$$

Since θ is small, we make the approximation

$$\sin\theta \approx \tan\theta = \frac{\partial y}{\partial x}$$

We use the partial derivative since y is a function of both x and t and we are considering a particular instant. When both sides of Eq. 16.17 are divided by Δx, the left side has the form

$$\frac{f(x + \Delta x) - f(x)}{\Delta x}$$

where $f(x) = \partial y/\partial x$. In the limit as $\Delta x \to 0$,

$$\frac{f(x + \Delta x) - f(x)}{\Delta x} = \frac{\partial f}{\partial x} = \frac{\partial^2 y}{\partial x^2}$$

Equation 16.17 becomes

$$\frac{\partial^2 y}{\partial x^2} = \frac{\mu}{F}\frac{\partial^2 y}{\partial t^2} \qquad (16.18)$$

Within the approximations made, this differential equation is essentially a state-

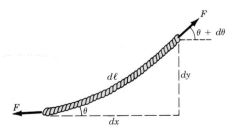

FIGURE 16.25 A length of string under the influence of a wave. The net transverse force produces the acceleration of the element.

ment of Newton's second law. Comparison with the wave equation (Eq. 16.16) shows that the wave velocity is

$$v = \sqrt{\frac{F}{\mu}} \qquad (16.19)$$

If the slope of the string ($\partial y/\partial x$) is not small, the displacements are not purely transverse and v would no longer be independent of the shape of a pulse. Furthermore, since $\sin\theta$ could not be replaced by $\tan\theta$, the differential equation would not be linear and linear superposition would not be valid.

SUMMARY

A **wave** is a disturbance that transports energy and momentum without the transport of matter. When two or more waves overlap in the same region, the resultant wave function is given by the principle of **linear superposition:**

$$y = y_1 + y_2 + \cdots + y_N$$

where the sum may be either a sum of scalars or a vector sum. Waves for which this principle is valid are said to be linear and satisfy the linear wave equation (Eq. 16.16).

The speed of a wave on a stretched string whose tension is F and whose linear mass density is μ is given by

$$v = \sqrt{\frac{F}{\mu}}$$

The **wave function** for a **harmonic wave** traveling along the $+x$ direction at speed v is

$$y = A\sin(kx - \omega t + \phi)$$

where the wave number $k = 2\pi/\lambda$ and $\omega = 2\pi/T$ is the angular frequency. The speed of the wave is given by

$$v = f\lambda = \frac{\omega}{k}$$

Two harmonic waves of equal frequency traveling in opposite directions can set up **standing waves.** In a system of finite size, such as a string fixed at both ends, the boundary conditions at the ends impose restrictions on the allowed frequencies of the **resonant** standing waves:

$$f_n = \frac{nv}{2L} \qquad (n = 1, 2, 3, \ldots)$$

When a harmonic wave propagates along a string, the instantaneous values of the kinetic energy and potential energy of a segment are equal. The **average power** transmitted by a harmonic wave of amplitude y_0 along a string of linear mass density μ is

$$P_{av} = \tfrac{1}{2}\mu(\omega y_0)^2 v$$

A wave that travels at speed v along the $+x$ direction without changing shape is described by a wave function of the form

$$y = f(x - vt)$$

Such a wave satisfies the linear **wave equation**

$$\frac{\partial^2 y}{\partial x^2} = \frac{1}{v^2} \frac{\partial^2 y}{\partial t^2}$$

In the case of mechanical waves, this equation may be derived by applying Newton's second law to the motion of an element of the medium.

ANSWERS TO IN-CHAPTER EXERCISES

1. $[v] = LT^{-1}$, $[F] = MLT^{-2}$, and $[\mu] = ML^{-1}$; thus, $LT^{-1} = (M^x L^x T^{-2x})(M^y L^{-y})$. On equating the powers of M, L, and T respectively, we obtain three equations:

$$0 = x + y \qquad 1 = x - y \qquad -1 = -2x$$

Thus, $x = \frac{1}{2}$ and $y = -\frac{1}{2}$.

2. The mass of a length L is $m = \rho V = \rho AL = \mu L$, so $\mu = \rho A = \rho \pi d^2/4$. Since $d_1 = 2d_2$, we have $\mu_1 = 4\mu_2$; also $F_1 = 0.5F_2$.

The ratio $v_2/v_1 = \sqrt{(F_2/F_1)(\mu_1/\mu_2)} = \sqrt{8}$.

3. Since the strings have the same tension, $v_2/v_1 = \sqrt{\mu_1/\mu_2}$. This is the ratio of the pulse lengths.

4. $y(x, t) = A \sin(2\pi x/\lambda - 2\pi t/T + \phi)$.

5. Comparing this equation with Eq. 16.8 we see that $k = 20$ rad/m and $\omega = 100$ rad/s. (a) $v = \omega/k = -5$ m/s. (b) The particle velocity is $\partial y/\partial t = 2 \cos(20x + 100t) = 2 \cos(4 + 30) = -1.7$ m/s.

QUESTIONS

1. Two pulses of identical shape overlap such that the displacement of the rope is momentarily zero at all points, as in Fig. 16.26. What happens to the energy at this time?

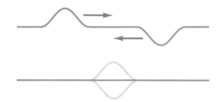

FIGURE 16.26 Question 1.

2. For linear superposition to be valid it is necessary that the amplitude of the wave be much less than the wavelength; that is, $A \ll \lambda$. Show that this implies $v \gg \partial y/\partial t$; that is, the wave speed is much greater than the particle velocity.

3. Some strings on a guitar, or in a piano, have round wire or metal tape wound around them. What purpose does this serve?

4. Is there any relationship between wave speed and the maximum particle speed for a wave traveling on a string? If so, what is it?

5. Is it possible to have a standing wave if the amplitudes of the two component waves are not equal?

6. In a standing wave on a string, is the energy density zero at the nodes?

7. If there were no dissipation of energy, how would the amplitude of circular waves on a pond decrease with distance from the source?

8. When a wave is transmitted from one medium to another, the frequency does not change. Why?

9. A pulse traveling in one rope is reflected at the boundary with another rope. If the reflected wave is not inverted, is the transmitted pulse shorter or longer than the original pulse?

10. (a) Does the interference of waves always involve the superposition of waves? (b) Does the superposition of waves always involve interference? (c) Do waves have to be periodic to interfere?

11. Why are the frets on a guitar not uniformly spaced?

12. Why do stringed instruments have hollow bodies? What role does the shape of the body play?

13. Why does the tone quality of a note played on a guitar depend on where the string is plucked?

EXERCISES

16.1 to 16.4 Properties of Waves; Pulses on Strings

1. (I) Calculate the range of wavelengths for each of the following radio broadcast bands: (a) the AM band from 550 kHz to 1600 kHz; (b) the FM band from 88 MHz to 108 MHz. The speed of propagation is 3×10^8 m/s.

2. (II) A 12-in. (30-cm) LP rotates at $33\frac{1}{3}$ rpm. At the rim the periodicity of the undulations in the groove is 1.2 mm. What is the frequency of the recorded signal?

3. (II) Consider the transverse wave depicted in Fig. 16.27. The wave velocity is $+40$ cm/s. Find: (a) the frequency; (b) the phase difference between points 2.5 cm apart; (c) how long it takes for the phase at a given position to change by $60°$; (d) the velocity of a particle at point P at the instant shown.

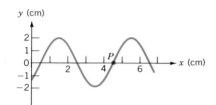

FIGURE 16.27 Exercise 3.

4. (I) A 12-in. (30-cm) LP rotates at $33\frac{1}{3}$ rpm. The frequency of a recorded signal is 10^4 Hz. (a) What is the distance between the peaks of the undulations on the vinyl if the needle is 14.5 cm from the center? (b) What is the wavelength of the recorded sound? The speed of sound is 340 m/s.

5. (I) An earthquake generates two types of seismic waves that propagate through the earth. The P waves have a characteristic speed of 8 km/s, whereas the S waves travel at 5 km/s. At an observation post these waves are detected with a time interval of 1.8 min. Assuming that the waves traveled in a straight line, how far away did the quake occur?

6. (I) A rope of length 3 m has a mass of 25 g. If the speed of transverse waves is 40 m/s, what is the tension in the rope?

7. (I) A stretched cord is 7.5 m long and under a tension of 30 N. If the wave speed is 20 m/s, what is the mass of the cord?

8. (I) When the tension in a stretched rope is 15 N, the wave speed is 28 m/s. What tension is needed to produce a wave speed of 45 m/s?

9. (I) Transverse waves travel along a stretched string. If the tension in a string is doubled (a) by what factor must the frequency change for the wavelength to be unchanged, and (b) how does the wave speed change?

10. (I) Figure 16.28 shows a triangular pulse on a rope. It is approaching a fixed end at 2 cm/s. (a) Draw the pulse at $\frac{1}{2}$-s intervals until it is completely reflected. (b) What is the particle speed on the leading edge at the instant depicted?

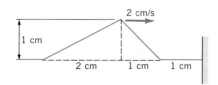

FIGURE 16.28 Exercise 10.

11. (I) A triangular pulse (Fig. 16.29) moving at 2 cm/s on a rope approaches an end at which it is free to slide on a vertical pole. (a) Draw the pulse at $\frac{1}{2}$-s intervals until it is completely reflected. (b) What is the particle speed on the trailing edge at the instant depicted?

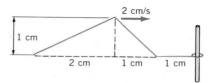

FIGURE 16.29 Exercise 11.

16.5 and 16.6 Traveling Waves

12. (I) The wave function for a traveling wave is given by

$$y(x, t) = \frac{5}{2 + (x - 2t)^2}$$

where both x and y are in centimeters and t is in seconds. Plot this function from $x = 0$ to 10 cm at (a) $t = 2$ s, and (b) $t = 3$ s.

13. (I) At $t = 0$, the shape of a traveling pulse is given by

$$y(x, 0) = \frac{2 \times 10^{-3}}{4 - x^2}$$

where x and y are in meters. What is the wave function for the traveling pulse if the velocity of propagation is 12 m/s in the negative x direction?

14. (I) Figure 16.30 shows a snapshot of a traveling wave taken at $t = 0.3$ s. The wavelength is 7.5 cm and the amplitude is 2 cm. If the crest P was at $x = 0$ at $t = 0$, write the wave function in the form of Eq. 16.8.

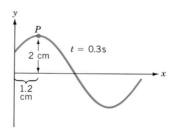

FIGURE 16.30 Exercise 14.

15. (II) The wave function for a traveling wave on a string is given by $y(x, t) = A \sin(kx - \omega t)$. (a) What is the slope of the string at an arbitrary position x and time t? (b) How is the maximum slope related to the wave speed and the maximum particle speed?

16. (II) The displacement due to a traveling wave is given by

$$y(x, t) = 3.2 \cos(0.2x - 50t)$$

where x and y are in centimeters. Plot $y(x, t)$ at $t = 0$ and $t = 0.1$ s. From your plots find: (a) the wave speed; (b) the distance between points whose phase differs by $2\pi/3$ rad.

17. (II) A transverse wave on a string is given by

$$y(x, t) = 2.4 \cos\left[\frac{\pi}{20}(0.5x - 40t)\right]$$

where both x and y are in centimeters. Find: (a) the maximum particle velocity; (b) the particle velocity at $x = 1.5$ cm at $t = 0.25$ s; (c) the maximum particle acceleration; (d) the acceleration at $x = 1.5$ cm and $t = 0.25$ s.

18. (II) The wave function for a traveling wave on a string is

$$y(x, t) = 0.03 \cos(2.4x - 12t + 0.1)$$

where both x and y are in centimeters. Find: (a) the frequency; (b) the wave speed; (c) the amplitude; (d) the particle velocity at $x = 15$ cm and $t = 0.2$ s; (e) The maximum particle acceleration.

19. (I) Which of the following functions represent traveling waves?

(a) $A \sin^2\left[\pi\left(t - \dfrac{x}{v}\right)\right]$ (b) $A \cos[(kx - \omega t)^2]$

(c) $A \sin[(kx)^2 - (\omega t)^2]$ (d) $Ae^{[-\sigma(x-vt)^2]}$

(e) $A(x + vt)^3$ (f) $Ae^{-\alpha t} \cos(kx - \omega t)$

20. (I) A traveling harmonic wave has a wavelength of 20 cm and a period of 0.02 s. Find the phase difference (a) between two points 8 cm apart; (b) at a given point but 0.035 s apart.

21. (II) The wave function of a wave is

$$y(x, t) = 0.02 \sin(0.4x + 50t + 0.8)$$

where x and y are in centimeters. Find: (a) the wavelength; (b) phase constant; (c) the period; (d) the amplitude; (e) the wave velocity; (f) the particle velocity at $x = 1$ cm and $t = 0.5$ s.

22. (I) A traveling harmonic wave function is

$$y = 0.04 \sin\left(\frac{x}{5} - 2t\right)$$

where x and y are in meters. Find: (a) the wavelength; (b) the period; (c) the wave velocity.

23. (II) A transverse wave traveling in the negative x direction has a wavelength of 2.5 cm, a period of 0.01 s, and an amplitude of 0.03 m. The displacement at $x = 0$ and $t = 0$ is $y = -0.02$ m and the particle velocity is positive. Write the wave function $y(x, t)$ in the form of Eq. 16.8.

24. (II) A transverse harmonic wave has an amplitude of 0.05 m, a wave number of 0.1 rad/m, and a wave velocity of 50 m/s in the negative x direction. At $x = 0$ and $t = 2$ s, the transverse displacement is $y = 1.25 \times 10^{-2}$ m and $\partial y/\partial t < 0$. Write an expression for the traveling wave function $y(x, t)$ in the form of Eq. 16.8.

25. (I) Expressed in terms of wave number and angular frequency, the equation for a traveling harmonic wave is

$$y(x, t) = A \sin(kx - \omega t)$$

Express this function in terms of: (a) wavelength and wave speed; (b) frequency and wave speed; (c) wave number and wave speed; (d) wavelength and frequency.

16.7 and 16.8 Standing Waves

26. (II) The displacement of a standing wave on a string is given by

$$y(x, t) = 4.0 \sin(0.5x) \cos(30t)$$

where x and y are in centimeters. (a) Find the frequency, amplitude, and wave speed of the component waves. (b) What is the particle velocity at $x = 2.4$ cm at $t = 0.8$ s?

27. (II) Two overlapping waves travel in opposite directions, each with a speed of 40 cm/s. They have the same amplitude of 2 cm and a frequency of 8 Hz. (a) Write the wave function for the resulting standing wave. (b) What is the distance between adjacent nodes? (c) What is the maximum displacement of a particle at $x = 0.5$ cm?

28. (II) A 60-cm-long guitar string has a linear mass density of 1.5 g/m. What is the tension required for the frequency of the second harmonic to be 450 Hz?

29. (II) A string fixed at both ends has consecutive standing wave modes for which the distances between adjacent nodes are 18 cm and 16 cm respectively. (a) What is the minimum possible length of the string? (b) If the tension is 10 N and the linear mass density is 4 g/m, what is the fundamental frequency?

30. (II) A string, with a linear mass density of 2.6 g/m, is fixed at both ends. It has consecutive standing wave modes with frequencies 480 Hz and 600 Hz. The tension is 12 N. Find: (a) the fundamental frequency; (b) the length of the string.

31. (I) One prong of a tuning fork that vibrates at 440 Hz is connected to a string ($\mu = 1.2$ g/m). A block of mass 50 g is suspended from the other end as shown in Fig. 16.22. For what length will the string resonate (a) at its fundamental frequency; (b) at the third harmonic?

32. (II) The amplitude of a standing wave on a string is 2 mm and the distance between adjacent nodes is 12 cm. Given that the mass linear density is 3 g/m and the tension is 15 N, write the wave function, $y(x, t)$, of the standing wave.

33. (II) Two wires have the same length and are under the same tension. Their radii are related according to $r_1 = 2r_2$, and the mass densities (kg/m³) according to $\rho_1 = 0.5\rho_2$. Compare their fundamental frequencies.

34. (II) Two strings are cut from the same roll. The tension in one is double that in the other (that is, $F_1 = 2F_2$), but its length is only one-third (that is, $L_1 = L_2/3$). Compare the fundamental frequencies.

35. (II) The wave function of a standing wave on a string is given by

$$y(x, t) = 0.02 \sin(0.3x) \cos(25t)$$

where x and y are in centimeters and t is in seconds. (a) Find the wavelength and wave speed of the component waves. (b) What is the length of the string if this function represents the third harmonic? (c) At what points is the particle velocity permanently zero?

36. An unfingered guitar string has a fundamental frequency of 320 Hz. What is the fundamental frequency when it is pressed against the fingerboard such that its length is reduced by one-third?

16.8 Energy Transport

37. (I) Transverse waves on a cord have an amplitude of 1.5 cm, a wavelength of 40 cm, and they travel at 30 m/s. If the linear mass density of the cord is 20 g/m, what power must be supplied?

38. (I) A mechanical oscillator supplies 3 W at a frequency of 30 Hz to a wire of length 15 m and mass 45 g. If the tension is 40 N, what is the amplitude of the waves that are generated?

39. (II) Transverse waves of amplitude 0.8 mm travel at 60 m/s along a rope of linear mass density 3.5 g/m. A 50-Hz oscillator is connected to one end. (a) What is the average power supplied by the oscillator? (b) What is the tension required to double the power at the same frequency?

16.9 The Wave Equation

40. (I) Find which of the following wave functions satisfy the wave equation:

$$\text{(a) } Ae^{[-\sigma(x - vt)^2]}; \qquad \text{(b) } A \ln[B(x - vt)]$$

41. (I) Show explicitly that the wave function for a standing wave,

$$y(x, t) = A \sin(kx) \cos(\omega t)$$

satisfies the wave equation.

PROBLEMS

1. (I) The tensile stress in a steel wire is 2×10^8 N/m^2. What is the speed of transverse waves in the wire? The density of steel is 7.8 g/cm^3.

2. (I) (a) Show that if the tension F in a string is changed by a *small* amount ΔF, the fractional change in frequency of a standing wave, $\Delta f/f$, is given by

$$\frac{\Delta f}{f} = 0.5 \frac{\Delta F}{F}$$

(b) A string vibrates in its fundamental mode at 400 Hz. If the tension is decreased by 3%, what is the new fundamental frequency? (c) What is the percentage change in tension required to change the fundamental frequency of a string from 260 Hz to 262 Hz?

3. (I) A short, light wire bent into the shape of a "V" is placed at the midpoint of a string fixed at both ends. Assuming that the string vibrates at its fundamental frequency f_1, at what amplitude A of the standing wave will the wire lose contact with the string?

4. (I) The G (196 Hz) string on a guitar is 64 cm long. Locate the position of the frets for the following notes: A (220 Hz); B (247 Hz); C (262 Hz); D (294 Hz).

5. (II) Consider a transverse wave of small amplitude on a string. Under the action of the wave, the string is moved upward, as shown in Fig. 16.31. (a) Show that the power supplied by the left side to the right side is

$$P = -F \frac{\partial y}{\partial x} \frac{\partial y}{\partial t}$$

(b) Use the expression $y(x, t) = A \sin(kx - \omega t)$ to find the average power transmitted along the rope. Compare your result with Eq. 16.16.

FIGURE 16.31 Problem 5.

6. (II) A rope of mass M and length L hangs vertically. Show that the time needed for a pulse to travel from the bottom end to the support is $T = 2\sqrt{L/g}$.

7. (I) Show that the instantaneous power transmitted through a rope by a traveling wave of amplitude A is a maximum when the displacement $y = 0$ and a minimum where $y = \pm A$.

8. (II) A rope of length L and linear mass density μ is under a tension F. The string is fixed at one end and free to slide on a frictionless pole at the other end. What are the normal mode (standing wave) frequencies? (*Hint:* The free end is an antinode.)

9. (I) A wire is stretched from L to $L + \Delta L$. Show that the wave speed of transverse waves is

$$v = \sqrt{\frac{Y}{\rho} \frac{\Delta L}{L}}$$

where Y is Young's modulus (Eq. 14.3) and ρ is the density of the wire.

10. (II) Show that for a standing wave on a string (Eq. 16.11) the average kinetic energy per unit length and the potential energy per unit length are given by

$$\left(\frac{dK}{dx}\right)_{av} = \mu(\omega A)^2 \sin^2 kx; \qquad \left(\frac{dU}{dx}\right)_{av} = F(kA)^2 \cos^2 kx$$

11. (II) (a) Consider the nth standing wave on a string of length L fixed at both ends in the form given in Eq. 16.11. Show that the total energy is

$$E_n = \mu(\omega_n A)^2 L$$

where μ is the linear mass density and ω is the angular frequency. (b) Show that the energy in one loop of a standing wave is

$$E = 2\pi^2 \mu A^2 f v$$

(Try Problem 10 first.)

12. (II) A guitar string of length 60 cm and mass 2 g has a tension of 200 N. It vibrates in its fundamental mode. The initial amplitude of 1 mm drops by 10% in 0.1 s and 50% of the associated decrease in energy is radiated as sound. Find the average radiated power. What happens to the other 50%? (Try Problem 11 first.)

13. (I) Show that the average power transmitted along a rope may be written in the form

$$P = \eta v$$

where η is the linear energy density (energy per unit length) and v is the wave velocity.

14. (II) A linear array of particles with equal masses m are connected by identical springs whose stiffness constant is k (Fig. 16.32). The equilibrium position of the nth particle is $x_n = na$, while s_n is its displacement from equilibrium. (a) Show that

$$m \frac{d^2 s_n}{dt^2} = k(s_{n+1} + s_{n-1} - 2s_n)$$

(b) Show that $s_n = A \sin(kx_n - \omega t)$ is a solution provided that

$$\omega^2 = \frac{4k}{m} \sin^2 \left(\frac{ka}{2}\right)$$

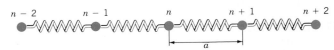

FIGURE 16.32 Problem 14.

(Hint: You will need trigonometric identities. See Appendix B.)

15. (I) (a) Show that the wave speed for transverse waves on a Slinky of mass m and spring constant k is $L\sqrt{k/m}$, where L, the stretched length, is much larger than the unstretched length. (b) Show that the time required for a pulse to travel the length of the Slinky is independent of the length.

CHAPTER 17

Sound

Major Points

1. **Sound waves** are longitudinal waves characterized by density or pressure fluctuations.
2. **Resonant standing waves** may be set up in **open** and **closed** pipes.
3. The **Doppler effect** is the change in the observed frequency of a wave when there is relative motion between the source and the observer.
4. The superposition of two waves of nearly equal frequency gives rise to **beats.**
5. The definition of the **intensity** of a wave.

An acoustic shock wave is formed when the speed of a body is greater than the speed of sound in the medium.

In this chapter we study some properties of sound waves in fluids, such as air or water. *Audible sound,* that our ears can detect, ranges in frequency from about 20 Hz to 20,000 Hz. Sound waves with frequencies below 20 Hz are called *infrasonic* and are produced during earthquakes, by thunder, and by the vibrations of heavy machinery or the tires of a car. *Ultrasonic* frequencies, above 20,000 Hz, can be heard by dogs, cats, and porpoises. Bats and porpoises, as well as underwater ''sonar'' devices, use ultrasonic waves to locate objects. In medicine, ultrasound is used to monitor fetal development (Fig. 17.1a) and acoustic shock waves are used to shatter kidney stones (Fig. 17.1b). Very high-frequency (10^9 Hz) ultrasound, produced by electrically stimulating a quartz crystal, is used in acoustic microscopy to yield sharp images (Fig. 17.1c).

17.1 THE NATURE OF A SOUND WAVE

In the equilibrium state the pressure and density in a fluid are uniform. However, the molecules are not at rest; they dart about randomly and collide frequently. In the presence of a sound wave, each tiny volume element of the gas undergoes periodic *longitudinal* vibrations. This orderly motion of an element is superimposed on the random motion of the molecules. The displacements of the elements give rise to a periodic fluctuation in air density and hence also in the pressure. The pressure fluctuations are of the order of 1 Pa (= 1 N/m²), whereas atmospheric pressure is about 10^5 Pa. One might compare these fluctuations to tiny ripples on the surface of a deep lake.

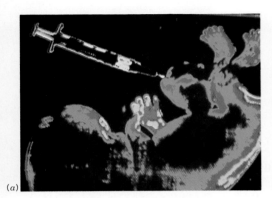

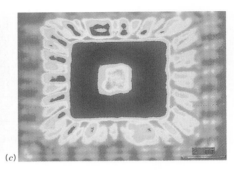

(a)   (b)   (c)

FIGURE 17.1 (*a*) Monitoring fetal development with ultrasound. (*b*) Shattering kidney stones with acoustic shock waves. (*c*) Acoustical microscopy provides detailed images of part of a microchip.

Figure 17.2 shows the cone of a loudspeaker at the end of an open pipe. When the cone moves forward, it produces a *compression,* that is, an increase in pressure, ΔP, above the equilibrium value P_0. When the cone moves backward, it produces a *rarefaction,* a decrease in pressure, $-\Delta P$, below P_0. The collisions between the molecules cause these compressions and rarefactions to propagate as a sound wave down the pipe. Keep in mind that it is the wave (that is, the disturbance from equilibrium), and not the molecules themselves, that travels down the pipe.

At a given instant there are points such as *b* and *d* toward which molecules converge and therefore increase the local pressure to a maximum of $P_0 + \Delta P_0$. Since the displacements on either side of *b* and *d* are in opposite directions, these two points must have zero displacement at this instant. On either side of points *a* and *c*, the molecules are moving away and so these points also have zero displacement. The local pressure at these points drops to a minimum of $P_0 - \Delta P_0$. In general, the points of maximum pressure deviation ($\pm\Delta P_0$) have zero displacement. In other words, pressure fluctuations are 90° out of phase with the displacements. The converse is also true: Zeroes in pressure fluctuation correspond to maxima or minima in the displacements.

We will show later (Section 17.5) that the speed of *longitudinal waves in a fluid* is given by

Speed of longitudinal waves in a fluid

(Fluid)
$$v = \sqrt{\frac{B}{\rho}} \qquad (17.1)$$

where B is the bulk modulus, defined in Eq. 14.7 as

$$B = -\frac{\Delta P}{\Delta V/V} \qquad (17.2)$$

where $\Delta V/V$ is the fractional change in volume produced by the change in pressure ΔP. The SI unit for B is N/m². The negative sign is needed to make B positive since a positive change in pressure results in a negative change in volume V. The bulk

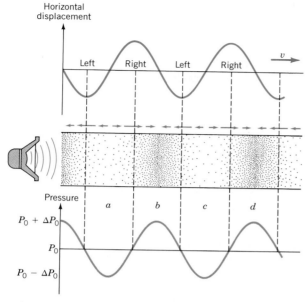

FIGURE 17.2 A loudspeaker produces compressions and rarefactions in the air in a tube. The longitudinal displacements (at a particular instant) are indicated by the arrows and are plotted in the upper graph. The deviations of the pressure above and below the atmospheric pressure are plotted in the lower graph. Note that a position of zero displacement corresponds to a maximum pressure deviation and vice versa.

modulus is an indication of the "springiness" of a compressible medium, just as the spring constant ($k = -dF/dx$) is for a spring. Notice that Eq. 17.2 has the same form as that for the speed of transverse waves on a string, $v = \sqrt{F/\mu}$.

EXAMPLE 17.1: Calculate the speed of longitudinal waves (a) in water, given that its bulk modulus is 2.1×10^9 N/m² and its density is 10^3 kg/m³, and (b) in air at 1 atm given that the bulk modulus is $B = 1.41 \times 10^5$ N/m² and its density is 1.29 kg/m³.

Solution: (a) From Eq. 17.2, we have

$$v = \sqrt{\frac{B}{\rho}} = \sqrt{\frac{2.1 \times 10^9 \text{ N/m}^2}{10^3 \text{ kg/m}^3}} = 1500 \text{ m/s}$$

(b) For air,

$$v = \sqrt{\frac{B}{\rho}} = \sqrt{\frac{1.41 \times 10^5 \text{ N/m}^2}{1.29 \text{ kg/m}^3}} = 330 \text{ m/s}$$

At a given temperature the speed of sound in air is independent of pressure. The variation in the speed of sound in air with *absolute* temperature T (measured in kelvins, K) is given approximately by

$$v \approx 20\sqrt{T}$$

To find T add 273° to the temperature in °C. For example, at 20 °C = 293 K, we find $v = 20\sqrt{293} = 342$ m/s.

EXERCISE 1. Porpoises can emit a sound signal of frequency 80 kHz. What is the wavelength of such a wave (a) in water, (b) in air at 20 °C?

In discussing the propagation of two-dimensional or three-dimensional waves, it is useful to introduce the concept of a wave front. For example, when a pebble is thrown into a pond, the waves that spread from the point of impact are circular (Fig. 17.3a). At a given instant, a continuous line that joins all the points with a given displacement, for example, a crest, forms a **wave front**. In general, a *wave front is a locus of points at which the wave function has the same phase.* Of special interest are the wave fronts emitted by a *point* source that emits waves uniformly in all directions. The wave fronts in this case are spherical surfaces with

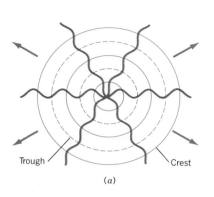

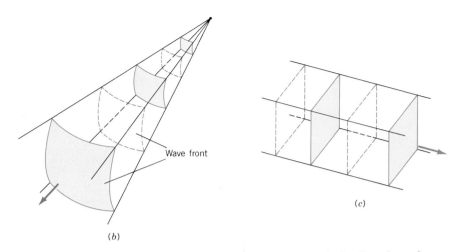

FIGURE 17.3 (*a*) Wave fronts spreading on a water surface are circles. (*b*) Small sections of spherical wave fronts spreading from a point source. (*c*) Far from a point source, the wave fronts become plane waves.

Many musical instruments depend on the phenomenon of resonance in an air column.

the source at the center. Figure 17.3*b* shows only part of such wave fronts. At great distances from a point source, the curvature of the fronts is small. They become essentially *plane* wave fronts, as shown in Fig. 17.3*c*.

17.2 RESONANT STANDING SOUND WAVES

In Chapter 16 we saw that musical instruments, such as guitars and violins, make use of resonant standing waves in strings. Resonant standing waves can also be produced in air columns, for example, in organ pipes, flutes, and other woodwind instruments, because sound waves are reflected both at a closed end and at an open end of a pipe. In an **open pipe** both ends are open, whereas in a **closed pipe** one end is closed. When a compressional pulse traveling along a pipe encounters a closed end it is reflected as a compression; a rarefaction is reflected as a rarefaction. A sound wave is also partly reflected and partly transmitted when the cross-sectional area of a pipe changes abruptly (recall the analogous situation with two strings of different mass densities). For example, when a rarefaction reaches an open end, the surrounding air rushes toward this region and creates a compression that travels back along the pipe. Similarly, when a compression reaches an open end, the air expands to form a rarefaction: A compression is therefore reflected as a rarefaction at an open end. Suppose that there is a periodic source, such as a loudspeaker, at an open end. If the time taken by each pulse to be reflected and to return is an appropriate multiple of the period of the source, a resonant standing wave will be produced.

There are several ways in which resonant standing waves can be excited in a pipe. The simplest is with a tuning fork or a loudspeaker connected to a signal generator. In some woodwind instruments a vibrating reed, or the vibrating lip of the player, is used. Air blown *across* the open end of a pipe will also set the air column into a normal mode of oscillation. (Try this with a soft-drink bottle.) This approach is used in the flute and in flue organ pipes. Although the actual motion of the air in an instrument is quite complicated, the following discussion is a good first approximation to reality.

CLOSED PIPES

In Fig. 17.4 the motion of the molecules is represented by solid arrows at $t = 0$ and by dashed arrows at $t = T/2$, where T is the period. At the closed end the displacement is permanently zero; that is, it is a displacement node. From the previous section we know that this corresponds to a pressure antinode. (At the instant depicted the pressure is a maximum at the closed end. Half a period later, the density and pressure will be a minimum.) The open end has to be at atmospheric pressure, so it is a pressure node and a displacement antinode. The fundamental mode occurs when $L = \lambda/4$, which means $f_1 = v/4L$. The higher harmonics are easily constructed if we keep in mind that the closed end is a displacement node, whereas the open end is a displacement antinode. (Traditionally the displacement is depicted rather than the pressure.) Figure 17.5 shows that for a closed pipe only the *odd* harmonics are possible:

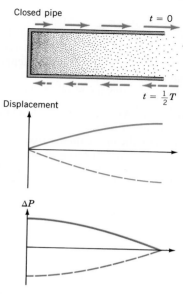

FIGURE 17.4 The fundamental mode in a closed pipe. The closed end is a displacement node and a pressure antinode. The solid curves depict the situation at $t = 0$, and the dashes at $t = T/2$.

$$(\text{Closed pipe}) \qquad f_n = \frac{nv}{4L} \qquad (n = 1, 3, 5 \ldots) \quad (17.3)$$

OPEN PIPES

The knowledge that each open end is a displacement antinode allows us to draw the various modes right away (see Fig. 17.6). Notice that the fundamental frequency of an open pipe is double that of a closed pipe of the same length. In an open pipe, *all* the harmonics are possible:

$$(\text{Open pipe}) \qquad f_n = \frac{nv}{2L} \qquad (n = 1, 2, 3 \ldots) \quad (17.4)$$

Standing waves in pipes decay very quickly after the source of excitation is removed. The interaction between the air in the pipe and the surrounding air also means that the above equations are not quite correct. It is found that pipes of the same length but of different diameters have slightly different resonant frequencies. The effective length is approximately the real length plus 0.61 of the radius of each open end.

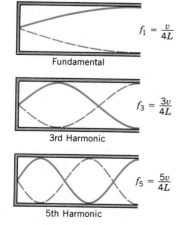

FIGURE 17.5 The first three resonant modes in a closed pipe. The solid and dashed curves represent the longitudinal displacements one-half period apart.

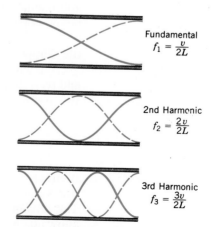

FIGURE 17.6 The first three resonant modes in an open pipe.

EXAMPLE 17.2: The length of an air column is varied by adjusting the water level in a pipe (Fig. 17.7). A vibrating tuning fork is placed at the open end. As the water is lowered, resonance is first heard when the length of the column is 18.9 cm and then when it is 57.5 cm. What is the frequency of the tuning fork? Take the speed of sound to be 340 m/s.

Solution: We are given the fundamental mode and the second harmonic of a closed pipe, as shown in Fig. 17.7. One could calculate the *approximate* frequency of the tuning fork by using Eq. 17.3. However, one avoids the problem of end corrections by noting that the difference in the measured lengths corresponds exactly to one half-wavelength of the sound. Thus,

$$\lambda = 2(57.5 \text{ cm} - 18.9 \text{ cm}) = 77.2 \text{ cm}$$

and the frequency is

$$f = \frac{v}{\lambda} = \frac{340 \text{ m/s}}{0.772 \text{ m}} = 440 \text{ Hz}$$

(You need not be concerned with end corrections in the exercises and problems unless the issue is specifically raised.)

EXERCISE 2. What would Eq. 17.3 yield for the frequency of the tuning fork in Example 17.2?

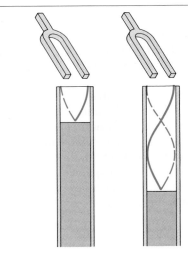

FIGURE 17.7 A tuning fork is at the open end of a pipe. The length of the air column is varied by adjusting the water level. The fundamental mode at one length has the same frequency as the second harmonic of a different length (one half-wavelength longer).

17.3   THE DOPPLER EFFECT

In 1842, J. C. Doppler published a paper in which he tried to relate the colors of stars to their motion. Although this idea was incorrect, he did analyze a similar phenomenon for sound waves. In 1845, C. Buys Ballot, tested that idea. He arranged to have one group of musicians stand at intervals along a railroad track while another group traveled in an open railcar. The job of each group was to estimate the pitch (frequency) of trumpet notes played by the other group. What they found is now a familiar phenomenon. When a car, with its horn blowing, moves past you quickly, the observed frequency of the note changes. As the car approaches, the frequency seems higher than normal and then suddenly drops below normal as the car moves away. The change in the observed frequency of a wave when there is relative motion between the source and an observer is called the **Doppler effect.**

We will use the following symbols: v = speed of sound, v_S = speed of source, and v_O = speed of observer. All of these are measured with respect to the ground. We assume that the air is at rest relative to the ground. When both source and observer are at rest, the frequency and the wavelength have their normal values f_0 and λ_0, and the speed of sound is then

$$v = f_0 \lambda_0$$

If either the source or the observer moves, the observed frequency changes to f'. The value of f' depends on the separate motions of the source and the observer. The reason is that the medium (air) acts as an "absolute" reference frame that allows us to distinguish whether it is the source and/or the observer that is moving. The wavelength is determined only by the motion of the source. Its value is the same in both the frame of the source and that of the observer. The following discussions will be confined to the effects heard along the line joining the source and the observer.

(a) Source at Rest, Observer Moves

Suppose that the observer O moves toward the source S at speed v_O, as shown in Fig. 17.8. The speed of the sound waves relative to O is $v' = v + v_O$, but the wavelength has its normal value, $\lambda_0 = v/f_0$. Thus, the frequency heard by O is

$$f' = \frac{v'}{\lambda_0} = \frac{v + v_O}{v} f_0$$

If O were moving away from S, the frequency heard by O would be $f' = [(v - v_O)/v]f_0$. Combining these two expressions we find

$$f' = \left(\frac{v \pm v_O}{v}\right) f_0 \tag{17.5}$$

(b) Source Moves, Observer at Rest

Suppose that the source S moves toward O, as in Fig. 17.9a. If S were at rest, the distance between crests would be $\lambda_0 = v/f_0 = vT_0$. However, in one period S moves a distance $v_S T_0$ before it emits the next crest. As a result the wavelength is modified, as shown in Fig. 17.9b. Directly ahead of S the effective wavelength (for both S and O) is

$$\lambda' = vT_0 - v_S T_0 = \frac{v - v_S}{f_0}$$

The speed of the sound waves relative to O is simply v; thus, the frequency heard by O is

$$f' = \frac{v}{\lambda'} = \frac{vf_0}{v - v_S}$$

If S were moving away from O, the effective wavelength would be $\lambda' = (v + v_S)/f_0$ and the apparent frequency would be $f' = vf_0/(v + v_S)$. Combining these two results, we have

$$f' = \left(\frac{v}{v \pm v_S}\right) f_0 \tag{17.6}$$

All four possibilities can be combined into one equation. We replace f_0 in one equation by f' from the other to find

$$f' = \left(\frac{v \pm v_O}{v \pm v_S}\right) f_0 \tag{17.7}$$

In any given situation, the appropriate signs are chosen as follows: Any relative motion of the source or observer toward the other tends to increase f'; any relative motion of one away from the other tends to decrease f'.

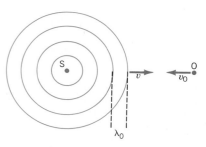

FIGURE 17.8 A stationary source produces spherical waves. The speed of the waves relative to the observer O, who is moving toward the source, is $(v + v_O)$.

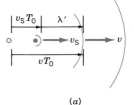

(a)

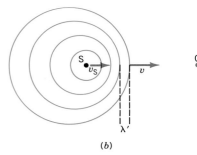

(b)

FIGURE 17.9 (a) In one period the source moves $v_S T_0$ while a crest, say, moves vT_0, where T_0 is the natural period of the source. The distance between crests, that is, the wavelength, is modified. (b) The wavelength in front of the moving source is less than normal, whereas in the rear, the wavelength is larger than normal.

EXAMPLE 17.3: A police car moves at 50 m/s in the same direction as a truck that has a speed of 25 m/s. The police siren has a frequency of 1200 Hz. What is the frequency heard by the truck driver when the police car is (a) behind the truck, or (b) ahead of the truck? Take the speed of sound to be 350 m/s.

Solution: (a) In Fig. 17.10a, the motion of the observer (truck) is away from the source, which tends to decrease the apparent frequency. In Eq. 17.7 the sign in the numerator is negative. The source is moving toward the observer, which tends to in-

crease the apparent frequency. The sign in the denominator is also negative. Thus,

$$f' = \left(\frac{v - v_O}{v - v_S}\right) f_0$$

$$= \left(\frac{325 \text{ m/s}}{300 \text{ m/s}}\right) 1200 \text{ Hz} = 1300 \text{ Hz}$$

(b) In Fig. 17.10b the observer's motion tends to increase the apparent frequency, whereas that of the source tends to decrease it. Consequently,

$$f' = \left(\frac{v + v_O}{v + v_S}\right) f_0$$

$$= \left(\frac{375 \text{ m/s}}{400 \text{ m/s}}\right) 1200 \text{ Hz} = 1125 \text{ Hz}$$

EXERCISE 3. Assume that the car and truck in Example 17.5 are moving in opposite directions. What is the frequency heard by the truck driver (a) when the car is approaching, and (b) after it has gone past the truck?

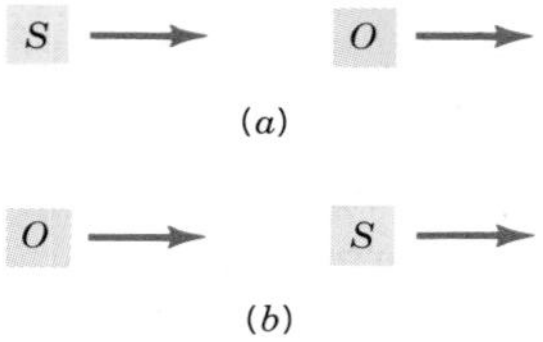

S →   O →

(a)

O →   S →

(b)

FIGURE 17.10 (a) The motion of the source tends to increase the observed frequency; the motion of the observer tends to decrease it. (b) The motion of the source tends to decrease the observed frequency; the motion of the observer tends to increase it.

17.4   INTERFERENCE IN TIME; BEATS

When two waves of slightly different frequencies are superposed, the resulting disturbance varies periodically in amplitude. Consider a specific position, which we take to be $x = 0$. For simplicity we assume the amplitudes are equal and that the phase constants are zero. The resultant wave function is*

$$y_T = y_1 + y_2 = A \sin \omega_1 t + A \sin \omega_2 t$$

$$= 2A \cos\left[2\pi \left(\frac{f_1 - f_2}{2}\right) t\right] \sin\left[2\pi \left(\frac{f_1 + f_2}{2}\right) t\right]$$

where we have used $\omega = 2\pi f$. This represents a wave function of frequency $f_{av} = (f_1 + f_2)/2$, whose amplitude is *modulated* at frequency $(f_1 - f_2)/2$. The individual functions and the resultant are shown in Fig. 17.11. Thus, when two tones of nearly equal frequencies f_1 and f_2 are sounded simultaneously, we hear the average frequency, f_{av}, but its loudness varies at the *beat frequency*, $|f_1 - f_2|$. Our ears

Beat frequency
perceive each peak in the envelope (dashed line), so the beat frequency is not $|f_1 - f_2|/2$—as one might infer from the term $\cos[2\pi(f_1 - f_2)t/2]$.

The phenomenon of beats is not restricted to mechanical waves. When radar signals are bounced off a moving target, the frequency of the reflected wave is Doppler shifted. When the reflected wave is combined with the original wave, the

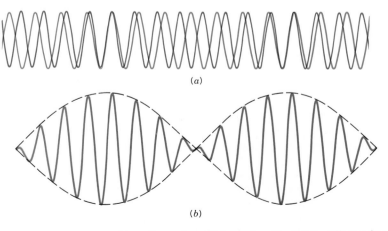

(a)

(b)

FIGURE 17.11 (a) The superposition of two sinusoidal signals with slightly different frequencies. (b) The resultant signal is sinusoidal but its amplitude fluctuates. The variation in intensity is perceived as beats with a frequency $|f_1 - f_2|$.

* $\sin A + \sin B = 2 \sin[(A + B)/2] \cos[(A - B)/2]$.

beat frequency gives an indication of the velocity of the object, for example a car. (The Doppler effect for electromagnetic waves, such as light and radar, is discussed in Section 39.9.)

17.5   VELOCITY OF LONGITUDINAL WAVES IN A FLUID

We now calculate the speed at which a longitudinal pulse propagates through a fluid. We will apply Newton's second law to the motion of an element of the fluid and from this we derive the wave equation. For convenience we assume that the fluid is confined to a tube of cross-sectional area A, as shown in Fig. 17.12. Under the action of a wave pulse, an element of thickness Δx at position x is moved to a new position $x + s$ and has a new thickness $\Delta x + \Delta s$. We assume that the equilibrium pressure of the fluid is P_0. Because of the disturbance, the pressure on the left side of the element becomes $(P_0 + p_1)$ and on the right side it becomes $(P_0 + p_2)$. Note that p_1 and p_2 are the *changes* in pressure caused by the pulse. If ρ is the equilibrium density, the mass of the element is $\rho A \Delta x$. (When the element moves its mass does not change, even though its volume and the density do change.) The net force acting on the element is $F = (p_1 - p_2)A$, and its acceleration is $a = \partial^2 s/\partial t^2$. Thus, Newton's second law applied to the motion of the element is

$$(p_1 - p_2)A = \rho A \Delta x \frac{\partial^2 s}{\partial t^2} \tag{17.8}$$

Next we divide both sides by Δx and note that in the limit as $\Delta x \to 0$, we have $(p_2 - p_1)/\Delta x \to \partial p/\partial x$. Equation 17.8 then takes the form

$$-\frac{\partial p}{\partial x} = \rho \frac{\partial^2 s}{\partial t^2} \tag{17.9}$$

The excess pressure p may be related to s via the bulk modulus B defined in Eq. 17.2: $B = -\Delta P/(\Delta V/V)$. For the element in Fig. 17.12, $V = A\Delta x$ and $\Delta V = A\Delta s$; therefore, $\Delta V/V = \Delta s/\Delta x$. The change in pressure ΔP has some "average" value p between p_1 and p_2, and so in the limit as $\Delta x \to 0$, the definition of the bulk modulus may be written as

$$p = -B \frac{\partial s}{\partial x} \tag{17.10}$$

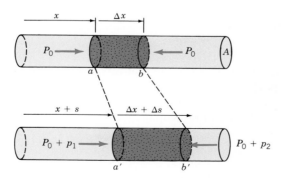

FIGURE 17.12 Under the influence of a longitudinal pulse, an element of length Δx at position x is moved to a new position $x + s$ and its length is changed to $\Delta x + \Delta s$. Note that p is the *change* in pressure relative to the equilibrium value P_0.

When this is used in Eq. 17.9, we obtain the wave equation

$$\frac{\partial^2 s}{\partial x^2} = \frac{\rho}{B}\frac{\partial^2 s}{\partial t^2} \qquad (17.11)$$

Comparison with the wave equation (Eq. 16.16) shows that the wave velocity is

(Fluid) $$v = \sqrt{\frac{B}{\rho}} \qquad (17.12)$$

This is the velocity of longitudinal waves within a gas or a liquid.

It is useful to relate the amplitude of the displacement fluctuations to the amplitude of the pressure fluctuations. For a harmonic wave, the displacement is given by

$$s = s_0 \sin(kx - \omega t) \qquad (17.13)$$

From Eq. 17.10, therefore,

$$p = -Bks_0 \cos(kx - \omega t)$$
$$= -p_0 \cos(kx - \omega t) \qquad (17.14)$$

Since $k = \omega/v$ and $B = \rho v^2$, we see that

$$p_0 = \rho \omega v s_0 \qquad (17.15)$$

Comparison of Eq. 17.13 with Eq. 17.14 shows that s and p are $\pi/2$ out of phase. This characteristic was deduced qualitatively for sound waves in Section 17.1.

17.6 SOUND INTENSITY

Consider a harmonic sound wave propagating along a tube of cross-sectional area A, as shown in Fig. 17.13. The quantity p is the *excess* pressure caused by the wave, and $\partial s/\partial t$ is the velocity of an element of the fluid. The instantaneous power supplied by the wave to the element is

$$P = Fv = pA\frac{\partial s}{\partial t}$$

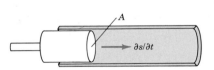

FIGURE 17.13 The motion of a piston, or a longitudinal pulse, produces an excess pressure p on one side of an element of air. The speed of the element is $\partial s/\partial t$.

Using Eq. 17.13 and Eq. 17.14, we have

$$P = p_0 A \omega s_0 \cos^2(kx - \omega t)$$

At any position, say $x = 0$, the average of $\cos^2 \omega t$ over one period is $\frac{1}{2}$; hence the average power transmitted by the wave is

$$P_{av} = \tfrac{1}{2}\rho A (\omega s_0)^2 v \qquad (17.16)$$

where we have used Eq. 17.15. Note that this has the same form as Eq. 16.16 for the power transported by a wave on a string.

The **intensity** I of a wave is defined as the energy incident per second per unit area normal to the direction of propagation:

$$I = \frac{\text{Power}}{\text{Area}} = \frac{P}{A} \qquad (17.17)$$

The SI unit of intensity is W/m². If we use $s_0 = p_0/\rho \omega v$ from Eq. 17.15 in Eq. 17.16, the average intensity of a sound wave is

$$I_{av} = \frac{p_0^2}{2\rho v} \qquad (17.18)$$

When expressed in terms of p_0, the intensity is proportional to the square of the pressure amplitude and is independent of frequency.

In the particular case of waves emitted by a point source, the radiated energy spreads uniformly over wave fronts that are spherical surfaces, as depicted in Fig. 17.14. Since the surface area of a sphere of radius r is $4\pi r^2$, the intensity at a distance r from a point source radiating with a power P is

(Point source) $$I = \frac{P}{4\pi r^2} \tag{17.19}$$

That is, $I \propto 1/r^2$, the intensity decreases as the inverse square of the distance from a point source.

EXERCISE 4. Express Eq. 17.18 for the average intensity in terms of s_0.

FIGURE 17.14 The energy emitted by a point source is spread over spherical wave fronts whose surface area is $4\pi r^2$. The intensity of the waves decreases as $I \propto 1/r^2$.

EXAMPLE 17.4: At 1 kHz, the minimum audible intensity is 10^{-12} W/m², whereas the maximum tolerable without pain is 1 W/m². Calculate the pressure and displacement amplitudes for: (a) the threshold of hearing; (b) the threshold of pain. The density of air is 1.29 kg/m³ and the speed of sound is 340 m/s.

Solution: From Eq. 17.18 the pressure amplitude is

$$p_0^2 = 2\rho v I_{av} \tag{i}$$

The displacement amplitude can be found from Eq. 17.15, $p_0 = \rho \omega v s_0$, where $\omega = 2\pi f = 6280$ rad/s.
(a) From (i) we have

$$p_0^2 = 2(1.29 \text{ kg/m}^3)(340 \text{ m/s})(10^{-12} \text{ W/m}^2) = 877 \times 10^{-12} \text{ Pa}^2$$

Thus $p_0 = 2.96 \times 10^{-5}$ Pa. This should be compared with the equilibrium air pressure, which is about 10^5 Pa. The displacement amplitude is

$$s_0 = \frac{p_0}{\rho \omega v} = \frac{2.96 \times 10^{-5} \text{ Pa}}{(1.29 \text{ kg/m}^3)(6280 \text{ rad/s})(340 \text{ m/s})}$$
$$= 1.07 \times 10^{-11} \text{ m}$$

This is an astonishing result when one realizes that the size of an atom is approximately 10^{-10} m! If the ear were any more sensitive we would constantly hear the flow of blood in nearby veins.
(b) $p_0^2 = 2(1.29 \text{ kg/m}^3)(340 \text{ m/s})(1 \text{ W/m}^2)$, from which we find $p_0 = 29.6$ Pa. This is still only a tiny fraction of the equilibrium pressure. From this p_0 we find $s_0 = 1.07 \times 10^{-5}$ m.

Intensity Level; The Decibel Scale

In the last example it was noted that the intensities the human ear can hear range from 10^{-12} W/m² to 1 W/m² (an enormous factor of 10^{12}!). The intensity of a sound is perceived by the ear as the subjective sensation of loudness. However, if the intensity doubles, the loudness does not increase by a factor of 2. Experiments first carried out by A. G. Bell showed that to produce an apparent doubling in loudness, the intensity of sound must be increased by a factor of about 10. Therefore, it is convenient to specify the **intensity level** β in terms of the decibel (dB), which is defined as

$$\beta = 10 \log \frac{I}{I_0} \tag{17.20}$$

where I is the measured intensity and I_0 is some reference value. If one takes I_0 to be 10^{-12} W/m², then the threshold of hearing corresponds to $\beta = 10 \log 1 = 0$ dB. At the threshold of pain, 1 W/m², the intensity level is

$$\beta = 10 \log \left(\frac{1 \text{ W/m}^2}{10^{-12} \text{ W/m}^2}\right) = 120 \text{ dB}$$

A list of the intensity levels of various sources is given in Table 17.1.

TABLE 17.1 INTENSITY LEVELS (dB)

Threshold of hearing	0
Leaves rustling	10
Quiet hall	25
Office	60
Conversation	60
Heavy traffic (3 m)	80
Loud classical music	95
Loud rock music	120
Jet engine (20 m)	130

EXAMPLE 17.5: The sound emitted by a source reaches a particular position with an intensity I_1. What is the change in intensity level when another identical source is placed next to the first? (There is no fixed phase relation between the sources.)

Solution: If the initial and final intensities are I_1 and I_2, then the two intensity levels are

$$\beta_1 = 10 \log \frac{I_1}{I_0}; \qquad \beta_2 = 10 \log \frac{I_2}{I_0}$$

The change in level is (note that $\log(A/B) = \log A - \log B$)

$$\beta_2 - \beta_1 = 10 \log \frac{I_2}{I_1}$$
$$= 10 \log 2 = 3 \text{ dB}$$

Thus, when the intensity doubles the intensity level changes by 3 dB. The response of the ear roughly corresponds to this logarithmic scale. The smallest change in level that can be detected by the human ear is about 1 dB.

EXAMPLE 17.6: A speaker emits 0.8 W of acoustic power. Assume that it behaves as a point source which emits uniformly in all directions. At what distance will the intensity level be 85 dB?

Solution: From Eq. 17.19 we know that the intensity of waves from a point source decreases as the inverse square of the distance r; that is,

$$I = \frac{P}{4\pi r^2} \qquad \text{(i)}$$

We must find the intensity corresponding to an 85-dB sound level:

$$85 = 10 \log \frac{I}{I_0}$$

Thus, $\log(I/I_0) = 8.5$, or

$$I = 10^{-12} \times 10^{8.5} = 10^{-3.5}$$
$$= 3.16 \times 10^{-4} \text{ W/m}^2 \qquad \text{(ii)}$$

Using (ii) in (i) we find

$$r^2 = \frac{P}{4\pi I}$$
$$= \frac{(0.8 \text{ W})}{4(3.14)(3.16 \times 10^{-4} \text{ W/m}^2)} = 201 \text{ m}^2$$

Thus, $r = 14.1$ m.

EXERCISE 5. The power output of an amplifier drops by 1.5 dB at low frequencies relative to 50 W at 1 kHz. What is the low-frequency power output?

17.7 FOURIER SERIES (Optional)

The sinusoidal wave functions we have employed so far are rarely encountered in practice. An almost purely sinusoidal wave is formed by the pressure variations produced by a tuning fork, as shown in Fig. 17.15a. In general, however, periodic wave functions have complex shapes. Figure 17.15b shows the pressure variations associated with a musical instrument playing a musical note of the same fundamental frequency as the tuning fork. Both would have the same apparent musical pitch, but the instrument has a "richer" and perhaps more pleasing sound. The richness arises from the fact that the wave function is the result of the superposition of a large number of harmonic waves of different amplitudes and frequencies. In 1807, Joseph Fourier showed that any reasonably well-behaved periodic function may be generated by the superposition of a sufficient number of sine or cosine functions. According to **Fourier's theorem,** the function is represented by the infinite sum

$$F(t) = \Sigma(a_n \sin n\omega t + b_n \cos n\omega t) \qquad (17.21)$$

where $\omega = 2\pi/T = 2\pi f$. The function has been decomposed into harmonic components whose frequencies are integer multiples of f, the frequency of the function. The Fourier coefficients a_n and b_n indicate the amplitude of the nth harmonic function. The process of determining these coefficients is called **Fourier analysis.**

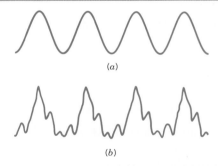

(a)

(b)

FIGURE 17.15 (a) A tuning fork produces a sinusoidal variation in pressure. (b) A hypothetical pressure fluctuation produced by a musical instrument.

Consider the periodic function depicted in Fig. 17.16. It is a square function for which $F(t) = +A$ between $t = 0$ and $t = T/2$, and $F(t) = -A$ between $t = T/2$ and $t = T$. The period of the function is T. (If the function were periodic in space, we might prefer to label the period by L or λ.) It can be shown that

$$F(t) = \frac{4A}{\pi}\left(\sin \omega t + \frac{1}{3} \sin 3\omega t + \frac{1}{5} \sin 5\omega t + \cdots\right)$$

In this particular case, only the sine functions are involved; furthermore, only the odd harmonics are present. The first

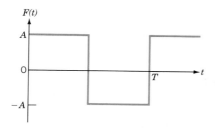

FIGURE 17.16 A "square wave" signal with a period T.

A useful method of presenting the results of the Fourier analysis of a function is by its *harmonic spectrum*, in which the relative amplitudes of the harmonic components are displayed. For example, Fig. 17.19a represents the output of a tuning fork—which has just one harmonic component, the fundamental. The square wave discussed earlier has only odd harmonics whose strength decreases monotonically as the frequency increases (Fig. 17.19b). A note from a musical instrument usually has a complex harmonic structure, such as that shown in Fig. 17.19c. Although two different instruments may sound the same note, their different harmonic structure allows us to easily identify them, for example, as a guitar and a piano.

three harmonic terms, each with its appropriate amplitude, are drawn in Fig. 17.17. When they are superimposed as in Fig. 17.18a we see that just three terms lead to a fairly good representation of $F(t)$. Figure 17.18b shows the effect of including ten terms.

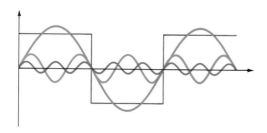

FIGURE 17.17 Three of the component signals used to synthesize a square wave.

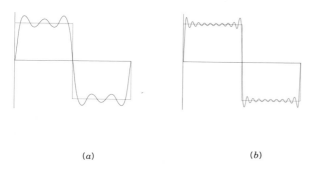

(a)   (b)

FIGURE 17.18 (a) The resultant of the three signals shown in Fig. 17.17 (b) When ten components are added, the resultant more closely resembles a square wave.

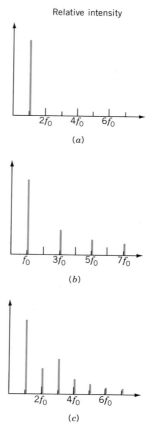

FIGURE 17.19 Sound spectra—displayed as harmonic components of the original signal. (a) A tuning fork has a single harmonic component. (b) The Fourier components of a square wave. (c) The harmonic components of a hypothetical musical instrument.

SUMMARY

The speed of a **longitudinal wave in a fluid** of density ρ is given by

$$v = \sqrt{\frac{B}{\rho}}$$

where B is the bulk modulus. **Sound waves** in air involve longitudinal oscillations of the molecules. These waves are characterized as pressure or density fluctuations.

The **resonant frequencies** of air columns in pipes are

(Open pipe) $f_n = \dfrac{nv}{2L}$ $(n = 1, 2, 3, \ldots)$

(Closed pipe) $f_n = \dfrac{nv}{4L}$ $(n = 1, 3, 5, \ldots)$

The apparent change in frequency when there is relative motion between a source of waves and an observer is called the **Doppler effect.** The observed frequency f' is related to the natural frequency f_0 by

$$f' = \left(\frac{v \pm v_O}{v \pm v_S}\right) f_0$$

where v is the speed of sound, v_O is the speed of the observer and v_S is the speed of the source. The signs are chosen by noting that any relative motion of source or observer toward the other tends to increase f'; any relative motion away from each other tends to decrease f'.

In a sound wave of angular frequency ω and speed v, the amplitude p_0 of the pressure deviation is related to the displacement amplitude s_0 by

$$p_0 = \rho \omega v s_0$$

The **intensity** I of a wave is defined as the power incident per unit area,

$$I = \frac{P}{A}$$

The **average intensity** transmitted by a sound wave is given by

$$I_{av} = \tfrac{1}{2}\rho(\omega s_0)^2 v = \frac{p_0^2}{2\rho v}$$

For a point source that radiates uniformly in all directions, the intensity at a distance r is

$$I = \frac{P}{4\pi r^2}$$

The **intensity level** β of a sound wave of intensity I is defined as

$$\beta = 10 \log \frac{I}{I_0}$$

where the reference intensity is $I_0 = 10^{-12}$ W/m^2. The unit of β is the decibel (dB).

ANSWERS TO IN-CHAPTER EXERCISES

1. Use $\lambda = v/f$. (a) $\lambda = 1.88$ cm; (b) $\lambda = 0.43$ cm.
2. The fundamental would yield

$$f_1 = \frac{v}{4L_1} = \frac{340 \text{ m/s}}{4 \times 0.189 \text{ m}} = 450 \text{ Hz}$$

   The third would yield

$$f_3 = \frac{3v}{4L_3} = 3 \times \frac{340 \text{ m/s}}{4 \times 0.575 \text{ m}} = 443 \text{ Hz}$$

3. (a) The motions of both the source and the observer tend to increase the observed frequency

$$f' = \left(\frac{v + v_O}{v - v_S}\right) f_0 = \left(\frac{375}{300}\right)(1200) = 1500 \text{ Hz}.$$

(b) The motions of both the source and the observer tend to decrease the observed frequency

$$f' = \left(\frac{v + v_O}{v + v_S}\right) f_0 = \left(\frac{325}{400}\right)(1200) = 975 \text{ Hz}.$$

4. From Eq. 17.15,

$$I = \frac{P}{A} = \tfrac{1}{2}\rho(\omega s_0)^2 v$$

5. Since $P \propto I$, we have

$$\beta = -1.5 = 10 \log(P_2/P_1),$$

so $P_2/P_1 = 10^{-0.15}$ or $P_2 = P_1(0.708) = 35.4$ W.

QUESTIONS

1. Does the speed of sound within the audible range of frequencies depend on wavelength? What evidence do you have in support of your answer?
2. If the temperature changes during an outdoor concert, would you expect the instruments to get out of tune?
3. Is it possible to measure temperature by using the vibrations of a tuning fork?
4. Suppose the interval between a lightning flash and the associated thunder is T seconds. The distance (in km) to the flash is numerically equal to $T/3$. Explain why.
5. The intensity of sound waves emitted by a point source actually decreases with distance somewhat faster than predicted by the inverse-square law. What is the reason for this?
6. Two sound waves have equal pressure amplitudes, but the frequency of one is twice that of the other; that is, $f_1 = 2f_2$. Compare (a) the displacement amplitudes; (b) the intensities.
7. What is the purpose of the flare at the end of a trumpet or a horn?
8. Why is your singing more impressive in a shower stall? Does the position of your mouth relative to the walls or ceiling have any affect?
9. A thin wine glass can be made to "sing" when a wet finger is rubbed along the rim. Why does this happen?
10. In some submersibles, divers work in an atmosphere in which helium replaces the nitrogen in ordinary air. Why do their voices sound unnaturally high.
11. A given sound wave has frequency f and displacement amplitude s_0. A second wave has half the frequency but double the amplitude. Compare their intensities.

EXERCISES

Unless otherwise specified, take the speed of sound in air to be 340 m/s and the density of air to be 1.29 kg/m³.

1. (I) Bats emit high-frequency sound to locate objects. The highest frequency emitted by a bat is 10^5 Hz. What is the wavelength of this signal?
2. (I) Dogs can hear sound with frequencies up to 35,000 Hz. What is the wavelength of such waves?
3. (I) In a medical ultrasound examination waves with a frequency of 4 MHz are used. If the speed of sound in tissue is 1500 m/s, what is the wavelength of these waves in tissue?
4. (I) There is an explosion on a boat. The sonar detector of another ship picks up the signal 3.2 s before the sailors on deck hear the sound. How far away was the boat from the ship? The speed of sound in water is 1500 m/s.
5. (I) (a) Calculate the speed of longitudinal sound waves in liquid mercury for which the bulk modulus is 2.8×10^{10} N/m² and the density is 13.6 g/cm³. (b) What would be the wavelength of a 1000-Hz sound wave in mercury?
6. (I) Calculate the speed of longitudinal waves in the following gases given data for 0 °C and a pressure of 1 atm: (a) oxygen for which the bulk modulus is 1.41×10^5 N/m² and the density is 1.43 kg/m³; (b) helium for which the bulk modulus is 1.7×10^5 N/m² and the density is 0.18 kg/m³.

7. (I) The speed of *longitudinal waves in a rod* is given by

$$v = \sqrt{\frac{Y}{\rho}}$$

where Y is Young's modulus. Calculate the speed of sound in a steel rod for which $Y = 2 \times 10^{11}$ N/m² and $\rho = 7.8$ g/cm³.

8. (II) The speed of longitudinal waves in a lead pipe is 1320 m/s. Consider a series of connected lead pipes of total length 100 m. If one end is tapped with a hammer, why will two taps be registered at the other end? Estimate the difference in time between the arrivals of the two sounds.

9. (I) The speed of *transverse waves in an infinite solid* is given by

$$v = \sqrt{\frac{S}{\rho}}$$

where S is the shear modulus. Calculate the speed of such waves in aluminum for which $S = 2.5 \times 10^{10}$ N/m² and $\rho = 2.7$ g/cm³.

10. (I) The speed of *longitudinal waves in an infinite solid* is given by

$$v = \sqrt{\frac{B + S/3}{\rho}}$$

where B is the bulk modulus and S is the shear modulus. Calculate the speed of such waves in copper for which $B = 1.4 \times 10^{11}$ N/m² and $S = 4.2 \times 10^{10}$ N/m². The density of copper is 8.92 g/cm³.

11. (II) Show that for a (longitudinal) wave in a fluid the pressure fluctuation p is related to the particle displacement s by $p = (B/v)(ds/dt)$, where B is the bulk modulus and v is the wave speed.

17.2 Resonance in Pipes

12. (I) The pressure variation in a standing sound wave has the form

$$p(x, t) = 4 \sin(5.3x) \cos(1800t) \text{ N/m}^2$$

Write the wave functions for the two component waves that make up this standing wave.

13. (I) A tuning fork with a frequency of 440 Hz is held at the open end of a tube as in Fig. 17.7. As the water level is lowered, what are the first and second lengths at which the air column resonates? (Ignore end corrections.)

14. (II) The fundamental frequencies for various open pipes are given in the table below:

L (cm):	18	35	52	76
f (Hz):	944	472	321	221

where L is the measured length. Plot a graph to determine the speed of sound.

15. (II) The 2nd harmonic of a string of length 60 cm and linear mass density 1.2 g/m has the same frequency as the 3rd harmonic of a closed pipe of length 1 m. Find the tension in the string.

16. (II) An air siren consists of a disk that has 40 uniformly spaced holes near its rim. Air from a small nozzle is blown through the holes as the disk rotates at 1200 rpm. (a) What is the frequency heard? (b) Would it sound as "pure" as the sound from a tuning fork? Explain.

17. (I) A straight and narrow open pipe is 20 m long. Estimate the values of the three lowest normal mode frequencies.

18. (I) A pipe has a fundamental frequency of 1 kHz at 20 °C. What would it be at 10 °C? (*Note:* The speed of sound in air varies according to $v \approx 20\sqrt{T}$ m/s where T is in kelvins.) Assume that the length is unchanged.

19. (I) A flute (which we treat as a pipe open at both ends) is 60 cm long. (a) What is the fundamental frequency when all the holes are covered? (b) How far from the mouthpiece should a hole be uncovered for the fundamental frequency to be 330 Hz?

20. (I) (a) What is the length of a closed organ pipe with a fundamental frequency of 25 Hz? (b) What is the length of an open organ pipe that has a fundamental frequency of 500 Hz?

21. (I) The speed of sound varies with temperature: At 20 °C it is 344 m/s, while at 5 °C it is 335 m/s. Consider an open pipe of length 30 cm. By how much does its fundamental frequency change when the temperature drops from 20 °C to 5 °C? Assume that the length is unchanged.

17.3 and 17.4 Doppler Effect; Beats

22. (I) The siren on a police car has a characteristic frequency of 1200 Hz. What is the frequency heard by a stationary observer if the car moves at 108 km/h (a) toward the observer; (b) away from the observer?

23. (I) A truck moving at 25 m/s emits a 400-Hz signal. Find the wavelength between the source and a stationary observer given that the truck is (a) approaching; (b) moving away.

24. (I) Repeat the previous exercise given that the source is stationary and that the observer moves at 40 m/s.

25. (I) A source emits sound with a characteristic frequency of 200 Hz. Calculate the observed frequency and the wavelength between source and observer in each of the following situations: (a) the source approaches a stationary observer at 40 m/s; (b) the observer approaches the stationary source at 40 m/s; (c) both the source and the observer move toward each other at 20 m/s relative to the ground.

26. (II) A battery-operated toy emits a signal at 1800 Hz. It moves in a circle of radius 1.2 m at 2.4 revolutions per second. What are the minimum and maximum frequencies heard by a stationary observer some distance away in the plane of the circle?

27. (II) A car moving at 40 m/s and a truck moving at 15 m/s travel along the same straight road. The car's horn has a natural frequency of 400 Hz. What is the change in the frequency observed by the truck driver as the car passes the truck? Assume that the car and truck are traveling (a) in the same direction; (b) in opposite directions.

28. (II) The siren of a police car moving at 40 m/s has a natural frequency of 600 Hz. A truck ahead of the car is moving at 20 m/s in the same direction. What is the frequency of the reflected sound heard by the police?

29. (I) A police car, whose siren has a natural frequency of 500 Hz, approaches a large wall at 30 km/h. A stationary observer detects the direct and the reflected waves. What is the beat frequency? (There are two possible answers.)

30. (I) Repeat the previous exercise given that the car moves away from the wall.

31. (II) A source of sound emits a signal at 600 Hz. This is observed as 640 Hz by a stationary observer as the source approaches. What is the observed frequency as the source recedes at the same speed?

17.5 Intensity, Intensity Level

32. (I) If a single person in the stands of a stadium shouts, the intensity level at the center of the field is 50 dB. What is the intensity level when 2×10^4 spectators are shouting from roughly the same distance?

33. (I) What is the power incident at an eardrum of area 0.4 cm², at the following intensity levels: (a) 120 dB (the threshold of pain); (b) 0 dB (the threshold of audibility)?

34. (II) The explosion of a firecracker in the air at a height of 40 m produces a 100-dB sound level at the ground below. What is the instantaneous total radiated power, assuming that it radiates as a point source?

35. (II) Two independent sound sources individually produce intensity levels of 80 dB and 85 dB at some point. What is the total intensity level at that point?

36. (I) (a) What is the intensity level for a sound of intensity 5×10^{-7} W/m²? (b) What is the intensity of a 75-dB sound wave?

37. (I) Find the intensity of each of the following sound waves in air: (a) a 600-Hz signal that has a displacement amplitude of 8 nm; (b) a 2-kHz signal with a pressure amplitude of 3.5 Pa.

38. (I) An amplifier has an 80-dB signal-to-noise ratio. What is the ratio of the power in the signal to that of the noise?

39. (I) Two 5-kHz sound waves differ in their intensity levels by 3 dB. What is the ratio of their displacement amplitudes?

40. (II) The intensities of two sounds of the same frequency differ by a factor of 1000. Find: (a) the difference in the intensity levels; (b) the ratio of the pressure amplitudes?

41. (II) A speaker is supplied with 40 W of electrical power at 1 kHz. Its efficiency in converting electrical power into acoustic power is 0.5%. Find the distance at which the sound is (a) painful (120 dB), (b) at the level of conversation (60 dB). Assume the speaker radiates uniformly as a point source.

42. (II) Figure 17.20 depicts a sound pulse, where the maximum longitudinal displacement $s_0 = 10^{-6}$ m and $d = 5$ cm. Find the speed of a "particle" (volume element) on both the leading and trailing edges.

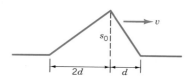

FIGURE 17.20 Exercise 42.

43. (II) The pressure variation in a sound wave is given by

$$p = 12 \sin\left(8.18x - 2700t + \frac{\pi}{4}\right) \text{ N/m}^2$$

Find: (a) the displacement amplitude; (b) the average intensity.

44. (I) The amplitude of the pressure variations in a 600-Hz sound wave in air is 0.3 N/m². What is the displacement amplitude?

45. (I) If the displacement amplitude of a harmonic sound wave is 2.4×10^{-9} m and the amplitude of the pressure variation is 4.2×10^{-2} N/m², what is the frequency of the sound?

PROBLEMS

1. (I) A point source of sound radiates at an acoustic power of 10^{-3} W at 240 Hz. At a distance of 4 m find: (a) the intensity; (b) the intensity level; (c) the pressure amplitude; (d) the displacement amplitude.

2. (I) A speaker vibrates at a frequency of 80 Hz and produces a pressure amplitude of 10 Pa 1 m from the speaker. Find: (a) the displacement amplitude 5 m from the speaker; (b) the intensity level (in dB) at 5 m. Assume the speaker radiates as a point source.

3. (I) The speed of sound in air is given approximately by $v \approx 20\sqrt{T}$, where T is the temperature in kelvins (K). (a) Show

that the fractional change in the speed of sound ($\Delta v/v$) caused by a fractional change in temperature ($\Delta T/T$) is

$$\frac{\Delta v}{v} = 0.5 \frac{\Delta T}{T}$$

(b) An organ pipe has a fundamental frequency of 400 Hz at 285 K. What is the fundamental frequency at 305 K? Assume the pipe length is unchanged. (c) Compare the percentage change in frequency to that of a semitone, which is about 6%.

4. (I) A pipe has two consecutive resonance frequencies of

607 Hz and 850 Hz. (a) Is the pipe open or closed? (b) What is the fundamental frequency? The speed of sound is not known.

5. (II) A stone is dropped into a well and the splash is heard 2.2 s later. What is the depth of the well?

6. (II) Figure 17.21 shows how the length of an air column can be continuously varied by raising or lowering a water reservoir. When a tuning fork is held at the open end, the air column first resonates when the measured length is 18.2 cm and then at 55.7 cm. The effective resonating length is the measured length plus $0.6r$, where r is the radius of the tube. (a) Find the radius of the tube. (b) If the speed of sound is 340 m/s, what is the frequency of the tuning fork?

FIGURE 17.21 Problem 6.

7. (I) Show that the wave function of a wave emitted by a point source has the form $y = (A/r) \sin(kr - \omega t)$. (*Hint:* Consider the variation of intensity with radial distance.)

8. (I) The musical note G is 7 semi-tones above C− whose frequency we take as 261.63 Hz. In the *diatonic* scale the ratio of the frequencies is $f_G/f_C = \frac{3}{2}$. In the *equally tempered* scale, the frequency changes by a factor of $2^{1/12}$ from one semitone to the next. What would be the beat frequency heard if the G's were played simultaneously?

9. (II) A guitar string of length 52 cm with a linear mass density of 2 g/m vibrates at its fundamental frequency of 400 Hz. When an open pipe resonates in its fundamental mode, one hears a beat frequency of 4 Hz. (a) What are the possible frequencies of the pipe? When the string is tightened, the beat frequency decreases. Find: (b) the original tension in the string, and (c) the length of the pipe.

10. (I) The acoustic wave impedance Z of a medium may be found by writing the intensity of a sound wave, with an angular frequency ω and a displacement amplitude A, in the form

$$I = \tfrac{1}{2}(\omega A)^2 Z$$

Show that for a fluid $Z = \sqrt{B\rho}$. (It is the change in Z at the boundary between two fluids that determines the phase of a reflected pulse.)

CHAPTER 18

Temperature, Thermal Expansion, and the Ideal Gas Law

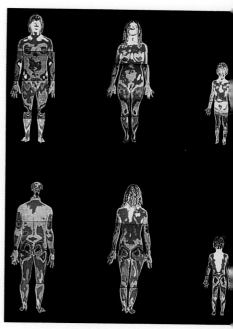

A color-coded scan of infrared radiation from the human body reveals small differences in skin temperature.

Major Points

1. The **Celsius, Fahrenheit,** and **Kelvin** temperature scales.
2. The concept of **thermal equilibrium** and the **zeroth law of thermodynamics.**
3. The **equation of state of an ideal gas.**
4. **Thermal expansion.**

The next four chapters deal with **thermodynamics.** The subject was given its name in 1854 by Lord Kelvin, who thought of it as the study of the "dynamical action of heat." Its origin lay in the study of the mechanical work performed by steam engines as a result of the flow of heat. In modern terms, thermodynamics deals with changes in the macroscopic variables that characterize a system, such as pressure, volume, and temperature, caused by **heat** exchanged with the surroundings and by **work** performed by the system on its surroundings. It leads to conclusions that are independent of the microscopic structure or composition of the system. This is the source of its generality and its power.

The **first law of thermodynamics** (Chapter 19) is a generalization of the conservation of energy to include heat. The **second law of thermodynamics** (Chapter 21) makes general statements regarding the efficiencies of heat engines, chemical equilibrium, the transmission of information, and the direction in which natural processes evolve. The **kinetic theory** of gases (Chapter 20) provides some insight into how the macroscopic behavior of a system is the result of the statistical behavior of a large number of particles. Fundamental to all aspects of thermodynamics is the concept of **temperature** with which we begin this chapter.

18.1 TEMPERATURE

The concept of temperature is based on our sense of hot and cold. However, our perception of the hotness or coldness of a body is deceptive. For example, a metal door knob feels colder than the wooden door to which it is attached, even though they are at the same temperature. In 1690 John Locke suggested a simple experiment to demonstrate the unreliability of our perception of hot and cold. Place one hand in hot water and the other in cold. Then place both in water at some intermediate temperature. This water will seem cool to one hand, but warm to the other. Clearly we require a more reliable means of establishing the temperature of an object.

A **thermometer** is an instrument that measures temperature. Any property of a substance or a device that changes when it is heated or cooled may be used as the basis of a thermometer. For example, the change in temperature may be defined to be proportional to the change in length of a column of liquid in a capillary tube, to the change in pressure of a gas held at constant volume, or to the change in electrical resistance of a wire.

Since temperature may be defined in terms of a variety of properties, it is not surprising that they lead to different values of temperature. Even in the case of a single property, say electrical resistance, two different materials do not necessarily respond in the same way when their temperature is raised. Therefore, not only does the value of temperature depend on the property chosen, it also depends on the particular substance.

Around 1595, Galileo built the first device to indicate changes in temperature. His "thermoscope," shown in Fig. 18.1a, consisted of a glass bulb with a long stem. Galileo first heated the air in the bulb (by the warmth of his hands) and then inverted the device with the stem into a bowl of colored liquid. As the air cooled, the liquid rose into the tube. Subsequent heating or cooling of the bulb caused the level in the tube to rise or fall. The thermoscope was a sensitive device, but as was realized only later, since the bowl was open to the air, the readings were affected by changes in atmospheric pressure. By the middle of the seventeenth century, thermometers were based on the expansion of a liquid, such as colored alcohol, within a thin glass stem (Fig. 18.1b). By 1724, D. G. Fahrenheit had perfected the technique of making uniform capillary tubes for this kind of thermometer.

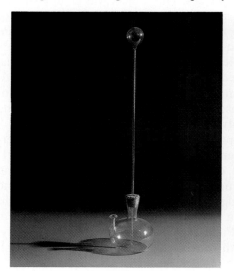 

FIGURE 18.1 (a) Galileo's thermoscope. (b) Thermometers made at the Academia del Cimento around 1650.

18.2 TEMPERATURE SCALES

In order to calibrate a thermometer, such as the liquid-in-glass type, we need to assign numerical values to the temperatures of two *fixed points*. In 1742, A. Celsius devised a scale based on the freezing point and the boiling point of water.

At the freezing point, liquid water and ice can coexist in equilibrium; that is, there is no tendency for one to transform into the other. At the boiling point, the liquid and the gas are in equilibrium. Since both of these points depend on the pressure, this is specified to be 1 atm. The positions of the liquid at these two temperatures are marked, and the distance between them is divided into equal intervals. In the **Celsius scale** there are 100 intervals or Celsius degrees (C°). Celsius chose the boiling point temperature to be zero and the freezing point to be 100! This bizarre choice was soon reversed, and the freezing point temperature was assigned the value zero degrees Celsius, 0 °C, and the boiling point, the value 100 °C. In the **Fahrenheit scale** there are 180 intervals, or Fahrenheit degrees (F°), between the freezing point and boiling point. The freezing point is at thirty two degrees Fahrenheit, 32 °F, and the boiling point is at 212 °F (see Fig. 18.2).

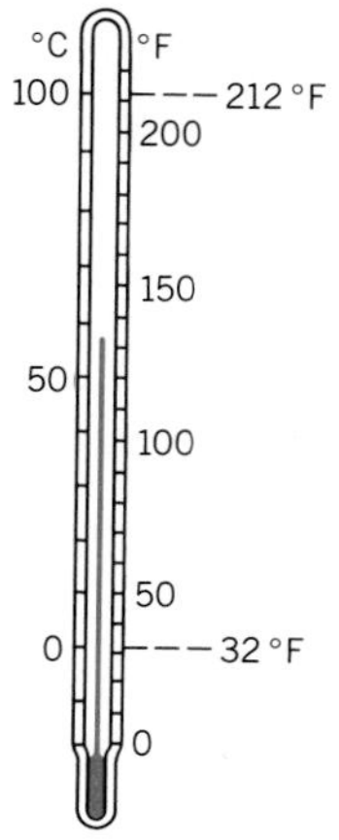

The apparently straightforward procedure outlined above is not without difficulties. The thermal expansion of a given liquid between, say, 10 °C and 20 °C, is not necessarily the same as that between, say, 60 °C and 70 °C. In other words, the thermal expansion is not uniform; it varies along the scale. Consequently, mercury and alcohol thermometers will agree at the two fixed points, but not necessarily at any other. When they are immersed in the same water bath, one thermometer may read 65.0 °C, while the other may read 64.8 °C. Another problem with such thermometers is that their range is limited. For example, mercury freezes at −39 °C. Clearly, we need a thermometer that can cover a large range and also serve as a laboratory standard. The constant-volume gas thermometer (Section 18.5) meets this requirement.

FIGURE 18.2 A liquid-in-glass thermometer.

EXERCISE 1. (a) Obtain an equation that allows one to convert a temperature t_C on the Celsius scale to the equivalent temperature t_F on the Fahrenheit scale. (b) What is 20 °C on the Fahrenheit scale?

18.3   THE ZEROTH LAW OF THERMODYNAMICS

The state of any system, such as a gas in a flask, may be specified by a certain number of macroscopic **state variables,** such as temperature, mass, pressure, and volume. In general, one would also have to include electrical and magnetic properties, chemical composition, and so on, but we ignore such complications. Suppose a system is isolated in a container with insulating walls that allow, ideally, no exchange of energy with the surroundings. In practice one might use thermal insulators such as Styrofoam or fiberglass wool. Consider, for example, a gas confined to a cylinder with a piston. If the piston is suddenly moved to compress the gas, the pressure will initially be greater near the piston than further along the cylinder. However, after some time the gas will reach an equilibrium state in which the pressure is uniform throughout the cylinder. Similarly, if the cylinder is heated at one end, the temperature will not be uniform. After the source of heat is removed, the system will ultimately reach a state in which all points have the same temperature. Initially, the state variables undergo changes, but after a sufficiently long time they stop changing. When all its state variables are constant in time, the system is said to be in **thermal equilibrium.** In this condition, the state variables have single values that characterize the whole system.

State variable

Thermal equilibrium

Now suppose two objects are placed on either side of an insulating wall within a container that isolates them from the environment. The wall prevents changes in one object from affecting the other. The state variables of the objects will change until each reaches thermal equilibrium. Next, the insulating wall is replaced by a thermal conductor, such as a thin metal sheet. The objects are now in good thermal contact, which means that they can exchange energy if one is hotter than the other. The state variables of both objects change to new values which, after a sufficiently long time, stay constant. The two systems are then in thermal equilibrium with each other. At this point, we find that the temperatures of the two objects are the same. *Two systems are in thermal equilibrium with each other if their temperatures are the same.* Although other state variables of the two systems are constant in time, thermal equilibrium does not necessarily mean that they have the same values for both systems. For example, the gases in two containers at the same temperature may have different pressures and volumes.

Consider three systems A, B, and C, which is a thermometer. Suppose A is in thermal equilibrium with C, and B is in thermal equilibrium with C. The question is whether A is also in thermal equilibrium with B. The answer is not obvious. For example, given that there is a mutual attraction between a magnet and two nails, it does not follow that the two nails attract each other. The results of experiments indicate that the answer is yes: A will be in thermal equilibrium with B. This fact is embodied in the **zeroth law of thermodynamics:**

> Two bodies in thermal equilibrium with a third are also in thermal equilibrium with each other.

The rather odd name of this law arises from the fact that the first and second laws of thermodynamics, to be discussed in Chapters 19 and 21, had already been formulated when its importance was realized.

According to the zeroth law, the two systems in thermal equilibrium do not have to be in actual thermal contact; it is sufficient for them to have the same stable temperature. This is an underlying assumption in the use of thermometers. Suppose a thermometer has the same reading when it is placed in contact with two separate systems, each of which is in thermal equilibrium. Then, if they are placed in thermal contact, the systems will remain in thermal equilibrium and register the same temperature.

18.4 THE EQUATION OF STATE OF AN IDEAL GAS

In 1662, Robert Boyle found that the volume of a gas kept at constant temperature is inversely proportional to the pressure: $V \propto 1/P$. Or, equivalently

(Boyle) $$PV = \text{constant}$$ (Fixed temperature)

Around 1800, J. Charles and J. L. Gay-Lussac independently found that, at a fixed pressure, the change in volume of a gas is proportional to the change in temperature. Figure 18.3a illustrates how the volume varies with the Celsius temperature t_C for different gases, or different amounts of a single gas. An interesting feature of this graph is that when the various lines are extrapolated, they all intersect the temperature axis at the same point, $-273.15\,°C$. (They must be extrapolated because all real gases liquefy above this temperature.) This point corresponds to the absolute zero of temperature in the Kelvin temperature scale, which will be defined in Section 18.5. The **Kelvin temperature** T is measured in kelvins (K) and is related to the Celsius temperature t_C by

$$T = t_C + 273.15$$

In terms of the Kelvin temperature, the law of Charles and Gay-Lussac is

(Charles and Gay-Lussac) $$V \propto T$$ (Fixed pressure)

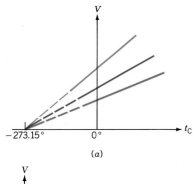

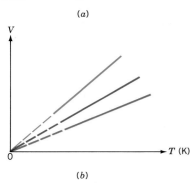

FIGURE 18.3 The volume of a sample of gas as a function of (a) the Celsius temperature t_C and (b) the Kelvin temperature $T\ (= t_C + 273.15)$. The different lines are for different gases.

and is shown in Fig. 18.3*b*. Gay-Lussac also found that at fixed volume, the change in pressure is proportional to the change in temperature. In terms of the absolute temperature, his result may be written as

(Gay-Lussac) $P \propto T$ (Fixed volume)

and is shown in Fig. 18.4. The results of all three investigators may be combined into one:

$$PV \propto T$$

The value of PV depends on the number of molecules N of the gas that are present. At a given pressure and temperature, $V \propto N$; at a given volume and temperature, $P \propto N$. Thus, $PV \propto N$. (Why not N^2?) Combining this with $PV \propto T$, we obtain the **equation of state for an ideal gas:**

(Ideal gas) $PV = NkT$ (18.1)

where k is a constant of proportionality called **Boltzmann's constant:**

$$k = 1.38 \times 10^{-23} \text{ J/K}$$

Instead of specifying the large number of molecules in a sample, it is often convenient to specify the number of moles. One **mole** (mol) of any substance contains as many elementary units (such as atoms or molecules) as the number of atoms in 12 g of the isotope carbon-12. (Isotopes are discussed in Chapter 43.) This number is called **Avogadro's number:**

$$N_A = 6.02 \times 10^{23} \text{ mol}^{-1}$$

Note that mol^{-1} stands for number per mole. The mass of one mole of a substance is called its molecular mass M, which may be measured in g/mol or in the SI unit kg/mol (see Table 18.1). For example, molecular oxygen (O_2) has a molecular mass of 32 g/mol, which means 32 g of this gas contain N_A molecules.

The number of molecules N in n moles is

$$N = nN_A$$

Thus, Eq. 18.1 is often written as

(Ideal gas) $PV = nRT$ (18.2)

where

$$R = kN_A = 8.314 \text{ J/mol·K}$$

is called the **universal gas constant.**

Equation 18.1 (or 18.2) is called an **equation of state** because it expresses a relationship between the state variables of the system. The equation is valid only for equilibrium states in which P, V, and T have well-defined values. In deriving Eq. 18.1 we assumed that the behavior of the gases could be extrapolated to the origin as a straight line. A gas that obeys Eq. 18.1 is called an **ideal gas.** At low temperatures, the behavior of real gases deviates from the linear relationships $P \propto T$ and $V \propto T$. In fact, the lines do not even pass through the origin. However, when a real gas is sufficiently dilute, and far above the temperature at which it liquefies, the ideal gas equation of state is approximately valid. For example, air at atmospheric pressure and room temperature behaves very nearly like an ideal gas.

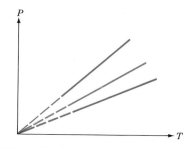

FIGURE 18.4 The variation of the pressure of various gases as a function of the Kelvin temperature T.

TABLE 18.1
MOLECULAR MASSES (g/mol)

Hydrogen	2.02
Helium	4
Nitrogen	28
Oxygen	32
Carbon dioxide	44

EXAMPLE 18.1: What is the volume of one mole of an ideal gas at 0 °C and 1 atm?

Solution: We first note that 0 °C = 273 K and 1 atm = 101 kPa. From Eq. 18.2, for one mole $n = 1$, we have

$$V = \frac{RT}{P} = \frac{(8.31 \text{ J/mol·K})(273 \text{ K})}{1.01 \times 10^5 \text{ N/m}^2}$$
$$= 22.4 \times 10^{-3} \text{ m}^3 = 22.4 \text{ L}$$

EXAMPLE 18.2: The absolute pressure in an automobile tire is 310 kPa at 10 °C. After a long drive the temperature rises to 30 °C. What is the new pressure?

Solution: Since the number of moles of gas is fixed, Eq. 18.1 tells us that

$$\frac{P_1 V_1}{T_1} = \frac{P_2 V_2}{T_2}$$

The volume of the tire does not change significantly; therefore, $V_1 = V_2$, and the above equation reduces to

$$\frac{P_1}{T_1} = \frac{P_2}{T_2}$$

After converting the temperatures to the absolute scale and inserting the numbers we find $P_2 = (303 \text{ K}/283 \text{ K})(310 \text{ Pa}) = 332$ Pa.

EXERCISE 2. A sample of gas at 20 °C has a volume of 0.6 L at a pressure of 0.8 atm. Find: (a) the number of moles; (b) the number of molecules.

18.5 CONSTANT-VOLUME GAS THERMOMETER (Optional)

In a constant-volume gas thermometer a gas is confined to a bulb that is placed in contact with the object, such as a liquid, whose temperature is to be measured (see Fig. 18.5). When the temperature of the gas changes, so does its pressure. If there is an increase in pressure, the level of mercury in tube A will fall. The reservoir R can be raised or lowered to restore the level to some fixed mark, so that the volume of the gas is held constant. The pressure is determined by measuring the difference in height between the levels of mercury in tubes A and B.

In order to form a temperature scale we need two fixed points. The boiling point and freezing point of water are not easy to determine accurately. In this gas thermometer, therefore, one fixed point is 0 K defined as the temperature (extrapo-

lated) at which $P = 0$. The second fixed point is the **triple point** of water—the point at which ice, liquid water, and water vapor coexist in equilibrium. This occurs at a unique temperature of 0.01 °C at a pressure of 4.58 mm Hg (610 Pa). The temperature of the triple point of water is defined to be 273.16 K on the absolute or Kelvin scale. Other temperatures are measured as follows.

We begin with a given amount of some gas, say nitrogen, in the bulb. The pressure in the bulb at the triple point temperature of water P_{tr} is recorded. The **absolute temperature** T at another point is defined in terms of the pressure P at that temperature by the linear relation

$$T = 273.16 \frac{P}{P_{tr}} \qquad (18.3)$$

If the amount of gas in the bulb is reduced, its pressure at the triple point will be lower. The temperature, according to Eq. 18.3, is also found to be slightly different. There are also slight discrepancies in the recorded temperatures when different gases are used. However, as Fig. 18.6 shows, the temperatures registered by the constant-volume gas thermometer tends to the same value as P_{tr} approaches zero, irrespective of the gas employed. As the gas in the bulb becomes very dilute, its properties approach that of an ideal gas because the molecules are farther apart on average and interact less frequently. The **ideal gas temperature** is defined in terms of the limiting value as P_{tr} tends to zero:

$$T = \lim_{P_{tr} \to 0} 273.16 \frac{P}{P_{tr}} \qquad (18.4)$$

Since the atoms of the inert gas helium interact very weakly, it has a very low liquefaction temperature. With helium as the working gas, the constant-volume gas thermometer can be used down to 1 K. The process of determining the ideal gas

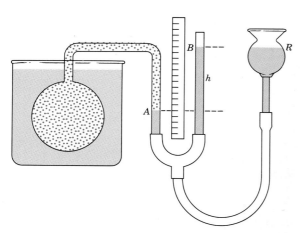

FIGURE 18.5 A constant-volume gas thermometer. The height of the reservoir R is adjusted so that the liquid in the left column is always at point A.

temperature is cumbersome, because the actual device is complex and is slow in responding to temperature changes. Nonetheless, this thermometer is an accurate and reproducible standard that can be used to calibrate other, more practical, thermometers.

The constant-volume gas thermometer is useful as a standard because all gases register the same temperature—provided they are dilute and far from their points of liquefaction. However, it still employs a particular substance, a gas. In Chapter 21 we will present another scale, the so-called thermodynamic scale, which is independent of the substance employed.

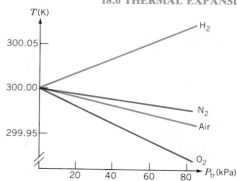

FIGURE 18.6 As the amount of gas in the bulb of a constant-volume thermometer is reduced, the temperature (defined by Eq. 18.3) decreases linearly for any gas. As the pressure at the triple point of water approaches zero, all gases lead to the same temperature.

18.6 THERMAL EXPANSION

Most materials expand when their temperature is increased. Railroad tracks, bridges, and clockwork mechanisms all have some means of compensating for thermal expansion. Figure 18.7a shows expansion slots in a bridge, and Fig. 18.7b shows what can happen to railroad tracks in hot weather. When a homogeneous object expands, the distance between *any* two points on the object increases. Figure 18.8 shows a block of metal with a hole in it. The expanded object is like a photographic enlargement. That is, the hole expands in the same proportion as the metal; it does not get smaller.

FIGURE 18.8 The thermal expansion of a plate with a hole is like a photographic enlargement—the hole also gets larger.

FIGURE 18.7 (a) Expansion slots in a bridge. (b) This railroad track buckled in hot weather.

A dish made by Corning can withstand great thermal stress.

TABLE 18.2
COEFFICIENTS OF
EXPANSION (20°C)

Linear α (10^{-6} K^{-1})	
Aluminum	24
Brass	18.7
Copper	17
Steel	11.7
Glass	9
Pyrex glass	3.2
Concrete	12

Volume β (10^{-4} K^{-1})	
Water	2.1
Ethyl alcohol	11
Mercury	1.8
Gasoline	9.5

The expansion of a solid may be discussed in terms of the change in any linear dimension. Consider a thin rod of original length L_0. It is found that the change in length ΔL is directly proportional to L_0 and to the change in temperature ΔT. This may be expressed as

$$\Delta L = \alpha L_0 \Delta T \qquad (18.5)$$

where α, measured in (°C)$^{-1}$ or K^{-1}, is called the **coefficient of linear expansion.** If Eq. 18.5 is written in the form

$$\alpha = \frac{\Delta L / L_0}{\Delta T} \qquad (18.6)$$

we see that the coefficient of linear expansion is the fractional change in length per unit change in temperature. The coefficient α is usually a function of temperature, so these equations are valid only for restricted values of ΔT. For some solids, such as wood, α also depends on direction within the material. Table 18.2 shows some average values in the vicinity of room temperature.

The difference in the coefficients of expansion of two metals can be used as a temperature-sensitive switch or a thermometer. Two metals having different expansion coefficients are welded together as a bimetallic strip (Fig. 18.9). As the temperature is increased, the strip bends toward the side with the lower expansion coefficient. The twisting of a bimetallic coil is used in some thermometers to control the rotation of the choke of a carburetor, in thermostats, and in electrical circuit breakers.

The thermal expansion of fluids is expressed in terms of the change in volume ΔV, which is found to be proportional to the change in temperature ΔT:

$$\Delta V = \beta V_0 \Delta T \qquad (18.7)$$

where V_0 is the original volume and β, measured in (°C)$^{-1}$ or K^{-1}, is called the **coefficient of volume expansion.** The coefficient β varies with temperature, and in the case of water is actually negative over a small temperature range (see Fig. 18.10). In the case of solids, $\beta = 3\alpha$, as is shown in Example 18.3.

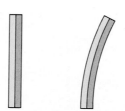

FIGURE 18.9 A bimetallic strip bends one way, or the other, when its temperature changes. The strip may be made into a coil.

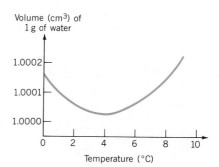

FIGURE 18.10 The volume of 1 g of water as a function of temperature. The highest density occurs at 4°C.

EXAMPLE 18.3: What is the relation between α and β for an isotropic solid—one whose properties do not depend on direction?

Solution: Consider a cube of side L. Since its volume is $V = L^3$, the rate of change of V with respect to L is $dV/dL = 3L^2$. Therefore, the differential change in volume dV associated with a differential change in length dL is

$$dV = 3L^2 dL$$

Dividing both sides by $V = L^3$, we have

$$\frac{dV}{V} = \frac{3dL}{L} \qquad \text{(i)}$$

Equations 18.5 and 18.7 may be written as $\Delta T = \Delta L/(\alpha L_0)$ and $\Delta T = \Delta V/(\beta V_0)$. For a given ΔT, we see from Eq. (i) that

$$\beta = 3\alpha$$

Note that for a given material, the change in volume of, say, a hollow cylinder, is the same as that of a solid cylinder.

EXAMPLE 18.4: The ends of a steel rod are fixed. What is the *thermal stress* in the rod when the temperature decreases by 80 K? Young's modulus for steel is $Y = 2 \times 10^{11}$ N/m^2.

Solution: When the temperature changes by ΔT the expected change in length is

$$\Delta L = \alpha L \Delta T \qquad \text{(i)}$$

where L is the original length. From the definition of Young's modulus, Eq. 14.5, we find that the stress in the rod is

$$\frac{F}{A} = Y\frac{\Delta L}{L} \qquad \text{(ii)}$$

Since the length of the rod is fixed, the ΔL associated with the stress in the rod must be equal and opposite to the ΔL that would have resulted from the change in temperature. From (i) and (ii) we find

$$\frac{F}{A} = -\alpha Y \Delta T$$

$$= -(11.7 \times 10^{-6} \text{ K}^{-1})(2 \times 10^{11} \text{ N/m}^2)(-80 \text{ K})$$

$$= 1.87 \times 10^8 \text{ N/m}^2$$

The rod is under a *tensile* stress. This value is about 2–3% of the breaking stress of most steels.

EXERCISE 3. Why do some glasses crack when hot water is poured into them?

The thermal expansion of water is interesting because it exhibits anomalous expansion between 0 and 4 °C. As Fig. 18.10 shows, between these two temperatures its volume decreases as the temperature rises. The maximum density of water is 1.000 g/cm^3 at 4.0 °C. This explains why water starts to freeze at the top of a lake. As the air temperature drops from some value above 4 °C, the cooler water at the top is denser and so it sinks. When the temperture of the water at the surface reaches 4 °C, this process stops. As the temperature decreases further, the cooler water at the surface has a lower density than that below, and therefore stays at the top. Finally it starts to freeze. If the water is deep enough, it remains liquid below the frozen top layer. This peculiar behavior of water allows fish and other aquatic life to survive through the winter.

At the atomic level, thermal expansion may be understood by considering how the potential energy of the atoms varies with distance. Recall from Section 8.8 that the equilibrium position of an atom will be at the minimum of the potential energy well if the well is symmetric. At a given temperature each atom vibrates about its equilibrium position and its average position remains at the minimum point. If the shape of the well is not symmetrical, as shown in Fig. 18.11, the average position of an atom will not be at the minimum point. When the temperature is raised the amplitude of the vibrations increases and the average position is located at a greater interatomic separation. This increased separation is manifested as expansion of the material.

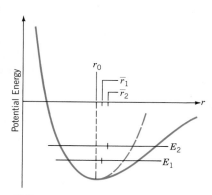

FIGURE 18.11 The potential energy of an atom. Thermal expansion arises because the "well" is not symmetrical about the equilibrium position r_0. As the temperature rises, the energy of the atom changes. The average position $\bar{r}$ when the energy is E_2 is not the same as that when the energy is E_1.

SUMMARY

The state of a system is characterized by a set of macroscopic variables such as pressure, temperature, and volume. When a system is in **thermal equilibrium** these quantities are constant. When two bodies are in thermal equilibrium with each other, they are at the same temperature.

According to the **zeroth law of thermodynamics** if bodies A and B are individually in thermal equilibrium with body C, then they are also in thermal equilibrium with each other.

The **equation of state** of N molecules or n moles of an *ideal gas* is

$$PV = NkT = nRT$$

where **Boltzmann's constant** k and the **universal gas constant** R are related by $R = kN_A$, where N_A is **Avogadro's number.** The temperature T is in kelvins.

When the temperature of a rod of length L is changed by ΔT its length changes by

$$\Delta L = \alpha L \Delta T$$

where α is the **coefficient of linear expansion.** When the temperature of an object of volume V is changed by ΔT, its volume changes by

$$\Delta V = \beta V \Delta T$$

where β is the **coefficient of volume expansion.**

ANSWERS TO IN-CHAPTER EXERCISES

1. (a) The temperature interval 100 C° corresponds to 180 F°, so $\Delta t_F = \frac{9}{5}\Delta t_C$. Since a temperature of 0 °C is the same as 32 °F, we have

$$t_F = \tfrac{9}{5}t_C + 32°$$

   (b) $t_F = (9/5)(20) + 32 = 68$ °F.

2. (a) The pressure is $P = (0.8)(1.01 \times 10^5 \text{ N/m}^2) = 8.08 \times 10^4$ N/m². From Eq. 18.2, the number of moles is

$$n = \frac{PV}{RT}$$

$$= \frac{(8.08 \times 10^4 \text{ N/m}^2)(0.6 \times 10^{-3} \text{ m}^3)}{(8.31 \text{ J/mol·K})(293 \text{ K})} = 0.02 \text{ moles}$$

   (b) We could use $N = nN_A$, or from Eq. 18.1, the number of molecules is

$$N = \frac{PV}{kT}$$

$$= \frac{(8.08 \times 10^4 \text{ N·m}^2)(0.6 \times 10^{-3} \text{ m}^3)}{(1.38 \times 10^{-23} \text{ J/K})(293 \text{ K})}$$

$$= 1.2 \times 10^{22} \text{ molecules}$$

3. The inner surface of the glass expands much more quickly than the outer surface. The resulting thermal stress in the glass may be great enough to cause a crack.

QUESTIONS

1. The lid on a tightly closed jar may be loosened by running hot water over it. Why does this work?

2. Is it possible for two bodies to be in thermal equilibrium with each other without being in physical contact?

3. When hot water is poured into an ordinary glass it is likely to crack. This does not happen with a Pyrex glass container. What is the difference between ordinary glass and Pyrex glass?

4. Why is it inadvisable to try to cool an overheated engine by letting cold water enter it or by spraying it with cold water?

5. A steel ball floats in mercury. Does it rise or sink as their common temperature is increased?

6. You are given a thermometer and asked to measure the outside air temperature. What precautions would you take?

7. For a given air temperature, it feels cooler when there is a breeze. Why is this?

8. When a cold mercury-in-glass thermometer is placed in hot water, the liquid column initially falls. Explain.

9. What are the disadvantages of using water in a liquid-in-glass thermometer?

10. One sometimes hears that the Fahrenheit scale is "more accurate." What is meant by this? How would you respond?

11. The sidewall of a tire indicates "35 psi max." After a long run on a hot day, the pressure is measured to be 38 psi. Should air be released from the warm tire?

12. Design a simple thermostat based on a bimetallic strip. How would you adjust it for different temperatures?

13. How does the value of a coefficient of linear expansion change when measured in $°F^{-1}$ rather than K^{-1}?

EXERCISES

18.2 Temperature Scales

1. (I) Convert the following temperatures to the Celsius scale: (a) a room at 70 °F; (b) an auto radiator at 195 °F; (c) normal body temperature of 98.6 °F.

2. (I) Convert the following temperatures to the Fahrenheit scale: (a) the melting point of lead at 327 °C; (b) the boiling point of hydrogen at −253 °C; (c) desert air at 45 °C.

3. (I) At what temperature are the numerical values on the Celsius and Fahrenheit scales identical?

4. (I) The length of the column in a mercury thermometer is 10 cm at 0 °C and 25 cm at 100 °C. Find: (a) the temperature that corresponds to 15 cm; (b) the length of the column at 70 °C.

5. (I) Some engineers find it convenient to use the Rankine scale, in which absolute zero is 0 °R and the degree intervals are those of the Fahrenheit scale. What is 0 °C on the Rankine scale?

6. (I) The temperature T on a thermometer is calibrated according to the relationship $T = (aR + b)$ °C, where R is the electrical resistance of a wire and a and b are constants. The resistance is 24 ohms at 0 °C and 35.6 ohms at 100 °C. Find: (a) the resistance at 60 °C; (b) the temperature when the resistance is 29 ohms.

18.4 Ideal Gas Law

7. (I) Show that at 0 °C and 1 atm, the number density of molecules in an ideal gas is 2.7×10^{19} molecules/cm^3.

8. (I) Show that the mass density ρ of n moles of a gas of particles with a molecular mass M within a volume V can be written as $\rho = nM/V$.

9. (I) Write the ideal gas law in terms of the mass density (measured in kg/m^3) of the gas. At 0 °C and 1 atm, find the density of the following: (a) nitrogen; (b) oxygen; (c) hydrogen.

10. (I) A cylinder contains 1 kg of oxygen at a pressure of 3 atm. (a) What would be the pressure if 1 kg of nitrogen replaces the oxygen at the same temperature? (b) What mass of nitrogen would produce a pressure of 2 atm at the same temperature?

11. (II) Two moles of helium gas are at 20 °C and a pressure of 200 kPa. (a) Find the volume of the gas. (b) If the gas is heated to 40 °C and its pressure reduced by 30%, what is the new volume?

12. (II) A log cabin has an internal volume of 220 m^3. (a) Find the mass of air contained at 20 °C and 1 atm. (b) If the cabin is warmed to 25 °C, find the new pressure, assuming that the cabin is airtight. (c) If the cabin were not airtight, what mass of air would escape when the temperature is raised? Take the "effective" molecular mass of molecules in air to be 29 g/mol.

13. (II) A car tire with an internal volume of 0.015 m^3 is filled to a gauge pressure of 200 kPa at 15 °C. After a long run, the pressure rises to 230 kPa. (a) What is the temperature of the air in the tire, assuming that the volume of the tire is unchanged? (b) What mass of air must be removed for the pressure to return to 200 kPa? (Add 1 atm ≈ 100 kPa to the gauge pressure to obtain the absolute pressure.) Take the "effective" molecular mass of molecules in air to be 29 g/mol.

14. (I) It is possible to achieve a pressure as low as 10^{-10} N/m^2. At this pressure, calculate the number of molecules of an ideal gas per cm^3 at 300 K.

15. (II) A cube of side 10 cm is filled with oxygen at 0 °C and 1 atm. The box is sealed and its temperature raised to 30 °C. What is the force exerted by the enclosed gas on each side of the box?

16. (II) (a) Two moles of an ideal gas are at a pressure of 100 kPa and have a volume of 16 L. What is the temperature? (b) The gas is heated at constant pressure until its volume is 32 L. What is the new temperature? (c) With the volume fixed at 32 L, the gas is heated until the temperature rises to 450 K. What is the new pressure?

17. (I) Two moles of nitrogen are at 3 atm and 300 K. (a) What is the volume of the gas? (b) It expands at constant temperature until the pressure drops to 1 atm. What is the new volume?

18.5 Constant-Volume Gas Thermometer

18. (I) In a constant-volume gas thermometer the pressure is 0.02 atm at 100 °C. Estimate: (a) the pressure at the triple point of water; (b) the temperature when the pressure is 0.027 atm.

19. (I) In a constant-volume gas thermometer the pressure in

the bulb at the triple point of water is 40 mm Hg. Estimate: (a) the pressure at 300 K; (b) the temperature when the pressure is 25 mm Hg.

18.6 Thermal Expansion

20. (I) A railroad is laid at 15 °C with steel tracks 20 m long. What is the minimum required separation between the ends of the tracks if the maximum temperature expected is 35 °C?

21. (I) A steel gauge block, which is used as a standard of length in machine shops, is 5.0000 cm long at 20 °C. Over what temperature range can it be used if an uncertainty of ±0.01 mm is acceptable?

22. (I) The scale on a steel meter stick is stamped at 15 °C. What is the error in a reading of 60 cm at 27 °C?

23. (I) A copper sphere of radius 2.000 cm is placed over a hole of radius 1.990 cm in an aluminum plate at 20 °C. At what common temperature will the sphere pass through the hole?

24. (I) A horizontal support beam in a house is made of steel and has a length 8 m at 20 °C. It extends from one wall to the other. During a fire its temperature rises to 80 °C. What is the increase in its length?

25. (I) The Eiffel tower, which is made of steel, is 320 m high at 20 °C. What is the change in its height over the range −20 °C to 35 °C?

26. (I) A steel I beam is installed at 20 °C between two fixed walls. What is the stress induced in the beam when the temperature increases to 35 °C? Young's modulus for steel is 2×10^{11} N/m².

PROBLEMS

1. (I) A vessel contains 5 moles of an ideal gas at 0 °C and 1 atm. It is heated at constant volume until its temperature is 100 °C. (a) What is the new pressure? (b) How many moles of gas should be allowed to escape for the pressure to return to 1 atm? (c) After the gas has escaped, the container is sealed and cooled to 0 °C. What is the pressure now?

2. (II) A hot air balloon has a volume of 1200 m³. The ambient air temperature is 15 °C and the pressure is 1 atm. To what temperature must the air in the balloon be raised for the balloon to just lift 200 kg (including the mass of the balloon)? Take the "effective" molecular mass of molecules in air to be 29 g/mol.

3. (I) Use the ideal gas law to show that the coefficient of volume expansion at constant pressure is $\beta = 1/T$.

4. (I) A spherical glass bulb is filled with 50 mL of water at 5 °C. The bulb has a neck of radius 0.15 cm that is initially empty. How high does the water rise in the neck at 35 °C? Include the effect of the expansion of the glass bulb.

5. (I) A pendulum clock has a rod with a period of 2 s at 20 °C. If the temperature rises to 30 °C, how much does the clock lose or gain in one week? Treat the rod as a physical pendulum pivoted at one end. (See Example 15.8.)

6. (II) Show that the fractional change in density of a liquid caused by a change in temperature is $\Delta\rho/\rho = -\beta\Delta T$ where β is the coefficient of volume expansion.

7. (II) A steel ball of radius 1.2 cm is in a cylindrical glass

beaker of radius 1.5 cm that contains 20 mL of water at 5 °C. What is the change in the water level when the temperature rises to 90 °C?

8. (I) A rectangular plate has a length L and a width W. The coefficient of linear expansion is α. Show that the change in area caused by a change in temperature ΔT is $\Delta A = 2\alpha A \Delta T$, where $A = LW$.

9. (I) A steel rod of radius 2 cm, whose ends are fixed, is subject to a tensile load of 15 kN. What is the stress if the temperature (a) falls by 10 C°; (b) rises by 10 C°? (Young's modulus for steel is 2×10^{11} N/m².)

10. (I) Concrete sections 18 m long are laid at 15 °C. (a) What is the required gap if the highest temperature expected is 35 °C? (b) What is the stress induced in the concrete if the temperature rises to 40 °C? (Young's modulus for concrete is 2×10^{10} N/m².)

11. (I) An air bubble has a radius of 2 mm at a depth of 12 m in water at temperature of 8 °C. What is its radius just below the surface where the temperature is 16 °C?

12. (I) A liquid completely fills a sealed container of fixed volume. Show that the pressure caused by a change in temperature ΔT is given by

$$\Delta P = \beta B \Delta T$$

where β is the coefficient of volume expansion and B is the bulk modulus.

CHAPTER 19

First Law of Thermodynamics

A thermogram records the heat radiated by two houses.

Major Points

1. The distinction between **heat** and **internal energy**.
2. The definition of the **specific heat** of a substance.
3. The **first law of thermodynamics** relates the *work* done by a system and the *heat* it exchanges with the environment to the change in its *internal energy*.
4. Changes in state associated with **quasistatic, isobaric, isothermal,** and **adiabatic** processes.
5. Heat is transferred by **conduction, convection,** and **radiation.**

When two bodies at different temperatures are placed in thermal contact, they ultimately reach a common temperature somewhere between the two initial temperatures. We say that **heat** has flowed from the hotter to the colder body. Heat, like light and force, impinges directly on our senses, so everyone has an intuitive understanding of the term. Yet there is a subtlety in its meaning that took many decades to clarify. Until the middle of the eighteenth century, the terms *heat* and *temperature* had essentially the same meaning. For example, thermometers were graduated in "degrees of heat." In 1760 Joseph Black was the first to make a clear distinction between temperature, as something measured on a thermometer, and heat, as something that flows from a hotter to a colder body to equalize their temperatures.

In the eighteenth century heat was believed to be an invisible, massless fluid called "caloric." According to this model, a body contained more caloric when it was hot than when it was cold. It was believed that caloric was a conserved quantity; it could neither be created nor destroyed. The caloric theory explained why two liquids, initially at different temperatures, always reach some intermediate temperature when they are mixed: Thermal equilibrium was said to be established by the flow of caloric from the hotter to the colder body. Heat conduction was explained by assuming that the particles of caloric repelled each other but were attracted to the particles of ordinary matter. Consequently, as caloric flowed into a body, it would spread throughout the volume. The mutual repulsion between the particles of caloric would also lead to thermal expansion.

However, the caloric theory failed to explain the generation of heat by friction—for example, when we rub our hands to stay warm. The American scientist

The caloric theory

FIGURE 19.1 Benjamin Thompson (1753–1814).

Benjamin Thompson (Fig. 19.1), who later became Count Rumford of Bavaria, had doubts about the material nature of caloric since it appeared to be weightless. In 1798, while supervising the boring of cannons, he was struck by the considerable heat generated. Water used to cool the metal had to be constantly replenished as it boiled away. According to the caloric theory the small chips of metal cut from the cannon could not retain their caloric and therefore released it to the water. Rumford noticed something that cast doubt on this explanation. Even when the tool was so dull that it no longer cut the metal, the water still got hot. Furthermore, he showed that the metal shavings had not lost any "ability to store caloric."* Clearly, the source of heat generated by friction in these experiments appeared to be inexhaustible and therefore could not be a material substance. Rumford's experiments showed that, far from being a conserved quantity, heat could be continually produced by mechanical work.

There is a strong tendency to think of heat as something stored in a system. The intuitive feeling that a body has something more when it is hot than when it is cold, is correct. But, as we shall see, this something is **internal energy,** not heat. The relationship between heat and internal energy is embodied in the first law of thermodynamics.

19.1 SPECIFIC HEAT

Joseph Black was the first person to realize that the rise in temperature of a body could be used to determine the quantity of heat absorbed by it. If a quantity of heat ΔQ produces a change in temperature ΔT in a body, its **heat capacity** is defined as

Definition of heat capacity

$$\text{Heat capacity} = \frac{\Delta Q}{\Delta T}$$

The SI unit of heat capacity is J/K. A commonly used (non-SI) unit of heat is the *calorie,* which used to be defined as the quantity of heat required to raise the temperature of 1 g of water from 14.5 °C to 15.5 °C. The modern definition of the calorie is in terms of the joule: 1 calorie = 4.186 J. The "energy values" quoted for foods are actually kilocalories (kcal or Cal). The *British thermal unit* (Btu) is the quantity of heat required to raise the temperature of 1 lb of water from 63 °F to 64 °F.

The quantity of heat ΔQ required to produce a change in temperature ΔT is proportional to the mass of the sample m, and to ΔT (for small ΔT). It also depends on the substance. These facts are incorporated in the equation

$$\Delta Q = mc\,\Delta T \tag{19.1}$$

where c is called the **specific heat** of the material. Equation 19.1 can be used to determine the heat transferred to or from a body. By expressing Eq. 19.1 in the form

Specific heat

$$c = \frac{1}{m}\frac{\Delta Q}{\Delta T} \tag{19.2}$$

we see that the specific heat is the heat capacity per unit mass. Its SI unit is J/kg·K (although the unit cal/g·K is often used). Specific heat is a property of a given substance, whereas heat capacity refers to a given sample of the material. As Table 19.1 shows, the specific heat of water, 1.00 cal/g·K, is large in comparison with the values for other substances.

It is sometimes convenient to work with the number of moles n of a substance

---

* By measuring their specific heat. See Section 19.1.

TABLE 19.1 SPECIFIC HEATS (20 °C AND 1 ATM)

	c (J/kg·K)	C (J/mol·K)
Aluminum	900	24.3
Copper	385	24.4
Gold	130	25.6
Steel/Iron	450	25.0
Lead	130	26.8
Mercury	140	28.0
Water	4190	75.4
Ice (−10 °C)	2100	38

rather than its mass. Equation 19.1 then becomes

$$\Delta Q = nC\,\Delta T \qquad (19.3)$$

where C is the **molar specific heat,** measured in J/mol·K (or cal/mol·K). Since $n = m/M$, where M is the molecular mass, we have

$$C = Mc \qquad (19.4)$$

Molar specific heat

The specific heat of a substance usually varies with temperature. In the case of water it varies by a few percent from 0 °C to 100 °C. The specific heat changes abruptly when the substance transforms from solid to liquid, or from liquid to gas. It also depends on the conditions under which the heat is applied. For example, the specific heat of a gas kept at constant pressure c_p is different from its specific heat at constant volume c_v. For air, $c_v = 0.17$ cal/g·K and $c_p = 0.24$ cal/g·K. For solids and liquids the difference is generally small, and in practice c_p is usually measured.

The Method of Mixtures

In the *method of mixtures,* first employed by Black, the specific heat of one body is determined by placing it in thermal contact with another body whose specific heat is known. Let us assume that the object whose specific heat is to be determined has mass m_1 and is at an initial temperature T_1. A liquid of mass m_2, at an initial temperature T_2, is in a cup within a thermally insulated container called a **calorimeter,** as in Fig. 19.2. The object is immersed in the liquid and the final equilibrium temperature T_f is recorded. Since there is no heat exchanged with the surroundings, the heat transferred to the colder body is equal to the heat transferred from the hotter body:

$$\Delta Q_1 + \Delta Q_2 = 0 \qquad (19.5)$$

Black based this assumption on the idea of "conservation of caloric." He was able to justify it by the consistency of the values he obtained for the specific heats of a variety of substances. We now realize that Eq. 19.5 is a particular example of the conservation of energy. In terms of the masses and specific heats of the two bodies, Eq. 19.5 becomes

$$m_1c_1\,\Delta T_1 + m_2c_2\,\Delta T_2 = 0 \qquad (19.6)$$

where $\Delta T_1 = T_f - T_1$ and $\Delta T_2 = T_f - T_2$. These changes in temperatures will have opposite signs. Calorimeters are used to measure the heats of combustion in chemical reactions. This is how the "energy values" of foods and fuels are determined. In practice, corrections have to be made to take into account the heat capacity of the calorimeter cup, as Example 19.1 illustrates.

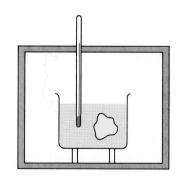

FIGURE 19.2 A calorimeter.

EXAMPLE 19.1: A steel ball of mass $m_1 = 80$ g has an initial temperature $T_1 = 200$ °C. It is immersed in $m_2 = 250$ g of water in a copper cup of mass $m_3 = 100$ g. The initial temperature of the water and cup is $T_2 = 20$ °C. Find the final temperature when the system reaches thermal equilibrium. The values of the specific heats are given in Table 19.1.

Solution: Equation 19.6 must be modified to take into account the calorimeter cup. Its change in temperature is the same as that of the water; that is, $\Delta T_3 = \Delta T_2$. When a term, $m_3 c_3 \Delta T_2$, is added for the cup, Eq. 19.6 takes the form

$$m_1 c_1 (T_f - T_1) + (m_2 c_2 + m_3 c_3)(T_f - T_2) = 0$$

Using the values in Table 19.1, we find $m_1 c_1 = (0.08$ kg$)(450$ J/kg $\cdot$ K$) = 36$ J/K. Similarly $m_2 c_2 = 1047$ J/K and $m_3 c_3 = 39$ J/K. Therefore,

$$(36)(T_f - 200) + (1047 + 39)(T_f - 20) = 0$$

We find $T_f = 25.8$ °C.

EXERCISE 1. What is the net transfer of heat to the water in Example 19.1?

19.2 LATENT HEAT

Black recognized that adding heat to a system does not always result in a change in its temperature. The temperature remains constant when a substance changes its **phase**—for example, from solid to liquid or from liquid to gas. He pointed out that if ice required just a small quantity of heat to melt, the spring thaw would occur quickly and lead to catastrophic flooding.

Consider a sample of ice at some arbitrary initial temperature, say -10 °C. As heat is applied to it at a steady rate, at first its temperature rises, as depicted in Fig. 19.3. However, when its temperature reaches 0 °C, the ice starts to melt and the temperature remains at 0 °C until all the ice has been converted to liquid water. The absorbed heat is not manifested as a rise in temperature. Measurements show that 1 kg of ice at 0 °C requires about 80 kcal to be fully converted to the liquid phase. Black called this "hidden" heat the **latent heat of fusion, L_f**. After all the ice has melted, the temperature rises steadily till it reaches 100 °C. At this point the liquid starts to convert to the gaseous phase and the temperature again stays constant. When all the water is in the form of vapor, the temperature again rises. At a pressure of 1 atm the **latent heat of vaporization, L_v**, of water is 540 kcal/kg. At other pressures, the temperature at which the liquid and gas phases are in equilibrium is different, and the value of the latent heat is also different. Table 19.2 lists some typical values of latent heats.

Consider a sample of mass m that changes its phase, as in Fig. 19.3. The heat exchanged with its surroundings is related to the **latent heat L** by

$$\Delta Q = mL \tag{19.7}$$

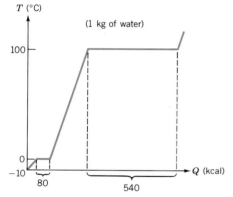

FIGURE 19.3 As heat is supplied to a sample of ice at 0°C its temperature does not change until all the ice has melted. Similarly, the temperature of water at 100°C does not change until all of it has been converted to steam.

TABLE 19.2 LATENT HEATS (AT 1 ATM)

	Melting Temperature	Latent Heat of Fusion (J/kg)	Boiling Temperature	Latent Heat of Vaporization (J/kg)
	°C		°C	
Aluminum	660	24.5×10^3	2450	$11,390 \times 10^3$
Copper	1083	134×10^3	1187	5065×10^3
Gold	1063	64.5×10^3	2660	1580×10^3
Lead	327	24.5×10^3	1750	870×10^3
Water	0	334×10^3	100	2260×10^3
	K		K	
Helium	3.5	5.23×10^3	4.2	20.9×10^3
Hydrogen	13.8	58.6×10^3	20.3	452×10^3
Nitrogen	63.2	25.5×10^3	77.3	201×10^3
Mercury	234	11.8×10^3	630	272×10^3

Latent heat is "hidden" in the sense that there is no temperature change, but the energy is not lost. When water condenses from the gaseous to the liquid phase, each kilogram releases the latent heat of vaporization. Similarly, when the liquid converts to the solid phase, each kilogram releases the latent heat of fusion.

In general, a phase of a substance is characterized by a certain arrangement of its molecules. Physically, the latent heat of fusion represents the work required to break the bonds between molecules in the solid phase and to allow them to move easily relative to each other in the liquid phase. The latent heat of vaporization is required to increase the separation between the molecules in the transition from the liquid to the gas phase. Other types of phase change are associated, for example, with changes in crystal structure and magnetization.

It should be noted that the latent heat of vaporization is required for a change of phase even when the temperature of a liquid is well below its normal boiling point. For example, when water evaporates at room temperature, the appropriate quantity of heat must be supplied by the surroundings. This is why water tends to cool a surface, such as our skin, from which it evaporates. The value of L_v is somewhat higher at temperatures below the boiling point. The value of the latent heat also depends on the pressure at which the transition takes place.

EXAMPLE 19.2: A 2-kg chunk of ice at $-10\ °C$ is added to 5 kg of liquid water at $45°C$. What is the final temperature of the system?

Solution: Three final states are possible: all ice, a mixture of ice and water at 0 °C, or all water. Before an equation can be written, the final state must be determined. We consider the following. The heat needed to raise the temperature of the ice to 0 °C is

$$\Delta Q_1 = m_i c_i\, \Delta T_1 = (2\ \text{kg})(2100\ \text{J/kg·K})(10\ \text{K}) = 42\ \text{kJ}$$

The heat then needed to convert all of the ice to the liquid phase is

$$\Delta Q_2 = m_i L_i = (2\ \text{kg})(334\ \text{kJ/kg}) = 668\ \text{kJ}$$

We compare this with the heat available if the water reaches the freezing temperature:

$$|\Delta Q_3| = m_w c_w |\Delta T_2| = (5\ \text{kg})(4190\,\text{J/kg·K})(45\ \text{K}) = 943\ \text{kJ}$$

Since $|\Delta Q_3| > (\Delta Q_1 + \Delta Q_2)$, all of the ice will melt. The final state is water at some temperature T_f. We may now write the equation

$$\Delta Q_1 + \Delta Q_2 + \Delta Q_3 + \Delta Q_4 = 0$$

where ΔQ_1 and ΔQ_2 have already been found, but ΔQ_3 is the heat lost by the water until it reaches T_f, and ΔQ_4 is the heat required to raise the temperature of the melted ice from 0 °C to T_f. Thus,

$$\Delta Q_1 + \Delta Q_2 + m_w c_w (T_f - 45°C) + m_i c_w(T_f - 0) = 0$$
$$42\ \text{kJ} + 668\ \text{kJ} + (21\ \text{kJ})(T_f - 45) + (8.38\ \text{kJ})T_f = 0$$

We find $T_f = 8.0°C$.

19.3   THE MECHANICAL EQUIVALENT OF HEAT

Rumford's experiments showed that a rise in temperature could be produced either by the absorption of heat or by mechanical work. This implied that work and heat are somehow equivalent. Around 1840, J. R. Mayer and J. P. Joule independently established a conversion factor between heat and work. Mayer noted that to produce a given change in temperature of a gas, the heat required at constant pressure is greater than that required at constant volume. He attributed this difference to the work done by the gas when it expands at constant pressure. His conclusion, expressed in modern units, was that the temperature increase produced by one calorie of heat could also be produced by about 4 J of mechanical work. Mayer's great contribution was slighted by his contemporaries because he had not actually performed experiments to arrive at his result and he expressed his ideas in language that was not considered "scientific."

Soon after electric motors and generators were invented in the 1830s, Joule (Fig. 19.4a) began performing experiments with them. In 1842 he used falling

FIGURE 19.4 (a) James Joule (1818–1889).

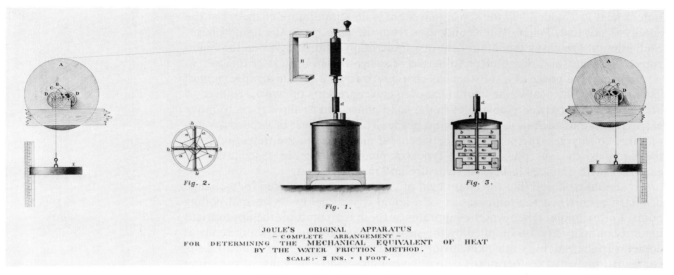

JOULE'S   ORIGINAL   APPARATUS
– COMPLETE   ARRANGEMENT –
FOR   DETERMINING   THE   MECHANICAL   EQUIVALENT   OF   HEAT
BY   THE   WATER   FRICTION   METHOD.
SCALE :- 3 INS. = 1 FOOT.

FIGURE 19.4 (*b*) The paddle wheel experiment.

weights to drive a generator. The electric current heated a wire immersed in water contained in an insulated vessel. Joule measured the quantity of heat produced and compared it with the mechanical work done by the weights. He deduced that 838 ft·lb of work produced the same rise in temperature as 1 Btu of heat. The determination of the precise value of this "mechanical equivalent of heat" became his lifelong passion. For the next four decades Joule performed a variety of experiments aimed at demonstrating the equivalence of heat and mechanical work.

Joule's most famous experiment is illustrated in Fig. 19.4*b*. It was first performed in 1845 and was repeated several times. He put water in an insulated container that had fixed vanes. A paddle with several vanes could rotate between the fixed vanes and thereby churn the water. The axle of the paddle was turned by the fall of two weights as shown in the figure. The 4-lb weights fell through 12 yards, were brought up again, and allowed to fall again a total of 16 times. Joule measured the small rise in temperature of the water (about 0.5 °F) and related it to the loss of potential energy of the weights. Since no heat could enter or leave the system, the temperature rise was due solely to the performance of mechanical work.

All of Joule's experiments led to the conclusion that the mechanical work required to produce a given change in temperature is in fixed proportion to the heat required for the same change in temperature. The intermediate step of electrical work did not affect that result. Joule's final result after 40 years of work was that 778 ft·lb of work are equivalent to 1 Btu of heat. This **mechanical equivalent of heat** is the source of the modern *definition* of the calorie: 1 calorie = 4.186 J. A change in the state of a system produced by the addition of 1 calorie of heat may also be produced by the performance of 4.186 J of work on the system.

Joule and Mayer had identified heat as another form of energy. This was a crucial step that led to the formulation of the principle of conservation of energy. Here is a modern definition of **heat:**

Definition of heat   Heat is energy transferred between two bodies as a consequence of a temperature difference between them.

In contrast, work is a mode of energy transfer in which the point of application of a force moves through a displacement and is not associated with a temperature difference. Both heat and work are "energy in transit" from one body to another during the operation of some process. Once the process stops, heat and work have no meaning. The experiments of Rumford showed that there is no "conservation of heat." There is only the conservation of energy, which includes all forms: mechanical, heat, electrical, magnetic, chemical, nuclear, and so on.

19.4   WORK IN THERMODYNAMICS

Thermodynamics is concerned with the work done by a system and the heat it exchanges with its surroundings. We are concerned only with work done by a system on its surroundings or on the system by the surroundings. That is, "external" work done by one system on another. We are not concerned with "internal" work done by one part of a system on another. Consequently, the boundaries of a given system must be clearly defined.

A useful concept in thermodynamics is that of a *heat reservoir*. This is a body with such a large heat capacity that large amounts of heat can flow into it, or out of it, without changing its temperature significantly. A large lake and the atmosphere are practical examples. In a steam engine, a boiler maintained at a constant temperature by a furnace serves as a heat reservoir.

Figure 19.5 shows a gas confined to a cylinder by a weight on a movable piston. Our system is the gas, whereas the cylinder and the piston form the environment. If the piston is allowed to move upward, the gas expands and does work on it. To calculate the work done by the gas, we assume that the process is **quasistatic.** In a quasistatic process the thermodynamic variables (P, V, T, n, etc.) of the system and its surroundings change infinitely slowly. Thus, the system is always *arbitrarily close to an equilibrium state,* in which it has a well-defined volume, and the whole system is characterized by single values of the macroscopic variables. To ensure that the piston moves very slowly, there must be some force, for example, provided by a weight, directed opposite to that due to the pressure. If the piston were to move suddenly, the rapid expansion would involve turbulence and the pressure would not be uniquely defined.

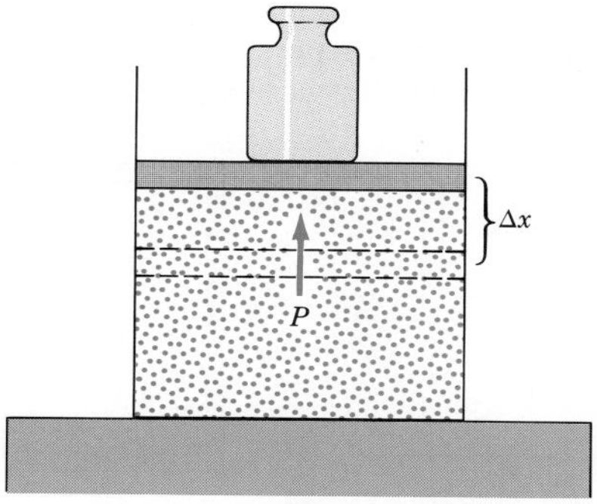

When the piston rises by dx, the work dW done by the gas is $dW = F\,dx = (PA)dx$ where A is the cross-sectional area of the piston. Since the change in volume of the gas is $dV = A\,dx$, the work may be expressed as

(Quasistatic) $$dW = P\,dV$$

As a quasistatic process evolves, P and V are always uniquely defined. This allows us to depict the process on a *PV* diagram such as Fig. 19.6. When the system is taken quasistatically from an equilibrium state i to another equilibrium state f, the total work done by the system is

$$W = \int_{V_i}^{V_f} P\,dV \qquad (19.8)$$

In Fig. 19.6 the work is represented by the area under the curve. If $V_f > V_i$, the work done *by* the gas is positive. If the volume decreases, the work done *by* the gas is negative. This may be interpreted as positive work done *on* the gas by the environment. The work done depends not only on the initial and final states but also on the details of the process, that is, the *thermodynamic path* between the states. Therefore, we need to know how the pressure varies with the volume. We will deal mainly with three basic cases: *Isobaric* work and *isothermal* work in this section, and *adiabatic* work in the next section.

Isobaric Work

In an **isobaric** process the expansion or compression occurs at *constant pressure.*

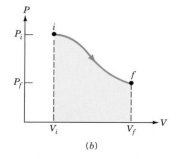

FIGURE 19.5 When a gas expands against an opposing force, the gas does work.

FIGURE 19.6 On a *PV* diagram, the work done by gas is the area under the curve, $W = \int P\,dV$.

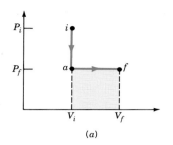

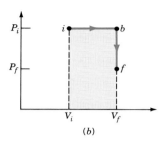

FIGURE 19.7 The work done by a gas depends on the path taken between the initial and final states. The work for the path in (b) is greater than that in (a).

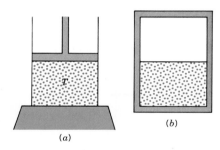

FIGURE 19.8 When quantity of gas is taken from a given initial state to a given final state, the heat transfer depends on the path taken. In (a) the gas is in contact with a heat reservoir and expands against a piston. In (b) the gas is in an insulated container and expands into a vacuum when a membrane is broken.

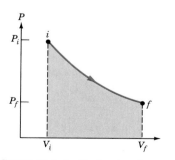

FIGURE 19.9 An ideal gas undergoes an isothermal expansion.

Thus, $W = \int P\, dV = P\int dV$; that is,

(Isobaric)
$$W = P(V_f - V_i)$$

Suppose we wish to calculate the work done when the system goes from an equilibrium state i to another equilibrium state f in Fig. 19.7a. Suppose we chose the path iaf. In segment ia the pressure of the gas is reduced at constant volume by cooling it. Since $dV = 0$, no work is done in this segment. In segment af, the gas expands at constant pressure and so the total work done by the gas is

$$\begin{aligned} W_{iaf} &= W_{ia} + W_{af} \\ &= 0 + P_f(V_f - V_i) \end{aligned}$$

An alternate route would be ibf in Fig. 19.7b. The gas first expands at constant pressure p_i, and then its pressure is reduced to p_f. Segment bf involves no work, whereas ib does. The total work done by the gas is

$$\begin{aligned} W_{ibf} &= W_{ib} + W_{bf} \\ &= P_i(V_f - V_i) + 0 \end{aligned}$$

Clearly the work done by a system depends on the details of the process that takes it from one equilibrium state to another. It makes no sense to speak of the "work in the system." In both of the above processes the system exchanges heat with its surroundings and its temperature varies. The quantity of heat transferred into or out of the system also depends on the thermodynamic path.

Consider an ideal gas confined to a cylinder by a piston. The walls of the cylinder are insulated but its base is in thermal contact with a heat reservoir at temperature T, as shown in Fig. 19.8a. If the piston is allowed to rise slowly, the gas will do work and absorb heat from the reservoir. For definiteness let us say the final volume is double the initial volume. Now consider the gas confined to part of the container by a thin membrane (Fig. 19.8b). Its initial temperature and volume are the same as those in Fig. 19.8a. The walls of the container are thermally insulated. If the membrane is punctured, the gas expands without doing any work. This is called a **free expansion.** It is an experimental fact that the temperature of an ideal gas does not change in such an expansion. The final state in this case is the same as that in the earlier quasistatic expansion, but no heat has been exchanged with the surroundings. These two cases illustrate that for given initial and final equilibrium states, the heat transfer into (or out of) a system depends on the thermodynamic path taken. For this reason it makes no sense to speak of the "heat content of a system."

Isothermal Work (Ideal Gas)

In an **isothermal** process, the system is kept in contact with a single heat reservoir at temperature T. The path followed in the PV diagram as the system expands at constant temperature is called an isotherm. A quasistatic expansion (in which the variables have well-defined values) will take the system from state i to state f along an isotherm, shown in Fig. 19.9. To evaluate this integral we need to know how the pressure varies with the volume. In the special case of an ideal gas we know from Eq. 18.2 that $PV = nRT$; therefore, $P = nRT/V$. Since T is constant, it may be taken out of the integral in Eq. 19.8:

(Isothermal, ideal gas)
$$W = nRT \int_{V_i}^{V_f} \frac{dV}{V} = nRT \ln\!\left(\frac{V_f}{V_i}\right)$$

where we have used the result $\int dx/x = \ln x$ and $\ln B - \ln A = \ln(B/A)$. The work depends on the ratio of the final volume to the initial volume.

EXAMPLE 19.3: Three moles of helium are initially at 20 °C and a pressure of 1 atm. What is the work done by the gas if the volume is doubled (a) at constant pressure, or (b) isothermally?

Solution: (a) From the ideal gas equation of state, the initial volume is

$$V_1 = \frac{nRT_1}{P_1}$$

$$= \frac{(3 \text{ mol})(8.31 \text{ J/mol·K})(293 \text{ K})}{(101 \text{ kPa})} = 0.072 \text{ m}^3$$

The final volume is $V_2 = 2V_1 = 0.144 \text{ m}^3$. The (isobaric) work done by the gas at constant pressure is

$$W = P_1(V_2 - V_1)$$

$$= (1.01 \times 10^5 \text{ N/m}^2)(0.144 \text{ m}^3 - 0.072 \text{ m}^3) = 7.27 \text{ kJ}$$

(b) The work done by an ideal gas under isothermal conditions is

$$W = nRT \ln \left(\frac{V_f}{V_i} \right)$$

$$= (3 \text{ mol})(8.31 \text{ J/mol·K})(293 \text{ K}) \ln 2 = 5.06 \text{ kJ}$$

The fact that the isothermal work is less than the isobaric work may be inferred by comparing the areas under the functions in Figs. 19.7b and 19.9.

EXERCISE 2. What is the final temperature of the gas in part (a) of Example 19.3?

19.5 FIRST LAW OF THERMODYNAMICS

Consider a system that consists of a gas enclosed by a piston in a cylinder. Suppose the system is taken quasistatically from an initial state P_i, V_i, T_i to a final state P_f, V_f, T_f. The system is allowed to reach equilibrium with a succession of reservoirs whose temperatures differ slightly. At each step the work done and heat exchanged are measured. We find that both the total work done W and the total heat transfer Q to or from the system depend on the thermodynamic path. However, the difference, $Q - W$, is the same for *all* paths between the given initial and final equilibrium states. This feature allows us to define a new function, called the **internal energy** U, such that the change in internal energy of the system is

$$\Delta U = Q - W \qquad (19.9) \qquad \text{First law of thermodynamics}$$

In this definition, Q is positive when heat *enters* the system and W is positive when work is done *by* the system on its surroundings. Equation 19.9 is called the **first law of thermodynamics.** It says that *the internal energy of a system changes when work is done on the system (or by it), and when it exchanges heat with the environment.** Notice that we can specify only the *change* in internal energy. The first law is valid for *all* processes, quasistatic or not. However, if friction is present, or the process is not quasistatic, the internal energy U is uniquely defined only at the initial and final equilibrium states.

The first law of thermodynamics is a generalization of the results of many experiments, starting with those of Mayer and Joule. It serves as a general definition of heat and establishes the existence of internal energy U as a *state function*—one that depends only on the thermodynamic state of the system. Heat can include radiation and other forms of energy released or deposited on the microscopic scale. To decide whether a given interaction involves heat or work, keep in mind that the transfer of heat can be prevented by a thermal insulator.

In the macroscopic approach of thermodynamics, there is no need to specify the physical nature of the internal energy. The experimental results are sufficient

---

* The internal energy of a system also changes when it exchanges particles with the environment. We ignore this aspect.

proof that such a function exists. However, it is worth noting that the internal energy is the sum of all possible kinds of energy "stored" in the system—mechanical, electrical, magnetic, chemical, nuclear, and so on. It does not include the kinetic and potential energies associated with the center of mass of the system. The kinetic and potential energies associated with the *random* motion of particles

Thermal energy

form a part of the internal energy called **thermal energy.** (However, some authors treat the terms internal energy and thermal energy as being synonymous.)

Just as the colloquial use of the term "work" differs from its definition in physics, the everyday use of the term "heat" does not always correspond to its meaning in the first law. It is common knowledge that a gas becomes warmer when it is compressed rapidly, as in a bicycle pump. This is often called "heating by compression." But since the compression occurs quickly, there is no chance for heat to flow into the system. In any case, where would the heat come from? In fact, the change in the internal energy of the air, manifested by the rise in temperature, comes from the work done by a person. The same change in temperature *could have* been produced by a flow of heat—the fact established by Mayer and Joule.

Confusion between the concepts of heat and internal energy arises from erroneous statements that refer to the "heat content" of a body. Even correct terms like "the heat capacity of a body" can mislead one to believe that heat is somehow stored in a system. This is *not* correct; it was the fatal flaw of the caloric theory. Thus, a "heat reservoir" does not possess a large quantity of heat!

The physical quantity possessed by a system is internal energy, which is the sum of *all* the kinds of energy in the system. As the first law indicates, U may be changed either by heat exchange or by work. The internal energy is a state function that depends on the equilibrium *state* of a system, whereas Q and W depend on the thermodynamic path between two equilibrium states. That is, Q and W are associated with *processes*. The heat absorbed by a system will increase its internal energy, only some of which is thermal energy. It is therefore incorrect to say that heat *is* the energy of the random motion. (This is how earlier scientists expressed themselves.)

Distinction between heat, work, and internal energy

Herein lies the subtlety of the concept of heat: How is it possible for heat to enter or to leave a system and yet not be stored in it? Perhaps an analogy with sound entering and leaving our bodies may help. We talk a lot each day, but we are not "full of sound"; we hear many noises, but our brains do not store sound waves! Sound energy enters and leaves our bodies, but at any given moment there is no sound stored.

19.6   APPLICATIONS OF THE FIRST LAW OF THERMODYNAMICS

We now apply the first law of thermodynamics to some simple situations.

(a) Isolated System

Consider first an isolated system for which there is no heat exchange and no work done on the external environment. In this case $Q = 0$ and $W = 0$, so from the first law we conclude

(Isolated) $\qquad\qquad \Delta U = 0 \quad$ or $\quad U = $ constant

The internal energy of an isolated system is constant.

(b) Cyclic Process

Engines operate in *cycles,* in which the system—for example, a gas—periodically returns to its initial state. In Fig. 19.10 the system goes from state a to state b via path I, for which $W_I > 0$, and returns to its initial state via path II, for which $W_{II} < 0$. The net work done by the system is the area enclosed by the curve. In a clockwise traversal the net work is positive. Since the system returns to its initial state, the change in internal energy in one complete cycle is zero; that is, $\Delta U = 0$. From the first law we see that

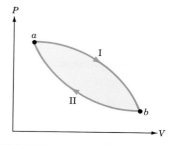

FIGURE 19.10 In a cyclic process the system returns to its initial state. The work done by the system equals the heat input.

(Cyclic) $$Q = W$$

The net work done by the system in each cycle, $W = W_I + W_{II}$, is equal to the net heat input per cycle. This result is of importance in the discussion of steam engines and diesel engines, for instance, in which the influx of heat is used to perform mechanical work.

(c) Constant-volume Process

In a constant-volume process, the volume of the system stays constant. Consequently, $W = 0$. From the first law we see that

(Constant volume) $$\Delta U = Q$$

All the heat entering the system goes into increasing the internal energy.

(d) Adiabatic Process

In an **adiabatic** process, the system does not exchange heat with its surroundings; that is, $Q = 0$. This may be accomplished in two ways. First, the system can be enclosed in a thermally insulated container. Second, the process could occur so quickly that there is not enough time for an appreciable quantity of heat to be exchanged with the surroundings. For example, the rapid compression stroke in a diesel engine is approximately adiabatic. The first law for an adiabatic process takes the form

(Adiabatic) $$\Delta U = -W \qquad (19.10)$$

When a gas expands against a piston, the gas does positive work. In an adiabatic expansion the internal energy decreases. This is usually manifested as a drop in temperature. Conversely, when a gas is compressed adiabatically, its internal energy increases and the temperature rises. This occurs when a bicycle pump is being used. In a diesel engine the volume of the air–fuel mixture is rapidly reduced by a factor of about 15. The rise in temperature is so great that the mixture spontaneously ignites.

(e) Adiabatic Free Expansion

We now consider what happens when a gas is allowed to expand adiabatically without doing any work. Figure 19.11 shows two vessels connected by a tube with a stopcock. Initially, one vessel is filled with gas while the other is evacuated. The system is thermally insulated; that is, $Q = 0$. When the stopcock is opened the gas quickly expands to fill the second chamber. The uncontrolled expansion is not quasistatic and cannot be depicted on a PV diagram. Since the gas does not

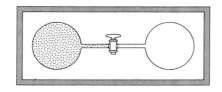

FIGURE 19.11 In an adiabatic free expansion gas initially in the left bulb is allowed to expand into the second (evacuated) bulb when the stopcock is opened.

expand against a piston, it does no work, $W = 0$. From the first law we conclude that

(Free expansion) $\qquad\qquad\qquad \Delta U = 0$

In an **adiabatic free expansion** *the internal energy of any gas (ideal or real) does not change.*

In the special case of an ideal gas, there is no change in temperature in an adiabatic free expansion. One can conclude that *the internal energy of an ideal gas depends only on temperature*, not on pressure or volume. Sensitive experiments show a small temperature change for a real gas at high pressures and low temperatures. This indicates that the internal energy of a real gas is also a function of pressure or volume.

EXAMPLE 19.4: A cylinder with a piston contains 0.2 kg of water at 100 °C. What is the change in internal energy of the water when it is converted to steam at 100 °C at a constant pressure of 1 atm? The density of water is $\rho_w = 10^3$ kg/m³ and that of steam is $\rho_s = 0.6$ kg/m³. The latent heat of vaporization of water is $L_v = 2.26 \times 10^6$ J/kg.

Solution: The heat transfer to the water is

$$Q = mL_v = (0.2 \text{ kg})(2.26 \times 10^6 \text{ J/kg}) = 4.52 \times 10^5 \text{ J}$$

The work done by the water when it expands against the piston at constant pressure is

$$W = P(V_s - V_w) = P\left(\frac{m}{\rho_s} - \frac{m}{\rho_w}\right)$$
$$= (1.01 \times 10^5 \text{ N/m}^2)\left(\frac{0.2 \text{ kg}}{0.6 \text{ m}^3} - \frac{0.2 \text{ kg}}{1000 \text{ m}^3}\right)$$
$$= 3.36 \times 10^4 \text{ J}$$

The change in internal energy is

$$\Delta U = Q - W = 452 \text{ kJ} - 33.6 \text{ kJ} = 418 \text{ kJ}$$

19.7   IDEAL GASES

The first law may be used to obtain information on the specific heats of an ideal gas. We will also obtain the equation of state for an ideal gas undergoing a quasi-static adiabatic process.

Specific Heats

The temperature of a system may be raised subject to a number of conditions. Of particular interest are the conditions of constant volume and constant pressure. When heat is added to any gas at constant volume, the work done by the gas is zero ($W = 0$), and so all the heat is used to increase the internal energy. If the heat is added when the pressure is kept constant, the volume of the gas increases. Because the gas does work during this expansion, for a given change in temperature it absorbs a greater quantity of heat than at constant volume. In the case of an ideal gas, one can obtain an expression for the difference between the specific heats at constant pressure and at constant volume.

If there are n moles of a gas in a system and the change in temperature is ΔT, the heat absorbed at constant volume (Eq. 19.3) is

$$Q_v = nC_v \Delta T \qquad\qquad (19.11)$$

where C_v is the molar specific heat at constant volume. Since the work done by the gas is zero, from the first law, $\Delta U = Q - W$, we see that

$$\Delta U = nC_v \Delta T \qquad\qquad (19.12)$$

This equation gives the change in internal energy of any gas, provided its volume is constant and it does not undergo a change of phase or composition. We have assumed that C_v stays constant as the temperature and pressure change. (This is not true for real gases.) However, in the case of an ideal gas the internal energy depends *only* on temperature, not on pressure or volume (as we noted for an adiabatic free expansion). Therefore, the equation $\Delta U = nC_v \Delta T$ applies to *any* process involving an ideal gas, not just to constant-volume processes.

Now suppose n moles of a gas are allowed to expand quasistatically at constant pressure. If the change in temperature is ΔT, the heat absorbed Q_p, is given by

$$Q_p = nC_p \Delta T \tag{19.13}$$

where C_p is the molar specific heat at constant pressure. The work done by the gas at constant pressure is $W = +P \Delta V$. From the first law, $\Delta U = Q - W$, we have

$$nC_p \Delta T = \Delta U + W \tag{19.14}$$

This equation applies to any gas (although in general we do not know ΔU).

We noted earlier that because the internal energy of an ideal gas depends *only* on temperature, not pressure or volume, Eq. 19.12 applies to *any* process involving an ideal gas. Hence Eq. 19.14 becomes

$$nC_p \Delta T = nC_v \Delta T + P \Delta V$$

The equation of state for an ideal gas is $PV = nRT$. At fixed pressure, this means $P \Delta V = nR \Delta T$, so the difference between the two specific heats is

(Ideal gas) $\qquad\qquad C_p - C_v = R \tag{19.15}$

Table 20.1 (in the next chapter) shows values of C_v and C_p. We see that the difference is close to 8.31 J/mol·K for various gases, as predicted by Eq. 19.15.

Adiabatic Quasistatic Process

The work done by an ideal gas during an isothermal expansion was treated in Section 19.4. Now, consider the quasistatic adiabatic expansion of an ideal gas in a thermally insulated container. In an adiabatic process there is no heat exchange with the surroundings, so $dQ = 0$. At any point in the process the equation $PV = nRT$ is valid, but the temperature varies as the process evolves.

The work done by the gas for an infinitesimal change in volume dV is $dW = P\,dV$. The temperature of the ideal gas will change by dT, which means that there will be a change in its internal energy given by Eq. 19.12, $dU = nC_v\,dT$. The first law, $dU = dQ - dW = 0 - dW$, takes the form

$$nC_v\,dT = -P\,dV \tag{19.16}$$

From the equation of state for an ideal gas, $PV = nRT$, we have

$$P\,dV + V\,dP = nR\,dT$$

Into this equation we substitute $n\,dT = -P\,dV/C_v$ from Eq. 19.16 and rearrange to find

$$P(C_v + R)dV + C_vV\,dP = 0$$

From Eq. 19.15 we know that $C_v + R = C_p$ for an ideal gas. With the definition

$$\gamma = \frac{C_p}{C_v}$$

the last equation becomes

$$\gamma \frac{dV}{V} + \frac{dP}{P} = 0$$

which, after integration, yields

$$\gamma \ln V + \ln P = \text{constant}$$

The constant depends on the initial conditions. From this equation we infer $\ln(PV^\gamma) = \text{constant}$, or equivalently

(Quasistatic, adiabatic) $PV^\gamma = \text{constant}$ (19.17)

This equation applies to a *quasistatic adiabatic process involving an ideal gas*. In the case of a quasistatic adiabatic expansion $P\,dV > 0$, so from Eq. 19.16 we see that $dU = nC_v\,dT < 0$: The temperature of the gas will fall. The ideal gas law, $PV = nRT$ (which applies to equilibrium states of any process), may be used to find the temperatures of the initial and final states.

A quasistatic adiabatic process is depicted as a **green** line in Fig. 19.12. The **red** lines are *isotherms*—which trace the paths of constant-temperature processes. The system starts in the state characterized by P_1, V_1, and T_1 and ends at state given by P_2, V_2, and T_2. From Eq. 19.17 we have

$$V^\gamma\,dP + \gamma PV^{\gamma-1}dV = 0$$

Thus, $dP/dV = -\gamma P/V$ for an adiabatic process. For an isothermal process, the equation $PV = nRT = \text{constant}$ leads to $dP/dV = -P/V$. Since $\gamma > 1$, we conclude that at a given point on a PV diagram, the slope of the adiabatic curve is steeper than that of an isothermal curve—as has been indicated in Fig. 19.12.

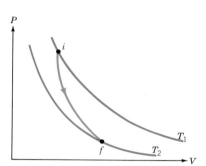

FIGURE 19.12 At a given point on a PV diagram, an adiabatic (solid) curve is steeper than the isothermal (dashed).

EXAMPLE 19.5: An ideal monatomic gas, for which $\gamma = 5/3$, undergoes a quasistatic expansion to one-third of its initial pressure. Find the ratio of the final volume to the initial volume if the process is (a) isothermal; (b) adiabatic.

Solution: (a) In an isothermal process, the ideal gas equation of state, $PV = nRT$, tells us that $P_1V_1 = P_2V_2$. Therefore,

$$\frac{V_2}{V_1} = \frac{P_1}{P_2} = 3$$

(b) In a quasistatic adiabatic process, $PV^\gamma = \text{constant}$. Therefore,

$$\frac{V_2}{V_1} = \left(\frac{P_1}{P_2}\right)^{\frac{1}{\gamma}}$$

Since $\gamma = 5/3$, we have $V_2/V_1 = (3)^{3/5} = 1.9$.

EXERCISE 3. If the initial temperature is 100 °C, what is the final temperature in part (b) of Example 19.5?

EXAMPLE 19.6: Find the work done by an ideal gas when its state changes adiabatically (a) from P_1 and V_1 to P_2 and V_2, and (b) from T_1 to T_2.

Solution: (a) Since we know the initial and final pressures and volumes we can apply Eq. 19.17 for a quasistatic adiabatic process: $P = K/V^\gamma$, where K is a constant. The work done by the gas in an infinitesimal change in volume, dV, is $dW = P\,dV$, so the total work done is

$$W = \int_{V_1}^{V_2} P\,dV = \int_{V_1}^{V_2} \frac{K}{V^\gamma}\,dV = \frac{K}{\gamma-1}\left(\frac{1}{V_1^{\gamma-1}} - \frac{1}{V_2^{\gamma-1}}\right)$$

If we use $K = P_1V_1^\gamma$, we find

(Adiabatic) $W = \dfrac{1}{\gamma-1}(P_1V_1 - P_2V_2)$

(b) Since $dQ = 0$ for an adiabatic change, the first law tells us $dU = -dW$. We also know that for *any* process the change in internal energy of an ideal gas is given by Eq. 19.12, $dU = nC_v\,dT$. Although the process need not be quasistatic, the result applies only to the initial and final equilibrium temperatures. The adiabatic work done by the gas is

(Adiabatic) $W = -nC_v\displaystyle\int_{T_1}^{T_2} dT = -nC_v(T_2 - T_1)$

Thus, if $W > 0$, then $T_2 < T_1$, which means positive work done by the gas leads to a drop in temperature.

EXERCISE 4. Show that the expressions for W found in parts (a) and (b) of Example 19.6 are equivalent.

19.8   SPEED OF SOUND (Optional)

In Eq. 17.1 the speed of sound in a gas was given as

$$v = \sqrt{\frac{B}{\rho}}$$

where ρ is the density and B is the bulk modulus defined in Eq. 14.7 as

$$B = -V \frac{dP}{dV}$$

When Newton first derived this expression for the speed of sound he assumed that the compressions and rarefactions occur isothermally. This led to a predicted value of about 280 m/s, which is well below the measured value of 330 m/s at 0 °C and 1 atm. In 1816, Simon Laplace suggested that the compressions and rarefactions associated with a sound wave occur adiabatically. The temperature increases in the compressions and decreases in the rarefactions. However, air is a poor heat conductor, and so heat does not have much chance to be transferred through a distance of half a wavelength within the time available. S. Poisson and J. Lagrange obtained a modified expression for the speed of sound in a gas by starting with the equation for an adiabatic process followed by an ideal gas, $PV^\gamma =$ constant.

If the pressure fluctuations occur adiabatically rather than isothermally, we must find the adiabatic bulk modulus. In Section 19.7 we saw that the equation $PV^\gamma =$ constant leads to

$$V \frac{dP}{dV} + \gamma P = 0$$

Therefore, using the definition given above, the *adiabatic* bulk modulus is

$$B = \gamma P$$

and the speed of sound, $v = \sqrt{B/\rho}$, is given by

$$v = \sqrt{\frac{\gamma P}{\rho}} \qquad (19.18a)$$

One can use the ideal gas law, $PV = nRT$, to express the pressure in terms of the mass density, $\rho = nM/V$, and the temperature: $P = \rho RT/M$, so

$$v = \sqrt{\frac{\gamma RT}{M}} \qquad (19.18b)$$

Note that M should be expressed in the unit kg/mol.

From Eq. 19.18b we see that at a given temperature, the speed of sound in a gas increases as the molecular mass decreases. For example, the speed of sound in helium ($M = 4$ g/mol) is much higher than the speed in air ($M = 29$ g/mol). An amusing consequence of this is heard when people inhale helium gas. The higher speed means that the resonant frequencies of the vocal cavity are higher. This makes them sound like the "Chipmunks."

EXAMPLE 19.7: Calculate the speed of sound in air at 0 °C at a pressure of 1 atm. Take $\gamma = 1.4$ and $\rho = 1.29$ kg/m³.

Solution: Substituting into Eq. 19.18a we have

$$v = \sqrt{\frac{(1.4)(1.01 \times 10^5 \text{ N/m}^2)}{1.29 \text{ kg/m}^3}} = 331 \text{ m/s}$$

This is in good agreement with the measured value at audible frequencies (below 20,000 Hz). At ultrasonic frequencies, where the wavelength is very short, the adiabatic approximation is not valid.

EXERCISE 5. From Eq. 19.18b we see that $v = A \sqrt{T}$, where A is a constant. Evaluate A for air ($M = 29$ g/mol).

19.9   HEAT TRANSPORT

Heat may be transported from one point to another by any of three possible mechanisms: *conduction*, *convection*, and *radiation*. We discuss each of these in turn.

Conduction

Figure 19.13a shows a rod whose ends are in thermal contact with a hot reservoir at temperature T_H and a cold reservoir at temperature T_C. The sides of the rod are covered with insulating material, so the transport of heat is along the rod, not through the sides. The molecules at the hot reservoir have greater vibrational energy. This energy is transferred by collisions to the atoms at the end face of the rod. These atoms in turn transfer energy to their neighbors further along the rod. Such transfer of heat through a substance is called **conduction.**

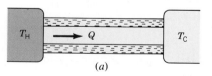

(a)

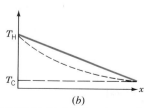

(b)

FIGURE 19.13 (a) Heat is conducted through an insulated bar whose ends are in thermal contact with two reservoirs. (b) In the steady state, the temperature varies linearly with distance along the bar.

TABLE 19.3 THERMAL CONDUCTIVITIES (W/m·K)

Aluminum	240
Copper	400
Gold	300
Iron	80
Lead	35
Glass	0.9
Wood	0.1–0.2
Concrete	0.9
Water	0.6
Glass wool	0.04
Air	0.024
Helium	0.14
Hydrogen	0.17
Oxygen	0.024

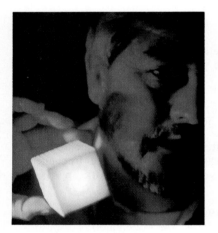

FIGURE 19.14 A cube of material used for the tiles on the Space Shuttle can be held even though it is at 1260 °C.

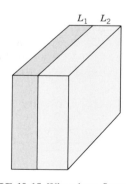

FIGURE 19.15 When heat flows through two slabs of different thicknesses and thermal conductivities, the R value is simply the sum of the individual R values.

The rate of transfer of heat through conduction, dQ/dt, is proportional to the cross-sectional area of the rod and to the temperature gradient, dT/dx, which is the rate of change of temperature with distance along the bar. In general,

$$\frac{dQ}{dt} = -\kappa A \frac{dT}{dx} \qquad (19.19)$$

The negative sign is used to make dQ/dt a positive quantity since dT/dx is negative (and κ is defined to be positive). The constant κ, called the **thermal conductivity,** is a measure of the ability of a material to conduct heat. The conductivities of various materials are listed in Table 19.3. A metal doorknob feels colder than the wooden door at the same temperature because the metal is a good thermal conductor and conducts heat away from the hand, whereas wood is a relatively poor conductor. The material used in the tiles for the Space Shuttle is an extremely poor conductor (see Fig. 19.14). In all solids and liquids heat is conducted by the transfer of vibrational energy of the atoms. In metals there are also a large number of so-called free electrons that move readily within the object. They account for the quick transfer of heat through metals. The insulating property (low thermal conductivity) of down and wool arise mainly from their ability to trap air.

Suppose that initially the temperature along the bar is uniform at the value T_c. At a later time, it varies somewhat like the dotted curve in Fig. 19.13b. After a long enough time, the system settles into a stable condition in which the temperature varies linearly with distance along the bar. In this condition, Eq. 19.19 may be written in the form

$$\frac{dQ}{dt} = \kappa A \frac{T_H - T_C}{L} \qquad (19.20)$$

Equation 19.20 is sometimes written as

$$\frac{dQ}{dt} = A \frac{\Delta T}{R} \qquad (19.21)$$

where $R = L/\kappa$ is called the **thermal resistance,** or **R value,** of the sample. The customary (non-SI) units used in this equation are Btu/h for dQ/dt, ft^2 for A, and F° for ΔT. Thus, the unit for R is ft^2·h·F/Btu! Given κ in SI units one can find the approximate R value (in the customary unit) for a 1-in. slab of a material by a conversion factor: $R = 0.14/\kappa$.

One benefit of the definition of R is that the effect of adding insulation is easily calculated. Consider two slabs of thickness L_1 and L_2 in contact as shown in Fig. 19.15. The temperature differences between their faces are ΔT_1 and ΔT_2. If no heat escapes through the sides, the heat flux is the same for both slabs. From Eq. 19.20, $\Delta T_1 = (L_1/\kappa_1 A)dQ/dt$ and $\Delta T_2 = (L_2/\kappa_2 A)dQ/dt$. Thus,

$$\Delta T_1 + \Delta T_2 = \frac{1}{A} \left(\frac{L_1}{\kappa_1} + \frac{L_2}{\kappa_2} \right) \frac{dQ}{dt}$$

Comparing this with Eq. 19.21, we see that the effective R value is simply the sum of the individual R values.

Convection

In the process of thermal conduction, atoms transfer their energy by colliding with their neighbors, but in a solid they only oscillate about their equilibrium positions. In liquid and gases, the atoms or molecules can move from point to point. The transfer of heat that accompanies mass transport is called **convection**. In *forced convection*, a fan or pump sets up fluid currents. For example, a fan blows air, or a

A glider can gain altitude by entering a rising column of warm air called a "thermal."

pump circulates water in a hot-water heating system in a house. *Free convection* occurs because the density of a fluid varies with its temperature. Hot air in contact with a radiator expands and therefore becomes less dense than the air around it. The hot air rises and is replaced by cooler air. The same process occurs when a liquid is heated, as depicted in Fig. 19.16. The warmer liquid at the base of the container rises and is replaced by cooler liquid from the top. There are large convection currents in our atmosphere. Gliders and birds use warm rising currents, called thermals, to gain altitude. The thermal insulation provided by still air in a double-glazed window is reduced by convection currents set up between the two panes.

Radiation

Radiation involves the transfer of heat without the action of an intervening medium. For example, we feel the heat radiated by hot bodies such as the sun or burning logs in a fireplace. Experimentally it is found that the power radiated from a body of area A at absolute temperature T is given by

$$\frac{dQ}{dt} = e\sigma A T^4 \qquad (19.22)$$

where $\sigma = 5.67 \times 10^{-8}$ W/m$^2\cdot$K^4. The quantity e, called the *emissivity,* depends on the nature of the surface. For example, $e \approx 0.1$ for a shiny metallic surface and $e \approx 0.95$ for a mat black surface.

A body both emits and absorbs radiant energy. If the body is at T_1, its radiated power is $e\sigma A T_1^4$. If the surroundings are at T_2, their radiation is proportional to T_2^4, so the rate of absorption by the body must also be proportional to T_2^4. When $T_1 = T_2$ there will be no net transfer of heat between the body and its surroundings, $dQ/dt = 0$. It follows immediately that the coefficient of the T_2^4 absorption term must also be $e\sigma A$. Therefore, a good emitter is also a good absorber. The net rate at which heat is radiated by the body is

$$\frac{dQ}{dt} = e\sigma A(T_1^4 - T_2^4) \qquad (19.23)$$

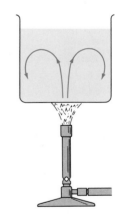

FIGURE 19.16 In convection, heat transfer accompanies the movement of a fluid.

FIGURE 19.17 People hidden behind some bushes are detected by a heat-sensing device.

Figure 19.17 shows how radiation emitted by a warm body can be used to detect a person hidden behind bushes.

EXERCISE 6.   (a) Why are both the inner and the outer surfaces of a Thermos flask very shiny? (b) Why does it help to remove air from the space between the inner and outer shells?

SUMMARY

Heat is a transfer of energy that occurs as a result of a temperature difference between bodies. Both heat and work are involved in processes; they *cannot* be stored in a system. The heat Q required to change the temperature of a mass m, or n moles, of a substance by ΔT is

$$Q = mc \, \Delta T = nC \, \Delta T$$

where c is the **specific heat** and C is the **molar specific heat.** When a mass m of a substance undergoes a change in phase it either absorbs or releases a quantity of heat

$$Q = mL$$

without a change in temperature. The **latent heat** L depends on the type of phase change.

The work done by a gas in a **quasistatic process** (in which the system is always close to an equilibrium state) is

(Quasistatic) $$W = \int_{V_i}^{V_f} P \, dV$$

This work depends on the thermodynamic path between the initial and final equilibrium states.

The **first law of thermodynamics** is a generalization of the conservation of energy to include heat. It says that the **internal energy** U of a system can be changed either by the input of **heat** Q or by work W done by the system on its surroundings:

$$\Delta U = Q - W$$

The negative sign is used to keep W, the work done *by* the system, a positive quantity. This is useful when we discuss heat engines.

In an **adiabatic** process there is no heat exchange with the environment; that is, $Q = 0$. Thus, $\Delta U = -W$. The change in internal energy is due solely to the work done by the system.

In an **adiabatic free expansion** a gas suddenly expands without doing work. Since both Q and W are zero, we have $\Delta U = 0$. The internal energy does not change. In the special case of an ideal gas, the temperature stays constant.

The change in internal energy of n moles of an ideal gas undergoing any process (not just at constant volume) is

$$\Delta U = nC_v \, \Delta T$$

For an ideal gas, the difference between the specific heats at constant pressure and at constant volume is

$$C_p - C_v = R$$

When an ideal gas undergoes a **quasistatic adiabatic** process, it obeys the relation

(Quasistatic adiabatic) $\quad\quad PV^\gamma = \text{constant}$

The ideal gas equation of state $PV = nRT$ may be used to find the temperature of the initial and final equilibrium states.

Heat transfer occurs by three mechanisms:

1. In **conduction** heat is transferred by collisions between molecules, and in the case of metals, by "free electrons."

2. In **convection** heat transport is associated with the movement of warm and cold parts of a fluid.

3. **Radiation** is the transfer of energy without an intervening medium. A warm body radiates to its cooler surroundings.

The rate of heat conduction along a rod of cross-sectional area A is

$$\frac{dQ}{dt} = -\kappa A \, \frac{dT}{dx}$$

where κ is the **thermal conductivity,** and dT/dx is the temperature gradient along the rod.

ANSWERS TO IN-CHAPTER EXERCISES

1. The transfer of heat to the water is

$$\Delta Q = m_2 c_2 \Delta T = (0.25 \text{ kg})(4190 \text{ J/kg·K})(5.8 \text{ K}) = 6.08 \text{ kJ}$$

2. From the ideal gas equation of state we have

$$T_2 = \frac{P_2 V_2}{nR}; \quad\quad T_1 = \frac{P_1 V_1}{nR}$$

Thus, $T_2 = \left(\frac{P_2}{P_1}\right)\left(\frac{V_2}{V_1}\right) T_1 = (1)(2)(293 \text{ K}) = 586 \text{ K}$.

3. The ideal gas equation of state may be applied to the initial and final states. Thus, $P_1 V_1 = nRT_1$ and $P_2 V_2 = nRT_2$, where $P_2 = P_1/3$ and $V_2 = 1.9V_1$. We see that $T_2 = (1.9/3)T_1 = 236$ K.

4. Note that $PV = nRT$ and that $C_p - C_v = R$ may be written in the form $(\gamma - 1) = R/C_v$.

5. From Eq. 19.18b,

$$v = \sqrt{\frac{\gamma RT}{M}}$$
$$= \sqrt{\frac{(1.4)(8.31 \text{ J/mol·K})}{29 \times 10^{-3} \text{ kg/mol}}} \sqrt{T} = 20\sqrt{T}$$

Note that T is in kelvins and v is in m/s.

6. (a) The mirror-like metallic coating minimizes both the absorption and the emission of radiant energy. (b) The removal of air minimizes heat transfer by conduction or convection.

QUESTIONS

1. Did Joule's paddle wheel experiment involve the transfer of heat to the water? If not, why did its temperature rise?

2. On hot desert roads, slightly porous canvas water bags are hung from the front bumpers of cars and trucks. What is the basis of this practice?

3. Why is ground frost more likely on clear (cloudless) nights?

4. Cooks often prefer copper pots to steel pots. Why?

5. (a) Can heat flow into a system of finite size without changing its temperature? (b) Can the temperature of a system change without an input of heat? Give examples.

6. (a) Is it possible to do mechanical work on a system, such as a fluid in a container, without changing its volume? If so, how? (b) Is it possible for a gas to do work without changing its volume?

7. The two processes depicted in Fig. 19.18 are quasistatic. In which case is the heat flow into the system greater?

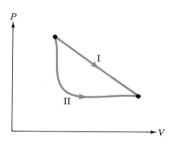

FIGURE 19.18 Question 7.

8. An ideal gas expands to twice its volume under the following conditions: (a) isothermal; (b) adiabatic; (c) isobaric.

Sketch each process on a PV diagram. In which case is the final temperature the greatest?

9. For the conditions mentioned in Question 8, in which case is the change in internal energy the greatest?

10. True/false: The expression $dW = p\,dV$ applies only to (a) an ideal gas, (b) a quasistatic process.

11. When an ideal gas undergoes a free expansion, its temperature is unchanged. Yet, when the gas undergoes an adiabatic expansion against a piston, its temperature drops. Why the difference?

12. Which of the following properties of an isolated system must be constant: (a) temperature; (b) pressure; (c) volume; (d) total energy?

13. Give an example in which the change in internal energy of a system is converted entirely into (a) heat; (b) work.

14. Aluminum foil used for cooking has a shiny side and a side with a mat finish. How would you wrap a potato to be baked?

15. Why does air escaping from a tire feel cool? Is this an example of free expansion?

16. What is the principal mechanism by which a fireplace heats a room? Is it better to have the front of the fireplace open or blocked with a glass partition?

17. Fiberglass insulation bats sometimes have aluminum foil on one side. What purpose does this serve?

18. On a very cold day, why is a wrench likely to stick to your skin?

19. Explain why we are comfortable in 15 °C air, but find swimming in 15 °C water unpleasant.

20. Is it possible for heat to flow from a system with low internal energy to another with higher internal energy? If so, explain why and give an example.

EXERCISES

19.1 Specific Heat; 19.2 Latent Heat

1. (I) One British thermal unit (Btu) raises the temperature of 1 lb of water by 1 F°. How many joules are equivalent to 1 Btu?

2. (I) When 400 J of heat are supplied to 150 g of a liquid, its temperature rises by 2.5 K. What is its specific heat?

3. (I) An 80-g steel ball at 180 °C is placed in a 90-g copper calorimeter that contains 500 g of water at 15 °C. What is the final temperature?

4. (I) A 250-g lead ball at 210 °C is placed in a 90-g aluminum calorimeter that contains 300 g of a liquid at 20 °C. If the final temperature is 30 °C, what is the specific heat of the liquid?

5. (I) A 0.5-kg steel electric kettle rated at 1200 W contains 0.6 kg of water at 10 °C. How long does it take the water to reach 90 °C? Assume that there are no losses.

6. (I) During light exercise a person generates heat at a rate of 600 kcal/h. If 60% of this heat is lost through evaporation of water, estimate the mass of water lost in 2 h. Take the latent heat of vaporization to be 2.45×10^6 J/kg.

7. (I) How much heat is needed to convert 80 g of ice initially at −10 °C to 60 g of water and 20 g of steam at 100 °C?

8. (II) A 70-g copper calorimeter contains 100 g of water. Two hundred grams of lead pellets at 200°C are added to the water. (a) What should be the initial temperature of the

water so that the final temperature is that of the room, 20 °C? (b) What would be the purpose of arranging this outcome?

9. (I) A nuclear power plant generates 500 MW of waste heat that must be carried away by water pumped from a lake. If the water temperature is to rise by 10 C°, what is the required flow rate in kg/s?

10. (I) Solar radiation supplies approximately 1 kW/m² at the earth's surface. A 3 m × 2 m solar collector is used to heat water. If the required temperature rise is 40 C°, what is the necessary flow rate of the water in kg/s? Assume 80% of the radiant energy is absorbed by the water.

19.3 Mechanical Equivalent of Heat

11. (I) A 1200-kg car is braked to a stop from 100 km/h. (a) How much kinetic energy is lost? (b) If 60% of this energy appears in the steel brake drums, whose total mass is 10 kg, what is their increase in temperature?

12. (I) A 0.5-kg hammer head strikes a 6-g steel nail 15 times at 2 m/s. Assume that 20% of the kinetic energy of the hammer serves to raise the temperature of the nail. What is the rise in temperature of the nail?

13. (II) A 20-g lead bullet at 30 °C and moving at 350 m/s embeds itself in a wooden block. (a) If 70% of the initial kinetic energy becomes internal energy of the bullet, what is its final temperature? (b) Does any of it melt? If so, how much?

14. (II) On his honeymoon in 1847, Joule took along a thermometer to measure the temperatures at the top and bottom of the falls at Chamonix in the French Alps. The falls are 120 m high. Assuming that all the kinetic energy of the water at the bottom becomes internal energy of the water, estimate the temperature rise. Do you think his measurement agreed with this estimate?

15. (II) In Joule's paddle wheel experiment the two blocks had a total mass of 3.6 kg and were allowed to fall 16 times through a height of 11 m. If the barrel contained 3.5 kg of water, what would be the expected rise in temperature of the water?

16. (II) Five kilograms of lead pellets are dropped from a height of 40 m into 50 kg of water. Estimate the rise in temperature of the water. Assume that the lead and the water are initially at the same temperature. Assume that 80% of the kinetic energy of the pellets goes into warming the water.

19.4 Work in Thermodynamics

17. (II) One kilogram of water at 0 °C and 1 atm freezes to ice at the same temperature. How much work is done by the water? The density of the liquid is 1000 kg/m³, and of ice it is 920 kg/m³.

18. (I) A sample of gas has a volume of 5 L at a pressure of 120 kPa. At this fixed pressure what is the work done by the gas if the volume is (a) doubled; (b) halved?

19. (II) An ideal gas expands at a constant pressure of 120 kPa from *a* to *b*; see Fig. 19.19. It is then compressed isother-

mally to point *c* where the volume is 40 L. Find the work done by the gas during (a) the expansion, (b) the compression.

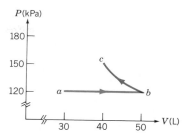

FIGURE 19.19 Exercise 19.

20. (II) Two moles of an ideal gas are initially at a pressure of 100 kPa. The gas is compressed isothermally at 0 °C to a pressure of 250 kPa. Find the work done by the gas.

19.5 and 19.6 The First Law of Thermodynamics

21. (I) A system absorbs 35 J of heat and in the process it does 11 J of work. (a) If the initial internal energy is 205 J what is the final internal energy? (b) The system follows a different thermodynamic path to the same final state and does 15 J of work, what is the heat transferred?

22. (I) While a gas is compressed from 1.2 L to 0.8 L at a constant pressure of 0.4 atm, it absorbs 400 J of heat. Find: (a) the work done by the gas; (b) the change in its internal energy.

23. (II) A gas is confined to a vertical cylinder by a piston of mass 2 kg and radius 1 cm. When 5 J of heat are added, the piston rises by 2.4 cm. Find: (a) the work done by the gas; (b) the change in its internal energy. Atmospheric pressure is 10⁵ Pa.

24. (II) A gas undergoes the cyclic process depicted in Fig. 19.20. In the process *abc*, the system absorbs 4500 J of heat. The internal energy at *a* is $U_a = 600$ J. (a) Determine U_c. The net heat absorbed during the complete cycle is 1000 J. For the process *c* to *a* find (b) the work done by the gas, and (c) the heat transfer.

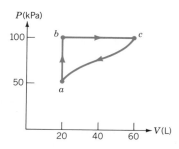

FIGURE 19.20 Exercise 24.

25. (II) When a gas undergoes a process depicted as the straight line from *a* to *c* in Fig. 19.21 the heat flow into the system is 180 J. (a) Find the work done from *a* to *c*. (b) If $U_a = 100$ J, find U_c. (c) What is the work done by the gas when it returns to a via b? (d) What is the heat transfer in the process cba?

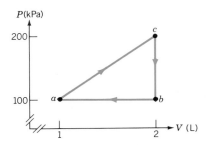

FIGURE 19.21 Exercise 25.

19.7 Ideal Gas

26. (II) (a) Find the work done by 20 g of oxygen ($M = 32$ g/mol) when it expands quasistatically and isothermally at 300 K from 0.12 m³ to 0.3 m³. (b) How much heat is absorbed?

27. (I) (a) The specific heat at constant volume of steam ($M = 18$ g/mol) is $c_v = 2.5$ kJ/kg·K. Find the specific heat at constant pressure. (b) The specific heat at constant pressure of air ($M = 29$ g/mol) is $c_p = 1$ kJ/kg·K. Find the specific heat at constant volume. Assume that both behave as ideal gases.

28. (I) A gas with a molecular mass of 32 g/mol has a specific heat at constant pressure $c_p = 0.918$ kJ/kg·K. Find the specific heat at constant volume.

29. (I) An airtight room has a volume of 80 m³. The air is initially at 0 °C and a pressure of 100 kPa. What is the change in temperature produced if 150 kJ of heat are absorbed by the air? Take $c_v = 0.72$ kJ/kg·K and $M = 29$ g/mol.

30. (I) A person produces heat at a rate of 60 W. What is the temperature increase in 30 min due to two people in an airtight room of volume 50 m³ initially at 0 °C and 10^5 Pa? Assume that all the heat is absorbed by the air. Take $c_v = 0.72$ kJ/kg·K and $M = 29$ g/mol.

31. (I) Two moles of air ($M = 29$ g/mol) at a fixed pressure of 1 atm are heated from 0 °C to 100 °C. Find: (a) the heat input; (b) the work done by the gas; (c) the change in its internal energy. Take $c_p = 1$ kJ/kg·K.

32. (II) One mole of an ideal gas is taken through the cyclic process depicted in Fig. 19.22. From a to b it undergoes an isothermal expansion. (a) Find the work done by the gas in each segment, ab, bc, and ca. (b) What is the net heat flow in a complete cycle?

33. (II) One mole of an ideal gas is first compressed isothermally at 350 K to 50% of its original volume. Next, 400 J of heat are transferred into it at fixed volume. Find: (a) the total work done by the gas; (b) the total change in its internal energy.

34. (I) An ideal gas expands adiabatically to twice its original

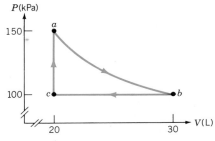

FIGURE 19.22 Exercise 32.

volume and in the process it does 400 J of work. (a) What is the change in its internal energy? (b) What is the heat transfer?

35. (II) Two moles of an ideal gas initially at a pressure of 150 kPa and a temperature of 20 °C expand quasistatically to twice the original volume. Find the work done and the change in internal energy, given that the expansion is: (a) isothermal; (b) adiabatic. Take $\gamma = 1.4$ and $C_v = 20.9$ J/mol·K.

36. (II) Helium gas initially at a temperature of 0 °C and a presure of 100 kPa is compressed quasistatically from 30 L to 20 L. What is the work required if the process is: (a) isothermal; (b) adiabatic? Take $\gamma = 5/3$.

37. (I) What is the quasistatic adiabatic work done by 2 moles of an ideal gas when its temperature changes from 15 °C to 90 °C? Take $C_v = 12.5$ J/mol·K.

38. (I) Two moles of an ideal gas initially at a pressure of 150 kPa and a temperature of 20 °C expand quasistatically at fixed pressure to twice the original volume. Find: (a) the work done by the gas; (b) the change in its internal energy. Take $C_v = 20.9$ J/mol·K.

39. (I) One mole of air ($\gamma = 1.4$) is initially at a pressure of 100 kPa and a temperature of 300 K. It expands adiabatically to five times the initial volume. What is (a) the final pressure, and (b) the final temperature?

40. (II) A diesel engine takes in air ($\gamma = 1.4$) at 20 °C and compresses it adiabatically to $\frac{1}{15}$ of its original volume (a compression ratio of 15:1). What is the final temperature of the air?

41. (I) Prove that for an ideal gas in a quasistatic adiabatic process the following equations are valid:

(a) $TV^{\gamma-1} = $ constant;   (b) $T^\gamma P^{1-\gamma} = $ constant

42. (II) Three moles of a monatomic ideal gas ($\gamma = 5/3$) expand adiabatically from an initial pressure of 200 kPa at a temperature of 10 °C to a final pressure of 50 kPa. Find: (a) the initial and final volumes; (b) the final temperature; (c) the work done by the gas.

43. (II) Two moles of a diatomic ideal gas ($\gamma = 1.4$) are initially at 17 °C. It is compressed adiabatically from a volume of 120 L to 80 L. Find: (a) the final temperature; (b) the initial and final pressures; (c) the work done by the gas.

19.8 The Speed of Sound

44. (I) Calculate the speed of sound at 300 K and 1 atm. (a) in oxygen ($M = 32$ g/mol, $\gamma = 1.4$) and (b) in helium ($M = 4$ g/mol, $\gamma = 1.66$).

45. (I) Carbon dioxide has a density of 1.98 kg/m³ at 0 °C and 1 atm. The constant $\gamma = 1.3$. (a) What is the speed of sound in this gas? (b) What is its adiabatic bulk modulus?

19.9 Heat Transport

46. (I) A 1.2 m × 1.4 m glass windowpane is 4 mm thick. What is the rate of heat transfer if the temperature difference across the pane is 30 °C?

47. (I) As one descends into the earth's crust, the temperature increases at the rate of 30 °C/km. What is the rate of heat flow per m² given that the conductivity of rock is 1 W/m·K?

48. (I) The sun may be treated as a body at 5800 K. Given that its radius is 7×10^8 m and $e = 1$, what is the total power radiated?

49. (I) The ends of an insulated copper rod of radius 2 cm and length 40 cm are maintained at 0 °C and 60 °C. (a) What is the rate of heat flow through the rod? (b) What is the temperature at a point 10 cm from the hot end when a steady state is reached?

50. (II) Two rods of copper and aluminum, each of length 50 cm and radius 1 cm, are placed in contact end to end as in Fig. 19.23. The sides of the rods are insulated. The other end of the copper rod is at 80 °C and that of the aluminum is at

FIGURE 19.23 Exercise 50.

10 °C. (a) What is the temperature at the junction? (b) What is the rate of heat conduction through the rods?

51. (II) One inch of Styrofoam has an R value of 6 in customary units. What thickness (in inches) of concrete would be needed to duplicate this figure?

PROBLEMS

1. (II) A steel cube has sides of length 10 cm. It is heated from 0 °C to 40 °C at a constant pressure of 100 kPa. Find: (a) the heat transferred to the cube; (b) the work done by the cube, (c) the change in the cube's internal energy. The coefficient of linear expansion is 11×10^{-6} K⁻¹ and the density is 7.8 g/cm³.

2. (I) A 2-kg chunk of ice at -10 °C is added to 1 kg of water at 8 °C. What is the final temperature of the system? Ignore the container.

3. (II) Hot fluid passes through a cylindrical pipe of length L with inner radius a and outer radius b; see Fig. 19.24. Show that the rate of heat conduction through the wall of the pipe is

$$\frac{dQ}{dt} = \frac{2\pi\kappa L \, (T_a - T_b)}{\ln(b/a)}$$

where T_a and T_b are the temperatures at the inner and outer surfaces, respectively. (*Hint:* Note that the temperature gradient may be written as dT/dr and that the heat flows through an area $2\pi rL$.)

FIGURE 19.24 Problem 3.

4. (II) A hot liquid is contained within a spherical shell of inner radius a and outer radius b. Show that the rate of heat transfer due to conduction is given by

$$\frac{dQ}{dt} = \frac{4\pi\kappa ab \, (T_a - T_b)}{b - a}$$

where T_a and T_b are the temperatures at the inner and outer surfaces, respectively.
(*Hint:* Note that the temperature gradient may be written as dt/dr and that the surface area of a sphere of radius r is $4\pi r^2$.)

5. (I) A "thermopane" window consists of two panes of glass of area A with a film of air in between; see Fig. 19.25. (a) Show that the rate of heat flow due to conduction is

$$\frac{dQ}{dt} = \frac{A(T_1 - T_2)}{2t_g/\kappa_g + t_a/\kappa_a}$$

where T_1 and T_2 are the inside and outside temperatures, and t_g and t_a are the thicknesses of each pane of glass and of the air film, respectively. The conductivity of glass is κ_g and that of air is κ_a. (b) The panes are 3.4 mm thick and the air film is 8 mm thick. If the inside and outside temperatures are 20 °C and -35 °C, respectively, what is the rate of heat flow if the area is 3.5 m²?

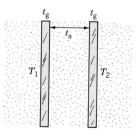

FIGURE 19.25 Problem 5.

6. (II) At low temperatures the molar specific heat of solids is given by Debye's law

$$C = \frac{kT^3}{T_D^3}$$

where $k = 1945$ J/mol·K and T_D is the *Debye* temperature characteristic of the solid. Calculate the heat input required to raise the temperature of 2.4 moles of aluminum ($T_D = 400$ K) from 10 K to 20 K.

7. (I) Show that the bulk modulus $B = -V(dP/dV)$ for an ideal gas in an isothermal process is $B = P$.

8. (I) An ideal gas is taken around the cycle *ABCA* shown in Fig. 19.26. Find: (a) the work done by the gas in each segment in terms of P_0 and V_0; (b) the net work done by the gas in each cycle; (c) the net heat input in each cycle. (d) Evaluate W and Q for 1.5 moles if A corresponds to 20 °C.

9. (II) A layer of ice of thickness y is on the surface of a lake. The air is at a constant temperature $-T$ and the ice–water interface is at 0 °C. Show that the rate at which the thickness increases is given by

$$\frac{dy}{dt} = \frac{\kappa T}{L\rho y}$$

where κ is the thermal conductivity of the ice, L is the latent heat of fusion, and ρ is the density of the ice.

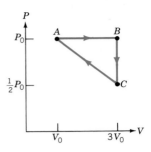

FIGURE 19.26 Problem 8.

10. (II) According to the *van der Waals* equation of state for one mole of a real gas, the pressure P and volume V at a temperature T are related by

$$\left(P + \frac{a}{V^2}\right)(V - b) = RT$$

where a and b are constants. Obtain an expression for the work done by the gas at a fixed value of T when the volume changes from V_i to V_f.

CHAPTER 20

Kinetic Theory

Major Points

1. The molecular model of an **ideal gas.**
2. **Kinetic theory** relates macroscopically observed quantities such as temperature and pressure to the behavior of molecules.
3. The **equipartition of energy** specifies how the total energy of a molecule is divided between translational, rotational, and vibrational *degrees of freedom.*
4. Kinetic theory and the equipartition of energy can be used to predict the values of the **specific heats** of an ideal gas.

Thermodynamics deals with macroscopically observed quantities, such as pressure, volume, and temperature. It does not require, nor does it yield, any information on microscopic details. This is the source both of its power and of its limitations. **Kinetic theory** attempts to provide insight into the underlying microscopic basis for the macroscopically observed behavior of gases. In kinetic theory, a gas is assumed to consist of a very large number of molecules that move about randomly and collide frequently. At atmospheric pressure and room temperature, 1 cm³ of gas has approximately 3×10^{19} molecules. Although one cannot apply Newton's laws to each molecule, one can relate the average values of certain quantities at the microscopic level to quantities, such as pressure and temperature, that are observable at the macroscopic level.

20.1 THE MODEL OF AN IDEAL GAS

In the kinetic theory of gases, the following assumptions are made about the molecules in a gas:

1. The gas consists of a *very large number of identical molecules moving with random velocities.*
2. The molecules have *no internal structure, so all their kinetic energy is purely translational.* (We will later incorporate the possibility of rotation or vibration of a molecule.)
3. The molecules *do not interact except during brief elastic collisions* with each

other and the walls of the container. This means that they exert strongly repulsive forces only at close range. The duration of each collision is much shorter than the time between collisions, so the potential energy associated with these forces is negligible.

4. *The average distance between the molecules is much greater than their diameters.* This means that they occupy a negligible fraction of the volume of the container. (From the incompressibility of liquids, we can infer that the molecules in a liquid are nearly as close together as possible. Since the density of the gas is about 10^{-3} of the density of the liquid, the assumption that the molecules are far apart is reasonable. The average separation between molecules in a gas is about 10 atomic diameters at atmospheric pressure and room temperature.)

In practice these assumptions are valid for real gases at low density and at temperatures well above the liquefaction (boiling) point. For simplicity, we also assume that there are no external forces, such as gravity. Also, the overall distribution in the speeds of the molecules is assumed not to vary in time—although the speeds of individual molecules change as a result of collisions.

20.2 KINETIC INTERPRETATION OF PRESSURE

In the seventeenth century, one explanation for the pressure exerted by a gas attributed it to repulsion between particles. Pressure was thought to be a static effect, as it is for liquids at rest. In this view, the absorption of "caloric" by a gas led to an increase in the repulsion, which in turn caused the pressure to rise. An alternative explanation was based on the idea that heat is associated with the motion of the particles. Robert Hooke was one of the first scientists to suggest that gas pressure is the result of the constant bombardment of the molecules against the walls of a container. In 1738, Daniel Bernoulli used this idea to derive Boyle's law. We will present a modified derivation given by Joule in 1848.

Consider a gas confined to a cube of side L, as in Fig. 20.1a. We wish to calculate the pressure exerted on a face perpendicular to the x axis. To simplify matters, we assume initially that the molecules do not collide with each other. We focus our attention on one molecule whose velocity has the x component v_{1x}. When it collides elastically with the wall, as in Fig. 20.1b, the y and z components of its velocity are unchanged, whereas the x component merely reverses its direction. The change in linear momentum of the molecule is

$$\Delta p_x = (-mv_{1x}) - (mv_{1x}) = -2mv_{1x}$$

The wall experiences an equal and opposite impulse. The molecule will strike the opposite wall and return after a time interval $\Delta t = 2L/v_{1x}$. (Collisions with other walls will not affect this time. Why?) A single molecule therefore exerts a series of impulses on the wall. The (time) average force exerted by this one molecule is

$$F_1 = \frac{\Delta p_1}{\Delta t_1} = \frac{2mv_{1x}}{2L/v_{1x}}$$
$$= \frac{mv_{1x}^2}{L}$$

The total average force exerted by all the molecules is the sum

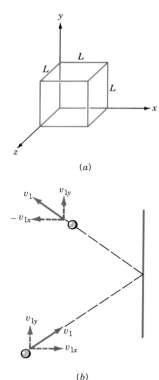

FIGURE 20.1 (a) A gas confined to a cube of side L. (b) When a gas molecule undergoes an elastic collision with a wall the component of its velocity perpendicular to the wall is reversed, but the component parallel to the wall is unchanged.

$$F = \sum F_i = \frac{m}{L} \sum v_{ix}^2 \qquad (20.1)$$

We cannot find $\sum v_{ix}^2$ since we do not know how the velocities of the molecules are distributed. However, we note that the average value of v_x^2 taken over all the N molecules in the gas (not the time average for any single molecule) is

$$\overline{v_x^2} = \sum \frac{v_{ix}^2}{N}$$

The speed of the ith molecule is given by $v_i^2 = v_{ix}^2 + v_{iy}^2 + v_{iz}^2$, so the average taken over all the molecules is

$$\overline{v^2} = \overline{v_x^2} + \overline{v_y^2} + \overline{v_z^2}$$

Since the motion of the molecules is completely random, there is no preferred direction. Thus, one would expect that

$$\overline{v_x^2} = \overline{v_y^2} = \overline{v_z^2} = \tfrac{1}{3}\overline{v^2}$$

Using $\sum v_{ix}^2 = N\overline{v_x^2} = N\overline{v^2}/3$ in Eq. 20.1, we find that the total average force is

$$F = \frac{mN\overline{v^2}}{3L}$$

and the pressure, $P = F/A$, exerted on the wall of area $A = L^2$, is

$$P = \frac{Nm\overline{v^2}}{3V} \qquad (20.2)$$

where $V = L^3$.

The **root mean square (rms)** speed of the molecules is defined as

$$v_{\text{rms}} = \sqrt{\overline{v^2}} \qquad (20.3)$$

In terms of the mass density, $\rho = Nm/V$, Eq. 20.2 becomes

$$P = \tfrac{1}{3}\rho v_{\text{rms}}^2 \qquad (20.4)$$

This equation is significant because the macroscopic variable pressure is expressed in terms of the rms speed of the molecules, a microscopic variable. Although this result has been derived for a cubic box, it does not depend on the shape of the container. Although collisions between pairs of molecules lead to the exchange of linear momentum between them, the force exerted on a wall averaged over all the molecules is not affected.

EXAMPLE 20.1: The speeds (in m/s) for 8 molecules are 2, 4, 5, 5, 8, 9, 11, and 15. Find: (a) their average speed; (b) their rms speed.

Solution: (a) The average speed is given by

$$v_{\text{av}} = \frac{1}{N}(v_1 + v_2 + v_3 + \cdots + v_N)$$

$$= \frac{(2 + 4 + 5 + 5 + 8 + 9 + 12 + 15)\text{m/s}}{8}$$

$$= 7.5 \text{ m/s}$$

(b) To find the rms speed we first find the average of v^2:

$$\overline{v^2} = \frac{1}{N}(v_1^2 + v_2^2 + v_3^2 + \cdots + v_N^2)$$

$$= \frac{(2^2 + 4^2 + 5^2 + 5^2 + 8^2 + 9^2 + 12^2 + 15^2)\text{m}^2/\text{s}^2}{8}$$

$$= 73 \text{ m}^2/\text{s}^2$$

Then,

$$v_{\text{rms}} = \sqrt{\overline{v^2}} = 8.54 \text{ m/s}$$

In general the average speed is not equal to the rms speed.

20.3   KINETIC INTERPRETATION OF TEMPERATURE

In thermodynamics the concept of temperature is merely the "reading on a thermometer." Kinetic theory provides insight into the physical basis of this concept. When Eq. 20.2 is written in the form

$$PV = \frac{2N}{3}\left(\frac{1}{2}m\overline{v^2}\right)$$

and compared with the ideal gas law, $PV = NkT$, we see that the **average kinetic energy** of a molecule is

Average kinetic energy
$$K_{av} = \tfrac{1}{2}mv_{rms}^2 = \tfrac{3}{2}kT \qquad (20.5)$$

Thus, for an ideal gas (or a real gas at low density and high enough temperature) *the absolute temperature is a measure of the average translational kinetic energy of the molecules.* This insight was one of the great successes of kinetic theory. The kinetic energy in Eq. 20.5 is associated with the *random* translational motion of the molecules. It does not include any orderly motion imposed, for example, by a wind.

From Eq. 20.5 we can see how the rms speed varies with temperature. The rms speed of the molecules is

$$v_{rms} = \sqrt{\frac{3kT}{m}} \qquad (20.6a)$$

Equation 20.6a may also be expressed in terms of the molecular mass M and the universal gas constant $R = kN_A$. The number of molecules N is related to the number of moles n ($N = nN_A$), and the total mass of the gas is $Nm = nM$. Combining these, we find (check it)

$$v_{rms} = \sqrt{\frac{3RT}{M}} \qquad (20.6b)$$

(In this equation the unit of M is kg/mol.) Thus, at a given temperature, lighter molecules move faster than heavier ones (they have the same average kinetic energy). This rms speed refers to the motion of the molecules *between* collisions. Notice that it is independent of pressure.

From Eq. 19.18b we know that the speed of sound is $v_s = \sqrt{\gamma RT/M}$. Since $\gamma \approx 1.3$ to 1.66 for many gases, we see that this is almost equal to the rms speed, $\sqrt{3RT/M}$. However, $v_s < v_{rms}$: We would not expect a sound wave to propagate faster than the molecules themselves.

EXAMPLE 20.2: Find: (a) the average translational kinetic energy of molecules in air at 300 K; (b) the rms speeds of the O_2 and N_2 molecules at this temperature.

Solution: (a) Although air consists of a mixture of gases, they all have the same average translational kinetic energy:

$$K_{av} = \tfrac{3}{2}kT = (1.5)(1.38 \times 10^{-23} \text{ J/K})(300 \text{ K})$$
$$= 6.21 \times 10^{-21} \text{ J}$$

(b) The molecular mass of O_2 is 32 g/mol = 32×10^{-3} kg/mol, so

the mass of one molecule is $m = M/N_A = 5.3 \times 10^{-26}$ kg. The rms speed is given by Eq. 20.6a:

$$v_{rms} = \sqrt{\frac{3kT}{m}}$$
$$= \sqrt{3(1.38 \times 10^{-23} \text{ J/K})(300 \text{ K})/(5.3 \times 10^{-26} \text{ kg})}$$
$$= 483 \text{ m/s}$$

The rms speed of N_2 molecules ($M = 28$ g/mol) works out to be 517 m/s.

EXERCISE 1. What is the rms speed of hydrogen molecules ($M = 2.02$ g/mol) at 300 K?

The relation between v_{rms} and temperature helps us understand why a gas confined to a cylinder warms up when it is rapidly compressed, even though there is no influx of heat. Consider a molecule that approaches the piston with a given speed. Because of the motion of the piston, the molecule will rebound with a greater speed. The average speed of the molecules increases because the piston does work on the gas. The increased average speed is manifested as a rise in temperature. Similarly when the gas expands, collisions with the receding piston cause the average speed of the molecules to drop. The gas does work on the piston and its temperature drops. In contrast, in an adiabatic free expansion into an evacuated chamber, the gas does no work. The average speed of the molecules does not change and, in the case of an ideal gas, the temperature stays constant.

EXERCISE 2. The rms speed of the molecules of a gas is 400 m/s at a certain temperature. The gas is put in a box on a train which travels at 40 m/s. By what factor does the temperature of the gas change?

20.4   SPECIFIC HEATS OF AN IDEAL GAS

The model of an ideal gas treats the particles as structureless hard spheres that can have only translational kinetic energy. This model is approximately valid for monatomic inert gases such as He, Ne, and Ar. The total energy of N molecules of such a gas is found from Eq. 20.5:

$$U = N(\tfrac{1}{2}mv^2)$$
$$= \tfrac{3}{2}NkT = \tfrac{3}{2}nRT \qquad (20.7)$$

where we have used the relations $R = kN_A$ and $N = nN_A$. We see that the internal energy of an ideal gas depends only on the temperature. We reached the same conclusion from the result of the experiment on adiabatic free expansion (Section 19.4).

According to Eq. 19.15, the difference in the molar specific heats of an ideal gas at constant pressure and at constant volume is

$$C_p - C_v = R$$

Equation 20.7 may be used to find the individual values of these specific heats for an ideal monatomic gas. For a given change in temperature the amount of heat absorbed by a gas depends on the process by which it changes its state. For example, for the same increase in temperature, it could go from state a to state b in Fig. 20.2 at constant volume or from a to c at constant pressure. Let us consider the transition at constant volume first.

When n moles of any gas are heated at constant volume, the gas does no work, $W = 0$. From the first law ($\Delta U = Q - W$) we see that all the heat goes into the internal energy: $Q_v = \Delta U$, where the subscript indicates that the heat is absorbed at constant volume. From Eq. 19.11, we know $Q_v = nC_v \Delta T$, where C_v is the molar specific heat at constant volume. Therefore, we have

$$\Delta U = nC_v \Delta T \qquad (20.8)$$

In general, this equation is restricted to a constant-volume process. However, the internal energy of an ideal gas is a function only of temperature. Thus, Eq. 20.8 gives the change in internal energy of an ideal gas for *any* process, even if the volume is not constant. Comparing Eq. 20.8 with Eq. 20.7 in the form $\Delta U =$

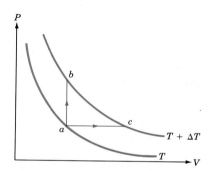

FIGURE 20.2 When a gas is heated at constant volume, it moves along the path ab. When it is heated at constant pressure, it moves along the path ac.

$\frac{3}{2}nR \, \Delta T$, we see that for an ideal monatomic gas the molar specific heat at constant volume is

(Monatomic ideal gas) $C_v = \frac{3}{2}R$ (20.9)

The numerical value is $C_v \approx 3$ cal/mol·K. Using Eq. 20.9 in $C_p - C_v = R$ we find

(Monatomic ideal gas) $C_p = \frac{5}{2}R$ (20.10)

The ratio of the specific heats is

$$\gamma = \frac{C_p}{C_v} = \frac{5}{3} \qquad (20.11)$$

From Table 20.1 it is clear that the predictions of kinetic theory for the molar specific heats and $C_p - C_v$ agree quite well for real monatomic gases. The prediction for the difference $C_p - C_v = R$ also holds for diatomic or polyatomic molecules, but the individual values of C_v and C_p do not agree with the above values for an ideal gas. The agreement for $C_p - C_v$ is not surprising since this difference represents the work term in the first law. The disagreement for diatomic molecules is not unexpected since in calculating the internal energy we ignored internal structure. If the molecules also rotate and vibrate, these motions will share in the available energy. Exactly how they do so is predicted by the equipartition theorem.

TABLE 20.1 MOLAR SPECIFIC HEATS (J/mol·K) AT 300 K AND 1 ATM.

	C_v	C_p	$C_p - C_v$	$\gamma = C_p/C_v$
Monatomic				
He	12.5	20.8	8.3	1.66
Ar	12.5	20.8	8.3	1.67
Diatomic				
H_2	20.4	28.8	8.4	1.41
N_2	20.8	29.1	8.3	1.40
O_2	21	29.4	8.4	1.40
Cl_2	25.2	34.0	8.8	1.35
Polyatomic				
CO_2	28.5	37	8.5	1.30
H_2O (100°C)	27.0	35.4	8.4	1.31

EXAMPLE 20.3: What is the heat input needed to raise the temperature of 2 moles of helium gas from 0 °C to 100 °C (a) at constant volume; (b) at constant pressure? (c) What is the work done by the gas in part (b)?

Solution: (a) The molar specific heat at constant volume is $C_v = 3R/2$. Thus,

$$Q_v = nC_v \, \Delta T = (2 \text{ mol})\left(\frac{3R}{2}\right)(100 \text{ K})$$

$$= (2 \text{ mol})(1.5 \times 8.31 \text{ J/mol·K})(100 \text{ K}) = 2.49 \text{ kJ}$$

(b) The molar specific heat at constant pressure is $C_p = 5R/2$. Thus,

$$Q_p = nC_p \, \Delta T = (2 \text{ mol})\left(\frac{5R}{2}\right)(100 \text{ K})$$

$$= (2 \text{ mol})(2.5 \times 8.31 \text{ J/mol·K})(100 \text{ K}) = 4.16 \text{ kJ}$$

(c) From the first law we know that the change in internal energy of the gas may be expressed as $\Delta U = Q_p - W$, and also as $\Delta U = Q_v - 0$ (since there is no work done at constant volume). Thus, the work done by the gas when it expands at constant pressure is

$$W = Q_p - Q_v = 1.67 \text{ kJ}$$

EXERCISE 3. A cylinder of volume 2.5 L contains 2 moles of helium ($M = 4$ g/mol) at 300 K. What is the internal energy of the (ideal) gas?

20.5   EQUIPARTITION OF ENERGY

In Section 20.5 we saw that the average translational kinetic energy of a molecule is given by

$$\tfrac{1}{2}m\overline{v^2} = \tfrac{3}{2}kT$$

and that $\overline{v_x^2} = \overline{v_y^2} = \overline{v_z^2} = v^2/3$. These results imply that

$$\tfrac{1}{2}m\overline{v_x^2} = \tfrac{1}{2}kT; \qquad \tfrac{1}{2}m\overline{v_y^2} = \tfrac{1}{2}kT; \qquad \tfrac{1}{2}m\overline{v_z^2} = \tfrac{1}{2}kT$$

The average translational kinetic energy associated with each component of velocity is equal to $\tfrac{1}{2}kT$. Since the molecule can move along three independent directions, we say that it has three translational degrees of freedom. A **degree of freedom** of a molecule is a way in which it can possess kinetic or potential energy. In the mathematical expression for the total energy, each degree of freedom appears as an independent term that involves the square of a position coordinate or a velocity component. According to Maxwell's theorem of the **equipartition of energy**:

> Each degree of freedom has an average energy $\tfrac{1}{2}kT$.

Degree of freedom

Equipartition of energy

If the molecules of a gas are structureless hard spheres, they have three translational degrees of freedom. As we saw earlier, this model is adequate for monatomic gases. However, it does not correctly predict the specific heat ratios of several gases. Let us see what happens when the idea of equipartition of energy is applied to a gas of diatomic molecules with a dumbbell shape as in Fig. 20.3. Suppose that initially all the molecules have only translational kinetic energy. Collisions between the molecules will convert some of the translational kinetic energy into rotational kinetic energy and vibrational energy. According to the equipartition theorem, this process will continue until the available energy is equally distributed, on average, among the various degrees of freedom.

If the interatomic separation is fixed, the molecule can still rotate about three mutually perpendicular axes (Fig. 20.3). Each rotational degree of freedom has kinetic energy

$$K_{\text{rot}} = \tfrac{1}{2}I\omega^2$$

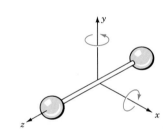

FIGURE 20.3 A dumbbell can rotate about two axes through its center of mass.

where I is the moment of inertia about the given axis. Although there appear to be three rotational degrees of freedom, better agreement with experiment is obtained if we assume there are just two rotational degrees of freedom. One can justify this by saying that the moment of inertia about the interatomic axis, I_z, is negligible in comparison with I_x and I_y.* If we take this approach, the rigid rotator has a total of five degrees of freedom, so its average energy is

$$U = 3(\tfrac{1}{2}kT) + 2(\tfrac{1}{2}kT) = \tfrac{5}{2}kT$$

The total energy of N molecules or n moles is

(Rigid rotator) $U = \tfrac{5}{2}NkT = \tfrac{5}{2}nRT$ (20.12)

Since $\Delta U = nC_v \Delta T$ and $C_p = C_v + R$, the molar specific heats and their ratio thus become

$$C_v = \tfrac{5}{2}R; \qquad C_p = \tfrac{7}{2}R; \qquad \gamma = \tfrac{7}{5} = 1.4 \qquad (20.13)$$

As Table 20.1 shows these are in reasonable agreement with values for N_2 and H_2, but not for Cl_2.

* A proper justification is provided by modern quantum mechanics.

FIGURE 20.4 A diatomic molecule has two vibrational degrees of freedom.

If one relaxes the condition of rigidity, then the atoms in a diatomic molecule can vibrate along the line joining them. The atoms may be thought of as being connected to each other by a spring, as in Fig. 20.4. (In reality the force is electrical.) The total vibrational energy is

$$E = \tfrac{1}{2}mv^2 + \tfrac{1}{2}kx^2$$

(Here, k is not Boltzmann's constant!) There are two vibrational degrees of freedom: one associated with the kinetic energy and the other with potential energy. With the addition of these contributions, the total energy of n moles of the nonrigid molecule is

(Nonrigid Rotator) $U = n(\tfrac{3}{2}RT + RT + RT) = \tfrac{7}{2}nRT$ (20.14)

The molar specific heats and their ratio are

$$C_v = \tfrac{7}{2}R; \qquad C_p = \tfrac{9}{2}R; \qquad \gamma = \tfrac{9}{7} = 1.29 \qquad (20.15)$$

In the case of Cl_2, the agreement between these values and the data is better than that predicted by Eq. 20.13 but still not completely satisfactory.

The Failure of Equipartition

Although the model of the rigid rotator appears to work well for some diatomic molecules, it does so only over a limited range of temperatures. Figure 20.5 illustrates the variation in the molar specific heat (at constant volume) of molecular hydrogen, H_2, over a wide range in temperatures. Below about 100 K, C_v is $3R/2$, which is characteristic of the three translational degrees of freedom. At room temperature (300 K) it is $5R/2$, which includes the two rotational degrees of freedom. It seems, therefore, that at low temperatures, rotation is not allowed. At high temperatures, C_v starts to rise toward the value $7R/2$ predicted for a nonrigid rotator. Thus, the vibrational degrees of freedom contribute only at these high temperatures; they appear to be "frozen" at room temperature. Problems also arise when the idea of equipartition is applied to solids.

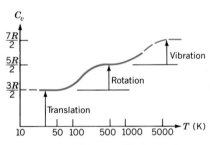

FIGURE 20.5 The specific heat of hydrogen as a function of temperature.

Solids

In crystalline solids, the atoms are arranged in a three-dimensional array, called a lattice. Each atom in a lattice can vibrate along three mutually perpendicular directions, each of which has two degrees of freedom. Thus, each atom has a total of six degrees of freedom. For n moles, the total energy should be

$$U = 3nRT$$

The volume of a solid does not change significantly with temperature. Thus, the work done to change the volume is nearly zero, and so there is little difference between C_v and C_p for a solid. The molar specific heat is expected to be (see Eq. 20.8)

$$C = \frac{1}{n}\frac{\Delta U}{\Delta T} = 3R$$

Its numerical value is $C \approx 25$ J/mol·K ≈ 6 cal/mol·K. This result was first found experimentally by Dulong and Petit in 1819. Figure 20.6 shows that the Dulong-Petit law is obeyed quite well at high (>250 K) temperatures. However, at low temperatures, the specific heat decreases. Again, the vibrational degrees of freedom seem to become "frozen." The failure of the equipartition theorem to predict the temperature dependence of the specific heats of gases and solids pointed to a major flaw in kinetic theory. The explanation of the variation in the specific heats of solids with temperature was provided by modern quantum theory.

According to quantum theory, the rotational and vibrational energies of a molecule are *quantized*. This means that only a certain set of discrete energy values, or *levels*, are allowed. Thus, a molecule can increase its rotational energy from one level to the next only if the separation, ΔE_{rot}, between the levels is less than kT—which is a measure of the available energy at a given temperature. At room temperature $\Delta E_{rot} < kT$, therefore, the rotational degrees of freedom are "active." For the vibrational energy levels, $\Delta E_{vib} > kT$ at room temperature, the vibrational degrees of freedom are therefore "frozen."

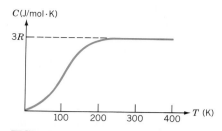

FIGURE 20.6 At low temperatures, the specific heat of a solid falls below the value $3R$ expected on the basis of classical physics.

20.6 MAXWELL-BOLTZMANN DISTRIBUTION OF SPEEDS (Optional)

Until now we have referred only to the average speed of the molecules. However, one might expect the speeds of individual molecules to vary over some considerable range. In 1859 James Clerk Maxwell showed that the speeds of the molecules in a gas of N particles of mass m are distributed according to the formula

$$f(v) = Av^2 e^{-mv^2/2kT} \qquad (20.16)$$

where $A = 4\pi N(m/2\pi kT)^{3/2}$. The function $f(v)$ is called the **Maxwell-Boltzmann distribution.** The name of L. Boltzmann (Fig. 20.7) is included because he proved that starting from any initial distribution of speeds, the collisions between the molecules lead to Maxwell's function as the *most probable* distribution. The number of particles with speeds in the range v to $v + dv$ is

$$dN = f(v) \, dv$$

FIGURE 20.7 Ludwig Boltzmann (1844–1906).

This number is represented as the shaded area in Fig. 20.8. The constant A in Eq. 20.16 comes from the condition that the total

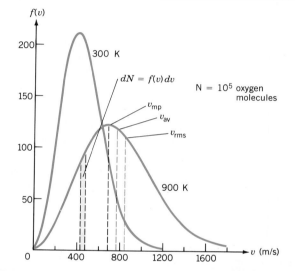

FIGURE 20.8 The Maxwell-Boltzmann speed distribution. The number of molecules dN within the range dv is given by $dN = f(v)dv$. As the temperature rises, the curve becomes broader and the peak shifts to higher speeds.

number of particles is N; that is,

$$N = \int_0^\infty f(v)dv$$

(see Problem 3). From Eq. 20.16 one can infer several features of the speed distribution. The results will be stated without proof and the derivations are left as problems. The peak of the curve is called the *most probable* speed v_{mp}, since the largest fraction of molecules have this speed. It may be located by using the condition $df/dv = 0$ for the maximum (or minimum) of a function. It is found (see Problem 2) that

$$v_{mp} = \sqrt{\frac{2kT}{m}} = 1.41 \sqrt{\frac{kT}{m}}$$

The *average* speed is (see Problem 4)

$$v_{av} = \sqrt{\frac{8kT}{\pi m}} = 1.59 \sqrt{\frac{kT}{m}}$$

The *root mean square* speed is (see Problem 4)

$$v_{rms} = \sqrt{\frac{3kT}{m}} = 1.73 \sqrt{\frac{kT}{m}}$$

which agrees with Eq. 20.6.

The curves in Fig. 20.8 are not symmetrical about the peaks. The minimum value of v is zero, but there is no limit (in classical physics) on the maximum speed. As the temperature increases, the peak shifts to higher speeds and the curve becomes broader. Also, for a fixed number of molecules, the height of the peak diminishes. The fraction of molecules with speeds above any given value increases.

The Maxwell-Boltzmann distribution helps explain the composition of our atmosphere. Although the average speed of molecules in the air is less than the escape speed from the earth, some fraction of the molecules will have speeds greater than the escape speed. Since $v \propto m^{-1/2}$, lighter elements have higher average speeds and a greater fraction of them have speeds exceeding the escape speed. As a consequence, hydrogen and helium escape from our atmosphere, whereas O_2 and N_2 do not. The escape speed from the moon is low enough that it has no atmosphere.

The distribution in the speeds of the molecules of a liquid is qualitatively similar to that for a gas. Thus, although there are strong cohesive forces tending to keep the molecules confined to the liquid, some molecules have enough kinetic energy to escape from the surface. This is called evaporation. When the molecules with the highest energies leave, the average energy of the remaining molecules is lower, which means that the temperature drops. This is why evaporation tends to cool a surface.

The Boltzmann Factor

The function $f(v)$ in Eq. 20.16 contains the factor $\exp(-E/kT)$, where $E = \frac{1}{2}mv^2$ is just the kinetic energy. In 1868, Boltzmann extended Maxwell's derivation to include the effect of an external force, such as the force of gravity. He showed that in a system in thermal equilibrium at temperature T, the number of particles N_i, with *total* energy E_i is given by

$$N_i = Ce^{-E_i/kT}$$

where C is a constant. The exponential term is called the **Boltzmann factor.** The ratio of the numbers of particles with energies E_1 and E_2 is therefore

$$\frac{N_1}{N_2} = \frac{e^{-E_1/kT}}{e^{-E_2/kT}} \tag{20.17}$$

This relation finds applications in many situations in physics and chemistry. In particular, the Maxwell-Boltzmann distribution may be derived by starting with the Boltzmann factor.

EXERCISE 4. Since evaporation tends to cool a surface, why does all the water in a puddle disappear?

20.7 MEAN FREE PATH (Optional)

In a paper published in 1857 by R. Clausius, many of the ideas in this chapter were discussed clearly for the first time. It was this work that stimulated Maxwell's own contributions. Clausius showed that the molecules in a gas move at high average speeds. He treated the molecules as point masses and therefore assumed that they collide only rarely. The Dutch scientist Buys-Ballot wondered why gaseous diffusion is so slow. For example, when a perfume bottle is opened at one corner of a room, the scent does not reach the other corner almost instantly—as the high average speed of the molecules might lead one to expect. Instead it spreads (diffuses) slowly throughout the available volume. From the slowness of the phenomenon of diffusion we infer that the paths of molecules are limited by collisions. Each molecule follows an erratic path, although to a good approximation the paths are straight between collisions, as illustrated in Fig. 20.9. In 1858 Clausius included the effect of collisions by introducing the concept of the **mean free path λ**, defined as the average distance traveled by a molecule between collisions.

FIGURE 20.9 The erratic path followed by a molecule as it collides with other molecules.

Let us assume that all the molecules are hard spheres with radius r. For simplicity we examine the motion of just one molecule and treat the others as if they were at rest. If the lines of motion of the centers of two molecules are separated by less than $2r$, as shown in Fig. 20.10*a*, there will be a collision. In effect the moving molecule is surrounded by a "sphere of action" of radius $2r$, and has an effective cross-sectional area of $\sigma = \pi(2r)^2 = \pi d^2$, where d is the diameter of the molecule (Fig. 20.10*b*). As the molecule moves at the average speed v from collision to collision, the sphere of action sweeps out a volume made up of cylindrical segments, as in Fig. 20.10*c*.

In a time Δt the total distance traveled is $v\Delta t$, and so the total volume swept out is $v\sigma\Delta t$. If $n_v = N/V$ is the number of molecules per unit volume, the number of molecules encountered in the above volume is $(n_v)v\sigma\Delta t$. If τ is the *mean time between collisions*, the number of collisions in Δt is $\Delta t/\tau$. We

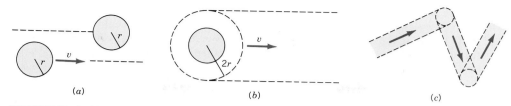

FIGURE 20.10 (*a*) Two molecules, each of radius *r*, will collide if their lines of motion are separated by less than 2*r*. (*b*) In effect, a molecule has a "sphere of action" of radius *d* = 2*r* and sweeps out a cylindrical volume, as shown in (*c*).

now equate the number of particles encountered to the number of collisions:

$$(n_v)v\sigma\Delta t = \frac{\Delta t}{\tau}$$

The mean free path is defined as the average distance between collisions and is related to the mean time between collisions by $\lambda = v\tau$. From the above equation we find

$$\lambda = \frac{1}{n_v\sigma} = \frac{1}{n_v\pi d^2}$$

where we have used $\sigma = \pi d^2$. When the motion of the other molecules is taken into account by including the Maxwell speed distribution, it is found that

$$\lambda = \frac{1}{\sqrt{2}\,n_v\pi d^2} \qquad (20.18)$$

The mean free path is reduced by a factor of $\sqrt{2}$.

EXAMPLE 20.4: Find: (a) the mean free path, and (b) the collision frequency for molecules in air at 300 K and 1 atm. Take the diameter of a molecule in air to be 0.3 nm.

Solution: (a) From Eq. 18.2 the volume of one mole is

$$V = \frac{nRT}{P} = \frac{(1\text{ mol})(8.31\text{ J/mol·K})(300\text{ K})}{1.01\times10^5\text{ N/m}^2}$$
$$= 0.0247\text{ m}^3$$

The number of molecules in one mole is N_A, so the number density is $n_v = N_A/0.0247\text{ m}^3 = 2.44\times10^{25}$ molecules per m³. Thus

$$\lambda = \frac{1}{\sqrt{2}\,n_v\pi d^2} \approx 1.02\times10^{-7}\text{ m}$$

Since the diameter of each molecule is 3×10^{-10} m, this is about 400 molecular diameters. In contrast, the average separation between molecules may be estimated as follows. Since *N* molecules occupy a volume *V*, each molecule occupies, on average, a volume *V/N*—which is the volume of a cube of side $(V/N)^{1/3}$. We take the length of the side of this cube as an estimate of the

separation between molecules:

$$\left(\frac{V}{N}\right)^{1/3} = (2.44\times10^{25})^{-1/3} = 3.5\times10^{-9}\text{ m}$$

This is about 11 molecular diameters.
(b) From Example 20.3 we know that the rms speed of oxygen molecules at 300 K is 483 m/s. The collision frequency is the number of collisions per second:

$$f = \frac{1}{\tau} = \frac{v}{\lambda}$$
$$= \frac{(483\text{ m/s})}{(1.2\times10^{-7}\text{ m})} = 4\times10^9\text{ s}^{-1}$$

20.8 THE VAN DER WAALS EQUATION; PHASE DIAGRAMS (Optional)

The ideal gas law, $PV = nRT$, holds for real gases at low densities and high temperatures. As the density increases, the size of the molecules and the interactions between them have to be taken into account.

First, because of the finite size of the molecules, the volume available for their motion is less than the volume of the container. Thus, we replace *V* by an effective volume $(V - b)$. The ideal gas pressure is modified to $P' = nRT/(V - b)$. The pressure at a given temperature is therefore greater than that for an ideal gas. Second, there must be attractive forces between molecules since gases condense under appropriate conditions. The long-range attractive interactions (van der Waals' forces) responsible for condensation are much weaker than other mechanisms of chemical bonding.

For a given molecule near a wall of the container, the net attractive force due to the other molecules will be directed toward the interior, away from the wall. The magnitude of this force on each molecule will be proportional to the density of the gas. The number of molecules near the wall is also proportional to the density. Therefore, the total force on all the molecules at the wall is proportional to the square of the density, ρ^2. Since this force is directed away from the wall, there is a decrease in pressure, $\Delta P \propto \rho^2$. Since $\rho \propto 1/V$, the decrease in *P* may be expressed as a/V^2, where a is a constant. The pressure P' is therefore reduced to $P'' = P' - a/V^2$.

When we use $P' = P'' + a/V^2$ in the expression $P' = nRT/(V - b)$, the gas law (for one mole) becomes the **van der Waals equation**

$$\left(P + \frac{a}{V^2}\right)(V - b) = RT \qquad (20.19)$$

The constants a and b depend on the gas. This equation, first proposed in 1873 by J. D. van der Waals, is remarkably successful in predicting the behavior of real gases.

Figure 20.11 illustrates isotherms in a PV diagram obtained from Eq. 20.19. Experimental data for such curves may be obtained by confining a gas in a cylinder with a movable piston and placing it in contact with a heat reservoir at some desired temperature. As the volume of the gas is changed isothermally, the pressure is recorded. At high temperatures the isotherms are the hyperbolas ($P \propto 1/V$) of an ideal gas. However, at lower temperatures, the curves become distorted. At a critical temperature, T_c, there is a point of zero slope called the **critical point**. Above T_c it is not possible to condense the gas merely by the application of pressure. Below T_c, the gas is called a vapor.

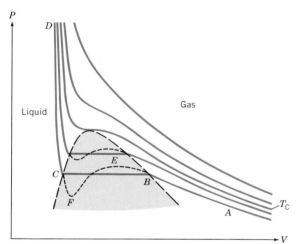

FIGURE 20.11 The PV diagram for a gas as predicted by the van der Waals equation. The solid curves are isotherms. At high temperatures, the isotherms are the hyperbolas of an ideal gas. At low temperatures, the behavior deviates markedly from that of an ideal gas. For example, the attractive forces between the molecules leads to liquefaction.

Consider what happens to a vapor as the volume is reduced isothermally, starting at point A. As V is reduced from A to B the pressure rises. At B, drops start to form—the gas begins to liquefy. As the volume is reduced further, the pressure does not change but the amount of liquid increases. Along BC the liquid and vapor are said to be in phase equilibrium. At point C all the vapor has condensed to liquid. Any attempt to reduce the volume further is hindered by the low compressibility of the liquid. This is why CD is so steep.

The van der Waals equation predicts the form of isothermal curves quite well except within the region of liquid–vapor phase equilibrium. However, the dotted line that arises from the equation is not entirely meaningless. Under carefully controlled conditions it is possible to change the state of a vapor along BE. It becomes a *supercooled* vapor—that is, the pressure is greater than that at which drops would normally appear at that temperature. This is an unstable condition, in which a slight disturbance will cause drops to form. It is also possible to move a liquid along CF, where it is *superheated*—that is, it remains a liquid above the boiling point at that pressure. This is also an unstable condition. Both supercooled vapors and superheated liquids are used in the detection of high-energy elementary particles. Since these states are unstable, the passage of a charged particle leads to the formation of drops or bubbles which show the path of the particle (see Fig. 44.16). The segment EF does not correspond to a real situation.

EXERCISE 5. In the van der Waals equation what are the units of a and b? (*Note:* Here V is the volume per mole with a unit m^3/mol.)

Phase Diagrams

Another useful way of displaying the behavior of a gas is a **phase diagram.** This is a plot of P versus T. Figure 20.12 is the phase diagram for water. It is not drawn to scale. The solid lines represent the points at which two phases are in equilibrium. The **line of vaporization** shows that the temperature at which a liquid boils depends on the pressure. At a pressure of one atmosphere, the boiling point is 100 °C. At lower pressures, the boiling point is lower. The line ends at the **critical point** C (see also Fig. 20.11). Clearly, a liquid can be converted to the vapor phase at a temperature well below the normal boiling point. The fact that the boiling point is lowered as the pressure decreases means that it takes longer to cook food properly at high altitudes. This is because the rates of chemical reactions depend on temperature.

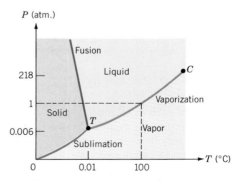

FIGURE 20.12 The phase diagram for water (not to scale). T is the triple point and C is the critical point.

The **line of fusion** represents the melting points. In the case of water, an increase in pressure results in a lowering of the melting point temperature. At a given temperature, ice can be made to melt by the application of pressure. This phenomenon plays a secondary role in the formation of a water film under ice skates. (The main reason for the melting is friction.)

The third line is the **line of sublimation.** At these points, which are at low pressures, the solid "sublimes"—it transforms directly to the vapor phase.

The three phase equilibrium lines intersect at the **triple point.** At this point the three phases can coexist in equilibrium. Recall that this property was used in the definition of temperature scales. The triple point temperature for water is 0.01 °C and the triple point pressure is 0.006 atm. In contrast, the values for CO_2 are about -57 °C and 5 atm. The sublimation of dry ice (solid CO_2) is often used for theatrical effect. The cool CO_2 gas itself is invisible, but water vapor condenses to form a fog.

SUMMARY

In the **kinetic theory of gases** a gas consists of a large number of molecules that move about randomly and collide elastically with each other and the walls of the container. It allows one to relate macroscopic quantities, such as pressure and temperature, to microscopic quantities, such as the speeds and energies of the molecules.

The pressure exerted by an ideal gas of density ρ is

$$P = \tfrac{1}{3}\rho v_{\text{rms}}^2$$

where $v_{\text{rms}} = \sqrt{\overline{v^2}}$, the **root mean square** speed of the molecules, is given by

$$v_{\text{rms}} = \sqrt{\frac{3RT}{M}} = \sqrt{\frac{3kT}{m}}$$

where M is the molecular mass (in kg/mol), R is the universal gas constant, k is Boltzmann's constant, and m is the mass of a single molecule.

The **average translational kinetic energy** of a molecule in a gas at temperature T is given by

$$K_{\text{av}} = \tfrac{1}{2}mv_{\text{rms}}^2 = \tfrac{3}{2}kT$$

The temperature of a gas is a measure of the average translational kinetic energy of the molecules.

According to the **equipartition theorem,** the average energy associated with each degree of freedom of a particle is $\tfrac{1}{2}kT$.

The **total energy** of N molecules or n moles of an ideal monatomic gas is

$$U = \tfrac{3}{2}NkT = \tfrac{3}{2}nRT$$

The **molar specific heats** of an ideal monatomic gas at constant volume and constant pressure are

$$C_v = \frac{3R}{2}; \qquad C_p = \frac{5R}{2}; \qquad \gamma = \frac{C_p}{C_v} = \frac{5}{3}$$

These values are modified for diatomic molecules when rotational and vibrational degrees of freedom are taken into account. However, the difference, $C_p - C_v = R$, is unchanged.

ANSWERS TO IN-CHAPTER EXERCISES

1. The mass of a hydrogen molecule is $m = M/N_A = 3.36 \times 10^{-27}$ kg. From Eq. 20.6a, the rms speed is

$$v_{rms} = \sqrt{\frac{3kT}{m}}$$

$$= \sqrt{\frac{3(1.38 \times 10^{-23} \text{ J/K})(300 \text{ K})}{(3.36 \times 10^{-27} \text{ kg})}}$$

$$= 1920 \text{ m/s}$$

2. The temperature is related to the kinetic energy of the *random* motion of the molecules and therefore it does not change.

3. $U = \frac{3}{2}nRT = (1.5)(2 \text{ mol})(8.31 \text{ J/mol·K})(300 \text{ K}) = 7.48$ kJ. The volume and molecular mass are not relevant.

4. The water continues to absorb heat from the surroundings so its temperature, and the speed distribution of the molecules, do not change.

5. The unit of a/V^2 must be the same as that of pressure, so the unit of a is the same as that of PV^2, which is $(\text{N/m}^2)(\text{m}^3/\text{mol})^2 = \text{N·m}^4/\text{mol}^2$. The unit of b is m^3/mol.

QUESTIONS

1. The ratio C_p/C_v for air is 1.4. Does this tell us anything about the atomic structure of the molecules?

2. A gas and its solid container are at the same temperature. Yet the gas molecules move freely, whereas those in the solid do not. Explain the difference.

3. Is it possible to transfer heat to a gas without raising its temperature? If so, how?

4. A gas is confined to a cylinder with a piston. Explain, in terms of the motions of molecules and of the piston, why the temperature of the gas rises in an adiabatic compression.

5. A container has a mixture of several noninteracting gases. According to *Dalton's law of partial pressures*, the total pressure is the sum of the pressures that would be exerted by each of the gases acting independently. Use kinetic theory to prove this statement.

6. A box contains 2 moles of a gas at temperature T. Does the pressure depend on the type of gas?

7. A box that contains a gas is placed on a scale. The gas molecules are in contact with the base of the box for only a very brief time. Why does the scale register the total weight, including all of the gas?

8. Suppose the collisions between molecules were not perfectly elastic. How would this become apparent?

9. Why does it take a shorter time to cook food, without burning it, in a pressure cooker?

10. In virtually all problems in kinetic theory, the gravitational force may be ignored. Why is this justified? Indicate one situation in which this force must be taken into account.

11. Why does the boiling point temperature of water decrease with altitude? (How is the boiling point defined?)

12. Does the mean free path of oxygen or nitrogen molecules affect the propagation of sound? If so, how?

13. A chemical reaction will occur only if the energy available is above a certain threshold "activation energy." How does the Maxwell-Boltzmann distribution account for the fact that reactions proceed faster as the temperature increases?

14. What property of water allows you to make a snowball?

15. Both diffusion and the propagation of sound waves in air involve collisions between molecules. Why is one process slow and the other fast?

EXERCISES

20.2 and 20.3 Kinetic Interpretation of Temperature and Pressure

1. (I) Calculate the rms speeds at 20 °C of atoms of: (a) helium (4 u); (b) neon (20 u); (c) radon (222 u).

2. (I) The temperature at the surface of the sun is estimated to be 5800 K. At this temperature, what is the rms speed of (a) hydrogen atoms (1 u), and (b) uranium atoms (238 u)?

3. (I) The temperature in the interior of the sun is about 2×10^7 K. What is (a) the average kinetic energy, and (b) the rms speed of protons?

4. (I) The escape speed from the surface of the earth is 11.2 km/s. Find at what temperature the following gases have this value as their rms speed: (a) N_2; (b) O_2; (c) H_2.

5. (I) The temperature at an altitude of 200 km in the upper atmosphere is 1200 K. At this temperature what would be the rms speed of: (a) helium atoms; (b) oxygen molecules?

6. (II) Naturally occurring uranium is a mixture of two isotopes ^{235}U and ^{238}U. (They have the same position in the periodic table but have different masses.) To produce "enriched uranium," with a greater fraction of ^{235}U that is required by nuclear reactors, the compound UF_6 is first formed and the gas allowed to diffuse through porous mate-

rial. The rate of diffusion depends on the rms speed. What is the ratio of the rms speeds for the molecules of the two compounds? The atomic mass of fluorine is 19 u.

7. (I) Neutrons produced in the fission of uranium have very high energies. A "moderator" such as graphite is used to slow them down to the thermal speeds characteristic of the temperature of the material. What is the rms speed of a "thermal neutron" at 300 K?

8. (I) The rms speed of a molecule is 500 m/s at 300 K. (a) At what temperature will the rms speed be twice as large? (b) Four times as large?

9. (I) In order to initiate a fusion reaction, deuterons (of mass 2 u) must have a temperature of about 10^9 K. At this temperature what is (a) the average kinetic energy, and (b) the rms speed?

10. (II) Two moles of nitrogen (N_2) are in a 6-L container at a pressure of 500 kPa. Find the average kinetic energy of one molecule.

11. (I) The temperature in intergalactic space is estimated to be 3 K. What is the rms speed of a hydrogen atom (1 u) at such a temperature?

12. (I) A 20-L flask contains 0.3 mole of oxygen gas ($M = 32$ g/mol.) at a temperature of 30 °C. (a) What is the average kinetic energy per molecule? (b) What is the pressure?

13. (II) (a) Find the rms speed of nitrogen molecules at 300 K. (b) If 2 moles of the gas are confined to a cube of side 15 cm, what is the pressure?

14. (I) What is the volume of a sample of an ideal gas at 0 °C and 1 atm in which the number of molecules is equal to the human population of the earth, which is about 5×10^9?

15. (II) Make an estimate of the average distance between molecules of oxygen (O_2) at a temperature of 0 °C and a pressure of 1 atm. (*Hint*: Consider one mole.)

16. (II) Through what vertical distance must an oxygen molecule (O_2) move so that the change in its gravitational potential energy is 10^{-3} of its average kinetic energy at 0 °C and 1 atm?

17. (I) By what factor does the rms speed of molecules in a gas increase when the temperature is raised from 0 °C to 100 °C?

18. (II) A box of volume V contains n_1 moles of oxygen (O_2) and n_2 moles of nitrogen (N_2). The *partial pressures* exerted by the gases are $P_1 = n_1RT/V$ and $P_2 = n_2RT/V$. (a) What is the ratio of their total masses if the partial pressures are equal? (b) What is the ratio of the partial pressures if the total masses are equal?

19. (I) The ionization energy (the energy needed to strip an electron from the atom) of helium is 4×10^{-18} J. At what temperature would the average kinetic energy of such atoms equal the ionization energy?

20.4 and 20.5 Specific Heats; Equipartition of Energy

20. (I) The heat capacity at constant volume of a sample of a monatomic ideal gas is 7.45 cal/K. Find: (a) the number of moles; (b) the total energy at 290 K.

21. (I) The heat capacity at constant volume of a sample of a monatomic gas is 35 J/K. Find: (a) the number of moles; (b) the internal energy at 0 °C; (c) the molar specific heat at constant pressure.

22. (I) For a given sample of gas, the difference in the heat capacities at constant pressure and at constant volume is 21.6 J/K. Find: (a) the number of moles; (b) the heat capacity at constant pressure given that the gas is diatomic (rigid dumbbell).

23. (II) One mole of an ideal monatomic gas is heated from 0 °C to 100 °C. Find the change in internal energy and the heat absorbed, given that the process takes place under the following condition: (a) constant volume; (b) constant pressure.

24. (I) The Dulong-Petit law ($C = 3R$) for the molar specific heat of a solid was initially used to infer the molecular mass M from the measured specific heat. Given that $c = 0.6$ kJ/kg·K for a particular solid, find M.

25. (I) One mole of an ideal monatomic gas is initially at 300 K. Find the final temperature if 200 J of heat are added as follows: (a) at constant volume; (b) at constant pressure.

26. (I) Find the internal energy of one mole of an ideal gas at 300 K, given that it is: (a) monatomic; (b) diatomic with no rotation; (c) diatomic with rotation and vibration.

27. (I) Given that the specific heat at constant volume of a monatomic gas is $c_v = 0.148$ kcal/kg·K, find the molecular mass of the gas (assumed to be ideal). What gas could this be?

20.7 Mean Free Path

28. (II) At a frequency of 20 kHz, what is the ratio of the wavelength of sound in air to the mean free path of O_2 molecules at 0 °C and 1 atm? Take the speed of sound to be 330 m/s and the diameter of the O_2 molecule to be 0.3 nm.

29. (II) The mean free path of O_2 molecules in a box is 9×10^{-8} m. Given that the number density is 2.7×10^{19} molecules/cm³, estimate the diameter of an oxygen molecule.

30. (II) At what pressure would the mean free path for an oxygen molecule be 1 mm at a temperature of 300 K? Take the diameter of a molecule to be 0.3 nm.

31. (I) Show that for an ideal gas the mean free path may be expressed in the form

$$\lambda = \frac{kT}{\sqrt{2}\,\pi d^2 P}$$

32. (I) The oxygen molecule (O_2) has a diameter of 3×10^{-10} m. At a pressure of 10^{-8} Pa (in a vacuum system) and a temperature of 300 K find: (a) the number of molecules per cm³; (b) the mean free path.

33. (I) The density of hydrogen atoms in intergalactic space is about 1 per m³. Assuming an atomic diameter of 10^{-10} m, estimate the mean free path.

PROBLEMS

1. (I) A collection of molecules has the following distribution of speeds:

Speed (m/s):	1	2	3	4	5	6
Number:	1	3	5	8	4	2

Find: (a) the average speed; (b) the rms speed; (c) the most probable speed.

2. (I) Consider the Maxwell speed distribution function, Eq. 20.16. By setting the derivative $df(v)/dv = 0$, show that the most probable speed is given by $v_{mp} = \sqrt{2kT/m}$.

3. (II) Use the Maxwell speed distribution function to show explicitly that the total number of particles is N, that is

$$\int_0^\infty f(v) \, dv = N$$

Note that $\int_0^\infty x^2 \exp(-ax^2)dx = \sqrt{\pi/16a^3}$.

4. (II) Given the Maxwell speed distribution function, the average value of v^n is

$$\overline{v^n} = \int_0^\infty \frac{v^n f(v) \, dv}{N}$$

where N is the total number of particles.
(a) Show that the average speed is

$$v_{av} = \sqrt{\frac{8kT}{\pi m}}$$

Note that $\int_0^\infty x^3 \exp(-ax^2)dx = 1/2a^2$.

(b) Show that the rms speed is

$$v_{rms} = \sqrt{\frac{3kT}{m}}$$

Note that $\int_0^\infty x^4 \exp(-ax^2)dx = \sqrt{9\pi/64a^5}$.

5. (II) A vessel contains 10^5 nitrogen (N_2) molecules at 300 K. Find the number of molecules with speeds within the following ranges: (a) 100 m/s to 110 m/s; (b) 330 m/s to 340 m/s; (c) 1000 m/s to 1010 m/s.

6. (II) Oxygen (O_2) gas at 0 °C and 1 atm is in a cubic box of side 40 cm. How many molecules strike each face per second?

7. (II) Show that according to the van der Waals equation, the critical point temperature and pressure are given by

$$T_c = \frac{8a}{27bR}; \qquad P_c = \frac{a}{27b^2}$$

(Note that the critical point is a point of inflection; that is, $dP/dV = d^2P/dV^2 = 0$. First use these conditions to show that the critical point volume is $V_c = 3b$.)

8. (I) One mole of a monatomic ideal gas has an initial volume of 1 L, an internal energy of 100 J and is at a pressure of 3 atm. (a) What is the initial temperature? (b) The gas expands quasistatically at constant pressure until its volume doubles. Determine the new temperature and internal energy. (c) The gas then cools at constant volume until its pressure is halved. Determine the new temperature and internal energy. Depict the process on a PV diagram.

CHAPTER 21

Entropy and the Second Law of Thermodynamics

One form of the second law of thermodynamics sets a limit on the efficiency that can be attained by heat engines, such as a steam engine.

Major Points

1. A **heat engine** uses part of the heat flowing from a hot reservoir to a cold reservoir to do work. A **refrigerator** requires the input of work to make heat flow from a cold reservoir to a hot reservoir.

2. According to the **second law of thermodynamics** perfect heat engines or perfect refrigerators are not possible.

3. The distinction between **reversible** and **irreversible** processes.

4. A **Carnot** cycle is a reversible cycle of operations that serves as a standard by which all other heat engines can be judged.

5. (a) The **entropy** of a system is a measure of the disorder of its constituent particles.
   (b) According to the second law of thermodynamics the entropy of an isolated system cannot decrease.

The first law of thermodynamics is a statement of the conservation of energy. It allows any process that conserves energy to occur. Yet several processes consistent with the first law do not occur in nature. For example, an ice cube put in water does not discharge heat to raise the temperature of the water. Heat always flows from a hot body to a cold body, never in the reverse direction. When a body is dropped to the ground, its initial potential energy is converted to kinetic energy and then, on impact, to internal energy of the body and its surroundings. This is manifested as a rise in temperature. We never witness internal energy from the environment converging onto the body and raising it off the ground. This process would not conflict with the conservation of energy, but it never happens. When a sugar cube is placed in a cup of tea, it dissolves and spreads throughout the liquid. An initially uniform distribution of sugar particles never spontaneously collects itself into a cube.

All the natural processes mentioned above proceed only in one direction; we never see the reversed order of events. Such seemingly trivial observations form the basis of the second law of thermodynamics. Among other things, the second law tells us something about how natural processes evolve. Our intuitive conviction of the correctness of this law is so firmly grounded in everyday experience that we laugh at the time-reversed order of events when a movie is run backward.

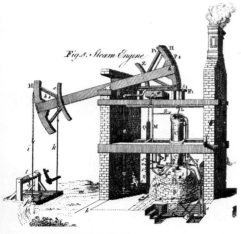

FIGURE 21.1 A Newcomen steam engine used to pump water.

The early history of thermodynamics was linked to the development of the steam engine. In one of its first applications, Newcomen used a steam engine to pump water from a mine in 1712 (Fig. 21.1). After significant improvements made by James Watt between 1763 and 1782, the steam engine supplied the motive force for the industrial revolution. Sadi Carnot, a young engineer in Paris, was impressed by the British pioneers. However, he noted that the development of the engine had been accomplished by native skill and ingenuity, rather than an understanding of the underlying principles of its operation. He saw the need to investigate the limits on the output of these engines.

Carnot's analysis of steam engines was guided by an analogy to a watermill. Water at some height is caught in the vanes and is discharged at a lower level. As the water falls, it causes the wheel to rotate, and this in turn may be used to do work. With this image in mind, Carnot established a basic requirement: The steam engine has to operate between two heat reservoirs whose "levels" are determined by their temperatures. In his view it was the *fall* of caloric from the hot to the cold reservoir that led to the work output. He assumed that the caloric itself was conserved. Despite this incorrect notion that heat is a conserved substance, Carnot formulated important theorems that laid the foundation for the subsequent development of thermodynamics. In particular, he devised an ideal thermodynamic cycle of operations that is useful in establishing limits on the efficiency of real engines.

The initial formulations of the second law of thermodynamics were in terms of the efficiencies of heat engines, such as the steam engine, which converts part of the heat absorbed from a reservoir into useful work. After a few decades it became a profound statement regarding the tendency of natural processes to evolve from a state of order to one of disorder. In this form it may be applied to a wide range of fields, such as chemistry, telecommunications, and microbiology.

One of James Watt's steam engines working at a coal mine in the 1790s. An analysis of such engines led to the initial formulation of the second law of thermodynamics.

21.1 HEAT ENGINES AND THE KELVIN-PLANCK STATEMENT OF THE SECOND LAW

A **heat engine** is a device that converts heat into mechanical work. Examples include steam engines, gasoline engines, and diesel engines. We will be concerned in particular with heat engines that operate in a repetitive cycle of processes. Such an engine has some "working substance" that is returned to its initial state at the end of each cycle. Steam engines use water, while gasoline and diesel engines use a mixture of fuel and air. One can make general statements regarding cyclical heat engines without having to specify the details of the processes.

Figure 21.2 is a schematic representation of a heat engine that operates between a hot reservoir at temperature T_H and a cold reservoir at temperature T_C. (Recall that a heat reservoir is a system whose temperature is not appreciably altered by a transfer of heat.) In each cycle the engine absorbs heat Q_H from the hot reservoir. Some of this heat is used to do work, W, and the remaining heat, Q_C, is discharged to the cold reservoir. It is convenient in the following discussions to include the sign of the heat transfer explicitly. Thus, heat entering the engine is $+|Q_H|$, and heat leaving the engine is $-|Q_C|$. In one complete cycle, the system returns to its initial state, so the internal energy of the working substance does not change. According to the first law, $\Delta U = Q - W = 0$, the net work done by the engine in a cyclic process is equal to the net heat influx:

$$W = |Q_H| - |Q_C| \tag{21.1}$$

The **thermal efficiency** ϵ of a heat engine is defined as the work *output* divided by the heat *input* ($W_{\text{OUTPUT}}/Q_{\text{INPUT}}$):

$$\epsilon = \frac{W}{|Q_H|} = 1 - \frac{|Q_C|}{|Q_H|} \tag{21.2}$$

where we have used Eq. 21.1. The engine would be 100% efficient ($\epsilon = 1$) only if $Q_C = 0$. In this case, all the heat input would be converted into work. However, as we will see, this is not possible, and the efficiency of even an "ideal" engine is less than 100%. A gasoline engine has an efficiency of about 20%, whereas for a diesel engine it is about 30%. Figure 21.3 shows a proposed plant that would use the difference in temperature between deep water and the warmer water close to the surface to generate electricity.

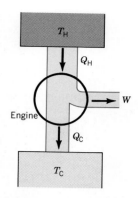

FIGURE 21.2 A heat engine extracts heat Q_H from a hot reservoir, does a quantity of work W, and discharges heat Q_C to a cold reservoir. In a complete cycle $W = |Q_H| - |Q_C|$.

Thermal efficiency

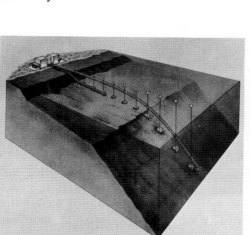

FIGURE 21.3 An OTEC (Ocean Thermal Energy Conversion) plant designed to use the difference in temperature between deep water and water near the surface of the ocean to generate electricity.

FIGURE 21.4 William Thomson, Lord Kelvin (1824–1907).

The Kelvin-Planck Statement of the Second Law

In 1851 Lord Kelvin (Fig. 21.4) made a statement that is now called the **Kelvin-Planck statement** of the **second law of thermodynamics.** We rephrase it this way:

> It is impossible for a heat engine that operates in a cycle to convert its heat input completely into work.

Note the qualifying phrase "in a cycle." It is possible to convert heat completely into work in an isothermal expansion of an ideal gas (Section 19.4). However, the system would not be left in its initial state; its volume would be greater and its pressure would be lower. The Kelvin-Planck statement of the second law says that Q_C is always nonzero; there must always be a reservoir to accept the heat discharged by the engine. Figure 21.5 illustrates a "perfect" (and impossible) heat engine. If the Kelvin-Planck statement were not true, it would be possible to use the vast internal energy of the ocean to power a ship without the necessity of a reservoir at a lower temperature.

EXERCISE 1. A heat engine has a thermal efficiency of 20%. It runs at 120 rpm and delivers 80 W. For each cycle find: (a) the work done; (b) the heat absorbed from the hot reservoir; (c) the heat discharged to the cold reservoir.

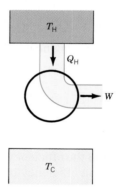

FIGURE 21.5 A perfect, and impossible, heat engine.

21.2 REFRIGERATORS AND THE CLAUSIUS STATEMENT OF THE SECOND LAW

Heat flows naturally from a hotter body to a colder body; it does not flow spontaneously from a colder to a hotter body. Based on this commonplace observation, in 1850 R. Clausius (Fig. 21.6) presented what is now known as the **Clausius statement** of the second law of thermodynamics:

> It is impossible for a cyclical device to transfer heat continuously from a cold body to a hot body without the input of work or other effect on the environment.

Figure 21.7 illustrates a "perfect," and impossible, refrigerator. It is possible to make heat flow from a cold reservoir to a hot reservoir, but a refrigerator is needed. A refrigerator, or a heat pump, is a heat engine working in reverse. Work is done *on* the device, which absorbs heat Q_C from a low-temperature reservoir (the contents of the refrigerator) and deposits a greater quantity of heat Q_H into a high-temperature reservoir (the air in the room), as shown in Fig. 21.8. A heat pump extracts heat from the cold outside air and transfers a greater quantity of

FIGURE 21.6 Rudolf Clausius (1822–1888).

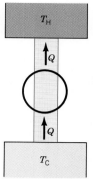

FIGURE 21.7 A perfect, and impossible, refrigerator.

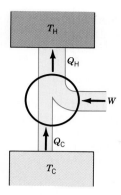

FIGURE 21.8 When work W is done on a refrigerator, it extracts heat Q_C from a cold reservoir and deposits a greater quantity of heat Q_H to a hot reservoir.

heat to the warmer air inside a room. Since the operation is cyclical, the internal energy of the engine does not change, so from the first law we have

$$|Q_H| = W + |Q_C| \qquad (21.3)$$

Refrigerators are rated according to their *coefficient of performance* (COP) which is defined as the ratio of the heat extracted from the cold reservoir Q_C to the work input:

(Refrigerator) $$\text{COP} = \frac{|Q_C|}{W} \qquad (21.4)$$ Coefficient of performance

A practical refrigerator has COP ≈ 5. The coefficient of performance of a heat pump is defined as

(Heat pump) $$\text{COP} = \frac{|Q_H|}{W}$$

EXERCISE 2. A refrigerator has a coefficient of performance of 4. It deposits 250 J per cycle into the hot reservoir. Find: (a) the heat extracted from the cold reservoir; (b) the work required.

21.3 EQUIVALENCE OF THE KELVIN-PLANCK AND CLAUSIUS STATEMENTS (Optional)

Although the Kelvin-Planck and Clausius statements appear to be unrelated, one can easily show that they are equivalent. We do so by showing that if one is assumed to be false, the other is violated.

First we assume that the Clausius statement is false and that a perfect refrigerator is possible. Let us combine this perfect refrigerator with an ordinary heat engine operating between the same two heat reservoirs. We adjust matters so that the heat $|Q_C|$ extracted by the refrigerator from the cold reservoir equals the heat discharged by the heat engine into this reservoir, as shown in Fig. 21.9a. Figure 21.9b shows that the combination is equivalent to a perfect heat engine that extracts a quantity of heat $(|Q_H| - |Q_C|)$ from the hot reservoir and converts this completely into work. This violates the Kelvin-Planck statement.

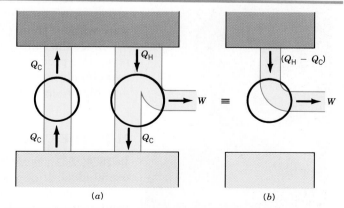

(a)

(b)

FIGURE 21.9 A perfect refrigerator combined with a real heat engine is equivalent to a perfect, and impossible, heat engine.

Next we assume that the Kelvin-Planck statement is false and that a perfect heat engine is possible. In Fig. 21.10a the perfect heat engine is combined with an ordinary refrigerator. The work output of the engine is used by the refrigerator to extract heat $|Q_C|$ from the cold reservoir and deposit heat $|Q_H|$ into the hot reservoir. As Fig. 21.10b shows, the combination is equivalent to a perfect refrigerator that transfers heat $|Q_C|$ from the cold to the hot reservoir. This violates the Clausius statement.

The Kelvin-Planck and Clausius statements apply to all types of engines and refrigerators without reference to details of the cyclical processes involved.

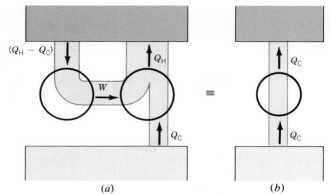

FIGURE 21.10 A perfect heat engine combined with a real refrigerator is equivalent to a perfect, and impossible, refrigerator.

21.4   REVERSIBLE AND IRREVERSIBLE PROCESSES

In a **quasistatic** process the state variables of a system change infinitely slowly so that the system is always arbitrarily close to thermal equilibrium. In practice, a quasistatic process does not have to be infinitely slow. Any system has some characteristic time that it requires to reach equilibrium from an initial nonequilibrium state. Provided the process takes much longer than this so called *relaxation time*, the process is effectively quasistatic.

Reversible process

In a **reversible** process the system may also be made to retrace its thermodynamic path back to the initial state. For a process to be reversible three conditions must be satisfied: (1) It must be quasistatic. (2) There must be no friction. (3) Any transfer of heat must occur at a constant temperature, or be associated with an infinitesimal temperature difference. Any process that does not satisfy these conditions is **irreversible.** All natural processes, which evolve only in one direction, are irreversible; examples were given in the introduction. A sudden expansion or compression is irreversible since the system must go through a succession of nonequilibrium states. Explosions, diffusion, conduction due to a finite temperature difference, and chemical reactions are also examples of irreversible processes. After an irreversible process the system cannot be returned to its initial state without some change in the surroundings.

Since the state variables have well-defined values in any quasistatic process, a reversible process can be represented on a PV diagram. The work done between two equilibrium states can be expressed in terms of the state variables. An irreversible process cannot be depicted on a PV diagram.

21.5   THE CARNOT CYCLE

In 1824, S. Carnot (Fig. 21.11) devised a reversible cycle of operations that is a useful ideal. It may involve any electrical, magnetic, or chemical cycle. We will assume that the working substance is an ideal gas confined to a cylinder with a frictionless piston. In 1834, E. Clapeyron simplified Carnot's original cycle and represented it on a PV diagram, as shown in Fig. 21.12. The Carnot cycle consists of two isothermal and two adiabatic processes:

FIGURE 21.11 Sadi Carnot (1796–1832).

1. The system starts at point a at temperature T_H. The gas undergoes an isothermal expansion from a to b while in contact with a hot reservoir at temperature T_H. During this process the internal energy of the ideal gas, which depends only on its temperature, does not change. The gas absorbs an amount of heat $|Q_H|$ and does an equal amount of work W_{ab} on the piston.

2. The reservoir is removed and the system is thermally isolated from its surroundings. The gas is allowed to undergo an adiabatic expansion ($Q = 0$) from b to c. The gas does positive work W_{bc} at the expense of its internal energy until the temperature drops to T_C.

3. The gas is placed in contact with a cold reservoir at temperature T_C and compressed isothermally from c to d. Negative work W_{cd} is done by the gas which discharges an equal quantity of heat $|Q_C|$ to the cold reservoir.

4. The last stage is an adiabatic compression from d to a during which the temperature rises to T_H. The adiabatic work done by the gas is the negative of that in stage 2, that is, $W_{da} = -W_{bc}$, because the magnitudes of the changes in internal energy are the same.

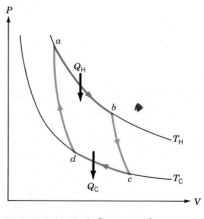

FIGURE 21.12 A Carnot cycle consists of two isothermal and two adiabatic operations.

Since the operation is in a closed cycle, the internal energy of the gas does not change. Consequently, the net work done by the gas on the piston is equal to the net heat input:

$$W = |Q_H| - |Q_C|$$

This work is represented by the area enclosed within the cycle $abcd$ in Fig. 21.12.

Efficiency of the Carnot Cycle

Let us find the efficiency of a heat engine operating in a Carnot cycle, which has an ideal gas as the working substance. The work done by an ideal gas in an isothermal expansion is given in Section 19.4. Thus, in the process a to b, the heat absorbed from the hot reservoir is

$$|Q_H| = nRT_H \ln\left(\frac{V_b}{V_a}\right) \tag{i}$$

whereas the heat discharged into the cold reservoir is

$$|Q_C| = nRT_C \ln\left(\frac{V_c}{V_d}\right) \tag{ii}$$

From the relation $PV^\gamma = \text{constant}$ for an adiabatic process, it can be shown that $TV^{\gamma-1} = \text{constant}$ (see Exercise 19.41). Using this result we have

$$T_H V_b^{\gamma-1} = T_C V_c^{\gamma-1}$$
$$T_H V_a^{\gamma-1} = T_C V_d^{\gamma-1}$$

The ratio of these two equations is $(V_b/V_a)^{\gamma-1} = (V_c/V_d)^{\gamma-1}$, which leads to

$$\frac{V_b}{V_a} = \frac{V_c}{V_d}$$

Thus the arguments of the logarithms in Eqs. (i) and (ii) are the same. The ratio of these two equations is therefore

(Carnot cycle)
$$\frac{|Q_C|}{|Q_H|} = \frac{T_C}{T_H} \tag{21.5}$$

We see that in the special case of a Carnot cycle, the ratio of the heat transfers is equal to the ratio of the Kelvin temperatures of the reservoirs.

The efficiency of a heat engine is $\epsilon = 1 - |Q_C|/|Q_H|$. So, using Eq. 21.5 for the Carnot engine, we see that the **Carnot efficiency** ϵ_C is

Efficiency of a Carnot engine

$$\epsilon_C = 1 - \frac{T_C}{T_H} \tag{21.6}$$

The Carnot efficiency depends only on the Kelvin temperatures of the reservoirs. The efficiency is always less than 100%, unless $T_C = 0$. Carnot showed that Eq. 21.6 applies to *any* reversible engine operating between the same two reservoirs, and that it sets an upper limit to the efficiency of a real (irreversible) engine.

Carnot's Theorem

Carnot presented the following theorem:

Carnot's theorem

> (i) All reversible engines operating between two given reservoirs have the same efficiency.
> (ii) No cyclical heat engine has a greater efficiency than a reversible engine operating between the same two temperatures.

One can use the Clausius statement to prove Carnot's theorem. Let us start with two reversible engines, one of which operates as a refrigerator (Fig. 21.13). The engine's work output is used to operate the refrigerator. We assume that all the heat transfers take place either at T_H or T_C. The efficiency of one engine is $\epsilon = W/|Q_H|$, and that of the other, now operating as a refrigerator, is $\epsilon' = W/|Q'_H|$. We begin by assuming that $\epsilon > \epsilon'$, that is,

$$\frac{W}{|Q_H|} > \frac{W}{|Q'_H|}$$

From this we infer that $|Q'_H| > |Q_H|$. From the figure we see that

$$W = |Q_H| - |Q_C| = |Q'_H| - |Q'_C|$$

Therefore, the combined system is equivalent to a perfect refrigerator that transfers a quantity of heat, $\Delta Q = |Q'_H| - |Q_H| = |Q'_C| - |Q_C|$, from the cold to the hot reservoir. This violates the Clausius statement unless $\Delta Q < 0$, which means heat flows the other way, or equivalently that $|Q'_H| < |Q_H|$. From this we conclude that

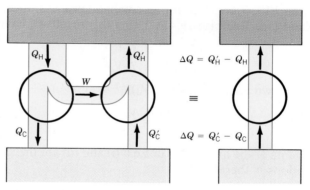

FIGURE 21.13 Two reversible engines, one operating as a refrigerator, are equivalent to a perfect, and impossible, refrigerator.

the condition $\epsilon > \epsilon'$ is not consistent with the Clausius statement of the second law. By interchanging roles of the engine and the refrigerator, this argument could be used to show that the assumption $\epsilon' > \epsilon$ is also untenable. The only condition consistent with the Clausius statement is $\epsilon = \epsilon'$.

We have shown that all reversible engines, for which the heat transfers take place only at T_H and T_C, have the same efficiency, which is equal to the Carnot efficiency. However, for an arbitrary reversible cycle with more than two reservoirs, all the heat transfers do not occur at the maximum and minimum temperatures. As a result, the efficiency of a reversible engine is either less than or equal to that of a Carnot engine operating between the same maximum and minimum temperatures.

For the second part of the theorem suppose that an irreversible engine supplies the work for a reversible engine that operates as a refrigerator. The efficiency of the irreversible engine is ϵ_{irrev}, whereas that of the reversible engine is ϵ_{rev}. The argument in the previous paragraph leads to the conclusion that $\epsilon_{\text{irrev}} > \epsilon_{\text{rev}}$ is not possible. But since the roles of the engine and refrigerator cannot be interchanged, we are unable to show that $\epsilon_{\text{rev}} > \epsilon_{\text{irrev}}$ is also untenable. We are left with the condition

$$\epsilon_{\text{irrev}} < \epsilon_{\text{rev}}$$

The efficiency of an irreversible engine is less than that of a reversible engine operating between the same two reservoirs. Any real engine is irreversible because of the presence of friction and heat conduction due to finite temperature differences.

EXAMPLE 21.1: A heat pump operating between reservoirs at $-5\ °C$ and $20\ °C$ requires an electrical input of 1.2 kJ in each cycle. (a) What is the maximum possible coefficient of performance? Find (b) the heat deposited into the hot reservoir in each cycle, and (c) the heat extracted from the cold reservoir.

Solution: (a) The coefficient of performance of a heat pump is

$$\text{COP} = \frac{|Q_H|}{W} = \frac{|Q_H|}{|Q_H| - |Q_C|}$$

If we use Eq. 21.5 for a Carnot engine, we find

$$\text{COP}_C = \frac{T_H}{T_H - T_C}$$

$$= \frac{293\ \text{K}}{293\ \text{K} - 268\ \text{K}} = 11.7$$

(b) $|Q_H| = \text{COP} \times W = (11.7)(1200\ \text{J}) = 14$ kJ.
(c) $|Q_C| = |Q_H| - W = 14\ \text{kJ} - 1.2\ \text{kJ} = 12.8$ kJ.

We see that for an input of 1.2 kJ, the ideal heat pump would discharge 14 kJ to the hot reservoir—which is the interior of a house.

EXERCISE 3. An ideal refrigerator is a Carnot engine operating in reverse with the same numerical values of Q_C, Q_H, and W. What is the coefficient of performance if the temperatures of the two reservoirs are 260 K and 300 K?

21.6 THE GASOLINE ENGINE (OTTO CYCLE)

The gasoline engine is the most common example of a heat engine. Its four-stroke cycle was devised in 1862 by A. Beau de Rochas, whereas N. Otto developed and built the first prototype in 1876. The **Otto cycle** actually consists of six steps, but only four "strokes" involve motion of the piston. The operation of a real engine involves irreversible processes, friction, heat loss, and a change in the working fluid because of the combustion of the fuel. We will discuss an idealized cycle in which the working fluid is considered to be an ideal gas and all processes are

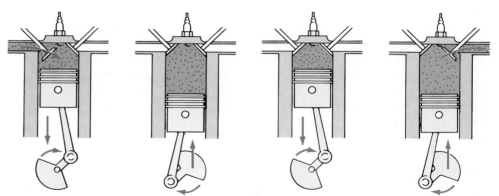

FIGURE 21.14 The four-stroke cycle in a gasoline engine: (*a*) intake stroke; (*b*) compression stroke; (*c*) ignition followed by the power stroke; (*d*) exhaust stroke.

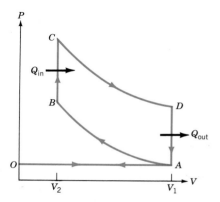

FIGURE 21.15 The idealized Otto cycle.

reversible. The combustion chamber consists of a piston in a cylinder with two valves at the top and a spark plug (see Fig. 21.14).

The six steps in the idealized Otto cycle may be plotted on a *PV* diagram (see Fig. 21.15).

1. *Intake stroke (O to A):* The cycle begins with the piston at the top of the cyclinder and the intake valve open to the atmosphere, at pressure P_0. As the piston moves down, it draws a mixture of gasoline and air into the cylinder until it reaches a volume V_1.

2. *Compression stroke (A to B):* The intake valve is closed and the piston moves upward until the volume reaches V_2. The fuel mixture is compressed rapidly, so the process may be considered to be adiabatic. Both the temperature and the pressure of the fuel rise significantly.

3. *Ignition (B to C):* Just before the piston reaches the top of its stroke, the spark plug ignites the already heated mixture. The explosion occurs so quickly that the piston does not move appreciably, so the volume stays constant at V_2. The temperature and pressure increase to very high values as heat Q_{in} enters the system.

4. *Power stroke (C to D):* As the piston is forced downward, it drives the crankshaft to which it is connected. The volume expands to V_1, while the temperature and pressure drop. This process is essentially adiabatic.

5. *Exhaust (D to A):* The piston does not move, but the exhaust valve opens, thereby allowing the gas to escape until the mixture in the cylinder reaches atmospheric pressure. The temperature drops as heat Q_{out} leaves the system.

6. *Exhaust stroke (A to O):* The piston moves upward, thereby forcing the remainder of the burned gases out of the piston. The volume drops to a small value (≈ 0). The exhaust valve closes and a new cycle begins.

The efficiency of the idealized Otto cycle is (see Example 21.2 below)

$$\epsilon = 1 - \frac{T_D}{T_C} = 1 - \frac{1}{r^{\gamma-1}}$$

where $r = V_1/V_2$ is called the *compression ratio*. The efficiency increases with the compression ratio. However, if the compression ratio is too high, the temperature and pressure of the gas after the compression stroke become high enough for the gas to ignite spontaneously. This premature ignition, which is heard as a knock or ping, is damaging to the engine. If one takes a typical value of $r = 8$ and $\gamma = 1.4$ (appropriate for air), we find for the idealized cycle that $\epsilon = 0.56$ or 56%. In practice, a value around 20% is attained.

The efficiency of a Carnot cycle operating between the same maximum temperature T_C and minimum temperature T_A is

$$\epsilon_C = 1 - \frac{T_A}{T_C}$$

This is greater than the efficiency of the Otto cycle since $T_A < T_D$. The reason is that in the Carnot cycle, all the transfer of heat occurs at only two temperatures, which is not the case in the Otto cycle. In the Otto cycle, Q_{in} and Q_{out} are not the simple heat transfers of the Carnot cycle. The input heat is released during the combustion, and the output heat is associated with the exchange of cool intake gases for the exhaust gases.

EXAMPLE 21.2: Calculate the efficiency of the (idealized) Otto cycle.

Solution: The heat input occurs during the ignition step from B to C. Since the volume does not change, there is no work done. Therefore, for n moles,

$$|Q_{in}| = nC_v(T_C - T_B)$$

where C_v is the molar specific heat. The heat transferred out during step D to A is

$$|Q_{out}| = nC_v(T_D - T_A)$$

The net work done in the cycle is $W = |Q_{in}| - |Q_{out}|$, so the efficiency, $\epsilon = W/|Q_{in}|$, is

$$\epsilon = 1 - \frac{|Q_{out}|}{|Q_{in}|} = 1 - \frac{T_D - T_A}{T_C - T_B}$$

For the (quasistatic) adiabatic processes A to B and C to D, we use $TV^{\gamma-1} = \text{constant}$, or $T = \text{constant}/V^{\gamma-1}$. (The constant is different for the two adiabatic processes.) Since $V_A = V_D = V_1$ and $V_B = V_C = V_2$, we find

$$\frac{T_D - T_A}{T_C - T_B} = \left(\frac{V_2}{V_1}\right)^{\gamma-1}$$

Therefore, with $r = V_1/V_2$, the efficiency is

$$\epsilon = 1 - \frac{1}{r^{\gamma-1}}$$

In a real engine, the efficiency is less than this value because of friction, heat loss, and the irreversible nature of the processes such as the combustion and the exhaust.

Another common internal combustion cycle was devised by R. Diesel. In this cycle the compression ratio is about 15. The air is first compressed by itself so that its temperature rises. The fuel, which is injected only after the compression, ignites spontaneously. There is no problem of preignition. The efficiency of diesel engines is higher ($\approx 25\%$) than gasoline engines, and they burn a less refined fuel. However, they are difficult to start in cold weather and the "power to weight" ratio is not as good as that for the gasoline engine.

21.7 ENTROPY

The zeroth law of thermodynamics leads to the identification of temperature as a state variable. The first law leads to the concept of internal energy. The second law of thermodynamics leads to another state function called **entropy.** The introduction of entropy broadens the scope of the second law of thermodynamics from heat engines to the evolution of natural processes, such as chemical reactions. We will introduce the concept of entropy by means of the Carnot cycle and later give it a physical interpretation.

From Eq. 21.5, we know that for a Carnot cycle

$$\frac{|Q_H|}{T_H} - \frac{|Q_C|}{T_C} = 0$$

It is now convenient to stop using the absolute values and explicit signs introduced in Section 21.1. However, we maintain the convention that heat entering a system is positive, whereas heat leaving a system is negative. The above condition then takes the form

$$\frac{Q_H}{T_H} + \frac{Q_C}{T_C} = 0$$

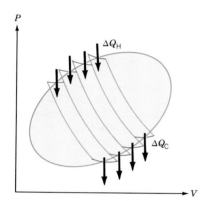

P

Consider an arbitrary reversible cycle, depicted as a closed curve in Fig. 21.16. We may approximate it by a series of Carnot cycles. The isotherms are the short lines, and the adiabatic curves of adjacent cycles overlap. For each cycle, we have $\Delta Q_H/T_H + \Delta Q_C/T_C = 0$. For a finite number of cycles,

$$\sum \frac{\Delta Q}{T} = 0$$

In the limit of infinitely many Carnot cycles, we have

(Reversible) $$\oint \frac{dQ_R}{T} = 0 \qquad (21.7)$$

FIGURE 21.16 A reversible cycle may be divided into a large number of Carnot cycles.

The circle on the integral sign indicates that it is taken around a closed path. The subscript R emphasizes that the cycle must be reversible. The quantity dQ_R represents an infinitesimal transfer of heat into or out of the system at temperature T, which varies along the path.

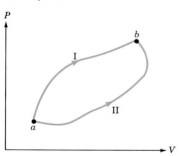

Equation 21.7 reminds us of Eq. 8.3, the condition for a force to be conservative. Consider two equilibrium states a and b, as shown in Fig. 21.17. The paths I and II indicate just two of many possible paths the system can follow from a to b. Equation 21.7 may be written as the sum of two terms:

$$\int_a^b \frac{dQ_R}{T} + \int_b^a \frac{dQ_R}{T} = 0$$

But, $\int_a^b = -\int_b^a$, and so

$$\int_a^b \frac{dQ_R}{T} = \int_a^b \frac{dQ_R}{T}$$
$$\quad\text{(I)} \qquad\qquad \text{(II)}$$

FIGURE 21.17 The quantity $\int dQ_R/T$ is the same for any path between the initial and final states a and b. This property allows us to define the entropy function.

The integral of dQ_R/T is independent of the path between the equilibrium states. It was just such a condition satisfied by a conservative force (see Eq. 8.2) that allowed us to define potential energy. In the present context we define an infinitesimal change in **entropy** as

$$dS = \frac{dQ_R}{T} \qquad (21.8)$$

For a finite change,

Change in entropy in a reversible process

$$\Delta S = S_f - S_i = \int_i^f \frac{dQ_R}{T} \qquad (21.9)$$

Equation 21.9 gives only the *change* in entropy; the initial value may be chosen arbitrarily. *The change in entropy depends only on the initial and final equilibrium states, not on the thermodynamic path.* Therefore, entropy is a state function, like internal energy. Although Eq. 21.9 is restricted to reversible processes, one can find the change in entropy for an irreversible process by imagining a suitable reversible path between the same initial and final equilibrium states, so that Eq. 21.9 may be used.

Before we connect the concept of entropy with the second law of thermodynamics, let us calculate the change in entropy in a few processes. We begin with a reversible process and an adiabatic free expansion of an ideal gas.

(a) Reversible Process (Ideal Gas)

The existence of the entropy function may also be inferred by considering a reversible process that takes an ideal gas from an initial to a final equilibrium state. According to the first law,

$$dQ = dU + dW$$

For an ideal gas, the change in internal energy is given by $dU = nC_v \, dT$ in any process (see Section 19.7). Since $PV = nRT$, the work done by the gas is $dW = P \, dV = nRT \, dV/V$. Thus,

$$dQ = nC_v \, dT + \frac{nRT \, dV}{V}$$

The contribution to the heat transfer from the second term on the right depends on the path between the initial and final states. That is, we need to know T as a function of V to calculate the heat transfer. However, when both sides are divided by T we find

$$\frac{dQ}{T} = \frac{nC_v \, dT}{T} + \frac{nR \, dV}{V} \tag{21.10}$$

The right side may now be integrated since each term contains only one variable. Thus, although dQ could not be integrated without knowledge of the path, the quantity dQ/T can be integrated. ($1/T$ is called an integrating factor.) Integrating Eq. 21.10 from an initial to a final equilibrium state we find

$$\int_i^f \frac{dQ}{T} = nC_v \ln\left(\frac{T_f}{T_i}\right) + nR \ln\left(\frac{V_f}{V_i}\right)$$

The integral depends only on the initial and final equilibrium states and not on the path. This property allows us to define the state function S such that, for an ideal gas,

(Ideal gas) $$\Delta S = nC_v \ln\left(\frac{T_f}{T_i}\right) + nR \ln\left(\frac{V_f}{V_i}\right) \tag{21.11}$$ Change in entropy of an ideal gas

The fact that S is a state function means that this expression for ΔS may be used even for an irreversible process involving an ideal gas between the same initial and final equilibrium states.

(b) Adiabatic Free Expansion (Ideal Gas)

Let us calculate the change in entropy of n moles of an ideal gas that undergoes an adiabatic free expansion from volume V_i to V_f. The gas is initially confined to part of a thermally insulated container with a thin membrane. It has a well-defined pressure, volume, and temperature. When the membrane is punctured, the gas rapidly expands to fill the whole container. In this uncontrolled expansion, the pressure, volume, and temperature do not have single, well-defined values characteristic of the whole system. Thus the process is irreversible.

Since there is no heat exchanged with the environment, $\Delta Q = 0$. We saw in Section 19.6 that in an adiabatic free expansion, the internal energy does not change; that is, $\Delta U = 0$. In the special case of an ideal gas, the temperature is also constant; that is, $\Delta T = 0$. However, it would not be correct to assume that $\Delta S = 0$.

To find the change in entropy of the gas, we must find a reversible path between the same initial and final equilibrium states. Since the temperature of an ideal gas does not change in an adiabatic free expansion, a suitable replacement would be a quasistatic isothermal expansion. From Eq. 21.11, with $T_f = T_i$ we see that

(Free expansion, ideal gas) $\Delta S_g = nR \ln \dfrac{V_f}{V_i}$

Since $V_f > V_i$, the entropy of the gas increases. Note that there is no change in the condition of the environment, so $\Delta S_e = 0$. The term *universe* in thermodynamics is applied to a system plus its surroundings. In the present case, the entropy change of the universe is

$$\Delta S_u = \Delta S_g + \Delta S_e > 0$$

EXERCISE 4. Two moles of an ideal gas undergo an adiabatic free expansion from an initial volume of 0.6 L to 1.3 L. What is the change in entropy of the gas?

EXERCISE 5. What is the change in entropy of a system that undergoes a reversible adiabatic process?

EXAMPLE 21.3: Figure 21.18 shows an insulated metal rod whose ends are in thermal contact with two reservoirs. When a steady state has been reached, during a certain time interval a quantity of heat $Q = 200$ J is transferred from the hot reservoir at $T_H = 60$ °C to the cold reservoir at $T_C = 15$ °C. What are the changes in entropy (a) of each reservoir; (b) of the universe?

FIGURE 21.18 In the steady state, the rod extracts a certain quantity of heat from the hot reservoir and deposits an equal quantity into the cold reservoir.

Solution: (a) The changes in entropy of each of these reservoirs are

$$\Delta S_H = -\frac{Q}{T_H} = \frac{-200 \text{ J}}{333 \text{ K}} = -0.6 \text{ J/K}$$

$$\Delta S_C = +\frac{Q}{T_C} = \frac{200 \text{ J}}{285 \text{ K}} = 0.7 \text{ J/K}$$

The net change in entropy of the two reservoirs is

$$\Delta S = \frac{Q}{T_C} - \frac{Q}{T_H} = +0.1 \text{ J/K}$$

(b) After a steady state has been reached, there is no net heat flow into the rod. The state of the rod does not change. In the irreversible process of heat transfer associated with a finite temperature difference, the entropy of the universe increases.

EXAMPLE 21.4: An ice cube of mass 400 g at temperature 0 °C (273 K) melts to water at 0 °C. Heat is supplied from the surrounding air which acts as a reservoir at slightly above 0 °C. The process takes place very slowly, so it is reversible. What is the change in entropy (a) of the ice when it has all melted, (b) of the reservoir, and (c) of the universe?

Solution: (a) The change in entropy as the ice is converted to liquid is $\Delta S = Q/T$, where $Q = mL$ and $L = 334$ J/kg is the latent heat of fusion and $T = 273$ K is the fixed temperature at which the heat is supplied. Thus,

$$\Delta S_i = \frac{mL}{T} = \frac{(0.4 \text{ kg})(334 \text{ J/kg})}{273 \text{ K}} = 488 \text{ J/K}$$

(b) The reservoir is at nearly the same temperature so its entropy change is

$$\Delta S_r = -\frac{mL}{T} = -488 \text{ J/K}$$

(c) The entropy change of the universe in this reversible process is $\Delta S_u = \Delta S_i + \Delta S_r = 0$. Note that if the air were at a higher temperature, the melting process would not be reversible. Furthermore, ΔS_u would then be positive.

EXAMPLE 21.5: A copper ball of mass $m = 0.5$ kg and specific heat $c = 390$ J/kg·K is at a temperature $T_1 = 90$ °C. The ball is thrown into a large lake at $T_2 = 10$ °C, which stays constant. Find the change in entropy of (a) the ball, (b) the lake, and (c) the universe.

Solution: Since the transfer of heat takes place with a finite temperature difference, this process is irreversible. However, we can imagine that the ball is successively placed in contact with a series of reservoirs whose temperatures differ slightly. We can compute the change in S for such a reversible process. (a) For an infinitesimal change in temperature, the heat transferred from the ball is $dQ = mc\,dT$. The related change in entropy is $dS = dQ/T = mc\,dT/T$. The temperature of the ball changes from T_1 to T_2, so

$$\Delta S_B = mc \int_{T_1}^{T_2} \frac{dT}{T} = mc \ln\left(\frac{T_2}{T_1}\right)$$
$$= (0.5 \text{ kg})(390 \text{ J/kg·K}) \ln\frac{283}{363} = -48.5 \text{ J/K}$$

Note that since $T_2 < T_1$, the change in entropy of the ball is negative: $\Delta S_B < 0$.

(b) The total heat inflow into the lake is equal to the loss from the ball:

$$\Delta Q = -\int_{T_1}^{T_2} mc\,dT = mc(T_1 - T_2)$$
$$= (0.5 \text{ kg})(390 \text{ J/kg·K})(363 \text{ K} - 283 \text{ K}) = 1.56 \times 10^4 \text{ J}$$

Since the heat transfer to the lake occurs at a single temperature, the change in entropy of the lake is

$$\Delta S_L = \frac{\Delta Q}{T_2} = \frac{1.56 \times 10^4 \text{ J}}{283 \text{ K}} = 55.1 \text{ J/K}$$

(c) The entropy change of the universe, $\Delta S_u = \Delta S_B + \Delta S_L = 6.6$ J/K, is greater than zero.

EXERCISE 6. What is the change in entropy of 300 g of water as its temperature increases from 10 °C to 25 °C? The specific heat of water is 4.19 kJ/kg·K.

21.8   ENTROPY AND THE SECOND LAW

In the examples in the last section, we found that the change in entropy is zero for reversible processes and greater than zero for irreversible processes. This leads to another formulation of the **second law of thermodynamics:**

$$\Delta S \geq 0 \tag{21.12}$$

Second law of thermodynamics in terms of entropy

In a reversible process the entropy of an isolated system stays constant; in an irreversible process the entropy increases.

Since all natural processes are irreversible, the entropy of a system plus its environment always increases. There may be a local decrease in entropy in one part of the universe, provided it is accompanied by a greater increase elsewhere.

One can show that Eq. 21.12 is consistent with, for example, the Clausius statement. Imagine a perfect refrigerator that transfers heat Q from a cold to a hot reservoir. Since it operates in a cycle, $\Delta S = 0$ for the refrigerator itself. The change in entropy of the reservoirs is $\Delta S = Q/T_H - Q/T_C < 0$. But a perfect refrigerator violates the Clausius statement. Therefore, we must have $\Delta S > 0$, in conformity with Eq. 21.12.

When Clausius introduced the entropy function, it served as a convenient calculational aid. It helped distinguish between reversible and irreversible processes. The physical meaning of entropy was elucidated only later, by L. Boltzmann. In the following sections we will discuss the implications of this formulation of the second law. We will see that the principle of increase of entropy is connected with the transition of a system from ordered to disordered states. This in turn will lead to a relation between entropy and probability.

21.9 THE AVAILABILITY OF ENERGY

According to the first law, all forms of energy have equal status. Electrical, gravitational, chemical, nuclear, and other forms of energy may, in principle, be fully converted from one form to another. The second law addresses a lack of such symmetry when dealing with heat. Although one can convert work completely into heat, it is a matter of experience that one cannot completely convert heat into work—at least not in a continuous cycle.

A car converts chemical energy into mechanical energy of motion and heat transferred to the road and air. When the car is brought to a stop, the mechanical energy is transformed into internal energy at the brakes and is then dissipated as heat into the atmosphere. Thus, ultimately, *all* the chemical energy of the fuel is converted into internal energy. Recall that internal energy does not include mechanical energy associated with the motion of the center of mass of a system. It does include thermal energy, which is the kinetic and potential energies associated with the random motion of the molecules. The energy of the center of mass involves "ordered" motion and can be used to do useful work. The thermal energy involves "disordered" motion, only part of which may be converted to useful work, provided a reservoir at a lower temperature is available. In any case, the efficiency of the conversion is always less than 100%.

Therefore, although complete conversions among electrical, chemical, mechanical, and other forms of energy are possible, the use of mechanical work to generate heat always means that some energy becomes unavailable to do work. This is because internal energy cannot be completely converted to other forms or fully used to do work. One often expresses this by saying that the original energy has become "degraded." The ordered kinetic energy of the center of mass motion of a car is "high-grade" energy because it can be fully used to do work. Once the object comes to a stop, friction has transformed it into random kinetic energy of the molecules, which is "low-grade" energy since it cannot be completely converted to work.

A similar situation occurs with the motion of molecules in air. In "stationary" air the molecules actually move about randomly at high speeds. This disordered motion is of no use in turning a windmill. When a wind blows, all the molecules acquire an additional small velocity component. It is this orderly component of the motion that turns the windmill and does work. To summarize:

1. The degradation of energy is associated with the transition of the system from an ordered to a disordered state.

2. In any natural (irreversible) process, some energy becomes unavailable to do useful work.

21.10 ENTROPY AND DISORDER

The second law says that an isolated system tends to evolve toward states of higher entropy. This principle of the increase of entropy may be connected with the transition of a system from ordered to disordered states. This leads us to identify entropy as a measure of disorder in a system. A highly ordered state has low entropy, whereas a disordered state has high entropy. Yet the growth of plants and animals clearly demonstrates evolution toward increasing order, and hence lower entropy. Although this seems to conflict with the second law, it does not. A local decrease in entropy of a system can occur at the expense of a greater increase in entropy in its environment.*

The second law, in the form of the principle of increase in entropy, is often called "time's arrow" because it tells us that in (irreversible) natural processes, an isolated system always evolves toward states of higher entropy. Any spontaneous change, such as the equalization of temperature, pressure, or concentration of particles, is always in the direction of increasing the entropy of the system.

One prediction of the second law of thermodynamics concerns the evolution of the universe. If we consider it to be an isolated system, it will evolve toward a state of thermodynamic equilibrium characterized by a uniform density, temperature, and pressure. Since there would be no temperature differences, no useful work could be performed. All physical and biological activity would cease. This gloomy state of affairs is usually called the "heat death." We need have no fear of this condition. The evolution of the sun into a "red giant" that will expand perhaps to the earth's orbit will occur a lot sooner.

* Evolution toward states of increasing order is a consequence of interatomic interactions that lead to lower *energy* states associated with more complex molecules. At a low enough temperature, the tendency to minimize the energy can be greater than the tendency to maximize the entropy. Thus, for example, the transition of liquid water to the more ordered crystalline state of ice is not forbidden by the second law. See P. W. Atkins, *The Second Law*. Freeman, San Francisco, 1986.

21.11 STATISTICAL MECHANICS (Optional)

The second law, $\Delta S \geq 0$, expresses the irreversibility of natural phenomena. Yet the laws of mechanics, which the molecules are assumed to obey, remain the same when time is reversed. They are said to be *time-reversal invariant*. For example, the initial and final velocities of balls in a collision could be interchanged, or the planets could orbit in the opposite sense, without violating any of the laws of mechanics. That is, the laws of mechanics do not rule out the order of events seen in a film run backward. Thus the second law of thermodynamics appears to be in conflict with the laws of mechanics. How do we reconcile the *reversibility* of the mechanics of individual events, with the *irreversible* behavior of large numbers of molecules?

The answer lies in the connection between entropy and probability. J. Clerk Maxwell suggested that the second law is a statistical truth, not absolutely valid, but very highly probable. Later, L. Boltzmann interpreted the tendency of systems to evolve from ordered states to disordered states to be the result of transitions from states of low probability to states of higher probability.

To see how order and disorder are connected with probability, consider a box that contains four coins. The state of the system may be characterized by the number of heads. If we are not concerned with specifying *which* coins are heads up, it is called a *macrostate*. There are five macrostates: Four heads, three heads, . . . , no heads. If the box is vigorously shaken, what is the probability of finding a given number of heads, say two? We assume that each coin has an equal probability of landing heads up or tails up. There are several possible ways

that the system could end up with two heads. If we take note of which coins are H or T, each configuration is called a *microstate*. Figure 21.19 shows the number of microstates that correspond to a given macrostate. A basic assumption in statistical mechanics is that *all microstates have equal probability*. The probability of finding a given macrostate is proportional to the number of microstates that correspond to it:

Probability of a given macrostate

$$= \frac{\text{Number of microstates for the given macrostate}}{\text{Total number of possible microstates}}$$

Macro State	Micro States		Probability
4H	HHHH	1	1/16
3H	HHHT HHTH HTHH THHH	4	1/4
2H	HHTT · · · · · TTHH	6	3/8
1H	HTTT · · · · TTTH	4	1/4
0H	TTTT	1	1/16

FIGURE 21.19 The probability that a given macrostate will occur depends on the number of microstates that are associated with it.

In Fig. 21.20 this probability is plotted versus the macrostate (the number of heads). After a limited number of trials, we would not expect to see three heads exactly 1/4 of the time. In 100 trials we may observe three heads 21 times or 27 times, for example. However, as the number of trials becomes large, say 1000, the predicted probabilities will be more closely approached.

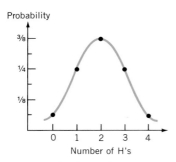

FIGURE 21.20 The probability that a given macrostate will occur based on the data in Fig. 21.19.

The macrostates with four heads or no heads are the most ordered, but the least probable. The most probable macrostate involves two heads and is the least ordered. Thus, given that each coin has no built-in bias for either heads or tails, the collection is most likely to be found in the macrostate with least order. As the number of coins is increased, the macrostates with some degree of order become relatively less probable.

Suppose we start with a thousand coins in any initial macrostate—even the extremely unlikely one with all heads up. After a sufficiently large number of trials, only the macrostate with equal numbers of heads and tails (or one with a tiny imbalance in the numbers) will be found (Fig. 21.21). Large departures from this state become less and less likely as the number of trials increases. Thus, the increase in entropy of a system is associated with the transition from order to disorder, or from states of low probability to those of high probability. *The most probable state is the one for which the entropy is a maximum.*

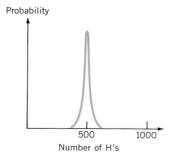

FIGURE 21.21 When the number of coins is large, states far from the state of greatest disorder have negligible probability.

We now return to the original question of the reversibility of the laws of mechanics and the irreversibility of natural processes. In the case of a gas, a microstate specifies the position and velocity of a single molecule. The macrostates are characterized by quantities such as pressure, volume, and temperature of the gas as a whole. Each microstate is equally probable, but each macrostate is not. Left to itself the gas will tend to increase its entropy. In the most probable state, with maximum entropy, the gas will fill the volume of its container. Loschmidt pointed to the following paradox. At any given moment the velocities of the molecules have certain values that, via collisions, will lead ultimately to states of higher entropy. But we can imagine that the velocities of all the molecules are reversed. Since all microscopic states are assumed to be equally probable, should not the entropy be just as likely to decrease as to increase? Boltzmann replied that the entropy can decrease momentarily, but the system will quickly return to a more probable state in which the motion is more random and the entropy is greater. In the case of a gas of 10^{21} particles, it is not impossible for all of them to be found in one half of a container. However, the probability of this occurrence is so small that we might have to wait 10^{100} years before encountering it!

21.12 ENTROPY AND PROBABILITY (Optional)

Although we cannot present Boltzmann's formulation of the relation between entropy and probability, we can show how the concept of probability might have been introduced. Equation 21.11 shows that the change in entropy when n moles of an ideal gas undergo an adiabatic free expansion from volume V_1 to V_2 is $S_2 - S_1 = nR \ln (V_2/V_1)$. Since $R = kN_A$ this may be written as

$$S_1 - S_2 = k \ln\left(\frac{V_1}{V_2}\right)^N \tag{21.13}$$

where $N = nN_A$ is the total number of molecules. We will now show how the argument of the logarithm is associated with probability.

Consider molecules in a container. The probability that one molecule will be found in the left half is simply 1/2. The probability that two molecules will be found in the same half volume is $(1/2)^2$. (This assumes that the molecules do not interact.) For N molecules the probability of finding all of them in the same half is $(1/2)^N$. In general, if the volume of the container is V_2, the probability of finding all N molecules in a particular volume V_1 is $(V_1/V_2)^N$.

Equation 21.13 gives us the clue to relating entropy and probability. The justification for the logarithm function is as follows. As we have just seen, the overall probability is the *product* of the probabilities of independent events. We also know that the entropy of two systems is the *sum* of their entropies. The logarithmic function converts the multiplicative property of probability to the additive property of entropy. If we choose $S_2 = 0$, we may define the entropy of a system in a given state as

$$S = k \ln W \tag{21.14}$$

where W is the probability of that state. The quantity W is proportional to the number of microstates that correspond to the given macrostate. The significance of this equation, first proposed by Boltzmann in 1877, is that it connects probabilities in the microscopic world of the molecules to the macroscopic variable S.

21.13 ABSOLUTE TEMPERATURE SCALE (Optional)

In 1851 Lord Kelvin realized an important implication of Eq. 21.5. This relationship is independent of the properties of the working substance in the Carnot cycle and thus could serve as a definition of an absoute temperature scale. The temperature T could be determined by taking the working substance through a Carnot cycle and carefully measuring the heat input and heat output. In order for the absolute thermodynamic scale to coincide with the absolute ideal gas scale, the temperature of the triple point of water is chosen to be $T_{tr} = 273.16$ K. Then from Eq. 21.5 any arbitrary temperature T is given by

$$T = 273.16 \frac{Q}{Q_{tr}}$$

where Q_{tr} is the heat transfer at the triple point and Q is the heat expelled to the reservoir at temperature T. This method is actually used below about 1 K where the ideal gas thermometer using helium does not function. Note that at the absolute zero of temperature, the heat discharged to the reservoir would be zero.*

---

* According to the third law of thermodynamics it is not possible to attain the absolute zero of temperature.

SUMMARY

A **heat engine** is a device that absorbs heat $|Q_H|$ from a hot reservoir and discharges a quantity of heat $|Q_C|$ to a cold reservoir and does an amount of work $W = |Q_H| - |Q_C|$.

The **thermal efficiency** ϵ of a heat engine is defined to be the *work output* divided by the *heat input:*

$$\epsilon = \frac{W}{|Q_H|} = 1 - \frac{|Q_C|}{|Q_H|}$$

The **second law of thermodynamics** can be expressed in several ways:

Kelvin-Planck: It is impossible for a heat engine that operates in a cycle to convert its heat input completely into work.

Clausius: It is impossible for a cyclical device to transfer heat continuously from a cold body to a hot body without the input of work or some other effect on the environment.

A **reversible** process proceeds quasistatically from the initial equilibrium state to the final equilibrium state There is no friction and no transfer of heat associated with a finite temperature difference.

A **Carnot engine** operates in a cycle that involves two isothermal and two adiabatic processes. The thermal efficiency of a Carnot engine operating between reservoirs at absolute temperatures T_H and T_C is

$$\epsilon_C = 1 - \frac{T_C}{T_H}$$

This is the maximum efficiency of any engine operating between the same two reservoirs. The efficiency of an irreversible engine operating between the same maximum and minimum temperatures is always lower.

Entropy is a state function of a system; it depends only on the equilibrium state of the system. The change in entropy between initial and final equilibrium states is

$$\Delta S = \int_i^f \frac{dQ_R}{T}$$

where dQ_R is an infinitesimal heat transfer that takes place reversibly. The change in entropy for any process, including an irreversible one, between given initial and final equilibrium states, is the same.

The **second law of thermodynamics** may be expressed in terms of entropy:

$$\Delta S \geqslant 0$$

The entropy change in an isolated system is either zero (for a reversible process) or greater than zero (for a real, irreversible process).

Entropy is a measure of the disorder in a system. The second law states that natural (irreversible) processes tend to evolve to states of greater disorder, or from states of low probability to states of high probability.

ANSWERS TO IN-CHAPTER EXERCISES

1. (a) At 2 cycles per second the engine delivers 80 W; thus, the work done per cycle is 40 J.
(b) $|Q_H| = W/\epsilon = (40 \text{ J})/(0.2) = 200$ J.
(c) $|Q_C| = |Q_H| - W = 160$ J.

2. Express the COP as

$$\text{COP} = \frac{|Q_C|}{|Q_H| - |Q_C|}$$

(a) $|Q_C| = 200$ J.
(b) $W = 250 - 200 = 50$ J.

3. We have

$$\text{COP} = \frac{|Q_C|}{|Q_H| - |Q_C|}$$

Using Eq. 21.5 for a Carnot cycle, we have $\text{COP}_C = T_C/(T_H - T_C) = 260/40 = 6.5$.

4. From Eq. 21.11:

$$\Delta S_g = nR \ln(V_f/V_i)$$

$$= (2 \text{ mol})(8.31 \text{ J/mol·K}) \ln (1.3/0.6) = 12.9 \text{ J/K}.$$

5. In any reversible adiabatic process $dQ = 0$ and T is well defined at each stage. Thus, $\Delta S = \int dQ/T = 0$.

6. The change in entropy is

$$\Delta S = \int \frac{dQ}{T} = mc \int_{T_1}^{T_2} \frac{dT}{T}$$

$$= mc \ln\left(\frac{T_2}{T_1}\right) = (0.3 \text{ kg})(4.19 \text{ kJ/kg·K}) \ln\left(\frac{298}{283}\right) = 65 \text{ J/K}$$

QUESTIONS

1. Is it possible to cool a room by leaving the refrigerator door open?

2. Are any natural processes reversible? If so, give an example.

3. Does the entropy of an ideal gas change when it is compressed adiabatically? Does your answer depend on whether or not the process is reversible?

4. Does the entropy of an ideal gas change when it is compressed isothermally? Does your answer depend on whether or not the process is reversible?

5. "Every blade of grass stands triumphant over the second law of thermodynamics." What is meant by this statement? Is there any validity to it?

6. There is a vast quantity of thermal energy in the oceans. Why has this not been used to power ships?

7. A system is taken from one equilibrium state to another by an irreversible process. Does the change in entropy depend on the details of the process?

8. Suppose you watch a film run backward. Name a few occurrences that would tell you time is flowing in the wrong direction.

9. For a given input of electrical energy, is there any advantage to heating a house with a heat pump rather than by direct electrical resistance heating?

10. The sun's radiation causes an iceberg to melt and thereby increases its entropy. The radiation also helps a plant to grow, thereby decreasing its entropy. Is there a contradiction here?

11. Is there any process for which the entropy change of the universe is zero? If so, give an example.

12. Is entropy a property of a substance or a particular sample of a substance?

13. When you blow on a bowl of hot soup, it cools and its entropy decreases. Discuss this phenomenon from the viewpoint of the second law.

14. Are all quasistatic processes reversible? Are all reversible processes quasistatic?

15. Hydrogen and oxygen combine to form water; water can be decomposed into hydrogen and oxygen. Are these processes reversible in the thermodynamic sense?

EXERCISES

21.1 and 21.2 Heat Engines; Refrigerators

1. (I) A heat engine with an efficiency of 25% does 200 J of work per cycle. Find: (a) the heat absorbed from the hot reservoir; (b) the heat expelled to the cold reservoir.

2. (I) In one cycle an engine absorbs 800 J from the hot reservoir and expels 550 J to the cold reservoir. If a cycle takes 0.4 s, what is the mechanical power output?

3. (I) A refrigerator absorbs 100 J per cycle from the cold reservoir and expels 125 J to the hot reservoir. It operates at 20 cycles/s. Find: (a) the electrical power required; (b) the coefficient of performance.

4. (I) A refrigerator with a coefficient of performance of 3.5 absorbs 80 J from the freezer compartment. (a) How much work does this require? (b) How much heat is expelled to the surroundings?

5. (I) A residential heat pump with a coefficient of performance of 4 requires 10 kWh of electrical energy during a certain period. How many kWh would be needed if the heating were accomplished with electrical resistance heaters?

6. (II) A power plant with an efficiency of 30% has a useful (electrical) output of 100 MW. Water from a river is pumped in to carry away 80% of the waste heat. What is the minimum necessary flow rate (kg/s) if the water temperature should not rise by more than 8 °C? The specific heat of water is 4190 J/kg.

7. (II) A gasoline engine has a 30 kW (≈40 hp) output and a thermal efficiency of 22%. Find: (a) the rate of heat input; (b) the rate of heat output. (c) If the heat of combustion of gasoline is 1.3×10^8 J/gal, what is the number of gallons used per hour?

8. (II) A heat pump extracts 6000 Btu/h from a house at 70 °F while the outside air is at 90 °F. Its power requirement is 1 kW. (a) What is its coefficient of performance? (b) At what rate does it discharge heat to the environment? (1 Btu/h = 0.293 W)

21.5 Carnot Engine

9. (I) A heat engine uses the 90 °C water from a geyser and the atmosphere at 10 °C as reservoirs. What is the maximum possible efficiency?

10. (I) A Carnot engine operates between reservoirs at 400 K and 300 K. It discharges 330 J to the lower temperature reservoir. What is the work done?

11. (I) It has been proposed that the temperature difference between water at the surface of a tropical ocean and cooler water at a depth of several hundred meters could be used in a heat engine. (a) If the two temperatures are 22 °C and 5 °C, find the maximum possible efficiency of such an engine. (b) If the useful output is 1 MW, what is the rate at which heat is discharged to the deep water?

12. (I) A Carnot engine operates between a hot reservoir at 600 K and a cold reservoir at 350 K. It produces 500 W of mechanical power. What is the rate of heat discharge into the cold reservoir?

13. (II) A house requires an average of 5 kW of heat input to maintain an inside temperature of 20 °C when the outside air is at 0 °C. (a) If the heating is accomplished by electrical resistance heaters, at 10 cents per kWh, how much does it cost to heat per day? (b) If an ideal heat pump were installed and operated as a Carnot engine in reverse, what would be the daily electricity bill?

14. (I) An ideal refrigerator, which is a Carnot engine operating in reverse, operates between a freezer at −5 °C and a room at 25 °C. In a certain period, it absorbs 100 J from the freezer. How much heat is rejected to the room?

15. (I) (a) How much heat is discharged by an ideal heat pump (Carnot engine in reverse) that extracts 1000 J of heat from outside air at 0 °C and discharges the heat output into a room at 20 °C? (b) How much work is required?

16. (I) A Carnot engine has a 35% efficiency. It expels 200 J into the lower temperature reservoir at 300 K. (a) What is the heat extracted from the hot reservoir? (b) What is the temperature of the hot reservoir?

17. (I) A Carnot engine operates between 200 °C and 20 °C and has a mechanical output of 360 W. If each cycle takes 0.2 s, find the heat (a) absorbed, and (b) expelled in each cycle.

18. (I) A heat pump operates between 0 °C and 25 °C. It has 60% of the maximum possible coefficient of performance. If it absorbs 100 J from the cold reservoir, find: (a) the work required; (b) the heat discharged.

19. (I) A Carnot engine operates between 280 °C and 40 °C. In one cycle it absorbs 1000 J from the hot reservoir. (a) What is the work done per cycle? (b) What would be its coefficient of performance if it were run as a heat pump?

20. (II) Which of the following has the greater effect on the efficiency of a Carnot engine: (a) A 5 K rise in the temperature of the hot reservoir? or (b) A 5 K decrease in the temperature of the cold reservoir?

21.7 Entropy (See Tables 19.1 and 19.2 for data)

21. (II) What is the change in entropy in each of the following processes: (a) 1 kg of water at 100 °C changes to steam at 100 °C and at constant pressure; (b) 1 kg of ice at 0 °C is converted to water at 0 °C; (c) 1 kg of water at 0 °C is heated to 100 °C.

22. (II) Find the change in entropy when 1 kg of ice at 0 °C is added to 1 kg of water at 100 °C in an insulated container.

23. (II) A 50-g ice cube at 0 °C melts (reversibly) to water at 0 °C. What is the change in entropy (a) of the ice; and (b) of the universe?

24. (I) A 100-g steel ball at 200 °C is thrown into a lake at 20 °C. What is the entropy change of (a) the ball? (b) the lake? (c) the universe?

25. (I) A 100-g lead ball at 100 °C is placed in 300 g of water at 20 °C in an insulated container. (a) What is the final equilibrium temperature? (b) What is the change in entropy (b) of the ball, (c) of the water, (d) of the universe?

26. (II) Heat is supplied to 1 kg of ice initially at −5 °C slowly changing it to water at +5 °C. What is the change in entropy of the sample?

27. (I) After a steady state has been reached, a metal rod conducts 1200 J from a 400 K reservoir to one at 250 K during a certain time interval. Find the change in entropy of: (a) the hot reservoir; (b) the cold reservoir; (c) the rod; (d) the universe.

28. (I) A 10-kg cannonball is projected at 50 m/s from a clifftop 60 m above sea level. What is the change in entropy of the universe after the ball lands in the sea? Assume the ball, the air, and the sea remain at 20 °C.

29. (II) Find the change in entropy for n moles of an ideal monatomic gas in the following processes:
    (a) the temperature changes from T_1 to T_2 at constant pressure

(b) the pressure changes from P_1 to P_2 at constant volume

30. (II) Find the change in entropy for n moles of an ideal monatomic gas in the following processes:
    (a) The temperature changes from T_1 to T_2 at constant volume
    (b) The volume changes from V_1 to V_2 at constant temperature

31. (I) A thin membrane divides an insulated container into two equal volumes. One side contains 2 mol of an ideal gas. (a) What is the change in entropy of the gas when the membrane is broken? (b) What is the change in entropy of the universe?

32. (I) A steam engine with 50% of the ideal Carnot efficiency takes in superheated steam at 250 °C and discharges steam at 105 °C. Its mechanical power output is 200 kW. For a period of 1 h find (a) how much heat it discharges; (b) the change in entropy of the hot reservoir; (c) the change in entropy of the cold reservoir.

33. (II) A Carnot engine operates between reservoirs at 300 K and 550 K. During each cycle, it absorbs 1000 J from the hot reservoir. (a) Calculate the changes in entropy during each of the four stages. (b) What is the net change in entropy for a complete cycle?

PROBLEMS

1. (I) One mole of an ideal monatomic gas is taken around the reversible cycle shown in Fig. 21.22. Find: (a) the heat input or output in each leg; (b) the work done in one cycle; (c) the efficiency.

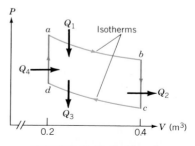

FIGURE 21.23 Problem 2

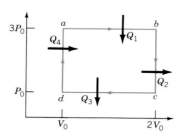

FIGURE 21.22 Problem 1

2. (I) One mole of an ideal monatomic gas is taken around the reversible cycle of Fig. 21.23. The isothermals are at 500 K and 300 K. Find the efficiency of the engine.

3. (II) Two moles of an ideal diatomic gas ($\gamma = \frac{7}{5}$) operate in the cycle of Fig. 21.24, where $T_a = 400$ K, $T_c = 250$ K, and $P_c = 100$ kPa. Find: (a) the work done per cycle; (b) the efficiency of the engine.

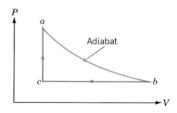

FIGURE 21.24 Problem 3

4. (I) Two moles of an ideal diatomic gas ($\gamma = \frac{7}{5}$) are taken around the cycle of Fig. 21.25. Find: (a) the heat absorbed or rejected in each segment; (b) the work done per cycle; (c) the efficiency.

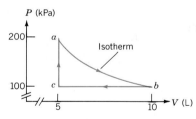

FIGURE 21.25 Problem 4

5. (II) The idealized *air–standard Diesel cycle* is shown in Fig. 21.26. Air enters the combustion chamber in an intake stroke (not shown). It is adiabatically compressed from a to b. At b fuel enters the system and starts to burn, essentially at constant pressure. This process continues until point c, when the gas expands adiabatically to point d. This is the power stroke. In the exhaust stroke, the combustion products cool at constant volume from d back to the initial point a. Show that the efficiency is

$$\epsilon = 1 - \frac{1}{\gamma}\left(\frac{T_d - T_a}{T_c - T_b}\right)$$

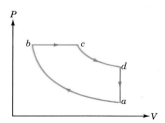

FIGURE 21.26 Problem 5

6. Show that an equivalent expression for the efficiency of the idealized air–standard Diesel cycle discussed in Problem 5 is

$$\epsilon = 1 - \frac{(V_c/V_a)^\gamma - (V_b/V_a)^\gamma}{\gamma[(V_c/V_a) - (V_b/V_a)]}$$

Evaluate the efficiency given that the compression ratio $V_a/V_b = 15$ and the expansion ratio $V_a/V_c = 5$. Take $\gamma = 1.4$.

7. (I) A Carnot engine that uses one mole of an ideal gas ($\gamma = \frac{5}{3}$) operates between 500 K and 300 K. The highest and lowest pressures are 500 kPa and 100 kPa. Find: (a) the net work done per cycle; (b) the efficiency.

8. (II) Draw the Carnot cycle on a T versus S diagram. Show that the net work done per cycle is $\oint T\, dS$, where the integral is taken around a complete cycle.

9. (I) Show that if the Kelvin-Planck statement were violated, that is, if a perfect engine were possible, the entropy of the universe could decrease.

10. (II) A heat engine operates between reservoirs at temperatures 800 K and 300 K. In one cycle it absorbs 1200 J of heat and does 300 J of work. Find: (a) the efficiency of the engine; (b) the change in entropy of each of the two reservoirs and of the universe ΔS_u. (c) How much work would be done by a Carnot engine that absorbed the same quantity of heat from the hot reservoir? (d) Show that the work done by the real engine is $T_C \Delta S_u$ less than that done by the Carnot engine, where T_C is the temperature of the cold reservoir.

11. (II) One mole of water is heated from 0 °C to 100 °C by bringing it into contact with a different number of reservoirs. Find the change in entropy of the universe given the following: (a) Only one reservoir at 100 °C is used. (b) The water is first brought to equilibrium with a reservoir at 50 °C and then put in contact with the reservoir at 100 °C. (c) The water is brought to equilibrium successively with reservoirs at the temperatures 25 °C, 50 °C, 75 °C, and 100 °C. (d) In practice how could the water be heated reversibly?

12. (II) Two Carnot engines operate in tandem: The heat output of one at an intermediate temperature T_i is used as the input of the other. (a) What is the overall efficiency? (b) Compare the result of (a) with the efficiency of a single engine, that is, when there is no intermediate stage.

13. (I) A freezer converts 0.5 kg of water at 10 °C to ice at −8 °C. If the coefficient of performance is 4, what is the electrical energy required? (See Table 19.1 for data on specific heats and the latent heat of fusion.)

14. (II) In a reversible adiabatic process the entropy of an ideal gas does not change (see in-chapter Exercise 5). Show explicitly that Eq. 21.11 is consistent with this statement.

CHAPTER 22

Electrostatics

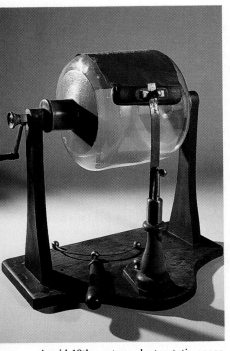

A mid-18th century electrostatic generator.

Major Points

1. The properties of **charge.** Its conservation and quantization.
2. The distinction between **conductors** and **insulators.**
3. **Coulomb's law:** The force between *static, point* charges.
4. The **principle of superposition:** The force between two particles is not affected by the presence of other particles.

On a dry day, a comb brushed through hair can pick up pieces of paper or deflect the stream of water from a tap. You may see flashes when a blanket is peeled off a sheet, or feel a shock on touching a doorknob after walking on a carpet. These are **electrical** effects. The orientation of a compass needle and the attraction between a magnet and a nail are **magnetic** effects.

A variety of electrical and magnetic phenomena were observed in ancient times. Thales of Miletus (ca. 600 B.C.) noted that when the mineral amber (fossilized tree resin) was rubbed by wool or fur, it could attract small pieces of straw or feathers. Aristotle commented on the ability of the electric torpedo fish to stun its prey and noted that people had also received shocks. Italian sailors in the 4th century A.D. were familiar with St. Elmo's fire, a glow seen at the tops of masts during thunderstorm activity. The poet Lucretius, writing in the 1st century B.C., described the mysterious power of lodestones, which were found in Magnesia, a region of Asia Minor. Lodestones differed from amber in that they attracted only iron and did not need to be rubbed. By the 11th century, Chinese and Arabic sailors were using floating lodestones as compasses.

In 1600, William Gilbert, physician to Queen Elizabeth I, was the first to distinguish clearly between electrical and magnetic phenomena. He coined the word electric, deriving it from *elektron*, which is Greek for amber. Gilbert showed that electrical effects were not confined to amber; many other substances could be electrified by rubbing. The first frictional electrical machine was made in 1663 by Otto von Guericke, famous for his demonstration of the pressure of the atmosphere (see Chapter 14). A ball of sulfur was electrified by placing a hand on it as it rotated on an axle (see Fig. 22.1). Later electrical machines, which could produce large and dangerous sparks, served as a source of amusement for many decades.

Virtually all the physical phenomena we experience, such as light, chemical reactions, the properties of materials, and the transmission of signals along nerve

FIGURE 22.1 The first electrical machine, made in 1663 by Otto von Guericke. The globe of sulfur was charged by the friction of a hand pressed against it as it rotated. Guericke showed that a charged feather could be suspended in the air by the repulsion of the globe.

fibers, are electrical in nature. In fact, the only nonelectrical force we experience every day is the force of gravity. The design and operation of radios, TVs, computers, motors, and X-ray machines are based on knowledge of the electrical interaction between charges. **Charge** is a property of matter that causes it to produce and experience electrical and magnetic effects. The subject of the electrical effects of charges at rest is called **electrostatics.** When both electrical and magnetic effects are present, the interaction between charges is referred to as **electromagnetic.** The everyday examples cited in this paragraph are actually electromagnetic in nature. For 200 years after the pioneering work of Gilbert, electricity and magnetism developed as separate subjects. We will maintain this distinction for the next few chapters.

22.1   CHARGE

When a glass rod is rubbed with a silk cloth, both become charged. One way of studying these charges is to use Styrofoam balls, which are light and able to hold charge well. Figure 22.2 shows two Styrofoam balls suspended close to each other. When one is touched by the rod and the other by the silk, the balls attract each other. When both balls are touched by the same object, either the rod or the silk, the balls repel each other.

From such evidence, Charles du Fay surmised in 1733 that there are two kinds of charge, which he called "electric fluid." Clearly, *like charges repel, and unlike charges attract.* The kind on glass was called "vitreous," while that on silk or amber was "resinous." Du Fay believed that these fluids were separated by the friction involved in rubbing. Around 1750, Benjamin Franklin proposed that during the process of rubbing, a single fluid flows from one body to the other. The body that gains fluid was designated as positively charged, while the one that loses fluid was negatively charged. (In this scheme, vitreous corresponds to positive.)

In our modern view, a neutral object possesses an equal number of positive and negative charges. In simplified terms, matter is made up of atoms (radius $\approx$ 10^{-10} m) each consisting of a tiny nucleus (radius $\approx$ 10^{-15} m) of positively charged **protons** and neutral **neutrons.** Around the nucleus, negatively charged **electrons** form clouds of various shapes. A neutral atom has an equal number of electrons and protons. An ion is an atom or molecule that has lost or gained one or more electrons.

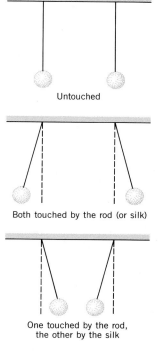

Untouched

Both touched by the rod (or silk)

One touched by the rod, the other by the silk

FIGURE 22.2 The phenomena of attraction and repulsion are easily demonstrated with two Styrofoam balls.

In the rubbing process, electrons or ions are transferred from one body to the other. This leaves a net positive charge on one body and a net negative charge on the other. The signs of the acquired charges depend on the electrical properties of the two materials and the condition of their surfaces. In fact, mere contact between two materials causes them to become charged; the rubbing merely enhances the effect. In some cases, a change from gentle rubbing to hard rubbing can change the signs of the charges acquired by the two bodies. This unpredictability in the signs is caused by minute quantities of dust, which are very difficult to remove.

The SI unit of charge is the coulomb (C). It is defined in terms of electrical current, which is the rate of flow of charge. The reason is that one can make precise measurements of the current flowing in wire, whereas the charge on a body tends to leak away (see Chapter 31). The coulomb is a very large amount of charge. A typical charge acquired by a rubbed body is 10^{-8} C, whereas a lightning bolt may transfer as much as 20 C between the earth and a cloud. When a body is charged by rubbing, only about 1 in 10^5 of the surface atoms gains or loses an electron. Even for very highly charged objects, only about one in 500 of the surface atoms have a net charge. Electrical effects arise from a very small imbalance in the usual neutral state of matter.

EXERCISE 1. What are the disadvantages of defining positive charge as that acquired by a glass rod?

Quantization of Charge

In the 17th and 18th centuries, both electrical charge and ordinary matter were considered to be continuous. However, during the 19th century, the simple rules that govern the chemical combinations of elements provided strong support for the notion that matter consists of atoms. Chemical evidence also led to the suggestion that molecules dissociate into ions, each of which carries a fixed charge that is a multiple of some basic unit of charge. Subsequent experiments have confirmed this suggestion. Electrical charge appears only in discrete amounts; it is said to be *quantized*. The quantum of charge, first directly measured in 1909 by R. A. Millikan (see Section 23.7), is approximately

The elementary charge

$$e = 1.602 \times 10^{-19} \text{ C}$$

Any charge q must be an integer multiple of this basic unit. That is, $q = 0, \pm e,$ $\pm 2e, \pm 3e$, etc. Although the mass of the proton is about 1800 times greater than that of the electron, their charges have the same magnitude:

$$q_e = -e \qquad q_p = +e$$

Note that the electron itself is *not* the charge: Charge is a property, like mass, of elementary particles, such as the electron.

Recent theories have postulated the existence of particles called *quarks* as the building blocks for the dozens of elementary particles presently known. The theory requires these quarks to carry fractions of the elementary charge: $\pm e/3$ or $\pm 2e/3$. Although there is much indirect evidence for their existence, no one has yet detected an isolated quark. It seems that e is still the smallest *isolated* charge in nature.

Conservation of Charge

Franklin went further with his one-fluid theory. In one experiment, two persons, A and B, stood on wax stands (to prevent the loss of charge). Person A was given the charge of a glass rod while B received the charge from the silk. When the knuckles of either A or B were brought close to those of a third person C, a spark was seen. However, if A and B touched each other before C approached, no spark was seen. Franklin concluded that the charges acquired by A and B were equal in magnitude and opposite in sign. By implication, the gain in fluid of the rod was equal to the loss in fluid of the silk, the total amount of fluid being unchanged. It was a significant realization: Charge is neither created nor destroyed; it is *transferred* from one body to the other. We now call this fact the **conservation of charge:**

> In an isolated system, the total charge stays constant.

Conservation of charge

By "isolated" we mean that there are no paths, such as wires or damp air, by which charges can leave or enter the system. To apply the law of conservation of charge, we add up the number of elementary charges before the interaction and then again after it. An example is a simple chemical reaction:

$$Na^+ + Cl^- \rightarrow NaCl$$
$$(+e) + (-e) = (0)$$

A sodium (Na) atom loses an electron and becomes a positive Na^+ ion. The chlorine (Cl) atom gains the electron and thereby becomes a negative Cl^- ion. The ions bond together to form the neutral sodium chloride (NaCl) molecule. Here is an example of radioactive decay:

$$n \rightarrow p + e^- + \bar{\nu}$$
$$(0) = (+e) + (-e) + (0)$$

In this case an uncharged neutron undergoes spontaneous decay into a proton, an electron, and a neutral particle called the antineutrino. The sum of the charges on the decay products is equal to the charge on the neutron, which is zero.

22.2   CONDUCTORS AND INSULATORS

In the early days of electrical researches, major advances could be made even by amateurs. Such was the case in 1729, when Stephen Gray charged a glass tube and noticed that cork plugs at the ends of the tube also became charged. The observation was significant because it showed that bodies can become charged without being rubbed. Gray succeeded in transferring the charge on a glass rod to an ivory ball hung from his window on a line of thread. Being curious as to how far the "fluid" could travel, he supported a line of thread a few hundred feet long with silk loops, but these soon broke. The experiment would not work when the silk loops were replaced by metal hooks. He concluded that the metal "carried away" the charge. To demonstate the transfer of charge through the human body, Gray suspended a small boy by silk threads and placed his feet in contact with a charging machine. The boy's fingers became charged and could attract small bodies or give shocks to others. Demonstrations like these became very popular (see Fig. 22.3).

FIGURE 22.3 An amusing and scientifically interesting demonstration that charge could travel from the feet of a boy to his fingers. This showed that the human body is a conductor.

Conductors, insulators, and semiconductors

Gray discovered that most substances can be classified into one of two groups. Those, such as metals or ionic solutions, that allow charge to flow freely are called **conductors.** Others, such as wood, rubber, silk, and glass, that do not allow the flow of charge are called **insulators.** (Thread is actually an insulator. Gray's experiments worked probably because moisture had been absorbed by the fibers.) A third group of materials, called **semiconductors,** includes silicon, germanium, and carbon. When very pure, they behave like insulators. However, with the careful addition of certain impurities, their ability to conduct can be controlled. Silicon and germanium are widely used in solid-state electronic circuits.

The mobility of charges in a substance can be characterized by a *relaxation time*. When a charge is placed on a small region on an object, the relaxation time tells us how quickly the charge diminishes at that point, or, equivalently, how long it takes the charges to reach their equilibrium positions. The relaxation time for copper is about 10^{-12} s, for glass it is 2 s, in the case of amber it is 4×10^3 s, and for polystyrene it is about 10^{10} s. These values differ by an enormous factor of 10^{22}! The relaxation time for copper indicates that any charge acquired by a metal spreads very quickly over the surface, as shown in Fig. 22.4a. On the other hand, one can find localized packets of charge on a good insulator, as in Fig. 22.4b. In order to transfer charge from an insulator to another body, one has to make contact with the insulator at many points.

At the atomic level one can understand the difference between conductors and insulators by considering what happens to the outer, loosely bound, valence electrons. In an insulator, such as NaCl, the valence electron in the Na atom is transferred to the Cl atom. The Na^+ and Cl^- ions form "ionic" bonds in which all electrons are bound to particular atomic sites. On the other hand, in a metallic conductor, about one electron per atom is free to roam throughout the body of the material. A metal consists of essentially stationary positive ions, usually arranged in a three-dimensional array called a *lattice*, immersed in a "sea" of **free electrons.** Conduction in a metal involves the movement of free electrons, which behave somewhat like the particles of a gas in a closed container. In an electrolytic solution (in which molecules are dissociated into oppositely charged ions), or in an ionized gas, both positive and negative charges move. Even in dry air there are enough ions to discharge a body in a few minutes.

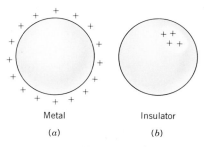

FIGURE 22.4 (*a*) Any net charge placed on a metal sphere spreads very quickly over its surface. (*b*) Net charge may be found in small regions on the surface of an insulator.

Metal
(a)

Insulator
(b)

22.3 CHARGING BY INDUCTION

Stephen Gray had shown that charge can be transferred to an object by conduction. In 1753, John Canton discovered that an insulated metal object can be charged even without connecting it to a charged body. The process of charging without contact is called **induction**. In Fig. 22.5a, A and B are two metal spheres on insulating stands and initially in contact. When a positively charged rod is brought close to A, the free electrons in the metal are attracted to the rod, and some move to the left side of A. This movement leaves unbalanced positive charge on the right surface of B. The rod has *induced* a charge separation. In Fig. 22.5b, the spheres are separated, with the rod *still* present. Finally, in Fig. 22.5c the two spheres carry equal and opposite charges.

A single metal sphere may also be charged by induction. When the positively charged rod is brought near, as in Fig. 22.6a, it induces a charge separation. Next, a connection is made to the ground—for example, through a water pipe. (The earth is a reasonably good conductor and acts as a nearly infinite reservoir of charge.) The symbol for ground is ⏚. As Fig. 22.6b shows, electrons will flow from the ground to neutralize the positive charge. In Fig. 22.6c, the ground connection has been broken. After the rod is taken away, as in Fig. 22.6d, we are left with a uniformly distributed negative charge on the sphere.

EXERCISE 2. Consider Fig. 22.5. (a) What would happen if the rod were removed before the spheres are separated? (b) Would the induced charges be equal in magnitude even if the spheres had different sizes? (c) Is there something slightly misleading about Fig. 22.5c?

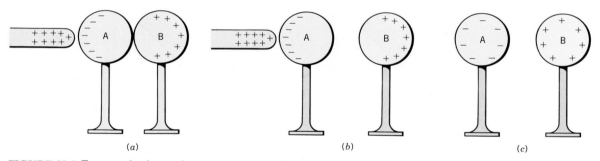

FIGURE 22.5 Two metal spheres given equal and opposite charges by the process of induction.

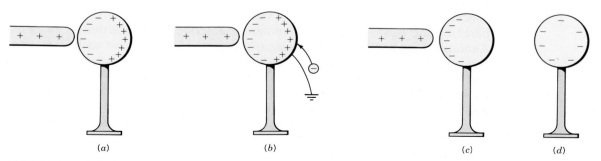

FIGURE 22.6 A single metal sphere charged by induction.

Pan

FIGURE 22.7 A leaf electroscope.

22.4 GOLD LEAF ELECTROSCOPE

An electroscope is a device that detects charge. A leaf electroscope, devised by Rev. Abraham Bennet in 1786, has one or two thin foils of gold or aluminum attached to a metal rod, as in Fig. 22.7. The rod is held in a transparent container with an insulating plug and has a metal pan or sphere at the exposed end. In Fig. 22.8a, a positively charged glass rod is brought near an uncharged electroscope. The electrons in the metal pan are attracted to the glass rod. Their movement results in an unbalanced positive charge on the leaves, which causes them to repel each other. If the rod were to be moved away, the leaves would simply fall back to the vertical position. In Fig. 22.8b, the rod has touched the pan and thereby given it a positive charge. Since we do not know exactly what fraction of charge on the rod has been transferred to the pan, the electroscope detects charge but does not actually *measure* it. The charged electroscope may also be used to indicate the sign of a charged object. When a positively charged object is brought close to the pan, the leaves move farther apart. A negatively charged object would cause the deflection of the leaves to decrease, as shown in Fig. 22.8c.

The electroscope can be used as a primitive detector of ionizing radiation, such as X-rays or high-energy particles. The container is filled with an appropriate gas and the leaves are charged. An incoming X-ray or ionizing particle can often break the atomic bonds in a molecule and thereby produce oppositely charged ions. Those ions whose charge is opposite to that on the leaves migrate to the leaves. Some of the charge on the leaves is neutralized, and so their deflection diminishes. Around 1900, Marie Curie used such a device in her early studies of radioactivity.

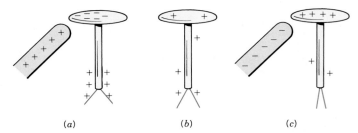

(a)   (b)   (c)

FIGURE 22.8 A charged leaf electroscope may be used to determine the sign of the charge on another body.

22.5 COULOMB'S LAW

Throughout most of the 18th century, scientists made notable conceptual advances in electricity, but their observations were qualitative. Progress was hampered by the lack of a quantitative law to describe the interaction of charges. Although there had been a few unsatisfactory attempts to obtain such a law, the first significant step was taken in 1766 by the chemist Joseph Priestley, the discoverer of oxygen. Sometime earlier, Franklin had conducted a simple experiment. He knew that an uncharged body is attracted to the outside of a charged metal cup and becomes charged if placed in contact with the cup. He suspended an uncharged cork ball inside a charged cup and was surprised that it did not experience any force. When the ball was put in contact with the inside of the cup and withdrawn, it had no charge. On Franklin's request, Priestley confirmed these results and drew an important conclusion. The only force law then known was the law of gravitation. Recall (Section 13.5) that the $1/r^2$ nature of this law means that the net force on a particle inside a uniform spherical shell is zero. By analogy with this result, Priestley deduced that the electrostatic force between charges must also vary as $1/r^2$. Although this was a reasonable inference, it was not entirely convincing. For example, the hollow conductor did not have to be spherically symmetric, as a hollow shell must in the case of gravity.

Credit for determining the force law for electrostatic charges goes to Charles Coulomb (Fig. 22.9) because he determined the law directly by experiment. He did this in 1785, almost exactly 100 years after Newton published his law of gravitation. In those days there was neither a unit of charge nor a reliable means of measuring it. Undaunted, Coulomb devised a simple scheme for keeping track of the sizes of the charges. He charged a small gold-plated pith ball and then placed an identical uncharged ball in contact with the first. If Q is the original charge, by symmetry each sphere acquires $Q/2$. By repeating this procedure, he could obtain various fractions of Q.

To measure force he used a torsion balance in which a dumbbell-like arrangement of a small charged metal sphere and a counterweight is suspended by a silk fiber, as shown in Fig. 22.10. When another charged sphere was brought close to the suspended one, the force between them could be inferred from the angle of twist. Coulomb found that for fixed charges q and Q, the force between them varies as the inverse square of their separation r; that is, $F \propto 1/r^2$. For a fixed separation, the force is proportional to the product of the charges; that is, $F \propto qQ$. These two results are combined in **Coulomb's law** for the electrostatic force between two point charges:

$$F = \frac{kqQ}{r^2} \qquad (22.1)$$

where k is a constant that depends on the system of units. In SI units its approximate value is

$$k \approx 9.0 \times 10^9 \text{ N} \cdot \text{m}^2/\text{C}^2$$

FIGURE 22.9 Charles A. Coulomb (1736–1806).

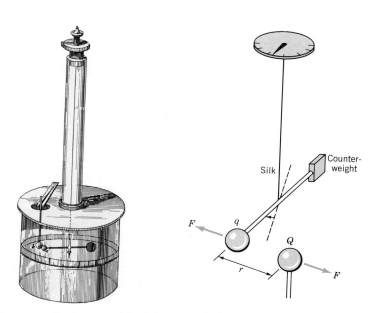

FIGURE 22.10 Coulomb's torsion balance. The electrical force between two spheres was determined by measuring the angle of twist in a silk fiber.

This constant is often written in the form

$$k = \frac{1}{4\pi\varepsilon_0}$$

where ε_0, called the **permittivity constant,** is

$$\varepsilon_0 = 8.85 \times 10^{-12} \text{ C}^2/\text{N} \cdot \text{m}^2$$

Although the use of $1/4\pi\varepsilon_0$ makes Coulomb's law look more complicated, it does simplify the appearance of other equations in electromagnetism.

The electrostatic force is *central* (it is directed along the line joining the particles) and *spherically symmetric* (it is a function only of r). The vector form of Coulomb's law is

$$\mathbf{F} = \frac{kqQ}{r^2} \hat{\mathbf{r}} \tag{22.2}$$

The unit vector $\hat{\mathbf{r}}$ has its origin at the ''source of the force.'' For example, to find the force on q, the origin of $\hat{\mathbf{r}}$ is placed at Q, as in Fig. 22.11. The signs of the charges must be explicitly included in Eq. 22.2. If F is the magnitude of the force (a positive scalar), then $\mathbf{F} = +F\hat{\mathbf{r}}$ means a *repulsion,* whereas $\mathbf{F} = -F\hat{\mathbf{r}}$ means an *attraction.*

Two points regarding Coulomb's law should be kept in mind. First, the charges are assumed to be *at rest.* As we will see later, charges in motion also produce and experience magnetic forces. Second, the charges are assumed to be on *point* particles. For charged bodies of finite size, as in Fig. 22.12, there is no well-defined value of the separation r. As is the case with the gravitational force, there is an exception. When the charge is distributed *uniformly over a spherical surface,* the force on a point charge outside the surface may be computed from Coulomb's law by treating the charge on the sphere as if it were concentrated at the center. Also, if the dimensions of two charged bodies are small in comparison to their separation, then Coulomb's law will provide an approximate value for the force between them. In all other cases, integration is required (see Section 23.5).

Principle of Superposition

Figure 22.13 shows a charge q_1 interacting with other charges. Electrostatic forces obey the **principle of linear superposition** (Section 13.3). Thus, to find the force on q_1, we first calculate the forces exerted by each of the other charges, one at a time. With the notation $\mathbf{F}_{AB}$ = Force on A due to B, the net force $\mathbf{F}_1$ on q_1 is simply the vector sum

$$\mathbf{F}_1 = \mathbf{F}_{12} + \mathbf{F}_{13} + \cdots + \mathbf{F}_{1N} \tag{22.3}$$

Note that the force $\mathbf{F}_{12}$ ($= -\mathbf{F}_{21}$) between q_1 and q_2 is not affected by the presence of the other charges, q_3 and q_4.

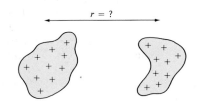

FIGURE 22.11 Coulomb's law applies to *point* charges. The unit vector $\hat{\mathbf{r}}$ has its origin at the ''source of the force.''

FIGURE 22.12 For charged bodies of arbitrary shape, there is no well-defined value for the separation r.

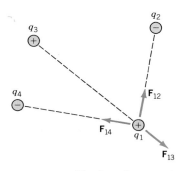

FIGURE 22.13 The force between two charges is not affected by the presence of other charges. The net force on q_1 is the vector sum of the forces exerted by the other charges, calculated one at a time.

Problem-Solving Guide for Coulomb's Law

Suppose we wish to find the net force on charge q_1. In applying Eq. 22.3, the vector form of Coulomb's law is not helpful because each unit vector ($\hat{\mathbf{r}}$) has a different origin. Instead, we use the following steps.

1. Decide whether the force due to a given charge is an attraction or a repulsion. Draw the force vector, with its tail at q_1, either toward or away from the other charge.

2. Find the magnitude of the force from Eq. 22.1—*ignoring* the signs of the charges. The positive sign of the magnitude is ensured by using the form

$$F = \frac{k|qQ|}{r^2}$$

3. Use Eq. 22.3 to find F_{1x} and F_{1y}. Your choice of coordinate axes will determine the signs of these components.

4. All vectors should be expressed in unit vector (**i j k**) notation unless otherwise indicated.

These steps are illustrated in the next example.

EXAMPLE 22.1: Find the net force on charge q_1 due to the three other charges in Fig. 22.14. Take $q_1 = -5 \ \mu$C, $q_2 = -8 \ \mu$C, $q_3 = 15 \ \mu$C, and $q_4 = -16 \ \mu$C.

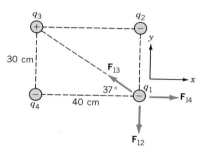

FIGURE 22.14 To find the net force on q_1 one must first calculate the magnitudes of the individual forces and then take components relative to a coordinate system.

Solution: The directions of the forces on q_1 and the coordinate axes are shown in the figure. The magnitude of the force on q_1 due to q_2 is

$$F_{12} = \frac{k|q_1 q_2|}{r^2}$$

$$= \frac{(9.0 \times 10^9 \text{ N} \cdot \text{m}^2/\text{C}^2)(5 \times 10^{-6} \text{ C})(8 \times 10^{-6} \text{ C})}{(3 \times 10^{-1}\text{m})^2}$$

$$= 4 \text{ N}$$

Similarly, we find $F_{13} = 2.7$ N and $F_{14} = 4.5$ N. (Check these.) The components of the net force are

$$F_{1x} = 0 - F_{13} \cos 37° + F_{14} = 2.3 \text{ N}$$
$$F_{1y} = -F_{12} + F_{13} \sin 37° + 0 = -2.4 \text{ N}$$

The net force on q_1 is $\mathbf{F}_1 = 2.3\mathbf{i} - 2.4\mathbf{j}$ N.

EXAMPLE 22.2: A point charge $q_1 = -9 \ \mu$C is at $x = 0$, while $q_2 = 4 \ \mu$C is at $x = 1$ m. At what point, besides infinity, would the net force on a positive charge q_3 be zero?

Solution: If q_3 is placed off the x axis, it will always experience some net force due to the two fixed charges. (Why?) At any point between q_1 and q_2, the forces on q_3 would be in the same direction, so this region is eliminated. For negative x, $\mathbf{F}_{31}$ and $\mathbf{F}_{32}$ are in opposite directions, which means there is a possibility that they will cancel. However, since $F \propto 1/r^2$, in order for the force due to the smaller charge (q_2) to balance the force due to the larger charge (q_1), the charge q_3 must be placed *closer to the smaller charge q_2*. This leaves the region $x > 1$ m on the x axis.

The force vectors are drawn in Fig. 22.15. The condition that the net force on q_3 must be zero is

$$\mathbf{F}_3 = \mathbf{F}_{31} + \mathbf{F}_{32} = -F_{31} \mathbf{i} + F_{32} \mathbf{i} = 0$$

or

$$F_{31} = F_{32}$$

Note that F_{31} and F_{32} are the (positive) *magnitudes* of the forces. The fact that q_1 is negative has *already* been taken into account in the way the vector $\mathbf{F}_{31}$ was drawn. From Coulomb's law

$$\frac{k|q_3 q_1|}{(1 + d)^2} = \frac{k|q_3 q_2|}{d^2}$$

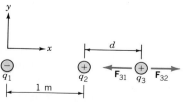

FIGURE 22.15 Locating the point at which the net force on q_3 is zero. The point is closer to the smaller charge.

Next we remove the common factor kq_3, substitute the given magnitudes of q_1 and q_2, and take the square root of both sides. This yields $3/(1 + d) = \pm 2/d$. Thus, $d = 2$ m or $-2/5$ m. Of the two solutions, $d = 2$ m is the proper answer to the given question.

EXERCISE 3. What is the significance of the discarded solution?

EXAMPLE 22.3: The electron and the proton in a hydrogen atom are 0.53×10^{-10} m apart. Compare the electrostatic and gravitational forces between them.

Solution: The magnitude of the electrostatic force is

$$F_E = \frac{ke^2}{r^2}$$

$$= \frac{(9.0 \times 10^9 \text{ N} \cdot \text{m}^2/\text{C}^2)(1.6 \times 10^{-19} \text{ C})^2}{(5.3 \times 10^{-11} \text{ m})^2}$$

$$= 8.2 \times 10^{-8} \text{ N}$$

The magnitude of the gravitational force is

$$F_G = \frac{Gm_e m_p}{r^2}$$

$$= \frac{(6.67 \times 10^{-11} \text{ N} \cdot \text{m}^2/\text{kg}^2)(9.11 \times 10^{-31} \text{ kg})(1.67 \times 10^{-27} \text{ kg})}{(5.3 \times 10^{-11} \text{ m})^2}$$

$$= 3.6 \times 10^{-47} \text{ N}$$

The ratio of the forces,

$$\frac{F_G}{F_E} = \frac{Gm_e m_p}{ke^2} = 4.4 \times 10^{-40}$$

is extremely small and does not depend on the separation r. When we deal with the electrical interaction between elementary particles, gravity may safely be ignored. One can also see why a charged comb can lift a piece of paper and thereby overcome the gravitational force exerted by the whole earth!

Coulomb's law has been found to be valid down to distances as small as 10^{-15} m. The exponent n in the denominator of $1/r^n$ has been determined to be 2 to within an uncertainty of $\pm 10^{-16}$ (see Section 24.3). The Coulomb interaction is the fundamental interaction between charges, and it therefore forms the basis of electromagnetism. We will see much later that one can infer the existence of magnetic forces from Coulomb's law by applying the theory of special relativity.

EXERCISE 4. By considering the phenomena of attraction and repulsion, comment on the possible validity of each of the following expressions for the force between two point charges q_1 and q_2 separated by a distance r:

$$(a) \ \frac{q_1^2 q_2}{r^2} \qquad (b) \ \frac{k(q_1 + q_2)}{r^2}$$

SUMMARY

Electrical **charge** is a property of matter that causes it to produce and to experience electric and magnetic effects. According to the **conservation of charge:** The net charge in an isolated system is constant. Charge is **quantized**—that is, it appears only in discrete packets. Any charge q is given by

$$q = \pm ne$$

where n is an integer and $e = 1.6 \times 10^{-19}$ C is the elementary charge.

A **conductor** is a material that allows the flow of charge. In a metal the charges that move are free electrons. In ionized gases and electrolytic solutions, ions of both signs are mobile. In an **insulator** all the charges are attached to specific sites, which means that charge cannot flow through an insulator. A **semiconductor** behaves like an insulator when very pure. Its ability to conduct can be controlled by the addition of certain impurities.

The magnitude of the electrostatic force between two *static, point* charges q and Q separated by a distance r is given by **Coulomb's law:**

$$F = \frac{k|qQ|}{r^2}$$

This is a *central* force (it acts along the line joining the two charges). It is also *spherically symmetric* (it is a function only of r). With the exception of a spherically symmetric charge distribution, Coulomb's law cannot be applied directly to a finite charge distribution.

The electrostatic force obeys the principle of **linear superposition,** which means that the force between two particles is not affected by the presence of other charges. This principle is used to find the net force exerted on a given particle by other charged particles.

ANSWERS TO IN-CHAPTER EXERCISES

1. The sign of the charge on a rubbed glass rod depends on the material used, such as silk or wool. Furthermore, the sign of the charge depends on how hard the rod is rubbed.

2. (a) The excess electrons on sphere A would flow back to B. (b) Yes. (c) The charges on conducting spheres placed close to each other would not be uniformly distributed.

3. If q_1 and q_2 had the same sign, the force on q_3 would be zero at a point 0.4 m to the left of q_2. Verify this for yourself.

4. (a) The sign of q_1 would not matter. If q_1 and q_2 were interchanged, the force would not be the same. (b) If the charges had equal and opposite signs, the force would be zero.

QUESTIONS

1. Protons in a nucleus are separated by a very small distance ($\approx 10^{-15}$ m). Why doesn't the nucleus fly apart because of the strong coulomb repulsion between the protons?

2. Since the electrostatic force is so much stronger than the gravitational force, why do we not have more direct and frequent encounters with it?

3. Can a metal object be charged by rubbing it? Explain why or why not. If it can, give details.

4. A typical charge produced by rubbing is 1 nC. Roughly how many elementary charges (e) does this involve?

5. A positively charged glass rod is brought close to a suspended metal needle. What can you say about the charge on the needle given that the needle is (a) attracted; (b) repelled?

6. How would you determine the sign of the charge on a body?

7. A thin strip of aluminum foil is attracted to a comb that is charged by passing it through hair. What happens after the foil touches the comb? Try it first and then explain what you see.

8. A point charge q is placed at the midpoint between two equal point charges of magnitude Q, as in Fig. 22.16. Is q in equilibrium? If so, is it stable or unstable equilibrium? Consider q to have (a) the same sign and (b) the opposite sign to

Q. (*Hint:* Consider small displacements away from the midpoint.)

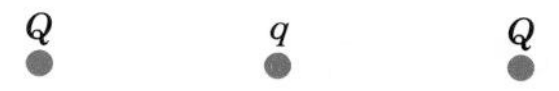

FIGURE 22.16 Question 8.

9. An uncharged metal sphere is brought close to a point charge. Does either body experience a force?

10. Two identical metal spheres are charged and placed close to each other. Can one calculate the force between them by using Coulomb's law with r as the separation between the centers? Explain why or why not.

11. A newspaper reports that a new elementary particle with a charge of 9.00×10^{-19} C has been discovered. What is your response?

12. In what ways does the conduction of heat differ from electrical conduction?

13. When a charged object is brought close to one end of an uncharged metal rod, some electrons in the rod flow from one end to the other. Since there is an enormous supply of electrons, why does the flow stop?

14. Why do electrostatic demonstrations tend to fail on humid days? Reconcile your answer with the fact that part of the invigorating effect of a shower is associated with charges on water drops.

15. Why is it not advisable to wipe a phonograph disc with a woolen cloth?

16. Trucks and cars used to have chains dangling onto the road. What purpose was this supposed to serve?

EXERCISES

1. (I) Three point charges lie on a straight line, as shown in Fig. 22.17. Find the resultant force exerted on (a) the -2 μC charge, (b) the 5 μC charge.

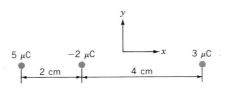

FIGURE 22.17 Exercise 1.

2. (I) Three point charges are held in the positions shown in Fig. 22.18. Take $q = 1$ nC. Find the resultant force exerted on (a) the charge $4q$; (b) the charge $-3q$.

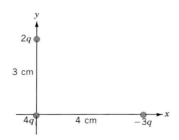

FIGURE 22.18 Exercise 2.

3. (I) Three point charges are held at the corners of an equilateral triangle as shown in Fig. 22.19. Take $Q = 2$ μC and $L = 3$ cm. What is the resultant force exerted on the charge (a) $3Q$, and (b) $-2Q$?

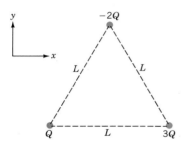

FIGURE 22.19 Exercise 3.

4. (I) Four point charges are located at the corners of a rectangle as shown in Fig. 22.20. Take $Q = 4$ nC. What is the resultant force on (a) $-2Q$, and (b) $-3Q$?

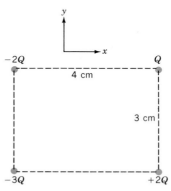

FIGURE 22.20 Exercise 4.

5. (I) A point charge $q_1 = 27$ μC is placed at $x = 0$ while $q_2 = 3$ μC is at $x = 1$ m. (a) At what point (other than infinity) would the net force on a third point charge be zero? (b) Repeat part (a) for $q_2 = -3$ μC.

6. (II) How much equal charge would be needed on the earth and on the moon for the electrostatic repulsion to balance the gravitational attraction?

7. (I) At what separation would the force between a proton and an electron be 1 N?

8. (I) A radioactive uranium nucleus has a charge $92e$. It can spontaneously decay into a thorium nucleus with charge $90e$ and a helium nucleus (α particle) of charge $2e$. Just after the decay, the helium and the thorium are 3×10^{-15} m apart. (a) What is the electrostatic force between them? (b) What is the acceleration of the α particle, whose mass is 6.7×10^{-27} kg?

9. (I) (a) The two protons in the H_2 molecule are 0.74×10^{-10} m apart. What is the electrostatic force between them? (b) In a NaCl crystal, the Na^+ and Cl^- ions are 2.82×10^{-10} m apart. What is the force between them?

10. (II) Two point charges, Q and $-2Q$, are located at the positions shown in Fig. 22.21. (a) What is the force on a charge q at the origin? (b) Where would you place a point charge $+2.5Q$ such that the net force on q is zero?

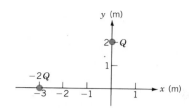

FIGURE 22.21 Exercise 10.

11. (II) Figure 22.22 shows five point charges on a straight line. The separation between the charges is 1 cm. For what values of q_1 and q_2 would the net force on each of the other three charges be zero?

FIGURE 22.22 Exercise 11.

12. (II) Each of two identical Styrofoam balls has a charge Q and a mass $m = 2$ g. They are suspended by threads of length $L = 1$ m, as shown in Fig. 22.23. Because of the mutual repulsion of the balls, the threads are at a 15° angle to the vertical. Find Q.

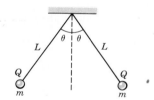

FIGURE 22.23 Exercise 12.

13. (II) Suppose the electrons and protons in 1 g of hydrogen could be separated and placed on the earth and the moon, respectively. Compare the electrostatic attraction with the gravitational force between the earth and the moon. (The number of atoms in 1 g of hydrogen is Avogadro's number N_A. There is one electron and one proton in a hydrogen atom.)

14. (II) (a) A point charge Q is at $x = 0$ and $9Q$ is at $x = 4$ m. Where should a third charge q be placed such that the net force on each of the three charges is zero? What is the value of q? (b) Repeat part (a) but replace $9Q$ by $-9Q$.

15. (I) Two small Styrofoam balls are 4 cm apart and repel each

other with a force of 0.2 N. If the charge on one ball is twice the other, find both charges.

16. (II) Two equal and opposite charges (± 1 nC) are separated by a distance $2d$ where $d = 1$ cm, as shown in Fig. 22.24. Find the force on a charge of 2 nC when it is placed at the point (a) A, (b) B, (c) C, and (d) D.

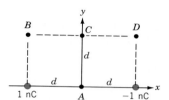

FIGURE 22.24 Exercise 16.

17. (I) In the quark model of elementary particles a proton consists of two "up" (u) quarks, each with charge $2e/3$, and a "down" (d) quark of charge $-e/3$. Suppose these particles lie equally spaced on a circle of radius 1.2×10^{-15} m, as in Fig. 22.25. Find the magnitude of the electrostatic force on each quark.

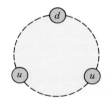

FIGURE 22.25 Exercise 17.

18. (I) A point charge $q_1 = 2 \ \mu$C is located at (2 m, 1 m) and a charge $q_2 = -5 \ \mu$C is at (-2 m, 4 m). Find the force exerted on q_2 by q_1.

19. (I) In a thundercloud there are two equal and opposite charges (± 40 C) separated by 5 km, as in Fig. 22.26. Assuming that they can be treated as point charges, what is the force between them?

FIGURE 22.26 Exercise 19.

PROBLEMS

1. (I) Three point charges, q_1, q_2, and q_3, lie at the corners of an equilateral triangle of side 10 cm. The forces between them are $F_{12} = 5.4$ N (attractive), $F_{13} = 15$ N (repulsive), and $F_{23} = 9$ N (attractive). Given that q_1 is negative, what are q_2 and q_3?

2. (II) Two equal point charges Q are located on the y axis at $y = a$ and $y = -a$. (a) What is the force on a charge q located at $(x, 0)$? (b) For what value of x is the force a maximum? Make a rough plot of $F(x)$, the force as a function of x. (c) When $x \gg a$, what is the form of $F(x)$? (*Hint:* Use the binomial expansion $(1 + z)^n \approx 1 + nz$ for small z.)

3. (II) A point charge $-Q$ is located at $(0, -a)$ and charge $+Q$ is at $(0, a)$. (a) Find the force on a charge q at $(x, 0)$. (b) At what point is the force a maximum?

4. (I) A point charge $-Q$ is at $(0, -a)$ and charge Q is at $(0, a)$. (a) Find the force on a charge q located at $(0, y)$, where $y > a$. (b) What is the form of $F(y)$ when $y \gg a$. (*Hint:* Use the binomial expansion $(1 + z)^n \approx 1 + nz$ for small z.)

5. (I) A charge Q is to be divided into two parts, q and $(Q - q)$, such that, for a given separation, the force between them is a maximum. What is the value of q? (*Hint:* In calculus what is the condition for a function to be a maximum?)

6. (II) Two small identical metal spheres are 3 cm apart and attract each other with a force of 150 N. They are momentarily connected by a wire. (a) Find the original charges if they now repel each other with a force of 10 N. (Assume the charge on each sphere is uniformly distributed.) (b) Find the original charges if the magnitude of the repulsive force is 150 N.

7. (I) Two 10-g copper spheres are separated by 10 cm. (a) How many electrons need to be removed from each sphere for the spheres to repel each other by a force of 10 N? (b) What fraction of the total number of electrons in each sphere does the answer to (a) represent? (*Hint:* The number of atoms in 63.5 g of Cu is Avogadro's number. There are 29 electrons in each Cu atom.)

8. (II) Eight charges of magnitude Q are located at the corners of a cube of side L; see Fig. 22.27. One corner is at the origin and the edges lie along the rectangular axes. Find the net force on the charge at $\mathbf{r} = L\mathbf{i} + L\mathbf{j} + L\mathbf{k}$ given the following: (a) all have the same sign; (b) the nearest neighbor of any charge has the opposite sign.

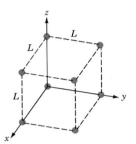

FIGURE 22.27 Problem 8.

9. (II) In the Bohr model of the hydrogen atom an electron orbits a stationary proton in a circular orbit of radius r. (a) Write Newton's second law for the circular motion and obtain an expression for the speed v. (b) Bohr imposed the condition that the angular momentum L of the electron could take on only discrete values given by $L = nh/2\pi$ where n is an integer and h is a constant (Planck's constant). Show that the radius of the nth allowed orbit is given by

$$r_n = \frac{n^2 h^2}{4\pi^2 kme^2}$$

(c) Evaluate r_n for $n = 1, 2, 3$.

10. (II) The sum of two point charges is $+8$ μC. When they are 3 cm apart, each experiences a force of 150 N. Find the charges given that the force is (a) repulsive; (b) attractive.

CHAPTER 23

The Electric Field

A lightning stroke is an awesome display of electrical activity.

Major Points

1. (a) The concept of an **electric field** is used to describe how charges interact.
   (b) The use of **field lines** to represent the field.
2. Under static conditions: (a) The electric field within a conductor is zero.
   (b) The electric field is perpendicular to the surface of a conductor.
3. The motion of charges in a static uniform field.
4. **Dipoles:** (a) The field produced by a dipole.
   (b) The torque and potential energy associated with a dipole in an electric field.

Coulomb's law, like Newton's law of gravitation, involves the concept of *action at a distance:* It simply states how the particles interact but provides no explanation of the mechanism by which the force is transmitted from one particle to the other. Newton himself was not comfortable with this aspect of his theory. An earlier attempt to explain how a charged body can "reach out" and affect another was made in 1600 by William Gilbert. He proposed that an electrified body released vapors, or "effluvia," when rubbed and thereby created an "atmosphere" around itself. As the effluvia returned to the parent body, light objects were swept up in the stream. He suggested that one could feel that effluvia as a tingling sensation if one's face were brought close to the electrified body. A different mechanism was proposed around 1650 by René Descartes, who thought of space as being filled with an invisible medium called the ether. He believed that a charged body set up vortex motions (whirlpools) in this ether, which then traveled to other bodies and exerted forces on them.

In the modern view, a charged particle does not emit an "atmosphere," nor does it require an intervening medium to interact with another charge. The modern description of the interaction between particles is based on the concept of a field, which was introduced in Chapter 13.

23.1 THE ELECTRIC FIELD

Consider two point charges separated by some distance. We know that the particles interact, but exactly how does one particle sense the presence of the other? We say that an electric charge creates an **electric field** in the space around it. A second charged particle does not interact directly with the first; rather, it responds to whatever field it encounters. In this sense, the field acts as an intermediary between the particles.

Let us examine the electric field created by a static point charge Q. One can obtain a map of the field by measuring the force experienced by a small test charge q_t at various points. Each point in space then has a unique force vector associated with it, as shown in Fig. 23.1. At a given point, the **electric field strength E** is defined as the force per unit charge placed at that point:

$$\mathbf{E} = \frac{\mathbf{F}}{q_t} \qquad (23.1)$$

The SI unit for electric field strength is newton per coulomb (N/C). The direction of **E** is that of the force experienced by a positive test charge. The test charge must be small enough not to disturb the charges that create the electric field being measured. In the particular case of the point charge Q, we have from Coulomb's law

$$\mathbf{F} = \frac{kq_t Q}{r^2} \hat{\mathbf{r}}$$

Therefore, from Eq. 23.1, the electric field created by the point charge Q is

$$\mathbf{E} = \frac{kQ}{r^2} \hat{\mathbf{r}} \qquad (23.2)$$

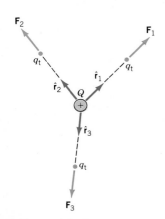

FIGURE 23.1 The electric field of the point charge Q can be mapped by measuring the force **F** on a test charge q_t placed at various points. The field strength at a given point is $\mathbf{E} = \mathbf{F}/q_t$.

The field strength **E** is a property of a point in space and depends only on the source of the field, Q. The field exists whether or not we choose to examine it with a test charge. Once the field strength is known, the force on any charge q can be found from

$$\mathbf{F} = q\mathbf{E} \qquad (23.3)$$

where **E** is the resultant field strength due to all the charges present, but *not* including q itself. If q is positive, the force on it is in the same direction as the field strength; if q is negative, the force on it is directed opposite to the field. Notice that Eq. 23.3 has the same form as $\mathbf{F} = m\mathbf{g}$, where **g** is the gravitational field strength (N/kg).

EXAMPLE 23.1: On a clear day there is an electric field of approximately 100 N/C directed vertically down at the earth's surface. Compare the electrical and gravitational forces on an electron.

Solution: The magnitude of the electrical force is

$$F_E = eE = (1.6 \times 10^{-19} \text{ C})(100 \text{ N/C}) = 1.6 \times 10^{-17} \text{ N}$$

and it is directed vertically upward, as shown in Fig. 23.2. The magnitude of the gravitational force is

$$F_g = mg = (9.1 \times 10^{-31} \text{ kg})(9.8 \text{ N/kg}) = 8.9 \times 10^{-30} \text{ N}$$

and points downward. Their ratio is

$$\frac{F_g}{F_E} = 5.6 \times 10^{-13}$$

The gravitational force on particles such as the electron and proton may be safely ignored in problems involving electric fields.

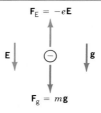

FIGURE 23.2 A charge held in equilibrium by electrical and gravitational forces.

EXERCISE 1. A point charge $q = 2\ \mu\text{C}$ experiences a force $\mathbf{F} = 5 \times 10^{-5}\mathbf{i}$ N. (a) What is the field strength **E**? (b) What happens to **E** if q is changed to 1 μC? (c) What happens to **F** if q is changed to $-2\ \mu$C? (d) What happens to **E** if q is changed to $-2\ \mu$C?

Since the principle of linear superposition is valid for Coulomb's law, it is also valid for the electric field. To calculate the field strength at a point due to a system of charges, we first find the individual field strengths $\mathbf{E}_1$ due to Q_1, $\mathbf{E}_2$ due to Q_2, and so on. For N point charges, the resultant field strength is the vector sum

$$\mathbf{E} = \mathbf{E}_1 + \mathbf{E}_2 + \cdots + \mathbf{E}_N = \sum \mathbf{E}_i \qquad (23.4) \qquad \text{Linear superposition}$$

where

$$\mathbf{E}_i = \frac{kQ_i}{r_i^2} \hat{\mathbf{r}}_i$$

Since each unit vector has its origin at a different charge, this equation can be inconvenient to use. It is usually easier to follow the approach in Chapter 22.

Problem-Solving Guide for the Electric Field

1. First draw the field vectors at the given location. (Their directions can be found by imagining that there is a positive charge at that point.)
2. Find the (scalar) magnitude of the field strength due to each charge. The signs of the charges must be ignored. This step may be ensured by writing the magnitude in the form

$$E = \frac{k|Q|}{r^2} \qquad (23.5)$$

3. Place the origin at the point at which $\mathbf{E}$ is being calculated. The choice of coordinate axes determines the signs of the components of the field strength $\mathbf{E}$.

EXAMPLE 23.2: A point charge $Q_1 = 20\ \mu\text{C}$ is at $(-d, 0)$ while $Q_2 = -10\ \mu\text{C}$ is at $(+d, 0)$. Find the resultant field strength at a point with coordinates (x, y) Take $d = 1.0$ m and $x = y = 2$ m.

Solution: The charges, the field vectors, and the coordinate system are shown in Fig. 23.3. The distances are $r_1 = \sqrt{(x + d)^2 + y^2} = \sqrt{13} = 3.6$ m and $r_2 = \sqrt{(x - d)^2 + y^2} = \sqrt{5} = 2.2$ m. The magnitudes of the fields are

$$E_1 = \frac{k|Q_1|}{r_1^2}$$

$$= \frac{(9.0 \times 10^9\ \text{N} \cdot \text{m}^2/\text{C}^2)(2 \times 10^{-5}\ \text{C})}{13\ \text{m}^2}$$

$$= 1.4 \times 10^4\ \text{N/C}$$

$$E_2 = \frac{k|Q_2|}{r_2^2}$$

$$= \frac{(9.0 \times 10^9\ \text{N} \cdot \text{m}^2/\text{C}^2)(10^{-5}\ \text{C})}{5\ \text{m}^2}$$

$$= 1.8 \times 10^4\ \text{N/C}$$

The components of the resultant field strength, $\mathbf{E} = \mathbf{E}_1 + \mathbf{E}_2$, are

$$E_x = E_{1x} + E_{2x} = E_1 \cos \theta_1 - E_2 \cos \theta_2$$
$$E_y = E_{1y} + E_{2y} = E_1 \sin \theta_1 - E_2 \sin \theta_2$$

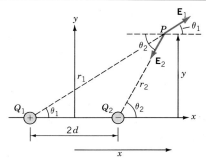

FIGURE 23.3 Finding the resultant field due to charges Q_1 and Q_2 at point P.

From Fig. 23.3, we see that $\sin \theta_1 = y/r_1$, $\sin \theta_2 = y/r_2$, $\cos \theta_1 = (x + d)/r_1$, and $\cos \theta_2 = (x - d)/r_2$. Therefore,

$$E_x = (1.4 \times 10^4\ \text{N/C}) \frac{3}{3.6} - (1.8 \times 10^4\ \text{N/C}) \frac{1.0}{2.2}$$

$$= 3.5 \times 10^3\ \text{N/C}$$

$$E_y = (1.4 \times 10^4\ \text{N/C}) \frac{2}{3.6} - (1.8 \times 10^4\ \text{N/C}) \frac{2}{2.2}$$

$$= -8.6 \times 10^3\ \text{N/C}$$

Finally, $\mathbf{E} = 3.5 \times 10^3 \mathbf{i} - 8.6 \times 10^3 \mathbf{j}$ N/C.

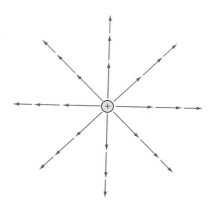

FIGURE 23.4 The electric field of a point charge represented by arrows drawn to scale.

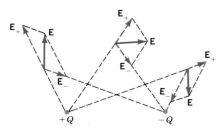

FIGURE 23.5 A few arrows to depict the electric field due to two equal and opposite charges.

23.2 LINES OF FORCE

Consider the electric field created by a positive point charge Q. The field strength at any point could be represented by an arrow drawn to scale. A map of the electric field might then look like Fig. 23.4. A few arrows for the field due to two equal and opposite point charges are shown in Fig. 23.5. When several charges are present, the use of arrows of varying lengths and orientations becomes confusing. Instead we represent the electric field by continuous **field lines** or **lines of force.** The lines emerge from a positive charge and enter a negative charge, as shown in Fig. 23.6. When a person with long hair touches a highly charged sphere, the hairs align along the field lines and spread out radially from the head thereby providing a dramatic picture of the field (Fig. 23.7). The field pattern may also be visualized by sprinkling grass seeds over the surface of a liquid, such as oil. When highly charged electrodes are immersed, the seeds orient themselves in the direction of the local field. The seed pattern and the corresponding field lines for two equal and opposite point charges are shown in Figs. 23.8a and 23.8b. The seed pattern and field lines for two point charges of the same sign are shown in Figs. 23.9a and 23.9b. (Keep in mind that we depict only a two-dimensional cross section of the three-dimensional field.)

Faraday, who introduced the use of field lines around 1840, considered the lines to be real and even endowed them with elastic properties: One can almost "feel" the lines either pulling the charges together or pushing them apart. Although in our modern view the field lines are *not* real, they do help us visualize the field, which *is* real.

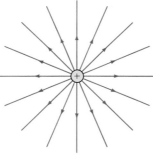

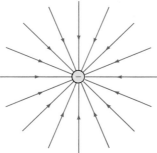

FIGURE 23.6 The electric field of a charge represented by continuous *field lines* or *lines of force*. The lines emerge from a positive charge and enter a negative charge.

FIGURE 23.7 When a person touches a highly charged object, the hairs on the head align along the field lines.

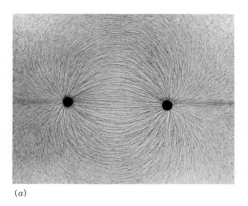
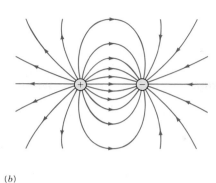

(a)   (b)

FIGURE 23.8 The field due to two equal and opposite charges. (a) The pattern of seeds sprinkled on a liquid. (b) The field lines.

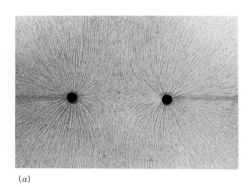
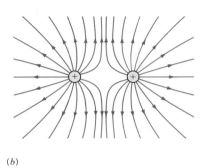

(a)   (b)

FIGURE 23.9 The field due to two equal charges of the same sign. (a) The seed pattern and (b) the field lines.

Field lines can also give us information on the field strength. Notice that the lines are crowded together where the field is strong and spread far apart where the field is weak. The field strength is proportional to the *density* of the lines—that is, the number of lines passing through a unit area normal to the field direction. Suppose N lines emerge from an isolated point charge. At a distance r from the charge the lines are spread over a spherical surface of area $4\pi r^2$. Thus, the density of lines is $N/4\pi r^2$, and decreases as $1/r^2$. This is exactly the way the field strength varies according to Eq. 23.5. In Fig. 23.10, the field is strong at A, while at B it is weaker. Since no lines cross the area at C, one might think that the field strength there is zero. However, we have drawn just a few lines for clarity. If the number were increased tenfold, some would surely pass through the area at C. The number of lines that emerge from a unit charge is a matter of choice; what really matters is the *relative* density of lines at various points.

Here is a summary of the properties of lines of force:

1. Electrostatic field lines always start on positive charges and end on negative charges.

2. The number of lines that originate from, or terminate on, a charge is proportional to the magnitude of the charge.

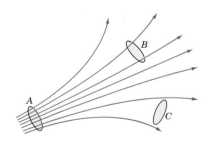

FIGURE 23.10 The field strength is related to the number of lines that cross unit area perpendicular to the field. It would be incorrect to assume that the field strength is zero at C: If the number of lines were increased tenfold some would pass through C. Only the *relative* density of lines is significant.

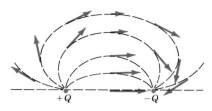

FIGURE 23.11 The direction of the field at a point is along the tangent to the field line.

3. The direction of the field at a point is along the *tangent* to the line of force, as shown in Fig. 23.11.

4. The field strength is proportional to the *density* of the lines, that is, the number of lines per unit area, intercepted by a surface normal to the field.

5. Lines of force never cross.

The techniques we have developed so far do not allow us to plot field lines accurately. (One approach is presented in Chapter 25.) We can, however, deduce the qualitative features of the field pattern due to two or more charges by considering the points raised in the next example.

EXERCISE 2. Why do lines of force never cross?

EXAMPLE 23.3: Sketch the field lines for two point charges $2Q$ and $-Q$.

Solution: The pattern of field lines can be deduced by considering the following points:

(a) *Symmetry:* For every point above the line joining the two charges there is an equivalent point below it. Therefore, the pattern must be symmetrical about the line joining the two charges.

(b) *Near field:* Very close to a charge, its field predominates. Therefore, the lines are radial and spherically symmetric.

(c) *Far field:* Far from the system of charges, the pattern should look like that of a single point charge of value $(2Q - Q) = +Q$. That is, the lines should be radially outward.

(d) *Null point:* There is one point at which $E = 0$ (see Example 22.2). No lines should pass through this point

(e) *Number of lines:* Twice as many lines leave $+2Q$ as enter $-Q$.

The resulting sketch is shown in Fig. 23.12. Notice that in this sketch the field line that goes straight from $2Q$ to $-Q$ has not been drawn because it would upset the proper ratio of lines as required in (e) above. This line has been included in Fig. 24.8.

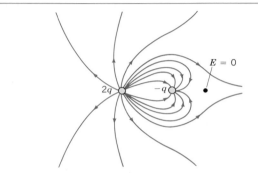

FIGURE 23.12 The field due to charges $2Q$ and $-Q$.

EXERCISE 3. Is the magnitude of the field strength constant along a line of force?

EXERCISE 4. A test charge is released in the field due to two point charges. Do the field lines indicate the possible paths traveled by the test charge?

23.3 ELECTRIC FIELD AND CONDUCTORS

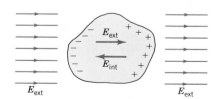

FIGURE 23.13 When a conductor is placed in an external field, the electrons are redistributed and produce an "internal" field. Under static conditions, the net field inside a conductor is zero.

When an external field E_{ext} is applied to a conductor, the free electrons move very quickly in a direction opposite to E_{ext}, as shown in Fig. 23.13. In doing so, they leave unbalanced positive charges on the right. The redistribution of charge creates an internal field E_{int} directed opposite to E_{ext}. Suppose that the net field inside the conductor, $E_{ext} - E_{int}$, is not zero. The free electrons will move under the influence of this net field, thereby increasing the internal field. This process will continue until electrostatic equilibrium is reached, that is, until the net internal field is reduced to zero.

Under static conditions, the net macroscopic field within the material of a homogeneous conductor is zero.

The word *macroscopic* (meaning large scale) is added because there are many complex fields between the electrons and nuclei. The point is that these fields do not add up to anything on a large scale. The average value over many atoms is zero. The word *homogeneous* is also important. When two metals—for example, Zn and Cu—are placed in contact, a separation of positive and negative occurs at the interface. There is an electric field across the interface even though the overall charge on the conductors is zero.

Now suppose that there is an electric field at an angle to the surface of a conductor. The source of the field could be either external or due to charges on the object itself. The free electrons will respond to the component parallel to the surface and quickly reduce it to zero, as Fig. 23.14 shows. *Under static conditions, the electric field at all points on the surface of a conductor is normal to the surface.*

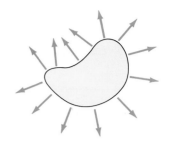

FIGURE 23.14 Under static conditions the electric field is normal to the surface of a conductor.

23.4 MOTION OF CHARGES IN UNIFORM STATIC FIELDS

We now consider the motion of charged particles in static uniform fields. In dealing with the motion of elementary particles, such as electrons and protons, in electric fields, the force of gravity may be neglected.* A particle of mass m and charge q in an electric field experiences a force $\mathbf{F} = q\mathbf{E}$. From the second law, $\mathbf{F} = m\mathbf{a}$, its acceleration is

$$\mathbf{a} = \frac{q\mathbf{E}}{m} \tag{23.6}$$

If the field is uniform, the acceleration is constant in magnitude and direction, so we may use the equations of kinematics for constant acceleration.

* We also assume that the speed of the particles is much less than the speed of light. This allows us to ignore corrections that arise from the theory of special relativity (Chapter 39).

EXAMPLE 23.4: A proton travels a distance of 4 cm parallel to a uniform electric field $\mathbf{E} = 10^3\mathbf{i}$ N/C, as shown in Fig. 23.15. If its initial velocity is 10^5 m/s, find its final velocity.

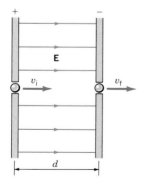

FIGURE 23.15 A proton is accelerated in the uniform field between two charged plates.

Solution: The magnitude of the acceleration of the proton is

$$a = \frac{eE}{m} = \frac{(1.6 \times 10^{-19} \text{ C})(10^3 \text{ N/C})}{1.67 \times 10^{-27} \text{ kg}}$$
$$= 9.6 \times 10^{10} \text{ m/s}^2$$

and it is directed toward the negative plate. The appropriate equation of kinematics is Eq. 3.12:

$$v_f^2 = v_i^2 + 2a\Delta x$$
$$= (10^5 \text{ m/s})^2 + 2(9.6 \times 10^{10} \text{ m/s}^2)(4 \times 10^{-2} \text{ m})$$
$$= 1.77 \times 10^{10} \text{ m}^2/\text{s}^2$$

Thus, $v_f = 1.3 \times 10^5$ m/s.

EXAMPLE 23.5: The *cathode ray tube* (CRT) is used in TVs, video display terminals, and electronic instruments such as the oscilloscope. Electrons are emitted from a thin, heated filament and collimated (made into a pencil-like beam) by two disks with holes (see Fig. 23.16). Their initial velocity is $v_0\mathbf{i}$. They pass between two plates of length ℓ that produce a uniform electric field, $\mathbf{E} = -E\mathbf{j}$. Within the field their acceleration is constant and so their path is parabolic, as in projectile motion near the earth's surface. After they leave the region between the plates, they travel in a straight line to a screen coated with a phosphorescent material, such as ZnS. As each electron strikes the screen, a tiny flash of light is emitted. Find: (a) its vertical position as it emerges from the plates; (b) the angle at which it emerges from the plates; (c) its final vertical displacement on the screen which is at a distance L from the ends of the plates.

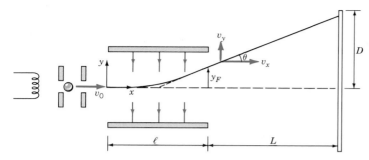

FIGURE 23.16 In a cathode ray tube electrons are emitted by a heated filament, accelerated by the field between two charged disks (with holes), and deflected by the field between two plates. When an electron strikes a coated screen, a flash of light is emitted.

Solution: (a) Since $q = -e$ and $\mathbf{E} = -E\mathbf{j}$, the acceleration is

$$\mathbf{a} = +\frac{eE}{m}\mathbf{j}$$

There is no acceleration in the x direction. The position coordinates within the plates are given by

$$x = v_0 t; \qquad y = \tfrac{1}{2}at^2$$

The electron spends a time $t = \ell/v_0$ between the plates, so its final vertical coordinate as it emerges from the plates is

$$y_F = \frac{1}{2}\frac{eE}{m}\left(\frac{\ell}{v_0}\right)^2 \qquad \text{(i)}$$

(b) The angle at which it emerges from the plates may be found from the components of its final velocity:

$$v_x = v_0; \qquad v_y = at = \frac{eE}{m}\frac{\ell}{v_0}$$

Thus,

$$\tan\theta = \frac{v_y}{v_x} = \frac{eE\ell}{mv_0^2} \qquad \text{(ii)}$$

(c) From Fig. 23.16, we see that $\tan\theta = (D - y_F)/L$. Thus,

$$D = y_F + L\tan\theta \qquad \text{(iii)}$$

23.5 CONTINUOUS CHARGE DISTRIBUTIONS

The expression $\mathbf{E} = (kq/r^2)\,\hat{\mathbf{r}}$ is valid only for point charges. In order to find the electric field due to a continuous distribution of charge, one must divide the charge distribution into infinitesimal elements of charge dq which may be considered to be point charges. The infinitesimal contribution to the total field produced by such an element, shown in Fig. 23.17, is

$$d\mathbf{E} = \frac{k\,dq}{r^2}\,\hat{\mathbf{r}} \qquad (23.7)$$

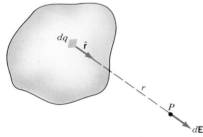

FIGURE 23.17 In order to calculate the field due to a finite charged distribution, one must first find the contribution $d\mathbf{E}$ to the total field from an infinitesimal charge element dq. The total field is the integral of all such contributions.

where $\hat{\mathbf{r}}$ has it origin at the charge element. According to the principle of superposition, the total field is the sum (integral) of all such contributions over the charge distribution; that is,

$$\mathbf{E} = k\int\frac{dq}{r^2}\,\hat{\mathbf{r}} \qquad (23.8)$$

In order to evaluate this integral, we must express dq in terms of r. The process is illustrated in the following examples.

EXAMPLE 23.6: A *thin insulating rod* of length L carries a uniformly distributed charge Q. Find the field strength at a point along its axis at a distance a from one end.

Solution: We must first find the contribution to the field from an infinitesimal length element dx, as shown in Fig. 23.18. Since $dq/Q = dx/L$, the charge on this element is $dq = \lambda dx$, where $\lambda = Q/L$ is the linear charge density. The magnitude of its contribution to the field at point P is

$$dE = \frac{k(\lambda\,dx)}{x^2}$$

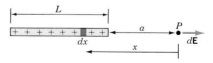

FIGURE 23.18 In finding the field due to a charged rod, the rod is divided into infinitesimal elements of length dx with charge $dq = (Q/L)\,dx$.

and its direction is to the right if λ is positive. The total field strength is the integral of dE over all elements from one end of the rod to the other:

$$E = k\lambda \int_a^{a+L} \frac{dx}{x^2}$$

$$= k\lambda \left[-\frac{1}{x}\right]_a^{a+L}$$

$$= k\lambda \left(\frac{1}{a} - \frac{1}{a+L}\right)$$

$$= \frac{kQ}{a(a+L)}$$

where $Q = \lambda L$. At large distances from the rod (that is, when $a \gg L$), we may neglect L in comparison with a. The expression then reduces to $E = kQ/a^2$. As we might expect, at large distances the field of the rod is nearly the same as that of a point charge. Such a consistency check should always be made when you obtain an algebraic expression.

EXAMPLE 23.7: What is the field strength at a distance R from an *infinite line of charge* with linear charge density λ C/m?

Solution: The magnitude of the contribution to the field at point P from the element of length $d\ell$ in Fig. 23.19 is

$$dE = \frac{k\lambda \, d\ell}{r^2} \qquad (i)$$

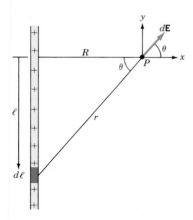

FIGURE 23.19 In finding the field due to an infinite line of charge we choose to express the variables ℓ and r in terms of θ.

To perform the integration we express the variables in terms of the angle θ. From the diagram we see that $r = R \sec \theta$ and $\ell = R \tan \theta$, from which we find $d\ell = R \sec^2 \theta \, d\theta$. Using these expressions in (i), we find

$$dE = \frac{k\lambda \, d\theta}{R} \qquad (ii)$$

For every element on the $+y$ axis, there is a symmetrically placed element on the $-y$ axis. The y components of the fields due to such pairs of elements will cancel. As a result, the resultant field is perpendicular to the line; that is, it points along the x axis. From Fig. 23.19 we see that $dE_x = dE \cos \theta$. The total field is the integral of this component from $-\theta_1$ to θ_2:

$$E = \frac{k\lambda}{R} \int_{-\theta_1}^{\theta_2} \cos \theta \, d\theta = \frac{k\lambda}{R} \left[\sin \theta\right]_{-\theta_1}^{\theta_2}$$

$$= \frac{k\lambda}{R} (\sin \theta_2 + \sin \theta_1)$$

For an infinite line, $\theta_1 = \theta_2 = \pi/2$; therefore,

$$E = \frac{2k\lambda}{R} \qquad (23.9)$$

Notice that this field decreases as $1/R$. Since the line is infinite the field never reduces to the $1/r^2$ behavior of a point charge. The field lines are radial at all points along the line of charge. The integration may also be performed with the distance along the line as the variable; see Problem 7.

EXAMPLE 23.8:   *nonconducting disk* of radius a has a uniform surface charge density σ C/m². What is the field strength at a distance y from the center along the central axis?

Solution: The circular symmetry of the disk indicates that an appropriate choice of element is a ring of radius x and width dx, as shown in Fig. 23.20. All points on this ring are equidistant from point P. Consider the component of the field parallel to the plane of the disk. Any contribution to this component from some point on the ring has an equal and opposite contribution from the diametrically opposite part of the ring. Thus, we can say that by symmetry there is no component of the field parallel to the plane of the disk, that is, $E_x = 0$.

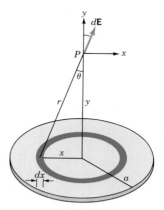

FIGURE 23.20 The field due to a uniformly charged disk. The charge element is a ring of radius x and thickness dx that carries a charge $dq = \sigma \, dA$, where dA is the area of the ring.

The y component of the field is

$$dE_y = dE \cos \theta = \frac{k \, dq}{r^2} \frac{y}{r}$$

where $r^2 = x^2 + y^2$ and $dq = \sigma \, dA = \sigma(2\pi x \, dx)$ is the charge on the elemental ring. Next we integrate over all rings to find the total field:

$$E = \pi k \sigma y \int_0^a \frac{2x\,dx}{(x^2 + y^2)^{3/2}}$$

$$= \pi k \sigma y \int_0^a \frac{d(x^2)}{(x^2 + y^2)^{3/2}}$$

$$= \pi k \sigma y \left[\frac{-2}{(x^2 + y^2)^{1/2}} \right]_0^a$$

$$= 2\pi k \sigma \left[1 - \frac{y}{(a^2 + y^2)^{1/2}} \right] \tag{i}$$

On obtaining such an expression you should ensure that it is dimensionally correct. In particular, the second term in the square brackets should be dimensionless. (Why?) As another check on the correctness of this expression, let us see if it reduces to the field of a point charge for large values of y. We will use the binomial expansion $(1 + z)^n \approx 1 + nz$, for small z (see Appendix B). The second term in the brackets may be written $y(a^2 + y^2)^{-1/2} = (1 + a^2/y^2)^{-1/2}$. When $y \gg a$, the binomial expansion yields

$$\left(1 + \frac{a^2}{y^2} \right)^{-1/2} \approx 1 - \frac{1}{2}\left(\frac{a^2}{y^2} \right) + \cdots$$

On substituting this into (i) and using $Q = \sigma \pi a^2$, we find $E \to kQ/y^2$, which is the field due to a point charge.

EXAMPLE 23.9: Find the field due to the following: (a) an *infinite sheet of charge* with surface charge density $+\sigma$; (b) two parallel infinite sheets with charge densities $+\sigma$ and $-\sigma$.

Solution: (a) We may use the result in Example 23.8 for the electric field due to a disk. In the limit as $a \to \infty$, the second term in Eq. (i) of Example 23.8 vanishes and we are left with $E = 2\pi k \sigma$, or

$$E = \frac{\sigma}{2\varepsilon_0} \tag{23.10}$$

where we have used $k = 1/4\pi\varepsilon_0$. The field is *uniform*; that is, it is constant in both magnitude and direction, as shown in Fig. 23.21.

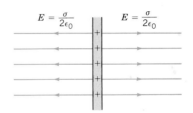

FIGURE 23.21 The field due to an infinite sheet of charge is uniform.

(b) Figure 23.22a shows two infinite parallel sheets with equal and opposite surface charge densities. The individual contributions to the field are dashed or dotted lines. In the region between the sheets the two fields are in the same direction and thus reinforce each other. In the other regions on the left and on the right, the fields cancel. The resultant field between the plates, shown in Fig. 23.22b, is

$$E = \frac{\sigma}{\varepsilon_0} \tag{23.11}$$

The magnitude of this uniform field is just twice that for a single infinite sheet of charge.

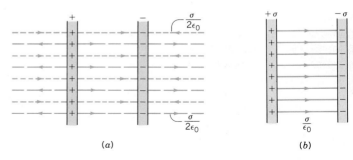

FIGURE 23.22 (a) The fields due to two infinite sheets with equal and opposite charge densities. (b) The resultant field.

23.6 DIPOLES

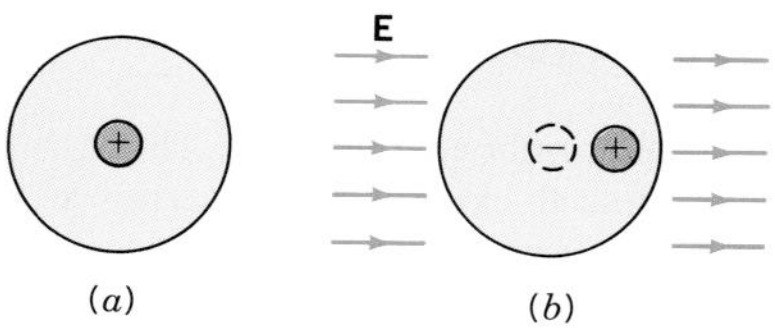

The arrangement of a pair of equal and opposite charges separated by some distance is called an **electric dipole**. Any molecule in which the centers of the positive and negative charges do not coincide may, to a first approximation, be treated as a dipole. Molecules such as HCl, CO, and H_2O have permanent dipoles and are called *polar* molecules. An electric field may also induce a charge separation in an atom or a nonpolar molecule. Figure 23.23a shows an atom as a positive point charge surrounded by a sphere of equal negative charge. When an external electric field is applied, these charges are displaced in opposite directions, as shown in Fig. 23.23b, thereby creating an *induced* dipole. Such an induced dipole vanishes when the external field is removed. We will consider the electric field

FIGURE 23.23 (a) In an atom the positive nucleus is located at the center of the negative charge distribution of the electrons. (b) When an external field is applied, an induced dipole appears.

created by a dipole and the interaction of a dipole with an external field or with other dipoles.

Field Due to a Dipole

Figure 23.24 shows a dipole consisting of charges Q at $(0, a)$ and $-Q$ at $(0, -a)$. We wish to find the electric field along the perpendicular bisector at a distance r from its center. At any point on the x axis, the fields due to the two charges have the same magnitude:

$$E_+ = E_- = \frac{kQ}{r^2 + a^2}$$

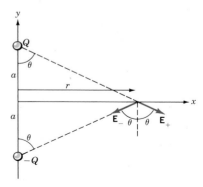

FIGURE 23.24 Calculating the field along the bisector of a dipole.

Since they are inclined at the same angle to the y axis, the x components cancel. The y component of the field is

$$
\begin{aligned}
E_y &= -(E_+ + E_-) \cos \theta \\
&= -\frac{2kQ}{(r^2 + a^2)} \frac{a}{(r^2 + a^2)^{1/2}} \\
&= \frac{-k2aQ}{(r^2 + a^2)^{3/2}}
\end{aligned}
$$

The **electric dipole moment p** is defined as the product of one of the charges and their separation:

$$\mathbf{p} = Q\mathbf{d} \tag{23.12}$$

where $d = 2a$. It is a vector that points from the negative to the positive charge, as shown in Fig. 23.25. The SI unit for the electric dipole moment is $C \cdot m$. When three charges are involved, as in a water molecule (Fig. 23.26), the net dipole moment is the vector sum of two dipole moments. At large distances from the dipole (that is, when $r \gg a$), we may ignore a in comparison with r, and so $(r^2 + a^2)^{3/2} \rightarrow r^3$. The expression for E_y then gives the magnitude of the resultant "far field" at a point along the perpendicular bisector:

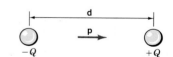

FIGURE 23.25 The dipole moment is defined to be $\mathbf{p} = Q\mathbf{d}$.

(Bisector) $$E = \frac{kp}{r^3} \qquad (r \gg a) \tag{23.13}$$

The resultant field falls off as the inverse *cube* of the distance, which is faster than for a single charge. The reason is that the field components partially cancel because the charges have opposite signs. In Exercise 5 below you are asked to show that the "far field" along the axis of a dipole is given by

(Axis) $$E = \frac{2kp}{r^3} \qquad (r \gg a) \tag{23.14}$$

The field lines for a dipole are displayed in Fig. 23.8b.

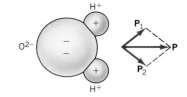

FIGURE 23.26 When more than one dipole is present, the net dipole moment is the vector sum of the individual moments.

EXERCISE 5. Calculate the field at a point on the y axis at a distance r from the center of the dipole. Use your result to derive Eq. 23.14.

Torque in a Uniform Field

Figure 23.27 shows a dipole oriented at an angle θ to a uniform electric field. The charges experience equal and opposite forces due to the field, so there is no net force on the dipole. However, the dipole does experience a torque. The torque

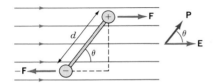

FIGURE 23.27 An electric dipole experiences a torque in an electric field.

due to each force about the center is $\tau_+ = \tau_- = r_\perp F$, where $r_\perp = (d/2) \sin \theta$. These torques are in the same sense, so the magnitude of the net torque is the sum

$$\tau = 2(qE)\left(\frac{d}{2} \sin \theta\right) = pE \sin \theta \qquad (23.15)$$

The torque due to the field tends to align the dipole moment along the field lines. The vector expression for the torque is

$$\tau = \mathbf{p} \times \mathbf{E} \qquad (23.16)$$

Potential Energy

We have seen that a dipole in an external electric field tends to align itself along the electric field. It therefore takes work to rotate the dipole. The work done by an external torque to rotate the dipole from θ_1 to θ_2, with no change in kinetic energy, is $W = \int -\tau d\theta$ (see Section 11.7). The negative sign appears because the external torque is opposite to the torque due to the field. From Eq. 23.15 we have

$$W_{\text{EXT}} = -\int_{\theta_1}^{\theta_2} pE \sin \theta \, d\theta = pE(-\cos \theta_2 + \cos \theta_1)$$

This external work is stored as electrical potential energy: $W_{\text{EXT}} = \Delta U = U_2 - U_1$. Since only changes in potential energy are physically significant, it is convenient to pick $U_1 = 0$ at $\theta_1 = \pi/2$ so that $\cos \theta_1 = 0$. The potential energy of a dipole in an external field then takes the form

$$U = -pE \cos \theta = -\mathbf{p} \cdot \mathbf{E} \qquad (23.17)$$

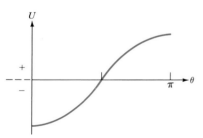

FIGURE 23.28 The potential energy of a dipole as a function of its orientation.

The potential energy as a function of the angle θ is shown in Fig. 23.28. The minimum potential energy occurs at $\theta = 0$, and the maximum at $\theta = \pi$. If the dipole is allowed to rotate, it oscillates about the direction of the field. If there is a mechanism that dissipates its mechanical energy, such as collisions with other molecules or radiation, the dipole will eventually occupy the state of lowest energy. That is, it will align itself along the field. We reached the same conclusion in the earlier discussion based on torque.

The large dipole moment of a water molecule (6.2×10^{-30} C · m) is an important property. For example, when a salt crystal is placed in water, the attraction between the charges of the polar water molecule and the Na$^+$ and Cl$^-$ ions is sufficient to break the ionic bond between these ions. Therefore, salt dissolves easily in water (see Fig. 23.29). The solubility of a liquid in water depends on whether its molecules are polar or nonpolar. If the substance is polar, its dipoles can combine with the dipoles of water in simple patterns. Oils consist of nonpolar molecules and do not mix with water. Microwave cooking depends on the response of water dipoles to an oscillating electric field that reverses direction with a high frequency (2.45×10^9 Hz). As the dipoles vibrate in response to the field, they generate thermal energy in the surrounding medium. Materials such as paper and glass, which have no dipoles that can respond to the field, do not become warm.

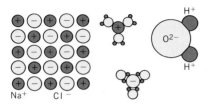

FIGURE 23.29 When a NaCl crystal dissolves, the Na$^+$ and Cl$^-$ ions are attached to water molecules.

The molecules of soaps and detergents are unusual. For example, a soap molecule has a long hydrocarbon chain in which one end is nonpolar, while the other end possesses a dipole moment (see Fig. 23.30). The nonpolar end mixes easily with (nonpolar) fatty acids, while the polar end is attracted by water. When the soapy water is washed away, the oil and grease are washed away with it.

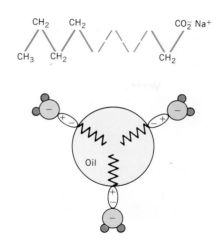

FIGURE 23.30 A soap molecule has a polar end and a nonpolar end. The nonpolar end combines with an oil drop, whereas the polar end combines with a water molecule.

23.7   DIPOLE IN A NONUNIFORM FIELD (Optional)

When a dipole is placed in a nonuniform electric field, the net force on it is not zero. In Fig. 23.31, the field is stronger at the position of the positive charge. If E_+ and E_- are the field strengths at the positive and negative charges, respectively, the net force on the dipole is

$$F = q(E_+ - E_-) = q\Delta E$$

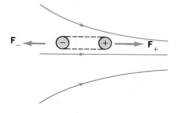

FIGURE 23.31 A dipole experiences a net force in a nonuniform field.

If the dipole is oriented along the x axis, we may write $p = q\Delta x$; then the force may be written as $F = p\Delta E/\Delta x$. In the limit as $\Delta x \to 0$, we have*

$$F_x = p \frac{dE}{dx} \qquad (23.18)$$

If dE/dx is positive, the force is in the $+x$ direction. We see that a neutral body may experience a net force in an electric field provided that it has a dipole moment and the field is nonuni-

---

* This expression may also be derived from $U = -\mathbf{p} \cdot \mathbf{E}$ and the relation $F_x = -dU/dx$.

form. This is how a charged comb can attract a neutral piece of paper. The field due to the comb induces a charge separation in the paper (see Fig. 23.32). Because the field due to the comb is not uniform, the positive and negative charges on the paper do not experience equal forces. The net force on the paper is directed toward the comb.

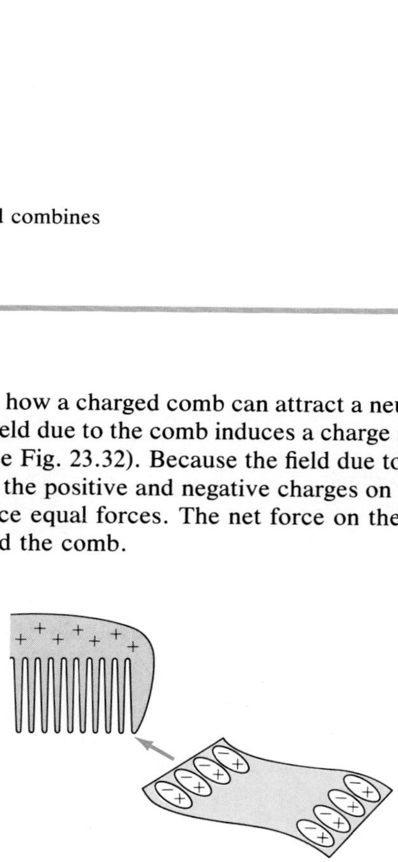

FIGURE 23.32 A charged comb induces dipoles in a piece of paper (an insulator). Since the field produced by the comb is nonuniform, the paper experiences a net attractive force.

Interaction between Dipoles

The interaction between induced dipoles in neutral atoms is responsible for the weak form of bonding called the **van der Waals** force. A random fluctuation in the charge distribution of one atom may result in a temporary dipole moment, say p_1. At large distances, the field along the axis due to this dipole is given by Eq. 23.14:

(Axis)
$$E_1 = \frac{2kp_1}{x^3}$$

As Fig. 23.33 shows, this field induces a dipole moment p_2 in a nearby atom. The second atom therefore experiences a force

$$F_2 = p_2 \frac{dE_1}{dx}$$

directed toward the first atom. The induced dipole moment is proportional to the external field strength; that is, $p_2 \propto E_1$. Since $dE_1/dx \propto 1/x^4$, we see that the force of interaction between the two dipoles varies as

$$F \propto \frac{1}{x^7}$$

The nonuniformity of the fields due to these dipoles results in a net attractive force between the uncharged atoms. This van der Waals force plays a role in the condensation of gases into liquids. The ease with which mica can be cleaved into planes is explained by the presence of this form of bonding rather than other, stronger, bonds.

FIGURE 23.34 Robert A. Millikan.

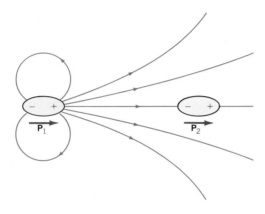

FIGURE 23.33 The field due to one dipole can induce a dipole in a nearby atom or molecule. As a result, there is an attractive force between them.

23.8 MILLIKAN OIL DROP EXPERIMENT (Optional)

In Section 22.1 we noted that charge is quantized; that is, it appears only in multiples of a basic unit, e. The experiment that first established this fact definitively was performed in 1909 by R. A. Millikan (Fig. 23.34). In Millikan's **oil drop experiment** an electric field is set up between two plates, as in Fig. 23.35. Oil from an atomizer (originally a simple perfume sprayer) is sprayed above the top plate and becomes charged by friction. A few drops fall through the hole in the top plate and can be viewed with a telescope. The potential difference between the plates is adjusted till a particular drop is held at rest. At this point the electrical and gravitational forces balance: $qE = mg$.

The mass of a drop is determined from the density of the oil and its radius, r. However, since the drops are very small ($r \approx 2.5 \times 10^{-5}$ cm) their radii must be measured indirectly. The field is switched off and the drops are allowed to fall subject to the viscous drag of the air, which is approximately given by Eq. 6.7: $F_D = \gamma v$, where v is the speed of the drop and γ is a constant that depends on the radius of the drop and the flow resistance (viscosity) of the air. By measuring the terminal speed, one can determine r and then m (see Problem 15). Finally, the charge is given by $q = mg/E$.

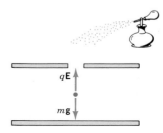

FIGURE 23.35 In Millikan's oil drop experiment, charged drops are suspended in the field set up by two plates.

After hundreds of measurements Millikan found that the charges on the drops were always integer multiples of the basic unit, $e = 1.602 \times 10^{-19}$ C. That is, charge is **quantized**. The delicacy of his work may be appreciated when one realizes that if a body charged by rubbing acquires even a tiny charge of 10^{-10} C, this involves a *billion* electronic charges. A few million here or there would not make much difference. Millikan was able to make measurements with drops that had a *single* electronic charge!

SUMMARY

An electric charge influences its surroundings by creating an **electric field.** The **electric field strength** at a point is defined as the force per unit charge on a test charge q_t placed at that point: $\mathbf{E} = \mathbf{F}/q_t$. The direction of $\mathbf{E}$ is that of the force on a positive charge. Once the field strength is known, the force on any charge q can be found from

$$\mathbf{F} = q\mathbf{E}$$

From Coulomb's law we find the electric field created by a point charge Q is

$$\mathbf{E} = \frac{kQ}{r^2}\,\hat{\mathbf{r}}$$

where the unit vector $\hat{\mathbf{r}}$ has its origin at the source charge Q. When several charges are present, the total field is given by the principle of linear superposition: $\mathbf{E} = \Sigma\mathbf{E}_i$. When the charge distribution is continuous, the field is found by integrating over the charge distribution:

$$\mathbf{E} = k\int \frac{dq}{r^2}\,\hat{\mathbf{r}}$$

Electric **field lines** are helpful in visualizing field patterns. They provide the following basic information: (a) The direction of the field is along the tangent to a line, and (b) the strength or magnitude of the field is proportional to the number of lines that cross a unit area perpendicular to the lines. Other properties of electric field lines and the approach to plotting them are outlined on page 451.

Two equal and opposite point charges $+q$ separated by a distance d form an **electric dipole.** The electric **dipole moment** is defined to be

$$\mathbf{p} = q\mathbf{d}$$

where the displacement $\mathbf{d}$, and hence also $\mathbf{p}$, points from the negative to the positive charge. In an external field, a dipole experiences a torque given by

$$\boldsymbol{\tau} = \mathbf{p} \times \mathbf{E}$$

This torque tends to align the dipole moment parallel to the field. The potential energy of a dipole in an external field is

$$U = -\mathbf{p} \cdot \mathbf{E}$$

The minimum of potential energy occurs when $\mathbf{p}$ is parallel to $\mathbf{E}$.

ANSWERS TO IN-CHAPTER EXERCISES

1. (a) $\mathbf{E} = \mathbf{F}/q = 25\mathbf{i}$ N/C. (b) Nothing. (c) $\mathbf{F}$ is in the opposite direction. (d) Nothing.
2. The direction of the field at a given point is along the tangent to a field line. The lines do not cross because the field cannot be in two directions at the same point.
3. No. For example, the field due to a point charge decreases as the inverse-square of the distance from the charge.
4. No. The charged particle experiences an acceleration that varies both in magnitude and in direction. Is there any case in which a test charge would travel along field lines?

5. The resultant field is

$$E = \frac{kQ}{(r - a)^2} - \frac{kQ}{(r + a)^2} = \frac{2kpr}{(r^2 - a^2)^2}$$

where $p = 2aQ$. When $r \gg a$, we obtain Eq. 23.14.

QUESTIONS

1. In Example 23.3 we found a point at which $E = 0$. Would a charge placed at this point be in stable equilibrium?

2. An electric field is created by a set of fixed charges. When an external charge is introduced into the region, the field lines are modified. Should we use the original lines or the modified ones to determine the direction of the force on the new charge?

3. Electrostatic field lines begin on positive charges and end on negative charges. What happens to the lines due to an isolated charge?

4. A point charge is placed at the center of an uncharged metal cube. Sketch the field lines within the cube in a plane parallel to a face and containing the charge.

5. Explain qualitatively why the field due to an infinite sheet of charge is uniform.

6. Four equal point charges are at the corners of a square. Where, besides infinity and the center of the square, is the field strength zero?

7. In what ways are Coulomb's law and Newton's law of gravitation similar? In what ways are they different? Consider the laws themselves and how they are applied.

8. Does the gravitational field ever display a dipole pattern? If so, give an example of how this can arise.

9. Name two fields that you encounter in everyday life that are (a) scalar, and (b) vector.

10. What is the work needed to rotate an electric dipole in a uniform electric field by 180° in each of the following cases: (a) 0 to 180°; (b) −90° to +90°? The angles are relative to **E**.

EXERCISES

23.1 The Electric Field

1. (I) What is the electric field strength needed to balance the weight of the following particles near the earth's surface: (a) an electron; (b) a proton?

2. (I) In fine weather there is a field of 120 N/C directed downward near the surface of the earth. (a) What is the electrical force on a proton in such a field? (b) What is the acceleration of the proton?

3. (I) A point charge $q_1 = 3.2$ nC experiences a force $\mathbf{F} = 8 \times 10^{-6}\mathbf{i}$ N. (a) What is the external electric field strength? (b) What would be the force on a point charge $q_2 = -6.4$ nC placed at the same point?

4. (I) A point charge $-4q$ is at $x = 0$ while a second charge is at $x = 1$ m. Besides infinity, where is the field strength zero given that the second charge is (a) $9q$, (b) $-q$?

5. (I) Four point charges are located at the corners of a square of side L, as shown in Fig. 23.36. Find the electric field strength at the point (a) A at the center, and (b) B.

6. (I) A point charge Q_1 is at the origin and $-Q_2$ is at $x = 2$ m. The field strength at $x = 1$ m is $10.8\mathbf{i}$ N/C and at $x = 3$ m, it is $-0.8\mathbf{i}$ N/C. Find Q_1 and Q_2.

7. (I) A droplet has a mass of 10^{-13} kg and a charge $+2e$. In what vertical electric field would the droplet be in equilibrium near the earth's surface?

8. (I) A point charge $q_1 = 3$ nC is at the origin and $q_2 = -7$ nC is at $x = 8$ cm. (a) Find the field strength due to q_1 at the position of q_2. (b) Find the field strength due to q_2 at the position of q_1. (c) What is the force exerted by q_1 on q_2? (d) What is the force exerted by q_2 on q_1?

9. (I) A point charge of -5 μC is at the origin. Find the electric field strength at the following points: (a) (2 m, -1 m); (b) (-2 m, 3 m).

10. (I) A point charge $Q_1 = -4$ μC is at (2 m, 1 m) while $Q_2 = +15$ μC is at (1 m, 4 m). Find the field strength at (3 m, 5 m).

11. (I) Three point charges are placed at the corners of an equilateral triangle as in Fig. 23.37. (a) Find the field strength due to the charges -2 μC and $+4$ μC at the origin. (b) What is the force on the -3 μC charge? (c) If the sign of the charge at the origin is changed, how is the field strength calculated in part (a) affected?

12. (II) A point charge Q_1 is at $x = 0$ and Q_2 is at $x = d$. What is the relation between these charges if the resultant field strength is zero at the following points: (a) $x = d/2$; (b) $x = 2d$; (c) $x = -d/2$?

13. (II) A point charge Q is at the origin. Show that the components of the electric field at a distance r are given by

$$E_\alpha = \frac{kQ\alpha}{r^3}$$

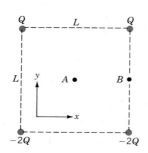

FIGURE 23.36 Exercise 5.

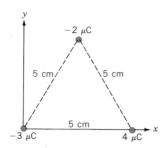

FIGURE 23.37 Exercise 11.

where $\alpha = x$, y, or z. (*Hint:* Note that $\mathbf{E} = E_x\mathbf{i} + E_y\mathbf{j} + E_z\mathbf{k} = (kQ/r^3)\mathbf{r}$. Use the dot product.)

14. (I) The radius of the proton is 0.8×10^{-15} m. (a) What is the field strength at its surface? (b) What is the field strength at a distance of 0.53×10^{-10} m—which is the position of the electron in the hydrogen atom?

15. (II) A point charge q is at $x = 0$, while $-q$ is at $x = 6$ m. Calculate the field strength as a function of x for both positive and negative values of x at 0.5-m intervals. Make a rough sketch of $E(x)$.

16. (II) A point charge $2q$ is at $x = 0$, while $-q$ is at $x = 6$ m. (a) Calculate the field strength as a function of x for both positive and negative values of x at 0.5-m intervals. (b) Where is $E = 0$? (c) Make a rough sketch of $E(x)$.

17. (I) A point charge of $2 \ \mu C$ is placed at the origin. There is an external uniform field $\mathbf{E} = 500\mathbf{i}$ N/C. What is the net force on a 5-μC charge placed at (3 m, 4 m)?

18. (I) Point charges $Q_1 = 25 \ \mu C$ and $Q_2 = -50 \ \mu C$ are placed on the y axis as shown in Fig. 23.38. A point charge $q = 2 \ \mu C$ is on the x axis. (a) Find the field strength due to Q_1 and Q_2 at the position of q. (b) If q is halved, what happens to the field calculated in part (a)? (c) What happens to the field found in part (a) if the sign of q is changed?

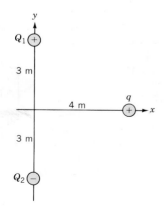

FIGURE 23.38 Exercise 18.

19. (II) Figure 23.39 shows a combination of charges called an electric quadrupole. Find the electric field strength (a) at point A at $(x, 0)$, and (b) at point B at $(0, y)$. (c) Show that in either case $E \propto 1/r^4$ for $r \gg a$, where r is the distance from the origin. (*Hint:* Use the binomial expansion $(1 + z)^n \approx 1 + nz$, for small z.)

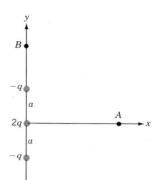

FIGURE 23.39 Exercise 19.

23.2 Lines of Force

20. (II) Three point charges are at the corners of an equilateral triangle. Two of the charges are equal to q and the third charge is $-q$. Sketch the field lines. (Is there a point where $\mathbf{E} = 0$?)

21. (II) Sketch the resultant field lines for a point charge placed in front of a uniformly and positively charged infinite plane. Take the point charge to be: (a) positive; (b) negative.

22. (II) Sketch the field lines for a pair of charges $+3q$ and $-q$.

23. (I) Sketch the field lines for a uniformly charged finite disk in a plane perpendicular to the plane of the disk that passes through the center. (Consider the shape of the lines close to the center and far from the disk.)

24. (II) Sketch the field lines for a pair of charges $+2q$ and $+q$.

25. (II) Two equal positive charges $+Q$ are placed at opposite ends of a diagonal of a square. Two negative charges $-Q$ are at the ends of the other diagonal. Sketch the field lines.

26. (II) A 16-μC charge is placed at the center of a spherical metallic shell that carries $-8 \ \mu C$. What are the charges on the inner and outer surfaces of the shell? Sketch the field lines both inside and outside the shell. (See Section 23.3.)

23.4 Motion of Charges

27. (I) An electron is accelerated from rest by a uniform field of magnitude 10^5 N/C. (a) How long does it take to reach $0.1c$, where $c = 3 \times 10^8$ m/s is the speed of light? (b) How far does it travel in this time? (c) What is its final kinetic energy?

28. (I) An electron in a TV tube is accelerated from rest to 5×10^6 m/s by a uniform electric field in a distance of 1.6 cm. What is the field strength?

29. (I) A proton is fired with an initial speed of 8×10^5 m/s in a direction opposite to a uniform field of 2.4×10^4 N/C. (a) How far does it travel before coming to rest? (b) How long does it take to come to rest?

30. (I) An electron enters the region between two horizontal plates with an initial velocity of $2 \times 10^6 \mathbf{i}$ m/s midway between two horizontal plates; see Fig. 23.40. The plates are 4 cm long and 1.6 cm apart. What is the magnitude of the maximum vertical electric field such that the electron does not strike either plate?

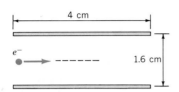

FIGURE 23.40 Exercise 30.

31. (I) Air breaks down—that is, the molecules rapidly become ionized and produce a spark—when the field strength is about 3×10^6 N/C. Given such a field, find: (a) how long it would take an electron starting from rest to acquire a kinetic energy of 4×10^{-19} J needed to knock out an electron from a molecule; (b) how far the electron would travel.

32. (II) A positron is a particle that has the same mass as the electron but a charge $+e$. An electron and a positron orbit around their center of mass. The radius of the orbit is 0.5×10^{-10} m. Find: (a) the speed of each particle; (b) the period.

33. (II) An electron is fired with an initial speed v_0 at 45° to the horizontal from the bottom plate of a parallel-plate arrangement as shown in Fig. 23.41. The plates are separated by 2 cm and are very long. What is the maximum value of v_0 for the electron not to hit the upper plate? Take $E = 10^3$ N/C.

FIGURE 23.41 Exercise 33.

34. (II) A uniform field $E = -10^5 \mathbf{j}$ N/C exists between two plates of length 4 cm, as shown in Fig. 23.42. A proton is fired at 30° to the x axis with an initial speed of 8×10^5 m/s. Find: (a) its vertical coordinate as it emerges from the plates; (b) the angle at which it emerges.

23.5 Continuous Charge Distributions

35. (I) The field on either side of an infinite sheet of charge of

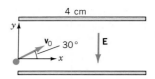

FIGURE 23.42 Exercise 34.

density σ C/m^2 is $E = \sigma/2\varepsilon_0$. Use this result to find the field in the four regions indicated in Fig. 23.43.

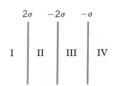

FIGURE 23.43 Exercise 35.

36. (II) A large, uncharged metal plate is placed with its plane perpendicular to the lines of a uniform field of value 1000 N/C. What is the surface charge density on each face of the plate?

37. (II) A point charge $q = 2$ μC is at a distance $d = 20$ cm from a uniformly charged infinite sheet with a surface charge density $\sigma = 20$ μC/m^2. (a) What is the force on the point charge? (b) At what point(s) is the resultant field strength zero?

38. (II) A rod of length 10 cm has a linear charge density 2 μC/m. What is the field strength along the axis at a distance 20 cm from the center?

39. (II) A disk of radius 4 cm has a uniform charge density 5 μC/m^2. What is the field strength at a point along its central axis at a distance 10 cm from the center?

40. (II) Two infinite lines of charge with equal linear charge densities λ C/m are placed along the x and y axes, as in Fig. 23.44. What is the electric field strength at an arbitrary point (x, y)?

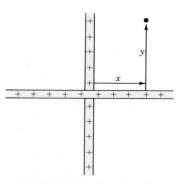

FIGURE 23.44 Exercise 40.

41. (II) Two finite rods of length L are uniformly, and oppositely, charged. They lie along the x and y axes with their ends at a distance d from the origin, as shown in Fig. 23.45. What is the field strength at the origin? Take $Q = 0.2 \ \mu C$, $L = 5$ cm, and $d = 1$ cm.

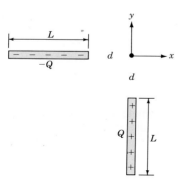

FIGURE 23.45 Exercise 41.

23.6 Dipoles

42. (I) A dipole for which $p = 3.8 \times 10^{-30}$ C · m is in a uniform field $E = 7 \times 10^4$ N/C. (a) What is the external work needed to rotate the dipole from alignment with the field to an angle 60° to the field? (b) When it is at 60°, what is the magnitude of the torque exerted by the field?

43. (I) A dipole consists of two point charges ± 2 nC separated by 4 cm. (a) What is the dipole moment? (b) What is the change in potential energy when the dipole rotates from alignment along a field $E = 10^5$ N/C to an orientation 90° to **E**?

44. (II) The water molecule has a dipole moment $p = 6.2 \times 10^{-30}$ C · m. Find the force on an ion of charge $+e$ at a distance of 0.5 nm: (a) along **p**; (b) normal to **p**. (Use the far field approximation.)

PROBLEMS

1. (I) Use the fact that the electrostatic field is conservative to show that the lines at the ends of two oppositely charged plates cannot end abruptly, as shown in Fig. 23.46. Sketch the proper lines. (*Hint:* A force is conservative if the work done around a closed path is zero: $\oint \mathbf{F} \cdot d\boldsymbol{\ell} = 0$.)

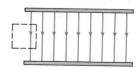

FIGURE 23.46 Problem 1.

2. (I) A circular ring of radius R has a linear charge density λ C/m; see Fig. 23.47. (a) What is the field strength along the axis at a distance x from the center? (b) For what value of x is the field a maximum? (c) How should the field strength vary for $x \gg R$? Verify that your function meets this criterion. (d) Sketch qualitatively the field strength as a function of x.

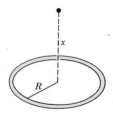

FIGURE 23.47 Problem 2.

3. (I) A dipole whose dipole moment is **p** lies along the x axis

$(\mathbf{p} = p\mathbf{i})$ in a nonuniform field $\mathbf{E} = C/x \ \mathbf{i}$. What is the force on the dipole?

4. (I) A uniformly charged rod with a linear charge density λ C/m is in the form of a circular arc of radius R, as shown in Fig. 23.48. (a) What is the field strength at the center? (b) Show that the field strength at the center of a uniformly charged semicircular rod is $2k\lambda/R$.

FIGURE 23.48 Problem 4.

5. (I) A point charge Q_1 is at $(-a, 0)$ and charge Q_2 is at $(a, 0)$. Sketch qualitatively the variation of the field strength along the x axis given: (a) $Q_1 = Q_2$; (b) $Q_1 = -Q_2$.

6. (II) The $1/r^2$ form of Coulomb's law implies the following: (i) The electric field is zero at all points inside a uniformly charged shell. (ii) The electric field outside a uniformly charged sphere can be found by treating the charge as being concentrated at the center. Use these facts to show that within a uniformly charged sphere of radius R having a volume charge density ρ C/m^3, the field strength increases linearly with the distance r from the center. That is, $E \propto r$ for $r < R$.

7. (II) Show that the field strength at a distance y along the perpendicular bisector of a uniformly charged rod of length L, as in Fig. 23.49 is given by

$$E = \frac{2kQ}{y(L^2 + 4y^2)^{1/2}}$$

(a) What is the form of this expression when $y \gg L$? (b) What is the form when $y \ll L$? (Refer to the table of integrals in Appendix C.)

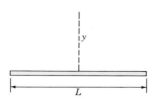

FIGURE 23.49 Problem 7.

8. (I) Use the results obtained in Example 23.5 to show that after emerging from the plates the electrons may be considered as having traveled in a straight line starting at the midpoint of the plates.

9. (II) Use the result $E = 2k\lambda/r$ for the field due to an infinite line of charge to obtain the result $E = \sigma/2\varepsilon_0$ for the field due to an infinite plane of charge with surface charge density σ. (Note: $\int dx/(a^2 + x^2) = (1/a) \tan^{-1}(x/a)$.)

10. (II) A charge $-q$ is in a circular orbit of radius R around an infinite line of charge with linear charge density λ C/m. The plane of the orbit is perpendicular to the line. Obtain an expression for the period.

11. (II) A dipole with a dipole moment p is pivoted freely at its midpoint. It lies in a uniform field. If its moment of inertia about the center is I, show that for small angular displacements, the dipole oscillates at a frequency

$$f = \frac{1}{2\pi} \sqrt{\frac{pE}{I}}$$

12. (II) A semi-infinite line of charge has a uniform charge density λ C/m. Find the field strength at a distance R from its end: (a) along the axis; (b) perpendicular to the axis. See Fig. 23.50.

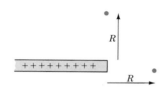

FIGURE 23.50 Problem 12.

13. (I) Two equal charges lie on the y axis, as shown in Fig. 23.51. (a) Find the field strength at the point $(x, 0)$. (b) What

is the form of $E(x)$ for $x \gg a$? (c) At what point is $E(x)$ a maximum?

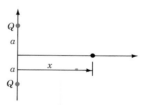

FIGURE 23.51 Problem 13.

14. (II) The field strength at a distance x from the center along the axis of a circular ring of radius R and total charge Q is given by

$$E(x) = \frac{kQx}{(x^2 + R^2)^{3/2}}$$

(a) Use the binomial expansion to obtain a simplified expression when $x \ll R$. (b) Show that a charge $-q$ would undergo simple harmonic motion for small displacements from the center along the axis. (c) Show that the angular frequency of the oscillation is

$$\omega = \sqrt{\frac{kqQ}{mR^3}}$$

15. (II) In Millikan's oil drop experiment, the drops are first held motionless by the application of a uniform field E. Next, the field is switched off and the drops are allowed to fall in air until they reach the terminal speed v_T. The fluid resistance is given by Stokes law, $F = 6\pi\eta r v_T$, where η is the coefficient of viscosity and r is the radius. The condition for falling at the terminal speed is

$$6\pi\eta r v_T = m_{eff} g$$

The effective mass of a drop is $m_{eff} = \frac{4}{3}\pi r^3 (\rho - \rho_A)$ where ρ is the density of the drop and ρ_A is the density of the air—which has a buoyant effect. Show that the charge on a drop is given by

$$q = \frac{18\pi}{E} \sqrt{\frac{\eta^3 v_T^3}{2(\rho - \rho_A)g}}$$

16. (II) An electron is fired at $\theta = 30°$ to the horizontal midway between two 4-cm-long horizontal plates separated by 1 cm, as shown in Fig. 23.52. Find the minimum and maximum values of the initial speed v_0 for the electron not to hit either plate.

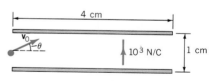

FIGURE 23.52 Problems 16 and 17.

17. (I) An electron is fired with an initial speed $v_0 = 3 \times 10^6$ m/s midway between two 4-cm-long horizontal plates, as shown in Fig. 23.52. For what initial angle will the electron emerge midway between the plates?

18. (I) Two uniformly charged rods, bent into circular arcs, have equal and opposite linear charge densities $\pm\lambda$ C/m. Their ends are placed together so that they form a semicircle of radius R as shown in Fig. 23.53. What is the electric field strength at the center?

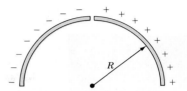

FIGURE 23.53 Problem 18.

CHAPTER 24

Gauss's Law

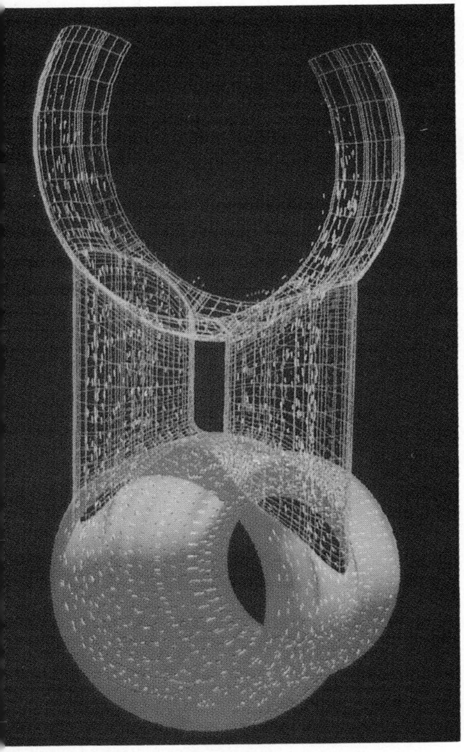

A computer-operated surface representing the hot gas manifold of the Space Shuttle's main engine.

Major Points

1. The concept of **electric flux.**
2. **Gauss's law** relates the flux through a closed surface to the net charge enclosed by it.
3. Any net charge on a conductor in electrostatic equilibrium resides on the surface.

In principle, the electrostatic field due to a continuous charge distribution can always be found by using Coulomb's law, but the integration required may be complex. In this chapter we present an alternative approach, based on the concept of lines of force, which, in some cases, can be much simpler. Although Michael Faraday had established the usefulness of field lines in visualizing the field, he did not express the idea in mathematical form. The mathematician Carl F. Gauss (Fig. 24.1) later put the concept of field lines into quantitative form. He built on the picture of lines "flowing" through a closed surface by introducing a quantity called **flux** and related it to the net charge enclosed by the surface. **Gauss's law** is a general statement about the properties of electric fields; it is not restricted to electrostatic fields as is Coulomb's law. When a charge distribution has sufficient symmetry, Gauss's law can provide an elegant way to determine the electrostatic field in a few simple steps.

24.1 ELECTRIC FLUX

One can think loosely of field lines crossing an area as being analogous to streamlines in a fluid flowing through an area. Gauss developed this analogy by defining a quantity called **electric flux.** Figure 24.2 shows a flat surface of area A set perpendicular to the lines of a uniform electric field. The electric flux Φ_E through this surface is defined as

$$\Phi_E = EA$$

The SI unit of electric flux is $N \cdot m^2/C$. Although the definition of flux does not involve lines of force, the electric flux through a given surface is proportional to the number of field lines passing through it. If the surface is inclined at some angle to the field, as in Fig. 24.3, the number of lines intercepted is determined by A_n,

FIGURE 24.1 Carl F. Gauss (1777–1855).

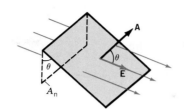

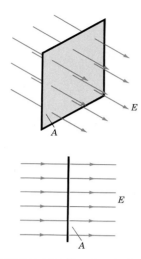

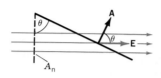

FIGURE 24.2 The electric flux through a flat surface of area A is $\Phi_E = EA$.

FIGURE 24.3 If the surface is at an angle to the field, the electric flux is $\Phi_E = EA \cos \theta$.

the projection of the area normal to the lines. Or, equivalently, the flux is determined by the component of **E** normal to the surface. Thus,

$$\Phi_E = EA_n = E_n A$$

The orientation of the surface may be specified by the vector **A** whose magnitude is equal to A and whose direction is perpendicular to the plane of the surface. The direction of **A** is still ambiguous; for the moment we choose it such that Φ_E is positive. Both of the expressions given above then reduce to

$$\Phi_E = EA \cos \theta$$

where θ is the angle between **A** and **E.** In terms of the scalar product, the flux associated with a uniform field is

(Uniform E) $$\Phi_E = \mathbf{E} \cdot \mathbf{A} \qquad (24.1)$$ Electric flux in a uniform field

Equation 24.1 has to be modified when the field is not uniform or the surface is not flat. In this case, the surface may be divided into tiny elements of area $\Delta \mathbf{A}$ that are essentially flat, as shown in Fig. 24.4. Even if the field is not uniform, it should not change significantly over each elemental area. The total flux through the surface is the sum

$$\Phi_E \approx \mathbf{E}_1 \cdot \Delta \mathbf{A}_1 + \mathbf{E}_2 \cdot \Delta \mathbf{A}_2 + \cdots + = \Sigma \mathbf{E}_i \cdot \Delta \mathbf{A}_i$$

In the limit as $\Delta A \to 0$, the approximate, discrete sum becomes an exact, continuous integral. Thus, the general definition of electric flux is

$$\Phi_E = \int \mathbf{E} \cdot d\mathbf{A} \qquad (24.2)$$

The right side of Eq. 24.2 is called a surface integral and is quite difficult to evaluate for an arbitrary surface or electric field. However, when the charge distribution has a high degree of symmetry, a careful choice of the surface over which the integral is performed can greatly simplify the calculations.

Figure 24.5 shows field lines passing through an (imaginary) closed surface. The direction of the vector $d\mathbf{A}$ at a given location is defined to be in the direction

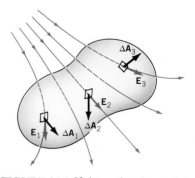

FIGURE 24.4 If the surface is not flat or the field is not uniform, one must sum the contributions to the flux from elements of area that may be treated as flat and over which the field has a single value.

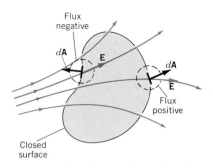

FIGURE 24.5 Flux leaving a closed surface is positive, whereas flux entering a closed surface is negative.

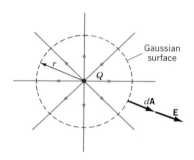

FIGURE 24.6 A spherical Gaussian surface around a point charge.

of the *outward* normal to the surface. Flux leaving a closed surface is positive, whereas flux entering a closed surface is negative. In Fig. 24.5, the net flux through the surface is zero since the number of lines that enter the surface is equal to the number that leave.

EXERCISE 1. The plane of a circle of radius 8 cm is at 40° to a uniform 600-N/C electric field. What is the flux through the circle?

24.2 GAUSS'S LAW

Consider a positive point charge Q, as in Fig. 24.6. From the symmetry of the situation, we see that the field strength will have the same value at all points on an imaginary sphere centered on the charge. Furthermore, each element of area (treated as a vector perpendicular to its plane) is parallel to the local field, and therefore $\mathbf{E} \cdot d\mathbf{A} = E \, dA$. The total flux through this closed *Gaussian* surface is

$$\Phi_E = \oint E \, dA = E \oint dA = E(4\pi r^2)$$

where E was taken out of the integral because the magnitude of the field is constant over the surface. The integral therefore reduces to a sum of the area elements, which is simply the surface area of a sphere, $4\pi r^2$. From Coulomb's law we know that $E = kQ/r^2$, so $\Phi_E = 4\pi kQ$. The factor of 4π can be eliminated by substituting $k = 1/4\pi\varepsilon_0$. The total flux then takes the form

$$\Phi_E = \frac{Q}{\varepsilon_0}$$

The flux through the closed surface is $1/\varepsilon_0$ times the charge enclosed by the surface. If we define the number of lines emerging from a point charge Q to be Q/ε_0, then we may say that the flux is *equal* to the number of lines.[*] The radius of the sphere does not appear in Φ_E because the radial dependence of the field ($E \propto 1/r^2$) exactly compensates for the increase in area ($A \propto r^2$). As a result, the number of lines passing through a large sphere is exactly equal to the number through a small sphere. Gauss's law for the electrostatic field can be formulated precisely because of the inverse square nature of the field.

Suppose the Gaussian surface has some arbitrary shape, as in Fig. 24.7. Although the integration required to determine the flux would be difficult to perform, one can see immediately that the number of lines, and hence also the flux, through this surface is exactly the same as that for a sphere. A proof of this point is provided in Section 24.4.

Consider the field due to two point charges, as in Fig. 24.8. The (positive) flux through surface S_1 around $2Q$ is twice the (negative) flux through surface S_2 around $-Q$. The net flux through surface S_3 around both charges is equal to that for the net charge $2Q - Q = +Q$ enclosed by S_3. Gauss's law expresses these results in the equation

Gauss's law

$$\oint \mathbf{E} \cdot d\mathbf{A} = \frac{Q}{\varepsilon_0} \qquad (24.3)$$

The net flux through a closed surface equals $1/\varepsilon_0$ times the net charge enclosed by the surface.

---

[*] Do not be concerned about a fractional number of lines. The electric flux is the real physical quantity, whereas the field lines are drawn merely to help us visualize the field.

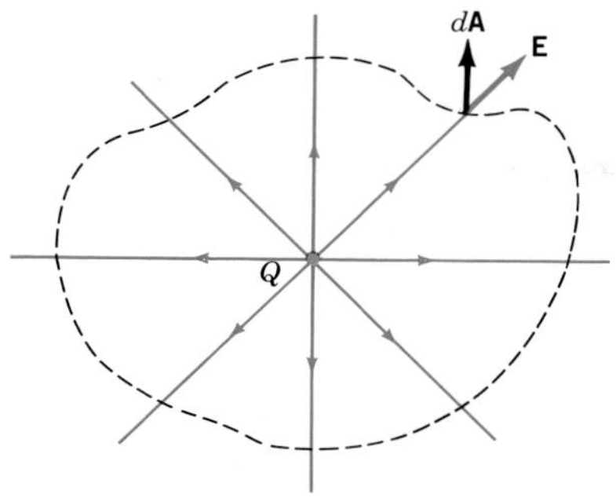

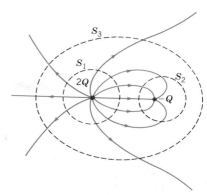

FIGURE 24.7 A Gaussian surface of arbitrary shape around a point charge. The flux through this surface is the same as that for a spherical surface around the charge.

FIGURE 24.8 The flux through a surface is determined by the *net* charge enclosed.

The circle on the integral sign indicates that the Gaussian surface must be closed. Note that Gauss's law does not depend on the exact location of the charges within the surface. The field that appears in Eq. 24.3 is the *total* field due to all charges, not just those within the Gaussian surface. If the enclosed charge is zero, it does *not* necessarily follow that **E** = 0 on the Gaussian surface. The field could be created by charges *outside* the surface, as shown in Fig. 24.5. However, the external charges do not contribute any *net* flux through the surface.

Problem-Solving Guide for Gauss's Law

Gauss's law in the integral form of Eq. 24.3 is useful in determining an electrostatic field whenever the charge distribution has enough symmetry to make the integration simple. In choosing a Gaussian surface, it helps to keep three points in mind:

1. Use the symmetry of the charge distribution to determine the pattern of the field lines.

2. Choose a Gaussian surface for which **E** is either parallel to $d\mathbf{A}$ or perpendicular to $d\mathbf{A}$.

3. If **E** is parallel to $d\mathbf{A}$, then the magnitude E should be constant over this part of the surface. The integral then reduces to a sum over area elements.

The approach is illustrated in the next few examples.

EXAMPLE 24.1: *spherical shell* of radius R has charge Q uniformly distributed over its surface. Find the field at points (a) outside, and (b) inside the shell.

Solution: (a) *Outside:* Since the charge distribution has spherical symmetry, the field is also spherically symmetric. The lines must point radially outward. Also, the field strength will have the same value at all points on any imaginary spherical surface concentric with the charged shell. This symmetry leads us to choose the Gaussian surface to be a sphere of radius $r > R$, as in Fig. 24.9. Any arbitrary element of area $d\mathbf{A}$ is parallel to the local **E**, so $\mathbf{E} \cdot d\mathbf{A} = E \, dA$ at all points on the surface. From Eq. 24.3

$$\oint E \, dA = E \oint dA = E(4\pi r^2) = \frac{Q}{\varepsilon_0}$$

Therefore,

$$E = \frac{Q}{4\pi\varepsilon_0 r^2} = \frac{kQ}{r^2} \qquad (24.4)$$

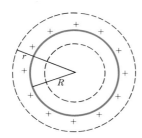

FIGURE 24.9 The Gaussian surface for a spherically symmetric charge distribution is a sphere.

For points outside the shell, the field is the same as that of a point charge at the center. This derivation is almost trivial in comparison to the integration that is required in a direct application of Coulomb's law. Recall that the analogous problem in the context of gravitation (the point mass theorem, Example 13.5) was an enormous obstacle to Newton.

(b) *Inside:* The field still has the same symmetry and so we again pick a spherical Gaussian surface, but now with radius r less than R. Since the enclosed charge is zero, Eq. 24.3 becomes

$$E(4\pi r^2) = 0$$

Since r can have any value, we conclude that $E = 0$ at *all* points inside a uniformly charged spherical shell. This result is a direct consequence of the inverse square nature of Coulomb's law.

EXAMPLE 24.2: A nonconducting *uniformly charged sphere* of radius R has a total charge Q uniformly distributed throughout its volume. Find the field (a) inside, and (b) outside, the sphere.

Solution: (a) *Outside:* For points outside the sphere the situation is identical to the previous example. Since the charge enclosed by a spherical Gaussian surface of radius $r > R$ is Q, we have

$$E(4\pi r^2) = \frac{Q}{\varepsilon_0}$$

from which

$$E = \frac{kQ}{r^2} \tag{24.5}$$

The field at points outside the sphere is the same as that of a point charge at the center. Notice that this result depends only on the spherical symmetry of the charge distribution, not on the fact that it is uniform.

(b) *Inside:* We choose a spherical Gaussian surface of radius $r < R$, as shown in Fig. 24.10. The charge enclosed by a sphere is proportional to its volume, $4\pi r^3/3$. Thus, the fraction of the total charge Q within the Gaussian surface is $(r^3/R^3)Q$. Using the same symmetry arguments as before, Gauss's law becomes

$$E(4\pi r^2) = \frac{(r^3/R^3)Q}{\varepsilon_0}$$

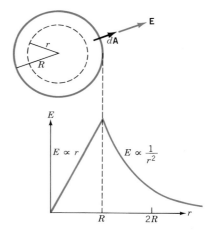

FIGURE 24.10 A spherical Gaussian surface within a uniformly charged sphere.

which yields (with $k = 1/4\pi\varepsilon_0$)

$$E = \frac{kQr}{R^3} \tag{24.6}$$

The electric field increases linearly with distance from the center. The variation of E with r is shown in Fig. 24.10. Note that at $r = R$, the two expressions for E agree; the field is continuous at the surface.

EXAMPLE 24.3: An *infinite line of charge* has a linear charge density λ C/m. Find the electric field at a distance r from the line.

Solution: The cylindrical symmetry tells us that the field strength will be the same at all points at a fixed distance r from the line. Since the line is infinite and uniform, for every charge element on the $+y$ axis in Fig. 24.11 there is a symmetrically located element on the $-y$ axis. The y components of the fields due to all such elements cancel in pairs. Thus, the field lines are directed radially outward, perpendicular to the line of charge.

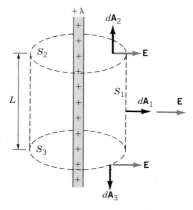

FIGURE 24.11 A cyclindrical Gaussian surface around an infinite line of charge. There is flux only through the curved surface.

The appropriate choice of Gaussian surface is a cylinder of radius r and length L. On the flat end faces S_2 and S_3, **E** is perpendicular to **A**, which means no flux crosses them. On the curved surface S_1, **E** is parallel to d**A**, so that $\mathbf{E} \cdot d\mathbf{A} = E\, dA$. The charge enclosed by the cylinder is $Q = \lambda L$. Applying Gauss's law to the curved surface, we have

$$E \oint dA = E(2\pi rL) = \frac{\lambda L}{\varepsilon_0}$$

Since $k = 1/4\pi\varepsilon_0$ we find,

$$E = \frac{2k\lambda}{r} \qquad (24.7)$$

This derivation is much simpler than the direct calculation based on Coulomb's law given in Example 23.7. Notice that the charges outside the Gaussian surface do not contribute to the net flux through the surface. However, they *do* contribute to the total field, and it is their presence that allows us to make the symmetry arguments.

EXAMPLE 24.4: Find the field due to an *infinite flat sheet of charge* with a uniform areal charge density σ C/m².

Solution: Since the charge is distributed over an infinite plane, all points equidistant from the sheet are equivalent. Thus the field must be constant in magnitude over any plane parallel to the sheet. The symmetry also indicates that the lines must be perpendicular to the plane. Therefore we choose a cylindrical Gaussian surface whose end faces are on either side of the sheet and at equal distances from it, as shown in Fig. 24.12. (It is sometimes called a "Gaussian pillbox.")

In this case, the curved surface of the cylinder has no flux passing through it ($\mathbf{E}_3$ is normal to $\mathbf{A}_3$). If the end faces have area A, the charge enclosed is σA. Applying Gauss's law we have

$$\oint \mathbf{E} \cdot d\mathbf{A} = E_1A_1 + E_2A_2 = \frac{\sigma A}{\varepsilon_0}$$

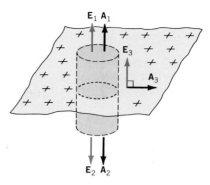

FIGURE 24.12 A Gaussian "pillbox" appropriate for an infinite sheet of charge. Flux passes through only the end faces.

But we know that $A_1 = A_2$ and that the field strengths must have the same value on both sides. The above expression becomes $2EA = \sigma A/\varepsilon_0$, which yields

$$E = \frac{\sigma}{2\varepsilon_0} \qquad (24.8)$$

As we found in Chapter 23, the field due to an infinite sheet is uniform and everywhere perpendicular to the sheet (see Fig. 24.13).

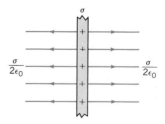

FIGURE 24.13 The field of an infinite sheet of charge is uniform.

24.3 CONDUCTORS

Gauss's law may be used to infer some interesting information about the charges and fields associated with conductors. When a net charge is added to a conductor, an electric field will be set up temporarily in the body of the conductor. The free electrons will redistribute themselves and within a tiny fraction of a second (approx. 10^{-12} s) the internal field will vanish. Thus, in electrostatic equilibrium, the field within a conductor is zero (see Section 23.5). In Fig. 24.14 we imagine a Gaussian surface just inside an arbitrary conductor. Since $E = 0$ at all points on this surface, there is no net flux through it. According to Eq. 24.3, the net charge enclosed by the Gaussian surface must also be zero. We conclude that *any net charge on a conductor must reside on its surface*.

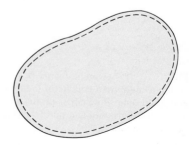

FIGURE 24.14 A Gaussian surface just inside a conductor.

EXAMPLE 24.5: Find the field due to an *infinite conducting plate* with a uniform surface charge density σ C/m^2.

Solution: From the symmetry arguments presented for the infinite sheet, we know that the field must be uniform and perpendicular to the plane. In Fig. 24.15 we set up a Gaussian pillbox as in the previous example. In this case, the field is zero within the conductor; therefore only one flat face has flux passing through it. If the area of the end face is A, from Gauss's law we have

$$EA = \frac{\sigma A}{\varepsilon_0}$$

which gives

$$E = \frac{\sigma}{\varepsilon_0} \qquad (24.9)$$

Note that we have used *more* than Gauss's law in deriving Eq. 24.9: We also used the fact that $E = 0$ inside a conductor.

The field in Eq. 24.9 is twice the value obtained for the infinite sheet of charge with the same surface charge density. In the case of a conductor all of the flux goes one way, whereas in the case of a sheet, the flux is divided between two directions. Although Eq. 24.9 was derived for a flat infinite conductor, it may be applied as a first approximation to any charged conductor without sharp points. Equation 24.9 implies that the field strength does not diminish with distance from the conductor. For a conductor of finite size this result would apply only to regions close enough to the surface for it to be considered flat.

The fields both inside and outside the conductor are produced by the superposition of two contributions. The charge within the pillbox produces a "local" field, $E_{\text{local}} = \sigma/2\varepsilon_0$, on *both* sides of the surface of the conductor, as shown in Fig. 24.16. Since the net field inside the conductor is zero, all *other* charges on its surface must produce a "far" field, $E_{\text{far}} = \sigma/2\varepsilon_0$, that exactly cancels the local field just inside the conductor. Just outside the conductor, the far field reinforces the local field to produce the resultant σ/ε_0.

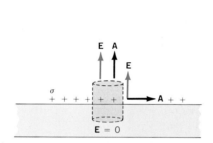

FIGURE 24.15 An infinite charged conducting plate. Since $E = 0$ inside the conductor, there is flux only through one end face of the Gaussian pillbox.

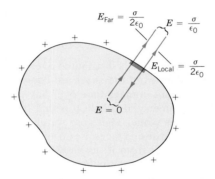

FIGURE 24.16 The field just outside a small area of a charged conductor is the sum of two contributions: A local field due to charges on the area and a far field due to all the other charges.

Cavity in a Conductor

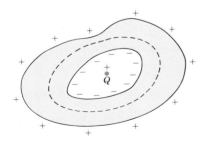

FIGURE 24.17 A point charge Q within a cavity in a conductor induces equal and opposite charges on the surface of the cavity and on the surface of the conductor.

Figure 24.17 shows a conductor with a cavity in which there is a point charge Q. Within the material of the conductor, $E = 0$ on any Gaussian surface we choose around the cavity, and so the net flux through this surface is zero. From Gauss's law the net charge enclosed by the surface must also be zero. This implies that there is an induced charge $-Q$ on the inside walls of the cavity. Since the conductor as a whole is neutral, its external surface acquires a charge $+Q$.

One way of confirming the values of the induced charges on the conductor is to repeat the *ice pail experiment* devised by Faraday. A charged metal ball is placed inside a hollow metal container. A metal cap is then placed over the top of the container. The ball induces total charges of equal magnitude and opposite signs on the inner and outer walls of the container, as shown in Fig. 24.18a. When the ball touches the inner wall, their charges are neutralized. This leaves a charge on the outer wall equal to that originally on the ball (see Fig. 24.18b).

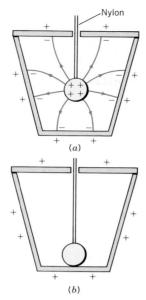

Nylon

(a)

(b)

FIGURE 24.18 In Faraday's ice pail experiment the charge on a metal ball is completely transferred to the outside of a metal pail.

The Cavendish Experiment

Gauss's law in electrostatics is essentially a restatement of Coulomb's law; either law can be derived from the other. By using Gauss's law, we showed that in electrostatic equilibrium any net charge on a conductor must reside on its surface. Thus, any charge detected on the inside of a conductor would indicate a violation of both Gauss's law and Coulomb's law. In 1771, Henry Cavendish placed a metal shell B inside a metal shell A that consisted of two hemispheres, as in Fig. 24.19. The shells were connected with a wire, and the outer shell A was given a charge. If charges within A felt any force, they would flow either to B or from it. The connection was then broken and A removed. Cavendish could detect no charge on B and concluded that the force law was $1/r^n$ where $n = 2 \pm 1/60$. Modern versions of this technique have reduced the uncertainty to 10^{-16}!

EXERCISE 2. A point charge q is at the center of a metal shell of radius R that has a net charge $-Q$. Find the field (a) inside the shell, and (b) outside the shell.

EXERCISE 3. Show that $E = 0$ within an empty cavity of *any* shape inside a conductor.

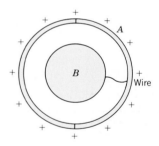

FIGURE 24.19 A test of Gauss's law and indirectly a test of Coulomb's law. A metal shell B is connected by a wire to shell A. When A is charged, no charge is detected on B, in agreement with Gauss's law.

24.4   PROOF OF GAUSS'S LAW (Optional)

From Fig. 24.7 we saw that the number of lines that pass through two arbitrary closed surfaces enclosing a point charge are the same. Since the concept of flux, rather than lines, is used in Gauss's law, one must show that the flux through two arbitrary surfaces is the same. Consider a cone of field lines that emerge from a point charge Q, as shown in Fig. 24.20. The size

of the cone is specified by the solid angle Ω which is defined by

$$\Omega = \frac{A_n}{r^2} = \frac{A \cos \theta}{r^2}$$

where $A_n = A \cos \theta$ is the projection of A perpendicular to the axis of the cone. The unit of solid angle is the steradian. A closed surface subtends a solid angle of 4π at any internal point.

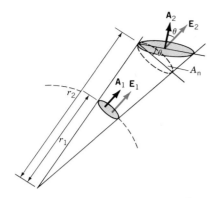

FIGURE 24.20 The flux within a given solid angle is constant.

(Consider the special case of a sphere whose surface area is $4\pi r^2$.) In Fig. 24.20 the cone intercepts an area A_1 of a spherical surface of radius r_1 and an area A_2 of an arbitrary surface. The areas can be of any shape. From Coulomb's law ($E \propto 1/r^2$) the ratio of the field strengths at the two areas is

$$\frac{E_2}{E_1} = \frac{r_1^2}{r_2^2} \qquad \text{(i)}$$

The solid angle of the cone may be expressed in terms of either area:

$$\Omega = \frac{A_1}{r_1^2} = \frac{A_2 \cos \theta}{r_2^2} \qquad \text{(ii)}$$

The flux through each surface is $\Phi_1 = E_1 A_1$ and $\Phi_2 = E_2 A_2 \cos \theta$. Using (i) and (ii) we see that $\Phi_1 = \Phi_2$. That is, the flux within a given solid angle is constant, independent of the shape or orientation of the surface.

SUMMARY

The picture of field lines intercepting an area is made quantitative by the concept of **electric flux, Φ_E**. For a flat surface of area $\mathbf{A}$ in a uniform field $\mathbf{E}$, the flux through the surface is defined as

$$\Phi_E = \mathbf{E} \cdot \mathbf{A} = EA \cos \theta$$

where the vector area $\mathbf{A}$ is perpendicular to the plane of the surface. If the field is not uniform or the surface is not flat, the flux is found from

$$\Phi_E = \oint \mathbf{E} \cdot d\mathbf{A}$$

Gauss's law is a general statement regarding electric fields. It relates the flux through a closed surface to the net charge Q enclosed by the surface:

$$\oint \mathbf{E} \cdot d\mathbf{A} = \frac{Q}{\varepsilon_0}$$

Note that the field $\mathbf{E}$ may include contributions from charges *not* enclosed by the closed surface. Of course the net flux due to such charges is zero. Gauss's law provides a simple and powerful way to determine electrostatic fields if the charge distribution has sufficient symmetry for the integration to be performed easily. In such cases, one can pick a Gaussian surface for which $\mathbf{E}$ at a given point is either parallel to, or perpendicular to, the local $d\mathbf{A}$.

One can make the following statements regarding the electric field associated with a homogeneous conductor in electrostatic equilibrium:

(a) Under static conditions, $E = 0$ everywhere inside the material of the conductor.

(b) Any excess charge resides only on the surface.

(c) $\mathbf{E}$ is perpendicular to the surface of a charged conductor.

ANSWERS TO IN-CHAPTER EXERCISES

1. The vector area of the circle is at 50° to the field. The area is $A = \pi r^2 = 0.02 \text{ m}^2$. Thus, the flux is

$$\Phi_E = EA \cos 50°$$
$$= (600 \text{ N/C})(0.02 \text{ m}^2)(0.643) = 7.7 \text{ N} \cdot \text{m}^2/\text{C}$$

2. Use a spherical Gaussian surface of radius r whose center coincides with that of the shell.
   (a) The enclosed charge is q so

$$E(4\pi r^2) = \frac{q}{\varepsilon_0}$$

which yields $E = kq/r^2$. (b) The net enclosed charge is $q - Q$. So, the field outside the shell is $E = k(q - Q)/r^2$. There are induced charges $\pm q$ on the inner and outer surfaces of the shell. Specifically, the charge on the outer surface is $q - Q$.

3. The flux through any Gaussian surface of arbitrary shape within the empty cavity is zero. The only value of the field consistent with this statement is zero. Does this result depend on whether the conductor has a net charge?

QUESTIONS

1. A positive point charge is placed at the center of an uncharged metal cube. Sketch the field lines within the cube. Is the induced charge on the cube uniformly distributed?

2. A charge Q is inside a cubic Gaussian surface. What information can one deduce from Gauss's law about the following: (a) the location of the charge; (b) the total flux through the surface; (c) the electric field at some pont on the surface?

3. The total flux through a Gaussian surface is zero. (a) What is the net charge enclosed? (b) Is $E = 0$ at all points on the surface?

4. Consider the three charges shown in Fig. 24.21. A Gaussian surface is also indicated. (a) Which charges contribute to the net flux through the Gaussian surface? (b) Which charges contribute to the field at a given point on the surface? (c) Write Gauss's law for the surface.

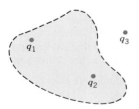

FIGURE 24.21 Question 4.

5. Charge is uniformly distributed around the circumference of a circle. Can one use Gauss's law to calculate the field? If so, at what points?

6. A point charge Q is placed off center inside a spherical metal shell. (a) Is the field inside the shell determined solely by the charge within the shell? (b) What is the net flux through a closed surface that is inside the shell and surrounds the charge? (c) Can one use Gauss's law to calculate the field at the surface?

7. (a) If the net charge enclosed by a surface is zero, does this imply that the field is zero at all points on the surface? (b) If the field is zero at all points on a surface, does this imply that the net charge enclosed is zero?

8. In Gauss's law, the electric field is determined by: (a) the charge within the Gaussian surface; (b) the charge on the Gaussian surface; (c) all charges that contribute to the field at any point on the Gaussian surface.

9. True/False: In order to use Gauss's law to determine the electric field, one must know the location of all the charges that contribute to the field.

10. Use Gauss's law to show that electrostatic field lines must begin or end on point charges.

11. What can Gauss's law tell us about the field due to a dipole?

12. An isolated metal shell has a charge $+Q$ on its inner surface of radius a and a charge $-Q$ on its outer surface of radius b. What can you deduce from this information?

13. An isolated metal shell has a uniform surface charge density $-\sigma$ on its inner surface of radius a and a uniform surface charge density $+\sigma$ on its outer surface of radius b. What can you deduce from this information?

14. A point charge Q is at the center of a spherical conducting shell. There is a point charge q outside the shell. (a) Does q experience a force? (b) Does Q experience a force? (c) If there is a difference in the forces experienced by the charges, reconcile your answer with Newton's third law.

EXERCISES

24.1 Electric Flux

1. (I) A circular plate has a radius of 12 cm. The plane of the plate is set at a 30° angle to a uniform field $\mathbf{E} = 450\mathbf{i}$ N/C, as shown in Fig. 24.22. What is the flux through the plate?

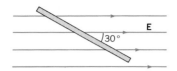

FIGURE 24.22 Exercise 1.

2. (I) A flat plate with dimensions 4 cm × 6 cm is set with its plane at 37° to a uniform electric field $\mathbf{E} = -600\mathbf{j}$ N/C, as in Fig. 24.23. What is the flux through the plate?

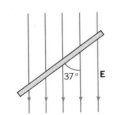

FIGURE 24.23 Exercise 2.

3. (I) A uniform electric field E is parallel to the central axis of a hemisphere of radius R, as shown in Fig. 24.24. What is the flux through the hemisphere?

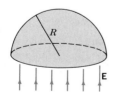

FIGURE 24.24 Exercise 3.

4. (I) A square plate of side 12 cm lies in the xy plane. What is the flux through the plate due to a uniform field $\mathbf{E} = 70\mathbf{i} + 90\mathbf{k}$ N/C?

24.2 and 24.3 Gauss's Law; Conductors

5. (I) Two charges $q_1 = 6$ μC and $q_2 = -8$ μC are within a spherical surface of radius 5 cm. What is the total flux through the surface?

6. (I) The flux through each face of a cubic Gaussian surface of side 10 cm is 3×10^4 N $\cdot$ m²/C. What is the net enclosed charge?

7. (I) A 60-μC charge is at the center of a cube of side 10 cm. (a) What is the total flux through the cube? (b) What is the flux through one face? (c) Would your answers to (a) or (b) change if the charge were not at the center?

8. (II) A point charge Q is placed at one corner of a Gaussian surface in the form of a cube of side L. What is the flux through each face?

9. (I) A spherical conductor of radius 8 cm has a uniform surface charge density 0.1 nC/m². Find the electric field: (a) at the surface; (b) at a distance 10 cm from the center.

10. (I) A 16-μC point charge is placed at the center of a spherical conducting shell that carries a uniformly distributed charge of -8 μC. (a) Find the field both inside and outside the shell. (b) What are the charges on the inner and outer surfaces of the shell? (c) Sketch the field lines.

11. (I) Show that the field strength at the surface of a uniformly charged spherical shell is $E = \sigma/\varepsilon_0$, where σ is the surface charge density.

12. (I) The electric field strength at all points of a spherical surface of radius 2 cm is 800 N/C directed radially inward. (a) What is the net enclosed charge? (b) Does the charge have to be a point charge at the center? If not, what other possibility is there?

13. (I) Two infinite and parallel sheets of charge have the same surface charge density σ C/m². What is the field (a) in the region between the sheets and (b) in the regions not between the sheets?

14. (I) A nonconducting infinite plate has a surface charge density σ C/m² on each face. It is placed parallel to a similar plate with $-\sigma$ C/m² on each face. Find the field (a) in the region between the plates, and (b) within the positive plate.

15. (I) Two infinite conducting plates are placed parallel to each other. They carry equal and opposite surface charge densities $\pm\sigma$ C/m². What is the net electric field (a) between the plates, (b) in the regions not between the plates?

16. (II) A cube of side L has one corner at the origin and its sides lie along the x, y, and z axes, respectively. There is a field given by $\mathbf{E} = (a + bx)\mathbf{i}$. (a) What is the net flux through the cube? (b) What is the net charge enclosed by the cube?

17. (I) A long, straight coaxial cable, shown in Fig. 24.25, has an inner wire of radius a with a surface charge density σ_1 and an outer cylindrical shell of radius b with σ_2 C/m². Find the relationship between σ_1 and σ_2 for the field strength to be zero outside the cable, that is, for $r > b$.

18. (I) A long, straight coaxial cable (Fig. 24.25) has an inner wire of radius a with a surface charge density σ and an outer cylindrical shell of radius b with $-\sigma$. Find the field in the regions (a) $a < r < b$, and (b) $r > b$.

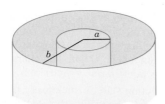

FIGURE 24.25 Exercises 17, 18, 19, and 20.

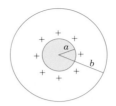

FIGURE 24.26 Exercises 21, 22, and 23.

19. (I) A long, straight coaxial cable (Fig. 24.25) has an inner wire of radius a that carries a linear charge density λ_1 C/m and an outer cylindrical shell of radius b that carries λ_2 C/m. (a) What is the relationship between λ_1 and λ_2 for the field outside the cable to be zero?

20. (I) A long, straight coaxial cable (Fig. 24.25) has an inner wire of radius a that carries a linear charge density λ C/m and an outer cylindrical shell of radius b that has a linear charge density $-\lambda$. Find the field in the regions (a) $a < r < b$, and (b) $r > b$.

21. (I) A positively charged metal sphere of radius a is at the center of a metal shell of radius b (Fig. 24.26). The spheres carry equal and opposite charges $\pm Q$. Find the electric field as a function of the distance r from the common center for (a) $a < r < b$, and (b) $r > b$.

22. (I) A positively charged metal sphere of radius a is at the center of a metal shell of radius b (Fig. 24.26). The spheres carry equal and opposite surface charge densities $\pm\sigma$ C/m^2. Find the electric field as a function of the distance r from the common center for (a) $a < r < b$, and (b) $r > b$.

23. (I) A positively charged metal sphere of radius a is at the center of a metal shell of radius b (Fig. 24.26). How should the surface charge densities be related for the field to be zero for $r > b$?

24. (II) A point charge Q is placed at the center of a conducting spherical shell of inner radius R_1 and outer radius R_2. (a) What are the surface charge densities on the inner and outer surfaces? (b) What is the field for $r < R_1$? (c) Find the field for $r > R_2$. (d) If the charge moves off center, can Gauss's law be used to find the field outside the shell?

PROBLEMS

1. (I) A nonconducting sphere of radius R has a uniform charge density ρ C/m^3 throughout its volume. Determine the electric field at a distance r from the center for (a) $r < R$; (b) $r > R$. Do your results agree at $r = R$?

2. (II) Repeat Problem 1 for the nonuniform density $\rho(r) = Ar$, where A is a constant. Express your answers in terms of the total charge Q. (*Hint:* The charge within a shell of thickness dr is $dq = \rho \, dV = \rho 4\pi r^2 dr$.)

3. (I) A conducting spherical shell has inner radius R_1 and outer radius R_2 with uniform charge densities σ C/m^2 on the inner and $-\sigma$ C/m^2 on the outer. (a) What can you say about the charge within the cavity? (b) What can you say about the net charge on the shell? (c) Find the field outside the shell.

4. (I) A conductor has a surface charge density σ C/m^2. Show that the force per unit area on the surface is $\sigma^2/2\varepsilon_0$ N/m^2. (*Hint:* The field at the surface has two contributions. Also, a static charge does not experience a force due to its own field.)

5. (I) Charge is uniformly distributed throughout an infinitely long cylinder of radius R. The density is ρ C/m^3. Find the electric field at a radius r when (a) $r < R$; (b) $r > R$. Do your results agree at $r = R$?

6. (I) A nonconducting sphere of radius R has a cavity of radius a at its center. The rest of the sphere has a uniform charge density ρ C/m^3 (Fig. 24.27). What is the electric field in the following regions: (a) $r > R$; (b) $a < r < R$? (*Hint:* The charge within a shell of thickness dr is $dq = \rho \, dV = \rho 4\pi r^2 dr$.)

7. (II) Consider a hydrogen atom to be a positive point charge e at the center of a uniformly charged sphere of radius R and with *total* charge $-e$. Determine the field as a function of the distance r from the nucleus.

8. (I) Gauss's law for the gravitational field is

$$\oint \mathbf{g} \cdot d\mathbf{A} = -4\pi Gm$$

where $\mathbf{g}$ is the gravitational field strength, G is the gravitational constant, and m is the mass enclosed by the Gaussian surface. Show that one can obtain Newton's law of gravitation from this expression. What is the significance of the negative sign?

9. (I) A metal sphere of radius a carrying a charge $+Q$ is at the center of an uncharged thick metal shell of inner radius b and outer radius c, as in Fig. 24.28. Determine the electric field in the following regions: (a) $a < r < b$; (b) $r > c$.

10. (II) A nonconducting sphere of radius R has a spherical cavity of radius a at its center; see Fig. 24.27. The charge density within the rest of the sphere varies according to $\rho = A/r$. Determine the electric field for $a < r < R$. (*Hint:* The charge within a shell of thickness dr is $dq = \rho \, dV = \rho 4\pi r^2 dr$.)

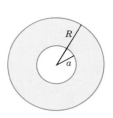
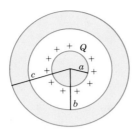

FIGURE 24.27 Problems 6 and 10.   **FIGURE 24.28** Problem 9.

11. (I) An infinite cylinder of radius R has a hole of radius a along its central axis as in Fig. 24.29. The rest of the cylinder has a uniform charge density ρ C/m³. Determine the electric field in the following regions: (a) $a < r < R$; (b) $r > R$.

12. (I) The density of charge within a nonconducting sphere of radius R varies according to $\rho = \rho_0(1 - r/R)$ C/m³ where ρ_0 is a constant and r is the distance from the center. Determine the field in the region $r < R$. (*Hint:* The charge within a shell of thickness dr is $dq = \rho \, dV = \rho 4\pi r^2 dr$.)

13. (I) An infinite nonconducting slab of thickness t has a uniform charge density ρ C/m³. Determine the electric field as a function of the distance from the central plane of symmetry.

14. (II) A sphere of radius R has a uniform charge density ρ C/m³ except for a spherical cavity of radius a as shown in Fig. 24.30. (a) Show that the field within the cavity is uniform. (*Hint:* The field at any point within the cavity is the sum of the field due to a sphere of radius R and a sphere of radius a with charge density $-\rho$ C/m³.) (b) What is the magnitude and direction of the field within the cavity?

FIGURE 24.29 Problem 11.

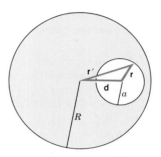

FIGURE 24.30 Problem 14.

CHAPTER 25

Electric Potential

This Cockcroft-Walton generator at Los Alamos laboratory produces a large potential difference to accelerate elementary particles called mesons.

Major Points

1. The definition of **electric potential.**
2. (a) Calculating the potential given the electric field.
   (b) Calculating the field strength given the potential function.
3. **Electrostatic potential energy** of point charges.
4. Electric field lines are perpendicular to **equipotential surfaces** and point toward lower potentials.

In mechanics we introduced the concept of potential energy and used it to formulate the law of conservation of energy. In Chapter 8 we discussed the fact that potential energy can be defined only for conservative forces. Also recall that the gravitational force, given by Newton's of gravitation, which has the form $\mathbf{F} = f(r)\hat{\mathbf{r}}$, is conservative. Since Coulomb's law has the same form, the electrostatic force is also conservative. This allows us to define an electrostatic potential energy, analogous to gravitational potential energy, and to apply the law of conservation of energy in the analysis of electrical problems.

This chapter is devoted to the concept of **electric potential,** which is closely related to potential energy. But whereas potential energy is a property of a *system* of particles (including any "test" particle), potential, like the electric field strength, is a property of a point in space and depends only on the *source* charges. The electric field strength gives the force per unit charge at a given point. The potential tells us the potential energy per unit charge. It is often easier to analyze a physical situation in terms of potential, which is a scalar, rather than the electric field strength, which is a vector.

25.1 POTENTIAL

The motion of a particle with positive charge q in a uniform electric field is analogous to the motion of a particle of mass m in the uniform gravitational field near the earth; see Fig. 25.1. To move a particle against the field requires work by an external agent—for example, you. If the external force is equal and opposite to

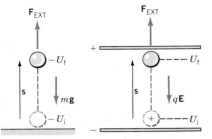

FIGURE 25.1 The motion of a point mass m in a gravitational field is analogous to the motion of a point charge q in an electric field. If the speed of the particle is constant, the change in potential energy is related to the work done by an external agent: $W_{EXT} = +\Delta U$.

Definition of electric potential

the force due to the field, the kinetic energy of the particle will not change. In this case, all the external work is stored as potential energy in the system:

(v constant) $$W_{EXT} = +\Delta U = U_f - U_i \qquad (25.1)$$

where U_f and U_i are the final and initial potential energies.

The gravitational potential energy function near the surface of the earth is $U_g = mgy$. One can obtain a function that does not depend on m by defining the *gravitational potential* as the potential energy per unit mass: $V_g = U_g/m = gy$. The SI unit of V_g is J/kg. The gravitational potential at a point is the external work needed to lift a unit mass from the zero level of potential ($y = 0$) to the given height, without a change in speed. A useful feature of the potential function is that it depends only the *source* of the field (the earth) through the value of the gravitational field strength g, and not on the value of the "test" mass, m.

When a charge q moves between two points in an electrostatic field, the change in **electric potential**, ΔV, is defined as the change in electrostatic potential energy per unit charge,

$$\Delta V = \frac{\Delta U}{q} \qquad (25.2)$$

The SI unit of electric potential is the volt (V), in honor of Alessandro Volta, inventor of the voltaic pile (the first primitive electric battery). Note that

$$1 \text{ V} = 1 \text{ J/C}$$

The quantity ΔV depends only on the field set up by the source charges, not on the test charge. Once the potential difference between two points is known, the external work needed to move a charge q, with no change in its speed, may be found from Eq. 25.1:

(v constant) $$W_{EXT} = q \, \Delta V = q(V_f - V_i) \qquad (25.3)$$

The sign of this work depends on the sign of q and the relative magnitudes of V_i and V_f. If $W_{EXT} > 0$, work is done by the external agent on the charge. If $W_{EXT} < 0$, work is done on the external agent by the field. In the latter case, in order to keep the speed constant, the external force acts opposite to the displacement of the charge.

From Eq. 25.3 we see that only changes in potential, rather than the specific value of V_i and V_f, are significant. One can choose the reference point at which the potential is zero at some convenient point such as infinity. In electronic circuits it is convenient to choose the ground connection to earth as the zero of potential. If $V_i = 0$, we may write $V_f = W_{EXT}/q$:

The potential at a point is the external work needed to bring a positive unit charge, at constant speed, from the position of zero potential to the given point.

Electric potential, measured in J/C, is analogous to gravitational potential, measured in J/kg. When the height of a particle is increased, its gravitational potential energy increases. Similarly, when a positive charge is moved to a point of higher potential, its electrostatic potential energy increases. If allowed to, positive charges tend to move "downhill" in potential, just as do ordinary masses. However, negative charges tend to move "uphill" in potential. In an external electric field, both positive and negative charges tend to decrease the electrostatic potential energy.

Although the concept of work done by an external agent is helpful in introducing potential energy, it is preferable to refer instead to the internal conservative forces within the system of interacting particles. From Eq. 8.4, the definition of potential energy in terms of the work done by the conservative force is $\Delta U = -W_c$. The negative sign tells us that positive work by the conservative force leads to a decrease in potential energy. In an electrostatic field, the (conservative) force on a test charge q is $\mathbf{F}_c = q\mathbf{E}$. Therefore, the change in potential energy, $dU = -dW_c$, associated with an infinitesimal displacement $d\mathbf{s}$, is

$$dU = -\mathbf{F}_c \cdot d\mathbf{s} = -q\mathbf{E} \cdot d\mathbf{s}$$

From Eq. 25.2 the infinitesimal change in potential associated with the displacement $d\mathbf{s}$ is

$$dV = \frac{dU}{q} = -\mathbf{E} \cdot d\mathbf{s} \qquad (25.4)$$

Figure 25.2 shows a curved path in a nonuniform field. The finite change in potential in going from point A to point B is the sum (integral) of these infinitesimal changes,

$$V_B - V_A = -\int_A^B \mathbf{E} \cdot d\mathbf{s} \qquad (25.5)$$

Since the electrostatic field is conservative, the value of this line integral depends only on the end points A and B, not on the path taken. The sign of the integral is determined (1) by the signs of the components of $\mathbf{E}$, and (2) by the direction of the path taken—which is indicated by the limits.

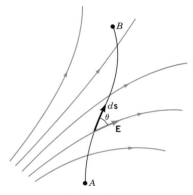

FIGURE 25.2 The change in potential in moving from point A to point B in an electrostatic field is $V_B - V_A = -\int \mathbf{E} \cdot d\mathbf{s}$ and is independent of the path taken.

25.2   POTENTIAL AND POTENTIAL ENERGY IN A UNIFORM FIELD

In a uniform field, $\mathbf{E}$ is constant, and therefore the integral in Eq. 25.5 may be written as $\int \mathbf{E} \cdot d\mathbf{s} = \mathbf{E} \cdot \int d\mathbf{s} = \mathbf{E} \cdot \Delta\mathbf{s}$. The finite change in potential ΔV associated with a finite displacement $\Delta\mathbf{s}$ takes the form

(Uniform E) $\qquad\qquad \Delta V = -\mathbf{E} \cdot \Delta\mathbf{s} \qquad (25.6a)$

Note that $\Delta\mathbf{s}$ and ΔV depend only on the initial and final positions, not on the path taken.

Figure 25.3 shows a uniform field $\mathbf{E} = E\mathbf{i}$. Let us find the change in potential in going from point A to point B, which are separated by a distance d along the lines. Since the electric field has only an x component, Eq. 25.6a reduces to $\Delta V = -E_x\Delta x$. If we write $E_x = E$ and $\Delta x = +x$, we have

$$V(x) - V(0) = -Ex \qquad (25.6b)$$

The potential decreases linearly along the x axis, as depicted in the graph of Fig. 25.3. Notice that the field lines point from high potential to low potential. Suppose now that the actual path in Fig. 25.3 is replaced by the two steps AC and CB. Since $\mathbf{E}$ is perpendicular to the displacement along BC, no work will be done on a test charge along this segment. Work is done only along the segment AC parallel to the field lines. Since only the component of the displacement along, or against, the field lines is significant, Eq. 25.6a is often written in the form

(Uniform E) $\qquad\qquad \Delta V = \pm Ed \qquad (25.6c)$

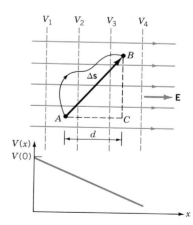

FIGURE 25.3 In a uniform field, the change in potential in moving from A to B is $\Delta V = -\mathbf{E} \cdot \Delta\mathbf{s}$. In a uniform field the potential decreases linearly with distance along the field lines.

where d is the magnitude of the component of the displacement along, or against, the field. The positive sign applies to a displacement *opposite* to the field. From Eq. 25.6c we see that an equivalent unit for electric field is V/m:

$$1 \text{ V/m} = 1 \text{ N/C}$$

Equipotentials

A relief map, as in Fig. 25.4, has contour lines that are formed by joining points of equal elevation. Usually, the contours are drawn for equal intervals in elevation, say 100 m. The lines are close together where the terrain is steep; they are far apart where the terrain slopes gradually. An **equipotential** is a surface that joins points of equal potential. In a two-dimensional plot, the surfaces are depicted as equipotential lines. The contour lines actually trace the gravitational equipotentials. In a similar fashion, one can draw electrical equipotentials.

In the uniform field of Fig. 25.3, each value of x has a particular value of V. Thus, the equipotential surfaces are flat planes, although they are depicted as straight dashed lines in Fig. 25.3. Note that the *electric field lines are perpendicular to the equipotentials and point from higher to lower potentials*, that is, "downhill" in potential. The fact that the field lines are perpendicular to the equipotentials is a general result. From Eq. 25.4 the change in potential associated with an infinitesimal displacement $d\mathbf{s}$ is $dV = -\mathbf{E} \cdot d\mathbf{s}$. If the displacement is along an equipotential, then $dV = 0$. Thus, $\mathbf{E} \cdot d\mathbf{s} = 0$, from which we conclude that $\mathbf{E}$ is perpendicular to $d\mathbf{s}$. No work is required to move a particle along an equipotential surface.

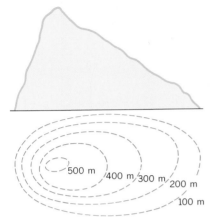

FIGURE 25.4 In a relief map, the contours join points of equal elevation (or gravitational potential).

EXERCISE 1. Figure 25.5 shows two points A and B in a uniform electrical field. A charge q moves from A to B. (a) Does the potential increase or decrease? (b) Does its potential energy increase or decrease? Consider both positive and negative values for q.

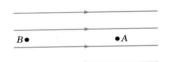

FIGURE 25.5 When a charge moves from A to B what happens to its potential and to its potential energy?

Motion of Charges

The motion of a charge in an electric field may be discussed in terms of the conservation of energy, $\Delta K + \Delta U = 0$. When we refer to the "potential energy of a charge," it is implied that the other charges are fixed in position. In terms of potential, the conservation law may be written as

$$\Delta K = -q \, \Delta V \qquad (25.7)$$

The sign of ΔK depends on the signs of both q and ΔV. For example, if $q > 0$, and the charge moves "downhill" in potential ($\Delta V < 0$), it will gain kinetic energy. It is often convenient to measure the energy of elementary particles, such as electrons and protons, in terms of a non-SI unit called the *electronvolt* (eV). When a particle with a charge of magnitude e moves through a potential difference of one volt, its kinetic energy changes by one electronvolt. From Eq. 25.7,

$$\Delta K = e \, \Delta V = (1.602 \times 10^{-19} \text{ C}) \, (1 \text{ V})$$

Thus,

$$1 \text{ eV} = 1.602 \times 10^{-19} \text{ J} \qquad (25.8)$$

In terms of this unit, chemical bonding energies are of the order of a few electronvolts per bond. The electrons in the beam of a cathode ray tube have approximately 10^4 eV.

EXAMPLE 25.1: A proton, of mass 1.67×10^{-27} kg, enters the region between two parallel plates a distance 20 cm apart. There is a uniform electric field of 3×10^5 V/m between the plates, as shown in Fig. 25.6. If the initial speed of the proton is 5×10^6 m/s, what is its final speed?

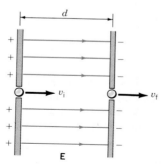

FIGURE 25.6 As a proton moves along the field lines, its potential energy decreases and its kinetic energy increases.

Solution: According to Eq. 25.7, the change in kinetic energy is

$$\tfrac{1}{2}mv_f^2 - \tfrac{1}{2}mv_i^2 = -q\,\Delta V \qquad \text{(i)}$$

Since the displacement is *along* the direction of the field lines, the change in potential is negative. From Eq. 25.6c,

$$\Delta V = -Ed = -6 \times 10^4 \text{ V}$$

From (i) we have

$$v_f^2 = v_i^2 - \frac{2q\,\Delta V}{m}$$

$$= (5 \times 10^6 \text{ m/s})^2 - \frac{2(1.6 \times 10^{-19} \text{ C})(-6 \times 10^4 \text{ V})}{1.67 \times 10^{-27} \text{ kg}}$$

$$= 36.5 \times 10^{12} \text{ m}^2/\text{s}^2$$

Thus, $v_f = 6 \times 10^6$ m/s.

EXERCISE 2. A charge q (<0) moves in an electric field. If it moves "uphill" in potential, does its kinetic energy increase or decrease?

25.3   POTENTIAL AND POTENTIAL ENERGY OF POINT CHARGES

We now consider how the potential varies in the vicinity of a point charge Q. The electric field is

$$\mathbf{E} = E_r\,\hat{\mathbf{r}} = \frac{kQ}{r^2}\,\hat{\mathbf{r}}$$

Since **E** is radial, as in Fig. 25.7, only the radial component of the displacement $d\mathbf{s}$ can contribute to $\mathbf{E} \cdot d\mathbf{s}$; thus, $\mathbf{E} \cdot d\mathbf{s} = E_r\,dr$. From Eq. 25.5 the change in potential in moving from A to B along any path is

$$V_B - V_A = -\int_A^B E_r\,dr = -\left[-\frac{kQ}{r} \right]_A^B$$

$$= kQ\left(\frac{1}{r_B} - \frac{1}{r_A} \right)$$

If we choose $V = 0$ at $r = \infty$, the potential at a distance r from Q is

$$V = \frac{kQ}{r} \qquad (25.9)$$

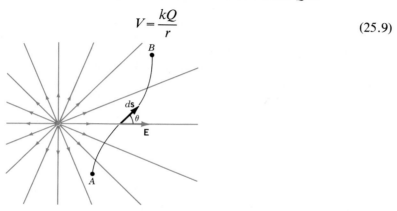

FIGURE 25.7 The change in potential from point A to point B is $V_B - V_A = -\int_A^B \mathbf{E} \cdot d\mathbf{s}$.

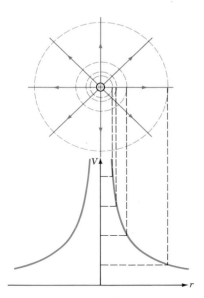

FIGURE 25.8 The potential function $V = kQ/r$ for a point charge. The dashed circles represent the equipotential surfaces (which are spheres centered on the charge).

This potential function, which depends only on the source charge Q, is plotted in Fig. 25.8. Since each value of r has a unique value of V, the equipotentials are spherical surfaces centered on the charge. In Fig. 25.8 the equipotentials are drawn as dashed circles. Near the charge the potential changes rapidly with distance, so the equipotentials are close together. The field lines (solid lines) are normal to the equipotentials and point toward lower values of potential. The field is strong where the equipotentials are closely spaced.

Potential of a System of Point Charges

In Chapter 23 it was pointed out that electric fields obey the principle of linear superposition. Since the potential function is derived from the electric field (as in Eq. 25.5), the potential function also obeys this principle. When several point charges are present, the total potential at some point is given by the *algebraic* sum of the potentials due to all the charges:

$$V = \sum \frac{kQ_i}{r_i} \tag{25.10}$$

The scalar nature of potential means that we need to keep track only of the signs of the charges in the above sum.

Figure 25.9 shows the total potential due to two equal and opposite point charges. The dashed curves are the individual potential functions whereas the solid curve is the total potential function that would be encountered by another charge brought into the region. Figure 25.10 is a two-dimensional plot of the equipotentials and field lines for two equal and opposite charges. Once the equipotentials have been obtained, the field lines are easily drawn perpendicular to them. Notice that at the midpoint in Fig. 25.9, $V = 0$ but $E \neq 0$.

Figure 25.11 is a two-dimensional plot of the equipotentials and field lines for two equal positive charges. Figure 25.12 shows the total potential due to two equal positive point charges. Notice that at the midpoint in Fig. 25.12, $E = 0$ but $V \neq 0$.

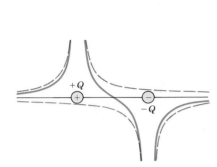

FIGURE 25.9. The dashed lines are the potentials due to each of two equal and opposite charges. The solid lines show the total potential.

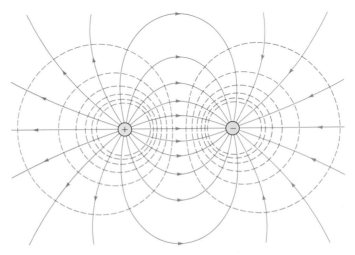

FIGURE 25.10 A two-dimensional view of the equipotentials (dashed lines) and the field lines (solid lines) for two equal and opposite charges.

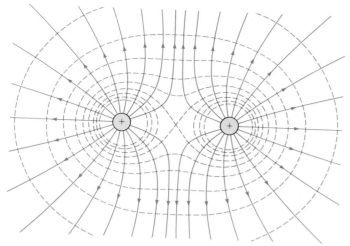

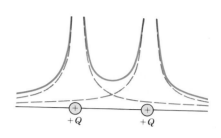

FIGURE 25.12 The dashed lines are the potentials due to each of two equal positive charges. The solid lines show the total potential.

FIGURE 25.11 A two-dimensional view of the equipotentials (dashed lines) and the field lines (solid lines) for two equal and positive charges.

Potential Energy of Point Charges

Consider a point charge q placed at a position where the potential is V. The potential energy associated with the interaction of this single charge with the charges that created V is

$$U = qV \qquad (25.11)$$

If the source of the potential is a point charge Q, the potential at a distance r from Q is $V = kQ/r$. Therefore, the potential energy shared by two charges q and Q separated by r is

$$U = \frac{kqQ}{r} \qquad (25.12a)$$

Implicit in Eq. 25.12a is the choice $U = 0$ at $r = \infty$, which allows the following interpretation:

> The potential energy of the system of two charges is the external work needed to bring the charges from infinity to the separation r without a change in kinetic energy.

When both charges have the same sign, their potential energy is positive: Positive work is needed to reduce their separation against their mutual repulsion. When the charges have opposite signs, the external work is negative. In this case, the external force has to prevent the particles from speeding up—which means that the external force is directed opposite to the displacement. Negative potential energy means that external work is required to separate the charges.

When calculating the total potential energy of a system of several charges, it is better to write Eq. 25.12a as

$$U_{ij} = \frac{kq_i q_j}{r_{ij}} \qquad (25.12b)$$

This form helps us not to double-count the contributions of the charges. Note that $U_{ij} = U_{ji}$ and that we do *not* include terms for which $i = j$. Since the potentials

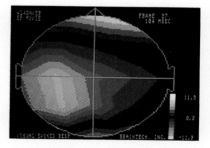

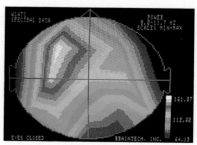

A color-coded map of equipotentials in the brain. These are "evoked" potentials measured about 0.1 s after a stimulus, such as a flash or a click. The upper display shows the presence of a tumor; the lower display is that of a person with epilepsy.

obey the principle of linear superposition, the total potential energy of a system is simply an algebraic sum and does not depend on how the charges are assembled.

EXAMPLE 25.2: Three point charges, $q_1 = 1\ \mu C$, $q_2 = -2\ \mu C$, and $q_3 = 3\ \mu C$ are fixed at the positions shown in Fig. 25.13a. (a) What is the potential at point P at the corner of the rectangle? (b) How much work would be needed to bring a charge $q_4 = 2.5\ \mu C$ from infinity and to place it at P? (c) What is the total potential energy of q_1, q_2, and q_3?

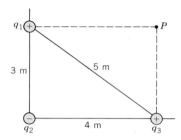

FIGURE 25.13 The potential energy of this system of charges is negative.

Solution: (a) The total potential at the point P is the scalar sum

$$V_P = V_1 + V_2 + V_3 = \frac{kq_1}{r_1} + \frac{kq_2}{r_2} + \frac{kq_3}{r_3}$$

With the given values,

$$V_1 = \frac{(9.0 \times 10^9\ \text{N·m}^2/\text{C}^2)(10^{-6}\ \text{C})}{4\ \text{m}} = 2.25 \times 10^3\ \text{V}$$

Similarly, $V_2 = -3.6 \times 10^3$ V, and $V_3 = 9 \times 10^3$ V. The total potential is $V_P = 7.65 \times 10^3$ V.
(b) The external work is $W_{\text{EXT}} = q(V_f - V_i)$. In this case, $V_i = 0$, so

$$W_{\text{EXT}} = q_4 V_P = (2.5 \times 10^{-6}\ \text{C})(7.65 \times 10^3\ \text{V}) = 0.19\ \text{J}$$

(c) The total potential energy of the three charges is the (scalar) sum

$$U = U_{12} + U_{13} + U_{23}$$
$$= \frac{kq_1q_2}{r_{12}} + \frac{kq_1q_3}{r_{13}} + \frac{kq_2q_3}{r_{23}}$$

We find, for example

$$U_{12} = \frac{(9.0 \times 10^9\ \text{N·m}^2/\text{C}^2)(10^{-6}\ \text{C})(-2 \times 10^{-6}\ \text{C})}{3\ \text{m}}$$
$$= -6 \times 10^{-3}\ \text{J}$$

Similarly, $U_{13} = +5.4 \times 10^{-3}$ J and $U_{23} = -13.5 \times 10^{-3}$ J. The

total potential energy is therefore $U = -1.41 \times 10^{-2}$ J. This negative potential energy means that external work is needed to *separate* the particles and place them at infinity.

EXERCISE 3. A point charge $q_1 = -2\ \mu C$ is located at $(-2\ \text{m}, 0)$ and charge $q_2 = 3\ \mu C$ is at $(4\ \text{m}, 3\ \text{m})$. Point A is $(0, 0)$ and point B is $(4\ \text{m}, 0)$. (a) Find the total potential at points A and B. (b) How much work would be required to move a point charge $q_3 = 5\ \mu C$ from A to B at constant speed?

EXAMPLE 25.3: In 1913, Niels Bohr proposed a model of the hydrogen atom in which an electron orbits a stationary proton in a circular path. Find the total mechanical energy of the electron given that the radius of the orbit is 0.53×10^{-10} m.

Solution: This problem is analogous to that of a satellite in orbit around the earth. The mechanical energy is the sum of the kinetic and potential energies, $E = K + U$. From Eq. 25.12a, the potential energy is

$$U = -\frac{ke^2}{r} \qquad \text{(i)}$$

To find the kinetic energy we must calculate the orbital speed, v. The centripetal force is provided by the coulomb attraction between the proton and the electron. From Newton's second law, $F = ma$,

$$\frac{ke^2}{r^2} = \frac{mv^2}{r}$$

Thus, the kinetic energy of the electron is

$$K = \tfrac{1}{2}mv^2 = \frac{ke^2}{2r} \qquad \text{(ii)}$$

Therefore, the total mechanical energy is

$$K + U = -\frac{ke^2}{2r} \qquad \text{(iii)}$$

$$= \frac{-(8.99 \times 10^9\ \text{N·m}^2/\text{C}^2)(1.60 \times 10^{-19}\ \text{C})^2}{(1.06 \times 10^{-10}\ \text{m})}$$

$$= -2.18 \times 10^{-18}\ \text{J} = -13.6\ \text{eV}$$

In Section 8.9 we saw that a negative total energy means that the orbiting particle is bound. The value 13.6 eV is in very good agreement with the experimental value of the ionization energy of the atom—the minimum energy required to remove the electron from its lowest orbit. The Bohr model is discussed further in Chapter 40.

25.4   ELECTRIC FIELD DERIVED FROM POTENTIAL

In Section 8.7 we saw how a conservative force can be derived from the derivative of the associated potential energy function, $F_x = -dU/dx$. Similarly, once the (scalar) potential function is known, one can determine the (vector) electric field. From Eq. 25.4 the change in potential associated with a displacement $d\mathbf{s}$ is

$$dV = -\mathbf{E} \cdot d\mathbf{s} = -E \, ds \cos \theta$$

Since $E_s = E \cos \theta$ is the component of $\mathbf{E}$ along $d\mathbf{s}$, the above equation may be written as $dV = -E_s \, ds$, from which we infer that

$$E_s = -\frac{dV}{ds} \qquad (25.13)$$

Since the direction of $d\mathbf{s}$ is arbitrary, Eq. 25.13 may be interpreted as follows: Any component of $\mathbf{E}$ may be found from the rate of change of V with distance in the chosen direction. There will be one direction for which this rate of change is a maximum. The full magnitude of $\mathbf{E}$ is given by this maximum value of the spatial derivative: that is, $E = -(dV/ds)_{\max}$. As Fig. 25.14 shows, the maximum occurs in the direction in which the equipotentials are most closely spaced.

In rectangular components the electric field is $\mathbf{E} = E_x\mathbf{i} + E_y\mathbf{j} + E_z\mathbf{k}$ and an infinitesimal displacement is $d\mathbf{s} = dx\mathbf{i} + dy\mathbf{j} + dz\mathbf{k}$. Thus,

$$dV = -\mathbf{E} \cdot d\mathbf{s} = -(E_x \, dx + E_y \, dy + E_z \, dz)$$

For a displacement in the x direction, $dy = dz = 0$ and so $dV = -E_x dx$. Therefore,

$$E_x = -\left(\frac{dV}{dx}\right)_{y,z \text{ constant}}$$

A derivative in which all variables except one are held constant is called a *partial* derivative and is written with a ∂ instead of d. The electric field is therefore

$$\mathbf{E} = -\frac{\partial V}{\partial x}\mathbf{i} - \frac{\partial V}{\partial y}\mathbf{j} - \frac{\partial V}{\partial z}\mathbf{k} \qquad (25.14)$$

The right side of Eq. 25.14 is called the *gradient* of V. There are no new rules of differentiation to learn, as the following example illustrates.

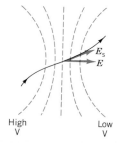

FIGURE 25.14 The electric field points from high potential to low potential. The component of the field along a displacement $d\mathbf{s}$ is $E_s = -dV/ds$. The field itself is normal to the equipotentials.

EXAMPLE 25.4: The potential due to a point charge is given by $V = kQ/r$. Find: (a) the radial component of the electric field; (b) the x component of the electric field.

Solution: (a) From Eq. 25.13 the radial component is given by

$$E_r = -\frac{dV}{dr}$$

$$= +\frac{kQ}{r^2}$$

This expression agrees with what we already know from Coulomb's law.

(b) In terms of rectangular components, the radial distance is $r = (x^2 + y^2 + z^2)^{1/2}$; therefore, the potential function $V = kQ/r$ is

$$V = \frac{kQ}{(x^2 + y^2 + z^2)^{1/2}}$$

To find the x component of the electric field, we treat y and z as constants. Thus,

$$E_x = -\frac{\partial V}{\partial x}$$

$$= +\frac{kQx}{(x^2 + y^2 + z^2)^{3/2}}$$

$$= +\frac{kQx}{r^3}$$

EXERCISE 4. Continue the calculation and show that the vector is $\mathbf{E} = (kQ/r^2)\hat{\mathbf{r}}$.

25.5   CONTINUOUS CHARGE DISTRIBUTIONS

The potential due to a system of discrete point charges is given by Eq. 25.10. The potential due to a continuous charge distribution may be found in two ways. The first is a direct calculation based on the contribution of an arbitrary charge element dq. In Fig. 25.15, the contribution to the potential at point P at a distance r from an infinitesimal (point) charge dq is

$$dV = \frac{k\,dq}{r}$$

The total potential at P is the integral of this over the charge distribution,

$$V = k \int \frac{dq}{r} \tag{25.15}$$

This equation implies the choice $V = 0$ at infinity. Equation 25.15 is generally not suitable for an infinite distribution of charge because the potential at infinity cannot be uniquely defined.

The second approach to calculating the potential is based on Eq. 25.5:

$$V_B - V_A = -\int_A^B \mathbf{E} \cdot d\mathbf{s} \tag{25.16}$$

If $\mathbf{E}$ is already known—for example, from Gauss's law—then this equation may be used to calculate ΔV. In this case, the potential can be chosen to be zero at any convenient point.

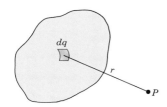

FIGURE 25.15 One way to find the potential due to a finite charge distribution is to integrate the contributions from infinitesimal charge elements dq, so $V = k \int dq/r$.

EXAMPLE 25.5: A nonconducting disk of radius a has a uniform surface charge density σ C/m². What is the potential at a point on the axis of the disk at a distance y from its center?

Solution: The symmetry of the disk tells us that the appropriate choice of element is a ring of radius x and thickness dx, as shown in Fig. 25.16. All points on this ring are at the same distance, $r = (x^2 + y^2)^{1/2}$, from the point P. The charge on the ring is $dq = \sigma\,dA = \sigma(2\pi x\,dx)$ and so the potential due to the ring is

$$dV = \frac{k\,dq}{r} = \frac{k\sigma(2\pi x\,dx)}{(x^2 + y^2)^{1/2}}$$

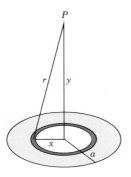

FIGURE 25.16 The appropriate charge element for a disk is a thin ring. Note that to find the potential one does not have to take components.

Since potential is a scalar, there are no components to worry about. Notice that there is only one variable, x, in this expression; the distance y is fixed. The potential due to the whole disk is the integral of the above expression:

$$\begin{aligned}
V &= 2\pi k\sigma \int_0^a \frac{x\,dx}{(x^2 + y^2)^{1/2}} \\
&= 2\pi k\sigma \left[(x^2 + y^2)^{1/2} \right]_0^a \\
&= 2\pi k\sigma [(a^2 + y^2)^{1/2} - y]
\end{aligned}$$

Let us see how this expression behaves at large distances, when $y \gg a$, or $a/y \ll 1$. We use the binomial theorem $[(1 + z)^n \approx 1 + nz$ for small z to expand the first term:

$$(a^2 + y^2)^{1/2} = y \left(1 + \frac{a^2}{y^2} \right)^{1/2}$$

$$\approx y \left(1 + \frac{a^2}{2y^2} + \cdots \right)$$

Substituting this into the expression for V we find

$$V = \frac{kQ}{y}$$

where $Q = \sigma\pi a^2$ is the total charge on the disk. At large distances, the potential due to the disk is the same as that of a point charge Q.

EXAMPLE 25.6: A shell of radius R has a charge Q uniformly distributed over its surface. Find the potential at a distance $r > R$ from its center.

Solution: The calculation of the potential from Eq. 25.15 is similar to that in Example 13.4 for gravitational potential energy. Here it is more straightforward to use the electric field, which we know from Gauss's law. At points outside a uniform spherical distribution the field is

$$\mathbf{E} = \frac{kQ}{r^2}\hat{\mathbf{r}}$$

The potential at a point for which $r > R$ may be found from Eq. 25.5. Since $\mathbf{E}$ is radial, $\mathbf{E} \cdot d\mathbf{s} = E_r\, dr$. Since $V(\infty) = 0$, we have

$$V(r) - V(\infty) = -\int_\infty^r \frac{kQ}{r^2}\, dr = -kQ\left[-\frac{1}{r}\right]_\infty^r$$

$$V = \frac{kQ}{r} \qquad (r > R) \qquad (25.17)$$

We see that the potential due to the uniformly charged shell is the same as that due to a point charge Q at the center. The result is valid for any *spherically symmetric* charge distribution since it may be treated as a series of concentric shells.

Equation 25.17 also applies to a conducting sphere since its charge resides only on the surface (see Section 24.3). Figure 25.17 shows a spherical conductor of radius R with a charge Q uniformly distributed over its surface. Under static conditions, $E = 0$ at points inside a conductor. From Eq. 25.16 we infer that $V_B = V_A$, which means that any two points in the conductor must have the same potential. Thus, *the potential has a fixed value at all points within the conducting sphere equal to the potential at the surface, $V = kQ/R$*. The variation of the electric field and the potential with r are shown in the figure.

EXAMPLE 25.7: A metal sphere of radius R has a charge Q. Find its potential energy.

Solution: We can find the potential energy by calculating the work needed to build up the charge to the final value. Suppose at some time that the charge on the sphere is q. From Eq. 25.15, its potential is $V = kq/R$. The external work required to bring an infinitesimal charge dq from infinity and to deposit it on the

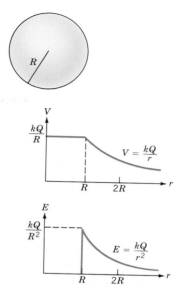

FIGURE 25.17 The variation of the potential and electric field for a spherical charged conductor.

sphere is $dW = V\, dq = (kq/R)dq$. The total work required to give the sphere a charge Q is therefore

$$W = \int_0^Q \frac{kq}{R}\, dq = \frac{kQ^2}{2R}$$

This is the potential energy of the charge distribution on the surface of the conducting sphere. It has the form $U = \frac{1}{2}QV$, where $V = kQ/R$ is the potential of the sphere. This expression should be compared with Eq. 25.11, $U = QV$. The factor of one-half arises because these two potential energies have different meanings. The expression $U = QV$ represents the potential energy associated with a *single* charge Q at a point where the potential due to other charges is V. It is the work needed to bring Q from infinity to the given position. The expression $U = \frac{1}{2}QV$ that we have just obtained is the potential energy of the whole *system* of charges. It is the work needed to bring the system of charges together.

25.6 CONDUCTORS

Figure 25.18 shows an empty cavity within a conductor in electrostatic equilibrium. It might be charged or it might be placed in an external electric field. Within the material of the conductor $\mathbf{E} = 0$; thus the change in potential $V_B - V_A = -\int_A^B \mathbf{E} \cdot d\mathbf{s}$ is zero between any two points in the material of the conductor, including the surface. Since the integral is zero for *any* path, including one through the cavity, we conclude that $\mathbf{E}$ must also be zero in the cavity. In general,

All points within and on the surface of a conductor in electrostatic equilibrium are at the same potential.

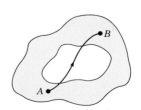

FIGURE 25.18 The field is zero inside an empty cavity in a conductor.

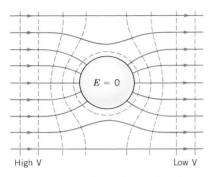

FIGURE 25.19 A conductor shields interior points from an external field.

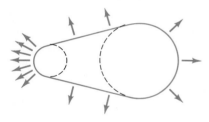

FIGURE 25.20 When two charged spheres are joined by a wire they are at the same potential.

For a displacement $d\mathbf{s}$ along the surface of a conductor, we have $dV = \mathbf{E} \cdot d\mathbf{s} = 0$, which means $\mathbf{E}$ is perpendicular to $d\mathbf{s}$. As we noted in Section 23.3, the field lines are perpendicular to the surface.

Figure 25.19 shows an uncharged spherical conductor in a uniform field. Far from the sphere, we would expect the field pattern to be unchanged: The field lines there are uniform, and the equipotentials are planes. Near the sphere, the equipotentials must be spherical and the field lines radial. The charges in the sphere redistribute themselves so as to ensure that these conditions are met. The fact that the field within the cavity is zero means that a conductor "shields" interior points from an external field. This property is useful when one wants to isolate equipment, or a signal-carrying cable, from external influences.

Suppose two charged metal spheres with radii R_1 and R_2 are connected by a long wire, as in Fig. 25.20. Charge will flow from one to the other until their potentials are equal, that is, $V_1 = V_2$. Since the spheres are far apart, their charges will be uniformly distributed, and the potential of each may be taken to be $V = kQ/R$. The equality of the potentials implies that

$$\frac{Q_1}{R_1} = \frac{Q_2}{R_2}$$

For a uniform surface charge density σ C/m², the total charge $Q = 4\pi R^2 \sigma$, so the above equation becomes

$$\frac{\sigma_1}{\sigma_2} = \frac{R_2}{R_1} \tag{25.18}$$

From Eq. 25.18 we infer that $\sigma \propto 1/R$: The surface charge density on each sphere is inversely proportional to the radius. This relationship allows us to make at least a qualitative statement regarding the charge distribution on a conductor of irregular shape, such as that in Fig. 25.21: The regions with the smallest radii of curvature have the greatest surface charge densities.*

In Section 24.3, it was shown that close to the surface of a conductor, the field strength is $E = \sigma/\varepsilon_0$. From Eq. 25.18 we infer that the field strength is greatest at sharp points on a conductor. If the field strength is great enough (about 3×10^6 V/m for dry air) it can cause an electrical discharge in air. The breakdown occurs because there are usually some molecules in the air that have been ionized (that is, electrons have been detached) by cosmic rays from space or by natural radioactivity in the soil. The electrons accelerate rapidly under the action of the electric field, collide with other molecules, and thereby create more ions. At this stage, the air loses its insulating properties and becomes a conductor. The result is a "corona discharge" that is accompanied by a visible glow. Examples include St. Elmo's fire and a glow sometimes seen around electrical transmission wires. To prevent corona discharge, high-voltage equipment has smooth surfaces with the largest possible radius of curvature.

In some instances sharp points are desirable. A lightning rod is designed to produce a continuous discharge that tends to neutralize the cloud above.** Airplanes have short wires that trail from wings and serve the same purpose. In field-ion microscopy, which is discussed later, very high electric fields are produced by extremely sharp needles.

FIGURE 25.21 On an arbitrary charged conductor the surface charge density is large where the radius of curvature is small.

---

* We assume that all parts of the surface are convex; that is, they bulge outward. See R. H. Price and R. J. Crowley, *Am. J. Phys. 53*: 843, (1985).

** However, see the special topic on Atmospheric Electricity.

The potential at the surface of a charged sphere is $V = kQ/R$ and the field strength is $E = kQ/R^2$. Thus, at the surface $V = ER$; so, for a given breakdown field strength, $V \propto R$. The potential of a sphere of radius 10 cm may be raised to 3×10^5 V before breakdown. On the other hand, a 0.05-mm dust particle can initiate a discharge at 150 V. Dust in grain silos or cement mills can easily become charged by friction and raised to this potential. The resulting electric discharges have led to several serious dust explosions in Canada and the United States.

SUMMARY

Electric potential is a scalar quantity that is related to work and potential energy in a conservative, electric field. The change in potential ΔV in moving from point A to point B is

$$V_B - V_A = \frac{W_{\text{EXT}}}{q} = \frac{\Delta U}{q}$$

where W_{EXT} is the work required to move the charge q from A to B at constant speed. $W_{\text{EXT}} = \Delta U$ is the associated change in potential energy. Like the electric field, potential is a function that depends on the *source* charges, not on any "test" charge. Only changes in potential are significant, so one can arbitrarily choose the point at which $V = 0$. Potential can also be related to the electric field:

$$V_B - V_A = -\int_A^B \mathbf{E} \cdot d\mathbf{s}$$

The integral does not depend on the path taken from A to B.

In a *uniform* field, the change in potential may be written

(Uniform field) $\qquad\qquad \Delta V = \pm Ed$

where $\pm d$ is the component of the displacement, from the initial to the final point, parallel to $\mathbf{E}$. The positive sign applies to a displacement *against* the field.

The potential at a distance r from a point charge Q is

$$V = \frac{kQ}{r}$$

It is implies that $V = 0$ at $r = \infty$. The sign of Q must be included. The potential due to a system of charges is given by the *algebraic* sum of the potentials due to individual charges. The potential due to a continuous charge distribution is given by Eq. 25.16 or

$$V = \int \frac{k \, dq}{r}$$

The potential function may be represented by equipotential surfaces. In a two-dimensional plot, the equipotentials are lines. The electric field is perpendicular to the equipotentials and points from higher to lower potential.

The **potential energy** associated with a single charge at a point where the potential due to other charges is V is given by

$$U = qV$$

In calculating the potential energy of a system of charges, care must be taken not to double-count contributions. Positive potential energy means that positive external work was required to bring the charges from infinite and to place them at their given positions. Negative potential energy means that positive external work must be done to *separate* the charges.

Once the scalar potential function V has been found, the component of the electric field in the direction of the displacement $d\mathbf{s}$ may be found from

$$E_s = -\frac{\partial V}{\partial s}$$

where s is usually x, y, z, or r.

For a homogeneous **conductor** in electrostatic equilibrium, the potential is the same at all points within the material and on the surface.

ANSWERS TO IN-CHAPTER EXERCISES

1. (a) The potential increases in a direction opposite to the field lines, $\Delta V > 0$. (b) $\Delta U = q \, \Delta V$. The potential energy of a positive charge would increase. The potential energy of a negative charge would decrease.

2. $\Delta K = -q \, \Delta V$. Since q is negative and ΔV is positive, it follows that $\Delta K > 0$.

3. (a) The total potential at any point is $V = kq_1/r_1 + kq_2/r_2$.

$$V_A = \frac{kq_1}{2 \text{ m}} + \frac{kq_2}{5 \text{ m}} = -3.6 \times 10^3 \text{ V}$$

$$V_B = \frac{kq_1}{6 \text{ m}} + \frac{kq_2}{3 \text{ m}} = 6 \times 10^3 \text{ V}$$

(b) The work needed to move q_3 from A to B is

$$W_{\text{EXT}} = q_3(V_B - V_A) = (5 \times 10^{-6} \text{ C})(9.6 \times 10^3 \text{ V}) = 48 \text{ mJ}$$

4. E_y and E_z have the same form as E_x. So,

$$\mathbf{E} = \frac{kQ}{r^3}(x\mathbf{i} + y\mathbf{j} + z\mathbf{k}) = \frac{kQ}{r^3}\mathbf{r}$$

Note that $\mathbf{r} = r\hat{\mathbf{r}}$.

QUESTIONS

1. (a) If the potential is zero at a point, what can you say about the electric field at that point? (a) If the field strength is zero at a point, what can you say about the potential function?

2. Does it make sense to speak of a "potential field"? If so, how would you represent it pictorially?

3. Points A and B have the same potential. In general would a net force be needed to move a charge from A to B? Would net work be required?

4. (a) Can the potential of a charged object be zero relative to ground? If so, explain how. (b) Is it possible for an uncharged body to be at a nonzero potential?

5. On a dry day, a spark between your finger and some object may involve several thousand volts. Why is this not dangerous when a mere 120 V at a wall outlet can be fatal?

6. Points A and B are at the same potential. What can be said about the field strengths at these points?

7. Why is the reception on a pocket radio better outside a car than inside it?

8. The surface of a metal object is an equipotential. Does this mean that the excess charge on it is uniformly distributed?

9. Is the equation $\Delta V = \pm Ed$ generally valid? Explain why or why not.

10. As one follows a given field line in the direction of the field, does the potential increase, decrease, or stay fixed?

11. Can two equipotential surfaces cross? Explain why or why not.

12. A circular ring of radius R has a charge $+Q$ uniformly distributed around its circumference. A negative point charge $-q$ starts at an arbitrary point on the axis and moves to the center of the ring. (a) Does the potential of the point charge increase or decrease? (b) Does the potential energy of the point charge increase or decrease?

13. Is it possible to move a charge in an electric field without doing work? If so, how?

14. What is the shape of an equipotential surface for an infinite line of charge?

15. A metal shell of radius 10 cm is charged till its potential is 70 V. (a) What is the potential at the center? (b) What is the electric field at the center?

16. Two charged metal spheres of radii R and $2R$ are temporarily placed in contact and then separated. At the surface of each sphere, which sphere has the greater value for the following: (a) charge density; (b) total charge; (c) potential; (d) electric field?

17. The electric field strength inside a charged hollow metal cube is zero, but the gravitational field strength inside a hollow cubic mass distribution is not. Why the difference?

EXERCISES

25.1 Potential

1. (I) A lightning flash may transfer up to 30 C of charge through a potential difference of 10^8 V. (a) How much energy does this involve? State your answer in eV. (b) For how long could this much energy light a 60-W bulb?

2. (I) A 12-V car battery is rated at 80 A · h, which is the charge it can transfer from one terminal to the other through an external circuit. (1A = 1 C/s.) (a) How much charge can the battery transfer? (b) How much energy can it provide, assuming that the potential difference between the terminals stays constant as it discharges?

3. (I) External work equal to 4×10^{-7} J is needed to move a -5-nC charge at constant speed to a point at which the potential is -20 V. What is the potential at the initial point?

4. (I) A uniform electric field is given by $\mathbf{E} = -180\mathbf{k}$ V/m. (a) What is the change in potential from $z_A = 5$ cm to $z_B = 15$ cm? (b) How far apart along the z axis are two points that differ in potential by 27 V?

5. (II) An electric field is given by $\mathbf{E} = 2x\mathbf{i} - 3y^2\mathbf{j}$ N/C. Find the change in potential from the position $\mathbf{r}_A = \mathbf{i} - 2\mathbf{j}$ m to $\mathbf{r}_B = 2\mathbf{i} + \mathbf{j} + 3\mathbf{k}$ m.

6. (II) Given the following electric fields, find the corresponding potential functions $V(x)$: (a) $\mathbf{E} = (A/x)\mathbf{i}$. Take $V = 0$ at $x = x_0$; (b) $\mathbf{E} = A \exp(-Bx)\mathbf{i}$. Take $V = 0$ at $x = 0$.

25.2 Potential and Potential Energy in a Uniform Electric Field

7. (I) Given that an electron starts from rest in a uniform field, what potential difference is needed to give it the following speeds: (a) 330 m/s (the speed of sound); (b) 11.2 km/s (the escape speed from the earth); (c) $0.1c$ (10% of the speed of light)?

8. (I) Repeat Exercise 7 for a proton.

9. (I) Suppose a 12-V car battery is used as a source of energy to accelerate particles. Find the speeds that would be attained by (a) an electron; and (b) a proton. Assume the particles start at rest.

10. (I) The gap in the spark plug of a car is 0.1 cm. What potential difference is needed to produce a spark given that the breakdown field strength of air is 3×10^6 V/m.

11. (I) Figure 25.22 shows two equipotential (dashed) surfaces such that $V_A = -5$ V and $V_B = -15$ V. What is the external work needed to move a -2 μC charge at constant speed from A to B along the indicated path?

12. (I) In Fig. 25.23 points A and B are 4 cm apart along the lines of a uniform field $\mathbf{E} = 600\mathbf{i}$ V/m. (a) Find the change in potential $V_B - V_A$. (b) What is the change in potential energy $U_B - U_A$ as a point charge $q = -3$ μC is moved from A to B?

13. (I) Two large parallel conducting plates carry equal and opposite charges and are separated by 5 cm. An 8-μC point

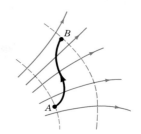

FIGURE 25.22 Exercise 11.

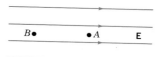

FIGURE 25.23 Exercise 12.

charge placed between them experiences a force of 2.4×10^{-2} N. Find the potential difference between the plates.

14. (I) What potential difference would be required to accelerate the following particles from rest to $0.1c = 3 \times 10^7$ m/s: (a) an alpha particle with a charge $2e$ and a mass 4 u; (b) a uranium nucleus with a charge $92e$ and a mass 235 u?

15. (I) In clear weather, at the earth's surface there is a uniform electric field of about 120 V/m directed vertically down. What is the potential difference between the ground and the following heights: (a) the top of the head of a person of height 1.8 m; (b) the top of the Sears tower whose height is 433 m?

16. (I) Two infinite parallel plates separated by 3 cm are connected to a 120-V battery. An electron starts at rest from the negative plate. (a) What is the electric field strength? (b) What is the work done by the field on the electron by the time it hits the positive plate? (c) What is the change in potential of the electron? (d) What is the change in potential energy of the electron?

17. (I) What is the work required to move a particle of mass 2×10^{-2} g and a charge of -15 μC through a change in potential of -6000 V and also to increase its speed from zero to 400 m/s?

25.3 Potential due to a Point Charge

18. (I) Two protons in a nucleus are 10^{-15} m apart. (a) What is their electrical potential energy? (b) If they were free to move and start from rest, find their speeds when they are 4×10^{-15} m apart.

19. (I) A uranium nucleus with a charge $+92e$ can spontaneously undergo fission into two fragments carrying charges $+48e$ and $+44e$. If the fragments are initially at rest $7 \times$

10^{-15} m apart, what is their total kinetic energy when they are infinitely far apart?

20. (I) (a) Find the potential due to the three charges in Fig. 25.24 at the lower left corner. (b) A charge -2 μC is placed at the lower left corner. What is its potential energy? (c) What is the potential energy of the system of four charges?

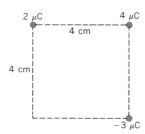

FIGURE 25.24 Exercise 20.

21. (I) Four point particles with charges 0.6 μC, 2.2 μC, -3.6 μC and $+4.8$ μC are placed at the corners of a square of side 10 cm. What is the external work needed to bring a charge of -5 μC from infinity to the center of the square? (Assume the speed of the -5 μC charge is kept constant.) What is the significance of the sign of your answer?

22. (I) Two charges Q and $-Q$ are held fixed at a separation of 4 m as shown in Fig. 25.25. Take $Q = 5$ μC. (a) What is the change in potential $V_B - V_A$? (b) A point particle of mass $m = 0.3$ g and charge $q = 2$ μC starts from rest at A. What is its speed at B?

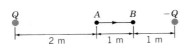

FIGURE 25.25 Exercise 22.

23. (I) Two point particles with equal charge Q are located as shown in Fig. 25.26. The points A and B are $(0, 4\text{ m})$ and $(0, 0)$, respectively. (a) Find the change in potential $V_B - V_A$. (b) If a point charge $-q$ of mass 3×10^{-8} kg is released from rest at A, what is its speed at point B? Take $q = Q = 5$ μC.

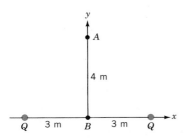

FIGURE 25.26 Exercise 23.

24. (I) At a distance r from a point charge Q, the field strength is 200 V/m and the potential is 600 V. Determine Q and r.

25. (I) A point charge $+4Q$ is at $x = 0$. At what point(s) on the x axis is the total potential zero if at $x = 1$ m there is a second charge equal to (a) $-Q$; (b) $-9Q$?

26. (I) In Fig. 25.27, the charges $Q_1 = 3$ μC, $Q_2 = -2$ μC and $Q_3 = 5$ μC are fixed. What is the external work needed to move a charge $q = -4$ μC at constant speed from point A at the center of the square to point B at the corner? What is the significance of the sign of your answer?

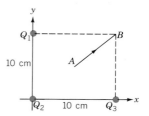

FIGURE 25.27 Exercise 26.

27. (I) A 5-μC point charge is placed at the origin in Fig. 25.28. Find the potential at the points (a) A, (b) B, and (c) C.

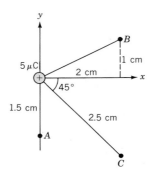

FIGURE 25.28 Exercise 27.

28. (I) Two point charges, -4 μC and $+6$ μC, are located as shown in Fig. 25.29. (a) What is the potential at the origin? (b) What is the external work required to bring a 2-μC charge at constant speed from infinity to the origin?

29. (I) In the quark model of elementary particles a proton consists of two up (u) quarks, each with charge $+2e/3$, and a down (d) quark with charge $-e/3$. Assuming that the quarks are equally spaced around a circle of radius 1.2×10^{-15} m, find the total electrostatic potential energy.

30. (I) A point charge $q_1 = -4$ μC is located at (3 cm, 0) and a charge $q_2 = 3.2$ μC is located at (0, 5 cm). Find: (a) the potential due to q_2 at the position of q_1; (b) the potential due to q_1 at the position of q_2; (c) the potential energy of the pair.

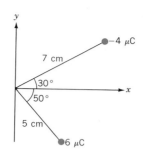

FIGURE 25.29 Exercise 28.

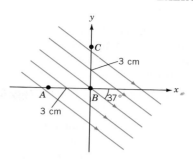

FIGURE 25.31 Exercise 35.

31. (I) Three point charges, $q_1 = 6\,\mu C$, $q_2 = -2\,\mu C$, and q_3, are located as shown in Fig. 25.30. For what value of q_3 will the potential at the origin be: (a) 0 V; (b) -400 kV?

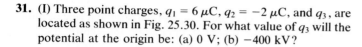
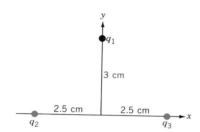

FIGURE 25.30 Exercise 31.

32. (I) A -10-μC point charge is located at (0, 3 cm) and a 6-μC point charge is located at (4 cm, 0). (a) What is the potential difference between the origin and the point (4 cm, 3 cm)? (b) How much external work would it take to bring a -2-μC point charge at constant speed from infinity to the origin?

33. (I) A uranium nucleus with a charge of $92e$ undergoes spontaneous fission into two fragments with equal charges. They are initially at rest and are separated by 7.4×10^{-15} m. (a) What is the initial potential energy? (b) What is the final kinetic energy of the fragments when they are infinitely far apart? (c) Assuming 30% of the kinetic energy of the fragments can be harnessed in a nuclear reactor, how many fissions per second are required for a power output of 1 MW?

34. (II) The field produced by an infinite sheet of charge in the yz plane with density σ C/m^2 is $\sigma/2\varepsilon_0\,\mathbf{i}$. (a) Write an expression for the potential $V(x)$ at a distance x from the sheet. Take $V = 0$ at a distance x_0. (b) What displacement Δx is associated with a potential difference of 20 V? Take $\sigma = 7$ nC/m^2.

35. (II) A uniform electric field of 400 V/m is directed at $37°$ below the x axis, as shown in Fig. 25.31. Find the changes in potential: (a) $V_B - V_A$; (b) $V_B - V_C$.

36. (II) An electron travels along the lines of a uniform electric field. Its initial speed is 8×10^6 m/s and its final speed after traveling a distance of 3 mm along the positive x axis is $3 \times$

10^6 m/s. (a) What is the potential difference between the two positions? (b) What is the electric field?

37. (II) Carbon dioxide (CO_2) is an example of a linear quadrupole, shown in Fig. 25.32. Find the potential at a point (a) $(x, 0)$, and (b) $(0, y)$ for $y > a$. In each case, show that $V \propto 1/r^3$ for $r \gg a$, where r is the distance from the origin.

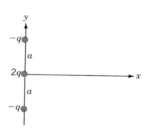

FIGURE 25.32 Exercise 37.

38. (II) (a) A 2-nC charge is at the origin. Find the distances at which the potential is 0.5 V, 1 V, 1.5 V, 2 V, 2.5 V, 3 V and 3.5 V. (b) Repeat part (a) for a negative charge (-2 nC) and negative potentials. (c) Place the charges 12 m apart. Draw circles to represent the equipotentials of the invididual charges. Indicate the points of intersection of the two sets of circles where the net potential is either 1 V or 0.5 V. Finally, join each set of points having the same potential. Compare the shapes of your equipotential curves with Fig. 25.10.

39. (II) Starting at a point 1 m away from a 2 nC charge, how far in the radial direction are the points at which the potential is (a) 1 V higher; (b) 1 V lower?

40. (II) An α particle of mass 6.7×10^{-27} kg and charge $+2e$ has an initial kinetic energy of 4.2 MeV. It is fired at a gold nucleus of charge $+79e$. Assuming that the nucleus stays at rest and that the α particle returns along its original path, find the distance of closest approach.

41. (II) A uranium nucleus with a charge of $92e$ can spontaneously decay into a thorium nucleus of charge $90e$ and an α particle of charge $2e$. The mass of the thorium is 234 u and that of the α particle is 4 u. Assume that just after the decay the decay products are 7.4×10^{-15} m apart and at rest. (a) What is the initial potential energy? (b) Find the final kinetic energy of the α particle assuming that the thorium stays at rest. (In Problem 1 this assumption is omitted.)

25.4 Electric Field Derived from Potential

42. (II) Two equal positive charges Q are at $(0, a)$ and $(0, -a)$, respectively. (a) Find the potential $V(x)$ at a point $(x, 0)$. (b) Use $V(x)$ to find the electric field along the x axis.

43. (II) Two equal positive charges Q are located at $(0, a)$ and $(0, -a)$. (a) Find the potential $V(y)$ at a point $(0, y)$ for $y > a$. (b) Use $V(y)$ to find the electric field along the y axis.

44. (II) A dipole consists of charges $-Q$ at $(-a, 0)$ and $+Q$ at $(a, 0)$. (a) What is the potential $V(x)$ at a point $(x, 0)$ for $x > a$? (b) From $V(x)$ find the electric field along the x axis.

45. (II) A sphere of radius R has a charge Q uniformly distributed throughout its volume. For $r < R$, the potential function is

$$V(r) = \frac{kQ(3R^2 - r^2)}{2R^3}$$

Find the radial component of the electric field from $V(r)$.

46. (II) The potential $V(r)$ at a perpendicular distance r from an infinite line of charge with density λ C/m is

$$V(r) = V(r_0) - 2k\lambda \ln\left(\frac{r}{r_0}\right)$$

where r_0 and $V(r_0)$ are constants. From $V(r)$, find the electric field.

47. (II) The potential at a point along the axis of a uniformly charged disk was found in Example 25.5. Use this expression to find the electric field strength along the central axis.

48. (II) A hypothetical potential function has the following form:

$$V(x, y, z) = 2x^3y - 3xy^2z + 5yz^3$$

What is the electric field?

25.5 Continuous Charge Distributions

49. (I) Assume the proton is a uniformly charged sphere of radius 10^{-15} m. Find the potential at the following points: (a) its surface; (b) the position of the electron in a hydrogen atom; that is, 5.3×10^{-11} m. (c) How would these results change if the proton were a spherical shell instead?

50. (II) A charge Q is uniformly spread around a ring of radius a. (a) Find the potential $V(y)$ along the central axis at a distance y from the center. Does your expression behave appropriately when $y \gg a$? (b) Use $V(y)$ to find the electric field strength along the central axis. Does your expression behave appropriately when $y \gg a$?

25.6 Conductors

51. (I) The breakdown field strength of dry air is 3×10^6 N/C. At this value, what would be the potential at the surface of a charged metal sphere of radius: (a) 0.01 mm; (b) 1 cm; (c) 1 m?

52. (I) The potential of a metal sphere of diameter 2 cm is 10^4 V relative to ground. (a) What is the surface charge density? (b) How many electrons were removed from the sphere? (c) What is the electric field strength at the surface?

53. (II) Two concentric spherical metal shells have radii a and b, respectively. The inner shell of radius a has charge Q, while the outer shell has charge $-2Q$. Sketch V and E as functions of r, the distance from the center.

PROBLEMS

1. (I) A uranium nucleus (charge $92e$, mass 238 u) at rest decays into a thorium nucleus (charge $90e$, mass 234 u) and an α particle (charge $+2e$, mass 4 u). Just after the decay the particles are at rest and separated by 7.4×10^{-15} m. Find the kinetic energy of each decay particle when they are infinitely far apart. Do not assume that the thorium stays at rest.

2. (I) A disk of radius b has a concentric hole of radius a. There is a uniform surface charge density σ. Find the potential at a point on the axis of the disk at a distance y from the center.

3. (I) In a NaCl crystal, Na^+ and Cl^- ions lie on a three-dimensional cubic array, as shown in Fig. 25.33. The nearest neighbor of any ion is at a distance 2.82×10^{-10} m. Find the potential energy of a Na^+ ion: (a) by including only the contributions of the six nearest neighbors; (b) by including the contributions of the twelve next-nearest neighbors.

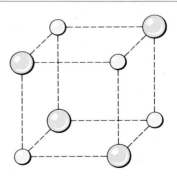

FIGURE 25.33 Problem 3.

4. (II) A beam of electrons is accelerated by 20 kV and bombards a 500-g tungsten target with 4×10^{16} electrons per second. Assuming that 30% of the energy of the electrons goes into heating the target, how long does it take for the temperature of the target to rise by 10 °C? (The specific heat of tungsten is 134 J/kg · K.)

5. (I) A metal sphere of radius R_1 has a charge Q_1. It is enclosed by a conducting spherical shell of radius R_2 that has a charge $-Q_2$; see Fig. 25.34. Determine: (a) the potential V_1 of the inner sphere; (b) the potential V_2 of the outer sphere; (c) the potential difference $V_1 - V_2$. (d) Under what condition is $V_1 = V_2$?

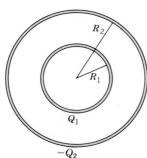

FIGURE 25.34 Problem 5.

6. (II) A balloon of radius R has a uniform surface charge density σ C/m². Show that the surface experiences an electrostatic force per unit area equal to $\sigma^2/2\varepsilon_0$ N/m². (*Hint:* Use the relation $F_r = -dU/dr$.)

7. (I) A coaxial cable has an inner wire of radius a with a linear charge density λ C/m surrounded by a cylindrical sheath of radius b carrying a linear charge density $-\lambda$. (a) Use the electric field ($E = 2k\lambda/r$) between the wire and the sheath to show that the potential difference between them is

$$V(b) - V(a) = -2k\lambda \ln\left(\frac{b}{a}\right)$$

(b) In a radiation detector, called a Geiger counter, that has a similar cylindrical geometry, $a = 3 \times 10^{-3}$ cm, $b = 2.5$ cm, and $\Delta V = 800$ V. What is the electric field strength at the surface of the inner wire?

8. (I) A rod of length L has a charge Q uniformly distributed along its length. Find the potential at a distance a from one end, along the axis of the rod, as in Fig. 25.35.

FIGURE 25.35 Problem 8.

9. (I) A rod of length L with a uniform linear charge density λ C/m lies along the x axis. Find the potential at a distance y from one end, perpendicular to the rod; see Fig. 25.36. The length of the rod is L. (See the table of integrals in Appendix C.)

FIGURE 25.36 Problem 9.

10. (II) A sphere of radius R has a charge Q uniformly distributed throughout its volume. Show that for $r < R$, the potential is

$$V(r) = \frac{kQ(3R^2 - r^2)}{2R^3}$$

(*Hint:* The electric field within a uniformly charged sphere is $E = kQr/R^3$. Evaluate $V(r) - V(R)$.)

11. (II) A nonconducting sphere of radius R has a total charge Q spread uniformly throughout its volume. Show that the potential energy of the sphere is

$$U = \frac{3kQ^2}{5R}$$

(*Hint:* First obtain an expression for the potential at the surface of a uniformly charged sphere of radius $r < R$. The charge within a thin shell from r to $r + dr$ is $dq = \rho(4\pi r^2\, dr)$. The work needed to bring an infinitesmal charge dq from infinity to a point with potential V is $V\, dq$.)

12. (II) (a) Show that the potential due to a dipole (see Fig. 25.37) with a dipole moment $p = 2aq$ at a distance r from its center is given by

$$V(\theta) = \frac{kp \cos \theta}{r^2}$$

where $r \gg a$. (*Note:* $r_- - r_+ \approx 2a \cos \theta$ and $r_+ r_- \approx r^2$.) (b) Use the above expression to find the components of the electric field:

$$E_r = -\frac{\partial V}{\partial r}; \qquad E_\theta = -\frac{1}{r}\frac{\partial V}{\partial \theta}$$

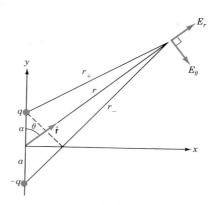

FIGURE 25.37 Problem 12.

13. (II) At large distances, the potential due to a dipole may be written in the form (see Problem 12)

$$V = \frac{k\mathbf{p} \cdot \mathbf{r}}{r^3}$$

Use $E_s = -\partial V/\partial s$, where $s = x$ or y, to show that if the dipole moment is directed along the $+x$ axis, then

$$E_x = \frac{kp(2x^2 - y^2)}{r^5}; \qquad E_y = \frac{3kpxy}{r^5}$$

14. (II) Show that the expression

$$\mathbf{E} = \frac{k}{r^3}[3(\mathbf{p} \cdot \hat{\mathbf{r}})\hat{\mathbf{r}} - \mathbf{p}]$$

yields the same results for E_x and E_y as Problem 13 for the field due to a dipole. (*Hint:* Assume $\mathbf{p} = p\mathbf{i}$. Express $\hat{\mathbf{r}} = \mathbf{r}/r$ in terms of x and y.)

15. (II) The potential energy of a dipole, whose dipole moment is $\mathbf{p}_2$, in the field $\mathbf{E}_1$ due to another dipole is $U = -\mathbf{p}_2 \cdot \mathbf{E}_1$. Use the expression for $\mathbf{E}$ in Problem 14 to show that the potential energy of the dipole–dipole interaction is

$$U = \frac{k}{r^3}[\mathbf{p}_1 \cdot \mathbf{p}_2 - 3(\mathbf{p}_1 \cdot \hat{\mathbf{r}})(\mathbf{p}_2 \cdot \hat{\mathbf{r}})]$$

Evaluate the interaction energy for two water molecules for which $p = 6.2 \times 10^{-30}$ C $\cdot$ m. Take $r = 0.4$ nm. Do this for four configurations of the dipole moments: (a) parallel side-by-side; (b) antiparallel side-by-side; (c) in line parallel; (d) in line antiparallel.

16. (II) The potential energy of a system of charges is given by

$$U = \sum \frac{kq_iq_j}{r_{ij}}$$

where the sum is taken over all *distinct* pairs; that is, there is no double-counting and $i \neq j$. Show that an equivalent expression is

$$U = \sum \tfrac{1}{2}q_iV_i$$

where the sum is over all charges in the system and V_i is the potential at the position of q_i produced by all charges, but *not* including q_i.

SPECIAL TOPIC: Electrostatics

Static electricity can be a nuisance—for example, when clothes stick together or when we receive tiny shocks on a dry day. It can also trigger dangerous events such as lightning strokes or explosions in grain silos and oil tankers. There are, however, several useful applications of electrostatics. Electrostatic charges are used in the separation of minerals, in spraying paint or chemicals, in separating grain from rodent droppings, in the coating of sandpaper or flocked wallpaper, and in priming auto bodies (Fig. 25.38). Here are a few applications discussed in more detail.

VAN DE GRAAFF ACCELERATOR

In Section 24.3 it was pointed out that the net charge on a conductor resides on its surface. This fact is used in a charged particle accelerator invented by Van de Graaff of MIT in 1932. In this device, charge is sprayed via corona discharge from a metal comb at high potential (2×10^4 V) onto a moving belt (Fig. 25.39). The insulating belt carries the charge into a large spherical dome supported by an insulating column. A second comb collects the charge from the belt and transfers it to the outer surface of the sphere. The charge and potential of the sphere rise until the breakdown field is reached at the surface. To increase the breakdown field, the whole apparatus is enclosed in a high-pressure (400-psi) vessel containing some gas. For a belt 50 cm wide moving at about 20 m/s, the charging current is $I = 1$ mA.

An ion source placed inside the sphere is at a high

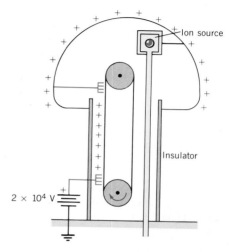

FIGURE 25.39 A Van de Graaff generator.

potential relative to ground. Therefore, charged particles, such as protons or ions, can be accelerated along a tube to collide with a target where they produce effects studied in nuclear physics, solid state physics, or medical applications.

PRECIPITATOR

In 1907, F. G. Cottrell invented a simple device to clean emissions from the smoke stacks in cement mills, steel smelters, power generating stations, and other chemical processing plants. In a typical precipitator, shown in Fig. 25.40, a short wire is maintained at a high potential (60 kV) relative to a grounded outer cylindrical conductor. Polluted gases enter at the bottom and pass through the strong electric field around the wire. There is a constant corona discharge from the wire into the surrounding air. Electrons accelerated by the high field produce further ionization in the particles of the pollutants. The resulting positively charged particles are attracted to the outer sheath, where they stick. This enables the preciptator to remove particles whose size is about 10 μm. Figure 25.41 shows the effectiveness of the technique. The cylinder has to be periodically shaken or flushed to remove the collected material. In commercial plants the central wire is negative. In household versions, the central wire is positive because it has been found that this polarity minimizes the production of ozone.

FIGURE 25.38 The particles of a paint primer are attracted to a car that is raised to a high potential.

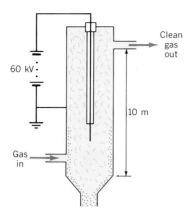

FIGURE 25.40 In an electrostatic precipitator, a large potential difference is maintained between a short central wire and the outer casing. Pollutant particles become ionized and attracted to the casing and collect there.

FIGURE 25.41 The emissions from a smoke stack are reduced by an electrostatic precipitator.

IMAGING

Perhaps the most widespread use of electrostatics is in the copying machines found in most modern offices. The process of electrostatic imaging was invented in 1935 by C. F.

Coulson. After several years of development, the first product was put on the market by the Xerox Corporation in 1948.

At the heart of the process is a material called a *photo-conductor*, which is an insulator in the dark. When exposed to light, it becomes a conductor because some electrons acquire enough energy to escape from their parent atoms and become free electrons. The photoconducting material is usually in the form of a thin layer (25 μm thick) of selenium or ZnO powder in plastic, spread over a conducting substrate. The essential steps in the copying process are outlined below:

1. A thin wire (0.015 cm) at high potential (7 kV) moves over the plate and applies a uniform layer of positive charge on the photoconducting layer via corona discharge (see Fig. 25.42a).

2. The photoconductor is then exposed to light reflected from the subject—for example, a typed page. The areas exposed to the light become conducting and allow the surface charge to drain through the grounded plate below (see Fig. 25.42b).

3. The photoconductor is now coated with "toner" particles, as in Fig. 25.42c. For example, glass beads (600 μm in diameter) may be coated with a monomolecular layer of some carbonized plastic or resin. The two materials become oppositely charged as they are shaken. Alternatively, a charged carbon particles (1 μm in diameter) or an aerosol is sprayed over the photoconductor. The negatively charged toner particles adhere to the positively charged regions.

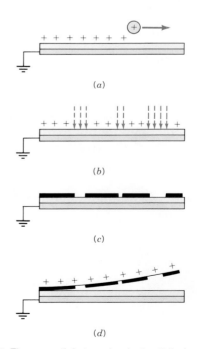

FIGURE 25.42 The essential steps in electrostatic imaging.

4. The latent image must now be transferred to a paper hard copy. Since the toner particles retain some negative charge, it is necessary to spray the paper with positive charge, as shown in Fig. 25.42*d*.

5. The image on the paper is fused by heat from a filament.

As you know, the whole process occurs in about one second.

FIELD-ION MICROSCOPE

E. W. Muller of The Pennsylvania State University invented the field-ion microscope (Fig. 25.43) in 1955. This device is used to study defects in semiconductors, thin films, and other surface structures. For this device, a very fine wire is etched with acid to produce a tip whose radius is about 0.05 μm. The tip is then inserted into a glass enclosure in which there is a high vacuum (10^{-9} mm Hg), and a large potential difference is applied between the (positive) tip and the (negative) enclosure. The field strength at the tip is about 4.5×10^8 V/m. Only certain metals, such as platinum, tungsten, and chromium, can withstand such high fields without disintegrating. Finally, a gas of inert atoms, such as He or Ne, is introduced into the enclosure. When a He atom comes close to the tip, it becomes ionized and the He$^+$ ion is accelerated toward a fluorescent screen on the other side of the enclosure. The pattern of dots on the screen (Fig. 25.44) reflects the arrangement of atoms on the surface of the tip. When the tip is cooled—for example, by liquid hydrogen—to minimize the thermal vibrations of the atoms, details as small as 2.5×10^{-10} m can be resolved.

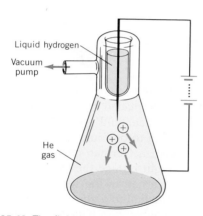

FIGURE 25.43 The field-ion microscope.

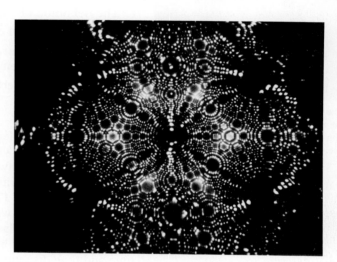

FIGURE 25.44 The tip of a needle as revealed in a field-ion microscope.

CHAPTER 26

Capacitors and Dielectrics

The "Cap Tree" at Fermilab reduces the demand on the local community's power supply when the accelerator is in use.

In the early days of electrical investigations there was no way of storing charge for long periods. Even when a charged body is placed on an insulated stand, the charge tends to leak away. The loss of "electrical fluid" (charge) was attributed to some form of evaporation, and so a way had to be found to "condense" charge without losing it. In 1745, E. G. von Kleist, a German clergyman, thought that enclosing electrified water in a glass jar might reduce the loss of charge. He put some water into a glass jar and immersed a nail in the water, as in Fig. 26.1a. Taking the jar in one hand, he connected the nail to a charging machine for some time and then disconnected it. Being an amateur, von Kleist made the mistake of not putting the jar on an insulated stand. When he touched the nail with the other hand, he got a tremendous shock. He later found that the jar could remain electrified for long periods provided it was not disturbed.

Others had difficulty reproducing von Kleist's feat because they followed the common procedure of keeping the jar insulated as it was being charged. Three months later, in 1746, Pieter van Musschenbroek, a professor at the University of Leiden, realized that it was necessary to hold the jar both during the charging process and also later in order to receive a shock. One shock was quite enough for him: "I thought it was all up with me!" (In the charging process the charged conductor on the inside (the water) induces an opposite charge on the outer conductor (the hand), which is connected to ground via the body—which is a conductor (see Fig. 22.6b). Later, the shock is felt when the charges pass from one hand to the other through the body. The discharge is much quicker than the charging process.)

Others soon realized that the water could be replaced by lead shot. Later, the lead shot and the hand were replaced by metal foil covering the inner and outer surfaces of the glass jar. Ben Franklin then substituted a flat pane of glass for the glass jar. Finally, the simplest device consisted of two flat metal plates separated by air.

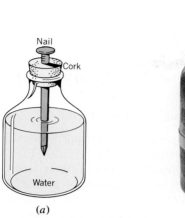

Nail
Cork
Water

(a)   *(b)*

FIGURE 26.1 (*a*) von Kleist charged the water by connecting the nail to an electrical machine. (*b*) A Leyden jar.

Von Kleist's accidental invention came to be known as the Leyden jar (Fig. 26.1*b*) and served as a basis of electrical researches for the next 50 years. It was the first "condenser," or what is now called a **capacitor,** a device that stores charge and electrical energy. Although they are relatively simple devices, capacitors play a vital role in the tuning circuits of radios, in electronic timing circuits, in electronic flashguns, and in other devices. Capacitors are used to smooth fluctuations in the output of power supplies in radios and TVs. Experiments in high-energy particle accelerators, and especially those in fusion power research, require extremely high instantaneous power levels—greater than the capability of all the generating stations in the United States! Therefore, to prevent overloading of the transmission lines in a region, huge banks of capacitors are slowly charged up and then rapidly discharged when needed.

26.1   CAPACITANCE

A capacitor consists of two conductors, called *plates,* separated by an insulator, such as air or paper. The plates may be given equal but opposite charges by connecting them to a battery, which is a device that maintains a constant potential difference between its terminals (see Fig. 26.2). In effect, the battery transfers charge from one plate to the other. The circuit symbol for a capacitor is ─┤├─ , whereas a battery is drawn as ─┤┠─, where the short line represents the negative terminal. The potential of each plate is the same as that of the terminal to which it is connected since there is no potential difference across a conductor (the wire and the plate) under static conditions. Therefore, the potential difference between the plates is the same as that between the terminals of the battery. When the battery is disconnected, the charges remain on the plates, held by their mutual attraction.

The magnitude of the charge Q stored on either plate of a capacitor is directly proportional to the potential difference V between the plates. Therefore, we may write

$$Q = CV \qquad (26.1)$$

where C is a constant of proportionality called the **capacitance** of the capacitor. The capacitance of a capacitor is a measure of its ability to store charge and

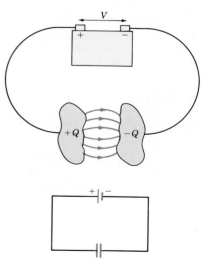

FIGURE 26.2 The plates of a capacitor are given equal and opposite charges by connecting them to a battery.

electrical energy. By expressing Eq. 26.1 in the form

$$C = \frac{Q}{V} \qquad (26.2)$$

we see that the capacitance tells us the quantity of charge a capacitor can store per unit potential difference between the plates. The SI unit of capacitance is the farad (F). From Eq. 26.2 we see that

$$1 \text{ farad} = 1 \text{ coulomb/volt}$$

In practice one farad is a very large value. As a result, values are usually quoted in picofarads (1 pF = 10^{-12} F) or microfarads (1 μF = 10^{-6} F). The capacitance of a capacitor depends on the *geometry* of the plates (their size, shape, and relative positions) and the *medium* (such as air, paper, or plastic) between them. The capacitance does *not* depend on Q or V individually. If the potential difference is doubled, the stored charge will also double, so their ratio is unchanged.

Parallel-Plate Capacitor

A common arrangement found in capacitors consists of two flat plates. If the plate separation is small, we may ignore the fringing fields at the ends, as shown in Fig. 26.3, and assume that the field is uniform, as in Fig. 26.4. The plates, which each have an area A and are separated by a distance d, are given opposite charges of the same magnitude Q. The charges lie on the inner surfaces of the plates. From Gauss's law applied to a conductor (Eq. 24.8), or by direct calculation (Eq. 23.10), we know that the field between the plates is given by

$$E = \frac{\sigma}{\varepsilon_0} = \frac{Q}{\varepsilon_0 A}$$

where $\sigma = Q/A$ is the surface charge density. From Eq. 25.6c, the potential difference in a uniform field is $V = Ed$. Therefore, the capacitance ($C = Q/V$) is

$$C = \frac{\varepsilon_0 A}{d} \qquad (26.3)$$

From Eq. 26.3 we see that an alternative unit for ε_0 is F/m:

$$\varepsilon_0 = 8.85 \times 10^{-12} \text{ F/m}$$

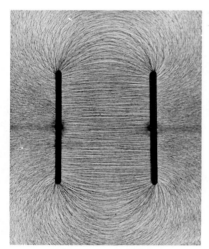

FIGURE 26.3 The electric field produced by two oppositely charged plates of finite size is not uniform.

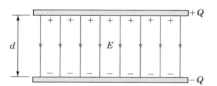

FIGURE 26.4 If the plate separation is small we may ignore the fringing fields and consider the field to be uniform.

One can understand how the capacitance depends on the geometrical quantities A and d as follows. The proportionality $C \propto A$ arises because for a given potential difference, more charge can be held on a larger plate. For a given potential difference V, the field $E = V/d$. Since $E = \sigma/\varepsilon_0$ and $Q = \sigma A$, we see that $Q = \varepsilon_0 VA/d$, which means $Q \propto 1/d$. That is, the amount of charge stored is inversely proportional to the plate separation. Therefore, $C \propto 1/d$ is also reasonable. Alternatively, consider how the potential difference varies for a given charge on the plates. We know that E is constant (ignoring fringe effects), and so $V (=Ed) \propto d$. As the plates are moved farther apart, the potential difference between them increases. Again, we find that $C (=Q/V) \propto 1/d$ is reasonable.

Practical parallel plate capacitors are made with two strips of metal foil separated by insulating plastic sheets. The sandwich is wrapped into a cylindrical form and encased, as in Fig. 26.5a. A different form of parallel plate capacitor, found in older radios, is shown in Fig. 26.5b. The area of overlap between two sets of semicircular plates can be varied by rotating one set. As you turn the tuning knob

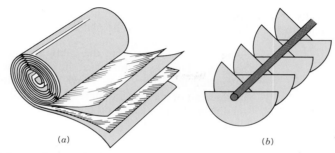

FIGURE 26.5 (a) A capacitor can be made by putting plastic between two metallic foils. (b) A variable capacitor. The capacitance depends on the area of overlap between two sets of plates; one is fixed, whereas the other can rotate.

of the radio, you are varying the capacitance. This in turn changes the frequency of the radio waves to which the radio responds (see Section 33.4).

EXAMPLE 26.1: A parallel-plate capacitor with a plate separation of 1 mm has a capacitance of 1 F. What is the area of each plate?

Solution: From Eq. 26.3,

$$A = \frac{Cd}{\varepsilon_0} = \frac{(1 \text{ F})(10^{-3} \text{ m})}{8.85 \times 10^{-12} \text{ F/m}}$$
$$= 1.13 \times 10^8 \text{ m}^2$$

This is approximately 10 km × 10 km! Clearly, the farad is a very large unit.

EXAMPLE 26.2: A parallel-plate capacitor has plates with dimensions 3 cm × 4 cm, separated by 2 mm. The plates are connected across a 60-V battery. Find: (a) the capacitance; (b) the magnitude of the charge on each plate.

Solution: (a) The area of the plates is $A = 12 \times 10^{-4}$ m². The capacitance is given by Eq. 26.3:

$$C = \frac{\varepsilon_0 A}{d}$$
$$= \frac{(8.85 \times 10^{-12} \text{ F/m})(1.2 \times 10^{-3} \text{ m}^2)}{2 \times 10^{-3} \text{ m}}$$
$$= 5.31 \text{ pF}$$

(b) The magnitude of the charge on each plate may be determined with Eq. 26.1, $Q = CV$. The value of the capacitance has been found in part (a), therefore,

$$Q = CV$$
$$= (5.31 \times 10^{-12} \text{ F})(60 \text{ V})$$
$$= 3.19 \times 10^{-10} \text{ C}$$

EXAMPLE 26.3: What is the capacitance of an *isolated sphere* of radius R?

Solution: If an isolated metal sphere of radius has a charge $+Q$, one can consider this charge as having been transferred from the ground—which is a reasonable conductor and acts as the "other plate." The potential of the sphere is $V = kQ/R$. Since $k = 1/(4\pi\varepsilon_0)$, the capacitance, $C = Q/V$, is

$$C = 4\pi\varepsilon_0 R \qquad (26.4)$$

We see that the capacitance depends on the radius, a geometric quantity. This result seems reasonable since a larger sphere requires a greater charge to raise it to a given potential. If we assume that the earth is a conducting sphere of radius 6370 km, Eq. 26.4 predicts that its capacitance would be 710 μF.

EXAMPLE 26.4: A *spherical capacitor* consists of two concentric conducting spheres, as shown in Fig. 26.6. The inner sphere, of radius R_1, has charge $+Q$. The charge on the outer shell of radius R_2 is $-Q$. Find its capacitance.

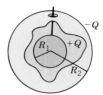

FIGURE 26.6 A spherical capacitor.

Solution: The potential difference between the spheres is determined by the electric field, $\Delta V = -\int \mathbf{E} \cdot d\mathbf{s}$. Since the field has only a radial component, the dot product is $\mathbf{E} \cdot d\mathbf{s} = E_r \, dr$, where $E_r = +kQ/r^2$. (Note that E_r is due solely to the charge on the *inner* sphere.) If we choose a path from the inner to the outer sphere, the potential difference is

$$V_2 - V_1 = - \int_{R_1}^{R_2} E_r\,dr = -\left[-\frac{kQ}{r} \right]_{R_1}^{R_2}$$

$$= kQ \left(\frac{1}{R_2} - \frac{1}{R_1} \right)$$

This happens to be a negative quantity (because of the direction in which the integral was evaluated), but we are interested only in its magnitude. From $C = Q/V$ we find

$$C = \frac{R_1 R_2}{k(R_2 - R_1)} \tag{26.5}$$

This expression can be related to the capacitance of an isolated sphere and to that of a parallel plate capacitor (see Problem 11).

EXAMPLE 26.5: A *cylindrical capacitor* consists of a central conductor of radius a surrounded by a cylindrical shell of radius b, as shown in Fig. 26.7. A coaxial cable used for transmission of TV signals has this geometry. Usually, the outer sheath is grounded and shields the signal in the inner wire from electrical disturbances. A nylon, or Teflon, sleeve separates the inner

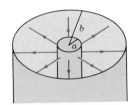

FIGURE 26.7 A cylindrical capacitor.

wire from the sheath. Find the capacitance of a length L assuming that air is between the plates.

Solution: We assume that the cable is long enough for us to neglect end effects. In this case, the electric field, which may be calculated by using Gauss's law, is the same as that for an infinite line of charge (Example 23.7 or 24.3). Thus, if λ C/m is the linear charge density on the central conductor, the electric field in the region $a < r < b$ is

$$E_r = \frac{2k\lambda}{r}$$

Note that the field is determined only by the charge on the *inner* conductor. Since the field has only a radial component, we have $\mathbf{E} \cdot d\mathbf{s} = E_r\,dr$. The potential difference between the conductors is given by

$$V_b - V_a = - \int_a^b E_r\,dr = -2k\lambda \int_a^b \frac{dr}{r}$$

$$= -2k\lambda \ln \frac{b}{a}$$

We need just the magnitude of the potential difference. The charge in a length L is $Q = \lambda L$, and so the capacitance is

$$C = \frac{2\pi\varepsilon_0 L}{\ln(b/a)} \tag{26.6}$$

The capacitance is proportional to the length of the cable and increases as the value of a approaches b. To understand this relation consider a fixed outer radius b and a given potential difference between the plates. As the radius of the inner conductor increases, the expression for $V_b - V_a$ shows that λ must increase since $\ln(b/a)$ decreases. A greater λ means that the total charge stored for the given potential difference increases—that is, the capacitance also increases.

26.2 SERIES AND PARALLEL COMBINATIONS

A capacitor is rated according to its capacitance and the maximum potential difference that can be applied without damaging the insulator between the plates. If the capacitance required for a particular application is not available, two or more capacitors may be connected to each other in different combinations. We wish to find the effective capacitance of two basic combinations.

In a **series** connection, two circuit elements are connected one after the other; they share *one* common terminal. Figure 26.8a shows two capacitors connected in series with a battery. The electric fields in the capacitors are in the same direction, so the potential difference across the combination is simply the sum of the individual potential differences; that is,

$$V = V_1 + V_2$$

The battery will transfer charge between plates a and d. Plates b and c will acquire induced charges that must be equal and opposite, since they are in effect part of a single conductor on which the net charge is zero. Thus, in a series connection, the magnitude of the charge on each capacitor is the same. The two capacitors are equivalent to a single capacitor C_{eq} (Fig. 26.8b). The charge on this equivalent capacitor is the same as that on each of the original capacitors, so $Q = Q_1 = Q_2$. Since $V = V_1 + V_2$ and $V = Q/C$, we have

$$\frac{Q}{C_{eq}} = \frac{Q}{C_1} + \frac{Q}{C_2}$$

From this we see that

$$\frac{1}{C_{eq}} = \frac{1}{C_1} + \frac{1}{C_2}$$

This logic is easily extended to several capacitors. For N capacitors in series, the equivalent capacitance is

(Series) $$\frac{1}{C_{eq}} = \frac{1}{C_1} + \frac{1}{C_2} + \cdots + \frac{1}{C_N} \qquad (26.7)$$

In a series combination, the effective capacitance is always less than that of the smallest capacitor.

In a **parallel** connection, the two circuit elements are connected side by side; they share *two* common terminals. Figure 26.9a shows two capacitors connected in parallel with a battery. The potentials of the left terminals must be the same in electrostatic equilibrium because they are connected by conducting wires—in which the field is zero. Similarly, the potentials of the right terminals are also the same. Consequently, the potential differences across both elements are the same; that is, $V = V_1 = V_2$. The total charge on each plate of the equivalent capacitance is the sum of the individual charges, so

$$Q = Q_1 + Q_2 = (C_1 + C_2)V$$
$$= C_{eq}V$$

where we have used $Q = CV$ and $V_1 = V_2 = V$. Clearly,

$$C_{eq} = C_1 + C_2$$

Extending this logic to N capacitors, we find

(Parallel) $$C_{eq} = C_1 + C_2 + \cdots + C_N \qquad (26.8)$$

In a parallel combination, the effective capacitance is always greater than any individual capacitance.

EXERCISE 1. Give a simple physical explanation for Eq. 26.8.

EXERCISE 2. (True/False): In a series connection, the largest potential difference appears across the smallest capacitor.

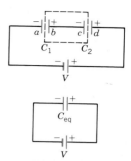

FIGURE 26.8 Two capacitors connected in series. The magnitudes of the charges on all the plates are the same. The equivalent capacitance is given by $1/C_{eq} = 1/C_1 + 1/C_2$.

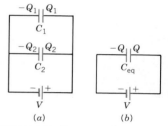

FIGURE 26.9 Two capacitors connected in parallel. The potential differences across the capacitors are the same. The equivalent capacitance is given by $C_{eq} = C_1 + C_2$.

EXAMPLE 26.6: For the circuit in Fig. 26.10a, find: (a) the equivalent capacitance; (b) the charge and potential difference for each capacitor.

Solution: When dealing with large combinations of capacitors, one should start by simplifying the combination(s) with the smallest number of elements in series or parallel.
(a) We start with C_2 and C_3, which are in parallel and are therefore equivalent to a 4-μF capacitor. This 4-μF capacitor is in series with C_1 and C_4 (see Fig. 26.10b). The total equivalent capacitance, Fig. 26.10c, is given by

$$\frac{1}{C_{eq}} = \frac{1}{6} + \frac{1}{4} + \frac{1}{12} = \frac{1}{2}$$

which yields $C_{eq} = 2 \ \mu$F.
(b) Since capacitors in series have the same charge,

$$Q_1 = Q_2 + Q_3 = Q_4$$

This is also the magnitude of the charge on the equivalent capacitance, $Q = C_{eq}V = (2 \ \mu\text{F})(48 \ \text{V}) = 96 \ \mu$C. Thus, $Q_1 = Q_4 = 96 \ \mu$C.

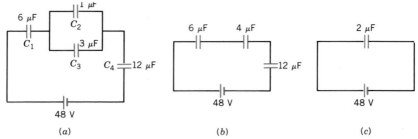

FIGURE 26.10 In calculating the equivalent capacitance of several capacitors, the problem can be broken down into intermediate steps.

To find the charges on the other two capacitors, we need their common potential difference $V_2 = V_3$. We first find the potential differences across C_1 and C_4:

$$V_1 = \frac{Q_1}{C_1} = \frac{96\ \mu C}{6\ \mu F} = 16\ V$$

$$V_4 = \frac{Q_4}{C_4} = \frac{96\ \mu C}{12\ \mu F} = 8\ V$$

Since the battery potential difference is 48 V, we must have

$V_2 = V_3 = 48\ V - (16\ V + 8\ V) = 24\ V$. Finally,

$$Q_2 = C_2 V_2 = (1\ \mu F)(24\ V) = 24\ \mu C$$
$$Q_3 = C_3 V_3 = (3\ \mu F)(24\ V) = 72\ \mu C$$

Notice that the sum $Q_2 + Q_3 = 96\ \mu C$, as it should.

EXERCISE 3. Given two capacitors for which $C_2 = 2C_1$, compare the charges and potential differences when they are connected: (a) in series; (b) in parallel.

26.3 ENERGY STORED IN A CAPACITOR

The energy stored in a capacitor is equal to the work done—for example, by a battery—to charge it. Suppose that at a particular time the magnitude of the charge on either plate is q and the potential difference between the plates is $V = q/C$. The work needed to transfer an infinitesimal charge dq from the negative plate to the positive plate is $dW = V\,dq = (q/C)dq$. (The charge moves through the wires, not across the gap between the plates!) The total work done to transfer charge Q is

$$W = \int_0^Q \frac{q}{C}\,dq = \frac{Q^2}{2C}$$

This work is stored as electrical potential energy, U_E. Since $Q = CV$, we have

Energy stored in a capacitor

$$U_E = \frac{Q^2}{2C} = \tfrac{1}{2}QV = \tfrac{1}{2}CV^2 \qquad (26.9)$$

Equation 26.9 represents the potential energy of the *system* of charges on both plates. The expression $U = qV$ in Eq. 25.11 is the potential energy associated with a *single* charge q at a potential V created by other charges in the vicinity. The factor of $\tfrac{1}{2}$ in the expression $\tfrac{1}{2}QV$ arises because the charge Q was not transferred in one step through a potential difference V. Both the charge and the potential difference were built up gradually to their final values.

EXAMPLE 26.7: Two capacitors, $C_1 = 5\ \mu F$ and $C_2 = 3\ \mu F$, are initially in parallel with a 12-V battery, as in Fig. 26.11a. They are disconnected and then reconnected as shown in Fig. 26.11b. Note carefully the numbering on the plates. Find the charges, potential differences, and energies stored (a) in the initial state, and (b) in the final state.

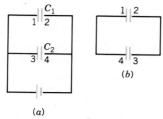

(a)

FIGURE 26.11 (a) Two capacitors are charged by placing them in parallel. (b) The battery is disconnected and the plates are reconnected.

Solution: (a) Since the capacitors are initially in parallel with the battery, their potential differences are equal to that of the battery:

$$\frac{Q_1}{C_1} = \frac{Q_2}{C_2} = 12 \text{ V}$$

Therefore, $Q_1 = 60 \ \mu C$ and $Q_2 = 36 \ \mu C$. The initial energies are

$$U_1 = \tfrac{1}{2}Q_1V_1 = \tfrac{1}{2}(60 \ \mu C)(12 \text{ V}) = 360 \ \mu J$$
$$U_2 = \tfrac{1}{2}Q_2V_2 = \tfrac{1}{2}(36 \ \mu C)(12 \text{ V}) = 216 \ \mu J$$

(b) When plate 1 (with charge 60 μC) and plate 4 (with charge $-36 \ \mu C$) are connected, there is a partial cancellation of charge.

That is, only the difference, 24 μC, is available for the capacitors to share. Denoting values after reconnection by primes, we have

$$Q_1' + Q_2' = 24 \ \mu C \qquad \text{(i)}$$

The potential differences across the capacitors are the same:

$$\frac{Q_1'}{C_1} = \frac{Q_2'}{C_2}$$

Thus

$$3Q_1' = 5Q_2' \qquad \text{(ii)}$$

Solving (i) and (ii) leads to $Q_1' = 15 \ \mu C$ and $Q_2' = 9 \ \mu C$. The potential differences ($V = Q/C$) are $V_1' = V_2' = 3$ V. The final energies are $U_1' = \tfrac{1}{2}Q_1'V_1' = 22.5 \ \mu J$ and $U_2' = \tfrac{1}{2}Q_2'V_2' = 13.5 \ \mu J$.

The initial total energy $U = 576 \ \mu J$ is much larger than the final total energy $U' = 36 \ \mu J$. The missing energy can be accounted for in two ways. First, in any real system there will be thermal losses in the connecting wires. Second, even if there is no resistance, the charges on the capacitors do not reach their final values instantaneously. The charge oscillates between the plates of the capacitors, like the water in a U-tube. As we will show in Chapter 34, oscillating charges produce electromagnetic radiation, such as light, heat, and radio waves. Thus, some of the "missing" energy is radiated away.

26.4 ENERGY DENSITY OF THE ELECTRIC FIELD

We know that the work required to bring two point charges from infinity to some finite separation is stored as potential energy. Since the charges themselves have not changed, where is this potential energy stored? To answer this question let us consider the energy stored in a capacitor. The capacitance of a parallel plate capacitor (Fig. 26.12) is $C = \varepsilon_0 A/d$, and the potential difference between its plates is $V = Ed$. The energy stored, $U_E = \tfrac{1}{2}CV^2$, may be written as

$$U_E = \tfrac{1}{2} \frac{\varepsilon_0 A}{d}(Ed)^2 = \tfrac{1}{2}\varepsilon_0 E^2 (Ad)$$

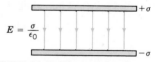

FIGURE 26.12 The energy of a capacitor is stored in the electric field. The energy density is $u_E = \tfrac{1}{2}\varepsilon_0 E^2$.

Since the volume between the plates where the field exists is Ad, the energy density, or the energy per unit volume (J/m³), is

$$u_E = \tfrac{1}{2}\varepsilon_0 E^2 \qquad (26.10)$$

Energy density of the electric field

Notice that there is no direct reference to the capacitor itself. From Eq. 26.10 we may conclude that the energy is stored in the electric field. Although Eq. 26.10 has been derived from a particular case, it is a generally valid expression for the **energy density of an electric field.**

The potential energy of a system of charges is associated with the change that has taken place in the electric field. For example, as a capacitor is charged, work is done to create the field between the plates. When two positive point charges are brought together, work is done to modify the electric field in the region surrounding them. When the charges are released, they fly apart. The increase in kinetic energy of the particles is paid for by a loss in the field energy.

In the 19th century, the electric field was thought of as some form of tension or deformation in a rarefied medium called the ether. We now realize that the ether does not exist. Electric fields can exist in a vacuum, far from the source charges. In this sense, they have an independent existence. The energy we receive in the form of light and heat from the sun is carried by electric and magnetic fields. As we will see in Chapter 34, not only is the field a repository of energy, but it can also transport momentum.

EXAMPLE 26.8: The "breakdown" field strength, at which dry air loses its insulating ability and allows a discharge to pass through it, is about 3×10^6 V/m. What is the energy density at this field strength?

Solution: From Eq. 26.10 the energy density at this critical field strength is

$$u_E = \tfrac{1}{2}(8.85 \times 10^{-12} \text{ C}^2/\text{N·m}^2)(3 \times 10^6 \text{ V/m})^2$$
$$= 40 \text{ J/m}^3$$

Since this corresponds to the breakdown field strength, it also represents the maximum energy density that can be attained with an electrostatic field in air.

EXAMPLE 26.9: Use Eq. 26.10 to derive the potential energy of a metal sphere of radius R with charge Q.

Solution: First, note that there is no electric field inside the conducting sphere. From Gauss's law, or a direct calculation, we know that the electric field at points outside the sphere is the same as that of a point charge Q at the center; that is,

$$E = \frac{kQ}{r^2} \qquad (r > R)$$

The volume of an imaginary shell of radius r and infinitesimal thickness dr, shown in Fig. 26.13, is $4\pi r^2 \, dr$. The energy in the electric field within this shell is

$$dU_E = u_E(4\pi r^2 \, dr)$$

$$= \tfrac{1}{2}\,\varepsilon_0 \left(\frac{kQ}{r^2}\right)^2 (4\pi r^2 \, dr)$$
$$= \frac{kQ^2}{2r^2} \, dr$$

We have used $k = 1/4\pi\varepsilon_0$. The total potential energy is

$$U_E = \frac{kQ^2}{2} \int_R^\infty r^{-2} \, dr = \frac{kQ^2}{2R}$$

This agrees with the derivation in Example 25.6. Clearly, one may regard the energy of the charged sphere as being stored in its electric field.

EXERCISE 4. Show that Eq. 26.9 leads directly to the potential energy of the conducting sphere.

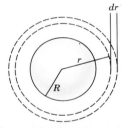

FIGURE 26.13 In order to calculate the energy stored in the electric field outside a charged metal sphere, we first find the energy in a spherical shell of radius r and thickness dr.

26.5   DIELECTRICS

When certain nonconducting materials, such as glass, paper, or plastic, are introduced between the plates of a capacitor, its capacitance increases. Michael Faraday, who first noticed this effect, called such materials **dielectrics.** Two simple experiments illustrate the effects of inserting a dielectric into a capacitor.

(i) Battery Not Connected

Figure 26.14a shows a capacitor with charges $+Q_0$ on its plates and a potential difference V_0 across them. The initial capacitance with vacuum between the plates is $C_0 = Q_0/V_0$. When a slab of dielectric is inserted to completely fill the space between the plates, as in Fig. 26.14b, the potential difference between them decreases by a factor κ, called the **dielectric constant:**

$$V_D = \frac{V_0}{\kappa} \qquad (26.11)$$

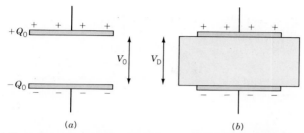

FIGURE 26.14 (a) Two plates carry equal and opposite charges $\pm Q_0$ and have a potential difference V_0 between them. (b) When a dielectric completely fills the space between the plates the potential difference decreases to $V_D = V_0/\kappa$ where κ is the dielectric constant.

From the relation $V = Ed$, we see that the electric field strength decreases by the same factor:

$$E_D = \frac{E_0}{\kappa} \qquad (26.12)$$

Electric field in a dielectric

Since the charge on each plate is unaffected (it has nowhere to go), the capacitance in the presence of the dielectric is $C_D = Q_0/V_D = \kappa C_0$.

(ii) Battery Connected

In Fig. 26.15a, the initial conditions are the same as in Fig. 26.14a. However, a battery will maintain the potential difference across the plates at V_0. When the dielectric is inserted to fill the space between the plates, as in Fig. 26.15b, the charge on the plates increases by a factor κ to $Q_D = \kappa Q_0$. Using $C_D = Q_D/V_0$, we again find $C_D = \kappa C_0$.

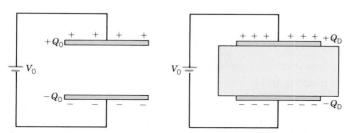

FIGURE 26.15 (a) This situation is similar to that in Fig. 26.14a except that the battery remains connected. (b) The potential difference is unchanged but the charge on the plates increases to $Q_D = \kappa Q_0$.

In either situation, the effect of inserting the dielectric is to increase the capacitance by a factor κ:

$$C_D = \kappa C_0 \qquad (26.13)$$

Typical values of κ are shown in Table 26.1. Although the term dielectric is almost invariably applied to insulating materials, this is somewhat misleading. Water, which has a high dielectric constant, is not an insulator.

In addition to increasing the capacitance for a given plate geometry, the use of

TABLE 26.1 DIELECTRIC CONSTANTS AND STRENGTHS

Material	Dielectric Constant	Dielectric Strength (10^6 V/m)
Air	1.00059	3
Paper	3.7	16
Glass	4–6	9
Paraffin	2.3	11
Rubber	2–3.5	30
Mica	6	150
Water	80	—

FIGURE 26.16 A tree-like pattern produced when a dielectric "breaks down"; that is, an electrical discharge passes through it.

a dielectric has other advantages. For example, a thin plastic sheet, or an oxide layer, allows the plates of a parallel-plate capacitor to be very close without danger of touching. Since $C \propto 1/d$, for a given capacitance, the size of the capacitor can be reduced. Another advantage of using a dielectric is its *dielectric strength*. This is the maximum electric field strength that can be applied to the material before it loses its insulating property and allows a discharge to pass through it (see Fig. 26.16). A dielectric increases the critical potential difference at which a spark would jump across the plates.

26.6   ATOMIC VIEW OF DIELECTRICS

The dielectric constant of a substance is a measure of the response of its charges to an external electric field. The charges can redistribute themselves in two ways. We noted in Section 23.6 that in an external electric field, an atom acquires an *induced* dipole moment. This is the only electrostatic response in a nonpolar substance. In a polar molecule, such as water, the centers of the positive and negative charge do not coincide. Consequently, the molecule possesses a permanent electric dipole moment. In the absence of an external field, the dipoles are randomly oriented, as in Fig. 26.17a. When an external field is applied, the torque on the dipoles tends to align them along the field lines, although the degree of alignment is not complete because of thermal agitation. At the end faces of the dielectric, there appear more charges of one sign than the other. The result is an effective charge separation across the width of the material, as shown in Fig. 26.17b. (This is in addition to the nonpolar contribution characteristic of all materials.)

Thus, whether the dielectric is polar or nonpolar, its end faces acquire induced charges whose sign is opposite to that of the adjacent capacitor plate. This effective charge separation is called **polarization.** Within the material, the induced charges create an *induced* electric field E_i that is directed opposite to the external field E_0, as shown in Fig. 26.18a. Consequently, as Fig. 26.18b shows, the net field

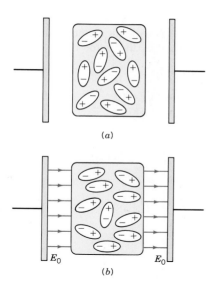

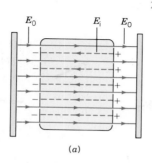

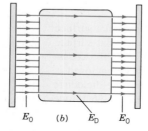

FIGURE 26.17 (a) In the absence of an external field, the dipoles in a polar substance are randomly oriented. (b) When an external field is applied, the dipoles tend to align along the field lines. In effect, the end faces of the dielectric become charged.

FIGURE 26.18 (a) The induced electric field E_i due to the surface charges on the dielectric is opposite to the external field E_0. (b) The net field within the dielectric is $E_D = E_0 - E_i$.

inside the dielectric, E_D, is reduced by a factor of κ:

$$E_D = E_0 - E_i = \frac{E_0}{\kappa} \qquad (26.14)$$

Because of the low density of molecules, one would not expect κ to be large for a gas. Since water has a polar molecule and is a liquid, it is relatively easy to reorient the molecules. This explains the large value of its dielectric constant.

EXERCISE 5. What is κ for (a) a metal, and (b) the substance in Fig. 26.18b?

EXAMPLE 26.10: A dielectric slab of thickness t and dielectric constant κ is inserted into a parallel plate capacitor with plates of area A, separated by distance d, as shown in Fig. 26.19. Assume that the battery is disconnected before the slab is inserted. What is the capacitance?

Solution: First one must find the potential difference between the plates. The electric field strengths in the air and in the dielectric are

$$E_0 = \frac{\sigma}{\varepsilon_0} = \frac{Q}{A\varepsilon_0}; \qquad E_D = \frac{E_0}{\kappa}$$

The total potential difference across the air spaces is $\Delta V_0 = E_0(d - t)$, while that across the dielectric is $\Delta V_D = E_D t = \sigma t / (\kappa \varepsilon_0)$. Thus, the total potential difference between the plates is

$$\Delta V = \frac{\sigma}{\varepsilon_0}\left[(d - t) + \frac{t}{\kappa}\right]$$

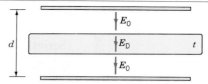

FIGURE 26.19 A slab of dielectric is inserted between two parallel plates. To calculate the capacitance one must first calculate the potential difference between the plates.

The capacitance is $C = Q/\Delta V = \sigma A/\Delta V$, which is

$$C = \frac{\varepsilon_0 A}{d + t(1/\kappa - 1)}$$

Notice that when $\kappa = 1$, $C = \varepsilon_0 A/d$, which we obtained earlier.

26.7 GAUSS'S LAW FOR DIELECTRICS (Optional)

We now consider how Gauss's law is modified when a dielectric is present, for example, in a parallel plate capacitor. The electric fields in vacuum and in the dielectric may be expressed in terms of the *free* charge density σ_f on the metal plates and the *bound* charge density σ_b on the surface of the dielectric: $E_0 = \sigma_f/\varepsilon_0$ and $E_i = \sigma_b/\varepsilon_0$. From Eq. 26.14

$$\frac{\sigma_f}{\varepsilon_0} - \frac{\sigma_b}{\varepsilon_0} = \frac{\sigma_f}{\kappa\varepsilon_0} \qquad (26.15)$$

Figure 26.20 shows a Gaussian pillbox with one flat surface within the metal plate (where $E = 0$) and the other flat surface within the dielectric, where the field is E_D. If A is the cross-sectional area of the cylinder, then the net charge enclosed is $\sigma_f A - \sigma_b A = Q_f - Q_b$. Only the flat face within the dielectric contributes to the flux in Gauss's law; thus, using Eq. 26.15, we find

$$\int \mathbf{E}_D \cdot d\mathbf{A} = \frac{Q_f - Q_b}{\varepsilon_0} = \frac{Q_f}{\kappa\varepsilon_0} \qquad (26.16)$$

From this equation we see that within a material, the electric field due to a charge is reduced by the factor $1/\kappa$.

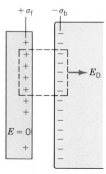

FIGURE 26.20 A Gaussian pillbox that extends from the metal to the dielectric encloses a net charge $(Q_f - Q_b)$, where Q_f is the free charge on the metal and Q_f is the bound charge on the face of the dielectric.

SUMMARY

A **capacitor** is a device that stores charge and electrical energy. It consists of two conducting plates separated by an insulator. If the plates carry charges $\pm Q$ and have a potential difference V, the capacitance of the capacitor is defined as the magnitude of the charge on one plate divided by the magnitude of the potential difference V between them:

$$C = \frac{Q}{V}$$

Capacitance depends on the size and shape of the plates and the material between them. It does *not* depend on Q or V individually.

When capacitors are connected in series or in parallel, the equivalent capacitance is found from

(Series) $$\frac{1}{C_{eq}} = \frac{1}{C_1} + \frac{1}{C_2} + \cdots + \frac{1}{C_N}$$

(Parallel) $$C_{eq} = C_1 + C_2 + \cdots + C_N$$

In effect, when a capacitor is charged, positive charge is transferred from a low potential to a high potential. The work done by the external agent is stored as electrical potential energy

$$U_E = \tfrac{1}{2}QV = \frac{Q^2}{2C} = \tfrac{1}{2}CV^2$$

The energy of the capacitor is actually stored in the electric field that is created between the plates. The **energy density of an electric field** (in free space) is given by

$$u_E = \tfrac{1}{2}\varepsilon_0 E^2$$

Suppose a capacitor has a capacitance C_0 when there is no material between the plates. When a dielectric material is inserted to completely fill the space between the plates, the capacitance increases to

$$C = \kappa C_0$$

where κ is called the **dielectric constant** of the material. An external electric field E_0 is reduced in a dielectric by the factor κ to the value $E_D = E_0/\kappa$.

ANSWERS TO IN-CHAPTER EXERCISES

1. The area to which charge flows is the sum of the areas of the plates. Think of a parallel-plate capacitor: $C \propto A$.

2. True. The charges on all the capacitors are the same and $V = Q/C$.

3. (a) In series the charges are the same. From $V = Q/C$ we infer that $V_2 = 0.5V_1$.

(b) In parallel, $V_1 = V_2$. Since $Q = CV$, it follows that $Q_2 = 2Q_1$.

4. From Example 26.2 we know that for an isolated sphere $C = R/k$. Thus,

$$U = Q^2/2C = kQ^2/2R.$$

5. (a) Since $E = 0$ within a metal, $\kappa = \infty$.

(b) From the density of lines we infer that $E_D = E_0/3$, so $\kappa = 3$.

QUESTIONS

1. When a battery is connected across a capacitor are the charges on the plates always equal and opposite, even for plates of different size?

2. Would two conductors have capacitance even if they did not have equal and opposite charges?

3. When a dielectric is inserted to fill the space between a charged parallel-plate capacitor, does the stored energy increase or decrease given that (a) the battery remains connected; or (b) the battery is first disconnected?

4. The potential difference across a capacitor is doubled. How does each of the following quantities change: (a) the capacitance; (b) the stored charge; (c) the stored energy?

5. Given a battery, how would you connect two capacitors, in series or in parallel, for them to store the greater: (a) total charge; (b) total energy?

6. A parallel plate capacitor with large plates is charged and then disconnected from the battery. As the plates are pulled apart, does the potential difference increase, decrease, or stay the same? How is the stored energy affected?

7. A parallel-plate capacitor is connected to a battery. Consider what happens as you move the plates closer together. (a) How are the charge, potential difference, and energy affected? (b) Would you do positive or negative work in moving the plates?

8. Repeat Question 7 for a charged parallel plate capacitor with the battery disconnected.

9. Give two reasons for using dielectrics in capacitors.

10. Water has a high dielectric constant. Why is it not frequently used in capacitors?

11. What is the difference between the dielectric constant and the dielectric strength of a material?

12. A metal sheet of negligible thickness is inserted between the plates of a parallel plate capacitor, as shown in Fig. 26.21. (a) How is the capacitance affected? (b) Does the position of the sheet matter? (c) What if the sheet has finite thickness?

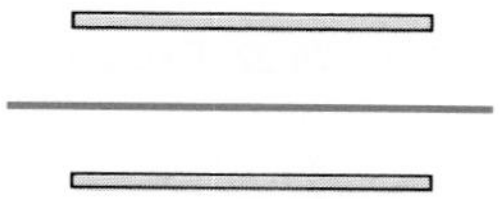

FIGURE 26.21 Question 12.

13. Would you expect the dielectric constant of a polar substance to depend on temperature? If so, would it increase or decrease with increasing temperature?

14. Show that $F/m = C^2/N \cdot m^2$.

EXERCISES

26.1 Capacitance

1. (I) A parallel-plate capacitor has circular plates of radius 6 cm separated by 2 mm. Find: (a) the capacitance; (b) the charge on each plate when a 12-V battery is connected.

2. (I) What is the capacitance per unit length of a long straight coaxial cable with an inner wire of radius 0.5 mm and an

outer sheath of radius 0.5 cm? (b) When a 24-V potential difference is applied, how much charge is stored on 2.5 m of the wire?

3. (I) A 240-pF parallel-plate capacitor has ±40-nC charges on its plates. The plates are separated by 0.2 mm. Find: (a) the area of each plate; (b) the potential difference between the plates; (c) the electric field between the plates.

4. (I) In a parallel-plate capacitor, the plates are separated by 0.8 mm. The plates carry charges ±60-nC and there is an electric field of magnitude 3×10^4 V/m between the plates. Find: (a) the potential difference; (b) the capacitance; (c) the plate area.

5. (I) Assume that the earth (radius 6400 km) is surrounded by a conducting sphere 50 km above the surface and that there is a constant electric field of 100 N/C directed vertically downward. (a) What is the surface charge density on the earth's surface? (b) What is the capacitance of the system? (c) Compare the answer to part (b) with the capacitance of the earth treated as an isolated sphere.

6. (I) A long, straight coaxial cable has an inner wire of radius $r = 1$ mm and an outer sheath of radius r_2. When a potential difference of 27 V is applied, the linear charge density on the inner wire is 4 nC/m. Find r_2.

7. (II) A capacitor consists of two interleaving sets of plates, as shown in Fig. 26.22. The plate separations and the effective area of overlap are shown in the figure. What is the capacitance of this arrangement?

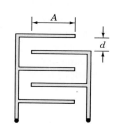

FIGURE 26.22 Exercise 7.

8. (I) A 24-pF parallel-plate capacitor has plates of area 0.06 m². (a) What potential difference would cause a spark to jump across the plates? The breakdown field strength for air is 3×10^6 V/m. (b) What would be the charge on the plates at this potential difference?

9. (I) When 10^{12} electrons are transferred from one plate to the other, the potential difference across an initially uncharged capacitor reaches 20 V. What is its capacitance?

10. (I) A 12-V battery is connected between the two concentric spheres of a spherical capacitor. The radii of the spheres are 15 cm and 20 cm. What is the charge on each sphere?

11. (II) A capacitor $C_1 = 4$ μF is connected across a 20-V battery. The battery is removed and the capacitor is reconnected to another capacitor $C_2 = 6$ μF. What are the final charges and potential differences for the capacitors?

12. (II) A variable capacitor has seven semicircular plates of radius 2 cm as in Fig. 26.23. The plate separation is 1 mm. Find the capacitance when the angle θ is: (a) zero; (b) 45°; (c) 135°.

FIGURE 26.23 Exercise 12.

13. (I) A spherical capacitor consists of an inner sphere of radius 3 cm and an outer shell of radius 11 cm. (a) What is the capacitance? (b) How many electrons would need to be transferred from one sphere to the other to create a potential difference of 5 V?

26.2 Series and Parallel Connections

14. (I) Given two capacitors, $C_1 = 0.1$ μF and $C_2 = 0.25$ μF, and a 12-V battery, find the charge and potential difference for each if they are connected (a) in series; (b) in parallel, with the battery.

15. (I) The three capacitors in Fig. 26.24a have an equivalent capacitance of 12.4 μF. Find C_1.

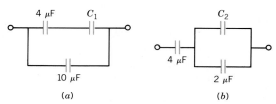

FIGURE 26.24 Exercises 15 and 16.

16. (I) The three capacitors in Fig. 26.24b have an equivalent capacitance of 2.77 μF. What is C_2?

17. (I) You are given four 10-μF capacitors. Put together a combination that has a capacitance of (a) 4 μF; (b) 2.5 μF.

18. (I) All the capacitors in Fig. 26.25 are identical, with $C = 1$ μF. What is their equivalent capacitance?

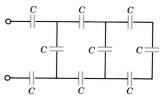

FIGURE 26.25 Exercise 18.

19. (II) Two capacitors, $C_1 = 2$ μF and $C_2 = 4$ μF, are connected in series with an 18-V battery. The battery is removed and the plates of like sign are connected. Find the final charge and potential difference for each capacitor.

20. (II) Two capacitors, $C_1 = 2 \ \mu F$ and $C_2 = 6 \ \mu F$, are in parallel with a 60-V battery. The battery is removed and plates of opposite sign are connected. Find the final charge and potential difference for each capacitor.

21. (II) A capacitor, $C_1 = 3 \ \mu F$, has an initial potential difference of 12 V and a second capacitor, $C_2 = 5 \ \mu F$, has an initial potential difference of 10 V. Find the final charges and potential differences for each if they are then connected with the following polarities for the plates: (a) like signs together; (b) unlike signs together.

22. (II) You are given three capacitors, $C_1 = 1 \ \mu F$, $C_2 = 2 \ \mu F$, and $C_3 = 4 \ \mu F$. How many distinct values of capacitance could these be used to produce? State the values.

26.3 and 26.4 Energy and Energy Density

23. (I) What capacitance would be required to store 100 MeV at a potential difference of 12 V across the plates?

24. (I) Given two 50-μF capacitors and a 20-V battery, find the total energy stored when the capacitors are connected: (a) in parallel; (b) in series, with the battery.

25. (I) A parallel-plate capacitor has plates of area 40 cm^2 separated by 2.5 mm. It is connected to a 24-V battery. Find: (a) the capacitance; (b) the energy stored; (c) the electric field; (d) the energy density in the electric field.

26. (I) A parallel-plate capacitor has a plate separation of 0.6 mm and a charge of magnitude 0.03 μC on each plate. If the electric field between the plates is 4×10^5 V/m, find: (a) the capacitance; (b) the energy stored.

27. (I) The plates of a 400-pF parallel-plate capacitor are separated by 1.2 mm. Find the energy density when a potential difference of 250 V is applied across the plates.

28. (I) Given two capacitors, $C_1 = 3 \ \mu F$ and $C_2 = 5 \ \mu F$, find the energy stored in each when they are connected: (a) in parallel; (b) in series, with a 20-V battery.

29. (II) Two capacitors, $C_1 = 2 \ \mu F$ and $C_2 = 5 \ \mu F$, are connected in series with a 20-V battery. The battery is removed and the plates of like sign are connected. Find the initial and final energies for each capacitor.

30. (II) Two capacitors, $C_1 = 2 \ \mu F$ and $C_2 = 5 \ \mu F$, are in parallel with a 40-V battery. The battery is removed and plates of opposite sign are connected. Find the initial and final energies for each capacitor.

31. (I) Consider the arrangement of capacitors shown in Fig. 26.26. Find the energy stored in (a) the 5-μF capacitor; (b) the 4-μF capacitor.

32. (I) A 5-pF parallel-plate capacitor has a potential difference of 25 V across its plates. The plate area is 40 cm^2. Find: (a) the energy stored; (b) the energy density in the field.

33. (I) A parallel-plate capacitor has plates of area A separated by a distance d. A metal block of thickness ℓ is inserted midway between the plates, as shown in Fig. 26.27. (a)

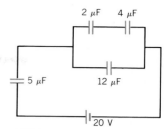

FIGURE 26.26 Exercise 31.

What is the capacitance? (b) What is the capacitance if the block is moved so that it touches one of the plates?

FIGURE 26.27 Exercise 33 and Problem 2.

34. (II) A parallel-plate capacitor with a plate separation d is connected to a battery with a potential difference V. The plates are pulled apart till the separation is $2d$. What is the change in each of the following quantities: (a) the potential difference; (b) the charge on each plate; (c) the energy stored in the capacitor?

35. (II) Repeat Exercise 34, with a charged capacitor but with the battery disconnected.

36. (I) Consider the combination of capacitors in Fig. 26.28. The energy stored in the 5-μF capacitor is 200 mJ. What is the energy stored in (a) the 4-μF capacitor, (b) the 3-μF capacitor?

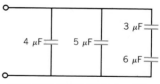

FIGURE 26.28 Exercise 36.

37. (I) The field strength at the tip of the needle in a field-ion microscope (see p. 503) is about 4.5×10^8 V/m. What is the energy density in such a field?

38. (I) In fair weather there is an electric field of 120 N/C directed vertically downward at the earth's surface. How much electrical energy is there in a cubic enclosure of side 10 m?

39. (I) The plates of a parallel-plate capacitor are separated by 1 mm. For what potential difference would the energy density be 1.8×10^{-4} J/m^3?

40. (I) A 15-pF parallel-plate capacitor is connected to a 48-V battery. The area of each plate is 80 cm^2. What is the energy density in the field?

26.5 Dielectrics

41. (II) The space between the plates of a parallel-plate capacitor is filled with two dielectrics of equal size, as shown in Fig. 26.29. What is the resulting capacitance in terms of κ_1, κ_2, and C_0, the capacitance with a vacuum between the plates?

FIGURE 26.29 Exercise 41.

42. (II) A parallel-plate capacitor is half-filled with a dielectric slab of constant κ_1, while the other half contains a slab of constant κ_2, as in Fig. 26.30. What is the resulting capacitance? Express your answer in terms of C_0, the capacitance with no dielectric.

FIGURE 26.30 Exercise 42.

43. (I) The plates of a parallel-plate capacitor carry surface charge density σ C/m^2, as shown in Fig. 26.31. A dielectric slab of thickness ℓ and dielectric constant κ is placed between the plates. Find (a) the potential difference, (b) the capacitance. Take $d = 1$ cm, $\ell = 0.3$ cm, $\sigma = 2$ nC/m^2, $\kappa = 5$, $A = 40$ cm^2.

FIGURE 26.31 Exercise 43.

44. (I) A 0.1-μF parallel-plate capacitor is connected to a 12-V battery. A dielectric slab ($\kappa = 4$) is inserted to fill the space between the plates. Find the additional charge that flows to the plates.

45. (II) A parallel-plate capacitor with a mica sheet as dielectric has a capacitance of 50 ρF. If the plate separation is 0.1 mm, find: (a) the plate area; (b) the maximum operating potential difference.

46. (I) A dielectric is inserted and completely fills the space between the plates of a capacitor. Determine the dielectric constant in each of the following cases: (a) The capacitance increases by 50%. (b) The potential difference decreases by 25%. (c) The stored charge doubles.

PROBLEMS

1. (I) A parallel-plate capacitor is filled with a material whose dielectric constant is κ. Show that the energy density in a dielectric is

$$\tfrac{1}{2} \kappa \varepsilon_0 E^2$$

where E is the field in the dielectric. Does this result depend on whether or not a battery is connected?

2. (I) A parallel-plate capacitor has plates of area A separated by a distance d and is connected to a battery with a potential difference V. A metal block of thickness ℓ is midway between the plates, as shown in Fig. 26.27. What is the work required to remove the block given that the battery remains connected?

3. (I) Repeat Problem 2 given that the battery is disconnected before the block is removed.

4. (I) Two identical parallel-plate capacitors are connected in series with a 12-V battery. The plate area is 16 cm^2 and the plate separation is 0.4 mm. (a) What is the charge and potential difference for each capacitor? (b) A dielectric slab ($\kappa = 5$) is inserted to completely fill one capacitor. What are the new values of the charge and potential difference for each capacitor?

5. (II) The pattern of connections between identical capacitors in Fig. 26.32 is continued indefinitely. What is the equivalent capacitance between the terminals a and b? (*Hint*: Since the pattern is infinite, the capacitance between the points a' and b' is the same as that between a and b.)

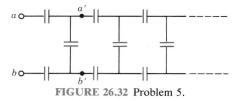

FIGURE 26.32 Problem 5.

6. (II) What is the equivalent capacitance of the combination shown in Fig. 26.33. Take $C_1 = 2$ μF, $C_2 = 4$ μF and $C_3 = 3$ μF. (*Hint*: Apply a potential difference across the terminals. This potential difference is the same for any path between the terminals. What is the relationship between the charges on the plates?)

7. (II) A parallel-plate capacitor has plates of area A that are separated by a distance d. The charges on the plates are $\pm Q$. What is the force between the plates given that the battery has been removed? Is the force attractive or repulsive? (*Hint*: Use $F_x = -dU/dx$.)

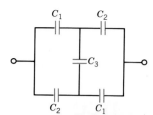

FIGURE 26.33 Problem 6.

8. (I) A parallel-plate capacitor of capacitance C is charged by a battery whose potential difference is V. The battery is disconnected and a dielectric with constant κ is inserted and completely fills the space between the plates. Find the energy stored in the capacitor.

9. (I) Repeat Problem 8, given that the battery remains connected.

10. (II) A cylindrical capacitor has a central conductor of radius a and an outer sheath of radius b. Show that when $b -$

$a \ll b$ the capacitance becomes that of a parallel-plate capacitor.

11. (II) A spherical capacitor consists of concentric spheres of radii R_1 and R_2. (a) Show that when $R_2 - R_1 \ll R_2$, the capacitance becomes that of a parallel-plate capacitor. (b) Show that in the appropriate limit the capacitance of the spherical capacitor reduces to that of an isolated sphere.

12. (II) (a) Find the energy density as a function of r for a cylindrical capacitor with an inner wire of radius a and an outer conductor of radius b. (b) What is the total energy stored in a length ℓ of the capacitor? (c) Compare your result with a calculation based on $\frac{1}{2}CV^2$ or $Q^2/2C$.

13. (I) A dielectric slab of dielectric constant κ is inserted between the plates of a parallel-plate capacitor. Show that the bound surface charge density σ_b on the dielectric and the free surface charge density σ_f on the capacitor plates are related by

$$\sigma_b = (\kappa - 1)\frac{\sigma_f}{\kappa}$$

CHAPTER 27

Current and Resistance

The glow produced when a current flows through the filament of an antique light bulb.

Major Points.

1. (a) The definition of **current** and **current density.**
   (b) The nature of a current in a wire.
2. (a) The definition of **resistivity;** its causes and temperature dependence.
   (b) The definition of **resistance;** its relation to resistivity.
3. **Ohm's law.** The limits to its validity.

The preceding chapters have dealt with charges at rest. We now go on to study effects associated with charges in motion, that is, electric currents. The central question is how the current through a wire depends on the potential difference applied across its ends. Although the concepts of potential difference and electric current were being slowly clarified during the 18th century, they could not be related for several reasons. First, there was no source of continuous current. Until 1800, the only way of producing an electric current in a wire was to discharge a Leyden jar through it. Of course, this produced only a transient effect. Second, it was not clear whether the conducting wire was merely a conduit for the electrical "fluid" or whether it played some more active role. Third, the lack of measuring instruments severely hampered the development of the subject. Investigators had to use their bodies, their tongues, and even their eyes as detectors of electrical current. Electroscopes could detect "electrification," but it was not clear exactly what quantity was being measured. (We now realize that the deflection of the leaf electroscope depends on the potential difference between the leaves and the outer casing.)

A major advance in electrical research came with the chance discovery of "animal electricity" by the Italian physiologist Luigi Galvani in 1780. Galvani was using an electrostatic generator to investigate the effects of electrical discharges through tissues. He had dissected a frog and happened to be touching a nerve with a scalpel when a nearby generator produced a spark. He was surprised to see the frog's muscles contract several times even though it was not connected to the machine. The tissue was responding to the discharge of charges that had been induced on it. But Galvani had not heard of induced charges, and instead of rediscovering electrostatic induction, he came across something far more signifi-

FIGURE 27.1 A collection of instruments used by Luigi Galvani (1737–1798) in his study of "animal electricity."

cant. He noted that a frog's leg, when hung by a nerve, twitched during flashes of lightning, and he decided to see if he could use the frog to detect fair weather electricity—which was known to exist. He attached a brass hook to the spinal cord of a frog and hung the frog on an iron railing. When nothing happened for some time, he grew impatient and offhandedly touched the hook to the railing. The muscles immediately displayed several contractions. The effect was seen again when the frog was placed on an iron table and the brass hook was put in contact with the table. He later found that other pairs of metals, such as Cu and Zn, also produced contractions. He published his results in 1791, and ascribed the phenomena to "animal electricity." Figure 27.1 is an illustration from his work.

The physicist Alessandro Volta (Fig. 27.2a) at the University of Pavia repeated these experiments and initially accepted the explanation in terms of animal electricity. He found that if the ends of two metal strips—for example, of silver and zinc—were joined, and the other ends placed on either side of the tongue, a definite taste and sensation was produced. (It seems his tongue did not twitch!) Volta actually used taste as a means of classifying the electrical properties of metals. In 1796, he discovered that plates of Cu and Zn become charged merely by being in contact. He finally realized that these effects depended on the use of different metals in contact and that the tissue merely provided a conducting path between them. In an effort to amplify the charges produced by contact, he piled disks of Cu and Zn one on top of the other but found no increase in the effect. Then came the crucial insight. The ability of electric torpedo fish and eels to produce electric shocks had been known since Greek times. Volta knew that the electric organs of these fish involved a laminated (layered) structure filled with fluid. He used this clue and separated the pairs of Cu and Zn disks with cardboard soaked in a saline solution or acid. With this arrangement, he was able to produce sparks repeatedly and to heat fine wires to incandescence. In 1799, he announced his invention of the "voltaic pile," shown in Fig. 27.2b. The voltaic pile allowed, for the first time, the production of continuous current. The subsequent development of the subject of electromagnetism could not have occurred without this invention.

FIGURE 27.2 (a) Alessandro Volta (1745–1827).

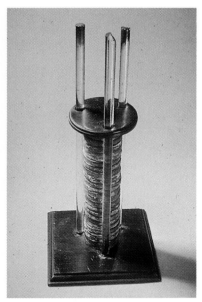

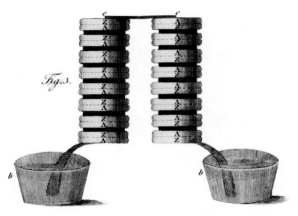

FIGURE 27.2 (*b*) A "voltaic pile."

Volta's pile in the Royal Institution.

27.1 CURRENT

Consider the flow of charge through a surface, as in Fig. 27.3*a*. If in time Δt a net charge ΔQ flows through the surface, the average electric current is defined as

$$I_{av} = \frac{\Delta Q}{\Delta t}$$

If the flow is not steady, the instantaneous **electric current** I is defined as

$$I = \frac{dQ}{dt} \tag{27.1}$$

An electrical current is the *rate* of flow of charge through a surface. The SI unit of current is the ampere (A). From Eq. 27.1 we see

$$1 \text{ A} = 1 \text{ C/s}$$

For almost all purposes, the flow of positively charged particles in one direction is equivalent to the flow of negatively charged particles in the opposite direction, as shown in Fig. 27.3*b*. (One exception is the Hall effect, Section 29.6.) Even though we know that it is really the electrons that move in a metal, for historical reasons based on Franklin's "one-fluid" theory, the direction of I is taken by convention to be that of the flow of *positive* charge. The most common example of electric currents occurs in wires. However, a beam of charged particles, such as electrons or protons moving in a vacuum, also constitutes a current. In ionized gases and in liquid electrolytes, both signs of charge contribute to the current.

In order for a current to flow in a wire, a potential difference must exist between its ends. One advantage of the conventional direction of current is that *current flows from high potential to low potential.* Thus the flow of electric current "downhill" in electrical potential is analogous to the flow of water down an inclined pipe. We may pursue the analogy a bit further. Just as a pump raises the gravitational potential of water, one might say that the job of a battery is to raise positive charges from a low potential (negative terminal) and to place them at a high potential (positive terminal). A current will flow continuously only through a closed loop, called an electrical circuit, that contains the battery and the wire, as in Fig. 27.4. When charge enters at one end of the wire, an equal amount leaves at the other end; the wire itself does not acquire any net charge.

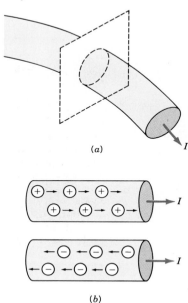

(*a*)

(*b*)

FIGURE 27.3 (*a*) A current is defined as the rate at which charge flows through a surface. (*b*) The current produced by positive charges moving in one direction is the same as the current due to an equal number of negative charges moving in the opposite direction.

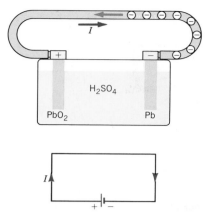

FIGURE 27.4 A current flows *through* a wire when a potential difference is applied *across* its ends.

EXERCISE 1. A current of 1 A flows through a wire. How many electrons pass a given point in 1 s?

The Electric Field within a Wire

A potential difference between two points on a wire implies the existence of an electric field along the wire. Let us consider the source of that field. When a wire is connected to the terminals of a battery, some charge flows between the terminals and the *surface* of the wire. The magnitude of the surface charge density diminishes with distance from each terminal, as depicted in Fig. 27.5a. We know that under static conditions, the field within a conductor is zero, and at the surface the field is perpendicular to the conductor. However, when a potential difference is applied across the ends of a wire, the conditions are no longer static. The electrostatic field due to the surface charges on the wire has a component *along* the wire. It is this electrostatic field *within* the wire that causes the current to flow through the wire. Figure 27.5b shows the pattern of the electrostatic field due to a short length of wire.

The Nature of a Current in a Wire

The path of a conduction electron in a current-carrying wire is quite erratic, as shown in Fig. 27.6a. The motion involves two distinct components. First, the conduction electrons behave somewhat like gas molecules in a container. They dart around randomly at high speeds and collide frequently with the basically stationary ions. The number that move in one direction is exactly balanced by the number that move in the opposite direction. Second, when a battery is connected, an electric field is created within the wire. This field causes a slight tendency in the electrons to move in one direction (opposite to the field) rather than the other. The motion of an electron resembles that of a steel ball rolling down a nail-studded incline (Fig. 27.6b). The imbalance in the electron flow, which amounts to only about 1 in 10^4 electrons, is what constitutes a current.

There is an analogy between a wind and an electric current. The molecules in air have random thermal velocities whose average magnitude is somewhat larger than the speed of sound, which is about 330 m/s. A pressure difference between two regions causes a net flow of molecules in one direction. The wind velocity,

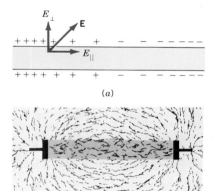

FIGURE 27.5 (a) When a wire is connected to a battery the surface of the wire becomes charged (although there is no net charge on the wire as a whole). It is the electrostatic field within the wire produced by these surface charges that "drives" the current. (b) The pattern of the electrostatic field due to a short length of wire.

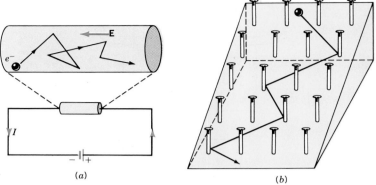

FIGURE 27.6 (*a*) The conduction electrons in a metal travel in zigzag paths as they collide with the positive ions of the lattice. When a battery is connected, the field within the wire causes a slight tendency of the electrons to move opposite to the field. (*b*) The motion of the electrons is analogous to that of a steel ball rolling down a nail-studded incline.

about 10 m/s, say, is much smaller than the random velocities. In a similar fashion, the conduction electrons in a wire have random thermal velocities up to about 10^6 m/s. When a potential difference is applied, they acquire a very slow drift velocity ($\approx 10^{-4}$ m/s) that is superimposed on the random thermal motion.

27.2   CURRENT DENSITY

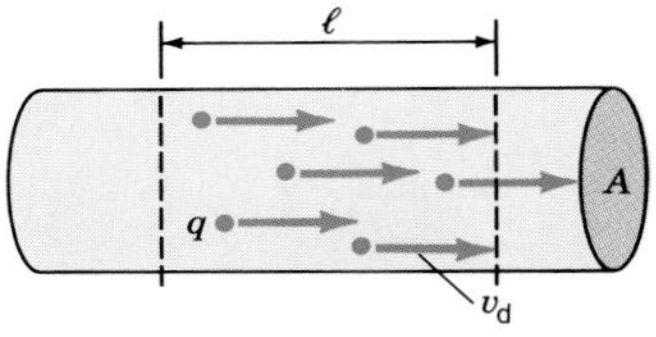

Since the random motion of the conduction electrons does not contribute to a current, we consider only the net effect of the electric field on the motion along the wire. Figure 27.7 shows particles of charge q moving with a drift velocity $\mathbf{v}_d$ along a wire. If there are n charges per unit volume, the total charge within a cylinder of length ℓ and area A is $\Delta Q = n(A\ell)q$. This charge takes a time $\Delta t = \ell/v_d$ to pass through the end of the cylinder. The current, given by $I = \Delta Q/\Delta t$, is

$$I = nAqv_d \tag{27.2}$$

The (average) **current density** is defined as the current per unit area:

$$J = \frac{I}{A} \tag{27.3}$$

The SI unit of current density is A/m². Whereas current is a scalar, current density is defined to be a vector in the direction of the drift velocity. From Eq. 27.2, we have

$$\mathbf{J} = nq\mathbf{v}_d \tag{27.4}$$

For negative charge carriers $\mathbf{J}$ is opposite to $\mathbf{v}_d$. Notice that the current I is a scalar quantity measured on a macroscopic scale; it is defined in terms of charge flowing through a surface. The current density $\mathbf{J}$ is a vector expressed in terms of microscopic quantities, and may vary from point to point. (If the current density is not uniform, the current through a surface is given by $I = \int \mathbf{J} \cdot d\mathbf{A}$.) An estimate of the magnitude of the drift velocity is obtained in the following example.

FIGURE 27.7 In order to calculate the current we ignore the random motion and consider only the small drift velocity acquired by the "free electron gas" as a whole.

EXAMPLE 27.1: A copper wire carries a current of 10 A. It has a cross-sectional area of 0.05 cm². Estimate the drift velocity of the electrons.

Solution: In order to find the carrier charge density needed in Eq. 27.4, we must first find the density of atoms. Recall that if M is the molecular mass of a substance, the number of elementary units (atoms or molecules) in M grams is Avogadro's number N_A (= 6.02×10^{23} mol⁻¹). For a mass m, the number of atoms N is therefore given by

$$\frac{N}{N_A} = \frac{m}{M} \qquad \text{(i)}$$

The mass density of the substance is $\rho = m/V$, where V is the volume. The number density of atoms, $n = N/V$, is related to the mass density ρ. Using $N = (m/M)N_A$ from (i) and $V = m/\rho$, we find

$$n = \frac{N}{V} = \frac{\rho N_A}{M}$$

For copper, $\rho = 8.9$ g/cm³ $= 8.9 \times 10^3$ kg/m³ and $M = 63.5 \times 10^{-3}$ kg/mol. With these numbers,

$$n = \frac{(8.9 \times 10^3 \text{ kg/m}^3)(6.02 \times 10^{23} \text{ atoms/mol})}{63.5 \times 10^{-3} \text{ kg/mol}}$$
$$= 8.5 \times 10^{28} \text{ atoms/m}^3$$

In Cu, each atom contributes one electron to the free electron gas and so the number we have just found is also the free electron density.

From Eq. 27.3 the magnitude of the current density ($J = I/A$) is

$$J = \frac{10 \text{ A}}{(5 \times 10^{-6} \text{ m}^2)} = 2 \times 10^6 \text{ A/m}^2$$

From Eq. 27.4 the drift speed is

$$v_d = \frac{J}{ne} = 1.5 \times 10^{-4} \text{ m/s}$$

This is the extremely small drift speed at which the electron gas as a whole moves along a wire.

EXERCISE 2. Since the drift velocity is so small, how do you reconcile this with the fact that a bulb lights almost instantly when the switch is turned on?

27.3 RESISTANCE

In 1729, Stephen Gray made the distinction between insulators and conductors. However, without adequate instruments there was no way of comparing the ability of different materials to conduct electricity. The first progress in classifying conductors came in remarkable experiments performed in 1772 by Henry Cavendish, who used his own body as a detector of shocks from a Leyden jar. For example, he passed the discharge through tubes of pure water and seawater. By adjusting the lengths of the tubes until the shocks were equally severe, he found that "a saturated solution of seawater conducts 720 times better than fresh water." He also tried holding known lengths of wire with both hands as a discharge passed through the wire. In this way he was able to compare how well different metals conduct.

Suppose that a current I flows through a conductor when a potential difference V is applied between two points. The **resistance** of the conductor between these points is defined as

$$R = \frac{V}{I} \qquad (27.5) \qquad \text{Definition of electrical resistance}$$

The SI unit of resistance is the ohm (Ω). From Eq. 27.5 we see that 1 Ω = 1 V/A. The resistance of an object tells us the potential difference that must be applied for a current of 1 A to flow through it. The resistance of a given sample depends on its geometry (size and shape) and the electrical properties of the conducting medium. It may also depend on V (or I).

The electric field within a wire accelerates the electrons. However, their velocities do not increase indefinitely because they collide with the array of positive ions that form the crystal lattice. As discussed in Section 27.1, the electrons acquire a slow drift velocity. At this velocity, the rate of work done by the electric field is equal to the rate of conversion of their kinetic energy into thermal energy. This steady drift velocity is analogous to the terminal velocity of a body falling through a viscous fluid (Section 6.5). The drift velocity is proportional to the electric field in the wire, that is, $\mathbf{v}_d \propto \mathbf{E}$. Since the current density is $\mathbf{J} = nq\mathbf{v}_d$, it follows that $\mathbf{J} \propto \mathbf{E}$. The relation between $\mathbf{J}$ and $\mathbf{E}$ is written as

Current density related to electric field

$$\mathbf{J} = \frac{1}{\rho}\mathbf{E} = \sigma\mathbf{E} \qquad (27.6)$$

where the constant ρ is called the **resistivity** of the medium and its SI unit is ohm · meter ($\Omega \cdot$ m). The **conductivity** is defined as $\sigma = 1/\rho$. A good electrical conductor has low resistivity (and high conductivity). Equation 27.6, which relates $\mathbf{J}$ to $\mathbf{E}$ at a given point within a medium, is actually a *definition* of ρ (or σ). This relationship is valid for conductors of any size and shape, including electrolytes and ionized gases. Some typical values of ρ are listed in Table 27.1.

TABLE 27.1 RESISTIVITIES AT 20°C

Material	Resistivity ($\Omega \cdot$ m)	Temperature Coefficient (C^{-1})
Mica	2×10^{15}	-50×10^{-3}
Glass	10^{12}–10^{13}	-70×10^{-3}
Hard rubber	10^{13}	
Silicon	2200	-0.7
Germanium	0.45	-0.05
Carbon (graphite)	3.5×10^{-5}	-0.5×10^{-3}
Nichrome	1.2×10^{-6}	0.4×10^{-3}
Manganin	44×10^{-8}	5×10^{-7}
Steel	40×10^{-8}	8×10^{-4}
Platinum	11×10^{-8}	3.9×10^{-3}
Aluminum	2.8×10^{-8}	3.9×10^{-3}
Copper	1.7×10^{-8}	3.9×10^{-3}
Silver	1.5×10^{-8}	3.8×10^{-3}

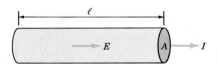

FIGURE 27.8 The current that flows through a wire when a potential difference is applied depends on the length and area of the wire and also on its resistivity.

Resistivity is a property of the material itself. Let us consider what role the size or shape of a conductor can play in determining the current. Figure 27.8 shows a wire of length ℓ and cross-sectional area A. We assume that the electric field in the wire is uniform. Therefore the potential difference across the ends of a wire of length ℓ is $V = E\ell$. Using $E = V/\ell$ in $J = I/A = (1/\rho)E$, we find

$$I = \frac{A}{\rho\ell} V \qquad (27.7)$$

By comparing the definition of resistance, $R = V/I$, with Eq. 27.7, which applies to the particular case of a wire, we find

$$R = \frac{\rho\ell}{A} \qquad (27.8)$$

The resistance of a wire is directly proportional to the length and inversely proportional to the area.

EXAMPLE 27.2: The radius of #8 copper wire is 1.63 mm. A potential difference of 60 V is applied across a 20-m length of this wire. Find: (a) its resistance; (b) the current; (c) the electric field.

Solution: (a) The resistivity of Cu, given in Table 27.1, is $\rho = 1.7 \times 10^{-8}\ \Omega \cdot$ m. The cross-sectional area is $A = \pi r^2$. Thus, from Eq. 27.8

$$R = \frac{\rho \ell}{A} = \frac{(1.7 \times 10^{-8} \ \Omega \cdot \text{m})(20 \ \text{m})}{(3.14)(1.63 \times 10^{-3} \ \text{m})^2}$$
$$= 0.04 \ \Omega$$

(b) The current is $I = V/R = 60 \ \text{V}/0.04 \ \Omega = 1500 \ \text{A}$.

(c) The electric field is $E = V/\ell = 60 \ \text{V}/20 \ \text{m} = 3 \ \text{V/m}$.

EXERCISE 3. A 20-m length of copper wire has a resistance of 0.4Ω. What is its diameter?

Temperature Dependence of Resistivity

Usually, the resistivity of a material depends on temperature. The resistivity ρ of a metal at a temperature T is expressed in terms of the resistivity ρ_0 at some reference temperature T_0:

$$\rho = \rho_0[1 + \alpha(T - T_0)] \qquad (27.9)$$

where α is the **temperature coefficient of resistivity,** measured in $°\text{C}^{-1}$. Equation 27.9 is valid only for a limited range of temperatures. Figure 27.9a shows how the resistivity of a typical metal varies with temperature. This behavior may be understood by considering three factors that contribute to the resistivity.

First, the electrons collide with the positive ions of the crystal lattice. These ions vibrate about their equilibrium positions. As the temperature increases, the amplitude of the vibrations increases, and consequently the flow of electrons is further impeded. Therefore, it is not surprising that the resistivity of a metal increases with temperature. The other two factors involve impurities, which are always present, and imperfections in the crystal lattice. The contributions to the resistivity from collisions with impurities and imperfections in the crystal are essentially independent of temperature. As a result, the resistivity of normal metals does not vanish even at $T = 0$ K.

Factors that influence electrical resistance

Two other types of material deserve mention. The resistivity of pure **semiconductors,** such as Si, Ge, and C, decreases as the temperature is increased, as shown in Fig. 27.9b. This is because more electrons become free to participate in the conduction process. More significant is the fact that the resistivity can be controlled by adding certain impurities to the pure material. It is this feature that allows the production of transistors and integrated circuits. In certain materials, called **superconductors,** the resistivity vanishes below a critical temperature T_c, indicated in Fig. 27.9c. Once a current is established in a superconductor, it persists indefinitely provided the low temperature is maintained. Semiconductors and superconductors are discussed further in Chapter 42.

EXERCISE 4. The resistance of a platinum resistance thermometer increases from 75 Ω to 80 Ω. What was the change in temperature?

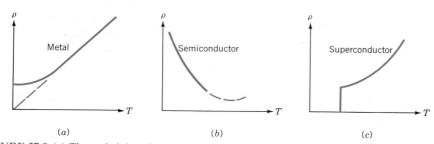

(a) (b) (c)

FIGURE 27.9 (a) The resistivity of a normal metal varies linearly with temperature over a broad range of temperatures. The finite resistivity at low temperatures is due to impurities and imperfections. (b) The resistivity of a semiconductor decreases as the temperature increase because more charge carriers are freed to take part in conduction. (c) The resistivity of a superconductor drop sharply to zero at a *transition* temperature that depends on the material.

FIGURE 27.10 Georg S. Ohm (1787–1854).

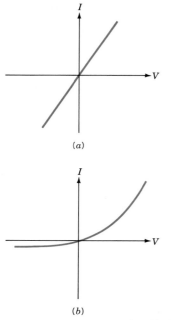

(a)

(b)

FIGURE 27.11 The *I-V* relationship for (a) an ohmic conductor and (b) for a junction diode, which is a nonohmic device.

27.4 OHM'S LAW

Equation 27.5, $R = V/I$, may be rewritten as

$$V = IR \qquad (27.10)$$

As it stands, this equation is merely a rewritten form of the definition of resistance. However, in those special cases in which R is a constant, *independent of V or I*, this equation also expresses a functional relationship, called Ohm's law, formulated by Georg Ohm in 1827 (Fig. 27.10). In modern terms, **Ohm's law** says that *the potential difference across a device is directly proportional to the current through it.*

Since V and I are macroscopically measured quantities, the equation $V = IR$ is called the macroscopic form of Ohm's law, provided R is constant. The relation $\mathbf{J} = \mathbf{E}/\rho$ is called the microscopic form of Ohm's law, provided ρ is constant, *independent of J or E*. In practice the condition of constant resistance (or resistivity) is satisfied by metals, provided the temperature is held constant. In certain instances (some alloys, carbon) Ohm's law is obeyed even when the temperature varies over a limited range. A material that obey's Ohm's law is said to be *ohmic*; otherwise it is *nonohmic*.

The *I-V* relationship for an ohmic device is a straight line, as in Fig. 27.11a. The *I-V* relationship for a *nonohmic* device, such as a junction diode (Fig. 27.11b), is not a straight line. The equation $R = V/I$ can be used as a definition of resistance at any point on such curves. But this does *not* mean that the object obeys Ohm's law. In fact, the resistance of a diode depends on the direction of current flow.

A **resistor** is a simple device that provides a specified amount of resistance in an electrical circuit. It can be made with a thin wire or a slab of ceramic. Since the resistivity of carbon is nearly constant over a wide range of temperatures, it is often used in the manufacture of resistors. We will assume that resistors obey Ohm's law in the form of Eq. 27.10. A resistor can be used to control the current flowing in a particular branch of a circuit. Two resistors placed in series can be used to divide a fixed potential difference, such as that due to a battery, into fractions needed by other elements, such as transistors. A variable "output" potential difference can be obtained by using a contact that slides over a single wire. Such a device is used as a volume control, for example, in a radio.

EXERCISE 5. What are the similarities and differences between electrical conduction and heat conduction?

27.5 POWER

Consider a stream of charged particles moving under the influence of an electric field. When a given charge q moves through a fixed potential difference V, its potential energy changes by $U = qV$. The rate at which the field delivers energy to the charge is the power delivered, which is $P = dU/dt = (dq/dt)V$, or

$$P = IV \qquad (27.11)$$

If the charged particles are electrons moving through a resistive medium, the electrical energy is converted into thermal energy. From the relation $V = IR$, the electrical power dissipated may also be written as

$$P = I^2R = \frac{V^2}{R} \qquad (27.12)$$

Recall that the SI unit of power is the watt (W). The dependence on the square of current was first demonstrated by James Joule, so the thermal loss in a resistor is often called Joule heating.

EXAMPLE 27.3: The heating element of a space heater is rated at 1000 W when operating at 120 V. (a) What is the current through it under normal conditions? (b) What would its power consumption be if the potential difference drops to 110 V?

Solution: (a) From Eq. 27.12,

$$I = \frac{P}{V} = \frac{1000 \text{ W}}{120 \text{ V}} = 8.33 \text{ A}$$

(b) We need to first find the resistance of the element:

$$R = \frac{V^2}{P} = \frac{(120 \text{ V})^2}{1000 \text{ W}} = 14.4 \text{ }\Omega$$

Since we assume the element obeys Ohm's law, its resistance will not change under the new operating conditions. Thus, the new power consumption is

$$P = \frac{V^2}{R} = \frac{(110 \text{ V})^2}{14.4 \text{ }\Omega} = 840 \text{ W}$$

27.6 MICROSCOPIC THEORY OF CONDUCTION (Optional)

A classical model for electrical conduction was proposed by P. K. Drude in 1900, shortly after the electron was discovered. He was able to relate the resistivity of a conductor to the motion of the electrons. In the following simplified version, we assume that a metal consists of a lattice of positive ions and a gas of free electrons. In the absence of an external electric field, the thermal velocities of the electrons are randomly oriented. The average thermal velocity is zero, which means that there is no net flow in any direction. When an electric field is applied, each electron experiences an acceleration $\mathbf{a} = -e\mathbf{E}/m$; therefore, over a time interval Δt, the change in velocity of an electron is

$$\Delta \mathbf{v} = -\frac{e\mathbf{E}}{m} \Delta t$$

Because of collisions with the lattice, $\Delta \mathbf{v}$ does not increase indefinitely. At each collision an electron gives up all the *excess* energy it gained from the field to vibrational energy of the ions. The collisions thus tend to destroy the orderly motion, so that the electrons retain only their random thermal velocities after each collision. With no electric field the paths of an electron between collisions are straight lines. In the presence of a field, the paths are parabolic, as depicted in Fig. 27.12. The times between collisions depend on the magnitude and direction of the velocity of the electron after each collision. The variation in

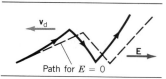

FIGURE 27.12 When a field is applied, the path of a conduction electron changes from the dashed lines to the solid curves.

FIGURE 27.13 The change in speed Δv acquired by an electron between collisions fluctuates in time. The change, averaged over all electrons, is called the drift speed v_d.

speed of a given electron might look like Fig. 27.13. Since the times between collisions span a range of values, we need an average taken over all electrons. This is called the mean time between collisions, τ. Using this time in our equation for $\Delta \mathbf{v}$, we see that the electron gas as a whole acquires an average drift velocity given by

$$\mathbf{v}_d = -\frac{e\mathbf{E}\tau}{m}$$

The constant τ is a property of the material. It does *not* depend on the electric field because the drift velocity is so much smaller than the thermal velocities ($v_d \ll v_{Th}$). From Eq. 27.4, $v_d = J/ne$, and therefore the current density is

$$J = \frac{ne^2\tau}{m} E = \frac{1}{\rho} E$$

where

$$\rho = \frac{m}{ne^2\tau} \tag{27.13}$$

The resistivity ρ is independent of $\mathbf{E}$, as is required by Ohm's law.

We may use Eq. 27.13 to obtain an estimate for τ. Using values appropriate for copper, we find

$$\tau = \frac{m}{ne^2\rho}$$

$$= \frac{(9.1 \times 10^{-31} \text{ kg})}{[(8.5 \times 10^{28} \text{ m}^{-3})(1.6 \times 10^{-19} \text{ C})^2(1.7 \times 10^{-8} \text{ }\Omega \cdot \text{m})]}$$

$$= 5 \times 10^{-14} \text{ s}$$

If the electrons are treated as an ideal gas, their rms speed is given by the equipartition theorem (Section 20.6), which says that $\frac{1}{2}mv^2 = \frac{3}{2}kT$. At 300 K, the rms speed, v_{rms}, would be about 10^5 m/s. Using this value, we can calculate the distance traveled between collisions, called the **mean free path:**

$$\lambda = v_{rms}\tau = 50 \times 10^{-10} \text{ m}$$

This may be compared with the interatomic spacing, which is about 2.5×10^{-10} m. In classical physics, one would expect the mean free path to depend on the interatomic spacing and the size of the atoms.

Although Drude's approach predicts that the resistivity is independent of the field strength, as is required by Ohm's law, there are problems with the theory. Classically, the thermal velocity varies with temperature according to $v_{Th} \propto \sqrt{T}$. Since $\tau = \lambda/v_{Th}$, Eq. 27.13 implies that the resistivity is proportional to $\sqrt{T}$. In fact, as Fig. 27.9a shows, the resistivity of metals is directly proportional to temperature for a wide range of temperatures (not including very low temperatures). As the temperature is lowered, the mean free path increases. At low temperatures λ can exceed 1 mm! Classical physics cannot explain how the electrons manage to avoid collisions with so many ions.

These problems were solved with the advent of quantum mechanics and a modern understanding of the statistical behavior of the free electron gas. First, as we will see in Chapter 41, electrons have wavelike properties. Thus, the very idea of a one-on-one collision between an electron and an ion is not correct. Second, the energies of the electrons are not distributed according to the equipartition theorem. In fact, only about 1 in 10^4 of the free electrons (those with the highest speeds, around 10^6 m/s) actually take part in the conduction process.

SUMMARY

Electric current is defined as the rate of flow of charge through a surface:

$$I = \frac{dQ}{dt}$$

When the current flows through an area A, the (average) **current density** is defined as

$$J = \frac{I}{A}$$

The current density at a point is a vector in the direction of motion of the charged particles:

$$\mathbf{J} = nq\mathbf{v}_d$$

where n is the number density of charge carriers, each of which has a charge q and moves with the drift velocity $\mathbf{v}_d$. The vector $\mathbf{J}$ can be related to the electric field that "drives" the current:

$$\mathbf{J} = \frac{1}{\rho}\mathbf{E} = \sigma\mathbf{E}$$

where ρ is the **resistivity** and $\sigma = 1/\rho$ is the **conductivity.** This equation is the microscopic form of Ohm's law *only* if ρ (or σ) is a constant, independent of $\mathbf{J}$ and $\mathbf{E}$.

If a current I flows through a device when a potential difference V is applied across it, the **resistance** is defined as the ratio

$$R = \frac{V}{I}$$

In the special case of a wire of length ρ and cross-sectional area A, the resistance is given by

$$R = \frac{\rho\ell}{A}$$

For metals, ρ varies with temperature.

The definition of resistance may be written in the form

$$V = IR$$

This equation applies to any conductor of any shape. It is *not* necessarily a statement of Ohm's law. This equation is called the macroscopic form of **Ohm's law** *only* if R is a constant, independent of V and I.

When a stream of charged particles moves through a potential difference V, the rate at which the field or an external agent does work, which is the **power** supplied, is

$$P = IV$$

From the relation $V = IR$ for a conductor, the power dissipated as thermal energy is

$$P = I^2R = \frac{V^2}{R}$$

ANSWERS TO IN-CHAPTER EXERCISES

1. From the definition of current, $\Delta q = I\,\Delta t = (1\ A)(1\ s) = 1\ C$. Since $e = 1.6 \times 10^{-19}\ C$, the number of electrons is $(1\ C)/(1.6 \times 10^{-19}\ C) = 6.3 \times 10^{18}$ electrons.

2. The information that an electron has entered or departed from the ends of a wire is transmitted as a disturbance in the electric field. This *disturbance* travels at the speed of light within the material. A mechanical analogy: Consider a tube full of marbles. When an extra marble is forced in at one end, another one pops out at the other end almost instantly.

3. From Eq. 27.8, we have $A = \rho\ell/R$ where $A = \pi d^2/4$. Thus,

$$d = \sqrt{\frac{4\rho\ell}{\pi R}} = 1.04\ mm$$

4. If we assume that the dimensions of the wire are unchanged, Eq. 27.9 may be rewritten in terms of resistance,

$$R = R_0(1 + \alpha\Delta T)$$

Thus,

$$\Delta T = \frac{R - R_0}{\alpha R_0} = 17\ K$$

5. Conduction electrons play a role in both electrical and heat conduction. However, heat is also conducted by means of collisions between the ions in a material that is an electrical insulator. Note the similarity between the equations for heat flow (Eq. 19.21) and charge flow (Eq. 27.7):

Heat flow: $\qquad \dfrac{\Delta Q}{\Delta t} = \dfrac{\kappa A \Delta T}{\ell}$

Charge flow: $\qquad \dfrac{\Delta q}{\Delta t} = \dfrac{\sigma A \Delta V}{\ell}$

Ohm's essential insight was that the relation *Flow rate = Driving mechanism/Resistance* applies as much to the flow of charge as it does to the flow of heat or of a fluid.

QUESTIONS

1. What is the effect of thermal expansion on the resistance of a wire? Does it tend to increase R or to decrease it?

2. Is Ohm's law valid only for wires?

3. If an appliance keeps "blowing" its own 15-A fuse, or an external 15-A fuse, would it be a good idea to switch to a 20-A fuse?

4. How does a current in a metal differ from that in an electrolyte? Can you give another example of an electrical current that differs qualitatively from a current in a metal?

5. Why is a light bulb most likely to burn out just as it is switched on? Why does it emit a bright flash just before it burns out?

6. What is meant by a "short circuit"? Illustrate your answer with a diagram.

7. Why is it possible for birds to perch on power transmission lines without being electrocuted?

8. Liquids exhibit greater resistance to flow as the temperature is lowered. Is the flow of charge in a wire affected in the same way?

9. Assuming that all other quantities are held constant, how does the drift velocity along a wire depend on each of the following: (a) length; (b) potential difference; (c) cross-sectional area; (d) current?

10. A copper and a silver wire of the same length and diameter carry the same current. In which is the electric field greater?

11. In what way(s) does the electron beam current in a TV tube differ from a current in a wire?

12. The expression $P = I^2 R$ indicates that the power increases with resistance, whereas $P = V^2/R$ seems to indicate the opposite. Reconcile these two apparently contradictory views.

13. A beam of electrons is accelerated without increasing its area. Does the current density change as the particles speed up?

14. What are the relative advantages, or disadvantages, of using a single strand of wire versus multiple strands with the same overall resistance?

EXERCISES

27.1 and 27.2 Current; Current Density

1. (I) The beam current in a color TV tube is 1.9 mA. The beam cross section is circular with radius 0.5 mm. (a) How many electrons strike the screen per second? (b) What is the current density?

2. (I) Protons in an accelerator are moving at 5×10^6 m/s and produce a beam current of 1 μA. If the radius of the beam is 1 mm, find: (a) the current density; (b) the number density of particles.

3. (I) A current of 200 mA flows in a silver wire of radius 0.8 mm. Find: (a) the drift speed of the electrons; (b) the electric field within the wire. The free electron density is 5.8×10^{28} m^{-3}.

4. (I) A 30-km length of electrical transmission cable made of copper wire of diameter 1 cm carries a 500-A current. Find: (a) the current density; (b) the electric field within the wire: (c) the drift speed; (d) how long it takes for a given electron to travel the length of the wire. The free electron density is 8.5×10^{28} m^{-3}.

5. (I) The starter motor on a car draws 80 A through a copper cable of radius 0.3 cm. (a) What is the current density? (b) Find the electric field within the wire.

6. (I) A 14-gauge copper wire of diameter 1.628 mm carries 15 A. Find: (a) the current density; (b) the drift velocity. The free electron density is 8.5×10^{28} m^{-3}.

7. (I) The electron in a hydrogen atom moves in a circle of radius 5.3×10^{-11} m at a speed of 2.2×10^6 m/s. What is the average current?

8. (II) The current in a wire varies as $I = 2t^2 - 3t + 5$ A. How much charge flows through a cross section of the wire between $t = 2$ s and 5 s?

9. (I) A 10-m length of aluminum wire has a diameter of 1.5 mm. It carries a current of 12 A. Find: (a) the current density; (b) the drift velocity; (c) the electric field in the wire. Aluminum has approximately 10^{29} free electrons per m^3.

27.3 Resistance

10. (I) When 100 V is applied across a wire of length 25 m and radius 1 mm, a current of 11 A flows through it. Find the resistivity of the material.

11. (I) A cylindrical rod of silicon has a length of 1 cm and a radius of 2 mm. What is the current when a potential difference of 120 V is applied across the ends?

12. (II) A wire of length ℓ and cross-sectional area A has a resistance R. What is the resistance if the same amount of material is used to make a wire twice as long?

13. (I) A cylindrical tube of length ℓ has an inner radius a and an outer radius b, as shown in Fig. 27.14. The resistivity is ρ. What is the resistance between the ends?

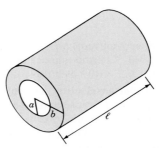

FIGURE 27.14 Exercise 13.

14. (I) A silver wire has a resistance of 1.20 Ω at 20 °C. What is its resistance at 35 °C? (Ignore changes in dimensions.)

15. (I) The resistance of a copper wire is 0.8 Ω at 20 °C. When it is placed in an oven, its resistance is 1.2 Ω. What is the temperature of the oven?

16. (I) A wire of length 4 m and diameter 0.8 mm has a resistance of 16 Ω at 20 °C. The resistance rises to 16.5 Ω at 35 °C. What is the temperature coefficient of resistivity?

17. (I) The resistances of a copper wire and an aluminum wire are equal. What is the ratio of the lengths, ℓ_{Cu}/ℓ_{Al}, if they have the same diameter?

18. (II) By combining a carbon and a nichrome resistor in series, one can devise an equivalent resistor whose resistance is independent of temperature. What percentage of the resistance should be contributed by carbon?

19. (I) The resistances of a copper wire and an aluminum wire are equal. What is the ratio of the diameters, d_{Cu}/d_{Al}, if they have the same length?

20. (I) The conductance G of a device is defined to be the inverse of its resistance, $G = 1/R$. Its SI unit is the mho (ohm^{-1}) or the siemens (S). What is the conductance of a device through which the current is 2 A when a potential difference of 60 V is applied?

21. (I) A wire of radius 2 mm and length 12 m has a resistance of 0.027 Ω, what is its resistivity? Can you identify the material?

22. (I) The resistance of a carbon rod is 0.6 Ω at 0 °C. What is the resistance at 30 °C?

23. (I) A wire of length 10 m and 1.2 mm diameter has a resistance of 1.4 Ω. What is the resistance if the length is 16 m and the diameter is 0.8 mm?

24. (I) A wire is connected to a 6-V battery. At 20 °C, the current is 2 A, whereas at 100 °C, the current is 1.7 A. What is the temperature coefficient of resistivity? Ignore changes in dimensions of the wire.

25. (I) A copper wire has a resistance of 1 Ω at 20 °C. At what temperature is the resistance 10% (a) higher; (b) lower?

27.4 and 27.5 Ohm's Law; Power

26. (I) A loudspeaker is connected to an audio amplifier with 18-gauge copper wire (diameter 1.024 mm) of total length 20 m. (a) What is the resistance of the wire? (b) If the speaker has a 4-Ω resistance, what percentage of the power delivered by the amplifier is dissipated in the wire? (For simplicity assume that the potential difference does not vary in time and that the speaker is a resistor.)

27. (I) A transmission line 200 km long has a resistance of 10 Ω and carries a current of 1200 A. What is the potential difference between two pilons separated by 200 m?

28. (I) According to a safety code, the maximum current allowed for 14-gauge copper wire (diameter 1.628 mm) is 15 A, and for 18-gauge wire (diameter 1.024 mm) it is 5 A. What would be the potential difference between the ends of a 10 m length of each type of wire at the maximum current?

29. (I) A 12-V car battery is rated at 80 A · h. (a) How much charge can it deliver? (b) For how long could it supply 25 W, assuming that the potential difference is fixed?

30. (I) A toaster operates at 120 V with a current of 7 A. It takes 30 s to complete its task. At 6¢ per kWh, what does it cost to toast a slice of bread?

31. (I) The diameter of 14-gauge Cu wire is 1.628 mm, while an 18-gauge wire has a diameter of 1.024 mm. Compare the electrical power losses when a current of 8 A flows in a 10-m length of each wire.

32. (I) The two headlights of a car require a total of 10 A supplied at 12 V. Given that the combustion of 1 L of gasoline releases 3×10^7 J and that the conversion to electrical power has an efficiency of 25%, how much gasoline is consumed in one hour for this purpose alone?

33. (I) A battery delivers 30 mW to an 8-Ω resistor. How many electrons leave the negative terminal in 1 min.?

34. (I) A 12-gauge copper wire of diameter 2.05 mm is used to supply 12 A to an appliance. What is the power dissipated in 20 m of the wire?

35. (I) An aluminum wire has a resistance of 1.8×10^{-3} Ω per meter and carries a 200-A current. What is the power dissipated by 10 km of such wire?

36. (II) A motor operating at 240 V draws 10 A while lifting a 2000-kg block vertically upward at a constant 2.5 cm/s. Find: (a) its horsepower; (b) the percentage efficiency of the conversion of electrical power to mechanical power.

37. (I) A power station supplies 100 kW to a load via cables whose total resistance is 5 Ω. Find the power loss in the cables if the potential difference across the load is (a) 10^4 V, (b) 2×10^5 V.

38. (II) A kettle operating at 120 V heats 1.5 L of water from 20 °C to 90 °C in 8 min. What is the current?

39. (II) Water is contained in a glass tube of radius 1 cm and length 20 cm. What potential difference would be required to heat the water by 30 °C in 4 min? The resistivity of water is 10^{-2} Ω · m.

40. (I) A three-way light bulb uses two filaments, singly or in series, to produce three power values: 41 W, 70 W, and 100 W when connected to a 120 V source. What are the two resistances?

41. (I) A three-way light bulb uses two filaments, singly or in parallel, to produce three power values: 50 W, 100 W, and 150 W when connected to a 120 V source. What are the two resistances?

PROBLEMS

1. (I) Nichrome is an alloy used in the heating element of a water heater that operates at 120 V. The "cold" resistance at 20 °C is 16 Ω. (a) If the radius of the wire is 1 mm, what is its length?; (b) What is the current at 200 °C?

2. (I) Consider the following data for a circuit element (such as a light bulb):

V (V):	2	4	6
I (A):	0.3	0.5	0.7

(a) What is the resistance when V = 5 V? (b) What would be the current when V = 0? (c) Does the element obey Ohm's law?

3. (II) A cylindrical tube of length L has inner radius a and outer radius b, as shown in Fig. 27.15. The material has resistivity ρ. Current flows radially from the inner to the outer surface. (a) Show that the resistance is

$$R = \frac{\rho}{2\pi L} \ln \frac{b}{a}$$

(b) What is the resistance of a carbon filament whose dimensions are a = 0.4 cm, b = 3 cm, and L = 30 cm? (*Hint:* Start with the equation $J = E/\rho$ and note that $E_r = -dV/dr$.)

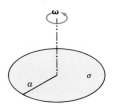

4. (II) A thick spherical shell has an inner radius a and an outer radius b. The material has resistivity ρ. Show that when a potential difference is applied between the inner and outer surfaces, the resistance is

$$R = \frac{(b - a)\rho}{4\pi ab}$$

Assume the current is radial at all points. (*Hint:* Start with $J = E/\rho$ and then use $E_r = -dV/dr$.)

5. (I) A nonconducting disk of radius a has a uniform surface charge density σ C/m². It rotates at angular velocity ω as depicted in Fig. 27.16. What is the current passing through a surface perpendicular to the plane of the disk? (*Hint:* First find the current associated with a ring of radius r and width dr.)

6. (I) A potential difference of 2 V is applied across the ends of a 30-m length of a silver wire of diameter 0.5 mm. Find: (a)

the drift speed; (b) the mean time between collisions; (c) the mean free path at 300 K. The free electron density is 5.8×10^{28} m^{-3}.

7. (I) An electroplating cell uses AgNO$_3$ to deposit silver (108 u) on an electrode. If a current of 0.2 A is equally shared by the Ag$^+$ and NO$_3^-$ ions, what mass of Ag is deposited in 10 min?

8. (II) A 600-kg electric car has a bank of twenty 12-V batteries in parallel, each rated at 100 A · h. The car is subject to a total resistance of 180 N at 60 km/h. For how long can the car travel (a) on level ground, (b) up a 10° incline? Assume that the potential difference stays constant.

9. (I) The ends of a copper and a steel wire, each 40 m long and of radius 1 mm, are connected together. A potential difference of 10 V is applied across the free ends. Find: (a) the power dissipated in each wire; (b) the electric field in each wire.

SPECIAL TOPIC: Atmospheric Electricity

Lightning inspires fear and awe in humans and was taken by the ancients to be a demonstration of the wrath of the Gods. In the mid-eighteenth century Ben Franklin established that lightning is actually an electrical phenomenon. In this essay we discuss some aspects of lightning and how thunderstorms develop. We begin with the less well-known fact there is an electric field in the atmosphere even in fair weather.

The Fair-Weather Field

On clear days there is an electric field of approximately 100 V/m directed downward at the surface of the earth. From the equation $E = \sigma/\varepsilon_0$, we infer that there is a negative surface charge density of about -10^{-9} C/m^2 on the ground. The field lines start at a positively charged layer at an altitude of about 50 km, the lower limit of a region called the ionosphere. The potential difference between the earth and this layer is about 3×10^5 V. (The field strength decreases with altitude so the potential difference is not 5 MV.)

Both the earth and the ionosphere are good conductors. Although dry air is a good insulator, the atmosphere does allow the flow of current because of the presence of oxygen and nitrogen ions that are created by cosmic rays, by natural radioactivity, and, at high altitudes, by photoionization due to ultraviolet and X rays from the sun. One would expect the vertical current through the air (3×10^{-12} A/m^2, or 1500 A globally) to result in the neutralization of the charges on the ground in about 10 min. We will see that it is the charge transfer associated with lightning that maintains the fair-weather field.

From the value of the fair-weather field, we see that the potential difference between two levels 2 m apart is about 200 V. Does this potential difference pose a threat to people? Could it be used as a source of electrical energy? No, because objects such as people, motors, and trees are good conductors, and all parts of a conductor in contact with the ground are at ground potential. As a result, the horizontal equipotentials are distorted, as shown in Fig. 27.17.

The fair-weather field strength can be measured with an instrument called a *field mill*. When a horizontal metal plate A is grounded, as in Fig. 27.18a, its top surface becomes negatively charged, the charges being held by the external field. If a second plate B suddenly covers A, as in Fig. 27.18b, the field lines no longer reach A, so the negative charge flows to ground through a measuring device. In practice, A and B are in the form of vanes, as in Fig. 27.18c. As the upper plate rotates, it alternately exposes and

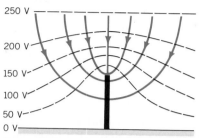

FIGURE 27.17 A metal rod inserted into the ground is at ground potential. The horizontal equipotentials become distorted as shown.

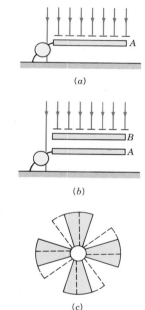

FIGURE 27.18 (*a*) The earth's field induces a charge on a grounded metal plate. (*b*) When the plate is shielded, the induced charge passes through the detection device. (*c*) In a field mill, the upper and lower plates are in the form of vanes. As the upper vane rotates, there is a pulsed current that can be amplified and recorded.

shields the lower plate from the field. The current from the lower plate is in the form of pulses that can be amplified and calibrated to yield the field strength.

The fair-weather field varies during each day. When measured at sea, away from other disturbances, it is sur-

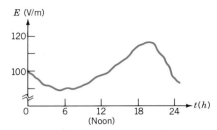

FIGURE 27.19 The fair-weather field varies with time in the same way around the earth.

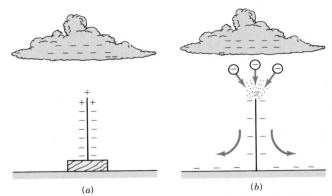

FIGURE 27.20 (*a*) There is a charge separation in an insulated metal rod. (*b*) If the rod is grounded, there is a constant current.

prising that the maximum field strength occurs at about 7 p.m. GMT (Greenwich Mean Time) and the minimum at 4 a.m. GMT at all points around the world (see Fig. 27.19). Because of the high conductivity of the earth and the ionosphere, local changes in charge density are dispersed very quickly around the globe.

Franklin's Kite Experiment

In 1750, Ben Franklin brought a grounded pointed conductor close to an electrified body and found that it would discharge the body more quickly than a blunt conductor. He surmised that if thunderclouds were electrified, perhaps they could also be harmlessly discharged, thereby preventing damage from lightning. First he had to demonstrate that the thundercloud is charged. His approach, expressed in modern terms, involved the following.

Figure 27.20*a* shows an insulated metal rod under a thundercloud, the base of which is negatively charged. The electric field, directed vertically upward, induces a separation of charge in the rod. Negative charges in the air neutralize some of the positive charge at the tip of the rod, which thus acquires an excess negative charge. A grounded conductor, such as a person, will elicit a spark on bringing a finger near the insulated charged rod. If instead the rod is grounded, as in Fig. 27.20*b*, negative charge continues to flow through the rod into the ground and a faint glow may be seen at the tip. This glow discharge is caused by electrons, which are accelerated by the strong electric field near the tip and then cause ionization of molecules. When the molecules recombine with other electrons, light is emitted. In the 4th century such a glow from the masts and rigging of ships was called St. Elmo's fire.

In order to show that a thundercloud is charged, Franklin proposed that a man in a sentry box might draw sparks from a tall insulated metal rod by bringing to it a grounded wire—held by a wax handle to protect the person. Scientists in France, who found out about this proposal, set up a 40-ft rod and in May 1752 obtained the expected sparks. Before he got news of their success, Franklin decided that a kite, with a pointed wire attached, would serve his purpose just as well. Accordingly, he flew a kite into an approaching thundercloud and held the twine with an insulating silk thread, taking care not to get it wet. Franklin had attached a key to the twine. When he saw the fine threads bristle, he realized that the twine was electrified and then drew a spark from the key with his knuckle. This experiment showed that a thundercloud is charged, and indirectly, that lightning is an electrical phenomenon. Both the kite and the sentry box experiments are very dangerous. A few months later, a Professor Richmann of St. Petersberg was trying to repeat the sentry box experiment and was killed instantly when the rod carried a lightning stroke.

Benjamin Franklin doing his kite experiment.

Thunderstorms

There are approximately 40,000 thunderstorms worldwide each day and about 100 lightning flashes each second. A *cumulonimbus* thunderstorm cloud begins as a relatively small cloud extending from an altitude of 2 to 5 km. The thundercloud is formed by a strong updraft of warm, humid air. The cloud grows rapidly in size and in a matter of minutes can reach an altitude of 10 to 15 km. Since pressure decreases with altitude, the moist air expands as it rises and its temperature decreases. The water vapor then condenses into drops and releases its latent heat. This makes the humid air warmer than the surrounding dry air, and so the humid air continues to rise at about 25 m/s. Near the top of the cloud, the dry surrounding air mixes with the updraft, causing evaporative cooling of the drops. Ice crystals are formed and grow into hailstones as they collide with water drops. When the hailstones become too heavy to be supported by the updraft, they start to fall, setting up a downdraft in a region of the cloud that does not overlap the updraft (see Fig. 27.21). Because hailstones usually melt

FIGURE 27.21 A thunderstorm cloud. The updraft of warm, moist air rises until ice crystals form. The ice crystals grow into hailstones that fall in the region of the downdraft.

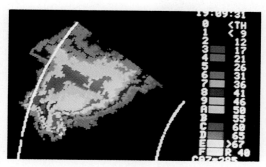

A storm, and the possibility of lightning, detected by Doppler radar. The frequency shift of the radar signal indicates the (color-coded) wind velocity in m/s.

before they reach the ground, there is a heavy rainfall. A thunderstorm cloud moves horizontally at about 30 km/h, gathering up fresh moist air. In the final stages of the development of the thundercloud, the downdraft prevails and a light rain falls.

Charge Separation

An important feature of a thundercloud is the appearance of large quantities of charge at the base and at the top of the cloud. Measurements of charge within the cloud, and of the electric field at the earth's surface, show that the lower part of the cloud is negatively charged ($N = -40$ C) whereas the upper part is positively charged ($P = +40$ C) (see Fig. 27.22). There is also a small positive charge ($p = +10$ C) at the base. The field under the thundercloud, which is opposite to the fair weather field, has a strength of about 10^4 V/m. The potential difference between the base of the cloud and ground is about 3 MV.

The mechanism by which the charge separation occurs is not clear. Here is one proposal. As a hailstone falls it is polarized by the existing fair weather field: Its base is positive, whereas its top is negative, as in Fig. 27.23a. A water droplet, or a tiny ice crystal, can collide with the base of the hailstone and acquire a positive charge. The lighter positively charged particle is carried up by the updraft, but the negatively charged hailstone continues to fall. The electric field is strengthened by this separation of charge and produces even greater polarization of the hailstones. The process involves "positive feedback." Ions already present in the air probably contribute to the effect.

An alternative mechanism involves the process of freezing. It is known that when there is a temperature difference across a sample of ice, the warm end becomes negatively charged. This occurs because the light H^+ ions are more mobile than OH^- ions. The H^+ ions migrate from the warm end in greater numbers, leaving an unbalanced neg-

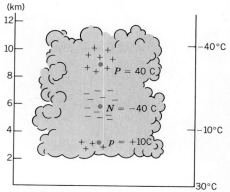

FIGURE 27.22 The charge separation in a thundercloud.

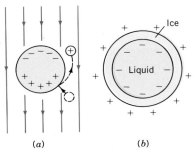

FIGURE 27.23 (*a*) A droplet can become charged when it collides with the lower positive side of a falling hailstone. (*b*) When a droplet of water starts to freeze, a shell of ice forms with the charges indicated.

ative charge. When a drop freezes, a thin layer of ice first forms on its surface. As the inner liquid freezes, it releases the latent heat of fusion, so the temperature on the inner surface of the skin is higher than on the outer surface. Because of this *thermoelectric* effect, the outer surface becomes positively charged (see Fig. 27.23*b*). A collision with another particle can burst the skin, allowing positively charged ice splinters to form. Being very light, they are carried by the updraft. The remainder of the drop is negatively charged and continues to fall.

Both of these explanations become untenable if it is established that the charge separation occurs before there is a downdraft. There is no universally accepted mechanism.

Lightning

Lightning is the most spectacular display of atmospheric electricity (see Fig. 27.24). It has been studied with high-speed cameras, by its radio-frequency emissions, by radar reflections, and by means of the electric field changes at ground level. A lightning *flash* consists of several distinct *strokes*. The process probably starts with a discharge within the cloud from the small positive base *p* to the higher negative charge *N*. This internal discharge lasts about 50 ms and is recorded on film as a faint glow. Half the flashes in a storm occur between clouds up to 10 km apart. Sometimes the discharge travels so far that it reaches clear air and becomes the proverbial "bolt from the blue."

A lightning stroke is initiated by a *step leader*. This is a highly ionized gas that carries mostly negative charge (see Fig. 27.25*a*). It is recorded on film as a bright spot that moves 50 m in 1 μs, pauses for 50 μs, and then takes another "step." Its average speed is about 2×10^5 m/s and the potential of the tip is -10^8 V relative to ground. When the tip is about 50 m from the ground, a *streamer* leaves the earth, usually from a sharp point (see Fig. 27.25*b*). When the streamer meets the step leader, a continuous conducting path is formed from the ground to the base of the cloud. The negative charge in this ionized channel drains very rapidly into the ground. The portion near the ground is discharged first, producing an intense light, as shown in Fig. 27.25*c*. As higher portions are discharged, the illumination travels upward in a *return stroke*, which has a diameter of about 20 cm (see Fig. 27.25*d*). Initially, the illumination wave front travels at nearly one-third the speed of light, but

FIGURE 27.24 A lightning flash.

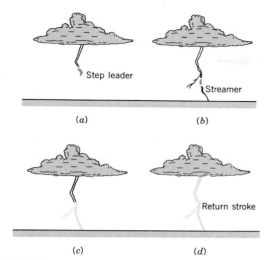

FIGURE 27.25 Four stages in the development of a lightning return stroke.

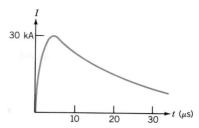

FIGURE 27.27 The variation in current during a return stroke.

it slows to $c/10$ near the base of the cloud. (Note that this is the speed at which the discharge propagates, not the speed of individual electrons.) The step leader transports -5 C in about 40 ms, which means that the average current is about 100 A.

After the return stroke, a small current continues to flow in the channel. After a pause, a *dart leader*, about 1 m wide, travels down to earth at 5×10^6 m/s and initiates a second return stroke (Fig. 27.26). The dart leader transports 1 C in 2 ms, for an average current of 500 A. A single flash, which lasts 0.3–0.5 s, involves four or five return strokes, each lasting 2 ms and separated by 50 ms. The individual strokes are visible as a flicker during the flash.

The variation of the current as a function of time is shown in Fig. 27.27. The current rises to 30 kA or more in about 2 μs, then decreases more gradually. (The maximum current can be estimated by placing small bars of cobalt steel near a long rod or transmission tower. After a lightning stroke the bar becomes magnetized and from the strength of the magnetization, the current can be estimated. Currents

as high as 60 kA have been recorded.) About 10 to 20 C of negative charge is transferred to the ground in 100 μs.

During the return stroke the temperature of the ionized channel reaches 30,000 K. The increase in pressure generates a shock wave that is heard as thunder up to 25 km away. The rumble that is usually heard is due to the different times of arrival of the sound from different regions of the stroke.

A single stroke can transfer 5 C through a potential difference 10^8 V in 10 μs. The energy involved is 5×10^8 J, and the power is 5×10^{13} W! The energy is dissipated in molecular excitation, creation of ions, kinetic energy of particles, and radiation. Despite the dramatic display of the lightning flash, about 80% of the charge transfer is due to point discharge, for example, from trees. The average current during a thunderstorm is 1.5 A.

Lightning Protection

Lightning causes much property damage and starts many forest fires. In the United States alone, 200 people are killed by lightning each year. Franklin believed that a grounded metal rod with a sharp tip drains away the charge in a thundercloud and thereby prevents a lightning stroke. This explanation of the action of a "lightning rod" is not correct. All objects with sharp points, such as leaves, exhibit point discharge. However, if a step leader does approach the area around the rod, the rod will be struck and will carry the discharge safely to ground. A lightning rod protects a region about itself, shown by the dashed lines in Fig. 27.28. In very rare instances a single sharp point will in fact discharge the cloud. For several years lightning was studied at the Empire State Building. It was found that a positive dis-

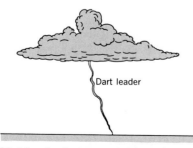

FIGURE 27.26 After the first return stroke, other return strokes are initiated by dart leaders.

FIGURE 27.28 A lightning rod "protects" the area indicated by the dashed lines.

charge (250 A) would initiate at the top of the building and rise to the cloud, without any return stroke. Thus, Franklin's idea works only from the tops of very high masts or buildings.

What should one do during a thunderstorm? First, avoid isolated trees or shacks that are obvious targets. Second, since water is a conductor, stop swimming. A discharge can be carried into a house through the water pipes or even via telephone wires. It is possible for the current to flow through the pipes and to electrocute someone taking a bath. Indeed, discharges through taps have been fatal. Therefore, during a thunderstorm, avoid using the telephone or taking a bath.

A lightning stroke can induce a rapid flow of charge in a grounded conductor. For example, a current of 100 A has been measured in a transmission wire 0.5 km from a stroke. When lightning strikes the ground, large currents flow through it. The potential difference between two points 1 m apart can result in a fatal current for animals or people. One can minimize the effect of such a "step voltage" by keeping one's feet together. When a stroke hits a tree, the current flows along the wet channels on the bark and may jump (flashover) from the trunk to a person who is either standing near or leaning on the tree. Thus, one should avoid taking shelter under an isolated tree.

CHAPTER 28

Direct Current Circuits

A worker makes a "routine" electrical connection during construction of the John Hancock Tower in Chicago.

Major Points

1. The definition of **emf.** The distinction between emf and potential difference.
2. **Kirchhoff's junction rule** and **loop rule.**
3. Series and parallel combinations of resistors.
4. The variation of charge and potential difference in **RC circuits**.

In this chapter we study currents and potential differences in circuits and describe some simple instruments used to measure these quantities. The discussion is restricted to direct currents (dc) that flow only in one direction. We start with steady dc currents in circuits that contain resistors, and then go on to time-varying dc currents in circuits that contain both a resistor and a capacitor.

When a current flows through a resistor, electrical energy is dissipated. One cannot have a circuit that consists solely of devices that dissipate electrical energy; there must also be some source of electrical energy. Such a device is called a source of **electromotive force,** abbreviated **emf.** Electrical circuits usually consist of several loops that contain circuit elements, such as resistors and capacitors, and sources of emf. Gustav Kirchhoff formulated two simple rules that form the basis of analyzing the currents and potential differences in electrical circuits. As we will see, they are actually statements of the conservation of charge and the conservation of energy.

28.1 ELECTROMOTIVE FORCE

To gain insight into the operation of a simple circuit, let us consider a mechanical analogy. Figure 28.1a shows a vertical belt, driven by a motor or a handcrank, that lifts balls through some height. Clearly it takes work to increase the gravitational potential energy of the balls. They roll on a horizontal section without changing their energy and then fall through a vertical tube filled with wire netting. As the balls fall at a constant (terminal) velocity, their potential energy is converted to thermal energy. At the bottom the balls roll back to the belt to repeat the round trip.

Now consider the motion of a hypothetical positively charged particle around a closed circuit that consists of a battery, a resistor R, and two wires, as in Fig.

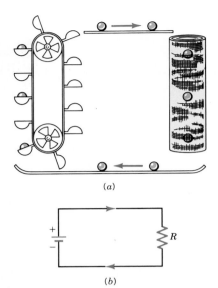

(a)

(b)

FIGURE 28.1 (*a*) A mechanical analogy of an electrical circuit. A mechanical device provides the energy to lift balls through some height. They subsequently fall at a constant speed through a tube filled with wire mesh, their potential energy being converted to thermal energy. (*b*) In an electrical circuit, a source of emf raises the potential energy of charges. This energy is dissipated as thermal energy in the resistor.

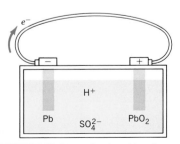

FIGURE 28.2 In a lead–acid cell electrons are continually transferred through the wire from the Pb terminal to the PbO$_2$ terminal.

28.1b. For convenience we assume the potential at the negative terminal of the battery is zero. When the particle enters at the negative terminal, the chemical action raises its electrical potential energy and places it at the positive terminal. As the particle moves through the wire (which we assume has zero resistance), the particle's energy does not change. When it encounters the resistor, the particle suffers many collisions with the ions of the lattice and moves through the wire at a small constant (drift) velocity. The electrical potential energy of the particle is converted to thermal energy in the resistor. Finally, the charge leaves the resistor with zero potential energy.

A battery must do work to separate positive and negative charges and to place them at the terminals against the repulsion of charges already there. A battery is an example of a source of electromotive force (emf). The term electromotive force is misleading; it was introduced by Volta who thought of it as a kind of "force" that causes current to flow. Since this is not the modern view, from now on we will use just emf. In modern terms, a source of emf converts some form of energy, such as chemical, thermal, radiant, or mechanical, into electrical potential energy. The **emf**, $\mathscr{E}$, of a device is defined as

$$\mathscr{E} = \frac{W_{ne}}{q} \tag{28.1}$$

An emf is the work per unit charge done by the source of emf in moving the charge around a *closed loop*.

The subscript "ne" emphasizes that the work is done by some nonelectrostatic agent, such as a battery or an electrical generator. The value of the emf depends on the particular physical process used to produce the charge separation and its associated potential difference. It is usually, but not always, a fixed property of a device.

One must be careful to distinguish between the concepts of emf and potential difference. A potential difference is associated only with a (conservative) electrostatic field. As we pointed out in Section 27.1, when a current flows in a wire, the "driving" electric field is produced by the distribution of charges on the terminals of the battery and on the surface of the wire. That distribution of charge is produced by a source of emf. An emf is always associated with some nonelectrostatic mechanism that provides the energy required to separate positive and negative charges. Thus a source of emf *converts* some form of energy into electrostatic potential energy.

The Production of a Current

Volta explained the action of the voltaic pile in terms of a potential difference set up by the contact of two metals. In his view the saline or acidic solution (called an electrolyte) served merely as a conductor. The English chemist, Humphrey Davy, correctly shifted attention to the interactions between the metals and the fluid electrolyte. A simplified explanation of the action of a lead–acid cell, found in car batteries, follows.

In a lead–acid cell, shown in Fig. 28.2, a Pb plate and a PbO$_2$ plate are immersed in an aqueous solution of sulfuric acid (H$_2$SO$_4$), which dissociates into positive hydrogen (H$^+$) and negative sulphate ions (SO$_4^{2-}$). When a wire is connected between the terminals, the following reactions take place. At the Pb plate,

$$\text{Pb} + \text{SO}_4^{2-} \rightarrow \text{PbSO}_4 + 2e^-$$

The two electrons freed in this reaction leave the Pb terminal and enter the wire. At the PbO_2 plate, two other electrons enter the terminal from the wire and the following reaction occurs:

$$PbO_2 + 4H^+ + SO_4^{2-} + 2e^- \rightarrow PbSO_4 + 2H_2O$$

Note that for every electron that leaves the Pb plate, another enters the PbO_2 plate; the wire itself does not acquire a net charge. Lead sulfate ($PbSO_4$) is deposited on both plates, and the acid is consumed. Electrons are continually transferred from the Pb plate, which acts as the negative terminal, to the PbO_2 plate, which acts as the positive terminal. The result is that a current flows in the external wire. A constant potential difference of 2.05 V is maintained between the plates. In a car battery six cells are placed in series for a total potential difference of about 12 V. Keep in mind that it is the *difference* in potential that is important; one can assign $V = 0$ to either terminal.

Terminal Potential Difference

A real source of emf, such as a battery, has internal resistance. When a current flows, there is a potential drop across this internal resistance. Let us find the potential difference between the terminals of a battery through which a current is flowing. In Fig. 28.3, the battery is treated as an ideal source of emf $\mathscr{E}$ with a resistor r in series with it. We start at point a and record the changes in potential energy of a unit positive test charge. As the charge moves through the battery from the negative to the positive terminal, the source of emf changes the potential of the charge by an amount numerically equal to $+\mathscr{E}$. In moving through the internal resistance, the potential of the unit charge decreases by Ir. The change in potential

$$V_{ba} = V_b - V_a = \mathscr{E} - Ir \qquad (28.2)$$

is called the **terminal potential difference.** Note that when *either* $I = 0$ or $r = 0$, we have $V_{ba} = \mathscr{E}$. Therefore, the emf of many sources may be measured by the "open-circuit" potential difference between the terminals. Unlike the emf, which is (usually) a fixed property of the source, the terminal potential difference depends on the current flowing through it. As a battery ages its internal resistance increases, and so, for a given output current, the terminal potential difference falls.

The fact that the emf is numerically equal to the terminal potential difference when $I = 0$ does not mean that the emf is the "same thing" as the potential difference. In a sense, the condition $I = 0$ represents a balance between two opposing tendencies: The charges tend to minimize their electrostatic potential energy and the source of emf tends to separate them, and thereby to minimize some other form of energy, such as chemical bond energy.

When a circuit is being analyzed, it is important to use correct terminology. One should say that current flows *through* a resistor when there is a potential difference *across* it. Furthermore, current is *not* "used up": The number of charges that leave one terminal of the battery is exactly equal to the number that enter via the other terminal. What they lose is potential energy, which is converted to thermal energy.

EXERCISE 1. Can you think of a practical situation in which the current through a battery is opposite to the direction shown in Fig. 28.3? What is the terminal potential difference in this case?

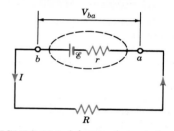

FIGURE 28.3 A battery is treated as an ideal source of emf $\mathscr{E}$ with an internal (series) resistance r. When a current flows in the direction shown, the terminal potential difference is $V_{ba} = \mathscr{E} - Ir$.

FIGURE 28.4 Gustav R. Kirchhoff (1824–1887).

FIGURE 28.5 The sum of the currents entering a junction must equal the sum of the currents leaving the junction, or $\Sigma I = 0$.

28.2 KIRCHHOFF'S RULES

The analysis of electrical circuits is aided by a *junction* rule and a *loop* rule first stated by G. R. Kirchhoff (Fig. 28.4). We begin by considering the junction rule. Often a circuit has two or more branches, as in Fig. 28.5. **Kirchhoff's junction rule** is

$$\Sigma I = 0 \tag{28.3}$$

The algebraic sum of the currents entering or leaving a junction is zero.

The sign assigned to a current entering a junction is opposite to that for a current leaving the junction. For the currents in Fig. 28.5 we may write $I_1 + I_2 - I_3 - I_4 = 0$. The junction rule is a statement of the conservation of charge: Charge is neither created nor destroyed at the junction, nor does it accumulate at this point.

In order to discuss the loop rule, consider the circuit in Fig. 28.6, which consists of a battery with internal resistance r in series with a resistor R. We assume that the resistance of the connecting wires is small enough to be ignored. Following the convention that current flows from high to low potential, we draw the current in the circuit as flowing from the positive to the negative terminal of the battery. For convenience the negative terminal is taken to be at zero potential. We will examine the potential energy changes of a hypothetical unit positive charge as it moves around the circuit. (Throughout its journey, the charge retains the small kinetic energy associated with the drift velocity.)

We start at A, just as the charge enters the negative terminal of the battery. At this point it has no potential energy. The source of emf increases the potential energy of the (unit) charge by $\mathscr{E}$. Between B and C, the charge does not lose any potential energy, assuming the wires have no resistance. At C it encounters the internal resistance and steadily loses potential energy, until it reaches the positive terminal of the battery at point D. The charge coasts with its drift speed from D to E, at which point it meets the resistor R and loses the rest of its potential energy as it passes through R. Finally, the charge travels from F back to A at the drift speed.

The charge must lose *all* the energy supplied by the battery by the time it reaches point F because of the conservative nature of the electrostatic field. Thus, in any round trip of the circuit, the charge returns to the same electrostatic potential. This fact is expressed as **Kirchhoff's loop rule:**

$$\Sigma V = 0 \tag{28.4}$$

The algebraic sum of the *changes* in potential around a closed loop is zero.

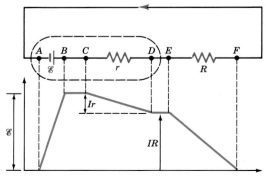

FIGURE 28.6 The sum of the *changes* in potential around a closed loop is zero: $\Sigma V = 0$. (Note V is now used for change in potential.)

This rule is simply a statement of the conservation of energy: The potential energy given a charge by the source of emf is lost in going through the resistors.

When adding up the changes in potential, we may trace a path around a loop either along or against the known (or assumed) direction of the current, as shown in Fig. 28.7. If we go along the current, the change in potential in going through a resistor (including any internal resistance of the source of emf) is negative. This follows from the fact that current flows "downhill" in potential. The sign of the change in potential in going through an ideal source of emf depends on the order in which the terminals are encountered. It does *not* depend on the direction of the current.

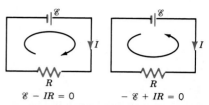

$$\mathscr{E} - IR = 0 \qquad -\mathscr{E} + IR = 0$$

FIGURE 28.7 In applying the loop rule one can trace the circuit in either sense.

EXERCISE 2. In the course of his experiments, Ohm was faced with the following problem. (a) When the external resistance of his circuit was large, the current through the circuit was proportional to the number of cells in the circuit; (b) when the external resistance was small, the current was independent of the number of cells. Explain these results.

28.3 SERIES AND PARALLEL CONNECTIONS

Resistors, like capacitors, can be connected in series and in parallel. We wish to find the equivalent resistance for each of these combinations. When two resistors, R_1 and R_2, are connected in series as in Fig. 28.8, the current is the same through both. Since the electric field in the wires is in the same direction, the potential difference across the combination is the sum of the individual potential differences:

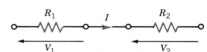

FIGURE 28.8 Two resistors in series have an equivalent resistance $R_{eq} = R_1 + R_2$.

$$V = V_1 + V_2 = I(R_1 + R_2) = IR_{eq}$$

The two resistors have an *equivalent* resistance $R_{eq} = R_1 + R_2$. Clearly, this logic can be extended to any number of resistances in series; that is,

(Series) $$R_{eq} = R_1 + R_2 + R_3 + \cdots + R_N \qquad (28.5)$$

The equivalent resistance for a series combination of resistors is simply the sum of the individual resistances. This result should be compared with that for capacitors in series (Eq. 26.7).

Figure 28.9 shows two resistors in parallel. The currents through the resistors are related by the junction rule applied at point a or b, $I - I_1 - I_2 = 0$, or

$$I = I_1 + I_2$$

We wish to find the equivalent resistance R_{eq} through which the same total current I would flow. By using Ohm's law, we may express the junction rule as

$$I = \frac{V_1}{R_1} + \frac{V_2}{R_2}$$

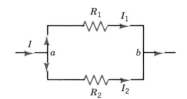

FIGURE 28.9 Two resistors in parallel have an equivalent resistance given by $1/R_{eq} = 1/R_1 + 1/R_2$.

The potentials at points a and b are unique, independent of the paths traveled by the charges. Therefore, the resistors have the same potential difference across them: $V_1 = V_2 = V$. The equivalent resistance is $R_{eq} = V/I$. Using this in the junction rule leads to

$$\frac{1}{R_{eq}} = \frac{1}{R_1} + \frac{1}{R_2}$$

Extending this logic to N resistors in parallel, we find

(**Parallel**) $$\frac{1}{R_{eq}} = \frac{1}{R_1} + \frac{1}{R_2} + \cdots + \frac{1}{R_N}$$ (28.6)

The equivalent resistance of a parallel combination of resistors is always less than that of the smallest resistance. This result should be compared with that for capacitors in parallel (Eq. 26.8).

EXAMPLE 28.1: Find the equivalent resistance of the combination of resistors shown in Fig. 28.10a.

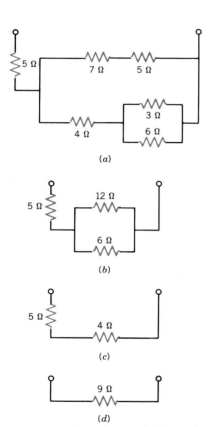

(a)

(b)

(c)

(d)

FIGURE 28.10 The calculation of the equivalent resistance for a combination of resistors is broken down into several steps.

Solution: In this type of problem it helps to start with the smallest identifiable series, or parallel, combination. The equivalent resistance of the 3-Ω and 6-Ω resistors in parallel is found from

$$\frac{1}{3\ \Omega} + \frac{1}{6\ \Omega} = \frac{1}{2\ \Omega}$$

Thus, these two resistors are equivalent to 2 Ω. When this 2-Ω resistance is added in series to the 4-Ω resistance, we obtain a 6-Ω resistance which is in parallel with a 12-Ω resistance, as

shown in Fig. 28.10b. Combining these two resistances in parallel,

$$\frac{1}{6\ \Omega} + \frac{1}{12\ \Omega} = \frac{1}{4\ \Omega}$$

we find an equivalent resistance of 4 Ω. Finally, we add this 4-Ω resistance to the 5 Ω in series with it (Fig. 28.10c) to find an equivalent resistance of 9 Ω for the whole combination (Fig. 28.10d).

EXAMPLE 28.2: Two resistors dissipate 60 W and 90 W, respectively, when each is connected separately to a 120-V source of emf. Find the power dissipated in each when they are connected in series with the 120-V source.

Solution: We first need to find the resistance of each resistor. From $P = V^2/R$ we have $R_1 = (120\ \text{V})^2/(60\ \text{W}) = 240\ \Omega$ and $R_2 = (120\ \text{V})^2/(90\ \text{W}) = 160\ \Omega$. When the resistors are connected in series, the current through them is the same and equal to $I = \mathscr{E}/(R_1 + R_2) = (120\ \text{V})/(400\ \Omega) = 0.3\ \text{A}$. The power dissipated in each resistor is

$$P_1 = I^2 R_1 = 21.6\ \text{W} \qquad P_2 = I^2 R_2 = 14.4\ \text{W}$$

The resistor with the higher wattage rating dissipates less power in this situation.

EXERCISE 3. The resistors in Example 28.2 are connected in parallel across the 120-V source of emf. (a) What is the current through the source? (b) What is the power dissipated in each resistor?

EXAMPLE 28.3: Whenever a real source of emf supplies power to an external load, some power is also dissipated in the internal resistance. A load resistance R is connected to a source of emf whose internal resistance is r, as in Fig. 28.11a. For what value of R will the power supplied to the load be a maximum?

Solution: From the loop rule, $\mathscr{E} - Ir - IR = 0$, we obtain the current $I = \mathscr{E}/(R + r)$. Therefore, the power dissipated in R is given by

$$P = I^2 R = \frac{\mathscr{E}^2 R}{(R + r)^2}$$

In order to find the maximum value of P, one could plot P as a function of R, as we have done in Fig. 28.11b. A better approach is to find the derivative of P with respect to R (treating r and $\mathscr{E}$ as constants). Thus,

$$\frac{1}{\mathscr{E}^2}\frac{dP}{dR} = \frac{1}{(R+r)^2} - \frac{2R}{(R+r)^3}$$

According to the rules of calculus, the value of R at which P is a maximum is found by setting $dP/dR = 0$.* As you can easily verify, this condition leads to the equation $(R + r) = 2R$, from which we obtain $R = r$. Thus, *the maximum transfer of power to the load occurs when the load resistance is equal to the internal resistance of the source of emf.* The source and load are said to be "matched." Such matching is important for the efficient transfer of power from, say, an audio amplifier to a loudspeaker—although the analysis is somewhat more complex because the current is not direct current (see Section 33.8).

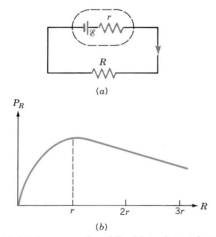

(a)

(b)

FIGURE 28.11 (*a*) A source of emf $\mathscr{E}$ with an internal resistance r connected to a load resistor R. (*b*) The power delivered to the load resistor as a function of R. The maximum power transfer occurs when $R = r$.

EXAMPLE 28.4: A battery with an emf of 20 V and an internal resistance of 1 Ω is connected to three resistors, as shown in Fig. 28.12. Find: (a) the terminal potential difference; (b) the current through and the potential difference across each resistor; (c) the power supplied by the emf; (d) the power dissipated in each resistor.

Solution: (a) To find the terminal potential difference we need the current through the battery. Since $\frac{1}{4} + \frac{1}{12} = \frac{1}{3}$, the equivalent resistance of R_2 and R_3 is 3 Ω. The equivalent resistance of the whole circuit is 6 Ω + 3 Ω + 1 Ω = 10 Ω. Thus, the current $I_1 = \mathscr{E}/R_{eq} = 2$ A, so the terminal potential difference is

$$V_{ba} = \mathscr{E} - I_1 r = 18 \text{ V}$$

(b) Since the current I_1 is known, the potential differences across r and R_1 are easily found. Thus, $V_r = I_1 r = 2$ V and $V_1 = I_1 R_1 = 12$ V. Since R_2 and R_3 are in parallel, it follows that $V_2 =$

* Strictly speaking one must also show that the second derivative, d^2P/dR^2, is negative.

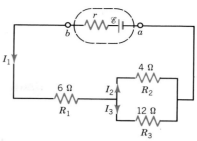

FIGURE 28.12 A source of emf with an internal resistance connected to three resistors. The current through the source is $I_1 = \mathscr{E}/R_{eq}$ where R_{eq} is the equivalent resistance for the whole circuit.

V_3. From the loop rule, we infer that the sum of the potential changes around the circuit must equal the emf. We have accounted for 14 V, so we are left with $V_2 = V_3 = 6$ V.

The currents through R_2 and R_3 are $I_2 = V_2/R_2 = \frac{3}{2}$ A and $I_3 = V_3/R_3 = \frac{1}{2}$ A. Notice that $I_1 = I_2 + I_3$ as is required by the junction rule. Always try to make this kind of check on the consistency of your calculations.
(c) The power supplied by the source of emf is $P = I_1\mathscr{E} = 40$ W.
(d) The power dissipated in each resistor is found by using either $P = IV$ or I^2R. We find

$$P_r = 4 \text{ W}; \qquad P_1 = 24 \text{ W}; \qquad P_2 = 9 \text{ W}; \qquad P_3 = 3 \text{ W}$$

The sum of these powers is 40 W, which is the power supplied by the source of emf. This is another check on the calculations.

EXAMPLE 28.5: The circuit in Fig. 28.13 has two loops and three sources of emf. (a) Determine the currents given that $r_1 = r_2 = 2 \text{ Ω}, r_3 = 1 \text{ Ω}, R_1 = 4 \text{ Ω}, R_2 = 3 \text{ Ω}, \mathscr{E}_1 = 15$ V, $\mathscr{E}_2 = 6$ V and $\mathscr{E}_3 = 4$ V. (b) What is the change in potential $V_A - V_B$?

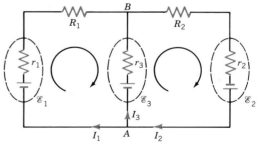

FIGURE 28.13 A two-loop circuit with three sources of emf. The directions in which each loop is traced is indicated with a curved arrow.

Solution: When we start to analyze a circuit such as that in Fig. 28.13, the currents may be drawn in arbitrary directions. An incorrect assumption will show up as a negative sign when numbers are used. The junction rule applied at point A gives

(assuming that current coming out is positive)

$$I_1 - I_2 + I_3 = 0$$

To apply the loop rule, we may trace each loop in a clockwise sense—which is indicated by a curved arrow in Fig. 28.13.

Left loop: $\mathcal{E}_1 - I_1 r_1 - I_1 R_1 + I_3 r_3 - \mathcal{E}_3 = 0$
Right loop: $\mathcal{E}_3 - I_3 r_3 - I_2 R_2 + \mathcal{E}_2 - I_2 r_2 = 0$

Note that these equations apply *only* to the circuit of Fig. 28.13. Since there are three independent equations, there can be no more than three unknowns.
(a) On inserting the given values into the loop equations, we obtain

Left loop: $15 - 2I_1 - 4I_1 + I_3 - 4 = 0$ (i)
Right loop: $4 - I_3 - 3I_2 + 6 - 2I_2 = 0$ (ii)

When we substitute $I_2 = I_1 + I_3$ from the junction rule into (ii), equations (i) and (ii) become, respectively,

$$11 - 6I_1 + I_3 = 0 \qquad \text{(iii)}$$
$$10 - 5I_1 - 6I_3 = 0 \qquad \text{(iv)}$$

These equations yield $I_1 = \frac{76}{41} = 1.85$ A and $I_3 = \frac{5}{41} = 0.12$ A. Finally, $I_2 = I_1 + I_3 = \frac{81}{41} = 1.97$ A.
(b) To determine $V_A - V_B$ we must *start* at B and add the changes in potential. Since the potentials are unique, we may take *any* path from B to A. (What fundamental property does this reflect?) Along the central branch we find

$$V_A - V_B = I_3 r_3 - \mathcal{E}_3 = (0.12)(1) - 4 = -3.78 \text{ V}$$

The negative sign means that V_A is less than V_B. (Try the other two paths.)

EXERCISE 4. Write the loop equation for the large loop in Fig. 28.13, that is, without $\mathcal{E}_3$ and r_3. Compare it to the sum of Eqs. (i) and (ii). What do you conclude?

EXERCISE 5. Consider the circuit in Fig. 28.14. Initially switches S_1 and S_2 are open. Specify whether the readings on the ammeter and the voltmeter increase or decrease under each of the following conditions: (a) S_1 is open, S_2 is closed; (b) S_1 is closed, S_2 is open; (c) S_1 and S_2 are closed.

EXAMPLE 28.6: Five resistors are connected as shown in Fig. 28.15. What is the equivalent resistance between points a and b?

Solution: In this example one cannot use techniques based on series and parallel connections to find the equivalent resistance. In approaching such a problem one needs to minimize the number of unknowns. For example, once we have used I_1 and I_2 for the currents through R_1 and R_2, and $I = I_1 + I_2$ for the current entering at junction a, the three other currents can be expressed, using the junction rule, in terms of these:

$$I_3 = I - I_1; \qquad I_4 = I - I_2; \qquad I_5 = I_1 - I_2$$

These values have already been indicated on the diagram. We will find the equivalent resistance from the equation $R_{eq} = V_{ba}/I$, where V_{ba} is the potential difference between points a and b.

On applying the loop rule to the left and right loops we find:

$$-I_1 R_1 - (I_1 - I_2)R_5 + (I - I_1)R_3 = 0 \qquad \text{(i)}$$
$$+(I_1 - I_2)R_5 - I_2 R_2 + (I - I_2)R_4 = 0 \qquad \text{(ii)}$$

Using (i) and (ii) one can express I_1 and I_2 in terms of I: $I_1 = \alpha_1 I$, and $I_2 = \alpha_2 I$, where the coefficients α_1 and α_2 are expressions involving the resistances.

The potential difference between points a and b is

$$V_b - V_a = -I_1 R_1 - I_2 R_2 = -(\alpha_1 R_1 + \alpha_2 R_2)I$$

Therefore, the equivalent resistance is $R_{eq} = V_{ba}/I = \alpha_1 R_1 + \alpha_2 R_2$. (You are invited to calculate α_1 and α_2 in Problem 16.)

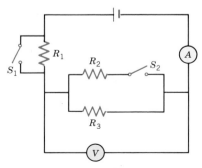

FIGURE 28.14 Exercise 5.

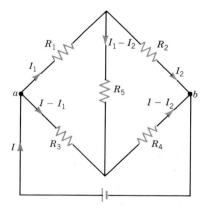

FIGURE 28.15 The equivalent resistance of this combination of resistors cannot be found by considering series and parallel combinations.

28.4   *RC* CIRCUITS

Thus far the currents we have dealt with have been steady. When a capacitor is included in a circuit, the current varies as a function of time for the period during which the capacitor is charging or discharging. When a capacitor is connected directly across the terminals of an ideal battery (no internal resistance), the capacitor becomes charged instantaneously. Similarly, if the terminals of a charged capacitor are connected by a wire, the capacitor is discharged instantaneously. We wish to determine how the charge on the capacitor and the current through a circuit vary as functions of time when there is resistance in the circuit.

(i) Discharge

Figure 28.16 shows a capacitor and a resistor in parallel with an ideal battery (no internal resistance) with emf $\mathscr{E}$. As long as the switch is closed, the potential difference across C and R is equal to the emf $\mathscr{E}$. The charge on the capacitor is $Q_0 = C\mathscr{E}$. When the switch is opened at $t = 0$, the capacitor starts to discharge through the resistor. From the loop rule we have

$$\frac{Q}{C} - IR = 0$$

where Q is the instantaneous value of the charge on the positive plate of the capacitor and I is the current through the wire. We note that the current I is equal to the rate at which the charge Q is *decreasing*; therefore, $I = -dQ/dt$. The loop rule becomes

$$\frac{dQ}{dt} = -\frac{Q}{RC}$$

This may be rewritten and integrated,

$$\int \frac{dQ}{Q} = -\frac{1}{RC} \int dt$$

to yield

$$\ln Q = -\frac{t}{RC} + k$$

where k is a constant of integration. We know that at $t = 0$, the charge $Q = Q_0$; therefore $k = \ln Q_0$. On taking the antilog, we find

$$Q = Q_0 e^{-t/RC} \tag{28.7}$$

This *exponential decay* function is plotted in Fig. 28.17. At a time

$$\tau = RC \tag{28.8}$$

called the **time constant,** the charge falls to $Q = Q_0 e^{-1} = 0.37 Q_0$, that is, 37% of its initial value. Another time of interest is the **half-life,** $T_{1/2}$, which is the time needed to drop to 50% of the initial value. Thus,

$$\tfrac{1}{2} Q_0 = Q_0 e^{-T_{1/2}/RC}$$

Taking the natural logarithm and rearranging, we find

$$T_{1/2} = RC \ln 2 = 0.693\tau \tag{28.9}$$

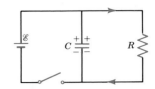

FIGURE 28.16 A circuit to study the discharge of a capacitor through a resistor.

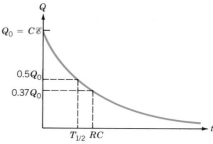

FIGURE 28.17 The exponential decay of the charge on a capacitor. $T_{1/2}$ is the half-life and τ is the time constant.

The current may be found from $I = -dQ/dt$ and Eq. 28.7:

$$I = I_0 e^{-t/RC} \tag{28.10}$$

where $I_0 = \mathscr{E}/R$ is the current at $t = 0$. The current has the same time dependence as the charge.

EXERCISE 6. Show that the time constant tells us how long it takes the charge to fall to $1/e$ or 37% of *any* starting value, not just the initial value, Q_0, at $t = 0$. (Similarly, it takes one half-life, $T_{1/2}$, to fall to 50% of any given initial value.)

EXERCISE 7. How many half-lives does it take for the charge to drop to 12.5% of its initial value?

(ii) Charging

We now consider how the charge on a capacitor increases when there is a resistor in series, as shown in Fig. 28.18. At $t = 0$, when the switch is closed, we assume that there is no charge on C and therefore no potential difference across it. The potential difference across R is $\mathscr{E}$, so the initial (maximum) current through the circuit is $I_0 = \mathscr{E}/R$. Applying the loop rule to the circuit (in the clockwise sense), we find

$$\mathscr{E} - IR - \frac{Q}{C} = 0$$

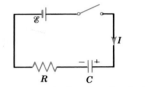

FIGURE 28.18 A charging circuit for a capacitor.

The sum of the potential difference across the capacitor and resistor is always constant, so $V_C + V_R = \mathscr{E}$. As charge flows to the capacitor, the potential difference across it increases, which means that the potential difference across R, and the current through the circuit, must decrease. In this circuit, the current I *increases* the charge on the capacitor, and therefore $I = +dQ/dt$. As the current decreases, we expect the rate at which the capacitor charges also to decrease. When the potential difference across C reaches $\mathscr{E}$, the potential difference across R is zero: The current stops flowing, and the charge on the capacitor has its maximum value $Q_0 = C\mathscr{E}$.

When $I = +dQ/dt$ is used in the loop rule, we find

$$C\mathscr{E} - Q = \frac{dQ}{dt} RC$$

After some rearrangement, integration of both sides,

$$\int \frac{dQ}{C\mathscr{E} - Q} = \frac{1}{RC} \int dt$$

leads to

$$-\ln(C\mathscr{E} - Q) = \frac{t}{RC} + k$$

where k is a constant of integration. At $t = 0$, $Q = 0$; therefore, $k = -\ln(C\mathscr{E})$. The above equation may therefore be written as (note that $\ln A - \ln B = \ln A/B$)

$$\ln \left(\frac{C\mathscr{E} - Q}{C\mathscr{E}} \right) = - \frac{t}{RC}$$

On taking the antilog and rearranging terms, we find

Capacitor charging

$$Q = Q_0(1 - e^{-t/RC}) \tag{28.11}$$

where $Q_0 = C\mathscr{E}$. This function is plotted in Fig. 28.19. In this case, the time constant $\tau = RC$ tells us how long it takes the charge to rise to $Q = Q_0(1 - e^{-1}) = 0.63Q_0$, that is, to 63% of its final value. The current, found from $I = +dQ/dt$, is

$$I = I_0 e^{-t/RC} \tag{28.12}$$

where $I_0 = \mathscr{E}/R$. Note that Eq. 28.12 has the same form as for the discharge of the capacitor.

In the discharge circuit of Fig. 28.16, the potential differences across R and C are equal, $V_C = V_R = \mathscr{E}$. In the charging circuit of Fig. 28.18, they are not equal. Indeed, the potential difference across the resistor, $V_R = IR$, decreases with time, whereas the potential difference across the capacitor, $V_C = Q/C$, increases with time. Their sum, $V_R + V_C = \mathscr{E}$, is constant.

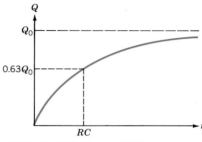

FIGURE 28.19 The increase in the charge on the capacitor as a function of time.

EXAMPLE 28.7: For the charging circuit of Fig. 28.18, take $\mathscr{E} = 200$ V, $R = 2 \times 10^5\ \Omega$, and $C = 50\ \mu$F. Find: (a) the time taken for the charge to rise to 90% of its final value; (b) the energy stored in the capacitor at $t = RC$; (c) the power loss in R at $t = RC$; (d) the total work done by the battery when the capacitor is fully charged ($t = \infty$); (e) the final energy stored in the capacitor; (f) the total energy loss in the resistor.

Solution: (a) We are looking for the time at which $Q = 0.9Q_0$, where $Q_0 = C\mathscr{E} = 0.01$ C. The time constant is $\tau = RC = 10$ s. From Eq. 28.11,

$$0.9Q_0 = Q_0(1 - e^{-t/10})$$

This becomes $\exp(-t/10) = 0.1$. Taking the natural logarithm gives $-t/10 = -2.3$, and so $t = 23$ s.
(b) In one time constant $Q = Q_0(1 - 1/e) = 0.632Q_0$. The energy stored in the capacitor is

$$U_C = \frac{Q^2}{2C} = \frac{(0.63 \times 0.01\ \text{C})^2}{10^{-4}\ \text{F}} = 0.4\ \text{J}$$

(c) The power loss in R is $P_R = I^2 R$. In one time constant $I =$ $0.37I_0$, where $I_0 = \mathscr{E}/R = 10^{-3}$ A. Thus,

$$P_R = I^2 R = (0.37 \times 10^{-3}\ \text{A})^2 (2 \times 10^5\ \Omega) = 2.7 \times 10^{-2}\ \text{W}$$

(d) The battery transfers an amount of charge Q_0 from one plate of the capacitor to the other and moves it through a potential difference $\mathscr{E}$. Thus the total work done by the battery is $Q_0\mathscr{E} = C\mathscr{E}^2 = 2$J.
(e) The final energy stored in the capacitor is

$$U_C = \frac{Q_0^2}{2C} = \frac{1}{2} C\mathscr{E}^2 = 1\ \text{J}$$

(f) The rate of energy loss in R is $P_R = dU_R/dt = I^2 R$. Thus, $dU_R = I^2 R\ dt$ where I is given by Eq. 28.12. The total energy loss is

$$U_R = \int_0^\infty P_R\ dt = \int_0^\infty I_0^2 R\ e^{-2t/RC}\ dt$$
$$= \tfrac{1}{2} C\mathscr{E}^2$$

This is a surprising result: The energy stored in the capacitor is exactly equal to the energy lost in the resistor. Of course, the sum $U_C + U_R$ is equal to the energy delivered by the source of emf.

28.5   DIRECT CURRENT INSTRUMENTS

An instrument that measures current is called an **ammeter,** and one that measures potential difference is called a **voltmeter.** Many of these meters are based on the **galvanometer.** This device, which we study in the next chapter, registers the deflection of a current-carrying coil suspended between the poles of a magnet (see Section 29.4). Potential differences are also measured with an oscilloscope, which registers the deflection of a beam of electrons. Modern digital meters use electronic circuits that can measure current and potential difference and display the values on a small screen.

The current through a resistor is measured by connecting an ammeter *in series* with it, as in Fig. 28.20a. The ammeter measures the current at *one* point in the circuit. The resistance of the ammeter should be small enough so that its insertion does not significantly alter the current in the circuit. The potential difference across a resistor is determined by a voltmeter connected *in parallel* with the resistor, as in Fig. 28.20b. The voltmeter measures the potential difference between *two* points in a circuit. The presence of the voltmeter results in a lower

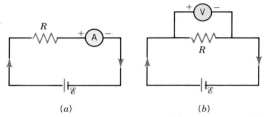

FIGURE 28.20 (*a*) An ammeter is connected *in series* with the resistor. (*b*) A voltmeter is connected *in parallel* with the resistor.

effective resistance and an increase in the current in the circuit. This disturbance is minimized if the voltmeter has a large resistance.

The coil of a galvanometer has a resistance in the range 10 Ω to 100 Ω and will produce a full-scale deflection for a current in the range 10 μA to 1 mA. (The most sensitive galvanometers can indicate currents less than 10^{-9} A.) With a typical resistance $R_G = 20$ Ω and 1 mA for full-scale deflection, the potential difference across the terminals is 20 mV. Such a galvanometer can be used as an ammeter for currents up to 1 mA and as a voltmeter up to 20 mV. To extend these ranges, the galvanometer may be combined with resistors either in series or in parallel, as the next example shows.

EXAMPLE 28.8: A galvanometer has a full-scale deflection for a current of 1 mA. The coil has a resistance of 20 Ω. Modify the instrument to produce the following: (a) an ammeter that can measure up to 500 mA; (b) a voltmeter capable of measuring 25 V.

Solution: (a) Since only 1 mA should flow through the galvanometer, a *shunt* resistor, R_{sh}, is connected in parallel with it, as shown in Fig. 28.21. Applying the junction rule, we see that the incoming current is

$$I = I_G + I_{sh}$$

Thus, $I_{sh} = 499$ mA. The potential differences across the galvanometer and the shunt are equal:

$$I_G R_G = I_{sh} R_{sh}$$

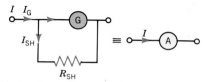

FIGURE 28.21 A galvanometer may be used as an ammeter by connecting a *shunt* resistance R_{sh} in parallel with it.

From this we find $R_{sh} = I_G R_G / I_{sh} = (1 \text{ mA})(20 \text{ Ω})/(499 \text{ mA}) = 0.04$ Ω. Since this resistance is in parallel with the 20 Ω of the galvanometer, the effective resistance of the ammeter is essentially 0.04 Ω.

(b) When 1 mA flows through the coil, the potential difference across it is only 20 mV. To measure 25 V, one must connect a *series* resistor, R_s, in series with the galvanometer, as shown in Fig. 28.22. From Ohm's law,

$$\begin{aligned} 25 &= I_G(R_G + R_s) \\ &= (1 \text{ mA})(20 \text{ Ω} + R_s) \end{aligned}$$

From which we find $R_s = 24{,}980$ Ω ≈ 25 kΩ. The effective resistance of the voltmeter will be 25 kΩ. Such an instrument is not adequate if the potential difference is being measured across, say, a 10-kΩ resistor.

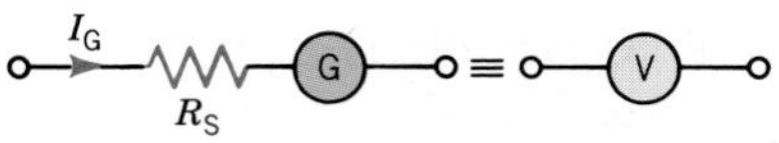

FIGURE 28.22 A galvanometer may be used as a voltmeter by connecting it to a *series* resistance R_s.

A commercial "multimeter" uses a single galvanometer to form several ranges for current, potential difference, and also to measure resistance. A multimeter is rated in terms of its *sensitivity*, which is specified, for example, as 1000 Ω/V. This value means that when the instrument is used as a voltmeter, the effective resistance is 1000 Ω times the maximum value on a given range. For example, on the 25-V range, the effective resistance would be $(1000\ \Omega/V)(25\ V) = 25\ k\Omega$. Since $I = V/R$, the sensitivity is simply the reciprocal of the current required by the galvanometer for full-scale deflection. With the given value of sensitivity the current in the galvanometer would be $1/1000\ V/\Omega = 10^{-3}\ A$.

Measurement of Resistance

An **ohmmeter** is an instrument designed to measure resistance. It contains an ammeter, a small battery, and a variable series resistance R_s, as shown in Fig. 28.23. The terminals are first "shorted" (connected directly to each other), and a small series resistor (not shown) is adjusted so that the deflection of the needle is full-scale. This procedure compensates for variations in the emf of the battery. The full-scale reading then corresponds to zero (external) resistance. When a resistor R is connected, the current is lower, so the deflection of the needle is less. Thus, the resistance scale is calibrated to increase from right to left.

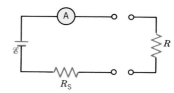

FIGURE 28.23 In an ohmmeter a source of emf is in series with an ammeter and a series resistance R_s. If the terminals are shorted, the ammeter registers the maximum deflection. When the resistor R is connected the deflection is lower.

EXAMPLE 28.9: An ammeter on the 0.1-mA scale has a resistance of 20 Ω. It is connected in series with a 1.5-V battery. (a) What is the series resistance required for full-scale deflection? (b) What external resistance will produce a half-scale reading?

Solution: (a) We consider the situation when the terminals are shorted (no external resistance). From the loop rule, with R_A as the resistance of the ammeter

$$\mathcal{E} = I_A(R_A + R_s)$$
$$1.5 = (0.1\ \text{mA})(20\ \Omega + R_s)$$

Thus, $R_s = 14{,}980\ \Omega \approx 15\ k\Omega$.
(b) When the external resistance R is added, the current is

$$I = \frac{\mathcal{E}}{R_A + R_s + R} = \frac{1.5\ V}{15\ k\Omega + R}$$

For a half-scale $I = I_A/2$. It is convenient to express I_A as (1.5 V)/(15 kΩ). We see at once that $2 \times 15\ k\Omega = (15\ k\Omega + R)$; thus, $R = 15\ k\Omega$. Thus, half-scale deflection occurs when the external resistance equals the internal resistance. When the resistance to be measured is small, the internal resistance needs to be reduced for improved sensitivity. The current through the ammeter will be larger and so its range must also be modified. Keep in mind that an ohmmeter sends current through the device whose resistance it is measuring and in fact may damage a sensitive device—for example, another galvanometer.

Wheatstone Bridge

An accurate method for measuring resistance was devised in 1843 by Charles Wheatstone, a pioneer in the development of the electric telegraph. Four resistors are connected in the form of a "bridge," shown in Fig. 28.24. The resistances R_1, R_2, and R_s are known, whereas R_x is the unknown to be measured. There are two ways the instrument may be used.

In the first method, R_1 and R_2 are fixed, and the calibrated or standard resistor R_s is changed until the galvanometer registers no current. In this condition, the bridge is said to be "balanced." (In such a "null" method, the value of the emf and the calibration of the galvanometer are not important. In practice, a resistor is used to limit the current through the galvanometer in the initial stages.) When the bridge is balanced, the points P and Q are at the same potential. Therefore, the potential differences across R_1 and R_s and across R_2 and R_x are equal:

$$I_1R_1 = I_2R_s$$
$$I_1R_2 = I_2R_x$$

Taking the ratio of these equations, we find

$$R_x = \frac{R_2}{R_1} R_s$$

The unknown resistance is determined in terms of the precision standard resistor.

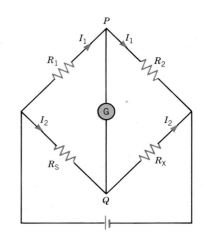

FIGURE 28.24 A Wheatstone bridge is used for precise measurements of resistance. The instrument is "balanced" when no current flows through the galvanometer.

In the second method, R_s is a fixed precision standard resistor and R_1 and R_2 are parts of a single continuous wire. At point P a sliding contact moves over the wire until balance is achieved. For a uniform wire, the resistance is proportional to the length, and so $R_x = (\ell_2/\ell_1)R_s$, is found from a ratio of the lengths on either side of the contact.

The Wheatstone bridge is capable of high precision. It is useful in measuring the resistance of a platinum resistance thermometer and of a strain gauge.

Potentiometer

A quick method of measuring the emf of a battery is to connect a voltmeter directly across its terminals. However, since the voltmeter has finite resistance, one inevitably measures the terminal potential difference rather than the true emf. A **potentiometer** is an instrument that allows one to compare an unknown emf with a standard emf. It is a null device, like the Wheatstone bridge.

A "working battery" supplies a steady current to a slide wire (usually 1 m long). A galvanometer is connected at one end to a switch and the other end to a sliding contact P that moves over the wire (Fig. 28.25). First the switch is thrown to connect the standard battery (which may be a cadmium cell, whose emf is 1.01826 V) of emf $\mathscr{E}_s$. The sliding contact is moved until the galvanometer registers no current. If the resistance of the wire between O and P in the lower loop is R_s, Kirchhoff's loop rule gives

$$\mathscr{E}_s - IR_s = 0$$

The standard battery is now replaced by the unknown emf and again P is moved until no current flows in the galvanometer. Now we have

$$\mathscr{E}_x - IR_x = 0$$

The resistances are proportional to the lengths of the wire between O and P. On eliminating I, we see that

$$\mathscr{E}_x = \frac{\ell_x}{\ell_s}\mathscr{E}_s$$

The ratio of the lengths yields the unknown emf in terms of $\mathscr{E}_s$. Since no current flows through the sources, the internal resistance is not important. The emf of the working battery is not relevant either, provided it is greater than both $\mathscr{E}_s$ and $\mathscr{E}_x$. Modern precision potentiometers can measure in steps of 10^{-6} V with an accuracy of $10^{-3}\%$.

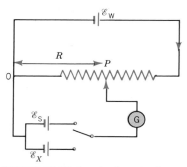

FIGURE 28.25 A potentiometer is used to compare an unknown emf $\mathscr{E}_x$ with a standard emf $\mathscr{E}_s$.

HISTORICAL NOTE:
The Electric Light

In the early 19th century, gas lighting was commonplace. By 1850 however, the arc light, produced by electrical discharges between two carbon rods held a small distance apart, was used in lighthouses, in railway stations, and for street lighting. When Thomas Edison, inventor of the phonograph, visited one of the first displays of arc lights in 1877, he was entranced and immediately thought of the possibility of installing electric lights in every home and office. However, arc lights were very bright and produced noxious fumes, which meant they were suitable only for large open spaces. He decided that the incandescent lamp, in which light is given off by an electrically heated filament, would produce a softer illumination and was therefore a better source of light—although such lamps had not been very successful thus far.

In 1878 Edison boldly set up the Edison Electric Light Co. and, on the strength of his reputation, got financial backing for it. Edison boasted that he would soon develop a complete distribution system for electricity that would run lights and electric motors, and serve other industrial purposes. Because of the publicity, gas company shares fell sharply both in the United States and in England. A committee of the British parliament, aided by prominent scientists, was appointed to look into the feasibility of Edison's plans.

An important feature of the gas distribution system already in use was that each consumer could switch the supply on or off. Any electrical system would clearly have to allow a similar "subdivision of the electric light." Arc lights had a resistance of about 5 Ω and required a current of 10 A. They were normally connected in a series circuit of about ten lamps. Of course, if one lamp was disconnected, all the lamps went out. In order to allow some lights to be on while others were off, they would have to be connected in parallel. But ten arc lights in parallel would require 100 A—which was already beyond the capability of any existing electrical generator. The few short-lived incandescent lamps that had been produced up to that time had resistances of about 0.5 Ω. If the power required by each incandescent light was about the same as that for an arc light, the current required would be much greater. Yet Edison was talking of thousands of lights! The transmission cables would have to handle very high currents without overheating, which meant that vast amounts of copper would be required. The British committee concluded that the "subdivision of the electric light" was an impossibility. But, as Napoleon once said, "Impossible is a word in the dictionary of fools." Edison felt much the same way. If gas could be subdivided, why not electricity?

A New Lamp

From the equation for power, $P = I^2R$, Edison realized that a given quantity of power could be delivered either by a large current to a low resistance, or, by a low current to a large resistance. To minimize the current in the transmission lines, he needed a lamp with a *high* resistance. His initial attempts with carbon filaments were unsuccessful because they tended to burn up (oxidize in the air that remained in the glass bulbs). Other materials met the same fate. In January 1879 he was able to borrow a new vacuum pump and with it attained the highest vacuum that had ever been achieved (10^{-6} atm). Nonetheless, after one year of work, his best effort was a Pt–Ir spiral whose resistance was 3 Ω. He then read an article in the July issue of *Scientific American* describing the carbon filament lamp of the Englishman Joseph Swan. It had worked only for a few hours. Edison had not tried carbon with the new vacuum pump, so he returned to carbon. By November 1879 he had developed a carbonized filament that had a resistance of 100 Ω. Over a period of about 15 months, he had tried over 1600 materials! Figure 28.26 shows an early filament lamp.

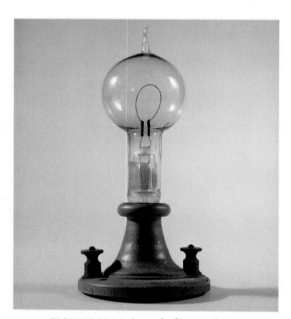

FIGURE 28.26 An early filament lamp.

A New Generator

It was shown in Example 28.3, that maximum transfer of power occurs when the internal resistance of the source of emf is equal to the external resistance. All previous electrical generators (which will be discussed in Chapter 31) had been designed with this principle in mind. Furthermore, their efficiency in converting mechanical energy to electrical energy was under 40% and each could supply only a few arc lights. With great insight, Edison saw a flaw in this approach. The efficiency of power transfer can be defined as the ratio

$$\text{Efficiency} = \frac{P_L}{P_S + P_L} = \frac{R_L}{R_S + R_L}$$

where $P_L = I^2 R_L$ is the power delivered to the load resistance R_L and $P_S = I^2 R_S$ is the power lost in the source resistance R_S. When $R_S = R_L$, fully one-half of the electrical power generated is lost in the generator itself. The power transferred is a maximum, but the *efficiency* is only 0.5 or 50%. Figure 28.27 shows how the efficiency of power transfer varies with the external resistance. The efficiency approaches 1 as $R_S \rightarrow 0$ or $R_L \rightarrow \infty$. Clearly, the greatest efficiency is obtained when the internal resistance of the source is as *small* as possible and the external resistance is as *large* as possible. With this new insight, Francis Upton, an electrical engineer hired by Edison, designed a new type of dc electrical generator that included many of the latest developments. It converted mechanical energy to electrical energy with an efficiency of 90% and produced a relatively constant 110 V even if the output current varied.

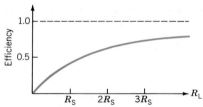

FIGURE 28.27 The efficiency of power transfer increases as the load resistance increases, or as the internal resistance of the source *decreases*.

A New Distribution System

Edison next devised a three-wire system for distributing electrical power that is still used to supply homes and offices—although nowadays it operates an alternating current rather than direct current as he had envisioned. Figure 28.28 shows a source of emf with three terminals. (Edison actually used two generators in series.) A potential difference of 110 V exists between the center terminal and each of the other two—one positive and the other negative. The 110-V loads are represented by two resistors R_1 and R_2. If only R_1, or R_2, is connected, current will flow through the ground wire and one of the other wires. When both resistors are connected, the current in the central ground wire, $I_g = I_2 - I_1$, is determined by the potential at the point P relative to ground. However, if $R_1 = R_2$, the potential at P will be midway between +110 V and −110 V—that is, it will be zero. Therefore, if the loads are "balanced," there will be *no* current in the ground connection. Instead of heating losses in all three transmission wires, there are losses only in the "hot" wires. A further advantage of the three-wire system is that two potential differences are available: 110 V for light duty, and 220 V for heavier-duty appliances, such as ovens and clothes dryers, which are connected between the two "hot" wires.

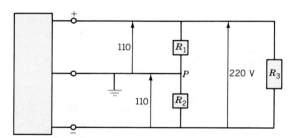

FIGURE 28.28 Edison's three-wire system for electrical power distribution. By having the load resistances R_1 and R_2 almost equal, the point P is kept close to the ground potential of the center terminal of the source. Consequently the power loss in the middle wire is very low.

SUMMARY

A source of emf converts some form of energy into electrical energy. The **emf** is defined as the work done per unit charge in moving the charge around a closed loop:

$$\mathcal{E} = \frac{W_{ne}}{q}$$

The subscript "ne" emphasizes that the work is done by some nonelectrostatic agent. When a source of emf produces a current I, the terminal potential difference is

$$V_{ba} = \mathcal{E} - Ir$$

where r is the internal resistance of the source. Clearly, $V_{ba} = \mathcal{E}$ when either $I = 0$ or $r = 0$. Therefore, the emf of a source can be measured by the "open circuit" potential difference.

(Kirchhoff's junction rule) $\Sigma I = 0$

The sum of currents at a junction is zero, or equivalently, the current entering the junction must equal the current going out.

The junction rule is an example of the conservation of charge.

(Kirchhoff's loop rule)

The algebraic sum of the *changes* in potential in going around a closed loop is zero.

Keep in mind that current flows "downhill" in potential. Therefore, if we go through a resistor along the current, the change in potential is negative. If we go against the current, the change is positive. The sign of the change in going through a source of emf does *not* depend on the direction of the current. The loop rule is an example of the conservation of energy.

Resistances can be connected in series or in parallel. The equivalent resistance in each case is

(Series) $\qquad R_{eq} = R_1 + R_2 + \cdots + R_N$

(Parallel) $\qquad \dfrac{1}{R_{eq}} = \dfrac{1}{R_1} + \dfrac{1}{R_2} + \cdots + \dfrac{1}{R_N}$

A resistor controls the rate at which a capacitor charges or discharges. For a time constant, $\tau = RC$, the equations for the discharging and charging circuits are

(Discharge) $\quad Q = Q_0 e^{-t/\tau}; \qquad I = I_0 e^{-t/\tau}; \qquad V_C = V_R = \mathcal{E}$

(Charge) $\quad Q = Q_0(1 - e^{-t/\tau}); \qquad I = I_0 e^{-t/\tau}; \qquad V_C + V_R = \mathcal{E}$

ANSWERS TO IN-CHAPTER EXERCISES

1. When a battery is being "charged" as in Fig. 28.29, the terminal potential difference is $V_b - V_a = V_{ba} = \mathcal{E} + Ir$.

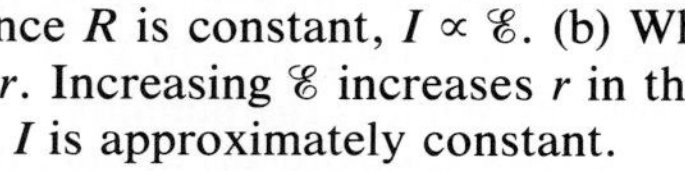

FIGURE 28.29 When a battery is being charged, the terminal potential difference is greater than the emf.

2. Consider the circuit of Fig. 28.3, in which as source of emf $\mathcal{E}$ and internal resistance r is connected to an external resistor R. The current is $I = \mathcal{E}/(R + r)$. (a) When $R \gg r$, we have $I \approx \mathcal{E}/R$. Since R is constant, $I \propto \mathcal{E}$. (b) When $R \ll r$, we have $I \approx \mathcal{E}/r$. Increasing $\mathcal{E}$ increases r in the same proportion, so that I is approximately constant.

3. (a) The equivalant resistance is found from $1/R_{eq} = 1/R_1 + 1/R_2 = 1/240 + 1/160$. We find $R_{eq} = 96\ \Omega$. The current is $I = \mathcal{E}/R_{eq} = 120\text{ V}/96\ \Omega = 1.25$ A. (b) 60 W and 90 W.

4. $\mathcal{E}_1 - I_1(r_1 + R_1) - I_2(R_2 + r_2) + \mathcal{E}_2 = 0$. This is simply the sum of the equations for the left and right loops. There are only two independent loop equations.

5. (a) The equivalent resistance of R_2 and R_3 is less than either of these two resistances. Thus, the resistance of the whole circuit decreases and the reading of the ammeter increases. Since the potential difference across R_1 increases, the potential difference across the parallel combination decreases. (b) R_1 is shorted out of the circuit, which leaves just R_3. The ammeter reading increases and the voltmeter reading increases to equal the terminal potential difference of the source. (c) R_1 is shorted out. The potential difference across the parallel combination increases to that of the source. The equivalent resistance decreases, so the ammeter reading goes up.

6. (a) Consider the charges at two times t_1 and t_2 ($> t_1$):

$$Q_1 = Q_0 e^{-t_1/\tau}; \qquad Q_2 = Q_0 e^{-t_2/\tau}$$

Taking the ratio of these equations we see that

$$Q_2 = Q_1 e^{-(t_2 - t_1)/\tau}$$

When $t_2 - t_1 = \tau$, we have $Q_2 = Q_1 e^{-1} = 0.37 Q_1$.

7. Since $0.125 = (1/2)^3$, it takes $3T_{1/2}$.

QUESTIONS

1. When operating separately, two bulbs are rated at 25 W and 100 W, respectively. Which bulb is closer to its normal brightness when they are connected in series?

2. What purpose does a fuse serve? Is there a difference in its function when used in a household wiring circuit and when it is included in an electronic circuit or loudspeaker?

3. (a) Would Kirchhoff's loop rule be valid even if part of the loop were traced through the air? (b) From your answer, what conclusion can you draw regarding the electric field in the space surrounding a circuit?

4. Consider two resistors with resistances R_1 and R_2, where $R_2 > R_1$. For a given battery potential difference, in which resistor will the power dissipated be greater when they are connected: (a) in series; (b) in parallel?

5. (a) Distinguish between an emf and a potential difference. (b) Are all potential differences created by emf's? Explain.

6. Eight D-cell batteries placed in series provide a 12-V potential difference. Could this arrangement be used to start a car? Explain why or why not?

7. Figure 28.30 shows two resistors connected to a source of emf. Initially switches S_2 and S_3 are open. How does the current through R_1 change when: (a) S_2 is closed, S_3 is open? (b) S_3 is closed, S_2 is open?

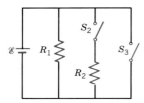

FIGURE 28.30 Question 7.

8. Repeat Question 7 for the circuit shown in Fig. 28.31.

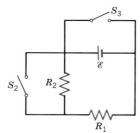

FIGURE 28.31 Question 8.

9. A source of emf $\mathcal{E}$ with an internal resistance r is connected to two resistors, as shown in Fig. 28.32. The switch is initially open. When the switch is closed, does the terminal potential difference across the source increase or decrease?

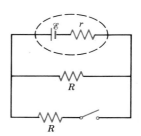

FIGURE 28.32 Question 9.

10. Describe how you would measure the internal resistance of a battery.

11. The circuit in Fig. 28.33 consists of an ideal source of emf, a resistor and two switches.
    (a) If S_1 is closed and S_2 is open, what are the potential differences across R and across the contacts of S_2?
    (b) If S_1 is open and S_2 is closed, what are the potential differences across R and across the contacts of S_1?
    (c) If both S_1 and S_2 are closed, what is the potential difference across R?
    (d) If both S_1 and S_2 are open, what are the potential differences across R and across the contacts of S_1 and S_2?

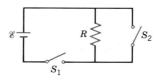

FIGURE 28.33 Question 11.

12. What are the advantages of the Wheatstone bridge over other methods of determining resistance? What factors affect the precision?

13. How would you modify the potentiometer to measure emf's much smaller than $\mathcal{E}_s$, the emf of the "standard" source?

14. Assume that the circuits in Fig. 28.34 have reached a steady-state condition. Take $\mathcal{E} = 10$ V, $R_1 = 5\ \Omega$, $R_2 = 10\ \Omega$, and $C = 40\ \mu$F. What is the potential difference across the resistors and the capacitor (a) in Fig. 28.34a and (b) in Fig. 28.34b?

15. The circuits in Fig. 28.35 have reached steady-state conditions. Find the potential difference across each resistor and capacitor in (a) Fig. 28.35a and (b) in Fig. 28.35b. Take $\mathcal{E} = 12$ V, $R_1 = 2\ \Omega$, $R_2 = 3\ \Omega$, $C_1 = 6\ \mu$F, and $C_2 = 3\ \mu$F

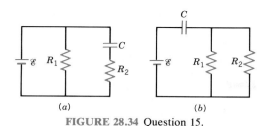

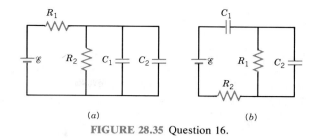

(a)

(b)

FIGURE 28.34 Question 15.

(a)

(b)

FIGURE 28.35 Question 16.

EXERCISES

28.1 Emf

1. (II) A battery has an emf $\mathscr{E}$ and an internal resistance r. It is connected to an external resistor R. When $R = 4\ \Omega$ the terminal potential difference is 9.5 V, and when $R = 6\ \Omega$ the terminal potential difference is 10 V. Find $\mathscr{E}$ and r.

2. (I) A car battery has a terminal potential difference of 12.4 V when there is no load. When the starter motor draws 80 A, the terminal potential difference drops to 11.2 V. What is the internal resistance of the battery?

3. (I) A battery is connected across a variable external resistance R. When the current is 6 A, the terminal potential difference is 8.4 V; when the current is 8 A, the terminal potential difference is 7.2 V. Find the emf and the internal resistance.

4. (I) An ideal battery is connected to an external resistor. When 2 Ω is added, the current drops from 8 A to 6 A. Find the value of the resistor and the emf of the battery.

5. (I) A 16-V battery delivers 50 W to a 4-Ω external resistor. (a) Find the internal resistance of the battery. (b) For what value of external resistor would the delivered power be 100 W?

6. (I) A 12.4-V battery whose internal resistance is 0.05 Ω is charged by an ideal external 14.2-V source of emf. Find: (a) the thermal power loss; (b) the rate at which electrical energy is being converted to chemical energy in the battery.

7. (I) When a battery with an emf of 12 V delivers 50 W to an external resistor, the terminal potential difference drops to 11.2 V. Find the internal resistance of the battery.

8. (II) A source of emf has $\mathscr{E} = 10$ V and an internal resistance of 1 Ω. When it is connected to an external resistor R, the power dissipated in the load is P. Find R, given that when it is increased by 50% the power changes as follows: (a) increases by 25%; (b) decreases by 25%.

28.2 Kirchhoff's Rules; 28.3 Series and Parallel Connections

9. (I) (a) Find the equivalent resistance of the combination of resistors shown in Fig. 28.36. (b) If a 10-V potential difference is applied across points a and b find the potential difference across the 4-Ω resistor.

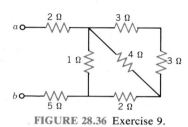

FIGURE 28.36 Exercise 9.

10. (I) The equivalent resistance of the combination shown in Fig. 28.37 is 16 Ω. What is R?

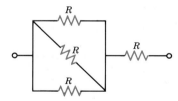

FIGURE 28.37 Exercise 10.

11. (I) Given three resistors with values 2 Ω, 3 Ω, and 4 Ω, how many different equivalent resistances could they be used to produce? State the values.

12. (I) Show two ways in which four equal resistances of value R can be combined to give an equivalent resistance R.

13. (I) What are the values of the minimum number of resistors needed to give the range 1 Ω to 10 Ω in integer steps? Is your set of values unique?

14. (I) Two batteries are connected in parallel, as shown in Fig. 28.38. $\mathscr{E}_1 = 1.53$ V, $r_1 = 0.05\ \Omega$, $\mathscr{E}_2 = 1.48$ V, and $r_2 = 0.15\ \Omega$. Find the open circuit potential difference between a and b and the internal power loss.

15. (I) A circuit in a home might look like Fig. 28.39. Calculate the current through each device. (The maximum current

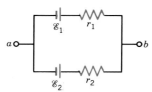

FIGURE 28.38 Exercise 14.

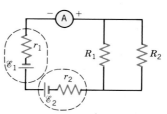

FIGURE 28.41 Exercise 18.

allowed for 14-gauge copper wire commonly used in household wiring is 15 A. The heater should be in a separate circuit.)

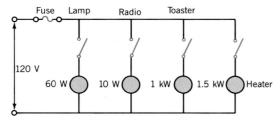

FIGURE 28.39 Exercise 15.

16. (I) Given three 5-Ω resistors rated at 10 W, find the maximum potential difference that can be applied if they are connected (a) all in series; (b) as in Fig. 28.40a.

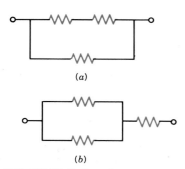

(a)

(b)

FIGURE 28.40 Exercises 16 and 17.

17. (I) Given three 4-Ω resistors rated at 20 W, find the maximum potential difference that can be applied if they are connected (a) all in parallel; (b) as in Fig. 28.40b.

18. (I) The current in the ammeter in Fig. 28.41 is 2 A. Find: (a) R_2; (b) the power dissipated in R_1 and R_2; (c) the terminal potential difference for each battery; (d) the power delivered by each battery. Take $\mathscr{E}_1 = 12$ V, $\mathscr{E}_2 = 6$ V, $R_1 = 3$ Ω, $r_1 = r_2 = 1$ Ω.

19. (I) Two ideal batteries are connected in series with two resistors, as in Fig. 28.42. Find: (a) the current in the circuit; (b) the power dissipated in each resistor; (c) the power delivered by each battery. Assume $\mathscr{E}_1 = 9$ V, $\mathscr{E}_2 = 6$V.

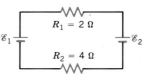

FIGURE 28.42 Exercise 19.

20. (I) In the circuit shown in Fig. 28.43, one point is grounded. (a) What is the potential at point P? (b) What is the terminal potential difference for each battery? (c) Find the power loss in R. Take $\mathscr{E}_1 = 5$ V, $\mathscr{E}_2 = 9.5$ V, $r_1 = 1$ Ω, $r_2 = 2$ Ω, $R = 1.5$ Ω.

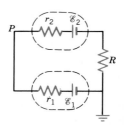

FIGURE 28.43 Exercise 20.

21. (I) A single battery is connected to three resistors, as shown in Fig. 28.44. Find the current in each resistor.

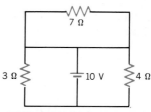

FIGURE 28.44 Exercise 21.

22. (II) In the circuit of Fig. 28.45, the ammeter reads 2 A and the voltmeter reads 4 V. Find the emf $\mathscr{E}$ and the resistance R.

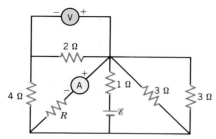

FIGURE 28.45 Exercise 22.

23. (II) Two batteries with the same emf $\mathscr{E}$ and internal resistance r are in parallel with a resistor R. For what value of R will the power loss in it be at maximum?

24. (II) A potential difference is applied across the resistors R_1 and R_2 in Fig. 28.46. R_L is a "load" resistance. What value of R_L would result in the maximum power dissipation in R_L?

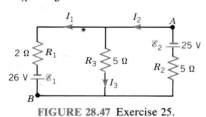

FIGURE 28.46 Exercises 24 and 48.

25. (II) For the circuit shown in Fig. 28.47, find: (a) the current and potential difference for each resistor; (b) the change in potential $V_A - V_B$.

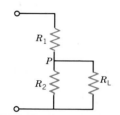

FIGURE 28.47 Exercise 25.

26. (II) For the circuit shown in Fig. 28.48, find the current and potential difference for each resistor, given the values: $R_1 = R_2 = 2\ \Omega$, $R_3 = 3\ \Omega$, $\mathscr{E}_1 = 12$ V, $\mathscr{E}_2 = 8$ V, and $\mathscr{E}_3 = 6$ V.

27. (II) For the circuit shown in Fig. 28.48, find the current and

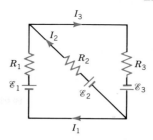

FIGURE 28.48 Exercises 26 and 27.

potential difference for each resistor, given the values: $R_1 = 4\ \Omega$, $R_2 = R_3 = 3\ \Omega$, $\mathscr{E}_1 = 12$ V, $\mathscr{E}_2 = 7$ V, and $\mathscr{E}_3 = 5$ V.

28. (II) For the circuit in Fig. 28.49, find the currents in the three branches. What is the change in potential $V_A - V_B$?

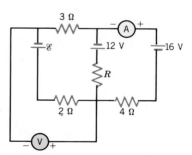

FIGURE 28.49 Exercise 28.

29. (II) In the circuit of Fig. 28.50, the ammeter reads 6.0 A and the voltmeter reads 14 V. Find the emf $\mathscr{E}$ and the resistance R.

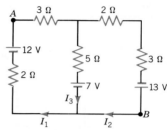

FIGURE 28.50 Exercise 29.

30. (II) What are the readings on the ammeter and the voltmeter in Fig. 28.51 when (a) the switch is open; (b) the switch is closed?

31. (II) When two resistors R_1 and R_2 are connected in parallel, they dissipate four times the power that they dissipate when they are in series with the same ideal source of emf. If $R_1 = 3\ \Omega$, find R_2.

32. (II) Consider the resistors in Fig. 28.52. They form neither

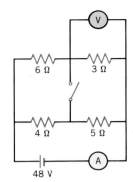

FIGURE 28.51 Exercise 30.

a series nor a parallel combination. (a) When a potential difference V is applied to the terminals, what is the potential difference across each resistor? (b) What is the effective resistance of the combination?

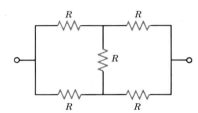

FIGURE 28.52 Exercise 32.

33. (II) In the circuit in Fig. 28.53, (a) what value of emf will deliver 6 W to R_3? (b) What would be the power loss in the other resistors? Take $R_1 = 2\ \Omega$, $R_2 = 4\ \Omega$, $R_3 = 2\ \Omega$.

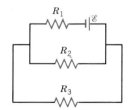

FIGURE 28.53 Exercise 33.

34. (II) In the circuit shown in Fig. 28.54, determine: (a) the values of the emf's; (b) the change in potential $V_b - V_a$.

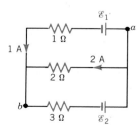

FIGURE 28.54 Exercise 34.

28.4 RC Circuits

35. (I) A 0.01-μF capacitor is connected in parallel with a battery and a resistor. When the battery is disconnected the charge on the capacitor falls to 25% of its initial value in 2 ms. What is the resistance?

36. (I) What is the time constant for the combination shown in Fig. 28.55?

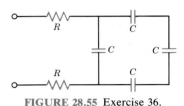

FIGURE 28.55 Exercise 36.

37. (I) In an RC charging circuit $R = 10^4\ \Omega$. If the charge on C rises from zero to 90% of its final value in 2 s, find C.

38. (II) In an RC charging circuit (Fig. 28.18), $\mathscr{E} = 200$ V, $R = 2 \times 10^5\ \Omega$, $C = 50\ \mu$F and $Q = 0$ at $t = 0$. Find: (a) the potential difference across C after one time constant; (b) the potential difference across R after one time constant; (c) the energy stored in C after 5 s; (d) the power loss in R after 5 s.

39. (II) In an RC discharge circuit (Fig. 28.16), $R = 2.5 \times 10^4\ \Omega$ and $C = 40\ \mu$F. The initial potential difference across C is 25 V. Find: (a) the charge on C and the current through R after one time constant; (b) the energy stored in C after one time constant; (c) the power loss in R at 0.5 s; (d) the rate at which energy is being lost by C at 0.5 s.

40. (II) (a) What is the initial rate of charging of C in an RC charging circuit (Fig. 28.18) given $Q = 0$ at $t = 0$? (b) If this rate were constant, how long would it take for C to reach its maximum charge?

41. (II) In an RC discharge circuit (Fig. 28.16), the current drops to 10% of the initial value in 5 s. (a) How long does it take to drop to 50%? (b) Express the current flowing at 10 s as a percentage of the initial current.

42. (II) An air-filled capacitor of capacitance 250 pF is in series with a 2×10^6-Ω resistor and a battery. A material with dielectric constant κ is inserted to fill the space between the plates. When the switch is closed, the current drops to 5% of it initial value in 0.02 s. What is the dielectric constant? (Assume there is no leakage current through the material.)

43. (II) The circuit in Fig. 28.56 includes a capacitor. (a) What is the current in each resistor after a steady state has been reached? (b) What is the charge on the capacitor?

28.5 Direct Current Instruments

44. A multimeter is rated at 20,000 Ω/V. (a) What is the current required for full-scale deflection? (b) What is the effective resistance of the meter when it is used as a voltmeter on the 50-V scale?

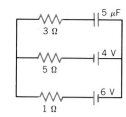

FIGURE 28.56 Exercise 43.

45. (I) A galvanometer has an internal resistance of 50 Ω and is rated at 1 mA for full-scale deflection. What are the series resistances required in Fig. 28.57 to use it as a voltmeter with different ranges. The maximum values for each range are indicated.

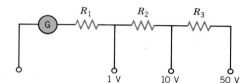

FIGURE 28.57 Exercise 45.

46. (I) A multimeter is rated at 20,000 Ω/V. (a) What current is needed by the galvanometer for full-scale deflection? If the resistance of the galvanometer is 40 Ω find the series of shunt resistance for a full-scale deflection of (b) 250 V; (c) 5 A.

47. (I) A galvanometer has a resistance of 20 Ω and registers a full-scale deflection for a current of 50 μA. Convert it into the following: (a) a voltmeter that reads 0 to 10 V; (b) an ammeter that reads 0 to 500 mA.

48. (II) A potential difference V is applied across the terminals of the "voltage divider" circuit in Fig. 28.46. Find the potential at P for the following values of the load resistance R_L: (a) zero; (b) infinity; (c) R_2; (d) 0.5 R_2.

49. (II) In the circuit of Fig. 28.58a, the internal resistances of the voltmeter and ammeter are $R_V = 1$ kΩ, $R_A = 0.1$ Ω. The resistor has a 10-Ω resistance. (a) What are the true values for the current through and potential difference across the resistor? (b) What are the current and potential difference as measured by the meters?

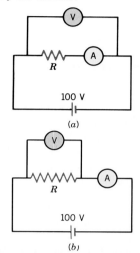

FIGURE 28.58 Exercises 49 and 50.

50. (II). Repeat Exercise 49 for the circuit in Fig. 28.58b.

PROBLEMS

1. (I) A galvanometer has an internal resistance of 20 Ω and requires 2 mA for full-scale deflection. What are the shunt resistances needed for the three ranges indicated in Fig. 28.59?

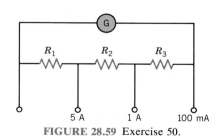

FIGURE 28.59 Exercise 50.

2. (II) The pattern of resistors of equal value shown in Fig. 28.60 is repeated indefinitely. Show that the effective resistance between terminals a and b is $(1 + \sqrt{3})R$. (*Hint*: Since the pattern is infinite, the resistance between points a' and b' is the same as that between a and b.)

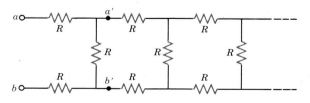

FIGURE 28.60 Problem 2.

3. (II) Twelve identical resistors form a cube, as shown in Fig. 28.61. Find the equivalent resistance between the points A and D. (*Hint*: Number the corners. Use the symmetry of the arrangement to locate points at the same potential and join them. Redraw the pattern in two dimensions.)

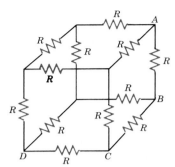

FIGURE 28.61 Problems 3 and 4.

4. (II) Repeat Problem 3 for the points (a) A and B, (b) A and C.

5. (I) A parallel-plate capacitor is filled with a material whose dielectric constant is κ and whose resistivity is ρ. Show that if the capacitor is charged and the battery removed, the charge will decay with a time constant $\tau = \varepsilon_0 \kappa \rho$.

6. (II) A capacitor can be used to vary the times between flashes of a small neon gas tube with the circuit shown in Fig. 28.62. When cool, the gas is a good insulater. The tube "fires" (ionizes and emits light) when the potential difference across it reaches the firing value V_f. Its resistance becomes very small and so the capacitor rapidly discharges through it. As the potential difference drops the gas cools down and becomes an insulator at the "extinction" potential difference V_e. At this stage the capacitor again starts to charge. The variation of the potential difference across the capacitor and the tube is shown. Note that

$$V_e = V_0(1 - e^{-t/RC}); \qquad V_f = V_0(1 - e^{-(t+T)/RC})$$

where t is an unknown time and T is the period of the flashing show that,

$$T = RC \ln \left(\frac{V_0 - V_e}{V_0 - V_f} \right)$$

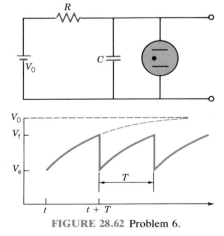

FIGURE 28.62 Problem 6.

7. (II) What is the value for the current in the galvanometer in the unbalanced Wheatstone bridge shown in Fig. 28.63. The resistance of the galvanometer is 20 Ω.

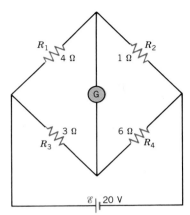

FIGURE 28.63 Problem 7.

8. (I) A 40 μF capacitor has an initial charge of 50 μC. It is discharged through an 8000-Ω resistor. Find: (a) the current at 10 ms; (b) the charge at 10 ms; (c) the power loss in the resistor at 10 ms. (d) How long does it take for the energy in the capacitor to drop to 10% of its initial value?

9. (I) In the circuit of Fig. 28.64, switch S_1 is initially closed and S_2 is open. (a) Find $V_a - V_b$; (b) After S_2 is also closed, what is $V_a - V_b$? (c) S_1 is opened and S_2 is left closed. What is the time constant for the capacitor discharge?

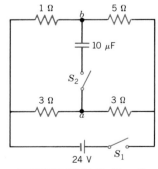

FIGURE 28.64 Problem 9.

10. (I) The circuit in Fig. 28.65 has three batteries and three loops. Find the current in each resistor.

11. (I) Find the current in each resistor in the circuit in Fig. 28.66.

12. (I) For the circuit in Fig. 28.67 find: (a) the initial current

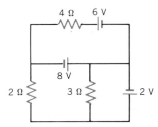

4 Ω 6 V

8 V

2 Ω   3 Ω   2 V

FIGURE 28.65 Problem 10.

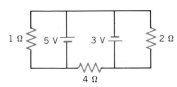

1 Ω   5 V   3 V   2 Ω

4 Ω

FIGURE 28.66 Problem 11.

through each resistor when the switch is closed; (b) the final steady-state current through each resistor; (c) the final energy stored in the capacitor; (d) the time constant when the switch is opened.

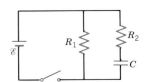

R_1 R_2

$\mathcal{E}$ C

FIGURE 28.67 Problem 12.

13. (II) The switch in Fig. 28.68 is closed at $t = 0$. (a) What are the currents through the resistors at (a) $t = 0$, and (b) $t = \infty$. (c) Show that the current charging the capacitor is given by

$$i_C = \frac{\mathcal{E}}{R_1} e^{-t/\tau}$$

where $\tau = R_1 R_2 C/(R_1 + R_2)$. (*Hint*: Write the loop equation for each loop and relate the currents via the junction rule. Obtain a differential equation for i_C and integrate it.)

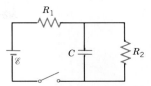

R_1

$\mathcal{E}$ C R_2

FIGURE 28.68 Problem 13.

14. (I) In an RC charging circuit (Fig. 28.18), $\mathcal{E} = 100$ V, $C = 80$ μF, and $R = 10^5$ Ω. Find: (a) the time needed for the energy in the capacitor to reach 50% of the final value; (b) the rate at which the capacitor is charging at 2 s; (c) the power loss in R at 2 s; (d) the total energy dissipated by the resistor from 0 to 10 s.

15. (II) Two batteries have emf's $\mathcal{E}_1$ and $\mathcal{E}_2$ and internal resistances r_1 and r_2. They are connected in parallel. Show that they are equivalent to a battery of emf $\mathcal{E}_{eq}$ where

$$\mathcal{E}_{eq} = \left(\frac{\mathcal{E}_1}{r_1} + \frac{\mathcal{E}_2}{r_2}\right)\left(\frac{1}{r_1} + \frac{1}{r_2}\right)^{-1}$$

(*Hint*: Consider the batteries connected to an external resistor R. Note that $\mathcal{E}_{eq} - Ir_{eq} = IR$.)

16. (II) Evaluate the coefficients α_1 and α_2 in Example 28.6 and thereby find the equivalent resistance between points a and b. Take $R_1 = 2$ Ω; $R_2 = 3$ Ω; $R_3 = 1$ Ω, $R_4 = 5$ Ω and $R_5 = 4$ Ω.

17. (II) Two batteries have the same emf $\mathcal{E}$ and internal resistance r. They are connected to an external resistor R. How should the batteries be connected, in series or in parallel, for the greater power loss in R given (a) $r < R$; (b) $r > R$?

CHAPTER 29

The Magnetic Field

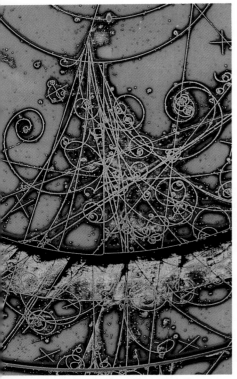

A false-color photo of the tracks of elementary particles in a bubble chamber. A magnetic field causes the paths to be curved.

Major Points

1. (a) The definition of the **magnetic field.**
   (b) The **magnetic force** on a moving charge or current element.
2. (a) The **torque on a current loop** in a magnetic field.
   (b) The definition of **magnetic dipole moment.**
3. The principle of the **galvanometer.**
4. The motion of charged particles in magnetic fields. The *velocity selector,* the *cyclotron,* and the *mass spectrometer.*
5. In the **Hall effect** a potential difference appears across a current-carrying strip placed in a magnetic field.

In Chapter 22 we briefly mentioned the ancient discovery of natural magnets, called lodestones, and their use as compasses in the 11th century. The first systematic study of magnets was made by Pierre de Maricourt in 1269. He used a magnetized needle to trace the "lines of force" around a spherical lodestone and found that they converged on two regions on opposite sides of the sphere, like the lines of longitude on the earth. By analogy, he called the regions magnetic *poles.* In 1600, William Gilbert extended this work and also made the important suggestion that the earth itself is a giant magnet. The end of a suspended bar magnet that points toward geographic north is called the "north-seeking" pole, or north pole for short. The north pole of one magnet attracts the south pole of another magnet and repels the north pole. Therefore, the *geographic* north pole is a *magnetic* south pole.

For over 200 years after the pioneering work of Gilbert, the subjects of electricity and magnetism developed independently. However by 1735, some connection between them was suspected when it was discovered that lightning could magnetize metal objects such as forks and spoons. Hans Christian Oersted (Fig. 29.1), a Danish professor, held a metaphysical belief in the unity of the "forces" in nature. In his view, all "forces," in particular electricity and magnetism, were interconnected. As early as 1813, he began thinking about how a magnetic effect might be produced from electricity. He, and others, tried various experiments. For example, a voltaic pile was suspended by a rope to see if it would orient like a compass.

In the spring of 1820, while preparing a lecture for advanced students, Oersted recalled that a compass needle fluctuates during a thunderstorm, particularly when lightning strikes. At the end of the lecture, he placed a compass needle below a fine platinum wire that was laid along the north–south axis. When a large current was switched on, he was thrilled to see the compass needle rotate from its normal alignment with the earth's magnetic field. Oersted had discovered that an *electrical* current can produce a *magnetic* effect. He later showed that a magnet exerts a force on a current-carrying wire. His results, published in July 1820, established that there is a link between electricity and magnetism.

Nowadays, magnets are used in meters, motors, loudspeakers, tape recording, computer memories, chemical analysis, the focusing of the electron beam in a TV tube, and a myriad of other applications. In addition to providing an aid to navigation, the earth's magnetic field protects us from the harmful effects of high-energy charged particles from outer space (see p. 580). In this chapter we discuss the forces exerted by magnetic fields on charged particles and on electric currents. In the next chapter we will discuss how such fields are produced by currents and charges in motion.

FIGURE 29.1 Hans Christian Oersted (1770–1851).

29.1   THE MAGNETIC FIELD

When iron filings are sprinkled around a bar magnet, they form a characteristic pattern (Fig. 29.2*a*) that shows how the influence of the magnet spreads to the surrounding space. Such patterns inspired Michael Faraday to introduce the concept of the magnetic field and the associated field lines (Fig. 29.2*b*). The **magnetic field, B,** at a point is along the tangent to a field line. The direction of **B** is that of the force on the north pole of a bar magnet, or the direction in which a compass needle points. The strength of the field is proportional to the number of lines passing through a unit area normal to the field. (Thus **B** is also called the magnetic flux density.)

Notice in Fig. 29.2 that the poles are not located at precise points but are

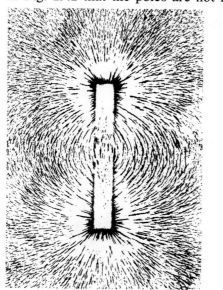

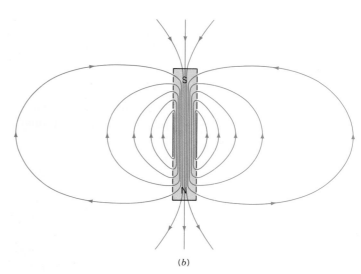

(a)   (b)

FIGURE 29.2 (*a*) The pattern of iron filings around a bar magnet. (*b*) The magnetic field lines for a bar magnet. The lines are closed loops.

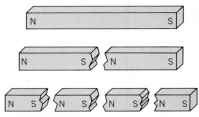

FIGURE 29.3 When a magnet is cut, one obtains two smaller magnets. The north pole or south pole cannot be isolated.

. . .	× × ×
. . .	× × ×
. . .	× × ×
B (out)	**B** (in)

FIGURE 29.4 The convention for depicting magnetic field lines (a) coming out and (b) going into the page.

rather ill-defined regions near the ends of the magnet. If one tries to isolate the poles by cutting the magnet, a curious thing happens: One obtains two magnets, as shown in Fig. 29.3. No matter how thinly the magnet is sliced, each fragment always has two poles. Even down to the atomic level, no one has found an isolated magnetic pole, called a *monopole*. For this reason, magnetic field lines form closed loops, as shown in Fig. 29.2b. Outside a magnet the lines emerge from the north pole and enter at the south pole; within the magnet they are directed from the south pole to the north pole. The usual convention used to denote lines going into, or coming out of, the page is shown in Fig. 29.4. The dot represents the tip of an arrow coming toward you (**B** is directed out of the page). The cross represents the tail of an arrow moving away (**B** points into the page).

Definition of the Magnetic Field

The definition of the electric field strength was quite simple. If **F** is the force on a *stationary* electric charge q placed in the field, then the electric field strength is **E** = **F**/q, or, the force per unit charge. Since an isolated pole is not available, the definition of the magnetic field is not as simple. Instead, we examine how an electric charge is affected by a magnetic field. We find the following:

(i) The force on a charged particle is directly proportional to its charge q and to its speed v; that is, $F \propto qv$.

(ii) If the particle's velocity **v** is at an angle θ to the **B** lines, we find $F \propto \sin \theta$.

Combining these two results, we have

$$F \propto qv \sin \theta$$

Obviously, the force must also depend on the strength of the field. The above proportionality is turned into an equation by defining the magnetic field strength B as the constant of proportionality:

$$F = qvB \sin \theta \qquad (29.1)$$

(iii) The direction of **F** is perpendicular to both **v** and **B**, as shown in Fig. 29.5.

All these results can be neatly incorporated in the definition of the vector product (Section 2.5):

$$\mathbf{F} = q\mathbf{v} \times \mathbf{B} \qquad (29.2)$$

Since **F** is always perpendicular to **v**, *a magnetic force does no work on a particle and cannot be used to change its kinetic energy.* The SI unit of magnetic field is the tesla (T). A conventional laboratory magnet can produce a field of about 1 T,

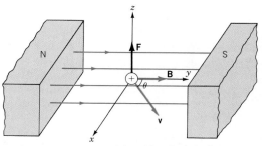

FIGURE 29.5 The magnetic force **F** on a particle with velocity **v** in a magnetic field **B** is perpendicular to both **v** and **B**.

whereas a superconducting magnet can produce over 30 T. Since the tesla is a large unit, a unit called the gauss (G) is often used. The conversion is

$$1 \text{ T} = 10^4 \text{ G}$$

The magnitude of the earth's magnetic field near the surface is about 0.5 G, whereas the field strength in the vicinity of a bar magnet may be 50 G.

EXAMPLE 29.1: An electron has a velocity $\mathbf{v} = 10^6\mathbf{j}$ m/s in a field $\mathbf{B} = 500\mathbf{k}$ G, as shown in Fig. 29.6. What is the force on the electron?

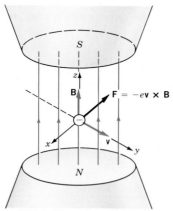

FIGURE 29.6 The magnetic force on an electron is $\mathbf{F} = -e\mathbf{v} \times \mathbf{B}$.

Solution: First, it is necessary to convert B to SI units. We obtain $B = 5 \times 10^{-2}$ T. Using $q = -e$ in Eq. 29.2, we have

$$\mathbf{F} = -e\mathbf{v} \times \mathbf{B} = (-1.6 \times 10^{-19} \text{ C})(10^6\mathbf{j} \text{ m/s}) \times (5 \times 10^{-2}\mathbf{k} \text{ T})$$
$$= -8 \times 10^{-15}\mathbf{i} \text{ N}$$

Note that the force is perpendicular to both the velocity and the magnetic field.

EXERCISE 1. At a certain location, the earth's magnetic field is horizontal and directed due north. What is the direction of the force on an electron moving (a) vertically up, (b) horizontally due east, (c) horizontally toward the southwest?

EXERCISE 2. At a certain location, the earth's magnetic field is directed north and downward at 60° to horizontal. What is the direction of the force on a proton moving (a) vertically up, or (b) horizontally eastward?

29.2 FORCE ON A CURRENT-CARRYING CONDUCTOR

When a wire is placed in a magnetic field, it experiences no force. The thermal velocities of the free electrons are randomly oriented and so the net force on them is zero. However, when a current flows, the electrons as a whole acquire a slow drift speed, v_d, and experience a magnetic force, which is then transmitted to the wire. (The mechanism by which this occurs is discussed in Section 29.8.) Consider a length ℓ of a straight wire of cross-sectional area A that is carrying a current I perpendicular to a uniform magnetic field, as shown in Fig. 29.7. If n is the number of conduction electrons per unit volume, the number in this length of wire is $nA\ell$. Each electron experiences a force ev_dB; therefore, the total force on the electrons in this section is

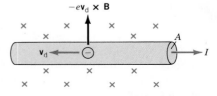

FIGURE 29.7 When a current flows in a wire, the magnetic force on the moving electrons is transmitted to the wire.

$$F = (nA\ell)ev_dB$$

From Eq. 27.2, the current $I = nAev_d$, so the above expression becomes $F = I\ell B$. When the current-carrying wire is not perpendicular to the field, the force on the wire is given by the vector expression

$$\mathbf{F} = I\boldsymbol{\ell} \times \mathbf{B} \tag{29.3}$$

where the vector $\boldsymbol{\ell}$ is defined to be in the direction in which the current is flowing. As Fig. 29.8 shows, the force is always normal to both the wire and the field lines. The magnitude of the force is

$$F = I\ell B \sin \theta \tag{29.4}$$

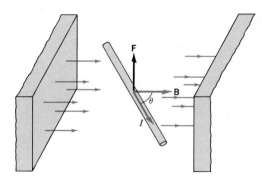

FIGURE 29.8 The magnetic force on a straight current-carrying wire in a uniform field is $\mathbf{F} = I\boldsymbol{\ell} \times \mathbf{B}$, where $\boldsymbol{\ell}$ is in the direction of the current.

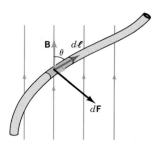

FIGURE 29.9 The magnetic force on an element of a current-carrying wire is $d\mathbf{F} = I\,d\boldsymbol{\ell} \times \mathbf{B}$.

where θ is the angle between the vector $\boldsymbol{\ell}$ and the **B** field. If the wire is not straight or the field is not uniform, the force on an infinitesimal **current element** $I\,d\boldsymbol{\ell}$, as in Fig. 29.9, is

$$d\mathbf{F} = I\,d\boldsymbol{\ell} \times \mathbf{B} \qquad (29.5)$$

The force on a wire is the vector sum (integral) of the forces on all current elements.

EXAMPLE 29.2: A straight wire of length 30 cm and mass 50 g lies along the east–west direction. The earth's magnetic field at this point is horizontal and has a magnitude of 0.8 G. For what current will the wire's weight be sustained by the field?

Solution: The earth's field points north (Fig. 29.10). In order for the magnetic force to be vertically upward, the current must flow from west to east. Since $\theta = 90°$, from Eq. 29.4 we have $F = I\ell B$. This force must equal the weight mg. Therefore,

$$I = \frac{mg}{\ell B} = 2.1 \times 10^4 \text{ A}$$

Although an ordinary copper wire would melt quickly if such a current flowed through it, a superconducting wire could sustain such a current.

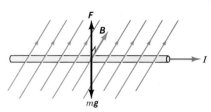

FIGURE 29.10 The wire and the magnetic field are in the horizontal plane. The magnetic force is vertically upward.

EXAMPLE 29.3: A straight wire lies along a body diagonal of an imaginary cube of side $a = 20$ cm, and carries a current of 5

A (see Fig. 29.11). Find the force on it due to a uniform field $\mathbf{B} = 0.6\mathbf{j}$ T.

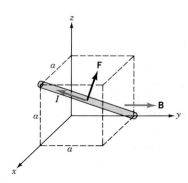

FIGURE 29.11 A length of straight wire lies along a body diagonal of a cube. The magnetic force is parallel to a face diagonal. Unit vector notation is useful in such cases.

Solution: In order to use Eq. 29.4, we would need to find the angle between $\boldsymbol{\ell}$ and **B**. We avoid this task by using unit vector notation. From Fig. 29.11 we see that

$$\boldsymbol{\ell} = a\mathbf{i} - a\mathbf{j} + a\mathbf{k}$$

Thus, the force is,

$$\mathbf{F} = I\boldsymbol{\ell} \times \mathbf{B} = IaB(\mathbf{i} - \mathbf{j} + \mathbf{k}) \times (\mathbf{j})$$
$$= IaB(-\mathbf{i} + \mathbf{k})$$

The force lies in the xz plane and has a magnitude

$$F = \sqrt{2}\, IaB = 0.85 \text{ N}$$

EXAMPLE 29.4: A wire is bent into a semicircular loop of radius R. It carries a current I, and its plane is perpendicular to a uniform magnetic field B, as shown in Fig. 29.12. Find the force on the loop.

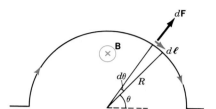

FIGURE 29.12 The force $d\mathbf{F}$ on a current element of length $d\ell$. By symmetry the x component of the total force on the semicircular loop is zero.

Solution: We must first consider the force on an arbitrary current element of length $d\ell = R\, d\theta$, as shown in Fig. 29.12. Since the current element is perpendicular to $\mathbf{B}$, the magnitude of the force on it is $dF = I\, d\ell\, B$ and its direction is radially outward. From the symmetry, we see that for every element on the right of the y axis, there is an equivalent element on the left. The x components of the forces on such elements will cancel in pairs.

The y component of the force on the element shown is

$$dF_y = dF \sin \theta = I\,(R\, d\theta)B \sin \theta$$

where we have used $d\ell = R\, d\theta$. The total force on the semicircle is

$$F_y = IRB \int_0^{\pi} \sin \theta\, d\theta = IRB \left[-\cos \theta \right]_0^{\pi}$$
$$= 2IRB$$

Notice that this would be just the force on a straight wire of length $2R$ joining the ends of the semicircle. In fact, the net force on a length of wire of any shape is equal to the force on a straight wire connecting the end points, as shown in Fig. 29.13 (see Problem 5). It follows immediately that in a *uniform* magnetic field, the net force on any *closed* current-carrying loop is zero.

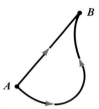

FIGURE 29.13 In a uniform field, the force on a curved wire joining points A and B is the same as that on a straight wire joining these points.

29.3 TORQUE ON A CURRENT LOOP

We have just seen that the net force on a current loop in a uniform magnetic field is zero. However, the loop may experience a net torque that tends to rotate it. Figure 29.14a shows a rectangular loop with sides a and c pivoted about a vertical axis. The loop is oriented at an angle to a uniform magnetic field $\mathbf{B} = B\mathbf{i}$. Figure 29.14b is a top view of the loop, looking down along the negative z direction. It is convenient to specify the angle θ between the *normal* to the plane of the loop and the field. The horizontal sections experience equal and opposite vertical forces that tend to stretch the loop. (Check this with the right-hand rule.) These vertical forces exert no torque since they lie in the plane of the loop. The forces on the vertical sections are also equal and opposite:

$$\mathbf{F}_1 = I(-c\mathbf{k}) \times (B\mathbf{i}) = -IcB\mathbf{j}$$
$$\mathbf{F}_2 = I(c\mathbf{k}) \times (B\mathbf{i}) = IcB\mathbf{j}$$

From Fig. 29.14b we see that these forces produce torques in the same sense about the central axis. The lever arm for each of these forces is $r_{\perp} = (a/2) \sin \theta$; thus, the net torque about the axis is

$$\tau = 2(IcB)\left(\frac{a}{2} \sin \theta\right) = IAB \sin \theta$$

where $A = ac$ is the area of the loop. For a loop with N turns, the torque is N times larger.

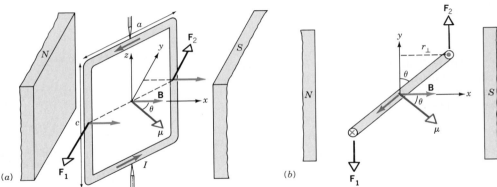

FIGURE 29.14 A current-carrying loop is pivoted in a magnetic field. The forces on the vertical sides produce a torque about the central axis. (*a*) A side view; (*b*) a top view. The magnetic moment μ is perpendicular to the plane of the loop.

A current loop, like a bar magnet, is a magnetic dipole (we will see why in Example 30.3). The **magnetic dipole moment** of a plane loop of any shape is defined as

$$\mu = NIA\hat{n} \qquad (29.6)$$

The SI unit of magnetic moment is A $\cdot$ m². The direction of the unit vector $\hat{n}$ is given by a right-hand rule: When you curve the fingers of your right hand along the current, the extended thumb points along the direction of $\hat{n}$, as shown in Fig. 29.15. The equation for the torque may now be written compactly as

$$\tau = \mu \times \mathbf{B} \qquad (29.7)$$

The tendency of the torque is to align the magnetic moment along the field—just the way a compass needle behaves.

The torque experienced by a current-carrying coil in a magnetic field is used in motors and meters. Equation 29.7 should be compared with Eq. 23.16 for the torque on an electric dipole in an electric field: $\tau = \mathbf{p} \times \mathbf{E}$. By analogy with Eq. 23.17, $U = -\mathbf{p} \cdot \mathbf{E}$, for the potential energy of an electric dipole in an electric field, we may write

$$U = -\mu \cdot \mathbf{B} \qquad (29.8)$$

for the potential energy of a magnetic dipole in the magnetic field. As we did for the electric dipole, we choose $U = 0$ when the angle between μ and $\mathbf{B}$ is 90°. (Note that we do not associate a potential energy with a single charge in a magnetic field.)

FIGURE 29.15 The direction of the magnetic moment μ is given by a right-hand rule: When the fingers are curved in the sense of the current, the extended thumb gives the direction of $\hat{n}$.

EXAMPLE 29.5: The square loop in Fig. 29.16*a* has sides of length 20 cm. It has 5 turns and carries a current of 2 A. The normal to the loop is at 37° to a uniform field $\mathbf{B} = 0.5\mathbf{j}$ T. Find: (a) the magnetic moment; (b) the torque on the loop; (c) the work needed to rotate the loop from its position of minimum energy to the given orientation.

Solution: In such problems it often helps to redraw the diagram from a different perspective, such as the top view shown in Fig. 29.16*b*.
(a) From Fig. 29.16*b* we see that $\hat{n} = -\sin 37°\mathbf{i} + \cos 37°\mathbf{j} =$

$-0.6\mathbf{i} + 0.8\mathbf{j}$. Note that the magnitude of $\hat{n}$ is one. The magnetic moment is

$$\mu = NIA\hat{n} = (5)(2 \text{ A})(0.2 \text{ m})^2(-0.6\mathbf{i} + 0.8\mathbf{j})$$
$$= -0.24\mathbf{i} + 0.32\mathbf{j} \text{ A} \cdot \text{m}^2$$

(b) The torque may be found by first calculating the forces on the sides parallel to the z axis. Instead, we use Eq. 29.7.

$$\tau = \mu \times \mathbf{B} = (-0.24\mathbf{i} + 0.32\mathbf{j}) \times (0.5\mathbf{j})$$
$$= -0.12\mathbf{k} \text{ N} \cdot \text{m}$$

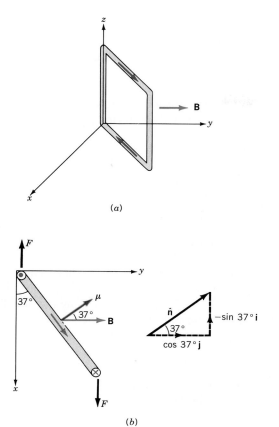

FIGURE 29.16 The torque on the rectangular loop may be found by first calculating the forces on the vertical sides or, as is done here, by determining the magnetic moment. (a) A side view; (b) a top view. The torque tends to align the magnetic moment along the field.

Looking in the direction of the negative z axis, as in Fig. 29.16b, we see that the (clockwise) torque tends to align $\boldsymbol{\mu}$ along $\mathbf{B}$.

(c) The potential energy of the loop is $U = -\mu B \cos \theta$ where $\mu = NIA = 0.4$ A·m^2 and $B = 0.4$ T. The position of minimum energy is $\theta = 0$. Thus, the external work, $W_{EXT} = +\Delta U$, needed to rotate it to the given orientation is

$$U_f - U_i = (-\mu B \cos 37°) - (-\mu B \cos 0°)$$
$$= (0.4)(0.5)(1 - 0.8) = 0.04 \text{ J}$$

The external work is positive since the dipole moment is rotated away from alignment with the field.

EXERCISE 3. What are the forces on the horizontal sections of the loop?

EXAMPLE 29.6: In the Bohr model of the hydrogen atom, an electron travels in a circular orbit about a stationary proton. The radius of the orbit is 0.53×10^{-10} m and the speed of the electron is 2.2×10^6 m/s. (a) What is the magnetic moment associated with the orbital motion? (b) Relate the magnetic moment to the orbital angular momentum of the electron.

Solution: (a) The average current associated with the electron's motion is $I = e/T$ where $T = 2\pi r/v$ is the period of the orbit. The magnetic moment of this current loop is

$$\mu = IA = \left(\frac{ev}{2\pi r}\right)(\pi r^2)$$

$$= \frac{evr}{2} \qquad \text{(i)}$$

On inserting the values into (i) we find a quantity called the *Bohr magneton*:

$$\mu_B = \tfrac{1}{2}(1.6 \times 10^{-19} \text{ C})(2.2 \times 10^6 \text{ m/s})(0.53 \times 10^{-10} \text{ m})$$
$$= 9.3 \times 10^{-24} \text{ A m}^2$$

This value serves as a convenient unit for the magnetic moments of atoms.

(b) The angular momentum of a particle in circular orbit is $L = mvr$. From (i) we have $\mu = evr/2$. Thus, in vector notation,

$$\boldsymbol{\mu} = -\frac{e}{2m}\mathbf{L} \qquad \text{(ii)}$$

The negative sign indicates that the vectors are in opposite directions.

29.4 THE GALVANOMETER

Volta's invention of the voltaic pile in 1800 was a great boon to electrical researchers because it could be used to produce a continuous current. Yet prior to 1820 the presence of an electric current could be detected only through the rise in temperature of a wire, by the chemical changes produced in a cell, or by the sensation the current produced when passed through the tongue. These methods did not lend themselves to precise *measurement* of current. Soon after Oersted's discovery, it was realized that the deflection of a compass needle could be used to measure current. J. Schweigger placed a compass needle at the center of a rectangular coil with many turns of wire (Fig. 29.17) and thereby greatly enhanced the magnetic effect of a current. An alternative design, perfected by d'Arsonval, is called a

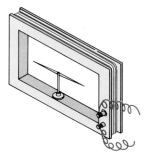

FIGURE 29.17 An early instrument to detect current. The magnetized needle is deflected from its normal alignment with the earth's field when a current flows in the coil.

moving-coil galvanometer and is still in use today. The current is passed through a coil with many turns of fine wire, suspended or pivoted in a magnetic field. When a current flows through the coil, the torque due to a uniform magnetic field is

$$\tau_B = \mu B \sin \theta$$

where θ is the angle between $\boldsymbol{\mu}$ and $\mathbf{B}$. The torque due to the field is opposed by the mechanical torque due to a coiled spring. This restoring torque obeys Hooke's law:

$$\tau_{sp} = \kappa \phi$$

where ϕ is the angle through which the coiled spring has been twisted. The loop comes to rest where these two torques balance:

$$NIAB \sin \theta = \kappa \phi$$

This equation may be used to determine the current, I. However, the factor $\sin \theta$ is a complication since it makes the scale nonlinear. To make the scale linear, the magnet's pole faces are made cylindrical, as in Fig. 29.18. This shape tends to make the field lines radial, rather than uniform. (A soft iron cylinder placed within the loop helps ensure this.) If the plane of the loop is always along the field lines—that is, the magnetic moment is always normal to them—then $\sin \theta = 1$ and

$$\phi = \frac{NAB}{\kappa} I$$

With this arrangement, the deflection is directly proportional to the current. A good galvanometer can register 1 μA, whereas the best can measure as low as 1 pA.

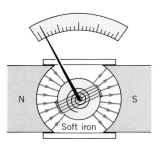

FIGURE 29.18 The basic movement in a D'Arsonval galvanometer. When a current flows, the coil experiences a magnetic torque and an opposing mechanical torque due to a spring.

HISTORICAL NOTE: The Electric Motor

Oersted's discovery of the magnetic field produced by a wire had an intriguing aspect: The magnetic force (whose direction was determined by a compass needle) acted along a tangent to a circle in the plane perpendicular to the wire (see Fig. 30.1*b*). This "circularity" of the magnetic force was quite unlike the central forces of gravitation or electrostatics, which act along the line joining the particles. In July 1821, Michael Faraday was able to demonstrate this new property. He reasoned that an "isolated" pole (roughly the end of a long bar magnet) would experience a force that would take it in a circular path around a current-carrying wire. This is shown on the left side of the apparatus in Fig. 29.19. The continuity of the current was ensured by liquid mercury. He also demonstrated that a current-carrying wire would move in a circle about one pole of a stationary magnet. This is shown on the right side of the apparatus. Faraday had produced the first dc electric motor! Faraday's apparatus brilliantly demonstrated the possibility of continuous mechanical motion produced by magnetic forces, but it was of no practical use.

Like the galvanometer, a simple modern **dc motor** employs the torque on a current-carrying coil in a magnetic field to do work. However, if the current in the rotating coil were maintained in the same direction, the torque on it would reverse direction just as the plane of the coil passes the dashed plane in Fig. 29.20*a*. Therefore, a device known as a **commutator,** which consists of two half-rings electrically connected to the coil (see Fig. 29.20*b*), is used to reverse the direction of the current at the proper time, after each half revolution. Current enters and leaves the coil through two brush contacts that

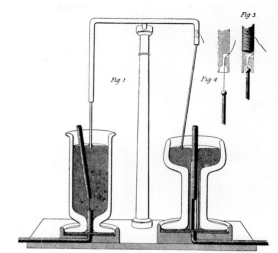

FIGURE 29.19 Michael Faraday's demonstration of continuous motion produced by magnetic force. This was the first motor.

slide on the commutator. One problem with this arrangement is that the torque drops to zero each time the current has to change direction. Thus, the output torque fluctuates. In practical designs, many turns of wire are wrapped around a soft-iron cylinder. The torque delivered by the motor is then greatly increased and relatively steady.

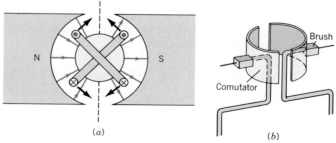

(*a*)   (*b*)

FIGURE 29.20 (*a*) When a loop rotates past the dashed line, the direction of the magnetic torque is reversed. (*b*) A commutator reverses the direction of the current in the loop so that the torque is always in the same sense.

29.5   THE MOTION OF CHARGED PARTICLES IN MAGNETIC FIELDS

As we saw at the beginning of this chapter, a moving charged particle experiences a force in a magnetic field. Magnetic fields are used to deflect and to focus the electron beam in a TV tube and to separate elementary particles produced in a high-energy particle accelerator (Fig. 29.21). The completely ionized gas (plasma) that is required in fusion power research is confined and controlled by magnetic fields. Even though the earth's magnetic field is relatively weak, it plays a part in protecting us from high-energy particles that bombard us from space (see p. 580). In this section we discuss some features of the motion of charged particles in magnetic fields.

Figure 29.22 shows two views of a positively charged particle with an initial velocity **v** perpendicular to a uniform magnetic field **B**. Since **v** and **B** are perpendicular, the particle experiences a force $F = qvB$ of constant magnitude directed perpendicular to **v**. Under the action of such a force, the particle will move in a circular path at constant speed. From Newton's second law, $F = ma$, we have

$$qvB = \frac{mv^2}{r} \tag{29.9}$$

FIGURE 29.21 Elementary particles move in circular paths when subjected to a magnetic field.

where r is the radius of the circle. From Eq. 29.9 we obtain

$$r = \frac{mv}{qB}$$

The radius of the orbit is directly proportional to the linear momentum of the particle and inversely proportional to the magnetic field strength. Since $r/v = m/qB$, the period of the orbit is

$$T = \frac{2\pi r}{v} = \frac{2\pi m}{qB} \tag{29.10}$$

The frequency, $f = 1/T$, is

$$f_c = \frac{qB}{2\pi m} \tag{29.11}$$

Because of its importance in the operation of a particle accelerator called the *cyclotron*, f_c is called the **cyclotron frequency.** Equations 29.10 and 29.11 lead to two important conclusions:

(i) Both the period and the frequency are independent of the speed of the particle.

(ii) All particles with the same charge-to-mass ratio, q/m, have the same period and cyclotron frequency.

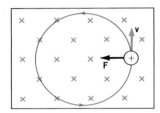

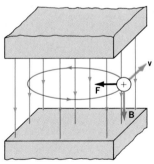

FIGURE 29.22 If a charged particle moves perpendicular to the field lines, its path is circular.

EXAMPLE 29.7: An electron with a kinetic energy of 10^3 eV moves perpendicular to the lines of a uniform field $B = 1$ G. (a) What is the period of its orbit? (b) What is the radius of the orbit?

Solution: (a) From Eq. 29.10 the period is

$$T = \frac{2\pi m}{eB}$$

$$= \frac{2(3.14)(9.11 \times 10^{-31} \text{ kg})}{(1.6 \times 10^{-19} \text{ C})(10^{-4} \text{ T})}$$

$$= 3.6 \times 10^{-7} \text{ s}$$

Note that we did not use the sign of the charge. Even in such a weak field, approximately that of the earth, the frequency of the orbit is large: $f = 1/T = 2.8$ MHz.

(b) In order to find the radius from Eq. 29.9 we must first find the speed. Since $K = \frac{1}{2}mv^2 = 1.6 \times 10^{-16}$ J, we find $v = \sqrt{2K/m} = 1.9 \times 10^7$ m/s. (When the speed of the particle is not much less than the speed of light, one must use a different expression for the kinetic energy. See Chapter 39.)

The radius of the circular path is

$$r = \frac{mv}{eB}$$

$$= \frac{(9.11 \times 10^{-31}\ \text{kg})(1.9 \times 10^7\ \text{m/s})}{(1.6 \times 10^{-19}\ \text{C})(10^{-4}\ \text{T})}$$

$$= 1.1\ \text{m}$$

EXAMPLE 29.8: A proton moves in a circle in radius 20 cm perpendicular to a field of magnitude 0.05 T. Find: (a) the magnitude of its momentum, and (b) its kinetic energy in eV.

Solution: (a) From Eq. 29.9 the linear momentum is

$$p = mv = qrB \qquad \text{(i)}$$
$$= (1.6 \times 10^{-19}\ \text{C})(0.2\ \text{m})(0.05\ \text{T})$$
$$= 1.6 \times 10^{-21}\ \text{kg} \cdot \text{m/s}$$

(b) The kinetic energy $K = mv^2/2$ is

$$K = \frac{p^2}{2m} = \frac{(qrB)^2}{2m} \qquad \text{(ii)}$$

$$= \frac{(1.6 \times 10^{-21}\ \text{kg} \cdot \text{m/s})^2}{2 \times 1.67 \times 10^{-27}\ \text{kg}}$$

$$= 7.7 \times 10^{-16}\ \text{J} = 4.8\ \text{keV}$$

EXERCISE 4. Two charged particles move perpendicular to a uniform magnetic field. The masses and charges are related as follows: $m_2 = 4m_1$ and $q_2 = 2q_1$. What is the ratio of the radii of the orbits given that the particles have (a) the same speed; or (b) the same kinetic energy?

Helical Motion

Let us consider the motion of a positive particle whose velocity has a component along the lines of a uniform magnetic field. We take the magnitudes of the components to be $v_\parallel$ parallel to **B** and $v_\perp$ perpendicular to **B,** as in Fig. 29.23a. The perpendicular component $v_\perp$ gives rise to a force $qv_\perp B$ that produces circular motion (as in Fig. 29.22), but the parallel component $v_\parallel$ is not affected. The result is the superposition of a uniform circular motion normal to the lines and a constant linear motion along the lines. These two motions combine to produce a spiral or *helical* path. The "pitch" of the helix is the displacement of the particle in the direction of the lines in one period,

$$d = v_\parallel T = v_\parallel \frac{2\pi m}{qB}$$

In a nonuniform field the radius of the path will vary. If the other quantities are fixed, then from Eq. 29.9 we see that $r \propto 1/B$, which means that the radius decreases as field strength increases. A more important effect is that the particle experiences a force that points toward the region of weaker field, as in Fig. 29.24. As a result, the component of the velocity along the B lines is not constant. If the particle is moving toward the region of stronger field, at some point it may be

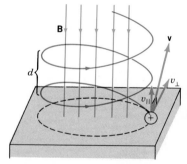

FIGURE 29.23 When a charged particle moves at an angle to the field it travels in a helical path.

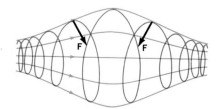

FIGURE 29.24 In a nonuniform magnetic field a charged particle experiences a force directed toward regions where the field is weaker. The direction of motion along the spiral path may be reversed. This is the principle of the "magnetic bottle."

The shape of these solar prominences demonstrates the existence of the sun's magnetic field.

stopped and made to reverse the direction of its travel (provided that v is not too large). This feature is used in the design of "magnetic bottles" that are used to confine high-temperature plasmas in fusion research. The plasma cannot be confined in an ordinary container because the plasma would rapidly cool down on contact with the walls. Another important example of magnetic confinement occurs in the magnetic field of the earth. Charged particles from space travel in spiral paths along the field lines from one pole to the other, as shown in Fig. 29.25. These trapped particles are confined to regions called the van Allen radiation belts (see p. 615).

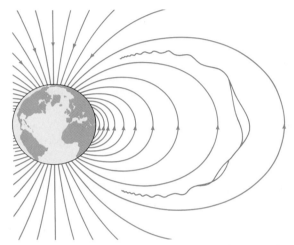

FIGURE 29.25 Protons and electrons from space are magnetically confined by the earth's magnetic field.

29.6 COMBINED ELECTRIC AND MAGNETIC FIELDS

When a particle is subject to both electric and magnetic fields in the same region, the total force on it is

Lorentz force

$$\mathbf{F} = q(\mathbf{E} + \mathbf{v} \times \mathbf{B}) \qquad (29.12)$$

This is called the **Lorentz force** after the Dutch physicist H. A. Lorentz. We have just seen that the path of a particle in a uniform magnetic field is a helix, and we know that the path in a uniform electric field is a parabola. When a particle is subject to both electric and magnetic fields, the motion is in general quite complex. However, the special cases in which the fields are either parallel or perpendicular to each other are useful and simple to analyze.

Velocity Selector

Figure 29.26 shows a region in which an electric field $\mathbf{E} = -E\mathbf{j}$ is perpendicular to a magnetic field $\mathbf{B} = -B\mathbf{k}$. Suppose a particle with charge $+q$ enters this region of crossed fields with initial velocity $\mathbf{v} = v\mathbf{i}$. The electric and magnetic forces are $\mathbf{F_E} = -qE\mathbf{j}$ and $\mathbf{F_B} = qvB\mathbf{j}$. These forces are in opposite directions and will cancel if they have the same magnitude. That is, $\mathbf{F_E} + \mathbf{F_B} = q(\mathbf{E} + \mathbf{v} \times \mathbf{B}) = 0$ if

$$\mathbf{E} = -\mathbf{v} \times \mathbf{B}$$

or, in terms of the magnitudes, $E = vB$. Therefore, if a stream of particles has a distribution of velocities, only those particles with speed

$$v = \frac{E}{B} \tag{29.13}$$

pass through the crossed fields undeflected. This arrangement of perpendicular electric and magnetic fields provides a convenient way of either measuring or selecting the velocities of charged particles.

EXERCISE 5. A negative charge moves from the top to the bottom of the page. There is a uniform magnetic field directed into the page. What should be the direction of the electric field for the charge to be undeflected?

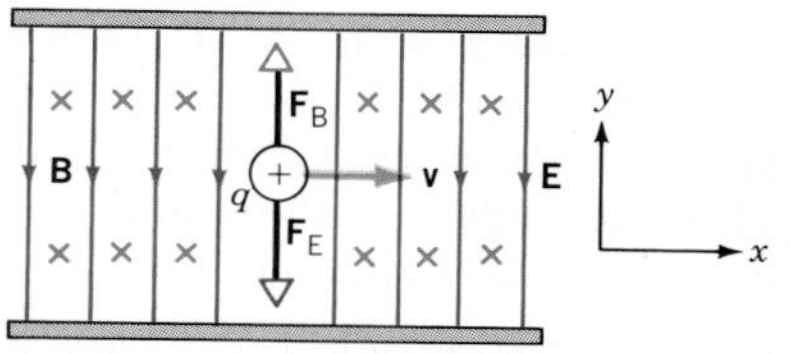

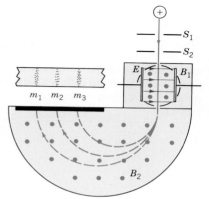

FIGURE 29.26 A velocity selector. In a region of perpendicular electric and magnetic fields only those particles that have a velocity that satisfies the condition $\mathbf{E} = -\mathbf{v} \times \mathbf{B}$ are not deflected.

Mass Spectrometer

A mass spectrometer is a device that separates charged particles, usually ions, according to their charge-to-mass ratios. If the charges are the same, the instrument can be used to measure the mass of the ions. An instrument made by K. T. Bainbridge in 1933 is shown in Fig. 29.27. A beam of charged particles is collimated (made into a pencil-like beam) by slits S_1 and S_2. The particles then enter a velocity selector in which the magnetic field is B_1 and the perpendicular electric field is E. Consequently, only those particles with speed $v = E/B_1$ enter the next section which has only a magnetic field B_2. The particles move in semicircular paths and strike a photographic plate. From Eq. 29.9 we know that the radius of the path is $r = mv/qB_2$. Substituting $v = E/B_1$, we find

$$\frac{m}{q} = \frac{B_1 B_2}{E} r \tag{29.14}$$

For a given value of q, the radius of the path is proportional to the mass. This technique is used to separate isotopes—atoms that have the same chemical properties but slightly different masses (see Chapter 43). Mass differences of about 0.01% can be detected with this instrument. The density of the traces on the photograph also yields the relative abundance of the particles. Mass spectroscopy is used routinely in chemical analysis—for example, in the detection of pollutants or impurities (see Fig. 29.28).

FIGURE 29.27 The Bainbridge mass spectrometer separates charged particles according to their charge-to-mass ration. They first pass through a velocity selector and then travel in a semicircular path in the magnetic field B_2. A photographic plate records the impacts of the particles.

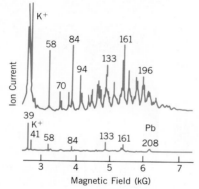

FIGURE 29.28 A mass spectrograph of air in a room before and after one cigarette was smoked. The peaks at 84 and 161 are due to nicotine. (W. D. Davis, *Environ. Sci. Technol.* **11**: 543, 1977)

EXAMPLE 29.9: In A. J. Dempster's mass spectrometer, shown in Fig. 29.29, two isotopes of an element with masses m_1 and m_2 are accelerated from rest by a potential difference V. They then enter a uniform field B normal to the magnetic field lines. What is the ratio of the radii of their paths?

Solution: The kinetic energy of a particle is given by $\frac{1}{2}mv^2 = qV$, so

$$v = \sqrt{\frac{2qV}{m}} \qquad (i)$$

From Eq. 29.9, we know that in the magnetic field $v = qrB/m$. On equating these expressions for v, we find that the radius is given by

$$r = \sqrt{\frac{2mV}{qB^2}} \qquad (ii)$$

Thus, if the charge q is the same, then $r_1/r_2 = \sqrt{m_1/m_2}$.

In practice, the radius of the path is held fixed and the accelerating potential difference is varied. This method can be used to isolate and to collect a particular isotope—for example, radioactive carbon-14 rather than the more abundant carbon-12.

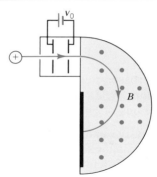

FIGURE 29.29 Dempster's mass spectrometer. The energy of the particles is determined by the accelerating potential difference which is adjusted so that all particles travel in a semicircle of fixed radius. The arrival of the particles is registered by an electrometer.

29.7 THE CYCLOTRON

A great deal of information regarding the properties of nuclei and elementary particles is obtained by bombarding atomic targets with high-energy particles. In 1932, J. Cockcroft and W. Walton in England produced the first "atom smasher" by bombarding a lithium target with protons that had been accelerated by 700,000 V. They adopted a brute force method in which they actually set up this large potential difference. In 1929, Ernest Lawrence, an American physicist, had thought of the possibility of accelerating particles repeatedly through a relatively small potential difference instead of in one giant step. He developed his device, called the **cyclotron**, in association with M. S. Livingston. The first cyclotron (built in 1930) and a later version (built in 1934) are shown in Fig. 29.30.

(a)   (b)

FIGURE 29.30 (a) The first prototype of the cyclotron could be held in the hand. (b) A 27-in. cyclotron made in 1934. Lawrence is on the right and Livingston is on the left.

The operation of the cyclotron is based on the fact that the orbital period of a particle in a magnetic field is independent of its speed. The cyclotron, depicted in Fig. 29.31, consists of two cylindrical dees, D_1 and D_2, (named after their shape) that are separated by a small gap and located in a uniform magnetic field. Charged particles, such as protons or alpha particles, are produced by an ion source at the center and are injected into one of the dees at some small speed. The apparatus is in a high vacuum to minimize losses due to collisions with air molecules. The magnetic field penetrates the dees and causes the particles to move in a circular path.

The dees are connected to a high-voltage supply that reverses its polarity with a period chosen to match the time it takes the particles to complete half a circle. The electric field associated with this potential difference is confined mainly to the gap. Just as the particles complete the first half-circle, D_1 is made positive and D_2 is made negative. Since this polarity accelerates the (positive) particles across the gap, they gain kinetic energy $\Delta K = qV$ and move in a larger orbit. In time $T/2$, they are back at the gap, but the potential difference has reversed its polarity, so they are again accelerated across the gap. This process is repeated for many traversals of the gap: The particles always speed up across the gap and travel in ever larger circles—*but always with the same period*. When the particles reach the maximum radius, a deflector plate channels them to the experimental area.

In practice there are complications. First, it is difficult to produce a uniform magnetic field over the large area of a modern cyclotron (radius about 2 m). Second, as a particle speeds up, its *relativistic* mass (Eq. 39.25) becomes significantly greater than the usual mass measured at low speeds. As a result, the alternating potential difference and the traversals across the gap are no longer in phase. For this reason heavier particles such as protons, rather than electrons, are used: For a given energy, the speed of a proton is much lower than that of an electron. Protons in a cyclotron attain a maximum energy of about 25 MeV. In the *synchrocyclotron* the frequency of the voltage supply steadily decreases to compensate for the mass increase. This machine can accelerate protons to energies up to 200 MeV.

In a *synchrotron,* both the magnetic field and the frequency are varied in such a way that the particles travel in an orbit of fixed radius. The particles accelerate as they pass between long tubes whose electric potential alternates at the appropriate frequency. At the Fermilab accelerator, shown in Fig. 29.32, protons travel in a circle of radius about 1 km and have final energies of about 1 TeV.

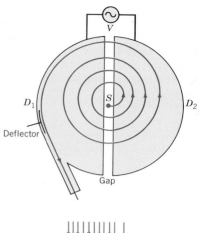
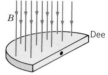

FIGURE 29.31 The operation of a cyclotron depends on the fact that the period of the circular motion is independent of the speed of the particle. An alternating potential difference applied between two cyclindrical dees is used to accelerate particles across the gap. A magnetic field is applied perpendicular to the dees so that the particles travel in semi-circular paths within each dee.

FIGURE 29.32 The Fermilab particle accelerator at Batavia, Illinois.

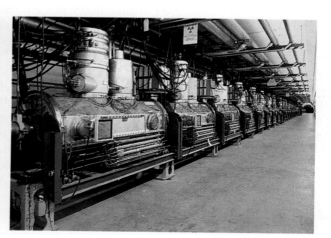

EXAMPLE 29.10: A cyclotron is used to accelerate protons from rest. It has a radius of 60 cm and a magnetic field of 0.8 T. The potential difference across the dees is 75 kV. Find: (a) the frequency of the alternating potential difference; (b) the maximum kinetic energy; (c) the number of revolutions made by the protons.

Solution: (a) From Eq. 29.11 the cyclotron frequency is

$$f_c = \frac{qB}{2\pi m}$$
$$= \frac{(1.6 \times 10^{-19}\ \text{C})(0.8\ \text{T})}{(6.28)(1.67 \times 10^{-27}\ \text{kg})} = 12\ \text{MHz}$$

(b) The maximum kinetic energy occurs when $r = 0.6$ m. Using the expression obtained in Example 29.8, we have

$$K_{max} = \frac{(qrB)^2}{2m}$$
$$= \frac{[(1.6 \times 10^{-19}\ \text{C})(0.6\ \text{m})(0.8\ \text{T})]^2}{3.34 \times 10^{-27}\ \text{kg}}$$
$$= 1.76 \times 10^{-12}\ \text{J} = 11\ \text{MeV}$$

(c) In each revolution the proton is accelerated twice, so its gain in energy is

$$\Delta K = 2qV$$
$$= 2(1.6 \times 10^{-19}\ \text{C})(7.5 \times 10^4\ \text{V}) = 2.4 \times 10^{-14}\ \text{J}$$

The total number of revolutions is simply $K_{max}/\Delta K = 73.5$ revs.

29.8   THE HALL EFFECT

In 1820 Oersted had shown that a magnet exerts a force on a current-carrying wire. In an important text published by J. C. Maxwell (whose work we will discuss in Chapter 34), he suggested that the magnetic force acts not on the electric current but on the conductor that carries it. He concluded that the distribution of current within the wire should be the same as if no field were present. These statements bothered Edwin Hall (Fig. 29.33), who in 1877 had just begun his graduate studies at Johns Hopkins University under Professor Henry Rowland. It was clear that the magnetic force is proportional to the current and is not affected by the size of the wire. In particular, the force vanishes when the current is zero. Maxwell's suggestion did not seem reasonable.

At that time electric current was taken to be the flow of one, or possibly two, fluids. (The discovery of the electron was still 20 years in the future.) Hall reasoned that in a magnetic field the "fluid" would be drawn to one side of the wire. As a result, the effective cross-sectional area of the wire would decrease, and therefore the resistance should increase. The idea was correct, but his instruments were not sensitive enough to detect any change in resistance. He then tried another line of attack. Perhaps, under the action of the magnetic field, the fluid would produce a "state of stress" on the conductor. A few years earlier, Rowland had actually detected a feeble potential difference across a current-carrying conductor in a magnetic field. He suggested that Hall repeat the experiment, but this time with a very thin gold leaf. In October 1879, Hall discovered the effect that now bears his name.

Figure 29.34 shows a flat metal strip, of width w and thickness t, in which a current I is flowing. A uniform magnetic field B is directed as shown in the figure. In 1879, Hall discovered that a potential difference is produced across the width of the strip. A charge $+q$ moving along the strip at its drift velocity $\mathbf{v}_d$ will experience an upward magnetic force $\mathbf{F}_B = qv_d B\mathbf{j}$. The upper face becomes positively charged, while the bottom face becomes negatively charged. These charges now produce an electric field directed downward—which results in a downward electric force $\mathbf{F}_E = -qE\mathbf{j}$. As more charges build up on the top and bottom faces, the electric force becomes strong enough to balance the magnetic force. The net

FIGURE 29.33 Edwin H. Hall (1855–1938).

force, $\mathbf{F} = q(\mathbf{E} + \mathbf{v} \times \mathbf{B})$, is zero when $\mathbf{E} = -\mathbf{v}_d \times \mathbf{B}$, that is, when $E = v_d B$. Associated with this "Hall field" is a potential difference across the width of the strip. This **Hall potential difference** is

$$V_H = Ew = v_d Bw \qquad (29.15)$$

From Eq. 27.2, the current is $I = nqv_d A$, where n is the carrier charge density and $A = wt$ is the cross-sectional area of the strip. If we substitute $v_d = I/nqwt$ into Eq. 29.15, we find

$$V_H = \frac{IB}{nqt} \qquad (29.16)$$

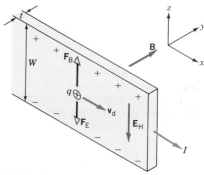

FIGURE 29.34 The Hall effect. When a magnetic field is applied perpendicular to a current-carrying strip, a potential difference appears across the width of the strip.

If the current consists of negative charges moving in the opposite direction, the polarity of V_H reverses. The polarity of the Hall potential difference tells us that the sign of the charges moving in a conductor is negative. Equation 29.16 may be used either to determine v_d if B is known or to measure B if v_d is known. An instrument called the Hall probe uses the Hall potential difference to measure magnetic fields.

From Eq. 29.16 we see that V_H is inversely proportional to the carrier charge density. If the other quantities are known, one can determine the carrier density. For metals, V_H is in the μV range, whereas for semiconductors it can be in the mV range. For metals such as Au, Cu, Ag, Pt, and Al, the sign of V_H is appropriate for negative charge carriers. However, for many years scientists were baffled by the results for Co, Zn, Pb, Fe, and the semiconductors Si and Ge. These materials appeared to have positive charge carriers. An explanation of this anomaly requires quantum mechanics.

EXAMPLE 29.11: A metal strip of width 1 cm and thickness 2 mm lies normal to a magnetic field. When it carries a current of 10 A, the Hall voltage is 0.4 μV. What is the magnitude of B? Take the carrier charge density to be 5×10^{28} m^{-3}.

Solution: From Eq. 29.16,

$$B = \frac{nqtV_H}{I}$$

$$= \frac{(5 \times 10^{28}\ \text{m}^{-3})(1.6 \times 10^{-19}\ \text{C})(2 \times 10^{-3}\ \text{m})(4 \times 10^{-7}\ \text{V})}{10\ \text{A}}$$

$$= 0.64\ \text{T}$$

EXERCISE 6. Show that the number density of charge carriers can be expressed in the form

$$n = \frac{J_x B_y}{q E_z}$$

where J is the current density.

The Nature of the Magnetic Force on a Current-carrying Wire

The Hall effect shows that Maxwell was wrong: The magnetic field acts directly on the moving charges that constitute the electric current. Since the positive ions that form the lattice of a conductor are stationary (except for thermal motion) they cannot experience a magnetic force. So how is the force transmitted to the wire? As we have seen, there is a separation of charge and an associated electrostatic field across the wire. The positive ions of the conductor experience the electrostatic force due to this field. Therefore, the "magnetic force on the wire" is really an electrostatic force on the lattice of positive ions in the conductor.

29.9   THE DISCOVERY OF THE ELECTRON (Optional)

The discovery of the electron, the first subatomic particle, came about through the study of electrical discharges in rarefied gases, which began in earnest around 1860. When a high potential difference is applied across a glass enclosure that contains a gas at low pressure (0.01 atm), the gas begins to glow, somewhat like a fluorescent tube. When the pressure is very low (10^{-3} mm Hg), the whole tube becomes dark and streams of faint blue light emanate from the negative electrode (cathode). Where they impinge on the enclosure, the invisible "cathode rays," which are responsible for the light, also cause the glass to fluoresce with a greenish or bluish glow. A mica cross placed between the cathode and the glass casts a sharp shadow on the glass. This showed that the "cathode rays" traveled in straight lines. In the 1880s several properties of these rays were discovered: (1) They were deflected by a magnetic field as if they were negatively charged. (2) They were emitted perpendicular to the surface of the cathode, unlike light, which is emitted in all directions. (3) They carried momentum (they could rotate a tiny paddle wheel) and energy (they could heat a body).

Some thought the rays were electromagnetic waves, like light or radio waves, while others believed the rays were charged particles. Nowadays we would consider the deflection in a magnetic field to be conclusive evidence that these were in fact charged particles. However, at that time it was not entirely clear whether electromagnetic waves could be deflected by a magnetic field. Heinrich Hertz tried to deflect the rays by an electric field by applying 22 V between two plates, but found no effect. When he tried a higher potential difference (about 500 V), there was an arc discharge across the plates, so he gave up this pursuit. Hertz's assistant, P. Lenard, found that the rays could penetrate thin (2 μm) metal foils and could travel through 1 cm of air. He proved that the rays could not be atoms because the foils did not allow the passage of hydrogen gas. (Hydrogen, the lightest element, was correctly assumed to have the smallest atoms.) If the rays were moving charged particles, then they should *produce* a magnetic field. Hertz was unable to detect this field. Most German physicists stuck to the opinion that the rays were waves of some sort.

In 1895, J. Perrin in France collected the rays in a cylinder and showed that they carried negative charge. When a magnetic field deflected them away from the collector, no charge was collected. This was clear evidence that the rays were in fact particles. In 1897, J.J. Thomson in Cambridge (Fig. 29.35) decided to nail down the identity of the "corpuscles," as he called them. His first important step was to show that the particles could be deflected by an electric field. He succeeded in doing this because he was able to achieve a better vacuum than Hertz could.

J.J. Thomson's apparatus is shown in Fig. 29.36. The rays were produced at the cathode C and accelerated to the anode A that has a small opening in it. They then passed through crossed E and B fields whose values were adjusted to leave the beam

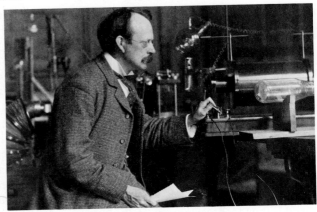

FIGURE 29.35 J. J. Thomson (1856–1940).

undeflected. The position of the beam was recorded as a spot on a screen coated with a phosphorescent material such as ZnS. From our discussion of the velocity selector, we know that the velocity of the particles is $v = E/B$. He found their speed to be about 3×10^7 m/s. Next, the particles were deflected by *either* the electric field or the magnetic field. We will discuss just the electric deflection. The path of the particles in a uniform electric field is a parabola. In Example 23.5 we showed that the angle at which they emerge from the plates is given by

$$\tan \theta = \frac{qE\ell}{mv^2}$$

Since $v = E/B$, this expression can be rearranged to give

$$\frac{q}{m} = \frac{E \tan \theta}{B^2 \ell} \tag{29.17}$$

All the quantities on the right side are easily measured.

Thomson tried several different gases in the tube and showed that the charge-to-mass ratio of the "corpuscles" had the same value, $q/m = 10^{11}$ C/kg, for each of them. Clearly, he was dealing with a single type of particle. This charge-to-mass ratio was 1000 times larger than that for the hydrogen ion, the lightest known. If the "corpuscle" had the same charge as the H$^+$ ion, its mass had to be 1000 times smaller. The ability of the "corpuscles" to penetrate foils and to travel quite far in air clearly indicated that these were very small particles. At a time when some scientists were still not convinced of the existence of atoms, Thomson realized that he was dealing with a "new form of matter," even smaller than an atom. The corpuscle was soon identified as the *electron* whose existence had been postulated from chemical evidence. The definitive measurement of the charge on the electron was made only in 1909 by R. A. Millikan (see Section 23.7). The role of the electron in the atom was clarified several years later. (We note parenthetically that the electrons were emitted by the cathode because of the bombardment by positive ions of the gas in the enclosure.)

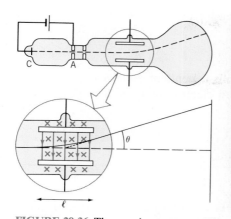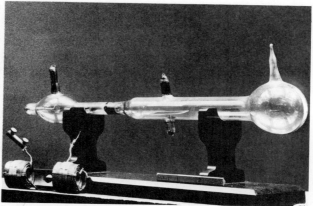

FIGURE 29.36 Thomson's apparatus. Electrons emitted by the cathode C were accelerated to the anode A. They then passed through crossed electric and magnetic fields and impinged on a coated screen.

SUMMARY

The **magnetic force on a charge** q with a velocity **v** in a magnetic field **B** is

$$\mathbf{F} = q\mathbf{v} \times \mathbf{B}$$

The magnitude of the force is $F = qvB \sin \theta$ and its direction is always perpendicular to both **v** and **B.** A magnetic force can accelerate a particle but cannot do work on the particle, and therefore cannot change its energy.

The **magnetic force on a straight wire** of length ℓ carrying a current I in a uniform magnetic field **B** is

$$\mathbf{F} = I\boldsymbol{\ell} \times \mathbf{B}$$

where $\boldsymbol{\ell}$ is in the direction of the current. If the wire is not straight or the field is not uniform, the force on an infinitesimal length $d\ell$ is

$$d\mathbf{F} = I\, d\boldsymbol{\ell} \times \mathbf{B}$$

The **magnetic dipole moment** of a coil of N turns of cross-sectional area A and carrying a current I is

$$\boldsymbol{\mu} = NIA\hat{\mathbf{n}}$$

where the unit vector $\hat{\mathbf{n}}$ is perpendicular to the plane of the coil and its direction is given by the right-hand rule. When a coil of magnetic moment $\boldsymbol{\mu}$ is placed in a uniform field **B,** it experiences a **torque**

$$\boldsymbol{\tau} = \boldsymbol{\mu} \times \mathbf{B}$$

that tends to align $\boldsymbol{\mu}$ parallel to **B.**

When a particle of charge q and mass m moves at speed v in a plane perpendicular to a uniform field, the particle moves in a circular path. From Newton's second law we have

$$qvB = \frac{mv^2}{r}$$

The period $T = 2\pi r/v$ and the **cyclotron frequency** f_c are given by

$$f_c = \frac{1}{T} = \frac{qB}{2\pi m}$$

Note that these quantities are not dependent on v or r.

A particle subject to both an electric field and a magnetic field, experiences the **Lorentz force:**

$$\mathbf{F} = q(\mathbf{E} + \mathbf{v} \times \mathbf{B})$$

ANSWERS TO IN-CHAPTER EXERCISES

1. (a) Eastward; (b) vertically downward; (c) vertically upward.

2. (a) Westward; (b) northward at 30° above the horizontal.

3. First we use $F = IlB \sin \theta$ for the top: $\theta = 53°$, so $F = (2 \text{ A})(0.2 \text{ m})(0.5 \text{ T})(0.8) = 0.16 \text{ N}$. From the right-hand rule, the force is upward. The force on the bottom is equal and opposite. Next we use $\mathbf{F} = I\boldsymbol{\ell} \times \mathbf{B}$ for the top:

$$\boldsymbol{\ell} = a \cos 37°\mathbf{i} + a \sin 37°\mathbf{j}$$

So,

$$\mathbf{F} = (2 \text{ A})(0.16\mathbf{i} + 0.12\mathbf{j} \text{ m}) \times (0.5\mathbf{j} \text{ T}) = 0.16\mathbf{k} \text{ N}$$

For the bottom, $\mathbf{F} = -0.16\mathbf{k}$ N.

4. (a) From (i), $v = qrB/m$, thus $r_2/r_1 = (m_2/m_1)(q_1/q_2) = 2$.
(b) From (ii), $K = (qrB)^2/2m$, thus

$$r_2/r_1 = (\sqrt{m_2/m_1})(q_1/q_2) = 1.$$

5. From right to left parallel to the page.

6. From Eq. 29.17: $n = IB/V_H qt$. Then we note that $I = J_x(tw)$; $V_H = E_z w$, and $B = B_y$, to obtain the given expression.

QUESTIONS

1. Show that 1 T = 1 N/A · m.

2. Can a magnetic field be used to speed up a charged particle? Explain why or why not.

3. In the equation $\mathbf{F} = q\mathbf{v} \times \mathbf{B}$, which pairs of quantities are always perpendicular? Which pairs can be at any angle?

4. How does the radius of the path of a charged particle in a magnetic field depend on its kinetic energy?

5. A beam of electrons passes undeflected through a region. What possible conclusions could be drawn regarding the existence of electric and magnetic fields?

6. The cosmic ray flux of charged particles approaching the earth at the equator tends to come from the west. What can be deduced from this fact?

7. A stationary charge is affected only by an electric field, whereas a moving charge may also experience a magnetic force. When a bar magnet moves rapidly past a stationary charge, as in Fig. 29.37, does the charge experience any force? If so, what is its nature?

8. Consider a charge in a magnetic field. Given q, $\mathbf{v}$, and $\mathbf{F}$, what can you determine about $\mathbf{B}$?

9. In a certain region the earth's field is horizontal and points due north. What is the direction of the force on an electron moving: (a) north; (b) east; (c) vertically up?

10. The field lines at the rim of a magnet curve as shown in Fig. 29.38. An electron is in orbit in a plane perpendicular to the lines. Show that the orbit is stable with respect to small displacements perpendicular to the plane of the orbit.

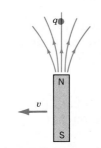

FIGURE 29.37 Question 7.

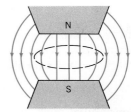

FIGURE 29.38 Question 10.

11. Would it be acceptable to define the direction of $\mathbf{B}$ as that of the force on a positive moving charge? Explain why or why not.

12. A charged particle enters a uniform magnetic field from a

field-free region. Can it complete a closed path in the field?

13. A strip of semiconductor is used as a Hall probe to measure magnetic field strength (Fig. 29.39). What is the sign of the potential difference $V_a - V_b$ if the charge carriers are (a) negative; or (b) positive?

14. Cosmic rays are high-energy charged particles that bombard the earth from space. The flux of cosmic ray particles reaching the earth is higher at the poles than at the equator. Why is this? (Think of the earth's field.)

15. If, in a given magnetic field, two particles have the same cyclotron angular frequency, what can one conclude?

16. Why does a bar magnet placed close to the side of a TV tube distort the picture on the screen?

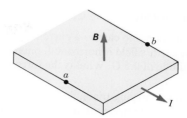

FIGURE 29.39 Question 13.

17. Is the maximum kinetic energy attained by particles in a cyclotron affected by the value of the potential difference across the dees? If so, how?

EXERCISES

29.1 Magnetic Force

1. (I) A proton moves at 3×10^7 m/s normal to a uniform field of 0.05 T. Find: (a) the radius of the path; (b) the period.

2. (I) An electron with a kinetic energy of 1 keV is fired normal to the lines of a field of magnitude 50 G. Find: (a) the radius of its path; (b) its acceleration; (c) its period.

3. (I) A proton moves in a circle of radius 10 cm normal to the lines of a field of 1.0 T. Find: (a) its linear momentum; (b) its kinetic energy in eV.

4. (I) A charge $q = 1 \ \mu$C moves with a speed of 10^6 m/s in a uniform field $\mathbf{B} = 500\mathbf{j}$ G. Find the force on it for each of the three directions of the velocity shown in Fig. 29.40. (Use **ijk** notation.)

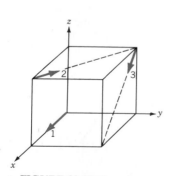

FIGURE 29.40 Exercise 4.

5. (I) When a positive charge moves in the xy plane at 30° to the $+x$ axis, it experiences a force in the $+z$ direction. When it moves with the same speed along $+y$ axis, the force is along the $-z$ axis and has the same magnitude as before. What is the direction of the magnetic field?

6. (II) A charge $q = -0.25 \ \mu$C has a velocity of 2×10^6 m/s at 45° to the x axis in the xz plane, as shown in Fig. 29.41. There is a magnetic field of magnitude 0.03 T. (a) If **B** is

along the $+z$ axis, what is the force on the charge? (b) If the force on the charge is 4×10^{-3} N along $+y$ axis, what is the direction of **B**?

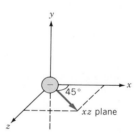

FIGURE 29.41 Exercise 6.

7. (I) A charge $q = -4 \ \mu$C has an instantaneous velocity $\mathbf{v} = 2 \times 10^6\mathbf{i} - 3 \times 10^6\mathbf{j} + 10^6\mathbf{k}$ m/s in a uniform field $\mathbf{B} = 2 \times 10^{-2}\mathbf{i} + 5 \times 10^{-2}\mathbf{j} - 3 \times 10^{-2}\mathbf{k}$ T. What is the force on the charge?

8. (II) When a charge $q = -2 \ \mu$C has an instantaneous velocity $\mathbf{v} = (-\mathbf{i} + 3\mathbf{j}) \times 10^6$ m/s, it experiences a force $\mathbf{F} = 3\mathbf{i} + \mathbf{j} + 2\mathbf{k}$ N. Determine the magnetic field, given that $B_x = 0$.

9. (I) An electron experiences a force $\mathbf{F} = (-2\mathbf{i} + 6\mathbf{j}) \times 10^{-13}$ N in a magnetic field $\mathbf{B} = -1.2\mathbf{k}$ T. What is the velocity of the electron given that $v_z = 0$?

10. (I) An electron moving at $10^6\mathbf{i}$ m/s in a magnetic field experiences a force of $4 \times 10^{-14}\mathbf{j}$ N. (a) What can you deduce about **B** with the given information? (b) Suppose the given force were the maximum possible in the field. What can you deduce in this case?

11. (II) When a proton has a velocity $\mathbf{v} = (2\mathbf{i} + 3\mathbf{j}) \times 10^6$ m/s, it experiences a force $\mathbf{F} = -1.28 \times 10^{-13}$ **k** N. When its velocity is along the $+z$ axis, it experiences a force along the $+x$ axis. What is the magnetic field?

29.2 Force on a Current-carrying Conductor

12. (I) A transmission line carries a current of 10^3 A from west to east. The earth's field is horizontal, points due north, and has a magnitude of 0.5 G. What is the force on one meter of the wire?

13. (I) The current-carrying triangular wire loop shown in Fig. 29.42 is in a uniform magnetic field $\mathbf{B}_1 = -B_1\mathbf{k}$. What is the force on each side of the loop?

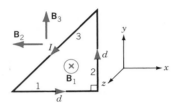

FIGURE 29.42 Exercises 13, 14, 15, and 24.

14. (I) The current-carrying triangular wire loop shown in Fig. 29.42 is in a uniform magnetic field $\mathbf{B}_2 = -B_2\mathbf{i}$. What is the force on each side of the loop?

15. (I) The current-carrying triangular wire loop shown in Fig. 29.42 is in a uniform magnetic field $\mathbf{B}_3 = +B_3\mathbf{j}$. What is the force on each side of the loop?

16. (II) A rod of length $\ell = 15$ cm and mass $m = 30$ g lies on a plane inclined at $37°$ to the horizontal, as shown in Fig. 29.43. A current enters and leaves the rod via light flexible wires which we ignore. For what current (magnitude and direction) will the rod be in equilibrium in a magnetic field $\mathbf{B} = 0.25\mathbf{j}$ T.

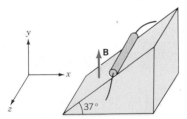

FIGURE 29.43 Exercise 16.

17. (I) A transmission line carries an 800-A current from east to west. The earth's magnetic field is 0.8 G due north but directed $60°$ below the horizontal. What is the force per meter on the line?

18. (I) Consider an 80-cm length of straight wire that carries a 3-A current in a 0.6-T uniform magnetic field. Find the force on the wire when the field is directed as shown in Fig. 29.44.

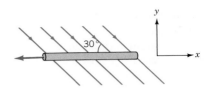

FIGURE 29.44 Exercise 18.

19. (I) A 45-cm length of straight wire carries a 6-A current along the $+z$ axis. It experiences a force 0.05 N in the $-x$ direction. Find the magnetic field given that it is directed: (a) perpendicular to the wire; (b) $30°$ to the $+z$ axis.

29.3 Torque on a Current Loop

20. (I) A rectangular loop has 25 turns with sides $a = 2$ cm and $c = 5$ cm, as shown in Fig. 29.45. What is the force on each side and the torque on the loop if the external field is 0.3 T and is directed: (a) parallel to the plane of the loop ($\mathbf{B}_1$); (b) normal to the plane of the loop ($\mathbf{B}_2$)? Take $I = 8$ A.

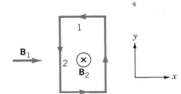

FIGURE 29.45 Exercise 20.

21. (II) A rectangular coil has 16 turns and sides of length $a = 20$ cm and $c = 50$ cm. The coil is pivoted about the z axis and its plane is at $30°$ to a magnetic field $\mathbf{B} = 0.5\mathbf{i}$ T, as shown in Fig. 29.46. (a) Find the force on each side. (b) What is the magnetic moment of the coil? (c) What is the torque on the coil? Take $I = 10$ A.

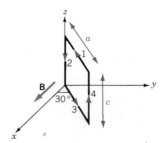

FIGURE 29.46 Exercise 21.

22. (II) The armature of a motor has 8 square coils of side 10 cm. They are all perpendicular to a radially directed magnetic field of magnitude 0.2 T, as shown in Fig. 29.47. If the

current is 10 A find: (a) the torque produced; (b) the mechanical power output at 1200 rpm.

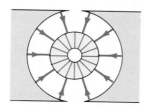

FIGURE 29.47 Exercise 22.

23. (I) A 5-A current flows in a circular loop of radius 2 cm. The axis of the loop is at 30° to a uniform field 0.06 T. What is the magnitude of the torque on the loop?

24. (I) Consider the triangular current loop in Fig. 29.42. Find: (a) the magnetic moment; (b) the torque in a magnetic field $\mathbf{B} = -B\mathbf{i}$.

25. (I) A circular loop of radius 4 cm carries a 2.8-A current. The magnetic moment of the loop is directed along $\hat{\mathbf{n}} = 0.6\mathbf{i} - 0.8\mathbf{j}$. The magnetic field is $\mathbf{B} = 0.2\mathbf{i} - 0.4\mathbf{k}$ T. Find: (a) the torque on the loop; (b) the potential energy of the loop.

29.4 Galvanometer

26. (I) A galvanometer has a square coil with sides of length 2 cm and 20 turns. When it is suspended in a radial magnetic field ($B = 400$ G) it registers a deflection of 30° for a current of 2 mA. What is the torsional constant of the suspension?

27. (I) A galvanometer has a square coil of 200 turns with sides of length 2.5 cm. The radial magnetic field strength is 500 G and normal to the vertical sides. If the torsional constant of the suspension is 2×10^{-8} N · m/degree, what is the angular deflection produced by a current of 10 μA?

29.5 to 29.7 Motion of Charged Particles; Cyclotron

28. (I) A proton is in a circular orbit of radius 20 cm perpendicular to a 0.8-T field. Find: (a) its speed; (b) its period; (c) its kinetic energy.

29. (I) What is the magnetic moment of the orbital motion of an electron in the hydrogen atom if its orbital angular momentum is 2.11×10^{-34} kg · m²/s?

30. (I) The mass of a deuteron is twice that of a proton, $m_d = 2m_p$, but the charge is the same. They both move normal to a uniform magnetic field. What is the ratio of the radii of their paths if they have the same value for the following: (a) linear momentum; (b) speed; (c) kinetic energy?

31. (I) Suppose that an electron and a proton both move perpendicular to the same uniform magnetic field. Find the

ratio of the radii of their orbits given that they have the same value for the following quantities: (a) speed; (b) kinetic energy.

32. (I) In a certain experiment protons (m_p, q_p) and α particles ($m_\alpha = 4m_p$; $q_\alpha = 2q_p$) must move with the same radius normal to a magnetic field whose strength can be changed. Find the ratio of the field strengths needed if they have the same value for the following: (a) speed; (b) linear momentum; (c) kinetic energy.

33. (I) An electron moving at 4×10^6 m/s enters a uniform field $B = 0.04$ T at 30° to the lines. What is the pitch of the helical path?

34. (I) A cosmic ray proton moves toward the earth at $0.1c$ along a radial line in the equatorial plane, that is, normal to the field lines. Suppose the earth's field strength in the region is 0.2 G. (a) What is the radius of the proton's path? (b) Is the deflection to the east or to the west?

35. (I) A proton moves in a circle of radius 3.2 cm perpendicular to a magnetic field of 0.75 T. Find: (a) the cyclotron frequency; (b) the kinetic energy; (c) the linear momentum.

36. (I) An alpha particle of mass 6.7×10^{-27} kg and charge $2e$ is accelerated from rest by a potential difference of 14 kV and enters a uniform magnetic field of 0.6 T normal to the lines. Find the radius of its path.

37. (II) The two isotopes of neon have masses 20 u and 22 u. Singly charged ions are accelerated from rest by a potential difference of 1 kV and then enter a uniform magnetic field of 0.4 T, normal to the lines. What is their separation after one half-revolution in a spectrometer?

38. (II) In a Bainbridge mass spectrometer (Fig. 29.27) ions pass through the crossed electric and magnetic field (B_1) of a velocity selector and are then deflected by a magnetic field (B_2) alone. Given that $E = 3 \times 10^5$ V/m and $B_1 = B_2 = 0.4$ T, calculate the difference in position on the photographic plate for singly charged ions of the carbon isotopes with masses 12 u and 14 u.

39. (I) An electron with a velocity $\mathbf{v} = 2 \times 10^6\mathbf{i}$ m/s enters a region in which the electric field is $\mathbf{E} = -200\mathbf{j}$ V/m. (a) What is the magnetic field required for the electron not to be deflected? (b) If the electric field is switched off, what is the radius of the path in the magnetic field?

40. (I) A proton is accelerated from rest along the $-x$ direction by a potential difference of 10 kV. It passes a region in which the electric field is $-10^3\mathbf{j}$ V/m. What magnetic field is required for the particle to experience no deflection?

41. (I) A proton makes 100 revs in a cyclotron and emerges at a radius of 50 cm with an energy of 10 MeV. Find: (a) the magnetic field in the cyclotron; (b) the potential difference across the dees; (c) the frequency of the voltage source.

42. (I) A cyclotron used to accelerate protons has a radius of 75 cm and a field of 0.9 T. Find: (a) the cyclotron angular frequency; (b) the maximum kinetic energy.

43. (II) Protons emerge from a cyclotron with a kinetic energy of 12 MeV. The alternating potential difference is 6×10^4 V and the magnetic field strength is 1.6 T. (a) What is the radius of the cyclotron? (b) How long, after they are emitted by the source, do the protons take to emerge?

44. (II) A charged particle is accelerated from rest along the x axis by a potential difference of 225 V and then enters a uniform field $\mathbf{B} = 10\mathbf{k}$ G. The radius of its path is 5 cm. What is the charge to mass ratio of the particle?

45. (II) An ion ($m = 1.2 \times 10^{-25}$ kg, $q = 2e$) is accelerated from rest by a voltage of 200 V and then enters a uniform field $B = 0.2$ T normal to the lines. Find the radius of its path.

29.8 The Hall Effect

46. (II) A metallic strip 0.1 cm thick and 1.6 cm wide carries a 15-A current in a field of 0.2 T normal to the width. The Hall potential difference is 6 μV. Find: (a) the drift speed of the carriers; (b) the conduction electron density.

47. (II) A Cu strip of thickness 0.25 cm carrying a current of 10 A is set normal to a magnetic field. The Hall potential difference is 1.2 μV. What is B? Take $n = 8.5 \times 10^{28}$ electrons/m^3.

48. (II) A 2-A current flows along a metal strip 0.1 mm thick, 0.8 cm wide. In a 0.8-T magnetic field normal to the width, the Hall potential difference is 1.4 μV. What is the conduction electron density?

PROBLEMS

1. (II) A dipole with a magnetic moment μ is freely pivoted at its center and has a moment of inertia I about this axis. (a) Show that for small angular displacements, the dipole will execute simple harmonic motion in a uniform magnetic field B. (b) What is the period of oscillation?

2. (II) A constant current flows through a metal rod of length ℓ and mass m that slides on frictionless rails, as shown in Fig. 29.48. A uniform magnetic field is directed into the page. If the initial speed of the rod is v_0, find: (a) the speed as a function of time; (b) the distance moved before coming to a stop.

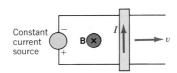

FIGURE 29.48 Problem 2.

3. (I) Figure 29.49 shows a curved wire carrying a current I from a to b in a uniform magnetic field B. Show that the net force on the curved wire is the same as that on a straight wire from a to b carrying the same current.

FIGURE 29.49 Problem 3.

4. (I) For a given length of wire carrying a current I, how many circular turns would produce the maximum magnetic moment?

5. (II) A disk of radius R has a uniform charge density σ C/m^2. It rotates about its central axis at ω rad/s with its axis normal to a uniform field B. (a) Find its magnetic moment. (b) Show that the torque on the disk is

$$\tau = \tfrac{1}{4}\sigma\omega\pi BR^4$$

(*Hint:* Divide the disk into rings of radius r and width dr.)

6. (II) An electron is orbiting a proton in a hydrogen atom. A weak magnetic field is turned on normal to the plane of the orbit. Show that if the radius of the orbit is unchanged, the angular velocity changes by

$$\Delta\omega = \pm\frac{eB}{2m}$$

7. (I) A metal rod of mass 10 g and length 8 cm is suspended on two springs, as shown in Fig. 29.50. The springs are extended by 4 cm. When a 20-A current flows through the rod it rises by 1 cm. Determine the magnetic field.

FIGURE 29.50 Problem 7.

8. (II) An electron has an initial velocity of 3×10^7 **i** m/s at the origin. (a) What magnetic field along the z axis will result in a circular motion of radius 2 cm? (b) How long would it take for the electron's velocity to be deflected by 30°? (c) What are the electron's x and y coordinates at the time found in (b)?

CHAPTER 30

Sources of the Magnetic Field

Major Points

1. (a) The magnetic field produced by a long, straight wire.
   (b) The magnetic force between two current-carrying wires.
2. The **Biot-Savart law** for the magnetic field due to an infinitesimal current element.
3. **Ampère's law** relates an integral of the magnetic field around a closed loop to the net current enclosed by the loop.

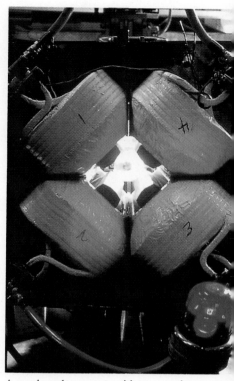

A quadrupole magnet, with two north poles and two south poles, is used to focus the beam in a particle accelerator.

In the last chapter we studied the forces exerted by a magnetic field on currents and charges in motion. We now go on to discuss how magnetic fields are produced by currents and charges in motion and to calculate them for some simple geometries of current-carrying wires. Oersted's work was described to the members of the Paris Academy of Sciences in September 1820. In a matter of weeks, Jean Biot and Felix Savart obtained an expression for the magnetic field produced by an infinitesimal current element in a wire. The Biot-Savart law is analogous to the expression for the electric field obtained from Coulomb's law for the force between point charges. Andre Marie Ampère, working independently, related an integral of the magnetic field around a closed loop to the net current enclosed by the loop. Ampère's law is analogous to Gauss's law in electrostatics. It is useful in determining the magnetic field due to a symmetric current distribution.

30.1 FIELD DUE TO A LONG, STRAIGHT WIRE

A current in a long, straight wire produces a magnetic field with circular field lines—as may be verified by sprinkling iron filings on a board normal to the wire, as in Fig. 30.1a, or with a compass needle, as in Fig. 30.1b. Biot and Savart investigated how the magnetic field strength B depended on the perpendicular distance R from the wire. Their method involved measuring the period of oscillation of a magnetized needle. In October 1820 they announced that B is inversely proportional to R, that is, $B \propto 1/R$. Although they could keep the current constant, they had no way to *measure* it accurately. It was later found that the field is directly proportional to the current I. In SI units, we express these results as

(Infinite wire)
$$B = \frac{\mu_0 I}{2\pi R}$$
(30.1)

Magnetic field due to a long, straight wire

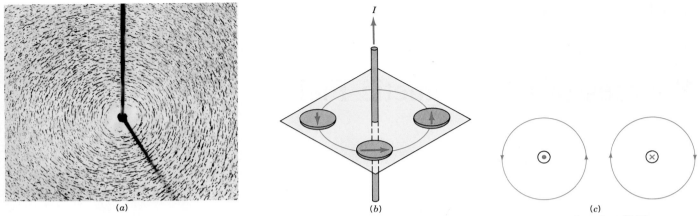

FIGURE 30.1 (*a*) The pattern of iron filings around a straight current-carrying wire. (*b*) The circular nature of the field lines can also be demonstrated with a compass. (*c*) A field line for a current coming out of the page (dot) and going into the page (cross).

where μ_0, called the **permeability constant,** is defined to have the value

$$\mu_0 = 4\pi \times 10^{-7} \text{ T} \cdot \text{m/A}$$

Figure 30.1*c* shows the convention for drawing currents that go into or come out of the page. Notice that the direction of the field is given by another right-hand rule: When the thumb points along the current, the curled fingers indicate the direction of the field. Figure 30.2 depicts the variation in the density of lines with distance from the wire.

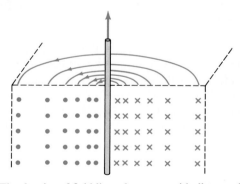

FIGURE 30.2 The density of field lines decreases with distance from the wire.

EXAMPLE 30.1: Two long, straight, parallel wires are 3 cm apart. They carry currents $I_1 = 3$ A and $I_2 = 5$ A in opposite directions, as shown in Fig. 30.3*a*. (a) Find the field strength at point P. (b) At what point, besides infinity, is the field strength zero?

Solution: (a) This problem is similar to that of finding the electric field strength due to two point charges. However, there is an important difference: The direction of the vector **B** is always *perpendicular* to the radial line from the current: It is along a tangent to an arc whose center is at the wire. In order to find the total field strength,

$$\mathbf{B}_T = \mathbf{B}_1 + \mathbf{B}_2$$

we first find the magnitudes B_1 and B_2 from Eq. 30.1. Thus,

$$B_1 = \frac{(2 \times 10^{-7} \text{ T} \cdot \text{m/A})(3 \text{ A})}{4 \times 10^{-2} \text{ m}} = 1.5 \times 10^{-5} \text{ T}$$

Similarly, $B_2 = 2 \times 10^{-5}$ T. These are quite weak fields, comparable to that at the earth's surface. Next we set up a coordinate system, as shown in the figure, and find the components. Since $\tan \theta = \frac{4}{3}$, the angle $\theta = 53°$.

$$B_{Tx} = B_2 \cos \theta = 1.2 \times 10^{-5} \text{ T}$$
$$B_{Ty} = B_1 - B_2 \sin \theta = -10^{-6} \text{ T}$$

Finally,

$$\mathbf{B}_T = (12\mathbf{i} - \mathbf{j}) \times 10^{-6} \text{ T}$$

(b) The null point, at which $\mathbf{B}_1 + \mathbf{B}_2 = 0$, must lie along the line joining the wires. (Why?) Between the wires the fields are in the same direction, so this region is eliminated. Below I_2, the fields are in opposite directions, but these points are closer to the

larger current. The fields can cancel above I_1 at some distance d, as shown in Fig. 30.3b. The condition

$$\Sigma B_x = -B_1 + B_2 = 0$$

yields $I_1/d = I_2/(3 + d)$, from which we find $d = 4.5$ cm.

EXERCISE 1. Two parallel wires carry currents into the page. $I_1 = 8$ A is at the origin, while $I_2 = 4.2$ A is at (4 m, 0). What is the resultant magnetic field at a point (4 m, 3 m)?

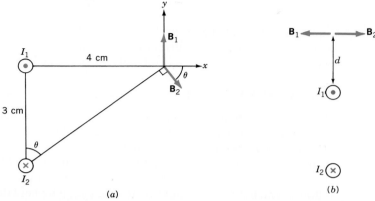

FIGURE 30.3 (a) The field vectors associated with currents I_1 and I_2 lie along the circular arcs centered on each wire. (b) The fields cancel at a point on the line joining the two wires. It is closer to the weaker current.

30.2   MAGNETIC FORCE BETWEEN PARALLEL WIRES

Oersted's demonstration that an electric current exerts a force on a magnetic needle did not, of course, prove that there is an interaction between two currents. (A magnet attracts two iron bars, but the bars by themselves do not attract each other.) In October 1820, Ampere demonstrated that two current-carrying wires do in fact exert forces on each other. Consider two long, straight wires that carry currents I_1 and I_2, as shown in Fig. 30.4. They are parallel and separated by a distance d. From Eq. 29.3, the force on a length ℓ_2 of wire 2 exerted by the field B_1 of wire 1 is $\mathbf{F}_{21} = I_2\boldsymbol{\ell}_2 \times \mathbf{B}_1$, where $\boldsymbol{\ell}_2$ is in the direction of I_2. Since $\boldsymbol{\ell}_2$ is perpendicular to $\mathbf{B}_1$, from Eq. 30.1 we find that the magnitude of the force is

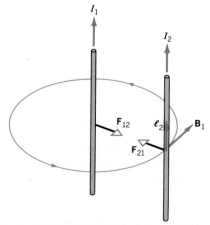

FIGURE 30.4 The equal and opposite forces exerted by two parallel current-carrying wires. The force is repulsive if the currents are in opposite directions.

$$F_{21} = I_2\ell_2 B_1 = I_2\ell_2 \frac{\mu_0 I_1}{2\pi d}$$

A similar expression is found for F_{12}. By applying the right-hand rule, we see that currents in the *same* direction *attract* each other. Conversely, currents in opposite directions repel each other. The *force per unit length* on either wire (for example, F_{21}/ℓ_2 on wire 2) is the same:

$$\frac{F}{\ell} = \frac{\mu_0 I_1 I_2}{2\pi d} \qquad (30.2)$$

Equation 30.2 is the basis for the definition of the ampere (A): If two long, parallel wires carrying the same current are 1 m apart and each unit length (1 m) experiences a force of 2×10^{-7} N, the current is defined to be 1 A.

30.3 BIOT-SAVART LAW FOR A CURRENT ELEMENT

Having determined the magnetic field for a long, straight wire, Biot and Savart next sought a more general expression for the field due to an infinitesimal length of any current-carrying wire. The mathematician Simon de Laplace pointed out to them that the result for the long wire implies that the field due to a current element should depend on the inverse square of the distance. One might infer this as follows. The form of the equation for the magnetic field due to a long, straight wire may be written as

$$B = \frac{2k'I}{R} \qquad (30.3)$$

where $k' = \mu_0/4\pi$. This has the same form as that for the electric field due to an infinite line of charge with linear charge density λ (see Example 23.7):

$$E = \frac{2k\lambda}{R} \qquad (30.4)$$

This result is obtained by integrating the contributions of infinitesimal charge elements $dq = \lambda\, d\ell$, each of which makes the following contribution to the electric field:

$$d\mathbf{E} = k\frac{\lambda\, d\ell}{r^2}\,\hat{\mathbf{r}} \qquad (30.5)$$

The radial unit vector $\hat{\mathbf{r}}$ is directed from the charge element to the point at which the field is being calculated (see Fig. 30.5a).

Biot and Savart wished to derive Eq. 30.3 by summing the contributions of *current elements,* $I\, d\ell$. Given Laplace's hint, they quickly concluded that the only possible form for the field due to such an element is $dB = k'(I\, d\ell/r^2)f(\theta)$. This differs in form from Eq. 30.5 by the angular factor $f(\theta)$. By doing further experiments they determined this function to be $f(\theta) = \sin\theta$ and in December 1820 announced the result:

$$dB = k'\frac{I\, d\ell}{r^2}\sin\theta$$

Unlike the equation for the electric field due to a charge element, this expression involves the angular factor, $\sin\theta$. Furthermore, the direction of the magnetic field is perpendicular to both the current element $I\, d\ell$ (treated as a vector in the direction of I) and the radial unit vector $\hat{\mathbf{r}}$ from the element to the point at which the field is being calculated.

In SI units and vector notation, the **Biot-Savart law** for the magnetic field due

The Aurora Borealis is produced by the interaction of electrons from the sun with the atmosphere. The green light comes from oxygen and the pink light comes from nitrogen (see also p. 616).

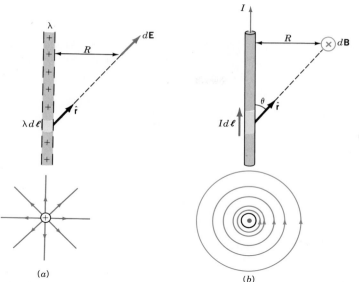

FIGURE 30.5 (*a*) The electric field *d***E** due to the charge element *dq* lies along the unit vector $\hat{\mathbf{r}}$.
(*b*) The magnetic field *d***B** due to the current element $I\,d\boldsymbol{\ell}$ is perpendicular to both $\hat{\mathbf{r}}$ and $d\boldsymbol{\ell}$.

to a current element $I\,d\boldsymbol{\ell}$, shown in Fig. 30.5*b*, is

$$d\mathbf{B} = \frac{\mu_0}{4\pi}\,\frac{I\,d\boldsymbol{\ell} \times \hat{\mathbf{r}}}{r^2} \qquad (30.6)$$

Biot-Savart law for a current element

The magnitude of the field is

$$dB = \frac{\mu_0}{4\pi}\,\frac{I\,d\ell\,\sin\theta}{r^2} \qquad (30.7)$$

Keep in mind that there is no such thing as an isolated current element; the element is always part of a circuit. We will now illustrate the use of this equation for a few simple geometries.

EXAMPLE 30.2: Find the field strength at a distance R from an *infinite straight wire* that carries a current I.

Solution: Figure 30.6 shows the infinitesimal contribution to the field, *d***B,** from an arbitrary current element. In order to integrate the contributions of all the elements, we use the angle α, measured from the perpendicular, as the variable. From the diagram we see that

$$|d\boldsymbol{\ell} \times \hat{\mathbf{r}}| = d\ell\,\sin\theta = d\ell\,\sin(\pi - \theta) = d\ell\,\cos\alpha$$

Since $\ell = R\tan\alpha$, we have $d\ell = R\sec^2\alpha\,d\alpha$. Furthermore, $r = R\sec\alpha$. On substituting these expressions into Eq. 30.6 and integrating we find

$$B = \frac{\mu_0 I}{4\pi R}\int_{-\alpha_1}^{\alpha_2}\cos\alpha\,d\alpha$$

$$= \frac{\mu_0 I}{4\pi R}(\sin\alpha_1 + \sin\alpha_2)$$

Notice that angles on opposite sides of the perpendicular are

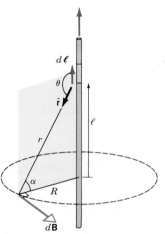

FIGURE 30.6 In calculating the magnetic field produced by a long straight wire we may use the angle α as the variable.

assigned opposite signs. This ensures that the contributions from either side of the wire add together rather than subtract.

For an infinite wire, the limits of integration are $-\pi/2$ to $+\pi/2$. Thus,

$$B = \frac{\mu_0 I}{2\pi R}$$

This, of course, is the original Biot-Savart result.

EXAMPLE 30.3: A *circular loop* of radius a carries a current I. Find the magnetic field along the axis of the loop at a distance z from the center.

Solution: Figure 30.7 shows the infinitesimal contribution to the field $d\mathbf{B}$ from an arbitrary current element $I\,d\boldsymbol{\ell}$. (Use the right-hand rule to confirm the direction of the field.) The field increment $d\mathbf{B}$ has components both along and perpendicular to the axis. However, if we consider the contributions of current elements that are diametrically opposite, we see that their components normal to the axis will cancel. Since $d\boldsymbol{\ell}$ and $\hat{\mathbf{r}}$ are perpendicular, it follows that $|d\boldsymbol{\ell} \times \hat{\mathbf{r}}| = d\ell$. Therefore, the component of $d\mathbf{B}$ along the axis is

$$dB_{\text{axis}} = dB \sin \alpha = \left(\frac{\mu_0 I\,d\ell}{4\pi r^2}\right)\left(\frac{a}{r}\right)$$

The total field strength is given by the integral of this expression over all elements. Since the only variable is ℓ, the integral reduces to a sum of length elements.

$$B_{\text{axis}} = \int dB_{\text{axis}} = \frac{\mu_0 I a}{4\pi r^3} \int_0^{2\pi a} d\ell$$

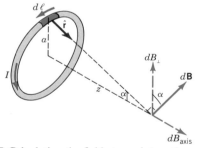

FIGURE 30.7 Calculating the field at a point on the axis of a current-carrying loop. By symmetry, the component of the field perpendicular to the axis is zero.

Since $\int d\ell = 2\pi a$ and $r^2 = a^2 + z^2$, we may express this result as

$$B_{\text{axis}} = \frac{\mu_0 I a^2}{2(a^2 + z^2)^{3/2}} \quad (30.8)$$

Note that this expression applies only to points on the *axis* of the loop. The overall field pattern is revealed by iron filings in Fig. 30.8.

At points far from the loop, that is, when $z \gg a$, Eq. 30.8 becomes

$$B_{\text{axis}} = \frac{2k'\mu}{z^3} \quad (30.9)$$

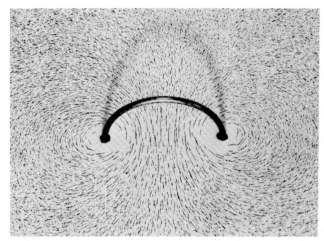

FIGURE 30.8 The pattern of iron filings for a current-carrying loop.

where $k' = \mu_0/4\pi$ and $\mu = I(\pi a^2)$ is the dipole moment of the loop. (Do not confuse the constant μ_0 with the dipole moment μ.) Equation 30.9 has the same form as Eq. 23.14 for the "far field" on the axis of an electric dipole:

$$E = \frac{2kp}{z^3}$$

This similarity in form of the far field expressions prompts us to refer to the current loop as a magnetic dipole. However, as Fig. 30.9 shows, the magnetic field pattern near the loop is quite different from that of the electric field due to a dipole. In fact, the fields are in opposite directions.

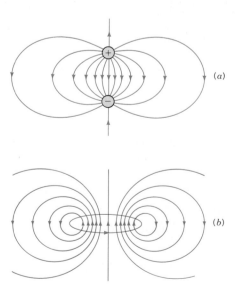

FIGURE 30.9 The field lines for (a) an electric dipole and (b) for a magnetic dipole. Although the far fields appear to be similar, the near fields are in opposite directions.

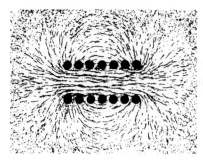

FIGURE 30.10 The pattern of iron filings for seven loops.

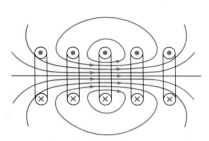

FIGURE 30.11 The field lines for five loops.

Figure 30.10 shows the pattern of iron filings for a coil with seven turns. Figure 30.11 shows the field lines for a coil with five turns. Notice that the field lines are always closed loops. Very close to each wire the lines are circular (not shown). Inside the coil the contributions from the different turns reinforce each other, so the field is strong. Near the axis it is fairly uniform. Outside the coil, the contributions from the various current elements tend to cancel, so the field is much weaker. The field outside the coil resembles that of a bar magnet. One end of the coil acts as a north pole, whereas the other end acts as a south pole. When the turns are tightly packed and their number becomes very large, the device is called a **solenoid.** The field within a long solenoid is quite uniform and strong, whereas it is essentially zero outside, as shown in Fig. 30.12. We will now calculate the field strength along the axis of a long solenoid.

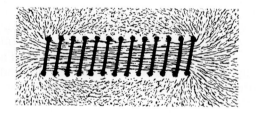

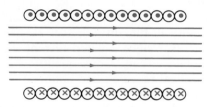

FIGURE 30.12 The field within a long solenoid is uniform. Outside, it is essentially zero.

EXAMPLE 30.4: A *solenoid* of length ℓ and radius a has N turns of wire and carries a current I. Find the field strength at a point along the axis.

Solution: Since the solenoid is a series of closely packed loops, we may use Eq. 30.8 for the field on the axis of a current-carrying loop. The solenoid is divided into current loops of width dz, as in Fig. 30.13a, each of which contains $n\,dz$ turns, where $n = N/\ell$ is the number of turns per unit length. The current within such a loop is $(n\,dz)I$. From the diagram we see that

$$z = a \tan \theta$$

from which

$$dz = a \sec^2 \theta\, d\theta$$

Thus, I in Eq. 30.8 is replaced by

$$nI\, dz = nIa \sec^2 \theta\, d\theta$$

After substituting for I and z in Eq. 30.8 and simplifying, we find

$$dB = \frac{\mu_0 n I a^3 \sec^2 \theta\, d\theta}{2(a^2 + a^2 \tan^2 \theta)^{3/2}}$$
$$= \tfrac{1}{2}\mu_0 nI \cos \theta\, d\theta$$

For a finite solenoid, the total field strength is

$$B = \tfrac{1}{2}\mu_0 nI \int_{\theta_1}^{\theta_2} \cos \theta\, d\theta$$
$$= \tfrac{1}{2}\mu_0 nI(\sin \theta_2 - \sin \theta_1) \qquad \text{(i)}$$

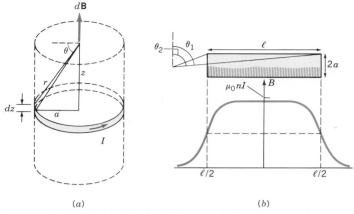

(a)   (b)

FIGURE 30.13 To calculate the field strength along the axis of a solenoid, the solenoid is divided into infinitesimal loops.

Figure 30.13*b* shows the variation in B along the axis of a solenoid for which $\ell = 10a$.

If the solenoid is infinite, $\theta_1 = -90°$ and $\theta_2 = 90°$, therefore, the field at the axis of the solenoid is

(Solenoid) $$B = \mu_0 nI \qquad (30.10)$$

We will show in the next section that this expression is also valid for points off the central axis; that is, the field is uniform.

EXERCISE 2. Show that the field strength at the end of a very long solenoid is close to $\mu_0 nI/2$.

30.4 AMPÈRE'S LAW

FIGURE 30.14 Andre Marie Ampère (1775–1836).

Ampère (Fig. 30.14) had several objections to the work of Biot and Savart. For example, he felt that their experiments were not precise enough for them to claim certainty in the $\sin \theta$ factor. He was uncomfortable with their use of "current elements," since isolated current elements do not exist; they are always part of a complete circuit. Consequently, he pursued his own line of experimental and theoretical research and obtained a different relation, now called Ampère's law, between a current and the magnetic field it produces.

Although Ampère's law can be derived from the Biot-Savart expression for $d\mathbf{B}$, we will not do so. Instead, we can make it plausible by considering the field due to an infinite straight wire. We know that the field lines are concentric circles. If Eq. 30.1 is written in the form

$$B(2\pi r) = \mu_0 I$$

we may interpret it as follows: $2\pi r$ is the length of a circular path around the wire, B is the component of the magnetic field *tangential* to the path, and I is the current through the area bounded by the path. Ampere generalized this result to paths and wires of any shape.

Figure 30.15 shows a current coming out of the page and an arbitrary closed path around it. For an infinitesimal displacement $d\ell$ along the path, the product of $d\ell$ and the component of $\mathbf{B}$ along $d\ell$ is $d\ell(B \cos \theta) = \mathbf{B} \cdot d\ell$. According to **Ampère's law** the sum (integral) of this product around a *closed* path is given by

$$\oint \mathbf{B} \cdot d\ell = \mu_0 I \qquad (30.11)$$

where I is the net current flowing through the surface enclosed by the path. The sense (clockwise or counterclockwise) in which the integral is to be evaluated is given by a right-hand rule: When the thumb of the right hand points along the

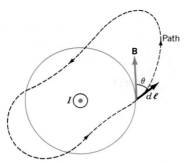

FIGURE 30.15 A current is coming out of the page. According to Ampère's law the integral $\oint \mathbf{B} \cdot d\boldsymbol{\ell}$ around any closed path that encloses the current is equal to $\mu_0 I$.

current, the curled fingers indicate the positive sense along the path. Equation 30.11 is valid only for *steady* currents and nonmagnetic materials, such as Cu. The enclosed current does not have to flow in a wire; a beam of charged particles also constitutes a current. The **B** that appears in Ampère's law is due to *all* currents in the vicinity, not just the current enclosed by the path.

In order to use Ampère's law to determine the magnetic field, it is necessary for the geometry of the current flow to possess sufficient symmetry so that the integral can be evaluated easily. One needs to know the field pattern and then to make a suitable choice for the path of integration.

EXAMPLE 30.5: An *infinite straight wire* of radius R carries a current I. Find the magnetic field at a distance r from the center of the wire for (a) $r > R$, and (b) $r < R$. Assume that the current is uniformly distributed across the cross section of the wire.

Solution: (a) From the symmetry of the situation we know that the field strength is the same at all points at a distance r from the center. We also know that the field lines are circular. Thus, we choose as the path of integration a circle of radius r whose center coincides with the center of the wire (Fig. 30.16a). At any point along the path, **B** is parallel to $d\boldsymbol{\ell}$, which means $\mathbf{B} \cdot d\boldsymbol{\ell} = B\, d\ell$. From Eq. 30.11,

$$\oint \mathbf{B} \cdot d\boldsymbol{\ell} = B \oint d\ell = \mu_0 I \qquad \text{(i)}$$

The magnitude of the field B is taken out of the integrand because it is a constant along the chosen path. The remaining integral is simply $2\pi r$ and the enclosed current is I. Thus, $B(2\pi r) = \mu_0 I$, and

$$B = \frac{\mu_0 I}{2\pi r} \qquad \text{(ii)}$$

This approach is much simpler than the integration required in Example 30.2

(b) The same symmetry considerations apply within the wire, so (i) is still applicable, provided the closed circular path is inside the wire. Only a fraction of the total current I flows through the path in Fig. 30.16b. This fraction is given by the ratio of the area enclosed by the path to that of the wire, that is $(\pi r^2/\pi R^2)I$. Equation (i) takes the form

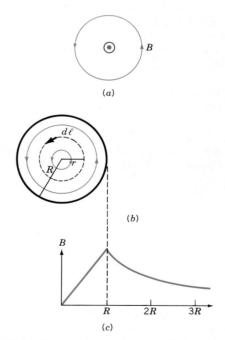

FIGURE 30.16 (a) The symmetry indicates that a suitable path of integration is a circular loop centered on the wire. (b) The path of integration is taken within the wire so only part of the total current is enclosed by the circular loop. (c) The variation of the field with distance from the center of the wire.

$$B(2\pi r) = \mu_0 \frac{r^2}{R^2} I$$

$$B = \frac{\mu_0 I r}{2\pi R^2} \qquad \text{(iii)}$$

Notice that at $r = R$, (ii) and (iii) yield the same result. Therefore, the magnitude of the magnetic field is continuous across the boundary of the wire. The field strength is plotted as a function of r in Fig. 30.16c.

EXAMPLE 30.6: An *ideal infinite solenoid* has n turns per unit length and carries a current I. Find its magnetic field.

Solution: The field outside an ideal infinite solenoid is zero (see Fig. 30.12). All points along an infinite solenoid are equivalent; we say that the solenoid has translational symmetry. The contribution to the total field inside the solenoid from each loop is directed along the axis, so we expect the field lines to be parallel to the axis. To take advantage of this geometry, we choose the rectangle *abcd* in Fig. 30.17 as the path of integration. The path integral consists of four segments:

$$\oint \mathbf{B} \cdot d\boldsymbol{\ell}$$

$$= \int_a^b \mathbf{B} \cdot d\boldsymbol{\ell} + \int_b^c \mathbf{B} \cdot d\boldsymbol{\ell} + \int_c^d \mathbf{B} \cdot d\boldsymbol{\ell} + \int_d^a \mathbf{B} \cdot d\boldsymbol{\ell}$$

Along *cd*, $\mathbf{B} = 0$ and so the third integral vanishes. $\mathbf{B}$ is also zero for those parts of *bc* and *da* outside the solenoid. Within the solenoid, $\mathbf{B}$ is perpendicular to $d\boldsymbol{\ell}$; thus, $\mathbf{B} \cdot d\boldsymbol{\ell} = 0$. For these two reasons the second and fourth integrals vanish. Finally along *ab*, $\mathbf{B}$ is constant (because of the translational symmetry) and parallel to $d\boldsymbol{\ell}$; so $\mathbf{B} \cdot d\boldsymbol{\ell} = B \, d\ell$. If the length of *ab* is L, the number of turns is nL and the current enclosed is nLI. Ampère's law now becomes

$$\oint \mathbf{B} \cdot d\boldsymbol{\ell} = B \int d\ell = \mu_0(nLI)$$

from which we find

$$B = \mu_0 n I$$

Once a suitable path is chosen for Ampere's law, the long integration of Example 30.4 is replaced by a one-line calculation!

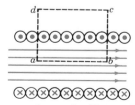

FIGURE 30.17 The appropriate path of integration for an infinite solenoid is a rectangular loop. Only the section within the solenoid contributes to the integral.

This result for B applies everywhere inside the (infinite) solenoid since it does not depend on the location of the segment *ab* in Fig. 30.17. Therefore, the magnetic field within the infinite solenoid is *uniform* throughout its cross section. This is *far* more than one can easily deduce from the Biot-Savart approach. Notice that the B within the path has contributions from current loops *not* enclosed by the path. Indeed, without these exterior contributions, we could not have claimed that $B = 0$ outside, or that the field lines inside the solenoid are parallel to the axis.

EXAMPLE 30.7: A *toroidal coil* (shaped like a doughnut) is tightly wound with N turns and carries a current I. We assume that it has a rectangular cross section, as shown in Fig. 30.18. Find the field strength within the toroid.

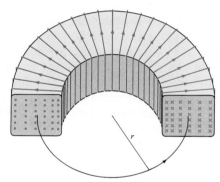

FIGURE 30.18 Within a toroid, the field is constant along a given circle of radius r.

Solution: The toroid is a rare example in which we know the shape of the field lines without approximation. The field lines are circular, and so we choose the path of integration to be a circle of radius r. If this path is outside the toroid, it will enclose no net current. From Ampère's law we have $\oint \mathbf{B} \cdot d\boldsymbol{\ell} = 0$. This result by itself does not allow us to conclude that $B = 0$. However, the circular symmetry tells us that B must be constant at all points of the circular path and parallel to $d\boldsymbol{\ell}$. Therefore, $\oint \mathbf{B} \cdot d\boldsymbol{\ell} = B \oint d\ell = B(2\pi r)$. Since $r \neq 0$, we conclude that $B = 0$ outside.

Within the toroid, $\mathbf{B}$ is parallel to $d\boldsymbol{\ell}$ and has the same magnitude at all points along the circular path. The current enclosed is NI, and so Ampère's law takes the form

$$\oint \mathbf{B} \cdot d\boldsymbol{\ell} = B \oint d\ell = \mu_0(NI)$$

Since $\oint d\ell = 2\pi r$, we find

$$B = \frac{\mu_0 N I}{2\pi r} \qquad (30.12)$$

The field is *not uniform*; it varies as $1/r$. Toroidal fields are used in research on fusion power (see p. 907).

EXAMPLE 30.8 (Optional): (a) What is the *magnetic field produced by a point charge q* moving at velocity **v**? (b) What is the force between two equal charges moving parallel to each other with the same velocity?

Solution: The Biot-Savart law refers the magnetic field produced by a current element, $I\,d\boldsymbol{\ell}$. Since $I = dq/dt$ we rewrite

$$I\,d\boldsymbol{\ell} = \frac{dq}{dt}\,d\boldsymbol{\ell} = dq\,\frac{d\boldsymbol{\ell}}{dt} = dq\,\mathbf{v}$$

where **v** is the velocity of the charge dq. From Eq. 30.6, we infer that the field due to a single charge q moving at velocity **v** is

$$\mathbf{B} = \frac{\mu_0}{4\pi} \cdot \frac{q\mathbf{v} \times \hat{\mathbf{r}}}{r^2} \qquad (30.13)$$

The magnetic field lines are circular, as shown in Fig. 30.19a. (b) In Fig. 30.19b, the repulsive electrostatic force between the charge is

$$F_E = \frac{kq^2}{d^2}$$

The magnitude of the magnetic force $\mathbf{F} = q\mathbf{v} \times \mathbf{B}$ exerted by one charge on the other is

$$F_B = qv\,\frac{k'qv}{d^2}$$
$$= \frac{k'q^2v^2}{d^2}$$

This force is attractive. Since $k' = \mu_0/4\pi$ and $k = 1/4\pi\varepsilon_0$ we find that $k' = k/c^2$, where c is the speed of light in vacuum.

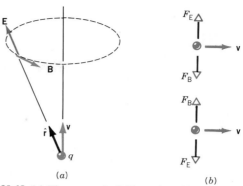

(a)   (b)

FIGURE 30.19 (a) The magnetic field produced by a charge q moving at velocity **v**. (b) Two charges moving side-by-side with the same velocity. The net force between them is less than when they are at rest (when there is no magnetic force).

(We will show in Chapter 34 that $c = 1/\sqrt{\mu_0\varepsilon_0}$.) Thus, the net force between the charges is

$$F = \left(1 - \frac{v^2}{c^2}\right)\frac{kq^2}{r^2} \qquad (30.14)$$

The net force on each of the particles moving with the same velocity is *less* than that when they are at rest. The significance of this result will be discussed in Chapter 39.

EXERCISE 3. Show that the electric and magnetic fields produced by a moving point charge are related according to $\mathbf{B} = (\mu_0\varepsilon_0)\mathbf{v} \times \mathbf{E}$.

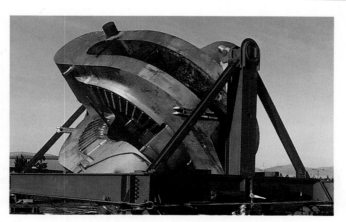

An unusual magnet at Lawrence Livermore Labs which is designed as part of a scheme to confine a hot plasma (ionized gas) in experiments for harnessing the energy released by the fusion of nuclei.

HISTORICAL NOTE: Electromagnets

In September 1820, Francois Arago placed an iron bar inside a current-carrying solenoid and found that the bar had become magnetized. This was the first electromagnet. In 1825 William Sturgeon used an iron bar in the form of a horseshoe. He varnished the bar (to provide an insulating layer) and wrapped several loops of bare wire around it, as in Fig. 30.20. Because the loops were far apart, this electromagnet could lift only a few ounces.

Joseph Henry, at Princeton University, greatly improved Sturgeon's design. Since insulated wire was not available, Henry used silk from his wife's petticoats to laboriously cover hundreds of meters of bare wire. The effort was worth it because the insulation allowed him to wrap many turns around a given iron core, as in Fig. 30.21a. His biggest electromagnet was capable of lifting 750 pounds. Henry offered considerable help and advice to Morse in the United States, and to Wheatstone in England, while they were setting up telegraph companies. The essential feature of the electric telegraph was dem-onstrated by Henry with the apparatus in Fig. 30.21b. When a current, from a distant source, is applied to the electromagnet, the suspended bar magnet rotates and strikes the bell. Nowadays, electromagnets are used, for example, in tape recording heads and to produce magnetic fields for research.

(a)

(b)

FIGURE 30.21 (a) An electromagnet made by Joseph Henry. (b) Henry's demonstration of the principle of the electric telegraph.

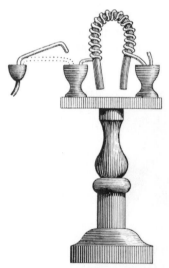

FIGURE 30.20 The first practical electromagnet made by William Sturgeon in 1825.

SUMMARY

The magnetic field at a distance R from an infinite (long) straight wire is

(Straight wire)
$$B = \frac{\mu_0 I}{2\pi R}$$

The magnetic force per unit length between two parallel wires carrying currents I_1 and I_2 and separated by a distance d is

$$\frac{F}{\ell} = \frac{\mu_0 I_1 I_2}{2\pi d}$$

The force is attractive if the currents are in the same direction.

The **Biot-Savart law** states that the magnetic field $d\mathbf{B}$ produced by a current element $I\,d\ell$ at a distance r from the element is given by

$$d\mathbf{B} = \frac{\mu_0}{4\pi} \frac{I\,d\ell \times \hat{\mathbf{r}}}{r^2}$$

The unit vector $\hat{\mathbf{r}}$ is directed from the current element to the point at which the field is being determined. When this equation is used to find the resultant field due to some current geometry, the components of the field should be referred to a coordinate system. The magnitude of the field is

$$dB = \frac{\mu_0}{4\pi} \frac{I\,d\ell\,\sin\theta}{r^2}$$

The magnetic field along the axis of an infinite (long) solenoid is

(Solenoid)
$$B = \mu_0 n I$$

where $n = N/\ell$ is the number of turns per unit length.

According to **Ampère's law,** the integral of the quantity $\mathbf{B} \cdot d\ell$ around a closed loop is related to the current flowing through the surface bounded by the loop:

$$\oint \mathbf{B} \cdot d\ell = \mu_0 I$$

Note that there are often contributions to $\mathbf{B}$ from currents *not* enclosed by the loop. This law may be used to find $\mathbf{B}$ if the geometry of the currents allows the choice of a path for which the integral may be easily evaluated.

ANSWERS TO IN-CHAPTER QUESTIONS

1. The field directions are drawn in Fig. 30.22. The magnitudes are $B_1 = 3.2 \times 10^{-7}$ T and $B_2 = 2.8 \times 10^{-7}$ T.

$$B_{Tx} = B_1 \cos 53° + B_2 = 4.72 \times 10^{-7}\text{ T}$$
$$B_{Ty} = -B_1 \sin 53° = -2.56 \times 10^{-7}\text{ T}$$
$$\mathbf{B} = (4.72\mathbf{i} - 2.56\mathbf{j}) \times 10^{-7}\text{ T}$$

2. In (i), set $\theta_1 = 90°$ and $\theta_2 = 0$.

3. Note that $\mathbf{E} = (kq/r^2)\hat{\mathbf{r}}$ where $k = 1/4\pi\varepsilon_0$.

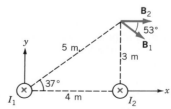

FIGURE 30.22 Finding the resultant field for two long, straight current-carrying wires.

QUESTIONS

1. Two long and straight current-carrying wires are perpendicular to each other. Describe the forces they exert on each other.

2. A weight suspended by a spring ensures that adjacent coils of the spring do not touch. What happens when a current passes through the spring?

3. A long, straight wire carries a current along the $+y$ axis. What is the direction of the force on a charge $+q$ located instantaneously on the $+x$ axis and whose velocity is: (a) away from the wire along the $+x$ axis; (b) parallel to the $+y$ axis; (c) parallel to the $-z$ axis?

4. Sketch the magnetic field lines for two long, straight, parallel wires in a plane perpendicular to the wires. The currents are in the same direction.

5. Sketch the magnetic field lines for two long, straight, parallel wires in a plane perpendicular to the wires. The currents are in opposite directions.

6. A metal tube carries a current along its length. What can you say about the magnetic field within the cavity of the tube, if the cross section is (a) circular, or (b) square?

7. From considerations of symmetry alone, what patterns are possible for the field lines due to an infinite straight wire?

8. From symmetry considerations alone, what patterns are possible for the field lines inside an infinite solenoid?

9. What are the dimensions of the quantity $1/\sqrt{\mu_0\varepsilon_0}$? What is its numerical value?

10. Three wires carry currents in the directions shown in Fig. 30.26. Write Ampère's law for a loop that encloses all three wires. Indicate the sense (clockwise or counterclockwise) in which the integration is to be performed.

11. A very long solenoid is wound with a single strand of wire. Is it possible for the magnetic field to be identically zero outside the solenoid? (*Hint:* The loops cannot be exactly perpendicular to the axis.)

EXERCISES

30.1 and 30.2 The Magnetic Field Due to a Long Straight Wire

1. (I) A long, straight wire and a rectangular loop lie in the same plane, as shown in Fig. 30.23. The dimensions and currents are indicated. Find the net force on the loop.

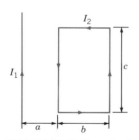

FIGURE 30.23 Exercise 1.

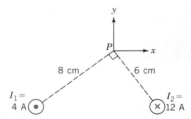

FIGURE 30.24 Exercise 2.

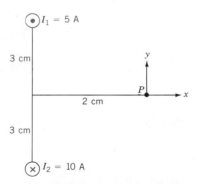

FIGURE 30.25 Exercise 3.

2. (I) Two long, straight wires carry currents with the magnitudes and directions shown in Fig. 30.24. (a) What is the total magnetic field at point P? (b) At what point is the total field strength zero? (c) What is the force per unit length between the wires?

3. (I) Two long, straight wires carry currents in the directions shown in Fig. 30.25. (a) Find the magnetic field strength at point P. (b) What force would 1 m of a third wire, carrying a 3-A current out of the page, experience if placed at P?

4. (I) A long, straight wire carries a 20-A current vertically upward. Where would its field cancel the earth's field—which is 0.5 G, horizontal, and points due north?

5. (I) A lightning bolt may carry a current of 5×10^3 A for 1 ms. Estimate its magnetic field strength at a perpendicular distance of 2 m.

6. (I) Three long, straight wires are at the corners of an equilateral triangle with sides of length $L = 6$ cm. They are parallel and carry currents as shown in Fig. 30.26. What is the force per unit length on the top wire?

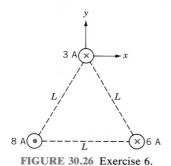

FIGURE 30.26 Exercise 6.

7. (I) A dc power line 20 m above the ground carries a 600-A current due north. If the horizontal component of the earth's field is 0.5 G due north, in what direction does a compass needle, on the ground directly below the line, point?

8. (I) A long, straight wire carries a 15-A current along the $+y$ axis. What is the force on an electron located instantaneously at $x = 6$ cm and moving with a speed of 10^6 m/s in the following directions: (a) away from the wire along the $+x$ axis; (b) parallel to the $+y$ axis; (c) parallel to the $+z$ axis?

30.3 The Biot-Savart Law

9. (I) A straight infinite current-carrying wire is bent into the form shown in Fig. 30.27. The curve is a semicircle of radius a. What is the field at the point P?

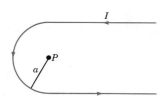

FIGURE 30.27 Exercise 9.

10. (I) Part of a long, flexible, current-carrying wire is made into a circular loop, while the rest of it lies in a straight line; see Fig. 30.28. What is the field strength at the center of the loop?

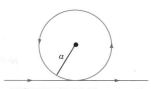

FIGURE 30.28 Exercise 10.

11. (I) A current loop consists of two concentric semicircles joined by radial sections, as shown in Fig. 30.29. What is the field strength at the center of the semicircles?

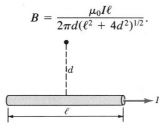

FIGURE 30.29 Exercise 11.

12. (II) A straight wire of length ℓ carries a current I (see Fig. 30.30). Show that the field strength along the perpendicular bisector at a distance d from its mid-point is

$$B = \frac{\mu_0 I \ell}{2\pi d(\ell^2 + 4d^2)^{1/2}}.$$

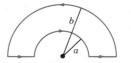

FIGURE 30.30 Exercise 12.

13. (II) A square coil of side ℓ carries a current I. Show that the field strength at its center is

$$B = \frac{2\sqrt{2}\mu_0 I}{\pi \ell}$$

14. (II) Consider the field due to a short segment of a wire that carries a 10-A current along the z axis. The segment has a length of 1 mm and is at the origin, as shown in Fig. 30.31. What is the field at points a, b, c, d, and e at the corners of a cube of side 2 cm?

15. (II) (a) Use the result of Example 30.3 to plot the field strength as a function of the distance from the center along the axis of a circular loop. Take $\mu_0 I/2a = 1G$. (b) At what point, in terms of a, is the field strength 50% of its value at the center?

16. (I) A tangent galvanometer consists of a large, circular coil with N tightly wound turns of radius R. Its plane is in the direction of the horizontal component of the earth's field

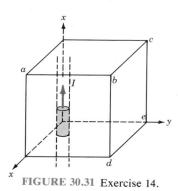

FIGURE 30.31 Exercise 14.

$\mathbf{B_e}$, as shown in Fig. 30.32. When a current flows in the coil, a small compass at its center rotates by some angle θ. Obtain an expression for the current in terms of θ.

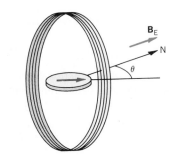

FIGURE 30.32 Exercise 16.

17. (I) Four long, parallel wires are located at the corners of a square of side 15 cm and carry currents as shown in Fig. 30.33. Find: (a) the resultant field at the center of the square; (b) the force on an electron moving at $4 \times 10^6 \mathbf{i}$ m/s as it passes the center.

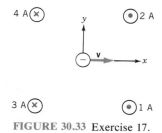

FIGURE 30.33 Exercise 17.

18. (I) Two long wires are parallel to the z axis and are located at $x = 0$, $y = \pm a$. (a) Find $B(x)$ at a point $(x, 0)$ in the xy plane, given that the wires carry equal currents in opposite directions. See Fig. 30.34. (b) At what point is B equal to 20% of its value at $x = 0$?

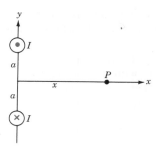

FIGURE 30.34 Exercise 18.

19. (II) (a) Repeat part (a) of Exercise 18 for currents in the same direction. (b) At what point is B a maximum?

20. (I) What current should flow in a circular loop of radius 4 cm so that the field at the center is equal in magnitude to that of the earth, 0.8 G?

21. (II) Two perpendicular straight wires join the ends of a semicircular loop of radius R, as shown in Fig. 30.35. If the current is I, what is the resultant field at the center of the circular section?

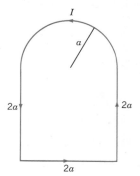

FIGURE 30.35 Exercise 21.

22. (II) The ends of a semicircular loop carrying a current I are connected to three wires that lie along the sides of a square (see Fig. 30.36). What is the field at the center of the circular section?

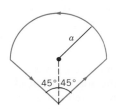

FIGURE 30.36 Exercise 22.

23. (I) Two semicircular loops of radii a and b have a common center and their ends are joined by straight wires, as shown in Fig. 30.37. (a) What is the resultant field at the center? (b) What is the magnetic moment of the loop? Take $a = 6$ cm, $b = 18$ cm, and $I = 4.5$ A.

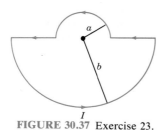

FIGURE 30.37 Exercise 23.

24. (II) Two infinite straight wires lie parallel to the x and z axes respectively, as in Fig. 30.38. One wire lies along the z axis while the other is located at $y = 10$ cm. Find the resultant magnetic field at the point P, $y = 5$ cm. The directions and values of the currents are indicated.

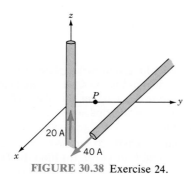

FIGURE 30.38 Exercise 24.

25. (I) How many turns of insulated copper wire are required to form a solenoid of length 25 cm and radius 2 cm if the current is 15 A and the field along the axis is 0.02 T. (Ignore end effects.)

26. (II) A length of copper wire of radius 1 mm is made into a closely wound coil with 60 turns of radius 10 cm. The ends of the wire, whose resistivity is 1.7×10^{-8} $\Omega \cdot$ m, are connected to a 1.5-V battery. What is the magnetic field at the center of the coil?

30.4 Ampère's Law

27. (II) Use Ampère's law to show that the field lines due to a magnet cannot end abruptly as depicted in Fig. 30.39.

28. (II) A metal tube has an inner radius a and outer radius b. It carries a current I uniformly spread over its cross-sectional area. Find the magnetic field at all points.

29. (II) An infinite metal plate of thickness t, as in Fig. 30.40, carries a uniform current density J. (a) Use the right-hand rule and symmetry to determine the direction of the field above and below the plate. (b) Determine the magnetic field at a distance a from the plate.

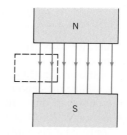

FIGURE 30.39 Exercise 27.

FIGURE 30.40 Exercise 29.

30. (I) A toroid with 240 turns has a square cross section (2 cm $\times$ 2 cm) with an inner radius of 3.6 cm. The current is 6 A. Find the magnetic field: (a) at the inner radius; (b) at the outer radius.

31. (I) A long, straight wire of radius 2 mm carries a 12-A current uniformly distributed throughout its cross section. At what points, both inside and outside the wire, is the field strength 25% the value at the surface of the wire?

32. (II) Use the Biot-Savart equation for the field due to an infinite wire to show that Ampère's law is valid for the loop shown in Fig. 30.41. The circular sections are connected by radial lines.

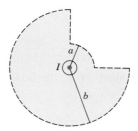

FIGURE 30.41 Exercise 32.

PROBLEMS

1. (I) A particle of mass m and charge q is in a circular orbit normal to an external field B. Show that the charge creates a magnetic field at the center of its orbit given by

$$\frac{\mu_0}{4\pi}\frac{q^2B}{mR}$$

2. (I) Helmholtz coils are two large circular coils with N tightly wound turns of radius R. The centers of the coils are separated by R, as shown Fig. 30.42. (a) Find the magnetic field along the line joining the centers as a function of x, the distance from the center of one coil. (b) Show that at the midpoint $B = (4/5)^{3/2}\mu_0 NI/R$. (c) Show that the field around the midpoint is approximately uniform. (*Hint:* Show that dB/dx and d^2B/dx^2 are both zero at $x = R/2$.)

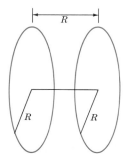

FIGURE 30.42 Problem 2.

3. (I) A nonconducting disk of radius R has a surface charge density σ C/m² and rotates about its central axis at ω rad/s. (a) What is the current dI in an elemental ring of width dr? (b) What is the magnetic field at the center of such a ring? (c) Show that the total magnetic field at the center of the disk is

$$B = \tfrac{1}{2}\mu_0 \sigma \omega R$$

4. (II) (a) Show that the magnetic field at a distance y from the center of a wire bent into a square of side ℓ carrying current I, as shown in Fig. 30.43, is

$$B = \frac{\mu_0 I\ell^2}{2\pi(y^2 + \ell^2/4)(y^2 + \ell^2/2)^{1/2}}$$

(b) Show that when $y \gg \ell$, the field is that of a dipole.

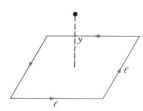

FIGURE 30.43 Problem 4.

5. (II) An infinitely long thin metal plate of width w carries a current I, as shown in Fig. 30.44. (a) By dividing the plate into infinitesimal strips and using the result for the field due to an infinite wire, show that the magnetic field at point P is

$$B_P = \frac{\mu_0 I}{\pi w}\tan^{-1}\left(\frac{w}{2D}\right)$$

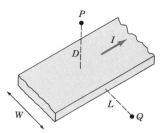

FIGURE 30.44 Problems 5 and 6.

6. (II) An infinitely long thin metal plate of width w carries a current I as shown in Fig. 30.44. Find the field at point Q at a distance L from the edge of the plate. (*Hint:* Divide the plate into infinitesimal strips and use the result for the field due to an infinite wire.)

7. (I) A solenoid of length 20 cm and radius 2 cm carries a current I. One end is at $x = 0$. (a) Find the field strength in terms of $\mu_0 nI$ along the axis at 1-cm intervals from $x = -5$ cm to 5 cm. (b) Plot your results and compare your plot with Fig. 30.13.

8. (a) Use the Biot-Savart law for an infinite wire to show that the field at any point inside a current-carrying tube is zero. (*Hint:* Show that the contributions due to elements intercepted by the pair of lines in Fig. 30.45 cancel.) (b) Use Ampere's law to obtain the same result.

FIGURE 30.45 Problem 8.

9. A long, straight, solid wire of radius R contains a cavity of radius r along its length. As shown in Fig. 30.46 the centers of the wire and of the cavity are a distance a apart. The current is uniformly distributed throughout the rest of the wire. (a) Show that the field in the cavity is uniform. (b) What is its value? (*Hint:* Use the superposition principle. Add the field due to a completely solid wire of radius R to that of a wire of radius a carrying a current in the opposite direction.)

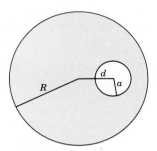

FIGURE 30.46 Problem 9.

10. A current I is uniformly distributed in one-half of a cylindrical tube of radius R, as shown in Fig. 30.47. What is the magnetic field at a point on the axis? Assume the cylinder is infinitely long.

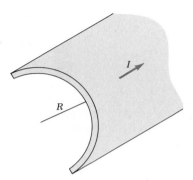

FIGURE 30.47 Problem 10.

SPECIAL TOPIC: The Earth's Magnetic Field

The use of lodestones as compasses dates from the 11th century. It was believed at that time that the compass pointed to the polestar under the influence of some extraterrestrial source. However, during a voyage in 1492, Columbus observed that his compass did not point to the polestar. His sailors became alarmed. They thought that this signified that they had entered a region where the laws of nature were different. Columbus assured them that the compass really orients toward a point beyond the polestar, which, he asserted, had moved slightly during the night! His reputation as an astronomer carried the day.

The beginning of a true understanding of the behavior of a compass came in 1544 when it was discovered that if an unmagnetized needle is initially balanced on a pivot, its north pole will dip below the horizontal after it is magnetized. In 1600, William Gilbert used a magnetized needle to map the region around a spherical lodestone. The similarity between this pattern of deflections and data concerning compass readings collected from various parts of the earth led him to suggest, correctly, that the earth itself is a giant magnet.

The magnetic field at the surface of the earth is essentially that of a magnetic dipole, as shown in Fig. 30.48. The field strength at the surface varies from 0.3 G to 0.6 G. The direction of the field at a given point on the surface is specified by the declination and the inclination. The *declination* is the angle between the horizontal component of the field and geographic north. The *inclination* is the angle of the field to the local horizontal. The points at which the inclination is ±90° are called the dip poles. Several points satisfy this condition.

The best fit to the observed field is obtained by placing a dipole of magnetic moment 8×10^{22} A · m² about 400 km from the center of the earth. The axis of the dipole is at an angle of 11.5° to the axis of rotation of the earth. The north and south magnetic poles lie on the axis of this "best-fit" dipole. The present north geographic pole is at 78.5° N 100° W, off Bathust Island in the Canadian Arctic. The locus of points for which the inclination is zero (the field is horizontal) is called the dip equator.

There are significant departures from a pure dipole field. On average, the strength of the nondipolar field is about 5% of the total field, although there can be much greater local anomalies. Ore deposits can contribute about 10^{-4} G. Figure 30.49 shows the nondipolar field, which is the difference between the actual field and the "best-fit" dipole field. The arrows show the horizontal component, whereas the contours indicate specific values of the vertical component in mG. Clearly any compass reading must be corrected with the help of such a chart. The total field can be fitted by combining the main dipole plus about eight radially oriented dipoles with various magnetic moments.

Variation in the Field

The earth's magnetic field is not constant in time. Its variation has components that cover a time scale from minutes to millions of years. The short-term changes are associated with disturbances caused by the "solar wind" (see below). Over the course of one day, the horizontal component of the field at a given point might vary as in Fig. 30.50. Such changes are caused by currents in the ionosphere and the magnetosphere (see below). "Magnetic storms," which last a couple of days, are caused by solar flares and lead to disruptions of radio communications.

Besides changing shape, the features of the nondipolar field tend to drift westward at about 0.2° per year, although some features drift eastward. For this reason, worldwide charts have to be made every few years. Satellite measurements now allow quicker compilation of data.

Using measurements made in London between 1580 and 1634, H. Gillibrand discovered that the declination at this location had gradually changed. Figure 30.51 shows the data over the past few centuries. Continuous measurements made between 1835 and 1955 showed that the magnetic dipole moment of the earth had decreased from about 8.5×10^{22} A · m² to 8×10^{22} A · m². If the current rate of decrease (0.05% per year) were to be maintained, the dipole field would vanish in about 2000 years. (The latest satellite measurements show an accelerated rate of 0.09%, which indicates the dipolar field would vanish in 1200 years.)

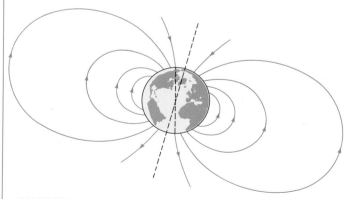

FIGURE 30.48 The magnetic field of the earth is essentially that of a dipole.

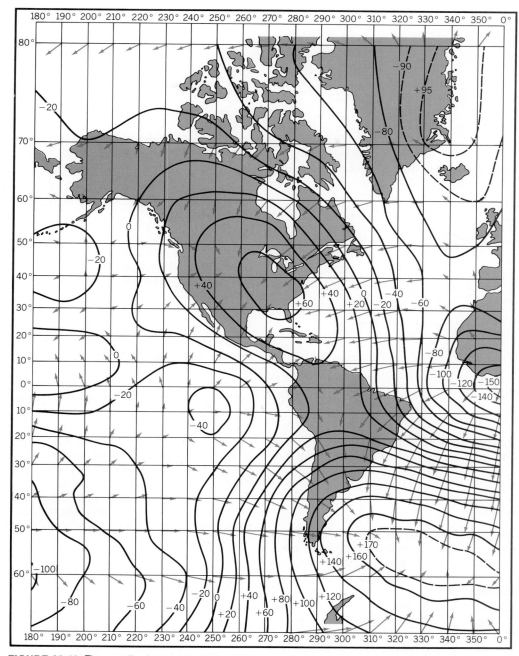

FIGURE 30.49 The nondipolar contribution to the earth's field. The arrows show the horizontal component, whereas the contours indicate specific values for the vertical component.

Archeomagnetism

Clay and rocks contain iron in the form of minerals such as magnetite. When they are heated and then cooled in the presence of an external field, they acquire a "thermal remnant magnetization" that can provide information on the history of the earth's field. Therefore, when pottery is baked or kilns are fired, they preserve a record of the field. If they have not been moved, firebricks provide information on both the magnitude and the direction of the field.

Archeological data covering a few thousand years, show that the north magnetic pole wanders up to 20° from

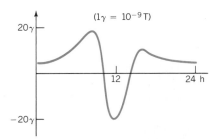

FIGURE 30.50 The magnetic field varies during each day.

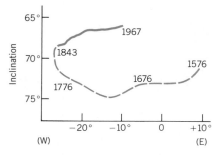

FIGURE 30.51 The variation in the declination and inclination at London over several centuries.

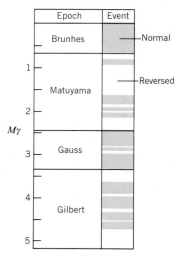

FIGURE 30.52 The reversals of the earth's magnetic field over a period of millions of years.

the geographical pole. However, the average position over a thousand years appears to coincide with the geographical pole.

Paleomagnetism

On a time scale of millions of years, volcanic lava flows, sedimentary rocks, and igneous rocks also preserve a record of the field. In addition to the thermal remnant magnetization mentioned above, sedimentary rocks can acquire magnetization in the following manner. As small grains (10 μm) settle in the presence of the field they are oriented along its lines. When they are later compacted, they preserve this orientation. Data collected from around the world, show that the main dipole field has reversed direction many times.

Figure 30.52 shows the pattern of reversals for the past 5 million years (My). Each *epoch*, lasting about 1 My, is characterized by a relatively stable direction punctuated by short-term (10^4 to 10^5 y) reversal *events*. The transition from one direction to the opposite takes about 5000 years. Instead of rotating continuously from one direction to the other, the main dipole field decreases to zero (probably leaving some nondipolar field) and then builds up in the

opposite direction. Measurements extended to 80 My show no preference for one or the other orientation—although there is a clear tendency to align with the earth's rotation axis.

Magnetic surveys of the seafloor have corroborated the evidence from rocks. There are relatively straight strips on the seafloor with magnetizations in opposite directions. The pattern, shown in Fig. 30.53, is symmetrical about a central line. When hot material emerges from the earth's interior, it cools and acquires a thermal remnant magnetization along the existing field. The alternating direction of the magnetization of the strips shows that the earth's field has reversed direction (the dates agree with those deduced from volcanic lava). This pattern incidently provided dramatic evidence that the seafloor is spreading at about 2.5 cm/year.

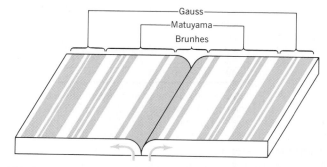

FIGURE 30.53 Sediments on the seabed are found to have a symmetrical pattern of strips with opposite magnetization. This finding confirmed that the earth's field has reversed direction and that the sea floor is spreading.

The Source of the Earth's Field

It is generally agreed that the earth's magnetic field is produced by currents in the outer, liquid part of the core (see Fig. 30.54). This region extends from 1000 km to 3000 km from the center. The earth cannot be an ordinary permanent magnet since the temperature of the core is high enough to destroy any "natural" magnetism such as that found in lodestones. The fact that the magnetic poles have a strong tendency to align with the axis of rotation, indicates that the earth's rotation is involved in the generation process. The existence of the nondipolar field shows, however, that the fluid motions are complex. In addition to the rotational motion, there are radial convection currents caused by the temperature difference between the hot inner core and the cooler mantle above the liquid. The exact mechanism by which the currents are generated is not clear. The field within the liquid core is strong (500 G) and has a complex shape. The predominantly dipolar field observed at the surface is only the small part that "leaks" through the mantle. A small instability, for example, associated with the rotation, or with an interaction at the core–mantle interface, might trigger a reversal of the field. Meteor showers and volcanic activity have also been proposed as triggers for the reversals.

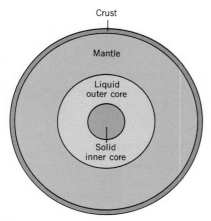

FIGURE 30.54 The earth's field is produced by currents in the fluid outer core. The temperature is higher than that at which permanent magnetism (as in a bar magnet) can exist.

The Van Allen Radiation Belts

When the first American satellite, *Explorer I,* was launched in January 1958, it detected an unexpectedly high flux of particles. Data from subsequent satellites led to the discovery of two belts of charged particles surrounding the earth, as shown in Fig. 30.55. These Van Allen belts, named after their discoverer, consist mainly of electrons and protons trapped by the earth's field. The doughnut-shaped regions

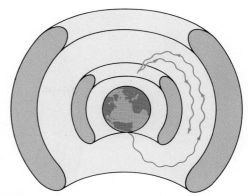

FIGURE 30.55 The Van Allen radiation belts contain charged particles trapped in the earth's field.

are not as sharply defined as the figure implies; actually the plasma extends to 40,000 km from the earth. The inner belt consists mainly of protons whose number is constant, whereas the outer belt consists of electrons whose number varies.

As was pointed out in Section 29.5, a charged particle moving in a nonuniform field will trace an ever-tightening spiral and at some point will reverse its direction of travel. The particles in the belts travel back and forth between the mirror points in about 0.25 s to 1 s. Some of the trapped particles "leak" out near the poles and produce auroral displays. The leakage of these high-energy particles may also explain the high temperatures found in the upper atmosphere. The source of the particles is not clear. The inner belt may arise from the radioactive decay of neutrons that have been produced in the upper atmosphere by cosmic rays. The outer belt may be replenished by the solar wind (see below).

The Magnetosphere

As satellites probed further into space, it was discovered that the earth's dipolar field is severely distorted. A constant stream of protons and electrons emitted by the sun bombards the earth. This *solar wind* is a hot ($T \approx 5 \times 10^5$ K), neutral plasma, with a density of about 5 particles/cm^3. The interaction between the solar wind and the earth's field compresses the field on the daylight side. The mechanism is as follows. When the particles encounter the earth's field, they are deflected to one side or the other, as depicted in Fig. 30.56. The magnetic fields created by the particles are opposite to the earth's field on one side of the path and in the same direction on the other side. The net effect is to confine the earth's field to the *magnetosphere*. The boundary between the solar wind and the field is called the *magnetopause*. The tail of the magnetosphere extends to

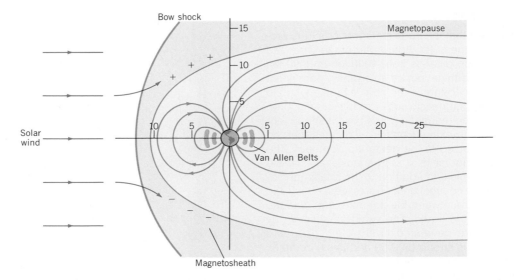

FIGURE 30.56 The earth's field seen on a large scale is far from that of a dipole. The distortion is created by the "solar wind," which is a stream of charged particles emanating from the sun.

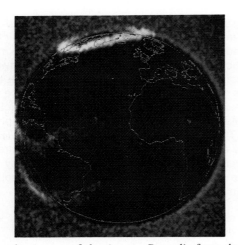

A false-color image of the Aurora Borealis from data obtained by a satellite about 3 earth radii from the North Pole. The wavelengths recorded were the 130.4 nm and 135.6 nm lines emitted by oxygen. The auroral displays result from the interaction of electrons in the solar wind with atoms in the ionosphere. The motion of the electrons is governed in a complex way by the earth's magnetic field. (See "The Dynamic Aurora," S.I. Akasofu, Scientific American, May 1989.)

over 2 million km on the night side. Notice that the field lines are parallel and oppositely directed on either side of the plane of symmetry.

The speed of the wind (400 km/s) relative to the earth is greater than that of sound waves that can propagate through it. Thus, the wind is supersonic and a "bow shock" front, about 10 km wide, forms when it encounters the earth's field. Between the magnetopause and the bow shock is the *magnetosheath* (about four earth radii thick) in which the field strength is about 25×10^{-9} T.

Evolution and Field Reversals

After the existence of the reversals of the earth's field was confirmed, it was suggested that they coincide with times of major evolutionary change. When the dipolar field vanishes, cosmic ray particles would not be deflected or trapped by the field. They could then strike the earth and cause drastic mutations in all living things. There is some statistical evidence that faunal extinctions do coincide with the reversals, but there is still controversy on this issue. For example, cosmic rays interact strongly with the atmosphere, which means any effects at the surface are, at best, associated with secondary particles produced by decay or collisions. There may also be some correlation between the earth's temperature and changes in the field strength over the past 10,000 years. If such correlations extend to tens of millions of years, they may play a role in the correlations between evolution and the field.

CHAPTER 31

Electromagnetic Induction

The lighting of the Las Vegas strip depends on the phenomenon of electromagnetic induction in the generators at nearby Hoover Dam.

Major Points

1. The production of **induced emf's.**
2. (a) **Faraday's law** relates the induced emf in a closed loop to the rate of change of magnetic flux through the loop.
   (b) **Lenz's law** is used to determine the direction of the induced emf.
3. The operation of ac and dc **generators.**
4. (a) **Induced electric fields** associated with time-varying magnetic fields.
   (b) **Motional emf's** produced by motion of a circuit relative to a magnetic field.

Oersted's discovery in 1820 of the magnetic effect produced by an electric current revealed a link between electricity and magnetism. Within weeks it was found that an iron bar was magnetized when placed inside a current-carrying solenoid. This demonstration that an electric current could produce magnetism encouraged many scientists to look for the inverse effect—an electric current produced by magnetism. As early as 1821 Michael Faraday made a note that he should try to "convert magnetism to electricity."

While on a short vacation in August 1830, the American physicist Joseph Henry placed a bar across the poles of an electromagnet and wrapped an insulated coil of wire around the bar (Fig. 31.1). The leads of the coil were connected to a galvanometer. When the current in the electromagnet was turned on, he observed a momentary deflection on the galvanometer—even though there was no electrical connection between the coil and the wires of the electromagnet. He had discovered that a current is induced (generated) in the coil when the magnetic field through it changes. Although Henry realized that he had "converted magnetism into electricity," his teaching duties prevented him from doing further work in this area and so the results were not immediately published. A year later, Michael Faraday (Fig. 31.2) independently made the same discovery—through essentially the same experiment.

The term **electromagnetic induction** encompasses two phenomena. The first involves a current that is induced in a conductor moving relative to magnetic field lines. This effect can be deduced from what we already know about the magnetic force on moving charges. The second, as in Henry's experiment, involves the generation of an electric field associated with a time-dependent magnetic field.

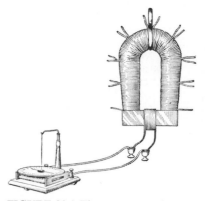

FIGURE 31.1 The apparatus with which Joseph Henry "converted magnetism to electricity."

FIGURE 31.2 Michael Faraday.

The induced electric field can produce an induced current in a conductor. Electromagnetic induction governs the operation of generators and transformers, and, as we will see in Chapter 34, it is vital to the propagation of electromagnetic waves, such as light, radio and TV signals, and X-rays.

31.1   ELECTROMAGNETIC INDUCTION

The essential features of electromagnetic induction can be demonstrated by some simple experiments.

(i) Change in Field Strength

Figure 31.3 shows a simple way to generate an electric current with a magnet and a loop of wire. When the magnet and the loop are stationary, nothing happens. When the north pole moves toward the loop (Fig. 31.3*a*), a current flows in the counterclockwise sense—as seen from the side on which the magnet is located. When the north pole moves away (Fig. 31.3*b*), there is a clockwise current. If the north pole and south pole are interchanged (Fig. 31.3*c*), the currents are reversed. These results are unchanged when the magnet is kept at rest and the loop is moved instead. The magnitude and direction of the induced current depend on the *relative* velocity of the coil and magnet.

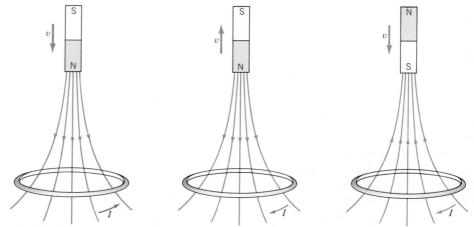

FIGURE 31.3 When a bar magnet moves relative to a loop of wire, there is an induced current in the loop.

Now consider two coils at rest, as shown in Fig. 31.4*a*. The "primary" coil is connected in series with a battery and a switch, whereas the "secondary" coil is connected to a galvanometer. When the switch in the primary circuit is closed, the meter in the secondary deflects *for an instant*. As long as the primary current stays constant, nothing happens. When the switch is opened, the meter again has a momentary deflection, but now in the opposite sense. This is essentially the experiment performed by Henry, and later by Faraday, who wrapped both the primary and the secondary coils around an iron ring, as shown in Fig. 31.4*b*.

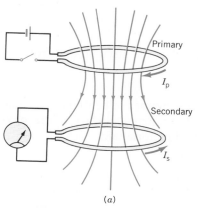

(a)   (b)

FIGURE 31.4 (a) When the current in the primary loop changes, an induced current appears in the secondary loop. (b) Faraday wrapped the primary and secondary coils on a circular iron ring to improve the magnetic coupling between the coils.

(ii) Change in Area

In Fig. 31.5 a circular coil made with flexible wire lies with its plane perpendicular to a uniform field which is constant in time. When opposite ends of a diameter are suddenly pulled apart, thereby reducing the area enclosed by the loop, an induced current is produced.

(iii) Change in Orientation

Now suppose that the field strength and the area of the coil are kept constant. When the plane of the coil is rotated relative to the direction of the field, as in Fig. 31.6, an induced current is produced as long as the rotation lasts.

31.2   MAGNETIC FLUX

In order to explain the above results we introduce the concept of **magnetic flux,** Φ_B, which is defined in the same way as electric flux, Φ_E (see Section 24.1). For a uniform magnetic field and a flat surface, as shown in Fig. 31.7, the magnetic flux intercepting the area is defined as

(Uniform **B**) $\Phi_B = BA \cos \theta = \mathbf{B} \cdot \mathbf{A}$ (31.1)

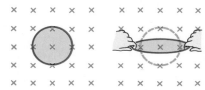

FIGURE 31.5 The plane of a coil is placed perpendicular to field lines. There is an induced current when the area of the coil is changed.

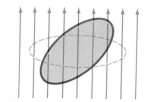

FIGURE 31.6 An induced current appears when a loop rotates in an external field.

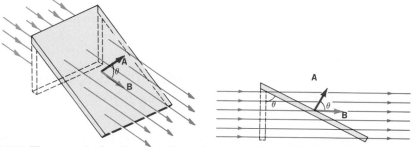

FIGURE 31.7 The magnetic flux through a flat surface in a uniform field depends on the projection of the area perpendicular to the field lines.

The SI unit of magnetic flux is the weber (Wb). From Eq. 31.1 we see that

$$1 \text{ T} = 1 \text{ Wb/m}^2$$

If the field is nonuniform or the surface is not flat, the flux is found from

Magnetic flux

$$\Phi_B = \int \mathbf{B} \cdot d\mathbf{A} \qquad (31.2)$$

As a matter of convenience we may choose the number of field lines to equal the magnetic flux. Keep in mind, however, that the magnetic flux is the physical quantity of interest; the field lines are merely a pictorial aid.

EXAMPLE 31.1: A square coil of side 20 cm is pivoted about the y axis. It is oriented as shown in Fig. 31.8a. The external field is $\mathbf{B} = 0.5\mathbf{i}$ T. What is the change in flux if the angle changes from 37° to 53°?

Solution: In this sort of problem it is often useful to get another perspective of the diagram, for example, a top view, as in Fig. 31.8b. The flux is

$$\Phi = BA \cos \alpha$$

where $\alpha = 90 - \theta$. As θ increases (α decreases) the flux intercepting the loop increases. The change in flux is

$$\Delta \Phi = BA(\cos \alpha_f - \cos \alpha_i) = 4 \times 10^{-3} \text{ Wb}$$

EXERCISE 1. What do the three types of change described in Section 31.1 have in common?

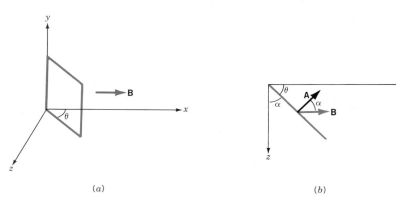

(a)   (b)

FIGURE 31.8 The flux through the coil changes when its orientation changes.

31.3 FARADAY'S LAW AND LENZ'S LAW

The generation of an electric current in a circuit implies the existence of an emf. Faraday stated that the induced emf in a circuit is proportional to the rate at which magnetic field lines cross the boundary of the circuit. Faraday's statement is nowadays expressed in terms of magnetic flux:

$$\mathscr{E} \propto \frac{d\Phi}{dt} \qquad (31.3)$$

> The induced emf along any closed path is proportional to the rate of change of magnetic flux through the area bounded by the path.

Note that the induced emf is not confined to a particular point; it is distributed around the loop. The derivative of Eq. 31.1 is

$$\frac{d\Phi}{dt} = \frac{dB}{dt} A \cos \theta + B \frac{dA}{dt} \cos \theta - BA \sin \theta \frac{d\theta}{dt}$$

The three terms represent the contributions from the rate of change of B, A, and θ, respectively, to the rate of change of flux. Each term is applicable to one of the experiments described in Section 31.1. The first term contributes when the field is explicitly time dependent, the second involves a change in the area of the path, and the third involves a change in orientation of the boundary of the path. In a given situation, more than one term may contribute.

EXERCISE 2. What is the physical significance of the negative sign in the last term of $d\Phi/dt$?

Lenz's Law

The Russian physicist H. F. Lenz was not satisfied with the way Faraday described the direction of the induced current in experiments that involved relative motion. In 1834, he proposed a simple rule to cover such cases. In Fig. 31.9a, as the north pole of the magnet approaches the loop, the side of the loop facing the magnet behaves like a north pole and repels the magnet. When the north pole moves away, as in Fig. 31.9b, the induced current flows in the opposite sense and the magnet is attracted. Lenz noted that in either case, the magnetic force exerted by the induced current opposes the relative motion. About 30 years later, J. C. Maxwell stated Lenz's rule in a more general way:

> The effect of the induced emf is such as to oppose the change in flux that produces it.

In Fig. 31.9a, as the magnet approaches, the (positive) flux through the coil increases. The induced current sets up an induced magnetic field, $\mathbf{B}_{ind}$, whose (negative) flux opposes this change. The direction of $\mathbf{B}_{ind}$ is opposite to that of external field, $\mathbf{B}_{ext}$, due to the magnet. In Fig. 31.9b, the induced current sets up an induced field whose (positive) flux opposes the decrease in flux of $\mathbf{B}_{ext}$. In this case, the induced magnetic field points in the *same* direction as the external field.

In 1851, von Helmholtz pointed out that Lenz's law is simply a consequence of the conservation of energy. Let us consider Fig. 31.9a. If the induced magnetic field were to reinforce the external field, this additional field would cause an increase in the induced current. The larger current would create a larger induced field, which in turn would lead to a larger induced current, and so on. Clearly, this runaway situation is not energetically possible. An external agent must supply the energy to create the induced emf.

In order to incorporate Lenz's law into Eq. 31.3, we need a sign convention for the induced emf. First we choose the direction of the vector area to make the initial flux positive. The right-hand rule, in this case with the thumb along $\mathbf{B}$ and the fingers curled around the loop, tells us whether clockwise or counterclockwise is the positive sense, as shown in Fig. 31.10. Figure 31.11 shows that the sign of

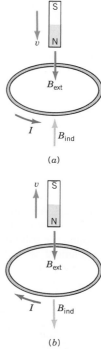

(a)

(b)

FIGURE 31.9 (a) When the flux through the loop increases, the flux due to the induced magnetic field opposes this increase. (b) When the flux through the loop decreases, the flux due to the induced magnetic field tries to maintain the flux through the loop.

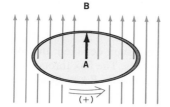

FIGURE 31.10 The vector **A** and the positive sense is determined by a right-hand rule with the thumb along the external field.

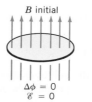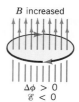

FIGURE 31.11 The sign of the induced emf is always opposite to that of the change in flux.

the emf is always opposite to the sign of the change in flux $\Delta\Phi$. This feature can be incorporated into Faraday's law by including a negative sign. The proportionality in Eq. 31.3 can now be turned into an equation with an appropriate constant of proportionality that depends on the system of units. In SI units, the constant is unity. Therefore, the modern statement of **Faraday's law** of electromagnetic induction is

Faraday's law

$$\mathcal{E} = -\frac{d\Phi}{dt} \qquad (31.4)$$

Suppose that the loop is replaced by a coil with N turns. If the flux through each turn is the same, each turn has the same induced emf. (This is strictly true only for a tightly wound toroid or infinite solenoid.) Since all these emf's are in the same sense, they are in series. Hence, the net emf induced in a coil with N turns is

$$\mathcal{E} = -N\frac{d\Phi}{dt} \qquad (31.5)$$

where Φ is the flux through *each* turn.

EXERCISE 3. For the situation depicted in Fig. 31.9, external work is needed to move the magnet against the attraction or repulsion of the loop. What happens to this work?

EXERCISE 4. For the situation discussed in Example 31.1, what is the direction of the induced current in the length of the loop that lies along the y axis?

EXAMPLE 31.2: An infinite solenoid has 10 turns/cm and a radius of 2 cm. A flat circular coil, of radius 4 cm and 15 turns, is placed around the solenoid with its plane perpendicular to the axis of the solenoid, as in Fig. 31.12. If the current in the solenoid drops steadily from 3 A to 2 A in 0.05 s, what is the emf induced in the coil?

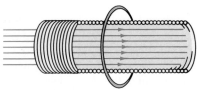

FIGURE 31.12 A loop outside a long solenoid through which the current is changing. There is an induced emf in the loop even though the magnetic field lines are confined within the solenoid.

Solution: From Eq. 30.11, the field within a long solenoid is $B = \mu_0 nI$. The flux through the coil is

$$\Phi = BA = \mu_0 nIA$$

where A is the cross-sectional area of the solenoid, not of the coil, because the field is confined to the solenoid. The induced

emf is

$$\mathcal{E} = -N\frac{\Delta\Phi}{\Delta t}$$

$$= -N\mu_0 nA\frac{\Delta I}{\Delta t}$$

Using the given values $n = 1000$ turns/m, $A = \pi(2 \times 10^{-2})^2$ m^2 and $\Delta I/\Delta t = -20$ A/s, we find $\mathcal{E} = +4.7 \times 10^{-4}$ V. Since the flux is dropping, the induced magnetic field is in the same direction as that of the solenoid. This example is noteworthy because the magnetic field strength outside the solenoid is zero. We will return to this point later.

EXAMPLE 31.3: A metal rod of length ℓ slides at constant velocity v on conducting rails that terminate in a resistor R. There is a uniform and constant magnetic field perpendicular to the plane of the rails, as shown in Fig. 31.13. Find: (a) the current in the resistor; (b) the power dissipated in the resistor; (c) the mechanical power needed to pull the rod.

Solution: (a) At the instant depicted, the rod is at a distance x from the end of the rails. The flux through the area enclosed by the rod and the rails is $\Phi = BA = B\ell x$. The magnitude of the induced emf is

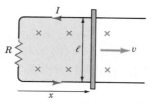

FIGURE 31.13 When a bar moves across conducting rails there is an induced current in the direction shown.

$$|\mathscr{E}| = \frac{d\Phi}{dt} = B\ell v$$

since $v = dx/dt$. The flux is increasing because the area is increasing. The induced emf opposes the increase in flux, which means that the induced magnetic field is opposite to the external field. Thus, the induced current in the circuit is counterclockwise. The magnitude of the current is

$$I = \frac{|\mathscr{E}|}{R} = \frac{B\ell v}{R} \qquad \text{(i)}$$

(b) The electrical power dissipated in the resistor is

$$P_{\text{elec}} = I^2 R = \frac{(B\ell v)^2}{R} \qquad \text{(ii)}$$

(c) Because of the induced current flowing in it, the rod experiences a force $\mathbf{F} = I\boldsymbol{\ell} \times \mathbf{B}$ due to the external field. The force $\mathbf{F}$ is directed opposite to $\mathbf{v}$. Thus, in order to keep the velocity constant, there must be an external agent that applies an equal and opposite force $F_{\text{ext}} = I\ell B$ toward the right. The mechanical power supplied by the external agent is

$$P_{\text{mech}} = \mathbf{F}_{\text{ext}} \cdot \mathbf{v} = \frac{(B\ell v)^2}{R} \qquad \text{(iii)}$$

Comparison of (ii) and (iii) shows that all the mechanical energy supplied by the external agent is converted to electrical energy and then to thermal energy.

EXAMPLE 31.4: A metal bar is moving at 2 cm/s over a U-shaped metal rail, as in Fig. 31.14. At $t = 0$, the external field is 0.2 T out of the page and is increasing at the rate of 0.1 T/s. Take $\ell = 5$ cm, and at $t = 0$, $x = 5$ cm. Find the induced emf.

Solution: In this example both the field strength and the area change. The flux is $\Phi = BA = B\ell x$ and $dx/dt = -v$. Therefore,

$$\frac{d\Phi}{dt} = B\frac{dA}{dt} + \frac{dB}{dt}A$$

$$= B\ell\frac{dx}{dt} + \frac{dB}{dt}A$$

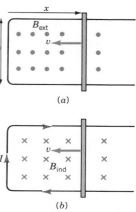

(a)

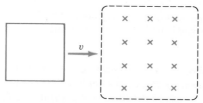

(b)

FIGURE 31.14 (a) The induced current is produced both by the motion of the bar and by the changing external magnetic field. (b) Since the flux through the loop is increasing, the induced magnetic field is directed opposite to the external field.

$$= B\ell(-v) + \frac{dB}{dt}A$$

$$= (0.2\text{ T})(5 \times 10^{-2}\text{ m})(-2 \times 10^{-2}\text{ m/s})$$
$$\quad + (0.1\text{ T/s})(25 \times 10^{-4}\text{ m}^2)$$

$$= +5 \times 10^{-5}\text{ V}$$

The net rate of change of flux is positive and the induced emf must oppose this increase. Thus the induced current is *clockwise* and the induced magnetic field is directed *into* the page.

EXERCISE 5. A rectangular loop moves at constant velocity perpendicular to a uniform magnetic field as shown in Fig. 31.15. Starting at the time at which it enters the field until it leaves the field, make plots of the variation in the flux through the coil and the emf induced in it as functions of time.

FIGURE 31.15 A rectangular loop moves at constant velocity perpendicular to a uniform field.

31.4   GENERATORS

An important application of electromagnetic induction is found in the generator. It consists of a coil with N turns rotating at angular velocity ω in a uniform external magnetic field. Figure 31.16 shows two views of just one turn. If we assume $\theta = 0$ at $t = 0$, then $\theta = \omega t$ and the flux is

$$\Phi = BA \cos(\omega t)$$

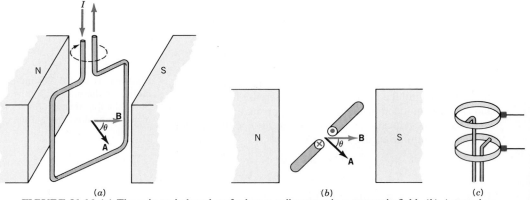

(a)   (b)   (c)

FIGURE 31.16 (a) There is an induced emf when a coil rotates in a magnetic field. (b) A top view of the rotating coil. (c) The current generated is fed to an external circuit via brush contacts that slide on two slip rings.

The induced emf is

$$\mathscr{E} = -N\frac{d\Phi}{dt} = NAB\omega \sin(\omega t)$$

which may be written as

$$\mathscr{E} = \mathscr{E}_0 \sin(\omega t) \tag{31.6}$$

As the coil rotates, the emf varies in a sinusoidal fashion; it *alternates* in sign, with an amplitude, or peak value,

$$\mathscr{E}_0 = NAB\omega \tag{31.7}$$

as shown in Fig. 31.17. Note that the peak emf occurs at the instant when the flux through the coil is zero. The alternating output of the coil is fed to two slip rings, as shown in Fig. 31.16c. When a circuit is connected to the generator, there is an alternating current (ac) that reverses direction periodically.

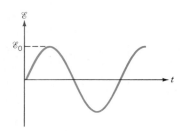

FIGURE 31.17 The sinusoidal alternating output emf of a coil rotating in a uniform magnetic field.

EXAMPLE 31.5: A square coil has 25 turns and sides of length 50 cm. It rotates at 120 rpm in a field of 400 G. At $t = 0$ the plane of the coil is normal to the lines. Find: (a) the peak value of the emf; (b) the emf at $\frac{1}{24}$ s.

Solution: First we must convert the angular frequency to rad/s and the field to teslas. Thus, 120 rev/min is equivalent to $\omega = 4\pi$ rad/s and $B = 4 \times 10^{-2}$ T.

(a) From Eq. 31.7 we have

$$\mathscr{E}_0 = NAB\omega$$
$$= (25)(0.5 \text{ m})^2(4 \times 10^{-2} \text{ T})(4\pi \text{ rad/s}) = 3.14 \text{ V}$$

(b) For the emf at a given instant we use Eq. 31.6:

$$\mathscr{E} = \mathscr{E}_0 \sin(\omega t)$$
$$= (3.14 \text{ V}) \sin(4\pi/24) = 1.57 \text{ V}$$

The first generators produced alternating current (ac) which was unsuitable for many types of experiment, or for running direct current (dc) motors. In 1834, William Sturgeon invented a simple device called a **commutator** that prevents the direction of the current from changing. It consists of two half-rings (attached to the coil) that make contact with metal brushes connected to the lead-in wires (Fig. 31.18a). When the current in the coil is zero and about to change direction, each brush switches from one ring to the other. As a result, the current in the external circuit does not change direction—although it is far from being constant in magnitude, as Fig. 31.18b shows. In 1841, Charles Wheatstone used multiple coils wrapped around a cylindrical form and a multielement commutator. The output of each coil was "tapped" only as it reached its peak emf. The fluctuations in the output were significantly reduced, as depicted in Fig. 31.18c.

The dc motor and the dc generator evolved separately into similar designs (a coil of many turns rotating in a magnetic field). Yet the fact that a generator could be run as a dc motor, and vice versa, escaped most engineers. At the Vienna exhibition of 1873, two giant generators were on display side by side. One was running (driven by a steam engine), while the other was stationary. A careless workman connected the output of the running generator to the terminals of the other one, which then started to rotate. It then became obvious that dc motors could be operated with the output from generators instead of large batteries. It is astonishing that this should have come as a surprise to engineers who had developed quite sophisticated machines.

The Back Emf of Motors

In Chapter 29 we saw that when a current is passed through a coil pivoted in a magnetic field it is subject to a torque and therefore rotates. This is the principle of the electric motor. As the coil rotates in the magnetic field there is an induced emf, similar to that of a generator, that opposes the external emf. The **back emf,** as it is called, is proportional to the angular speed ω of the motor. When the motor is first turned on, the coil is at rest, and so there is no back emf. The "start-up" current can be quite large because it is limited only by the resistance of the coil. As the rotation rate increases, the rise in the back emf reduces the current—which depends on the net emf. If there is no load attached to the motor, the angular speed increases until the input energy just balances the frictional and resistive losses. At this stage the current is quite small.

When a load is applied (that is, the motor is made to do mechanical work), the angular speed decreases, which in turn reduces the back emf. As a result the current increases. The additional power supplied by the external source of emf is converted to mechanical power by the motor. If the load is too large, the back emf is reduced still further. The larger current that results may cause the motor to "burn out."

EXERCISE 6. The lights in a house may temporarily dim when the refrigerator turns on. Why does this occur?

31.5   THE ORIGINS OF THE INDUCED EMF

So far, we have referred to the "induced emf" without concern for the mechanism by which such an emf is generated. In Chapter 28, emf was defined as the work done per unit charge by a source of emf as the charge moves around a closed loop:

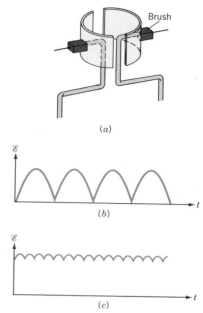

FIGURE 31.18 (a) A spit-ring commutator ensures that the sign of the emf at the brush contacts does not change. (b) The emf supplied by a single coil with a commutator. (c) When many coils are used, the output fluctuations are greatly reduced.

Back emf

Assembling a new turbine generator at the Grand Coulee Dam.

$$\mathscr{E} = \frac{W_{ne}}{q} = \frac{1}{q} \oint \mathbf{F} \cdot d\boldsymbol{\ell} \qquad (31.8)$$

where the subscript "ne" emphasizes that the work is done by a nonelectrostatic force. In the presence of both electric and magnetic fields, the total force on a charged particle is given by the Lorentz force, Eq. 29.12,

$$\mathbf{F} = q(\mathbf{E} + \mathbf{v} \times \mathbf{B}) \qquad (31.9)$$

Therefore, the induced emf may be written as

$$\mathscr{E} = \oint (\mathbf{E} + \mathbf{v} \times \mathbf{B}) \cdot d\boldsymbol{\ell} \qquad (31.10)$$

Equation 31.10 allows us to identify the factors that contribute to the induced emf. The first term, $\oint \mathbf{E} \cdot d\boldsymbol{\ell}$, involves an **induced electric field.** We will see that it is associated with a time-dependent magnetic field. The second term, $\oint (\mathbf{v} \times \mathbf{B}) \cdot d\boldsymbol{\ell}$, involves motion relative to a magnetic field and is called a **motional emf.** We see that Faraday's law encompasses two quite distinct phenomena. Both may be present in a given situation, but we will consider them separately.

31.6 INDUCED ELECTRIC FIELDS

When there is no relative motion between the source of the magnetic field and the boundary of the path around which the emf is evaluated, only the first term in Eq. 31.10 is present. Furthermore, since the path is not moving, only the explicit time dependence of the magnetic field contributes to the change in flux. For a *uniform* magnetic field directed perpendicular to the plane of the area of the path, the flux is $\Phi = BA$ and $d\Phi/dt = A\, dB/dt$. Faraday's law, $\mathscr{E} = -d\Phi/dt$, becomes

$$\mathscr{E} = \oint \mathbf{E} \cdot d\boldsymbol{\ell} = -A \frac{dB}{dt} \qquad (31.11)$$

Thus Eq. 31.11 may be interpreted as follows:

> There is an **induced electric field** in *any* closed path, whether in matter or in empty space, through which the magnetic field is changing.

Figure 31.19 shows the induced electric field associated with a time-dependent magnetic field of a solenoid. The induced electric field differs from an electrostatic field in two ways. First, the *induced electric field* lines are closed loops, whereas *electrostatic field* lines always begin and end on charges. Second, the induced electric field is a nonconservative field since its line integral around a closed path is not zero.

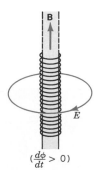

$$\left(\frac{d\phi}{dt} > 0\right)$$

FIGURE 31.19 As the magnetic field within the solenoid is changing, there is an induced electric field whose lines are closed circles.

EXAMPLE 31.6: The current in an ideal solenoid of radius R varies as a function of time. Find the induced electric field at points (a) inside, and (b) outside the solenoid. Express the results in terms of dB/dt.

Solution: In order to evaluate the integral in Eq. 31.11 we choose a path of integration that exploits the symmetry of the situation. The value of the induced electric field will be the same at all points on any circular loop concentric with the solenoid. Such a loop is a convenient path of integration. Whether the loop is inside or outside the solenoid, we have

$\mathbf{E} \cdot d\boldsymbol{\ell} = E\, d\ell$ since $\mathbf{E}$ is parallel to $d\boldsymbol{\ell}$. For a loop of radius r, the integral becomes

$$\oint \mathbf{E} \cdot d\boldsymbol{\ell} = E \oint d\ell = E(2\pi r)$$

(a) When $r < R$, the flux through the loop is $\Phi = BA = B(\pi r^2)$. Thus, from Eq. 31.11,

$$E(2\pi r) = -(\pi r^2) \frac{dB}{dt}$$

$(r < \;$ 　　　　　　　　　　　　　(i)

The induced electric field strength increases linearly with distance from the center.

(b) When $r > R$, the flux through the loop is $\Phi = B(\pi R^2)$. Thus,

$$E(2\pi r) = -(\pi R^2)\frac{dB}{dt}$$

$(r > R)$ $\qquad\qquad E = -\frac{R^2}{2r}\frac{dB}{dt} \qquad\qquad$ (ii)

Outside the solenoid, the induced electric field strength is inversely proportional to the distance from the center. In Fig. 31.20, the field is plotted as a function of r. We have assumed dB/dt is negative. You should confirm the direction of E by using Lenz's law.

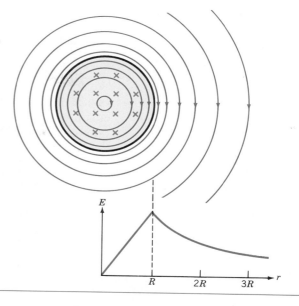

FIGURE 31.20 The variation of the induced electric field with distance from the center of a solenoid.

The physical significance of Eq. 31.11 is sometimes stated as follows: An induced electric field is created by a changing magnetic field. Consider a loop outside a solenoid as in the above example and Example 31.2. With this interpretation, it is difficult to explain how an induced electric field is created at the position of a loop where $B = 0$ at all times! In order to avoid such a paradox, one should keep in mind that *fields are created by electric charges, not by other fields.* When the current is changing in a solenoid, the accelerated motion of the charges is responsible for *both* the changing magnetic field and the induced electric field. For this reason, we say that an induced electric field is "associated with" a changing magnetic field. Equation 31.11 expresses the mathematical relationship between these fields.

31.7 MOTIONAL EMF

When the magnetic field is constant in time, there is no induced electric field. Using the second term in Eq. 31.10, we write Faraday's law as

$$\mathcal{E} = \oint(\mathbf{v} \times \mathbf{B}) \cdot d\ell = -\frac{d\Phi}{dt} \qquad (31.12)$$

Figure 31.21 shows a conducting rod of length ℓ moving at constant velocity $\mathbf{v}$ perpendicular to a uniform magnetic field directed into the page. An electron will experience a magnetic force $\mathbf{F}_B = -e\mathbf{v} \times \mathbf{B}$ directed downward along the rod. As a result, electrons migrate to the lower end and leave unbalanced positive charge at the top. This redistribution of charge sets up an electrostatic field $\mathbf{E}_0$ directed downward. The electrostatic force, $\mathbf{F}_E = -e\mathbf{E}_0$, is directed upward. The system quickly reaches an equilibrium state in which these two forces on an electron balance. That is, $\mathbf{E}_0 + \mathbf{v} \times \mathbf{B} = 0$, and there is no further motion of charge along the rod. The final magnitude of the electrostatic field is $E_0 = vB$. The potential difference associated with this electrostatic field is given by

$$V_b - V_a = E_0\ell = B\ell v$$

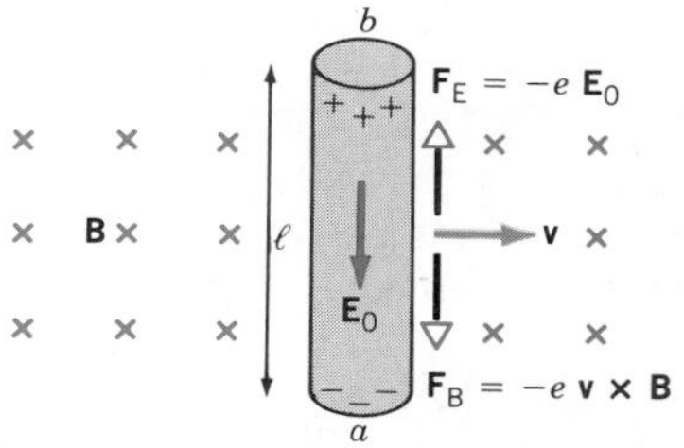

FIGURE 31.21 A metal rod moving perpendicular to magnetic field lines. There is a separation of charge and an associated electrostatic potential difference set up.

Note that $V_b > V_a$. The charge separation and its associated potential difference were created by an emf in the rod. Since there is no current flowing, the "terminal potential difference" is equal to the **motional emf**:

Motional emf

$$\mathscr{E} = B\ell v$$

We see that when the rod moves at constant velocity $\mathbf{v}$ on conducting rails, as in Fig. 31.13, it acts as a source of emf. Since $\mathbf{v}$ and $\mathbf{B}$ are perpendicular, the induced emf is $\mathscr{E} = vB\ell$, which agrees with the calculation in Example 31.3 based on the rate of change of flux.

EXAMPLE 31.7: In a *homopolar generator* a conducting disk of radius R rotates at angular velocity ω rad/s. Its plane is perpendicular to a uniform and constant magnetic field $\mathbf{B}$, as shown in Fig. 31.22. What is the emf generated between the center and the rim?

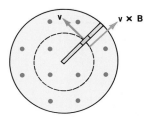

FIGURE 31.22 In a homopolar generator, a conducting disk rotates perpendicular to a magnetic field. To find the motional emf generated, the disk is divided into elementary rods.

Solution: A disk may be treated as a collection of radially oriented rods. The magnitude and polarity of the induced emfs are thus the same as for a single rod, but the current generated is much larger with a disk. Consider a small segment of width dr at a distance r from the center. The speed of the segment is $v = \omega r$. The electrons in the segment are subject to the magnetic force $\mathbf{F} = -e\mathbf{v} \times \mathbf{B}$ directed radially inward. Since $\mathbf{v}$ is perpendicular to $\mathbf{B}$, we have $|\mathbf{v} \times \mathbf{B}| = vB$. The length element is $d\ell = dr\hat{\mathbf{r}}$, and therefore

$$(\mathbf{v} \times \mathbf{B}) \cdot d\ell = vB\,dr = \omega Br\,dr$$

From Eq. 31.16, the total emf between the center and the rim is

$$\mathscr{E} = \int_0^R \omega Br\,dr = \tfrac{1}{2}\omega BR^2 \qquad \text{(ii)}$$

With the given directions of $\mathbf{v}$ and $\mathbf{B}$, the center is at a higher potential than the rim. If sliding contacts are placed at these two points, a constant dc current will flow in an external resistor.

If we had started to evaluate the emf of a rotating disk by considering the flux, a paradoxical situation would have arisen. Since the total flux through the disk is not changing, this result appears to be in conflict with the equation $\mathscr{E} = -d\Phi/dt$. This is not so. In order to apply Faraday's law, it is necessary to choose an appropriate loop that *incorporates the motion of the disk*. In the present case, the loop is a triangular sector of the circle, as shown in Fig. 31.23. One radial line (OP) is fixed while the other (OQ) rotates with the disk at ω. The area of the sector is $dA = \tfrac{1}{2}(R\,d\theta)R$ and the flux through it is $d\Phi = B\,dA = \tfrac{1}{2}R^2B\,d\theta$. The rate of change of flux is $d\Phi/dt = \tfrac{1}{2}BR^2\,d\theta/dt = \tfrac{1}{2}\omega BR^2$, which is the expression we found above. The direction of the emf is found from Lenz's law.

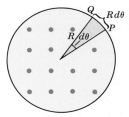

FIGURE 31.23 Although the total flux through a rotating disk does not change, one can use the equation $\mathscr{E} = -d\Phi/dt$ provided one chooses a loop that incorporates the motion of the disk.

The moving rod in Example 31.3 acts as a source of emf. In general, a source of emf converts some form of energy into electrostatic energy and does work on charges. Since magnetic forces do no work we must consider more closely the source of the work.

The velocity of a given electron has a component v along the motion of the rod and a drift velocity v_d along the rod, as shown in Fig. 31.24. The magnetic forces associated with these components are evB downward and ev_dB opposite to the motion of the rod. The power ($P = \mathbf{F} \cdot \mathbf{v}$) delivered by the magnetic force associated with the drift motion is $+(evB)v_d$ and the power associated with the horizontal motion is $-(ev_dB)v$. The net power supplied by the magnetic forces is zero, as it should be. Since the electrons are constrained to stay within the rod, they are also subject to a force due to the rod. In the steady state the two horizontal forces on each charge balance: $F_{rod} = ev_dB$. This equation applies to the charges within the rod. In order to keep the rod itself moving at constant velocity, there must be an *external* agent that applies a force toward the right to balance the magnetic force $I\ell B$ (which is the sum of the magnetic forces on all the electrons). The required energy must be supplied by the external agent. The magnetic field acts, in a sense, as an intermediary in the transfer of the energy from the external agent to the rod.

The "force due to the rod," F_{rod}, arises from the Hall effect. As the electrons move along the rod, they are subject to a magnetic force to the left. As the sides of the rod become oppositely charged, a Hall electric field is created across the rod (from right to left). It is the electrostatic force due to this Hall field that leads to F_{rod}.

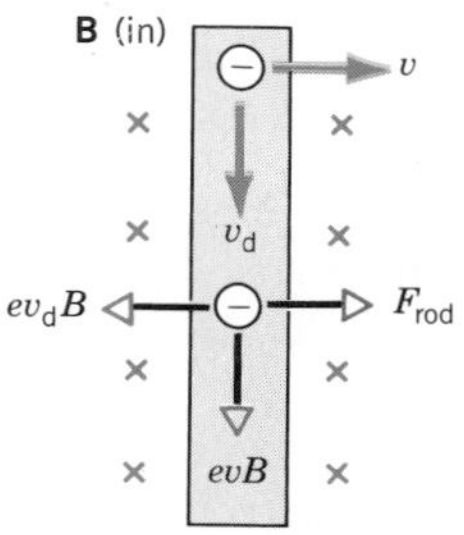

FIGURE 31.24 A conducting rod moving normal to a magnetic field. The velocity of each electron has a component in the direction of the rod's motion and a drift velocity along the rod. The force due to the rod, F_{rod}, is an electric force associated with a Hall field across the rod.

31.8 EDDY CURRENTS

Figure 31.25a shows a bar magnet approaching a conducting plate. Since the flux through any loop on the plate is changing, currents will be induced in the counterclockwise sense. If the magnet moves parallel to the plate, as in Fig. 31.25b, the nonuniformity of the field means that regions ahead of the magnet experience an increasing flux, while those behind experience a decreasing flux. Ahead of the magnet the currents flow counterclockwise, and behind it they flow clockwise. Such currents induced in bulk material are called *eddy currents* because they resemble the shape of eddies in a fluid.

In Example 31.3 we discussed what happens when a rod moves across a uniform magnetic field. If the rod is replaced by a conducting plate, as in Fig. 31.26, the induced currents are distributed throughout the plate. When only part of the plate is in the field, the currents in this portion will experience a force opposite to the direction of motion. This retarding force may be used to damp the oscillations of a chemical balance or the coil in a galvanometer. Eddy currents may also be used as part of the braking system of a train. An electromagnet on a railcar travels close to a rail. When the current in the magnet is turned on, large eddy currents are induced in the rail. The magnetic force exerted on these currents by the magnet is in the forward direction. From Newton's third law, the reaction force on the train is in the backward direction. The forces due to induced eddy currents are also used in the speedometers of cars.

As eddy currents flow through the body of a conductor, they generate thermal energy. This method of heating is used in smelting and in a refining process for semiconductors. Eddy currents generated in copper pots can also be used for "induction cooking".

In Fig. 31.27a a magnet is suspended at the rim of a conducting disk that is rapidly rotating about a vertical axis. The eddy currents induced in the disk produce a force tending to drag the magnet in the direction of motion of the rim. There is also a repulsive force on the magnet. The repulsive force that results from

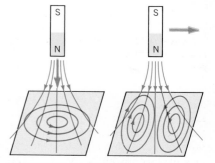

FIGURE 31.25 When a bar magnet moves relative to a conducting plate, eddy currents are induced in the plate.

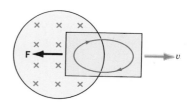

FIGURE 31.26 A conducting plate pulled across field lines. The net magnetic force on the induced eddy currents is opposite to the direction of motion.

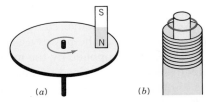

FIGURE 31.27 (*a*) A magnet suspended above a rapidly spinning metal disk induces eddy current in the disk. The magnetic forces are such that the magnet tends to be dragged along with the disk and is also repelled. (*b*) An iron bar is placed within a solenoid (to enhance the magnetic field produced by the solenoid) and a metal ring is placed at the end, as shown. When the current in the solenoid changes rapidly, the ring flies off.

eddy currents may be demonstrated as follows. A solenoid is wrapped around an iron bar (to intensify the field), as in Fig. 31.27*b*. A copper ring is placed at the edge of the solenoid. When a (rapidly changing) current is switched on, the ring is repelled (according to Lenz's law in his own form) and flies vertically upward. Instead of a ring one could use a flat plate. The repulsive force that results from eddy currents is used in magnetic levitation and propulsion of trains (see p. 659).

Faraday working in his laboratory at the Royal Institution.

HISTORICAL NOTE: The Search for Electromagnetic Induction

In the 18th century, discharges of Leyden jars were used to heat wires and to induce chemical changes in ionic solutions. These were examples in which electricity evolved heat and chemical change. Of course, it was known that the application of heat could initiate a chemical reaction and that a chemical reaction would produce heat. Volta's pile and other galvanic cells showed that chemical changes could produce electricity. In 1822 Thomas Seebeck discovered that the application of heat at the junction of two metals produces an electric current. Such evidence reinforced a belief among many scientists that all the "forces in nature" were interconnected. Recall that such a belief had motivated Oersted in his search for a connection between electricity and magnetism. A year later, in 1821, Francois Arago showed that an iron bar is magnetized when placed inside a current-carrying solenoid. Having seen that electricity (that is, current) produces magnetism in an iron bar, it was natural to search for the inverse effect: an electrical current produced by magnetism.

In addition to the metaphysical belief in the "unity of the forces in nature," there was another motivation in the search for the induction of currents. It was known that a charged object could induce charges in a nearby conductor and that a bar magnet could induce a temporary magnetization in an iron nail. Several scientists wondered whether an electric current could induce a current in a nearby conductor. The story of the search for electromagnetic induction is interesting because the effect was seen in different forms, but not recognized. Even when it was recognized, the discovery was not publicized.

In 1821, Ampère showed that a current-carrying solenoid behaves like a bar magnet and that two current-carrying wires exert magnetic forces on each other. He concluded that all magnetic effects were due to electric currents and developed a theory of magnetism in terms of current elements interacting through central forces. Yet, the exact nature of the currents in a magnet was not clear. They could have been microscopic "molecular" currents or macroscopic currents flowing in circular paths about the axis of the magnet.

In contrast to Ampère's mathematically sophisticated approach, Faraday relied on physical intuition and easily visualized models. He was particularly struck by the "circularity" of the lines of force around a current-carrying wire. In September 1821 he demonstrated this feature beautifully and incidently invented the electric motor (see the Historical Note in Chapter 29). Faraday was not impressed by the central forces in Ampère's theory, or by the idea that magnetism is produced by currents. He performed some subtle experiments aimed at disproving these ideas. For example, he showed that the "poles" of a current-carrying solenoid are not exactly in the same place as in a bar magnet. Ampère was forced to abandon the notion of macroscopic currents. In an effort to save his theory, he quickly devised an explanation of Faraday's experiments in terms of microscopic currents. Other scientists were not pleased at the way Ampère had so easily modified his theory to suit new experimental results.

In 1822 Ampère repeated an earlier (unsuccessful) experiment designed to clarify the issue of the nature of the currents. A copper ring was suspended inside a coil with many turns, and the poles of a magnet were placed across a point on the rim as in, Fig. 31.28. When the current was turned on in the coil, the ring rotated by some angle. When the current was turned off, the ring returned to its original position.* He concluded that the nonmagnetic copper ring had acquired a "temporary magnetization" because of steady microscopic induced currents. He did not bother to determine the direction of the currents.

Arago's Disk

In 1824 there was another very interesting discovery. Francois Arago, a colleague of Ampère, found that the oscillations of a suspended bar magnet are damped in the presence of a conducting sheet. The next year he showed that a rapidly spinning magnet could set a Cu disk rotating and a rapidly spinning disk could set a magnetic needle rotating. Arago also suspended a solenoid electromagnet over a rotating disk and observed its deflection. Ampère took this merely as confirmation of his ideas that currents were the ultimate basis of magnetism.

In London, J. Babbage and W. Herschel pursued Arago's work. They suspended a magnet above spinning disks made of various metals, as in Fig. 31.29a, and noted that the deflection of the magnet depended on the metal. For example, it was greater with a copper disk than with a lead disk. (The conduc-

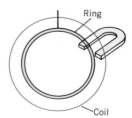

FIGURE 31.28 A copper ring suspended in the plane of a coil. When a current was turned on in the coil, Ampere noted that the ring rotated.

*There was a macroscopic induced current around the ring when the current was turned on. The ring should have immediately returned to its equilibrium position when the induced current fell to zero, but the suspension probably did not provide sufficient restoring torque. The accounts of this experiment are not clear as to exactly what was observed.

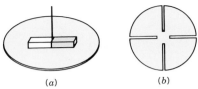

FIGURE 31.29 (*a*) A magnet suspended above a rapidly spinning disk experiences a torque. The size of the torque depended on the conductivity of the disk—but the connection was not made at the time. (*b*) When the disk had radial slots cut into it, the magnet was unaffected. (In this case the paths of the induced currents were severely limited.)

tivity of Cu is greater than that of Pb.) There was no deflection for nonmetallic disks. Babbage and Herschel also decided that the disk acquired temporary induced magnetism. Next they cut radial slits (Fig. 31.29*b*) and observed the deflection decrease as the number of slits increased. This effect was explained as being due to the reduction of the magnetization caused by the insertion of air gaps. The puzzle of Arago's disk was not solved, and interest in it slowly waned.

The relationship between the deflection of the suspended magnet and the conductivity pointed to the existence of induced currents in the disks. This was emphasized by the fact that the slits interrupted the flow of the currents. Also, the induced current in Arago's suspended solenoid was large enough for it to be set rotating! In writing about his own 1822 experiment, and those of Babbage and Hershel, Ampère referred explicitly to "little electric currents." In other words he understood fully that currents had been induced.

Ampère had all the evidence he needed to "discover" electromagnetic induction, but he did not. There are two reasons. First, it would have been very inconvenient for him to admit the existence of macrocurrents because his explanation of Faraday's experiments had locked him into the model of microcurrents. Second, he, along with everyone else, believed that a steady current should induce a *steady* current. Blinded by his preconceived notion as to what he should find and his desire to preserve his theory, he saw all that was necessary but got little out of it. This is a striking example of the fact what one observes depends very much on one's theory or viewpoint.

Meanwhile Faraday had also searched for induced currents over several years. When he heard of Ampère's experiment with the copper ring he tried to repeat it. A slip in the translation into English resulted in an unsuccessful experiment because he used a copper disk instead of a ring. (The moment of inertia of the disk was much larger than that of the ring.) In 1828 he suspended a ring and inserted a bar magnet into it. He then tried to detect the induced currents with other magnets. (If the magnet had been thrust quickly into the ring, what would he have seen?) Every one of these experiments could have led to discovery, but at this stage the experimental arrangements were not sensitive enough.

It is worth recording the unfortunate luck of J. D. Colladon. In 1825 he brought up a powerful magnet to a solenoid with many turns. In order to shield the galvanometer from any direct effects of the magnet, he placed it in the next room. He was too careful. By the time he went to check the deflection of the needle, the transient effect had, of course, ended.

In August 1830, quite independent of the events in Europe, Joseph Henry observed the "conversion of magnetism into electricity," but it seems he did not have the time to fully pursue or to immediately publish his discovery. He displayed an extraordinarily callous attitude to a discovery of major importance. Nonetheless, he did observe something new that Faraday had missed. We discuss this in the next chapter.

Without knowledge of Henry's discovery, Faraday returned to the problem in 1831 with a burst of astonishing creativity and certainty. He not only solved the puzzle of Arago's disk but, with the homopolar generator (Fig. 31.30), he went on to produce a *continuous* induced current—the trophy that had eluded everyone for a decade. Before knowing the full details of Faraday's work, Ampère quickly had his 1822 experiment published. Others also tried to establish claims of priority—with the exception of Arago, whose disk was the most spectacular demonstration of induced currents. When the tensions subsided, Ampère acknowledged that he had not realized the essential factor of time in electromagnetic induction.

The three simple experiments used to introduce this chapter seem so straightforward and obvious. That tidy presentation is distilled from trials that spanned a decade. The most brilliant theoretical and experimental minds were either unable, or unwilling, to recognize the underlying principle.

FIGURE 31.30 Faraday's homopolar generator with which he was able to produce a *continuous* induced current.

SUMMARY

In a uniform magnetic field the **magnetic flux** through a plane area **A** is given by

$$\Phi_B = \mathbf{B} \cdot \mathbf{A}$$

If the area is not flat or the field is not uniform, the flux is given by

$$\Phi_B = \int \mathbf{B} \cdot d\mathbf{A}$$

Faraday's law of electromagnetic induction relates the induced emf in a closed loop to the rate of change of flux through it:

$$\mathscr{E} = -\frac{d\Phi_B}{dt}$$

The negative sign takes into account the direction of $\mathscr{E}$, which is given by **Lenz's law:** The induced emf opposes the change in magnetic flux that produces it. In a circuit, the induced field produced by the induced current may either oppose or reinforce the external field.

In general, emf is defined as the work done by a nonelectrostatic agent in carrying a unit charge around a closed loop:

$$\mathscr{E} = \frac{W_{ne}}{q} = \oint \frac{\mathbf{F} \cdot d\boldsymbol{\ell}}{q}$$

The Lorentz force on a charge is given by $\mathbf{F} = q(\mathbf{E} + \mathbf{v} \times \mathbf{B})$. Thus,

$$\mathscr{E} = \oint (\mathbf{E} + \mathbf{v} \times \mathbf{B}) \cdot d\boldsymbol{\ell}$$

The first term is associated with an **induced electric field** and arises when the magnetic field depends explicitly on time. The second term arises when a conductor moves relative to a magnetic field. This **motional emf** arises from the magnetic force on a moving charge.

ANSWERS TO IN-CHAPTER EXERCISES

1. In each case there is a change in the number of field lines passing through the loop.
2. An increase in θ, as defined in Fig. 31.7, leads to a decrease in flux.
3. It is converted to thermal energy.
4. The flux through the loop increases. The flux due to the *induced* magnetic field must oppose this increase. Thus, the component of the induced magnetic field along the x axis is opposite to the external field. From the right-hand rule the induced current in the loop is *downward* on the y axis.
5. As Fig. 31.31 shows the flux is $\Phi = B\ell x = B\ell vt$. The flux increases linearly with time. The induced emf, $\mathscr{E} = -d\Phi/dt = -B\ell v$, is constant. When the loop is completely within the field, the flux is constant and the emf is zero. As the loop emerges, the flux decreases linearly with time and the direction of the induced emf is opposite to initial direction.
6. The "start-up" current is large enough for there to be a significant potential drop along the household wiring. The potential difference across the lights is temporarily lower than normal. The refrigerator should be in a separate circuit if possible.

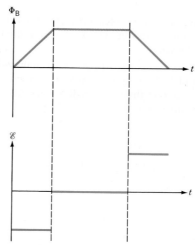

FIGURE 31.31 Since the loop moves at constant speed, the flux varies linearly with time and the induced emf has constant values. When the loop is entirely within the field, the flux is constant, so the induced emf is zero.

QUESTIONS

1. What is the difference between a magnetic field and a magnetic flux?

2. Consider a long, straight wire that passes through the center of a ring. If the current in the wire changes, is there an induced emf in the ring in either of the two geometries depicted in Fig. 31.32? (a) in Fig. 31.32a the wire lies along a diameter; (b) In Fig. 31.32b the wire lies along the axis of the ring.

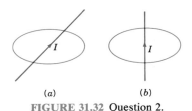

(a)   (b)
FIGURE 31.32 Question 2.

3. A bar magnet lies along the axis of a circular ring at some distance from the center, as shown in Fig. 31.33. Is there an induced emf in the ring if the ring rotates about its central axis?

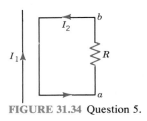

FIGURE 31.33 Questions 3 and 4.

4. A light metallic ring is released from above a vertical bar magnet, as shown in Fig. 31.33. Describe qualitatively the motion of the ring.

5. The current, I_1, in a long, straight wire changes in time. The current I_2 induced in the nearby loop, shown in Fig. 31.34, flows from a to b in the resistor. If a voltmeter is connected between a and b, what will it read?

FIGURE 31.34 Question 5.

6. A magnet moves at constant velocity along the axis of a

stationary loop, as in Fig. 31.35. Make a qualitative graph of the variation in (a) the flux, and (b) in the induced emf, as functions of time.

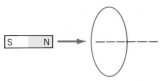

FIGURE 31.35 Question 6.

7. A rod of length d and a rectangular loop of width d are released together and fall in a uniform field, as shown in Fig. 31.36. Is there any difference in their motion? (Assume the loop is never totally within the field.)

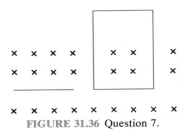

FIGURE 31.36 Question 7.

8. Two coils are wrapped around cylindrical forms, as shown in Fig. 31.37. One coil is in series with a battery, a switch, and a variable resistance. The other is connected to an ammeter. State the direction of the induced current measured by the ammeter (x to y, or y to x) in the following circumstances: (a) when the switch is first closed; (b) with the switch closed, the resistance is decreased; (c) with the switch closed, the coils are moved apart.

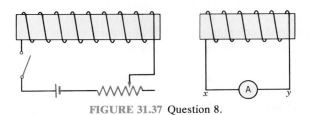

FIGURE 31.37 Question 8.

9. A solenoid is used as an antenna in an AM pocket radio. On what principle is the design of such an antenna based?

10. A bar magnet is dropped into a long, vertical, copper pipe. Describe qualitatively its motion. Ignore air resistance.

11. In what way(s) is an induced emf different from the emf due to a battery?

12. A flat loop and a long, straight wire lie in the same plane, Fig. 31.38. If the current in the wire decreases suddenly, in which sense (clockwise or counterclockwise) is the induced current in the loop?

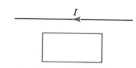

FIGURE 31.38 Question 12.

13. A metal box acts as a shield against external electric fields. Does it also serve as a shield for external magnetic fields? Would there be a difference between static and time-varying magnetic fields?

14. When an electric motor, such as an electric drill, is being used, does it also act as a generator? If so, what is the consequence of this?

15. A suspended magnet is oscillating freely in a horizontal plane. The oscillations are strongly damped when a metal plate is placed under the magnet. Explain why this occurs.

16. If an aluminum plate is moved rapidly through the region between the poles of an electromagnet, it experiences a strong retarding force. However, if slots are cut into it, as shown in Fig. 31.39, the force is greatly diminished. Why?

FIGURE 31.39 Question 16.

17. Two coils face each other with their axes coincident. Is it

possible for an induced emf to exist in one coil if the current in the other coil is instantaneously zero?

18. A small, flat coil is flipped over (by 180°) in a uniform magnetic field. (a) Does the number of lines through the coil change? (b) Does the flux through the coil change?

19. A metal bar moves over conducting rails perpendicular to a magnetic field, as in Fig. 31.40. There is a motional emf generated in the rod. Does a stationary voltmeter connected as in Fig. 31.40a or Fig. 31.40b register a reading?

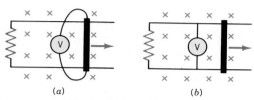

(a)   (b)

FIGURE 31.40 Question 19.

20. A thin metallic ring lies below a coil connected to a battery and a switch, as in Fig. 31.41. When the switch is opened, is the ring attracted to, or repelled by, the coil?

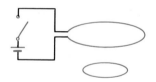

FIGURE 31.41 Question 20.

21. In a certain region the earth's magnetic field points vertically down. When a plane flies due north, which wingtip is positively charged?

22. Two loops lie side by side on a table. If a clockwise current suddenly starts to flow in one, what is the sense of the induced current in the second?

EXERCISES

31.2 Magnetic Flux

1. (I) The plane of loop with dimensions 12 cm × 7 cm is initially perpendicular to a uniform 0.2-T magnetic field. Find the change in flux through the loop if it is turned through 120 about an axis perpendicular to the field lines.

2. (I) The plane of a circular loop of radius 6 cm makes an angle of 30° with a uniform magnetic field of magnitude 0.25 T. (a) What is the flux through the loop? (b) If the field reverses in direction, what is the change in flux?

31.3 Faraday's Law

3. (I) A solenoid has 10 turns/cm and carries a 4-A current. A circular loop with 5 turns of area 8 cm² lies within the solenoid with its axis at 37° to the axis of the solenoid. Find the magnitude of the average induced emf if the current increases by 25% in 0.1 s.

4. (I) A metal rod of length ℓ = 5 cm moves at constant speed v on rails of negligible resistance that terminate in a resistor $R = 0.2 \ \Omega$, as shown in Fig. 31.42. A uniform and constant

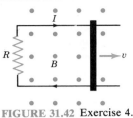

FIGURE 31.42 Exercise 4.

magnetic field $B = 0.25$ T is normal to the plane of the rails. The induced current is $I = 2$ A and flows in the direction shown. Find: (a) the speed v; (b) the external force needed to keep the rod moving at v.

5. (II) A coil of resistance 3 Ω has 25 turns of area 8 cm². Its plane is perpendicular to a field whose variation in time is given by $B(t) = 0.4t - 0.3t^2$ T (a) What is the flux as a function of time? (b) What is the induced current at 1 s?

6. (I) The plane of a circular coil with 15 turns of radius 2 cm lies at 40° to a uniform field of 0.2 T. Find the magnitude of the induced emf if the field increases linearly with time to 0.5 T in 0.2 s.

7. (I) A solenoid of length 30 cm has 240 turns of radius 2 cm. A tightly wound coil with 12 turns of radius 3 cm is at the center of the solenoid. The axes of the coil and solenoid coincide. Find the emf induced in the coil if the current in the solenoid varies according to $I(t) = 4.8 \sin(60\pi t)$ A.

8. (I) The antenna on a radio receiving an AM radio station that broadcasts at 800 kHz consists of a coil of 120 turns of radius 0.6 cm. The coil experiences an induced emf due to the oscillating magnetic field of the radio wave. If the field is given by $B(t) = 10^{-5} \sin(2\pi ft)$ T, find the emf induced in the coil. Assume the magnetic field is directed along the axis of the coil.

9. (I) A circular loop of diameter 10 cm is placed on a horizontal table. A uniform magnetic field of magnitude 0.2 T is directed vertically up at $t = 0$. The field varies according to $B = 0.2 - 12.5t$ T. (a) What is the change in flux through the loop from 0 to 20 ms? (b) What is the induced emf? (c) What is the sense of the induced current (clockwise/counterclockwise) looking down from above the coil?

10. (I) A rectangular loop 25 cm × 40 cm moves at a constant 20 m/s with its plane normal to a uniform magnetic field of magnitude 0.18 T (Fig. 31.43). The resistance of the loop is 1.2 Ω. Given that only one side of the loop is inside the field, find: (a) the induced emf; (b) the force exerted on the loop by the field; (c) the electrical power dissipated; (d) the mechanical power required to move the loop at constant velocity.

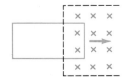

FIGURE 31.43 Exercise 10.

11. (I) A metal rod slides at a constant 30 m/s over frictionless rails separated by 24 cm, as shown in Fig. 31.44. A 0.45-T uniform magnetic field is directed out of the page. Assume the resistance of the rod is 2.7 Ω and that the rails have negligible resistance. Find: (a) the current flowing in the rails; (b) the magnetic force acting on the rod; (c) the mechanical power needed to keep the rod moving at constant velocity; (d) the electrical power dissipated.

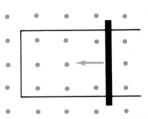

FIGURE 31.44 Exercise 11.

12. (I) A flat circular coil has 80 turns of diameter 20 cm with a total resistance of 40 Ω. The plane of the coil is perpendicular to a uniform field. At what rate should the field change for the thermal power dissipated in the coil to be 2 W?

13. (II) A coil of radius 5 cm has 20 turns of Cu wire of diameter 1 mm. The plane of the coil is perpendicular to a field that changes at the rate of 0.2 T/s. What is the power loss in the coil? The resistivity of Cu is 1.7×10^{-8} Ω·m.

14. (II) A conducting loop is formed with two springs ($k = 2$ N/m) and a rod of length $\ell = 30$ cm and mass $m = 20$ g, as shown in Fig. 31.45. A uniform magnetic field of 0.4 T is directed perpendicular to the plane of the loop. At $t = 0$, the rod is released with the springs extended by $A = 10$ cm. (a) Write an expression for the induced emf, $\mathscr{E}(t)$. (b) What is the maximum value of the emf, and when does it occur for the first time?

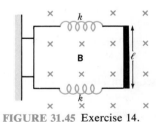

FIGURE 31.45 Exercise 14.

15. (II) A rectangular loop of mass m, width ℓ, and resistance R falls vertically through a uniform horizontal field **B,** as shown in Fig. 31.46. (a) Show that the loop reaches a terminal velocity $v_T = mgR/(B\ell)^2$. (b) Show that at v_T the rate at which gravitational energy is being lost is equal to the electrical power dissipation.

16. (I) A coil has N turns of area A and total resistance R. It is connected to a galvanometer and its plane is normal to a uniform magnetic field B. In a short time interval, the coil is flipped over (turned through 180°). Show that the charge

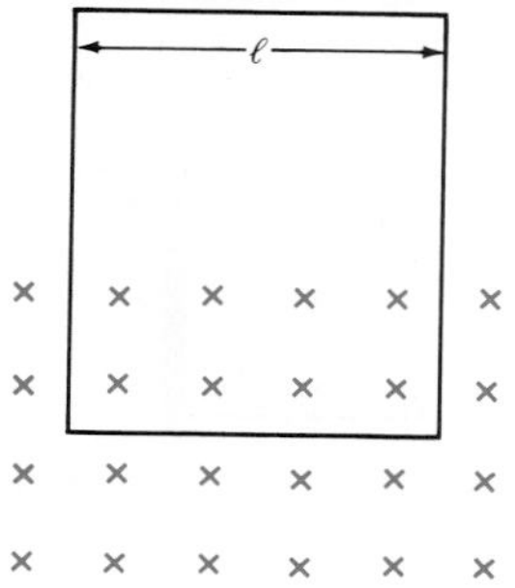

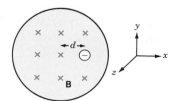

FIGURE 31.46 Exercise 15.

the solenoid is changing according to $B = Ct$. Obtain an expression for the electric force on the electron.

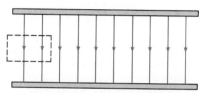

FIGURE 31.47 Exercise 23.

that flows through the galvanometer is $Q = 2NAB/R$.

17. (II) A magnetic field given by $B(t) = 0.2t - 0.5t^2$ T is directed perpendicular to the plane of a circular coil containing 25 turns of radius 1.8 cm and whose total resistance is 1.5 Ω. Find the power dissipation at 3 s.

18. (II) Wire is braided on an elastic circular loop that is oriented with its plane perpendicular to a uniform field of magnitude 0.32 T. At a particular moment the radius is 6 cm and is expanding at 20 cm/s. What is the magnitude of the induced emf?

31.4 Generator

19. (I) A square coil of side 8 cm has 180 turns and rotates in a uniform field of magnitude 0.08 T. If the peak emf is 12 V, what is the angular velocity of the coil?

20. (I) An enterprising store owner decides to use the large (2 m × 3 m) revolving door at the entrance as a generator. She wraps 100 turns around the perimeter of the door. A constant stream of customers keeps the doors rotating at 0.25 rev/s. If the horizontal component of the earth's field is 0.6 G, what is the maximum emf?

21. (I) A coil of cross-sectional area 40 cm² consists of 100 turns and has a resistance of 4.5 Ω. It rotates at 120 rpm with its axis perpendicular to a 0.04-T field. Find: (a) the maximum emf generated; (b) the maximum magnetic torque experienced by the coil.

22. (II) A square coil has 25 turns of side 5 cm and a resistance of 2.5 Ω. It rotates at 120 rpm about a vertical axis in a horizontal magnetic field of 0.04 T. At $t = 0$, the plane of the coil is perpendicular to the field. At $t = 0.1$ s find: (a) the induced emf; (b) the torque needed to keep the coil rotating; (c) the mechanical power needed to keep the coil rotating; (d) the electrical power dissipated.

31.6 Induced Electric Fields

23. (II) An electron is at a distance d from the axis of a solenoid, as shown in Fig. 31.47. The uniform magnetic field in

24. (II) Use Faraday's law (Eq. 31.11) to show that the electric field lines between the plates of a capacitor cannot end abruptly at the edge of the plates, as is shown in Fig. 31.48.

FIGURE 31.48 Exercise 24.

25. (II) The current in a long solenoid varies according to $I(t) = 4 + 6t^2$ A. The solenoid has 800 turns/m and a radius of 2 cm. At $t = 2$ s find the magnitude of the induced electric field at the following distances from the central axis: (a) 0.5 cm; (b) 4 cm.

26. (II) A long solenoid with 20 turns/cm has a radius of 2.4 cm. The magnitude of the induced electric field 2 cm from the axis is 5×10^{-3} V/m. At what rate is the current in the solenoid changing?

31.7 Motional Emf

27. (I) An airliner with a wingspan of 45 m flies at 300 m/s in a region in which the vertical component of the earth's field is 0.6 G. (a) What is the potential difference between the wingtips?; (b) What would a voltmeter moving with the plane read if its leads were connected to the wingtips?

28. (II) The propeller on an airplane is 1.5 m from tip to tip. It rotates at 1800 rpm in a plane perpendicular to the horizontal component of the earth's field, which is 0.6 G. What is the emf induced between a tip and the center?

29. (II) A Faraday disk dynamo (homopolar generator) of radius 20 cm generates 1.2 V in a 0.08-T magnetic field directed perpendicular to the plane of the disk. What is the frequency of rotation in rpm?

PROBLEMS

1. (I) A long, straight wire carries a constant current I. A metal rod of length ℓ moves at velocity v relative to the wire, as shown in Fig. 31.49. What is the potential difference between the ends of the rod? (Note that the field is not uniform.)

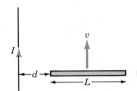

FIGURE 31.49 Problem 1.

2. (I) A long, straight wire carries a constant current $I = 15$ A. A metal rod of length $\ell = 40$ cm moves at constant velocity on rails of negligible resistance that terminate in a resistor $R = 0.05\ \Omega$, as shown in Fig. 31.50. Find the induced current in the resistor given $a = 1$ cm, $d = 5$ cm, and $v = 25$ cm/s. (*Hint:* First consider the flux through an infinitesimal strip of width dx at a distance x from the wire.)

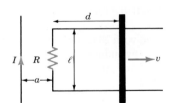

FIGURE 31.50 Problem 2.

3. (II) A metal rod of mass m and of length ℓ slides on frictionless rails of negligible resistance which terminate in a resistor R, as shown in Fig. 31.51. A uniform magnetic field is directed perpendicular to the plane of the rails. The initial velocity of the rod is v_0. There is no external agent applying a force to the rod. (a) Show that

$$v(t) = v_0 e^{-t/\tau}$$

where $\tau = mR/(B\ell)^2$. (b) Show that the distance traveled before coming to rest is $v_0\tau$. (c) Show that the total electrical energy lost is $\frac{1}{2}mv_0^2$.

4. (I) A triangular loop travels at a constant velocity v. Its plane is perpendicular to a uniform magnetic field; see Fig. 31.52. Obtain an expression, as a function of time, for the induced emf before the whole loop enters the field.

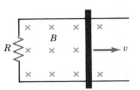

FIGURE 31.51 Problem 3.

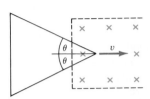

FIGURE 31.52 Problem 4.

5. (II) An elasticized conducting band is around a spherical balloon. Its plane passes through the center of the balloon. A uniform magnetic field of magnitude 0.4 T is directed perpendicular to the plane of the band. Air is let out of the balloon at 100 cm³/s at an instant when the radius of the balloon is 6 cm. What is the induced emf in the band?

6. (II) A metal bar of mass m, length ℓ, and resistance R slides down a pair of frictionless rails of negligible resistance inclined at angle θ to the horizontal; see Fig. 31.53. A uniform magnetic field is directed vertically upward. (a) Obtain an expression for the current induced in the rod. Assume that the rails have no resistance. (b) Show that the rod attains a terminal speed given by

$$v_T = \frac{mgR \sin\theta}{(B\ell \cos\theta)^2}$$

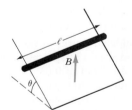

FIGURE 31.53 Problem 6.

7. (II) A flat loop and a long straight wire lie in the same plane; see Fig. 31.54. The current in the wire varies according to $I = I_0 \sin(\omega t)$. Find the emf induced in the loop. (*Hint:* First consider the flux through an infinitesimal strip of width dx at a distance x from the wire.

8. (I) A rod of mass m and resistance R slides on frictionless and resistanceless rails a distance ℓ apart that include a source of emf $\mathcal{E}_0$; see Fig. 31.55. The rod is initially at rest.

FIGURE 31.54 Problem 7.

(a) Show that the rod reaches a terminal speed, v_T. (b) What is v_T?

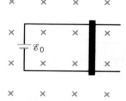

FIGURE 31.55 Problem 8.

9. (II) Figure 31.56 shows a square loop of side L perpendicular to the uniform field of a solenoid. (a) Show that at any point on a side the component of the induced electric field along the side is $\frac{1}{4}L\,dB/dt$. (b) Evaluate $\oint \mathbf{E} \cdot d\boldsymbol{\ell}$ around the loop.

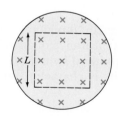

FIGURE 31.56 Problem 9.

10. (I) A square coil has 40 turns of side 4 cm, and a total resistance of 2.5 Ω. Its plane is perpendicular to a uniform magnetic field that varies in time according to $B = B_0 \exp(-t/\tau)$, where $B_0 = 0.2$ T and $\tau = 50$ ms. (a) What is the induced current in the coil? (b) Show that the total charge that flows through it is NAB_0/R. (*Note:* $\int \exp(ax)dx = (1/a) \exp(ax)$.)

11. (II) A *betatron* is a machine that uses an induced electric field to speed up electrons that travel in a circular path within a doughnut-shaped cavity (Fig. 31.57). The magnetic field is nonuniform and varies in time. (a) Write Newton's second law, $F = ma$, for the circular motion of an electron given that B_{orb} is the magnetic field at the position of the orbit of radius r. Show that $mv = erB_{orb}$. (b) If B_{av} is the average value of the field over the area within the orbit, show that the magnitude of the induced electric field is given by $|E| = (r/2)dB_{av}/dt$. (c) Apply Newton's second law in the form $F = d(mv)/dt$ to the electric force on the electron to show that $B_{orb} = B_{av}/2$. When this condition is met the electron is held in a fixed orbit even as its speed increases.

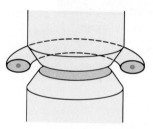

FIGURE 31.57 Problem 11.

CHAPTER 32

Inductance and Magnetic Materials

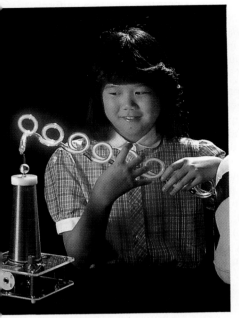

An oscillating potential difference induced in the "Telsa coil" is large enough to cause the gas in the tube to glow although there is no contact.

Major Points

1. The definitions of **self-inductance** and **mutual inductance.**
2. The rise and decay of current in an ***LR* circuit.**
3. (a) The **energy** stored in an inductor.
   (b) The **energy density** of the magnetic field.
4. (a) **Undamped oscillations** in an *LC* circuit.
   (b) **Damped oscillations** in an *RLC* circuit.
5. **Ferromagnetism, paramagnetism,** and **diamagnetism.** (Optional)

In the last chapter we saw that when the magnetic flux due to one coil changes, there is an induced emf in a nearby coil. The appearance of an induced emf in one circuit due to changes in the magnetic field produced by a nearby circuit is called **mutual induction.** The response of the circuits is characterized by their *mutual inductance.*

Faraday believed that an induced current should also appear in a circuit when the flux due to its own magnetic field changes, but he was unable to detect this effect. Although Joseph Henry (Fig. 32.1) had lost his claim to priority in the discovery of electromagnetic induction, he belatedly resumed his investigations. He noticed that when the current in the windings of an electromagnet was interrupted, bright sparks appeared at the switch contacts. As we will see, this confirmed Faraday's belief. Faraday somehow overlooked Henry's report but was told of a similar observation made by a William Jenkin in 1834. The appearance of an induced emf in a circuit associated with changes in its own magnetic field is called **self-induction.** The corresponding property is called *self-inductance.* A circuit element, such as a coil, that is designed specifically to have self-inductance is called an **inductor.** Its circuit symbol is ℓℓℓℓ. In this chapter we investigate the behavior of circuits containing inductors. A most important property of circuits that contain both inductance and capacitance is that they can undergo electrical oscillations and display the phenomenon of resonance. These properties are important in the generation and reception of electromagnetic waves, such as radio and TV signals. Mutual inductance is important in the operation of transformers, which are discussed in the next chapter.

FIGURE 32.1 Joseph Henry (1797–1878).

32.1 INDUCTANCE

Figure 32.2a shows a coil in series with a switch and a battery. When the switch is first closed, the growing current creates a changing magnetic field. As the flux through the coil changes, there is an induced emf that opposes this change. This phenomenon of **self-induction** appears in any circuit; the coil merely enhances the effect. In this case the *self-induced emf* tries to prevent the rise in the current. As a result, the current does not reach its final value instantly, but instead rises gradually as depicted in Fig. 32.2b. When the switch is opened, the flux rapidly decreases. This time the self-induced emf tries to maintain the flux; it has the same polarity as the battery. When the current in the windings of an electromagnet is shut off, the self-induced emf can be large enough to produce a spark across the switch contacts. (This was Henry's discovery.)

If a coil has N turns and the flux Φ has the same value for each turn, then the induced emf is $\mathscr{E} = -N \, d\Phi/dt$. Since N is fixed, we may write the emf as

$$\mathscr{E} = -\frac{d(N\Phi)}{dt} \tag{32.1}$$

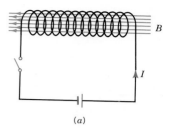

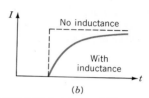

FIGURE 32.2 (a) A coil in series with a battery. When the switch is closed, the induced emf in the coil opposes the change in flux through it. (b) The current in the coil rises gradually.

The quantity $N\Phi$ is called the **flux linkage** through the coil. The SI unit of flux linkage is the weber (Wb), the same as that of flux. Since self-induction and mutual induction occur simultaneously, both contribute to the flux and to the induced emf in each coil. For example, in Fig. 32.3, the flux linkage through coil 1 is the sum of two terms:

$$N_1\Phi_1 = N_1(\Phi_{11} + \Phi_{12}) \tag{32.2}$$

where Φ_{11} is the flux in coil 1 due to its own current I_1, and Φ_{12} is the flux in coil 1 due to I_2, the current in the second coil. The flux Φ_{11} is associated with self-induction, and Φ_{12} with mutual induction. The net emf induced in coil 1 due to changes in I_1 and I_2 is

$$\mathscr{E}_1 = -N_1 \frac{d}{dt}(\Phi_{11} + \Phi_{12})$$

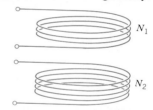

FIGURE 32.3 Two coils placed close to each other. Each coil has self-inductance and the pair has mutual inductance.

In the last chapter, we ignored the first (self-induction) term and expressed the second (mutual induction) term simply as $\mathscr{E} = -N \, d\Phi/dt$. There, it was understood that $\mathscr{E}$ and Φ in one coil are associated with the I in a *different* coil.

EXERCISE 1. Write the corresponding expression for $\mathscr{E}_2$.

Self-inductance

It is convenient to express the induced emf in terms of the current in a circuit rather than the magnetic flux through it. If no magnetic materials are present, the magnetic field produced by a coil, and hence also the flux, are directly proportional to the current flowing through it. Thus, the first term in Eq. 32.2 may be written as

$$N_1\Phi_{11} = L_1 I_1 \tag{32.3}$$

where L_1 is a constant of proportionality called the **self-inductance** of coil 1. The SI unit of self-inductance is the henry (H). The self-inductance of a circuit depends on its size and its shape. From Eq. 32.1, the self-induced emf in coil 1 due to changes in I_1 takes the form

$$\mathscr{E}_{11} = -L_1 \frac{dI_1}{dt} \tag{32.4} \qquad \text{Self-induced emf}$$

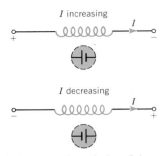

FIGURE 32.4 The polarity of the self-induced emf is determined by the *rate of change* of the current.

Figure 32.4 shows that the polarity of the self-induced emf depends on the *rate of change* of the current, not on its magnitude or its direction.

Mutual Inductance

The flux produced by coil 2 is proportional to I_2. Thus, the flux linkage through coil 1 due to I_2 (the second term in Eq. 32.3) may be written as

$$N_1\Phi_{12} = MI_2 \qquad (32.5)$$

where the constant of proportionality M is called the **mutual inductance** of the two coils. (At this stage we should have written M_{12}. However, it can be shown that $M_{12} = M_{21} = M$.) The SI unit of mutual inductance is the henry (H). The mutual inductance of two circuits depends on their sizes, their shapes, and their relative positions. Intuitively one would expect the mutual inductance to be greater when the coils are near each other and oriented so that the maximum amount of flux from one coil intercepts the other. The emf induced in coil 1 due to changes in I_2 takes the form

$$\mathscr{E}_{12} = -M\frac{dI_2}{dt} \qquad (32.6)$$

The total emf induced in coil 1 due to changes in I_1 and I_2 is $\mathscr{E}_1 = \mathscr{E}_{11} + \mathscr{E}_{12}$. Since most problems are concerned with only self-inductance or mutual inductance, but not both, the subscripts are often dropped.

EXERCISE 2. Show that 1 H = 1 Wb/A = 1 V · s/A.

EXAMPLE 32.1: A *long solenoid* of length ℓ and cross-sectional area A has N turns. Find its self-inductance. Assume that the field is uniform throughout the solenoid.

Solution: From Eq. 30.11 we know that the field within a long solenoid is $B = \mu_0 nI$, where $n = N/\ell$ is the number of turns per unit length. The flux through each turn (actually Φ_{11}) is

$$\Phi = BA = \mu_0 nIA$$

From Eq. 32.3, the self-inductance is

$$L = \frac{N\Phi}{I} = \mu_0 n^2 A\ell$$

As we see, the self-inductance depends on the geometrical properties of the circuit. (Note that $A\ell$ is the volume of the solenoid.) This result may be compared to the capacitance of two parallel plates ($C = \varepsilon_0 A/d$), which depends on the geometry of the capacitor. Both expressions are only approximately true. In the case of the capacitor we ignored the fringing field. In the case of the solenoid, we ignored the fact that the field strength drops toward each end. The self-inductance of a real solenoid will be less than the value computed from the above expression.

EXAMPLE 32.2: A *coaxial cable* is often used to carry electrical signals, for example, from an antenna to a TV set. As Fig. 32.5 shows, it consists of an inner wire of radius a that carries a current I upward, and an outer cylindrical conductor of radius b that carries the same current downward. Find the self-induc-

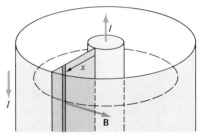

FIGURE 32.5 In determining the flux in the region between the inner wire and the outer cylinder one must take into account that the field is not uniform.

tance of a coaxial cable of length ℓ. Ignore the magnetic flux within the inner wire.

Solution: We again use Eq. 32.3, $N\Phi = LI$, to determine L. First it is necessary to find the flux through a cross section of the space between the conductors. Our calculation is approximate because we consider only the field in the region between the conductors. (It would be exact if the inner wire were replaced by a hollow conductor.) From Eq. 30.1 we know that field produced by the inner wire at a distance x ($>a$) from its center is

$$B = \frac{\mu_0 I}{2\pi x}$$

To determine L it is necessary to evaluate the flux through a rectangular loop oriented perpendicular to the field lines between the wire and the outer cylinder. Since the field is nonuniform, we must first find the flux through an infinitesimal strip of width dx and area $dA = \ell\, dx$. Thus,

$$d\Phi = B\, dA = \frac{\mu_0 I \ell}{2\pi} \frac{dx}{x}$$

The total flux through the loop is

$$\Phi = \frac{\mu_0 I \ell}{2\pi} \int_a^b \frac{dx}{x}$$
$$= \frac{\mu_0 I \ell}{2\pi} \ln \frac{b}{a}$$

From Eq. 32.3 the self-inductance of the coaxial cable is

$$L = \frac{\Phi}{I} = \frac{\mu_0 \ell}{2\pi} \ln \frac{b}{a}$$

The dependence of L on the ratio b/a is reasonable: As b gets larger or a becomes smaller, the flux enclosed increases. This expression should be compared with the capacitance of the cable; see Example 26.5.

EXAMPLE 32.3: A circular coil with a cross-sectional area of 4 cm² has 10 turns. It is placed at the center of a long solenoid that has 15 turns/cm and a cross-sectional area of 10 cm², as in Fig. 32.6. The axis of the coil coincides with the axis of the solenoid. What is their mutual inductance?

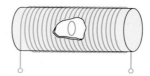

FIGURE 32.6 A small coil placed within a solenoid.

Solution: Let us refer to the coil as circuit 1 and the solenoid as circuit 2. The field in the central region of the solenoid is uniform, so the flux through the coil is

$$\Phi_{12} = B_2 A_1 = (\mu_0 n_2 I_2) A_1$$

where $n_2 = N_2/\ell = 1500$ turns/m. From Eq. 32.5, the mutual inductance is

$$M = \frac{N_1 \Phi_{12}}{I_2}$$
$$= \mu_0 n_2 N_1 A_1$$
$$= (4\pi \times 10^{-7}\,\text{T} \cdot \text{m/A})(1500\,\text{m}^{-1})(10)(4 \times 10^{-2}\,\text{m}^2)$$
$$= 7.54 \times 10^{-6}\,\text{H}$$

Notice that although $M_{12} = M_{21}$, it would have been much more difficult to find Φ_{21} because the field due to the coil is quite nonuniform. Furthermore, since we needed only the flux at the central region of the solenoid, this expression for M is more accurate than the equation for the self-inductance of a solenoid. (Recall that in that case we had to ignore end effects.)

32.2   *LR* CIRCUITS

In Section 32.1 it was pointed out that self-inductance in a circuit prevents the current from changing abruptly. We now consider how the current rises or falls as a function of time in a circuit containing an inductor and a resistor in series. We assume that the ideal inductor has no resistance. The actual resistance of a real inductor is considered as part of the external resistor.

Rise

Figure 32.7a shows an inductor in series with a resistor, a battery with an emf $\mathcal{E}$, and a switch. At $t = 0$, the switch is closed and the current starts to flow in the direction shown. The current is increasing, $dI/dt > 0$, and therefore the polarity of the induced emf in the inductor is opposite to that of the battery. That is, $\mathcal{E}_L = -L\, dI/dt = V_a - V_b > 0$. From Kirchhoff's loop rule we have

$$\mathcal{E} - IR - L\frac{dI}{dt} = 0 \qquad (32.7)$$

To solve this differential equation we let $y = \mathcal{E}/R - I$, which means that $dy/dt = -dI/dt$. Using this in Eq. 32.7, we find $dy/dt = -(R/L)y$, or

$$\frac{dy}{y} = -\frac{R}{L}\, dt$$

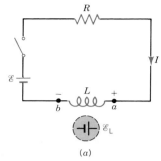

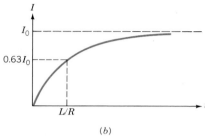

FIGURE 32.7 (a) A resistor and an inductor in series with an ideal source of emf. (b) When the switch is closed, the current rises gradually.

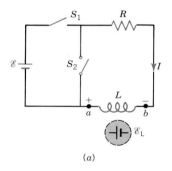

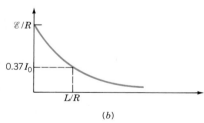

FIGURE 32.8 (a) In order to study the decay of the current in an LR circuit, switch S_2 is closed just before switch S_1 is opened. (b) The current decays exponentially.

Integration of both sides leads to

$$\ln y = -\frac{R}{L} t + \ln y_0$$

where, for convenience, the integration constant has been written as $\ln y_0$, where y_0 is the value of y at $t = 0$. The antilog of this equation is

$$y = y_0 e^{-Rt/L}$$

At $t = 0$ the current $I = 0$, and therefore $y = \mathcal{E}/R - I$ has the value $y_0 = \mathcal{E}/R$. Returning to the original variables we find

$$I = I_0(1 - e^{-t/\tau}) \tag{32.8}$$

where $I_0 = \mathcal{E}/R$ is the final value of I at $t = \infty$. The quantity

$$\tau = \frac{L}{R} \tag{32.9}$$

is called the **time constant.** In one time constant, the current rises to $(1 - e^{-1})I_0 = 0.63I_0$, as drawn in Fig. 32.7b. (Compare Eq. 32.8 with Eq. 28.11 for the charging of a capacitor.)

Decay

We now consider what happens when the battery is suddenly removed without breaking the continuity of the circuit. In Fig. 32.8a switch 2 is closed first and switch 1 is opened a very short time later. The current through R and L starts to decrease, so $dI/dt < 0$. The induced emf in the inductor tries to maintain the current, so its polarity is the same as that of the battery. That is, $\mathcal{E}_L = -L dI/dt = V_a - V_b > 0$. According to the loop rule traced along the direction of the current,

$$-IR - L\frac{dI}{dt} = 0$$

This may be rearranged and integrated

$$\int \frac{dI}{I} = -\frac{R}{L} \int dt$$

to yield

$$\ln I = -\frac{R}{L} t + \ln I_0$$

At $t = 0$, $I = I_0 = \mathcal{E}/R$. Taking the antilog we find

$$I = \frac{\mathcal{E}}{R} e^{-Rt/L}$$

or

$$I = I_0 e^{-t/\tau} \tag{32.10}$$

The variation in the current is drawn in Fig. 32.8b. In this case the time constant $\tau = L/R$ is the time at which the current falls to $1/e$ or 37% of its initial value $I_0 = \mathcal{E}/R$. (Compare Eq. 32.10 with Eq. 28.7 for the discharge of a capacitor.)

EXERCISE 3. Show that the unit of $\tau = L/R$ is the second.

32.3   ENERGY STORED IN AN INDUCTOR

The battery that establishes the current in an inductor has to do work against the opposing induced emf. The energy supplied by the battery is stored in the inductor. To obtain an expression for the energy stored in the inductor we consider the circuit in Fig. 32.7a. In Kirchhoff's loop rule, Eq. 32.7, it is convenient to make a temporary change in the symbol for current from I to i, so

$$\mathscr{E} = iR + L\frac{di}{dt} \qquad (32.11)$$

If we multiply each term in Eq. 32.11 by i and rearrange them, we find

$$i\mathscr{E} = i^2R + Li\frac{di}{dt}$$

The product $i\mathscr{E}$ is the power supplied by the battery and i^2R is the power dissipated in the resistor. The last term represents the rate at which energy is being supplied to the inductor; that is,

$$\frac{dU_L}{dt} = Li\frac{di}{dt}$$

The total energy stored when the current has risen from 0 to I is found by integration:

$$U_L = \int_0^I Li\,di$$

which yields

$$U_L = \tfrac{1}{2}LI^2 \qquad (32.12) \qquad \text{Energy stored in an inductor}$$

This result should be compared with the expression for the energy stored in a capacitor, $U_C = \tfrac{1}{2}Q^2/C$.

EXAMPLE 32.4: A 50-mH inductor is in series with a 10-Ω resistor and a battery with an emf of 25 V. At $t = 0$ the switch is closed. Find: (a) the time constant of the circuit; (b) how long it takes the current to rise to 90% of its final value; (c) the rate at which energy is stored in the inductor; (d) the power dissipated in the resistor.

Solution: (a) The time constant is $\tau = L/R = 5 \times 10^{-3}$ s.
(b) We need the time taken for I to reach $0.9I_0 = 0.9\mathscr{E}/R$. From Eq. 32.8,

$$0.9I_0 = I_0(1 - e^{-t/\tau})$$

From this we find that $\exp(-t/\tau) = 0.1$, which may be expressed as $(-t/\tau) = \ln(0.1)$. Thus,

$$t = -\tau \ln(0.1) = 11.5 \times 10^{-3} \text{ s}$$

(c) The rate at which energy is supplied to the inductor is

$$\frac{dU_L}{dt} = +LI\frac{dI}{dt}$$

From Eq. 32.8, $dI/dt = +\mathscr{E}/L\ e^{-Rt/L}$. Therefore,

$$P_L = \frac{dU_L}{dt} = I\mathscr{E}e^{-t/\tau}$$

We now substitute for I again from Eq. 32.8 to obtain

$$P_L = \frac{\mathscr{E}^2}{R}[e^{-t/\tau} - e^{-2t/\tau}]$$

(d) The power dissipated in the resistor is

$$P_R = I^2R = I_0^2R(1 - 2e^{-t/\tau} + e^{-2t/\tau})$$

EXERCISE 4.   What is the rate at which the battery is supplying energy? How is your answer related to P_R and P_L in parts (c) and (d)?

Energy Density of the Magnetic Field

We may regard the energy of the inductor as being stored in its magnetic field. Let us consider the specific case of a solenoid. From Example 32.1 we know that the self-inductance of a solenoid is $L = \mu_0 n^2 A\ell$. From Eq. 30.11 we know that the field strength (assumed to be uniform) is $B = \mu_0 nI$. Equation 32.12 may be expressed in the form

$$U = \frac{1}{2} LI^2 = \frac{B^2}{2\mu_0} A\ell$$

Since $A\ell$ is the volume of the solenoid, the energy per unit volume, $u = U/\text{volume}$, is

Energy density of a magnetic field

$$u_B = \frac{B^2}{2\mu_0} \tag{32.13}$$

This is the **energy density of a magnetic field** in free space. Although Eq. 32.13 has been obtained from a special case, the expression is valid for any magnetic field. Compare it with Eq. 26.10 for the energy density of an electric field, $u_E = \frac{1}{2}\varepsilon_0 E^2$. In both cases, the energy density is proportional to the square of the field strength.

EXAMPLE 32.5: The breakdown electric field strength of air is 3×10^6 V/m. A very large magnetic field strength is 20 T. Compare the energy densities of these fields.

Solution: The electric field energy density is

$u_E = \frac{1}{2}\varepsilon_0 E^2$
   $= (0.5)(8.85 \times 10^{-12} \text{ C}^2/\text{N} \cdot \text{m}^2)(3 \times 10^6 \text{ V/m})^2$
   $\approx 40 \text{ J/m}^3$

The energy density for the magnetic field is

$$u_B = \frac{B^2}{2\mu_0}$$
$$= \frac{(20 \text{ T})^2}{2 \times 4\pi \times 10^{-7} \text{ N/A}^2}$$
$$= 3.2 \times 10^8 \text{ J/m}^3$$

Clearly magnetic fields are an effective means of storing energy without breakdown of the air. However, it is difficult to produce such large fields over large regions.

EXAMPLE 32.6: Use the expression for the energy density of the magnetic field to calculate the self-inductance of a toroid with a rectangular cross section; see Fig. 32.9.

Solution: The magnetic field within a toroid was found by using Ampère's law in Example 30.7:

$$B = \frac{\mu_0 NI}{2\pi r}$$

Figure 32.9 shows a thin cylindrical ring within the toroid of height h, radius r, and thickness dr. The volume of the cylinder is

$$dV = h(2\pi r \, dr)$$

From Eq. 32.13, $u_B = B^2/2\mu_0$, so the energy in this elemental volume is

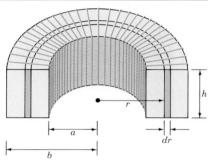

FIGURE 32.9 Since the field within a toroid is not uniform one must first obtain an expression for the energy in a cylindrical volume of radius r and thickness dr.

$$dU = u_B \, dV = \frac{\mu_0 N^2 I^2 h}{4\pi} \frac{dr}{r}$$

The total energy within the toroid is

$$U = \int dU = \frac{\mu_0 N^2 I^2 h}{4\pi} \int_a^b \frac{dr}{r}$$
$$= \frac{\mu_0 N^2 I^2 h}{4\pi} \ln \frac{b}{a}$$

On comparing this with the expression $U = \frac{1}{2}LI^2$ we see that the self-inductance of the toroid is

$$L = \frac{\mu_0 N^2 h}{2\pi} \ln \frac{b}{a}$$

You are asked to derive this expression in Problem 6 by calculating the flux through the toroid.

32.4 *LC* OSCILLATIONS

The ability of an inductor and a capacitor to store energy leads to the important phenomenon of electrical oscillations. Figure 32.10*a* shows a capacitor with initial charge Q_0 connected to an ideal inductor having no resistance. All the energy in the system is stored in the electric field: $U_E = Q_0^2/2C$. At $t = 0$, the switch is closed and the capacitor starts to discharge (see Fig. 32.10*b*). As the current increases, it sets up a magnetic field in the inductor, and so part of the energy is stored in the magnetic field, $U_B = \frac{1}{2}LI^2$. When the current reaches its maximum value I_0, as in Fig. 32.10*c*, all the energy is in the magnetic field: $U_B = \frac{1}{2}LI_0^2$. The capacitor now has no energy, which means $Q = 0$. Thus, $I = 0$ when $Q = Q_0$, and $Q = 0$ when $I = I_0$. The current now starts to charge the capacitor, as in Fig. 32.10*d*. In Fig. 32.10*e*, the capacitor is fully charged, but with a polarity opposite to its initial state in Fig. 32.10*a*. The process just described will now repeat itself till the system reverts to its original state. Therefore the energy in the system oscillates between the capacitor and the inductor. As the block-spring system shown in the diagram

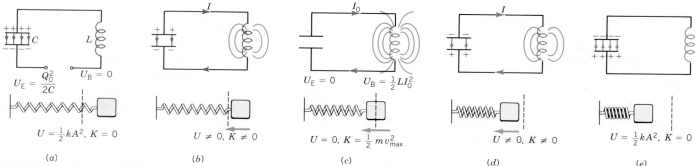

FIGURE 32.10 The oscillations in an *LC* circuit are analogous to the oscillation of a block at the end of a spring. The figure depicts one-half of a cycle.

suggests, the current and the charge in fact undergo simple harmonic oscillations. We will pursue the analogy later.

Consider the situation depicted in Fig. 32.10*b* and redrawn in Fig. 32.11. The current is increasing ($dI/dt > 0$), which means the induced emf in the inductor has the polarity shown, so $V_b < V_a$. According to Kirchhoff's loop rule,

$$\frac{Q}{C} - L\frac{dI}{dt} = 0$$

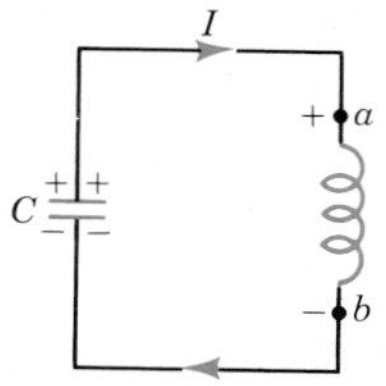

In order to relate the current in the wire to the charge on the capacitor, we note that the current causes Q on the capacitor to decrease, so $I = -dQ/dt$. With this, the loop rule becomes

$$\frac{d^2Q}{dt^2} + \frac{1}{LC}Q = 0$$

FIGURE 32.11 At the instant depicted the current is increasing, so the polarity of the induced emf in the inductor is as shown.

This has the same form as Eq. 15.5 for simple harmonic oscillation: $d^2x/dt^2 + \omega^2 x = 0$. The charge therefore oscillates with a **natural angular frequency**

$$\omega_0 = \frac{1}{\sqrt{LC}}$$

(32.14)   Natural angular frequency

and, in general, its variation with time is given by

$$Q = Q_0 \sin(\omega_0 t + \phi)$$

where Q_0 is the maximum value (amplitude) of Q and ϕ is a phase constant. Since $Q = Q_0$ at $t = 0$, we have $\sin \phi = 1$, and so $\phi = \pi/2$. Since $\sin(\theta + \pi/2) = \cos \theta$, the charge is given by

$$Q = Q_0 \cos(\omega_0 t) \tag{32.15}$$

The current, $I = -dQ/dt$, is given by

$$I = I_0 \sin(\omega_0 t) \tag{32.16}$$

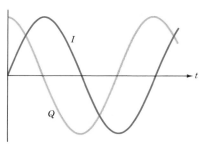

where $I_0 = \omega_0 Q_0$. These two functions are plotted in Fig. 32.12. To understand why I and Q are 90° out of phase, consider the loop rule, which requires the potential differences across C and L to be equal at all times; that is, $Q/C = L \, dI/dt$. Thus, if $Q = 0$, then $dI/dt = 0$, which means I must be a maximum or a minimum. This confirms the analysis based on the energy exchanges between C and L.

FIGURE 32.12 Both the current in the inductor and the charge on the capacitor vary sinusoidally. They are one-quarter of a cycle out of phase.

It is helpful to pursue the analogy between LC oscillations and the mechanical oscillation of a block attached to a spring. For a spring, the displacement x is determined by $F = ma$, where $F = -kx$ and $a = d^2x/dt^2$. Therefore, $d^2x/dt^2 + (k/m)x = 0$ and the natural angular frequency is $\omega_0 = \sqrt{k/m}$. Note first that x corresponds to Q in the differential equation. Second, the equation $F = m \, dv/dt$ has the same form as $\mathscr{E} = -L \, dI/dt$. Thus, L is analogous to m: Mass is a measure of resistance to change in velocity and L is a measure of resistance to change in current. Finally, on comparing $\omega_0 = 1/\sqrt{LC}$ with $\omega_0 = \sqrt{k/m}$ we see that $1/C$ is analogous to k. The constant $k = F/x$ tells us the (external) force needed to produce a unit displacement, whereas $1/C = V/Q$ tells us the potential difference needed to store a unit charge. Table 32.1 illustrates further analogies. Each part of Fig. 32.10 is related to the corresponding stage of the mechanical oscillator.

TABLE 32.1 ANALOGIES BETWEEN MECHANICAL AND ELECTRICAL QUANTITIES

Mechanical:	x	v	m	$\frac{1}{2}mv^2$	k	$\frac{1}{2}kx^2$	F	$P = Fv$
Electrical:	Q	I	L	$\frac{1}{2}LI^2$	$\dfrac{1}{C}$	$\dfrac{1}{2}\dfrac{Q^2}{C}$	V	$P = VI$

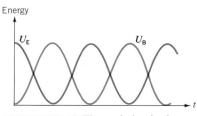

Energy

The energy stored in the capacitor is $U_E = Q^2/2C$ and that in the inductor is $U_B = \frac{1}{2}LI^2$. In the mechanical analog, the spring potential energy corresponds to U_E, whereas the kinetic energy corresponds to U_B. From Eqs. 32.15 and 32.16, the total energy is

$$U = U_E + U_B = \frac{Q_0^2}{2C} \cos^2(\omega_0 t) + \frac{LI_0^2}{2} \sin^2(\omega_0 t)$$

The variation in both U_E and U_B is depicted in Fig. 32.13. Since $I_0 = \omega_0 Q_0$ and $\omega_0 = 1/\sqrt{LC}$, it is easy to show that U is a constant:

$$U = \frac{Q_0^2}{2C} = \frac{1}{2} LI_0^2$$

FIGURE 32.13 The variation in the energy in the capacitor and the inductor.

EXAMPLE 32.7: In an *LC* circuit, as in Fig. 32.10, $L = 40$ mH, $C = 20\ \mu$F, and the maximum potential difference across the capacitor is 80 V. Find: (a) the maximum charge on C; (b) the angular frequency of the oscillation; (c) the maximum current; (d) the total energy.

Solution: (a) $Q_0 = CV_0 = (2 \times 10^{-5}\ \text{F})(80\ \text{V}) = 1.6 \times 10^{-3}\ \text{C}$.

(b) The angular frequency is

$$\omega_0 = \frac{1}{\sqrt{LC}}$$

$$= \frac{1}{\sqrt{(4 \times 10^{-2}\ \text{H})(2 \times 10^{-5}\ \text{F})}} = 1120\ \text{rad/s}$$

(c) The maximum current is

$$I_0 = \omega_0 Q_0 = (1120\ \text{rad/s})(1.6 \times 10^{-3}\ \text{C}) = 1.79\ \text{A}$$

(d) The total energy is simply the initial energy of the capacitor,

$$U = \frac{Q_0^2}{2C} = 6.4 \times 10^{-2}\ \text{J}$$

The foregoing analysis of *LC* oscillations is not realistic for two reasons. First, any real inductor has resistance. This will be taken into account in the next section. A second, more subtle, problem is that even if the resistance were zero, the total energy of the system would *not* stay constant. It is radiated away from the system in the form of electromagnetic waves (which are discussed in Chapter 34). In fact, radio and TV transmitters depend on this radiation! Nonetheless, our analysis has shown that the system is a simple harmonic oscillator and we have obtained the natural frequency of oscillation.

32.5   DAMPED *LC* OSCILLATIONS

We now consider what effect a resistor has on *LC* oscillations. Figure 32.14 shows a resistor, an inductor, and a capacitor in series. We assume that the capacitor has an initial charge Q_0 and that the switch is closed at $t = 0$. At the instant depicted, the current is increasing, and so the polarity of the induced emf in the inductor is as shown. From the loop rule we have

$$\frac{Q}{C} - IR - L\frac{dI}{dt} = 0$$

Since $I = -dQ/dt$, the loop rule may be written as

$$L\frac{d^2Q}{dt^2} + R\frac{dQ}{dt} + \frac{Q}{C} = 0 \tag{32.17}$$

This equation has the same form as Eq. 15.21 for damped harmonic motion; that is,

$$m\frac{d^2x}{dt^2} + \gamma\frac{dx}{dt} + kx = 0 \tag{32.18}$$

Notice that R is analogous to the damping constant γ. By analogy with the expressions in Section 15.5, we write down the solution to Eq. 32.18:

$$Q = Q_0 e^{-Rt/2L}\cos(\omega't + \delta) \tag{32.19}$$

The **damped angular frequency** is

$$\omega' = \sqrt{\omega_0^2 - \left(\frac{R}{2L}\right)^2} \tag{32.20}$$

where $\omega_0 = 1/\sqrt{LC}$ is the natural angular frequency.

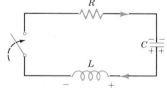

FIGURE 32.14 An *LRC* series circuit. At the instant depicted it is assumed that the current is increasing.

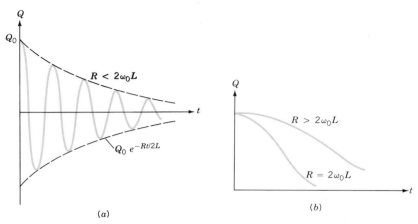

FIGURE 32.15 (a) When $R < 2\omega_0 L$, the system is underdamped and it oscillates with an exponentially decaying amplitude. (b) When $R = 2\omega_0 L$, the system is critically damped; when $R > 2\omega_0 L$, the system is overdamped.

The behavior of the system depends on the relative values of ω_0 and $R/2L$. When $R/2L < \omega_0$, or equivalently, when $R < 2\omega_0 L$, the system is *underdamped* and the charge varies according to Eq. 32.19. The amplitude of the oscillations decays exponentially as shown in Fig. 32.15a. When $R = 2\omega_0 L$, the system is *critically damped*. In this case there is no oscillation, and the charge falls to zero in the shortest time. Finally, when $R > 2\omega_0 L$, the system is *overdamped*. The last two cases are shown in Fig. 32.15b. In the next chapter we will discuss the response of an *LRC* series circuit to an external sinusoidally varying emf.

EXAMPLE 32.8: In an *RLC* series circuit $L = 20.0$ mH, $C = 50\ \mu$F, and $R = 6.0\ \Omega$. Find: (a) the time taken for the amplitude to fall to half the initial value; (b) the damped angular frequency; (b) the number of oscillations in 20 ms.

Solution: (a) From Eq. 32.19 we see that when the amplitude falls to half the initial value, $0.5 = \exp(-Rt/2L)$; therefore, the half-life of the oscillations is

$$T_{1/2} = \frac{2L}{R} \ln 2 = 4.6 \times 10^{-3} \text{ s}$$

(b) The natural angular frequency is $\omega_0 = 1/\sqrt{LC} = 10^3$ rad/s and the damped angular frequency is

$$\omega' = \sqrt{\omega_0^2 - \left(\frac{R}{2L}\right)^2}$$

$$= \sqrt{(10^3 \text{ rad/s})^2 - \left(\frac{6\ \Omega}{4 \times 10^{-2} \text{ H}}\right)^2}$$

$$= 990 \text{ rad/s}$$

(c) The period of the oscillations is $T' = 2\pi/\omega' = 6.4$ ms. Therefore, in 20 ms there are $20/6.4 = 3.1$ oscillations.

EXERCISE 5. For what value of R would the system be critically damped?

32.6 MAGNETIC PROPERTIES OF MATTER (Optional)

A bar magnet strongly attracts an iron nail, but other materials are only weakly attracted, and some are actually repelled. We may use the response of material to the nonuniform field of a bar magnet to broadly classify magnetic materials. When a **ferromagnetic** sample is placed in a nonuniform field, it is attracted to the stronger region of the field (Fig. 32.16a). A **paramagnetic** material is also attracted to the region of stronger field but only weakly. A **diamagnetic** material is weakly repelled by the mag-

net. As shown in Fig. 32.16b it tends to move to regions of weaker field. Some examples of these materials are given below:

Ferromagnetic: Fe, Ni, Co, Gd, and Dy; alloys of these and other elements and oxides, such as CrO_2, EuO, and Fe_3O_4 (magnetite)

Paramagnetic: Al, Cr, K, Mg, Mn, and Na

Diamagnetic: Cu, Bi, C, Ag, Au, Pb, and Zn

When a material is placed in an external magnetic field B_0,

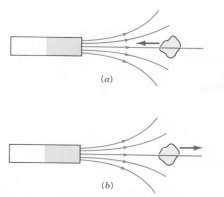

FIGURE 32.16 (*a*) A ferromagnetic or paramagnetic material is attracted to a magnet. (*b*) A diamagnetic material is repelled by a magnet.

the resultant field within the material is different from B_0. The field due to the material itself, B_M, is directly proportional to B_0:

$$B_M = \chi_m B_0 \qquad (32.21)$$

where χ_m is called the **magnetic susceptibility.** Thus, the total field within the material is

$$
\begin{aligned}
B &= B_0 + B_M \\
&= (1 + \chi_m)B_0 \\
&= \kappa_m B_0 \qquad (32.22)
\end{aligned}
$$

where $\kappa_m = 1 + \chi_m$, called the **relative permeability,** plays a role similar to that of the dielectric constant. Both χ_m and κ_m are dimensionless numbers.

In a paramagnetic material, the field increases, which means that the magnetic susceptibility is positive. Typical values are $\chi_m \approx 10^{-5}$ and depend on temperature. Within a diamagnetic material, the field is reduced, which means that the susceptibility is negative. Typical values are $\chi_m \approx -10^{-5}$ and do not depend on temperature. The susceptibility of a ferromagnetic material is dependent on temperature, the value of the external field B_0, and the (magnetic) history of the particular sample. (We will see why later.) Typical values of χ_m range from 10^3 to 10^5.

Atomic Moments

The magnetic properties of matter are mainly associated with the motions of electrons. The moving electrons in an atom set up atomic currents which have magnetic dipole moments and produce magnetic fields. In the semiclassical Bohr model of the hydrogen atom, an electron orbits a stationary proton. In Example 29.6, we showed that the magnetic moment (μ) associated with the orbital motion is related to the *orbital angular momentum* (*L*) by

$$\mu = \frac{eL}{2m}$$

This relation turns out to be true even in quantum mechanics. One rule of quantum mechanics (which we will discuss in Chapter 41) is that angular momentum is **quantized,** that is, it appears only in integer multiples of a basic unit: $L = n\hbar = 0, \hbar, 2\hbar, \ldots$, where $\hbar = h/2\pi$ and h is called Planck's constant. On substituting $L = \hbar$ into the above equation, we find

$$\mu_B = \frac{e\hbar}{4m} \qquad (32.23)$$

The value of this quantity, called the **Bohr magneton,** is $9.27 \times 10^{-24} \ \text{A} \cdot \text{m}^2$. In most substances, the orientations of the angular momenta differ from atom to atom. As a result, the average value of the dipole moment taken over all atoms is zero. However, there is another source of magnetism.

According to quantum mechanics each electron has a certain *spin angular momentum*. Although the picture is not correct, one can imagine the electron spinning about an internal axis and thereby setting up internal currents. The magnetic moment associated with the spin is equal to the Bohr magneton. In many atoms and ions, the spin angular momenta are paired in opposite directions. As a result there is no net dipole moment. In some instances, one or two electrons are not paired and the atom acquires a permanent dipole moment. Let us see how this view helps us to understand the behavior of different magnetic materials.

DIAMAGNETISM

In a diamagnetic material the atoms have no permanent dipole moment. When an external field is applied, the orbital motion of the electrons is modified such that the change in dipole moment is directed opposite to the external field. As a result, the net field is lower than the external field. This behavior may be understood in terms of Lenz's law: The change in flux is opposed by the induced field. In the absence of a mechanism to dissipate the energy, the change in the electron currents persists even after the external field has reached a constant value. The diamagnetic effect, which is present in all materials, is very weak and is often masked by paramagnetic or ferromagnetic effects. A superconductor is a perfect diamagnet; its susceptibility is $\chi = -1$. That is, the external field is completely excluded from the interior of the superconductor.

PARAMAGNETISM

In a paramagnetic material, the atoms or ions have a permanent dipole moment. However, they interact only weakly and are randomly oriented in the absence of an external field. When an external field is applied, the dipoles tend to align along the field, but the alignment is counteracted by thermal agitation. Consequently, the alignment is not complete unless the external field is very high and the temperature is low. The partial alignment of the dipole moments enhances the external field. (Notice that this behavior is opposite to that of electric dipoles in a dielectric. In that situation, the alignment of dipoles results in a lower net internal field.)

The energies associated with the magnetic field and thermal agitation are estimated as follows. The energy required to turn a dipole through 180° is $2\mu B$. In a field of 1 T and $\mu = \mu_B = 9.27 \times 10^{-24}$ A·m², $2\mu B \approx 1.9 \times 10^{-23}$ J. The energy associated with the thermal agitation is approximately $kT \approx 6 \times 10^{-21}$ J at 300 K. Thus, the thermal energy is 200 times larger.

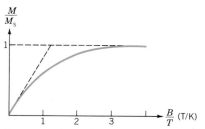

FIGURE 32.17 The magnetization of a paramagnetic sample as a function of B/T where B is the magnetic field and T is the absolute temperature.

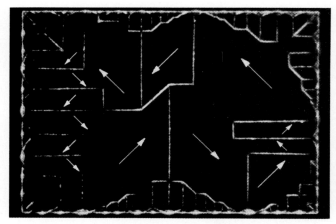

FIGURE 32.19 Domain walls made visible by iron filings.

Figure 32.17 shows how the magnetization M, defined as the magnetic moment per unit volume, varies with the ratio of the external field B over the absolute temperature T. The initial linear portion is adequately described by **Curie's law:**

$$M = C \left(\frac{B}{T} \right) \qquad (32.24)$$

where C is a constant. Note that even for a large field of 1 T and a temperature of 300 K, the horizontal coordinate is well within the linear range. As the field is increased or the temperature is lowered, the magnetization increases until it reaches the saturation value M_s, at which point all the dipoles are perfectly aligned with the field.

Ferromagnetism

In a ferromagnetic material, each atom has a magnetic moment that arises from the spin of one or two electrons. The moments on neighboring atoms tend to align parallel to each other via an interaction that can be explained only with quantum mechanics. In practice, the moments align perfectly only within small magnetic **domains** whose linear dimension is about 1 mm. Each domain contains about 10^{16} atoms. Although the alignment within each domain is perfect, the domains are randomly oriented, as shown in Fig. 32.18. They are separated by walls a few atoms thick in which the direction of magnetization gradu-

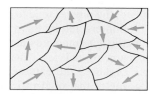

FIGURE 32.18 The domains in an unmagnetized sample are randomly oriented.

ally changes from one orientation to another. If ferromagnetic particles are sprinkled on the surface, the domain walls become visible under a microscope (see Fig. 32.19). The powder tends to collect at the walls, where the field is strongly nonuniform.

The formation of the domains may be understood qualitatively by considering the energies involved. Suppose all the moments in a crystal were aligned perfectly, as in Fig. 32.20. The end faces would then consist solely of north or south poles. The field due to these poles would be directed from the north to the south pole. The dipoles within the material would find themselves aligned opposite to this field. But this is the orientation with the highest energy. The spin–spin interaction between atoms lowers the energy of adjacent dipoles, but as the domains grow, the "bulk" magnetic energy just discussed increases. The sizes and random orientation of the domains represent a situation in which the total energy of the system is a minimum.

When an external field is applied, the domains respond in two ways. For weak fields, the domains whose moments are

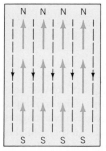

FIGURE 32.20 If all the domains were aligned perfectly as shown, there would be a magnetic field directed from the north poles at one end to the south poles at the other. This orientation of the domains is one in which the "bulk" magnetic energy is highest. It is not a stable configuration.

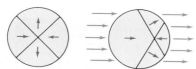

FIGURE 32.21 When an external magnetic field is applied, domains aligned parallel to the field grow at the expense of others. Domains also tend to align with the field.

aligned parallel to the field grow at the expense of the others. At stronger fields, the domains also rotate to align with the external field. Both effects are illustrated in Fig. 32.21.

The magnetization of a permanent magnet can be destroyed by dropping it or striking it sharply. The domains are jolted out of alignment. If the temperature is raised, the saturation magnetization decreases. Above the **Curie temperature** T_C, ferromagnetism vanishes and the substance becomes paramagnetic (see Fig. 32.22). The Curie temperatures of the five ferromagnetic elements are: Fe (1043 K), Co (1404 K), Ni (631 K), Gd (289 K), and Dy (85 K).

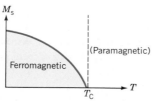

FIGURE 32.22 As the temperature of a ferromagnet is increased, its saturation magnetization decreases. Above the Curie temperature, the sample becomes paramagnetic.

Hysteresis

We now consider what happens when an external field is applied to an initially unmagnetized ferromagnet. In Fig. 32.23 we compare the total field B to the external field B_0. (Note that the scale for B is 1000 times larger!) As B_0 is increased from zero, B increases along the curve ab. After the initial steep rise, B approaches the saturation field B_s (2.1 T for Fe, 1.6 T for Permalloy) relatively slowly. When B_0 is reduced, the B field does not return along the original curve, but instead follows curve bcd. When B_0 has been reduced to zero, there is still a remnant field B_r, at point c, produced by the magnetization of the sample. This is what makes a permanent magnet. Once most of the domains have rotated to align along the field, they do not return to their original orientations. Their response "lags" behind the change in B_0. This lagging effect is called **magnetic hysteresis.** (Recall that we mentioned *elastic hysteresis* in connection with rolling friction.) As B_0 is reversed in direction, B reaches zero at point d. The field B_c, called the coercive force, tells us how difficult it is to destroy the magnetization of a sample. As B_0 is made more negative, the domains begin to align in the opposite

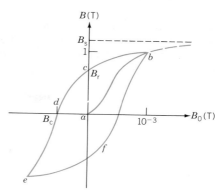

FIGURE 32.23 A hysteresis curve: The total field B is plotted as a function of the external field B_0. Note the difference in the scales.

direction, till they reach point e. When B_0 is reduced to zero and increased in the original direction, curve efb is traced. The area of the hysteresis loop is equal to the work required to take the material through a complete cycle. When the external field is produced by an alternating current, the reversals of the field result in the production of thermal energy.

Notice that the relative permeability, $\kappa_m = B/B_0$, is not constant but depends on both B_0 *and* the previous history of the sample. At the saturation field for Fe, $\kappa_m = 5000$, whereas for Permalloy it is 25,000.

Materials that have a large B_c are called magnetically "hard." Their hysteresis loop looks like Fig. 32.24a. Such materials are suitable for permanent magnets needed for speakers and moving-coil meters. Those, like Fe, that have a small B_c are magnetically "soft" and have hysteresis loops like that in Fig. 32.24b. The small area of the loop means that heat dissipation is minimized. Iron is useful in transformers, electromagnets, magnetic tape, and computer diskettes.

In order to demagnetize an object—for example, a magnetic tape recording head or a watch—it must be taken through a series of hysteresis cycles with ever-decreasing values of the

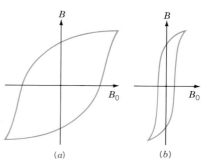

FIGURE 32.24 (a) A magnetically "hard" material suitable for permanent magnets. (b) A magnetically "soft" material suitable for electromagnets and magnetic recording.

external field (Fig. 32.25). An oscillating magnetic field is produced by passing an ac current through a coil. The coil is first placed near the magnetized object and then gradually moved away.

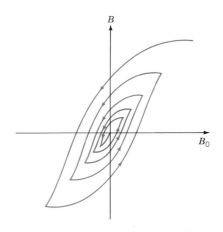

FIGURE 32.25 In order to demagnetize a sample it must be taken through many hysteresis cycles as the external field is decreased.

SUMMARY

When two current-carrying coils are close to each other, the magnetic flux in coil 1 has two contributions:

$$\Phi_1 = \Phi_{11} + \Phi_{12}$$

where Φ_{11} is the flux in coil 1 due to its own current I_1, and Φ_{12} is the flux in coil 1 due to the magnetic field produced by current I_2 of coil 2. When the current in the coils change, the net emf in coil 1 is $\mathcal{E}_1 = \mathcal{E}_{11} + \mathcal{E}_{12} = -d(N_1\Phi_1)/dt$. The self-induced emf in coil 1 is given by

$$\mathcal{E}_{11} = -L_1 \frac{dI_1}{dt}$$

where the **self-inductance** L_1 is found from

$$N_1\Phi_{11} = L_1 I_1$$

The emf induced in coil 1 due to changes in I_2 is

$$\mathcal{E}_{12} = -M \frac{dI_2}{dt}$$

where the **mutual inductance** M is found from

$$N_1\Phi_{12} = M I_2$$

If a given problem involves only self-inductance *or* mutual inductance, the subscripts may be dropped provided the meanings of the terms are kept clear.

In a circuit containing an inductor and a resistor, the inductor prevents the current from changing abruptly. When the current is first switched on, it rises according to

$$I = I_0(1 - e^{-t/\tau})$$

where the **time constant** is

$$\tau = \frac{L}{R}$$

When the battery is disconnected, the current decreases according to

$$I = I_0 e^{-t/\tau}$$

The **energy** stored in an inductor is

$$U_L = \tfrac{1}{2} LI^2$$

This energy is stored in the magnetic field. The **energy density in a magnetic field** is given by

$$u_B = \frac{B^2}{2\mu_0}$$

In an LC circuit, the oscillations of the charge on C are simple harmonic

$$Q = Q_0 \sin(\omega_0 t + \phi)$$

where the natural angular frequency is $\omega_0 = 1/\sqrt{LC}$. In the presence of a resistance, the angular frequency shifts to a lower value and the amplitude of the oscillations decreases.

ANSWERS TO IN-CHAPTER EXERCISES

1. $\mathcal{E}_2 = -N_2 \dfrac{d}{dt}(\Phi_{21} + \Phi_{22})$.

2. Use Eqs. 32.5 and 32.6.

3. From Exercise 2 we know that 1 H = 1 V · s/A. From $V = IR$ we know that 1 Ω = 1 V/A. Thus, the unit of L/R is 1 H/Ω = 1 s.

4. The power supplied by the battery is

$$P = I\mathcal{E} = I_0^2 R(1 - e^{-t/\tau})$$

This equals the sum, $P_L + P_R$, of the power supplied to the inductor and the resistor.

5. For critical damping $R = 2\omega_0 L = 40$ Ω.

QUESTIONS

1. Precision resistors are often made of wire wound around a ceramic core. How can the self-inductance be minimized?

2. Two circular coils are near each other. How should they be oriented such that their mutual inductance is (a) a maximum; (b) a minimum?

3. Are the relations $L = N\Phi/I$ and $L = -\mathcal{E}/(dI/dt)$ equally general? If not, which is preferable as a definition of L? Explain why you eliminated the other possibility.

4. For a solenoid of finite length, is the self-inductance per unit length different at the center than at the ends? If so, at which point is it greater?

5. A circuit has a high inductance. Devise a scheme that allows the current to be reduced to zero quickly, yet safely.

6. Show that $1/\sqrt{LC}$ has dimensions of T^{-1}.

7. Can there be an induced emf in an inductor even if the current through it is zero?

8. A real inductor has some resistance. Can the potential difference across the terminals of the inductor be (a) greater than the induced emf; (b) less than the induced emf?

9. (Optional) What is the effect on the inductance of a solenoid when an iron core is inserted? Does the inductance have a single value?

10. A coil surrounds a long solenoid. The magnetic field due to the solenoid is essentially zero at the position of the coil. Do the coil and solenoid have mutual inductance?

11. Is it possible to have mutual inductance without self-inductance? How about self-inductance without mutual inductance?

12. In an LR circuit does the value of the battery emf affect the time it takes to reach a given current? If not, what effect does it have?

13. Can one have an inductor without resistance? How about a resistor without inductance?

14. How does changing the radius of a coil that is around a toroid affect their mutual inductance?

15. Why is k analogous to $1/C$ rather than to C?

EXERCISES

32.1 Inductance

1. (I) (a) A solenoid of length 15 cm has 120 turns of radius 2 cm. Ignoring end effects, what is its self-inductance? (b) At what rate should the current through it change to produce a self-induced emf of 4 mV?

2. (I) The self-induced emf in a solenoid of length 25 cm and radius 1.5 cm is 1.6 mV when the current is 3 A and increasing at the rate 200 A/s. (a) What is the number of turns? (b) What is the magnetic field within the solenoid at the given instant? Ignore end effects.

3. (I) A solenoid has 500 turns and its self-inductance is 1.2 mH. (a) What is the flux through each turn when the current is 2 A? (b) What is the induced emf when the current changes at 35 A/s?

4. (I) Find the induced emf in an inductor L when the current varies according to the following functions of time: (a) $I = I_0 \exp(-t/\tau)$; (b) $I = at - bt^2$; (c) $I = I_0 \sin(\omega t)$.

5. (I) A solenoidal inductor has 50 turns. When the current is 2 A, the flux through each turn is 15 μWb. What is the induced emf when the current changes at 25 A/s?

6. (I) The self-induced emf in an inductor with 60 turns is 7.2 mV when the current changes at 16 A/s. What is the flux through each turn when the current is 4.5 A?

7. (I) When the current in an inductor changes at 128 A/s, the self-induced emf is 12 V. What is the self-inductance?

8. (I) A coaxial cable has a wire of radius 0.3 mm surrounded by a sheath of radius 4 mm. What is the self-inductance of 18 m of the cable?

9. (II) A flat circular coil with 5 turns of radius 2.4 cm is placed around a solenoid of length 24 cm that has 360 turns of radius 1.7 cm. The axis of the coil is at 10° to the axis of the solenoid. What is the mutual inductance?

10. (I) Two coils have a mutual inductance of 40 mH. What is the magnitude of the emf induced in coil 2 when the current in coil 1 changes at 25 A/s?

11. (I) A solenoid with a cross-sectional area of 8 cm² has 20 turns/cm. A second coil of 40 turns is closely wound around the solenoid. (a) What is the mutual inductance? (b) If the current in the solenoid changes according to $I = 3t - 2t^2$ A, what is the magnitude of the induced emf in the second coil at $t = 2$ s?

12. (I) Coil A has 5 turns of area 2.4 cm² and coil B has 6 turns of area 0.5 cm². The planes of the coils coincide. When the current in coil A is 2 A it produces an essentially uniform field of 10 μT over the area of coil B. Find: (a) the mutual inductance; (b) the emf induced in coil A when the current in B changes at 40 A/s?

13. (II) A toroid with N_1 turns has a rectangular cross section; see Fig. 32.9. The inner radius is a, the outer radius is b, and the height is h. A coil with N_2 turns is wrapped around the toroid. Find: (a) the flux through the coil; (b) their mutual inductance. (See Example 30.7.)

14. (I) A coil of radius 2 cm has 12 turns. It is placed inside a solenoid of radius 2.5 cm that has 20 turns/cm. The axis of the coil is at 60° to the axis of the solenoid. Find their mutual inductance. Ignore end effects.

15. (II) Consider two solenoids with the following specifications: $L_1 = 20$ mH, $N_1 = 80$ turns; $L_2 = 30$ mH, $N_2 = 120$ turns, and $M = 7$ mH. At a certain instant the current in coil 1 is 2.4 A and is increasing at 4 A/s, the current in coil 2 is 4.5 A and increasing at 1.8 A/s. Find the magnitude of: (a) Φ_{11}; (b) Φ_{12}; (c) Φ_{21}; (d) $\mathscr{E}_{11}$; (e) $\mathscr{E}_{12}$; (f) $\mathscr{E}_{21}$.

32.2 *LR* Circuits

16. (II) An ideal battery with an emf $\mathscr{E}$ is connected at $t = 0$ to an ideal resistanceless inductor L. (a) How does the current vary as a function of time? (b) Obtain the result of part (a) from Eq. 32.8, by using the expansion $e^x \approx 1 + x$, which is valid for small x.

17. (I) In the circuit shown in Fig. 32.26, S_2 is open and S_1 is closed at $t = 0$. Find: (a) the current after 50 ms; (b) the emf in the inductor after 50 ms; (c) the time needed for the current to reach 80% of its final value.

18. (I) In Fig. 32.26, S_1 has been closed for a long time. At $t = 0$, S_2 is closed and S_1 is opened. (a) At what time does the potential difference across the resistor drop to 12.5% of its initial value? (b) What is the emf in the inductor at the time found in (a)?

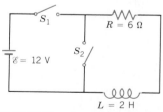

FIGURE 32.26 Exercises 17, 18, 19, 29, and 30

19. (I) In Fig. 32.26, S_2 is open and S_1 is closed at $t = 0$. (a) What is the initial rate of change of the current? (b) At what time does this rate drop to 50% of the initial value? (c) How long would it take for the current to reach its final value if the initial rate of change were maintained?

20. (I) (a) In an *LR* circuit the current falls to 25% of its initial value in 0.05 s. If $L = 6$ mH, what is R? (b) In an *LR* circuit, the current rises to 40% of its final value in 0.02 s. If $R = 10$ Ω, what is L?

21. (I) A coil has a resistance of 2 Ω and a self-inductance of 40

mH. The current is 6 A and is changing at 25 A/s. What is the potential difference across the coil given that the current is (a) increasing; (b) decreasing?

22. (I) In an *LR* circuit, the current rises to 40% of its final value in 40 ms. (a) How long does it take to reach 80% of the final value? (b) If $R = 12 \ \Omega$, what is *L*?

23. (I) In an *LR* circuit, $L = 120$ mH and $R = 15 \ \Omega$. The switch is closed at $t = 0$. (a) How long does it take for the current to rise to 50% of its final value? (b) What percentage of the final current is attained in five time constants?

24. (II) Consider the circuit shown in Fig. 32.27. Find the three currents: (a) when the switch is first closed; (b) after the currents have reached steady values; (c) when the switch is first opened (after being closed for a long time). (d) What is the potential difference across R_2 for case (c)?

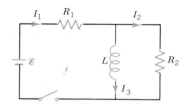

FIGURE 32.27 Exercise 24, Problem 10

25. (II) A solenoid of length 18 cm and radius 2 cm is formed with a tightly wrapped single layer of copper wire (diameter 1.0 mm, resistivity $1.7 \times 10^{-8} \ \Omega \cdot$ m). Estimate its time constant?

32.3 Energy

26. (I) (a) A solenoid has a self-inductance of 1.5 H. What is the energy stored when the current is 20 A? (b) A long solenoid with 120 turns produces a flux of 4×10^{-5} Wb (through a plane normal to the axis) when $I = 1.5$ A. What is the energy stored? Ignore end effects.

27. (I) (a) The earth's magnetic field strength is approximately 1 G near the surface. What is the magnetic energy density? (b) A solenoid of length 10 cm and radius 1 cm has 100 turns. What current would produce the energy density found in (a)? Ignore end effects.

28. (I) A coaxial cable has an inner wire of radius $a = 0.5$ mm and an outer sheath of radius $b = 2$ mm. If the current is 2 A what is the energy stored per meter along the cable?

29. (II) In Fig. 32.26, S_2 is open and S_1 is closed at $t = 0$. After a period of one time constant find: (a) the power loss in the resistor; (b) the rate at which energy is being stored in the inductor; (c) the power supplied by the battery.

30. (II) In the circuit of Fig. 32.26, S_2 is open and S_1 is closed at $t = 0$. Take $L = 25$ mH, $R = 60 \ \Omega$ and the battery emf to be 40 V. At $t = 1$ ms find: (a) the self-induced emf in the

inductor; (b) the power dissipated in the resistor; (c) the power supplied to the inductor; and (d) the power delivered by the battery.

31. (II) A resistor $R = 5 \ \Omega$ is in series with an inductor $L = 40$ mH and a 20-V battery. The switch is closed at $t = 0$. At what time will the power loss in R equal the rate at which energy is being stored in L? Also express your answer in terms of the number of time constants.

32. (I) An inductor stores 1.2 J when the current through it is 4 A. What is the self-inductance?

33. (I) A solenoid with 300 turns has a length of 20 cm and a radius of 1.8 cm. For what current will the energy density within the solenoid be 8 mJ/m³?

34. (I) A toroidal coil is in series with a 60-Ω resistor and a 24-V battery. The switch is closed at $t = 0$. Given that the current reaches 180 mA in 2 ms, find: (a) the self-inductance of the toroid; (b) the final energy stored in the toroid.

35. (II) A solenoid has a cross-sectional area A and a length ℓ. (a) If it carries a current I, obtain an expression for the energy density within the solenoid. (b) Equate the total energy of the solenoid to $\frac{1}{2} LI^2$, and thereby determine the self-inductance. Compare your result with that obtained in Example 32.1.

36. (I) The current in an inductor $L = 160$ mH varies according to $I = 2.5 \sin (150 \ t)$ A. Find: (a) the induced emf $\mathcal{E}$ at 1.2 ms; (b) the instantaneous power supplied to the inductor at $t = 1.2$ ms.

32.4 and 32.5 *LC* and Damped *LC* Oscillations

37. (II) A capacitor $C = 10 \ \mu$F has an initial charge of 60 μC. It is connected across an inductor $L = 8$ mH at $t = 0$. (a) What is the frequency of oscillation? (b) What is the maximum current through L? (c) What is the first time at which the energy is equally shared by C and L?

38. (I) In an *LC* circuit, the capacitor $C = 25$ nF, takes 10^{-4} s to lose all 20 μC of its initial charge. (a) What is L? (b) What is the maximum energy stored in L?

39. (I) In the tuning circuit of an AM radio, the inductance is 5 mH. Over what range must the capacitance vary for the circuit to detect the AM band from 550 kHz to 1600 kHz?

40. (II) In a series *RLC* circuit, $R = 20 \ \Omega$, $L = 4$ mH and $C = 20 \ \mu$F. (a) What is the damped angular frequency? (b) For what resistance will the system be critically damped?

41. (II) In an *RLC* circuit $R^2 \ll 4L/C$. Show that the total energy stored in C and L is given by

$$U \approx \frac{Q_0^2}{C} e^{-Rt/L}$$

42. (II) In an *RLC* circuit $L = 40$ mH and $C = 0.01 \ \mu$F. (a) What is the natural angular frequency ω_0? (b) For what resistance would the damped angular frequency be 0.1% lower than ω_0?

PROBLEMS

1. (I) Find the effective self-inductance of two inductors connected as follows: (a) in series; (b) in parallel. Ignore their mutual inductance.

2. (II) Two solenoids with self-inductances L_1 and L_2 and a mutual inductance M are connected in series. Show that their effective self-inductance is $L_{eff} = L_1 + L_2 \pm 2M$. Why are there two possible signs?

3. (I) The centers of two parallel long wires of radius a are separated by a distance d. They carry equal currents in opposite directions. Show that if the flux within the wires is neglected, the self-inductance per unit length is $L = (\mu_0/\pi)$ $\ln[(d - a)/a]$.

4. (I) Consider a long, straight wire of radius a. What is the contribution to its self-inductance per unit length from the flux within the wire? (See Example 30.5.)

5. (I) A long, straight wire and a rectangular loop lie in the same plane, as shown in Fig. 32.28. What is their mutual inductance?

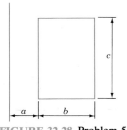

FIGURE 32.28 Problem 5

6. (I) Find the self-inductance of the toroid in Fig. 32.9 by calculating the flux through it. (See Example 32.6.)

7. (II) (a) Show that in an underdamped RLC oscillation, the fraction of energy lost per cycle is

$$\frac{|\Delta U|}{U} \approx \frac{2\pi}{Q}$$

where $Q = \omega L/R$ is called the Q-factor. (*Hint*: Use Eq. 32.19 to calculate the total energy at two times one period apart. The cosine will have the same value. Then use $e^x \approx$

1 + x for small x.) (b) If the energy loss per cycle is 2%, what is Q? (c) For the Q found in part (b), suppose $R = 0.5$ Ω, and $L = 18$ mH. What is C?

8. (I) In an LR circuit, $L = 20$ mH and $R = 9$ Ω. These elements are connected in series with a 60-V battery at $t = 0$. Find the time required for each of the following quantities to reach 50% of their maximum values: (a) the current; (b) the power dissipation in the resistor; (c) the energy stored in the inductor.

9. (I) A solenoid consists of a wire of radius a wrapped in a single layer on a paper cylinder of radius r. Show that the time constant is

$$\tau = \frac{L}{R} = \frac{\mu_0 \pi a r}{4\rho}$$

where ρ is the resistivity.

10. (II) In the circuit of Fig. 32.27 show that after the switch is closed at $t = 0$, I_3 is given by

$$I_3 = \frac{\mathcal{E}}{R_1}(1 - e^{-t/\tau})$$

where $\tau = L(R_1 + R_2)/R_1 R_2$.

11. (II) Prove that in an LR circuit (Fig. 32.8a) all the energy stored in the inductor is dissipated as thermal energy in the resistor.

12. (II) Given that the charge on the capacitor in an RLC circuit varies as $Q(t) = Q_0 e^{-Rt/2L} \cos(\omega' t)$, obtain an expression for the current as a function of time. Show that if $R/2L \ll \omega'$, then the magnitude of the current can be written in the form

$$I(t) \approx A(t) \sin(\omega' t + \delta)$$

where $A(t) = \omega' Q_0 e^{-Rt/2L}$ and $\tan \delta = R/(2L\omega')$.

13. (II) Consider two solenoids, one within the other. Write expressions for L_1 and L_2 and two expressions for M. Assuming all the flux from one intercepts the other, show that $M = \sqrt{L_1 L_2}$.

14. Show that the average power loss in an underdamped RLC circuit, as in Fig. 32.14, is

$$P_{av} = \frac{\omega_0^2 Q_0^2 R}{2} e^{-Rt/L}$$

SPECIAL TOPIC: Magnetic Levitation and Propulsion

In May, 1990, the *train à grende vitesse* (TGV) reached a record speed of 515 km/h. Yet trains that employ friction between wheels and rails ordinarily cannot travel faster than about 300 km/h without losing traction. Another approach to high-speed transportation is the hovercraft—a vehicle supported on a cushion of air and propelled by an aircraft engine. However, a hovercraft is noisy and there are pollution problems. Another alternative employs electromagnetic forces both to lift and to propel a train. Such trains are quiet and can travel at speeds up to 500 km/h supported by **magnetic levitation.** The propulsion systems for such trains are based either on the *induction motor* or on the *synchronous motor.*

INDUCTION MOTORS

In the period between 1835 and 1885, electric motors operated mainly on direct current. There were no satisfactory motors that could operate on alternating current until the induction motor was invented by N. Tesla in 1888. To understand the principle of an induction motor, first consider a pivoted compass needle. If the south pole of a bar magnet is moved in a circle, as in Fig. 32.29, the north pole of the needle will be attracted to the magnet and set into rotation. Tesla devised a method for creating a rotating magnetic field without actual mechanical motion. This is accomplished by passing ac currents to two or more pairs of poles. Figure 32.30 shows a *stator*, which is usually made from laminated steel, with protruding poles around which coils are wrapped. Alternating currents are passed through the coils *AA'* and *BB'*, but they are shifted in phase by 90°, as shown in Fig. 32.31a. For example, when the current in one pair is at a maximum, the current in the other pair is zero. In this example, which employs a two-phase current and two pairs of poles, the resultant magnetic field rotates once per cycle of the current (see Fig. 32.31b). Thus, if the

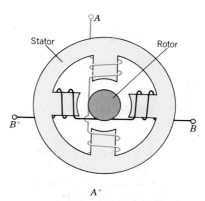

FIGURE 32.30 A rotating magnetic field is produced by supplying two (or more) sets of coils with ac currents that are not in phase.

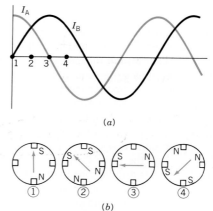

FIGURE 32.31 (*a*) Two currents 90° out of phase. (*b*) The direction of the magnetic field at four times.

frequency of the currents is 60 Hz, the magnetic field rotates at 60 rev/s.

The central *rotor* is not connected electrically to the stator. In one form of rotor called a squirrel cage, shown in Fig. 32.32a, copper rods are inserted into holes in a laminated iron cylinder. The rods are connected together by end rings. Suppose that the rotor is initially at rest and the two-phase current is applied to the stator. Since the rotor moves relative to the (rotating) magnetic field, large currents are induced in the rods. The direction of the induced

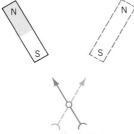

FIGURE 32.29 A compass needle follows the rotation of a magnet.

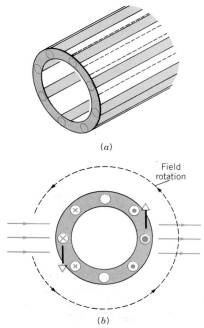

(a)

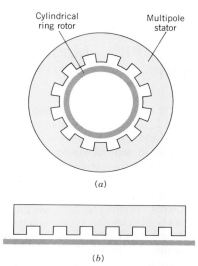

Cylindrical ring rotor

Multipole stator

(a)

(b)

FIGURE 32.33 (a) The squirrel cage may be replaced by a continuous ring. (b) In the linear induction motor the stator and rotor are "opened up" and laid flat.

(b)

FIGURE 32.32 (a) A squirrel cage. (b) The induced currents experience forces as shown.

currents can be found from the magnetic force $\mathbf{F} = q\mathbf{v} \times \mathbf{B}$, where $\mathbf{v}$ is the velocity of a rod relative to the field.

Figure 32.32b shows the direction of the induced currents and the forces they experience due to the stator field. The sizes of the dots and crosses indicate the relative magnitudes of the currents. These forces on the rods cause the rotor to turn in the sense of rotation of the field. The rotor speeds up until it almost reaches the synchronous angular speed Ω_s, at which point there would be no relative motion between the rotor and the field. Consequently, there would be no induced currents and no torque on the rotor. In practice, in order to overcome frictional losses, the angular speed remains slightly below the synchronous value. When a load is applied to the rotor, its angular speed decreases. The induced currents, and therefore the torque generated, increase.

The induction motor also works if the copper rods in the rotor are replaced by a continuous cylindrical ring, as shown in Fig. 32.33a. Its efficiency is only about 10% of a normal induction motor. The rotational form of the induction motor could be used for propulsion of a vehicle, but there would be inevitable frictional losses in gears and wheels. The linear induction motor avoids these problems.

LINEAR INDUCTION MOTOR

Imagine the motor of Fig. 32.33a cut along a radial line and laid flat, as in Fig. 32.33b. The rotor becomes a flat con-

ducting strip laid on a bed of concrete, and is called the reaction rail. The vehicle carries a series of magnets through which an alternating current passes. To a first approximation, the magnetic field has a sinusoidal variation in space (see Fig. 32.34a). As a function of time, the magnetic

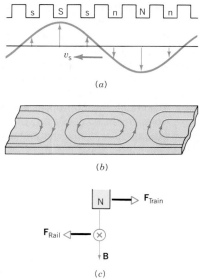

(a)

(b)

(c)

FIGURE 32.34 (a) By varying the currents in the electromagnets, the magnetic field travels from the front to the rear of the train. (b) The currents induced in the track. (c) The magnetic force on the induced currents is toward the rear. The force on the magnet (on the train) is in the forward direction.

field is made to travel from the front to the rear of the train. If λ is the separation between like poles, and f is the frequency of the current, the speed of the field relative to the train is $v_s = \lambda f$, the synchronous speed.

As the magnetic field sweeps across the reaction rail, eddy currents are induced in the rail (Fig. 32.34b). The forces on the induced currents due to the field are in the same direction as the motion of field—that is, toward the rear of the train. From Newton's third law, the forces on the magnets are in the forward direction (Fig. 32.34c). As the train speeds up, the relative speed between the magnetic field and the reaction rail decreases, so the propulsive force diminishes. The propulsive force would be zero at the synchronous speed.

Although several models have worked, there are problems with implementing the linear induction motor for high-speed intercity travel. The air gap between the train and the reaction rail is only about 3 cm and therefore does not allow much clearance for pitching or rolling motions of the train. Furthermore, about 5 MW of electrical power has to be picked up by a contact sliding on a railside cable. This would be a major problem at 400 km/h.

LINEAR SYNCHRONOUS MOTOR

In the linear synchronous motor powerful electromagnets on a train are supplied with direct current. As Fig. 32.35 shows, three wires, embedded in a concrete track, are connected to an alternating current supply of frequency f. The direction of the current under each pole is adjusted (by a suitable choice of phase constant) so that the force on the magnet is always in the forward direction. The spacing between poles must match the repeat distance λ of the track currents. Figure 32.36 shows how the currents in the wires change as the train moves forward. The synchronous speed is $v_s = \lambda f$. When a synchronous motor is used, it is necessary to vary the frequency of the currents to match the speed of the train. Since this is difficult to accomplish, other means, such as a linear induction motor, are employed to get the train's speed up to v_s. At any given time, only short segments of

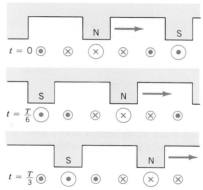

FIGURE 32.36 The current in each wire is adjusted so that the force on each pole is in the forward direction.

the track, about 5 km long, need to be hooked up to the (60 Hz) transmission lines. The train can be slowed down by changing the phase of the currents.

A given train may use 50 magnets of cross section 0.5 m × 1.5 m spaced 60 cm apart. The currents in the field coils, which are superconducting, can be as high as 5 × 10^5 A. For a thrust of 4 × 10^4 N, a current of only 250 A is required in the guideway.

MAGNETIC LEVITATION

There are two ways in which magnetic forces can be used to support the weight of a train. One employs the attraction between electromagnets on the train and an iron rail, as in Fig. 32.37. This approach is inherently unstable, because if the magnet gets closer to the rail, the attractive force increases. Careful electronic feedback controls are needed to adjust the current in the electromagnets. The other approach uses the repulsion between a magnet and the eddy currents it induces in a conductor. In practice, superconducting electromagnets are used.

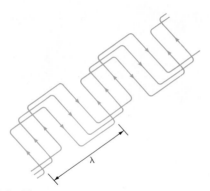

FIGURE 32.35 The pattern of wires embedded in a concrete track used in the linear synchronous motor.

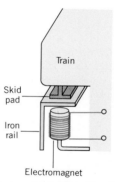

FIGURE 32.37 A train may be levitated by the attraction between an electromagnet and an iron rail.

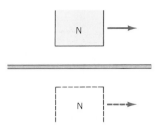

FIGURE 32.38 When a pole moves over a metal track it experiences a repulsive force (due to eddy currents) as if there were an "image" pole.

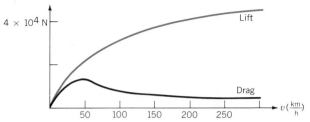

FIGURE 32.40 The variation in the lift and drag forces produced by a superconducting electromagnet in the Canadian MAGLEV project.

When a magnet moves relative to a conducting plate, the eddy currents induced in the plate result in both lift and drag forces on the magnet. The drag force arises because of the heat dissipation associated with the eddy currents. More significant is the presence of repulsive forces between the magnet and the conducting plate. It can be shown that the conducting plate acts like a "mirror." That is, the repulsive force exerted by the eddy currents can be calculated by imagining an "image" magnet underneath the plate (see Fig. 32.38).

Lift and drag forces are evident even with a long, straight, current-carrying wire that moves perpendicular to its length. The variation in the component of the wire's magnetic field perpendicular to the plate is shown in Fig. 32.39a. As the wire moves, the time variation in this vertical component is shown in Fig. 32.39b. The induced current just below the wire produces a lift force. The variation of the lift and drag force with speed is similar for wire and for a

magnet. Figure 32.40 illustrates the forces for a superconducting electromagnet used in the Canadian MAGLEV project. The magnet, 0.3 m × 1 m in cross section, carried a current of 4×10^5 A. Whereas the lift force increases steadily with speed, the drag force reaches a maximum at fairly low speeds and then decreases as the speed increases. This feature is, of course, very useful. The drag at high speeds is therefore mainly aerodynamic. The low-speed eddy current drag may be reduced by using a thicker plate (about 3 cm) for the first kilometer or so from a station and then reducing the thickness to the normal 1 cm. In any case, wheels are needed to support the train at slow speeds. Figure 32.41 shows a magnetically levitated train currently being operated in Japan.

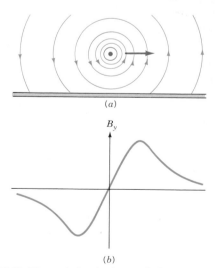

(a)

(b)

FIGURE 32.39 The variation in the vertical component of the magnetic field of a single current-carrying wire.

FIGURE 32.41 A magnetically levitated train in Japan.

CHAPTER 33

Alternating Current Circuits

The generator at the 47-MW hydroelectric plant at Snettisham, Alaska.

Major Points

1. The distinction between *instantaneous*, *peak*, and *root mean square* (*rms*) values of current, potential difference, and power.

2. The use of **phasors** to determine phase relationships between current and potential difference.

3. *RLC* series circuit:
    (a) **Ohm's law** for ac circuits.
    (b) **Impedance.**
    (c) **Resonance.**

4. The operation of a **transformer.**

Until now, we have discussed only circuits with *direct current* (dc)—which flows in only one direction. However, many sources produce an *alternating current* (ac) that changes direction periodically. The ac generator, discussed in Section 31.4, is used in power stations, bicycle dynamos, alternators in cars, and other devices. Every instrument or appliance plugged into a wall outlet is powered by an ac source of emf. The transmission and reception of radio or TV signals involve currents that vary sinusoidally in time. In audio systems, the outputs of phono cartridges, magnetic tape, and microphones are ac signals, which are boosted by an amplifier and fed to loudspeakers.

In this chapter we study the response of resistors, inductors, and capacitors to an applied alternating emf. We first treat each of these circuit elements individually and then study a series combination. The series *RLC* circuit is particularly important because the current in it exhibits resonance as the frequency of the ac source is varied.

33.1 SOME PRELIMINARIES

From the discussion of the ac generator in Section 31.4, we know that both the emf and the current produced vary sinusoidally in time. In this chapter lowercase letters are used for the *instantaneous* values of current and potential difference. It will be convenient for us to assume that the instantaneous current i always has the form

$$i = i_0 \sin(\omega t) \tag{33.1}$$

Instantaneous current

where $\omega = 2\pi f$ is the angular frequency in rad/s and f is the frequency in hertz (Hz) of the ac source of emf. The amplitude i_0 is called the *peak* value of the current. Two quantities, such as current and potential difference, are said to be *in phase* if they reach their peak values at the same time. As we will see, the instantaneous potential difference across a circuit element will not, in general, be in phase with the current through it; the peak values occur at different times. Thus, the instantaneous potential difference across the terminals of the source is written as

$$v = v_0 \sin(\omega t + \phi) \tag{33.2}$$

where v_0 is the peak value and ϕ is the phase angle between the current and the potential difference. (We assume that the source has no internal resistance, so the terminal potential difference is equal to the emf.) Our initial problem is to determine ϕ for a resistor, an inductor, and a capacitor connected individually to an ac source.

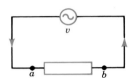

The sign of the instantaneous potential difference across a resistor (v_R), an inductor (v_L), or a capacitor (v_C), is determined by the way it is defined. Figure 33.1 shows a circuit element connected to the ac source, for which the circuit symbol is ⊙. The direction of the instantaneous current is indicated. If V_a and V_b are the potentials at points a and b, then any one of v_R, v_L, or v_C is defined to be $V_a - V_b$. We next discuss the response of R, L, and C individually to an ac source of emf.

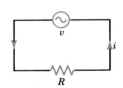

FIGURE 33.1 A circuit element, R, L, or C, connected to an ac source with an instantaneous potential difference v across its terminals. When the current flows in the direction shown, the potential at a is greater than the potential at b.

33.2   A RESISTOR IN AN AC CIRCUIT; ROOT MEAN SQUARE VALUES

In Fig. 33.2 a resistor is connected to an ideal ac source of emf whose instantaneous terminal potential difference is v. According to Kirchhoff's loop rule, $v - v_R = 0$ where $v_R = iR$ is the potential difference across the resistor. From Eq. 33.1, the instantaneous potential difference across the resistor is

$$v_R = i_0 R \sin(\omega t) = v_{0R} \sin(\omega t) \tag{33.3}$$

FIGURE 33.2 A resistor connected to an ac source.

so the peak value is

$$v_{0R} = i_0 R \tag{33.4}$$

Comparing Eqs. 33.1 and 33.3, we see that the current and potential difference are in phase ($\phi = 0$), as shown in Fig. 33.3.

The instantaneous power, p, dissipated in the resistor is

$$p = i^2 R = i_0^2 \sin^2(\omega t) R$$

In order to find the average power loss, one cannot use the average value of i, since this is obviously zero for each complete cycle. Physically we know that the average power loss is not zero because the energy dissipation is independent of the direction of the current. Therefore, we need to find the average value of i^2 over one complete cycle. This may be found by using the trigonometric identity $\sin^2\theta = \frac{1}{2}(1 - \cos 2\theta)$. The average of $\cos 2\theta$ over a complete cycle is zero, which leaves the $\frac{1}{2}$.* Thus

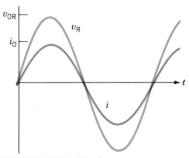

FIGURE 33.3 The instantaneous current i and the instantaneous potential difference v_R are in phase.

$$(i^2)_{av} = \frac{i_0^2}{2}$$

* More formally, the average value of a function $F(t)$ over a period T is given by $F_{av} = (1/T) \int_0^T F(t)\, dt$. In the present case $F(t) = i_0^2 \sin^2(\omega t) = i_0^2(1 - \cos 2\omega t)/2$.

The square root of this average (mean) is called the **root mean square (rms) current**, I (see Fig. 33.4):

$$I = \sqrt{(i^2)_{av}} = \frac{i_0}{\sqrt{2}} = 0.707i_0 \qquad (33.5a)$$

The uppercase letters I, V, and P, will be used for rms values of current, potential difference, and power. The rms potential difference is related to the peak potential difference in a similar fashion:

$$V = \sqrt{(v^2)_{av}} = \frac{v_0}{\sqrt{2}} = 0.707v_0 \qquad (33.5b)$$

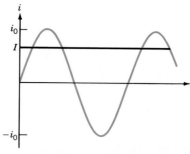

FIGURE 33.4 The rms current I is related to the instantaneous peak current i_0 by $I = i_0/\sqrt{2} = 0.707i_0$.

Since the rms values of the current and potential difference differ from their instantaneous values only by the factor $\sqrt{2}$, Eq. 33.4 may also be written as

$$V_R = IR \qquad (33.6)$$

This, of course, has the usual (dc) form of Ohm's law if R is constant. The average power, also called the **rms power**, is $P = p_{av} = (i^2)_{av}R$, so from Eq. 33.5a we have

(RMS power) $\qquad\qquad P = I^2R \qquad\qquad (33.7)$

We retain the usual (dc) form of the equation for electrical power provided the rms values of current or potential difference are used. The rms current I is the equivalent dc current that would produce the same average power loss as the alternating current. Equation 33.7 may also be written as $P = V_R^2/R = IV_R$.

EXAMPLE 33.1: A light bulb is rated at 100 W rms when connected to a 120-V rms wall outlet. Find: (a) the resistance of the bulb; (b) the peak potential difference of the source; (c) the rms current through the bulb.

Solution: (a) We are given $P = 100$ W and $V = 120$ V. The resistance is given by

$$R = \frac{V_R^2}{P} = 144 \ \Omega$$

(b) From Eq. 33.6 the peak potential difference of the source is

$$v_0 = \sqrt{2}\ V = 170 \ V$$

The potential difference fluctuates between -170 V and $+170$ V.

(c) Since $P = IV_R$, we have

$$I = \frac{P}{V_R}$$
$$= \frac{(100 \ W)}{(120 \ V)} = 0.833 \ A$$

EXERCISE 1. For the numbers in Example 33.1, what is the peak value of (a) the current, and (b) the instantaneous power?

33.3 AN INDUCTOR IN AN AC CIRCUIT

Figure 33.5 shows an inductor connected to an ac source. From the loop rule, we know that $v - v_L = 0$, where $v_L = L\ di/dt$ is the instantaneous potential difference across the inductor. (We use v_L rather than the induced emf $\mathscr{E}_L = -L\ di/dt$, so that the equations for R, L, and C will have the same form.) From Eq. 33.1, we find that the rate of change of current is

$$\frac{di}{dt} = i_0\omega\ \cos(\omega t)$$

so

$$v_L = L\frac{di}{dt} = v_{0L}\ \cos(\omega t) \qquad (33.8)$$

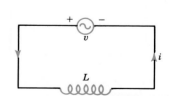

FIGURE 33.5 An inductor connected to an ac source.

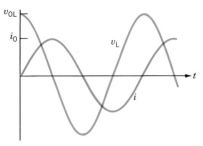

FIGURE 33.6 The instantaneous potential difference across the inductor, v_L, *leads* the instantaneous current, i, by 90°.

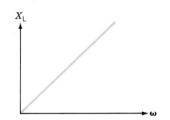

FIGURE 33.7 The reactance of an inductor is directly proportional to the angular frequency ω.

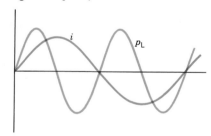

FIGURE 33.8 The average power delivered to an inductor is zero.

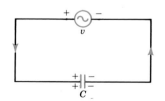

FIGURE 33.9 A capacitor connected to an ac source.

where the peak value is

$$v_{0L} = i_0 \omega L \tag{33.9}$$

Since $\cos(\omega t) = \sin(\omega t + 90°)$, the phase angle is $\phi = +90°$, which means that v_L *leads* i by 90°. The potential difference v_L reaches its peak value one quarter cycle before the corresponding peak in the current, as shown in Fig. 33.6, because v_L is determined not by the current, but instead by its rate of change di/dt. For example, v_L has its peak value when $i = 0$, because at this instant di/dt has its maximum value. Equation 33.9 may be written in a form that resembles Ohm's law:

$$v_{0L} = i_0 X_L \quad \text{or} \quad V_L = I X_L \tag{33.10}$$

where the quantity

$$X_L = \omega L \tag{33.11}$$

is called the **reactance** of the inductor. The SI unit of reactance is the ohm. The reactance of a circuit element is a measure of its opposition to the flow of ac current. As Eq. 33.10 suggests, reactance plays a role in an ac circuit that is similar to that of resistance in a dc circuit. The reactance of an element tells us the ac potential difference that must be applied to produce unit ac current through the circuit at a given frequency.

One can see why the form of X_L is reasonable by noting the following. In Section 32.4 we saw that inductance is analogous to mechanical inertia—that is, L is a measure of the opposition to change in current. Thus, in an ac circuit it is not surprising that $X_L \propto L$. Since the rate of change of current, di/dt, is proportional to ω, the potential difference across the inductor, $v_L = L \, di/dt$, is also proportional to ω. A larger angular frequency results in a greater opposing induced emf and a smaller current in the circuit. Thus, $X_L \propto \omega$ is also reasonable (see Fig. 33.7).

The instantaneous power supplied to the inductor is

$$p = i v_L = i_0 v_{0L} \sin(\omega t) \cos(\omega t)$$

By using the identity $\sin 2\theta = 2 \sin \theta \cos \theta$, we see that the average power over a complete cycle is zero (since the average of $\sin 2\theta$ is zero), as shown in Fig. 33.8. The energy stored by the inductor in one quarter period is returned to the source in the next quarter period.

EXERCISE 2. Is Eq. 33.10 also valid for the instantaneous current i and instantaneous potential difference v_L? Explain why or why not.

33.4   A CAPACITOR IN AN AC CIRCUIT

Figure 33.9 shows a capacitor connected to an ac source of emf. The charge on the positive plate of the capacitor is q. The current in the circuit (*not* through the capacitor itself!) is charging the plates, so $i = +dq/dt$ or $dq = i \, dt$. Thus,

$$q = \int i \, dt = \int i_0 \sin(\omega t) \, dt$$
$$= -\frac{i_0}{\omega} \cos(\omega t) + \text{constant} \tag{33.12}$$

The constant depends on the initial conditions and we take it to be zero. According to the loop rule, $v - v_C = 0$, where $v_C = q/C$ is the instantaneous potential difference across the capacitor. Using Eq. 33.12, we have

$$v_C = -\frac{i_0}{\omega C} \cos(\omega t) = -v_{0C} \cos(\omega t)$$

where the peak value is

$$v_{0C} = i_0 \frac{1}{\omega C} \qquad (33.13)$$

Since $-\cos(\omega t) = \sin(\omega t - 90°)$, the phase angle is $\phi = -90°$. This means that the potential difference across the capacitor v_C *lags* the current by 90°. As shown in Fig. 33.10, v_C reaches its peak value one quarter cycle after the peak in the current. In order to understand this phase relationship, start at time a in Fig. 33.10, where $v = v_C = 0$—which means $q = 0$ as well. As the source potential difference v rises, the charge on the capacitor increases to maintain the condition $v = v_C$. When v reaches its peak value at time b, there is no further need for charge to flow, and thus $i = 0$. As v starts to decrease, charge flows from the capacitor back to the source—which means that the current has reversed. (Note also that $i = dq/dt = C\, dv/dt$. Thus, i has its maximum value when $v = 0$ because dv/dt is a maximum at this time.)

Equation 33.13 may be written in the form of Ohm's law:

$$v_{0C} = i_0 X_C \quad \text{or} \quad V_C = IX_C \qquad (33.14)$$

where

$$X_C = \frac{1}{\omega C} \qquad (33.15)$$

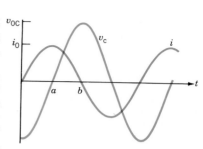

FIGURE 33.10 The instantaneous potential difference across the capacitor, v_C, *lags* the instantaneous current, i, by 90°.

is the reactance of the capacitor. To understand the form of Eq. 33.15 we first note that $X_C = \infty$ when $\omega = 0$. This is reasonable since a capacitor does not allow the steady flow of dc current. (Of course, a time-dependent dc current flows in the circuit when a capacitor is being charged or discharged.) To understand why $X_C \propto 1/\omega$, as depicted in Fig. 33.11, we note that the time required to reach a given peak potential difference is determined by the frequency of the source. As the frequency rises, charge has to flow to or from the capacitor in a shorter time. Thus, as the frequency increases, the current increases, which implies that the reactance decreases. (Also note that $i = C\, dv/dt$ and that $dv/dt \propto \omega$. That is, the current $i \propto \omega$.) Next let us see why $X_C \propto 1/C$. For a given potential difference, a large capacitor stores more charge than a small capacitor. Consequently, at any instant the current will be larger with the larger capacitor. Thus, $X_C \propto 1/C$ makes sense intuitively.

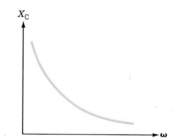

FIGURE 33.11 The reactance of a capacitor is inversely proportional to the angular frequency ω.

The instantaneous power supplied to the capacitor is

$$p = iv_C = -i_0 v_{0C} \sin(\omega t) \cos(\omega t)$$

As is the case with the inductor, the average power is zero. The energy stored by the capacitor in each quarter period is returned to the source in the next quarter period, as shown in Fig. 33.12.

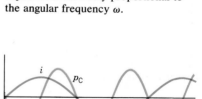

FIGURE 33.12 The average power delivered to a capacitor is zero.

33.5 PHASORS

The phase relationships between current and potential difference are easily determined for a single capacitor or inductor. When several such elements are combined in a circuit, we require more powerful analytical tools. One technique involves **phasors**. A phasor is a rotating vector that is used to represent quantities that vary sinusoidally in time. For example, the function $i = i_0 \sin(\omega t)$ may be represented by a phasor $\mathbf{i}_0$ that rotates counterclockwise at the angular frequency ω. As shown in Fig. 33.13a, the length of the phasor is the peak value i_0. As the phasor rotates, its component along the "vertical" axis indicates the variation of the instantaneous current. Figure 33.13b shows how the function $v = v_0 \sin(\omega t + \phi)$

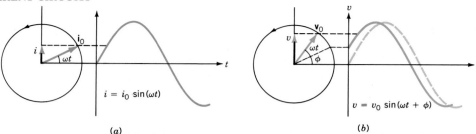

FIGURE 33.13 (*a*) As the current phasor $\mathbf{i}_0$ rotates, its vertical component is the instantaneous current i. (*b*) The potential difference phasor $\mathbf{v}_0$ is shifted by a phase angle ϕ relative to the current phasor.

may be generated. The position of the phasor $\mathbf{v}_0$ at $t = 0$ is determined by the phase angle ϕ. Notice that phasors may be used for quantities such as current and potential difference that are not vectors themselves.

Figure 33.14 illustrates the phase relationships between current and potential difference that were found in the last sections. In all cases $i = i_0 \sin(\omega t)$ and $v = v_0 \sin(\omega t + \phi)$. A *positive* phase angle means that the potential difference *leads* the current. In summary, $\phi = 0$ for the resistor, $\phi = +\pi/2$ for the inductor, and $\phi = -\pi/2$ for the capacitor. In the next section we will see how phasors help us to analyze circuits with more than one element.

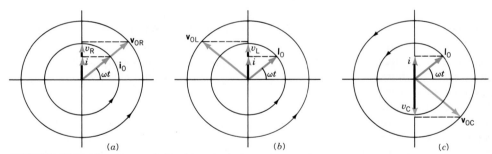

FIGURE 33.14 The phase relationships between the instantaneous current and the potential difference for (*a*) a resistor, (*b*) an inductor, and (*c*) a capacitor.

33.6 *RLC* SERIES CIRCUIT

We are now ready to consider a circuit consisting of a resistor, an inductor, and a capacitor in series with an ac source, as shown in Fig. 33.15. Our problem is to determine the instantaneous current and its phase relationship to the applied alternating potential difference v. The instantaneous current, $i = i_0 \sin(\omega t)$, is the *same* at all points in the circuit. (Why?) At the instant depicted we assume that the current is increasing. According to the loop rule, the instantaneous potential differences are related by

$$v - v_R - v_L - v_C = 0$$

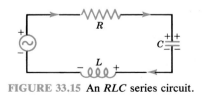

FIGURE 33.15 An *RLC* series circuit.

Each term in the sum $v = v_R + v_L + v_C$ has a different phase with respect to the current. In order to relate v to i, we must first find the (vector) sum of the potential difference phasors:

$$\mathbf{v}_0 = \mathbf{v}_{0R} + \mathbf{v}_{0L} + \mathbf{v}_{0C}$$

The "vertical" component of $\mathbf{v_0}$ gives the instantaneous value v. The potential difference phasors for each element and the current phasor, $\mathbf{i_0}$, are drawn in Fig. 33.16a, where we have assumed that v_{0L} is greater than v_{0C}. Thus, in Fig. 33.16b, $|\mathbf{v_{0L}} + \mathbf{v_{0C}}| = (v_{0L} - v_{0C})$. By Pythagoras' theorem, the magnitude of the sum is given by

$$v_0^2 = v_{0R}^2 + (v_{0L} - v_{0C})^2$$
$$= i_0^2[R^2 + (X_L - X_C)^2]$$

This may be written in the form of Ohm's law,

$$v_0 = i_0 Z \quad \text{or} \quad V = IZ \tag{33.16}$$

where V is the rms value of the potential difference applied by the source. The quantity

$$Z = \sqrt{R^2 + (X_L - X_C)^2} \tag{33.17}$$

is the **impedance** of the series circuit. The SI unit of impedance is the ohm. The impedance of a circuit determines the ac current that will flow for a given applied ac potential difference.

Since $\mathbf{v_{0R}}$ is always parallel to $\mathbf{i_0}$, the angle ϕ in Fig. 33.16b is the phase angle between $\mathbf{v_0}$ and $\mathbf{i_0}$. From the diagram we see that $\tan \phi = (v_{0L} - v_{0C})/v_{0R}$, which reduces to

$$\tan \phi = \frac{X_L - X_C}{R} \tag{33.18}$$

A positive phase angle indicates that the driving potential difference v is ahead of the current i by ϕ.

EXERCISE 3. Twice in every cycle the instantaneous potential difference applied by the source v is zero, but the current in the circuit is not zero. How is this possible?

EXERCISE 4. Are the peak potential differences related by $v_0 = v_{0R} + v_{0L} + v_{0C}$? Explain why or why not.

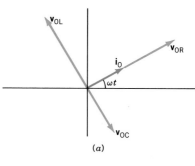

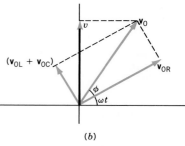

FIGURE 33.16 (a) The current phasor and the potential difference phasors for the resistor, the capacitor and the inductor. (b) The potential difference phasors are related by $\mathbf{v_0} = \mathbf{v_{0R}} + \mathbf{v_{0C}} + \mathbf{v_{0L}}$. We have assumed that $v_{0L} > v_{0C}$.

EXAMPLE 33.2: An ac source of emf with frequency 50 Hz and a peak potential difference of 100 V is in an *RLC* series circuit with $R = 9 \, \Omega$, $L = 0.04$ H, and $C = 100 \, \mu$F. Find: (a) the impedance; (b) the phase angle; (c) the peak potential difference across each element.

Solution: (a) The angular frequency is $\omega = 2\pi f = 100\pi$ rad/s. The reactances of L and C are

$$X_L = \omega L = 4\pi = 12.6 \, \Omega$$

$$X_C = \frac{1}{\omega C} = \frac{100}{\pi} = 31.8 \, \Omega$$

From Eq. 33.17 the impedance is

$$Z = \sqrt{R^2 + \left(\omega L - \frac{1}{\omega C}\right)^2}$$
$$= \sqrt{81 + (19.2)^2} = 21.2 \, \Omega$$

(b) From Eq. 33.18, the phase angle is given by

$$\tan \phi = \frac{X_L - X_C}{R} = \frac{-19.2}{9} = -2.13$$

Thus, $\phi = -64.8°$, which means that the potential difference of the source lags the current.

(c) The peak current through the circuit,

$$i_0 = v_0/Z$$
$$= \frac{100 \text{ V}}{21.2 \, \Omega} = 4.72 \text{ A}$$

is the same for all elements. The peak potential difference across each is given by

$$v_{0R} = i_0 R = 42.5 \text{ V}$$
$$v_{0L} = i_0 X_L = 59.5 \text{ V}$$
$$v_{0C} = i_0 X_C = 150 \text{ V}$$

Clearly, $v_0 \neq v_{0R} + v_{0L} + v_{0C}$.

33.7   *RLC* SERIES RESONANCE

For a given value of the rms potential difference V applied by the source, the rms current I is given by Eq. 33.16,

$$I = \frac{V}{Z} = \frac{V}{\sqrt{R^2 + (X_L - X_C)^2}}$$

where the impedance Z is a function of frequency. As the frequency of the source is varied, the impedance reaches a minimum value ($Z = R$) when $X_L = X_C$, that is, when $\omega L = 1/\omega C$. This condition defines the **resonance (angular) frequency**

$$\omega_0 = \frac{1}{\sqrt{LC}} \qquad (33.19)$$

This is the same as the natural angular frequency of oscillation of an LC circuit with no resistance (see Section 32.4). Figure 33.17 shows that as the angular frequency ω of the source is varied, the rms current I displays resonance behavior. At $\omega = \omega_0$, I reaches a maximum value given by

$$I_{max} = \frac{V}{R} \qquad (33.20)$$

The width of the resonance curve depends on the resistance, becoming sharper as the resistance is reduced. From Eq. 33.18 we see that when $X_L = X_C$, $\tan \phi = 0$, which means that $\phi = 0$: At the resonance frequency, the instantaneous current and potential difference are in phase.*

EXERCISE 5. How are the potential differences across L and C related at resonance?

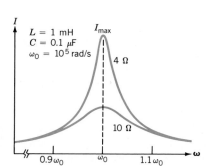

FIGURE 33.17 As the angular frequency ω of the source is varied, the (rms) current displays a resonance at the natural frequency ω_0. The width of the resonance peak increases as the resistance increases.

33.8   POWER IN AC CIRCUITS

The instantaneous power delivered by the source of emf is

$$p = iv = i_0 v_0 \sin(\omega t) \sin(\omega t + \phi)$$
$$= i_0 v_0 [\sin^2(\omega t) \cos\phi + \sin(\omega t) \cos(\omega t) \sin\phi]$$

Only the first term contributes to the power loss because the average of $\sin^2(\omega t)$ over one cycle is $\frac{1}{2}$, whereas the average of $\sin(\omega t) \cos(\omega t) = \frac{1}{2} \sin(2\omega t)$ is zero. Therefore, the average power is

$$p_{av} = \tfrac{1}{2} i_0 v_0 \cos \phi$$

From the phasor diagram of Fig. 33.16b, we see that $v_0 \cos \phi = v_{0R} = i_0 R$. Expressed in terms of rms values, the average, or **rms power**, $P = p_{av}$, delivered by the source is

Rms power

$$P = IV \cos \phi = I^2 R \qquad (33.21)$$

As we might have expected, power is dissipated only in the resistor. Note that V in Eq. 33.21 is the rms value of the potential difference of the *source* and not the potential difference across the resistor alone.

---

* Note that the resonance in the current occurs at exactly the natural frequency and not at the damped frequency as in the amplitude resonance shown in Fig. 15.17. The reason is that current is analogous to velocity, not amplitude. See Table 32.1.

The quantity $\cos \phi$ is called the **power factor.** Looking at $p_{av} = \frac{1}{2}i_0 v_0 \cos \phi$, we might say that only the component of $\mathbf{v}_0$ along $\mathbf{i}_0$; that is, $v_0 \cos \phi$, contributes to the average power loss. If $\cos \phi = 1$, the rms power has its maximum value, $P = IV$. In this case, the source "sees" a purely resistive load. If $\cos \phi = 0$, the load is either purely inductive or purely capacitive and there is no net transfer of energy from the source to the circuit.

The rms power, $P = I^2 R = (V/Z)^2 R$, delivered by the source is

$$P = \frac{V^2 R}{R^2 + \left(\omega L - \dfrac{1}{\omega C} \right)^2} \qquad (33.22)$$

As Fig. 33.18 shows, the rms power also displays a resonance at the natural angular frequency $\omega_0 = 1/\sqrt{LC}$, where it reaches its maximum value, $P_{max} = V^2/R$.

The resonance of an RLC series circuit is used in the detection of radio and TV signals. The sharpness of the resonance is an indication of the selectivity of the radio or TV receiver—its ability to reject signals whose frequency is close to ω_0. Typical values for Q in electronic circuits range from about 40 to 100.

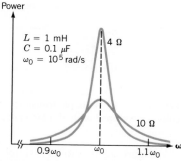

FIGURE 33.18 The (rms) power delivered by the source also displays a resonance.

EXAMPLE 33.3: In an RLC series circuit $R = 50 \ \Omega$, $C = 80 \ \mu$F, and $L = 30$ mH. The 60-Hz source has an rms potential difference of 120 V. Find: (a) the rms current and potential difference for each element; (b) the power factor; (c) the rms power delivered by the source; (d) the resonance frequency; (e) the peak values of current and potential difference for each element at the resonance frequency.

Solution: (a) The rms current $I = V/Z$ is the same for all elements. We first need to find the impedance. The reactances are $X_L = \omega L = (120\pi)(3 \times 10^{-2}) = 11.3 \ \Omega$ and $X_C = 1/\omega C = 1/(120\pi)(8 \times 10^{-5}) = 33.2 \ \Omega$. The impedance is

$$Z = \sqrt{R^2 + (X_L - X_C)^2} = 54.6 \ \Omega$$

Therefore, $I = V/Z = (120 \text{ V})/(54.6 \ \Omega) = 2.2$ A.

The rms potential difference across each element is

$$V_R = IR = 110 \text{ V}$$
$$V_L = IX_L = 24.9 \text{ V}$$
$$V_C = IX_C = 72.8 \text{ V}$$

Notice that $V \neq V_R + V_L + V_C$. An ac voltmeter reads rms values. Thus, the sum of its readings across the three elements will not equal the reading across the source.
(b) To determine the power factor $\cos \phi$, we first need to find ϕ.

$$\tan \phi = \frac{X_L - X_C}{R}$$
$$= \frac{11.3 \ \Omega - 33.2 \ \Omega}{50 \ \Omega} = -0.438$$

Thus, $\phi = -23.6°$. The power factor is $\cos(-23.6°) = 0.916$.
(c) The rms power delivered by the source is

$$P = IV \cos \phi = I^2 R = 242 \text{ W}$$

(d) The resonance frequency is

$$f_0 = \frac{\omega_0}{2\pi} = \frac{1}{2\pi \sqrt{LC}} = 103 \text{ Hz}$$

(e) At the resonance frequency $Z = R$; therefore, the peak current is $i_0 = v_0/R = \sqrt{2} \ V/R = (170 \text{ V})/(50 \ \Omega) = 3.4$ A. The reactances of L and C are equal at the resonance frequency:

$$X_L = X_C = \sqrt{\frac{L}{C}} = 19.4 \ \Omega$$

The peak potential differences are

$$v_{0R} = i_0 R = 170 \text{ V}; \qquad v_{0L} = v_{0C} = i_0 \sqrt{\frac{L}{C}} = 65.8 \text{ V}$$

EXERCISE 6.   What is the rms power delivered by the source at the resonance frequency?

33.9   THE TRANSFORMER (Optional)

The transformer is a device that can raise or lower the amplitude of ac potential differences. Transformers are used at several stages in the distribution of electrical power. In order to minimize thermal losses in the transmission cables, electrical power is transmitted at high voltage (perhaps 500 kV rms). For reasons of safety and simplicity in design, power is delivered to commercial and residential consumers at low voltage (typically 120 V rms). Many electronic circuits that obtain power from a wall outlet require a transformer. The "ignition coil" in a car is a transformer. Transformers are also helpful in isolating monitoring equipment connected to patients from interference caused by other devices or circuits.

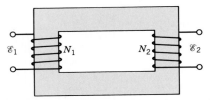

FIGURE 33.19 A simple transformer. The primary and secondary coils are wrapped around a laminated soft-iron core.

Figure 33.19 shows a simple transformer that consists of two coils wound around a soft iron core. The *primary* coil, connected to the ac source, has N_1 turns, while the *secondary* coil, connected to the load, has N_2 turns. A soft iron core serves to increase the flux and also to ensure that all the flux from one coil intercepts the other. The emf's that appear at the primary and secondary are

$$\mathscr{E}_1 = -N_1 \frac{d\Phi}{dt} \qquad \mathscr{E}_2 = -N_2 \frac{d\Phi}{dt}$$

where Φ is the flux through one turn. Taking the ratio of these we find

$$\frac{\mathscr{E}_1}{\mathscr{E}_2} = \frac{N_1}{N_2} \qquad (33.23)$$

The ratio of the emf's in the primary and secondary is equal to the "turns ratio." By adjusting the ratio of the number of turns in the coils, the primary emf may be "stepped up" or "stepped down." If the resistance of the wires in the primary circuit is negligible, the potential difference v_1 across the primary coil is equal to the source emf $\mathscr{E}_1$ at all times. Similarly, $\mathscr{E}_2 = v_2$.

With no load resistance at the secondary, there is no net transfer of energy. The ac source connected to the primary "sees" a purely inductive load, and only a small "magnetizing" current flows in the primary windings to establish a (varying) flux in the core. When a load resistance R_L is connected, as in Fig. 33.20, there is an induced current i_2 in the secondary. From Lenz's law, i_2 tends to oppose the flux changes in the core. This in turn would tend to lower the value of the emf of the primary. However, the emf of the primary must at all times be equal to the emf of the source (which we take to be unaffected). There-fore, to counteract the opposing emf associated with i_2, the primary current increases by an amount i_1. (i_1 is 180° out of phase with i_2.) From Eq. 32.5 we have $N_1\Phi_{12} = Mi_2$ and $N_2\Phi_{21} = Mi_1$. Assuming there is no leakage of flux, $\Phi_{12} = \Phi_{21}$, and so

$$\frac{i_2}{i_1} = \frac{N_1}{N_2} \qquad (33.24)$$

Since i_1 is much larger than the magnetizing current, i_1 is in effect the total current in the primary.

When a load is present in the secondary, the current in the primary does not have the value appropriate for a purely induc-

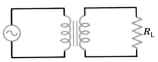

FIGURE 33.20 A schematic representation of a transformer transferring power from an ac source to a load resistor.

tive load. As a result, the power factor is no longer zero. From the conservation of energy, the extra energy supplied by the source of emf appears in the load connected to the secondary. Combining Eqs. 33.23 (with $\mathscr{E} = v$) and 33.24, we see that

$$i_1 v_1 = i_2 v_2 \qquad (33.25)$$

In practice, the efficiency of power transfer can be as high as 99%. The use of a soft iron core limits hysteresis losses (Section 32.6) and flux leakage. Eddy current heating is minimized by making the core with iron plates that are separated by an insu-lating material such as shellac or an oxide layer. Such *lamina-tions* greatly increase the resistance in the path of the induced eddy currents. Since the induced emf's are unchanged, the power loss is considerably reduced.

Impedance Matching

Another useful feature of a transformer arises because of the following circumstance. The rms values of the potential differ-ences across the primary and secondary are $V_1 = I_1Z_1$ and $V_2 = I_2Z_2$. If we express Eq. 33.25 in terms of these rms values, we have

$$\frac{Z_2}{Z_1} = \left(\frac{N_2}{N_1}\right)^2 \qquad (33.26)$$

Therefore, the primary current is given by

$$I_1 = \frac{V_1}{Z_1} = \frac{V_1}{(N_1/N_2)^2 Z_2}$$

The primary source "sees" an effective impedance $Z_1 = (N_1/N_2)^2 Z_2$. In other words, the transformer also "transforms" the impedance of the secondary. This feature is useful for the maximum transfer of power from a source of emf.

The transformer allows one to present the source with the optimum impedance. The transformer then transfers the power to the load with near ideal efficiency. Transformers are used to match the impedance of the output stage of an audio amplifier to the impedance of the loudspeakers. A similar kind of matching is required for a wave to be transmitted from one medium to another, for example, from solid to liquid. The maximum ampli-tude of the transmitted wave occurs when the "acoustic impe-dances" of the media are matched.

EXAMPLE 33.4: An 8-Ω speaker, rated at 20 W rms, is con-nected via a transformer to an amplifier whose output impe-dance is 1 kΩ. Find: (a) the turns ratio required; (b) the currents and potential differences in the secondary; (c) the current and potential difference in the primary.

Solution: (a) From Eq. 33.26, we require a step-down transformer with a turns ratio of

$$\frac{N_2}{N_1} = \sqrt{\frac{8}{1000}} \approx 0.09$$

(b) The power in the secondary is $P_2 = I_2^2 R_2$, and therefore $I_2 = \sqrt{20/8} = 1.6$ A. The potential difference across the secondary is $V_2 = I_2 R_2 = 12.8$ V.

(c) Given that the transformer has 100% efficiency, we can use Eq. 33.25 to find the current and potential difference in the primary from the corresponding values in the secondary:

$$I_1 = \sqrt{\frac{N_2}{N_1}} I_2 \approx (0.09)(1.6 \text{ A}) = 0.14 \text{ A}$$

$$V_1 = \sqrt{\frac{N_1}{N_2}} V_2 \approx \left(\frac{1}{0.09}\right)(12.8 \text{ V}) = 140 \text{ V}$$

Notice that $I_1 V_1 = 20$ W.

SUMMARY

In an ac circuit, the **instantaneous** current through the circuit and the instantaneous potential difference across the terminals of the source are given by

$$i = i_0 \sin(\omega t); \qquad v = v_0 \sin(\omega t + \phi)$$

where i_0 and v_0 are the peak values and ϕ is the phase angle by which the potential difference leads the current. **Root mean square (rms)** values of current and potential difference are given by

$$I = \frac{i_0}{\sqrt{2}}; \qquad V = \frac{v_0}{\sqrt{2}}$$

In an *RLC* series circuit the current and emf are related by an equation that has the same form as **Ohm's law:**

$$v_0 = i_0 Z; \qquad V = IZ$$

where Z is the **impedance** of the circuit:

$$Z = \sqrt{R^2 + (X_L - X_C)^2}$$

The inductive reactance is $X_L = \omega L$, and the capacitive reactance is $X_C = 1/\omega C$. The phase angle ϕ is given by

$$\tan \phi = \frac{X_L - X_C}{R}$$

When ϕ is positive, the potential difference leads the current.

As the driving frequency is varied, the current in the series *RLC* circuit displays a resonance. The rms current I reaches a maximum value $I_{max} = V/R$, when $X_L = X_C$, which occurs at the **resonance angular frequency**

$$\omega_0 = \frac{1}{\sqrt{LC}}$$

This is the same as the natural angular frequency of oscillation in an *LC* circuit.

The average or **rms power** delivered by the source of emf is

$$P = I^2 R = IV \cos \phi$$

The quantity $\cos \phi$ is called the power factor.

A **transformer** consists of a primary and a secondary coil wrapped around a common soft iron core. The potential differences across the primary and secondary depend on the number of turns in each:

$$\frac{V_2}{V_1} = \frac{N_2}{N_1}$$

The device transfers power with near ideal efficiency; thus

$$I_1 V_1 = I_2 V_2$$

ANSWERS TO IN-CHAPTER EXERCISES

1. (a) The peak current is $i_0 = \sqrt{2}\,I = 1.18$ A.
   (b) The instantaneous peak power is $p_0 = i_0^2 R = (1.18$ A$)^2$ $(144\ \Omega) = 200$ W. Note that $p_0 = 2P$, where P is the rms power.

2. False, the instantaneous current and potential difference are not in phase. Equation 33.10 refers to the *peak* values or the *rms* values.

3. The potential difference across the capacitor is determined by the *charge* on its plates, and the potential difference across the inductor depends on the *rate of change* of the current, not its value.

4. The peak potential differences have different phases relative to the current.

5. Since $X_L = X_C$, the potential differences are equal.

6. $P = I_{max}^2 R = (i_0^2/2)\,R = (3.4^2$ A$^2/2)\,(50\ \Omega) = 288$ W.

QUESTIONS

1. Why does a capacitor behave as a short circuit at high frequencies and as an open circuit at low frequencies?

2. Why is an inductor sometimes referred to as a "choke" coil? What does it "choke"?

3. In what situations is it preferable to use an ac source rather than a dc source? When is dc preferable to ac?

4. The average current supplied by an ac source to a circuit is zero, yet the average power supplied is not zero if there is resistance in the circuit. Explain why.

5. Can an ac source be connected to a circuit and yet deliver no power to it? If so, under what circumstances?

6. Four unmarked wires emerge from a transformer. What steps would you take to determine the turns ratio?

7. A transformer has a primary coil designed to operate at 120 V and 60 Hz. Yet, it could be damaged by 50 V dc. Why?

8. Ac generators are rated in volt-amperes (V · A) rather than watts (W). Why is this?

9. A bulb designed to operate at 120 V (rms) is connected in series with an inductor, a capacitor, and a 120-V (rms) ac source. Is it possible for the bulb to have its normal brightness?

10. Can a power factor be negative? If so, what would this imply regarding the power supplied by the source?

11. Why is electrical power from a power station transmitted at very high potential difference?

12. Electric utilities generally prefer that the electrical load presented by a consumer has a power factor of one. Why is this?

13. True or false: (a) Above the resonance frequency, the potential difference leads the current. (b) A positive power factor means that the potential difference leads the current.

14. If the impedance of a circuit decreases as the frequency increases, is the phase angle positive or negative?

15. In an *RLC* series circuit can the rms potential difference across either *L* or *C* exceed the rms potential difference of the source?

16. Consider the circuit in Fig. 33.21. The frequency of the source is constant. How is the brightness of the bulb affected as the capacitance is varied?

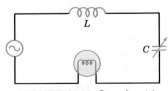

FIGURE 33.21 Question 16.

EXERCISES

33.2 to 33.4 Resistors, Inductors, and Capacitors

1. (I) An inductor $L = 40$ mH is connected to a source for which the peak potential difference is 120 V and the frequency is 60 Hz. (a) Find the peak current. (b) If the peak potential difference is unchanged, at what frequency would the peak current be 30% of the value found in (a)?

2. (I) A 50-μF capacitor is connected to a 70-V (peak) source of frequency 50 Hz. (a) Find the peak current. (b) Given the same peak potential difference, at what frequency would the peak current be 30% greater than that in (a)?

3. (I) Given a capacitor $C = 0.1$ μF and an inductor $L = 10$ mH, find the frequency at which: (a) $X_L = X_C$; (b) $X_L = 5$ X_C; (c) $X_C = 5$ X_L.

4. (I) The reactance of an inductor is 37.7 Ω at 60 Hz. It is connected to a 50-Hz source with a 120 V (rms) potential difference. What is the peak current?

5. (I) A 50-μF capacitor is connected to a 60-Hz source that provides 24 V (rms). Find: (a) the peak charge on the capacitor; (b) the peak current in the wires.

6. (II) A 72-mH inductor is connected to a source with a peak potential difference of 50 V and a frequency of 120 Hz. Find: (a) the peak current; (b) the current when the potential difference has its (positive) peak value; (c) the current when the potential difference is one-half the (positive) peak value (there are two answers); (d) the instantaneous power delivered to the inductor at 1 ms.

7. (II) A 108-μF capacitor is connected to a source that operates at 80 Hz with a peak potential difference of 24 V. Find: (a) the peak current; (b) the current when the potential difference has its peak value; (c) the current when the potential difference has one-half the (positive) peak value (there are two answers); (d) the instantaneous power delivered to the capacitor at 1 ms.

8. (I) A 6-μF capacitor has a reactance of 11 Ω. (a) What would be the reactance of a 0.2 mH inductor at the same frequency? (b) At what frequency would the reactances be equal?

9. (II) An ideal inductor $L = 80$ mH is connected to a source whose peak potential difference is 60 V. (a) If the frequency is 50 Hz, what is the current at 2 ms? What is the instantaneous power delivered to the inductor at this time? (b) At what frequency would the peak current be 1.8 A?

10. (II) A source whose peak potential difference is 72 V is connected to a capacitor $C = 80$ μF. (a) If the frequency is 50 Hz, what is the current at 2 ms? What is the instantaneous power delivered to the capacitor at this time? (b) At what frequency would the peak current be 4 A?

33.6 and 33.7 *RLC* Series Circuits; Resonance

11. (I) A resistor and a capacitor are in series with an ac source. The impedance $Z = 10.8$ Ω at 390 Hz and $Z = 18.8$ Ω at 200 Hz. Find R and C.

12. (I) In a circuit in which a resistor and inductor are in series, $Z = 28.3$ Ω at 100 Hz and $Z = 22.9$ Ω at 75 Hz. Find R and L.

13. (I) In an *RLC* series circuit, the source has an rms potential difference $V = 60$ V and a frequency of $250/\pi$ Hz, while

$R = 50$ Ω and $C = 10$ μF. If the peak potential difference across R is 25 V, find L. (There are two possible values.)

14. (I) A real coil, which may be treated as an inductor and a resistor in series, is connected in series with a capacitor and an ac source. The rms potential difference across the coil is 45 V and $V_C = 60$ V at a frequency of $200/\pi$ Hz. If $C = 25$ μF, and $R = 50$ Ω, find L.

15. (I) In an *RLC* series circuit, the rms potential difference provided by the source is $V = 120$ V, and the frequency is $f = 200/\pi$ Hz. Given that $L = 0.2$ H, $C = 20$ μF, and $V_R = 50$ V, find: (a) I; (b) R; (c) V_L; (d) V_C.

16. (I) A voltmeter across L and C in Fig. 33.15 reads 80 V (rms). Given that $L = 0.2$ H, $C = 50$ μF, the rms potential difference provided by the source is 120 V, and its frequency is $200/\pi$ Hz, find: (a) X_L; (b) X_C; (c) I; (d) R; (e) the rms power delivered by the source.

17. (I) A 100-Ω resistor is connected in series with a 25-μF capacitor and an inductor. The rms potential difference across the source is 240 V and its frequency is $800/\pi$ Hz. Given that $V_R = 80$ V, find: (a) Z; (b) L; (c) ϕ.

18. (I) At a certain frequency in an *RLC* series circuit, $X_L = 20$ Ω and $X_C = 8$ Ω. The resonance frequency is 2000 Hz. Find L and C.

19. (I) A resistor ($R = 10$ Ω) and an inductor ($L = 40$ mH) are connected in series with a capacitor. The source potential difference is 120 V rms at 60 Hz. The rms potential difference across the resistance is 30 V. (a) What is C? (b) What is the natural frequency, f_0?

20. (I) A series ac circuit has the following components: $R = 25$ Ω, $L = 320$ mH, and $C = 18$ μF. The peak potential difference across the source is 170 V and the frequency is 60 Hz. Find: (a) the impedance: (b) the rms current: (c) the phase angle.

21. (II) In an *RLC* series circuit take $R = 40$ Ω, $L = 20$ mH, $C = 60$ μF. Find at what frequency the potential difference leads the current by 30°.

22. (II) When an ac source with a peak potential difference of 48 V is connected to a series *RLC* circuit, the peak current is 2 A. The capacitor is $C = 10$ μF, the frequency is 50 Hz, and the current leads the potential difference by 45°. Find (a) the resistance, and (b) the inductance.

23. (II) An inductor ($L = 3$ mH), a resistor ($R = 8$ Ω), and a capacitor ($C = 10$ μF) are connected in series with an ac source whose rms potential difference is 25 V. Find: (a) the natural frequency f_0; (b) the frequencies at which the rms current is 50% of the value at f_0.

24. (II) A coil with an inductance of 80 mH is connected in series with a 120-Ω resistor and a capacitor. The frequency of the ac source is 600 Hz. What value(s) of C would result in an impedance of 200 Ω?

33.8 Power in Alternating Current Circuits

25. (I) An ac generator operates at 90 Hz and has an rms potential difference of 100 V. It is connected to a series circuit with components $R = 20\ \Omega$, $C = 80\ \mu F$, and $L = 9$ mH. Find: (a) the power factor; (b) the average power delivered by the generator.

26. (I) In an *RLC* series circuit, the source has a potential difference of 120 V (rms). The impedance is 110 Ω and the resistance is 40 Ω. Find: (a) the rms power delivered by the source; (b) the power factor.

27. (I) In an *RLC* series circuit, the source has a peak potential difference of 200 V and a frequency of $50/\pi$ Hz. Take $R = 15\ \Omega$, $C = 200\ \mu F$, and $L = 0.2$ H. Find: (a) X_L and X_C; (b) the phase angle; (c) the average power delivered by the source; (d) the power factor.

28. (I) A motor draws 8 A from a 120-V (rms) source. If its average power consumption is 800 W, what is its power factor?

29. (I) In an *RLC* series circuit, the 100-V (rms) source operates at 60 Hz. The resistor is 24 Ω and the phase angle is +53°. What is the rms power delivered by the source?

30. (II) For the function shown in Fig. 33.22, find (a) the average potential difference and (b) the rms potential difference.

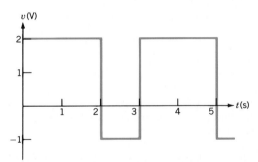

FIGURE 33.22 Exercise 30.

31. (II) Show that the rms power delivered to an *RLC* series circuit can be written as $P = (V \cos \phi)^2/R$, where V is the rms potential difference provided by the source.

32. (II) A 1-kW heater (purely resistive) operates from a 120-V source at 60 Hz. (a) What inductor in series with the heater would reduce the delivered power by one-half? (b) What would then be the phase angle?

33. (II) The instantaneous current in an *RLC* series circuit is given by $i = 0.06 \sin(320t)$ A. The three components are $R = 24\ \Omega$, $L = 18$ mH, and $C = 70\ \mu F$. Write an expression for the instantaneous potential difference across the terminals of the source.

33.10 Transformers

34. (II) An ideal 5 : 1 step-down transformer supplies 40 kW (rms) at 240 V (rms) to a building. If the transmission line (connected to the primary) has a total resistance of 1.2 Ω, what is the rms power loss in the line?

35. (II) Figure 33.23 shows a simple power transmission system. An ac generator produces 15 A (rms) at 300 V (rms). A step-up transformer boosts this potential difference and the power is transmitted along lines with a total resistance of 20 Ω. A step-down transformer supplies a load resistance R_L. What is the rms power loss in the lines if the potential difference of the generator is boosted to (a) 5 kV; (b) 20 kV?

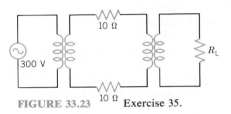

FIGURE 33.23 Exercise 35.

36. (II) A 5 : 1 step-down transformer has 120-V (rms) across the primary and is 90% efficient. The primary current is 2 A (rms) and lags the potential difference by 12°. (a) What is the input power? (b) What is the output power? (c) If the secondary circuit has a power factor of 0.75, what is the secondary rms current?

37. (II) A step-down transformer has 600 V across the primary and 120 V across the secondary. The secondary has 80 turns. (a) What is the number of turns in the primary? (b) If a load resistance $R_L = 10\ \Omega$ is in the secondary, what is the primary current?

38. (I) An ideal transformer has 400 turns in the primary coil and 50 turns in the secondary. When the potential difference across the primary is 120 V (rms), the rms current is 2.4 A. Determine the rms current and potential difference for the secondary.

PROBLEMS

1. (I) Show that the rms power delivered to an *RLC* series circuit may be written as

$$P = \frac{\omega^2 R V^2}{\omega^2 R^2 + (\omega^2 - \omega_0^2)^2 L^2}$$

2. (II) An expression for the rms power delivered to an *RLC* series circuit is given in Eq. 33.22. At what angular frequency is this power a maximum? (*Hint*: Find $dP/d\omega$.)

3. (I) For the function shown in Fig. 33.24, determine (a) the average value and (b) the rms value of the potential difference.

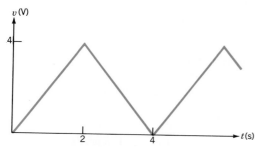

FIGURE 33.24 Problem 3.

4. (II) A resistor, an inductor, and a capacitor are in parallel with an ac source, as shown in Fig. 33.25. (a) What is the relationship between the instantaneous potential differences across the elements? (b) How are the instantaneous currents related? (c) Use the phasor diagram for the currents drawn in the diagram to show that the impedance is

$$Z = \frac{1}{\sqrt{1/R^2 + (1/X_L - 1/X_C)^2}}$$

(d) What is the resonance frequency at which, in this case, the impedance is a maximum?

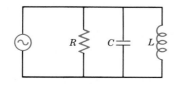

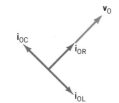

FIGURE 33.25 Problem 4.

5. (I) What is the rms potential difference for the "saw tooth" function shown in Fig. 33.26?

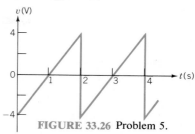

FIGURE 33.26 Problem 5.

6. (II) The impedance of a series *RLC* circuit is a minimum at the resonance frequency f_0. A given higher value of Z occurs at a frequency f_L below f_0 and at f_H above f_0, as shown in Fig. 33.27. Show that f_0 is the geometric mean of the other frequencies; that is,

$$f_0 = \sqrt{f_L f_H}$$

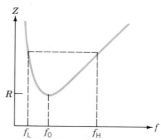

FIGURE 33.27 Problem 6.

7. (I) Figure 33.28 shows a circuit that acts as a crude filter. If the input potential difference contains a range of frequencies, the output consists mainly of the higher or the lower frequencies. Show that

$$\frac{V_{out}}{V_{in}} = \frac{R}{(R^2 + \omega^2 L^2)^{1/2}}$$

Take $R = 10\ \Omega$, $L = 25$ mH, and $V_{in} = 100$ V (rms). Find V_{out} for the following frequencies of the input potential difference: (a) 40 Hz; (b) 400 Hz; (c) 4000 Hz. Is this a "low-pass" or a "high-pass" filter?

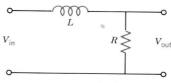

FIGURE 33.28 Problem 7.

8. (I) Repeat Problem 7 for the filter circuit drawn in Fig. 33.29. Show that

$$\frac{V_{out}}{V_{in}} = \frac{\omega L}{(R^2 + \omega^2 L^2)^{1/2}}$$

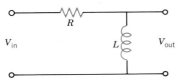

FIGURE 33.29 Problem 8.

9. (I) An ac source provides an RLC series circuit with a current $i = 4 \sin(377t)$ A and a potential difference $v = 160 \sin/(377t + \phi)$ V. Given that $R = 12.5$ Ω, $X_C = 52$ Ω, and $\phi < 0$, find L.

10. (II) (a) Show that the peak charge Q_0 on the capacitor in a series RLC circuit is $Q_0 = v_0/\omega Z$; where v_0 is the peak potential difference of the source, ω is the angular frequency and Z is the impedance. **(b)** Show that the maximum value of Q_0 occurs at an angular frequency given by

$$\omega_{max} = \sqrt{\omega_0^2 - \frac{R^2}{2L^2}}$$

11. (I) In the RLC series circuit of Fig. 33.15, take $R = 8$ Ω, $L = 40$ mH, $C = 20$ μF, the peak potential difference of the source, $v_0 = 100$ V, and $f = 200/\pi$ Hz. Find the peak potential difference across **(a)** R, C, and L individually; **(b)** R and C combined; **(c)** C and L combined.

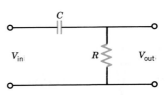

FIGURE 33.30 Problem 12.

12. (I) Figure 33.30 shows a simple filter circuit. The input ac potential difference, V_{in}, is across R and C, whereas the output potential difference, V_{out}, is that across R. Show that the ratio V_{out}/V_{in} is

$$\frac{V_{out}}{V_{in}} = \frac{1}{\sqrt{1 + 1/\omega^2 R^2 C^2}}$$

Plot this ratio for $\omega = 0, 0.5, 1, 1.5$, and 2 in units of $1/RC$. Is this a "high-pass" or a "low-pass" filter?

13. (I) Repeat Problem 12 for the filter circuit in Fig. 33.31. Show that in this case

$$\frac{V_{out}}{V_{in}} = \frac{1}{\sqrt{1 + \omega^2 R^2 C^2}}$$

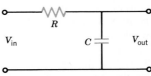

FIGURE 33.31 Problem 13.

CHAPTER 34

Maxwell's Equations; Electromagnetic Waves

A camper enjoys visible and infrared electromagnetic waves emitted by the campfire.

Major Points

1. Maxwell's modification of Ampère's law: The **displacement current.**
2. **Electromagnetic waves:** (a) Faraday's law and the Ampère–Maxwell law lead to a *wave equation* for electric and magnetic fields.
   (b) Electromagnetic waves are transverse waves that propagate at the speed of light.
3. An electromagnetic wave transports **energy.** The **Poynting vector** indicates the intensity of the wave.
4. Electromagnetic waves transport **linear momentum** and exert **radiation pressure** on a surface.

By about 1820, experimental and theoretical work had established that light is a transverse wave.* The precise nature of the waves, how they are produced, and how they interact with matter, were still unanswered questions. In 1845, Faraday demonstrated that a magnetic field produces a measurable effect on a beam of light passing through glass. This prompted him to speculate that light involves oscillations of electric and magnetic field lines, but his limited mathematical ability prevented him from pursuing this idea. Another clue that pointed to a connection between electromagnetism and light came from an unrelated experiment.

During the 19th century, two systems of units were used in electromagnetism: Electrostatic units were based on Coulomb's law for the force between charges, and electromagnetic units were based on an analogous expression for the force between magnetic poles. The ratio of the units of charge in these two systems is $1/(\varepsilon_0\mu_0)^{1/2}$ and has the dimensions of speed. In 1856, W. Weber and R. Kohlrausch experimentally determined this ratio to be 3.11×10^8 m/s. This value was almost exactly equal to the speed of light, 3.15×10^8 m/s, as measured by A. Fizeau in 1849.

A young admirer of Faraday, James Clerk Maxwell (Fig. 34.1), believed that the closeness of these two numbers was more than just coincidence and decided to

FIGURE 34.1 James Clerk Maxwell (1831–1879).

* How this came about is discussed in Chapter 37.

develop Faraday's bold hypothesis. He made a subtle, yet momentous, modification to Ampère's law (Eq. 30.11), and in 1865 predicted the existence of electromagnetic waves that propagate at the speed of light. The conclusion was irresistible that light itself is an electromagnetic wave. Maxwell's theory was a grand synthesis of the hitherto separate subjects of optics and electromagnetism. It was a triumph of the mind equal to that of Newton, two centuries earlier. Its experimental verification by H. Hertz in 1887 and its commercial exploitation by M. G. Marconi and others, led to our present radio, TV, and satellite communications.

34.1 DISPLACEMENT CURRENT

In Section 30.4 we saw that according to Ampère's law,

$$\oint \mathbf{B} \cdot d\boldsymbol{\ell} = \mu_0 I$$

the line integral of $\mathbf{B} \cdot d\boldsymbol{\ell}$ around a closed loop is equal to $\mu_0 I$, where I is the current flowing through a surface bounded by the loop. In 1861 Maxwell discovered an inconsistency in this statement since the "surface bounded by the loop" is not uniquely specified. Suppose a current is charging a capacitor, as in Fig. 34.2. At any point along the wire, we can find the magnetic field by using Ampère's law. Maxwell considered a single loop but two different surfaces. The flat surface of Fig. 34.2a, is an obvious choice, and it leads to the usual expression for the field due to a straight wire. The surface of Fig. 34.2b is bounded by the same loop, but it encloses one plate of the capacitor. Since there is no electric current flowing through this surface, Ampère's law implies that the field along the loop is zero—which is clearly false. In order to remove this inconsistency, Maxwell proposed that a new type of current, which he called the **displacement current**, I_D, can be associated with the nonconductor between the plates. Thus Ampère's law should be written as

$$\oint \mathbf{B} \cdot d\boldsymbol{\ell} = \mu_0(I + I_D)$$

The name *displacement current* arose from Maxwell's mechanistic conception of the ether—the medium through which electromagnetic waves were supposed to travel. By analogy with the polarization in a dielectric, he envisioned a real displacement of the particles of the ether when it was stressed by an electric field. He later abandoned this picture, but the name has stuck.

In order to obtain an expression for the displacement current, Maxwell noted that the electric field strength between the plates is increasing because of the charge buildup. For a parallel plate capacitor, $E = Q/(\varepsilon_0 A)$, from which we see that the rate of change of charge on the plates is $dQ/dt = \varepsilon_0 A dE/dt$. In this case, the electric flux is simply $\Phi_E = EA$; therefore

$$\frac{dQ}{dt} = \varepsilon_0 \frac{d\Phi_E}{dt} \tag{34.1}$$

Note that dQ/dt is equal to the conduction current I through the wire. It took Maxwell's genius to realize that the changing electric flux must also be associated with a magnetic field. Since the value of the magnetic field must be the same for either choice of surface, the magnitude of the displacement current between the plates must be the same as the conduction current in the wire:

$$I_D = \varepsilon_0 \frac{d\Phi_E}{dt}$$

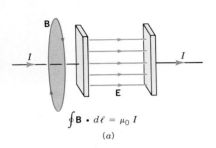

$$\oint \mathbf{B} \cdot d\boldsymbol{\ell} = \mu_0 I$$
(a)

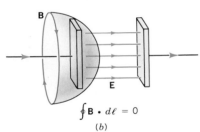

$$\oint \mathbf{B} \cdot d\boldsymbol{\ell} = 0$$
(b)

FIGURE 34.2 In evaluating the line integral ∫ **B** · *d*s around a closed loop, the surface through which the current flows may or may not include the plate of the capacitor. Ampère's law does not give consistent answers for these two choices.

With Maxwell's modification, Ampère's law becomes

$$\oint \mathbf{B} \cdot d\ell = \mu_0 \left(I + \varepsilon_0 \frac{d\Phi_E}{dt} \right) \tag{34.2}$$

Depending on the choice of surface, either the first or the second term on the right leads to the correct value for B around a loop. In some instances—for example, a leaky capacitor—a loop may enclose both a conduction current and a displacement current, in which case both terms contribute to the field. Maxwell's modification of Ampère's law is an excellent example of a discovery of great practical importance that was made on purely theoretical grounds.

EXAMPLE 34.1: Use the Ampère–Maxwell law to find the magnetic field between the circular plates of a parallel-plate capacitor that is charging. The radius of the plates is R. Ignore the fringing field.

Solution: Since we may take the electric field to be uniform, we pick a circular loop of radius r normal to the $\mathbf{E}$ lines, as in Fig. 34.3. In order to invoke symmetry, we assume that the wires are long and straight and are attached to the centers of the plates. From the circular symmetry of the system we infer that $\mathbf{B}$ has the same value at all points on the circular loop. Furthermore, $\mathbf{B}$ is always parallel to $d\ell$. Thus,

$$\oint \mathbf{B} \cdot d\ell = B \oint d\ell = B(2\pi r)$$

The electric flux is $\Phi_E = E(\pi r^2)$, so Eq. 34.2 becomes

$$B(2\pi r) = \mu_0 \varepsilon_0 (\pi r^2) \frac{dE}{dt}$$

$$(r < R) \qquad B = \frac{1}{2} \mu_0 \varepsilon_0 \left(\frac{dE}{dt} \right) r$$

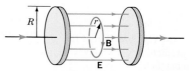

FIGURE 34.3 To evaluate the B field associated with a changing electric field between two circular capacitor plates, we choose a circular loop.

Note that $B \propto r$, just as it is inside a current-carrying wire (see Fig. 30.16).

EXERCISE 1. Obtain an expression for B when $r > R$.

34.2   MAXWELL'S EQUATIONS

With the inclusion of Maxwell's contribution, we now display all the fundamental equations in electromagnetism. There are just four:

Gauss	$\oint \mathbf{E} \cdot d\mathbf{A} = \dfrac{Q}{\varepsilon_0}$	(34.3)
Gauss	$\oint \mathbf{B} \cdot d\mathbf{A} = 0$	(34.4)
Faraday	$\oint \mathbf{E} \cdot d\ell = - \dfrac{d\Phi_B}{dt}$	(34.5)
Ampère–Maxwell	$\oint \mathbf{B} \cdot d\ell = \mu_0 \left(I + \varepsilon_0 \dfrac{d\Phi_E}{dt} \right)$	(34.6)

The first form of Gauss's law (Eq. 34.3) relates the electric field to electric charges. For the electrostatic field, whose lines begin and end on charges, it is equivalent to Coulomb's law. However, Eq. 34.3 is a more general statement; it also applies to induced electric fields for which the lines are closed loops. The second form of Gauss's law (Eq. 34.4) tells us that magnetic field lines always form closed loops because there are no magnetic monopoles. Faraday's law says

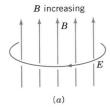

B increasing

(a)

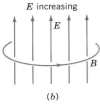

E increasing

(b)

FIGURE 34.4 (a) The direction of the electric field predicted by Faraday's law when the magnetic flux is increasing. (b) The direction of the magnetic field predicted by the Ampère–Maxwell law when the electric flux is increasing.

that an induced electric field is associated with a changing magnetic flux. Note that with the thumb along the magnetic field, the fingers of the right hand indicate the sense in which the integral is to be evaluated. The negative sign in Eq. 34.5 means that the induced electric field is in the opposite sense, as shown in Fig. 34.4a. According to the Ampère–Maxwell law, a magnetic field is produced by a conduction current I and may also be associated with a changing electric flux. In this case, the right-hand rule and the electric field determine the sense of the integration. The positive sign in Eq. 34.6 means that the magnetic field is in the same sense as that of the integral (see Fig. 34.4b).

These four equations are collectively called **Maxwell's equations.** (Maxwell actually presented twenty equations. In 1885 Oliver Heaviside reduced them to just four by using the vector notation he had developed.) These four equations, supplemented by the Lorentz force equation, $\mathbf{F} = q(\mathbf{E} + \mathbf{v} \times \mathbf{B})$, and the conservation of charge, describe all the electromagnetic phenomena and devices we encounter.*

34.3 ELECTROMAGNETIC WAVES

In Chapter 16 we saw that a wave traveling along the x axis with a wavespeed v satisfies the wave equation:

$$\frac{\partial^2 y}{\partial x^2} = \frac{1}{v^2} \frac{\partial^2 y}{\partial t^2}$$

Maxwell was able to show that time-dependent electric and magnetic fields also satisfy the wave equation. This led to the most significant outcome of Maxwell's theory, his prediction of the existence of electromagnetic waves.

From Faraday's law we infer that an E field accompanies a changing B field, and from the Maxwell–Ampère law that a B field accompanies a changing E field. It is this coupling of changing electric and magnetic fields that is the basis of electromagnetic waves. In Section 34.8 we will show that in free space, far from the source of the fields, the fields satisfy **Maxwell's wave equations:**

$$\frac{\partial^2 E}{\partial x^2} = \mu_0 \varepsilon_0 \frac{\partial^2 E}{\partial t^2} \tag{34.7}$$

$$\frac{\partial^2 B}{\partial x^2} = \mu_0 \varepsilon_0 \frac{\partial^2 B}{\partial t^2} \tag{34.8}$$

On comparing these with the standard wave equation, we see that the wave speed is

$$c = \frac{1}{\sqrt{\mu_0 \varepsilon_0}} \tag{34.9}$$

When the values $\mu_0 = 4\pi \times 10^{-7}$ H/m and $\varepsilon_0 = 8.85 \times 10^{-12}$ F/m are inserted, we find

$$c = 3.00 \times 10^8 \text{ m/s}$$

This is the speed of light in vacuum! One can hardly avoid the suggestion that light itself is an electromagnetic wave. Not only did Maxwell provide a theoretical foundation for the remarkable result of Weber and Kohlrausch, but he also unified the hitherto separate subjects of optics and electromagnetism.

---

* A modification to Eq. 34.6 is needed when a magnetic material is present.

The simplest plane wave solutions to Eqs. 34.7 and 34.8 are

$$E = E_0 \sin(kx - \omega t)$$
$$B = B_0 \sin(kx - \omega t)$$

From these equations we see that at any point, E and B are *in phase*. The electric and magnetic fields in a plane electromagnetic wave are perpendicular to each other and also perpendicular to the direction of propagation, as shown in Fig. 34.5. They are *transverse* electromagnetic waves. The magnitudes of the fields are related by (see Section 34.8)

$$E = cB \qquad (34.10)$$

There are two common ways of representing an electromagnetic plane wave. In the first, shown in Fig. 34.6, the length of a vector varies sinusoidally. For a plane wave moving in the x direction, the value of E or B is the same at all points on any yz plane. In the second method, illustrated in Fig. 34.7, the density of the field lines indicates the varying strengths of the fields. If a straight wire is placed parallel to the E field, an oscillating current is produced in it. If a coil is placed with its plane normal to B, the changing magnetic flux will induce an oscillating current. Both straight-wire and loop antennas are used for radio and TV reception.

According to the thinking of the 19th century, the constants μ_0 and ε_0 referred to properties of the *ether,* the medium through which the electromagnetic waves were assumed to propagate. This is not our present thinking. The ether does not exist and electromagnetic waves do not require any medium in which to propa-

FIGURE 34.5 In a plane electromagnetic wave the electric and magnetic fields are perpendicular to each other and to the direction of propagation.

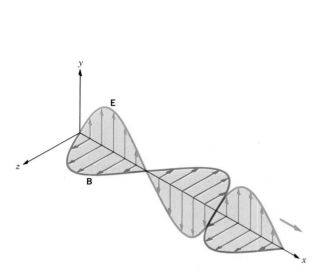

FIGURE 34.6 One representation of an electromagnetic wave traveling along the $+x$ direction. In a plane wave the fields have the same value at all points on any yz plane. The variation in the field strengths is indicated by the varying sinusoidal function.

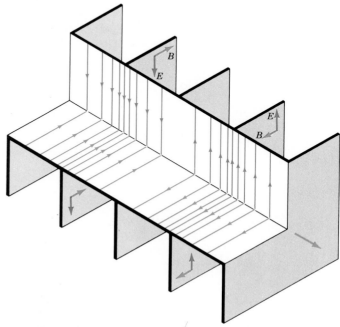

FIGURE 34.7 A representation of a plane electromagnetic wave in which the variation in the field strengths is depicted by the density of field lines.

gate. However, when they travel through a substance, the fields do interact with charges in the atoms. The strength of the interaction is related to the permittivity, ε, and the permeability, μ, of the substance. As a result, the speed of light is reduced from c to $1/\sqrt{\mu\varepsilon}$.

34.4   ENERGY TRANSPORT AND THE POYNTING VECTOR

The light that we see and the heat that we feel in the sun's rays show that electromagnetic waves transport energy. We now find an expression for the rate at which energy is transported by an electromagnetic wave. The energy densities for the electric and magnetic fields in free space are given in Eq. 26.10 and Eq. 32.13, respectively:

$$u_{\text{E}} = \frac{1}{2}\, \varepsilon_0 E^2; \qquad u_{\text{B}} = \frac{B^2}{2\mu_0} \tag{34.11}$$

Since $E = cB = B/\sqrt{\mu_0\varepsilon_0}$ for an electromagnetic wave, the instantaneous values of these energy densities are equal. The **total energy density,** $u = u_{\text{E}} + u_{\text{B}}$, is therefore

Energy density of an electromagnetic wave

$$u = \varepsilon_0 E^2 = \frac{B^2}{\mu_0} = \sqrt{\frac{\varepsilon_0}{\mu_0}}\, EB \tag{34.12}$$

Consider two planes, each of area A, a distance dx apart, and normal to the direction of propagation of the wave, as in Fig. 34.8. The total energy in the volume between the planes is $dU = u(A\, dx)$. The rate at which this energy passes through a unit area normal to the direction of propagation is

$$S = \frac{1}{A}\frac{dU}{dt} \tag{34.13}$$

Since the energy is carried by the fields, which move at speed c, it is also transported at this speed. Thus, $dU/dt = uA\, dx/dt = uAc$, and so $S = uc$. Using the last expression in Eq. 34.12 and $c = 1/\sqrt{\mu_0\varepsilon_0}$, we find

$$S = uc = \frac{EB}{\mu_0} \tag{34.14}$$

Notice that the energy flow is perpendicular to both **E** and **B.** In 1884 J. H. Poynting neatly incorporated this fact into Eq. 34.14 by using the recently developed vector analysis. The **Poynting vector** is defined as

$$\mathbf{S} = \frac{\mathbf{E} \times \mathbf{B}}{\mu_0} \tag{34.15}$$

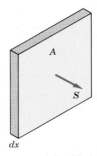

FIGURE 34.8 The energy contained between two planes of area A and set a distance dx apart is $dU = u(A\, dx)$, where u is the energy density of the electromagnetic wave.

The magnitude of **S** is the intensity, that is, instantaneous power that crosses a unit area normal to the direction of propagation. The direction of **S** is the direction of the energy flow. In an electromagnetic wave, the magnitude of **S** fluctuates rapidly in time. A more useful quantity is the average intensity of the wave, which is the average value of S. At any point in space (say, $x = 0$) the product EB in Eq. 34.14 is $E_0B_0 \sin^2(\omega t)$. The average of $\sin^2(\omega t)$ over one period is $\frac{1}{2}$. Thus the **average intensity** is

$$S_{av} = u_{av}c = \frac{E_0B_0}{2\mu_0} \qquad (34.16)$$

The quantity S_{av}, measured in W/m², is the average power incident per unit area normal to the direction of propagation. The average intensity of a plane wave does not diminish as it propagates.

EXAMPLE 34.2: A radio station transmits a 10-kW signal at a frequency of 100 MHz. For simplicity, assume that it radiates as a point source. At a distance of 1 km from the antenna, find: (a) the amplitudes of the electric and magnetic field strengths, and (b) the energy incident normally on a square plate of side 10 cm in 5 min.

Solution: (a) The energy of waves emitted by a point source spreads over ever-expanding spheres. The surface area of a sphere of radius r is $4\pi r^2$, so the intensity of the waves at a distance r is

$$\text{(Point source)} \qquad S_{av} = \frac{\text{Average power}}{4\pi r^2} \qquad (i)$$

Since $E = cB$, S_{av} may be written in terms of E_0:

$$S_{av} = \frac{E_0^2}{2\mu_0 c} \qquad (ii)$$

Equating (i) and (ii),

$$\frac{10^4 \ W}{(4\pi)(10^6 \ m^2)} = \frac{E_0^2}{2(4\pi \times 10^{-7} \ H/m)(3 \times 10^8 \ m/s)}$$

we find $E_0 = 0.775$ V/m.

The amplitude of the magnetic field is

$$B_0 = \frac{E_0}{c} = 2.58 \times 10^{-9} \ T$$

(b) From Eq. 34.13, the energy incident normally on an area A in time Δt is

$$\Delta U = S_{av}A\Delta t$$
$$= \frac{(10^4 \ W)}{(4\pi)(10^6 \ m^2)}(0.01 \ m^2)(300 \ s)$$
$$= 2.4 \times 10^{-3} \ J$$

EXERCISE 2. A plane electromagnetic wave of frequency 25 MHz travels in free space along the $+z$ direction. At a particular point in space and time, $\mathbf{E} = -5\mathbf{i}$ V/m. What is $\mathbf{B}$ at this point?

EXAMPLE 34.3: When a wire is connected to a battery, an electric field exists in the space surrounding the circuit (see Fig.

34.9). In addition, the current in the wire generates a magnetic field. Use the Poynting vector to show that the rate at which energy enters the wire is equal to the power loss due to heating.

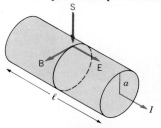

FIGURE 34.9 The electric and magnetic fields associated with a current flowing in a wire. The Poynting vector for these fields is directed toward the wire.

Solution: Let us consider a long straight wire that has a radius a and an electrical resistance R. We assume it carries a constant current I. In order to calculate the Poynting vector we need to find the electric and magnetic field strengths at the surface of the wire. If the potential difference across a length ℓ is $V = IR$, then the (constant) electric field along the wire, and hence at its surface, is

$$E = \frac{V}{\ell} = \frac{IR}{\ell} \qquad (i)$$

From Example 30.5, we know that the magnetic field at the surface of the wire is

$$B = \frac{\mu_0 I}{2\pi a} \qquad (ii)$$

Since the electric and magnetic fields at the surface of the wire are perpendicular, as shown in Fig. 34.9, it follows that $|\mathbf{E} \times \mathbf{B}| = EB$. The magnitude of the Poynting vector, Eq. 34.14, is

$$S = \frac{EB}{\mu_0} = \left(\frac{1}{\mu_0}\right)\left(\frac{IR}{\ell}\right)\left(\frac{\mu_0 I}{2\pi a}\right)$$
$$= \frac{I^2 R}{2\pi a \ell} \qquad (iii)$$

But $A = 2\pi a\ell$ is the surface area of the length of the wire, so (iii) may be written in the form

$$SA = I^2R$$

We see that the rate at which electromagnetic energy is deposited in the wire (SA) by the fields at the surface is equal to the rate of dissipation (I^2R).

34.5 MOMENTUM AND RADIATION PRESSURE

An electromagnetic wave transports linear momentum. We state, without proof, that the **linear momentum** carried by an electromagnetic wave is related to the energy it transports according to

Linear momentum carried by an electromagnetic wave

$$p = \frac{U}{c} \qquad (34.17)$$

If the wave is incident in the direction perpendicular to a surface and is completely absorbed, then Eq. 34.17 tells us the linear momentum imparted to the surface. If surface is perfectly reflecting, the momentum change of the wave is doubled. Consequently, the momentum imparted to the surface is also doubled, that is, $p = 2U/c$.

The force exerted by an electromagnetic wave on a surface may be related to the Poynting vector. If we use Eq. 34.17 in Newton's second law, $F = \Delta p/\Delta t$, we obtain $F = (1/c)(\Delta U/\Delta t)$. From Eq. 34.13 we have $\Delta U/\Delta t = SA$; therefore $F = SA/c$. The **radiation pressure** (force/area) at normal incidence is

(Pressure) $$\frac{F}{A} = \frac{S}{c} = u \qquad (34.18)$$

where we have used $S = uc$ from Eq. 34.14. The radiation pressure is equal to the energy density (N/m² = J/m³). At a perfectly reflecting surface the pressure on the surface is doubled.

The fact that light exerts a force is illustrated in Fig. 34.10, which shows a tiny particle suspended by the radiation pressure of light from a laser. The radiation pressure of sunlight causes dust particles in the tail of a comet to be deflected away from the sun, as shown in Fig. 34.11a. It has been proposed that spacecraft with giant "sails," as in Fig. 34.11b, could be propelled by the radiation pressure of sunlight.

FIGURE 34.10 A tiny particle suspended by the pressure of light from a laser.

(a)

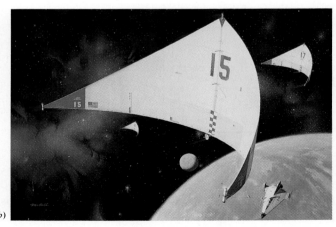

(b)

FIGURE 34.11 (a) The dust particles in a comet are deflected by the pressure of sunlight. The other tail consists of much smaller ions. (b) A "solar sail" may use the pressure of sunlight to transport minerals in space.

EXAMPLE 34.4: The intensity of solar radiation at the surface of the earth is 1 kW/m². A roof has a square solar panel of side 10 m that absorbs the radiation completely. What is the average force exerted on the panel?

Solution: From Eq. 34.18, the average force is

$$F_{av} = \frac{S_{av}A}{c} = \frac{(1 \times 10^3 \text{ W/m}^2)(100 \text{ m}^2)}{3 \times 10^8 \text{ m/s}}$$
$$= 0.5 \times 10^{-3} \text{ N}$$

Despite its tiny value, the pressure due to radiation was detected in 1899 by P. Lebedev in Russia and in 1901 by E. L. Nicholls and G. F. Hull in the United States.

34.6   HERTZ'S EXPERIMENT

When Maxwell's work was published in 1867 it did not receive immediate acceptance. Many physicists were skeptical about the concept of displacement current and the existence of electromagnetic waves. The experiment that conclusively demonstrated the existence of electromagnetic waves was performed by Heinrich Hertz (Fig. 34.12) in 1887. Maxwell had shown that electromagnetic radiation is produced when a charge *accelerates,* or, equivalently, when the current in a wire changes with time. One way to produce electromagnetic waves is by connecting an ac source to two rods, as in Fig. 34.13. As the polarity of the rods alternates in time, the charges oscillate and therefore radiate. The fields near the source are complicated, but at distances large compared with the wavelength of the radiation, the *radiation* fields vary as shown in Fig. 34.13. The electric field lines are drawn as closed loops while the magnetic field lines are normal to the page. Notice that the direction of **B** changes with that of **E** so that the Poynting vector, **S** = **E** × **B**/μ_0, always points radially outward, confirming that energy is being transported away.

The equipment used by Hertz is shown in Fig. 34.14. He used an *LC* circuit in which the inductor was the secondary of a transformer and the capacitor consisted of two metal balls connected to flat plates. When the current in the primary was switched off, a large potential difference was induced in the secondary. This caused the air between the balls to become ionized, temporarily making it a conductor. As a result there occurred a few cycles of damped *LC* oscillations,

FIGURE 34.12 Heinrich Hertz (1857–1894).

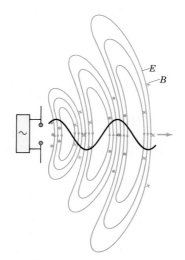

FIGURE 34.13 Electromagnetic waves may be generated by connecting two rods to an oscillator. The accelerating charges in the rods produce a radiation field as shown. The electric field is drawn in closed loops, whereas the magnetic field is normal to the page.

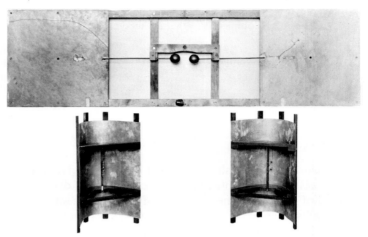

FIGURE 34.14 Some of the equipment used by Hertz.

which caused the charges in the air gap to oscillate and thereby to emit radiation. Hertz used a concave metal shield to focus the emitted waves on a single wire loop, which also terminated in an air gap. The size of the receiving loop and its air gap were adjusted so that the system would be in resonance with the emitted radiation. The magnetic field of the radiation induced an emf large enough to cause tiny sparks to jump across the gap in the receiving loop. Hertz went on to prove that the radiation exhibited all the phenomena we associate with waves, such as reflection, refraction, and interference. In particular, by reflecting the waves off a metal surface he set up standing waves. The wavelength ($\approx$ 33 cm) could be determined from the distance between the nodes or the antinodes whereas the frequency was known from the LC circuit. From $v = f\lambda$ he was able to deduce the speed of the waves to be about 3×10^8 m/s—which of course is the speed of light.

34.7 THE ELECTROMAGNETIC SPECTRUM

Electromagnetic waves span an immense range of frequencies, from very long wavelength radio waves, whose frequency is around 100 Hz, to extremely high energy γ rays from space, with frequencies around 10^{23} Hz. The electromagnetic spectrum, shown in Fig. 34.15, covers approximately 100 octaves. (The audible sound spectrum covers about nine octaves.) There is no theoretical limit to the high end. With the exception of the visible part of the spectrum, the boundaries between the classifications given below are not as sharp as Fig. 34.15 implies. The classifications are based roughly on how the waves are produced and/or detected.

Visible Light

The visible part of the electromagnetic spectrum covers roughly one octave, from 400 to 700 nm. An approximate range of wavelengths is associated with each color: violet (400–450 nm), blue (450–520 nm), green (520–560 nm), yellow (560–600 nm), orange (600–625 nm), and red (625–700 nm). As electrons undergo transitions between energy levels in an atom, light is produced at well-defined

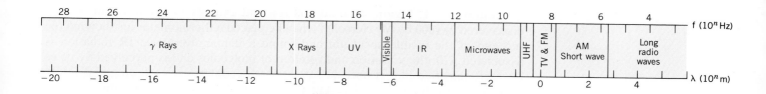

FIGURE 34.15 The electromagnetic spectrum. The divisions between the various regions are less abrupt than the diagram indicates.

wavelengths. Light covering a continuous range of wavelengths is produced by the random accelerations of electrons in hot bodies. Our sense of vision and the process of photosynthesis in plants have evolved within the range of those wavelengths of sunlight that our atmosphere does not absorb, which is between 300 nm and 1100 nm.

Ultraviolet Radiation

In 1801 J. W. Ritter, who was studying the blackening of silver chloride in various regions of the spectrum, found the strongest effect occurred beyond the violet. The ultraviolet (UV) region extends from 400 nm to about 10 nm. It plays a role in the production of vitamin D in our skins and leads to tanning. In large or prolonged doses, UV radiation kills bacteria and can induce cancer in humans. Glass absorbs UV radiation and hence can provide some protection against the sun's rays. If the ozone in our atmosphere did not absorb the UV below 300 nm, there would be a large number of cell mutations, especially cancerous ones, in humans. For this reason, the depletion of the ozone in our atmosphere by chlorofluorocarbons (CFCs) is now a matter of international concern. In some atoms, the absorption of UV is followed by the emission of longer wavelength visible light. This phenomenon, called fluorescence, is the process underlying the use of "blacklights."

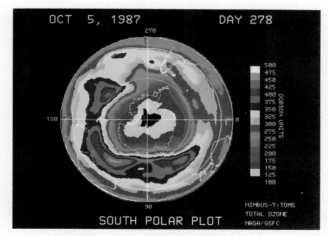

The ozone concentration over Antarctica was monitored by measuring the intensity of reflected UV radiation in the absorption bond of ozone.

Infrared Radiation

The infrared (IR) region starts at 700 nm and extends to about 1 mm. It was discovered in 1800 by M. Herschel when he placed a thermometer just beyond the red end of the visible spectrum and found a temperature rise. IR radiation is associated with the vibration and rotation of molecules and is perceived by us as heat. IR-sensitive film is used in satellites for geophysical surveying and in the detection of the hot exhaust gases of a rocket launch. Since it permits the scanning of the human body for minute temperature variations, IR is used in the early detection of tumors—which are warmer than the surrounding tissue (see p. 357). Snakes and "night vision" instruments (see Fig. 19.17) can detect IR radiation emitted by the warm bodies of animals.

Microwaves

Microwaves cover wavelengths from 1 mm to about 15 cm. Microwaves up to about 30 GHz (1 cm) may be generated by the oscillations of electrons in a device called a klystron. In the microwave ovens used in kitchens, the radiation has a frequency of 2450 MHz. Modern intercity communications, such as numerical data, phone conversations, and TV programs, are often carried via a cross-country network of microwave antennas. Microwaves have also been focused on cancerous tissue to raise the local temperature to about 115 °F. While normal cells are able to dissipate the thermal energy quickly, the cancerous cells have relatively poor circulation and are thereby destroyed.

Radio and TV Signals

These signals span the range from 15 cm to 2000 m. Dipoles, such the familiar "rabbit ears," are used for both transmission and reception. For AM radio, a coil

FIGURE 34.16 A radio telescope is used for communications and for radio astronomy. It is not dependent on clear weather, as is an optical telescope.

is usually used for reception because the wavelength is so much larger than is practical for an electric dipole. For UHF TV, the coil is used because the wavelengths are so small. Radio telescopes (Fig. 34-16b) are used to communicate with satellites and to study a range of wavelengths reaching us from space.

X Rays

X rays, discovered in 1895 by W. Roentgen, are adjacent to the UV and extend from 1 nm to 0.01 nm. X-ray machines produce these waves by the rapid deceleration of electrons that bombard a heavy metal target. This type of radiation spans a range of frequencies and is called *bremsstrahlung* or "braking radiation." X rays are also produced by electronic transitions between the energy levels in an atom. Since the sizes of atoms and their spacing in crystals fall in this domain, X rays are used to study the atomic structure of crystals or molecules, such as DNA (see X ray diffraction, Section 38.7). Besides their diagnostic and therapeutic use in medicine, X rays are used to detect tiny faults in machinery. With the advent of scientific satellites, X-ray astronomy has become an important tool in studying the universe.

γ Rays

Gamma rays, which produce effects similar to those of X rays, were first identified by P. Villiard in 1900 as part of the radioactive emission of some materials. Whereas X rays are produced by electrons, γ rays are usually produced within the nucleus of an atom and are extremely energetic by atomic standards. They cover the range from 0.01 nm down, or, equivalently, from 10^{20} Hz up.

34.8 DERIVATION OF THE WAVE EQUATION (Optional)

Mathematical manipulation of Faraday's law and the Ampère–Maxwell law leads directly to a wave equation for the electric and magnetic fields, but this approach is beyond our scope. Instead, we will assume E and B vary in a certain way, consistent with Maxwell's equations, and then show that electromagnetic waves are a consequence of the application of Faraday's law and the Ampère–Maxwell law. To simplify matters we consider only free space, where there is no charge or conduction current. The fields are produced by charges in some distant region of no interest to us.

Figure 34.17 shows two plane wavefronts traveling along the +x axis. The electric field is along the y axis, and the magnetic field is along the z axis. Each field is uniform on any yz plane and varies only along the x axis. First we apply Faraday's law, Eq. 34.5, to the small rectangular loop in the xy plane. The line integral consists of four parts. For the top and bottom sides **E** · d𝓵 = 0 since **E** is perpendicular to d𝓵. The contribution of the other two sides is

$$\oint \mathbf{E} \cdot d\boldsymbol{\ell} = E_{y2}\Delta y - E_{y1}\Delta y$$

Recall that with the thumb along the direction of the magnetic field, B_z, the fingers of the right hand determine the sense in which the integral is to be evaluated. Strictly speaking, we need to integrate over the area of the loop in order to find the mag-

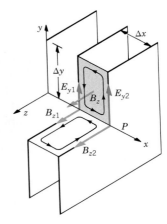

FIGURE 34.17 Two plane wavefronts with their associated electric and magnetic fields. We apply Faraday's law to the shaded rectangle in the xy plane and the Ampère–Maxwell law to a similar rectangle in the xz plane.

netic flux through it. We can do just as well for our purposes by taking the value of B_z at the center as the "average" value for the whole area. This approach is valid if the distance Δx between the wavefronts is much less than the wavelength. The magnetic flux through the loop is $\Phi_B = B_z \Delta x \Delta y$ and its rate of change is

$$\frac{d\Phi_B}{dt} = \frac{\partial B_z}{\partial t} \Delta x \Delta y$$

The partial derivatives are used because we are concerned with the explicit time dependence of these quantities at a fixed point in space. After dividing both sides of Faraday's law (Eq. 34.5) by $\Delta x \Delta y$, we find

$$\frac{(E_{y2} - E_{y1})}{\Delta x} = -\frac{\partial B_z}{\partial t}$$

As $\Delta x \rightarrow 0$, this equation takes the form

(Faraday's law) $\dfrac{\partial E_y}{\partial x} = -\dfrac{\partial B_z}{\partial t}$ (34.19)

An observer at P will note that B_z is decreasing in time, or $\partial B_z / \partial t < 0$. Thus from Eq. 34.19 it follows that $\partial E_y / \partial x > 0$, that is, $E_{y2} > E_{y1}$, as drawn in Fig. 34.17. Physically this is a statement of Lenz's law: The magnetic flux through the loop is decreasing in time, so the induced emf (given by the line integral) must oppose the change.

A similar argument may be used for the Ampère–Maxwell law applied to a loop in the xz plane. We take two distinct values for B_z in the integral $\oint \mathbf{B} \cdot d\boldsymbol{\ell}$ but use the average value of E_y for the electric flux. Again only two sides contribute to the line integral. Equation 34.6 yields

$$(-B_{z2} + B_{z1})\Delta z = \mu_0 \varepsilon_0 \frac{\partial E_y}{\partial t} \Delta x \Delta z$$

The sense of the integration is determined by E_y and the right-hand rule. Dividing by $\Delta z \Delta x$ and taking the limit as $\Delta x \rightarrow 0$, we find

(Ampère's law) $\dfrac{\partial B_z}{\partial x} = -\mu_0 \varepsilon_0 \dfrac{\partial E_y}{\partial t}$ (34.20)

By taking the appropriate derivatives of Eq. 34.19 and Eq. 34.20 (see Exercise 3 below), it is straightforward to obtain Maxwell's wave equations, Eqs. 34.7 and 34.8.

If we substitute the functions $E = E_0 \sin(kx - \omega t)$ and $B = B_0 \sin(kx - \omega t)$ into Eq. 34.19, we find

$$kE_0 \cos(kx - \omega t) = \omega B_0 \cos(kx - \omega t)$$

Thus $E_0 = (\omega/k)B_0 = cB_0$. At any point, E and B are in phase and their magnitudes are related by $E = cB$.

EXERCISE 3. Take the space derivative of Eq. 34.20 and then use Eq. 34.7 to obtain Eq. 34.8. (Drop the subscripts on E and B.)

SUMMARY

In a region where the electric field is changing, there is a **displacement current** given by

$$I_D = \varepsilon_0 \frac{d\Phi_E}{dt}$$

where Φ_E is the electric flux. Ampère's law must be modified to include the displacement current.

By combining the Ampère–Maxwell law and Faraday's law, one can show that the electric and magnetic fields obey the **Maxwell's wave equation,** for example,

$$\frac{\partial^2 E}{\partial x^2} = \mu_0 \varepsilon_0 \frac{\partial^2 E}{\partial t^2}$$

In a plane electromagnetic wave the electric and magnetic fields are perpendicular to each other and to the direction of propagation. These fields oscillate in phase and the waves propagate in free space with a speed

$$c = \frac{1}{\sqrt{\mu_0 \varepsilon_0}}$$

which is equal to the speed of light in free space. The instantaneous values of the field are related according to

$$E = cB$$

The intensity of an electromagnetic wave may be found from the **Poynting vector**

$$S = \frac{E \times B}{\mu_0}$$

which shows that the energy flows perpendicular to both **E** and **B**. The average intensity of a plane electromagnetic wave is given by

$$S_{av} = \frac{E_0 B_0}{2\mu_0}$$

where E_0 and B_0 are the amplitudes of the fields.

The **linear momentum** transported by an electromagnetic wave is given by

$$p = \frac{U}{c}$$

where U is the energy absorbed by a surface. If the waves are completely reflected, the momentum transfer is doubled.

The **radiation pressure** exerted by an electromagnetic wave incident normal to a surface and completely absorbed by it is given by

(Pressure) $$\frac{F}{A} = \frac{S}{c} = u$$

where $u = \varepsilon_0 E^2 = B^2/\mu_0$ is the energy density of the wave. If the waves are completely reflected, the pressure is doubled.

ANSWERS TO IN-CHAPTER EXERCISES

1. For a circular loop, the line integral of B is still $B(2\pi r)$ but the enclosed electric flux is $\Phi_E = (\pi R^2 E)$. Using these expressions in Eq. 34.2 (with $I = 0$) yields

$$B = \mu_0 \varepsilon_0 \frac{R^2}{2r} \cdot \frac{dE}{dt}$$

Check that at $r = R$ this is consistent with the expression in Exercise 34.1.

2. The magnitude of **B** is $B = E/c = 1.67 \times 10^{-8}$ T. The direction of **E** × **B** must be in the $+z$ direction. Since $(-\mathbf{i}) \times (-\mathbf{j}) = +\mathbf{k}$ we see that **B** is along the $-y$ direction.

3. From Eq. 34.19 we have

$$\frac{\partial}{\partial x}\left(\frac{\partial E}{\partial x}\right) = -\frac{\partial}{\partial x}\left(\frac{\partial B}{\partial t}\right)$$

which is equivalent to

$$\frac{\partial^2 E}{\partial x^2} = -\left(\frac{\partial}{\partial t}\right)\left(\frac{\partial B}{\partial x}\right)$$

When $\partial B/\partial x$ from Eq. 34.20 is used in the above equation, we obtain Eq. 34.7.

QUESTIONS

1. An empty dish gets hot in an ordinary oven, but in a microwave oven it may not. Why not?

2. How can a loop antenna be used to locate the source of a clandestine radio transmission?

3. Can a conduction current and displacement current coexist in the same region? If so, give an example.

4. While a capacitor is charging, is there a displacement current in the connecting wires?

5. A radio station broadcasts the sound of a singer. Outline, in simple terms, how the sound of her voice reaches your ears.

6. If a TV set or an FM radio lacks an antenna, it can "pull in" several stations if you touch the antenna terminals. Why does this occur?

7. What is the direction of the Poynting vector between the plates of a capacitor that is being charged?

8. Some phenomena associated with electromagnetic radiation do not depend on its frequency. Name two.

9. A microwave oven contains radiation with a frequency of 2450 MHz. What is the wavelength?

10. Why is the radiation pressure exerted by a given electromagnetic wave greater for a reflecting surface than for an absorbing surface?

11. Do ac power transmission lines emit electromagnetic waves?

12. (a) Why is it inadvisable to use a metal pot in a microwave oven? (b) Microwave ovens are prone to "dead spots" where food does not cook properly. What might be the origin of this phenomenon?

13. The magnetic field of an electromagnetic wave is given by $B = (2 \times 10^{-6}) \cos[\pi(0.04x + 10^7 t)]$ T. Is this a wave in free space?

14. Is it possible to produce a standing electromagnetic wave? How might this be accomplished?

15. A TV viewer and a spectator at a stadium both watch a baseball being struck. Do they both see and hear the contact at the same time? If not, how do differences arise?

16. A plane electromagnetic wave travels horizontally from east to west. If, at some point, **B** is vertically downward at one instant, what is the direction of **E**?

17. Could a powerful laser be used to propel a spacecraft? If so, how?

18. In what sense is the term $\varepsilon_0 d\phi/dt$ (a) like an electrical current, and (b) not like an electrical current?

19. If magnetic monopoles were discovered, which of Maxwell's equations would need to be modified?

EXERCISES

34.1 Displacement Current

1. (I) (a) Show that the unit of $1/\sqrt{\mu_0 \varepsilon_0}$ is m/s. (b) Show that the unit of EB/μ_0 is W/m^2.

2. (I) Show that the following equations are dimensionally correct: (a) $E = cB$. (b) $I_D = \varepsilon_0 d\Phi_E/dt$. (c) Pressure $= S/c$.

3. (I) A parallel plate capacitor has circular plates of radius 2.5 cm separated by 3 mm. If the potential difference between the plates changes at 5×10^4 V/s, what is the displacement current?

4. (I) A parallel plate capacitor has circular plates of radius 2 cm separated by 1.4 mm. At a given instant, the current in the long, straight connecting wires is 3 A. (a) What is the displacement current between the plates? (b) At what rate is the potential difference between the plates changing?

5. (I) Show that the displacement current through an air-filled parallel plate capacitor may be expressed as $I_D = C\, dV/dt$, where V is the potential difference across the capacitor.

6. (I) A parallel plate capacitor has circular plates of radius 2 cm separated by 2.4 mm. The potential difference between the plates is increasing at the rate of 8 kV/s. What is the displacement current between the center and a distance equal to half the radius from the center? (Assume that the electric field between the plates is uniform.)

7. (II) The circular plates of a parallel plate capacitor have a radius of 2 cm and are separated by 4 mm. They are connected to a 60-Hz ac source with a peak potential difference of 120 V. Find the peak magnetic field halfway from the center to the edge of the plates.

8. (II) Show that the magnetic field at a distance r from the center of a parallel plate capacitor with circular plates of radius R is given by

$$B(r) = \frac{\mu_0 I_D}{2\pi r} \qquad (r > R)$$

(b) Express $B(r)$ in terms of I_D for $r < R$.

9. (II) A parallel plate capacitor has circular plates of radius 2 cm separated by 2.4 mm. The current in the long straight connecting wires is 20 mA. Find the magnetic field at the following radial distances from the center of the plates: (a) 0.5 cm; (b) 5 cm.

34.3 Electromagnetic Waves

10. (I) A plane electromagnetic wave propagates in free space along the negative z direction. At a certain point, the electric field vector is $-21\mathbf{i}$ V/m. What is the magnetic field vector?

11. (I) The magnetic field in a plane electromagnetic wave is given by

$$B_y = 2 \times 10^{-7} \sin(0.5 \times 10^3 x + 1.5 \times 10^{11} t) \text{ T}$$

(a) What is the wavelength and frequency of the wave? (b) Write an expression for the electric field vector.

12. (I) The components of the electric field of a plane electromagnetic wave are given by $E_z = E_0 \sin(ky + \omega t)$, $E_x = E_y = 0$. Write an expression for **B**.

34.4 Energy Transport and the Poynting Vector

13. (I) The average energy density in a sinusoidal electromagnetic wave is 10^{-7} J/m^3. Find the amplitude of (a) the electric field, and (b) the magnetic field.

14. (I) The electric field of a plane wave is given by

$$E_y = 50 \sin[\pi(0.8x - 2.4 \times 10^8 t)] \text{ V/m}$$

Find: (a) The average energy density; (b) the amplitude and direction of the magnetic field; (c) the average Poynting vector.

15. (II) Summed over all wavelengths, the average intensity of solar radiation at the earth's surface is 1 kW/m². (a) What is the average energy density at the earth's surface? (b) Estimate the solar energy incident in 1 h on the earth's surface?

16. (II) A distress beacon (a point source) emits a single wavelength with an average 25-W output. Find the amplitudes of the electric and magnetic fields at the following points: (a) at a search plane at a distance of 25 km; (b) at a geosynchronous satellite at an altitude of 34,000 km.

17. (II) At a distance of 6 m from a monochromatic (single wavelength) point source, the amplitude of the electric field is 10 V/m. Find: (a) the amplitude of the magnetic field; (b) the average power output of the source.

34.5 Momentum and Radiation Pressure

18. (I) The intensity of solar radiation incident at the level of the upper atmosphere of the earth is 1.34 kW/m². What is the force exerted by this radiation on a satellite solar panel of area 100 m²? Assume normal incidence and complete absorption.

19. (I) Find the force exerted on a 5-cm² plate by radiation from the following lasers: (a) a 1-mW helium-neon laser; (b) a 1-kW carbon dioxide laser. Assume that the beam, which has a cross section of 10 mm², is incident normally and completely absorbed.

20. (I) A 10⁴-W radio antenna transmits at 98 MHz. Assuming it radiates as a point source, find the radiation pressure at a distance of 20 km.

21. (I) At a distance of 100 m from a point source, the amplitude of the magnetic field is equal in magnitude to 0.1% of the earth's field, that is, about 10^{-7} T. Estimate the power output of the transmitter?

22. (I) A 1-kW continuously operating laser is used as a "light rocket" to propel a 100-kg spacecraft. The cross-sectional area of the beam is 20 mm². What is its acceleration?

23. (I) At the threshold of detection, an FM receiver can pick up a signal for which $E_0 = 2 \; \mu$V/m. (a) What is the intensity of the electromagnetic wave? (b) At what distance would a 10 kW point source produce this intensity?

24. (I) (a) At what distance from a 100-W point source is the amplitude of the magnetic field 10^{-8} T? (b) What is the amplitude of the electric field at this point?

25. (I) The intensity of a plane electromagnetic wave is 5 W/m². It is incident on a perfectly reflecting surface. Find: (a) the radiation pressure; (b) the force exerted on a panel 60 cm × 40 cm set perpendicular to the direction of wave propagation.

26. (I) Assume that a 60-W light bulb radiates at a single wavelength and as a point source. At a distance of 10 m, find the amplitudes of (a) the electric field, and (b) the magnetic field.

27. (I) The intensity of solar radiation at the earth's surface is 1 kW/m². What would be the power entering an eye through a pupil of diameter 0.5 cm?

28. (I) A solar panel converts sunlight into electrical energy with an efficiency of 18%. The intensity of the sun's radiation at the earth's surface is 1 kW/m². What is the area needed to generate 10 kW of electrical energy?

29. (I) Show that the instantaneous magnitude of the Poynting vector in free space may be written in the form

$$S = \frac{c}{2} \left(\varepsilon_0 E^2 + \frac{B^2}{\mu_0} \right)$$

30. (II) A bulb radiates 120 W as a point source. At a distance of 10 m find: (a) the average intensity; (b) the average energy density; (c) the average force on a perfectly reflecting plate of area 1 cm² set perpendicular to the radiation.

PROBLEMS

1. (I) A capacitor with parallel circular plates of radius a and a distance d apart is being charged. Find: (a) B at its edge, and (b) the Poynting vector at the edge. (c) Show that the input power into the capacitor is $\varepsilon_0 \pi da^2 E(dE/dt)$. (Ignore the fringing electric fields.)

2. (I) The magnetic field of an 800-kHz AM radio signal has an amplitude of 4×10^{-10} T. If the wave is detected by a flat coil with 20 turns of radius 6 cm, what is the peak induced emf? Assume the magnetic field is directed along the axis of the coil.

3. (I) A straight wire has a length of 6 m and a radius of 0.5 mm. Its electrical resistance is 0.8 Ω and there is a potential difference 24 V across its ends. (a) What is the power loss due to Joule heating? (b) What is the Poynting

vector at the surface? (c) Show that the electromagnetic power entering the wire is equal to the value found in (a).

4. (I) A laser beam of intensity S and cross-sectional area A is completely absorbed by a particle of mass m for a period Δt. Show that the change in speed of the particle is $\Delta v = SA \; \Delta t / mc$.

5. (I) Plane electromagnetic waves of intensity S are incident normally on a flat surface. Only a fraction f of the incident energy is absorbed. What is the radiation pressure?

6. (II) A dust particle in the tail of a comet has a radius R. Its density is 1.2 g/cm³. It is subject both to the gravitational attraction of the sun and the force due to its radiation pressure. For what value of R will these be equal in magnitude? Assume complete absorption. The solar power output is 3.8×10^{26} W.

7. (II) A cylindrical coil used as an AM antenna has 250 turns of diameter 1.5 cm. Find the peak emf induced by a 10^4-W station (treated as a point source) broadcasting at 800 kHz at a distance of 2 km. The axis of the coil is parallel to the direction of the magnetic field of the wave.

8. (II) Radiation is incident at an angle θ to a flat surface, as shown in Fig. 34.18. Show that the radiation pressure is $u_{av} \cos^2 \theta$. Assume all the radiation is absorbed.

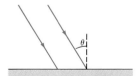

FIGURE 34.18 Problem 8.

9. (II) A leaky capacitor with circular plates of radius $R = 12$ cm has a capacitance of 5 μF and an effective resistance of 4×10^5 Ω. At $t = 0$, the potential difference across the plates is zero but is increasing at the constant rate of 2000 V/s. (a) Find the displacement current I_D. (b) At what time is I_D equal to the conduction current?

10. (II) A *radiometer* consists of two disks of radius 1.2 cm connected by a light rod of length 10 cm that is suspended at its midpoint by a fine thread; see Fig. 34.19. One disk is perfectly absorbing whereas the other is perfectly reflect-

ing. The torque required to rotate the thread is given by Hooke's law, $\tau = \kappa \theta$, where the constant $\kappa = 10^{-11}$ N·m/deg. What is the equilibrium deflection when the sun's radiation (1 kW/m²) is incident normally on the disks?

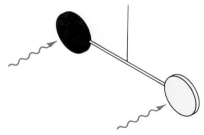

FIGURE 34.19 Problem 10.

11. (I) A plane electromagnetic wave with an intensity of 220 W/m² is incident normal to a flat plate of radius 30 cm. If the plate absorbs 60%, and reflects 40%, of the incident radiation, what is the momentum transferred to it in 5 min?

12. (I) It has been proposed that solar radiation could be used to propel spacecraft. Suppose a perfectly reflecting solar "sail" of area 10^3 m² is set normal to the sun's radiation of intensity 1 kW/m². Find: (a) the force on the sail; (b) if the craft has a mass of 10^3 kg, how long would it take to reach 1 m/s starting from rest? (Ignore the gravitational attraction of the sun.)

CHAPTER 35

Light: Reflection and Refraction

A double rainbow.

Major Points

1. In **geometrical optics,** light is considered to travel in **rays.**
2. (a) The **law of reflection.** (b) **Snell's law** and the definition of refractive index. (c) Total internal reflection.
3. **Huygens' principle** for the construction of wavefronts.
4. **Dispersion** and the spectrum produced by a prism.
5. The distinction between **real** and **virtual** objects or images.
6. Images formed by **concave** and **convex** spherical mirrors. (a) The **mirror formula.** (b) **Principal ray diagrams.**

Light is the principal means by which we gain knowledge of the world. Consequently, the nature of light has been the source of one of the longest debates in the history of science. In the 17th century Descartes and Newton considered light to be a stream of particles, whereas Huygens proposed that light is a disturbance in a medium called the "ether." Huygens knew that two beams of light can cross without affecting each other and could not imagine how a stream of particles could do the same without colliding. In the last chapter we saw that light is a transverse electromagnetic wave consisting of oscillating electric and magnetic fields. Furthermore, these waves can propagate in a vacuum. This wave theory of light underlies the next four chapters on optics.

Although all problems concerning the propagation of light can be solved with Maxwell's electromagnetic theory, such an approach is often not necessary. Our present interest is in knowing what happens when light meets the boundary between two media, such as glass and air. One can analyze the effects of mirrors and lenses by considering light to travel in rays and by using simple geometry. This approach is called **ray optics.**

35.1 RAY OPTICS

The light from a projector in a smoky cinema, or sunlight shining through trees on a misty morning, appears to travel in straight lines. We also know that on a clear sunny day, objects cast sharp shadows. Thus it is natural to treat the propagation of light in terms of **rays.** A ray is equivalent to a very narrow beam of light, and it

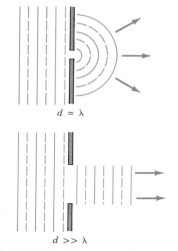

FIGURE 35.1 The passage of waves of wavelength λ through an opening of size d. When $d \approx \lambda$ the waves spread in the region to the right. The bending of the rays is called diffraction. When $d \gg \lambda$, the waves travel in the original direction.

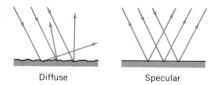

Diffuse   Specular

FIGURE 35.2 (a) In diffuse reflection off a rough surface, the reflected light travels in all directions. (b) In specular reflection off a smooth surface, the reflected rays travel in the same direction.

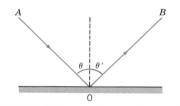

FIGURE 35.3 According to the law of reflection, the angle of incidence, θ, is equal to the angle of reflection, θ'.

indicates the path along which the energy of the wave travels. Rays are drawn perpendicular to the wavefronts. In a homogeneous medium, the rays are straight lines. This fact allowed the mathematician Euclid and the astronomer Ptolemy to use the power of geometry to analyze problems in optics. **Geometrical optics** is the study of the behavior of straight-line rays at the interface between two media by the use of simple geometrical constructions.

However, we know that light is an electromagnetic wave and that, in general, waves do not travel in straight lines. For example, water waves passing through a small opening in a barrier, as in Fig. 35.1a, spread into the region behind the barrier. This phenomenon, called **diffraction** (Chapter 37), is significant when the size of the aperture, d, is comparable to the wavelength of the wave, that is, when $d \approx \lambda$. When the aperture is much larger than the wavelength, that is, when $d \gg \lambda$, part of each wavefront is removed, but the remaining waves continue to travel mostly in the original direction, as shown in Fig. 35.1b. In this case the border of the shadow region is relatively sharp. In geometrical optics the bending of rays at the edges of apertures and obstacles is ignored. This is a reasonable approximation when the size of an apparatus is much larger than the wavelength of visible light, which is less than 1 μm.

35.2 REFLECTION

Suppose parallel light rays are incident at some angle on the boundary between two media, such as glass and air. In general, some light is reflected and the rest is either transmitted or absorbed. If the surface is rough, as in Fig. 35.2a, the reflected rays travel in random directions. As a result the object can be seen from any position. Such *diffuse* reflection is evident from the dull appearance of the object. When the surface is highly polished, as in Fig. 35.2b, the direction of the reflected rays is simply related to that of the incoming rays. Such *specular* reflection occurs when the size of the surface irregularities is smaller than the wavelength of the incident light. In specular reflection, the reflected beam can be seen only in one direction.* Even in diffuse reflection, each ray is specularly reflected at the tiny part of the surface it strikes. But since all these tiny parts are randomly oriented, the reflected rays do not have a common direction.

The Law of Reflection

Consider a ray incident on a flat surface, as in Fig. 35.3. The direction of the reflected ray is given by the law of reflection, which was known to Hero of Alexandria in the 2nd century B.C. The **law of reflection** states: *The angle of incidence, θ, is equal to the angle of reflection, θ'.* These angles are conventionally measured with respect to the normal to the plane. Around 1000 A.D., an Arab scholar named Alhazen pointed out that the incident ray, the normal to the plane, and the reflected ray, all lie in the same plane, which we call the *plane of incidence*.

---

* Cover all but a small portion of a flat mirror and place it on a sidewalk at night. Next, try to locate the reflection of a street lamp.

EXAMPLE 35.1: Two mirrors, M_1 and M_2, are placed in contact at an angle of 120° as shown in Fig. 35.4. A ray is incident at 50° to the normal to M_1. In what direction does the light leave M_2?

Solution: From the law of reflection, the angle of reflection at M_1 is also 50°, so the angle made by the reflected ray to the plane of M_1 is 40°. In triangle ABC, the angle at C is 180° − 40° − 120° = 20°. The angle of incidence to M_2 is 70°, so this is also the angle of reflection.

EXERCISE 1. What is the total angle through which the original ray is deflected?

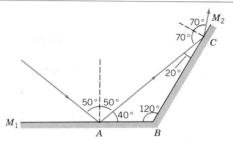

FIGURE 35.4 A ray reflected off two mirrors at an angle to each other.

A ray in a horizontal plane reflected by two vertical mirrors placed at right angles, as in Fig. 35.5a, is returned in exactly the opposite direction. A useful arrangement, consisting of three mutually perpendicular mirrors, is called a *corner reflector*. Any ray approaching the corner reflector is returned in the opposite direction. This property is used in the reflectors on cars or bicycles. Signals reflected by the corner reflectors in the LAGEOS satellite, Fig. 35.5b, were used as part of a geological survey of the earth. A corner reflector placed on the moon, Fig. 35.5c, has been used to determine its distance to an accuracy of 15 cm.

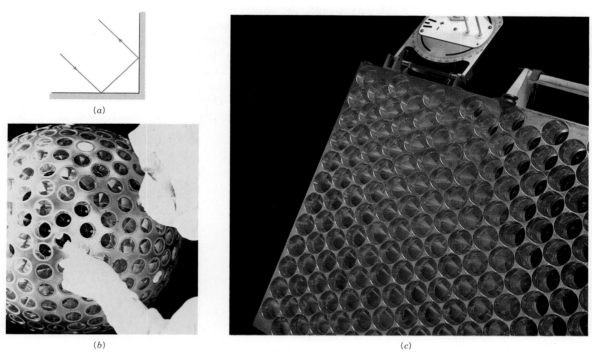

FIGURE 35.5 (a) The direction of a ray incident on two mirrors perpendicular to each other is reversed. (b) The Lageos satellite was covered in corner reflectors. By timing the return of laser pulses from different locations, minute movements of the continents could be detected. (c) A corner reflector placed on the moon allowed its distance to be measured with an uncertainty of only 15 cm.

Huygens' Principle

In 1678 C. Huygens enunciated a "principle" that is useful in predicting the propagation of wavefronts. He proposed that when a light pulse is emitted by a source, the nearby particles of the "ether" are set into motion. The light propagates because this motion is communicated to the neighboring particles. Therefore, each particle acts as a source of secondary *wavelets*. For example, particles at wavefront *AB* in Fig. 35.6 produce secondary wavelets, which, at a later time, form the new wavefront *CD*. In order to explain the rectilinear propagation of rays, he asserted that only the wavelets in the forward direction are strong. The wavelets that spread to the sides were quietly ignored as being "too feeble to be seen."* In modern terms, **Huygens' principle** for the construction of wavefronts is:

> Each point on a wavefront acts as a source of secondary wavelets. At a later time, the envelope of the leading edges of the wavelets forms the new wavefront.

In this statement Huygens' "particle" has been replaced by a mathematical "point." One might think in terms of particles for water waves or sound, but not for light waves, which can propagate in a vacuum. Huygens used his principle to obtain the law of reflection as is shown below.

Figure 35.7 shows plane wavefronts approaching a flat surface at angle θ to the surface and being reflected at θ' to the surface. Since the rays are perpendicular to the wavefronts, the angles θ and θ' are also the angles made by the rays to the *normal* to the surface. When edge *A* of wavefront *AB* meets the surface, it starts to produce its secondary wavelet. The same happens as each successive point of *AB* strikes the surface. In the time that the wavelet from *B* reaches the point *C,* the wavelet from *A* has expanded to point *D*. The line *DC* forms the reflected wavefront. Since the speeds of the incident and reflected waves are identical, $AD = BC$. Triangles *ACD* and *ACB* are both right angled and have a common hypotenuse. We conclude that $\theta = \theta'$, which is the law of reflection.

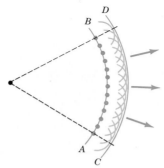

FIGURE 35.6 According to Huygens' principle each point on a wavefront acts as a source of secondary wavelets. The wavefront at a later time is formed from the envelope of the wavelets.

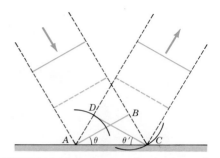

FIGURE 35.7 In the time that it takes point *B* of wavefront *AB* to reach the surface, the wavelet from *A* has expanded to *D*. From the geometry of triangles *ABC* and *ADC* we conclude that $\theta = \theta'$.

* Some justification for this procedure was first provided in the 19th century.

35.3   REFRACTION

A drinking straw appears to be bent when it is partially immersed in water (see Fig. 35.8); a magnifying lens can focus the sun's rays or make objects seem larger; sunlight passing through a prism produces a beautiful spectrum of colors. These and many other optical effects are caused by **refraction,** which is the bending of rays at the boundary between two media. As shown in Fig. 35.9, the directions of the incident ray and the refracted ray are specified by the *angle of incidence* θ_1 and the *angle of refraction* θ_2, both measured with respect to the normal to the boundary. Around 130 A.D., Ptolemy measured these angles for the boundary between air and water and suggested that the ratio θ_1/θ_2 is constant, but this is not correct. The correct relationship between the angle of incidence and the angle of refraction was found experimentally around 1621 by Willebrord Snell, a Dutch mathematician. Although he did not use the following terms, in effect he found that the ratio of the sine of the angle of incidence to the sine of the angle of refraction is constant, that is, $\sin \theta_1/\sin \theta_2 = $ constant.

In 1678 Huygens derived Snell's result as follows. Let us take the speed of light in the two media to be v_1 and v_2, with $v_1 > v_2$. In Fig. 35.10 the angles made with the boundary by the incident and refracted wavefronts are θ_1 and θ_2, respectively. In a short time interval Δt, the wavelet from point B of wavefront AB travels a distance $v_1\Delta t$ to point B', where $BB' = v_1\Delta t = AB' \sin \theta_1$. In this time, the wavelet from A travels a distance $v_2\Delta t$ to point A' in the second medium, where $AA' = v_2\Delta t = AB' \sin \theta_2$. The new wavefront, $A'B'$, is tangent to the wavelets from wavefront AB. The ratio BB'/AA' yields

$$\frac{\sin \theta_1}{\sin \theta_2} = \frac{v_1}{v_2} \tag{35.1}$$

According to this equation $\theta_1 > \theta_2$ when $v_1 > v_2$. Therefore the ray bends *toward* the normal when it enters a medium in which the *wave velocity is lower*.

Equation 35.1 is usually expressed in terms of the refractive index of each medium. The **refractive index,** n, of a medium is defined as the ratio of the speed of light in vacuum, c, to the speed v in the medium,

$$n = \frac{c}{v} \tag{35.2} \qquad \text{Refractive index}$$

FIGURE 35.8 A straw partially immersed in a liquid appears bent because of the refraction of light at the surface of the liquid.

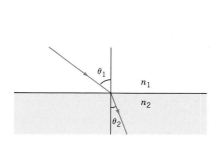

FIGURE 35.9 The angle of incidence θ_1 and the angle of refraction θ_2 are related by Snell's law. Note that the direction of the ray may be reversed without affecting the relationship $n_1\sin\theta_1 = n_2\sin\theta_2$.

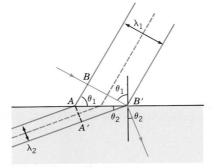

FIGURE 35.10. Refraction is explained by Huygens' principle. The speed of the waves is lower in the medium of higher refractive index. In one period, the wavelet from A travels one wavelength in the medium, λ_2 and the wavelet from B travels λ_1, where $\lambda_2 > \lambda_1$.

If Eq. 35.1 is rewritten in terms of the refractive indices, we obtain the modern form of **Snell's law:**

Snell's Law

$$n_1 \sin \theta_1 = n_2 \sin \theta_2 \qquad (35.3)$$

When $n_2 > n_1$, it follows that $\theta_2 < \theta_1$, that is, on entering a medium with a *higher refractive index* the ray bends *toward* the normal. Equation 35.3 is also valid when the direction of the ray is reversed. Thus the positions of a point source of light and the eye can be interchanged.

The frequency of the wave, which is determined by the source, is the same on both sides of the boundary. One can see this by noting that the number of crests that approach the boundary per second must equal the number that leave per second. Otherwise the waves would pile up at the boundary—an effect never observed. If the wavelength in vacuum is λ_0, and that in the medium is λ_n, then $v = f\lambda_n$ and $c = f\lambda_0$. Using this in Eq. 35.2 we find

FIGURE 35.11 The wavelength of light in a medium is less than its wavelength in vacuum: $\lambda_n = \lambda_0/n$. The frequency of the light is unchanged as it crosses the boundary.

$$\lambda_n = \frac{\lambda_0}{n} \qquad (35.4)$$

As Fig. 35.11 shows, the wavelength in the medium is shorter than the wavelength in vacuum.

EXAMPLE 35.2: Light of wavelength 600 nm in air is incident at an angle of 35° to the normal of a plate of heavy flint glass of refractive index 1.6. Assume the refractive index of air is 1. Find: (a) the angle of refraction, (b) the wavelength of the light in the glass, (c) the speed of light in the glass.

Solution: (a) From Eq. 35.3,

$$(1) \sin 35° = 1.6 \sin \theta_2$$

Thus $\sin \theta_2 = \sin 35°/1.6 = 0.358$, and so $\theta_2 = 21°$.
(b) The wavelength in glass is given by Eq. 35.4:

$$\lambda_n = \frac{\lambda_0}{n} = \frac{600 \text{ nm}}{1.6} = 355 \text{ nm}$$

(c) The speed of light in the glass is

$$v = \frac{c}{n} = \frac{3.00 \times 10^8 \text{ m/s}}{1.6} = 1.88 \times 10^8 \text{ m/s}.$$

EXAMPLE 35.3: A ray traveling in a medium of refractive index n_1 enters a flat plate of glass of refractive index n_2 at an angle α to its normal and exits into the original medium. Show that it emerges parallel to its original direction.

Solution: On entering the top surface of the glass in Fig. 35.12, the ray is refracted at angle β to the normal. According to Snell's law

$$n_1 \sin \alpha = n_2 \sin \beta \qquad (i)$$

The refracted ray strikes the lower surface also at β to the normal and emerges at γ to the normal. Thus,

$$n_2 \sin \beta = n_1 \sin \gamma \qquad (ii)$$

Comparing (i) and (ii) we see that $\alpha = \gamma$. The emergent ray is parallel to the incident ray but is displaced laterally as indicated in Fig. 35.12.

EXERCISE 2. Green light traveling in glass ($n = 1.5$) emerges into air at 40° to the normal to the glass–air boundary. The wavelength in air is 546 nm. (a) What is the angle of incidence in the glass? (b) What is the frequency in glass?

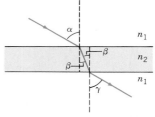

FIGURE 35.12 When a ray travels through a plate of uniform thickness, it emerges parallel to the original direction.

When a ray encounters a series of slabs of increasing refractive index, as shown in Fig. 35.13, its path tends increasingly toward the normal. When the variation in the index is continuous, the path is a smooth curve. Thus the path of a ray in a nonhomogeneous medium is not a straight line. The density of our atmosphere decreases with height, and as a result the refractive index also decreases.

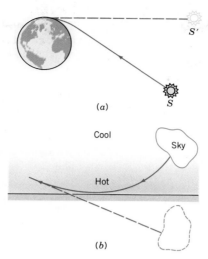

(a)

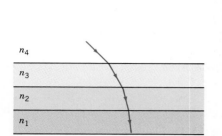

FIGURE 35.13 A ray encounters slabs of increasing refractive index. Its path tends increasingly toward the normal.

FIGURE 35.14 (a) Since the refractive index of the atmosphere decreases with altitude, light from the setting sun is refracted as shown. (b) On a hot day the refractive index of the air at ground level may be lower than that above. The mirage is actually an image of the sky.

For this reason one can see the sun at sunset, even after it has really fallen below the horizon. The real object, S, appears to be at S', as shown in Fig. 35.14a. Another phenomenon associated with a varying refractive index is the *mirage*. On a hot day, the air at ground level is less dense than that above, which means that the refractive index increases with height up to some level. As a result, rays close to the ground curve upward, as shown in Fig. 35.14b. The "water" on a hot road is really an image of the sky. The shimmering is caused by random fluctuations in air density.

35.4   TOTAL INTERNAL REFLECTION

Figure 35.15 shows the boundary between two media with refractive indices n_1 and n_2, where $n_2 > n_1$. A ray approaching the boundary from the medium of higher refractive index is refracted away from the normal. For small angles of incidence, there is both a reflected ray and a refracted ray. However, at some critical angle of incidence, θ_c, the refracted ray emerges parallel to the surface. For any angle of incidence greater than θ_c the light is totally reflected back into the medium of higher refractive index. This is called **total internal reflection** and was first noted by Kepler in 1604. The value of θ_c may be found from Snell's law by setting $\theta_2 = \theta_c$ and $\theta_1 = 90°$:

$$n_2 \sin \theta_c = n_1 \qquad (35.5)$$

If the medium with the lower refractive index is air, we may take $n_1 = 1$. Then, for water ($n = 1.33$) we find $\theta_c = 48.5°$, and for glass ($n \approx 1.5$) we find $\theta_c \approx 42°$.

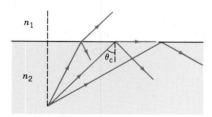

FIGURE 35.15 Light from a point source in a medium of refractive index n_2 meets the boundary to a medium of lower refractive index n_1. When the angle of incidence reaches a critical value, there is no refracted ray. The light suffers total internal reflection.

EXAMPLE 35.4: Kepler used total internal reflection in a glass block to deflect a beam of light as shown in Fig. 35.16. For an angle of incidence i at the top surface, what is the minimum refractive index needed for total internal reflection at the vertical face? The surrounding medium is air, for which $n = 1$.

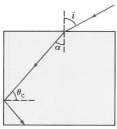

FIGURE 35.16 A ray enters a glass block and undergoes total internal reflection at the vertical face.

Solution: At the top surface the angle of refraction is found from Snell's law:

$$n \sin \alpha = \sin i \tag{i}$$

Total internal reflection will occur at the vertical face if the angle of incidence is greater than the critical angle, which is determined by

$$n \sin \theta_c = 1 \tag{ii}$$

From the diagram we see that $\theta_c = 90° - \alpha$ and so $\sin \theta_c = \cos \alpha$. Thus,

$$n \cos \alpha = 1 \tag{iii}$$

We square both sides of equations (i) and (iii) and add them to find

$$n^2 \cos^2 \alpha + n^2 \sin^2 \alpha = 1 + \sin^2 i$$

Therefore,

$$n = (1 + \sin^2 i)^{1/2} \tag{iv}$$

EXERCISE 3: Suppose that the block is immersed in water ($n = 4/3$), with its top face just above the surface. The angle $i = 45°$. What would be the minimum refractive index of the glass in this case?

Total internal reflection has several applications. Figure 35.17 shows a 45° prism and a ray incident normal to a short face. Since the angle of incidence at the long face is greater than θ_c (= 42°), the light undergoes total internal reflection— with almost 100% efficiency. Even the best metallic mirrors reflect only about 95% of the incident energy. Figure 35.18 shows two prisms used as part of the optical system of binoculars to invert (turn upside down) and to rectify (switch left and right) the image.

FIGURE 35.17 Light undergoes total internal reflection within a 45° prism.

FIGURE 35.18 The optical system of binoculars uses total internal reflection in two prisms to invert the image and to switch left and right so that the eye sees a normal field of view.

Glass and plastics can be made into thin fibers, between 10 μm and 50 μm thick. If a ray enters one end of a fiber at the appropriate angle, it will undergo total internal reflection and travel down the fiber without much loss through the sides, as shown in Fig. 35.19. A bundle of fibers closely packed, and with the

FIGURE 35.19 Light undergoes total internal reflection as it travels along an optical fiber.

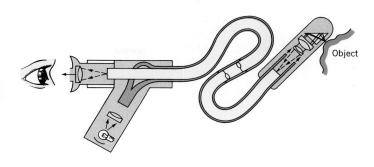

FIGURE 35.20 A ''coherent'' bundle of optical fibers can be used to view internal organs without surgery.

relative positions maintained, is a valuable tool in medicine. Such a ''coherent'' bundle may be used to view internal organs, such as the stomach, without major surgery (Fig. 35.20). It is important to coat each fiber with a material of different refractive index to prevent losses that would otherwise occur when two fibers come into contact. Optical fibers are also fast replacing wires for land-based telephone communication (Fig. 35.21).

FIGURE 35.21 A single optical fiber can carry as much information as a large bundle of wires.

35.5   THE PRISM AND DISPERSION

In general, the refractive index of any medium is a function of wavelength. Figure 35.22 shows a typical variation in the refractive index of a glass within the visible region (400–700 nm). A particular small range of wavelengths is perceived by us as a single color. Red, which has the longest wavelength, has a lower refractive index than violet, which has the shortest wavelength. When a beam of white light, which is a mixture of all the visible wavelengths, is incident at an angle to a glass surface, it is dispersed into a spectrum of colors. If the glass consists of two parallel surfaces, the rays that emerge from the second surface are parallel to the incident ray, as shown in Fig. 35.23. The nonparallel sides of a triangular prism serve to increase the angular separation between the colors and to make the emerging rays

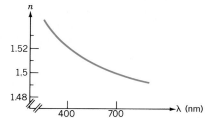

FIGURE 35.22 A typical dispersion curve. The refractive index decreases as the wavelength increases.

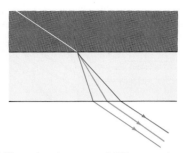

FIGURE 35.23 If overlapping rays of different colors pass through a slab of uniform thickness, the emerging rays are dispersed but they are parallel.

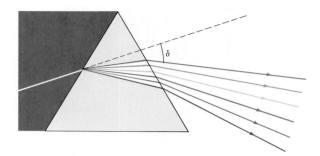

FIGURE 35.24 A triangle prism disperses white light, which is a mixture of all colors, and makes the emerging beam divergent. The angle δ, called the angle of deviation, depends on the color.

nonparallel, as shown in Fig. 35.24. Each wavelength has its own angle of deviation, δ, relative to the original ray.

A prism spectroscope, shown in Fig. 35.25, is a device used to analyze the light emitted by various sources. Light from a source, *S*, passes through a fine slit, and is collimated (made into a parallel beam) by a lens. When the beam passes through the prism, the various wavelengths are refracted by differing amounts. The emerging light is examined with a telescope, *T*. The source may be a salt placed in a flame, or more commonly, an electrical discharge passed through a low-density gas. The spectrum emitted by an element consists of a series of wavelengths that are observed in the telescope as colored lines (see p. 828). The line spectrum of an element is a signature that may be used to identify it.

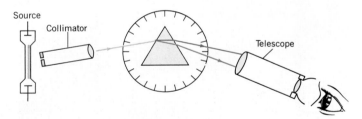

FIGURE 35.25 A prism spectroscope. Light from a source *S* is made into a parallel beam by a collimator *C*. The dispersed light is examined with a telescope *T*.

EXAMPLE 35.5: Obtain an expression for the refractive index of a prism in terms of the *minimum angle of deviation*.

Solution: Figure 34.26 shows a ray of a single wavelength incident on a prism. The angle of deviation, δ, of the beam depends on the angle *i* at which the beam strikes the face of the prism. It can be shown that the *minimum* value of the angle of deviation occurs when the ray goes through the prism symmetrically. That is, when the angle at which it emerges is equal to the angle of incidence.

The apex angle of the prism is φ. At each face the direction of the ray changes by (*i* − *r*) and so the total deviation is

$$\delta_{min} = 2(i - r)$$

Since the angle between the normals, *PR* and *QR*, to the faces is

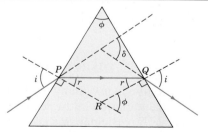

FIGURE 35.26 At the minimum angle of deviation a ray passes symmetrically through the prism.

the same as that between the faces, we see from triangle *PQR* that φ = 2*r*.

Snell's law is $\sin i = n \sin r$, where $i = \frac{1}{2}\delta_{min} + r$. Therefore, $n = \sin i/\sin r$ becomes

$$n = \frac{\sin\left(\dfrac{\phi + \delta_{min}}{2}\right)}{\sin\left(\dfrac{\phi}{2}\right)} \qquad (35.6)$$

This expression can be used to find n by measuring δ_{min}. By setting the prism for minimum deviation with known wavelengths, one can plot a *dispersion curve* such as that in Fig. 35.22.

35.6 IMAGES FORMED BY PLANE MIRRORS

The earliest references to "looking glasses" (mirrors) occur in *Exodus* 38.8 (ca. 1200 B.C.) and in *Job* 35.18 (ca. 600 B.C.). They were made from polished bronze, an alloy of copper and tin. The Romans improved these mirrors by increasing the proportion of tin to produce an alloy called speculum. (This is the origin of the term "specular reflection.") The Chinese may have also done this around 400 B.C. There is a legend that Archimedes used mirrors to set fire to the Roman fleet of Marcellus at Syracuse (ca. 212 B.C.). Although this may not be true, the feasibility of the plan was demonstrated in 1973 (see Fig. 35.27). Modern mirrors are made by evaporating aluminum onto a smooth surface.

FIGURE 35.27 In 1973, Greek historian I. Sakkas lined up 70 soldiers with flat copper shields and directed them to reflect sunlight to a rowboat 50 m from shore. The boat soon caught fire. Archimedes is reputed to have used this method to set fire to a Roman fleet.

Figure 35.28 shows a cone of rays diverging from a point source S and being reflected at a plane mirror. S is called a **real object** because the light rays actually emerge from it. When the reflected rays are extrapolated backward, they meet at point S'. The brain considers rays that enter the eye to have traveled in straight lines. Therefore, we interpret S' as the source of the reflected light. S' is the **image** of S in the mirror. An image is the real, or apparent, origin of light rays after they have been reflected or refracted at the boundary between two media. Since the light does not really come from S', but only appears to do so, it is called a **virtual image.** A virtual image cannot be displayed on a screen placed at the position of the image.

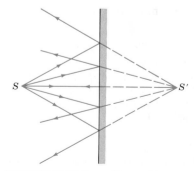

FIGURE 35.28 Rays diverge from a real object and appear to diverge from a virtual object.

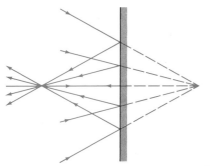

FIGURE 35.29 Rays converge to a real image and appear to converge to a virtual object.

TABLE 35.1

Rays diverge from a real object.
Rays converge to a real image.
Rays appear to converge to a
   virtual object.
Rays appear to diverge from a
   virtual image.

Figure 35.29 shows a cone of rays converging toward the mirror. (As we will see later, this can be accomplished with a curved mirror or a lens.) After being reflected, the rays intersect at the **real image,** S'. A real image can be displayed on a screen placed at the position of the image. One can associate the source of the rays (that is, before they strike the mirror) with a single point by extending them behind the mirror to intersect at S. Since the light does not really come from S, this point is called a **virtual object.** The characteristics of the rays associated with each of the four cases are summarized in Table 35.1.

The formation of the image of an extended object in a plane mirror is illustrated in Fig. 35.30. Rays diverge in all directions from each point on the object, but we choose to examine the behavior of just a small cone. Each point on the object has a corresponding image point. Consider a given point on the object, such as the tail of the arrow. This object is at a distance p from the mirror, and its image is at a distance q. *OMA* and *IMA* are right-angled triangles. From the law of reflection we infer that the angles *OAM* and *IAM* are equal. Thus triangles *OMA* and *IMA* are congruent, which means $p = q$: *the object distance is equal to the image distance.* The image in a plane mirror has the same size as the object. (If your eye is closer to one or the other, you will not perceive them as being equal.) Notice that for a given position of the eye, only part of the mirror is required to form a complete image.

A reflection in a plane mirror appears to interchange left and right. Figure 35.31a shows a person facing a mirror, whereas in Fig. 35.31b the person faces to one side. In either case, when the person lifts the right arm, the image lifts its "left" arm. We insist on applying the terms left and right to the image because we are conditioned to looking at other people this way. The operation performed by the mirror is best illustrated by considering a rectangular coordinate system. We see from Fig. 35.32 that the x and y axes, parallel to the plane of the mirror, maintain their direction in space, whereas the z axis, perpendicular to the plane of the mirror, reverses its direction.

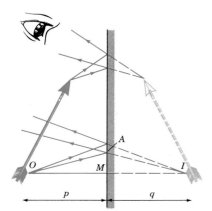

FIGURE 35.30 Rays used to form the image of an arrow in a plane mirror. The object distance p equals the image distance q.

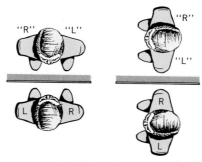

FIGURE 35.31 When a person stands in front of a plane mirror it seems that left and right have been interchanged.

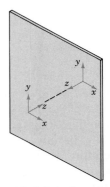

FIGURE 35.32 A flat mirror does not alter the x and y axes in the plane of the mirror but reverses the direction of the z axis, perpendicular to the mirror.

EXAMPLE 35.6: An object is placed between two perpendicular plane mirrors. How many images are seen?

Solution: In Fig. 35.33 we use the tail of an arrow as the object. Its images in mirrors M_1 and M_2 are I_1 and I_2 as shown. In these cases, light reaches the eye after one reflection. In addition, light can enter the eye after being reflected by both mirrors. The rays leading to the formation of image I_3 are shown. Notice that I_3 would be an image of I_1 if M_2 were extended downward. Similarly, I_3 is an image of I_2 in the "virtual mirror" formed by extending M_1 to the left.

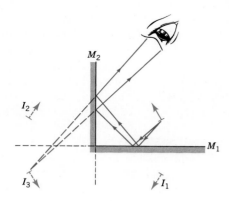

FIGURE 35.33 An object placed between two mirrors at 90° will produce three images. The image I_3 is produced after two reflections, as shown. One can quickly locate images by imagining that the mirrors extend into the "virtual" region (dashed lines).

EXAMPLE 35.7: What is the minimum length of mirror required for a person to see his or her full height? Assume that the eyes are a distance a below the top of the head and a distance b above the feet.

Solution: Rays enter the eye from the feet and the top of the head after reflection at the mirror, as shown in Fig. 35.34. We know that the angle of incidence is equal to the angle of reflection. Light from the feet reaches the eye after reflection at point B, which is $b/2$ above the floor. Light from the top of the head reaches the eye after reflection at point A, located $a/2$ below the top of the head. The person's total height is $a + b$, and the required length of mirror is $a/2 + b/2$, which is 50% of the full height of the person.

EXERCISE 4: (a) Does the person's horizontal distance from the mirror matter? (b) Does the vertical position of the mirror matter?

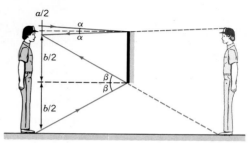

FIGURE 35.34 A length of mirror half the height of a person is sufficient for him to view his full height.

35.7   SPHERICAL MIRRORS

In this section we discuss the formation of images by mirrors whose surfaces are spherical. In a **concave** mirror, shown in Fig. 35.35a, the central section of the reflecting surface is recessed whereas in a **convex** mirror, shown in Fig. 35.35b, it

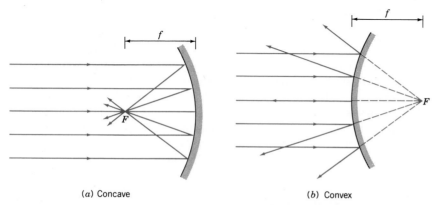

(a) Concave   (b) Convex

FIGURE 35.35 (a) After reflection in a concave mirror, rays parallel to the central axis converge to a real focal point. (b) After reflection in a convex mirror, rays parallel to the central axis diverge from a virtual focal point.

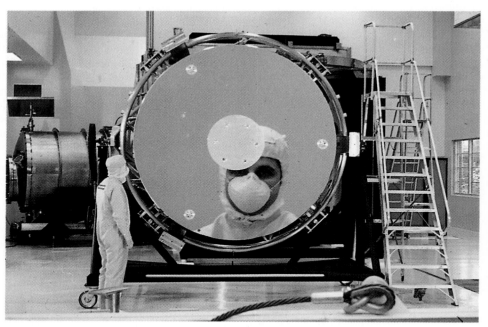

A magnified image in the mirror of the Hubble space telescope.

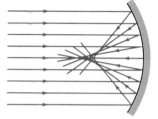

FIGURE 35.36 When parallel rays are incident at different distances from the axis of a spherical mirror, the rays do not converge to a single point.

bulges outward. In order to simplify our discussion, we consider only rays that are close to the central, or *principal,* axis and that make small angles with it. Such rays are called *paraxial rays.* In practice this condition may be achieved by using a mirror whose size is much smaller than the radius of curvature of the surface.

When a beam of parallel rays strikes a spherical mirror, each ray is reflected according to the law of reflection. Within the paraxial approximation, parallel rays reflected by a concave mirror converge toward, and pass through, a **real focal point,** *F,* as shown in Fig. 35.35*a.* Parallel rays reflected by a convex mirror appear to diverge from the focal point *F* behind the mirror as in Fig. 35.35*b.* Since the rays do not actually pass through this point, it is a **virtual focal point.** In either case, the distance from the mirror to the focal point is the **focal length,** *f.*

In general, parallel rays reflected by a concave mirror cross the principal axis at different points, as shown in Fig. 35.36. Consequently, the image of a source infinitely far away is blurred. This phenomenon, called **spherical aberration,** is a consequence of the spherical geometry of the mirror. It was known to the Greeks that a *parabolic* mirror will focus incoming parallel rays to a unique point. Rays from a point source placed at the focal point will form a beam of parallel rays after reflection. This property is used, for example, in search lights. Since spherical mirrors are much easier to make, they are still used in many applications, so we confine our attention to this type.

There is a simple relationship between the focal length and the radius of curvature of a spherical mirror. In Fig. 35.37, C is the center of curvature and R is the radius of curvature of a concave mirror. A ray is incident at point P, at a distance h above the principal axis, and makes an angle α to the radial line CP. The angles in the figure are greatly exaggerated. Since CP is normal to the surface, the reflected ray must also be at angle α to CP. The exterior angle PFA is the sum of the two interior angles FCP and CPF. Since all angles are assumed to be small, we may use the approximation $\tan \alpha \approx \alpha$; therefore $\alpha \approx h/R$ and $2\alpha \approx h/f$, which means

$$f = \frac{R}{2} \tag{35.7}$$

The focal length f is one-half the radius of curvature R of the mirror—a result first obtained in 1591 by G. Della Porta.

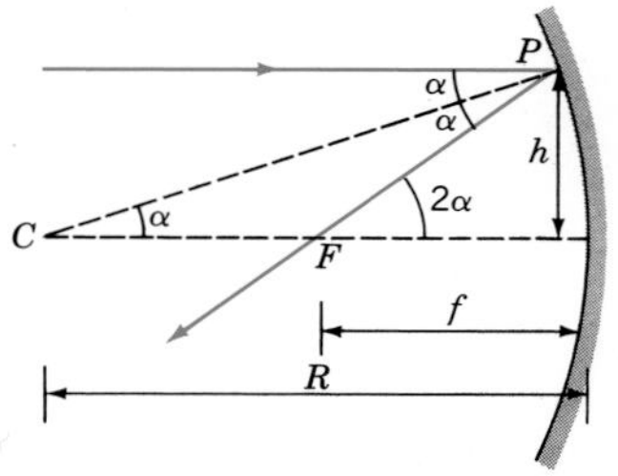

FIGURE 35.37 The focal length, f, of a spherical mirror is one half the radius of curvature, R.

Ray Diagrams

When an object is at infinity, its image is at the focal point. A simple way of locating the image of an object at an arbitrary position was devised in 1735 by Robert Smith. It is called a **ray diagram.** For clarity, rays are often drawn far from the principal axis or at a large angle to it. However, the results obtained from such ray diagrams are valid only in the paraxial approximation. Real images are drawn with solid lines whereas virtual images are drawn with dashed lines. Any two of the following *principal* rays are sufficient to locate the image.

1. A ray *along a radius* reverses its path upon reflection.

Principal rays for spherical mirrors

2. A ray *parallel to the axis* passes through the focal point after reflection.

3. A ray *through the focal point* is reflected parallel to the axis.

4. A ray *incident at the center* of the mirror makes the same angle with the axis after reflection.

Each of these rays is illustrated in Fig. 35.38. Note that for a convex mirror (Fig. 35.38b), the rays need to be extrapolated either forward or backward through the mirror.

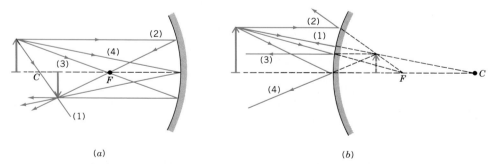

(a)   (b)

FIGURE 35.38 Principal rays used to form an image (a) in a concave mirror, and (b) in a convex mirror. Only two rays are needed to locate an image.

The Mirror Formula

Instead of using a ray diagram to locate the image, we can develop an equation that relates the **object distance,** p, and the **image distance,** q, to the focal length. We consider in detail only the case of a concave mirror. In the paraxial approximation, all the rays from the point object O in Fig. 35.39 will reach the image point I. The images of points off the axis but close to O will also be located at unique points close to I. In Fig. 35.39 we consider an arbitrary ray that, after reflection, crosses the principal axis at I. For small angles $\tan \theta \approx \theta$, so we have

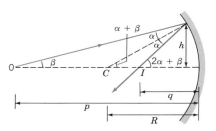

FIGURE 35.39 A ray from object point O on the central axis crosses the axis at the image point I. The object distance, p, and the image distance, q, are related by the mirror formula.

$$\beta \approx \frac{h}{p}; \qquad \alpha + \beta \approx \frac{h}{R}; \qquad 2\alpha + \beta \approx \frac{h}{q}$$

Since $(2\alpha + \beta) = 2(\alpha + \beta) - \beta$, we find

$$\frac{h}{q} = \frac{2h}{R} - \frac{h}{p}$$

Since $f = R/2$, this is equivalent to

$$\frac{1}{p} + \frac{1}{q} = \frac{1}{f}$$

The derivation for a convex mirror is much the same and is left as an exercise. The result is $1/p - 1/q = -1/f$. The negative signs indicate that both the image and the focal point are virtual. To avoid having to remember two equations, we use a single **mirror formula** and supplement it with a sign convention:

Mirror formula

$$\frac{1}{p} + \frac{1}{q} = \frac{1}{f} \tag{35.8}$$

SIGN CONVENTION (Mirrors): p, q, and f are positive (real) on the left and negative (virtual) on the right. (The light approaches from the left.)

Linear Magnification

In general, the size of the image is not the same as that of the object. The **transverse** (or **linear**) **magnification,** m, is defined as the ratio of the image height y_I to the object height y_O, that is, $m = y_I/y_O$. From Fig. 35.40 we see that $\tan \alpha = y_O/p = y_I/q$. Therefore,

Transverse (Linear) magnification

$$m = \frac{y_I}{y_O} = -\frac{q}{p} \tag{35.9}$$

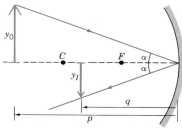

FIGURE 35.40 The transverse (linear) magnification, m, is defined as the ratio of the image height y_I to the object height, y_O: $m = y_I/y_O$.

When m is positive, the image is *erect;* when m is negative, the image is *inverted*. If $|m| > 1$, the image is *enlarged;* if $|m| < 1$, the image is *reduced*. The image produced by a concave mirror may be erect or inverted. A convex mirror always produces a virtual, erect, and reduced image of a real object. (See Figs. 35.38b and 35.41b.)

EXAMPLE 35.8: An object of height 1.2 cm is placed 2 cm from a spherical mirror whose radius of curvature is 8 cm. Find the position and size of the image given that the mirror is (a) concave, and (b) convex.

Solution: (a) We are given the focal length $f = R/2 = 4$ cm and the object distance $p = 2$ cm. The position of the image is found by applying the mirror formula, $1/p + 1/q = 1/f$. On inserting the given values, we find

$$\frac{1}{2 \text{ cm}} + \frac{1}{q} = \frac{1}{4 \text{ cm}}$$

Therefore $q = -4$ cm. The negative sign means that the image is virtual, to the right of the mirror. From Eq. 35.9 the transverse magnification is

$$m = -\frac{q}{p} = +2$$

Since m is positive, the image is erect. Since its magnitude is greater than one, the image is enlarged. The size of the image is $y_I = 2(1.2 \text{ cm}) = 2.4$ cm. The ray diagram in Fig. 35.41a (which is not to scale) is formed with principal rays 1 and 3 from the table on page 711.

(b) The procedure for the convex mirror is the same as above except that now $f = -4$ cm. From the mirror formula we have

$$\frac{1}{2 \text{ cm}} + \frac{1}{q} = -\frac{1}{4 \text{ cm}}$$

which yields $q = -4/3$ cm. The image is again virtual. The transverse magnification is

$$m = -\frac{q}{p} = +\frac{2}{3}$$

The image is erect and reduced. The ray diagram in Fig. 35.41b (which is not to scale) uses principal rays 2 and 4 from the table.

EXAMPLE 35.9: An object of height 2 cm is located 12 cm from a spherical mirror. The erect image is of height 3.2 cm. What type of mirror is it?

Solution: We are given $p = 12$ cm. The image is erect so m is

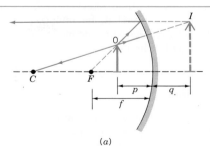

(a)

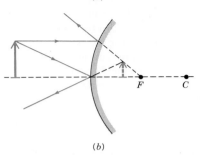

(b)

FIGURE 35.41 (a) An object closer to a concave mirror than its focal length results in a virtual, erect, and magnified image. (b) A concave mirror producing a virtual, erect, and reduced image of a real object.

positive and its value is $m = +3.2/2 = 1.6$. Since $m = -q/p$ we see that $q = -1.6p = -19.2$ cm; the image is virtual. From the mirror equation we have

$$\frac{1}{12 \text{ cm}} - \frac{1}{19.2 \text{ cm}} = \frac{1}{f}$$

which yields $f = +32$ cm. The positive sign means that the mirror is concave. Figure 35.41a is again an appropriate ray diagram.

EXERCISE 5. An object is located 15 cm from a spherical mirror whose focal length has a magnitude of 10 cm. Determine the position and transverse magnification of the image given that mirror is (a) concave and (b) convex.

35.8   THE SPEED OF LIGHT (Optional)

The first attempt to measure the speed of light was made by Galileo in 1635. One night, he and an assistant each carried lanterns to points about 1 km apart. Galileo briefly exposed his lantern. On seeing the light, the assistant flashed a similar signal back. By measuring the time interval for the round trip and knowing the distance traveled, he intended to compute the speed of light. The results were inconclusive since they were hopelessly masked by the uncertainties due to reaction times.

Roemer's Method

In 1676 Olaf Roemer, a Danish astronomer working in Paris, presented his measurements of the period of one of the moons of Jupiter. The moon, Io, has an average period of just under 42.5 h. This was measured by noting the time at which the moon either entered or emerged from the shadow of Jupiter—that is, at the beginning or the end of its eclipse. Roemer discovered a systematic variation in this period during the course of each year. Figure 35.42 shows the Earth and Jupiter in orbit around the Sun and the moon of Jupiter in orbit about that planet. Roemer found that the period had its average value when the Earth was closest or farthest from Jupiter, at points A and B. At times when the Earth was moving away from Jupiter, from C to D, the period was longer than average and when the Earth approached Jupiter, from E to F, the period was shorter.

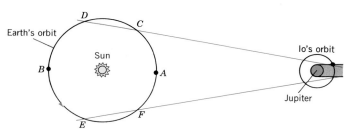

FIGURE 35.42 By timing the eclipses of a moon of Jupiter, Roemer showed that the speed of light is finite. (The diagram is not to scale.)

Roemer ascribed this variation to the changes in the distance the light has to travel from the moon to the Earth. For example, suppose the moon first emerges from the shadow when the Earth is at C. Within the next 42.5 h, the Earth will have moved to D. Therefore, when the moon emerges after one period, the light must travel the extra distance CD. (For simplicity we ignore the motion of Jupiter, whose orbital period is about 12 y.) For a single orbit of Io, the extra time involved is a matter of seconds. However, over an interval of several months, the discrepancy between the predicted time of an eclipse, based on the average period, and the actual time it occurs amounts to several minutes.

In September 1676, Roemer announced to fellow astronomers that the eclipse of November 9 would occur at 5 h 34 m 45 s, a full 10 min later than the time inferred from observations made in August. After the prediction had proven to be correct, Roemer explained that the speed of light was not infinite, as many had believed, but instead was finite. He estimated that it took 22 min for light to travel the diameter of the Earth's orbit. (In fact, it takes closer to 16.5 min.) Some time later, Huygens used the best-known value for the radius of the Earth's orbit and calculated the speed of light to be 2.1×10^8 m/s. Later measurements, based on other types of astronomical observations, yielded values closer to 3×10^8 m/s.

Fizeau's Method

The first terrestrial measurement of the speed of light was made in Paris by A. Fizeau in 1849. It involved a "time-of-flight" technique like that of Galileo. Light from a source S (Fig. 35.43) was reflected by a partly silvered plate P and passed through a slot in a wheel with n teeth around its rim. The light then reflected off a mirror M, at a distance d, and passed through the slotted wheel and through P to the observer E. The wheel was set into rotation. Let us say that the light from the source passed through slot 1. For low angular velocities, the light reflected by M was blocked by the tooth between slots 1 and 2 and the observer saw nothing. The angular speed was then increased until the light reappeared. At this angular speed, the time required for the light to make the return trip to M, $\Delta t = 2d/c$, exactly equals the time needed for the tooth between slot 1 and slot 2 to get out of the way. If the period of rotation is T, this time interval is $\Delta t = T/2n$. (Why is there a factor of 2?) Fizeau used a wheel with $n = 720$ teeth and placed the mirror M on a hill in Montmartre, at a distance of 8.633 km. The light first reappeared at a rotational rate of 12.6 rev/s, which corresponds to a period of 1/12.6 s. Equating the two expressions for Δt yields $c = 4nd/T = 3.13 \times 10^8$ m/s.

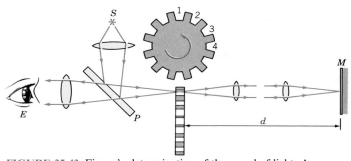

FIGURE 35.43 Fizeau's determination of the speed of light. A beam passed through a rapidly spinning toothed wheel and was reflected by a distant mirror M. Knowing the rate of rotation of the wheel and the distance to the mirror, the speed of light could be calculated.

A similar approach was used by A. A. Michelson in the early 1920s. He replaced the wheel by an eight-sided rotating mirror (Fig. 35.44). The source was on Mt. Wilson and the distant mirror was on Mt. Baldy approximately 35 km away. The distance between the mirrors was determined by surveying techniques to within 0.3 cm! Light that reflected off the distant mirror could enter the telescope only if the one face of the rotating mirror was in the correct position. Several hundred individual measurements led to value of 2.99796×10^8 m/s. For our purposes, the approximate value $c = 3 \times 10^8$ m/s is usually good enough.

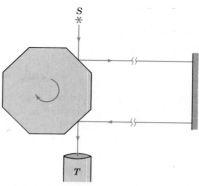

FIGURE 35.44 In A. A. Michelson's method light reflected off a rapidly rotating eight-sided mirror before being reflected by a distant mirror. Light emitted by source S could enter the telescope T only if a face of the mirror was in the correct position.

HISTORICAL NOTE: Newton's Prism Experiment

It was known in Greek times that when sunlight passes through a prism or a transparent spherical vessel full of water, a display of colors, similar to that of a rainbow, is produced. Ancient philosophers and medieval thinkers did not associate light itself with color. Light was believed to be a substance that is white in its purest form. Colors were assumed to be the result of the addition of differing amounts of "darkness" to light. Up until the 17th century, Descartes and others accepted the idea that white is the "natural" or "primitive" form of light. In their view, colors were produced when light is modified as it passes through a medium. Descartes, who considered light to be a stream of particles ("corpuscles"), assigned them a rotational motion on entering a medium: One sense produced the sensation of red, while the opposite produced blue. The colors in between came from a mixture of these two.

Newton's interest in optics was generated by Descartes' book *Dioptrice*. In 1662, at the age of twenty, and while still an undergraduate at Cambridge, he commenced his own investigations. In February 1672 he presented a paper which began as follows:

In the year 1666, I procured me a triangular glass prism, to try therewith the celebrated phaenomena of colours. And in order thereto, having darkened my chamber, and made a small hole in my window shuts, to let in a convenient quantity of the sun's light, I placed my prism at its entrance, that it might thereby be refracted to the opposite wall. It was at first a very pleasing divertisement, to view the vivid and intense colours produced thereby; but after a while applying myself to consider them more circumspectly, I became surprised to see them in an oblong form; which according to the received laws of refraction, I expected should have been circular.

Other scientists were not at all surprised at the oblong form of the spectrum in Fig. 35.45; it was what they usually saw. Descartes, Hooke, and others had directed their attention to the colors produced. But to Newton these were merely "a pleasing divertisement"; he was struck instead by something that everyone else had missed. How could the oblong shape be explained by a single value for the refractive index—an assumption implicit in the "received laws of refraction"?

He passed the beam through different parts of the prism, for example, near the apex or near the base, and found that the amount of material through which the light passed did not change the colors. He then placed a second prism so that it reversed the refraction of the first. When the colors were recombined, they again produced white. Next came a simple, yet crucial, experiment.

Newton used a hole to select single colors from the first prism and then passed them one at a time through a second prism, as shown in Fig. 35.46. His experimental setup ensured that each color entered the second prism at the same angle of incidence. First, he found that the second prism did not modify each "pure" color; red remained red, green remained green, and so on. Second, as different colors were selected, he discovered that they were projected onto different parts of a screen. This implied that *each color has its own refractive index*, red having the lowest and blue the highest. It was clear that color is an intrinsic property of a given ray.

Newton concluded that white light, far from being pure, consists of a mixture of all colors "mixed in due proportion." The prism merely separates the different colors because each has its own refractive index. (Earlier values for n were only approximate values, most likely for the middle of the spectrum.) The significance of this finding was not appreciated by Newton's contemporaries.

Hooke and others accepted the experimental results, but not that they clearly demonstrated, as Newton claimed, that white light is a mixture of all colors. They believed Newton was interpreting his results in terms of the corpuscular theory and

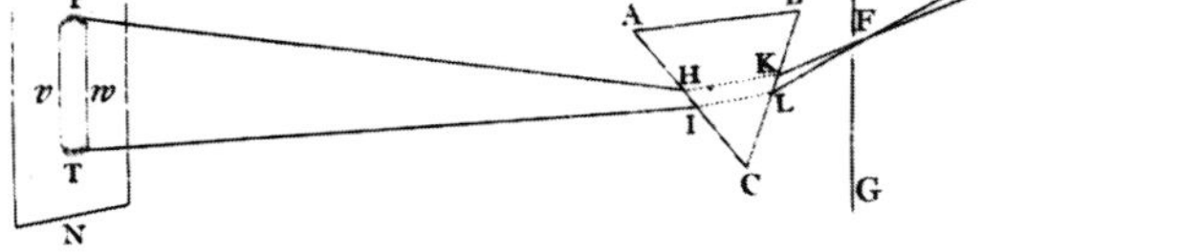

FIGURE 35.45 A diagram from Newton's *Opticks*. He was struck by the oblong form of the spectrum.

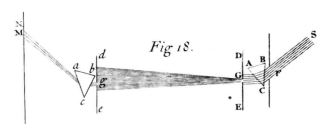

FIGURE 35.46 A "crucial" experiment from Newton's *Opticks*. Single colors selected from the first prism did not change color when they were passed through a second prism. With a fixed angle of incidence, each color was projected to a different point on the screen—which indicated that each color has a different refractive index.

treated his conclusion about white light as only an unproven hypothesis. Hooke suggested that the experiment might also be interpreted in terms of a wave theory. Newton agreed, but insisted that this was beside the point: The findings were independent of the wave or particle nature of light. In a letter to Hooke, he actually improved the wave theory by suggesting that the "bigness" (wavelength) of a wave could be associated with its color. Huygens claimed that colors are "qualities of things, not of light itself." Newton countered that nonluminous objects illuminated with light of just one given color will appear to have that color. He also pointed out that when objects are illuminated with white light, they reflect some colors more than others, and thereby appear to have an apparent color determined by the relative intensities of the "pure" colors.

Our account of Newton's prism experiment might create the impression that this was the fruit of one miraculous afternoon. In fact Newton developed his ideas over a period of about four years (1661–1665), and the prism experiment itself covered a period of about 18 months. His results were certainly not accepted as conclusive. The vehement criticism that greeted his ideas caused Newton to withdraw from the scientific community. It also set the stage for a lifetime of uneasy tension with Hooke—so much so, that Newton withheld publication of his *Optiks* until after the death of Hooke in 1703.

The spectrum of sunlight.

Newton's home in Woolsthrope, with descendents of the legendary apple tree.

SPECIAL TOPIC: The Rainbow

The rainbow has intrigued thinkers for centuries. The search for an explanation of this phenomenon is an interesting example of the development of science. The primary rainbow, with red at the top and violet at the bottom, is a common phenomenon. Occasionally, a secondary rainbow, with the order of colors reversed, is also seen above the primary rainbow. Aristotle recorded only four colors—red, yellow, green, and blue—and noted that the red was the purest. He suggested that the colors are produced by reflection at the drops in a cloud. Robert Grosseteste (ca. 1235) believed that the colors were produced by refraction in a whole cloud followed by reflection by another that acts as a screen. Although the explanation is far-fetched, it did propose a combined action of refraction and reflection. In 1267 Roger Bacon pointed out that the "screen" moves with the observer and therefore each person sees a different rainbow. Since rainbows are seen in sprays near the ground—for example when oars splash in water—he emphasized the role of individual drops of water (rather than a whole cloud). Bacon also noted that the sun, the observer, and the center of the bow lie along a line. He made the first quantitative measurement which showed that the angular radius of the primary rainbow is 42°. In 1275, Witelo showed that a spectrum is produced when sunlight passes through a spherical vessel filled with water or through a hexagonal prism. It was now clear that refraction and reflection by individual drops was the key factor.

In 1304 Theodoric of Freiburg developed an explanation of the rainbow that is remarkably close to our present view. He used a spherical vessel filled with water as a model of a drop to show that the significant reflection occurs at the inner surface of the drop. By raising and lowering the vessel he saw the colors of both the primary and the secondary rainbows. He associated the primary rainbow with a refraction as the light enters a drop, a reflection at the back surface, and finally a refraction as the light emerges from the drop (Fig. 35.47). The ludicity of his thinking and his appeal to a model and experiment are commendable. With knowledge no greater than that available to the Greeks, he did remarkably well.

Although Theodoric had the basic idea correct, he could not explain why the rainbow is limited to a small angular range, or why the order of the colors is reversed in the secondary rainbow. An almost complete explanation was given by Descartes in 1635. Knowing the correct law of refraction, which Theodoric did not, Descartes traced the paths of a large number of rays incident at different angles. He discovered that the angle θ between the incident light and the light emerging from a drop, as shown in Fig. 35.47, reaches a maximum of 42° for the primary bow. (The angle of deviation, $\delta = 180° - 42° = 138°$, is therefore a *mini-*

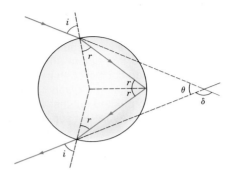

FIGURE 35.47 In the formation of the primary rainbow, each light ray undergoes two refractions and an internal reflection in water drops.

mum.) As a result, no rays can enter the eye at greater angles. He also discovered that the emerging rays tend to bunch up in the range $\theta = 40°$ to 42°, as shown in Fig. 35.48. This bunching explains why the light is most intense in this range. Descartes had solved the geometrical problem of the rainbow, but the explanation for the colors had to await Newton's prism experiment and his discovery of the nature of white light.

In modern terms the rainbow is caused by the disper-

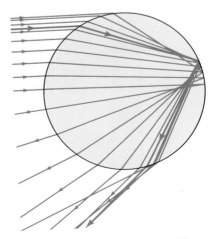

FIGURE 35.48 A parallel beam of monochromatic rays incident at different points on a drop. The emerging rays tend to "bunch up" as shown. The thicker ray undergoes minimum deviation. The light is most intense in this direction.

sion of light in water. White light from the sun enters a water drop and is dispersed into a spectrum, as shown in Fig. 35.49. After being reflected, the light is further dispersed as it emerges from the drop. From Fig. 35.47 we see that the change in direction at each refraction is $(i - r)$ and at the reflection it is $(\pi - 2r)$. The total angular deviation, δ, is therefore

$$\delta = 2(i - r) + (180° - 2r)$$

As Descartes discovered, the light coming from each drop is most intense at the angle of minimum deviation for each wavelength. The minimum angle of deviation varies from $(180° - 40.2°)$ for violet to $(180° - 42.1°)$ for red, as indicated in Fig. 35.50. To find this angle we calculate $d\delta/di$ and set it equal to zero. When this condition is met, δ is relatively constant for small changes in i. In other words the rays "bunch up" at the minimum angle of deviation. Using the expression given above, we find

$$\frac{d\delta}{di} = 2 - \frac{4dr}{di} = 0$$

By taking the derivative of Snell's law, $\sin i = n \sin r$, we find that

$$\cos i = n \cos r \frac{dr}{di} = \frac{1}{2} n \cos r$$

Next we use $\cos^2 r = 1 - \sin^2 r$ and Snell's law to find (see Problem 11)

$$\cos i = \sqrt{\frac{n^2 - 1}{3}}$$

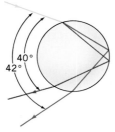

FIGURE 35.49 Newton showed that dispersion is responsible for the colors of the rainbow.

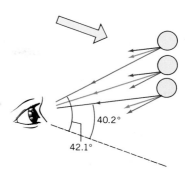

FIGURE 35.50 The angles (to the incoming sunlight) at which each color is most intense varies from 40.2° for violet to 42.1° for red. Light of only one wavelength enters the eye from any one drop.

This equation specifies the angle of incidence that leads to the minimum angle of deviation δ_m.* Here are some typical values:

$n_V = 1.3435$ $i = 58°48'$ $r = 39°33'$ $\delta_m = 180 - 40°36'$

$n_Y = 1.333$ $i = 59°23'$ $r = 40°12'$ $\delta_m = 180 - 42°02'$

$n_R = 1.3311$ $i = 59°31'$ $r = 40°21'$ $\delta_m = 180 - 42°22'$

Recall that Aristotle had mentioned that the red in the primary rainbow is the "purest" color. This is indeed true. Each color of the spectrum is spread through a range of angles. Basically just one color reaches the eye from any particular drop—the color that is at minimum deviation and therefore the most intense. However, there is considerable overlap between colors. With the exception of the red, the other colors seen are not the primary colors of a prism spectrum.

A secondary rainbow, associated with two internal reflections, is occasionally seen (see Problem 14). The order of colors is reversed. Supernumerary bows, which consist of reddish and greenish bands below the primary rainbow, are rarely seen. Their origin, which lies in the interference of light, was first explained by Young in 1803, using the wave theory.

* On evaluating the second derivative, $d^2\delta/di^2$, one finds that it is a positive quantity, and thus δ is a minimum rather than a maximum.

SUMMARY

Huygen's principle states that each point on a wavefront acts as a source of secondary wavelets. The wavefront at a later time is found by drawing the envelope of the wavelets.

The **refractive index**, n, of a medium in which the speed of light is v is

$$n = \frac{c}{v}$$

When light passes from a medium of refractive index n_1 to a medium of index n_2, its direction of propagation changes. According to **Snell's law of refraction,**

$$n_1 \sin \theta_1 = n_2 \sin \theta_2$$

where the angles are measured with respect to the normal to the boundary between the media.

When light travels from a medium of higher refractive index to one of lower refractive index, it bends away from the normal. At the **critical angle** given by

$$n_1 \sin \theta_1 = n_2$$

the angle of refraction is 90°. At greater angles of incidence the light is completely reflected into the first medium. This is **total internal reflection.**

In the paraxial approximation, the object distance, p, and image distance, q, are related to the focal length, f, of a spherical mirror by the **mirror formula:**

$$\frac{1}{p} + \frac{1}{q} = \frac{1}{f}$$

The sign convention on p. 712 helps to distinguish between real and virtual objects and images. Images may be located geometrically through the use of a **principal ray diagram.**

The **transverse (linear) magnification** of an image is given by

$$m = \frac{y_I}{y_O} = -\frac{q}{p}$$

where y_O and y_I are the object and image heights respectively.

ANSWERS TO IN-CHAPTER QUESTIONS

1. Extend the line of the ray incident on M_1 forward (through M_1) and extend the ray reflected off M_2 backward to intersect the plane of M_1. The total angle is 40° + 60° + 20° = 120°.

2. (a) From Snell's law, (1.5) $\sin \theta_1 = (1.0) \sin 40°$, we find $\theta_1 = 25.4°$.
   (b) The same as in air, that is, $f = c/\lambda_0 = 5.5 \times 10^{14}$ Hz.

3. Eq. (iii) becomes $n \cos \alpha = 4/3$ and (iv) becomes $n = (16/9 + 0.5)^{1/2} = 1.51$.

4. (a) No. (b) Yes, the bottom of the mirror must be at a height $b/2$.

5. (a) Given that $p = 15$ cm and $f = 10$ cm, from $1/p + 1/q = 1/f$ we find $q = 30$ cm. The magnification is $m = -1$.
   (b) With $p = 15$ cm and $f = -10$ cm, we find $q = -6$ cm and $m = +0.4$.

QUESTIONS

1. What kind of vanity mirror (concave or convex) is used for putting on makeup? Where is the face with respect to the focal point?

2. True or false: (a) A concave mirror always produces a real image. (b) A convex mirror always produces a virtual image.

3. Draw a large hemispherical mirror. By carefully drawing rays incident parallel to the central axis, show that reflections close to and reflections far from the axis do not converge to a single point.

4. A decorated hemispherical bowl is filled with water. Does the pattern at the bottom appear larger or smaller than when the bowl is empty? Explain.

5. Does a flat slab of glass display dispersion?

6. Why does dust on an optical fiber result in a loss of light?

7. Mirrors in amusement arcades can make you look fat or thin. How do they do this?

8. As you approach a concave mirror with your eye on the axis, discuss what you see when (a) $2f > p > f$; and (b) $p < f$.

9. Why are convex mirrors used in rear-view mirrors on trucks and in stores as part of the security system? What advantage do they have over plane mirrors?

10. Can you use two plane mirrors to see the back of your head? If so, draw a ray diagram. (Can the mirrors be parallel?)

11. A diamond ring displays more "fire" (vivid colors) than any other gemstone. Does the high refractive index of diamond explain this? If not, what does?

12. When light passes from one medium to another, the wavelength changes. Does the color also change? Explain why or why not.

13. Do light rays always travel in straight lines? If not, give an example.

14. Does the variation in the density of air with height affect the propagation of a sound wave? If so, how?

15. When a broad beam of light crosses a boundary at an angle, the intensity of the wave changes. Show how this occurs in a diagram.

16. Why do wavefronts of ocean waves tend to approach a beach parallel to the shore?

17. Write the word AMBULANCE so that it will appear normal when viewed in the rear-view mirror of a car. Verify your answer.

18. If it is possible, under what condition will the image in a concave mirror be (a) real; (b) virtual; (c) erect; (d) inverted; (e) magnified; (f) reduced?

19. Repeat Question 18 for a convex mirror.

20. How is the focal length of a spherical mirror affected when it is immersed in water?

21. True or false: When an object is closer to any spherical mirror than the magnitude of its focal length, the image is always: (a) virtual; (b) erect; (c) magnified.

22. Why does a stick partly immersed in water appear bent? Draw a ray diagram to support your answer.

23. What arrangement of mirrors would produce the multiple images of Ann Margaret shown in Fig. 35.51?

FIGURE 35.51 Question 23.

24. Consider the layered structure of Fig. 35.14b. Once the direction of a ray has become horizontal, why should it start to turn upward? (You need to examine the behavior of a vertical wavefront.)

EXERCISES

35.2 Reflection

1. (I) Show that when a plane mirror rotates by θ, the reflected ray rotates by 2θ.

2. (I) Two vertical plane mirrors are placed at 90° to each other. Show that after two reflections any horizontal ray is returned in a direction antiparallel to its original direction.

3. (II) An object is placed between two vertical mirrors that are 60° to each other as in Fig. 35.52. How many images are formed? Sketch the positions and the orientations of the images.

4. (I) A parallel beam is incident at a corner of a prism as shown in Fig. 35.53. Show that the angle θ between the two reflected rays is twice the apex angle, ϕ.

5. (II) Three mirrors are mutually perpendicular. Show that

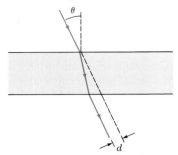

FIGURE 35.54 Exercise 11.

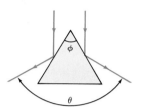

FIGURE 35.52 Exercise 3.

FIGURE 35.53 Exercise 4.

after three reflections, any ray is returned opposite to its original direction. (*Hint:* Express the direction of the ray in **ijk** notation.)

35.3 Refraction

6. (I) Light that has a wavelength of 450 nm in water ($n = 1.33$) has a wavelength of 400 nm in another medium. (a) What is the refractive index of the medium? (b) What is the speed of light in the medium?

7. (I) When a ray of light in air is incident on a material of refractive index 1.4, the angle of refraction is 32°. What is the angle between the reflected and refracted beams?

8. (I) Light traveling in air ($n = 1$) is incident on a flat surface of refractive index $n = 1.52$. For what angle of incidence are the refracted and reflected rays perpendicular?

9. (I) A diver 3 m under water ($n = 1.33$) directs a beam at 30° to the perpendicular to the water–air interface. At what horizontal distance from the diver should a person in a boat, whose eyes are 1 m above the water, be located to see the light?

10. (II) Around A.D. 150, Claudius Ptolemy published a table of angles of incidence and of refraction for the air–water boundary.

i:	10	20	30	40	50	60	70	80
r:	8	15.5	22.5	29	35	40.5	45.5	50

Ptolemy suggested that i/r is constant, which is clearly not true. Plot $\sin i$ vs $\sin r$ to obtain an estimate of the refractive index.

11. (II) Show that if a ray strikes a flat glass plate of thickness t and refractive index n, at a small angle of incidence θ, the lateral deviation of the ray (see Fig. 35.54) is approximately $d \approx t\theta(n - 1)/n$, where θ is in radians.

35.4 Total Internal Reflection

12. (I) A ray traveling in a transparent medium suffers total internal reflection at its interface with water ($n = 1.33$). The

critical angle is 68°. What is the speed of light in this medium?

13. (I) A point source of light is 2 m below the surface of a lake. Calculate the radius of the circle at the surface through which light can emerge into the air.

14. (I) A liquid of unknown index n_2 is placed on a hemisphere of known index n_1 as shown in Fig. 35.55. A ray enters the hemisphere in a radial direction. (a) How can one determine n_2 with this arrangement by measuring the angle θ? Are there any restrictions on the value of n_2? (b) Relate θ to n_2.

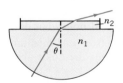

FIGURE 35.55 Exercise 14.

15. (II) A ray traveling in vacuum enters a long fiber of refractive index 1.5; see Fig. 35.56. Show that the ray undergoes total internal reflection for any angle of incidence.

FIGURE 35.56 Exercise 15.

35.5 Prism and Dispersion

16. (I) A ray is incident normal to one face of a 30°–60°–90° prism of refractive index 1.5, as shown in Fig. 35.57. In what direction does light emerge from the bottom surface?

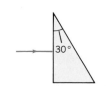

FIGURE 35.57 Exercise 16.

17. (I) The angle of minimum deviation for a 60° prism is 41°. What is the speed of light in the prism?

18. (II) A prism ($n = 1.6$) has an apex angle of 60°. Find the minimum angle of deviation if it is immersed in water ($n = 1.33$).

19. (I) Two rays are incident on a prism as shown in Fig. 35.58. Assuming that both refracted rays experience total internal reflection at the bottom face, draw the rays that emerge from the other side. Of what use is this arrangement?

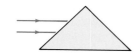

FIGURE 35.58 Exercise 19.

20. (I) The *dispersive power* of a medium is defined as $(n_B - n_R)/(n_Y - 1)$, where the subscripts serve to distinguish the refractive indices for blue, red, and yellow light. Evaluate this quantity given that $n_R = 1.611$, $n_Y = 1.620$, and $n_B = 1.633$.

21. (I) (a) What is the minimum refractive index needed for a 45° prism as in Fig. 35.59 to exhibit total internal reflection? (b) What would be the minimum value if the prism were immersed in water?

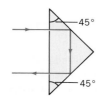

FIGURE 35.59 Exercise 21.

22. (I) An equilateral prism has a refractive index of 1.6. At what angle of incidence does a ray undergo minimum deviation?

23. (II) A ray is incident at 45° to the normal at the middle of one face of a 60° prism of refractive index 1.5. Trace the path of the ray and determine the angle of deviation between the incident ray and the ray emerging from the other face.

24. (II) Show that for a thin prism, as in Fig. 35.60, the angle of deviation is $\delta = (n - 1)\phi$. (Use $\sin \theta \approx \theta$.)

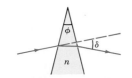

FIGURE 35.60 Exercise 24.

25. (II) A ray is obliquely incident on a prism of refractive

index n as shown in Fig. 35.61. Show that the maximum value of α for which there is an emergent ray along face AC is given by $\cos \alpha = n \sin(\phi - \theta_c)$, where θ_c is the critical angle for total internal reflection.

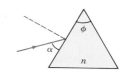

FIGURE 35.61 Exercise 25.

35.7 Spherical Mirrors

26. (I) A concave mirror has a radius of curvature of 40 cm. Determine the image position and linear magnification for the following object positions: (a) 15 cm, (b) 60 cm. Draw a ray diagram in each case.

27. (I) A convex mirror has a radius of curvature of 40 cm. Determine the image position and the linear magnification for the following object positions: (a) 15 cm, (b) 40 cm. Draw a ray diagram for each case.

28. (I) An object of height 2 cm is 40 cm from a spherical mirror. The erect virtual image is of height 3.6 cm. (a) What type of mirror is it? (b) Locate the image. (c) What is the focal length?

29. (I) An object is 60 cm from a concave mirror. The size of the real image is 40% of the size of the object. What is the radius of curvature of the mirror?

30. (I) The image in a concave mirror of focal length 30 cm is magnified 2.5 times. Where is the object given that the image is (a) erect, (b) inverted? Draw a ray diagram for each case.

31. (I) A convex mirror of focal length 30 cm produces an image with a magnification of 0.4. Where is the object?

32. (I) An object 22 cm from a concave mirror produces a real image with a magnification of -3.2. What is its focal length? Draw a ray diagram.

33. (I) The image size in a convex mirror, whose radius of curvature is 16 cm, is one-third of the (real) object's size. Locate (a) the object and (b) the image.

34. (I) The image of an object placed 3.2 cm from a convex mirror has a transverse magnification of $+0.4$. Find: (a) the location of the image; (b) the focal length.

35. (II) A concave mirror produces an image 40% larger when a real object is 20 cm from the mirror. Determine the possible focal lengths of the mirror.

36. (II) A real object is 60 cm from a concave mirror. Find the radius of curvature of the mirror given the following: (a) The real image is 40% smaller. (b) The real image is 25% larger. (c) The virtual image is 80% larger.

37. (II) A spherical object of radius r is at a very large distance d from a concave mirror of focal length f. (a) Show that the radius of the image is approximately rf/d. (See Fig. 35.40. Where is the image located?) (b) The concave mirror of the Mt. Palomar reflecting telescope has a focal length of 16.8 m. What is the radius of the image of the Moon? Take $r = 1.74 \times 10^6$ m.

35.8 Speed of Light

38. (I) How long does light take to reach us from (a) the moon; (b) the sun?

39. (I) A light-year is the distance traveled by light in one year. (a) What is a light-year in meters? (b) Express the distance between the earth and the sun in terms of light-years.

40. (I) Romer found that the time interval between successive eclipses of the moon of Jupiter changed by a maximum of 22 min in a 6-month period. Given that the radius of the earth's orbit is 1.5×10^8 km, what value for the speed of light would he have obtained?

41. (I) The distance between the mirrors in Michelson's experiment was 35 km. What was the minimum rotational frequency (rev/s) of the eight-sided mirror such that light would enter the telescope?

42. (II) In Fizeau's experiment, the light traveled 4 km each way. If the wheel had 360 teeth, what was its minimum rotational frequency (rev/s)?

PROBLEMS

1. (I) A point source is 10 cm below the surface of a pond. What is the depth of the image if it is viewed in air at $\theta = 5°$ to the normal? Assume the image is at the intersection of the normal from the source to the surface and the refracted ray extended backwards; see Fig. 35.62.

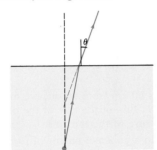

FIGURE 35.62 Problem 1.

2. (II) A person of height 2 m stands at a distance of 4 m from the edge of a pool of depth 2.5 m and width 4 m. A coin lies at the bottom at the far side; see Fig. 35.63. To what depth must the pool be filled for him to see the coin?

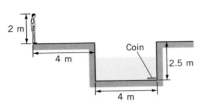

FIGURE 35.63 Problem 2.

3. (I) A beam of light is incident on a flat glass slab of thickness t at an angle of incidence i, as in Fig. 35.64. Show that the lateral displacement, d, of the beam passing through the slab is given by

$$d = \frac{t \sin(i - r)}{\cos r}$$

where r is the angle of refraction.

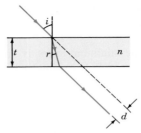

FIGURE 35.64 Problems 3 and 7.

4. (I) Overlapping rays of red and blue light are incident at 30° to the normal to a flat plate of thickness 2.4 cm. The indices of refraction are $n_R = 1.58$ and $n_B = 1.62$. (a) What is the angle between the refracted rays after they enter the plate? (b) What is the lateral separation (the perpendicular distance) between the rays that emerge on the other side?

5. (I) A point source is 10 cm below the surface of a pond ($n = 1.33$). Draw rays with angles of incidence θ and $\theta + 2°$ for $\theta = 0°, 10°, 20°, 30°, 40°,$ and $45°$. Calculate the directions of the refracted rays and draw them as well. Extend each pair of refracted rays backward to locate the image. (The curve joining the image points is called a *caustic*. It arises when spherical waves meet a plane boundary or when plane waves meet a spherical boundary.)

6. (I) The *longitudinal* magnification of a mirror is defined as $m_L = dq/dp$, where dp and dq are infinitesmal changes in the object and image distances, respectively. Show that

$$m_L = -\frac{q^2}{p^2}$$

7. (II) Figure 35.64 shows a ray of light incident at an angle i on a glass slab of thickness t and refractive index n. Show that the lateral displacement, d, is given by

$$d = t \sin i \left(1 - \frac{\cos i}{\sqrt{n^2 - \sin^2 i}} \right)$$

8. (II) A glass cylinder of index n is surrounded by a sheath of index n'. The surrounding medium has index n_0 (see Fig. 35.65). (a) Show that the maximum angle θ at which light will undergo total internal reflection is given by

$$n_0 \sin \theta = \sqrt{n^2 - n'^2}$$

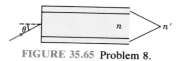

FIGURE 35.65 Problem 8.

(b) What is special about the condition $n_0 = n' = 1$?

9. (II) In Fig. 35.66 light travels from point A to point B after reflection in the mirror. According to *Fermat's principle* a light ray travels between two points along a path that takes the least time. (a) Show that the time taken is

$$t = \frac{(x^2 + a^2)^{1/2}}{c} + \frac{[(L - x)^2 + b^2]^{1/2}}{c}$$

(b) By setting $dt/dx = 0$, show that the angle of incidence is equal to the angle of reflection.

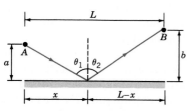

FIGURE 35.66 Problem 9.

10. (II) In Fig. 35-67 a light ray travels from point A in a medium of refractive index n_1 to point B in a medium of refractive index n_2. The ray strikes the boundary at a horizontal distance x from A. (a) Show that the time taken is

$$t = \frac{(a^2 + x^2)^{1/2}}{v_1} + \frac{[b^2 + (L - x)^2]^{1/2}}{v_2}$$

where v_1 and v_2 are the speeds of light in the two media. (b) Use *Fermat's principle* (stated in Problem 9) to find the value of x for which t is a minimum. By setting $dt/dx = 0$ and expressing $\sin \theta_1$ and $\sin \theta_2$ in terms of the given vari-

ables obtain Snell's law:

$$\frac{\sin \theta_1}{v_1} = \frac{\sin \theta_2}{v_2}$$

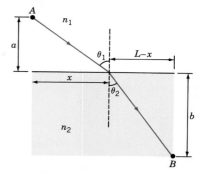

FIGURE 35.67 Problem 10.

11. (II) (a) Show that the angular deviation for the primary rainbow is $\delta = \pi + 2i - 4r$, where $\sin i = n \sin r$; see Fig. 35.68. (b) Show that δ has a minimum value of $180° - 42°$. See the special topic on the rainbow. Assume the refractive index is $n = 4/3$.

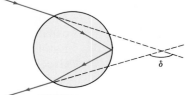

FIGURE 35.68 Problem 11.

12. (II) The secondary rainbow is formed when the incident light undergoes two internal reflection as shown in Fig. 35.69. (a) What is the angle of deviation δ in this case? (b) Show that the condition for minimum deviation is

$$\cos^2 i = \frac{n^2 - 1}{8}$$

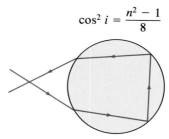

FIGURE 35.69 Problem 12.

CHAPTER 36

Lenses and Optical Instruments

A reflecting telescope on Mt. Hamilton, California.

Major Points

1. (a) Ray diagrams for converging and diverging lenses.
   (b) The **thin lens formula.**
2. The **angular magnification** produced by a single lens.
3. The optics of the compound **microscope** and of **telescopes.**
4. The **eye:** Correction of nearsight and farsight.

An optical **lens** is a sample of some transparent material, such as glass, that usually has spherical or cylindrical surfaces. The term lens is derived from the Latin for the lentil bean. It is unclear when lenses were first produced but the Greeks knew that a transparent spherical bowl filled with water could concentrate sunlight. Also, a character in Aristophanes' *Comedy of the Clouds* (423 B.C.) suggests the use of a "burning glass" to set a document on fire. Lenses were first used as spectacles in northern Italy around 1285. Kepler, in 1611, provided the foundation for the development of optical instruments. He was the first to analyze the relationship between objects and their images in optical systems, and was able to design telescopes even though he knew only Ptolemy's approximate law of refraction, $n_1\theta_1 = n_2\theta_2$. In this chapter we study the formation of images by lenses. We discuss the optics of microscopes and telescopes and how the problems of nearsight and farsight are corrected.

36.1 LENSES

In a **converging lens** (Fig. 36.1*a*), the central section is thicker than the rim. Such a lens causes rays parallel to the principal axis to be focused to a real focal point. In a **diverging lens** (Fig. 36.1*b*), the central section is thinner than the rim. Such a lens causes a parallel beam to appear to diverge from a virtual focal point. Note that both lenses have two focal points, at equal distances from the lens.

The present discussion is restricted to *thin* lenses whose thickness is much less than their diameters. This means that we may ignore the sideways displacement of a ray that is incident at an angle to the axis (Fig. 36.2*a*). Instead we assume that a ray that passes through the center is undeviated, as in Fig. 36.2*b*. Another problem arises from the dispersion of light in glass. As shown in Fig. 36.3, different colors have different focal points. This problem, called *chromatic*

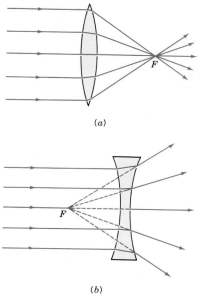

(a)

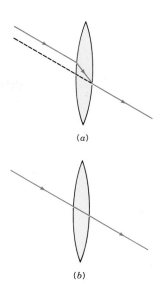

(a)

(b)

(b)

FIGURE 36.1 (*a*) A converging lens focuses rays parallel to the central axis to real focal point. (*b*) For a diverging lens, the focal point is virtual.

FIGURE 36.2 In the thin lens approximation we ignore the sideways displacement of a ray as in (*a*) and assume that a ray passes through the center undeviated as in (*b*).

aberration, can be corrected with a second (diverging) lens made of glass that has different dispersive properties. As is the case with spherical mirrors, a lens also suffers from *spherical aberration,* that is, even a monochromatic parallel beam is not brought to a unique focal point (see Fig. 36.4). This problem is minimized by dealing only with paraxial rays. We will ignore aberrations and assume that our thin lenses focus parallel rays to a unique focal point.

Spherical and chromatic aberration

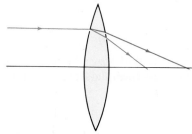

FIGURE 36.3 Because of dispersion, rays of different colors are focused to different points. This defect is called chromatic aberration.

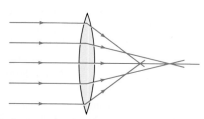

FIGURE 36.4 As a result of spherical aberration, rays incident at different distances from the central axis are focused to different points.

Principal Ray Diagrams

Ray diagrams are useful in locating images due to lenses. Any two of the following *principal* rays are sufficient to locate the image produced by a lens:

1. A ray *passing through the center* of the lens is undeviated.

2. A ray directed *parallel to the axis* passes through a focal point.

Principal rays for lenses

3. A ray directed *toward, or away from, a focal point* emerges parallel to the axis.

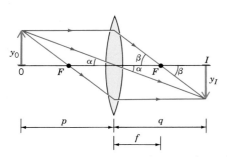

FIGURE 36.5 The principal rays used to locate an image in a converging lens.

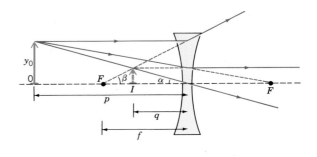

FIGURE 36.6 The principal rays used to locate an image in a diverging lens.

These three rays are illustrated in Fig. 36.5 and Fig. 36.6. Note that in the case of a diverging lens the rays do not actually pass through the focal point. They must be extrapolated either forward or backward.

The Thin Lens Formula

In the case of a thin lens, it is straightforward to relate the **object distance** p, the **image distance** q, and the **focal length** f. We assume that all rays are paraxial and make the approximation $\tan \theta \approx \theta$. From Fig. 36.5 we see that for a converging lens

$$\alpha = \frac{y_O}{p} = \frac{y_I}{q}; \qquad \beta = \frac{y_O}{f} = \frac{y_I}{q - f}$$

Therefore, $y_I/y_O = q/p = (q - f)/f$. From this we find $1/p + 1/q = 1/f$. (Supply the missing steps.) From Fig. 36.6, we see that for the diverging lens

$$\alpha = \frac{y_O}{p} = \frac{y_I}{q}; \qquad \beta = \frac{y_O}{f} = \frac{y_O - y_I}{q}$$

When the expressions for y_I/y_O are equated, we find $1/p - 1/q = -1/f$. (Check it.) As we did for mirrors, we use just one **thin lens formula** and supplement it with a sign convention:

Thin lens formula

$$\frac{1}{p} + \frac{1}{q} = \frac{1}{f} \qquad (36.1)$$

SIGN CONVENTION (Lenses): p is positive (real) on the left, negative (virtual) on the right. q is positive on the right, negative on the left. f is positive for a converging lens, negative for a diverging lens. (Light travels from left to right.)

As is the case with mirrors, the **transverse,** or **linear magnification,** m, is defined as the ratio of image height to object height. From Fig. 36.5 and Fig. 36.6 we see that

Transverse (Linear) magnification

$$m = \frac{y_I}{y_O} = -\frac{q}{p} \qquad (36.2)$$

The negative sign signifies that the image is inverted if both p and q have the same sign.

Problem Solving Guide for Lenses

This guide is meant to assist you in problems that involve two lenses. We assume that the object distance to the first lens and the distance between the lenses is specified. This is *not* always so. For example, the position of the final image may be specified and you may be asked to "work backward" to locate the object. The order in which the steps are taken may then be different.

THIN LENS FORMULA

1. Calculate the position of the image I_1 in the first lens, L_1. This first image acts as the *object* for the second lens, L_2.

2. Determine the object distance for the second lens and then calculate the position of the second image, I_2.

3. Keep in mind the sign convention for lenses. In particular, if I_1 is placed to the right of L_2, it acts as a *virtual* object, so p_2 would be negative.

PRINCIPAL RAY DIAGRAM

1. The values obtained above allow you to chose a suitable scale for a ray diagram. Turn ordinary lined paper sideways to give you a ready-made grid.

2. Draw each lens as a single line with triangles at the ends to indicate whether it is converging or diverging (see Fig. 36.7). This allows you to easily erase the triangles and "extend" the lens if the ray diagram requires it.

3. Use only principal rays to locate images. If an image is located to the right of the second lens, use only *dashed* lines beyond L_2 (see Fig. 36.8). The lines from a lens to a virtual image are also dashed (see Fig. 36.10).

4. In order to draw the rays to locate the second image I_2, you may choose *any* two rays emerging from I_1 that are convenient and useful. They need *not* be related to those that were used to form I_1.

5. The light travels from left to right. Do not draw any arrowheads pointing toward the left.

EXAMPLE 36.1: (a) A small object is placed 16 cm from a converging lens that has a focal length of 12 cm. Locate the image and determine its transverse magnification. (b) An object of height 0.8 cm is located 25 cm from a diverging lens of focal length −16 cm. Locate the image and its height.

Solution: (a) We are given $p = 16$ cm and $f = 12$ cm. From the lens formula we obtain

$$\frac{1}{16 \text{ cm}} + \frac{1}{q} = \frac{1}{12 \text{ cm}}$$

which yields $q = 48$ cm. The transverse magnification is

$$m = -\frac{q}{p} = -\frac{48 \text{ cm}}{16 \text{ cm}} = -3$$

The image is real (q is positive), inverted (m is negative), and magnified ($|m| > 1$). Figure 36.5 is an appropriate ray diagram (not to scale).

(b) From the lens formula we obtain

$$\frac{1}{25 \text{ cm}} + \frac{1}{q} = -\frac{1}{16 \text{ cm}}$$

which yields $q = -9.75$ cm. The magnification is

$$m = -\frac{q}{p} = -\frac{-9.75 \text{ cm}}{25 \text{ cm}} = +0.39$$

The height of the image is (0.8 cm)(0.39) = 0.31 cm. The image is virtual, erect, and reduced, as shown in Fig. 36.6.

EXERCISE 1. A small object is located 6 cm from a converging lens of focal length 12 cm. Locate the image and determine its transverse magnification. Draw a principal ray diagram.

EXAMPLE 36.2: A converging lens of focal length 4 cm is placed 12 cm ahead of a diverging lens of focal length −2 cm. A small object is 8 cm in front of the converging lens. Find: (a) the position of the final image; (b) the transverse magnification of the final image.

Solution: (a) We must first find the image distance for the converging lens, L_1, in Fig. 36.7. Using $p_1 = 8$ cm and $f_1 = 4$ cm in the lens formula gives

$$\frac{1}{8 \text{ cm}} + \frac{1}{q_1} = \frac{1}{4 \text{ cm}}$$

from which we find $q_1 = 8$ cm. Since q_1 is positive, the image I_1 due to the first lens is real and to the right of L_1. In drawing the ray diagram to locate I_1 only two rays have been used. (Add the third ray.)

Since I_1 is to the *left* of L_2, the image of the first lens acts as a *real* object for the second lens. With $p_2 = +4$ cm and $f_2 = -2$ cm in the lens formula,

$$\frac{1}{4 \text{ cm}} + \frac{1}{q_2} = -\frac{1}{2 \text{ cm}}$$

we find the image distance $q_2 = -1.33$ cm. The negative sign means that this image is virtual, to the left of L_2.

(b) The overall transverse magnification is the product of the individual magnifications:

$$m = m_1 m_2$$
$$= \left(-\frac{8}{8}\right)\left(\frac{1.33}{4}\right) = -\frac{1}{3}$$

The final image is virtual, inverted, and reduced.

Solution: (a) This problem is similar to the previous example but the drawing of the ray diagram is less straightforward. The location of the image I_1 in the first lens is found from the lens formula with $p_1 = 5$ cm and $f_1 = 4$ cm. Thus,

$$\frac{1}{5 \text{ cm}} + \frac{1}{q_1} = \frac{1}{4 \text{ cm}}$$

from which we find $q_1 = 20$ cm. The first image is real.

In the ray diagram of Fig. 36.8, rays 1 and 2 suffice to locate I_1. Note that they are continued beyond L_2 only as dashed lines, since they are refracted by this lens. Since I_1 is to the right of L_2, it acts as a *virtual object* for this lens, so $p_2 = -8$ cm. From the lens formula applied to the second lens we have

$$-\frac{1}{8 \text{ cm}} + \frac{1}{q_2} = \frac{1}{7 \text{ cm}}$$

which yields $q_2 = +3.7$ cm. Since q_2 is positive, the image is real.

In the ray diagram, ray 1, which is parallel to the axis after emerging from L_1, will converge to the focal point F_2' of the second lens. Our problem is to find a second ray to locate the second image I_2. To do this, we note that of the many rays that formed I_1, one must have passed through the center of L_2 undeviated. Having already located I_1, we can draw a (solid) ray from the center of L_2 to the tip of I_1. The intersection of ray 3 with the ray through F_2' locates the final image I_2. Ray 3 may then be extended backward to meet L_1 to make the diagram appear complete. (Actually, one more line can be added. Do so.)

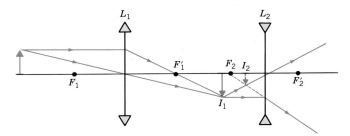

FIGURE 36.7 The image I_1 formed by the first lens, L_1, acts as a real object (since it is to the left) for the second lens, L_2.

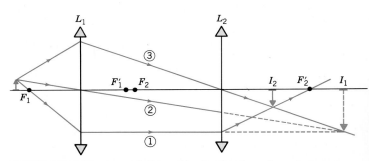

FIGURE 36.8 The image I_1 formed by the first lens, L_1, acts as a virtual object (since it is to the right) for the second lens, L_2.

EXAMPLE 36.3: A converging lens L_1 of focal length 4 cm is 12 cm ahead of a second converging lens L_2 of focal length 7 cm. A small object is 5 cm in front of L_1. Find the position of the final image.

EXERCISE 2. What is the overall transverse magnification of the two-lens combination in Example 36.3? Describe the nature of the final image.

36.2   THE SIMPLE MAGNIFIER

A small coin held at arms length can seem to be as large as the moon. The reason for this is that the apparent size of an object depends on the angle it subtends at the eye. This in turn determines how much of the retina is stimulated. In order to see detail on an object, we bring it as close as possible to the eye. The closest one can comfortably bring an object to an eye is called the *least distance of distinct vision* and is taken to be 25 cm for a normal eye. It may be as low as 10 cm for children and is larger for older people. Thus, 25 cm is used as a convenient reference value. The maximum angle an object can subtend occurs when it is at the least distance of distinct vision. This position also corresponds to maximum resolution since the retinal image is the largest it can be. From Fig. 36.9, we see that

$$\alpha_{25} = \frac{y_O}{0.25} \tag{36.3}$$

where y_O is in meters.

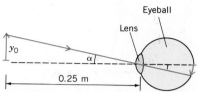

FIGURE 36.9 The apparent size of an object is determined by the angle it subtends at the eye. The maximum possible angle occurs when the object is at the least distance of distinct vision—taken to be 25 cm.

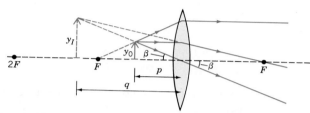

FIGURE 36.10 A real object placed closer to a converging lens than its focal length results in an enlarged virtual image.

Figure 36.10 shows that when an object is closer to a converging lens than its focal length, the image is erect, virtual, and enlarged. The angle subtended by the image is $\beta = y_I/q = y_O/p$. The **angular magnification,** M, of single lens, called a *simple magnifier,* is defined as the ratio of the angle subtended by the image produced by the lens to the angle subtended when the object itself is at 25 cm:

$$M = \frac{\beta}{\alpha_{25}}$$

On inserting the expressions for α_{25} and β, we find

$$M = \frac{0.25}{p} \tag{36.4}$$

where p is in meters. We assume that the distance between the eye and the lens is negligible. It is easier on the eye to have the image at infinity, as shown in Fig. 36.11, because then the normal eye is relaxed. The object is then at the focal point,

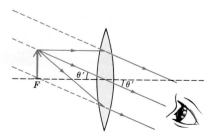

FIGURE 36.11 When an object is placed at the focal point of a converging lens, the image is at infinity.

FIGURE 36.12 One of the small single-lens microscopes used by van Leewenhoek.

that is, $p = f$, so Eq. 36.4 becomes

$$M_\infty = \frac{0.25}{f} \qquad (36.5)$$

This shows that the angular magnification increases as f gets smaller.

The first description of the use of a converging lens to produce a magnified image was presented in the 13th century by Roger Bacon, who suggested it as an aid for reading. Around 1670, Antony van Leeuwenhoek used tiny *single* lenses with focal lengths of about 1.5 mm to make detailed studies of insects and even bacteria.* He was able to make lenses with nonspherical surfaces to greatly improve the resolution of the images. Nine of the approximately 500 simple microscopes he made still exist (see Fig. 36.12). The best has a magnification of about 275—an astonishing figure given that a modern microscope achieves about 1000! With his best lens, he could have seen detail as small as 1 μm. However, these tiny simple microscopes place great demands on the eyesight of the observer. An alternative is the compound microscope.

36.3 THE COMPOUND MICROSCOPE

A microscope that consisted of a converging lens and a diverging lens originated in Holland around 1590 and was first used for scientific purposes by Galileo in 1610. The use of two converging lenses was suggested by Kepler in 1611, and this arrangement forms the basis for the **compound microscope.** An important event in the history of microscopy was the publication of the *Micrographia* by Robert Hooke in 1665. It contained detailed illustrations of his observations using the compound microscope shown in Fig. 36.13. Indeed, it was his drawings of cloth fibers that stimulated Leeuwenhoek, originally a draper, to make his own observations.

This microscope, similar to Hooke's, was made by C. Cock in the 1680s and belonged to George III.

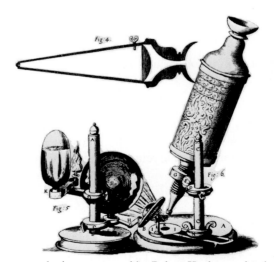

FIGURE 36.13 The compound microscope used by Robert Hooke consisted of two plano-convex lenses set about six inches apart.

* See *The Single Lens*, B. J. Ford, Heinemann, London, 1985. This book discusses how Leeuwenhoek single-handedly founded the subject of microbiology.

In a compound microscope the first lens is the **objective** O and the second is the **eyepiece** E, as shown in Fig. 36.14. The function of the objective is to place an enlarged image of the object at a point closer to the eyepiece than its focal length, f_E. The eyepiece then acts as a simple magnifier. The focal length of the objective, f_O, is quite small, of the order of 5 mm. This allows the instrument to be placed close to the object under study. As we shall see, it also leads to a greater overall magnification. The focal length of the eyepiece is approximately 15 mm. The distance between the focal points of the objective and the eyepiece is called the *optical tube length*, ℓ, and is usually fixed at 16 cm. The distance d between the lenses is $d = \ell + f_O + f_E$. The numbered arrows in Fig. 36.14 (which is not drawn to scale) signify the following:

1. The object just beyond the focal point of the objective ($p_O > f_O$).

2. The real, enlarged image produced by the objective ($q_O > \ell + f_O$). This image acts as a real object for the eyepiece.

3. The final virtual image produced by the eyepiece.

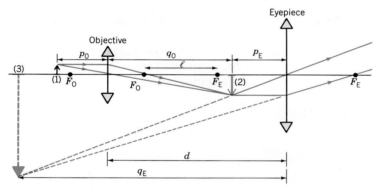

FIGURE 36.14 In a compound microscope the objective lens produces a real image within the focal length of the eyepiece, which acts as a simple magnifier. (The diagram is not to scale.)

The angular magnification of the microscope is again defined to be

$$M = \frac{\beta}{\alpha_{25}}$$

where α_{25} is the angle subtended by the object at 0.25 m and β is the angle subtended by the image in the instrument. From Eq. 36.5 we know that $\alpha_{25} = y_O/0.25$, and from Fig. 36.14 we see that $\beta = y_2/p_E$. Furthermore, $y_2/q_O = -y_O/p_O$, where the negative has been inserted because the image (2) is inverted. Combining these factors, we find the **angular magnification,** or **magnifying power,** of the compound microscope is

$$M = -\frac{q_O}{p_O} \cdot \frac{0.25}{p_E} \qquad (36.6)$$

Angular magnification of a compound microscope

This form for M is not convenient since q_O/p_O cannot be directly measured or specified. If the image due to the objective coincides with the focal point of the eyepiece, then $p_E = f_E$. In this case the final virtual image is at infinity and can be viewed by a normal relaxed eye. Since $1/q_O + 1/p_O = 1/f_O$, and $q_O = \ell + f_O$, the ratio q_O/p_O equals ℓ/f_O. (Check it.) Equation 36.6 then takes the form

$$M_\infty = -\frac{\ell}{f_O} \cdot \frac{0.25}{f_E} \tag{36.7}$$

Clearly, shorter focal lengths produce greater magnifications. Although $M_\infty < M$ (see Example 36.4 below), the slight loss in magnifying power is compensated for by the greater comfort of the eye. Figure 36.15 shows the optical system of a modern microscope.

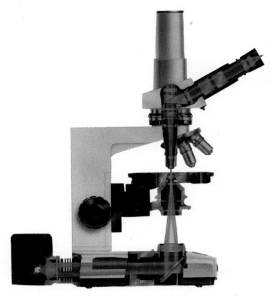

FIGURE 36.15 The optical system of a modern microscope.

EXAMPLE 36.4: A microscope has an objective of focal length 5 mm and an eyepiece of focal length 20 mm. The optical tube length is 15 cm and the final image is at 40 cm from the eyepiece. Find: (a) the position of the object; (b) the overall magnification.

Solution: The distance between the lenses is $d = \ell + f_O + f_E = 17.5$ cm.
(a) Since we have the image distance for the eyepiece, the object distance may be found from $1/p_E + 1/q_E = 1/f_E$ with $f_E =$ 2 cm and $q_E = -40$ cm. This leads to $p_E = 40/21 = 1.90$ cm.

The image distance for the objective is $q_O = d - p_E = 15.6$ cm. Finally, $1/p_O + 1/q_O = 1/f_O$ leads to $p_O = 0.517$ cm. Note that this is slightly greater than f_O.
(b) From Eq. 36.6 the overall angular magnification is

$$M = -\left(\frac{15.6}{0.517}\right)\left(\frac{25}{1.90}\right) = -397$$

EXERCISE 3. What is the magnification if the image is at infinity?

36.4 TELESCOPES

A **telescope** is an optical instrument used to view objects far away. Roger Bacon mentioned the idea of using a concave mirror and a magnifying lens "to make distant objects appear close to the eye," but it is uncertain whether he actually built such a telescope. The use of a diverging lens was later suggested by della Porta around 1590. Perhaps because of this suggestion or a chance discovery of his own, in 1608 the Dutch spectacle maker Hans Lippershey constructed the first telescope with a combination of a converging lens and a diverging lens.

When Galileo heard of this instrument in June 1609, he immediately made one for himself and turned it skyward (see Fig. 36.16). His discovery of surface structure on the moon conflicted with the long-held view that the celestial bodies were "perfect." His discovery in 1610 of the moons of Jupiter also cast doubt on the Aristotelian belief that the earth is at the center of the universe. When Kepler expressed disbelief in this discovery, Galileo sent him a telescope to see for himself. This stimulated Kepler's own work in optics and in that same year he proposed that two converging lenses could form a telescope. The first such **astronomical telescope** was constructed nearly 50 years later by Huygens. We will discuss this type first.

The objects viewed with a microscope are close to the objective lens. In contrast, the objects viewed with an astronomical telescope, such as the moon, are usually at infinity. As shown in Fig. 36.17, the objective O produces a real image within the focal length of the eyepiece E, which then acts as a simple magnifier. The angle subtended by the object at the unaided eye is α. The image in the telescope subtends an angle β, so the angular magnification of the telescope is

$$M = \frac{\beta}{\alpha} \tag{36.8}$$

Since the angles are small, $\alpha \approx h/f_O$ and $\beta \approx h/p_E$, and Eq. 36.8 becomes

$$M = -\frac{f_O}{p_E} \tag{36.9}$$

If the final image is also at infinity, the focal points of the objective and the eyepiece coincide, thus $q_O = f_O$ and $p_E = f_E$. The angular magnification in this case is

$$M_\infty = -\frac{f_O}{f_E} \tag{36.10}$$

For large magnification, we need $f_O \gg f_E$. This requirement means that the astronomical telescope is rather long. Also, since it produces an inverted image, it is inconvenient for terrestrial work. (A third lens may be used to produce an upright final image.)

In the **terrestrial** or **Galilean telescope,** shown in Fig. 36.18, the eyepiece is a diverging lens. If the focal points of the two lenses coincide, the final image is at

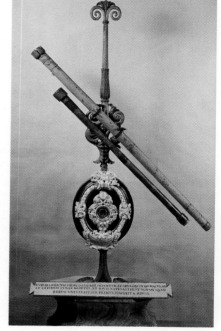

FIGURE 36.16 Telescopes used by Galileo.

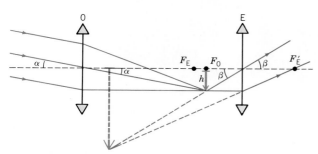

FIGURE 36.17 The objective of an astronomical telescope produces a real image within the focal length of the eyepiece, which then acts as a simple magnifier.

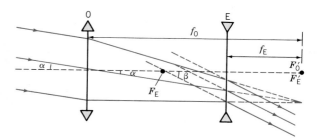

FIGURE 36.18 A terrestrial or Galilean telescope uses a diverging lens as the eyepiece.

infinity, as depicted in the figure. The expressions for the overall magnification are the same as those for the astronomical telescope. This arrangement is used in opera glasses since the image is upright and the separation between the lenses is not large.

Newton believed (erroneously) that chromatic aberration in lenses could not be eliminated. He made a **reflecting telescope** (Fig. 36.19) that consisted of a concave mirror and a small plane mirror that reflected light to the eyepiece, as shown in Fig. 36.20. The concave mirror provides a real image within the focal length of the eyepiece, which then acts as a simple magnifier. Since lenses of large diameter (needed to collect as much light as possible) are difficult to make and tend to become distorted under their own weight, modern telescopes usually employ a parabolic mirror. The telescope on Mount Palomar has a mirror 200 inches in diameter (Fig. 36.21). The angular magnification is given by Eq. 36.10 where f_O would be the focal length of the mirror.

FIGURE 36.19 A replica of Newton's reflecting telescope.

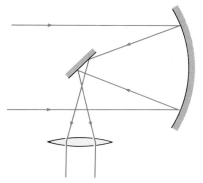

FIGURE 36.20 In a reflecting telescope a mirror replaces the objective of a (refracting) astronomical telescope. The eyepiece serves as a simple magnifier.

FIGURE 36.21 The mirror of the telescope at Mount Palomar, California, is made of 14.5 tons of Pyrex glass and has a diameter of 200 inches (5.08 m).

36.5 THE EYE

The human eye, shown in Fig. 36.22, is a wonderfully refined optical instrument. When light enters the eye, most of the refraction takes place at the outer membrane, called the *cornea*. The remaining refraction is due to the *crystalline lens,* which is immersed in two liquids (called humours) as shown. The size of the *pupil,* through which light enters the lens, is controlled by the *iris* (the colored part). The light is then focused on the *retina,* which contains light-sensitive *rods* and *cones* and the information is carried by the *optic nerve* to the brain. The eye is able to focus objects at different distances by varying the focal length of the crystalline lens. This process, called accommodation, is accomplished by contracting or relaxing the *ciliary muscles,* which change the radii of curvature of the two surfaces of the lens. In what follows, we consider a "reduced" eye, which consists of a single lens that replaces the cornea and the crystalline lens.

The farthest distance at which an object can be focused by the unaided eye is the **far point.** The closest distance that an object can be brought to the eye and be focused without difficulty is the **near point.** For a normal eye, the far point is at infinity and the near point is taken to be 25 cm.

When a person is farsighted, the eyeball is too short compared to the focal length of the relaxed eye. Objects at infinity are focused behind the retina. This condition, called **hyperopia,** is shown in Fig. 36.23a. The near point for such an eye may or may not be the normal 25 cm. A second kind of farsight is called **presbyopia,** shown in Fig. 36.23b. As a person gets older, the eye muscles get weaker or the lens hardens. The near point becomes greater than 25 cm, which makes it difficult to focus nearby objects. The person must hold a printed page at arm's length to read it. Both of these conditions are helped by a converging lens.

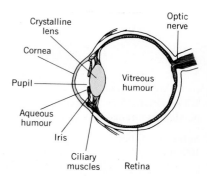

FIGURE 36.22 The human eye. Most of the refraction of light entering the eye occurs at the cornea. The focal length of the crystalline lens is adjusted by contracting or relaxing the ciliary muscles.

This photograph clearly shows the cornea and the crystalline lens.

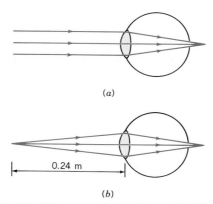

(a)

0.24 m

(b)

FIGURE 36.23 (a) Hyperopia: Parallel rays are focused to a point behind the retina. (b) Presbyopia: Light from objects near the eye are focused behind the retina.

When someone is nearsighted, the eyeball is too long compared to the focal length of the relaxed eye. Objects at infinity are focused in front of the retina, as shown in Fig. 36.24. This condition, called **myopia,** may be compensated for with a diverging lens.

Eyeglasses are specified in terms of the **power** of the lens. This is defined as the reciprocal of the focal length in meters:

$$\text{Power} = \frac{1}{f} \tag{36.11}$$

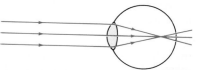

FIGURE 36.24 Myopia: Light from infinity is focused to a point in front of the retina.

The unit of power is the dioptre (D). A converging lens has a positive power, whereas a diverging lens has a negative power. Since we are able to vary the focal length of the crystalline lens, the range in focal lengths is expressed in terms of its *power of accommodation.*

EXAMPLE 36.5: An eye has a near point at 1 m. What lens is needed to change this to 25 cm?

Solution: When the object is at 25 cm, it must appear to be at 100 cm (Fig. 36.25). That is, the spectacle lens must produce a virtual image at this distance (1 m). Note that the rays that enter the eye are diverging. From the lens formula, $1/p + 1/q = 1/f$, we have

$$\frac{1}{25 \text{ cm}} - \frac{1}{100 \text{ cm}} = \frac{1}{f}$$

Thus $f = 100/3 = 33.3$ cm.

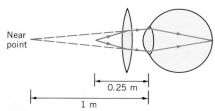

FIGURE 36.25 Rays that form the virtual image of the object at 25 cm appear to diverge from the person's near point.

EXAMPLE 36.6: (a) An eye has a far point at 2 m. What lens is needed to move this to infinity? (b) If the near point is 25 cm with these glasses, where is it without them?

Solution: (a) The object, which is at infinity, must appear to be at 2 m. That is, the rays reaching the eye must diverge from the far point, as in Fig. 36.26a. With $p = \infty$ and $q = -2$ m, the lens formula yields $f = -2$ m.
(b) An object at 25 cm will appear to be closer because of the diverging lens, as shown in Fig. 36.26b. With $p = 25$ cm and $f = -2$ m, we find $q = -22$ cm. Note that the image produced by the lens is virtual. The diverging lens shifts the near point from 22 cm to 25 cm from the eye. For close-up work, it is more comfortable to remove the glasses.

EXAMPLE 36.7: A normal eye has a diameter of 2 cm. What is its power of accommodation?

Solution: A normal eye can focus from 25 cm to infinity. The diameter of the eyeball is equal to the focal length for objects at infinity (Fig. 36.27a). Thus $1/f_1 = P_1 = 1/(0.02 \text{ m}) = 50$ D. For an object at the near point of 25 cm, the image distance is still 2

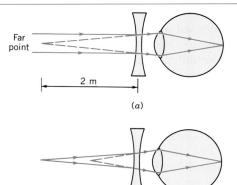

FIGURE 36.26 (a) Light from infinity appears to diverge from the person's far point. (b) With a diverging lens in an eyeglass, a person's near point is moved farther from the eye.

cm (Fig. 36.27b). The focal length is given by

$$\frac{1}{0.25 \text{ m}} + \frac{1}{0.02 \text{ m}} = \frac{1}{f_2} = P_2$$

Thus $P_2 = 54$ D. The power of accommodation is $(P_2 - P_1) = 4$ D.

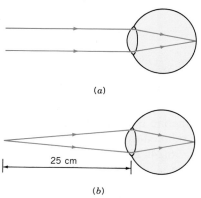

FIGURE 36.27 A normal eye can focus on the retina light from (a) infinity, and (b) the near point at 25 cm.

36.6 SPHERICAL BOUNDARIES (Optional)

We now examine the behavior of paraxial rays at a spherical boundary between two media. A surface will be classified as either convex or concave when viewed from the medium of lower refractive index. Thus, if $n_2 > n_1$ in Fig. 36.28, the surface is convex; if $n_2 < n_1$, the surface is concave. Only the convex surface will be treated in detail.

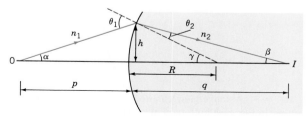

FIGURE 36.28 A ray from a point object at O in a medium of refractive index n_1 has an image point, I, in a medium of refractive index n_2.

Consider an arbitrary ray from the point object O. It strikes the boundary at a distance h from the axis and at an angle θ_1 to the normal (radial line) in medium 1. After refraction it travels at an angle θ_2 to the normal in medium 2. In the paraxial approximation, all rays from O will pass through the image point I. For small angles we may set $\sin \theta \approx \theta$, so Snell's law becomes

$$n_1 \theta_1 = n_2 \theta_2 \qquad (36.12)$$

From Fig. 36.28 we see that $\theta_1 = (\alpha + \gamma)$ and $\theta_2 = (\gamma - \beta)$, where $\alpha \approx h/p$, $\beta \approx h/q$, and $\gamma \approx h/R$. When these values for θ_1 and θ_2 are inserted into Snell's law we find

$$n_1 \left(\frac{h}{p} + \frac{h}{R} \right) = n_2 \left(\frac{h}{R} - \frac{h}{q} \right)$$

which, after some rearrangement, yields

$$\frac{n_1}{p} + \frac{n_2}{q} = \frac{n_2 - n_1}{R} \qquad (36.13)$$

This equation (valid in the paraxial approximation) relates the object distance p, in the medium of index n_1, and the image distance q, in the medium of index n_2, to the radius of curvature R of the surface. If similar derivations are carried out for other situations, we find that all possible cases can be handled with Eq. 36.13 and a sign convention.

> SIGN CONVENTION (Boundaries): p is positive (real) on the left of the surface, q is positive (real) on the right of the surface. R is positive if the center of curvature is on the right. R is negative if the center of curvature is on the left. (Light travels from left to right.)

Note that this convention for R is the opposite of the convention for spherical mirrors. This difference arises because the

direction of the reflected light is reversed, whereas that of the refracted light is not.

EXAMPLE 36.8: A point source O is inside a glass cylinder of refractive index $n = 1.5$, as shown in Fig. 36.29. It is 3 cm from a concave surface whose radius of curvature is 2 cm. Locate the image.

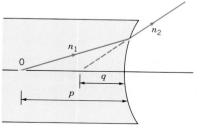

FIGURE 36.29 A ray from an object in glass emerging into air. The radius of curvature of the spherical surface is positive.

Solution: For the first medium (glass), $n_1 = n$, and for the second medium, (air) $n_2 = 1$. The object is in the glass. The radius of curvature of the concave surface is $R = +2$ cm, since the center of curvature is on the right. From Eq. 36.13, we have

$$\frac{n_1}{p} + \frac{n_2}{q} = \frac{n_2 - n_1}{R}$$

or

$$\frac{1.5}{3 \text{ cm}} + \frac{1}{q} = \frac{1 - 1.5}{2 \text{ cm}}$$

Therefore, $q = -1.33$ cm. The image is virtual, to the left of the boundary.

36.7 LENS MAKER'S FORMULA (Optional)

Until now we have treated the focal length of a lens as a quantity that may be determined by measuring the object and image distances. In this section we will relate the focal length of a lens to the radii of curvature of its two surfaces.

Figure 36.30 shows a double convex lens of refractive index n whose surfaces have radii of curvature R_1 and R_2. We take $n = 1$ for air. As a ray traverses the lens, it crosses two boundaries. Depending on where the object O is located, the image after the first refraction is either real or virtual. Let us assume that the image O' is virtual and at a distance q'. From Eq. 36.13 (in the paraxial approximation) we have

$$\frac{1}{p} - \frac{n}{q'} = \frac{n - 1}{R_1}$$

Note that the center of curvature of the first surface is on the

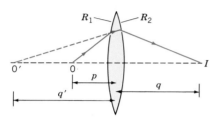

FIGURE 36.30 The real object at O has a virtual image due to the first surface at O', which then acts as a real object for the second surface. In this case, R_1 is positive and R_2 is negative.

right. For the second surface, O' acts as a real object. (It is to the left of the surface and rays are diverging from it.) If we ignore the thickness of the lens in comparison with R_1 and R_2, q' is the new object distance. We now have

$$\frac{n}{q'} + \frac{1}{q} = \frac{1 - n}{R_2}$$

Adding these two equations we find

$$\frac{1}{p} + \frac{1}{q} = (n - 1)\left(\frac{1}{R_1} - \frac{1}{R_2}\right) \qquad (36.14)$$

The quantity on the right-hand side may be identified as the reciprocal of the focal length:

(Thin lens) $\qquad \frac{1}{f} = (n - 1)\left(\frac{1}{R_1} - \frac{1}{R_2}\right) \qquad (36.15)$

Note that this derivation is valid only for a *thin lens in the paraxial approximation.*

EXAMPLE 36.9: A converging lens has surfaces with radii 2 cm and 3 cm as shown in Fig. 36.31. What is its focal length? The refractive index of the glass is 1.5.

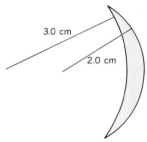

FIGURE 36.31 A converging lens with surfaces of different curvature. In this case both radii of curvature are negative.

Solution: For both surfaces, the center of curvature lies on the left; thus $R_1 = -3$ cm and $R_2 = -2$ cm. From Eq. 36.15,

$$\frac{1}{f} = (1.5 - 1)\left(-\frac{1}{3 \text{ cm}} + \frac{1}{2 \text{ cm}}\right)$$

$$= \frac{0.5}{6 \text{ cm}}$$

Thus $f = +12$ cm.

EXERCISE 4. Repeat Example 36.9 with the lens turned around.

SUMMARY

In the paraxial approximation, the object and image distances from a thin lens are related by the **thin lens formula**

$$\frac{1}{p} + \frac{1}{q} = \frac{1}{f}$$

The sign convention to distinguish between real and virtual objects and images is stated on *p.* 728.

The **transverse (linear) magnification** of a single thin lens is

$$m = -\frac{q}{p}$$

The **angular magnification** of a hand magnifier, or simple microscope, is

$$M = \frac{\beta}{\alpha_{25}} = \frac{0.25}{p}$$

where α_{25} is the angle subtended by the object at 0.25 m from the eye, and β is the angle subtended by the virtual image in the lens. The quantity p is the object distance in meters.

When the final image is at infinity, the angular magnification of a **compound microscope** is

$$M_\infty = -\frac{\ell}{f_O}\cdot\frac{0.25}{f_E}$$

where f_O and f_E are the focal lengths of the objective and eyepiece, respectively, and ℓ, the tube length, is the distance between the focal points.

When the final image is at infinity, the angular magnification of an **astronomical telescope** is

$$M_\infty = -\frac{f_O}{f_E}$$

where f_O and f_E are the focal lengths of the objective and the eyepiece, respectively. The expression for the Galilean telescope is the same, except it is positive since the final image is erect.

When light traveling in a medium of refractive index n_1 is refracted at a **spherical boundary** of radius of curvature R, the object and image distances are related according to

$$\frac{n_1}{p} + \frac{n_2}{q} = \frac{n_2 - n_1}{R}$$

where n_2 is the refractive index of the second medium. The sign convention is stated on p. 739.

The focal length f of a thin lens of refractive index n may be expressed in terms of the radii of curvature of the two surfaces:

$$\frac{1}{f} = (n - 1)\left(\frac{1}{R_1} - \frac{1}{R_2}\right)$$

It is assumed the lens is in air ($n = 1$). The sign convention is on p. 739.

ANSWERS TO IN-CHAPTER EXERCISES

1. With $p = 6$ cm and $f = 12$ cm, the lens formula is

$$\frac{1}{6\text{ cm}} + \frac{1}{q} = \frac{1}{12\text{ cm}}$$

Thus $q = -12$ cm. The transverse magnification is $m = -q/p = +2$. The image is virtual, erect and magnified. See Fig. 36.10 for the ray diagram.

2. The overall magnification is

$$m = m_1 m_2 = \left(-\frac{20}{5}\right)\left(\frac{3.7}{8}\right) = -1.85$$

The final image is real, inverted, and magnified.

3. From Eq. 36.7 we find

$$M_\infty = -\left(\frac{15}{0.5}\right)\left(\frac{25}{2}\right) = -375$$

which is slightly lower than M found in part (b) of Example 36.3.

4. In this case $R_1 = +2$ cm and $R_2 = +3$ cm, so

$$\frac{1}{f} = (1.5 - 1)\left(\frac{1}{2\text{ cm}} - \frac{1}{3\text{ cm}}\right)$$

which yields $f = +12$ cm, as before.

QUESTIONS

1. In a hand magnifier held close to the eye, the object and the image subtend approximately the same angle at the eye. So what purpose does the lens serve?

2. How would you determine the focal length of a diverging lens?

3. Why is it difficult for our eyes to form a clear image under water? Why do goggles help?

4. Does the focal length of a telescope affect the size of the image of an object?

5. Indicate as many ways as you can in which the diameter of a lens affects the image it produces.

6. How does the focusing of a camera lens differ from that of the crystalline lens in the eye?

7. What happens to the focal length of a lens when it is immersed in water? Discuss both converging and diverging lenses.

8. Can a virtual image be photographed?

9. Is it possible for a diverging lens to produce a real image? If so, explain how.

10. Can a converging lens produce an inverted, virtual image? If so, explain how.

11. Two thin plano-convex lenses are placed in contact. Compare the overall focal lengths when the plane surfaces touch and when the curved surfaces touch (see Fig. 36.32).

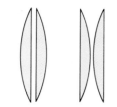

FIGURE 36.32 Question 11.

12. Which of the two orientations of a plano-convex lens shown in Fig. 36.33 is likely to produce a sharper image of an object at infinity? Explain why.

13. For a converging lens, specify the condition for the image to be each of the following (if possible): (a) real; (b) virtual; (c) erect; (d) inverted; (e) magnified; (f) reduced.

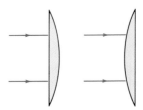

FIGURE 36.33 Question 12.

14. Repeat the above question for a diverging lens.

15. Can a virtual object produce a real image? If so, sketch a ray diagram to show how.

16. (a) Given a converging lens, where would you place an object to produce an image of the same size? (b) Could this be accomplished with a diverging lens? If so, how?

17. Why is it inadvisable to leave small water drops on a car in bright sunlight?

18. How could you make a lens to focus sound waves? Consider waves in air and in water.

19. A thin transparent plastic bag has the shape of a converging lens when inflated with air. How does it behave optically when placed in water?

20. When an astronomical telescope used to view the moon is adjusted for a relaxed normal eye, both the object and the image are at infinity. Why does the moon appear larger when viewed through the instrument?

21. An optical element is constructed with two surfaces of equal radius of curvature; see Fig. 36.34. Draw a ray diagram that shows the path of parallel rays incident from the left.

FIGURE 36.34 Question 21.

EXERCISES

36.1 Thin Lenses

1. (I) An object of size 2 m is 4 m from a converging lens. What is the size of the image on a screen given that the focal length is (a) 5 cm; (b) 20 cm?

2. (I) What is the size of the image produced by a converging lens of focal length 2 m used to take a picture of (a) the moon; (b) the sun?

3. (I) An object of size 1 cm is projected onto a screen 2 m

from a thin lens and forms a 5 cm image. (a) What is the position of the object, and (b) what is the focal length of the lens?

4. (I) A lens produces a virtual image, enlarged four times and located 16 cm in front of the lens (on the same side as the object). (a) Where is the object? (b) What is the focal length?

5. (I) A lens produces a real image, one third the size of the object and located 6 cm behind the lens (not on the side of the object). (a) Where is the object? (b) What is the focal length?

6. (I) A 35 mm slide (36 mm wide) is to be projected to fill a screen. What lens is needed if the screen is 2 m wide and 7 m from the lens?

7. (I) A simple camera has a single converging lens of focal length of 50 mm. How far from the film is the lens when the object being photographed is (a) 2 m from the lens, and (b) 0.5 m from the lens?

8. (II) A converging lens has a focal length of 15 cm. For what two object positions is the size of the image twice as large as the object?

9. (I) The image of an object located 12 cm from a converging lens has a linear magnification of $-2/3$. (a) Where is the image located? (b) What is the focal length of the lens?

10. (I) A converging lens of focal length 35 cm produces a magnified image 2.5 times the size of the object. What is the object distance if the image is (a) real, or (b) virtual?

11. (I) A converging lens of focal length 20 cm produces a reduced image that is 40% of the size of the object. Locate the object given that the image is (a) real, or (b) virtual.

12. (II) The focal length of a diverging lens is -20 cm. Locate the object, given that the image is (a) virtual, erect, and 20% of the size of the object; (b) real, erect, and 150% of the size of the object.

13. (II) A converging lens ($f_1 = 10$ cm) is separated by 10 cm from a diverging lens ($f_2 = -15$ cm). An object is 20 cm in front of the first lens. Locate the final image.

14. (II) Two converging lenses of focal lengths 10 cm and 20 cm are separated by 15 cm. An object is located 12 cm in front of the 10-cm lens. Where is the final image?

15. (II) An object is placed 40 cm in front of a converging lens of focal length 8 cm. Another converging lens of focal length 12 cm is placed 20 cm behind the first lens. Find the position and transverse magnification of the final image. Draw a principal ray diagram.

16. (I) Two thin lenses of focal lengths f_1 and f_2 are in contact. Show that their effective focal length is $f \approx f_1 f_2 / (f_1 + f_2)$.

17. (II) A converging lens of focal length $f_1 = 10$ cm is placed in contact with a diverging lens of focal length f_2. Given that the focal length of the combination is 14 cm, find f_2.

36.2 and 36.3 Simple Magnifier and Microscope

18. (I) A gemstone is located 5.7 cm from a hand magnifier of focal length 6 cm. Find: (a) the angular magnification; (b) the location of the image.

19. (I) A detail on a stamp is 1 mm wide. A converging lens of focal length 4 cm is used to produce a virtual image 40 cm from the lens (which is close to the eye). Find: (a) the size of the image produced by the lens; (b) the angular magnification.

20. (II) (a) Show that if the image in a simple magnifier is at the normal near point (0.25 m), then Eq. 36.4 becomes

$$M = 1 + \frac{0.25}{f}$$

where f is in meters. Assume the lens is close to the eye. (b) What is the maximum focal length that could produce an angular magnification of $2.4\times$ for a normal eye?

21. (II) The focal length of a hand magnifier is 10 cm. (a) Where should an object be placed for the maximum angular magnification (assume a normal eye)? (b) For the condition described in (a), if the object size is 2 mm, what is the image size?

22. (I) The angular magnification of a microscope is $400\times$ when the final image is at infinity. The optical tube length is 16 cm and the focal length of the objective is 5 mm. What is the focal length of the eyepiece?

23. (II) The focal length of the objective of a microscope is 8 mm and that of the eyepiece is 3 cm. The distance between the lenses is 17.5 cm. Find the angular magnification if the final image is at 40 cm from the eyepiece.

24. (II) In a microscope the focal lengths of the objective and eyepiece are 6 mm and 2.4 cm respectively. The object is located 6.25 mm from the objective. Find the angular magnification if the final image is at infinity.

36.4 Telescopes

25. (I) The objective of an astronomical telescope has a 60 cm focal length. The distance between the lenses is 65 cm. What is the angular magnification when the instrument is adjusted for a relaxed normal eye?

26. (I) A reflecting telescope has a mirror of focal length 180 cm and an eyepiece whose focal length is 5 cm. What is the angular magnification when the final image is at infinity?

27. (I) When viewed with a relaxed eye, the angular magnification of a Galilean telescope is $8\times$. What is the focal length of the eyepiece if the focal length of the objective is 36 cm?

28. (I) An astronomical telescope has an objective of focal length 5 m and an eyepiece of focal length 10 cm. What is the angular magnification if the final image is (a) at infinity, or (b) 40 cm from the eyepiece?

29. (I) The lenses in an astronomical telescope are separated by 65 cm. If the final image is at infinity and has an angular magnification of $25\times$, find the focal lengths of the lenses.

30. (I) A Galilean telescope is 15 cm long and has an objective of focal length 20 cm. If the final image is at infinity, what is its angular magnification?

31. (I) The 200-in. (5.1-m) diameter mirror of the Mt. Palomar telescope has a focal length of 16.8 m. If the image is examined with an eyepiece of focal length 3.5 cm, what is the angular magnification if the final image is at infinity?

32. (II) An astronomical telescope used to view the moon has an objective of focal length 1.8 m and an eyepiece of focal length 11 cm. What is the angular magnification if the final image is 40 cm from the eye?

33. (II) A Galilean telescope consists of a converging lens of focal length 24 cm and a diverging lens of focal length −8 cm separated by 16 cm. The object is 12 m away. (a) Where is the final image? (b) What should be the separation between the lenses for the final image to be at infinity?

36.5 The Eye

34. (I) Where is the near point of a person whose eyeglass prescription is +2.8 D?

35. (I) A person can focus objects clearly only between 15 cm and 40 cm. What eyeglasses would you prescribe?

36. (I) A person can focus objects clearly only between 40 cm and 4.0 m. What eyeglasses would you prescribe?

37. (I) What eyeglasses would you prescribe for persons with the following conditions: (a) a near point of 34 cm; (b) a far point of 34 cm?

38. (II) An unaided eye has a far point at 2 m. If the near point is at 28 cm with eyeglasses, where is it without the eyeglasses?

39. (II) A person requires a lens of +1.5 D in order to read the paper at 25 cm. A few years later, the paper must be held at 40 cm with the same glasses. What is the new prescription required for normal reading vision?

40. (II) A nearsighted person has eyeglasses with a −2 D prescription. (a) Where is the far point without eyeglasses? (b) If the near point is at 20 cm without eyeglasses, where is it when they are put on?

36.6 and 36.7 Spherical Boundaries; Lens Maker's Formula

41. (I) The curved surface of a thin plano-convex lens ($n = 1.5$) has a 12 cm radius of curvature. Locate the image of an object in air at infinity given that the surface that faces the object is (a) curved; (b) flat.

42. (I) A thin glass lens ($n = 1.5$) has a convex surface of radius of curvature 12 cm. What should be the radius of the other surface so that the focal length is (a) +16 cm; (b) −40 cm?

43. (II) Locate the image of a small object produced by a glass sphere ($n = 1.5$) of radius 4 cm given that the object is located in air at (a) infinity; (b) 20 cm from the center of the sphere.

44. (II) A rod of glass ($n = 1.5$) of length $2R = 16$ cm has a convex surface with radius of curvature $R = 8$ cm and a flat surface (see Fig. 36.35). Locate the final image of a small object placed at the following distances from the convex surface: (a) 24 cm; (b) 6 cm.

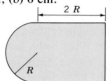

FIGURE 36.35 Exercise 44.

45. (II) A thin biconvex lens (as in Fig. 36.1a) is made of glass ($n = 1.5$) with surfaces of radii of curvature 12 cm and 16 cm. An object is located 20 cm from the lens. What are the position and linear magnification of the image?

46. (II) A thin biconcave lens (as in Fig. 36.1b) made of glass ($n = 1.5$) has two surfaces with radii of curvature 12 cm and 16 cm. An object is located 20 cm from the lens. What are the position and linear magnification of the image?

PROBLEMS

1. (I) A converging lens ($f = 4$ cm) is 12 cm in front of a second converging lens ($f = 7$ cm). Locate the final image and the transverse magnification for the following object distances from the first lens (a) 5 cm, (b) 12 cm. Draw a ray diagram for each case.

2. (I) A converging lens ($f = 10$ cm) is 30 cm in front of a diverging lens ($f = -5$ cm). Locate the final image and transverse magnification when the object distance from the first lens is 20 cm. Draw a ray diagram.

3. (I) An object of height 2 cm is 20 cm from a converging lens ($f = 10$ cm), which is located at a distance 12 cm in front of a diverging lens ($f = -15$ cm). Find the location of the final image and its transverse magnification. Draw a ray diagram.

4. (I) A diverging lens of focal length −15 cm is 12 cm in front of a converging lens of focal length 14 cm. An object is located 25 cm in front of the diverging lens. (a) Locate the final image. (b) What is the transverse magnification of the final image? Draw a ray diagram.

5. (II) A point source and a screen are a fixed distance D apart. There is a converging lens of focal length f between them. (a) Show that there are two positions of the lens for which a clear image is produced. (b) Show that the distance between the two possible positions of the lens is given by $d = \sqrt{D(D - 4f)}$.

6. (I) The Newtonian form of the lens equation is

$$xx' = f^2$$

where x and x' are the distances of the object and image from the first and second focal points. Prove this relation.

7. (II) An astronomical telescope is used to view an object of size 4 cm at a distance of 20 m. The focal lengths of the objective and the eyepiece are 80 cm and 5 cm, respectively. The final image is 25 cm from the eyepiece. (a) What is the size of the final image? (b) What is the angular magnification? (In Eq. 36.9 replace f_O by q_O. Draw a ray diagram to see why.)

8. (I) A point source is 15 cm from a converging lens of focal length 10 cm. A plane mirror is 10 cm behind the lens. Locate the final image.

9. (I) A hemispherical block of glass ($n = 1.5$) of radius 3 cm has a circular spot at its center (see Fig. 36.36). When viewed vertically from above, where is the image of the spot?

FIGURE 36.36 Problem 9.

10. (I) You are given a converging lens of focal length f. How is it possible to double the width of a parallel beam by using a second lens that is (a) converging, or (b) diverging? Specify

the focal length of the second lens and the distance between the lenses. Draw ray diagrams.

11. (II) A thin lens made of a material with a refractive index n_1 is surrounded by a medium of refractive index n_2. Show that the focal length f is given by

$$\frac{1}{f} = \left(\frac{n_1}{n_2} - 1\right)\left(\frac{1}{R_1} - \frac{1}{R_2}\right)$$

where R_1 and R_2 are the radii of the curvature of the surfaces.

12. (I) For a certain type of glass, the refractive indices for red and blue light are $n_B = 1.62$ and $n_R = 1.58$. What is the difference in focal lengths for these colors in a symmetrical converging lens with surfaces of radius of curvature 10 cm.

13. (I) The f *number* on a camera lens is the ratio of the focal length to the diameter of the aperture. The common values are: $f/1.4$, 2.0, 2.8, 4.0, 5.6, 8, 11, 16. By what factor does the amount of light passing through the lens change (a) from $f2.0$ to $f2.8$, and (b) from $f5.6$ to $f8$?

14. (I) An air-filled lens has thin plastic walls with radii of curvature 12 cm and 16 cm. What is the focal length of the "air lens" in water ($n = 1.33$). Ignore the plastic. (See Problem 11.)

CHAPTER 37

Wave Optics (I)

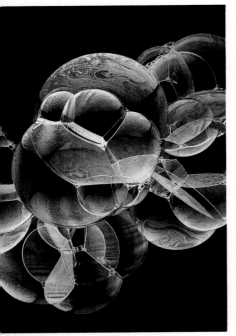

The colors in soap bubbles are produced by the interference of light.

Major Points

1. The basic features of **interference** and **diffraction.**
2. **Young's double-slit** experiment demonstrates the wave nature of light.
3. To exhibit an interference pattern, two waves must be **coherent.**
4. Interference in **thin films.**
5. The **Michelson interferometer** uses interference fringes to measure distances with great precision.

Although in the last two chapters we ignored the wave nature of light, there are many optical phenomena in which this aspect must be taken into account. First let us review the basic features of **interference,** introduced in Chapter 17. Figure 37.1 shows two pulses of a rope approaching each other. They have the same shape and amplitude. When a crest and a trough overlap, they momentarily cancel to produce zero displacement, as in Fig. 37.1a. This is called *destructive* interference; it occurs because the medium is simultaneously under the influence of two equal but opposite disturbances. When two crests overlap, as in Fig. 37.2b, they reinforce each other so that the amplitude of the resultant is double that of either pulse. This is *constructive* interference.

Sound waves always interfere when they are superposed because they are characterized by a scalar wave function (the pressure fluctuations). On the other hand, the displacement of a rope and the electric field in an electromagnetic wave are examples of vector wave functions. In order for interference to occur for such waves, the oscillations must lie along the same line. For example, suppose that a

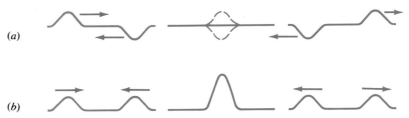

(a)

(b)

FIGURE 37.1 (a) Destructive interference between two pulses. (b) Constructive interference between two pulses.

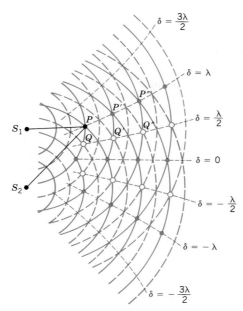

FIGURE 37.2 Circular wavefronts emitted by two sources in phase. The solid arcs are crests and the dashed arcs are troughs. The full circles denote points of constructive interference, where the path difference is $\delta = m\lambda$. The open circles denote points of destructive interference, where $\delta = (m + 1/2)\lambda$.

rope lies along the (horizontal) x axis. Vertical displacements along the y axis cannot interfere destructively with horizontal displacements along the z axis.

We can hear the sound of a distant bell or a rifle shot even though there may be a hill that prevents us from seeing the source. Similarly, two people standing beside adjacent walls near the corner of a building can talk to each other even though there is an obstruction between them. Such phenomena demonstrate that, in general, waves do not travel in straight lines. The bending of rays or wavefronts at the edge of an opening, or an obstruction, is called **diffraction.** Diffraction occurs whenever the physical limits on the length of a wavefront are either decreased or increased. For example, a barrier with an opening allows only part of the incoming wavefronts to continue and spread beyond the barrier (see Fig. 35.1). On the other hand, sound waves traveling down a pipe spread from an open end because the restriction has been removed.

37.1 INTERFERENCE

A ripple tank allows us to extend our study of interference to two dimensions. In Fig. 37.2, two plungers, S_1 and S_2, vibrate in phase on the surface of the water. Each produces expanding circular wavefronts of equal amplitude. The solid fronts are the crests, and the dashed fronts are the troughs. For simplicity we assume that the amplitude of the waves does not diminish with distance. When a crest from one source meets a trough from the other, the surface of the water has zero displacement. Such points of permanently destructive interference are indicated by open circles in Fig. 37.2. At other points, two crests or two troughs overlap to produce oscillations with twice the amplitude of either wave. These points of permanently constructive interference are indicated by solid circles in the diagram.

The positions of the points of constructive or destructive interference are easily obtained. The difference in the distances from S_1 and S_2 to a given point is called the **path difference,** δ:

$$\delta = r_2 - r_1 \qquad\qquad (37.1)$$

At all points along the central line, the path difference is zero since the waves have traveled equal distances. The waves are in phase and therefore interfere constructively. At a point such as P, where a crest meets a crest, the wave from S_2 has traveled an extra distance of one wavelength λ; hence $\delta = S_2P - S_1P = \lambda$. The waves are also in phase at this point. The same path difference applies to points P' and P''. (These and successive points trace a curve that has the shape of a hyperbola.) In general, constructive interference will occur wherever the path difference is an integer multiple of whole wavelengths:

(Maxima) $\qquad\qquad \delta = m\lambda; \qquad\qquad m = 0, \pm1, \pm2, \ldots \quad (37.2)$

At a point such as Q, where a crest meets a trough, one wave has traveled one half wavelength farther than the other, that is, $\delta = S_2Q - S_1Q = \lambda/2$. The waves are 180° out of phase and interfere destructively. The same path difference applies to points Q' and Q''. They trace a (hyperbolic) curve along which the water is calm. In general, there is destructive interference wherever the path difference is an odd number of half wavelengths:

(Minima) $\qquad\qquad \delta = \left(m + \dfrac{1}{2}\right)\lambda; \qquad m = 0, \pm1, \pm2, \ldots \quad (37.3)$

Note that although the interference pattern of maxima and minima is stationary, the waves are continually moving forward (see Fig. 37.3).

If the sources are loudspeakers driven by the same signal generator, the interference pattern at low frequencies can easily be detected by ear. As one walks parallel to the line joining the speakers, the loudness will rise and fall. (This is best done outdoors to avoid reflections off walls.) The spotty acoustics in some concert halls also arise from interference effects between the direct sound and sound reflected off the walls and ceiling. Similarly, interference between radio signals that have traveled different paths can lead to poor reception. One signal may have traveled to the receiving antenna in a straight line, whereas another may have been reflected off a building or a nearby airplane. The path difference may be just that required for destructive interference.

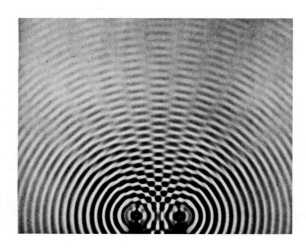

FIGURE 37.3 An interference pattern on a ripple tank. Notice the lines along which the water is calm (the locus of points of destructive interference).

EXAMPLE 37.1: Two speakers S_1 and S_2 are separated by 6 m and emit sound waves in phase. Point P in Fig. 37.4 is 8 m from S_1. What is the minimum frequency at which the intensity at P is (a) a minimum, (b) a maximum? Take the speed of sound to be 340 m/s.

Solution: Finding the minimum frequencies is equivalent to finding the maximum wavelengths that satisfy the conditions for constructive or destructive interference. The distance from S_2 to P is found from Pythagoras' theorem: $S_2P = (6^2 + 8^2)^{1/2} = 10$ m. The path difference from the waves from S_1 and S_2 to P is $\delta = S_2P - S_1P = 10 \text{ m} - 8 \text{ m} = 2$ m and is a fixed quantity. (a) From Eq. 37.3 we see that the maximum wavelength is associated with $m = 0$ and $\lambda = 2\delta = 4$m. The corresponding frequency is $f = v/\lambda = (340 \text{ m/s})/(4 \text{ m}) = 85$ Hz. (b) In Eq. 37.2 the value $m = 0$ is not acceptable. (Why?) Hence $m = 1$ for the maximum wavelength, so $\lambda = \delta = 2$ m. The associated frequency is $f = v/\lambda = (340 \text{ m/s})/(2 \text{ m}) = 170$ Hz.

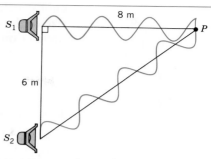

FIGURE 37.4 Two speakers emit sound waves in phase. The path difference to point P is fixed. The waves may interfere constructively or destructively depending on the wavelength.

37.2   DIFFRACTION

Diffraction phenomena are also easily demonstrated with water waves. The extent to which diffraction modifies the rectilinear propagation of waves depends on the relative size of the wavelength and the opening (or obstruction). When the width, d, of the opening or obstruction is much larger than the wavelength ($d \gg \lambda$), as in Fig. 37.5, the obstructed parts of the wavefronts are missing, but the remaining parts continue in the original direction. This is the condition assumed in geometrical optics. As d gets smaller, the waves start to spread into the regions just behind the barrier, as in Fig. 37.6. The wavefronts in the "shadow" region are arcs. When the size of the opening becomes comparable to the wavelength ($d \approx \lambda$) as in Fig. 37.7, the diffracted wavefronts are circular. For (three-dimensional) plane waves, the diffracted wavefronts are spherical. Although a small obstruction also produces diffraction, this is not easily observed.

Diffraction is easily explained with Huygens' construction (Section 35.2). Each point on the approaching wavefronts acts as a source of secondary wavelets. When the fronts reach the aperture or obstruction, only the wavelets from the

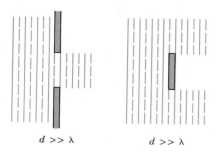

$d \gg \lambda$ $d \gg \lambda$

FIGURE 37.5 Straight wavefronts pass through an aperture. When the size, d, of an aperture is much larger than the wavelength, λ, the wavefronts remain straight.

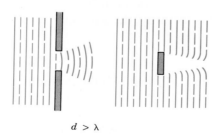

$d > \lambda$

FIGURE 37.6 When the size of the aperture is comparable to the wavelength, the wavefronts spread into the region behind the opening or obstacle. This is called diffraction.

$d \approx \lambda$

FIGURE 37.7 When the size of an aperture is approximately equal to the wavelength, the diffracted wavefronts are nearly circular (or spherical).

FIGURE 37.8 Diffraction is easily explained with Huygens' principle. Each point of the opening acts as a source of secondary wavelets. When the opening is large, the envelope of the wavelets is nearly straight.

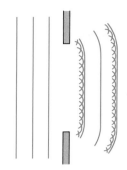

FIGURE 37.9 Thomas Young (1773–1829).

FIGURE 37.11 A set of fringes produced by double slits.

unobstructed region can contribute to the wavefronts on the right side. When the size of the opening is comparable to the wavelength, there is essentially just one wavelet on the right, as shown in Fig. 37.7. When d is much larger than λ, the envelope of the large number of wavelets produces almost straight (or plane) wavefronts, as in Fig. 37.8.

37.3 YOUNG'S EXPERIMENT

The wave nature of light was demonstrated in an historic experiment, performed in 1802, by Thomas Young (Fig. 37.9). Since it is not possible to get two ordinary sources to emit light in phase, he used a pinhole to let sunlight into a room and placed two closely spaced narrow slits at some distance from the pinhole. In such an arrangement, the cylindrical waves emerging from the slits are in phase since they are always derived from the same (essentially) plane wavefronts, as shown in Fig. 37.10. On a screen he observed bright and dark bands called interference *fringes* (Fig. 37.11). Such fringes cannot be explained by the particle model of light, which was advocated by Newton and generally accepted during the 18th century.

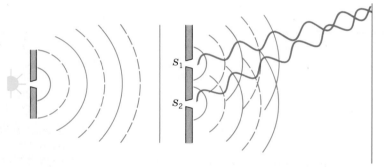

FIGURE 37.10 In Young's experiment a small slit is used to select light from a small region of a source. The expanding wavefronts become nearly plane at the double slits S_1 and S_2. Although the phase of the source may change, the changes take place simultaneously at both slits, which therefore always remain in phase.

In order to derive an expression for the position of the fringes, we assume the light has a single wavelength, λ, and that the separation between the slits is d. An arbitrary point P on the screen in Fig. 37.12 may be either bright or dark, depending on the path difference between the waves from the slits, S_1 and S_2. If the screen is far away, the outgoing rays are almost parallel, and the path difference, $\delta = S_2A$, is

$$\delta = r_2 - r_1 \approx d \sin \theta$$

From Eqs. 37.2 and 37.3 the positions of the bright and dark fringes are therefore given by

(Maxima)
$$d \sin \theta = m\lambda \tag{37.4}$$
$$m = 0, \pm 1, \pm 2, \ldots$$

(Minima)
$$d \sin \theta = \left(m + \frac{1}{2}\right) \lambda \tag{37.5}$$

Note that $\delta = d \sin \theta$ is valid only when the rays from the slits may be treated as being very nearly parallel. The integer m is referred to as the *order* of the fringe.

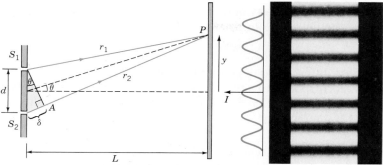

FIGURE 37.12 If the screen is very far from the slits, the path difference to point P is approximately $\delta \approx d \sin \theta$, where d is the slit separation.

EXAMPLE 37.2: Calculate the spacing between the bright fringes of yellow light of wavelength 600 nm. The slit separation is 0.8 mm, and the screen is 2 m from the slits.

Solution: Since the distance L to the screen is large compared with d, we can write $\sin \theta \approx \tan \theta = y_m/L$, where y_m is the position of the mth fringe relative to the center of the pattern. Thus the position of the bright fringes is given by

$$y_m = \frac{m \lambda L}{d}$$

The spacing between these fringes is

$$\Delta y = y_{m+1} - y_m = \frac{\lambda L}{d}$$

$$= \frac{(6 \times 10^{-7} \text{ m})(2\text{m})}{8 \times 10^{-4} \text{ m}}$$

$$= 1.5 \times 10^{-3} \text{ m} = 1.5 \text{ mm}$$

The spacing between the dark fringes is the same.

EXERCISE 1. A pair of slits separated by 0.8 mm are illuminated with light containing two wavelengths, 450 nm and 680 nm. What is the separation between the 6th-order bright fringes on a screen 3.2 m from the slits?

Under what conditions can an interference pattern be observed? In the previous section we required the sources to be in phase. Actually, they need only to have a constant phase relationship. The position of the central peak ($m = 0$) of the interference pattern depends on the phase difference between the sources. Also, the frequencies of the sources must be the same, otherwise the phase relationship at a given point will fluctuate in time and no steady interference will be seen.

Sources that emit waves of the same frequency and have a constant phase relationship are said to be **coherent.**

Coherence is easily achieved for sound waves or for radio waves by connecting the speakers, or transmitters, to the same oscillator. In Young's arrangement, light reaching both slits comes from a single (point) source. Any change in phase at the source occurs simultaneously at both slits—which therefore remain in phase. In Section 37.7 we discuss in more detail the coherence of light waves.

37.4   INTENSITY OF THE DOUBLE-SLIT PATTERN

In the earlier discussion of two-source interference, only the positions of the minima and maxima were found. We now obtain an expression for the distribution of intensity of two coherent sources that are in phase. The approach is valid for any type of wave, but our primary interest here is in light. The wave function in this case is the electric field. We assume that the slits are narrow enough for diffraction to spread light from each slit uniformly over the screen. Thus, the amplitudes of the fields at any point on the screen will be equal. At a given point on the screen the fields due to S_1 and S_2 are

$$E_1 = E_0 \sin(\omega t); \qquad E_2 = E_0 \sin(\omega t + \phi)$$

where the phase difference ϕ depends on the path difference $\delta = r_2 - r_1$. Since one wavelength λ corresponds to a phase change of 2π, a distance δ corresponds to a phase change ϕ given by $\phi/2\pi = \delta/\lambda$. If the screen is far from the slits, $\delta \approx d \sin \theta$ (see Fig. 37.12), therefore

$$\phi = \frac{2\pi\delta}{\lambda} = \frac{2\pi d \sin \theta}{\lambda} \tag{37.6}$$

The resultant field is found from the principle of superposition:

$$E = E_1 + E_2 = E_0 \sin(\omega t) + E_0 \sin(\omega t + \phi) \tag{37.7}$$

By using the trigonometric identity $\sin A + \sin B = \sin[(A + B)/2] \cos[(A - B)/2]$, we obtain

$$E = 2E_0 \cos\left(\frac{\phi}{2}\right) \sin\left(\omega t + \frac{\phi}{2}\right) \tag{37.8}$$

The amplitude of the resultant wave is $2E_0 \cos(\phi/2)$. The intensity of a wave is proportional to the square of the amplitude, so from Eq. 37.8 we have

$$I = 4I_0 \cos^2\left(\frac{\phi}{2}\right) \tag{37.9}$$

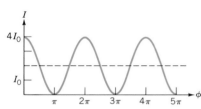

FIGURE 37.13 The intensity in a double-slit pattern as a function of the phase difference, ϕ. It is assumed that the slits are so fine that diffraction due to each slit would illuminate the screen uniformly.

where $I_0 \propto E_0^2$ is the intensity due to a *single* source. This function is plotted in Fig. 37.13. The maxima occur when $\phi = 0, 2\pi, 4\pi, \ldots = 2m\pi$. At these points $I = 4I_0$; that is, the intensity is four times that of a single source. The minima ($I = 0$) occur when $\phi = \pi, 3\pi, 5\pi, \ldots = (2m + 1)\pi$.

(Maxima) $\phi = 2m\pi$, $d \sin \theta = m\lambda$

$$m = 0, \pm 1, \pm 2, \ldots$$

(Minima) $\phi = (2m + 1)\pi$, $d \sin \theta = (m + 1/2)\lambda$

These agree with Eqs. 37.4 and 37.5.

EXERCISE 2. Calculate the intensity of the double-slit pattern for (a) $\phi = \pi/4$ and (b) $\phi = \pi/2$.

37.5   THIN FILMS

The colors in soap bubbles, in oil patches on a road, and in peacock feathers are due to the interference of light waves that have been reflected from the two surfaces of a thin film. Before we consider thin film interference, we must discuss an aspect of the reflection of light. Recall from Section 17.2 that when a pulse traveling in a light string encounters a heavy string, the reflected pulse suffers a 180° change of phase. In the case of light waves, the factor that determines whether or not a phase inversion occurs at a boundary is the refractive index. *When light encounters a medium of higher refractive index, the reflected wave suffers a phase change of π.*

Figure 37.14 shows a thin film of uniform thickness t (a few wavelengths) and refractive index n, with air ($n = 1$) on either side. To simplify our treatment we

Phase change on reflection

ignore refraction and consider only waves that travel essentially perpendicular to the film surfaces. (The angles in the figure are exaggerated.) Since the speed of light in the material is $v = c/n$, the wavelength in the film is $\lambda_F = \lambda/n$, where λ is the wavelength in air. When a wave arrives at point A, it is partly reflected and partly transmitted. At B and C there are also reflected and transmitted waves. We ignore the remaining weaker rays and consider just the rays moving upward from A and C—which have approximately the same amplitude. These rays are coherent since they are derived from the same incident ray. In Fig. 37.14 a phase change of π occurs for the reflection at A, but not at B. Thus the reflection process contributes π to the phase difference between the two rays. This is equivalent to a shift of half a wavelength.

The path difference between the rays is $ABC = 2t$, since we assume the rays are perpendicular to the surfaces. The two rays are already half a wavelength out of step because of the phase change on reflection at A. Thus when $2t = \lambda_F/2$, $3\lambda_F/2$, $5\lambda_F/2$, and so on, the rays will again be in phase and interfere constructively. In general, the film will appear brighter than normal in the reflected light when $2t = (m + 1/2)\lambda_F$. If $2t = m\lambda_F$, the rays interfere destructively and the film appears dark. These two conditions are valid only for this particular situation (a film with media of lower refractive index on both sides). Each case must be worked out from scratch by taking into account the presence of phase changes on reflection.

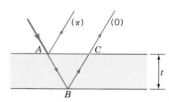

FIGURE 37.14 Light reflected at the top and bottom surfaces of a thin film. At the top surface there is a phase change of π. If the rays are essentially normal to the film, the path difference is $2t$.

The Nature of the Color in Thin Films

Suppose white light is incident normally on a uniform thin film, as shown in Fig. 37.15a. For simplicity, we assume that the intensities of all the colors are the same, and depict this fact by bars of equal length. Suppose that the thickness of the film is such that the yellow-green component ($\sim$550 nm) suffers complete destructive interference in the reflected light, as in Fig. 37.15b. The other wavelengths entering the film undergo various phase changes as they travel through the film. When they recombine with the waves reflected at the top surface, the phase difference between each pair of waves will be $\phi = 2\pi\delta/\lambda_F + \pi$, where $\delta = 2t$ and the term π takes into account the phase inversion at the top surface. The intensities of the other wavelengths may be found from Eq. 37.9 for two-source interference. We would obtain something like Fig. 37.15b.

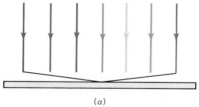
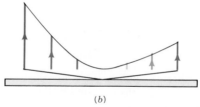
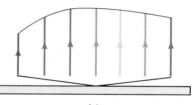

FIGURE 37.15 (a) White light incident at a point on a film is taken to be a mixture in which all colors are equally represented. (b) The thickness of the film is such that there is destructive interference for the green. (c) The film thickness is such that the green is enhanced in reflection.

When the green is missing, the apparent color of the film is magenta; when red is missing, the apparent color is cyan; and if blue is missing, the apparent color is yellow. When constructive interference enhances yellow-green in the reflected light, the reflected intensities of the adjacent wavelengths are not significantly weaker, as depicted in Fig. 37.15c. Thus, the colors seen in thin films are not the pure (monochromatic) colors of a prism spectrum. They are called *subtractive* colors, since the apparent color is determined primarily by what is *missing* from the reflected light.

Lens Coatings

When light is incident normally on the boundary between air ($n = 1.0$) and glass ($n = 1.5$), about 4% of the energy is reflected and 96% is transmitted. Thus, a camera with 6 lenses has 12 air–glass interfaces, which means that only $(0.96)^{12} = 0.61$ or 61% of the incident energy is transmitted. The loss due to reflection is minimized by coating each lens surface with a thin film. The thickness of the film is chosen so that in the reflected light there is destructive interference for yellow (550 nm), which is in the middle of the visible spectrum. Cancellation in the reflected light adds to the transmitted light. If the refractive index of the film is equal to the geometric mean of the indices for air and glass, that is, $n_F = (n_{air} n_g)^{1/2}$, the amplitudes of the two reflected waves in Fig. 37.14 are equal. This would ensure complete cancellation at this wavelength. In practice MgF_2 ($n = 1.38$) is often used even though it does not meet this criterion; its durability is a more important consideration.

EXAMPLE 37.3: White light is incident normally on a lens ($n = 1.52$) that is coated with a film of MgF_2 ($n = 1.38$). For what minimum thickness of the film will yellow light of wavelength 580 nm (in air) be missing in the reflected light?

Solution: Since the index for glass is higher than that of the film, there is a phase change for the reflection at the film–glass interface, as shown in Fig. 37.16. There is no net phase difference introduced by the reflections. Thus, in this case the condition for destructive interference in the reflected light is

$$2t = \left(m + \frac{1}{2}\right)\frac{\lambda}{n}$$

Notice that this differs from the expression given in the previous section. The minimum thickness occurs when $m = 0$; thus

$$t_{min} = \frac{\lambda}{4n} = \frac{5.5 \times 10^{-7} \text{ m}}{(4)(1.38)} = 99.6 \text{ nm}$$

Although the condition for destructive interference applies only to one wavelength, the reflection for other wavelengths is also reduced. The combination of the red and violet reflected light gives such a coated lens a purple hue. The net effect of a single coating is to reduce the overall reflected energy of white

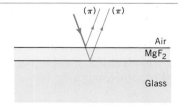

FIGURE 37.16 A thin film of MgF_2 ($n = 1.38$) on a glass lens ($n = 1.5$). Both reflected rays suffer a phase inversion. In order to produce destructive interference in the reflected light, the minimum thickness of the film is $\lambda_F/4$ (so that the path difference is one-half wavelength in the film).

light from 4% to about 1%. Multiple coatings of different thicknesses are also used to eliminate reflection at several wavelengths.

EXERCISE 3. If a coating has reduced the loss at each reflection to 1%, what is the percentage of light transmitted in the 6-element lens system?

EXERCISE 4. At what minimum thickness (other than zero) is there constructive interference in the reflected light?

Fringes of Equal Thickness

Soap bubbles and oil films on a road do not have uniform thickness. The thickness of the film at any given point determines whether the reflected light has a maximum or minimum intensity. When white light is used, each wavelength has its own fringe pattern. At a given point on the film, one wavelength may be enhanced and/or another wavelength suppressed. This is the source of the colors in soap bubbles and oil films on the road.

A wedge-shaped film of air may be produced by placing a sheet of paper or a hair between the ends of two glass plates, as in Fig. 37.17. With flat plates, one sees a series of bright and dark bands, each characteristic of a particular thickness (Fig. 37.18a). If the plates are not flat, the fringes are not straight; each is a locus of points with the same thickness. If one plate is known to be flat, the fringes display the irregularities of the other, as shown in Fig. 37.18b. The pattern shows where the plate needs to be polished for it to be made "optically flat."*

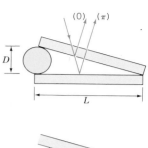

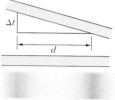

FIGURE 37.17 A wedge-shaped film of air formed by two plates separated at one end by a hair or fine wire.

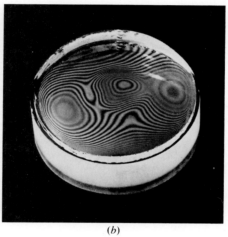

(a)   (b)

FIGURE 37.18 (a) When both plates forming an air film are flat, the fringes are straight and uniformly spaced. (b) If either plate is not "optically flat," the fringes are loci of points of equal thickness of the air film.

EXAMPLE 37.4: A wedge-shaped film of air is produced by placing a fine wire of diameter D between the ends of two flat glass plates of length $L = 20$ cm, as in Fig. 37.17. When the air film is illuminated with light of wavelength $\lambda = 550$ nm, there are 12 dark fringes per centimeter. Find D.

Solution: As indicated in Fig. 37.17, only one of the reflected rays suffers a phase inversion. At the thin end of the wedge, where the thickness is less than $\lambda/4$, the two rays interfere destructively. This region is dark in the reflected light. The condition for destructive interference in the reflected light is

$$2t = m\lambda \qquad m = 0, 1, 2, \ldots$$

The change in thickness between adjacent dark fringes is $\Delta t = \lambda/2$. The horizontal spacing between fringes $d = 1/12$ cm $= 8.3 \times 10^{-4}$ m. From Fig. 37.17 we see that $D/L = \Delta t/d$, so

$$D = \frac{\lambda L}{2d}$$

$$= \frac{(5.5 \times 10^{-7} \text{ m})(0.2 \text{ m})}{16.6 \times 10^{-4} \text{ m}}$$

Thus $D = 6.6 \times 10^{-5}$ m

---

* We do not need to consider the reflection at the top surface of the upper thick glass plate because the lateral separation between the two reflected rays in Fig. 37.14 is too great for both to enter the pupil of the eye.

Newton's Rings

When a lens with a large radius of curvature is placed on a flat plate, as in Fig. 37.19, a thin film of air is formed. When the film is illuminated with monochromatic light, circular fringes, called **Newton's rings,** can be seen with the unaided eye or with a low-power microscope (Fig.37.20). An important feature of Newton's rings is the dark central spot. Newton tried polishing the surfaces to get rid of it. The dark spot was also initially puzzling to Young. It implied that the light wave suffers a phase inversion on reflection at a medium with a higher refractive index. Young tested this idea by placing oil of sassafras between a lens of crown glass and a plate of flint glass. The refractive index of the oil is between the values for these two glasses. Since both reflections occur at a medium with a higher refractive index, they should both suffer a phase inversion and therefore be in phase. This is precisely what happened: The central spot became bright—and undoubtedly gave Young much satisfaction.

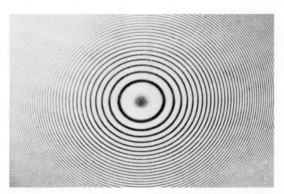

FIGURE 37.20 Newton's rings. The fringes are not equally spaced. Note the central dark spot.

EXAMPLE 37.5: In an experiment on Newton's rings the light has a wavelength of 600 nm. The lens has a refractive index of 1.5 and a radius of curvature of 2.5 m. Find the radius of the 5th bright fringe.

Solution: If R is the radius of curvature of the lens, then from Fig. 37.21 we see that $r^2 = R^2 - (R - t)^2$, where r is the radius of a fringe and t is the thickness of the film. Since t is very small, we may drop terms in t^2 to obtain

$$r^2 \approx 2Rt \qquad \text{(i)}$$

In order to find r, we must first find t. The condition for a bright fringe is

$$2t = \left(m + \frac{1}{2}\right)\lambda_F \qquad \text{(ii)}$$

We note that $n = 1$ for the air film (the index for the glass is

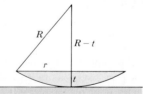

FIGURE 37.21 The radius r of a fringe may be related to the radius of curvature R of the lens and the thickness t of the air film.

irrelevant) and that $m = 4$ for the fifth bright fringe. Thus, from (ii)

$$t = \frac{(4.5)(6 \times 10^{-7})}{2} = 1.35 \times 10^{-6} \text{ m}$$

Substituting this into (i), we find $r = \sqrt{2Rt} = 2.6 \times 10^{-3}$ m.

37.6   MICHELSON INTERFEROMETER

An **interferometer** is a device that uses interference to make precise measurements of distances in terms of the wavelength of the light. Around 1880, A. A. Michelson (Fig. 37.22a) invented the elegant and versatile instrument shown in Fig. 37.22b. Light from an extended monochromatic source S is partly reflected and partly transmitted by a glass plate P that is "half-silvered" on one face. Approximately half the incident light goes to a mirror M_1, is reflected there, and then travels through P to reach the observer at O. The light from S that is transmitted by P is reflected by the mirror M_2 and then reaches the observer after being reflected by P. PM_1 and PM_2 are called the "arms" of the interferometer. The compensator plate C makes the distances traveled in glass the same for both beams. M_2' is the image of M_2 in the silvered surface of P.

The system is equivalent to an air film. If light travels slightly different distances to the mirrors, the resulting phase difference may lead to constructive or destructive interference. If the mirrors are exactly perpendicular, the "film" is of uniform thickness, and one observes circular fringes.* If the mirrors are not perpendicular, the film is a wedge and one sees straight fringes.

Michelson's interferometer is useful because one mirror may be moved on a finely threaded screw, so that the thickness of the film is continuously adjustable. If M_1 moves back by $\lambda/4$, a path difference of $\lambda/2$ is added for the light traveling in this arm. Thus, at any given point on the fringe pattern, a bright fringe is replaced by a dark fringe and vice versa. By counting the number of fringes that move past cross hairs in the telescope, one can determine the distance traveled by one mirror to within a fraction of the wavelength of light! Michelson measured the length of what was then the standard meter in terms of the wavelength of nearly monochromatic light from cesium. This later led to the modern definition of the standard meter in terms of the wavelength (see Section 1.4). An interferometer may also be used to determine the refractive index of a gas, as the next example shows.

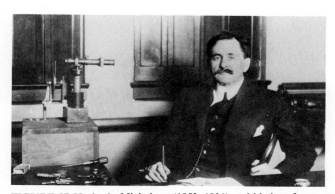
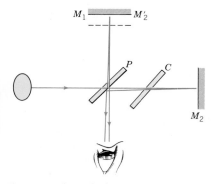

FIGURE 37.22 A. A. Michelson (1852–1931) and his interferometer. The system is equivalent to a thin film. If the mirrors are not exactly perpendicular, one observes the straight fringes of a wedge film (Fig. 37.18a).

---

* In calculating the path differences, one must take into account the fact that the light from the extended source reaches the film at various angles.

EXAMPLE 37.6: One arm of a Michelson interferometer contains a transparent cylinder of length $L = 1.5$ cm (Fig. 37.23). The cylinder is evacuated, and the cross hairs of the telescope are centered on a particular bright fringe with light of wavelength 600 nm (in vacuum). When a gas is introduced into it, fourteen fringes move past the cross hairs. What is the refractive index of the gas?

Solution: The light travels through the cylinder twice. The number of wavelengths within the distance $2L$ is $2L/\lambda_0$, where λ_0 is the wavelength in vacuum. When the gas is introduced, the wavelength changes to $\lambda = \lambda_0/n$, where n is the refractive index. The number of wavelengths in the same distance is $2L/\lambda = 2nL/\lambda_0$. A shift from one fringe to the next implies that the path difference has changed by one wavelength. Therefore,

$$\frac{2(n-1)L}{\lambda_0} = \Delta m$$

from which we find

$$n = \frac{\lambda_0 \Delta m}{2L} + 1$$

$$= \frac{(14)(6 \times 10^{-7}\ \text{m})}{0.03\ \text{m}} + 1 = 1.00028$$

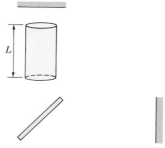

FIGURE 37.23 When a gas is introduced into a transparent cylinder in one arm, the shift in the fringes may be used to calculate the refractive index of the gas.

37.7 COHERENCE (Optional)

To gain a deeper understanding of coherence for light waves, we must consider the mechanism of light emission. An atom emits light when it makes a transition from an excited state to one of lower energy. This process typically takes about 10^{-8} s and is random in the sense that one cannot predict when a particular atom will radiate. Since the frequency of visible light is around 5×10^{14} Hz, a **wavetrain** of about 5×10^6 wavelengths (3 m) is emitted in this time.

Let us consider the interference between the wavetrains from two independent sources (Fig. 37.24a). At any point on a screen, the phase difference associated with the fixed path difference is constant. At any given instant there is a particular phase difference between the sources themselves, but it lasts only as long as the emission process, that is, 10^{-8} s. One interference pattern based on one value of ϕ will be replaced by a shifted pattern, based on another value of ϕ, 10^{-8} s later. For a collection of atoms, ϕ fluctuates randomly. Consequently, there is no fixed phase relationship and therefore no steady interference pattern. For this reason, different regions of a single extended source, such as a gas discharge tube or the incandescent wire in a light bulb, are also incoherent. The intensity at any point on the screen is the sum of the intensities due to the individual sources.

The size of the largest region of a source that still produces an interference pattern is an indication of its **spatial coherence.** Spatial incoherence arises from the random nature of the phases and directions of the emission events. An ordinary light source has poor spatial coherence. Therefore in order to produce an interference pattern, a pinhole has to be used to sample light from just a tiny region—which then acts approximately as a point source. An extended source at a great distance also acts

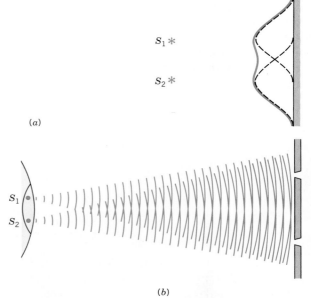

FIGURE 37.24 (a) With two independent sources, the intensity is simply the sum of the individual intensities. (b) Two separate points on an extended source act as independent sources and therefore do not maintain a constant phase relationship. However, at a distant screen the wavefronts from both points are nearly plane waves, which means that both slits stay in phase—even though this phase fluctuates.

like a point source. Suppose two slits are placed far enough (many wavelengths) from the source, as in Fig. 37.24b. The spherical wavefronts from the points S_1 and S_2 become nearly plane waves traveling in almost the same direction when they reach the slits. Wavetrains from different atoms reach the slits with various phases. However, even though the phase of the plane wave is changing rapidly, it is always the same for both slits. Therefore, the slits are always in phase. A laser (see the Special Topic for Ch. 40) has extraordinary spatial coherence. One can place double slits at opposite edges of the beam and still obtain an interference pattern. Even though the waves going through the two slits are emitted from different atoms, they are in phase.

Since the emission process for a given atom is of short duration, the wavetrains have a finite length, ℓ_c, called the **coherence length**. The **temporal coherence** of the waves is indicated by the **coherence time**, $\tau_c = \ell_c/c$. Figure 37.25 illustrates the difference between spatial and temporal coherence.

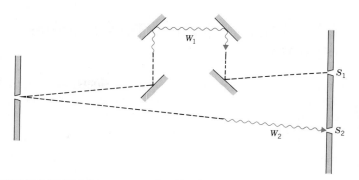

FIGURE 37.26 Mirrors are used to lengthen the path of the waves traveling to one slit. If the extra distance is greater than the coherence length, there will be no interference pattern.

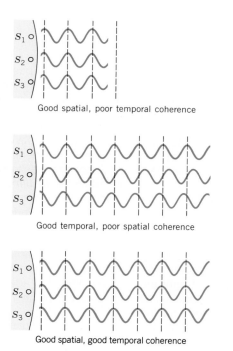

Good spatial, poor temporal coherence

Good temporal, poor spatial coherence

Good spatial, good temporal coherence

FIGURE 37.25 Good spatial (or lateral) coherence means that different points on an extended source are coherent. Good temporal (or longitudinal) coherence means that the wavetrains from each point source are long.

A steady interference pattern is produced only when a single wavetrain is divided into two parts that travel different distances and are then superposed. Suppose we use mirrors to lengthen the path of the waves heading for one of the slits, as in Fig. 37.26. If the extra distance is greater than ℓ_c, or the extra

time taken is greater than τ_c, there is no overlap between the two parts W_1 and W_2, and no interference is observed. The coherence length of light from an ordinary source, such as a sodium discharge lamp ($\lambda = 590$ nm) is about 3 mm. This is much less than the 3 m we mentioned for a single atom because of the random motion of the atoms and collisions between them. The best conventional sources are low-density discharges in cesium or potassium gas, for which ℓ_c is between 20 cm and 30 cm. The common helium–neon gas laser has a coherence length of about 20 cm. However, special lasers can have a coherence length of 30 km!

In the discussion of two *coherent* point sources in Section 37.4, we first used the principle of superposition to find the resultant wavefunction

(Coherent) $$E = E_1 + E_2$$

and then found the intensity from the amplitude of the resultant. Let us consider the intensity at the screen due to two *incoherent* sources. The phase difference ϕ at any given position on the screen is not constant but fluctuates randomly in time. Consequently, there is no steady interference pattern. Rewriting Eq. 37.9 in the form $I = 2I_0 (1 + \cos \phi)$, we note that the time average of $\cos \phi$ is zero and we are left with the first term, $I = 2I_0$. In general, the resultant intensity at any point on the screen is simply the sum of the intensities of the two sources acting independently:

(Incoherent) $$I = I_1 + I_2$$

Notice that in Fig. 37.13, the quantity $2I_0$ is the average value of the intensity taken over many fringes. The total energy arriving at the screen is the same for both types of sources, but it is redistributed by interference when the sources are coherent.

HISTORICAL NOTE: Two Theories of Light

The nature of light was the source of much debate during the 17th century. Descartes treated light as a stream of particles, or "corpuscles," to which the principles of mechanics could be applied. He considered the reflection of a beam of light to be analogous to the elastic collision of a tennis ball with a flat surface. To explain refraction when light enters glass from air he assumed that the component of the particle's velocity perpendicular to the boundary is greater in the glass. Newton supported the corpuscular hypothesis, but nevertheless thought the description of reflection was naive. He pointed out that a beam of light initially in glass is also partly reflected when it encounters a vacuum. What could the corpusles collide with?

Christiaan Huygens rejected the corpuscular hypothesis (see the introduction to Chapter 35). Instead, he proposed that light is a (longitudinal) pulse or disturbance in a medium called the "ether." His was not a wave theory in the modern sense: Huygens rejected the notion that light involves any form of periodicity (such as wavelength or frequency).

The phenomena of reflection and refraction of light could be accounted for in terms of either the wave theory or the corpuscular theory. The phenomena of interference and diffraction sharpened the debate. In 1665 Robert Hooke described the colors he saw in thin strips of mica and thin films of liquid placed between glass plates. By applying pressure to the plates with his hands, he discovered that the color in a given region depends on the thickness of the film. Hooke used a kind of "wave" theory to explain the colors qualitatively in terms of the interference of pulses reflected from the top and bottom surfaces of the film. Like Huygens, he did not associate any periodic behavior (frequency or wavelength) with the light pulses. He could not analyze the phenomenon any further because he did not know how to determine the thicknesses of such thin films. Newton did.

Newton investigated the colors in thin films between 1666 and 1672. In one arrangement, first used by Hooke, a lens with a long radius of curvature was placed on a flat plate to form a wedge-shaped film of air (Fig. 37.19). When the film is illuminated, a series of alternately bright and dark concentric rings are seen (Fig. 37.20). When he used different colors of a prism spectrum, he noted that the rings either expanded or contracted. He found that if the thickness t of the film corresponds to the first bright ring, then the other bright rings occur at $3t$, $5t$, $7t$, etc. The dark rings occur at $2t$, $4t$, $6t$, etc. From the fact that rings with red light were larger than those with blue, he inferred that red "corpuscles" of light were bigger than blue ones. The rings clearly showed that light involves a periodic phenomenon. However, neither Hooke nor Newton thought of associating the "bigness" of the corpuscles with wavelength.

An Italian Jesuit, F. M. Grimaldi, discovered another interesting phenomenon of light. In a work published posthumously in 1665, he described several experiments that showed that light does not travel in straight lines. He placed a thin opaque strip at some distance from a point source and found that the shadow was bordered by colored bands. When he replaced the strip with an aperature, he again found bands beyond the region of the geometrical shadow. The bending of light rays into the region of the geometrical shadow he called diffraction.

Even though the evidence of his own experiments (the periodicity in the rings) and those of Grimaldi pointed to the wave nature of light, Newton preferred to advocate the corpuscular theory. His main objection to the wave theory was that light appears to travel in straight lines, whereas waves, such as those in air and water, spread noticeably throughout the region behind an obstacle. He considered the diffraction of light to be too small an effect to warrant a wave theory. In his view, a light ray is *refracted* slightly as it passes near the surface of an object because the density of the ether is lower there. Newton "explained" the bending but completely failed to address the question of the colored bands. It is interesting to note that Huygens, who is often considered to be one of the original proponents of the wave theory of light, in which diffraction finds a clear explanation, did not even mention diffraction in his 1690 treatise on optics. His concern was to demonstrate that his "wave" theory could explain the *rectilinear* propagation of light! He too did not discuss the colors in thin films.

Since both Newton and Huygens chose to ignore experimental evidence with which they were uncomfortable, there was little hope for a wave theory of light. Newton acknowledged that the wave theory was not entirely unreasonable, but his stumbling block was that he did not realize how small the wavelengths of light are. Unfortunately, his advocacy of the corpuscular model stifled further investigation into the nature of light for over a century.

The first serious challenge to the corpuscular theory came from Thomas Young, who was the first to clearly enunciate the principle of superposition of waves. Young thought that the dark rings in Newton's experiment were produced by a process analogous to the phenomenon of beats: When two sound waves of nearly equal frequency are superposed, they can momentarily cancel to produce zero intensity. Like Huygens, he could not imagine how particles might display such behavior. Young used Newton's own data on the rings to calculate the wavelengths of visible light. He is best known for the historic double-slit experiment (Section 37.3), performed in 1802, that clearly demonstrated the wave nature of light.

SUMMARY

The conditions for constructive and destructive **interference** of waves from two sources may be expressed in terms of the path difference δ:

(Constructive) $\delta = m\lambda$

$$m = 0, \pm 1, \pm 2, \cdots$$

(Destructive) $\delta = \left(m + \dfrac{1}{2}\right)\lambda$

The **phase difference**, ϕ, associated with a **path difference**, δ, is

$$\frac{\phi}{2\pi} = \frac{\delta}{\lambda}$$

In Young's **double slit experiment**, $\delta = d \sin \theta$, so the positions of the fringes are determined by

(Maxima) $d \sin \theta = m\lambda$

$$m = 0, \pm 1, \pm 2, \cdots$$

(Minima) $d \sin \theta = \left(m + \dfrac{1}{2}\right)\lambda$

The **intensity** of the interference fringe pattern for the double slit is

$$I = 4I_0 \cos^2\!\left(\frac{\phi}{2}\right)$$

where I_0 is the intensity (assumed to be uniform across the screen) due to a single source.

The condition for constructive or destructive interference in **thin films** must be worked out each time with the following kept in mind:

1. Light undergoes a phase change of π when it is reflected at a medium of higher refractive index.

2. The wavelength λ within the medium of refractive index n is $\lambda = \lambda_0/n$, where λ_0 is the wavelength in air.

ANSWERS TO IN-CHAPTER EXERCISES

1. From Eq. (i) of Example 37.1 we see that

$$\Delta y_m = \frac{m\,\Delta\lambda\,L}{d}$$

$$= \frac{(6)(1.3 \times 10^{-7}\ \text{m})(3.2\ \text{m})}{(8 \times 10^{-4}\ \text{m})}$$

$$= 3.12\ \text{mm}.$$

2. We use Eq. 37.9.

(a) $I = 4I_0 \cos^2(\pi/8) = 3.41 I_0$
(b) $I = 4I_0 \cos^2(\pi/4) = 2 I_0$

3. The percentage of light transmitted is $(0.99)^{12} = 88.6\%$

4. In this case, the condition for constructive interference is

$$2t = \frac{m\lambda}{n}$$

The minimum thickness is $t = \lambda/2n = 199$ nm.

QUESTIONS

1. When a transmitter is obstructed from view by a mountain, it is possible to receive an AM radio signal but not an FM signal. Why?

2. Suppose a double-slit experiment was conducted under water. Would the pattern change? If so, how?

3. Why are interference effects more easily observed in *thin*

rather than in thick films? (Consider the lateral separation of the rays reflected from the two surfaces.)

4. Why is the spacing between Newton's rings not constant?

5. An oil film on water has a whitish perimeter where its thickness is much less than the wavelength of light in the film. What can you deduce about the oil from this observation?

6. True/false: In order for two waves to be coherent they must have the same (a) phase; (b) wavelength; (c) direction of propagation.

7. In a double-slit experiment a thin plate covers one slit and thereby introduces a 90° phase lag. What is the effect on the pattern on the screen?

8. Will two flashlight bulbs produce an interference pattern if they are sufficiently small? Explain.

9. As the upper region of a vertical soap film gets thinner, why does it appear dark in the reflected light (and clear in transmitted light), as shown in Fig. 37.27?

10. What is the purpose of the first slit in Young's double-slit arrangement?

11. Suppose one of the slits in a double-slit experiment were twice as wide as the other. How would the pattern be affected?

12. When light enters a different medium its wavelength

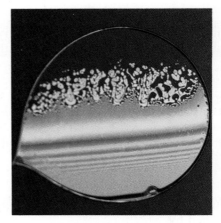

FIGURE 37.27 Question 9.

changes. Does its color also change? Explain why or why not.

13. Why does an overflying aircraft produce a disturbance in FM or TV reception?

EXERCISES

37.1 and 37.3 Interference and the Double-Slit Experiment

1. (I) In a double-slit experiment using light of wavelength 490 nm, the 6th-order bright fringe is 38 mm from the central fringe on a screen located 2.2 m from the slits. What is the slit separation?

2. (I) Two narrow slits separated by 0.4 mm are illuminated with light containing two wavelengths, 480 nm and 650 nm. What is the separation between the 2nd-order bright fringes on a screen 2.0 m from the slits?

3. (I) A double-slit pattern is observed on a screen 2 m from the slits. Given that the light is incident normally and has a wavelength of 450 nm, what is the minimum slit separation for a point 3.2 mm from the center to be (a) a minimum; or (b) a maximum?

4. (I) Light of wavelength 546 nm from a mercury source illuminates two slits separated by 0.32 mm. What is the distance between the 2nd- and 3rd-order dark fringes if the screen is placed 1.8 m from the slits?

5. (I) In a double-slit pattern, the distance between the fourth bright fringes on either side of the central maximum is 7 cm. If the slit separation is 0.2 mm and the distance to the screen is 3.0 m, what is the wavelength of the light?

6. (I) In a double-slit arrangement the 3rd-order bright fringe is 16 mm from the center on a screen 2 m from the slits. If

the wavelength is 590 nm, find (a) the slit separation, (b) the separation between like fringes.

7. (I) Two narrow slits are 0.2 mm apart. The 4th-order dark fringe is 0.7° from the central bright fringe. What is the wavelength of the light?

8. (I) The 10th-order bright fringe of wavelength 560 nm in a double-slit pattern overlaps the 9th-order dark fringe of another wavelength. Find the other wavelength.

9. (I) A pair of slits separated by 0.24 mm is illuminated by light containing two wavelengths, 480 nm and 560 nm. The pattern is viewed on a screen at a distance of 1.2 m from the slits. What is the first position relative to the central peak at which the maxima exactly overlap?

10. (I) A double slit is illuminated with yellow light (589.0 nm) emitted by sodium vapor. The eighth dark fringe is 6.5 mm from the central maximum. The screen is 1.2 m from the slits. What is the slit separation?

11. (II) Two line sources transmit microwaves of wavelength 3 cm in phase. How far apart should they be in order to produce a 10° angular separation between the first and second bright fringes on one side of the central peak?

12. (I) In a double-slit experiment there are 7 dark fringes per centimeter on a screen 2 m from the slits. What is the slit separation if $\lambda = 510$ nm?

13. (II) A thin glass plate placed in front of the upper slit in a

double-slit arrangement introduces a 270° phase lag. Assume light of wavelength 600 nm illuminates the slits, which are separated by 0.5 mm, and that the screen is 2.4 m from the slits. By how much, and in which direction, does the central fringe shift?

14. (II) A speaker, which emits a 200-Hz signal is 8 m from a microphone. They are equidistant from a wall. What is the minimum distance to the wall such that the sound that reaches the microphone directly, and that which is reflected off the wall, interfere constructively. Take the speed of sound to be 340 m/s. (There is no phase change on reflection.)

15. (II) Two speakers are 1 m apart and emit sound of frequency 1000 Hz in phase. A listener walks along a line that is parallel to the line joining the speakers and 8 m away from their midpoint. What is the separation between the central maximum and the first minimum in loudness? Take the speed of sound to be 340 m/s.

16. (II) Two speakers, S_1 and S_2, are a distance d apart, as shown in Fig. 37.28. They emit sound of frequency 95 Hz in phase. What is the minimum value of d for which the intensity is zero at each of the following points: (a) P; (b) Q? Assume the intensity does not decrease with distance from the speakers. Take the speed of sound to be 340 m/s.

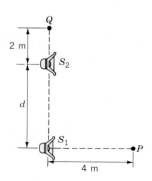

FIGURE 37.28 Exercises 16 to 18.

17. (II) For the arrangement of two speakers shown in Fig. 37.28, assume S_1 is π rad out of phase with S_2. The frequency is 500 Hz. What is the minimum value of d for which the intensity at P is a maximum? Take the speed of sound to be 340 m/s.

18. (II) For the arrangement of speakers shown in Fig. 37.28, take $d = 2$ m. What is the lowest frequency for which the intensity at P is (a) a maximum; and (b) a minimum? The speed of sound is 340 m/s and the speakers are in phase.

19. (II) Two point sources S_1 and S_2 separated by distance d, as in Fig. 37.29, emit waves at the same wavelength ($\lambda \ll d$). What is the condition on the distance x such that point P is a point of destructive interference given that (a) S_1 and S_2 are in phase; (b) S_1 and S_2 are π rad out of phase?

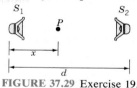

FIGURE 37.29 Exercise 19.

20. (II) Signals received by two microwave antennas that are 80 cm apart are fed to the same amplifier midway between the antennas. In order to receive a strong signal at a wavelength of 3 cm, the output from the upper antenna has to be delayed by 5 rad. In which direction is the source—assuming it is very far away.

21. (II) Plane waves are incident on a pair of slits at angle α, as shown in Fig. 37.30. (a) What is the path difference between the rays leaving at angle θ as shown? (The screen is far away.) (b) At what value of θ is the central peak located? (c) What is the minimum value of α for which the intensity at the center of the screen is a minimum?

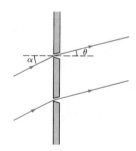

FIGURE 37.30 Exercise 21.

22. (II) A lens of focal length f is used to focus the light emerging from two slits onto a distant screen that is located in the focal plane of the lens, as in Fig. 37.31. Show that the positions of the minima are given by $y_m = (2m + 1)f\lambda/2d$.

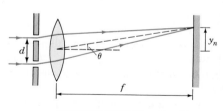

FIGURE 37.31 Exercise 22.

23. (II) A bright fringe in Young's double-slit experiment is 1.5 cm from the center of the pattern. The light has a wavelength of 600 nm and falls on a screen 1.4 m from the slits, whose separation is 0.4 mm. How many dark fringes are there between the center and the bright fringe at 1.5 cm?

24. (II) By using light reflected from a mirror, as in Fig. 37.32, a single thin line source may be used to produce fringes. Over what distance on the screen are fringes visible? Take $d = 0.4$ mm, $L = 3$ cm and $\lambda = 600$ nm.

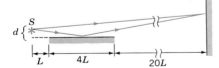

FIGURE 37.32 Exercise 24.

37.4 Intensity of Double-slit Pattern

25. (I) Show that the intensity of a double-slit pattern, Eq. 37.9, may be written in the form

$$I = 4I_0 \cos^2\left(\frac{\pi dy}{\lambda L}\right)$$

where y is the distance from the center of the pattern.

26. (I) At what distance from the center of a double-slit pattern does the intensity first fall to 50% of the central maximum? Assume the slit separation is 0.2 mm, the wavelength is 560 nm, and the screen is 1.6 m from the slits.

27. (I) Two narrow slits separated by 0.6 mm are illuminated by light of wavelength 480 nm. The pattern is observed on a screen 1.25 m from the slits. What is the intensity (relative to that of a single slit) of the pattern at a point 0.45 mm from the center?

28. (I) What would be the intensity (relative to that of a single slit) at the center of a double-slit pattern if a plastic sheet in front of one slit produced a phase shift of $\pi/2$ radians?

29. (II) Light of wavelength 627 nm illuminates two slits. What is the minimum path difference between the waves from the slits for the resultant intensity to fall to 25% of the central maximum?

30. (II) The intensity of the double-slit interference pattern is given by Eq. 37.9 for two slits in phase. Plot $I(\theta)$ vs θ in steps of 0.1 rad up to 1.5 rad. Take $d = 2\lambda$.

37.5 Thin Films

31. (I) A coating of MgF$_2$ ($n = 1.38$) on glass ($n = 1.6$) is 8.3×10^{-5} cm thick. If white light is incident normally, which visible wavelengths are missing in the reflected light?

32. (I) White light is incident normally on a film ($n = 1.4$) of thickness 90 nm deposited on glass ($n = 1.5$). What is the phase difference between rays reflected from the top and bottom surfaces for each of the following wavelengths: (a) 400 nm, (b) 550 nm, and (c) 700 nm?

33. (I) Light of wavelength 600 nm illuminates a glass ($n = 1.5$) wedge submerged in water ($n = 1.33$). If the distance between successive bright fringes is 2 mm, find (a) the change in thickness of the glass between these fringes, and (b) the angle of the wedge.

34. (I) Two glass plates of length 12 cm separated by a fine wire form an air wedge. Light of wavelength 480 nm is incident normally on the film. If there are 6 dark fringes per centimeter, estimate the radius of the wire.

35. (I) White light is incident normally on a uniform film of water ($n = 1.33$) lying on a glass plate ($n = 1.6$). Find the minimum possible thickness of the film given that in the reflected light: (a) 550 nm is enhanced, or (b) 550 nm is missing.

36. (II) A total of 42 dark fringes (excluding the central spot) are seen in a Newton's rings experiment. The wavelength used is 640 nm and the largest ring has a diameter of 2.2 cm. Find: (a) the thickness of the film at the last fringe; (b) the radius of curvature of the lens.

37. (II) When oil is introduced into the space between the lens and the plate in Newton's arrangement, the radius of the 8th dark ring decreases from 1.8 cm to 1.64 cm. What is the refractive index of the oil? Assume the refractive index of the glass is greater than that of the oil.

37.6 Michelson Interferometer

38. (I) When one mirror of a Michelson interferometer moves 0.08 mm, the pattern shifts by 240 dark fringes. What is the wavelength of the light?

39. (I) When a transparent sheet of thickness 2 μm is inserted into one arm of a Michelson–Morley interferometer, there is a fringe shift of 5 fringes. If the wavelength used is 600 nm, what is the refractive index of the sheet?

PROBLEMS

1. (I) White light is incident normally on a film ($n = 1.6$) in air. In the reflected light only 504 nm and 672 nm are missing. (a) What is the minimum possible thickness of the film? (b) Which wavelengths are most strongly reflected?

2. (I) White light is incident normally on a film of oil ($n = 1.2$) on water. In the reflected light, 544 nm is missing and 680

nm is particularly bright. (a) What is the minimum possible thickness of the film? (b) What other visible wavelengths experience either constructive or destructive interference?

3. (I) Light of wavelength 600 nm illuminates two narrow slits separated by 0.3 mm. When a plastic sheet is placed in front of the upper slit, it introduces a phase lag of 9 $\pi/2$ rad. If the pattern is observed on a screen 4 m away, by how much, and in which direction, does it shift?

4. (I) An oil film ($n = 1.2$) is on a slab of glass ($n = 1.5$). When white light is incident normally, the wavelengths 406 nm; and 522 nm are missing in the reflected light. (a) What is the minimum possible thickness of the film? (b) Which wavelengths are enhanced in the reflected light?

5. (I) White light is incident normally on a film of thickness 900 nm and refractive index 1.5 which is in air. In the reflected light which wavelengths (a) are missing, and (b) experience constructive interference?

6. (I) A wedge film of length $L = 12$ cm and height $h = 20$ μm has a refractive index of 1.5 (see Fig. 37.33). The wedge is in air and illuminated with light of wavelength 490 nm. How far from the thin end is the 20th bright fringe located?

FIGURE 37.33 Problem 6.

7. (I) In an experiment on Newton's rings the plano-convex lens has a radius of curvature of 3 m. Light of wavelength 600 nm is used. What is the number of bright fringes observed within a radius of 0.8 cm?

8. (I) At a certain point on one side of the central peak in a double-slit pattern the intensity is 50% of the central maximum when light of wavelength 400 nm is used. For what wavelength would the intensity at the same point be 64% of the maximum?

9. (I) Two point sources are 2 m apart and emit sound waves of 300 Hz in phase. A person walks along a line parallel to the line joining the sources at a distance of 10 m from the midpoint. At what distances from the central maximum is the loudness (a) a maximum, and (b) a minimum? Take the speed of sound to be 340 m/s.

10. (II) The yellow doublet emitted by sodium has wavelengths 589.0 nm and 589.6 nm. As one mirror is moved in the Michelson interferometer, the fringes periodically appear and disappear. (a) Why does this occur? (b) How far must the mirror move between successive points of maximum contrast between bright and dark?

11. (I) A thin plastic sheet ($n = 1.6$) placed in front of one slit in a double-slit arrangement results in a shift of the central bright fringe to a location previously occupied by the 12th bright fringe. Given that the light has a wavelength of 650 nm, what is the minimum thickness of the sheet?

12. (II) Two point sources S_1 and S_2 are in phase. They are a distance d apart along a line perpendicular to a screen, as in Fig. 37.34. (a) What is seen on the screen? (b) Assuming $d \ll L$, find the position, y_m, of the mth maximum relative to the center O.

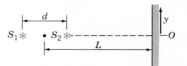

FIGURE 37.34 Problem 12.

13. (I) An evacuated cylinder of length $L = 4.0$ cm is placed in one arm of a Michelson–Morley interferometer. As air is slowly allowed to enter it, there is a shift of 40 fringes with light of wavelength 600 nm. Find the refractive index of air.

14. (II) A ray is incident obliquely on a thin film of thickness t in air, as shown in Fig. 37.35. Some light is reflected at the first surface (ray 1) and some is reflected at the second surface (ray 2). Show that the condition for destructive interference is $2nt \cos \theta = m\lambda$. (*Hint*: Consider the phase difference between the two rays.)

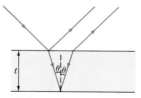

FIGURE 37.35 Problem 14.

15. (II) When white light is incident normally on a thin film in air, 550 nm is missing from the reflected light. Assuming the film has the minimum possible thickness, find the phase differences between the two interfering beams for (a) 400 nm, and (b) 700 nm. Estimate the factors by which the reflected intensities of (c) 400 nm, and (d) 700 nm are reduced relative to that for constructive interference.

CHAPTER 38

Wave Optics (II)

Major Points

1. The location of the minima in the **single-slit diffraction** pattern.
2. **Rayleigh's criterion** for the resolution of images.
3. The location of the principal maxima of a **grating.**
4. The use of **phasors** to determine the intensity patterns for multiple slits and a single slit.
5. The Bragg condition in **X-ray diffraction** (optional).

Opal consists of layers of tightly packed silica spheres (100 nm in diameter). The crystal acts as a three-dimensional grating. The color seen at a given region of the (nonuniform) surface depends on its orientation.

FIGURE 38.1 Jean Augustin Fresnel (1788–1827).

Thomas Young's ideas on light were daring and imaginative, but he did not provide the kind of rigorous mathematical theory that scientists of the day expected. This, coupled with his arrogant attitude, explains why he had little success in convincing the British scientific establishment of the validity of his wave theory of light. A few years later, a young Frenchman, J. A. Fresnel (Fig. 38.1), independently performed Young's experiments and did develop a mathematical theory of interference, diffraction, and other phenomena. He showed that the significant quantity in determining the amount of diffraction is the size of the aperture, or obstacle, relative to the wavelength. Newton had objected that the bending of light is too small for it to be a wave. Fresnel explained this as being a consequence of the smallness of the wavelength of light.

Because of Fresnel's work, the Paris Academy chose diffraction as the subject of its prize essay for 1818 and Fresnel submitted a paper. S. Poisson, a member of the selection committee, was hostile to the wave theory. In an attempt to scuttle it, he pointed out an "absurd" consequence of Fresnel's theory. All the diffracted waves from the edge of a circular obstacle should arrive in phase at the center of the shadow. Thus, there should be a bright spot at this point. Soon after, Fresnel and Arago demonstrated the existence of the "Poisson spot," as shown in Fig. 38.2. To his chagrin, Poisson had unwittingly devised a severe test that led to elegant confirmation of Fresnel's theory! After this the wave nature of light was generally accepted. It paved the way for Maxwell's beautiful and profound electromagnetic theory of light.

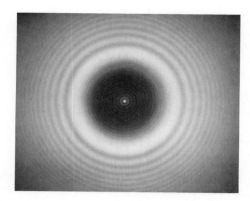

FIGURE 38.2 The Poisson spot clearly demonstrated that Fresnel's wave theory of light was correct.

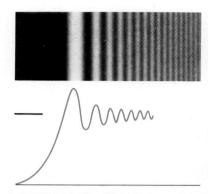

FIGURE 38.3 The diffraction pattern near an obstruction (the short black line) is called Fresnel diffraction.

38.1   FRAUNHOFER AND FRESNEL DIFFRACTION

Diffraction patterns are usually classified into two categories depending on where the source and screen are placed. When either the source or the screen is near the aperture or obstruction, the wavefronts are spherical and the pattern is quite complex. This is called **Fresnel diffraction,** and one case is illustrated in Fig. 38.3. Some light enters the region of the geometrical shadow, and fringes are observed near the edges of the obstruction. Because of its complexity, this pattern will not be discussed further. When both the source and the screen are at a great distance from the aperture or obstruction, the pattern is simpler to analyze. The incident light is in the form of plane waves and the rays leaving the opening are parallel. This is called **Fraunhofer diffraction** and is discussed below.

38.2   SINGLE-SLIT DIFFRACTION

Figure 38.4 shows plane wavefronts (parallel rays) incident on a slit of width a. For convenience we divide the slit into an even number (12) of point sources. (They are really *line* sources perpendicular to the paper.) In the forward direction all the secondary wavelets will be in phase, so there is a central bright region at $\theta = 0$. In general, when the path difference between two rays is $\lambda/2$, they interfere destructively. In the situation depicted in the figure, this occurs for the pairs (1) and (7); (2) and (8), (3) and (9), and so on. Thus when $\frac{1}{2} a \sin \theta = \lambda/2$, there is complete destructive interference. Suppose now that the slit is divided into four regions, each with many sources (Fig. 38.5). The corresponding sources in AB and BC will cancel in pairs when $\frac{1}{2} a \sin \theta = \lambda$. The same is true for CD and DE. By continuing this process of dividing the opening, we find that there is complete destructive interference whenever

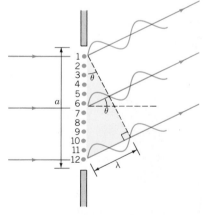

FIGURE 38.4 A single slit is treated as a series of point sources. When the path difference between the first source (1) and the last source (12) is one wavelength, there is destructive interference between the pairs 1 and 7, 2 and 8, etc.

(Minima) $a \sin \theta = m\lambda \qquad m = 1, 2, 3, \ldots$ (38.1)

This equation looks like Eq. 37.4 for the maxima in two-source interference. Make sure you understand the differences between them. Note that $m = 0$ is not included in Eq. 38.1 since $\theta = 0$ corresponds to the central *maximum*, not to a

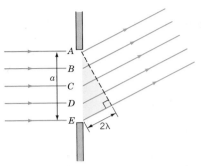

FIGURE 38.5 The slit is divided into four segments. If the path difference between waves from the top and the bottom is 2λ, there is destructive interference between sources in the regions AB and BC and between those in CD and DE.

minimum. The positions of the secondary maxima are *approximately* given by $a \sin \theta \approx (m + \frac{1}{2})\lambda$, where $m = 1, 2, 3, \ldots$ (see Section 38.6).

When $a \gg \lambda$, we see the usual uniform illumination in the shape of the slit. As the slit width is reduced, the illumination starts to spread out and dark bands become visible, as shown in Fig. 38.6. When the slit width is comparable to the wavelength, the central maximum becomes very wide and the screen far away may be uniformly illuminated.

The patterns produced by obstructions, such as hair or blood cells, are similar and may be used to infer the sizes of such small objects. Diffraction also plays a part in the pattern of radiation from a source of finite size. For example, a large loudspeaker tends to beam high frequencies in the forward direction but the low frequencies are uniformly radiated. A small tweeter helps disperse the high frequencies better.

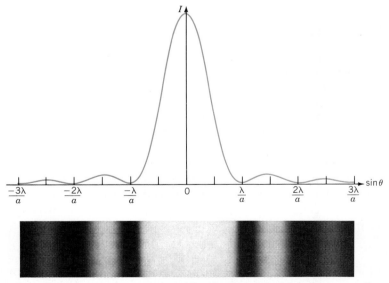

FIGURE 38.6 A wide slit produces a narrow diffraction pattern, whereas a narrow slit produces a large diffraction pattern.

EXAMPLE 38.1: Light of wavelength 600 nm is incident normally on a slit of width 0.1 mm. (a) What is the angular position of the first minimum? (b) What is the position of the second-order minimum on a screen 3 m from the slit?

Solution: From Eq. 38.1, the angular position of the first-order ($m = 1$) minimum is given by

$$\sin \theta_1 = \frac{\lambda}{a}$$

$$= \frac{6 \times 10^{-7} \text{ m}}{10^{-4} \text{ m}} = 6 \times 10^{-3}$$

Thus $\theta_1 = 0.34°$.

(b) If y is the distance from the center of the screen, then $\sin \theta \approx \tan \theta = y/L$, where L is the distance from the slit to the screen. Thus, for second order ($m = 2$) we have

$$y \approx L \sin \theta = L \left(\frac{2\lambda}{a} \right)$$

$$= \frac{(3 \text{ m})(2)(6 \times 10^{-7} \text{ m})}{10^{-4} \text{ m}} = 3.6 \text{ cm}$$

EXERCISE 1. At what value of a will there be no diffraction minimum observed?

Interference and Diffraction Combined

In our discussion of Young's double-slit experiment, the diffraction pattern due to each slit was ignored. When the screen is far away, the slits produce diffraction patterns that essentially overlap. As a result one sees the diffraction pattern of a single slit, but with maxima of higher intensity. The diffraction pattern determines the overall distribution of light on the screen. If the interference equation predicts a maximum at an angle for which the diffraction pattern has a minimum, the screen is dark. Similarly, an interference minimum will eliminate that part of a diffraction peak that it overlaps. We say that the interference pattern has a single-slit diffraction envelope.

EXAMPLE 38.2: In a double-slit experiment the slits are 0.25 mm wide and their centers are separated by 1 mm. Which interference maxima are missing?

Solution: The missing orders in the interference pattern occur when an interference maximum, given by

(Interference maxima) $d \sin \theta = m\lambda$; $m = 0, 1, 2,$ (i)

coincides with a diffraction minimum given by (with a temporary change in notation)

(Diffraction minima) $a \sin \theta = M\lambda$; $M = 1, 2, 3,$ (ii)

Note that a is the width of each slit and d is the separation between the slits. Dividing (i) by (ii) we obtain $d/a = m/M$. When the ratio $d/a = k$, an integer, the interference peaks given by $m = kM$ are missing. In the present example, $d = 4a$, and so the interference orders $m = 4, 8, 12, \ldots$ are missing, as shown in Fig. 38.7.

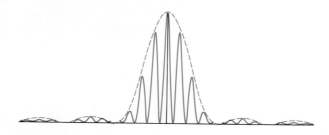

FIGURE 38.7 The actual pattern produced by a pair of slits is a double-slit interference pattern with a diffraction envelope due to single-slit diffraction.

38.3 THE RAYLEIGH CRITERION

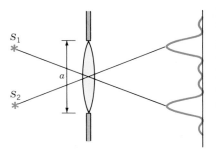

FIGURE 38.8 Two noncoherent point sources can be easily resolved if their diffraction patterns do not overlap.

Plane wavefronts passing through an aperture, such as a lens, undergo diffraction. Thus the image of a point source is not a point, but is instead a (Fraunhofer) diffraction pattern. The *resolution* of any optical system, its ability to produce sharp images, is limited by diffraction. Figure 38.8 shows light from two noncoherent point sources S_1 and S_2, passing through a lens. Each source produces its own diffraction pattern. For a circular aperture, it can be shown that the position of the first minimum in each diffraction pattern is given by

$$\sin \theta = \frac{1.22\lambda}{a} \qquad (38.2)$$

This differs from Eq. 38.1 for a rectangular slit by the factor 1.22.

When the angular separation between the sources is large, the diffraction patterns are far apart on the screen or photographic film. Each image is as shown in Fig. 38.9a. As the separation between the sources is reduced, the diffraction patterns start to overlap. Lord Rayleigh (Fig. 38.10) proposed a criterion for judging whether the images are resolved. According to **Rayleigh's criterion:** Two images are just resolved when the central maximum of one pattern coincides with the first minimum of the other. The appearance of the images and the associated diffraction patterns are shown in Fig. 39.9b. For small angles, $\sin \theta \approx \theta$, so the critical angular separation between the sources is

$$\theta_c = \frac{1.22\lambda}{a} \qquad (38.3)$$

where a is the diameter of the circular aperture. If the separation is reduced further, it is not possible to say that there are two distinct sources (Fig. 38.9c).

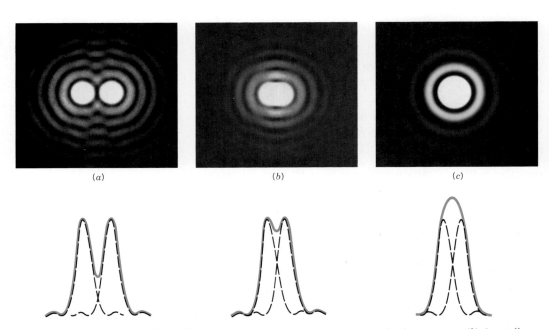

(a)   (b)   (c)

FIGURE 38.9 (a) The diffraction pattern of a point source and a circular aperture. (b) According to Rayleigh's criterion, two sources can just be resolved if the central maximum of one diffraction pattern coincides with the first minimum of the other. (c) If the separation between the sources is reduced further, they can no longer be resolved.

FIGURE 38.10 Lord Rayleigh (1842–1919).

EXAMPLE 38.3: The optical telescope at Mount Palomar has a diameter of 200 inches (5.08 m). At a wavelength of 550 nm, what is the minimum resolvable detail on the moon? The distance to the moon is 3.84×10^8 m.

Solution: According to Eq. 38.3, the critical angle is

$$\theta_c = \frac{(1.22)(5.5 \times 10^{-7}\ \text{m})}{5.08\ \text{m}} = 1.32 \times 10^{-7}\ \text{rad}$$

This is about 0.03 second of arc. With this angle, the separation between two resolvable points on the moon would be $s = L\theta_c$ where L is the distance to the moon. Therefore,

$$s = L\theta_c = (3.84 \times 10^8\ \text{m})(1.3 \times 10^{-7}) = 50\ \text{m}$$

In practice the resolution is limited to about 1 second of arc by atmospheric turbulence and optical aberrations in the mirror. The 2.4-m diameter mirror of the Hubble space telescope (Fig. 38.11) is expected to operate closer to the diffraction limit since it is in orbit above the atmosphere at an altitude of 600 km.

EXERCISE 2. A spy satellite in orbit at an altitude of 200 km has a mirror of diameter 50 cm. Assuming that it is limited only by diffraction, what is the closest distance between two bodies on the earth's surface for them to be resolved? Take $\lambda = 400$ nm.

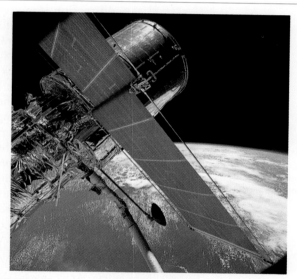

FIGURE 38.11 The Hubble space telescope, operating far above the earth's atmosphere, should have a resolution limited only by diffraction.

38.4   GRATINGS

A **grating** consists of thousands of very fine slits or grooves cut into a glass plate (in which case, the untouched parts act as the slits). We assume the slits are so fine that the single-slit diffraction pattern illuminates the screen uniformly. In the discussion of single-slit diffraction, we imagined the opening to consist of a large but indefinite number of point (line) sources. The point (line) sources in the grating are a small, but finite, distance d apart.

When the path difference between rays 1 and 2 in Fig. 38.12a is λ, they interfere constructively. The same holds for 2 and 3, and so on. Any path difference equal to an integral number of wavelengths would also lead to constructive

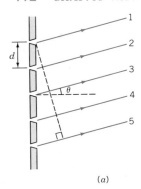

(a)

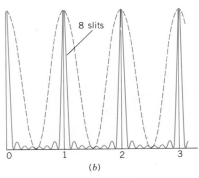

8 slits

(b)

FIGURE 38.12 (a) Interference due to multiple slits. (b) Changes in the interference pattern as the number of slits is increased. In a grating with thousands of slits, the principal maxima are very sharp and the secondary maxima are not visible. (See also Fig. 38.15.)

interference. The path difference between rays from adjacent slits is $\delta = d \sin \theta$, so the positions of the *principal maxima* are given by

(Principal maxima) $$d \sin \theta = m\lambda \qquad m = 0, 1, 2, 3 \qquad (38.4)$$

These are called principal maxima since the waves from *all* the slits are in phase.

Figure 38.12b shows how the pattern changes as the number of slits increases. Eight slits lead to principal maxima in the same positions as in the double-slit pattern, but there are six secondary peaks. (See also Fig. 38.15.) For a grating with thousands of slits, the principal maxima are sharp lines and the secondary peaks are not visible.

We may understand this sharpening qualitatively as follows. A small change in θ away from the condition in Eq. 38.4 is associated with a small path difference between adjacent slits. For two slits the resultant amplitude changes only slowly with θ. In the case of a grating with the same value of d, the path difference between adjacent slits is the same, but as θ varies the path difference between the first and the last slit quickly reaches λ. Following the logic used for single-slit diffraction, the contributions of the first slit and the one in the middle will then cancel, and so on for each pair of slits. It is for this reason that the principal maxima are narrow. (The variation in the intensity due to multiple slits is calculated in Section 38.5.)

Gratings are extremely important in the analysis of light emitted by atoms and molecules. Unlike the continuous spectrum of a hot body, such as a heated filament or the sun, the light emitted by a low-density gas through which an electrical discharge is passing consists of a series of discrete wavelengths. These appear as line spectra when passed through a grating. A grating acts somewhat like a prism but has much better resolution. However, the intensity of a given color is much less than with a prism because each wavelength is spread over many orders. A significant advantage offered by the use of a grating is that wavelengths can be determined. It was only after J. von Fraunhofer had learnt how to make fine gratings, around 1823, that wavelengths of light emitted by various sources could be determined precisely.

EXAMPLE 38.4: Light of wavelength 550 nm is incident normally on a grating that has 400 lines per mm. At what angle does the second-order principal maximum occur?

Solution: The spacing between the lines is

$$d = \frac{1 \text{ mm}}{400} = 2.5 \times 10^{-6} \text{ m}$$

From Eq. 38.4 with $m = 2$, the angular position of the second-order maximum is given by

$$\sin \theta_2 = \frac{2\lambda}{d}$$
$$= \frac{(2)(5.5 \times 10^{-7} \text{ m})}{2.5 \times 10^{-6} \text{ m}} = 0.44$$

Thus $\theta_2 = 26.1°$.

EXERCISE 3. (a) Locate the third- and fourth-order maxima. (b) What is the total number of maxima observable?

38.5   MULTIPLE SLITS

In Section 37.4 we found the intensity distribution of the interference pattern due to two fine slits by combining the wavefunctions trigonometrically. This procedure is not convenient when there are three or more sources. Instead, we use the technique of phasors introduced for ac circuits (Chapter 33). Recall that a **phasor** is a vector used to represent a physical quantity that varies sinusoidally in time. Here the physical quantity is the electric field of the light wave, $E = E_0 \sin(\omega t)$.

Phasors

The phasor $\mathbf{E}_0$ rotates at the angular frequency of the wave, and its magnitude is equal to the amplitude of the field. The projection of a phasor onto the "vertical" axis represents the variation of the physical quantity as a function of time. To find the result of superposing several waves, we must first find the vector sum of the phasors. The instantaneous value of the total field is the vertical component of the resultant phasor.

We assume that the slits are so narrow that diffraction spreads the light uniformly over the screen. That is, each slit contributes a wave of amplitude E_0 at the screen. If the screen is far from the sources, the outgoing rays are nearly parallel. The phase difference, ϕ, between the fields from adjacent slits is related to the path difference, $\delta \approx d \sin \theta$:

$$\phi = \frac{2\pi\delta}{\lambda} = \frac{2\pi d \sin \theta}{\lambda} \tag{38.5}$$

where d is the slit separation.

Three Slits

Let us consider the interference pattern for three identical coherent narrow slits. At a given point on the screen, the fields originating from adjacent slits have a phase difference ϕ:

$$E_1 = E_{01} \sin(\omega t)$$
$$E_2 = E_{02} \sin(\omega t + \phi)$$
$$E_3 = E_{03} \sin(\omega t + 2\phi)$$

where $E_{01} = E_{02} = E_{03} = E_0$. In order to find the magnitude of the total (resultant) field, we represent the fields by phasors. At some time t, the first phasor $\mathbf{E}_{01}$ is at angle ωt to the horizontal axis, $\mathbf{E}_{02}$ is at an angle ϕ to $\mathbf{E}_{01}$, and $\mathbf{E}_{03}$ is at an angle ϕ to $\mathbf{E}_{02}$, as shown in Fig. 38.13. The resultant is $\mathbf{E}_{0T} = \mathbf{E}_{01} + \mathbf{E}_{02} + \mathbf{E}_{03}$. Note that the angles between the phasors refer to the *phase* relationship between the fields; the field vectors themselves are assumed to lie along the same direction in *space*.

To develop the intensity pattern we draw phasor diagrams for various phase differences as in Fig. 38.14. The **principal maxima** occur when all the waves at the screen are in phase, that is, when $\phi = 0, 2\pi, 4\pi$, and so on. The amplitude of the resultant phasor in these cases is $E_{0T} = 3E_0$, so the intensity is $I_T = 9I_0$, where $I_0 \propto E_0^2$ is the intensity due to a single slit. When $\phi = 2\pi/3, 4\pi/3, 8\pi/3$, and so on, the phasor diagram closes on itself, so the intensity is zero. Notice that $\phi =$

How many orders do you see in this grating spectrum?

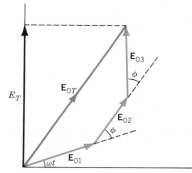

FIGURE 38.13 A phasor diagram for three slits. The phase difference between adjacent phasors is ϕ, but the actual field vectors lie along the same direction in *space*.

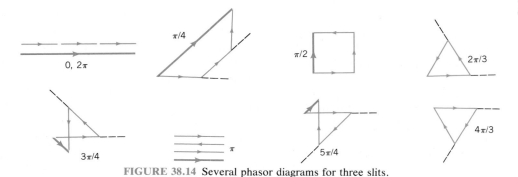

FIGURE 38.14 Several phasor diagrams for three slits.

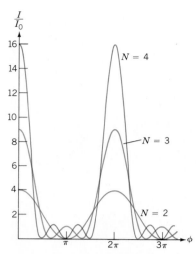

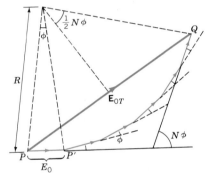

FIGURE 38.15 The interference pattern for three slits with the two-slit and four-slit pattern included for comparison.

FIGURE 38.16 A phasor diagram for N slits. It can be used to determine the intensity as a function of ϕ (see Problem 11).

$6\pi/3 = 2\pi$ corresponds to a principal maximum. To sum up:

Principal maxima:	$\phi = 2m\pi$	$m = 0, 1, 2, 3,$
Minima:	$\phi = \dfrac{2p\pi}{3}$	$p = 1, 2, 4, 5,$ $(p \neq 3, 6, 9, \ldots)$

The intensity pattern for three slits is drawn in Fig. 38.15. The pattern for two and four slits, with the same value for d, is included for comparison. Notice that between each pair of minima of the three-source pattern, **secondary maxima** appear at $\phi = \pi, 3\pi, 5\pi$, and so on. At a secondary maximum, two phasors are in opposite directions, so the resultant intensity is just I_0, for a single source. Notice that the principal maxima become sharper and higher as the number of slits increases.

N Slits

We now consider N equally spaced, coherent narrow slits. With a slit separation d, the phase difference between *adjacent* slits is still given by Eq. 38.5. A phasor diagram is shown in Fig. 38.16. The **principal maxima** occur when all the waves at the screen are in phase, that is, when $\phi = 0, 2\pi, 4\pi$, and so on. The amplitude of the resultant phasor in these cases is $E_{0T} = NE_0$, so the intensity of the principal maxima is $I_T = N^2 I_0$, where $I_0 \propto E_0^2$ is the intensity due to a single slit. When $N\phi = 2\pi, 4\pi$, and so on, or equivalently when $\phi = 2\pi/N, 4\pi/N$, and so on, the phasor diagram closes on itself and the intensity is zero. To sum up:

Principal maxima:	$\phi = 2m\pi$	$m = 0, 1, 2, 3,$	
Minima:	$\phi = \dfrac{2p\pi}{N}$	$p = 1, 2, 3, 4, 5,$ $(p \neq N, 2N, 3N, \ldots)$	(38.6)

Note that for the minima, p takes on all integer values except $N, 2N, 3N$, and so on, since the values $\phi = 2\pi, 4\pi$, and so on, correspond to the principal maxima. The first minimum ($p = 1$) beside the central ($\theta = 0$) principal maximum occurs at $\phi = 2\pi/N$, or from Eq. 38.5 when

$$\text{(First minimum)} \qquad \sin \theta = \frac{\lambda}{Nd} \qquad (38.7)$$

Clearly, as the number of slits, N, increases, the principal maxima become narrower, although their positions do not change relative to a two-source pattern. To obtain an accurate distribution of intensity using phasor diagrams, it is necessary to draw a large number, which is cumbersome. One can, however, derive an analytic expression for the intensity due to multiple slits (see Problem 11).

Figure 38.17 shows the *Pave Paws* radar installation in Cape Cod, which can monitor hundreds of targets simultaneously. There are 1800 antennas on each of two faces of a pyramid-like structure. Although the antennas are fixed, the phase relationship between adjacent antennas is varied electronically so that the emitted beam (which consists of 5 ns pulses) scans through a range of 120° within a few microseconds. The relationships between phase and source separation also apply to waves being *received* by an array of antennas. Thus, Eq. 38.7 may be used to estimate the resolution of such an array.

FIGURE 38.17 A two-dimensional array of microwave antennas in the *Pave Paws* radar installation used to monitor hundreds of targets simultaneously. By electronically varying the phase between the antennas, a narrow beam is made to sweep through 120° within a few microseconds.

EXAMPLE 38.5: The *Very Large Array* in Socorro, New Mexico consists of several radiotelescopes that can be moved on tracks (Fig. 38.18). Suppose one straight segment 10.8 km long has nine telescopes equally separated from each other and that the incoming signal from space has a wavelength of 21 cm. What is the minimum angular separation between two sources that the array can resolve?

Solution: The angular spread from a peak in reception sensitivity to the next minimum is given by Eq. 38.7. In the present case $d = 1.2$ km and $N = 9$, so

$$\sin \theta = \frac{\lambda}{Nd} = \frac{0.21 \text{ m}}{9 \times 1.2 \times 10^3 \text{ m}} = 1.9 \times 10^{-6}$$

For such a small angle $\sin \theta \approx \theta$. So the angular width is $2\theta \approx 3.8 \times 10^{-5}$ rad or 7.8 second of arc. Let us compare Eq. 38.7 with that for a single circular aperture, Eq. 38.3.

If the factor of 1.22 is ignored, we see that an array of N small antennas separated from each other by d has about the same resolution as a single antenna of diameter $a = Nd$. Signals received by radiotelescopes in different countries are sometimes combined to obtain resolution even better than the VLA.

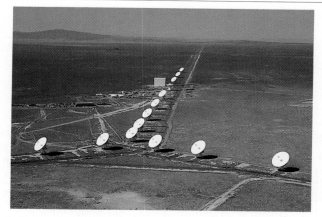

FIGURE 38.18 The Very Large Array in Socorro, New Mexico, has movable radio telescopes arranged in a Y formation. The signals from the antennas are fed to a central processing station. The system has a maximum resolution equivalent to that of a single radio telescope 37 km in diameter.

38.6   INTENSITY OF SINGLE-SLIT DIFFRACTION

In Section 38.2 we found only the positions of the minima for the single-slit diffraction pattern. We now calculate the variation in intensity across the pattern and also locate the secondary maxima. The phasor analysis for N slits can be carried over to the case of a single slit of width a. The slit is divided into a large but indefinite number of coherent line sources, each of which produces a wavelet of tiny amplitude. The only phase difference we can calculate is that between the

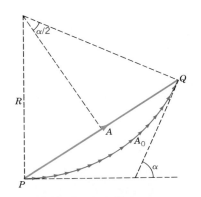

FIGURE 38.19 The phasor diagram for a single slit. The contributions from the infinitesimal sources in the slit form a continuous arc of length A_0. The resultant is the chord A.

waves from the top and bottom edges of the slit. When the screen is far away, we may treat the outgoing rays as parallel, so the path difference for the two extreme rays is $\delta = a \sin \theta$, and the phase difference is

$$\alpha = \frac{2\pi a \sin \theta}{\lambda} \tag{38.8}$$

In the forward direction ($\theta = 0$, $\alpha = 0$) all the phasors are aligned. Let us take the amplitude of the resultant in this case to be A_0. At some arbitrary angle θ, the discrete set of lines of Fig. 38.16 is replaced in Fig. 38.19 by a continuous arc of length A_0. The line PQ is the resultant amplitude A. From the figure we see that

$$A = 2R \sin \left(\frac{\alpha}{2} \right)$$

and

$$A_0 = R\alpha$$

Eliminating R we have

$$A = \frac{A_0 \sin \left(\frac{\alpha}{2} \right)}{\frac{\alpha}{2}} \tag{38.9}$$

The intensity ($I \alpha A^2$) is

$$I = I_0 \frac{\sin^2 \left(\frac{\alpha}{2} \right)}{\left(\frac{\alpha}{2} \right)^2} \tag{38.10}$$

where I_0 is the intensity at $\theta = 0$. The intensity pattern and a few phasor diagrams are shown in Fig. 38.20. Note that the length of the curved sections of all the phasor diagrams is fixed at A_0. The phasor diagram closes on itself when $\alpha = 2\pi$, 4π, 6π, and so on. So from Eq. 38.8 the intensity is zero when $\alpha = 2m\pi$ or

(Minima) $$a \sin \theta = m\lambda \qquad m = 1, 2, 3, \ldots \tag{38.11}$$

This agrees with the analysis in Section 38.2. Let us now consider the intensities.

At $\alpha = \pi$, the resultant is the diameter of the semicircle whose length is A_0, and so $\pi A/2 = A_0$, or $A = 2A_0/\pi$. The intensity at $\alpha = \pi$ is thus $I \alpha A^2 = 4/\pi^2 I_0 \approx 0.4I_0$.

At $\alpha = 3\pi$, the phasor diagram completes 1.5 circles. The diameter is A and the total length of the curve is A_0; therefore $(3\pi/2)A = A_0$, or $A = (2/3\pi)A_0$. The intensity at this point is $I = (4/9\pi^2) I_0 \approx 0.045I_0$. This is close, but not exactly equal, to the intensity of the first secondary maximum.

The positions of the secondary maxima are (see Problem 9)

(Secondary maxima) $$\alpha = 2.86\pi, \qquad 4.92\pi, \qquad 6.94\pi, \ldots$$

Notice that these are close, but not quite equal, to 3π, 5π, 7π, and so on. The intensities for these values of α are found from Eq. 38.10:

$$I = 0.047I_0; \qquad 0.017I_0; \qquad 0.008I_0, \ldots$$

The first secondary peak has an intensity of only 4.7% relative to the central peak.

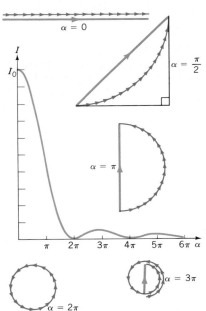

FIGURE 38.20 The intensity pattern and a few phasor diagrams for single-slit diffraction.

EXAMPLE 38.6: Light of wavelength 600 nm is incident normally on a slit of width 0.1 mm. What is the intensity at $\theta = 0.2°$?

Solution: From Eq. 38.8,

$$\alpha = \frac{2\pi a \sin \theta}{\lambda}$$

$$= \frac{(2\pi)(10^{-4} \text{ m})(3.5 \times 10^{-3})}{6 \times 10^{-7} \text{ m}}$$

$$= 3.67 \text{ rad}$$

On substituting into Eq. 38.10, we find the intensity is

$$I = I_0 \frac{\sin^2(1.84)}{(1.84)^2} = 0.27 I_0$$

EXERCISE 4. Use a phasor diagram to calculate the intensity at $\alpha = 5\pi$. Confirm your calculation by substituting into Eq. 38.10.

38.7 RESOLVING POWER OF A GRATING (Optional)

An important property of a grating is its ability to resolve nearly equal wavelengths. This depends on the width of each principal maximum and the difference between the wavelengths. Consider light consisting of a mixture of two wavelengths λ and $\lambda + \Delta\lambda$ incident normally on a grating. According to Rayleigh's criterion, two principal maxima will be just resolved when the mth-order principal maximum of $\lambda + \Delta\lambda$ coincides with the first minimum on one side of the principal maximum of λ of the same order, as drawn in Fig. 38.21.

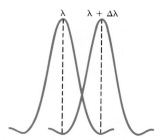

$\lambda \qquad \lambda + \Delta\lambda$

FIGURE 38.21 According to Rayleigh's criterion, two lines produced by a grating are resolved when a principal maximum of one wavelength coincides with the first minimum away from the maximum of the other wavelength.

From Eq. 38.4 the position of the mth-order principal maximum is

(Principal maxima) $d \sin \theta = m(\lambda + \Delta\lambda)$ (38.12)

To find the first minimum just above the mth-order principal maximum for λ, use Eq. 38.6 with $p = mN + 1$, which yields $\phi = 2(mN + 1)\pi/N$. Since $\phi = 2\pi d \sin \theta/\lambda$ from Eq. 38.5, we have

(Minimum) $d \sin \theta = (mN + 1)\dfrac{\lambda}{N}$ (38.13)

Since a minimum and a maximum coincide according to Rayleigh's criterion, we equate the conditions in Eq. 38.12 and 38.13:

$$m(\lambda + \Delta\lambda) = (mN + 1)\frac{\lambda}{N}$$

From this equation we obtain an expression for the **resolving power**, $R = \lambda/\Delta\lambda$, of the grating:

$$R = \frac{\lambda}{\Delta\lambda} = Nm \qquad (38.14)$$

The resolving power increases with the number of slits N and the order m of the principal maximum. Owing to the overlapping of spectra from adjacent orders, m is usually restricted to 2 or 3.

EXAMPLE 38.7: (a) What is the resolving power required to resolve the two sodium lines at 589.0 nm and 589.6 nm? (b) If a grating is 2 cm wide, how many lines per millimeter are needed to resolve these wavelengths in 3rd order?

Solution: (a) The difference in the wavelengths is $\Delta\lambda = 0.6$ nm and we take the average wavelength for λ. Thus the resolving power needed is

$$R = \frac{\lambda}{\Delta\lambda} = \frac{589.3 \text{ nm}}{0.6 \text{ nm}}$$

$$= 982$$

(b) The number of slits N is related to the width w of the grating by $w = Nd$, where d is the slit separation. Thus $R = Nm = wm/d$, or

$$d = \frac{wm}{R} = \frac{(2 \text{ cm})(3)}{982}$$

$$= 0.0061 \text{ cm}$$

The number of lines per millimeter is $1/(0.061 \text{ mm}) = 16.4$ lines per mm.

38.8   X-RAY DIFFRACTION (Optional)

In 1895, W. C. Roentgen was studying cathode rays (which we now know to be electrons) in a gas discharge tube. On one occasion he noticed that some nearby paper that had been coated with barium platinocyanide fluoresced (like watch dials that glow after being exposed to light). The fluorescence occurred even when the tube and paper were separated by a screen of black paper. Roentgen thought that the fluorescence was caused by hitherto unknown **X rays.** He soon discovered that they emanated from the spot on the glass at which the electrons impinged. William Crookes, who had designed the type of discharge tube used by Roentgen, had also noticed that film placed near the tube would fog up. However, instead of investigating this further, he complained to the Ilford film company. ''Chance'' discoveries are made only by minds prepared to make them.

The nature of the X rays remained a mystery for many years, although the possibilities had been narrowed to neutral particles or electromagnetic waves of very short wavelength ($\approx$0.1 nm). In 1912, Max von Laue suggested that their wave nature could be established if they exhibited diffraction. Since optical gratings are far too coarse, he suggested that the regular array of atoms in a crystal would act as a three-dimensional grating. In response to this suggestion, Friedrich and Knipping passed a narrow beam of X rays through thin strips of various crystals such as NaCl and ZnS, as shown in Fig. 38.22. A photographic plate was used to record the transmitted beam. They found a symmetrical arrangement of spots, which clearly indicated that some directions were favored more than others—just as for light passing through an ordinary grating. This established the wave nature of X rays.

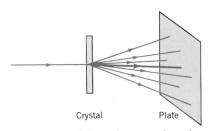

FIGURE 38.22 X rays passed through a crystal produce a pattern characteristic of the crystal structure.

In 1913, W. H. Bragg and his son W. L. Bragg provided the following analysis. Each atom in a crystal absorbs the incident radiation and re-emits it in all directions. In general, the ''scattered'' waves interfere destructively. The Braggs reasoned that the atoms could be considered to lie in various planes, each of which acts like a mirror (Fig. 38.23a). A given set of parallel planes would have a specific density of atoms and interplane spacing d (Fig. 38.23b). Consider rays reflected from two adjacent planes, as shown in Fig. 38.24a. When the path difference $ABC = 2d \sin \theta$ is an integral number of wavelengths, the rays are in phase and therefore interfere constructively:

$$2d \sin \theta = n\lambda \qquad (38.15)$$

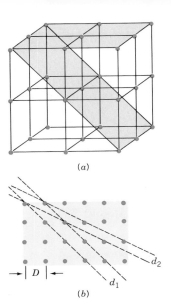

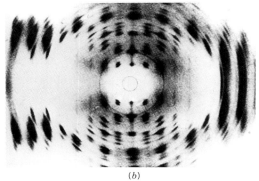

FIGURE 38.23 (a) A beam of X rays is reflected off the surface of a crystal. (b) The atoms form planes with various spacings.

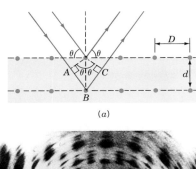

FIGURE 38.24 (a) When the spacing between the atomic planes and the angle of incidence satisfy the Bragg condition, $2d\sin\theta = n\lambda$, the reflected waves are strong. (b) An X-ray diffraction pattern of DNA.

where θ is the angle to the planes. This **Bragg condition** applies to *all* the planes parallel to the two shown in the figure. Thus we expect particularly strong reflections when this condition is satisfied. Figure 38.24b shows an X-ray diffraction pattern of a crystal. Such patterns can be used to infer the arrangement of the atoms in a crystal. Note that the Bragg condition looks quite similar to Eq. 38.5 for a diffraction grating. However, Eq. 38.15 involves a factor of 2, and the angle θ is defined differently.

38.9 POLARIZATION (Optional)

The phenomena of interfrence and diffraction provide evidence only that light has wave characteristics. Initially, both Young and Fresnel thought in terms of longitudinal waves. The realization that light is a transverse wave arose from the study of its polarization. To understand the meaning of this term consider a mechanical example. Figure 38.25a shows a rope vibrating in a vertical plane; and in Fig. 38.25b it vibrates in a horizontal plane. In each case the wave is said to be **linearly polarized** along the direction of oscillation. Polarization can be manifested only by transverse waves. The polarization of light was discovered through the study of double refraction.

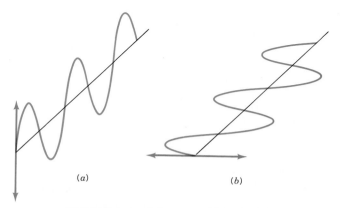

(a)

(b)

FIGURE 38.25 A linearly polarized wave.

In 1669, E. Bartholinius looked at a small object through a crystal of Iceland spar (calcite) and discovered two refracted images, as shown in Fig. 38.26. The *ordinary* (O) rays that form one image obey Snell's law, whereas the *extraordinary* (E) rays that form the other image do not. For example, if light is incident normally on the surface, the O ray continues undeviated, but the E ray travels at some angle to the surface (Fig. 38.27). Quartz, sugar solution, and ice also exhibit this phenomenon of **double refraction** or **birefringence.**

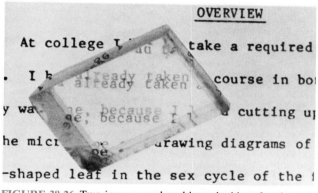

FIGURE 38.26 Two images produced by a doubly refracting or birefringent crystal.

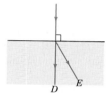

FIGURE 38.27 When light is incident normally on a birefringent crystal, the ordinary (O) ray behaves as usual; however, the extraordinary (E) ray does not obey Snell's law.

Huygens passed the O and E rays from one crystal through another and found that by rotating the second crystal relative to the first, he could turn an O ray into an E ray and vice versa. He noted that the two waves must propagate at different speeds but was unable to offer any real explanation in terms of his (longitudinal) wave theory. Newton suggested that rays of light had "sides," like the poles of a magnet. Depending on how they were oriented relative to the structure of the crystal, the light would propagate as either an O or an E ray. This "sidedness" of light was later called **polarization**—a result of Newton's reference to magnets!

In an electromagnetic wave, electric and magnetic fields oscillate perpendicular to the direction of propagation. Such waves can be produced by a dipole antenna connected to a radio frequency source (Fig. 38.28). The electrons in the wire oscillate and radiate, as we discussed in Chapter 34. The intensity of the radiation is zero along the axis of the dipole and a maximum perpendicular to it. Furthermore, at any given point of reception, the electric field vector oscillates only along a

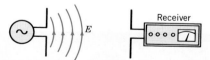

FIGURE 38.28 The radiation from a dipole (two rods connected to an oscillator) is polarized.

single direction. For electromagnetic waves, the direction of the electric field is defined to be the direction of polarization. The wave is said to be *linearly polarized* and is represented as shown in Fig. 38.29. The polarization is easily detected by connecting a second antenna to a receiver. When the dipoles are parallel, the electric field of the first induces currents in the second and a signal is registered. When the dipoles are perpendicular, there is no signal. The polarization of TV and radio broadcasts necessitates proper positioning of the receiving antennas.

FIGURE 38.29 The electric field of a linearly polarized wave.

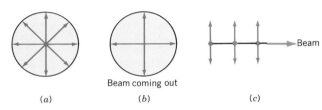

FIGURE 38.30 (*a*) In an unpolarized wave the direction of the electric field fluctuates. The electric field in an unpolarized wave may be decomposed into perpendicular components. In (*b*) the beam is normal to the page, whereas in (*c*) it travels to the right.

An atom emits electromagnetic radiation only for a brief time, typically 10^{-8} s. The electric field associated with the wave has a definite direction in space. However, the emission from another atom is completely independent, and so its **E** vector will point in another direction. The net effect is that the direction of **E** fluctuates randomly in time, as shown in Fig. 38.30*a*. The arrows depict the directions of **E** at different instants as we look into the beam. The beam is *unpolarized*. Since any electric field may be resolved along any two mutually perpendicular axes, an unpolarized beam may also be represented by two perpendicular arrows, as in Fig. 38.30*b* or Fig. 38.30*c*. Keep in mind that these two components are not coherent; they have no fixed phase relationship. We now consider some other ways in which polarized light is produced.

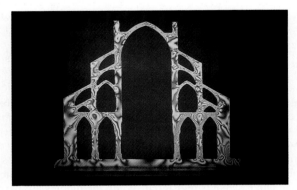

Plastic becomes birefringent when it is subjected to stress. Here, a stressed arch is placed between polarizing sheets. The amount of light transmitted at a given point on the arch depends on whether $\Delta n \cdot t$ is $m\lambda$ or $m\lambda/2$ where Δn is the difference in the refractive indices of the E and O rays at that point. Closely spaced bands indicate high stress. (See p. 358 of D. Falk, D. Brill, and D. Stork, Seeing the Light, Wiley, New York, 1988.)

Polarization by Double Refraction

Liquids and amorphous solids (such as glass) are called *isotropic* because their properties do not depend on direction. In particular, the speed of light is the same in all directions. Double refraction occurs in optically *anisotropic* crystals. The arrangement of atoms in an anisotropic crystal is such that the speed of light in a given direction depends on its state of polarization. Along the so-called *optic axis*, the *O* and *E* rays have the same speed, and therefore the same refractive index. The index for the *E* ray depends on its direction relative to the optic axis and reaches a maximum normal to this axis. When an unpolarized beam passes through a calcite crystal, the *E* and *O* rays are linearly polarized along perpendicular directions, as shown in Fig. 38.31.

FIGURE 38.31 The ordinary and extraordinary rays emerging from a birefringent crystal are polarized.

Polarization by Reflection

In 1808, a French engineer E. Malus happened to look through a calcite crystal at sunlight reflected off a window of the Luxembourg Palace. He saw one image instead of the usual two. He soon realized that light can be polarized by reflection. At a certain angle of incidence, the **polarizing angle** θ_p, the reflected ray is linearly polarized. In 1815, David Brewster found that at the polarizing angle of incidence, the reflected and the refracted rays are perpendicular; that is $i + r = 90°$, as in Fig. 38.32.

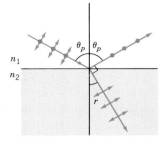

FIGURE 38.32 When the angle of incidence equals the polarizing angle θ_p, both the reflected and refracted rays are polarized as shown.

In terms of electromagnetic wave theory, we may resolve the fields of the unpolarized beam into components parallel to and normal to the plane of incidence—defined by the incident ray and the normal to the surface. The process of reflection does not occur precisely at the surface, but only after a wave has penetrated some distance into a material. As the wave interacts with the charges within the material, their oscillations produce both the reflected wave and the refracted wave. In general the reflected ray will have both components of E. Suppose, however, that the component in the plane of incidence were perpendicular to the refracted ray. This component would lie along the direction of the reflected ray. But such a longitudinal oscillation is contrary to electromagnetic theory, and so only the component normal to the plane of incidence is observed. Polarization by reflection is not an efficient method since only about 8% of the incident light energy is reflected at the polarizing angle, θ_p.

The polarizing angle may be found as follows. From Snell's law,

$$n_1 \sin \theta_p = n_2 \sin r$$

where r is the angle of refraction. We know that at the polarizing angle θ_p, the reflected and refracted rays are perpendicular. Since $\theta_p + r = 90°$, we have $\sin r = \cos \theta_p$. Using this in Snell's law, we find

$$\tan \theta_p = \frac{n_2}{n_1} \qquad (38.16)$$

This is called **Brewster's law.** For air $n_1 = 1$ and for glass $n_2 = 1.5$, so $\tan \theta_p = 1.5$, or $\theta_p = 57°$.

Polarization by Selective Absorption

Suppose an unpolarized beam of microwaves is incident on a grid of vertical wires or metal strips. The component of the electric field along the direction of the wires sets up macroscopic oscillating currents. These currents lead to joule heating and consequent absorption of energy from the wave. The oscillating currents also produce radiation in directions other than the original one. We say that the incoming wave is *scattered*. Since macroscopic currents cannot flow perpendicular to the wires, the field component normal to the wires is mostly transmitted. Therefore, the beam that emerges from the grid is polarized perpendicular to the wires.

In 1852, W. B. Herapath discovered that a properly oriented crystal of quinine sulphate periodide would completely absorb polarized light. The substance was extremely fragile and was not exploited till 1928, when E. H. Land, an undergraduate at Harvard, devised a way of embedding microscopic crystals of this substance in plastic sheets. He aligned the crystals by stretching the plastic while applying heat.

Nowadays, long-chain molecules of polyvinyl alcohol are used instead of Herapath's crystal. When the sheets are immersed in iodine, the chains become electrically conducting. An unpolarized beam that passes through a polaroid sheet emerges linearly polarized parallel to the transmission axis (which is normal to the axis of alignment of the molecules), as shown in Fig. 38.33. If a second sheet is placed in the polarized beam with its axis inclined to the first, only the component of the field along the transmission axis, $E \cos \theta$, is transmitted.

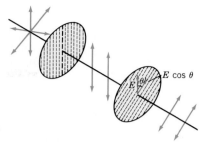

FIGURE 38.33 Unpolarized light passing through a polarizer is linearly polarized. A second polarizer transmits only the component of the electric field along its polarizing axis.

Since intensity is proportional to the square of the amplitude, the intensity of the transmitted light is given by

$$I = I_0 \cos^2 \theta \qquad (38.17)$$

where I_0 is the transmitted intensity at $\theta = 0$. This is called the law of Malus. In polaroid glasses the transmission axis is vertical. They absorb the horizontal component of the reflection off horizontal surfaces, such as roads and water, and thereby considerably reduce glare.

Polarization by Scattering

When an unpolarized electromagnetic wave traveling along the z axis is incident on a gas (Fig. 38.34), the electrons in each atom are set into oscillation in the xy plane. The atoms absorb the energy and then reemit dipole radiation in all directions—except, that is, along each dipole axis. The wave has been scattered. An observer receiving the radiation along a direction perpendicular to the incoming beam, as in Fig. 38.34, will not see an E field along the x axis, since this would imply a longitudinal component in the wave. Thus, the scattered wave is linearly polarized. (There is no oscillation along the direction of propagation of the original wave, the z axis.) Sunlight scattered off molecules in the atmosphere may also be polarized, which is why, on a clear day, polaroid glasses make certain regions of the sky appear black. It has been claimed that bees use this polarization of skylight to navigate.

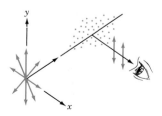

FIGURE 38.34 When light is scattered perpendicular to its original direction it is polarized as shown.

SUMMARY

The bending of rays, or the change in direction of propagation of wavefronts at the edge of an opening or an obstacle, is called **diffraction.** The positions of the minima of a **single-slit diffraction** pattern are given by

(Minima) $$a \sin \theta = m\lambda \qquad\qquad m = \pm 1, \pm 2, \ldots$$

where a is the slit width. Note that $m \neq 0$.

With a *circular aperture* of diameter a, **Rayleigh's criterion** for the resolution of the diffraction patterns due to two point sources yields the condition

$$\theta_c = \frac{1.22\lambda}{a}$$

The positions of the **principal maxima** of a **grating** are given by

$$d \sin \theta = m\lambda \qquad\qquad m = 0, \pm 1, \pm 2, \ldots$$

where d is the spacing between lines.

The variation in intensity in an interference pattern due to several sources may be determined by the use of **phasors** to represent the electric field. The phase difference between adjacent phasors is determined by the path difference from adjacent sources to the screen. The intensity at a given point is proportional to the square of the amplitude of the resultant phasor.

The variation in intensity in the diffraction pattern of a single slit of width a is given by

$$I = I_0 \frac{\sin^2 (\alpha/2)}{(\alpha/2)^2}$$

where $\alpha = 2\pi a \sin \theta / \lambda$ and I_0 is the intensity at $\theta = 0$.

ANSWERS TO IN-CHAPTER EXERCISES

1. The maximum possible value of θ is 90°. From

$$\sin \theta = \frac{\lambda}{a} = 1$$

we find $a = \lambda$. Note that if $a < \lambda$, $\sin \theta > 1$ and no minimum is observed.

2. Following Ex. 38.3, we have

$$s = L\theta_c = \frac{1.22L\lambda}{a}$$

thus

$$s = \frac{(1.22)(2 \times 10^5 \text{ m})(4 \times 10^{-7} \text{ m})}{(0.5 \text{ m})} = 0.2 \text{ m}$$

3. (a) From $\sin \theta_m = m\lambda/d$ we find $\theta_3 = 41°$ and $\theta_4 = 61.6°$.
(b) Since $\sin \theta_5 = 5\lambda/d > 1$, which is not possible, the fifth order is not observed. The total number of maxima is 9.

4. (a) The phase diagram completes 2.5 circles. The diameter is A and the circumference is A_0; thus

$$\left(\frac{5\pi}{2}\right)(A) = A_0$$

which means $A = 2A_0/5\pi$. On squaring we find

$$I = \frac{4I_0}{25\pi^2} = 0.016I_0$$

This agrees with Eq. 38.10.

QUESTIONS

1. Look at a distant street lamp at night with your eyelids nearly touching. Explain the fringes that appear.

2. Explain why the order of colors produced by a prism is reversed compared to that produced by a grating.

3. In a pinhole camera, there is an optimum size for the pinhole. Why does the sharpness of the image deteriorate when the aperture is (a) increased, or (b) decreased?

4. What is the source of the colors seen on the surface of compact discs or long-playing records?

A compact disc.

5. Is it possible that no minimum is recorded in a single-slit diffraction pattern? If so, under what condition?

6. Taking into account both interference and diffraction in a double-slit experiment, what is the effect of varying (a) the wavelength, (b) the slit separation, and (c) the slit width?

7. Distinguish between interference and diffraction. (a) Can one have diffraction without interference? (b) Can one have interference without diffraction? Give examples to support your answer.

8. Why does dust on a camera lens result in a loss in sharpness of the image on film?

9. In principle could one build a microscope to examine the structure of atoms with visible light?

10. Would a color filter that allows just one color to pass through it improve the resolution of a microscope? If so, for what reason(s)? Which color would produce the best resolution?

11. In a grating what is the effect of changing (a) the total number of lines, (b) the number of lines per centimeter, and (c) the width of the grating?

12. What is the shape and orientation of the aperture that produces the diffraction pattern shown in Fig. 38.35?

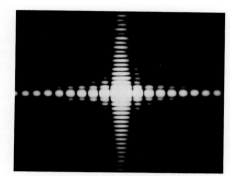

FIGURE 38.35 Question 12.

13. How would you test whether a pair of tinted sunglasses is made of a Polaroid sheet?

14. When unpolarized light passes through two crossed polarizing sheets, the transmitted intensity is zero. Is it possible to increase the transmitted intensity by using a third polarizing sheet? If so, how?

15. Why do you think it is necessary to adjust the orientation of "rabbit ear" antenna to obtain good reception of FM radio or TV signals?

16. Can two beams with perpendicular polarizations exhibit interference?

EXERCISES

38.2 Single-Slit Diffraction

1. (I) Light of wavelength 680 nm is incident normally on a slit of width 0.06 mm. The pattern is observed on a screen 1.8 m away. (a) What is the width of the central peak? (b) What is the distance between the first- and second-order minima?

2. (I) A loudspeaker with a flat diaphragm is fed a 5-kHz signal. What is the angular width of the central diffraction peak if the speaker's diameter is (a) 8 cm or (b) 30 cm? The speed of sound is 340 m/s.

3. (I) When light of wavelength 589 nm emitted by sodium vapor illuminates a single slit, the width of the central diffraction peak on a screen is 3 cm. What would be the width if the 436 nm line from mercury vapor were used instead?

4. (I) The green line emitted by mercury vapor, which has a wavelength of 546 nm, illuminates a single slit. The width of

the central diffraction peak is 8 mm on a screen 2 m from the slit. What is the slit width?

5. (I) In a single-slit diffraction pattern, the distance between the first and second minima is 3 cm on a screen 2.80 m from the slit. Find the slit width given that the wavelength of the light is 480 nm.

6. (I) A doorway is 76 cm wide. (a) For what sound frequency is the width equal to four wavelengths? (b) Assuming normal incidence, at what angle is the first diffraction minimum? The speed of sound is 340 m/s.

7. (I) In a double-slit experiment, the slit width is 0.15 mm and the slit separation is 0.6 mm. How many bright fringes are there in the central diffraction maximum?

38.3 Rayleigh Criterion

8. (I) A pinhole camera with a circular aperture of radius 0.5

mm is used to photograph a distant point source that emits at 550 nm. What is the width of the central diffraction peak on the film, which lies 22 cm from the hole?

9. (I) A pinhole camera has a hole of diameter 0.8 mm. The film is 20 cm from the hole. Two point sources are 16 m from the hole. Assuming the light has a wavelength of 600 nm, what is the minimum separation between the sources for them to be resolved?

10. (I) A dish of radius 1 m is used to receive 3-cm wavelength microwaves. At a distance of 20 km, what is the minimum separation between two point sources if they are to be resolved?

11. (I) What is the minimum separation between two point sources on the moon for them to be resolved according to the Rayleigh criterion (a) by an eye with pupil diameter 5 mm; (b) by a telescope of diameter 4.5 m? Take λ = 550 nm.

12. (I) A satellite at an altitude of 180 km has a telescope of diameter 30 cm. What is the smallest detail it can resolve on the surface of the earth at an ultraviolet wavelength of 280 nm? Ignore the atmosphere.

13. (I) Two tiny objects are 25 cm from an eye. What is the smallest separation between the objects if they are to be resolved when the pupil diameter is 3 mm? The wavelength of the light is 500 nm.

14. (I) A camera lens has an aperture of diameter 1.5 cm. At what distance could it resolve the headlights of a car that are 2 m apart? Assume λ = 550 nm.

15. (I) A star cluster is located at a distance of 10^{16} m. What is the smallest distance between two sources that can be resolved by each of the following: (a) The optical telescope at Mount Palomar, which has a diameter of 200 inches (5.08 m) and operates at 500 nm. (b) The radio telescope at Arecibo, Puerto Rico (Fig. 38.36), which has a diameter of 1000 feet (305 m) and operates at 21 cm? Assume that both are limited solely by diffraction.

16. (I) The tail lights of a car are 1.8 m apart and emit light of wavelength 650 nm. What is the maximum distance at which they can be resolved by the following: (a) an eye whose pupil diameter is 5 mm; (b) a telescope of diameter 2.8 m. Assume that each is limited only by diffraction.

38.4 Gratings

17. (I) A grating with 300 lines/mm is used to analyze the light from a hydrogen discharge tube that emits wavelengths 410.1 nm and 656.3 nm. What is the angular separation between these lines (a) in first order, (b) in second order? (c) Is there an overlap of second and third orders?

18. (II) Light is incident on a transmission grating at an angle ϕ to the normal. Show that Eq. 38.4 for the principal maxima must be replaced by

$$d(\sin \phi \pm \sin \theta) = m\lambda$$

Why is there a $\pm$ sign?

19. (I) How many complete orders are formed for the visible range 400–700 nm by a grating that has 6000 lines/cm?

20. (I) What is the angular separation in second order between the sodium doublet lines, 589.0 nm and 589.6 nm, produced by a grating with 5000 lines/cm?

21. (I) A grating places a spectral line of wavelength 640 nm at 11° in first order. At what angle is a line of wavelength 490 nm observed in second order?

22. (I) A grating is 2.8 cm wide. A spectral line of wavelength 468 nm is observed at 21° in second order. What is the number of lines in the grating?

38.5 Multiple Sources

23. (I) The electric fields at a point due to three sources are given by $E_1 = E_0 \sin(\omega t)$, $E_2 = E_0 \sin(\omega t + \phi)$ and $E_3 = E_0 \sin(\omega t + 2\phi)$. Use the phasor method to find the magnitude and phase of the resultant (relative to E_1) for the following phase differences, ϕ: (a) 30°; (b) 60°; (c) 90°; (d) 120°.

24. (I) The resultant of two phasors of equal amplitude is given by $E = 16 \sin(\omega t + 50°)$, where 50° is the angle between the resultant and one phasor. What are the individual amplitudes and the phase difference between the phasors?

25. (I) Five coherent point sources are separated from each other by 25 m along a line, as in Fig. 38.37. They emit waves of frequency 100 MHz and equal amplitude. What is the minimum phase difference between adjacent sources for the resultant amplitude to be zero at a distant point on the central axis?

FIGURE 38.36 Exercise 15.

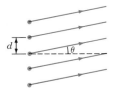

FIGURE 38.37 Exercise 25.

26. (II) A series of point sources are equally spaced at a distance d apart along a line, as in Fig. 38.37. Each source is α radians ahead in phase of the one above. What is the angular position of the central peak?

38.6 Intensity of Single-Slit Diffraction

27. (I) Hydrogen gas in a discharge tube emits a red line at 656.2 nm. The light passes through a single slit of width 0.08 mm. (a) At what angle is the first minimum? (b) What is the intensity (relative to the central peak) at half the angle found in (a)?

28. (I) A single slit of width 0.06 mm diffracts light of wavelength 523 nm to a screen 3.4 m away. What is the intensity (relative to the central peak) at a point 2 cm from the center of the central peak?

29. (I) If the width of a single slit is doubled, show that the intensity at the center of the central diffraction peak increases by a factor of four.

38.7 Resolving Power of a Grating

30. (I) The sodium D lines have wavelengths 589.0 nm and 589.6 nm. If these lines are to be resolved in 1st order, what is the required width of a grating with 300 lines/mm?

31. (I) A grating of width 2.8 cm has 4200 lines/cm. What is the minimum difference in wavelength that can be resolved in second order at 550 nm?

32. (I) It is necessary to resolve two spectral lines whose wavelengths are 586.32 nm and 586.85 nm. A grating is 3.2 cm wide. (a) What is the resolving power required? (b) How many rulings must the grating have to resolve these lines in second order?

38.8 X-Ray Diffraction

33. (I) X rays of wavelength 0.14 nm are incident on a crystal's atomic planes that are 0.32 nm apart. At what angle to the planes is the first-order diffracted beam?

34. (I) Certain atomic planes of a crystal are spaced 0.28 nm apart. The first-order Bragg diffraction maximum occurs at 15° to the planes. Find: (a) The wavelength of the X rays, (b) the angle at which the second-order Bragg maximum occurs.

35. (I) Monochromatic X rays are incident on certain atomic planes of a crystal that are 0.28 nm apart. The second-order Bragg diffraction maximum is observed at 19.5°. What is the wavelength of the radiation?

36. (I) When X rays of wavelength 0.13 nm are directed toward a crystal, the first-order Bragg diffraction maximum occurs at 9° to certain atomic planes. What is the spacing between the planes?

38.9 Polarization

37. (I) Unpolarized light of intensity I_0 falls on a polarizer and an analyzer whose transmission axis is rotated by 60° relative to the polarizer. What is the transmitted intensity?

38. (I) Unpolarized light of intensity I_0 is incident on crossed polarizers. (The transmission axes are at 90°.) A third polarizing plate is placed between the first two with its axis oriented at 45°. What is the final transmitted intensity?

39. (I) Show that the critical angle θ_c for total internal reflection and the polarizing angle θ_p are related by

$$\cot \theta_p = \sin \theta_c$$

40. (II) When light travels in a medium and is reflected at the interface with air, the critical angle for total internal reflection is found to 38°. What is the polarizing angle?

41. (I) Sunlight is reflected off the calm surface of a pond. At what angle above the horizon should the sun be located for the reflected light to be linearly polarized?

42. (I) Two polarizers are set for maximum transmission of unpolarized light. Through what angle should one be rotated for the transmitted intensity to decrease to 40%?

43. (I) A slab of flint glass ($n = 1.6$) is immersed in water ($n = 1.33$). What is the polarizing angle for reflection off the glass–water interface?

44. (II) A source of light is submerged under water ($n = 1.33$). Is there any angle of incidence at which light internally reflected at the water–air interface is linearly polarized? If so, what is this angle?

45. (I) A beam of light is incident at the polarizing angle on a plate of flint glass ($n = 1.6$). What is the angle of refraction?

PROBLEMS

1. (I) The spaces between the rulings on a metal ruler can be used as a reflection grating, as shown in Fig. 38.38. If light is incident at angle α to the plane of the ruler, what is the condition for a principal maximum at angle θ?

2. (I) Light of wavelength 450 nm is incident normally on three narrow slits spaced 0.5 mm apart. The interference is observed on a screen 3.6 m away. (a) Locate the first off-

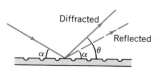

FIGURE 38.38 Problem 1.

center principal maximum; (b) the first and second secondary maxima; (c) the first two minima.

3. (I) Four narrow slits are illuminated by light of wavelength 450 nm. The slit separation is 0.08 mm and the interference is observed on a screen 3.6 m away. Locate (a) the first off-center principal maximum; (b) the first two minima.

4. (I) Show that the angular separation between the nth order principal maxima in a grating for wavelengths λ and $\lambda + \Delta\lambda$ is

$$\Delta\theta = \frac{n\,\Delta\lambda}{d\cos\theta}$$

where d is the grating spacing.

5. (II) An array of N line sources lies along a line as in Fig. 38.37. Show that the angular half-width $\Delta\theta$ of the principal maximum at θ is (for small $\Delta\theta$)

$$\Delta\theta \approx \frac{\lambda}{Nd\cos\theta}$$

where d is the spacing between sources. Evaluate this expression at $\theta = 0$ for $N = 32$, $d = 70$ m, and $\lambda = 21$ cm. (*Hint:* Consider a maximum at θ and a minimum at $\theta + \Delta\theta$. See Eq. 38.13.)

6. (I) The *dispersion* of a grating is defined as $d\theta/d\lambda$. Use Eq. 38.4 to show that the dispersion may be expressed as

$$\frac{d\theta}{d\lambda} = \frac{\tan\theta}{\lambda}$$

7. (I) Three polarizing plates are initially aligned for maximum transmission of unpolarized light of intensity I_0. The transmission axis of the second is rotated by $30°$ relative to the first and that of the third is rotated by $60°$ (in the same sense) relative to the first. What is the transmitted intensity?

8. (I) Four radio transmitters are separated by 100 m along a north–south line and emit waves of wavelength 600 m. How can the central peak be directed along $11°$ N of E?

9. (II) Show that the condition that determines the angular positions of the secondary maxima in a single-slit diffraction pattern is

$$\tan\left(\frac{\alpha}{2}\right) = \frac{\alpha}{2}$$

(Set the derivative of Eq. 38.10 equal to zero). (b) Plot $\tan(\alpha/2)$ and $\alpha/2$ vs. α (α in radians). The intersections of the functions are the solutions of the above (transcendental) equation. What is the first value of α?

10. (II) Light of wavelength 600 nm is incident normally on a single slit of width 0.5 mm. At what angle in the diffraction pattern does the intensity drop to 25% of the value at the center of the central peak? (*Hint:* First find α from Eq. 38.10. You could obtain a graphical solution or employ "trial and error.")

11. (II) In order to find an expression for the intensity distribution due to N sources, use Fig. 38.16 to obtain expressions for the phasors $\mathbf{E}_0$ and $\mathbf{E}_{0T}$ in terms of ϕ and R. Express E_{0T} in terms of E_0; then use the fact that the intensity is proportional to the square of the field to show that

$$I = I_0 \frac{\sin^2\left(\dfrac{N\phi}{2}\right)}{\sin^2\left(\dfrac{\phi}{2}\right)}$$

where $I_0 \propto E_0^2$ is the intensity due to one source.

SPECIAL TOPIC: Holography

Holography is a two-step process for the recording and viewing of images without the use of lenses. The basic principles were first outlined by D. Gabor in 1948. An ordinary photograph records only the intensities of the waves emanating from different parts of a scene. That is, the information on the amplitudes of the waves is retained but the information on the relative phases of the waves from different regions is lost. Gabor devised a technique, for which he received the Nobel prize in 1971, whereby both the amplitudes and the phases of the wavefronts are preserved on a photographic plate that is called a **hologram** ("whole record"). Whereas a photograph "projects" a three-dimensional object onto a two-dimensional format, a hologram preserves the information on the three-dimensional nature of the object.

THE PRINCIPLE OF THE HOLOGRAM

Figure 38.39a shows two monochromatic plane waves that overlap and interfere. For later reference, we call AA the *reference* wave and BB the *object* wave, which travels at an angle θ to AA. The points of constructive and destructive interference form lines perpendicular to the plane of the page. Thus, a thin photographic plate, P, records a series of straight bright and dark fringes (Fig. 38.39b). The interference pattern preserves information on both the amplitudes and the relative phase of the two wavefronts.

When the exposed plate P is illuminated by the monochromatic coherent reference wave AA, it acts like a grating. In a normal grating, Fig. 38.39c, the incoming waves are either transmitted or obstructed, and one observes several orders in the diffracted waves. The profile of the fringes recorded on the plate is sinusoidal, as shown in Fig. 38.39d. This results in a gradual variation in the transmitted intensity, which in turn means that there is only *one* (first-order) diffracted wave on either side of the (zero-order) forward wave (Fig. 38.40). Furthermore, the diffracted waves make the same angle θ with the reference wave as did the original object wave BB. Thus, one of the first-order diffracted wavefronts is an exact reconstruction of the original wavefront BB. If the plate had been illuminated with wavefront BB, the wavefront AA would have been reconstructed. In this case, either wave can serve as the "reference wave."

Now consider interference of a plane reference wave with spherical wavefronts that are either emitted by a point source or are waves scattered off a point object (Fig. 38.41). The interference fringes are recorded on a thin plate P parallel to the reference wavefronts. The interference pattern does not look like a dot; it is not a photographic image. The fringes, which look similar to Newton's rings (Fig. 37.20), form arcs whose center lies at the foot of the normal drawn

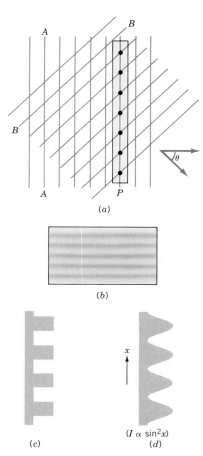

FIGURE 38.39 Two plane waves traveling at angle θ to each other as in (a) produce an interference pattern on the plate P as in (b). In a normal transmission grating light is either transmitted or obstructed as in (c). The variation in intensity of the pattern in the exposed plate is sinusoidal as in (d).

from the point to the plate, as indicated in Fig. 38.42a. The exposed plate is called a **Gabor zone plate.** When the plate is illuminated by the coherent plane wavefronts of the reference wave, it acts as a grating (with variable spacing). Whereas a straight-line grating diffracts incoming plane waves either "up" or "down," the circular lines of the Gabor zone plate diffract the incoming beam either inward (toward the central axis) or outward, away from the axis, as in Fig. 38.42b. In this sense, the zone plate acts like both a converging lens and a diverging lens. Because the transmitted intensity varies gradually along the grating, there is again

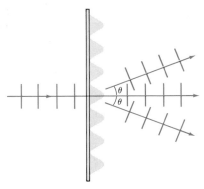

FIGURE 38.40 When the exposed plate is illuminated with a coherent beam, there are two diffracted beams.

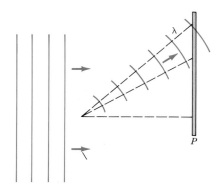

FIGURE 38.41 Interference between a plane reference wave and spherical wave emitted by a point source.

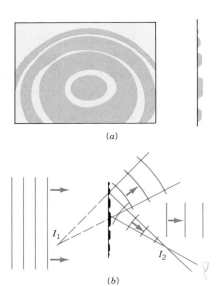

(a)

(b)

FIGURE 38.42 The fringes obtained from this setup resemble Newton's rings. When the exposed plate is illuminated by the plane reference beam, there are two diffracted beams. One produces the virtual image I_1 and the other produces the real image I_2.

just one first-order wavefront on either side of the forward wave. One first-order wave appears to diverge from the position of the original point object. Therefore, the original spherical waves emanating from the object have been reconstructed by the passage of the reference beam through the hologram. What we observe is a virtual image, I_1, in exactly the position of the original object—long after the object has been removed! The other diffracted wave converges to form a real image, I_2.

Two point sources, plus a reference beam, will produce two overlapping fringe patterns on the plate. Upon reconstruction with the reference wave, one observes two virtual point images. Any finite-sized object is a collection of point sources from which spherical waves are either emitted or scattered. When the hologram is illuminated by the coherent reference waves, each point on the object will be reproduced in exactly its original position. Since the reconstructed waves are an exact replica of the original waves from the object itself, a virtual image is observed with all the three-dimensional perspective of the original object! (The real image can be viewed or photographed, but it has a peculiar geometric property that makes it difficult to use.)

PROPERTIES OF A HOLOGRAM

Looking through a hologram is like looking through a window into a room. By moving one's head, one sees different views of the contents. For example, the hologram displays parallax: An object obscured by something in front of it may become visible when the observer's head is moved. Even a "stereo" pair of photos (such as those in a Viewmaster) provides a three-dimensional perspective from only one viewing direction. The hologram provides a range of viewpoints. As in real life, in order to make the images of near or far objects sharp, we have to refocus our eyes or a camera.

When a lens is used to form an image, each object point corresponds to a single image point. In a hologram, the information regarding each object point (its interference pattern) is spread over the *whole* plate. Therefore, even part of a hologram will reproduce the complete object, although the intensity and resolution will be poorer. One sees a different perspective from different parts of the hologram and a somewhat restricted view (as one would through a smaller window).

In order to produce a hologram of a three-dimensional object, light emanating from different parts of the object must be coherent. This means that the object must be smaller than the coherence length. Gabor was forced to use a flat transparency to demonstrate holography because of the poor coherence of his light source (a mercury discharge tube). Furthermore, with the "in-line" geometry of Fig. 38.40, the real image and the undeviated forward reference beam make it difficult to see the virtual image. The practical development of holography had to await the invention of the laser, with its extraordinary coherence, in 1960. Even a commonly available HeNe laser has a 15–20 cm coherence length.

In 1962, E. Leith and J. Upatnieks devised an "off-axis" technique for producing holograms (Fig. 38.43). The beam from the laser is divided into two parts by a partially reflecting slab of glass (called a beam splitter). The beams are broadened by lenses (a pair of lenses can serve as a beam expander). The reference beam goes directly to the plate, P. The object beam reflects off the object before reaching the plate. When the developed plate is later illuminated by the reference beam, the virtual image is not obstructed by the real image.

The fringes in a hologram are so fine that they are not visible to the naked eye. The film emulsion must be able to resolve 2000 lines/mm. (Conventional film is rated at under 100 lines/mm.) This type of film is not sensitive and requires an exposure time of 10 s with a low-power HeNe laser. During this time, the whole apparatus must not move by more than a fraction of a wavelength (about $\lambda/10$). One way to minimize exposure times is to use a pulsed laser of very high power.

Holograms produced just by developing the film have clear parts that transmit light and dark sections that do not. They are called absorption holograms. It is possible to bleach the hologram so that the dark parts are replaced by a transparent silver salt whose refractive index differs from that of the originally clear areas. When the reference beam passes through, its amplitude is not affected, but the phase

along the points of the wavefront are changed differently. The plate is called a phase hologram. Since much more light is deviated into the diffracted waves, the images are brighter.

Leith and Upatnieks were part of a team that was developing "coherent radar." In a modern "side-looking" radar, a coherent microwave beam ($\lambda \approx 1$ cm) is transmitted toward the ground by an aircraft that travels on a very straight path. The reflected signals from different parts of the terrain are mixed with a reference signal in the aircraft and displayed on an oscilloscope. The interference pattern (which changes with time) is recorded on a movie camera. Later, when the film is illuminated with laser light, it produces an image of the terrain in very fine detail. The technique has obvious applications in geophysical research and reconnaissance.

White Light Holograms

The plates used in the above discussions are relatively thin. In 1962, Y. N. Denisyuk produced holograms with thick photographic emulsions. In his method, the reference wave and the object wave reach the emulsion from opposite directions. The loci of constructive and destructive interference form nearly plane surfaces parallel to the surface of the plate, as in Fig. 38.44. The planes are separated by $\lambda/2$ (as in a standing wave) and there are about 50 planes in an emulsion 20 μm thick. The volume hologram thus consists of a set of reflecting planes like those in a crystal. Only when the Bragg condition ($2d \sin \theta = n\lambda$) is satisfied will there be a strong diffracted wave. The wavelength of the reference wave used to illuminate the hologram does not have to equal the wavelength of the reference wave used to form the hologram. The image produced by a given wave-

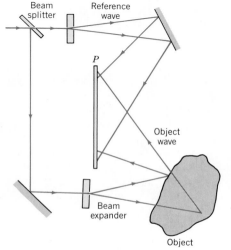

FIGURE 38.43 The off-axis method of producing a hologram.

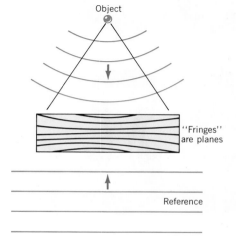

FIGURE 38.44 The production of a white-light hologram creates a set of reflecting planes like those in a crystal.

length appears at a unique angle. It was realized later that a volume hologram could be viewed even with incoherent white light, such as ordinary sunlight! At a given viewing angle, only one wavelength will satisfy the Bragg condition, and as the angle is changed, the color of the image changes.

Holographic Interferometry

Suppose a hologram of a rod is recorded. Next a small weight is placed on the rod and a second hologram is recorded on the same film. The rod has been slightly deformed, so the second hologram differs from the first. When the composite hologram is illuminated, the two images interfere with each other and produce a broad fringe pattern that shows where the object was deformed. The technique is sensitive to a displacement or deformation that is only a fraction of the laser wavelength used. Strains, the growth of plants, and defects in tires can be detected in this way. Holographic interferometry is also used in the study of aerodynamic flow—for example, to examine the wake of a bullet. First a hologram is made in still air. Next, when the bullet passes through the same region, a short high-power laser pulse produces a second hologram on the same plate. The variation in the air density results in a different hologram. When the plate is illuminated, the shock wave and turbulent eddies are clearly seen (Fig. 38.45). The

FIGURE 38.45 Shock waves made visible by holographic interferometry.

same technique can be used to study the vibrational modes of any structure, such as a bar, a musical instrument, or a loudspeaker. The exposure is made long enough to cover many cycles of vibration. Since the object spends a large fraction of time at the extreme positions of the vibrations, these positions result in greater reflected intensity. The interference between the waves from the extreme positions outlines the pattern of vibrations, as shown in Fig. 38.46.

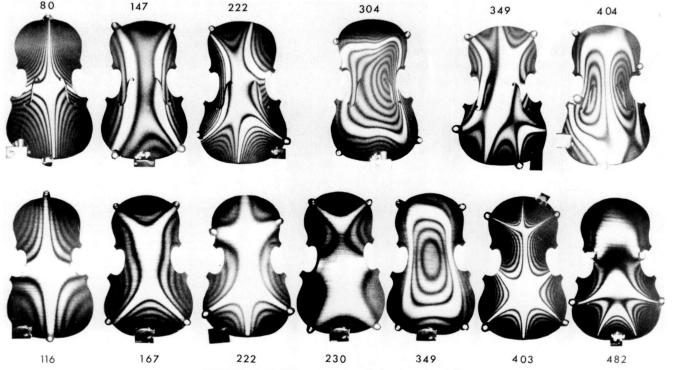

FIGURE 38.46 Different modes of vibration of a violin.

360° Holograms

Holograms can also be used to obtain a 360° view of an object. The object is first photographed from all angles by an ordinary camera. (The object may rotate while the camera is fixed.) The film frames are then used as objects for the holographic process. For each frame, only a thin strip (1 mm) of the hologram plate is exposed. The final hologram, which contains many pictures of the object, is made into a cylinder, and a line source is placed at its center. As one walks around, one sees the different views of the object—front, sides, and back. This approach can be used to make a "moving" hologram. In one example, as you walk around the hologram, you see a woman wink and then blow you a kiss (Fig. 38.47). These are not true holograms since they display only horizontal parallax; one cannot look "above" or "below."

Another way of producing a 360° view is shown in Fig. 38.48. A small object is placed inside a cylinder with a concave mirror at its base. The slightly divergent reference beam enters through the other end. The reference beam and the light reflected from the object are recorded on a film wrapped within the cylinder.

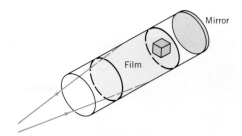

FIGURE 38.48 Producing a 360° hologram.

There are many other uses for holograms: information storage, check-out counters, many forms of nondestructive testing (of tires for example), determination of particle sizes in the air and in liquids, or pattern recognition. Gabor's original dream was to use holography to improve the resolution of the electron microscope. If a hologram is made with 0.1 nm X rays and reconstructed with visible light, a magnification of greater than 10^6 is possible. This is not "empty" magnification; resolution of about 0.1 nm might be possible, but this has not yet been realized.

FIGURE 38.47 "The Kiss."

CHAPTER 39

Special Relativity

When two nuclei coalesce to form a single nucleus—a process called fusion—energy is released. This is an example of the equivalence of mass and energy. The glowing plasma (hot ionized gas) is part of an experiment to harness this energy.

Major Points

1. (i) The **principle of relativity:** All physical laws have the same form in all inertial frames.
   (ii) The **constancy of the speed of light:** The speed of light in free space is the same in all inertial frames.

2. The **relativity of simultaneity.**

3. The phenomena of **time dilation** and **length contraction.**

4. The relativistic **Doppler effect** for light.

5. (a) The **Lorentz transformation** of space and time coordinates. (b) The addition of velocities.

6. **Mass–energy equivalence.**

7. The relativistic definitions of **linear momentum** and **kinetic energy.**

39.1 INTRODUCTION

The work of Young and Fresnel in the early 1800s served to establish that light is a transverse wave. Since mechanical waves require a medium in which to propagate, it was generally accepted that light also required a medium. This medium, called the **ether,** was assumed to pervade all matter and space in the universe. From Maxwell's wave equation,

$$\frac{\partial^2 E}{\partial x^2} = \frac{1}{c^2} \frac{\partial^2 E}{\partial t^2}$$

it was inferred that the speed of light should equal c only with respect to the ether. This meant that the ether was a "preferred" or "absolute" reference frame: A particle at rest in the ether frame would be at "absolute" rest, and motion with respect to the ether would be "absolute" motion.

The ether had bizarre properties. For example, in order for light to propagate at such an enormous speed, the ether had to be extremely rigid, yet it did not impede the motion of bodies. For a substance so crucial to optics and electromagnetism, it was embarrassingly elusive. Despite the many peculiar properties assigned to the ether, no one could detect its ghostly presence.

In 1879, A. A. Michelson read an article by Maxwell that mentioned the problems of detecting the ether and took it as a challenge. He developed his interferometer (Section 37.10) and in 1881 used it to try to detect the earth's motion relative to the ether. The results were not conclusive so an improved version was tried in 1887 with the assistance of E. W. Morley.

39.2 THE MICHELSON–MORLEY EXPERIMENT

Michelson and Morley wanted to detect the speed, v, of the earth relative to the ether. If the earth *were* moving relative to the ether, there should be an "ether wind" blowing at the same speed v relative to the earth but in the opposite direction. They needed to compare the times of travel of beams of light parallel to and perpendicular to the earth's direction of motion relative to the ether. Since the time of travel determines the phase of a wave, they compared the phases of the beams by viewing the interference fringes seen in the telescope of the interferometer.

We assume the arms PM_1 and PM_2 of the interferometer (Fig. 39.1) are of equal length, L_0, and are oriented along and perpendicular to the earth's motion. The arm PM_1 lies along the earth's motion, so the speed of light relative to the apparatus is either $c - v$ or $c + v$. The time for the round trip from P to M_1 and back is therefore

$$T_1 = \frac{L_0}{(c - v)} + \frac{L_0}{(c + v)} = \frac{(2L_0/c)}{(1 - v^2/c^2)} \qquad (39.1)$$

On the other hand, if the resultant direction of light is along PM_2, perpendicular to the motion, the speed of light relative to the apparatus is $(c^2 - v^2)^{1/2}$, as shown in Fig. 39.2. The time for the round trip from P to M_2 and back is

$$T_2 = \frac{2L_0}{(c^2 - v^2)^{1/2}} = \frac{(2L_0/c)}{(1 - v^2/c^2)^{1/2}} \qquad (39.2)$$

To obtain a simple expression for the difference between these two times, we use the binomial expansion $(1 + x)^n \approx 1 + nx + \ldots$, for small x. If v is of the order of the earth's orbital speed (≈ 30 km/s), then $v \ll c$. With $x = (v/c)^2$ we find (see Exercise 1 below)

$$\Delta T = T_1 - T_2 = \left(\frac{L_0}{c}\right)\left(\frac{v}{c}\right)^2$$

This time difference gives rise to a phase difference that causes either constructive or destructive interference at the crosshairs of the telescope.

If the apparatus is now rotated by 90°, the roles of the mirrors interchange, so

$$\Delta T' = T_1' - T_2' = -\left(\frac{L_0}{c}\right)\left(\frac{v}{c}\right)^2$$

The rotation should cause the fringe pattern to shift by an amount determined by the change

$$\Delta T - \Delta T' = \frac{2L_0}{c}\left(\frac{v}{c}\right)^2 \qquad (39.3)$$

Using $v = 30$ km/s, the expected shift was about 0.4 fringe. Even though Michelson and Morley were able to detect shifts smaller than 1/20 of a fringe, they found nothing. They concluded that the speed of the earth relative to the ether had to be less than 5 km/s. Similar experiments have been repeated many times since then. The latest upper limit for v is 5 cm/s.

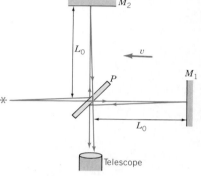

FIGURE 39.1 Michelson's interferometer. The arms PM_1 and PM_2 are set along and perpendicular to the earth's motion. If the earth is moving to the right, there is an "ether wind" flowing toward the left at speed v relative to the earth.

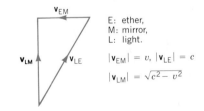

FIGURE 39.2 The velocity of light relative to the mirror, $\mathbf{v}_{LM}$ is the sum of $\mathbf{v}_{LE}$, the velocity of light relative to the ether, and $\mathbf{v}_{EM}$, the velocity of the ether relative to the mirror: $\mathbf{v}_{LM} = \mathbf{v}_{LE} + \mathbf{v}_{EM}$. In terms of the magnitudes, $|\mathbf{v}_{LM}| = (c^2 - v^2)^{1/2}$.

The analysis of this experiment involved the expressions $c \pm v$ and $(c^2 - v^2)^{1/2}$ for the speed of light relative to the earth, but the null result indicated that $v = 0$. This implies that all three expressions for the speed of light reduce to the value c.

EXERCISE 1. Use the binomial expansion in Eqs. 39.1 and 39.2 to obtain the expression $\Delta T = (L_0/c)(v/c)^2$.

The Contraction Hypothesis

The Michelson–Morley null result left scientists bewildered. The earth's motion relative to the ether, and hence indirectly the ether itself, could not be detected in this way. In 1889, G. F. FitzGerald suggested that perhaps the arm parallel to the earth's motion contracts by just the right amount to make $\Delta T = T_1 - T_2 = 0$. By comparing Eqs. 39.1 and 39.2 we see that the natural length L_0 in Eq. 39.1 should therefore be replaced by a "contracted" length, L, given by

$$L = L_0(1 - v^2/c^2)^{1/2} \tag{39.4}$$

This idea also occurred to H. A. Lorentz in 1892. He accounted for the null result by assuming that a contraction occurred because the electrical forces within a body were modified in the direction of motion through the ether.

39.3 COVARIANCE

Besides the questions regarding the ether, there were other theoretical problems in electromagnetism. According to the Galilean principle of relativity, the laws of mechanics have the same *form* in all inertial frames. For example, Newton's second law, $F = ma$, in one frame has the same form, $F' = ma'$, in another. We say that the laws of mechanics are **covariant**—they retain their form—with respect to the Galilean transformation (Section 4.8). In fact, the Galilean transformation equations

$$x' = x - vt; \qquad t' = t$$

show that $a' = a$, which means $F' = F$. Quantities that have the same value in two frames moving relative to each other are said to be **invariant.** As another example of covariance, note that the law of conservation of linear momentum in one frame, $m_1u_1 + m_2u_2 = m_1v_1 + m_2v_2$, has the same form, $m_1u_1' + m_2u_2' = m_1v_1' + m_2v_2'$ in another, even though the velocities have different values in different frames. This is a satisfying state of affairs: Mechanics on a ship moving at constant velocity is the same as it is on land. However, when the same considerations are applied to the laws of electromagnetism, problems arise. We consider three.

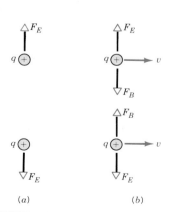

FIGURE 39.3 (a) In a frame in which two equal charges are at rest, they experience only an electric repulsion. (b) In a frame in which both charges have the same velocity, they also experience a magnetic attraction.

THREE PROBLEMS

1. Consider two equal point charges q moving with the same velocity. In a frame moving with the charges, they are at rest (Fig. 39.3a) and experience only an electrostatic repulsion $F' = F_E$. In our "laboratory" frame, in which the charges move at speed v (Fig. 39.3b), each charge creates a magnetic field. The force between the charges is therefore reduced by the magnetic attraction F_B, that is, $F = F_E - F_B$. The force between the charges depends on the frame of reference employed. This clearly conflicts with the classical invariance of force in Newtonian mechanics.

2. When the Galilean transformation is applied to Maxwell's wave equation, its form changes completely. So if the Galilean transformation equations are

correct, Maxwell's equations are valid in only one special frame—that of the ether. However, there is no evidence that Maxwell's equations are restricted in this way.

3. Consider a short wire moving at constant velocity across the pole face of a magnet. In the magnet's frame (Fig. 39.4a), the magnet is at rest and the wire moves at velocity $+v$. An observer in this frame says that a charge q in the wire experiences a magnetic force. In the wire's frame (Fig. 39.4b), the wire is at rest and the magnet has a velocity $-v$. Since the charges are at rest in the wire's frame, an observer in this frame would say that the charge q is subject to an electric force. We know experimentally that it is only the *relative* motion of the wire and the source of the magnetic field that matters. Yet merely switching from one inertial frame to another requires a change from a magnetic field to an electric field. Although both observers agree on the phenomenon, they use different laws to describe it.

As a student, Albert Einstein (Fig 39.5) was aware of these and other problems. Indeed, as a boy of 16, he conceived of an intriguing question: What would one see if one travels with a beam of light? One should see a *stationary* sinusoidal variation in space of the electric and magnetic fields that constitute the wave. But this is not an acceptable solution of Maxwell's wave equation—which requires a wave moving at c. Could the laws for the traveler be different from those for an observer at rest?

Although by 1904 Einstein had found out about the Michelson–Morley experiment through the work of Lorentz, this experiment did *not* play a significant role in the formulation of his theory. Recall from Eq. 39.3 that the Michelson–Morley result depended on the second-order quantity $(v/c)^2$. The results of earlier optical experiments, involving only first order in v/c, and the problems regarding the covariance of the laws of electromagnetism were sufficient for him to develop his theory.

Einstein had to make a choice. If the Galilean transformation and the laws of mechanics were correct, then Maxwell's equations had to be reformulated. If Maxwell's equations were correct, then the laws of mechanics were not exactly correct, even though no exception had yet been encountered. The striking success of Maxwell's theory made it improbable that it was incorrect, so he decided that the Galilean transformation and the laws of mechanics had to be modified. Einstein believed that there must exist some powerful "universal principle" that would guide him to the "true" laws of physics. In June 1905, in a paper entitled "On the Electrodynamics of Moving Bodies," Einstein introduced the special theory of relativity. Here is the opening passage:

> It is known that Maxwell's electrodynamics—as usually understood at the present time—when applied to moving bodies, leads to asymmetries which do not appear to be inherent in the phenomena. Take for example, the reciprocal electrodynamic action of a magnet and a conductor. The observable phenomenon here depends only on the relative motion of the conductor and the magnet, whereas the customary view draws a sharp distinction between the two cases in which either the one, or the other of these bodies, is in motion. Examples of this sort, together with the unsuccessful attempts to discover any motion of the earth relative to the "light medium," suggest that the phenomena of electrodynamics as well as of mechanics possess no properties corresponding to the idea of absolute rest.

He then went on to state the two postulates on which the theory is based. These are the "universal principles" he sought.

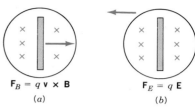

FIGURE 39.4 (a) The charges in a rod moving across the face of a magnet experience a magnetic force. (b) If the rod is at rest and the magnet moves in the opposite direction, the charges in the rod experience only an electric force.

$F_B = q\,v \times B$ $F_E = q\,E$
(a)   (b)

FIGURE 39.5 Albert Einstein (1879–1955).

39.4 THE TWO POSTULATES

Based on the problems discussed above, Einstein made two assumptions that are the starting points of the theory. He would take them to be true and study their logical consequences. The results or predictions could be tested experimentally. The **two postulates** in the **theory of special relativity** are:

1. THE PRINCIPLE OF RELATIVITY: All physical laws have the same form in all inertial frames.

2. THE PRINCIPLE OF THE CONSTANCY OF THE SPEED OF LIGHT: The speed of light in free space is the same in all inertial frames. It does not depend on the motion of the source or the observer.

Both postulates are restricted to inertial frames. This is why the theory is *special*. (In 1916 Einstein included accelerated frames in the *general theory of relativity*.)

The principle of relativity extends the concept of covariance from mechanics to *all* physical laws. Since all inertial frames are equally suitable for formulating physical laws, there is no "absolute" or "privileged" frame. The mysterious ether was blown away. This postulate immediately accounts for the Michelson–Morley result because no experiment can ever give the "absolute" velocity of an inertial frame. The first postulate is both a restriction and a guide in the formulation of any physical theory.

The second postulate may be deduced from the first when it is applied to Maxwell's wave equation—whose only solution is a wave moving at speed c. As we saw, it also follows as an interpretation of the Michelson–Morley null result. Lorentz had arrived at this statement theoretically via complex reasoning and arbitrary hypotheses. In contrast, Einstein simply *asserted* that this is the way things are and offered no "explanation." (This is reminiscent of Newton's approach to the law of gravitation.)

The idea that the speed of light does not depend on the motion of either the source or the observer is difficult to accept. If light is a wave, we expect its speed to be measured with respect to some medium—but there is no ether. If light consists of particles (as some had suggested), its speed should be measured with respect to the source. The evidence indicates that this is not correct. It is important to realize that the theory of special relativity does not speculate either on the nature of light itself, or how it propagates through space. These two models of light, the wave and the particle, are enormously fruitful but can hardly claim to be "true" descriptions. We simply have not come up with anything better. Consequently, do not worry about trying to "understand" the second postulate by visualizing some physical process. The point to keep in mind is that *all the experimental consequences have confirmed its correctness.*

39.5 SOME PRELIMINARIES

In special relativity an **event** is something that occurs at a single *point* in space at a single *instant* in time. Suppose there is a post beside a track along which a train is moving, as shown in Fig. 39.6. We cannot say "The train passed the post at 12 noon," since it took some time to go past. We can specify two events: The front coincides with the post at t_1 and the rear coincides with the post at t_2. The events are the coincidences.

An **observer** is either a person, or an automatic device, with a clock and a meter stick. Each observer can record events *only in the immediate vicinity*. Each has to rely on colleagues at other locations to record the times of distant events.

FIGURE 39.6 A train passing a post beside the track. In the first event the front of the train coincides with the post. The second event is the coincidence of the rear of the train and the post.

An observer may see or photograph a distant event, but these observations do not count as a record of the event.

A **reference frame** is a whole set of observers uniformly distributed in space as in Fig. 39.7. All observers in a given reference frame agree on the position and the time of an event. Only one observer would actually be close enough to record it, but the data are communicated to the others at a later time. We will use the notation $S, S', S'', \ldots$ to denote different inertial frames. The frame in which an object, for example a clock or a rod, is at rest is called its **rest frame.** We will use the notation shown in the table in the margin.

In special relativity it is extremely important to define precisely what is meant by the time in a given reference frame. This requires a careful procedure for the synchronization of clocks.

x	Position coordinate of an event, a *point* in space.
$\Delta x = x_2 - x_1$ $= L$	A space *interval*, a length
t	Time coordinate of an event, an *instant* in time
$\Delta t = t_2 - t_1$ $= T$	A time *interval*, a period

Synchronization of Clocks

Suppose synchronized clocks are to be placed at different locations along a line. We could first bring them together, synchronize them, and then move them to their desired locations. However, we already suspect, from the work of Lorentz, that the time kept by a clock may depend on its motion. Einstein proposed that the clocks be placed equidistant from each other, as in Fig. 39.8. Each location has an antenna for receiving and transmitting signals. To check the equality of the distances, B emits a light flash and notes that the reflections from A and C arrive back at the same instant, 2 seconds later. C can do the same with B and D, and so on down the line. Thus it takes 1 s for light to travel from one position to the next. All observers agree that A will emit a flash when his clock strikes 12 noon. Clock B is set at 1 s past noon, C is set 2 s past noon, and so on. At the appointed instant, the flash is emitted from A and triggers each of the other clocks in turn. All the clocks are now synchronized.

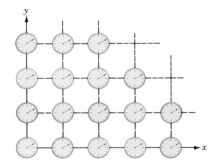

FIGURE 39.7 A reference frame is assumed to consist of many observers uniformly spread throughout space. Each observer has a meter stick and a clock to make measurements only in the immediate vicinity.

FIGURE 39.8 In order to synchronize four equally spaced clocks, a signal is sent out by clock A to trigger the other clocks—each of which has been set ahead by the amount of time it takes light to travel from A to the given clock.

39.6   RELATIVITY OF SIMULTANEITY

If two events occur in close proximity at the same time, they are clearly simultaneous. How can we determine whether two events at different locations are simultaneous? Suppose an observer O is equidistant from points A and B on a platform (frame S), as in Fig. 39.9. If two firecrackers explode at A and B simultaneously (according to frame S), observer O will receive the flashes at the same time. This is our definition of simultaneity: *Two events at different locations are simultaneous if an observer midway between them receives the flashes at the same instant.*

The explosions also leave marks at points A' and B' on a train (frame S') that is moving at constant velocity along the platform. As Fig. 39.10 shows, an observer O' who is midway between A' and B' receives the flash from B before the flash from A. Thus, observers in S' will not judge the two explosions to be simultaneous:

Spatially separated events that are simultaneous in one frame are not simultaneous in another, moving relative to the first.

This effect, called the **relativity of simultaneity,** is entirely reciprocal: Events simultaneous for O' will not be so for O. Since events that are simultaneous but at different locations in frame S are not simultaneous in frame S', the time *interval* between events will be different in the two frames. As we show next, the measurement of length is also affected by the relativity of simultaneity.

To measure the length of a stationary rod, we place a meter stick beside it and record the positions of its ends. What does one do if the rod is moving? We would not think of recording the position of the front at one time and that of the rear a little later! The positions must be recorded simultaneously. Figure 39.11 shows a rod $A'B'$ whose length measured at rest is L'. It moves at speed v relative to frame S. To measure the length of the rod in the frame S, we require *two* spatially separated observers A and B. They take great trouble to record the positions of the ends simultaneously. (As a practical matter, all the observers in frame S could take data at $t = 0$. Only two of them, A and B, happen to record the positions of the ends A' and B'.) However, observers in frame S', moving with the rod, will complain that the measurements by A and B were not made at the same instant. Consequently, observers in frames S and S' will not agree on the length of the rod.

It is the relativity of simultaneity that leads to the length contraction introduced by FitzGerald and Lorentz. They believed that there is a physical contraction caused by the modification of electrical forces between atoms. The theory of special relativity arrives at the same result from a very different viewpoint—a profound analysis of the process of measurement.

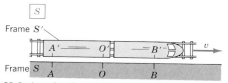

FIGURE 39.9 A train, frame S', moves relative to a platform, frame S. At a certain instant in S, explosions occur at A and B and leave marks at points A' and B' on the train. Observer O is midway between A and B and observer O' is midway between A' and B'.

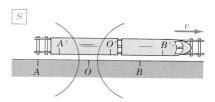

FIGURE 39.10 The flash from the explosion at B reaches O' before the flash from the explosion at A. Although the explosions are simultaneous for observer O, they are not simultaneous for observer O'.

FIGURE 39.11 To measure the length of a moving rod, observers at A and B locate its ends simultaneously. However, for observers in the frame of the rod, the two measurements are not simultaneous.

39.7 TIME DILATION

In order to determine how the relative motion of two frames affects the measured time interval between two events, we consider the "light clock" shown in Fig. 39.12. A pulse of light is emitted from a source A', reflected by a mirror M at a distance L_0, and detected by B', which is close to A'. The time interval between the events of emission and detection in the rest frame S' of the clock is

$$T_0 = \Delta t' = \frac{2L_0}{c} \qquad (39.5)$$

A **proper time,** T_0, is the time *interval* between two events as measured in the rest frame of a clock. In this frame both events occur at *the same position.*

Now let us find the time interval recorded in the frame S, in which the clock has velocity v. The time interval Δt in frame S is measured by *two* observers A and B at different positions. From Fig. 39.13 we see that

$$\left(c \cdot \frac{\Delta t}{2}\right)^2 = L_0^2 + \left(v \cdot \frac{\Delta t}{2}\right)^2$$

which leads to

$$T = \Delta t = \frac{2L_0}{c} \cdot \left(\frac{1}{\sqrt{1 - v^2/c^2}}\right) \qquad (39.6)$$

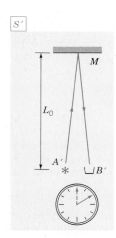

FIGURE 39.12 A "light clock." The time taken for light to travel from the source A' to the detector B' is $2L_0/c$ *in the frame in which the clock is at rest.* The emission and detection events occur at the same place in this frame.

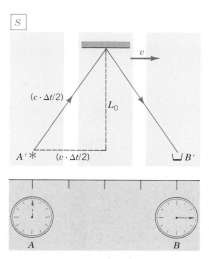

FIGURE 39.13 In a frame in which the clock is moving, the emission and detection events occur at *two* different positions. The time interval recorded is greater than that recorded in the rest frame of the clock.

Note that we have used c as the speed of light in both frames—in accord with the second postulate. The expression in brackets is given a special symbol:

$$\gamma = \frac{1}{\sqrt{1 - v^2/c^2}} \qquad (39.7)$$

Values of γ for a range of values of v/c are given in Table 39.1. Comparing Eq. 39.5 and Eq. 39.6 we see that

$$T = \gamma T_0 \qquad (39.8)$$

Since $\gamma > 1$, the time interval T measured in frame S (by two clocks) is greater than the proper time, T_0, registered by the clock in its rest frame S'. This effect is called **time dilation:**

Two spatially separated clocks, A and B, record a greater time interval between two events than the proper time recorded by a *single* clock that moves from A to B and is present at both events.

Time dilation is an entirely reciprocal effect. If Δt were the proper time (an interval) for a clock in S, then two observers in S' would measure $\Delta t' = \gamma \Delta t$. If this effect were not reciprocal, it would be a way to distinguish between inertial frames—in contradiction to the first postulate. Note that time dilation applies to any periodic phenomenon, be it electronic, mechanical, or biological.

TABLE 39.1*

$\dfrac{v}{c}$	γ
0.6	5/4
0.8	5/3
0.98	5
0.995	10
0.9965	12
0.9992	25

* Some values of γ have been rounded off.

EXAMPLE 39.1: Two clocks A and B are at rest and 100 m apart in frame S. Clock A', at rest in frame S', moves with velocity $0.6c$ along the line joining A to B. How long does it take to get from A to B according to observers in S and S'?

Solution: For convenience we assume that clocks A and A' register zero when they coincide, as shown in Fig. 39.14a. In frame S the time taken for A' to coincide with B (Fig. 39.14b) is

$$T = \Delta t = \frac{100 \text{ m}}{1.8 \times 10^8 \text{ m/s}} = 5.55 \times 10^{-7} \text{ s}$$

Clock A' records the proper time for the journey. (Why?) From Eq. 39.8,

$$T_0 = \left(\frac{1}{\gamma}\right) T$$

$$= \left(\frac{4}{5}\right) (5.55 \times 10^{-7} \text{ s})$$

$$= 4.44 \times 10^{-7} \text{ s}$$

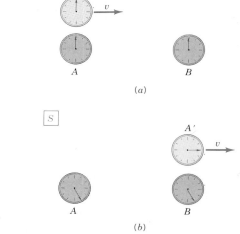

FIGURE 39.14 (a) Clocks A and A' are synchronized when they coincide. (b) When clock A' coincides with clock B, their readings do not agree. Clock B registers a later time.

The reality of time dilation was verified in an experiment performed by B. Rossi and D. B. Hall in 1941 and repeated in simplified form by D. H. Frisch and J. H. Smith in 1963. An elementary particle, the muon (μ), decays into other particles. With N_0 muons at $t = 0$, the number remaining at a later time t is

$$N = N_0 e^{-t/\tau}$$

where $\tau = 2.2$ μs is called the mean lifetime. Among other sources, muons are produced in the upper atmosphere from the bombardment of cosmic ray protons. The experiment involved the comparison of the number (N_1) of muons detected at an altitude of 2000 m with the number (N_2) detected at sea level, as shown in Fig. 39.15. Muons, whose speed was known to be $v = 0.995c$, should take 6.7 μs to travel from the mountaintop to sea level. From the above equation, the ratio of the numbers should be

$$\frac{N_2}{N_1} \approx e^{-3} = 0.05$$

Instead it was found that N_2/N_1 was 0.7. That is, instead of detecting just 5% at sea level, they found that 70% had survived! Since this value corresponds to $0.7 = \exp(-0.3)$, it meant that the moving muons lived 10 times longer than their cousins that decay at rest in the laboratory. This is precisely what time dilation would predict. For the given speed, $\gamma = 10$ and τ is the proper time. In our laboratory frame the lifetime would be $\tau' = \gamma\tau = 10\tau$.

EXERCISE 2. At what speed would time dilation lead to a result 10% greater than the proper time?

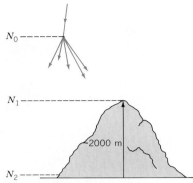

FIGURE 39.15 A number N_0 of muons are produced at some altitude. At the peak of a mountain there are N_1 remaining, whereas at sea level there are N_2 remaining. Many more muons "survive" at sea level than is predicted by classical physics.

39.8   LENGTH CONTRACTION

Consider a rod AB at rest in frame S, as shown in Fig. 39.16. The distance between its ends is its proper length L_0:

> The **proper length,** L_0, of an object is the space interval between its ends measured in the rest frame of the object.

An observer O' in frame S', which moves at velocity v relative to frame S, can measure the rod's length by recording the interval between the times at which O' passes A and B. Figures 39.16 and 39.17 show the two events in frames S and S', respectively. The measurements in the two frames are

$$\text{Frame } S: \quad L_0 = \Delta x = v\,\Delta t$$

$$\text{Frame } S': \quad L = \Delta x' = v\,\Delta t'$$

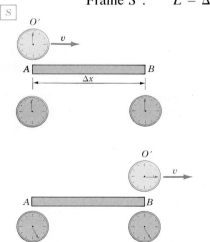

FIGURE 39.16 The observer O' can determine the length of the rod, which is at rest in frame S, by measuring the time taken to travel from A to B.

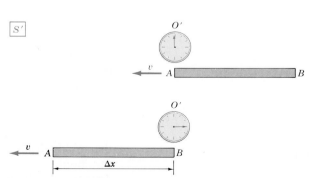

FIGURE 39.17 In frame S', in which observer O' is at rest, the rod moves in the opposite direction. It has a contracted length, less than its proper length in its rest frame, S.

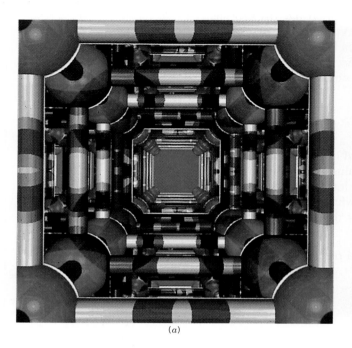

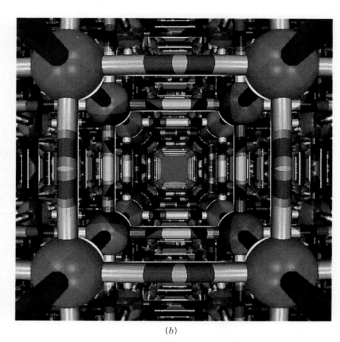

(a)   (b)

Computer generated graphics show the visual appearance of a three-dimensional lattice of rods and balls moving toward you at various speeds. (a) The normal view at rest. (b) Even at 0.5 c the rods appear straight. (c) At 0.95 c the rods appear bent. (d) At 0.99 c, the lattice appears severely distorted.

where Δt is the time interval in frame S between the measurements of the ends, and $\Delta t'$ is the proper time measured by O'. From Eq. 39.8, we have $\Delta t = \gamma \Delta t'$; therefore,

$$L = \frac{1}{\gamma} L_0 \tag{39.9}$$

Since $\gamma > 1$, we see that $L < L_0$. That is, the length measured in a frame moving relative to the rod is less than its proper length. This effect, called **length contraction**, is reciprocal: A rod at rest in frame S' will have a contracted length in frame S. Note that only lengths parallel to the direction of motion are affected; lengths perpendicular to v are unaffected.

It is misleading to state that the moving rod "appears contracted." A single observer viewing the rod would not, in general, see or photograph the contraction. Careful analysis of the times of travel from various parts of a moving box shows that a single observer photographs the body in a rotated position and with a distorted shape as in Fig. 39.18. If a photograph of a train at rest looks like Fig. 39.19a, then a photograph of the train moving at 0.9c would look like Fig. 39.19b.

The muon decay experiment can also be interpreted in terms of length contraction. In the land frame, S, the mountain is 2000 m high—its proper length. In the muon's frame, S', the mountain is moving toward it at $v = 0.995c$. In frame S', the mountain's contracted length is $L' = L_0/\gamma = 200$ m (see Fig. 39.20). This distance would be covered in $\Delta t' = L'/v = 0.67$ μs, which is only one-third of the mean lifetime.

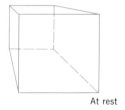

At rest

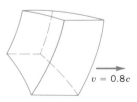

$v = 0.8c$

FIGURE 39.18 A view of a box at rest and a view of the box photographed by a single observer shows a distorted shape.

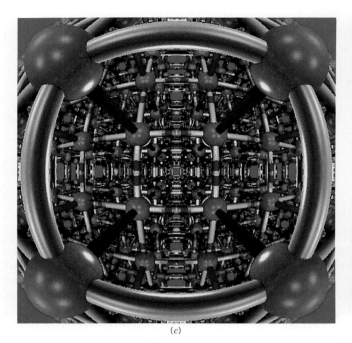

(c)

(d)

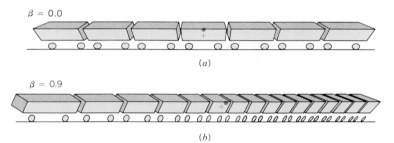

β = 0.0

(a)

β = 0.9

(b)

FIGURE 39.19 (a) The appearance of a train at rest as seen in a photograph. (b) The appearance of that train as it moves at 0.9c past a camera.

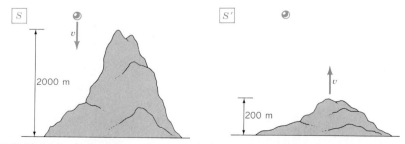

S

v

2000 m

S'

v

200 m

FIGURE 39.20 (a) In the land frame the mountain is 2000 m high and the muons are moving downward. (b) In the muon frame, the mountain is only 200 m high and is moving upward.

EXAMPLE 39.2: A platform has a proper length of 200 m. An engine takes 0.5 μs to travel from one end of the platform to the other, as measured by the train driver. What is the relative speed?

Solution: We are given the proper length of the platform, $L_0 = 200$ m, and the proper time, $T_0 = 0.5$ μs measured by the train driver. In the frame of the train, the platform has a contracted length

$$L = \frac{L_0}{\gamma} \qquad \text{(i)}$$

If v is the relative speed, then it is also true that

$$L = vT_0 \qquad \text{(ii)}$$

We equate (i) and (ii) and square both sides to obtain

$$L_0^2(1 - v^2/c^2) = v^2T_0^2$$

On rearranging this expression we find

$$v^2 = \frac{c^2}{1 + c^2T_0^2/L_0^2}$$

The given values of T_0 and L_0 lead to $v = 2.4 \times 10^8$ m/s $= 0.8c$.

39.9 THE RELATIVISTIC DOPPLER EFFECT

In the classical Doppler effect for sound waves (Section 17.3), the observed frequency depends differently on the velocities of the source and the observer. The underlying reason is that for sound there is a medium (the air) that serves as an "absolute" reference frame. In contrast, for light there is no absolute frame: The relativistic Doppler effect for light depends only on the *relative* velocity between source and observer. Figure 39.21 shows a source at O', the origin of frame S', that emits flashes with a period $\Delta t' = T_0$, and moves at velocity v relative to frame S. If one flash is emitted when O' coincides with the origin O in frame S, at what time will the next flash arrive at the *same* point? Observer B in frame S, who coincides with O' when the next flash is emitted, will record a dilated time interval $\Delta t = \gamma\Delta t' = \gamma T_0$ between the flashes. In this time interval, O' has moved a distance $d = v\Delta t = v\gamma T_0$, so this flash takes a time d/c to reach O. Thus the time of arrival at O is

$$T = \Delta t + \frac{d}{c} = \gamma\left(1 + \frac{v}{c}\right)T_0$$

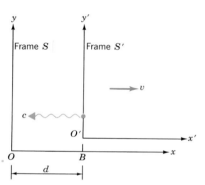

FIGURE 39.21 A source of the origin of frame S' emits a pulse toward the origin of frame S.

Since $\gamma = (1 - v/c)^{-1/2}(1 + v/c)^{-1/2}$, we find

$$T = \sqrt{\frac{c + v}{c - v}}\ T_0 \qquad (39.10)$$

The observed frequency, $f = 1/T$, is

(Longitudinal) $$f = \sqrt{\frac{c - v}{c + v}}\ f_0 \qquad (39.11)$$

This equation applies when the source and observer move *away* from each other. When they approach, the sign of v is changed. Since the signal travels along the direction of motion, the effect is called the *longitudinal* Doppler effect. If the signals are detected perpendicular to the direction of motion, the *transverse* Doppler effect includes only the effect of time dilation, $T = \gamma T_0$, or

(Transverse) $$f = \frac{1}{\gamma}f_0 \qquad (39.12)$$

39.10 THE TWIN PARADOX

Nothing in the theory of relativity catches the imagination more than the so-called twin paradox. Twin A stays on earth while twin B travels at high speed to a nearby star. When B returns, they both find that A has aged more than B. The paradox arises because of the apparent symmetry of the situation: In B's frame, it is A that leaves and returns, so one should also find that B has aged more than A. How can both points of view be correct? The early opponents of relativity believed this paradox pointed to a hidden flaw in the theory. We follow a discussion first presented in 1911 by P. Langevin.

Twin A is at the origin of frame S, while B is at the origin of frame S'. To simplify matters we assume B has already reached his cruising speed v as he passes A at $t = t' = 0$. On the return trip, they both record the time at which B flies past A. This procedure allows us to avoid having to deal with the initial and final acceleration phases of the motion. We have to calculate the duration of the journey as recorded separately by A and by B. To accomplish this, twin B emits N pulses of period T_0 on the outward trip and the same number on the return trip. According to B, the duration of each part of the journey at constant speed is therefore NT_0. For A, who makes his observations at a single point, we need the formula for the Doppler effect (Eq. 39.10). The measurements are presented below for each phase of B's journey. To make the results concrete we have taken $N = 20$, $T_0 = 5$ min, $v = 0.6c$, which means $\gamma = 5/4$.

(B)   (A)

Outward trip:

$$\Delta t_1' = NT_0 = 100 \text{ min} \qquad \Delta t_1 = NT_0 \sqrt{\frac{c+v}{c-v}} = 200 \text{ min}$$

Return trip:

$$\Delta t_2' = NT_0 = 100 \text{ min} \qquad \Delta t_2 = NT_0 \sqrt{\frac{c-v}{c+v}} = 50 \text{ min}$$

Total: $T' = 2NT_0 = 200$ min $T = \gamma(2NT_0) \quad = 250$ min

On the outward trip A receives N pulses with an expanded period, while on the return trip he receives N pulses with a shortened period. The times T and T' recorded by A and B respectively for the total trip are related by the time dilation equation $T = \gamma T'$. Thus A has aged more than B.

The paradox is partially resolved once we realize that the situation is *not* symmetrical. Twin B's turnaround involves acceleration (or, equivalently, he switches from one inertial frame to another). It is B who experiences the forces due to the firing of rockets, whereas A's existence is uneventful. A calculation in A's frame for the acceleration phase leads to a correction of the time calculated earlier, but does not affect the general conclusion that A has aged more than B. To resolve the paradox fully we would need to show that B agrees with A's calculation for the period of acceleration. This requires the general theory of relativity, which treats accelerated frames. A detailed calculation confirms the above discussion.

Although the predictions of special relativity are fully verified, there was still the need to erase any doubts regarding the twin paradox by using macroscopic clocks. In 1971, J. C. Hafele and R. E. Keating flew around the earth, once eastward and once westward, on commercial jets. They compared the time kept on the plane by four ceasium atomic clocks (that can measure time to 10^{-9} s) with that kept by identical clocks on the ground. Although their results included effects due to gravitation, the results were consistent with special relativity.

EXERCISE 3. Supply the algebraic steps that reduce $(\Delta t_1 + \Delta t_2)$ to $\gamma(2NT_0)$.

EXERCISE 4. Suppose A emits pulses with period T_0. B again divides his journey into two equal parts of 100 min according to *his* clock. Show that B receives (a) 10 pulses on the outward trip, and (b) 40 pulses on the return trip.

39.11   THE LORENTZ TRANSFORMATION

The laws of electromagnetism are not covariant with respect to the Galilean transformation $x' = x - vt$, $t' = t$. This transformation also conflicts with the principle of the constancy of the speed of light. A derivation of the correct *Lorentz transformation* of coordinates is given in Section 39.15. For the present we merely state it and discuss some of its consequences. We assume frame S' moves at velocity v along the x axis of frame S, as in Fig. 39.22. The y and z coordinates of an event will be the same in both frames, that is, $y' = y$ and $z' = z$. The x and t coordinates are related according to the **Lorentz transformation:**

$$x' = \gamma(x - vt) \tag{39.13}$$

$$t' = \gamma\left(t - \frac{vx}{c^2}\right) \tag{39.14}$$

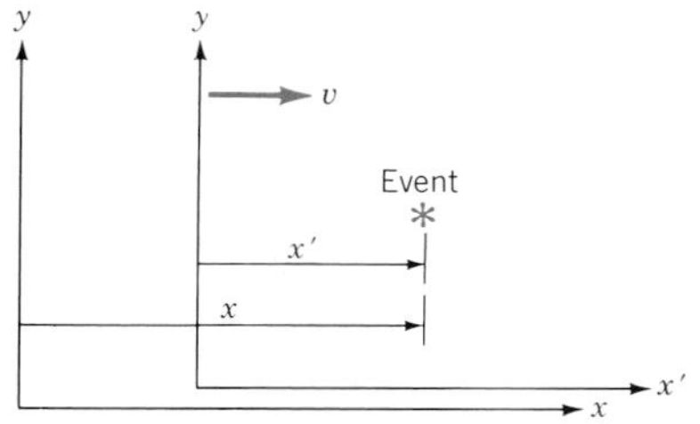

FIGURE 39.22 The x coordinates of an event as measured in frames S and S' (moving at speed v relative to S) are related according to the Lorentz transformation: $x' = \gamma(x - vt)$.

These equations relate the position and time coordinates of *single* events, such as flashes or coincidences, as measured in two inertial frames. Notice that in the limit as $v \to 0$, the quantity $\gamma \to 1$, that is, these equations reduce to the Galilean transformation at low speeds. Equation 39.14 shows that the time t' measured in frame S' depends on *both* t and x. Space and time have become inseparably mixed into something called **spacetime.** This is a consequence of the fact that the speed of light, a ratio of a space interval over a time interval, is a *universal* constant—it is the same in all inertial frames.

The coordinates of an event in spacetime are often written in the form:

$$x_1 = x; \qquad x_2 = y; \qquad x_3 = z; \qquad x_4 = ct$$

The "timelike" coordinate x_4 now has the same unit as the "spacelike" coordinates x_1, x_2, and x_3. With the definition $\beta = v/c$, the Lorentz transformation equations become

$$x_1' = \gamma(x_1 - \beta x_4) \tag{39.15}$$
$$x_4' = \gamma(x_4 - \beta x_1) \tag{39.16}$$

These two equations are identical in form. They beautifully illustrate the space–time symmetry of the Lorentz transformation. Since x_4 now has the same status as x_1, it is often associated with the "fourth" dimension.

Since only relative velocity is significant, a mere sign change yields the inverse transformation:

$$x = \gamma(x' + vt') \tag{39.17}$$

$$t = \gamma\left(t' + \frac{vx'}{c^2}\right) \tag{39.18}$$

The Lorentz transformation equations lead immediately to time dilation and length contraction as the next example shows.

EXAMPLE 39.3: Obtain equations for (a) time dilation, and (b) length contraction from the Lorentz transformation.

Solution: (a) Consider a clock that is at rest in the frame S'. Say $\Delta t'$ is the time interval between two events, for example, "tick" and "tock." Since the events occur at the same place in S', we have $\Delta x' = 0$. From Eq. 39.18 we obtain

$$\Delta t = \gamma \Delta t'$$

(b) Suppose a rod has a length $\Delta x'$ in its rest frame S'. Observers in S record the positions of its ends at the same time, so $\Delta t = 0$. From Eq. 39.13 we see that

$$\Delta x = \frac{1}{\gamma} \Delta x'$$

Notice that neither time dilation nor length contraction depend on the sign of v.

EXAMPLE 39.4: Figure 39.23a shows a train (frame S') of proper length $L_0 = 9$ km, moving at velocity $v = 0.8c$ relative to a platform (frame S). At $t' = 0$, observers A' and B' at the ends of the train fire bullets that make marks at A and B on the platform. According to observers in S, what is (a) the time interval and (b) the space interval between the shots?

Solution: (a) With $\Delta t' = 0$ and $\Delta x' = x'_B - x'_A = L_0$, the interval between the shots in frame S is given by Eq. 39.18, where $\gamma = 5/3$:

$$t_B - t_A = + \frac{\gamma v L_0}{c^2}$$

$$= \frac{(5/3)(2.4 \times 10^8 \text{ m/s})(9 \times 10^3 \text{ m})}{9 \times 10^{16} \text{ m}^2/\text{s}^2}$$

$$= 40 \ \mu\text{s}$$

Suppose that the clocks at A and A' both read zero when these points coincide, so $t_A = t'_A = 0$. In frame S, the clock at B reads $t_B = \gamma v L_0/c^2$, which means that the bullets are not fired simultaneously: The bullet from the *rear* was fired *first*. According to observers in frame S', the clocks in S are *out of synchronization*. Since frame S is moving to the left relative to frame S', the trailing clock leads. That is, the clock at B is set ahead of the clock at A.

(b) Figure 39.23*b* shows the version of events in frame S, in which the train has a contracted length $L_0/\gamma = 5.4$ km. At $t = 0$, the mark is made at A. The front of the train has not yet reached B. At $t = \gamma v L_0/c^2$, the front reaches B after moving a distance vt_B. The distance between A and B is

$$L = \frac{L_0}{\gamma} + v \left(\frac{\gamma v L_0}{c^2} \right) = \gamma L_0$$

The marks are *farther* apart than the proper length L_0. This does not contradict our findings on length contraction because the ends of the train were not recorded simultaneously in frame S.

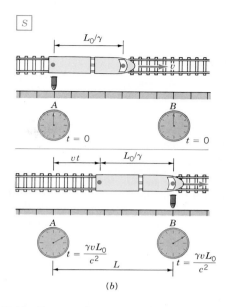

FIGURE 39.23. (Repeat). (*b*) In the platform frame, the train has a contracted length and the bullet from the rear is fired first.

EXERCISE 5. Obtain the result $L = \gamma L_0$ of part (b) of Example 39.4 directly from the Lorentz transformation equations.

EXAMPLE 39.5: In frame S, events A and B occur at different positions. Furthermore, event B occurs after event A. Is it possible for event B to precede event A in another frame S' moving at constant velocity relative to frame S? If so, does this mean an effect can precede its cause?

Solution: In frame S the *space interval* between the events is $\Delta x = x_B - x_A$ and the *time interval* between them is $\Delta t = t_B - t_A$. From Eq. 39.14, the time interval in frame S' is given by

$$\Delta t' = \gamma \left(\Delta t - \frac{v \Delta x}{c^2} \right)$$

We see that if $\Delta t < v \Delta x/c^2$, then $\Delta t' < 0$, which means that the order of *independent* events may be reversed.

Suppose, however, that the events are connected, in the sense that one is caused by, or is a consequence of, the other. This would require a body, or a signal, to travel from A to B. Since $\Delta x/\Delta t = v$, we may rewrite the above expression as $\Delta t' = \gamma \Delta t (1 - v^2/c^2)$. The condition $\Delta t' < 0$ thus implies that $v^2 > c^2$. As we will see later, it is not possible for the speed of a particle to exceed c, and therefore it is not possible for the order of connected events to be reversed: An effect cannot precede its cause.

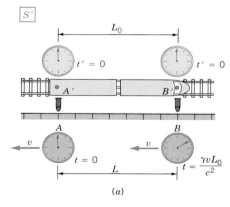

FIGURE 39.23 A train (S') moves at speed v relative to a platform (S). Two bullets are fired simultaneously according to observers on the train. (*a*) Observers on the train find the clocks in S are out of synchronization.

39.12   THE ADDITION OF VELOCITIES

Suppose that a particle has velocity u'_x relative to frame S', which is itself moving at velocity v along the x direction of frame S, as shown in Fig. 39.24. Classically,

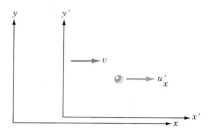

FIGURE 39.24 A particle moves with velocity u'_x, relative to frame S', which is itself moving at velocity v relative to frame S. The velocity of the particle relative to S is *not* simply the sum of v and u'_x.

the velocity of the particle relative to S would be $u_x = u'_x + v$. In the present theory, the particle velocities are given by $u_x = dx/dt$ and $u'_x = dx'/dt'$. From Eq. 39.17 and Eq. 39.18 we have

$$dx = \gamma(dx' + v\,dt') = \gamma\,dt'(u'_x + v) \tag{39.19}$$

$$dt = \gamma\left(dt' + \frac{v\,dx'}{c^2}\right) = \gamma\,dt'\left(1 + \frac{vu'_x}{c^2}\right) \tag{39.20}$$

Taking the ratio of these equations we find

$$u_x = \frac{u'_x + v}{1 + vu'_x/c^2} \tag{39.21}$$

When both u'_x and v are very small compared to c, this expression reduces to the classical result, $u_x = u'_x + v$. If the particle is a pulse of light, then $u'_x = c$. The velocity of the pulse relative to S is therefore

$$u_x = \frac{c + v}{1 + cv/c^2} = c$$

The addition of any velocity to the speed of light still yields the speed of light. This is, of course, specified in the second postulate. The speed of light is an absolute limit. As we will see later, the speed of a material particle may approach, but never equal, c.

EXAMPLE 39.6: Two rockets, A and B, approach each other with speeds of $0.995c$ relative to the earth (E) frame, as in Fig. 39.25. What is the velocity of A relative to B?

Solution: It is easier to keep track of signs if we express the law of addition of velocities in notation similar to that used in Sec. 4.6. That is, v_{AB} is the velocity of A relative to B. Equation 39.21 then takes the form

$$v_{AB} = \frac{v_{AE} + v_{EB}}{1 + v_{AE}v_{EB}/c^2} \tag{39.22}$$

$$= \frac{0.995c + 0.995c}{1 + 0.995^2}$$

We find $v_{AB} = 0.9999c$. The velocity of one particle relative to the other is less than c and not $1.99c$, as one would expect classically.

FIGURE 39.25 Two rockets approach each other with the same speed relative to earth.

39.13 MOMENTUM AND ENERGY

The principle of relativity requires the form of any physical law to be the same in any inertial frame. Maxwell's equations are already covariant with respect to the Lorentz transformation. However, the laws of mechanics do not retain their form under a Lorentz transformation, so they must be modified. As we will see, the relativistic definitions of linear momentum and kinetic energy are not the same as in classical mechanics. We begin with the profound relationship between mass and energy discovered by Einstein.

Mass–Energy Equivalence

Imagine an isolated box of length L, as in Fig. 39.26, with a source of light S at one end and a detector D at the other end. The mass of the box plus detector is M. In Section 34.5, we saw that when light waves transport an energy E, they also transport linear momentum, $p = E/c$. Thus, if the source emits a pulse of light, the box will recoil with some velocity v. From the conservation of linear momentum

$$\frac{E}{c} = Mv$$

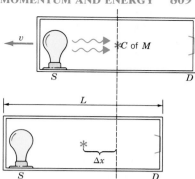

If $v \ll c$, it takes a time $\Delta t = L/c$ for the pulse to reach D. When the light is absorbed, the box experiences an impulse that brings it to rest. In this time the box moves a small distance

$$\Delta x = v\Delta t = \frac{EL}{Mc^2}$$

The net result of the emission followed by the absorption has been to shift the box by Δx.

This is perplexing: No *internal* process can ever move the center of mass of a system. If we are to preserve this idea, we have to assume that the pulse transfers some mass m from the source to the detector, through a displacement L. If the center of mass is fixed, then

$$-M\Delta x + mL = 0$$

FIGURE 39.26 (a) A pulse of light is emitted by a bulb at one end of a box, which recoils in the opposite direction. (b) After the pulse has been absorbed by the detector at the other end, the box comes to rest at a different position.

Comparing this equation with that for Δx obtained earlier, we see that $m = E/c^2$. Einstein reached a remarkable conclusion:

> *If a body gives off the energy E in the form of radiation, its mass diminishes by E/c². . . . The mass of a body is a measure of its energy content.*

Since radiation can be transformed into thermal, electrical, chemical, and other kinds of energy, it follows that the *inertial mass of a body changes whenever it gains or loses any kind of energy.* Thus, in any chemical reaction that produces heat or light, the total mass of the constituents is not constant. The conservation of mass is replaced by the conservation of **mass–energy.** The equation

$$E = mc^2 \qquad (39.23) \qquad \text{Mass-energy equivalence}$$

is perhaps the most famous equation in physics. Einstein himself considered it to be the most significant outcome of special relativity.

Linear Momentum

We saw in Section 39.12 that velocities do not add as in classical mechanics. As a result, if the conservation of linear momentum is to be valid in any inertial frame (as is required by the principle of relativity), a new definition of linear momentum is required. The relativistic definition of **linear momentum** is

$$p = mv \qquad (39.24) \qquad \text{Relativistic linear momentum}$$

where the **relativistic mass,** m, is

$$m = \gamma m_0 = \frac{m_0}{\sqrt{1 - v^2/c^2}} \qquad (39.25)$$

The quantity m_0 is the **rest mass** of the particle—the mass measured in its rest frame. When $v \ll c$, γ approaches 1, and $p \approx m_0 v$, which is the classical expression. Equation 39.25 is sometimes interpreted to mean that the mass of a particle increases with velocity. This can be misleading since the relativistic form of Newton's second law is not $F = ma$, and the relativistic equation for kinetic energy is not $K = \frac{1}{2}mv^2$.

Rest energy: m_0c^2

The quantity m_0c^2 is called the **rest energy** of the particle. The relativistic **kinetic energy** of a particle is defined as the difference between the **total energy**, $E = mc^2$, and the rest energy:

Relativistic kinetic energy

$$K = E - m_0c^2 = m_0c^2\,(\gamma - 1) \tag{39.26}$$

As v approaches c, the factor γ, and hence also the kinetic energy, approaches infinity. The attainment of this value would involve an infinite amount of work. Therefore, *it is impossible to accelerate a particle with a finite rest mass to the speed of light.* For low velocities ($v/c \ll 1$) we may expand γ in a binomial series, $\gamma \approx 1 + \frac{1}{2}\,(v/c)^2 + \frac{3}{8}\,(v/c)^3 + \ldots$, so the kinetic energy takes the form

$$K = m_0c^2\left[\frac{1}{2}\left(\frac{v}{c}\right)^2 + \frac{3}{8}\left(\frac{v}{c}\right)^4 + \cdots\right]$$
$$\approx \tfrac{1}{2}\,m_0v^2$$

At low speeds, the kinetic energy is roughly given by the first term, $\frac{1}{2}\,m_0v^2$, which is the classical kinetic energy of the particle. The remaining terms may be considered to be corrections to this expression.

The rest energy, m_0c^2, represents the sum of all the "internal" energies of the body. This includes electrical energy, nuclear energy, thermal energy, and so on. When we speak of the mass of a body, we could just as well speak of its energy content. In other words, *mass and energy are equivalent.* We are familiar with the concept of mass as a measure of the inertia of a body, independent of its temperature, potential energy, and so on. However, according to special relativity, the mass of a body changes when its temperature changes. When a spring is compressed, the added elastic energy also increases its mass. The quantity $E = mc^2$ was rightly called the *total* energy because it encompasses *all* forms of energy associated with the body. If the mass of a system were to decrease by Δm, an amount of energy $\Delta E = \Delta m\,c^2$ would be released. This is what occurs in the processes of nuclear fission and fusion (Fig. 39.27).

One can relate the relativistic momentum $p = \gamma m_0v$ of a particle to its energy $E = \gamma m_0c^2$ by eliminating v. This leads to (see Problem 1)

$$E^2 = p^2c^2 + m_0^2c^4 \tag{39.27}$$

For particles with zero rest mass, the second term vanishes. We are left with $E = pc$, the relation for light. If $E \gg m_0c^2$, then $E \approx pc$ for a particle whose speed is close to that of light.

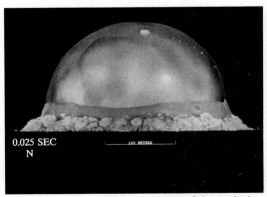

0.025 SEC
N

FIGURE 39.27 A nuclear explosion is convincing evidence of the equivalence of mass and energy.

EXAMPLE 39.7: An electron has a kinetic energy of 2 MeV. Find (a) its total energy, (b) its linear momentum, and (c) its speed.

Solution: (a) The rest energy of the electron is

$$E_0 = m_0 c^2 = (9.11 \times 10^{-31} \text{ kg})(3 \times 10^8 \text{ m/s})^2$$
$$= 8.20 \times 10^{-14} \text{ J} = 0.511 \text{ MeV}$$

The total energy is

$$E = K + m_0 c^2 = 2.51 \text{ MeV} = 4.02 \times 10^{-13} \text{ J}$$

(b) The linear momentum is given by Eq. 39.27:

$$p^2 c^2 = E^2 - m_0^2 c^4 = (16.1 \times 10^{-26} \text{ J}) - (6.72 \times 10^{-27} \text{ J})$$
$$= 15.4 \times 10^{-26} \text{ J}$$

Thus $p = 2.27 \times 10^{-21}$ kg · m/s.

(c) On comparing $E = \gamma m_0 c^2 = 2.51$ MeV with $m_0 c^2 = 0.511$ MeV we see that $\gamma = 4.91$. We rewrite $1/\gamma^2 = 1 - v^2/c^2$ in the form

$$\frac{v^2}{c^2} = 1 - \frac{1}{\gamma^2}$$

to find $v/c = 0.98$, or $v = 0.98c$.

39.14 RELATIVITY AND ELECTROMAGNETISM (Optional)

The fact that c is an unattainable speed for a material particle resolves Einstein's boyhood question regarding what he would see if he were to ride along with an electromagnetic wave. He would not see a stationary sinusoidal variation of electric and magnetic fields because he could *never* catch up with a light wave. We will briefly discuss the issue raised in the quote from the 1905 paper. Recall that Einstein was uneasy about the use of an electric field or a magnetic field depending on one's choice of reference frame.

Figure 39.28*a* shows a positive charge q moving at velocity u relative to a stationary wire that carries a current I. For simplicity we assume that the current in the wire arises from

both positive and negative charges moving with opposite velocities, $\pm v$. In the frame of the wire, the charge q experiences a magnetic force toward the wire, but no net electric force. In the frame in which the charge q is at rest (Fig. 39.28*b*), it does not experience any magnetic force. In this frame, the positive charges in the wire move more slowly than v, whereas the negative charges move faster than v. Electric charge is an invariant in special relativity. Hence, because of length contraction, the negative charge density is greater than the positive charge density. The wire has a net negative charge in the rest frame of charge q. Consequently, charge q experiences an electric force toward the wire. We see that an electrostatic field in the rest frame of charge q transforms into a magnetic field in another frame.

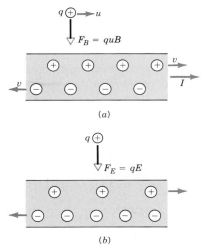

FIGURE 39.28 (*a*) A charge moving relative to a wire. The positive and negative charges in the wire are equally spaced and have equal and opposite velocities. (*b*) In the frame of the charge q, the positive and negative charges in the wire have different speeds. The different factors for length contraction mean that the negative charge density is greater than the positive charge density.

39.15 DERIVATION OF THE LORENTZ TRANSFORMATION (Optional)

The Lorentz transformation equations may be derived by using either postulate. We will use the principle of the constancy of the speed of light. In Fig. 39.29 frame S' moves at velocity v along the x direction of frame S. When the origins coincide, a pulse of light is emitted. By the second postulate, the speed of light is the same in both frames, and so the position of the pulse at a later time is given by

$$x = ct; \qquad x' = ct' \qquad (39.28)$$

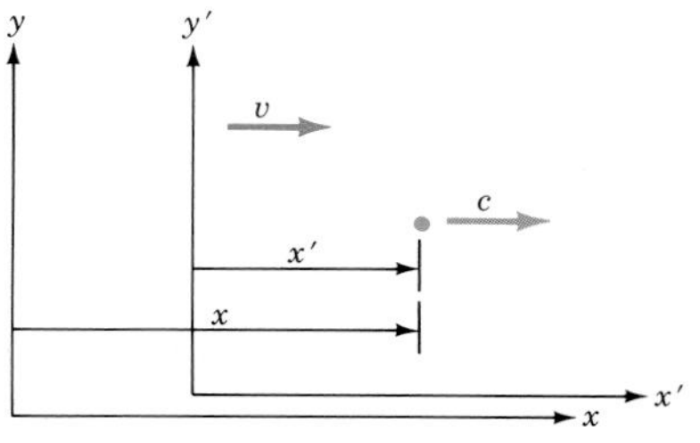

FIGURE 39.29 According to the second postulate, a pulse of light moves at c relative to *both* frames S and S'—which is moving at speed v relative to frame S.

We assume that the position coordinate transforms according to

$$x' = Ax + Bt$$

(The use of terms only to the first power in x and t can be justified by assuming that all points in space and time are equivalent. We omit the details.) The position of O' in frame S' is $x' = 0$, but in frame S it is $x = vt$. Thus $0 = A(vt) + Bt$, which yields $B = -Av$. The equation for x' therefore becomes

$$x' = A(x - vt) \tag{39.29}$$

By a similar argument applied to O but with the sign of v changed, we find

$$x = A(x' + vt') \tag{39.30}$$

We now use Eq. 39.28 in Eqs. 39.29 and 39.30 to obtain

$$ct' = A(ct - vt)$$
$$ct = A(ct' + vt')$$

These two equations lead to

$$A = \frac{1}{(1 - v^2/c^2)^{1/2}}$$

To obtain the transformation equation for the time coordinate, we substitute x' from Eq. 39.29 into Eq. 39.30:

$$x = A[A(x - vt) + vt']$$

Solving for t' then gives

$$t' = A\left(t - \frac{vx}{c^2}\right) \tag{39.31}$$

The derivation for t is much the same. The conventional notation replaces A by γ.

39.16   POLE–BARN PARADOX (Optional)

A pole vaulter and a farmer get into an argument. The athlete's pole and the farmer's barn have the same rest length L_0. They agree that the athlete (frame S') will run toward the barn (frame S) at a relative speed of $0.8c$ ($\gamma = 5/3$). The farmer says that the pole will have a contracted length and so will easily fit into the barn. The athlete claims that the barn will be contracted and so the pole cannot possibly fit inside. The resolution of this paradox lies in the relativity of simultaneity.

The farmer shuts the front (F) and rear (R) doors at $t_F = t_R = 0$. At this instant we assume that the tip B' of the pole coincides with the rear door, as in Fig. 39.30b. In frame S, the pole has a length $3L_0/5$. Therefore, the trailing tip A' must have coincided with the front door at an earlier time $t = -(2L_0/5)/v = -L_0/2c$. shown in Fig. 39.30$a$. So the farmer says the pole is trapped.

In frame S' the doors do not close simultaneously. The times t_F' and t_R' for the closing of F and R, respectively, are related by Eq. 39.14:

$$t_R' - t_F' = \gamma\left[(t_R - t_F) - \frac{v(x_R - x_F)}{c^2}\right]$$

where $t_R = t_F = 0$ and $x_R - x_F = L_0$. For convenience we again set the time of the closing of the rear door to zero, so that $t_R' = 0$, as in Fig. 39.31a. The above equation becomes

$$t_F' = \frac{\gamma v L_0}{c^2} = \frac{4L_0}{3c}$$

Thus in frame S', the front door is closed *after* the rear door, as shown in Fig. 39.31b.

At $t_R = t_R' = 0$, when B' coincides with R, the farmer says A' is inside the barn and the front door is closed. The athlete says that A' is outside the barn but the front door is still open. By the time t_F' that F is closed in frame S', the barn has moved a distance $vt_F' = 16L_0/15$. Herein lies the resolution of the paradox: Observers in both frames agree that B' is to the left of R when it closes and that A' is to the right of F when it closes. They merely disagree on where A' is when F closes.

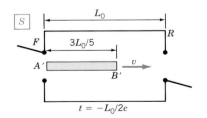

$t = -L_0/2c$

A' coincides with F

(a)

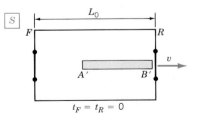

$t_F = t_R = 0$

B' coincides with R, F and R close

(b)

FIGURE 39.30 (a) The rear of the contracted pole coincides with the front door of the barn. (b) The front of the pole coincides with the rear door of the barn.

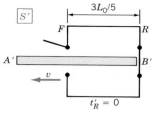
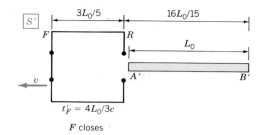

B' coincides with R, R closes F closes

(a)   (b)

FIGURE 39.31 (a) The front of the pole coincides with the rear door of the contracted barn. (b) The pole has crashed through the rear door when the front door of the barn is closed.

SUMMARY

The **two postulates of special relativity** are:

1. All physical laws have the same form in all inertial frames.
2. The speed of light in free space is the same in all inertial frames. It is independent of the motion of the source or of the observer.

Because of the **relativity of simultaneity,** events that are simultaneous for observers in one frame are not simultaneous for observers in another frame moving relative to the first.

The time interval T_0 recorded by a clock O' in its rest frame S' is called the **proper time.** The time interval T, recorded by *two* clocks A and B in frame S is greater than the proper time:

$$T = \gamma T_0$$

where

$$\gamma = \frac{1}{\sqrt{1 - v^2/c^2}}$$

This effect is called **time dilation.**

The length L_0 of a rod in its rest frame S' is called its **proper length.** The length L measured by observers in frame S is smaller:

$$L = \frac{L_0}{\gamma}$$

This effect is called **length contraction.**

The **Doppler effect** for light depends only on the *relative* motion of the source S and the observer O. If S and O approach each other while S emits light of frequency f_0, the frequency f' observed by O is

$$f' = \sqrt{\frac{c + v}{c - v}}\, f_0$$

If S and O recede from each other, the sign of v is changed.

The **Lorentz transformation** from coordinates (x, t) to (x', t') where frame S' moves at $+v$ along the x axis of frame S is

$$x' = \gamma(x - vt);$$

$$t' = \gamma\left(t - \frac{vx}{c^2}\right)$$

In this case $y' = y$ and $z' = z$. The inverse transformation, from (x', t') to (x, t), is obtained by changing the sign of v.

The relativistic **addition of velocities** is done as follows: The velocity of A relative to B, v_{AB} is given by

$$v_{AB} = \frac{v_{AC} + v_{CB}}{1 + v_{AC}v_{CB}/c^2}$$

where v_{AC} is the velocity of A relative to C and v_{CB} is the velocity of C relative to B.

The **linear momentum,** p, of a particle moving at velocity v is

$$p = mv$$

where the **relativistic mass** $m = \gamma m_0$ and m_0 is the **rest mass.** The **total energy,** E, of a particle is

$$E = mc^2$$

This equation expresses the equivalence of mass and energy. The **kinetic energy** of a particle is

$$K = E - m_0c^2 = (\gamma - 1)m_0c^2$$

The total energy and linear momentum are related according to

$$E^2 = p^2c^2 + m_0^2c^4$$

ANSWERS TO IN-CHAPTER EXERCISES

1. Since $v \ll c$, $(1 - v^2/c^2)^{-1} \approx 1 + v^2/c^2$, and $(1 - v^2/c^2)^{-1/2} = 1 + v^2/2c^2$. The difference between these expressions is $v^2/2c^2$. On multiplying by $(2L_0/c)$ we obtain the desired expression.

2. A 10% increase means $\gamma = 1.1$. From the expression for γ we find $(v/c)^2 = 1 - 1/\gamma = 1 - 1/1.1 = 0.091$. Thus $v = 0.3c$.

3. Note that

$$\left[\frac{(c + v)}{(c - v)}\right]^{1/2} + \left[\frac{(c - v)}{(c + v)}\right]^{1/2} = \frac{2c}{(c^2 - v^2)^{1/2}} = 2\gamma.$$

4. (a) On the outward trip the received period is

$$\Delta t' = \left[\frac{(c + v)}{(c - v)}\right]^{1/2} T_0 = 10 \text{ min.}$$

In 100 min there are 10 pulses.
(b) On the return trip,

$$\Delta t' = \left[\frac{(c - v)}{(c + v)}\right]^{1/2} T_0 = 2.5 \text{ min.}$$

In 100 min there are 40 pulses.

5. From Eq. 39.17, $\Delta x = \gamma(\Delta x' + v\Delta t')$. Since $\Delta t' = 0$ and $\Delta x' = L_0$, the result follows immediately.

QUESTIONS

1. Would it serve any purpose to repeat the Michelson–Morley experiment at different times of the year? If so, explain why.

2. Are any parts of the Galilean transformation still valid in special relativity?

3. What is the distinction between Galileo's principle of rela-

tivity and Einstein's principle of relativity?

4. Two events occur at the same position but at different times in one frame. Can these two events be simultaneous in another frame moving at constant velocity relative to the first?

5. In a transatlantic phone conversation, one person says it is 3 P.M. while the other says it is 7 P.M. Is this an example of lack of synchronization of clocks?

6. Reword the statement "Moving rods appear contracted" to avoid misconception.

7. It is sometimes stated that "moving clocks run slow." What happens to them?

8. Suppose that the speed of light were only 30 m/s. List a few everyday experiences that would change.

9. Why is it not possible for an electron or a proton to travel at the speed of light?

10. The average lifetime of a human is about 70 y. Does this mean that the upper limit to the distance a human can travel away from the earth is about 70 light-years?

11. In principle, would it be possible for parents on a galactic vacation trip to return younger than their children, who stayed at home?

12. Do you think a perfectly rigid body is a valid concept in special relativity? Explain why or why not.

13. In what ways is the Doppler effect for sound similar to the Doppler effect for light? In what ways are they different?

14. Is it possible for a person to move 200 y into the future? Could the person return to tell friends about what was found?

15. Under what condition is the equation $p = E/c$ valid for an electron or a proton?

16. Can the kinetic energy of a particle be $K = \frac{1}{2} m_0 c^2$?

17. True/false: According to special relativity, the linear momentum of a particle can approach, but never equal, $p = m_0 c$.

18. You hear your friends state that according to Einstein's theory "everything is relative." Convince them otherwise by making a list of quantities that according to special relativity are (a) relative, that is, whose value depends on the reference frame, and (b) invariant, that is, whose value is the same for all inertial frames.

EXERCISES

39.7 and 39.8 Time Dilation and Length Contraction

1. (I) A moving meter stick is measured to be 80 cm. What is its speed?

2. (I) Use the binomial expansion, $(1 + x)^n \approx 1 + nx$, for $x \ll 1$, to show that when $v \ll c$, (a) $\gamma \approx 1 + v^2/2c^2$. (b) $1/\gamma \approx 1 - v^2/2c^2$.

3. (I) A rod moving at $0.6c$ relative to the laboratory frame has a length of 1.2 m in this frame. What is the proper length of the rod?

4. (I) A star is 10 light-years (ly) from earth. At what speed should a rocket move relative to the earth–star frame so that the distance measured in the rocket frame is 3 ly?

5. (I) Use the results in Exercise 2 for $v \ll c$ to obtain an expression for (a) $(T - T_0)/T_0$, where $T = \gamma T_0$; and (b) $(L - L_0)/L_0$, where $L = L_0/\gamma$.

6. (I) A clock travels at constant velocity relative to frame S for one year as measured in its rest frame. How much time does it "lose" compared to clocks in S if the speed is (a) $0.1c$; (b) $0.998c$?

7. (I) How fast must a clock (frame S') move relative to frame S in order for it to "lose" one second in one year as measured in S? (See Exercise 2.)

8. (II) The mean lifetime of muons at rest is 2.2 μs. At what speed relative to frame S would they travel 400 m (as measured in S) before decaying?

9. (I) A train traveling at $0.8c$ takes 5 μs to pass an observer on the platform. (a) What is the time interval measured in the train's frame? What is the length of the train according to observers (b) on the train, and (c) on the platform?

10. (I) An atomic clock is flown at 400 m/s between two points 200 km apart on the earth's surface. What is the discrepancy between it and clocks on earth, given that it was initially in synchronization? (See Exercise 2.)

11. (I) At what speed will the measured rate of a clock be 50% of its rate measured at rest?

12. (II) The star Alpha Centauri is 4.2 ly from earth. If a spaceship travels at $0.98c$, how long does the journey take according to (a) the astronauts, and (b) observers in the earth–star frame? (c) What is the distance between the earth and the star in the spaceship's frame?

13. (II) Spaceship B overtakes spaceship A at a relative speed of $0.2c$. Observers in A measure the length of B to be 150 m. (a) What is the proper length of B? How long does it take B to pass a given point on A as measured by observers (b) in A, and (c) in B?

14. (II) A star is 80 light years from earth. At what speed must a spacecraft travel for the journey to be completed within the 70-year lifetime of an astronaut?

15. (I) A train moving at $0.6c$ relative to the ground has a measured length of 320 m in the ground frame. How long does it take to pass a tree as measured (a) in the ground frame, and (b) in the train frame?

16. (I) A train moves at $0.6c$ relative to a platform. The travelers measure the length of the platform to be 1.2 km. (a) What is the proper length of the platform? How long does it take the front of the train to move from one end of the platform to the other (b) in the platform frame, and (c) in the train frame?

17. (I) A spacecraft moves past a space station at $0.98c$. Observers on the station measure the craft's length to be 120

m. How long does it take the craft to pass a given point on the station (a) in the station frame? and (b) in the spacecraft frame?

18. (II) Alpha Centauri is 4.2 ly from earth. (a) At what speed relative to the earth–star frame would astronauts measure this distance to be 3.6 ly? (b) At what speed would the journey take 24 y according to the astronauts? (c) At the speed calculated in (b), what would be the duration of the journey in the earth–star frame?

19. (II) A craft travels the 500 km distance between two cities at 0.2c. (a) What is the duration of the trip according to the pilot? (b) What is the distance traveled according to the pilot?

20. (II) Pions have a mean lifetime 2.6×10^{-8} s when at rest. If they have a speed of 0.8c in the laboratory frame find:
(a) The mean lifetime measured in the laboratory frame.
(b) The distance traveled in one lifetime in the lab frame.
(c) The distance traveled in the laboratory in one lifetime as measured in the particle's rest frame.

21. (II) Muons have a mean lifetime of 2.2×10^{-6} s when at rest. They are produced at an altitude of 10 km and travel at 0.995c toward the earth. Find
(a) The mean lifetime measured on earth.
(b) The time taken to reach ground level in the earth frame.
(c) The time taken to reach ground level in the particle's frame.

39.10 Doppler Effect

22. (I) An astronaut moving at 0.6c relative to earth has a normal heartrate of 72 beats per min. What is the rate received on earth if the spacecraft is (a) approaching, and (b) receding?

23. (II) A speed trap uses 3-cm wavelength radar waves. What is the Doppler shift in frequency for a car moving at 108 km/h toward the source as measured by the policeman?

24. (I) A stellar speeder is given a ticket for going through a red (700 nm) light. The speeder pleads that the light appeared green (500 nm). How fast was he traveling?

25. (I) A galaxy moves away from the earth at 0.2c. What is the natural wavelength of a spectral line whose wavelength measured in a laboratory is 600 nm?

26. (II) A space center on earth tracks an approaching spaceship whose speed is 0.1c. If the radar signal has a frequency of 1000 MHz, what is the frequency of the reflected signal measured on earth?

27. (II) A 2-cm wavelength radar signal is reflected off a car whose speed is 40 m/s. The reflected signal is mixed with the transmitted signal. What is the beat frequency if the car (a) approaches, and (b) recedes?

39.11 Lorentz Transformation

28. (I) The origins of frames S and S' coincide at $t = t' = 0$. Frame S' has a velocity 0.6c along the +x direction. A bomb explodes at $x' = 400$ km at $t' = 0.01$s. Where and when does this occur in frame S?

29. (I) Two flashes occur simultaneously in frame S' but are separated by 480 km. What are the space and time intervals in frame S if S' moves at 0.6c in the + x direction?

30. (I) A train (frame S') of proper length 1.2 km moves at 0.98c along a platform (frame S). Observers at the ends of the train fire bullets onto the platform simultaneously in frame S'. Find the distance between the marks on the platform measured in frame S.

31. (I) A flasher on a spaceship moving at 0.8c toward the earth has a period of 0.01 s. What is the distance traveled by the ship between flashes as measured in the earth frame?

32. (II) A train (frame S') of proper length 3.2 km moves at 0.6c relative to a platform (frame S). At $t = t' = 0$, two light pulses are emitted in opposite directions from the center, as shown in Fig. 39.32. At what times do the pulses reach the ends of the train, A' and B' (a) in frame S', and (b) in frame S?

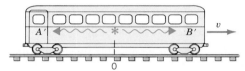

FIGURE 39.32 Exercise 32.

33. (II) When it is 10^8 m away in the earth's frame (S), a rocket (frame S') traveling at 0.8c toward the earth emits a flash. On receipt, it is immediately transmitted back to the rocket. How long does it take the flash return to the rocket according to the earth frame?

34. (II) A red flash occurs at the origin of frame S at $t = 0$. A green flash occurs at $x = 6 \times 10^4$ m at $t = 0.16$ ms. A spacecraft (frame S') is moving in the +x direction. (a) At what speed will the flashes occur simultaneously in S'? (b) What would be the effect if the speed were greater than that found in (a)?

35. (II) A red flash and a green flash occur simultaneously in frame S. The red flash occurs at the origin and the green flash at 240 m. Frame S' moves at 0.995c along the +x axis. Find the time interval and the space interval between the flashes as measured in S'.

36. (II) The *spacetime interval*, Δs, between two events is defined by the equation

$$(\Delta s)^2 = c^2(\Delta t)^2 - (\Delta x)^2 - (\Delta y)^2 - (\Delta z)^2$$

Show that this spacetime interval is an invariant, that is, $(\Delta s')^2 = (\Delta s)^2$.

37. (II) A detector (frame S') moves away from the origin of frame S at speed v along the +x axis. When it is at a distance $x = L$ from the origin of S, a flash is emitted at the origin. How long does it take the flash to reach the detector according to observers (a) in S, and (b) in S'?

39.12 Addition of Velocities

38. (I) Two protons approach each other, each moving at

0.960c relative to the laboratory frame. What is their relative speed?

39. (I) In the earth frame, starship A is chasing starship B at 0.8c while B's velocity is 0.6c. What is the velocity of A relative to B?

40. (I) A spaceship moving at 0.7c relative to the earth fires a missile at 0.1c relative to the spaceship. What is the velocity of the missile relative to earth if the missile is fired (a) forward; (b) backward?

41. (II) Relative to earth, spaceship A moves at 0.6c and is chasing B, whose speed is 0.8c. Spaceship A fires a missile at 0.3c relative to itself. (a) Does the missile hit B? (b) If the answer to (a) is negative, what is the minimum required speed of the missile relative to spaceship A?

42. (II) Two rockets approach each other with the same speed relative to the earth. What is this speed if their relative speed is 0.5c?

39.13 Energy and Momentum

43. (I) The sun radiates energy at the rate of 3.9×10^{26} W. Its mass is 2×10^{30} kg. (a) By how much does the mass decrease per second? (b) If this rate were constant, what would be the lifetime of the sun?

44. (I) What is the momentum of a proton moving at 0.998c?

45. (I) What is v/c for an electron whose kinetic energy is (a) 10^4 eV in a TV tube; (b) 10^7 eV in the tube of the Stanford Linear Accelerator?

46. (I) An electron moves at 0.998c. Find its (a) kinetic energy, and (b) its linear momentum.

47. (I) Calculate the energy required to accelerate an electron from (a) 0.6c to 0.8c; and (b) 0.995c to 0.998c.

48. (I) Find the speed of a particle whose kinetic energy equals (a) its rest energy; (b) eleven times its rest energy.

49. (I) The earth's surface receives 1 kW/m^2 of radiant energy from the sun. If your surface area were 0.5 m^2 and oriented normal to the radiation, how much weight would you gain by sunning yourself for a whole year? (Ignore all other factors and assume all the radiation is absorbed.)

50. (I) Prove that the quantity $E^2 - p^2c^2$ is an invariant, that is, it has the same value in all inertial frames.

51. (I) The total energy of a particle with rest mass m_0 is three times its rest energy. Find (a) its momentum and (b) its speed.

52. (I) The total energy consumption per year in the U.S. is about 3×10^{16} J. Assuming a 0.1% conversion efficiency, how much mass would be needed to provide this?

53. (II) In the Bohr model of the hydrogen atom, the electron travels at 2.2×10^6 m/s. What is the percentage error involved in using the classical expression for kinetic energy? (*Hint:* See the expansion below Eq. 39.26.)

54. (II) At what speed is the momentum of a particle 1% higher than the classical value?

55. (I) At what speed would the relativistic value for the linear momentum of a particle be twice the classical value?

56. (II) The total energy of an electron is 50 MeV. What is $\beta = v/c$?

57. (II) (a) According to classical physics, what is the potential difference needed to accelerate an electron from rest to 0.9c? (b) With the potential difference calculated in (a), what would the speed of the electron be according to special relativity?

58. (II) At what speed is the kinetic energy of a particle 1% higher than the classical value? (*Hint:* See the expansion below Eq. 39.26.)

59. (II) A proton has a kinetic energy of 40 GeV. What is (a) its speed, and (b) its linear momentum?

60. (II) An electron with a total energy of 10 GeV travels 3.2 km down an accelerator tube. (a) What is the length of the tube in the electron's frame? How long does it take in the frame (b) of the electron and (c) of the tube?

PROBLEMS

1. (I) Use $p = \gamma m_0 v$ and $E = \gamma m_0 c^2$ to prove that $E^2 = p^2c^2 + m_0^2c^4$.

2. (I) A particle is subject to a constant force F along the direction of its motion. Starting with the expression $F = dp/dt$, show that its acceleration is $dv/dt = F/\gamma^3 m_0$. Note that as v increases, the acceleration decreases.

3. (I) Prove the following expression for the linear momentum p of a particle of rest mass m_0 in terms of its kinetic energy K:

$$ p = \sqrt{2m_0K + \left(\frac{K}{c}\right)^2} $$

4. (I) Light moves at speed c/n in a medium of refractive index n. Show that if light travels "downstream" in water that flows at speed v relative to the laboratory, then the **speed of light relative to the laboratory is** $(c/n) [(1 + nv/c)/ (1 + v/nc)]$.

5. (I) An electron moves at 0.9995c. At what speed would a proton have (a) the same momentum, and (b) the same kinetic energy?

6. (I) A spaceship (frame S') is 100 m long and moves at 0.995c in the $+x$ direction of frame S. It has a source of light at its rear end and a mirror at the front. At $t = t' = 0$, when the rear coincides with the origin of frame S, a flash is emitted. When does the reflected pulse reach the rear (a) in the ship frame; (b) in frame S. (c) At what posi-

tion in frame S does the pulse reach the rear end?

7. (I) A light beam travels at θ' to the x' axis of frame S', which moves at velocity v in the $+x$ direction of frame S. If θ is the angle measured to the x axis, show that

$$\cos\theta = \frac{\cos\theta' + \beta}{1 + \beta\cos\theta'}$$

where $\beta = v/c$. [*Hint*: Use the Lorentz transformation and note that $\cos\theta = dx/(c\,dt)$]. (b) Given $\beta = 0.9$, evaluate θ for $\theta' = 30°$, $60°$, and $90°$. Why is this called the *headlight effect*?

8. (I) Frame S' moves at speed v along the $+x$ axis of frame S. The speed of a particle is measured to be u in S and u' in S'. Show that the y components of the velocity are related according to

$$u_y = \frac{u_y'}{\gamma(1 + vu_x'/c^2)}$$

A similar relation holds for the z component.

9. (I) The number N of muons remaining at time t is given by $N = N_0 e^{-t/\tau}$, where $\tau = 2.2\ \mu s$ is the mean lifetime in the rest frame and N_0 is the number at $t = 0$. Suppose $N_0 = 1000$ and that the muons move at speed $v = 0.98c$ relative to the earth. After they travel 3 km relative to the earth frame calculate the number remaining (a) in the rest frame, and (b) in the earth frame.

10. (I) A wave in one frame is also a wave in any other. All observers agree that a crest is a crest, a trough is a trough, and so on. That is, the *phase of the wave is an invariant*. Suppose that the wavefunction in frame S is $y = A\sin(kx - \omega t)$ and in frame S' it is $y' = A\sin(k'x' - \omega't')$. Frame S' moves at speed v in the $+x$ direction of frame S. Use the Lorentz transformation and the invariance of phase to show that (ω, k) transform like (x, t), that is,

$$k = \gamma(k' + v\omega'/c^2); \qquad \omega = \gamma(\omega' + vk')$$

(*Hint*: You will obtain an equation of the form $Ax - Bt = 0$ that must be true for any value of x and t).

11. (II) Frame S' moves at velocity $+v$ relative to frame S. (a) A rod at rest in frame S' is at an angle θ' with respect to the x' axis (Fig. 39.33). Use the Lorentz transformation equations to show that the angle θ with respect to the x axis is given by $\tan\theta = \gamma\tan\theta'$. (b) If a rod of proper length L_0 is at the origin of frame S' and makes an angle $\theta' = \theta_0$ with

respect to the x' axis, show that the length in frame S is

$$L = L_0(1 - \frac{v^2}{c^2}\cos^2\theta_0)^{1/2}$$

12. (I) The speed of a star receding from the earth can be determined by the Doppler shift of light emitted by the star. Show that if $v \ll c$, the fractional shift in wavelength is

$$\Delta\lambda/\lambda \approx v/c$$

The received wavelength is longer than the emitted wavelength, so the spectrum as a whole is shifted toward the red end.

13. (II) Two trains run on parallel tracks. Each has a proper length of 1 km. Train A has a velocity of $0.6c$ while train B has a velocity of $0.8c$ relative to the ground. How long does it take the faster train to fully pass the slower one (from the time when the front of B coincides with the rear of A to the time when the rear of B coincides with the front of A) according to observers in (a) the ground frame, (b) the frame of the slower train?

14. (II) A light at the front of a spaceship of proper length 1 km flashes and a light at the rear flashes 4 μs later in the spaceship frame. The ship moves at $0.98c$ along the $+x$ direction in frame S. (a) What are the space and time intervals measured in S? (b) What is the time interval measured by a single observer in S who sees the ship approaching?

15. (II) Two trains, each of proper length 1 km, approach each other on parallel tracks at $\pm0.6c$ relative to the platform. What is the time interval between the meeting of the front ends and the meeting of the rear ends (a) in the platform frame, (b) in the frame of one train?

16. (II) A particle travels at speed u' at angle θ' to the x' axis of frame S', which moves along the $+x$ axis of frame S at speed v, as in Fig. 39.34. Show that the angle at which the particle travels relative to the x axis is given by

$$\tan\theta = \frac{\tan\theta'}{\gamma\left(1 + \dfrac{v}{u'\cos\theta'}\right)}$$

(*Hint*: Consider the transformations of u_x and u_y. See Problem 8 and Eq. 39.12.)

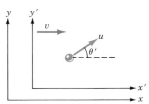

FIGURE 39.34 Problem 16.

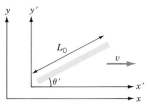

FIGURE 39.33 Problem 11.

17. (II) A particle of rest mass m_0 moves at speed u along the $+x$ axis in frame S. The linear momentum of the particle is $p_x =$

$\gamma(u)m_0u$ and its energy is $E = \gamma(u)m_0c^2$ where $\gamma(u) = (1 - u^2/c^2)^{1/2}$. Frame S' moves at velocity v along the $+x$ axis of frame S. Show that the linear momentum and energy in frame S' are related to the values in S according to

$$E' = \gamma(E - vp_x); \qquad p'_x = \gamma\left(p_x - \frac{vE}{c^2}\right)$$

where $\gamma = (1 - v^2/c^2)^{-1/2}$. Thus E and p_x transform in exactly the same way as x and t. [*Hint:* The speed of the particle in frame S' is $u' = (u - v)/(1 - uv/c^2)$.]

CHAPTER 40

Early Quantum Theory

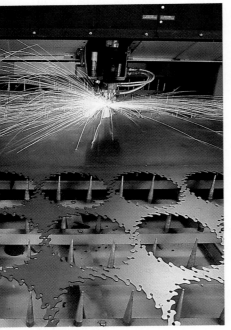

The development of the laser was based on ideas introduced by quantum theory. Here, a laser is being used as a precise cutting tool.

Major Points

1. The failure of classical physics to explain the spectrum of **blackbody radiation.**
2. **Planck's radiation law** and **Einstein's quantum hypothesis.**
3. Einstein's introduction of the **photon** to explain the **photoelectric effect.**
4. In the **Compton effect** a photon is scattered by an electron.
5. **Balmer's formula** for the spectrum of hydrogen.
6. **Rutherford's nuclear model** of the atom.
7. **Bohr's theory** of the hydrogen atom.

The theory of special relativity was the first part of the revolution in 20th-century physics. It showed that classical mechanics is not correct when particles move at high speed. Classical physics also does not provide adequate explanations for issues such as the line spectrum emitted by atoms in a gas discharge tube or the structure of the atom itself. The second part of the revolution was the formulation of the theory of quantum mechanics between about 1900 and 1930. Its origin lay in the study of the radiation emitted by hot bodies and in the photoelectric effect, in which light ejects electrons from a surface. The explanation of these phenomena required the introduction of the idea that energy appears only in discrete amounts; it is said to be *quantized*. In 1911, E. Rutherford proposed that an atom consists of a tiny nucleus surrounded by electrons. Two years later, Niels Bohr combined Rutherford's model and the idea of quantization to explain the origin of the visible spectrum of the hydrogen atom. Other developments in quantum theory are discussed in the next chapter.

40.1 BLACKBODY RADIATION

It is a familiar fact that hot bodies emit radiation, which we experience as heat. As the temperature of an object is raised, it first glows a dull red, then changes to orange-yellow, and finally becomes "white hot." Classically, such thermal radiation is produced by the accelerations of electrons and the oscillations of molecules. It was noted in the late 18th century that a variety of objects placed in a hot oven all glow with the same apparent color. That is, at a given temperature, the distribution of thermal radiation among the various wavelengths is the *same for all bodies*.

When an object is placed in a hot furnace, it absorbs energy until it reaches the temperature of the furnace. Since radiation continues to be incident on it, the object must also emit radiation to remain in thermal equilibrium. By definition a **blackbody** is an ideal system that absorbs all the radiation incident on it (sooty carbon absorbs about 97%). Since a blackbody is the perfect absorber, it must also be the ideal emitter (why?). In practice a cavity with a tiny opening, as in Fig. 40.1, acts as a blackbody since any radiation that enters the cavity is unlikely to reemerge and is ultimately absorbed. The radiation emitted by the opening is called blackbody or **cavity radiation.** *The spectrum of cavity radiation is independent of the material in the walls of the cavity.*

Figure 40.2 shows typical curves for the distribution of energy among the wavelengths. The **spectral energy density,** $u_\lambda(T)$, is defined so that $u_\lambda(T)d\lambda$ is the energy per unit volume of the cavity in the wavelength interval λ to $\lambda + d\lambda$. The unit of $u_\lambda(T)$ is $(J/m^3)/m$. When the temperature is raised, the total energy per unit volume increases and the peak shifts to shorter wavelengths. The wavelength at which the energy density is a maximum is related to the temperature by **Wien's displacement law:**

$$\lambda_{max} T = 2.898 \times 10^{-3} \text{ m} \cdot \text{K} \tag{40.1}$$

In 1896, W. Wien put forward an expression for the spectral energy density, now called **Wien's radiation law**

$$u_\lambda(T) = A\lambda^{-5}e^{-B/\lambda T} \tag{40.2}$$

where A and B were to be experimentally determined constants. This expression was found to agree well with data for wavelengths from 0.7 μm to 6 μm, as indicated in Fig. 40.3.

However, in June 1900, Lord Rayleigh pointed out that Wien's radiation law implies that at long wavelengths the energy density does not increase with temperature, which is not reasonable. He proposed another expression, now called the **Rayleigh–Jeans law,** which he thought would be suitable at long wavelengths:

$$u_\lambda(T) = CT\lambda^{-4} \tag{40.3}$$

where $C = 8\pi k$ and k is Boltzmann's constant. (Rayleigh had missed a factor of 2 in C, which was added later by James Jeans.) In September 1900, measurements

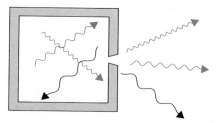

FIGURE 40.1 A cavity with a tiny opening absorbs any radiation entering it, so it acts like a blackbody. The radiation emitted by the opening when the walls of the cavity are hot is characteristic of a blackbody.

The spectrum of radiation emitted by a hot furnace depends on the temperature, not on the material in the walls.

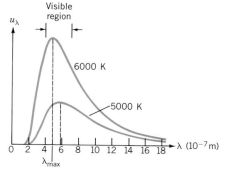

FIGURE 40.2 The spectrum of blackbody radiation at two temperatures. At the higher temperature, more radiation is emitted and the peak shifts to shorter wavelengths.

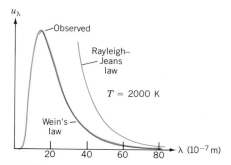

FIGURE 40.3 Wien's radiation law is quite successful at short wavelengths but not at long wavelengths. The Rayleigh–Jeans law works well at long wavelengths ($\approx 15\mu m$) but is a complete failure at shorter wavelengths.

between 12 μm and 18 μm confirmed Rayleigh's prediction. Indeed, the deviations from Wien's radiation law were up to 50% in this wavelength range. However, as Fig. 40.3 shows, the Rayleigh–Jeans law is totally inadequate at short wavelengths.

Planck's Radiation Law

Max Planck (Fig. 40.4), a specialist in thermodynamics, had been working on the problem of cavity radiation for several years. He was impressed by the fact that the spectrum of cavity radiation was a *universal property*—independent of the nature of the material in the walls. He noted that the entropy of any system, such as the radiating oscillators in the walls of a cavity, must be a maximum when the system reaches thermodynamic equilibrium (Section 21.8). In March 1900 he used a simple condition to maximize the entropy and from this derived Wien's radiation law—which is valid for short wavelengths. Later, in order to obtain Rayleigh's law, which is valid at long wavelengths, he had to use a different condition. Planck then simply combined these two conditions into one and obtained a new radiation formula, which he presented on October 19, 1900:*

$$u_\lambda = \frac{A\lambda^{-5}}{e^{B/\lambda T} - 1} \tag{40.4}$$

FIGURE 40.4 Max Planck (1858–1947).

where A and B are constants. At short wavelengths the -1 can be ignored in comparison with the exponential to yield Wien's radiation law. At long wavelengths, the exponential can be expanded, $\exp(B/\lambda T) \approx 1 + B/\lambda T + \cdots$. When this is substituted into Eq. 40.4, it yields the form of the Rayleigh–Jeans law. That same night it was confirmed that this equation fit all available data perfectly!

Planck now felt obliged to justify the way he had dealt with entropy in arriving at Eq. 40.4. He turned to the statistical approach developed by L. Boltzmann (Section 21.11). In order to calculate the entropy, Planck had to determine the number of ways a given total energy could be distributed among a fixed number of oscillators in the cavity walls. If the energy were treated as a continuous variable, there would be an infinite number of ways in which to distribute it. So instead, to facilitate the counting process, Planck divided the *total* energy of the oscillators into energy "elements" of size ε. He found that he could obtain the form of Eq. 40.4 provided he set $\varepsilon = hf$, where f is the frequency and h is a constant. The value of Planck's constant is

Planck's quantum hypothesis

$$h = 6.626 \times 10^{-34} \text{ J} \cdot \text{s}$$

Planck's radiation law was

Planck's radiation law

$$u_\lambda = \frac{8\pi hc\lambda^{-5}}{e^{hc/\lambda kT} - 1} \tag{40.5}$$

This function correctly yields the complete spectrum of cavity radiation.

---

* Planck considered the entropy S to be a function of energy U and noted that S is a maximum if $dS/dU = 0$ and $d^2S/dU^2 < 0$. He showed that the condition $d^2S/dU^2 \propto (-1/fU)$, where f is the frequency, led to Wien's law. However, Rayleigh's law required $d^2S/dU^2 \propto (-1/U^2)$. He combined these into $d^2S/dU^2 = -a/[U(bf + U)]$, where a and b are constants, and derived Eq. 40.4.

At this stage Planck did not realize the true significance of what he had done. Neither did anyone else. He thought of the discrete "energy elements" as merely a calculational aid in determining the entropy of the oscillators. Indeed, he tried for many years to incorporate the constant h into the framework of classical physics.

Einstein's Quantum Hypothesis

Classically, there is no restriction on the amount of energy that an oscillator can emit or absorb. In 1906, Einstein proved that Planck's radiation law could be derived only if the energy of *each individual* oscillator (rather than the *total* energy of all the oscillators) is quantized in steps of hf. Thus, according to **Einstein's quantum hypothesis,** the energy of an oscillator can take on only values that are integer multiples of hf. In the nth "level," the energy is

$$E_n = nhf \qquad n = 0, 1, 2, 3, \ldots \quad (40.6)$$

Einstein's quantum hypothesis

Einstein's hypothesis implies that an oscillator can emit or absorb radiation *only in multiples of hf*. The spacing between the energy levels depends on the frequency.

EXAMPLE 40.1: The peak in the radiation from the sun occurs at about 500 nm. What is the sun's surface temperature, assuming that it radiates as a blackbody?

Solution: From Wien's displacement law, Eq. 40.1, we have

$$T = \frac{2.898 \times 10^{-3} \text{ m} \cdot \text{K}}{500 \times 10^{-9} \text{ m}}$$
$$= 5800 \text{ K}$$

In comparison, the temperature of the filament of an incandescent bulb is about 2000 K.

EXERCISE 1. The temperature of a person's skin is 34°C. What is the wavelength at which the maximum radiation occurs?

EXAMPLE 40.2: A block of mass 0.2 kg oscillates at the end of a spring ($k = 5$ N/m) with an amplitude of 10 cm. What is its "quantum number" n?

Solution: In order to apply Einstein's hypothesis, $E_n = nhf$, we must first calculate the energy. The energy of a simple harmonic oscillator is

$$E = \tfrac{1}{2} kA^2 = \tfrac{1}{2} (5 \text{ N/m})(0.1 \text{ m})^2$$
$$= 0.025 \text{ J}$$

From Eq. 15.7, the frequency of the oscillation is

$$f = \frac{1}{2\pi}\sqrt{\frac{k}{m}} = 0.80 \text{ Hz}$$

From $E_n = nhf$ we find

$$n = \frac{1}{2}\frac{kA^2}{hf} = \frac{(0.025 \text{ J})}{(6.63 \times 10^{-34} \text{ J} \cdot \text{s})(0.80 \text{ Hz})}$$
$$\approx 10^{32}$$

The change in energy ($\Delta E = hf$) between levels n to $n - 1$ is insignificant in comparison to the total energy. Thus in such a macroscopic system we would not expect to experience the quantization of energy. For atomic systems, however, quantization assumes great importance.

Note that the physical quantity "energy" remains a continuous variable, that is, it can assume any value over a continuous range. It is the energy of the possible states of a bound *system* that is quantized.

For a few years, Einstein's quantum hypothesis was more or less ignored since few scientists were concerned with the "dull" problem of cavity radiation. They focused their attention on more exciting fields such as relativity and models of the atom. The most profound revolution in physics had begun, but hardly anyone noticed! However, by 1908 most physicists had become aware of the drastic disagreement between the prediction of classical physics (the Rayleigh–Jeans law) and the radiation curve at short wavelengths. It was dubbed the "ultraviolet catastrophe." Although Planck had introduced the constant h in 1900, the idea that the energy of an oscillator is *really* quantized came from Einstein. Planck himself accepted this idea only around 1910.

Ultraviolet catastrophe

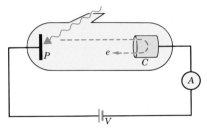

FIGURE 40.5 Light strikes a plate *P* in an evacuated chamber. The emitted photoelectrons are collected by the cylinder *C*, whose potential can be varied to be either positive or negative relative to *P*. When the retarding potential reaches a critical value, the stopping potential, even the most energetic electrons are turned back. The current in the ammeter, *A*, drops to zero.

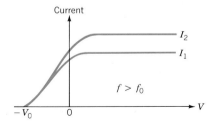

FIGURE 40.6 At positive accelerating potential differences, the maximum current is determined by the intensity of the radiation. However, the stopping potential does not change with the intensity.

40.2 THE PHOTOELECTRIC EFFECT

The work of Young and Fresnel in the early 19th century had converted scientists from the corpuscular theory to the wave theory of light. Maxwell's beautiful theory of 1865 predicted that light is an electromagnetic wave and Hertz's experiment in 1887 was its crowning glory (Section 34.6). Ironically, the very experiment that demonstrated that light is an electromagnetic wave also produced the first evidence of its corpuscular nature! Hertz noticed that the sparks at the detector loop jumped more easily when the loop's electrodes were illuminated by the light of the emitting electrodes. In 1888, Hallwachs found that when a zinc plate is illuminated with ultraviolet light, it becomes positively charged, and in 1899, J. J. Thomson showed that electrons were expelled from the plate. For some alkali metals, such as Na, K, and Cs, visible light can produce this emission of electrons, now called the **photoelectric effect.**

In 1902, P. Lenard conducted the experiment depicted in Fig. 40.5. Monochromatic light illuminates a plate *P* in a glass enclosure. A battery maintains a potential difference between *P* and a metal cylinder *C*, which collects the photoelectrons. When the collector is positive with respect to the plate, the electrons are attracted to it and the ammeter registers a current. At some value of the potential difference, all the emitted electrons are collected. Increasing the accelerating potential difference has no effect on the current, as is shown in Fig. 40.6. When the polarity is reversed, the electrons are repelled and only the most energetic ones reach the collector, so the current falls. When the retarding potential difference reaches a critical value, the current drops to zero. At this **stopping potential,** V_0, only those electrons with the maximum kinetic energy reach the collector:*

$$\tfrac{1}{2} mv_{\text{max}}^2 = eV_0 \qquad (40.7)$$

Lenard found that the number of photoelectrons (inferred from the maximum current in Fig. 40.6) is proportional to the light intensity, even at very low intensities. However, for very weak light, one might expect to wait a long time before the electrons absorb enough energy to escape from the material. In fact, the delay is less than 3×10^{-9} s. The absence of a threshold intensity is puzzling. Furthermore, the maximum kinetic energy of the electrons depends on the light source and the plate material, but *not* on the intensity of the source. Certain combinations of light sources and plate materials exhibit no photoelectric effect. According to the wave theory, photoemission should occur at any frequency, provided the intensity is high enough. Only *one* feature, the increase in the number of photoelectrons with intensity, can be understood in terms of classical physics.

The Photon

In March 1905, Einstein published a paper on cavity radiation. He was uneasy about a basic inconsistency in Planck's approach. Planck had treated the total energy of the oscillators as consisting of *discrete* ''elements'' but had assumed that the energy of the radiation is *continuous*. While he conceded that Maxwell's wave theory is extremely successful in dealing with interference, diffraction, and other properties of electromagnetic radiation, Einstein noted that optical observa-

---

* There is a correction to this equation if the emitter and collector are made of different metals.

tions refer to values averaged over time, not to instantaneous values. The wave theory may not apply to the individual events of absorption and emission. Einstein obtained an expression for the entropy of the radiation in terms of the volume of the cavity and noted that the *form* of this function was similar to that for the entropy of a system of gas particles (Eq. 21.13). This prompted him to propose that radiation behaves as if it were composed of a collection of discrete *energy quanta* of magnitude

$$E = hf \qquad (40.8)$$ Energy of a photon

where f is the frequency of the radiation. The name **photon** was given to these light quanta by G. N. Lewis in 1926. Einstein pictured a wavefront as consisting of billions of photons. He assumed that the energy was not spread uniformly over a wavefront, but concentrated in bundles, localized in space. (The modern view of the photon is not quite so straightforward.)

The Photoelectric Equation

Einstein immediately applied the idea of light quanta to the photoelectric effect. In the process of photoemission *a single photon gives up all its energy to a single electron.* As a result, the electron is ejected instantaneously. The intensity of light of a given frequency is determined by the number of photons incident. Increasing the intensity will increase the number of ejected electrons. The *maximum* possible kinetic energy, $\frac{1}{2}mv_{\max}^2$, of the photoelectrons is determined by the energy of each photon, hf:

$$hf = \tfrac{1}{2}mv_{\max}^2 + \phi \qquad (40.9)$$

where the **work function,** ϕ, is the *minimum* energy needed to extract an electron from the surface of the material. More tightly bound electrons will emerge with kinetic energies less than the maximum. From Eq. 40.9 we infer that there is a **threshold frequency,** f_0, given by

$$hf_0 = \phi \qquad (40.10)$$

Below f_0 there will be no photoemission. By using Eqs. 40.7 and 40.10 in Eq. 4.9 we find **Einstein's photoelectric equation:**

$$eV_0 = h(f - f_0) \qquad (40.11)$$

Not only did Einstein explain all the known facts, but he also predicted (1) the existence of a threshold frequency, and (2) that a plot of V_0 versus f should produce a straight line with a slope h/e, independent of the nature of the material.

R. A. Millikan (who measured the elementary charge e) was uncomfortable with the concept of the photon. In 1906, he began a series of experiments aimed at disproving Einstein's equation. However, after nearly a decade of work, and contrary to his expectations, he proved the validity of Einstein's equation in 1914. Figure 40.7 shows typical data on how the stopping potential depends on the frequency of the light. The slope of the line is correctly predicted by Eq. 40.11.

It is noteworthy that Einstein did *not* build on, or extend, Planck's idea of the quantum; he worked from his own ideas in statistical thermodynamics. In the 1905 paper, he arrived at the equation $E = Cf$, where C is a constant, by using Wien's radiation law, which is accurate only at high frequencies. It was only in the following year that Einstein realized the connection to Planck's theory and that $C = h$.

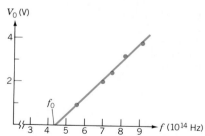

FIGURE 40.7 Millikan's data verified Einstein's equation for the photoelectric effect.

EXAMPLE 40.3: Ultraviolet light of wavelength 207 nm causes photoemission from a surface. The stopping potential is 2 V. Find: (a) the work function in eV; (b) the maximum speed of the photoelectrons.

Solution: (a) From Eq. 40.11,

$$\phi = \frac{hc}{\lambda} - eV_0$$

$$= \frac{(6.626 \times 10^{-34} \text{ J} \cdot \text{s})(3 \times 10^8 \text{ m/s})}{(2.07 \times 10^{-7} \text{ m})} - (1.6 \times 10^{-19}\text{C})(2 \text{ V})$$

$$= 6.4 \times 10^{-19} \text{ J} = 4 \text{ eV}$$

(b) From $\frac{1}{2} mv_{\text{max}}^2 = eV_0$ we have

$$v_{\text{max}} = \sqrt{\frac{2eV_0}{m}} = 8 \times 10^5 \text{ m/s}$$

The classical equation for kinetic energy is satisfactory provided $v < 0.1c$.

EXERCISE 2. For the values in Example 40.3 find: (a) the threshold wavelength; (b) the stopping potential when $\lambda = 250$ nm.

40.3 THE COMPTON EFFECT

The idea that the energy of a system is quantized required a wrenching adjustment away from classical physics. The striking successes in blackbody radiation and the photoelectric effect were still not sufficient to convince some scientists of the validity of the quantum concept. Although by 1910 Planck had accepted the quantization of the energy levels of an oscillator, he (and others) strongly rejected the idea that the radiation itself is quantized.

In 1923, A. H. Compton found further evidence for the photon concept while he was studying the scattering of X rays by graphite. Classically, the charges should oscillate at the frequency of the incoming radiation and re-radiate at the same frequency. Compton found that the scattered radiation had two components: one at the original wavelength (0.071 nm) and a second at a longer wavelength. The value of the shifted wavelength depended on the scattering angle but not on the material of the target.

Compton analyzed his results in terms of a collision between a photon and an electron. Since the energy of the X ray photon (≈ 20 keV) was very much larger than the binding energy of an atomic electron, the electrons could be treated as being "free." Classically we know that an electromagnetic wave carries momentum given by $p = E/c$ where E is the energy (Section 34.5). Since the energy of a photon is hf, its momentum is

$$p = \frac{hf}{c} = \frac{h}{\lambda} \tag{40.12}$$

The collision is depicted in Fig. 40.8a. The incoming photon is deflected by an angle θ and the electron, initially at rest, moves off at angle ϕ. From the conservation of energy, we have

$$\frac{hc}{\lambda} = \frac{hc}{\lambda'} + K \tag{40.13}$$

where $K = (\gamma - 1)m_0c^2$ is the relativistic kinetic energy of the electron after the collision. (We have omitted the rest energy of the electron from both sides.) The conservation of linear momentum for the two components yields

$$\sum p_x: \frac{h}{\lambda} = \frac{h}{\lambda'} \cos \theta + p \cos \phi \tag{40.14}$$

$$\sum p_y: 0 = \frac{h}{\lambda'} \sin \theta - p \sin \phi \tag{40.15}$$

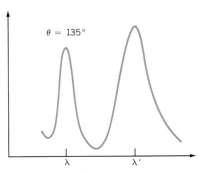

FIGURE 40.8 Compton scattering. (a) A photon of wavelength λ is scattered by an electron, which moves off at angle θ. The scattered photon has a longer wavelength λ. (b) The Compton shift at $\theta = 135°$.

where $p = mv = \gamma m_0 v$. Manipulation of Eqs. 40.13 to 40.15 (see Problem 7) leads to

$$\lambda' - \lambda = \left(\frac{h}{m_0 c}\right)(1 - \cos\theta) \qquad (40.16) \quad \text{Compton shift}$$

where m_0 is the rest mass of the electron. The quantity $h/m_0 c = 0.00243$ nm is called the **Compton wavelength.** One of the results is shown in Fig. 40.8b. The unshifted peak is due to a scattering process in which the whole atom is involved and is explainable in classical terms. The Compton effect persuaded most physicists to accept the concept of the photon.

EXAMPLE 40.4: X rays of wavelength 0.24 nm are scattered through an angle of 40° as they pass through a block of carbon. What is the wavelength of the scattered rays?

Solution: From Eq. 40.16, the shift in wavelength is

$$\lambda' - \lambda = (0.00243 \text{ nm})(1 - \cos 40°)$$
$$= 0.00057 \text{ nm}$$

The wavelength of the scattered rays is $\lambda' = \lambda + \Delta\lambda = 0.24057$ nm. By using shorter wavelength X rays, the fractional shift $\Delta\lambda/\lambda$, can be made larger.

EXERCISE 3. What is the kinetic energy of the recoiling electron?

40.4 LINE SPECTRA

We saw in Section 40.1 that a hot body emits a continuous spectrum of radiation. The emission from a rarefied gas that is very hot, or through which an electrical discharge passes, consists of sharp lines. Each element has a characteristic set of lines that may be used to identify it. Indeed, rubidium, cesium, helium, thorium, and indium were discovered in the 1860s through the study of spectra. Classical physics could neither explain these line spectra nor fit them into any scheme.

The visible spectrum of hydrogen consists of four lines: 410.12 nm, 434.01 nm, 486.07 nm, and 656.21 nm, as shown in Fig. 40.9. In 1884, J. J. Balmer, a Swiss mathematics teacher, found that these wavelengths (in nm) could be represented by a single formula. **Balmer's formula** was

$$\lambda_m = 364.56 \frac{m^2}{m^2 - 4} \qquad m = 3, 4, 5, 6. \quad (40.17) \quad \text{Balmer's formula}$$

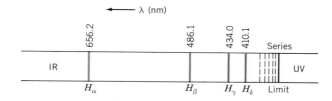

FIGURE 40.9 The visible spectrum of hydrogen consists of four lines.

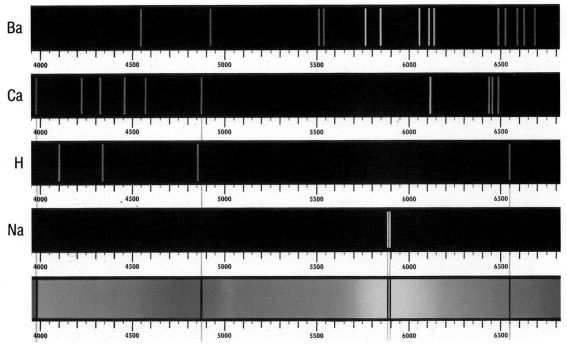

Line spectra for various elements.

By inserting the integer values for m, the wavelengths were reproduced with an error no greater than 1 in 40,000! By 1890, J. R. Rydberg had discovered similar formulas for the spectra of the alkali elements Li, Na, K, and Cs. He also suggested that the formula be rewritten as the difference between two terms. For hydrogen

Rydberg's formula

$$\frac{1}{\lambda} = R \left(\frac{1}{2^2} - \frac{1}{m^2} \right) \tag{40.18}$$

where

$$R = 1.09737 \times 10^7 \text{ m}^{-1}$$

is now called the Rydberg constant.

In 1908, Ritz discovered a "combination principle": The frequency of a line in the spectrum of a given element could be expressed as a simple combination (sum or difference) of the frequencies of two other lines in the same spectrum. At last, line spectra were falling into place, but it took another five years before these regularities were given some theoretical foundation.

40.5   ATOMIC MODELS

In the 19th century, there was considerable chemical and physical evidence for the existence of atoms, but nothing was known about their structure. In 1904, J. J. Thomson suggested that an atom consists of a positively charged sphere in which electrons were embedded. He tried, without success, to relate the normal mode frequencies of various electron configurations to line spectra. Although he made several qualitative deductions, the model was not satisfactory.

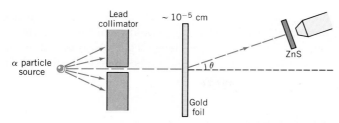

FIGURE 40.10 In the experiment of Geiger and Marsden, high-energy alpha particles were scattered by a thin gold foil. The scattered particles were detected as flashes on a ZnS screen.

In 1909, Ernest Rutherford asked two of his assistants, Hans Geiger and Ernst Marsden, to study the scattering of alpha particles (doubly ionized helium atoms) by a very thin gold foil (Fig. 40.10). Since the positive charge in the Thomson atom was spread throughout the atom, most of the scattered beam was expected to be about 3° wide, with perhaps a few scattered particles up to 20°. In fact, Geiger and Marsden found that about one in eight thousand of the α particles was scattered through angles larger than 90°. Rutherford was stunned:

It was quite the most incredible thing that has ever happened to me in my life. It was almost as incredible as if you fired a 15 inch shell at a piece of paper, and it came back and hit you.

Since the mass of an α particle is about 7000 times that of an electron, the electrons play a negligible part in the scattering. The mass of the positive part of the gold atom is about 50 times that of the α particle. If the positive part were spread uniformly over the whole atom ($r \approx 10^{-10}$ m) as Thomson suggested, it could not cause such large deflections. Rutherford concluded that each deflection was caused by a *single* strong interaction. This meant that the positive part of the atom had to be concentrated in an extremely small volume ($r \approx 10^{-14}$ m), which we now call the **nucleus.** The deflection of an α particle would depend on how closely it approached the nucleus, as shown in Fig. 40.11, with the force of repulsion given by Coulomb's law. Occasionally there would be a "head-on" collision in which the α particle simply reverses its direction of motion. This special case allows us to obtain an estimate of the size of the nucleus.

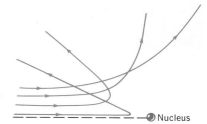

FIGURE 40.11 According to Rutherford's nuclear model, the alpha particles were scattered by the Coulomb force of a tiny particle (the nucleus) rather than a large sphere, as in Thomson's model of the atom. Each alpha particle experienced a single strong collision.

EXAMPLE 40.5: An alpha particle moving with speed 2×10^7 m/s makes a head-on collision with a gold (Au^{197}) nucleus that carries a charge of $79e$. What is the distance of closest approach? Take $m_\alpha = 6.7 \times 10^{-27}$ kg, $q_\alpha = 2e$, and assume that the gold nucleus remains at rest.

Solution: This problem is easily solved with the conservation of energy. The initial energy of the system is just the kinetic energy of the α particle. (Since $v < 0.1c$, the classical expression for kinetic energy is roughly correct.) At the distance of closest approach r_0, the α particle is momentarily at rest, and therefore the system has only potential energy. The initial and final energies are

$$E_i = \tfrac{1}{2} m_\alpha v^2; \qquad E_f = \frac{kq_\alpha q_{Au}}{r_0}$$

Equating these two we find $r_0 = 2.7 \times 10^{-14}$ m. This would be an upper limit to the size of the nucleus.

Rutherford's work on α particle scattering established the existence of the nucleus; he did not concern himself with how the electrons were distributed in an atom. In 1904, H. Nagaoka had proposed that the electrons formed rings (like those around Saturn), but he could not explain how such a system would be stable. A "planetary model," in which the electrons orbit the nucleus, would be mechanically stable. However, according to Maxwell's theory, an accelerating electron emits radiation. The loss in energy would mean that the electron would spiral, within about 10^{-8} s, into the nucleus. This obviously does not occur. Moreover, the radiation would cover a continuous range in frequency, in contrast to the line spectra actually observed.

40.6   THE BOHR MODEL

After Niels Bohr obtained his doctorate in 1911, he worked under Rutherford for a while (see Fig. 40.12). He believed that Planck's constant was the key to a successful atomic model and was encouraged by the following dimensional analysis. If m and e are the mass and charge of the electron, and k is the constant in Coulomb's law, then the quantity h^2/mke^2 has the dimensions of length and its numerical value agreed roughly with the known size of atoms. However, Bohr made limited progress until he found out about the Balmer–Rydberg formula in July 1912. He then focused his efforts on obtaining this result theoretically. In 1913, he presented a model of the hydrogen atom, which has one electron. Bohr stated two postulates.

1. The electron moves only in certain circular orbits, called *stationary states*. This motion can be described classically.

FIGURE 40.12 Ernest Rutherford (left) (1871–1937) and Niels Bohr (1885–1962).

This postulate does not explain why an atom is stable; it merely asserts that it is. Figure 40.13 shows an electron of mass m and charge $-e$, moving at speed v in a stable circular orbit of radius, r, around a nucleus of charge $+e$. The centripetal force is provided by the Coulomb attraction between the electron and the nucleus. From Newton's second law we have

$$\frac{mv^2}{r} = \frac{ke^2}{r^2} \qquad (40.19)$$

The total mechanical energy of the electron is

$$E = K + U = \tfrac{1}{2}mv^2 - \frac{ke^2}{r}$$

From Eq. 40.19 we find $K = ke^2/2r$, therefore

$$E = -\frac{ke^2}{2r} \qquad (40.20)$$

FIGURE 40.13 In the Bohr model of the hydrogen atom an electron is in a circular orbit around a single proton.

Bohr knew that Maxwell's theory could not explain the spectrum of blackbody radiation or the photoelectric effect. So he abandoned the ideas that the accelerating electron must radiate and that an oscillator radiates at the (mechanical) frequency of oscillation. The second postulate was

2. Radiation occurs only when an electron goes from one allowed orbit to another of lower energy. The radiated frequency is

$$hf = E_m - E_n \qquad (40.21)$$

where E_m and E_n are the energies of the two states.

At this point we deviate from Bohr's initial work and present a simplified approach. There is no justification for the first postulate about stationary orbits—a postulate Bohr had to make since atoms do not spontaneously collapse. To address this fact we need a quantum condition that restricts the allowed values of the orbital radius. Near the end of Bohr's first paper he made a passing comment that the angular momentum of the orbits is quantized, but he did not take this idea seriously. It was realized only later (c. 1915) that this is a *fundamental* aspect of quantum theory, and so it serves as our "third" postulate:

3. The angular momentum of the electron is restricted to integer multiples of $h/2\pi \ (= \hbar)$:

$$mvr = n\hbar \tag{40.22}$$

When $v = n\hbar/mr$ from this equation is equated to $v = \sqrt{ke^2/mr}$ from Eq. 40.19, we find the radius of the nth orbit is

$$r_n = \frac{n^2\hbar^2}{mke^2} \tag{40.23}$$

From Eq. 40.20, the total energy of the nth orbit is

$$E_n = -\frac{mk^2e^4}{2\hbar^2}\left(\frac{1}{n^2}\right) \tag{40.24}$$

Combining this with Eq. 40.21 immediately leads to Rydberg's formula:

$$f = Rc\left(\frac{1}{n^2} - \frac{1}{m^2}\right) \tag{40.25}$$

where

$$R = \frac{mk^2e^4}{4\pi c\hbar^3} \tag{40.26}$$

Bohr had derived the correct form of Rydberg's equation (Eq. 40.25). More importantly, he had expressed the empirical number R in terms of fundamental constants. When their values were inserted, the calculated value of R was within 6% of the then-accepted value.

Bohr's theory may be applied to other single electron systems such as He$^+$ or Li^{++}, provided the nuclear charge is replaced by Ze, where Z is the atomic number. The energy of the nth level is given by Eq. 40.24. Expressed in terms of electronvolts, we find

$$E_n = -\frac{13.6Z^2}{n^2} \text{ eV} \tag{40.27}$$

The energy level diagram for hydrogen ($Z = 1$) is shown in Fig. 40.14. Each state is characterized by the integer n, which is called the principal quantum number. When the atom is unexcited, the electron is in the ground state with $n = 1$. The energies of the various levels are $E_1 = -13.6$ eV, $E_2 = -13.6/2^2 = -3.4$ eV, $E_3 = -13.6/3^2 = -1.51$ eV, and so on. Since we have chosen the potential energy to be zero at $r = \infty$, the total energies are negative, as is required for the electron to be in a bound state. The electron may be raised to a higher level by a collision with another electron or by absorbing a photon. Note that the *photon energy must correspond exactly to the energy difference between the two states* involved in the transition. If the photon energy is large enough, the electron may be ejected from the atom. This is called *ionization*. From Fig. 40.14 we see that the ionization energy for hydrogen is 13.6 eV. The Bohr theory is remarkably accurate in this prediction.

The electron may return from an excited state to the ground state in one step or via intermediate levels. A single frequency is emitted in the first case, but there

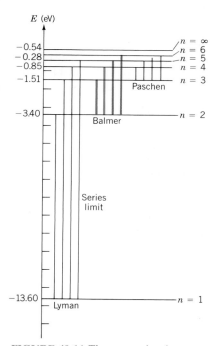

FIGURE 40.14 The energy level diagram for hydrogen. Light is emitted or absorbed when an electron makes a transition between two levels.

are two or more in the second. Balmer's series corresponds to transitions from higher levels to $n = 2$. Transitions to level $n = 1$ form the Lyman series; those to $n = 3$ form the Paschen series. For each series there is a maximum possible frequency called the series limit. This corresponds to a transition from $n = \infty$ to the lowest level of the series.

EXAMPLE 40.6: According to the Bohr theory, what is the radius of the ground state orbit of the hydrogen atom?

Solution: From Eq. 40.23,

$$r_1 = \frac{\hbar^2}{mke^2}$$

$$= \frac{1.05 \times 10^{-34}\ \text{J} \cdot \text{s}}{(9.11 \times 10^{-31}\ \text{kg})(9 \times 10^9\ \text{N} \cdot \text{m}^2/\text{C}^2)(1.60 \times 10^{-19}\ \text{C})^2}$$

$$= 5.29 \times 10^{-11}\ \text{m}$$

Note that the radius of the nth orbit is simply related to r_1:

$$r_n = n^2 r_1$$

EXAMPLE 40.7: An electron is in an excited state for which $n = 3$. (a) What is the highest frequency that can be radiated? (b) What other frequencies are possible?

Solution: (a) From the second postulate we have $f = \Delta E/h$. The maximum frequency will be emitted in a direct transition to the ground state, $n = 1$. From the energy level diagram we see that $E_1 = -13.6$ eV and $E_3 = -1.51$ eV. Therefore

$$f_{31} = \frac{(E_3 - E_1)}{h}$$

$$= \frac{(+12.1\ \text{eV})}{(4.14 \times 10^{-15}\ \text{eV} \cdot \text{s})}$$

$$= 2.92 \times 10^{15}\ \text{Hz}$$

(b) Instead of $f = \Delta E/h$, we could also use Eq. 40.25. Thus,

$$f_{32} = Rc \left(\frac{1}{2^2} - \frac{1}{3^2} \right)$$

$$= (1.09 \times 10^7\ \text{m}^{-1})(3.00 \times 10^8\ \text{m/s}) \left(\frac{5}{9} \right) = 4.58 \times 10^{14}\ \text{Hz}$$

$$f_{21} = Rc \left(\frac{1}{1^2} - \frac{1}{2^2} \right)$$

$$= (1.09 \times 10^7\ \text{m}^{-1})(3.00 \times 10^8\ \text{m/s}) \left(\frac{3}{4} \right) = 2.48 \times 10^{14}\ \text{Hz}$$

EXAMPLE 40.8: In Section 20.3 we showed that the average kinetic energy of a particle in a gas at temperature T is $\frac{3}{2} kT$, where k is Boltzmann's constant. At what temperature would this be equal to the energy needed to make a transition from the ground state to $n = 2$?

Solution: The energy needed is $13.6 - 3.4 = 10.2$ eV $= (10.2$ eV$)(1.6 \times 10^{-19}$ J/eV$) = 1.63 \times 10^{-18}$ J. We set this equal to the thermal energy:

$$\Delta E = \frac{3}{2} kT$$

from which

$$T = \frac{2\Delta E}{3k}$$

$$= \frac{2(1.63 \times 10^{-18}\ \text{J})}{3(1.38 \times 10^{-23}\ \text{J/K})}$$

$$= 7.87 \times 10^4\ \text{K}$$

It would be quite difficult to excite the hydrogen atom purely by thermal collisions. An electrical discharge through the gas is usually employed.

EXERCISE 4. What is the shortest wavelength possible in the Balmer series?

Bohr's theory correctly predicted the frequencies of the spectrum of hydrogen and other one-electron systems. However, it provided no information regarding the relative intensities of the lines or on the spectra of multi-electron atoms. As we will see in the next chapter, Bohr's theory has been replaced by quantum mechanics. The second and third postulates remain valid, but the picture of an electron in well-defined orbits is not correct. Nonetheless, Bohr's approach is a good example of how a scientist pulls together different strands in developing a remarkably successful theory.

40.7 WAVE–PARTICLE DUALITY OF LIGHT

In this chapter we have presented evidence that light behaves like a particle. However, Young's double-slit experiment provides clear evidence for the wave nature of light. How are we to reconcile these two apparently contradictory facts?

Suppose that the light intensity in the double-slit experiment is reduced to such an extent that only one photon is present in the system at any time. Depending on the size of the apparatus, this still involves a rate of perhaps 10^9 photons/s. With just one slit, one observes (on a photographic plate) a single-slit diffraction pattern. The light intensity recorded at a given point is proportional to the number of photons that arrive at that point. The two-slit experiment with very weak light was first performed by G. I. Taylor in 1909. After a long enough wait (up to 3 months!), the usual interference fringes were recorded. When both slits are open, the total number of photons reaching the screen doubles. Yet at some points the intensity is more than double that of the single-slit pattern, and at other points it is zero. This is strange behavior for "particles."

If light consists of particles, then it seems reasonable to assume that each photon has to go through either one slit or the other. Suppose that the slits are alternately closed and opened such that at any given instant one slit is open, while the other is closed. If this is done, the interference pattern is lost. In fact, any attempt to find out through which slit each photon passes causes the interference pattern to disappear.

We might try to explain the interference by saying that each photon somehow splits into two parts and therefore passes through both slits. Thus each photon would interfere only with itself. However, if the energy of the photon is halved, $E = hf$ implies that its wavelength ($\lambda = c/f$) is doubled. This would lead to a fringe spacing twice as large as that actually observed.

Young's double-slit experiment, which was so crucial in establishing the wave nature of light as opposed to the corpuscular model, turns out to be not so definitive after all. The same ambiguous result accompanied Hertz's demonstration of Maxwell's wave theory. It was conclusive proof of the wave nature of light and yet, by revealing the photoelectric effect, it was also a good demonstration of its particle nature.

As we saw earlier, the equation $E = hf$ for a photon leads to $p = h/\lambda$. Both of these equations involve a mixture of particle and wave concepts. E is the energy of a light quantum, whereas f is the frequency of a wave. p is a quantity we normally associate with particles, whereas λ is a property of waves. It seems that wave theory is appropriate for the *propagation* of light but that quantum theory is needed for the *interaction* of light with matter. So light that is a wave suddenly pulls itself together into a localized quantum when it encounters matter? Not quite. Everyday language is simply inadequate for describing many subatomic events.

Light exhibits a **wave–particle duality.** Depending on the experiment performed, it will behave either as a particle, or as a wave. This does not mean that light "really is" either a particle or a wave: These are our simple models that have proven to be very helpful. At low frequencies—for example, in radio waves—we can detect only billions of photons, and so the model of the radiation as a continuous wave is adequate. In the optical region, different experiments—for example, on interference or the photoelectric effect—require either the wave or the particle model. At high frequencies, such as X rays, we tend to observe only single-photon events, although it is still possible to demonstrate the wave nature of X rays through crystal diffraction (Section 38.8).

The concept of the wave–particle duality of light was introduced by Einstein. In his 1905 paper that introduced the photon, Einstein had used only Wien's radiation law to show that short wavelength radiation has properties similar to those of a gas of particles. The Rayleigh–Jeans law, which is valid for long wavelengths, was of course based on the wave nature of radiation. In yet another analysis of cavity radiation in 1909, Einstein proved (in effect) that the complete spectrum of cavity radiation, given by Planck's law, requires *both* the particle and the wave models.

We might make a connection between our inability to pin down the nature of light and the principle of the constancy of the speed of light. If light were a wave, its speed would be measured with respect to a medium—but there is no ether. If it consisted of particles, its speed would be measured with respect to the source—but this is not what is observed. According to special relativity, the photon moves at c simply because it has zero rest mass. We might conclude that light is *neither* a wave nor a particle! It is indeed remarkable that these simple models have been so fruitful.

40.8   BOHR'S CORRESPONDENCE PRINCIPLE

Classical physics is very successful in dealing with a large number of phenomena. Bohr felt that when a newer, more general, theory is proposed, its predictions should reduce to the classical results when the appropriate limit is taken. This requirement that the results of a new theory correspond, in the limit, to classical physics is called the **correspondence principle.** For example, Planck's radiation law reduces to the classical Rayleigh–Jeans formula when $h \to 0$. In the special theory of relativity, the Lorentz transformation reduces to the Galilean transformation when $v \ll c$. In his second derivation for the hydrogen atom Bohr used the correspondence principle.

Bohr compared the second postulate with Rydberg's formula to obtain

$$E_n = \frac{Rch}{n^2} \tag{40.28}$$

The (*mechanical*) frequency of the orbital motion is $\nu = v/2\pi r$ while from Newton's second law Eq. 40.19, we have $v^2 = (ke^2)/mr = 2E/m$. Thus, for the nth orbit, we find

$$\nu_n = \frac{(2E_n^3/m)^{1/2}}{\pi ke^2 n^3} \tag{40.29}$$

To proceed further Bohr invoked his correspondence principle. Here it meant that in the limit of large quantum numbers n (say, $n = 10^4$), the *radiated* frequency, f, should be the same as the *mechanical* frequency, ν, as predicted by Maxwell's theory. From the second postulate, $f = \Delta E/h$, and Eq. 40.28, the radiated frequency in the transition from n to $n - 1$ is

$$f = Rc\left[\frac{1}{(n-1)^2} - \frac{1}{n^2}\right]$$
$$= Rc\left[\frac{2n-1}{n^2(n-1)^2}\right]$$

As $n \to \infty$, this becomes

$$f \approx \frac{2Rc}{n^3}$$

Equating this to ν_n in Eq. 40.29 leads to Eq. 40.26 for R—which was a major success of the theory.

SUMMARY

The spectrum of the radiation emitted by a small opening to a cavity is independent of the material in the walls of the cavity. The spectrum is described by **Planck's radiation law.**

According to **Einstein's quantum hypothesis** the energy of an oscillator is quantized in steps of hf, where f is the frequency and h is Planck's constant. The energy of the nth level is

$$E = nhf$$

The energy of electromagnetic radiation of frequency f is also quantized in units of hf. Each quantum of energy is called a **photon.**

In the photoelectric effect a single photon, of frequency f, interacts with a single electron and ejects it from a material. The maximum kinetic energy of the electrons may be found from the **stopping potential** V_0: $\frac{1}{2}mv_{max}^2 = eV_0$. According to **Einstein's photoelectric equation,**

$$hf = eV_0 + \phi$$

where ϕ is the **work function**—the minimum energy needed to extract an electron from the surface. The photoelectric effect will not occur below the **threshold frequency,** f_0:

$$hf_0 = \phi$$

In the **Compton effect,** a photon is scattered by a free electron. The change in wavelength of the photon is given by

$$\Delta\lambda = \left(\frac{h}{m_0 c}\right)(1 - \cos\theta)$$

where θ is the angle through which the photon is deflected.

The **Bohr model** of the hydrogen atom successfully accounts for the line spectra of the hydrogen atom. It is also applicable to other one-electron systems. The energy levels can be derived with the following postulates:

(1) The electron moves only in certain stable circular orbits called stationary states.

(2) Radiation occurs only when an electron jumps from one orbit to another. The frequency being given by

$$hf = E_m - E_n.$$

(3) The angular momentum of an electron is quantized according to

$$mvr = n\hbar$$

The energy levels of the electron in the hydrogen atom are given by

$$E_n = -\frac{13.6}{n^2}\text{ eV}$$

An electron can be excited into a higher state by a collision with another electron or by the absorption of a photon of the correct frequency.

ANSWERS TO IN-CHAPTER EXERCISES

1. From Wien's displacement law, with $T = 307$ K, we find $\lambda_{max} = 9.4$ μm. This lies in the infrared region.

2. (a) Since $\phi = hc/\lambda_0$, we have $\lambda_0 = hc/\phi = 310$ nm.
   (b) From Eq. 40.11,
   $$V_0 = (hc/\lambda - \phi)/e$$
   $$= 0.97 \text{ V}$$

   Note that ϕ must be in joules.

3. From Eq. 40.13, the kinetic energy of the electron is
   $$K = hc\left(\frac{1}{\lambda} - \frac{1}{\lambda'}\right) = 1.97 \times 10^{-18} \text{ J} = 12.3 \text{ eV}$$

4. In this case the electron makes a transition from $n = \infty$ to $n = 2$. From Eq. 40.25 we have $1/\lambda = R(1/4 - 1/\infty)$, which means that $\lambda = 4/R = 365$ nm.

QUESTIONS

1. Could a sufficiently powerful AM radio signal produce a photoelectric effect?

2. (a) When a surface is illuminated with monochromatic light, why is there a maximum kinetic energy for photoelectrons? (b) For a given frequency greater than the threshold frequency, why is there a range of kinetic energies of the emitted electrons?

3. When light with a continuous range of frequencies passes through a sample of hydrogen gas at room temperature, only the Lyman series (see Fig. 40.14) is observed in the absorption spectrum. Why?

4. If the intensity of light is fixed, does the number of photoelectrons depend on frequency?

5. The existence of a photoelectric work function is not contrary to classical physics. Since the work function is equal to hf_0, why isn't the existence of a cutoff frequency also acceptable classically?

6. What easily observed phenomenon is described by the following: (a) The Stephan–Boltzmann law? (b) Wien's displacement law?

7. In what way(s) are the photoelectric effect and the Compton effect (a) similar, (b) different?

8. Why does the Compton effect not occur with visible light?

9. What effect, if any, would the temperature of a metal have on the photoelectric effect?

10. Light from stars may appear reddish or bluish. What information would one infer from this observation?

11. Why is it difficult to produce an incandescent bulb with a visible spectrum similar to sunlight?

12. Show that the unit of Planck's constant is the same as that of angular momentum.

13. Ultraviolet rays are responsible for tanning and sunburn. Why doesn't visible light have the same effects?

14. Would a hotter filament in a light bulb be more efficient in converting electrical energy to light? Explain why or why not.

15. According to Bohr's second postulate the frequency f of emitted light is given by $\Delta E = hf$, where ΔE is the difference in energy between two levels. Can this equation be exactly true? (Think of conservation of linear momentum.)

16. An electron in a hydrogen atom is in its ground state. (a) What happens when radiation with a frequency greater than $(E_3 - E_1)/h$ but less than $(E_4 - E_1)/h$ is incident? (b) What happens if a beam of electrons with a kinetic energy greater than $(E_3 - E_1)$ but less than $(E_4 - E_1)$ is used?

17. What experimental evidence did Bohr use to formulate his theory?

18. Bohr's first postulate abandons *two* features of classical radiation theory. One was mentioned explicitly. What is the other?

19. Which aspects of Bohr's model of the hydrogen atom are (a) classical, and (b) nonclassical?

20. Hydrogen has only one electron, yet one observes many spectral lines. Explain why.

21. What is the maximum possible kinetic energy of a beam of electrons such that collisions with hydrogen atoms are elastic?

22. Show how Fig. 40.6 is modified if the intensity is kept fixed but the frequency is varied ($f > f_0$).

23. Since Eq. 40.8 involves e^4, why does changing e to Ze lead to Z^2 in Eq. 40.11? Trace the steps.

24. Explain the physical basis of Ritz's combination principle (Section 40.1).

25. Suppose the electron in the hydrogen atom starts at the $n = 4$ level. How many possible lines could be observed?

26. In the Compton effect why is $\Delta\lambda$ independent of the material? Why is it independent of λ?

27. In the Compton effect, why is it preferable to use short wavelengths for the incident radiation?

EXERCISES

40.1 Blackbody Radiation

1. (I) What is the wavelength of the peak in blackbody radiation at the following temperatures: (a) The 3 K cosmic background radiation that is a remnant of the "big bang" that created the universe, (b) a tungsten filament at 3000 K, and (c) a fusion reaction at 10^7 K?

2. (I) (a) The peak in the radiation from the sun occurs at 470 nm. What is the surface temperature of the sun? (b) What would be the surface temperature of a star whose thermal radiation peaked at 350 nm?

3. (I) For what range of temperatures does the wavelength of the peak in blackbody radiation vary through the visible range (400 nm–700 nm)?

4. (I) The net loss due to radiation of a blackbody at temperature T is

$$R = \sigma(T^4 - T_0^4) \text{ W/m}^2$$

where $\sigma = 5.67 \times 10^{-8} \text{ W} \cdot \text{m}^{-2} \cdot \text{K}^{-4}$, and T_0 is the temperature of the surroundings. Estimate the net radiated intensity for the following:

(a) A hot coal at 2000 °C in a room at 20 °C.
(b) A person with a skin temperature of 34 °C in air at 10 °C.
(c) The earth's surface at 22 °C radiating into space at −270 °C.

5. (I) Given that the sun's surface temperature is 5760 K. Find the total power radiated into space (taken to be at 0 K). The sun's radius is 6.96×10^8 m. (See Exercise 4.)

6. (I) A heater filament has a radius of 2 mm and a length of 20 cm. If its temperature is 2000 K, what is the net radiated power? (See Exercise 4, and set $T_0 = 0$ K.)

7. (I) What is the wavelength of the peak in the blackbody radiation of a body at 300 K?

8. (I) A CO_2 molecule vibrates at 5.1×10^{13} Hz. What is the separation between adjacent energy levels in eV?

40.2 Photoelectric Effect

9. (I) A radio station transmits 40 kW at 100 MHz. How many photons per second does it emit?

10. (I) (a) Show that the energy, E, of a photon (in eV) can be written in the form

$$E = \frac{1240}{\lambda}$$

where the wavelength λ is in nanometers. (b) What is the range in energy of photons in the visible region from 400 nm to 700 nm?

11. (I) The work function for potassium is 2.25 eV. A beam with a wavelength of 400 nm has an intensity of 10^{-9} W/m^2. Find (a) the maximum kinetic energy of the photoelectrons, (b) the number of electrons emitted per meter squared per second from the surface assuming 3% of the incident photons are effective in ejecting electrons.

12. (I) The minimum intensity that the eye can detect is about 5×10^{-13} W/m^2. If the pupil diameter is 5 mm find: (a) the power needed, and (b) the number of photon/s required at 500 nm.

13. (I) The threshold wavelength for cesium is 686 nm. If light of wavelength 470 nm illuminates the surface, what is the maximum speed of the photoelectrons?

14. (I) Find the energy (in eV) of photons of the following wavelengths or frequencies: (a) visible light at 550 nm; (b) an FM radio wave at 100 MHz; (c) an AM radio wave at 940 kHz; (d) an X ray at 0.071 nm.

15. (I) (a) The dissociation energy of CO is 11 eV. What is the minimum frequency of radiation that could break this bond? (b) The maximum wavelength of radiation capable of dissociating the O_2 molecule is 175 nm. What is the binding energy in eV?

16. (I) The C–C bond has a dissociation energy of 2.8 eV. What is the longest wavelength of radiation that could break this bond? To what part of the spectrum does it belong?

17. (I) The intensity of solar radiation incident on the earth's atmosphere is 1.34 kW/m^2. Assuming it is monochromatic at 550 nm (yellow), how many photons/m$^2 \cdot$ s does this involve?

18. (I) A continuous wave helium–neon laser produces 1 mW at a wavelength of 632.8 nm. How many photons/s does it emit?

19. (I) The work function for lithium is 2.3 eV. (a) What is the maximum kinetic energy of photoelectrons when the surface is illuminated with light of wavelength 400 nm? (b) If the stopping potential is 0.6 V, what is the wavelength?

20. (I) Radiation of wavelength 200 nm is incident on mercury, which has a work function of 4.5 eV. What is (a) the maximum kinetic energy of the ejected electrons, and (b) the stopping potential?

21. (I) When radiation of wavelength 350 nm is incident on a surface, the maximum kinetic energy of the photoelectrons is 1.2 eV. What is the stopping potential for a wavelength of 230 nm?

22. (I) When violet light of wavelength 420 nm illuminates a surface, the stopping potential of the photoelectrons is 2.4 V. What is the threshold frequency for this surface?

23. (II) A 100-W bulb coverts 5% of the electrical energy input to visible light. Assume the light has a wavelength of 600 nm and the bulb is a point source. (a) What is the number of photons emitted per second? (b) If the eye can detect 20 photons/s, at what distance would the bulb be visible? Take the pupil diameter to be 3 mm.

24. (II) When a metal is illuminated with light of frequency f, the maximum kinetic energy of the photoelectrons is 1.3 eV. When the frequency is increased by 50%, the maximum kinetic energy increases to 3.6 eV. What is the threshold frequency for this metal?

25. (II) (a) What is the frequency of a photon whose energy is twice the rest energy of an electron? (b) What would be the linear momentum of the photon?

26. (II) With a pupil diameter of 5 mm, the eye can detect 8 photons/s at 500 nm. What is the required power of a point source at the distance of (a) the moon; (b) Alpha-centauri, 4.2 light-years away?

27. (II) The following data on wavelengths and stopping potentials were obtained from an experiment on the photoelectric effect.

λ (nm):	500	450	400	350	300
V_0(V):	0.37	0.65	1.0	1.37	2.0

Plot a graph and from it determine (a) h/e; (b) the threshold frequency.

40.3 Compton Effect

28. (I) A beam of X rays with an energy of 30 keV undergoes Compton scattering. A scattered photon emerges at 50° relative to the incoming beam. (a) Find the modified wavelength. (b) What is the kinetic energy of the scattered electron?

29. (I) An X-ray beam has an energy of 40 keV. Find the maximum possible kinetic energy of Compton scattered electrons.

30. (I) A 0.071-nm wavelength X ray is scattered by a carbon target. It suffers a 0.02% shift in wavelength. At what angle to its original direction does it emerge?

31. (I) The wavelength of a photon is equal to the Compton wavelength. What is its energy?

32. (I) A 30 keV beam of X rays is Compton scattered through 37°. (a) What is the shift in wavelength? (b) What is the energy of the scattered photon?

33. (II) The fractional shift experienced by a beam of Compton-scattered radiation is $\Delta\lambda/\lambda = 0.03\%$. What is the energy of the incident photon if it is scattered through 53°?

34. (I) An X ray of wavelength 0.08 nm is scattered by 70° by a block of carbon. (a) What is the Compton shift in wavelength? (b) What is the kinetic energy of the scattered electron?

35. (I) X rays with an energy of 50 keV are scattered by 45°. Find the frequency of the scattered photons.

36. (I) A beam of X rays of wavelength 0.08 nm undergoes Compton scattering by a target. Calculate the shift in wave-

length if the scattered photon is deflected by (a) 30°, (b) 90°, (c) 150°.

40.6 The Bohr Model

37. (I) (a) A gas of hydrogen atoms in their ground state is bombarded by electrons with kinetic energy 12.5 eV. What emitted wavelengths would you expect to see? (b) What if the electrons were replaced by photons of the same energy?

38. (I) (a) Find the three longest wavelengths of the Paschen series (to $n = 3$) for the hydrogen atom. In what part of the spectrum do they lie? (b) What is the shortest wavelength in this series?

39. (I) What is the maximum wavelength that can ionize a hydrogen atom in the ground state? In what region of the electromagnetic spectrum does this wavelength lie?

40. (I) Calculate the frequency of the orbit of the electron in the ground state of the hydrogen atom. If radiation were classical, to what part of the spectrum would this belong?

41. (I) The electron in the hydrogen atom is in the $n = 2$ state. What is its (a) potential energy, (b) kinetic energy?

42. (I) (a) Determine the first four energy levels of Li^{++} ion ($Z = 3$). (b) What are the wavelengths of the three highest frequency emissions possible with these four levels?

43. (I) Calculate the radii of the first three states of the hydrogen atom.

44. (I) (a) What are the first three energy levels of the He^+ ion ($Z = 2$). (b) What is the energy required to remove the electron from this ion?

45. (I) Consider an electron in the ground state of the hydrogen atom. Determine its (a) speed, (b) linear momentum, and (c) acceleration.

46. (II) An electron orbits a nucleus with a charge Ze. Show that the radius of the nth level is given by $r_n = n^2 r_1/Z$, where $r_1 = \hbar^2/mke^2$.

47. (II) An electron orbits a nucleus with a charge Ze. Show that the speed for the nth level is given by $v_n = 2.2 \times 10^6 Z/n$ m/s.

48. (II) In a muonic atom, the electron is replaced by a particle called the muon that has the same charge as the electron but with a mass 207 times larger. By what factor does each of the following quantities change in comparison with an ordinary "one electron" atom: (a) the energy levels? (b) the radii of the orbits?

49. (II) An electron orbits a nucleus with a charge Ze. Show that the energy of the nth level is given by Eq. 40.27.

PROBLEMS

1. (I) In a Compton-scattering experiment, the scattered photon has an energy of 130 keV and the scattered electron's kinetic energy is 45 keV. Find (a) the wavelength of the incident photons, (b) the angle θ through which the photon is scattered, and (c) the angle ϕ at which the electron moves off.

2. (I) Show that Wien's radiation law, Eq. 40.2, leads to Wien's displacement law, Eq. 40.1. (*Hint:* What is the condition for λ_{max}?)

3. (I) By considering the special case of a one-dimensional collision, show that a free electron cannot completely absorb a photon. (Show that linear momentum and energy cannot be simultaneously conserved.) The "free" electrons in a metal are still bound to the material. An atom or crystal as a whole satisfies the momentum conservation without taking much energy.

4. (I) Show that the fractional energy loss of a Compton-scattered photon is given approximately by $\Delta E/E = -\Delta\lambda/\lambda$.

5. (I) The two protons in the hydrogen molecule are 0.074 nm apart and rotate about their center of mass. The total angular momentum is quantized in units of $nh/2\pi$. (a) What is the moment of inertia, I? (b) If the angular momentum $I\omega_n$ is quantized, find ω_n. (c) Where does $f_{n+1} - f_n$ lie in the electromagnetic spectrum?

6. (I) The electron in a hydrogen atom makes a transition from the $n = 5$ level to $n = 1$. Find the recoil speed of the atom.

7. (II) Derive Eq. 40.16 for the Compton effect. (*Hint:* First use Eqs. 40.14 and 40.15 to eliminate ϕ and obtain an expression for p^2. Second, use $E^2 = p^2c^2 + m_0^2c^4 = (K + m_0c^2)^2$ and Eq. 40.13 to obtain another expression for p^2.)

8. (II) (a) In Planck's radiation law, set $x = hc/\lambda kT$. By taking the derivative with respect to x show that the wavelength at which the maximum occurs is given by the equation $5 - x = 5e^{-x}$. The solution to this equation is $x = 4.965$. (b) Show that Planck's radiation law leads to Wien's displacement law.

9. (II) The total intensity, R, radiated from the surface of a blackbody is found by multiplying the integral of the energy density over all wavelengths, $U = \int u_\lambda \, d\lambda$, by $c/4$, that is, $R = Uc/4$. Derive the Stefan–Boltzmann law $R = \sigma T^4$. Where the constant $\sigma = 2\pi^5 k^4/15c^2h^3$. Set $x = hc/\lambda kT$ and note that

$$\int_0^\infty \frac{x^3}{(e^x - 1)}dx = \frac{\pi^4}{15}.$$

10. (II) Positronium consists of an electron and a positron (a positive electron) orbiting about their common center of mass. Use the Bohr model to show that the energy levels are given by $E_n = -6.8 \text{ eV}/n^2$.

SPECIAL TOPIC: Lasers

In 1917 Einstein published a paper in which he discussed the thermodynamic equilibrium between cavity radiation and the matter in the walls of the cavity. He assumed that the atoms could occupy a discrete set of energy levels. Let us consider two atomic states of energies E_1 and E_2, as in Fig. 40.15. The ratio of the numbers in the levels at temperature T is given by the Boltzmann factor (Eq. 20.17):

$$\frac{N_2}{N_1} = e^{-(E_2 - E_1)/kT} \qquad (40.30)$$

In thermal equilibrium, $N_2 < N_1$; that is, the higher state is less populated.

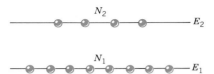

FIGURE 40.15 In thermal equilibrium the relative number of particles in the two energy levels is given by the Boltzmann factor: $N_2/N_1 = \exp[-(E_2 - E_1)/kT]$.

Radiation enters the problem as follows. Even at thermodynamic equilibrium, there is an ongoing absorption and emission of radiation. First consider an atom in state 1. If an incoming photon has the correct frequency ($hf = E_2 - E_1$), it will be absorbed and raise the atom to level 2, as in Fig. 40.16a. This process of **absorption** depends on the energy density ρ of the radiation at this frequency and on N_1. The number of such $1 \rightarrow 2$ transitions per unit time is $N_1 B_{12}\,\rho$, where B_{12} is a measure of the probability of the transition taking place. The probability that an atom in the upper level will drop to E_1, as in Fig. 40.16b, is given by A_{21}. The numbers of such **spontaneous emissions** depends on N_2, but not on the presence of the external radiation. The number of $2 \rightarrow 1$ transitions per unit time is $A_{21}N_2$.

In thermodynamic equilibrium one would expect the rate of upward transitions to equal the rate of downward transitions, that is, $N_2 A_{21} = N_1 B_{12}\rho$. Thus the energy density of the radiation has the form $\rho = (A_{21}/B_{12})(N_2/N_1)$. When Eq. 40.30 is substituted, the function resembles Wien's radiation law rather than Planck's radiation law—which perfectly describes the energy density of cavity radiation.

This difficulty led Einstein to propose another mechanism by which an atom can interact with radiation. In Fig. 40.17, the atom is in state 2. An incoming photon of the correct frequency causes the atom to "resonate" in some fashion and induces it to drop to level 1. In this case there are *two* outgoing photons of the same frequency. Einstein also showed that the stimulated photon must move off in the same direction as the incoming photon. This process of **stimulated emission** has a certain probability B_{21}. The number of such transitions per unit time, which depends on the radiation density and on N_2, is $N_2 B_{21}\rho$.

With the inclusion of stimulated emission, the requirement that the rate of emission balance the rate of absorption now reads

$$N_1 B_{12}\rho = N_2(A_{21} + B_{21}\rho)$$

When the Boltzmann factor for N_2/N_1 is used, we find that ρ has exactly the form of Planck's law provided $B_{21} = B_{12}$ and A_{21} is related to B in a simple way. This astonishingly simple derivation of Planck's law showed that the process of stimulated emission is necessary for the system to attain thermodynamic equilibrium. Under normal circumstances the absorption process dominates because there is a larger number of atoms in the lower level.

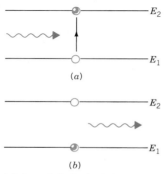

FIGURE 40.16 (a) A particle in the lower state absorbs a photon and makes a transition to the higher state. (b) The particle falls from the higher to the lower state in the process of spontaneous emission.

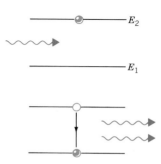

FIGURE 40.17 In the process of stimulated emission, an incoming photon causes a particle in the upper level to fall to the lower level. The photon that is emitted is coherent with the original photon and the two photons move off in the same direction.

The idea of using stimulated emission to amplify microwave radiation occurred to several people in the early 1950s. In the spring of 1951, C. H. Townes thought of a device that would accomplish this and in 1953, with co-workers, he successfully operated the first **maser** (*micro*wave *a*mplification by *s*timulated *e*mission of *r*adiation) using energy levels in the ammonia molecule. In 1958, Townes and A. Schawlow proposed a way of producing stimulated emission at optical frequencies, and in 1960, T. H. Maiman, operated the first ruby **laser** (*l*ight amplification by *s*timulated *e*mission of *r*adiation).

The Ruby Laser

The red color of ruby (Al_2O_3) is caused by a small number of Cr^{3+} impurities. The relevant energy levels of this ion are shown in Fig. 40.18. E_1 is the ground state and E_3 is a short-lived (10^{-8} s) excited state, whereas E_2 corresponds to a long-lived (3×10^{-3} s) **metastable** state. The atom decays readily from E_3 to E_2, but not from E_2 to E_1. Maiman placed a ruby crystal in the form of a rod within a coiled discharge tube, as in Fig. 40.19. A flash, covering a range of wavelengths around the required value of 550 nm, raises the

Maiman and his laser.

Cr^{3+} ions to E_3. From this state they decay quickly to E_2. If the flash is intense enough, it is possible to have more atoms in the metastable state than in the ground state. This process of **optical pumping** creates a nonequilibrium condition called **population inversion** in which $N_2 > N_1$.

Let us emphasize two conditions necessary for the operation of a laser:

1. Recall that $B_{21} = B_{12}$, that is, the probabilities for absorption and stimulated emission are equal. The **population inversion** ($N_2 > N_1$) allows the stimulated emission to occur more frequently than absorption.

2. The requirement of a **metastable state** allows stimulated emission to occur before spontaneous emission and makes population inversion a practical possibility.

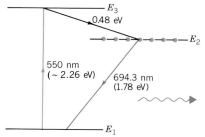

FIGURE 40.18 The energy levels for the operation of the ruby laser. The level marked E_2 is a metastable state. For laser action to occur, the population of this state must be greater than that of the ground state, E_1.

A stray (spontaneously emitted) photon will stimulate an atom in the state E_2 to emit a photon. The two photons have the *same frequency* and travel in the *same direction*. These two photons can then stimulate two other atoms to emit two more photons, and so on, as shown in Fig. 40.20. This process occurs simultaneously in various directions. In practice the ends of the rod are coated with aluminum so that they act as mirrors and are made parallel to an accuracy of less than 1' of arc! One end is slightly leaky—it transmits 1% of the light. With this arrangement, only those photons that travel along the axis of the rod will be reflected back and forth many times. The stimulated radiation builds in intensity only in this direction until, ultimately, a short pulse of nearly unidirectional and monochromatic radiation is emitted through the leaky mirror. The ruby laser emits only short pulses (several pulses per flash of the discharge tube). Furthermore, in this three-level scheme, the "lasing" leaves the atom in the ground state. It therefore takes a great deal of input power to produce a population inversion.

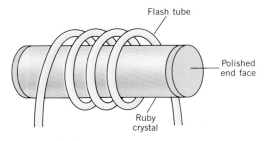

FIGURE 40.19 In the ruby laser the crystal is "optically pumped" by a flash tube wrapped around it. This produces the population inversion necessary for laser action.

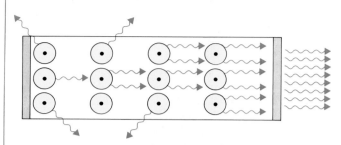

FIGURE 40.20 The ends of the laser have mirrors that cause the photons to reflect back and forth within the cavity. Although initially the stimulated photons are emitted in all directions, only those that travel parallel to the axis increase streadily. The laser light leaks through one mirror that is slightly transparent.

The Gas Laser

In 1960, A. Javan et al. operated the first continuous-wave laser using a mixture of He and Ne gases in a discharge tube. Collisions between the electrons and ions raise the He atoms to a metastable state at $E_1 = 20.61$ eV above the ground state (see Fig. 40.21). It so happens that Ne has a metastable state at nearly the same energy, $E_2 = 20.66$ eV. Rather than decaying to its own ground state by emitting a photon, the He atoms can transfer the energy to Ne atoms during collisions. The small difference of 0.05 eV is supplied by the kinetic energy of the atoms. The Ne atoms are also raised to E_2 by collisions with electrons, but the He helps considerably in populating this state. This four-level scheme is more efficient (a 15 W input produces a 1 mW beam output) than the three-level scheme because atoms in state $E_3 = 18.70$ eV decay very quickly to state E_4. It is therefore easier to maintain a population inversion between states E_2 and E_3. The laser light emitted is at 632.8 nm. Let us now consider some properties of laser light.

1. *The beam is unidirectional:* The beam that emerges from a typical laser has a divergence of about 1′ of arc. (Dif-

fraction always produces some divergence.) Thus the beam diameter increases by about 1 mm per meter of travel. This also implies that the light consists of nearly plane waves and that the intensity decreases slowly with distance.

2. *The beam intensity is high:* A powerful searchlight can produce about 1 kW of radiation, whereas a continuously operating CO_2 laser can produce 10 kW. Operating in pulses of duration 10^{-12} s, a neodymium–glass laser can produce an instantaneous power of 10^9 W! Consider the low (1 mW) power output of a continuous He–Ne laser at 632.8 nm. Depending on the beam size, its intensity is about 100 W/m². Thermal radiation from a blackbody at a temperature of 4580 K would have its peak at this wavelength. The radiated intensity within the specified range would be about 25 mW/m². Thus, within its spectral range, even the relatively low-power He–Ne laser is 4000 times brighter than sunlight (a good reason NEVER to look into a laser beam).

3. *Laser light is nearly monochromatic:* Although perfectly monochromatic light does not exist, laser light comes close to this ideal. Every spectral line from an atom has a natural range in wavelength or frequency. The effects of collisions between the atoms, and the Doppler effect, cause further broadening of the lines. A fine line from an ordinary gas discharge tube may have a spread in wavelengths, a *linewidth*, of ±0.01 nm, while the very best have ±0.0005 nm. In contrast, the linewidth of a He-Ne laser can be as small as ±10^{-6} nm.

   The narrowness of the laser output arises from a resonance effect. The radiation being reflected back and forth between the mirrors forms resonant standing wave modes with sharply defined frequencies. So, within the (Doppler and collision broadened) linewidth, the laser output consists of a few very sharp ($\Delta f < 10^3$ MHz) mode frequencies, as shown in Fig. 40.22. There are several ways of getting the laser to operate in a single mode, but we will not discuss them.

4. *Laser light is coherent:* It was pointed out earlier that a photon produced by stimulated emission travels in the

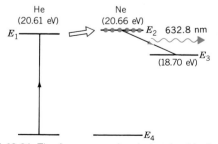

FIGURE 40.21 The four energy levels involved in the He-Ne laser. The metastable state, E_2, of the neon atoms is populated by collisions with electrons in the state E_1 of the helium atoms.

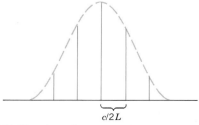

FIGURE 40.22 The sharp frequency of laser light is associated with resonant modes that are set up in the laser cavity.

same direction as the original photon. Just as significant is the fact that the two photons are exactly *in phase*, and have the *same polarization*. This leads to the remarkable coherence of laser light (see Section 37.7). The *spatial coherence* of laser light means that two points on opposite sides of the beam are coherent. Laser light also has great *temporal coherence*.

The coherence time, τ_c, is the maximum time for which two points on a wavetrain have a fixed phase relation. It is also an estimate of the lifetime of the upper level involved in a transition. Any wavetrain of finite length may be considered to be the result of superposing of waves of infinite length (each with a single frequency) but having a spread in frequencies Δf. Analysis shows that the linewidth and the coherence time are related according to

$$\Delta f = \frac{1}{\tau_c}$$

Thus a narrow linewidth implies that a beam has long *temporal coherence*. The **coherence length,** $\ell_c = c\tau_c$, is an indication of the length of a wavetrain. Typically, for a single atom $\tau_c \approx 10^{-8}$ s, and so $\ell_c = c\tau_c = 3$ m. In a gas discharge, Doppler and collision broadening increases the measured linewidth. One of the sharpest lines from cadmium has $\Delta\lambda = \pm 0.001$ nm, which leads to $\ell_c = 25$ cm if $\lambda = 500$ nm. In contrast, the coherence length of a laser line may be over 30 km!

A laser beam being used to determine particle size and concentration a flame.

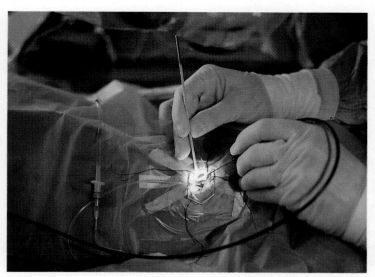

Laser eye surgery.

CHAPTER 41

Wave Mechanics

Major Points

1. **De Broglie's hypothesis** that particles have wave properties.
2. **Schrödinger's wave equation** is used to predict the behavior of matter waves.
3. The **wave function** tells us the **probability** of finding a particle within a given region.
4. A particle can penetrate into a classically forbidden region and can **tunnel** through a potential barrier.
5. The **Heisenberg uncertainty principle:** One cannot measure certain pairs of quantities simultaneously to arbitrary precision.

Images of silicon atoms produced by a scanning tunneling electron microscope. The magnification is about 10 million. The topmost atoms (top). The "dangling" bonds that reach up from the top atoms (top center). Bonds between the top atoms and the second layer (bottom center). The second layer (also shows "sideways" bonds) at a depth of 0.9 nm (bottom).

Bohr's theory was successful in explaining the spectrum of hydrogen, and it began to explain the stability of atoms. However, the theory applied only to one-electron systems; it could not predict the relative intensities of spectral lines, or explain why, with increased resolution, some lines were found to consist of two or more finer lines. In 1916, A. Sommerfeld refined Bohr's theory by incorporating special relativity and the possibility of elliptical orbits. With the addition of two new quantum numbers, the Bohr–Sommerfeld theory accounted for many features of spectra and showed how the periodic table is built up in a systematic way. Nonetheless, the rules that were used had no proper foundation, and by the early 1920s the theory reached the limits of its explanatory power. It was clear that radical reform was needed in quantum theory.

41.1 DE BROGLIE WAVES

In 1924, Louis de Broglie (Fig. 41.1) put forward an astounding proposition as his doctoral thesis. Its genesis lay in the metaphysical notion that "nature is symmetrical." De Broglie noted that light, which for a century had been successfully treated as a wave, had recently displayed particle characteristics in the photoelectric and Compton effects. He recalled that in 1909 Einstein had shown that a complete description of cavity radiation (Planck's law) requires *both* the particle and the wave aspects of radiation. De Broglie guessed that a similar wave–particle duality might apply to material particles. That is, matter may also display wave behavior. Relativity and quantum theory had shown that classical physics is inadequate in several areas, so there was no great need to rely on classical concepts when dealing with the submicroscopic world of the atom.

De Broglie used a combination of quantum theory and special relativity to propose that the wavelength, λ, associated with a particle is related to its linear momentum, $p = mv$, by

$$\lambda = \frac{h}{p} \qquad (41.1)$$

Note that this equation also applies to a photon: $p = E/c = hf/c = h/\lambda$. The physical significance of the "matter wave" was not clear, but he gained courage from the following demonstration. In the Bohr model, the angular momentum of the electron is quantized:

$$mvr = \frac{nh}{2\pi} \qquad (41.2)$$

When de Broglie's equation, $p = mv = h/\lambda$, is used, Eq. 41.2 becomes

$$2\pi r = n\lambda \qquad (41.3)$$

This looks like the condition for a standing wave! De Broglie had given Bohr's arbitrary postulate a clear interpretation: Only those orbits that can fit an integral number of wavelengths around the circumference are allowed (see Fig. 41.2). Einstein helped to publicize De Broglie's hypothesis—although it struck many as being nonsense.

41.2   ELECTRON DIFFRACTION

Evidence for the wave nature of electrons existed, without being recognized, even before de Broglie had presented his thesis. C. L. Davisson, who had been studying the scattering of electrons off nickel surfaces, reported the curious result that the reflected intensity depended on the orientation of the sample. After de Broglie's hypothesis was publicized, W. Elsasser suggested that this finding could involve diffraction of de Broglie waves, but Davisson did not pay much attention. Fortunately, an accident in his vacuum system forced Davisson to heat the target to remove an oxide coating. This process converted the original polycrystalline sample into almost a single crystal.

In 1926, Davisson and L. Germer resumed the experiment. Electrons were produced by a heated filament, accelerated by a potential difference, V, and then directed at the Ni target, as shown in Fig. 41.3. They found that the electrons were reflected primarily in certain directions that depended on their speed. If the electrons had interacted with the atoms on a one-to-one basis, this would have produced random scattering. The pronounced reflections implied that the electrons were interacting with an *array* of atoms—much like the reflection of X rays by atomic planes produces diffraction patterns (Section 38.10). The electrons were also exhibiting wave behavior!

When a particle of mass m and charge q is accelerated from rest by a potential difference V, its kinetic energy is given by $K = p^2/2m = qV$ (assuming that its speed is much less than the speed of light). Since $p = \sqrt{2mqV}$, the de Broglie relation, $\lambda = h/p$, takes the form

$$\lambda = \frac{h}{\sqrt{2mqV}} \qquad (41.4)$$

FIGURE 41.1 Louis de Broglie (b.1892).

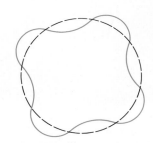

FIGURE 41.2 A standing wave set up around the perimeter of a circle. This picture was used by de Broglie to explain the quantization of angular momentum in Bohr's theory.

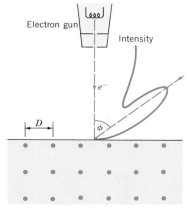

FIGURE 41.3 Electrons are directed at a nickel crystal. The reflected beam shows a strong angular dependence, which indicates the electrons are being diffracted by the atomic planes.

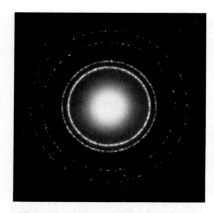

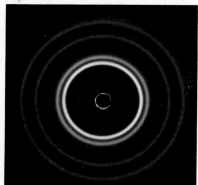

FIGURE 41.4 (*a*) A diffraction pattern produced by 0.071 nm X-rays passing through an aluminum foil. (*b*) The diffraction pattern produced when electrons pass through an aluminum foil. The circular patterns arise because the foil consisted of many randomly oriented tiny crystals.

For example, if $V = 150$ V, the de Broglie wavelength of an electron is about 0.1 nm, which is roughly the interatomic spacing in a crystal. By an analysis similar to that for X rays (Section 38.10), it is found that the angular positions of the diffraction maxima are given by

$$D \sin \phi = n\lambda \qquad (41.5)$$

where D is the spacing between atoms, which in the case of nickel is 0.215 nm. One pair of experimental values was $V = 54$ V and $\phi = 50°$. Equation 41.4 predicted $\lambda = 0.167$ nm whereas from Eq. 41.5 (with $n = 1$) he found $\lambda = 0.165$ nm. De Broglie's hypothesis was conclusively verified!

In 1927, G. P. Thomson and A. Reid passed a beam of 30-keV electrons through thin films of celluloid and gold, which consisted of tiny, randomly oriented crystals. In this arrangement there are always some crystals oriented such that Eq. 41.5 is satisfied. As a result, diffraction rings were recorded on a photographic plate, thereby confirming that electrons exhibit wave behavior. In Fig. 41.4 we compare the diffraction rings produced by X rays and by electrons. Wavelike behavior is exhibited by all elementary particles. For example, Fig. 41.5 shows the diffraction of neutrons by a polycrystalline sample of iron.

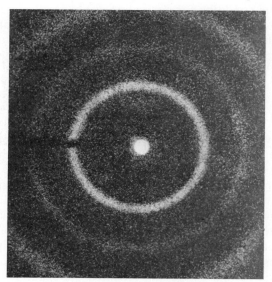

FIGURE 41.5 A diffraction pattern produced by 0.07-eV neutrons passing through a polycrystalline sample of iron.

EXAMPLE 41.1: What is the de Broglie wavelength of (a) an electron accelerated from rest by a potential difference of 54 V, and (b) a 10 g bullet moving at 400 m/s?

Solution: (a) From Eq. 41.5, the de Broglie wavelength is

$$\lambda = \frac{h}{p} = \frac{h}{\sqrt{2meV}}$$

$$= \frac{(6.626 \times 10^{-34} \text{ J} \cdot \text{s})}{\sqrt{(2 \times 9.11 \times 10^{-31} \text{ kg})(1.6 \times 10^{-19} \text{ C})(54 \text{ V})}}$$

$$= 0.167 \text{ nm}$$

(b) The de Broglie wavelength is

$$\lambda = \frac{h}{mv} = \frac{(6.63 \times 10^{-34} \text{ J} \cdot \text{s})}{(10^{-2} \text{ kg})(400 \text{ m/s})}$$

$$= 1.66 \times 10^{-34} \text{ m}$$

This value is far smaller than the size of a single nucleus, which is about 10^{-14} m. There is no chance of observing wave phenomena, such as diffraction, with macroscopic objects.

EXERCISE 1: Calculate the de Broglie wavelength of a particle moving at 10^6 m/s given that it is (a) an electron, and (b) a proton.

41.3 SCHRÖDINGER'S WAVE EQUATION

When Erwin Schrödinger (Fig. 41.6) first heard of de Broglie's hypothesis, he thought it was rubbish. However, when he realized that Einstein took this idea seriously, he decided to look for an equation to describe these matter waves. Schrödinger reasoned that just as geometrical optics is only an approximation to wave optics, classical (particle) mechanics may be just an approximation to a more correct wave mechanics. Although we cannot present Schrödinger's approach, a simplified discussion can be based on the wave equation (Eq. 16.15):

$$\frac{\partial^2 y}{\partial x^2} = \frac{1}{v^2}\frac{\partial^2 y}{\partial t^2} \tag{41.6}$$

FIGURE 41.6 Erwin Schrödinger (1887–1961).

In keeping with de Broglie's analysis of the Bohr atom, let us consider only standing wave solutions in one dimension. Recall from Section 16.7, that the form of a standing wave is

$$y(x,\ t) = \psi(x)\ \sin(\omega t) \tag{41.7}$$

When this is substituted into the wave equation, we find

$$\frac{d^2\psi}{dx^2} = -\frac{\omega^2}{v^2}\ \psi \tag{41.8}$$

We have used the ordinary derivative since $\psi(x)$ is a function only of x. Since $\omega/v = k = 2\pi/\lambda$ and $p = h/\lambda$, we have

$$\frac{\omega^2}{v^2} = \frac{p^2}{\hbar^2}$$

From the expression for the total energy, $E = p^2/2m + U$, where U is the potential energy, we see that $p^2 = 2m(E - U)$. Thus

$$\frac{\omega^2}{v^2} = \frac{2m(E - U)}{\hbar^2} \tag{41.9}$$

Equation 41.8 therefore becomes

$$\frac{d^2\psi}{dx^2} + \frac{2m}{\hbar^2}\ (E - U)\psi = 0 \tag{41.10}$$ Schrödinger wave equation

This is the one-dimensional **time-independent Schrödinger wave equation.** The wavefunction $\psi(x)$ represents *stationary* states of an atomic system for which E is constant in time.

How can a *continuous* description, such as that implied by a wave, lead to *discrete* quantities, such as the energy levels of the hydrogen atom? Recall that the continuous system of a string tied at both ends vibrates only at certain frequencies. The classical wave equation leads to discrete modes when we apply the *boundary conditions,* which state that the displacement of the string must be zero at the fixed ends. In wave mechanics, both ψ and $d\psi/dx$ must also be *continuous* functions. If, for example, $d\psi/dx$ were to have a discontinuity, then $d^2\psi/dx^2$ would be infinite and the solutions to Eq. 41.10 would be unphysical.

When Schrödinger applied his equation to the hydrogen atom (for which $U = -ke^2/r$), he found that the mathematics and the appropriate boundary conditions lead naturally to the discrete energy levels of the Bohr model. At about the same time (1925) W. Heisenberg developed a different form of quantum mechanics that was later shown to be equivalent to that of Schrödinger.

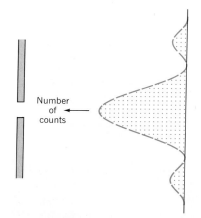

FIGURE 41.7 When electrons pass through a narrow slit, the number of counts display the usual diffraction pattern for a single slit.

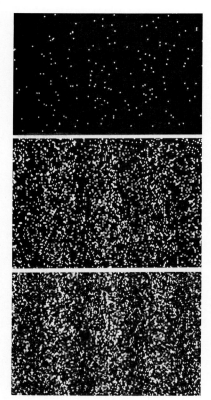

FIGURE 41.8 A double-slit interference pattern due to electrons recorded on a TV screen. Initially the dots seem to appear randomly located. However, after a large number of electrons has arrived, the pattern is quite clear.

41.4 THE WAVE FUNCTION

Schrödinger's success in tackling several problems confirmed that the new wave mechanics was an important advance. But how was the "wave associated with the particle" to be interpreted? De Broglie had tentatively suggested that the wave might represent the particle itself, or perhaps it might play some sort of guiding role in the particle's motion. Schrödinger believed that a particle is really a group of waves, a *wave packet,* somewhat like a fuzzy powderpuff.

A short time later Max Born proposed an interpretation of the wave function that has now been generally accepted. He was guided by Einstein's view that the intensity of a light wave at a given point (which is proportional to the square of the amplitude of the wave) is a measure of the number of photons that arrive at that point. In other words, the wave function for the electromagnetic field determines the *probability* of finding a photon. By analogy, Born suggested that the square of the wave function* tells us the probability per unit volume of finding the particle:

$$\psi^2 dV = \text{Probability of finding the particle within a volume } dV$$

The quantity ψ^2 is called the **probability density.** In one dimension, $\psi^2(x)dx$ is proportional to the probability of finding the particle within the interval x to $x + dx$. We are more likely to observe the particle wherever $\psi^2(x)$ is large and less likely to observe it where $\psi^2(x)$ is small. Thus the wave function represents an abstract *wave of probability.* Since the particle has to be found somewhere, the sum of all the probabilities along the x axis has to be one:

$$\int_{-\infty}^{\infty} \psi^2(x)dx = 1$$

A wave function that satisfies this condition is said to be **normalized.**

The probability interpretation of the wave function can be illustrated by a simple experiment in which a beam of electrons passes through a single slit, as in Fig. 41.7. The beam is so weak that only one electron at a time passes through the slit. A large number of densely packed detectors records exactly where each electron goes. (Equivalently, one could count the number of electrons that arrive at each position in a specified time interval.) The quantity $\psi^2(x)$ tells us what fraction of the total counts are recorded at position x. The initial pattern of counts is random. However, after several thousand counts, the number of counts displays the familiar single-slit diffraction pattern, thereby providing startling confirmation of these strange ideas! Figure 41.8 shows an interference pattern produced by electrons that passed through two slits.

Classical physics and special relativity are based on the principle of *determinism:* Given the initial position and velocity of a particle and all the forces acting on it, one can accurately predict its future course. The exact position of the particle can, at least in principle, be determined. The statistical interpretation of the wave function says that one can predict only the *probability* that a particle will be observed at a given position. It is no longer possible to predict exactly where a single particle will be detected. Quantum mechanics correctly predicts *average* values of physical quantities, not the results of individual measurements.

---

*Since ψ can be a complex number, the proper expression is $\psi\psi^*$ where * denotes the complex conjugate.

41.5   APPLICATIONS OF WAVE MECHANICS

Particle in a Box

Let us see how these new ideas are applied to a particle of mass m that bounces back and forth in a one-dimensional box of side L, as in Fig. 41.9. We assume the box is impenetrable: The potential energy is zero within the box and infinite at the walls. This is a somewhat artificial example, but it illustrates several important ideas. It is a first step in solving problems such as the motion of a conduction electron in a metal or a proton trapped in a nucleus.

Classically, the particle may be found anywhere from $x = 0$ to $x = L$ with equal probability. In the wave-mechanical approach we have to assign a wave function to the particle. Since the particle cannot penetrate the walls, $\psi = 0$ for $x < 0$ and $x > L$. The requirement that the wave function be continuous leads to the boundary condition

$$\psi(x) = 0 \text{ at } x = 0 \text{ and } x = L$$

With $U = 0$, the Schrödinger wave equation becomes

$$\frac{d^2\psi}{dx^2} + k^2\psi = 0$$

where $k = \sqrt{2mE}/\hbar$. The solution to this equation is $\psi(x) = A \sin(kx + \phi)$. From the boundary condition $\psi = 0$ at $x = 0$, it follows that $\phi = 0$. From the condition that $\psi = 0$ at $x = L$ we find $\sin(kL) = 0$, which means that $kL = n\pi$, where n is an integer. Thus, the wave function that satisfies the boundary conditions has the form of a standing wave:

$$\psi(x) = A \sin\left(\frac{n\pi x}{L}\right); \qquad n = 1, 2, 3, \ldots \quad (41.11)$$

Since $k = 2\pi/\lambda = n\pi/L$, the wavelength of the nth standing wave is $\lambda = 2L/n$. When this is equated to de Broglie's equation $\lambda = h/mv$, we find $v = nh/2mL$. Since n takes on only integer values, the speed is quantized. The particle's energy, which is purely kinetic, is $\frac{1}{2}mv^2$, is thus also quantized:

$$E_n = \frac{n^2h^2}{8mL^2}; \qquad n = 1, 2, 3, \ldots \quad (41.12)$$

The boundary conditions have led to a set of quantized energy levels, which are depicted in Fig. 41.10. Notice that the particle cannot have zero energy. The lowest value occurs at $n = 1$, and is called the **zero-point energy.** It is present for any particle that is confined to a region in space and exists even at 0 K—in stark contrast to the classical notion that everything should be at rest at 0 K.

The wave functions for the first few levels are shown in Fig. 41.11a. The

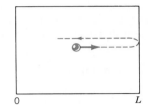

FIGURE 41.9 A particle confined to a box bounces back and forth. The walls are impenetrable; they define a region of infinite potential energy.

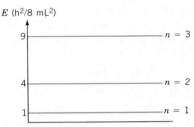

FIGURE 41.10 The quantized energy levels of a particle in an infinite potential well.

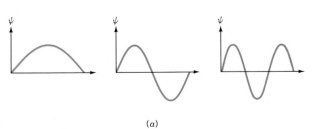 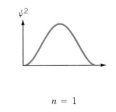 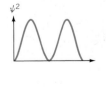

$n = 1$ $\qquad$ $n = 2$ $\qquad$ $n = 3$

(a) $\qquad\qquad\qquad\qquad\qquad\qquad$ (b)

FIGURE 41.11 (a) The first three wave functions for the particle in a box. (b) The probability densities for the first three states.

probability densities, $\psi^2(x)$, shown in Fig. 41.11b, are zero at certain points; we would never observe the particle at these locations. This appears to contradict our everyday experience, but fortunately the correspondence principle (Section 40.8) resolves the problem, as we will see in Example 41.3.

EXAMPLE 41.2: An electron is trapped within an infinite potential well of length 0.1 nm. What are the first three energy levels?

Solution: From Eq. 41.12, the energy levels are given by

$$E_n = \frac{n^2 h^2}{8mL^2}$$

$$= \frac{n^2(6.63 \times 10^{-34} \text{ J} \cdot \text{s})^2}{(8 \times 9.11 \times 10^{-31} \text{ kg})(10^{-10} \text{ m})^2}$$

$$= n^2(6.03 \times 10^{-19}) \text{ J} = 37.7 n^2 \text{ eV}$$

Thus the first three energies are $E_1 = 37.7$ eV, $E_2 = 151$ eV, and $E_3 = 339$ eV.

EXERCISE 2. What is the wavelength of the photon emitted when the electron makes a transition from the $n = 2$ level to the ground state?

EXAMPLE 41.3: Consider a 10^{-7}-kg dust particle confined to a 1-cm box. (a) What is the minimum speed possible? (b) What is the quantum number n if the particle's speed is 10^{-3} mm/s?

Solution: (a) From Eq. 41.12, the minimum allowed energy is E_1. Thus, $\frac{1}{2}mv^2 = h^2/8mL^2$, from which we find

$$v = \frac{h}{2mL} = \frac{6.63 \times 10^{-34} \text{ J} \cdot \text{s}}{2 \times 10^{-7} \text{ kg} \times 10^{-2} \text{ m}}$$

$$= 3.32 \times 10^{-25} \text{ m/s}$$

Even with its zero-point energy, the dust particle is essentially at rest—which agrees with our classical expectations.
(b) To find the quantum number n we equate the kinetic energy to E_n:

$$\frac{n^2 h^2}{8mL^2} = \frac{1}{2}mv^2$$

When $v = 10^{-6}$ m/s is substituted, we find $n \approx 10^{23}$! The quantized nature of energy in the transitions from n to $n - 1$ is not observable on a macroscopic scale. Also, the wave function goes through many oscillations between $x = 0$ and $x = L$. The peaks and dips in the probability function are so very close together that, in effect, the probability becomes uniform. This is just the classical result, as we would expect from the correspondence principle.

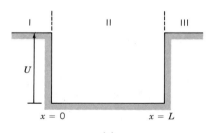

(a)

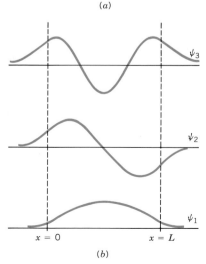

(b)

Finite Potential Well

We now consider a particle within a potential well of finite depth, U, that extends from $x = 0$ to $x = L$. We take $U = 0$ at the bottom of the well, as in Fig. 41.12a. Classically, if the energy of the particle is less than U (that is, $E < U$) the particle cannot enter the regions $x < 0$ and $x > L$. However, according to quantum mechanics, the wave function does not vanish outside the walls. Within region II where $U = 0$, the Schrödinger wave equation is

$$\frac{d^2\psi}{dx^2} + k^2\psi = 0$$

where $k = \sqrt{2mE}/\hbar$, and the wavefunction is sinusoidal:

$$\psi_{\text{II}} = C \sin(kx)$$

However, ψ is not zero at $x = 0$ and $x = L$. In the regions outside the well, $U > E$, and so the wave equation may be written in the form

$$\frac{d^2\psi}{dx^2} = K^2\psi$$

FIGURE 41.12 (a) A finite potential well of depth U. (b) The first three wave functions for a particle in a finite potential well. The wave functions decay exponentially in the classically forbidden regions where $U < E$.

where $K^2 = 2m(U - E)/\hbar^2$. The general solution to this equation is

$$\psi = Ae^{Kx} + Be^{-Kx}$$

In region III, ψ must approach zero as $x \to \infty$, so the appropriate function is

$$\psi_{III} = Be^{-Kx}$$

In region I, ψ must approach zero as $x \to -\infty$, so the appropriate choice is

$$\psi_I = Ae^{Kx}$$

In order to complete the solution, we must match the functions within the well with those outside. That is, we must apply the boundary conditions. For example,

$(x = 0)$ $\qquad\qquad \psi_I = \psi_{II}, \qquad$ and $\qquad \dfrac{d\psi_I}{dx} = \dfrac{d\psi_{II}}{dx}$

Boundary conditions

Similar conditions hold for ψ_{II} and ψ_{III} at $x = L$.

A few wave functions have been sketched in Fig. 41.12b. The fact that the wave functions are not zero outside the well means that there is a finite probability of finding the particle outside the well, in the region that would be forbidden classically. This surprising result is manifested in the phenomenon of tunneling, discussed next.

Barrier Penetration; Tunneling

Let us consider what happens when a particle with energy E encounters a potential energy barrier of height U ($> E$), as in Fig. 41.13. The wave function of the approaching particle, in the region where $U = 0$, is sinusoidal. Classically, the particle would be reflected. However, as we saw for a finite potential well, according to wave mechanics the wave function of the particle decays exponentially within the barrier region. If the thickness of the barrier is not great, the wave function may not decay completely to zero on the other side. In this case, there would again be a sinusoidal wave function of small amplitude. This means that there is a small, but finite, probability that the particle will **tunnel** through the barrier! Such tunneling occurs in a device called the tunnel diode, in the emission of α particles from radioactive nuclei (Section 43.3) and in superconducting Josephson junctions (see p. 885). It is also used in the scanning tunneling electron microscope.

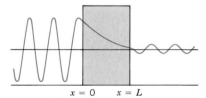

FIGURE 41.13 A particle with energy less than the height of a potential barrier has a probability of tunneling through the barrier.

The scanning tunneling microscope invented in 1981 by George Bennig and H. Rohrer (see p. 859).

41.6   HEISENBERG UNCERTAINTY PRINCIPLE

The fact that particles have wave characteristics has an important implication. The wavelength of a wave can be specified precisely only if the wave extends over many cycles, as in Fig. 41.14a. But if a matter wave is spread out in space, the position of the particle is poorly defined. To reduce the uncertainty in the position of the particle, Δx, one can superpose many wavelengths to form a reasonably well-localized *wave packet,* as shown in Fig. 41.14b. From $\lambda = h/p$ we see that a spread in wavelengths, $\Delta\lambda$, means that the wave packet involves a spread in momentum, Δp. According to the **Heisenberg uncertainty principle,** the uncertainties in position and in momentum are related by

$$\Delta x \Delta p \geq h \qquad (41.13)$$

It is not possible to measure both the position of a particle and its linear momentum simultaneously to arbitrary precision. This inability has nothing to do with experimental skill or equipment; it is a fundamental restriction imposed on us by nature. For a wave packet, the uncertainty relation is an intrinsic property, independent of the measuring apparatus.

W. Heisenberg (Fig. 41.15) arrived at the above relationship in 1927 through a simple yet profound analysis of the process of measurement. Here is a simplified version. Suppose we wish to locate an electron by bouncing a photon off it. One would not expect to locate it more precisely than the wavelength of the light used for observation. Thus the uncertainty in the position of the electron is at least $\Delta x = \lambda$. The photon may transfer some, or almost all, of its linear momentum to the electron. Thus, the uncertainty in the momentum of the electron is roughly equal to the original momentum of the photon, $\Delta p = h/\lambda$. The product of these uncertainties is $\Delta x \Delta p \approx h$, which is Eq. 41.13. If we try to reduce Δx by using light of shorter wavelength, the momentum of the photon increases, and so does Δp, the uncertainty in the electron's momentum. One can measure *either* the position or the momentum precisely, but one cannot measure both simultaneously. In this example the very act of measurement disturbs the system under study. One cannot speak of the system as if it were an isolated entity; there is always an inevitable interaction between the observer and what is being observed.

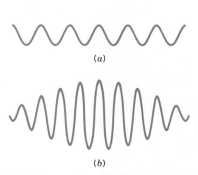

(a)

(b)

FIGURE 41.14 (a) The wavelength of a wave that extends over many cycles is well defined. The position of the wave is not well defined. (b) When waves of different wavelengths are superposed, they can form a localized wave packet but the wavelength is not well defined.

FIGURE 41.15 Werner Heisenberg (1901–1976).

Another derivation of the uncertainty relation is possible. Consider the diffraction of electrons by a single slit, as shown in Fig. 41.16. From Eq. 38.1 we know that the position of the first minimum is given by

$$\sin \theta = \frac{\lambda}{a} = \frac{\lambda}{\Delta y}$$

As the electron wave goes through the slit, the uncertainty in the lateral position is the slit width Δy. The uncertainty in the momentum in the y direction must be at least $p \sin \theta$, where θ corresponds to the first minimum. We may say $\Delta p_y > p \sin \theta$. Combining this with $p = h/\lambda$ we find

$$\Delta p_y \Delta y > h$$

A finer slit would locate the particle more precisely but lead to a wider diffraction pattern—that is, to a greater uncertainty in the transverse momentum.

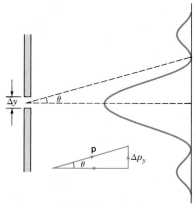

FIGURE 41.16 When an electron passes through a slit, the uncertainty in the vertical coordinate is the slit width and the uncertainty in the y component of its momentum may be estimated from the location of the first diffraction minimum.

EXAMPLE 41.4: What is the minimum uncertainty in the position of each of the following particles if the speed is measured to an uncertainty of 0.1%? (a) An electron moving at 4×10^6 m/s. (b) A 10-g bullet moving at 400 m/s.

Solution: (a) The uncertainty in momentum is

$$\Delta p = m\Delta v = (9.11 \times 10^{-31} \text{ kg})(4 \times 10^3 \text{ m/s})$$
$$= 3.64 \times 10^{-27} \text{ kg} \cdot \text{m/s}$$

From the Heisenberg uncertainty relation, the uncertainty in the position is

$$\Delta x \approx \frac{h}{\Delta p}$$
$$= \frac{6.63 \times 10^{-34} \text{ J} \cdot \text{s}}{3.64 \times 10^{-27} \text{ kg} \cdot \text{m/s}} = 18.2 \ \mu\text{m}$$

(b) In this case, $\Delta p = (0.01 \text{ kg})(0.4 \text{ m/s}) = 4 \times 10^{-3} \text{ kg} \cdot \text{m/s}$. The uncertainty in position is $\Delta x \approx h/\Delta p = 1.67 \times 10^{-31}$ m. This value is less than the diameter of a single proton. The uncertainty principle presents no practical restriction on determining the position of the bullet.

The Heisenberg uncertainty principle also applies to other pairs of variables. Among the most important are energy and time:

$$\Delta E \Delta t \geq h \qquad (41.14)$$

Heisenberg uncertainty relation for energy and time

To minimize the uncertainty in measuring the energy of a system, one must observe it for as long as possible. Since $\Delta E = h\Delta f$, we see that if an electron stays in an excited atomic state for a long time before making a transition to the ground state, the frequency of the photon emitted is sharply defined. If the lifetime of the upper state is short, the frequency of the emission is less well defined. From this version of Heisenberg uncertainty principle we may also infer that *the energy of a system can fluctuate from the value set by the conservation of energy*—provided the fluctuation occurs within the time interval specified by Eq. 41.14.

41.7 WAVE–PARTICLE DUALITY

Let us reconsider Young's double-slit experiment, but now conducted with electrons. The electrons may be detected by an array of counters. Each click suggests that the electron is a particle, but as we saw in Fig. 41.8, the overall pattern of clicks suggests that electrons behave like waves. Let us say that ψ_1 is the wave function that applies to the passage of an electron through slit S_1 whereas ψ_2 applies to slit S_2. When just one slit, say S_1, is open, the distribution of the electrons on a detecting screen will be given by ψ_1^2. When both slits are open, the distribution will display the familiar interference fringes. In this case the wave function (the probability amplitude) is $\psi = \psi_1 + \psi_2$. If we represent the wavefunctions at the screen as phasors (Section 38.5) differing in phase by ϕ, as in Fig. 41.17, the probability density, $\boldsymbol{\psi} \cdot \boldsymbol{\psi} = \psi^2$, is

$$\psi^2 = |\boldsymbol{\psi}_1 + \boldsymbol{\psi}_2|^2 = \psi_1^2 + \psi_2^2 + 2\psi_1\psi_2 \cos \phi$$

FIGURE 41.17 If the wave functions at the screen are treated as phasors, the resultant wave function is given by $\boldsymbol{\psi} \cdot \boldsymbol{\psi}$ where $\boldsymbol{\psi} = \boldsymbol{\psi}_1 + \boldsymbol{\psi}_2$.

The last term represents the interference between the two waves. The fact that that square of the *sum* of the probability amplitudes leads to the correct result implies that the electron is in a superposition of both states as it propagates through the apparatus.

Suppose we try to find out through which slit each electron passes. In order for us to detect an electron, it has to interact with something. For example, we could bounce a photon off it or have it collide with another electron. Whichever is done, the electron is found to pass through either one *or* the other slit. However, our intrusion causes the interference pattern to vanish. Just as we have verified the particle aspect, the wave aspect disappears!

Complementarity principle

Bohr noted that any given experiment reveals either the wave aspect or the particle aspect. According to his **complementarity principle:** A complete description of matter and radiation requires *both* particle and wave aspects. That is, the wave and particle pictures complement each other.

Quantum mechanics has caused a complete upheaval in our understanding of how nature operates. It is not a "reasonable" theory and cannot be even tenuously connected to our daily experiences. Even those who had the original ideas, such as Planck, Einstein, and Schrödinger, never accepted later developments in the theory. Schrödinger was sorry he ever had anything to do with it. Although he later became a champion of quantum mechanics, Bohr himself did not accept the existence of photons until about 1925. As Einstein put it: "I look upon quantum mechanics with admiration and suspicion." He proposed several ingenious experiments to bypass the restrictions imposed by the uncertainty principle, but Bohr always managed to find a subtle flaw in Einstein's reasoning. Above all, he rejected the idea that nature operates on pure chance and once remarked that "God does not play at dice!" Ironically, Einstein himself was the first to use the notion of probability in atomic transitions (see p. 840). The probability interpretation of the wave function is now generally accepted, and quantum mechanics is a foundation stone of physics.

SUMMARY

According to de Broglie's hypothesis, material particles exhibit wave behavior. The **de Broglie wavelength** of a particle, including the photon, with linear momentum p is given by

$$\lambda = \frac{h}{p}$$

Matter waves are governed by **Schrödinger's wave equation,** which in one dimension is

$$\frac{d^2\psi}{dx^2} + \frac{2m}{\hbar^2}(E - U)\psi = 0$$

where E is the total energy and U is the potential energy. The wave functions, ψ, that are the solutions to this equation indicate the **probability** of finding a particle within a volume dV:

$$\text{Probability} = \psi^2 dV$$

The wave function must be **normalized:** The integral of the probability over all space must equal 1: $\int \psi^2 dV = 1$. Also, the wave function must satisfy the *boundary conditions* appropriate for the situation. In particular ψ and $d\psi/dx$ must be continuous.

The allowed energy levels of a particle of mass m confined to an impenetrable one-dimensional box of length L are

$$E_n = \frac{n^2 h^2}{8mL^2} \qquad\qquad n = 1, 2, 3, \ldots$$

According to wave mechanics, a particle can penetrate into a region for which $U > E$, although this would be forbidden in classical physics. Thus a particle can **tunnel** through a potential barrier.

According to the **Heisenberg uncertainty principle** one cannot determine the position and linear momentum of a particle simultaneously to arbitrary precision. The uncertainties are related by

$$\Delta x \Delta p \geq h$$

These uncertainties are inherent in nature and *not* the result of inadequate equipment. This principle may also be considered to be the consequence of the inevitable interaction between an observer and what is being observed. Another form of the uncertainty principle relates energy and time:

$$\Delta E \Delta t \geq h$$

In order to minimize the uncertainty in the energy of a particle one must take the maximum possible time to measure it. This version allows the nonconservation of energy, provided the fluctuation in E occurs within $\Delta t \approx h/\Delta E$.

ANSWERS TO IN-CHAPTER EXERCISES

1. (a) $\lambda = h/mv = (6.63 \times 10^{-34} \text{ J} \cdot \text{s})/(9.11 \times 10^{-31} \text{ kg})(10^6 \text{ m/s})$ $= 0.73$ nm. (b) With $m = 1.67 \times 10^{-27}$ kg, $\lambda = 0.4$ pm.

2. $hf = hc/\lambda = \Delta E = (151 - 37.7) \times 1.6 \times 10^{-19}$ J. We find $\lambda = 11$ nm.

QUESTIONS

1. What do de Broglie waves and electromagnetic waves have in common that distinguishes them from other types of waves?

2. Would you expect de Broglie waves to exhibit a Doppler effect?

3. In what ways is Bohr's model of the hydrogen atom not compatible with quantum mechanics?

4. Compare the de Broglie wavelengths of an electron and a proton if they have (a) the same speed, and (b) the same energy.

5. What evidence would you present to skeptical friends in support of the contention that matter exhibits wave properties?

6. The wave function gives us information regarding only probabilities, yet the predictions of wave mechanics are quite precise. Reconcile these two statements.

7. Distinguish between Bohr's correspondence principle and Bohr's complementarity principle.

8. If one uses a cold thermometer to measure the temperature of warm water in a glass, the reading will not be accurate. Is this an example of Heisenberg's uncertainty principle? Explain why or why not.

EXERCISES

41.1 De Broglie Waves; 41.2 Electron Diffraction

1. (I) Use the classical expression relating linear momentum to kinetic energy to show that the de Broglie wavelength of an electron with kinetic energy K is given by

$$\lambda = \frac{1.23 \text{ nm}}{\sqrt{K}}$$

where K is in electron volts.

2. (I) An electron is accelerated from rest by a potential difference V. Show that its de Broglie wavelength, in nanometers (nm), is given by

$$\lambda = \sqrt{\frac{1.5}{V}}$$

where V is in volts. Assume the kinetic energy is given by the classical expression.

3. (I) An electron is accelerated from rest by a potential difference of 120 V. What is its de Broglie wavelength? (See the previous exercise.)

4. (I) Calculate the de Broglie wavelengths of (a) an electron and (b) a photon if the kinetic energy of the electron is equal to the energy of the photon, which is 2 eV.

5. (I) Find the de Broglie wavelength of a proton moving at (a) 10^3 m/s, and (b) 10^6 m/s.

6. (I) A thermal neutron, which has a kinetic energy of 0.04 eV at 300 K, plays an important role in the fission of uranium in a nuclear reactor. What is its de Broglie wavelength of such a neutron?

7. (II) A 1-g pellet moves at 10 m/s. For what slit width would the first diffraction minimum be at 0.5°? Is this a practical experiment?

8. (I) A photon and an electron each has a de Broglie wavelength of 5 nm. Compare their energies in eV.

9. (I) For what energy (in eV) is the wavelength of a photon (a) 10^{-10} m, (b) 10^{-15} m?

10. (I) At what speed would the de Broglie wavelength of an electron equal that of yellow light, which is 600 nm?

11. (I) Through what potential difference must a proton be accelerated for it to have a de Broglie wavelength of 0.1 pm?

12. (I) The electrons in the Davisson–Germer experiment were accelerated by a potential difference of 65 V. What was the corresponding angle ϕ (for the first order peak) in Fig. 41.3?

13. (I) An electron with an energy of 80 eV enters a region where there is a potential well of depth -20 eV, as shown in Fig. 41.18. Calculate its de Broglie wavelength (a) outside the well, and (b) inside the well.

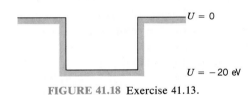

FIGURE 41.18 Exercise 41.13.

14. (I) The resolving power of a microscope, which tells us the smallest detail that can be distinguished, is approximately equal to one wavelength. At what speed is the de Broglie wavelength of an electron equal to 0.1 nm, the approximate size of an atom?

15. (II) (a) At what speed would the de Broglie wavelength of an electron equal the Bohr radius, which is 0.053 nm? (b) Compare the speed found in part (a) with the speed of the electron in the ground state, as predicted by the Bohr model.

16. (II) An electron is attracted to a proton that is held at rest. Assuming the electron starts from rest at infinity, find its de Broglie wavelength when it is 0.1 nm from the proton.

17. (II) Thermal neutrons, whose kinetic energy is 0.04 eV,

pass through two slits separated by 0.1 mm. What is the expected separation between like fringes on a screen 2 m from the slits?

41.5 Applications of Wave Mechanics

18. (I) A proton is trapped in a one-dimensional infinite potential well of length 10^{-14} m. (a) What are the first two energy levels? (b) What is the frequency of the photon emitted when the proton makes a transition from the upper level to the ground state? In what part of the electromagnetic spectrum does it lie?

19. (I) An electron moves within a one-dimensional infinite potential well of length 0.1 nm. (a) Calculate the energies of the ground state and the first excited state. (b) What is the wavelength of the photon emitted when the electron falls from the excited state to the ground state?

20. (I) An electron in an infinite potential well has an energy of 5 eV in the $n = 4$ level. What is the width of the well?

21. (I) What is the photon energy required to transfer an electron from the ground state to the $n = 3$ level in an infinite potential well of width 0.2 nm. To which part of the electromagnetic spectrum does the photon belong?

22. (I) What is the minimum speed of an electron in an infinite potential well of width 0.1 mm?

23. (II) The ground state energy of an electron in an infinite potential well is 20 eV. (a) What is the energy of the first excited level? (b) What is the length of the well?

24. (I) Suppose an electron were trapped with an infinite potential well whose length is 10^{-14} m, which is an approximate size of a nucleus. (a) Calculate the ground state energy of the electron. (b) Given that nuclear energies are of the order of tens of MeVs, what can you say about the possibility of electrons being within the nucleus?

41.6 Heisenberg Uncertainty Principle

25. (I) The position of the electron in the ground state of the hydrogen atom has an uncertainty of 0.1 nm. What is the uncertainty in its linear momentum?

26. (I) An electron is in an infinite potential well of length 0.2 nm. What is the uncertainty in its linear momentum?

27. (I) The lifetime of an excited state is 10^{-8} s. What is the uncertainty in (a) the energy, and (b) the frequency of the photon emitted?

28. (I) A proton is confined to a nucleus of radius 2×10^{-14} m. (a) Estimate the uncertainty in its linear momentum. (b) If the linear momentum were equal to the uncertainty found in (a), what would be the kinetic energy in MeV?

PROBLEMS

1. (II) Use the relativistic expressions for kinetic energy and linear momentum to show the following: (a) $\lambda \approx h/\sqrt{2m_0K}$, if $K \ll m_0c^2$, and (b) $\lambda \approx hc/K$ if $K \gg m_0c^2$.

2. (I) What is the de Broglie wavelength of an electron with an energy of 200 MeV? Use relativistic expressions and ignore the rest energy (0.5 MeV) of the electron.

3. (I) (a) Calculate the linear momentum of the electron in the ground state of the hydrogen atom according to Bohr's model. (b) If the uncertainty in the momentum is $\Delta p = 2p$, find the uncertainty in the position and compare it with the Bohr radius.

4. (I) Consider the wavefunction $\psi = A \sin(n\pi x/L)$ for a particle in a one-dimensional box of length L. Use the normalization condition $\int \psi^2 dx = 1$ to show that $A = \sqrt{2/L}$.

5. (II) Consider the ground state wave function for a particle in an infinite potential well that extends from $x = 0$ to $x = L$. What is the probability of finding the particle from $x = L/4$ to $3L/4$?

6. (I) A finite potential well extends from $x = 0$ to $x = L$, as in Fig. 41.12a. Show that the boundary conditions at $x = 0$ lead to the relation $C = AK/k$. (The notation is the same as that in the text accompanying Fig. 41.12.)

7. (I) An impenetrable box extends from $x = -L/2$ to $x = L/2$. What are the normalized wave functions for the three lowest energy states?

8. (I) A particle of energy E approaches a region where the potential suddenly rises to U as shown in Fig. 41.19. The probability of reflection is given by the reflection coefficient

$$R = \left[\frac{k_1 - k_2}{k_1 + k_2}\right]^2$$

where $k_1 = \sqrt{2mE}/\hbar$ and $k_2 = \sqrt{2m(E - U)}/\hbar$. Evaluate R for $E = 1.5U$.

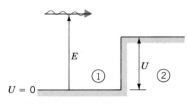

FIGURE 41.19 Problem 41.8.

9. (II) The potential energy of a simple harmonic oscillator is given by $U = \frac{1}{2} m\omega^2 x^2$. Show that $\psi = Ae^{-Bx^2}$ is a solution of the Schrödinger wave equation where $E = \hbar\omega/2$ is the energy of the state. What is B?

10. (II) A conduction electron in a metal may be treated as a particle trapped in a three-dimensional box of side L. The energy is determined by three quantum numbers, n_1, n_2, and n_3:

$$E = \frac{h^2}{8mL^2}(n_1^2 + n_2^2 + n_3^2)$$

(a) List the number of distinct values of the quantum numbers that would correspond to the ground state? (b) Repeat part (a) for the first excited state.

11. (II) The expectation value of a function $f(x)$ is given by

$$\langle f(x)\rangle = \int_{-\infty}^{\infty} f(x)\psi^2\, dx$$

Show that for a particle in the nth state of a one-dimensional impenetrable box of length L,

$$\langle x^2\rangle = \left(\frac{1}{3} - \frac{1}{2n^2\pi^2}\right)L^2$$

(See Problem 4.)

12. (II) An electron with energy E approaches a barrier of height U ($>E$) and length L, as in Fig. 41.13. The transmission coefficient, T, is found from the ratio of the probabilities at the two faces of the barrier. Show that

$$T \approx e^{-2KL}$$

where $K^2 = 2m(U - E)/\hbar^2$. (This expression is approximate because we ignore the "internal" reflection at the second face of the barrier.) Evaluate T for $L = 0.1$ nm, $U = 100$ eV and $E = 50$ eV.

SPECIAL TOPIC: Electron Microscopes

During the 1920s, when the cathode-ray oscilloscope was being developed, it was shown that the paths of electrons passing through a short magnetic deflection coil could be described by an equation analogous to the thin lens formula. The coil could therefore be used as a lens to focus a beam of electrons. The "focal length" of a magnetic lens depends on the strength of the magnetic field—which is controlled by the current in the coil. The analogy between "electron optics" and ray optics prompted engineers Max Knoll and Ernst Ruska to build the first **electron microscope** in 1931. The following year they heard of de Broglie's hypothesis (which had been published in 1925!) and realized that, in principle, such an instrument could exceed the resolving power of an optical microscope—which is limited by diffraction to about 200 nm at a wavelength of 400 nm.

An electron accelerated from rest by a potential difference of 40 kV has a de Broglie wavelength of about 0.006 nm. This might lead us to expect that an electron microscope would have a resolving power of about 0.003 nm—which is much smaller than the typical size of atoms (0.1–0.3 nm). In fact, this value is not attained. "Electron lenses" are usually magnets consisting of coils inside a soft-iron casing. The pole pieces are separated by a gap of a few millimeters, as shown in Fig. 41.20. The nonuniform fields produced by these magnets cannot be as finely shaped as the surface of a glass lens. As a result, spherical aberration (see Fig. 36.4) is a major problem. Nevertheless, as we will see, the electron microscope does provide resolution at the atomic scale. There are actually three types of electron microscope. We discuss them in turn.

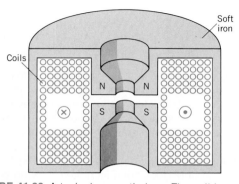

FIGURE 41.20 A typical magnetic lens. The coil is encased in soft iron with the pole pieces separated by a few millimeters.

TRANSMISSION ELECTRON MICROSCOPE

In the **transmission electron microscope** (TEM), first made in the 1930s by Ernst Ruska, electrons are emitted by a hot tungsten wire and then accelerated by a potential difference of 50–100 kV. The path of the beam is controlled by three lenses (see Fig. 41.21). The *condenser* lens produces a nearly parallel beam that is incident on the specimen. The *objective* lens produces a magnified image, which acts as the object for the *projector* lens. This lens produces further magnification and projects the final image onto a fluorescent screen or photographic plate. The system must be maintained at a "high vacuum" of about 10^{-5} mm Hg (10^{-7} atm), and the path of the beam must be kept steady to within 0.2 nm for a few seconds while a photograph is being taken.

Since the de Broglie wavelength of an electron depends on its speed, the accelerating potential difference must be stabilized to 1 part in 10^5. Nonetheless, the energies of the electrons emitted by the electron gun span a range of about 1 eV, which cannot be eliminated. Thus, the associated range in wavelengths of the electrons results in chromatic aberration (see Fig. 36.3). The effects of aberrations are reduced by limiting the angular spread of the beam with small circular apertures and by keeping the beam close to the central axis. Unfortunately, this procedure also limits the resolving power of the instrument and reduces the beam current from the 150 μA produced by the electron gun to about 10 μA passing through the specimen.

In a TEM contrast between adjacent areas is produced because the electrons are scattered (deflected away from the original direction of travel) by differing amounts in different areas. The specimen must be very thin so that the electrons do not lose energy as they pass through it. A spread in energies of the emerging electrons would result in a spread in wavelengths and additional chromatic aberration. Biological samples are first embedded in plastic, which is then sliced by a microtome to a thickness of about 20 nm. A typical resolving power of a TEM is 0.5 nm, while the best attain about 0.2 nm—corresponding to a magnification of $10^6\times$. Figure 41.22 shows an image of a crystal lattice produced by a TEM.

SCANNING ELECTRON MICROSCOPE

The **scanning electron microscope** (SEM) was first made in the mid-1930s by Max Knoll. In this device, shown in Fig. 41.23, a beam of electrons accelerated by 10–40 kV scans the surface of the specimen in a raster pattern (like the lines

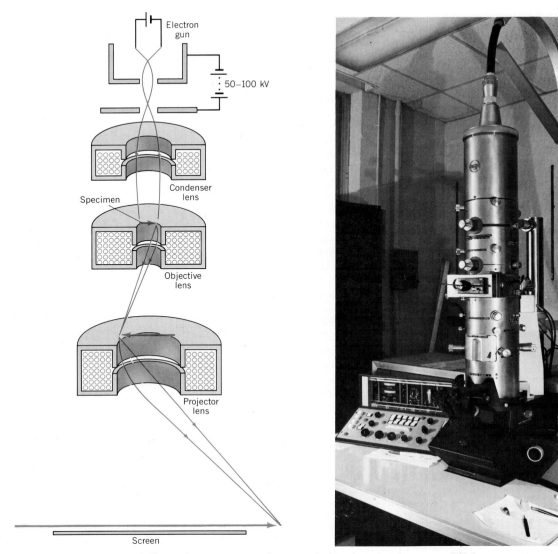

FIGURE 41.21 The main components of a transmission electron microscope (TEM). Beam-limiting apertures are not shown. There is often another "intermediate" lens between the objective and the projector. The motion of the electrons within the magnets actually involves a spiral.

on a TV screen). When the beam impinges on the specimen, some electrons are backscattered, others knock out low-energy ($\approx$50 eV) secondary electrons from the outer shells of atoms, while still others produce X rays. One or more of these can be detected and serve as the "signal." Even the current through the specimen can be monitored. The beam current reaching the specimen is only 10 pA and the secondary electron current is about 1 pA, so considerable amplification is necessary.

A scan does not produce an image in the usual sense; instead, it is a kind of map of the object. Each position of the beam corresponds to a point on the screen of a cath-

ode-ray tube. The scan of the specimen is synchronized with the scan on the screen and the signal is used to control the brightness of the display. The magnification is determined by the ratio of the picture point size to the beam point size. Since the beam size can be varied from 10 nm to 1 μm, an SEM can produce an enormous range of magnifications from 15$\times$ to about $10^5\times$. The highest resolving power is about 10 nm (still a factor of ten poorer than the TEM).

Although the SEM does not match the resolving power of the TEM, the image produced by an SEM has a large depth of field (the range of object distances for which the image is reasonably in focus). For example, with a beam

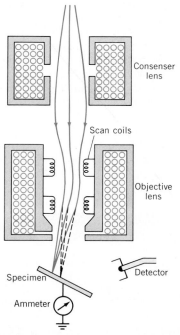

FIGURE 41.22 A color-coded TEM image of a high-temperature superconductor $Y_1Ba_2Cu_3O_{7-x}$. Yttrium atoms are black, barium atoms are yellow, and copper atoms are red. Oxygen atoms are not seen because of their low atomic number.

To understand how the contrast in an SEM image is produced, consider Fig. 41.25, which shows the beam impinging on an uneven surface. The high-energy backscattered electrons travel roughly normal to each face. Only those directed at the electron detector are registered. Thus

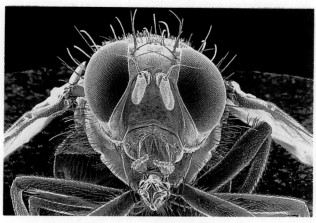

FIGURE 41.24 The image of a Mediterranean fruit fly produced by a scanning electron microscope has a great depth of field.

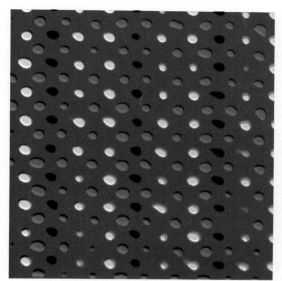

FIGURE 41.23 In the scanning electron microscope the beam cross section is reduced by two lenses. Two coils within the objective lens are used to scan the surface of the specimen.

faces 1 and 3 would appear dark, face 4 would appear bright and face 2 would have an intermediate brightness. Although the low-energy secondary electrons emerge in all directions, they can be attracted to the detector by maintaining it at about +10 kV. In this case, a signal will be registered even as the beam scans faces 1 and 3. If the specimen current is used as the signal, then all faces contribute. By combining these various signals, the contrast of the final image may be adjusted.

The electron beam in an SEM has been used to blast

size of 50 nm (at 1000×), the depth of field is 50 μm—which is 100 times greater than that of an optical microscope at the same magnification. As a result, the SEM produces an almost three-dimensional effect (see Fig. 41.24).

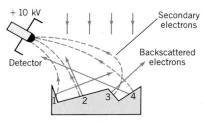

FIGURE 41.25 If just the backscattered electrons are detected, faces 1 and 3 will appear dark. If the detector is placed at a potential of +10 kV relative to the sample, the low-energy secondary electrons are also attracted to it. Then as the beam scans the surface, faces 1 and 3 will also contribute a signal.

holes in a pinhead with enough precision to form letters. The block of text shown in Fig. 41.26 is just 1 μm across. The letters are so small that all 29 volumes of the *Encyclopaedia Brittannica* could be inscribed on the pinhead!

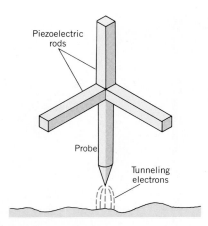

FIGURE 41.27 The tungsten probe in a scanning tunneling microscope is held about 0.1 nm above the surface it scans. Its position is controlled by a tripod of piezoelectric rods.

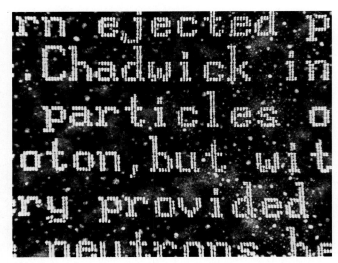

FIGURE 41.26 A block of text produced by the beam of a scanning electron microscope. The letters are so small that a whole encyclopedia could be inscribed on a pinhead!

SCANNING TUNNELING MICROSCOPE

In the transmission and scanning electron microscopes, the electron paths may be calculated classically. The wave nature of the electrons affects only the resolution of the image. In contrast, the **scanning tunneling microscope** depends on the quantum mechanical phenomenon of tunneling through a potential barrier (Section 41.5). This device was invented in 1981 by Gerd Bennig and Heinrich Rohrer—who shared the Nobel prize with Ruska in 1986. A tungsten probe with a very fine tip (even as small as one atom!) is held between 0.1 nm and 1 nm above a conducting surface. When a small potential difference is applied between the probe and the surface, an electron tunneling current flows through the vacuum between the tip and the surface.

The position of the probe is controlled (to within 10^{-5} nm!) by a tripod made of three mutually perpendicular *piezoelectric* rods, as shown in Fig. 41.27. (A piezoelectric crystal changes its size when a potential difference is applied to it.) As the probe slowly scans the surface, its vertical position is adjusted so that the tunneling current, and therefore the height above the surface, stays constant. The probe therefore traces the topography of the surface. The

"image" is built up either on a fluorescent screen or on a chart recorder.

In Fig. 41.13 we saw that a particle's wave function decays exponentially within a potential barrier. The amplitude of the transmitted wave (which determines the tunneling current) depends on the width of the barrier. The exponential dependence of the tunneling current on the separation between the probe and the surface results in exquisite sensitivity: When the vertical position of the probe is changed by just 0.1 nm, the tunneling current changes by a factor of 100. Vertical resolution is an astonishing 0.001 nm—much smaller than the size of a single atom! The best horizontal resolution achieved so far is about 0.1 nm. A motion of 0.1 nm across a sample is displayed as 1 cm on a screen or chart, so the overall magnification is $10^8\times$. Figure 41.28 shows an image produced by a scanning tunneling microscope (See also p. 851).

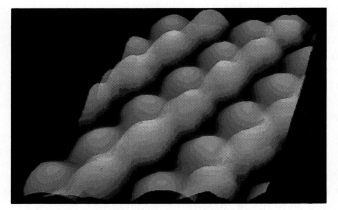

FIGURE 41.28 An STM image of Ga As: Ga is blue; As is red.

CHAPTER 42

Atoms and Solids

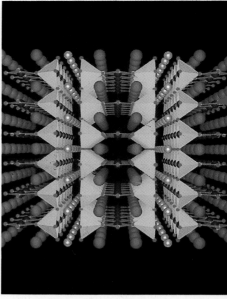

A computer-generated model of a high-temperature superconductor. Yttrium atoms are silver, barium atoms are green, copper atoms are blue, and oxygen atoms are red.

Major Points

1. **Four quantum numbers** are used to specify the state of an electron in an atom.
2. According to the **Pauli Exclusion Principle** no two electrons in an atom can have the same four quantum numbers.
3. The formation of the **periodic table** by the use of the four quantum numbers and the exclusion principle.
4. The **band theory of solids** explains the difference between the electrical conductivities of metals, insulators, and semiconductors.

The theory of quantum mechanics has been applied with great success to a wide variety of phenomena. It has provided insight into the structure and behavior of atoms, molecules, nuclei, and solids. We begin this chapter by briefly illustrating its application to the hydrogen atom. The solution of this problem, which is in three dimensions, requires the introduction of three quantum numbers to specify the states of the electron. These are the **principal quantum number,** n, the **orbital quantum number,** ℓ, and the **orbital magnetic quantum number,** m_ℓ. In addition, it turns out that particles, such as the electron, have an intrinsic spin angular momentum that is specified by a **spin magnetic quantum number,** m_s. The states of electrons in all atoms can be specified with these four quantum numbers. An important constraint on these numbers is the **Pauli Exclusion Principle,** which states that no two electrons in an atom can have the same four quantum numbers. We will see that the four quantum numbers and the exclusion principle allows us to systematically build up the periodic table and to explain several of its features in terms of the electron configurations of atoms. Finally, we discuss the formation of energy bands in solids and how they are used to explain the different electrical conductivities of metals, insulators, and semiconductors.

42.1 QUANTUM NUMBERS FOR THE HYDROGEN ATOM

When the Schrödinger wave equation is applied to the hydrogen atom, for which $U = -ke^2/r$, three quantum numbers arise naturally from the analysis. We will simply state the results. The energy levels are given by

$$E_n = -\frac{mk^2e^4}{2\hbar^2n^2} = -\frac{13.6}{n^2} \text{ eV} \tag{42.1}$$

Principal quantum number, n

The energies depend only on the **principal quantum number**, n, which varies from 1 to ∞. The values agree with Bohr's result.

The magnitude of the orbital angular momentum, L, of a state is determined by the **orbital quantum number**, ℓ:

Orbital angular momentum, **L**

$$L = \sqrt{\ell(\ell + 1)}\hbar \qquad (42.2)$$

where the maximum value of ℓ is restricted by the value of n:

Orbital quantum number, ℓ

$$\ell = 0, 1, 2, \ldots, (n - 1)$$

Notice that the lowest allowed value of angular momentum is $L = 0$, not $L = \hbar$ as in the Bohr theory. The value $L = 0$ corresponds to states for which the wave function is spherically symmetric, having no unique axis of rotation.

In order to specify the direction of the angular momentum vector, **L**, we need to set up a preferred axis. This may be accomplished by applying an external magnetic field along the z axis. The component of the orbital angular momentum along this axis is also quantized:

$$L_z = m_\ell \hbar \qquad (42.3)$$

Orbital magnetic quantum number, m_ℓ

where the values of the **orbital magnetic quantum number, m_ℓ**, are restricted to

$$m_\ell = 0, \pm 1, \pm 2, \ldots, \pm \ell$$

The angular momentum vector can point only in certain directions such that its z component assumes values given by Eq. 42.3. This phenomenon is called **space quantization**. The term "magnetic" is used because when the atom is placed in an external magnetic field, each value of m_ℓ corresponds to a different energy. A given spectral line can then split into several lines—a phenomenon called the Zeeman effect.

Each value of ℓ is associated with $2\ell + 1$ values of m_ℓ. Thus if $\ell = 2$, then $L = \sqrt{6}\hbar$ and the five allowed values of m_ℓ are 0, ± 1, and ± 2. The corresponding values of L_z are shown in Fig. 42.1.

All states with a given value of n are said to form a **shell**, whereas states with a given value of ℓ form a **subshell**. The designations are indicated in Table 42.1. The first four letters for the subshells are historical relics of terms (sharp, principal, diffuse, and fundamental) that were used to describe spectral lines. States are designated by the shell and subshell: $n\ell$. For example, for $n = 1$, $\ell = 0$ the state is $1s$; for $n = 2$ and $\ell = 1$, it is $2p$, etc. A state such as $3f$ is not possible since with $n = 3$, the maximum value of ℓ is 2.

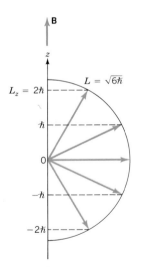

FIGURE 42.1 The orientation of the angular momentum vector is quantized. The z component can assume the values $L_z = 0, \pm \hbar, \pm 2\hbar$.

TABLE 42.1

n	Shell	ℓ	Subshell
1	K	0	s
2	L	1	p
3	M	2	d
4	N	3	f
5	O	4	g
6	P	5	h
.	.	.	.

According to quantum mechanics one can determine the values of L and L_z, but one cannot determine either L_x or L_y. To see this, suppose that the electron moves only in the *xy* plane, which means that $z = 0$ and $p_z = 0$. These precise values violate the uncertainty relation $\Delta p_z \Delta z > \hbar$. A geometrical interpretation of this restriction is shown in Fig. 42.2. The **L** vector precesses (it traces a cone) about the *z* axis, while the component L_z stays constant. The time-averaged values of the *x* and *y* components are zero.

The angle made by the vector **L** to the *z* axis is given by

$$\cos \theta = \frac{L_z}{L} = \frac{m_\ell}{\sqrt{\ell(\ell + 1)}} \qquad (42.4)$$

Notice that θ cannot be zero (why?). As ℓ becomes large, the change in θ, or L_z, from one value to the next becomes smaller. In the classical limit of very large quantum numbers, L_z or θ can assume an essentially continuous range of values. This is in agreement with Bohr's correspondence principle.

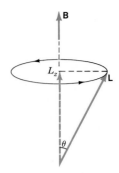

FIGURE 42.2 As the angular momentum vector precesses, the *z* component stays constant. The *x* and *y* components cannot be determined.

EXAMPLE 42.1: What are the allowed values of θ for $\ell = 2$?

Solution: With $\ell = 2$, we have $L = \sqrt{6}\hbar$ and $m_\ell = 0, \pm 1, \pm 2$. Therefore, from Eq. 42.4,

$$\cos \theta = \frac{m_\ell}{\sqrt{6}} = 0; \pm \frac{1}{\sqrt{6}}; \pm \frac{2}{\sqrt{6}}$$

from which we find $\theta = 90°, 65.9°, 35.3°, 114.1°,$ and $144.7°$.

EXERCISE 1. What is the minimum angle θ between L_z and L for $\ell = 100$?

42.2 SPIN

When the spectral line of sodium at 589.3 nm is examined under high resolution, it is found to consist of two finer lines at 589.0 nm and 589.6 nm. The Schrödinger wave equation cannot account for this *fine structure*, which is present in many spectral lines. Furthermore, the Schrödinger wave equation does not correctly predict the number of new lines that appear when the atom is placed in a magnetic field (the Zeeman effect).

In 1924, Wolfgang Pauli proposed that these problems would be solved if there existed a fourth quantum number that could assume only two values. This led S. A. Goudsmit and E. Uhlenbeck to suggest that each electron has an intrinsic *spin angular momentum* that can assume the values $\pm \frac{1}{2}\hbar$. They pictured the electron as a charged sphere spinning about an internal axis, as shown in Fig. 42.3. Although this (classical) picture is convenient, it is not correct. In 1929, when Paul Dirac incorporated relativity into quantum mechanics, he found that the spin quantum number arose naturally from the analysis. All one can say is that the electron has an intrinsic property, called spin for purely historical reasons, that manifests itself according to the following rules. The magnitude of the **spin angular momentum, S,** of the electron is determined by its **spin quantum number,** $s = \frac{1}{2}$:

$$S = \sqrt{s(s + 1)}\hbar = \frac{\sqrt{3}}{2}\hbar \qquad (42.5)$$

In a magnetic field, the *z* component can assume only two values

$$S_z = m_s \hbar \qquad (42.6)$$

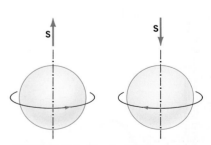

FIGURE 42.3 Classically the electron is pictured as a spinning sphere with a spin angular momentum that can assume two orientations, "spin up" and "spin down."

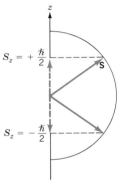

FIGURE 42.4 The z component of the spin angular momentum S can assume the values $S_z = \pm\hbar/2$.

where the **spin magnetic quantum number,** $m_s = \pm\frac{1}{2}$ (see Fig. 42.4). The introduction of spin doubles the number of states allowed for each value of n.

The splitting of spectral lines into doublets is explained through an effect called *spin–orbit coupling*. As we will see in Section 42.6, the electron has an intrinsic magnetic moment that is proportional to its spin. In the reference frame of an orbiting electron, it appears that the positive nucleus is moving and therefore produces a magnetic field. The two possible orientations of the spin magnetic moment relative to this field have slightly different energies. The energy associated with the sodium spectral line is 2.1 eV, while the energy difference between the two lines in the fine structure is 2.1 meV.

EXERCISE 2. What are the allowed angles of **S** relative to the z axis?

42.3 WAVE FUNCTIONS FOR THE HYDROGEN ATOM

In this section we consider the simplest wave functions for the hydrogen atom that are obtained from Schrödinger's wave equation. The wave function for the ground state, $n = 1$, $\ell = 0$, is

Ground state wave function

$$\psi_{1s}(r) = \sqrt{\frac{1}{\pi r_0^3}}\, e^{-r/r_0} \tag{42.7}$$

where $r_0 = \hbar^2/mke^2 = 0.529$ nm is called the *Bohr radius* (see Example 40.6). The probability density, given by ψ^2, is a maximum at $r = 0$ and decreases exponentially as r increases. We see that the Bohr model of the atom as mostly empty space is not correct. Instead of speaking of well-defined orbits, we sometimes refer to the distribution of ψ^2 in space as the "electron cloud."

The actual probability of finding the electron within a volume dV is $\psi^2 dV$. It is convenient to define the **radial probability density,** $P(r)$, such that $P(r)dr$ is the probability of finding the electron within the spherical shell from r to $r + dr$. The volume of a shell of radius r and thickness dr is $dV = 4\pi r^2 dr$. Therefore,

$$\text{Probability} = \psi^2 dV = \psi^2(4\pi r^2 dr) = P(r)dr$$

So the radial probability density is

$$P(r) = 4\pi r^2 \psi^2$$

From Eq. 42.7, for the $1s$ state we find

$$P_{1s}(r) = \frac{4r^2}{r_0^3}\, e^{-2r/r_0} \tag{42.8}$$

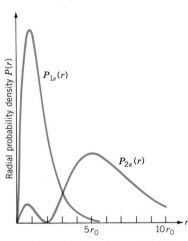

FIGURE 42.5 The radial probability densities for the $1s$ and $2s$ state of hydrogen.

This function is plotted in Fig. 42.5. Notice that the most probable value of r is r_0, which is equal to the first allowed radius in the Bohr model. However, in contrast to Bohr's picture, the electron may be found at values of r greater or smaller than r_0; the atom does not have a sharp boundary.

The wave function for the first excited state, $n = 2$, $\ell = 0$, is

$$\psi_{2s}(r) = \sqrt{\frac{1}{32\pi^2 r_0^3}}\left(2 - \frac{r}{r_0}\right) e^{-r/2r_0}$$

The corresponding radial probability density,

$$P_{2s}(r) = \left(\frac{r^2}{8 r_0^3}\right)\left(2 - \frac{r}{r_0}\right)^2 e^{-r/r_0} \tag{42.9}$$

is also shown in Fig. 42.5. Notice that P_{2s} has two peaks and that there is considerable overlap between the 1s and 2s functions. The most probable value of r for the 2s state is $r \approx 5r_0$. All s state wavefunctions ($\ell = 0$) depend only on r; they are spherically symmetric. The wave functions for states with $\ell \neq 0$ also include angular factors.

EXAMPLE 42.2: Calculate the most probable value of r for the electron in the 1s state of hydrogen.

$$\frac{dP}{dr} = (2r)e^{-2r/r_0} + r^2 \left(- \frac{2}{r_0}\right) e^{-2r/r_0}$$

$$= 2r \left(1 - \frac{r}{r_0}\right) e^{-2r/r_0} = 0$$

Solution: The most probable value of r corresponds to the maximum value of $P(r)$ in Fig. 42.5. One must calculate dP/dr and set it equal to zero. From Eq. 42.8, we find

Since the exponential goes to zero only at $r = \infty$, we find $r = r_0$. Thus, the electron is most likely to be found at a radial distance equal to the Bohr radius.

42.4 X RAYS AND MOSELEY'S LAW

By the mid-19th century, chemists recognized several groups or families of elements with similar properties, such as the alkalis, the halogens, and the noble gases. In 1871, D. Mendeleev formed the periodic table with the 62 elements then known. He arranged them in order of increasing atomic weight according to regularities in their physical and chemical properties. When gaps appeared in his arrangement, he was able to predict properties of the missing elements—such as atomic weight, boiling point, density, and color. Within a few years, elements such as Ge, Ga, and Sr were discovered and placed in the gaps.

It was difficult to place certain pairs of elements properly. For example, the atomic weights of copper and nickel were too close for them to be properly assigned. As another example, potassium is highly reactive (an alkali), whereas argon is inert (a noble gas). They belong in different groups, with Ar appearing before K. However, the atomic weight of argon is slightly greater than that of potassium. A major challenge for quantum theory was to account for the structure of the periodic table. But first, the proper basis for building up the table had to be found. This was provided by the work of H. G. J. Moseley (Fig. 42.6).

FIGURE 42.6 H. G. J. Moseley.

Intensity

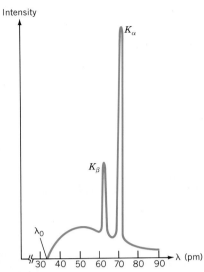

FIGURE 42.7 The X-ray emission when 35 keV electrons bombard a molybdenum target.

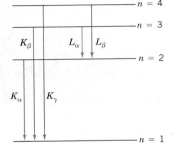

FIGURE 42.8 The characteristic X-ray lines are labeled according to the lower level (shell) in the transition.

We saw in Section 38.10 that when a heavy metal target is bombarded with high-energy (30–50 keV) electrons, it emits X rays. The radiation involves both a continuous and a line spectrum, as shown in Fig. 42.7. The *continuous* spectrum, which starts at some minimum wavelength λ_0, arises from the rapid deceleration of the electrons when they enter the target—it is called *bremsstrahlung,* or "braking radiation." The existence of a minimum wavelength (or maximum frequency) is further evidence in favor of the photon concept: The highest frequency photon is emitted when an electron loses all its energy in one step. By equating the energy, eV, of the electron to the energy $hf_0 = hc/\lambda_0$ of the photon, we find

$$\lambda_0 = \frac{hc}{eV} \tag{42.10}$$

The minimum wavelength depends on the electron energy, but not on the target material.

The *line* spectrum depends on the element used as target. These *characteristic* X rays are produced when an electron knocks out an atomic electron from one of the inner levels. The ejected electron leaves a vacancy, which is then filled by an electron falling from a higher level. In the process a high-energy photon is emitted. If the transitions are to the $n = 1$ level, the X rays are labeled K_α, K_β, . . . If they are to the $n = 2$ level, they are labeled L_α, L_β, . . . (see Fig. 42.8).

In 1913, Moseley noted that the characteristic lines shifted systematically as the target material was changed. He plotted the square root of the frequency of the K_α line versus the atomic number Z (the position in the periodic table) for many elements. The straight line he obtained is shown in Fig. 42.9. Moseley concluded:

We have here a proof that there is in the atom a fundamental quantity, which increases by regular steps as we pass from one element to the next. This quantity can only be the charge on the central nucleus.

Thus atomic numbers, rather than atomic weights, should be used to systematically build up the periodic table.* The evidence was so clear that he proposed that the positions of Ni and Co in the periodic table be interchanged, even though the atomic weight of Co is slightly greater than that of Ni. He identified gaps at $Z = 43, 61, 72,$ and 75. These elements were subsequently found.

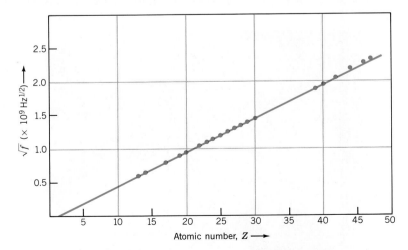

FIGURE 42.9 A plot of the square root of the frequency of the K_α lines versus atomic number using Moseley's data.

* If $\sqrt{f}$ is plotted versus atomic weight, the points do not lie so clearly on a straight line.

As Fig. 42.9 shows, Moseley's plot did not pass through the origin. Let us see why. Once one of the two electrons in the $n = 1$ level is ejected, an electron in the next highest level will drop to the lower state to fill the vacancy and in the process it emits the K_α frequency. For this electron the electric field due to the nucleus is screened by the remaining electron in the $n = 1$ level. Moseley estimated that the "effective" nuclear charge for the K_α transition is $(Z - 1)e$. (This is in agreement with the intercept of the line in Fig. 42.9.) Thus **Moseley's law** for the frequency of the K_α line is

$$\sqrt{f_{K_\alpha}} = a(Z - 1) \tag{42.11}$$

where a is a constant that can be related to Bohr's theory. After Moseley's death, a French chemist, G. Urbain, who had worked on sorting out the rare earth elements, commented that Moseley's law "established in a few days the conclusions of my efforts of 20 years of patient work."

EXERCISE 3. A target is bombarded with 30-keV electrons. What is the minimum wavelength in the continuous X-ray spectrum?

42.5 PAULI EXCLUSION PRINCIPLE AND THE PERIODIC TABLE

The four quantum numbers n, ℓ, m_ℓ, and m_s, may be used to classify the states of electrons in all atoms, although the energy associated with a given set of values depends on the atom. The question naturally arises as to why all electrons in an atom do not fall to the ground state. A study of the classification of spectral lines led W. Pauli (Fig. 42.10) in 1925 to make an important statement, now called the **Pauli Exclusion principle***:

No two electrons in an atom can have the same four quantum numbers n, ℓ, m_ℓ, and m_s.

With the aid of the exclusion principle one can see how electrons fill **shells** (n) and **subshells** (ℓ). For each value of ℓ there are $(2\ell + 1)$ values of m_ℓ. Since $m_s = \pm\frac{1}{2}$, each subshell can accommodate $2(2\ell + 1)$ electrons. Table 42.2 enumerates the available states, which in any atom are filled in order of increasing energy. In general, for a given value of n, the energy of the states increases with ℓ. Thus a $4s$ state is lower than a $4p$ state, which in turn is lower than a $4d$ state. However, the

FIGURE 42.10 Wolfgang Pauli (1900–1958).

TABLE 42.2

n	ℓ	m_ℓ	m_s	Shell	Subshell	Number in Subshell	Number in Shell
1	0	0	$\pm\frac{1}{2}$	K	$1s$	2	2
2	0	0	$\pm\frac{1}{2}$	L	$2s$	2	
	1	0, ±1	$\pm\frac{1}{2}$		$2p$	6	8
3	0	0	$\pm\frac{1}{2}$	M	$3s$	2	
	1	0, ±1	$\pm\frac{1}{2}$		$3p$	6	8
	2	0, ±1, ±2	$\pm\frac{1}{2}$		$3d$	10	
4	0	0	$\pm\frac{1}{2}$	N	$4s$	2	
	1	0, ±1	$\pm\frac{1}{2}$		$4p$	6	18

* The Pauli exclusion principle applies not just to electrons, but to any system of particles that have half-integer spin, $\hbar/2$, $3\hbar/2$, $5\hbar/2$, etc. Particles with integer values of spin, $\hbar$, $2\hbar$, $3\hbar$, etc., do not obey this principle.

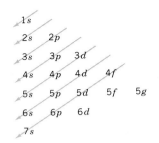

FIGURE 42.11 A simple, but approximate, mnemonic for the filling of subshells.

4s state is lower in energy than the 3d state, and the 5s state is lower than the 4d. Therefore the 4s subshell fills before the 3d subshell, and 5s fills before 4d. A useful, but *approximate,* mnemonic that tells us the order in which the subshells are first filled is shown in Fig. 42.11.

The periodic table is organized into groups (the columns) and periods (the rows). The number of elements in the six complete periods are 2, 8, 8, 18, 18, and 32. To see how these numbers arise, refer to the last column in Table 42.2. Notice that each number is associated with a completely filled (closed) subshell. In order to obtain the second 18 and the 32, one must keep in mind the filling order indicated in Fig. 42.11.

The ground state electron configurations are shown in the periodic table in Appendix D. The number of electrons in a subshell is indicated by a superscript. For example, $2p^3$ means that there are three electrons in this $\ell = 1$ subshell. We now briefly consider how the electron configuration of an atom helps to explain its chemical properties.

Each period begins with a chemically active alkali element and ends with a chemically inert noble element. In the last column are the **noble elements:** He, Ar, Ne, Kr, and Xe. All of these have completely filled subshells and the energy gap to the next higher level is large. These configurations are extremely stable, and so these elements do not react readily with other elements. In Fig. 42.12 we see that the ionization energy of each noble element is higher than that of the neighboring elements. In the first column are the **alkali elements:** Li, Na, K, Rb, Cs, and Fr. Each of these has a single, loosely bound electron, called the valence electron, outside a closed shell or subshell. These elements are highly reactive. As Fig. 42.13 shows, their atomic radii are consistently greater than those of neighboring elements. The **halogens,** F, Cl, Br, I, and At, are one electron short of a complete p subshell. It is energetically favorable for the valence electron in an alkali to transfer to a halogen atom to complete its shell. Sodium chloride, NaCl, is an example of such a compound. Properties of the **transition elements** and the **rare earths** can also be explained in terms of their electron configurations—which are characterized by incomplete inner subshells.

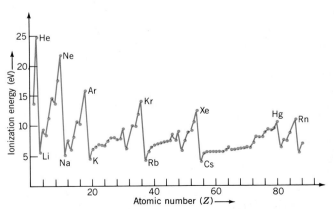

FIGURE 42.12 The ionization energy of atoms as a function of atomic number. The noble elements have consistently high values.

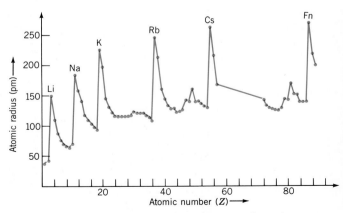

FIGURE 42.13 The radii of atoms as a function of atomic number. The radii of the alkali elements are consistently greater than those of neighboring elements.

42.6 MAGNETIC MOMENTS

In classical physics, an electron moving in an orbit is equivalent to a current loop. The orbital magnetic dipole moment, $\boldsymbol{\mu}_\ell$, is related to the orbital angular momentum, $\mathbf{L}$ (see Example 29.6):

$$\boldsymbol{\mu}_\ell = -\frac{e\mathbf{L}}{2m} \qquad (42.12)$$

Orbital magnetic moment

Quantum mechanics predicts precisely the same relation. In an external magnetic field, $\mathbf{B} = B_z\mathbf{k}$, the potential energy of the dipole is (Eq. 29.8),

$$U = -\boldsymbol{\mu} \cdot \mathbf{B} = -\mu_{\ell z}B_z \qquad (42.13)$$

Classically, the vector $\boldsymbol{\mu}$ may assume any orientation with respect to the field. The energy of the electron can therefore have any value between the parallel and antiparallel orientations. Thus, each atomic level should broaden and one should see a continuous spreading in spectral lines. In fact, each line splits into a finite number of discrete lines. This phenomenon (the Zeeman effect) implies that the energies of the states in the magnetic field, and therefore the orientations of the atomic moments, cannot assume any arbitrary value. From Eq. 42.13 it follows that the z component of the angular moment is quantized, as we noted in Eq. 42.3. The z component of the orbital magnetic moment is

$$\mu_{\ell z} = -\frac{e}{2m}L_z$$
$$= -\mu_\mathrm{B}m_\ell$$

where the quantity

$$\mu_\mathrm{B} = \frac{e\hbar}{2m} = 9.27 \times 10^{-24} \text{ J/T}$$

is called the **Bohr magneton.** It is a convenient unit for atomic magnetic moments.

The Stern–Gerlach Experiment

In 1921, Otto Stern and Walter Gerlach performed an experiment that demonstrated space quantization. They directed a beam of neutral silver atoms between the poles of a magnet and deposited them onto a glass plate, as in Fig. 42.14. In the absence of a field the beam produced a line on the plate (Fig. 42.15a). Next they applied the extremely nonuniform field of the magnet. The net force on a dipole in a nonuniform field is given by (see Eq. 8.18)

$$F_z = -\frac{dU}{dz} = \mu_z\frac{dB}{dz}$$

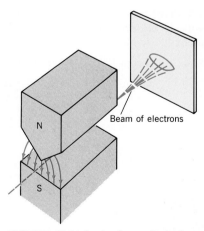

FIGURE 42.14 In the Stern–Gerlach experiment a beam of neutral silver atoms from an oven pass through an extremely nonhomogeneous magnetic field. The net force on the atoms depends on the orientation of the dipole moment.

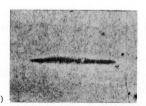 

(a)   (b)

FIGURE 42.15 The results of Stern and Gerlach. (a) The trace seen when there is no magnetic field. (b) The two distinct traces seen in the presence of the inhomogeneous field indicate that the magnetic moment can have only two orientations—a phenomenon called space quantization.

Classically, μ_z may assume a continuous range of values, so the beam should merely spread out. Instead, Stern and Gerlach found the two distinct traces on the plate shown in Fig. 42.15b. They interpreted this to mean that μ_z could point either up (along $+z$) or down (along $-z$). They assumed that this effect was associated with the quantization of L_z. It later became clear that one cannot assign this phenomenon to m_ℓ since this quantum number can assume $2\ell + 1$ values, which is always an *odd* number. Although space quantization had been verified, the number of traces was not correctly explained until 1925, when the concept of spin was introduced.

The relationship between the spin angular momentum **S** and the spin magnetic moment $\boldsymbol{\mu}_s$ is

Spin magnetic moment

$$\boldsymbol{\mu}_s = -\frac{e}{m}\mathbf{S} \qquad (42.14)$$

On comparing this with Eq. 42.12, $\boldsymbol{\mu}_\ell = -e\mathbf{L}/2m$, we see that the spin angular momentum is twice as effective in creating a magnetic moment as the orbital angular momentum. This difference emphasizes that one should not picture the electron as a spinning charge. The component of the magnetic moment along the z axis may assume two values given by

$$\mu_{sz} = -2m_s\mu_{\mathrm{B}}$$

where μ_{B} is the Bohr magneton. We can now offer the correct explanation of the Stern–Gerlach experiment.

In any atom, both the spin and the orbital angular momentum of each electron contribute to the total magnetic moment of the atom. The Stern–Gerlach experiment depends on this net magnetic moment of the atom. Of the 47 electrons in the silver atom, 46 form closed shells with zero angular momentum and zero magnetic moment. It is only the spin of the last electron, in the $\ell = 0$ state, that contributes to the angular momentum and the magnetic moment of the atom.

EXAMPLE 42.3: In an external magnetic field $\mathbf{B} = 0.5\mathbf{k}$ T, what is the separation between adjacent energy levels of an atom whose magnetic moment is equal to one Bohr magneton?

Solution: From Eq. 42.13 we see that the energy is given by

$$U = -\mu_z B_z = -\mu_{\mathrm{B}} m_\ell B_z$$

Between adjacent levels $\Delta m_\ell = \pm 1$, so

$$\Delta U = \mu_{\mathrm{B}} B_z$$
$$= 4.64 \times 10^{-24} \text{ J} = 2.9 \times 10^{-8} \text{ eV}$$

This energy difference should be compared with the energies of visible lines, which are about 2 eV.

42.7 BAND THEORY OF SOLIDS

The band theory of solids helps explain the different electrical conductivities of conductors, insulators, and semiconductors. To understand the formation of energy bands in a solid, let us consider the sodium atom, which has 10 electrons in a closed shell and a single electron in the 3s state. In isolated atoms the energy levels are sharply defined. Now suppose that two sodium atoms are brought close to each other so that their electron wave functions overlap. As a result of the interaction between the electrons, it turns out that each single state of the isolated atom splits into two states with different energies. As Fig. 42.16a shows, the degree of splitting increases as the interatomic separation decreases. Similarly, if five atoms are placed in close proximity, then each original energy level splits into five new levels as shown in Fig. 42.16b. The same process occurs in a solid, where there are roughly 10^{28} atoms/m³: The energy levels associated with each state of the isolated atom spread into essentially continuous **energy bands** separated from each other by energy gaps, as in Fig. 42.17. The bands derived from the lower energy atomic states are narrower since there is less overlap among the associated wave functions.

In a sodium atom, the 1s and 2s subshells have two electrons each and the 2p subshell has six electrons. They are all completely filled. When N atoms are brought together to form a solid, each level of the isolated atom splits into N new levels, each of which can accommodate two electrons with opposite spins. The atoms form a single system in which the Pauli exclusion principle permits only one electron to occupy each quantum state. Therefore in the solid, the 2N levels in the 1s band, the 2N levels in the 2s band and the 6N levels in the 2p band are completely filled. However, the 3s level in a sodium atom has just one electron instead of the two that are allowed. Therefore, the corresponding 3s band is only half-filled, as indicated in Fig. 42.17. Similar analysis can be applied to other solids to predict their electrical properties.

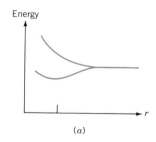

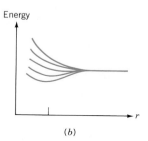

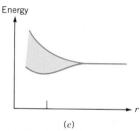

FIGURE 42.16 (*a*) As two atoms are brought closer together, a single atomic level splits into two states with different energies. (*b*) A single atomic level splits into five levels when five atoms are in close proximity. (*c*) In a crystal each atomic level splits into an essentially continuous band of energies.

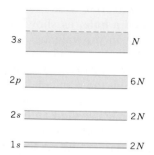

FIGURE 42.17 The band structure of sodium. If N atoms are brought together, each atomic level splits into N new levels. If the atomic state was fully occupied the corresponding band is full. Since the 3s state in sodium has just one electron, the corresponding 3s band in the solid is only half full.

Conductor

In a **conductor,** the highest occupied band is only partially filled, as shown in Fig. 42.18*a*. The electrons do not fall to the bottom of the band because they are prevented from doing so by the exclusion principle. Instead, they fill the available levels up to some maximum level, called the **Fermi energy** E_F, which is 3–8 eV above the bottom of the band. The electrons in this partially filled **conduction** band can respond to an external electric field because there are many nearby unoccupied levels available. This is why the electrical conductivity of a metal is good. Also, at room temperature ($kT \approx 0.025$ eV), the electrons near the Fermi level can be thermally excited to the unoccupied levels. The speed of an electron at the Fermi level is typically 10^6 m/s, so the classical idea that all motion ceases at $T = 0$ K is clearly false.

Insulator

In an **insulator** all the states in the highest occupied band are filled, as shown in Fig. 42.18*b*. This filled **valence band** is separated from the higher, unoccupied

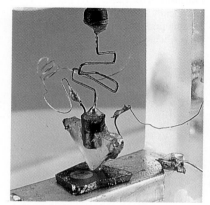

The first transistor, built in 1947.

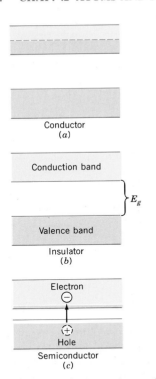

Conductor
(a)

Conduction band

E_g

Valence band

Insulator
(b)

Electron

Hole

Semiconductor
(c)

FIGURE 42.18 (a) In a conductor the highest occupied band is only partly filled. Electrons at the highest levels are able to absorb electrical or thermal energy to make transitions to nearby states within the band. (b) In an insulator the highest occupied band is completely full. There are no nearby states to which electrons can make transitions. (c) The band structure of a semiconductor is similar to that of an insulator, but the band gap is relatively small. An electron in the lower (valence) band can be thermally or electrically excited to the upper (conduction) band. In this process it leaves behind a "hole."

conduction band by a gap of about 5–8 eV. Thus, at room temperature electrons cannot be thermally excited to the higher band. The presence of the gap also means that the electrons cannot gain energy from an external electric field because there are no nearby unoccupied levels to which they can be transferred. Consequently, current does not flow in an insulator. If the band gap is greater than 3.2 eV, photons in the visible region (1.8–3.2 eV) will not be absorbed. The material— for example, sodium chloride—will be transparent to visible light.

Semiconductor

The band structure in a **semiconductor** is similar to that in an insulator, but the gap is much smaller (Fig. 42.18c). For example, it is 0.7 eV in Ge and 1.1 eV in Si. At room temperature, a few electrons can be thermally excited from the valence band to the conduction band. The density of electrons in the conduction band is about 10^{15} m^{-3}, which is much smaller than the 10^{28} m^{-3} typical of a conductor. As the temperature is increased, the number of conduction electrons increases, and as a consequence the conductivity also increases.

When an electron makes a transition from the valence band to the conduction band, it leaves a vacancy, called a **hole,** as shown in Fig. 42.18c. If an external field is applied, another electron in the valence band can move to fill the hole, thereby leaving a hole in *its* original position. This new hole can be filled by yet another electron, and so on. As this process continues, the hole migrates through the solid. The total current through the semiconductor arises from the motion of electrons in the conduction band and of holes in the valence band. A pure material in which these two processes occur is called an **intrinsic semiconductor.**

The conductivity of a semiconductor may be increased by the addition of selected impurities, a process called doping. When conduction is associated with impurities added to a pure semiconductor, the material is called an **extrinsic semiconductor.** Consider, for example, a Ge crystal. Each atom contributes four valence electrons to form covalent bonds with its neighbors (Fig. 42.19a). When an impurity, such as arsenic or phosphorus, which has five valence electrons, is added to the Ge crystal, four of these electrons form bonds, but the fifth is only loosely bound to the remaining positive P or As ion. The closely spaced energy levels of these electrons are just below the conduction band (0.01 eV for Ge, 0.05

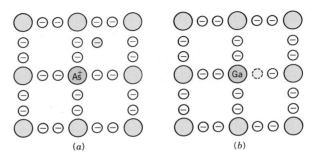

(a)   (b)

FIGURE 42.19 (a) When an element with five valence electrons enters a Ge or Si crystal one of its electrons is only loosely bound to the impurity atom. The impurity is called a donor atom. (b) When a trivalent atom is the impurity, it creates a "hole" that can accept electrons from other sites. The impurity is called an acceptor atom.

eV for Si), as shown by the gap E_d in Fig. 42.20. Electrons from these levels can easily be thermally excited into the conduction band. Since the arsenic atom contributes electrons, it is called a **donor** atom. The density of the impurity atoms is usually about 10^{21} m^{-3}, so the density of conduction electrons increases by a factor of about $10^{21}/10^{15} = 10^6$. Although holes still contribute to conduction, the (negative) electrons are the majority charge carriers, so the material is called an *n*-type semiconductor.

If gallium is the impurity, its three valence electrons form bonds with the neighboring Ge atoms, but a hole is left in one bond, as shown in Fig. 42.19*b*. The impurity atoms contribute a set of levels just above the valence band, as shown by the gap E_a in Fig. 42.20. Since the trivalent impurity accepts electrons from other sites, it is called an **acceptor** atom and the levels it contributes are called acceptor levels. In this case the majority charge carriers are (positive) holes, so the doped material is called a *p*-type semiconductor. Electrons can be thermally excited from the valence band into these levels, but in this case they are the minority carriers.

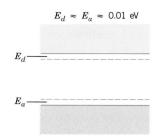

FIGURE 42.20 The energy levels contributed by the donor atoms are close to the bottom of the conduction band, whereas those contributed by the acceptor atoms are just above the top of the valence band.

42.8   SEMICONDUCTOR DEVICES (Optional)

In this section we discuss a few semiconductor devices, such as diodes, transistors, and solar cells, that are formed by the suitable combination of *n*-type and *p*-type semiconductors.

Junction Diode

In a *pn* **junction diode,** a *p*-type semiconductor is separated from an *n*-type semiconductor by a region, called the junction, that is about 1 μm thick (see Fig. 42.21*a*). The density of either holes or electrons in the junction is low, and its electrical resistance is high. To describe the operation of the diode, we consider only the motion of electrons; the discussion for holes is similar.

The density of conduction electrons in the *n*-type region is far greater than in the *p*-type region. Thus, electrons diffuse from the *n*-type to the *p*-type material, thereby setting up a *diffusion current, I_{diff}.* The diffusion current does not completely drain the electrons from the *n*-type region because this side of the junction develops a positive charge and the side

adjacent to the *p*-type region develops a negative charge. The electric field associated with these charges sets up a potential energy barrier that opposes the diffusion current. Only those electrons with energy greater than that of the barrier can surmount it. Electrons are also thermally excited in the *p*-type region, and some of them drift toward the junction where they can "slide downhill" in energy. This results in a very small *drift current, I_{drift},* which does not depend on the size of the potential hill. When equilibrium is established, the diffusion current (due to the majority carriers that surmount the energy barrier) is balanced by the drift current (due to minority carriers that are thermally excited).

Now consider what happens when an external potential difference is applied to the terminals of the diode. If the *p*-type region is connected to the negative terminal of the battery and the *n*-type region to the positive terminal, the external electric field is in the same direction as the internal field. The diode is said to be in *reverse bias* (Fig. 42.21*b*). The energy of the electrons in the *p*-type region is raised relative to those in the *n*-type region, so the potential barrier becomes larger. The diffusion current decreases, since the number of electrons in the *n*-type

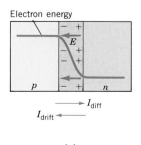
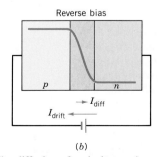
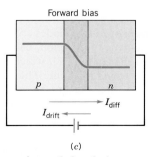

(a)   (b)   (c)

FIGURE 42.21 A *pn* junction diode. The diffusion of majority carriers sets up an internal electrical field across the junction. The potential energy of the electrons is indicated. The height of the potential hill is reduced when the diode is in forward bias and increased when it is in reverse bias.

region that are able to surmount the barrier diminishes. At high negative bias only the small drift current, produced by minority carriers, flows in the external circuit. This current due to the minority carriers is independent of the applied potential difference: It depends only on the number of conduction electrons thermally generated and the rate at which they drift toward the junction.

When the polarity of the battery is such that the *p*-type region is positive and the *n*-type region is negative, the external electric field is opposite to the internal field, which means that the potential barrier is lowered. The diode is in *forward bias* (as in Fig. 42.21*c*). The drift current is unchanged, but more electrons in the *n*-type region can surmount the smaller hill, so the diffusion current increases. If the external field is large enough, the net field will be directed from the *p*-type region to the *n*-type. In this case electrons are accelerated across the junction and a large current flows in the external circuit.

The current–voltage characteristic of a junction diode is shown in Fig. 42.22. Note that this device is highly nonlinear; it does not obey Ohm's law. When the diode is forward biased it has a low resistance, and when it is reverse biased, it has a high resistance. It is this property that makes it useful, as we see in the next section.

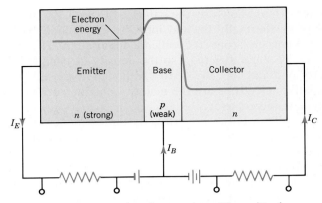

FIGURE 42.23 An *npn* junction transistor. The emitter-base junction is in forward bias (the potential hill is small) whereas the base-collector junction is in reverse bias (the potential hill is large).

holes, travel before recombining. Since the base is only weakly *p*-type, any electrons that enter the base from the emitter have a good chance of reaching the B–C junction—at which they merely slide downhill. Once on the collector side, they are quickly thermalized and no longer have the energy to diffuse backward and surmount the barrier. Thus, nearly the full emitter current flows through the collector connection to the external circuit.

The transistor is used as an amplifier. To understand how, consider the diode I–V characteristic in Fig. 42.24. When the diode is in forward bias, a small change in current is associated with a small change in potential difference across the diode.

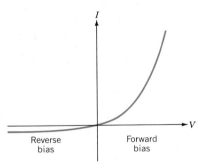

FIGURE 42.22 The current–voltage characteristic of a junction diode. The small current that flows when the diode is in reverse bias is due to the movement of thermally generated minority carriers across the junction.

Junction Transistor

In an *npn* **junction transistor,** two diodes are set back-to-back as in Fig. 42.23. A weakly *p*-type region, called the **base** (B), is sandwiched between the **emitter** (E), which is strongly *n*-type, and the **collector** (C), which is *n*-type. The E–B junction has a small (≈ 0.5 V) forward bias, whereas the B–C junction has a reverse bias of about 20 V. These applied potential differences appear mainly across the junctions. The E–B junction has a relatively low resistance, whereas the B–C junction has a high resistance. The B–C junction passes only the small drift current (the few thermally generated electrons from the *p*-type base region that "slide downhill"). The base thickness ($\approx 2 \times 10^{-5}$ m) is less than the average distance that minority electrons, or

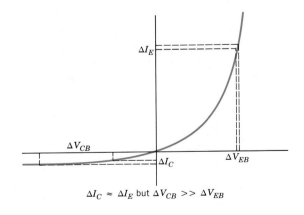

$\Delta I_C \approx \Delta I_E$ but $\Delta V_{CB} >> \Delta V_{EB}$

FIGURE 42.24 The transistor can serve as an amplifier because almost the whole emitter current passes through the collector. The current is "transferred" from a low resistance to a high resistance. Thus the same change in current produces a larger change in potential difference across the base-collector junction.

However, when the diode is in reverse bias, a similar change in current is associated with a very large change in potential difference. Suppose a varying potential difference is applied in series with the battery connected to the E–B junction. A small change in emitter current, ΔI_E, involves only a small change in the emitter–base potential difference, ΔV_{EB}, since the emitter–base resistance, R_{EB}, is small. If the change in the collector current is equal to the change in emitter current, that is, $\Delta I_C \approx \Delta I_E$, then the change in potential difference across the B–C junction, ΔV_{BC}, will be large because the resistance, R_{BC}, is large. The same current flows through a larger resistor but the potential difference is amplified, and so, therefore, is the power. Of course the power is ultimately derived from the battery connected to the base–collector junction.

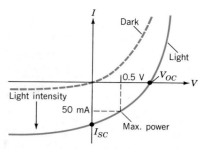

FIGURE 42.26 The I–V characteristic of a photovoltaic device. As the light intensity increases the curve moves downward. I_{SC} is the short-circuit current and V_{OC} is the open-circuit voltage.

Photovoltaic Devices

A **photovoltaic device** uses light energy to set up an emf. It is used as a light sensor and for electrical power generation in *solar cells*. A photovoltaic device consists of a thick *n*-type region covered by a thin *p*-type layer, as shown in Fig. 42.25. Incoming photons with sufficient energy can create electron–hole pairs. The extra holes in the *n*-type region and the extra electrons in the *p*-type region lead to a large percentage increase in minority carriers in both regions. Minority carriers created close to the junction have a good chance of reaching the junction—at which point they simply "slide downhill." To enhance this effect, the thickness of the *p*-type region ($\approx 10^{-5}$ m) must be less than the average distance (≈ 0.2 cm) that minority carriers, electrons or holes, travel before recombining. When the device is illuminated, the drift current due to minority carriers increases, while the (opposite) diffusion current due to majority carriers is unchanged.

decreases as the illumination intensity increases as shown in Fig. 42.27. The short-circuit current, I_{SC} (measured without an external load), increases with illumination intensity. The open circuit potential difference, V_{OC}, is measured with $I = 0$. At some point on the knee of the curve, the product of I and V is a maximum. This point corresponds to the maximum power that can be generated. Typical values at this point are 50 mA and 0.5 V, which means that the power generated is 25 mW. For a silicon cell, the efficiency (= electrical power out/optical power in) is about 12%. Although solar cells are used in satellites and electronic calculators, their efficiency needs to be improved before they can provide an alternate of source of power for the home.

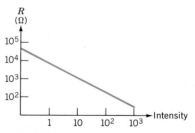

FIGURE 42.27 The resistance of a photoelectric device as a function of light intensity.

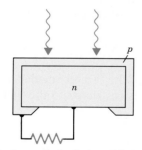

FIGURE 42.25 A photovoltaic device with an *n*-type semiconductor covered by a thin *p*-type semiconductor. The operation of the device depends on the creation of electron–hole pairs close to the junction.

Figure 42.26 shows the I–V curve for a photovoltaic device. The potential difference V is across the external load. In the dark, one obtains the usual junction diode characteristic (dashed curve). When the junction is illuminated, the curve shifts downward (solid curve). The resistance of the device

Light-Emitting Diode

In a light-emitting diode (LED), both the *n*-type and *p*-type regions are heavily doped. A large forward bias is applied, so that the external field is greater than the internal field. The electrons from the *n* side travel to the *p* side and recombine with holes, thereby producing photons. Thus, the LED emits light when a current flows through it. Some semiconductors, such as GaAs, emit light in the visible region. If the opposite faces of the device are made parallel (by cleaving along atomic planes) it is possible to generate laser action.

SUMMARY

When Schrödinger's wave equation is applied to the hydrogen atom, three quantum numbers arise from the mathematics. First, the energy levels are determined by the **principal quantum number** n:

$$E_n = -\frac{mk^2e^4}{\hbar^2 n^2}; \qquad n = 1, 2, 3, \ldots$$

Second, the angular momentum L is given by the **orbital quantum number** ℓ:

$$L = \sqrt{\ell(\ell + 1)}\, \hbar; \qquad \ell = 0, 1, 2, \ldots, (n-1)$$

Third, the component of L along the z axis is given by the **orbital magnetic quantum number**:

$$m_\ell = 0, \pm 1, \pm 2, \ldots, \pm \ell.$$

The electron has an intrinsic **spin angular momentum S** whose value is determined by the **spin quantum number,** $s = \frac{1}{2}$: $S = \sqrt{3/2}\, \hbar$. The z component of S can assume only the values

$$S_z = m_s \hbar$$

where $m_s = \pm \frac{1}{2}$ is the **spin magnetic quantum number.**

The **radial probability density** $P(r)$ is defined such that $P(r)dr$ is the probability of finding an electron within the shell from r to $r + dr$. It is related to the **probability density** ψ^2 by $P(r) = 4\pi r^2 \psi^2$.

X rays are high-energy photons that are produced when electrons strike a target. The continuous spectrum is produced during the rapid deceleration of the electrons. The line spectrum is produced when an atomic electron makes a transition to a lower state. The frequency of an X-ray line is characteristic of the element and is given by **Moseley's law.**

According to the **Pauli exclusion principle** *no two electrons in an atom can have the same four quantum numbers.* With the aid of this principle one can systematically build up the periodic table. The chemical properties of an element can be explained in terms of the configuration of the electrons in the atom.

The **magnetic moment** associated with the spin of an electron is given by

$$\boldsymbol{\mu}_s = -\frac{e}{m}\, \mathbf{S}$$

This is twice as large as the contribution of the orbital angular momentum. The z component of the spin magnetic moment is restricted to the values

$$\mu_{sz} = \pm \frac{e\hbar}{2m}$$

When atoms are brought together to form a solid, the energy levels broaden into **energy bands.** In an insulator the highest occupied band (called the valence band) is completely occupied and the gap to the next highest band (the conduction band) is large. In a semiconductor, the gap between the valence and conduction band is small compared to that in an insulator. Electrons can be thermally excited from the valence band to the conduction band. In a metal, the highest band is only partly full. Thus, there are a large number of states available to the electrons.

The conductivity of a semiconductor, such as Si or Ge with four valence electrons, can be drastically altered by **doping** it with impurities. Donor atoms have five valance electrons and make the material an **n-type** semiconductor. Acceptor atoms have three valence electrons and make the material *p*-type. Diodes and transistors are made by combining these two types of semiconductor.

ANSWERS TO IN-CHAPTER EXERCISES

1. The minimum value of θ occurs when $m_\ell = +\ell$. Thus,

$$\cos \theta = \frac{\ell}{\sqrt{\ell(\ell + 1)}} = \frac{100}{\sqrt{100(101)}}$$

From which we find $\theta = 5.7°$.

2. From Fig. 42.4, we see that

$$\cos \theta = \frac{S_z}{S} = \pm \frac{1}{2} \sqrt{\left(\frac{1}{2}\right)\left(\frac{3}{2}\right)} = \frac{\pm 1}{\sqrt{3}}$$

Thus $\theta = 54.7°$ or $125.3°$.

3. From Eq. 42.10,

$$\lambda_0 = \frac{hc}{eV}$$

$$= \frac{(4.14 \times 10^{-15} \text{ eV} \cdot \text{s})(3 \times 10^8 \text{ m/s})}{(3 \times 10^4 \text{ eV})}$$

$$= 41.4 \text{ pm}$$

QUESTIONS

1. Why do *three* quantum numbers appear naturally in the wave mechanical solution of the hydrogen atom?

2. Suppose that the Pauli exclusion principle were not valid. What would be some consequences of this?

3. Give examples to illustrate the relationship between the electron configuration in an atom and its chemical properties.

4. Why was a nonuniform magnetic field used in the Stern–Gerlach experiment?

5. How does the existence of a cutoff wavelength in the X-ray spectrum support the validity of the photon concept?

6. Would the result of a Stern–Gerlach experiment allow one to determine whether the angular momentum of an atom arises from orbital or spin contributions?

7. The Bohr model is totally inadequate in dealing with electrons in the higher energy states of a complex atom. Why does it do reasonably well in predicting the wavelengths of K_α lines?

8. The spectral lines emitted by a gas in a discharge tube become broader as the gas pressure is increased. Suggest a reason for this phenomenon.

9. The characteristic X-ray lines show a systematic shift with atomic number. Do the visible spectra of elements show a similar shift? If so, explain its features.

10. The resistivity of a metal increases with temperature, but the resistivity of an intrinsic semiconductor decreases. Explain this difference.

11. Can hydrogen atoms produce X rays? Can helium ions? Explain why or why not.

12. What aspects of Bohr's model reappear in the quantum mechanical solution of the hydrogen atom? Which features are missing?

13. How could one determine whether or not a given atom has net angular momentum?

14. Why are the rare earth elements chemically similar? How could one detect the presence of a given rare earth in a sample?

15. What role does the Pauli exclusion principle play in the electrical conduction of metals?

16. Is there a distinction between a "free" electron and a "conduction" electron? If so, what is it?

17. Why are the lowest energy levels in a solid narrower than the higher levels?

18. What is a "hole"? How does it contribute to electrical conduction in a semiconductor?

19. Would a study of the absorption of radiation by a semiconductor allow one to determine the band gap? If so, how?

20. What is a photoconductor? Explain how it operates.

EXERCISES

42.1 Quantum Numbers for the Hydrogen Atom; 42.2 Spin

1. (I) What is the orbital angular momentum of an electron in (a) a $3p$ state, and (b) a $4f$ state?

2. (I) The orbital angular momentum of an electron is 3.65×10^{-34} J · s. What is its orbital quantum number?

3. (I) Make a table of all the states in the $4d$ subshell.

4. (I) What are the possible values of L_z for an electron in a p subshell?

5. (I) The electron in a hydrogen atom is in an $n = 2$ state. What are the possible values of (a) L_z, and (b) the angle θ between $\mathbf{L}$ and L_z?

6. (I) List all the possible values of m_ℓ for a state with $n = 3$.

7. (I) In a given state, the maximum possible value of the magnetic quantum number is $m_\ell = 4$. What can you say about the value of (a) ℓ, and (b) n?

8. (II) (a) List the possible values for the quantum numbers ℓ and m_ℓ for the electron in He$^+$ in the $n = 3$ state. (b) What is the energy of the electron?

9. (I) (a) What is the energy of the electron in Li^{++} in the $n = 2$ state? (b) What are the possible values for the quantum numbers ℓ and m_ℓ?

10. (I) An electron is in a state for which $\ell = 4$. What is the minimum possible value of the angle between the vector **L** and L_z?

11. (I) An electron has an orbital angular momentum equal to 2.583×10^{-34} J $\cdot$ s. What is the maximum possible value for the z component, L_z, of its orbital angular momentum?

12. (II) Count the number of possible states for each value of n from $n = 1$ to $n = 5$. Can you find a simple formula that relates the number of states to n?

13. (II) It is possible to determine L and L_z exactly, but not L_x and L_y. Show that the x and y components satisfy the condition

$$\sqrt{L_x^2 + L_y^2} = (\sqrt{\ell(\ell + 1) - m_\ell^2})\hbar$$

14. (II) One version of the Heisenberg uncertainty principle relates the z component of the angular momentum to the angular position ϕ of **L** in the xy plane: $\Delta L_z \Delta \phi \approx \hbar$. (a) If L_z is known exactly, what can one infer about $\Delta \phi$? (b) What does your answer to part (a) imply about the components L_x and L_y?

42.3 Wave functions of the Hydrogen Atom

15. (I) The electron in a hydrogen atom is in the ground state, ψ_{1s}. Calculate the radial probability density $P_{1s}(r_0)$, where r_0 is the Bohr radius.

16. (I) For the first excited state of the hydrogen atom, calculate the radial probability density $P_{2s}(r_0)$, where r_0 is the Bohr radius.

17. (I) For an electron in the 1s state, calculate the radial probability density, $P(r)$, at (a) $r_0/2$, (b) r_0, and (c) $2r_0$, where r_0 is the Bohr radius.

18. (I) If the electron in the hydrogen atom is in the 2s state, what is the radial probability density, $P(r)$, at (a) $r_0/2$, (b) r_0, and (c) $2r_0$, where r_0 is the Bohr radius?

19. (II) Prove that the ground state wavefunction, ψ_{1s}, is normalized. That is, show

$$\int_0^\infty \psi^2 \, dV = 1$$

Note that the volume of a spherical shell of radius r and thickness dr is $dV = 4\pi r^2 dr$. (You will need to integrate by parts.)

20. (II) Show that the probability of finding an electron in the 1s

state within a sphere of radius r_0 centered at the nucleus is 0.32.

42.4 X Rays and Moseley's Law

21. (I) When electrons bombard a metal target, the shortest wavelength X rays produced have a wavelength of 0.05 nm. What is the accelerating potential difference?

22. (I) Electrons are accelerated by a potential difference of 25 kV and strike a metal target. What is the minimum wavelength of the X rays produced?

23. (I) Find the constant a in Moseley's law (Eq. 42.11) from Fig. 42.9.

24. (I) Show that the minimum wavelength of the X rays generated when electrons accelerated by a potential difference V strike a target is given by

$$\lambda_0 = \frac{1240 \text{ nm}}{V}$$

where V is in volts.

25. (I) For molybdenum, the wavelength of the K_α line is 0.71 nm and of the K_β line it is 0.63 nm. Use this information to find the wavelength of the L_α line.

26. (I) The wavelength of the K_α line for molybdenum ($Z = 42$) is 0.71 nm. Use this information to predict the wavelength of the K_α line (a) for silver ($Z = 47$), and (b) for iron ($Z = 26$).

27. (I) The energy of the $n = 2$ state in molybdenum is $E_2 = -2870$ eV. Given that the wavelengths of the K_α and K_β lines are 0.71 nm and 0.63 nm, respectively, determine the energies, E_1 and E_3.

42.5 Pauli Exclusion Principle and the Periodic Table

28. (I) The electron configuration of an atom is [Ar]$3d^3 4s^2$, where [Ar] stands for the configuration of the argon atom. Identify the element.

29. (I) List the quantum numbers for an oxygen atom in the ground state.

30. (II) Suppose that the electron did not have a spin quantum number but that the Pauli exclusion principle is still valid. Build up the periodic table for the first 15 elements. Which elements would you classify as "noble"?

42.6 Magnetic Moments

31. (I) The electron in the hydrogen atom has an orbital quantum number $\ell = 3$. What is the orbital magnetic moment?

32. (I) The orbital angular momentum of the silver atom is zero. (a) In the presence of a uniform magnetic field of magnitude 0.4 T along the z axis, what are the energies of the two possible orientations of the spin? (b) What is the frequency of the photon that would allow a transition from one level to the other?

33. (I) Sodium, which has a single electron in the 3s state, emits doublet lines at 589.0 nm and 589.6 nm in making transitions to the ground state (which consists of a single level).

(a) What is the difference in energy between the excited states? (b) What is the effective magnetic field experienced by the electron?

34. (II) A horizontal beam of neutral silver atoms traveling horizontally at 400 m/s passes through a vertical nonuni- form field for which $dB/dz = 120$ T/m. The mass of an atom is 1.8×10^{-25} kg. (a) What is the acceleration of an atom? (b) If the field extends for 20 cm along the horizontal direc- tion, what is the magnitude of the vertical deflection of the beam as it emerges from the field?

PROBLEMS

1. (I) Show that the most probable value of r for an electron in the $2s$ state of hydrogen is at $r \approx 5.2r_0$.

2. (I) The radial part of the wavefunction for the $2p$ state in hydrogen is

$$\psi_{2p}(r) = Cre^{-r/2r_0}$$

where C is a constant. Show that the most probable value of r is $4r_0$.

3. (I) Show that the probability of finding the electron in the $1s$ state in hydrogen within a sphere of radius $2r_0$ is 0.76.

4. (I) The average value of the radial coordinate is given by

$$r_{av} = \int_0^\infty r\psi^2 dV$$

Show that the average value for the $1s$ state in hydrogen is $1.5r_0$. Note that the volume of a shell of radius r and thick- ness dr is $dV = 4\pi r^2\, dr$.

5. (II) Electrons are accelerated from rest by a potential dif- ference of 40 kV and are used to bombard a metal target. Calculate the de Broglie wavelength. (You must use the relativistic expressions for kinetic energy and momentum.)

SPECIAL TOPIC: Superconductivity

PROPERTIES OF SUPERCONDUCTORS

Zero Electrical Resistance

The electrical resistivity of a metal arises from the interactions of the conduction electrons with impurities, defects, and the vibrating ions of the crystal lattice (see Chapter 27). As the temperature is lowered, the amplitudes of the lattice vibrations diminish, so one would expect the resistivity also to decrease gradually toward a small, but finite, value determined by the impurities and defects. This behavior is manifested by many materials. However, in 1911, H. Kamerlingh Onnes discovered that as the temperature of a specimen of mercury was reduced, its resistance suddenly dropped to an extremely small value at 4.15 K (see Fig. 42.28). The metal had made a transition to a new **superconducting** state. The resistivity of a superconductor is at least a factor of 10^{-12} less than that for an ordinary conductor. We can usually take it to be zero. An electrical current induced in a superconductor has been observed to flow for several years without any applied potential difference—provided the temperature is maintained below the **critical temperature, T_c**.

Over two dozen elements and thousands of alloys and compounds have been found to become superconducting. Semiconductors, such as Ge and Si, become superconducting only if very high pressure is applied. It is noteworthy that some of the best conductors, such as Ag, Au, and Cu, do not become superconductors. Also, elements with magnetic properties, such as Fe and Co, are not superconductors. Among the elements, niobium has the highest critical temperature of 9.26 K. Until 1986, the highest critical temperature that had been observed was 23.3 K in Nb_3Ge.

In addition to their use in electromagnets, there are other applications of superconductors. The resistive losses in transmission lines amount to about 10% of the power supplied. This heat dissipation would be eliminated by superconducting lines. Persistent currents around a hole in a superconductor may be used as a memory device. And since the transition to the superconducting state can occur within a very small temperature range (it is 10^{-3} K for Sn), a superconductor may be used as a detector of radiation. The device, called a bolometer, can have a sensitivity of 10^{-12} W!

Meissner–Ochsenfeld Effect

Consider a *perfect* conductor, placed in a magnetic field, as in Fig. 42.29a. As the external field is removed, induced currents would be set up to maintain the flux through the specimen at its original value, as shown in Fig. 42.29b. That is, the flux would be trapped or "frozen" in a perfect conductor. In 1933, W. Meissner and R. Ochsenfeld placed a sample of lead in a weak magnetic field, as in Fig. 42.30a, and started to cool it. They found that as the material made the transition to the superconducting state, the flux was

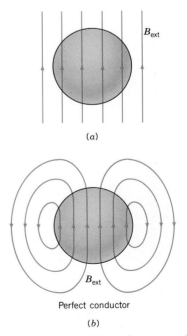

(a)

(b)

FIGURE 42.29 (a) A perfect conductor in a magnetic field. (b) When the external field is removed, the lines within the perfect conductor are "trapped" or "frozen."

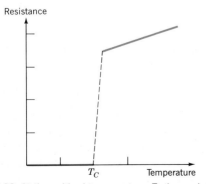

FIGURE 42.28 At the critical temperature T_c the resistivity of a superconductor goes to zero.

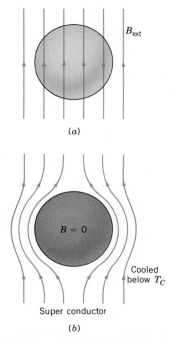

(a)

$B = 0$

Cooled
below T_C

Super conductor

(b)

FIGURE 42.30 (*a*) A magnetic field applied to a superconductor above its transition temperature. (*b*) When the material is cooled below T_c the flux is excluded from the superconducting material.

expelled, as shown Fig. 42.30*b*. (The exclusion of flux occurs for a long thin cylinder or a sphere but not for a flat plate.) This exclusion of magnetic flux from the interior of a superconductor is called the **Meissner–Ochsenfeld effect.** It shows that a superconductor is characterized by more than just perfect conductivity; it also displays perfect diamagnetism (see Section 32.6). The Meissner–Ochsenfeld effect may be demonstrated by first placing a small magnet on a sample of the material. When the sample is cooled below its critical temperature, the small magnet is seen to rise off the surface, as shown in Fig. 42.31.

FIGURE 42.31 The exclusion of magnetic flux by a superconducting material results in the suspension of a small magnet above a "high temperature" superconductor.

Flux exclusion and persistent currents are useful in making superconducting solenoids. The efficiency of an ordinary electromagnet is essentially zero: Once the magnetic field has been set up, all the electrical energy supplied is dissipated as heat in the windings. If the above procedure is repeated with a superconducting ring, the flux is expelled from the superconducting material, but the persistent induced currents maintain the flux within the hole. Energy is required only to maintain the temperature below the critical value.

Critical Magnetic Field

When the magnetic field applied to a superconductor exceeds a **critical field,** B_c, the flux penetrates into the material. The temperature dependence of the critical field is shown in Fig. 42.32. Below the critical field, a **Type I** superconductor is divided into normal and superconducting regions. As the temperature is increased, the superconducting regions diminish in size and finally disappear at T_c. Similarly, when the field is increased, the superconducting regions disappear at the critical field, which is typically 0.04 T at 0 K. At 0 K, roughly 10^{-7} eV/atom of magnetic energy is needed to destroy the superconducting state, whereas at zero field, roughly 5×10^{-4} eV/atom of thermal energy is needed.

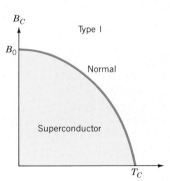

FIGURE 42.32 The variation in the critical field at which superconductivity is destroyed as a function of temperature. The curve is for a Type I superconductor.

In 1962, it was realized that there is a second class of superconductor, now called **Type II.** These materials are characterized by two critical fields, as shown in Fig. 42.33. Above the first critical field, B_{c1}, flux penetrates the superconductor in thin filaments called **fluxoids** or **vortices,** as shown in Fig. 42.34. The *core* of each filament is in the *normal* state: The flux through each filament is maintained by persistent currents that flow around the perimeter. The flux through each filament is called the **flux quantum:**

$$\phi_0 = \frac{h}{2e} = 2.07 \times 10^{-15} \text{ Wb}$$

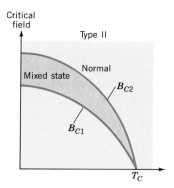

FIGURE 42.33 A Type II superconductor has two critical fields. When the field is between the values B_{c1} and B_{c2} the material is in the mixed state.

FIGURE 42.34 A Type II superconductor in the mixed state. The flux penetrates in the form of thin (nonsuperconducting) filaments but the resistivity of the sample is still zero. The filaments are made visible because iron filings tend to collect at the ends.

As the external field changes, the number of fluxoids also changes. The material is said to be in a **mixed state** (the resistivity is still zero). At the second critical field, B_{c2}, the flux penetrates completely and the sample reverts to the normal state. In $Nb_3Sn(T_c = 18.1$ K) it is found that $B_{c1} = 0.02$ T and $B_{c2} = 22$ T.

Superconductivity is also destroyed when the current density exceeds a critical value. Only Type II materials are suitable for superconducting solenoids. Current densities as high as 10^6 A/m^2 have been achieved in Type II materials, such as NbTi and Nb_3Sn, which are used for super-

conducting magnets in accelerators, fusion reactors, medical imaging, and magnetic levitation.

High-Frequency Properties

Although the dc resistance of a superconductor is zero, the ac resistance increases with frequency. It reaches the normal value at about 5×10^{11} Hz, in the microwave region. In the visible region (around 10^{15} Hz) the electromagnetic properties of the superconductor are the same as those of the normal state. For example, the material does not change its appearance as it undergoes the transition to the superconducting state. Also, the absorption of radiation increases sharply at microwave frequencies. Such behavior reminds us of the energy gap in the band structure of insulators and semiconductors. Recall that incident photons can be absorbed only if their energy exceeds the band gap energy.

THE BCS THEORY OF SUPERCONDUCTIVITY

Although several theories described the macroscopic electromagnetic and thermodynamic properties of superconductors, the development of a theory to explain superconductivity from a microscopic basis was slow in coming. An important clue came from the **isotope effect.** When different isotopes of a superconducting element are tested, the critical temperature is found to vary according to $T_c \propto 1/M^{1/2}$, where M is the atomic mass. Recall that the frequency of oscillation of a block at the end of a spring varies with mass in the same way (Eq. 15.8). This isotope effect thus provided an important clue that lattice vibrations are somehow involved in superconductivity.

In 1950, H. Frohlich proposed a mechanism by which conduction electrons could "couple" to the lattice of positive ions. At very low temperatures, the amplitudes of the thermal vibrations of the ions are significantly reduced in comparison with the values at room temperature. When an electron passes between the positive ions, they are attracted to it. The distortion in the lattice then travels away as a sound wave. Another electron, some distance away, is attracted to the increased positive charge density as the wave passes by. In effect, the two electrons interact via the exchange of sound waves. Frolich suggested that if the electrons are far enough apart, the attractive interaction might be greater than their coulomb repulsion—which is heavily screened by the presence of other electrons.

In 1956, L. Cooper showed that it is possible for two conduction electrons near the Fermi level to form a bound state for even a very weak attractive interaction. In 1957, J. Bardeen, L. Cooper, and J. Schriefer (Fig. 42.35) developed Cooper's idea into the first microscopic theory of superconductivity. They showed that in the ground state at 0 K, with no external field and no current flowing, all the electrons form **Cooper pairs,** in which two electrons have opposite

FIGURE 42.35 J. Bardeen (left), L. Cooper, and J. Schriefer (right) receiving the 1972 Nobel prize for their theory of superconductivity. Bardeen received the Nobel prize in 1956 for his work on the transistor.

momenta and spin. Under any other conditions, some electrons are unpaired. The distance between the electrons in a Cooper pair is about 10^{-6} m, which is about 200 times larger than the interatomic distance. This means that the wave function for each Cooper pair spreads over many other pairs.

The conduction electrons in a normal conductor occupy all the states up to the Fermi level. When a Cooper pair forms, the energy of the electrons is lower by an amount equal to the binding energy, which is about 10^{-3} eV. Since the electrons are either bound in a pair or free, there is a range of energies that they cannot assume. As a result, an **energy gap** appears about the Fermi level. The gap is a function of temperature. At 0 K it is $E_g = 3.5kT_c$, and goes to zero at T_c.

When a current flows in a normal conductor, any non-zero total linear momentum of the free electrons is transferred to the lattice via the electron–ion collisions. That is, the energy gained from the electric field is lost to the thermal vibrations of the lattice. The electrical resistivity of a superconductor is zero because a Cooper pair loses no energy to the lattice. This is because collisions with the lattice ions do not transfer enough energy to break the pair—the energy input must be greater than the energy gap. The energy needed to break up the pairs may be supplied by increasing the electrical current or the temperature.

One might say that when a current flows in a normal conductor the free electrons behave like people dancing to rock music on a crowded floor. Their random motion leads to many collisions. In contrast, when a current flows in a superconductor, all Cooper pairs have the same net momentum. In the superconducting state, the paired electrons are like choreographed dancers who move in step without collisions.

We noted above that the wave function for a Cooper pair spreads over large distances. Consider going around a superconducting loop and returning to the initial point. In order for the wave function to be single-valued, its phase can change only by $2\pi n$, where n is an integer. As a result, it can be shown that when a superconducting ring is placed in a magnetic field, the flux through the loop must be an integral multiple of the flux quantum mentioned above, that is, $\phi = n\phi_0 = nh/2e$.

JOSEPHSON JUNCTIONS

Consider two superconductors separated by a thin (1 nm) insulating barrier. In 1962, B. D. Josephson suggested that Cooper pairs can tunnel through such a thin barrier. Since each pair carries a charge $-2e$, the tunneling creates a *supercurrent*—in the absence of any applied potential difference. This is called the **dc Josephson effect.** The supercurrent flows to equalize the density and linear momenta of the Cooper pairs on either side of the barrier. Once the supercurrent exceeds a certain critical value, a potential difference appears across the junction. If a magnetic field is applied perpendicular to the supercurrent, the value of the critical supercurrent drops to zero whenever the flux through the junction is a multiple of the flux quantum, that is, $n\phi_0$ (see Fig. 42.36).

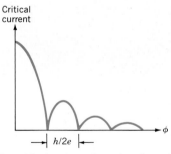

FIGURE 42.36 The critical current through a Josephson junction varies periodically with the flux through the junction.

Josephson also predicted that a dc potential difference V applied across the junction produces an ac supercurrent of frequency $f = 2eV/h$. (Note that $2e/h = 483.6$ MHz/μV.) This **ac Josephson effect** is used to measure potential difference to great accuracy.

Figure 42.37 shows two Josephson junctions connected in parallel. The supercurrent splits at junction A. The current at junction B is determined by the interference between the wave functions for the pairs that travel along the two paths. Since the phase difference depends on both the path difference and the magnetic flux through the loop, the current through the loop varies periodically with the external field. Such a device, called a **squid** (superconducting

FIGURE 42.37 A squid is a superconducting quantum interference device. The wave functions for the Cooper pairs that travel the two paths have a phase difference when the currents meet again at point *B*. The phase difference depends on the external magnetic field.

quantum interference device), may be used to detect extremely weak magnetic fields, even those produced by brain activity (Fig. 42.38).

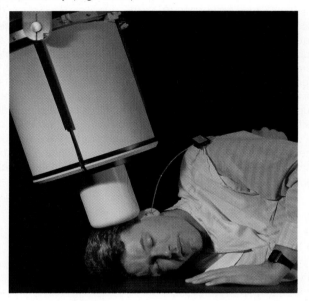

FIGURE 42.38 Squids can detect extremely weak magnetic fields. Here they are used to study the fields produced by the human brain.

HIGH-TEMPERATURE SUPERCONDUCTIVITY

Before the 1980s, most work on superconductors involved metallic alloys and a few organic compounds. In 1983, K. A. Muller and J. G. Bednorz decided to test metallic oxides, called ceramics, instead. In December 1985, they found that a compound of Ba–La–Cu–O became superconducting at 35 K—a full 12 K above the previous record set in 1973. P. Chu confirmed the result and then replaced La by Y. In February 1987, he found that $YBa_2Cu_3O_{7-x}$ has a transition temperature above 90 K! Not only was this a spectacular increase in T_c, but it occurred above the temperature of liquid nitrogen (77 K). Prior to this discovery, it was necessary to cool with expensive liquid helium, whereas liquid nitrogen is easier to work with and cheaper than beer or milk. It was soon found that yrttium can be replaced by several transition or rare earth elements. In January 1988, Paul Grant announced zero resistivity in a five-element compound, Tl–Ca–Ba–Cu–O at 125 K.

These new high-temperature superconductors may lead to a revolution equal to that associated with the transistor. However, these ceramics are too brittle to be made into wires for transmission or electromagnets, although they can be deposited as thin films for small-scale applications, such as electronic components. The critical current densities are only 1% of the best type II materials, although 10^5 A/cm^2 has been recorded along certain lattice directions in thin films. A typical critical field is 0.01 T. Although these ceramics show great promise, much developmental work remains to be done before we will see commercial applications. There is as yet no theory to explain high-temperature superconductivity.

This wire of high-temperature superconducting material was produced in May 1990 by the Argonne National Laboratory. It has a maximum current density of 100 Å/cm².

CHAPTER 43

Nuclear Physics

Major Points

1. The **binding energy** of a stable nucleus.
2. **Radioactivity** involves the emission of α particles, β particles, and γ rays.
3. The **radioactive decay law.**
4. In **nuclear reactions** nuclei are transformed from one type to another.
5. In the process of **fission** a heavy nucleus splits into two lighter fragments.
6. In the process of **fusion** two light nuclei combine to form a heavier nucleus.

Light emitted as a result of fusion in a tiny fuel pellet in the NOVA inertial confinement project.

Nuclear physics had its origin in the discovery of radioactivity in February 1896. This in turn was triggered by Roentgen's discovery of X rays in November 1895. Roentgen had noticed that electrons striking the glass in a discharge tube caused it to fluoresce and also to emit a new kind of radiation. It was not clear whether these "X rays" always accompany fluorescence. Several scientists tried to find out by wrapping a photographic plate in black paper and placing a fluorescent material on top. After the material was stimulated by sunlight for a few hours, the plate was developed. They found nothing.

H. Becquerel tried this technique with crystals of potassium uranyl sulphate. The sun appeared only intermittently on February 26 and 27, so he put the plates, with the crystals on top, into a drawer. Since the sun still had not appeared after several days, he decided to develop the plates on March 1, expecting to see only feeble images. Instead, the silhouettes of the crystals appeared with great intensity. Clearly, the action of the crystals had continued in the dark. Since fluorescence lasts only a short time, and these samples had been mostly in the dark, there had to be some cause other than X rays for the strong image. He soon discovered that even nonfluorescent salts of uranium emitted these new invisible rays, which pointed to uranium as the active agent.

Late in 1897, Marie Curie discovered that thorium was also **radioactive** (a term she coined). She and her husband, Pierre Curie (Fig. 43.1), used chemical techniques to isolate two new radioactive elements: polonium (in July 1898) and radium (in December 1898). Several other radioactive elements were found in the next few years.

Radioactivity was found to be unaffected by the temperature, the pressure, or the chemical state of the substance. The heat generated by a small sample of radium placed in a lead container (1 g of radium released 0.1 cal/h) was estimated

FIGURE 43.1 Marie and Pierre Curie in their laboratory.

to be 10^5 times greater per atom than that from any known chemical reaction. It was clear that radioactivity was produced by some unknown process within single atoms and was not the result of interaction between atoms. In 1911, when Rutherford's scattering experiment (Section 40.5) revealed the existence of the nucleus, he identified it as the source of radioactivity. We next consider some of the facts that have since been learned about the nucleus.

43.1   THE STRUCTURE OF THE NUCLEUS

The nucleus of an atom consists of protons and neutrons, which are collectively referred to as **nucleons.** (The hydrogen nucleus is a single proton.) The **atomic number,** Z, which uniquely specifies an element, is the number of protons in the nucleus. The naturally occurring elements cover the range from $Z = 1$ (hydrogen) to $Z = 92$ (uranium), whereas elements with atomic numbers up to $Z = 107$ have been briefly created by artificial means. The **atomic mass number,** $A = N + Z$, is the total number of nucleons in a nucleus. A nucleus with given numbers of protons and neutrons is called a **nuclide** and is denoted by

Atomic number, Z

Mass number, A

$$_Z^A X$$

where X is the chemical symbol. The subscript Z may be dropped since the chemical symbol identifies the element uniquely, but is often retained for convenience as in $_8^{16}O$, $_6^{12}C$, and $_7^{14}N$.

Isotopes

The **isotopes** of a given element are atoms whose nuclei have the same number of protons, but different numbers of neutrons. For example, naturally occurring carbon consists of 98.9% $_6^{12}C$ and 1.1% $_6^{13}C$, but other isotopes exist from $_6^{11}C$ to $_6^{16}C$. Since the chemical properties of an element are determined by its atomic electrons, isotopes are chemically identical even though their nuclei have different masses.

The masses of atoms are nearly integer multiples of the mass of the hydrogen atom. This is because the mass of the electron is very small compared to that of the proton, and the mass of the neutron is approximately the same as that of the proton. Thus, the atomic mass number, A, is approximately equal to the mass of an atom expressed as a multiple of the proton mass.

Atomic masses are specified in terms of the **unified mass unit** (u). The mass of a neutral atom of the carbon isotope $_6^{12}C$ is defined to be *exactly* 12 u. The neutral atom is chosen because its mass is easier to measure and the contributions of the electrons are easily taken into account.

Unified mass unit

$$1\ u = 1.66056 \times 10^{-27}\ kg = 931.5\ MeV/c^2$$

The second expression arises from the equation $E = mc^2$ and is derived in the next example.

EXAMPLE 43.1: (a) Obtain a value of the unified mass unit from Avogadro's number. (b) Express the unified mass unit in terms of its energy equivalent.

Solution: (a) One mole of ^{12}C has a mass of 12 g and contains Avogadro's number, N_A, of atoms. By definition, each ^{12}C atom has a mass of 12 u. Thus, 12 g corresponds to $12N_A$ u, which means

$$1\ u = \frac{1\ g}{N_A} = \frac{1\ g}{6.022045 \times 10^{23}}$$
$$= 1.66056 \times 10^{-27}\ kg$$

(b) From the equation $E = mc^2$, the energy equivalent of 1 u is

$$E = (1.66056 \times 10^{-27}\ kg)(2.9979 \times 10^8\ m/s)^2$$
$$= 1.4924 \times 10^{-9}\ J = 931.5\ MeV$$

Thus, 1 u = 931.5 MeV/c^2.

The masses of the proton, the neutron, and the electron are

$$m_p = 1.67264 \times 10^{-27} \text{ kg} = 1.007276 \text{ u} = 938.28 \text{ MeV}/c^2$$
$$m_n = 1.6750 \times 10^{-27} \text{ kg} = 1.008665 \text{ u} = 939.57 \text{ MeV}/c^2$$
$$m_e = 9.109 \times 10^{-31} \text{ kg} = 0.000549 \text{ u} = 0.511 \text{ MeV}/c^2$$

The atomic masses that appear in the periodic table are weighted averages over the various isotopes of each element. For example, Cl has two isotopes with approximate masses 35 u (75.4%) and 37 u (24.6%). Thus the atomic mass listed is $35(0.754) + 37(0.246) = 35.5$ u. The mass number A of an isotope is numerically equal to the atomic mass, in u, rounded to the nearest integer.

Rutherford's analysis of α particle scattering (Section 40.5) established that the nucleus has a radius of about 10^{-14} m. More recent experiments involve the scattering of electrons, protons, and neutrons. Electrons are particularly suitable since they are not affected by the nuclear interaction. If their energy is over 200 MeV, their de Broglie wavelength is less than the size of a nucleus, so they can probe the details of the nuclear charge distribution. Such experiments indicate that many nuclei are approximately spherical and that the radius R is approximately related to the mass number by

$$R \approx 1.2A^{1/3} \text{ fm} \tag{43.1}$$

Radius of a nucleus

where 1 fermi (fm) $= 10^{-15}$ m. Since the volume, V, of a sphere is proportional to R^3, we see from Eq. 43.1 that $V \propto A$, the number of nucleons. It appears that the nucleons pack closely together like molecules in a liquid drop.

EXAMPLE 43.2: What is the mass density of a typical nucleus, say $^{16}_{8}O$?

Solution: The volume of a sphere is

$$V = \frac{4}{3}\pi R^3 = \frac{4\pi}{3}(1.2)^3 A = 1.16 \times 10^{-43}\text{m}^3$$

The mass of an oxygen atom ($A = 16$) is 16 u (including the electrons, but their mass makes a negligible contribution). The density is

$$\rho = \frac{m}{V} = \frac{(16 \text{ u})(1.6606 \times 10^{-27} \text{ kg/u})}{(1.16 \times 10^{-43}\text{m}^3)}$$
$$= 2.3 \times 10^{17} \text{ kg/m}^3$$

This is over 10^{14} times the density of water! Since $m \propto A$ and $V \propto A$, the density $\rho = m/V$, is independent of A; it is approximately the same for all nuclei. Such incredible densities are also found in neutron stars.

43.2   BINDING ENERGY AND NUCLEAR STABILITY

The existence of a stable nucleus means that the nucleons are in a bound state. Since the protons in a nucleus experience strong electrical repulsion, there must exist a stronger attractive force that holds the nucleus together. The **nuclear force** is a short-range interaction that extends only to about 2 fm. (In contrast, the electromagnetic interaction is a long-range interaction.) An important feature of the nuclear force is that it is essentially the same for all nucleons, independent of charge.

The **binding energy** (BE) of a nucleus is the energy required to completely separate the nucleons. The origin of the binding energy may be understood with the help of the mass–energy relation, $\Delta E = \Delta mc^2$, where Δm is the difference between the total mass of the separated nucleons and the mass of the stable nucleus. The mass of a stable nucleus is less than the sum of the masses of its nucleons. The binding energy of a nuclide $^A_Z X$ is

$$BE = [Zm_H + Nm_n - m_X]c^2 \tag{43.2}$$

Binding energy

where m_n is the mass of the neutron. The quantities m_H and m_X are the masses of the *neutral atoms*, because it is these atomic masses that are listed in tables and the masses of the electrons in Zm_H and m_X cancel.

EXAMPLE 43.3: What is the binding energy of the helium nucleus, $^4_2$He?

Solution: The mass of the neutral helium atom is 4.002604 u. Therefore,

$$\Delta m = 2m_H + 2m_n - m_{He}$$
$$= (2 \times 1.007825) + (2 \times 1.008665) - 4.002604$$
$$= 0.030376 \text{ u}$$

Since 1 u = 931.5 MeV/c^2, the binding energy is

$$BE = \Delta mc^2 = 28.3 \text{ MeV}$$

The average binding energy per nucleon is BE/A = 28.3 MeV/4 = 7.1 MeV.

EXERCISE 1. Calculate the average binding energy per nucleon for $^{12}_6$C.

The **average binding energy per nucleon,** BE/A, for stable nuclei is plotted in Fig. 43.2. The curve reaches a maximum of about 8.75 MeV in the vicinity of $^{56}_{26}$Fe and then gradually falls to 7.6 MeV for $^{238}_{92}$U. If each of the A nucleons in a nucleus were to interact with all of the remaining $(A - 1)$ nucleons, there would be $A(A - 1)/2$ distinct pairwise interactions. The binding energy would increase approximately as A^2 and BE/A would be proportional to A. However, as Fig. 43.2 shows, the binding energy per nucleon stays roughly constant above $A = 30$. From this we infer that each nucleon interacts only with its nearest neighbors.

Of the approximately 1500 nuclides known, only about 260 are stable, while the rest are radioactive. Figure 43.3 shows a plot of N versus Z for the stable nuclei. For mass numbers up to about $A = 40$, we see that $N \approx Z$. For larger values of Z, the (short-range) nuclear force is unable to hold the nucleus together against the (long-range) electrical repulsion of the protons unless the number of neutrons exceeds the number of protons. At Bi ($Z = 83$, $A = 209$), the *neutron excess* is $N - Z = 43$. There are no stable nuclides with $Z > 83$.

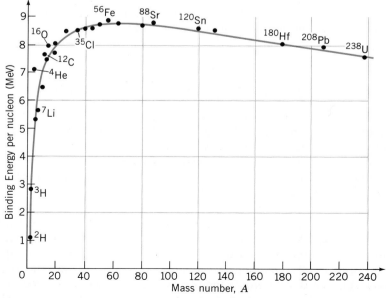

FIGURE 43.2 The average binding energy per nucleon as a function of atomic number A. Notice that $^4_2$He, $^{12}_6$C, and $^{16}_8$O are particularly stable. The maximum value occurs at $^{56}_{26}$Fe.

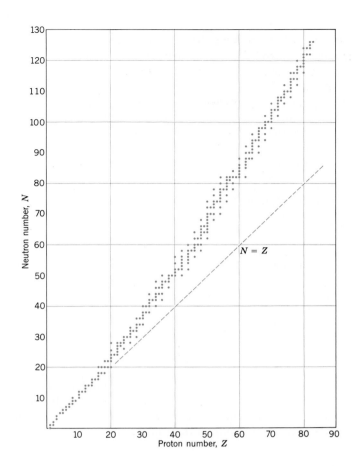

FIGURE 43.3 A plot of the neutron number N versus the atomic number Z for stable nuclides. The curve upward indicates that heavier nuclei need a greater proportion of neutrons to balance the electrical repulsion of the protons.

43.3   RADIOACTIVITY

In 1899, E. Rutherford classified radioactive emissions according to their charges and penetrating power. The positively charged *α* **particles** are stopped by a single sheet of paper or about 5 cm of air. In 1908, Rutherford and Royds identified alpha particles as doubly ionized helium atoms ($^4_2$He). The negatively charged *β* **particles** can penetrate several meters through air. They were identified as electrons by Becquerel in 1900. Also in 1900, P. Villard discovered *γ* **rays,** which can penetrate several centimeters of lead. Gamma rays were later identified as electromagnetic waves with wavelengths shorter than X rays. Most radioactive elements emit either α particles or β particles; a few emit both.

In order to clarify the distinction between the three types of emission, consider a hypothetical experiment. A radioactive sample is placed in a lead block, as in Fig. 43.4, and the emissions are subjected to a magnetic field. The γ rays are undeflected, whereas the α particles are only slightly deflected. The β particles experience large deflections in the opposite sense and are spread out in space.

A significant step in nuclear physics came with the understanding of the mechanism of radioactivity. In 1902, Rutherford and F. Soddy proposed that

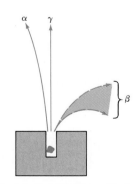

FIGURE 43.4 The three kinds of radioactive emission. Alpha particles are positive helium nuclei, beta particles are negative electrons, and gamma rays are high-energy photons.

radioactivity involves the distintegration of atoms. As a result one element is actually "transmuted" into another. They were excited but cautious because this sounded too much like alchemy. It meant the abandonment of the belief that atoms always retain their identities—the idea on which chemistry had been developed for nearly a century.

A nucleus X is theoretically unstable with respect to α or β decay if its mass is greater than the sum of the masses of the products Y + α or Y + β. A few nuclides between $A = 140$ to 190 undergo α decay, but most α decays occur with nuclides above $A = 200$. Almost all nuclides with $Z < 83$ are stable with respect to α decay.

EXERCISE 2. What can you conclude from the fact that the β particles experience a range of deflections in a magnetic field?

Alpha Decay

When a nucleus emits an α particle ($^4_2$He), the charge of the resulting nucleus is lower by $2e$ and its mass number is lower by four. For example,

$$^{226}_{88}\text{Ra} \rightarrow ^{222}_{86}\text{Rn} + ^4_2\text{He}$$

The energy released in any decay event is called the **disintegration energy, Q.** For α decay it is given by

Disintegration energy for α decay

(α decay) $\qquad\qquad Q = (m_X - m_Y - m_{He})c^2$

where X is the *parent* nucleus and Y is the *daughter* nucleus. The masses are of the *neutral atoms* since the masses of the electrons cancel. For the α decay of radium, $Q = 4.87$ MeV. This energy appears as kinetic energy of the α particle (about 4.8 MeV) and of the recoiling radon nucleus (about 0.1 MeV).

The energies of α particles have discrete values characteristic of the nucleus. This implies that nuclei have excited energy levels that are discrete. Thus, nucleus X may emit an α particle and may be left in the ground state of nucleus Y, or it may first reach an excited state of Y and then drop to the ground state by emitting a γ ray. The difference between the two α particle energies is just the γ ray energy. The above calculation for Q was based on the assumption that both nuclei are in their ground states, so that only an α particle is emitted. Other α particles emitted by radium have energies of 4.6 MeV and 4.2 MeV, as shown in Fig. 43.5.

It may seem strange that a nucleus emits the α particle, a set of four nucleons, rather than individual neutrons or protons. A particle can be emitted only if the total mass of the products is less than that of the parent nucleus. Neutron or proton emission is very rare (it occurs in $^{17}_7$N and $^{87}_{35}$Br) because the mass of (Y + n) or of (Y + p) is greater than the mass of the parent nucleus X. The large binding energy (7.1 MeV per nucleon) of the $^4_2$He nucleus reduces the sum of the masses of the products (Y + α) just enough to make α decay possible.

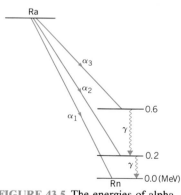

FIGURE 43.5 The energies of alpha particles are well defined and depend on whether the daughter nucleus is in the ground state or an excited state. Gamma rays are emitted when an excited daughter nucleus makes a transition to a lower state.

Alpha Particle Tunneling

Since α decay is energetically possible, why haven't all nuclei that can emit α particles already done so? The reason is the existence of a potential barrier. To understand this imagine the α particle bouncing back and forth within the nucleus. The short-range attractive nuclear force is represented by a rectangular potential well, whereas the potential energy associated with the electrical repulsion varies as $1/r$. The resultant potential energy function experienced by the α particle is shown in Fig. 43.6. The height of the peak is about 30 MeV. Classically, the α particle would be trapped unless its energy exceeds 30 MeV. Yet α particles emitted by nuclei have energies within the range 4–9 MeV. According to quantum mechanics, the α particle simply tunnels through the potential barrier (Section 41.5). The probability of emission, which determines the decay rate, is determined by the height and width of the barrier and the mass of the tunneling particle. This explanation, given in 1928, was one of the first major successes of quantum mechanics.

EXERCISE 3. Show that the disintegration energy for the α decay of $^{226}_{88}\text{Ra}$ is 4.87 MeV. Refer to Appendix E.

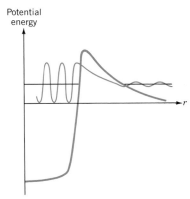

Beta Decay

Beta decay can involve the emission of either electrons or positrons. A positron is a form of antimatter, which will be discussed in the next chapter. For the moment we treat it simply as a positive electron. Since neither electrons nor positrons exist inside a nucleus, the β particles are created at the time of emission. When a nucleus emits a β particle, the charge on the daughter nucleus is $(Z + 1)e$ or $(Z - 1)e$, but the mass number does not change:

$$^{14}_{6}\text{C} \rightarrow \, ^{14}_{7}\text{N} + e^- + \, ?$$
$$^{13}_{7}\text{N} \rightarrow \, ^{13}_{6}\text{C} + e^+ + \, ?$$

FIGURE 43.6 The potential energy function of an alpha particle and a nucleus. The rectangular well arises from the attractive nuclear force, whereas the hill arises from the Coulomb repulsion. Although the energy of the alpha particle is less than the barrier height, it is able to tunnel through the barrier.

The question mark signifies a problem. The disintegration energies for β^- decay and β^+ decay (see Problem 6) are

(β^- decay) $Q = (m_X - m_Y)c^2$
(β^+ decay) $Q = (m_X - m_Y - 2m_e)c^2$

Disintegration energy for β decay

where the masses are of the neutral atoms. In the β^- decay of $^{14}_{6}\text{C}$ the disintegration energy is

$$Q = (14.00234 \text{ u} - 14.00397 \text{ u})(931.5 \text{ MeV/u}) = 156 \text{ keV}$$

With some allowance for the recoil of the ^{14}N nucleus, the energy of the β particle should be close to 156 keV. Yet when the energies of β particles are measured, the curve shown in Fig. 43.7 is obtained. Only a tiny fraction of the β particles have energies near the maximum kinetic energy ($= Q$). In the 1920s, it seemed that the law of conservation of energy did not hold for β decay. In 1930, W. Pauli suggested that there might exist a very light (perhaps massless) neutral particle that carries away the missing energy. It would interact very weakly with matter and therefore be almost undetectable. E. Fermi, who called it a **neutrino,** developed a theory of β decay that was quite successful in explaining the energy spectrum and other features. Fermi showed that the neutrino interacts via a new kind of force called the **weak interaction.** A neutrino interacts so weakly with ordinary matter that it can pass through the earth without a single interaction. It was detected in 1956 by F. Reines and C. L. Cowan.

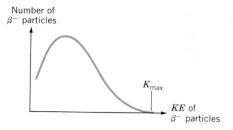

FIGURE 43.7 The energy distribution of β^- particles.

In β^- decay a neutron in the nucleus is transformed into a proton, an electron, and an antineutrino:

$$n \rightarrow p + e^- + \bar{\nu}$$

This process is also observed for free neutrons. The question mark in the β^+ decay is a neutrino.

Gamma Decay

Gamma rays, which are high-energy photons, are produced when nuclei make transitions between different energy levels. Gamma rays are usually emitted soon after α or β decay or after a nucleus has been left in an excited state as a result of a collision. Their energies, which range from 1 keV to a few MeV, can be measured by absorption, by crystal diffraction (up to 1 MeV), or by the energy of Compton-scattered electrons. Since they have discrete energies, γ rays are useful in determining the energy levels of stable nuclei.

43.4   THE RADIOACTIVE DECAY LAW

Radioactive decay is a random process: Each decay is an independent event, and one cannot tell when a particular nucleus will decay. When a given nucleus decays, it is transformed into another nuclide, which may or may not be radioactive. When there is a very large number of nuclei in a sample, the rate of decay is proportional to the number of nuclei, N, that are present:

Radioactive decay law

$$\frac{dN}{dt} = -\lambda N$$

where λ is called the **decay constant.** This equation may be expressed in the form $dN/N = -\lambda\, dt$ and integrated:

$$\int_{N_0}^{N} \frac{dN}{N} = -\lambda \int_0^t dt$$

to yield

$$\ln\left(\frac{N}{N_0}\right) = -\lambda t$$

where N_0 is the initial number of parent nuclei at $t = 0$. The number that survive at time t is therefore

$$N = N_0 e^{-\lambda t} \tag{43.3}$$

This function is plotted in Fig. 43.8.

The time required for the number of parent nuclei to fall to 50% is called the **half-life,** $T_{1/2}$, and may be related to λ as follows. Since

$$0.5 N_0 = N_0 e^{-\lambda T_{1/2}}$$

we have $\lambda T_{1/2} = \ln 2 = 0.693$. Therefore,

$$T_{1/2} = \frac{0.693}{\lambda} \tag{43.4}$$

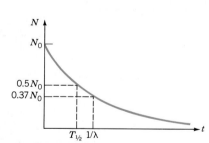

FIGURE 43.8 The number of radioactive nuclei in a sample as a function of time. The half-life is the time required for the number to fall to 50% of any initial value (not just $t = 0$).

It takes one half-life to drop to 50% of *any* starting value. The half-life for the decay of the free neutron is 12.8 min. Other half-lives range from about 10^{-20} s to 10^{16} y.

Since the number of atoms is not directly measurable, we measure the **decay rate,** $R = -dN/dt$. On taking the derivative of Eq. 43.3 we find

Decay rate

$$R = \lambda N = R_0 e^{-\lambda t} \tag{43.5}$$

where $R_0 = \lambda N_0$ is the initial decay rate. The decay rate is characterized by the same half-life. The SI unit for the decay rate is the becquerel (Bq), but the curie

(Ci) is often used in practice:

$$1 \text{ Bq} = 1 \text{ decay/s}; \qquad 1 \text{ Ci} = 3.7 \times 10^{10} \text{ Bq}$$

EXAMPLE 43.4: What is the initial decay rate of 1 g of radium 226? Its half-life is 1620 y, and its molecular mass $M = 226$ g/mole.

Solution: First we must find the initial number of atoms, N_0, in the sample of mass m. Since the number of moles is $n = N_0/N_A$, where N_A is Avogadro's number, the mass of the sample is $m = nM = (N_0/N_A)M$. Therefore,

$$N_0 = \frac{mN_A}{M} = \frac{(1 \text{ g})(6.02 \times 10^{23} \text{ atoms/mol})}{(226 \text{ g/mol})}$$
$$= 2.66 \times 10^{21} \text{ atoms}$$

From Eq. 43.4, we have $\lambda = 0.693/T_{1/2}$, where $T_{1/2} = 1620$ y $= 5.11 \times 10^{10}$ s. Thus from Eq. 43.5 the initial decay rate is

$$R_0 = \lambda N_0 = \frac{0.693 \, N_0}{T_{1/2}}$$
$$= 3.6 \times 10^{10} \text{ Bq} = 0.97 \text{ Ci}$$

The curie was originally defined as the decay rate of radon in equilibrium with 1 g of radium.

Radioactive Dating

The radioactive decay of the isotope ^{14}C, which has a half-life of 5730 y, has been an invaluable aid in determining the age of archeological specimens. The abundance of this isotope relative to the usual ^{12}C is $^{14}C/^{12}C = 1.3 \times 10^{-12}$. A living organism, such as an animal or a tree, exchanges CO_2 with the environment, so the ratio of the isotopes in the living organism is the same as it is in the atmosphere. When the organism dies, it ceases this exchange and the relative amount of ^{14}C decreases by decay. By determining the carbon content of a sample and measuring its activity, one can determine when the organism died.

The supply of ^{14}C in the atmosphere is constantly being replenished by the bombardment of cosmic rays via the reaction $n + N \rightarrow C + p$. Thus, although the ^{14}C concentration does vary over long periods, this method can still be used to about 40,000 y. Corroboration for carbon dating has been found from the growth rings on very old bristlecone pine trees and ice cores drilled in Antarctica. For geological time scales, similar methods based, for example, on the decay of ^{238}U are used.

EXAMPLE 43.5: (a) What is the decay rate of 1 g carbon in a living organism? (b) A sample with 10 g of carbon registers a decay rate of 30 decays/min. How old is it?

Solution: (a) The number of ^{12}C atoms in the sample is

$$N_0 = mN_A/M = \frac{(1 \text{ g})(6.02 \times 10^{23} \text{ atoms/mol})}{(12 \text{ g/mol})}$$
$$= 5.0 \times 10^{22} \text{ atoms}$$

From the relative abundance quoted above, we find the number of ^{14}C atoms is $(1.3 \times 10^{-12})(N_0) = 6.5 \times 10^{10}$. From Eq. 43.5 the initial decay rate is

$$R_0 = \lambda N_0 = \left(\frac{0.693}{T_{1/2}}\right) N_0$$
$$= 14 \text{ decays/min} = 0.25 \text{ Bq}$$

(b) For the 10 g sample there are 0.5 decays/s, which means

$R = 0.05$ Bq for each gram. From part (a) we know that $R_0 = 0.25$ Bq for each gram. From Eq. 43.5

$$\frac{R}{R_0} = e^{-\lambda t}$$

we obtain (fill in the steps)

$$t = \frac{1}{\lambda} \ln(R_0/R)$$
$$= T_{1/2} \frac{\ln(R_0/R)}{0.693}$$

where $T_{1/2} = 0.693/\lambda$ is the half-life. Using $T_{1/2} = 5730$ y and $R_0/R = 0.25/0.05 = 5$, we find $t = 13,300$ y.

EXERCISE 4. How many half-lives would it take for the decay rate to drop to 15% of its initial value?

43.5 NUCLEAR REACTIONS

In 1902, Rutherford and Soddy showed that radioactivity involves the spontaneous transmutation of atoms. In 1919, they discovered that α particles could combine with nitrogen nuclei to produce protons and an oxygen isotope:

$$^4_2\alpha + {}^{14}_7N \rightarrow {}^{17}_8O + {}^1_1p$$

This was the first *artificially induced* transmutation of one element into another. In a sense, the dream of alchemists had come true! Since then, much information about nuclei has been obtained by bombarding them with particles such as protons, neutrons, electrons, and α particles. A **nuclear reaction** in which a collision between particle a and nucleus X produces nucleus Y and particle b is represented as

$$a + X \rightarrow Y + b \tag{43.6}$$

The reaction is sometimes expressed in the shorthand notation X(a, b)Y. Reactions are subject to the restrictions imposed by the conservation of charge, energy, linear momentum, and angular momentum. In addition, the total number of nucleons remains unchanged.

The **reaction energy,** Q, is determined by the mass difference between the initial set of particles and the final set:

Reaction energy

$$Q = \Delta mc^2 = (m_a + m_X - m_Y - m_b)c^2 \tag{43.7}$$

where the masses are of the neutral atoms. If $Q > 0$, the reaction is said to be *exothermic*. The energy released generally goes into kinetic energy of the products and γ rays due to transitions between excited states of Y. If $Q < 0$, the reaction is *endothermic*. In this case there is a certain threshold energy of the incoming particle, below which the reaction will not take place (see Problem 10). The special case $Q = 0$ corresponds to elastic scattering, denoted by X(a, a)X. Although a and X might exchange energy, the total kinetic energy does not change.

In 1932, J. Cockcroft and E. Walton completed the first "atom-smasher" that could accelerate protons up to 0.6 MeV (Fig. 43.9). With 0.125 MeV protons and a lithium target, they observed the reaction

$$p + {}^7_3Li \rightarrow {}^4_2He + {}^4_2He$$

for which the reaction energy is $Q = 17.3$ MeV. Thus, although the incoming protons had energies of only 0.125 MeV, the two outgoing α particles had a total energy of over 17 MeV. This reaction served as the first direct experimental check of the mass–energy relation $\Delta E = \Delta mc^2$. The Cockcroft–Walton experiment was a significant milestone in nuclear physics because the energies of the incoming particles were controlled. It was soon followed by the development of the cyclotron and the Van de Graaff accelerators.

Prior to 1934, all known radioactive isotopes were heavier than lead ($Z = 82$). In that year F. Joliot and I. Joliot-Curie bombarded an aluminum target with α particles and discovered that positrons were emitted for several minutes after the source of α particles had been removed. The phosphorus isotope produced in the reaction

$$^{27}_{13}Al + \alpha \rightarrow {}^{30}_{15}P + n$$

was found to undergo β^+ decay:

$$^{30}_{15}P \rightarrow {}^{30}_{14}Si + \beta^+ + \nu$$

FIGURE 43.9 The equipment used by J. Crockcroft and E. Walton in the first nuclear reaction produced by protons whose energy could be controlled.

Such *artificially induced radioactivity* has had great practical consequences. Various radioactive isotopes of elements important in biological and chemical processes can be produced and used as "tracers" in analyzing the sequence of events in complex reactions. Stable nuclides can be transformed into radioactive nuclei when bombarded with neutrons. Each **neutron-activated** nucleus undergoes β decay, which can be used in the analysis of samples too small for other methods. An important consequence of the study of artificially induced radioactivity was the discovery of fission.

43.6 FISSION

Enrico Fermi (Fig. 43.10) realized that since neutrons have no charge, they could induce artificial radioactivity more easily than protons or α particles. Furthermore, a neutron absorbed by a nucleus with charge Ze would place it in an excited state from which it might undergo β^- decay. This would result in a nucleus with charge $(Z + 1)e$. With this in mind, Fermi bombarded uranium ($Z = 92$) with neutrons in an attempt to produce "transuranic" elements. His chemical analysis led him to suggest that a radioactive element with $Z = 93$ (Np) or $Z = 94$ (Pu) had been created.

Two nuclear chemists, O. Hahn and F. Strassmann, pursued Fermi's work and in 1938 isolated a radioactive element which they could not separate chemically from barium ($Z = 56$). They believed it to be an isotope of radium ($Z = 88$). The β^- decay product of this element was chemically similar to lanthanum ($Z = 57$) and they assumed it was actinium ($Z = 89$). Irene Joliot-Curie and P. Savitch, however, insisted that their (different) chemical technique showed that the decay

FIGURE 43.10 Enrico Fermi (1901–1954).

product was actually a lanthanum isotope. Hahn and Strassmann then confirmed that they could separate the radioactive product from another isotope of radium, but not from barium. They were still reluctant to state that the radioactive product was in fact a barium isotope. Since all previous nuclear reactions had involved only small changes in mass number, they could not imagine how a nucleus whose mass number is approximately half that of uranium could be in the products.

Within weeks, Lise Meitner (Fig. 43.11), who had previously collaborated with both Hahn and Strassmann, and her nephew, Otto Frisch, explained what had happened. They proposed a mechanism based on a "liquid-drop" model of the nucleus put forward by Niels Bohr in 1936. In this model the nucleons are assumed to move about freely and randomly within the nucleus and to interact only with their nearest neighbors, like molecules in a drop. At the surface, the nucleons experience a net inward force. When a spherical uranium nucleus absorbs a neutron, it becomes unstable and undergoes oscillations. The shape of the "drop" can become distorted, as in Fig. 43.12a. If a neck forms, as in Fig. 43.12b, the short-range nuclear force between the two parts of the "dumbbell" is greatly reduced. However, the (long-range) electrical repulsion between these two parts is only slightly diminished. As a result, the nucleus splits into two roughly equal fragments, as in Fig. 43.12c. This process was named **fission** after the process of cell division in biology.

When a neutron is captured by a $^{235}_{92}U$ nucleus, it creates a short-lived ($\approx 10^{-14}$ s) **compound nucleus**, $^{236}_{92}U^*$, in an excited state (indicated by the asterisk), which then undergoes fission. For example,

$$n + {}^{235}_{92}U \rightarrow {}^{236}_{92}U^* \rightarrow {}^{140}_{54}Xe + {}^{94}_{38}Sr + 2n + Q$$

The primary fission fragments have an excess of neutrons and therefore release an average of 2.5 "prompt" neutrons. The final steps are several β^- and γ decays to stable end products. For example,

$$^{140}_{54}Xe \rightarrow {}^{140}_{55}Cs \rightarrow {}^{140}_{56}Ba \rightarrow {}^{140}_{57}La \rightarrow {}^{140}_{58}Ce \text{ (stable)}$$
$$\quad\; 16 \text{ s} \qquad 66 \text{ s} \qquad 13 \text{ d} \qquad 40 \text{ h}$$

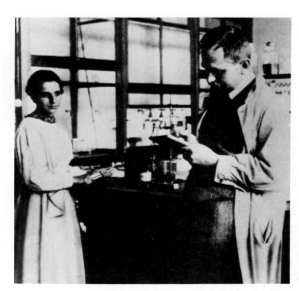

FIGURE 43.11 Lise Meitner (1878–1968) and Otto Hahn (1879–1968) in their laboratory in Berlin.

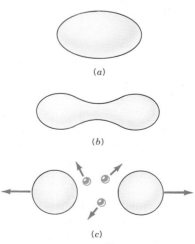

FIGURE 43.12 When a spherical $^{235}_{92}U$ nucleus absorbs a thermal neutron it undergoes oscillations. When its shape distorts into the form of a dumbbell, the electrical repulsion overcomes the nuclear force and the nucleus undergoes fission.

The half-life of each product is also indicated. The actual distribution of fission products of $^{236}_{92}U$ is shown in Fig. 43.13. The peaks (7%) are centered at about $A = 95$ and $A = 140$. There are hardly any fragments (0.01%) of equal mass number $A = 118$.

The energy released in each fission event may be estimated as follows. The binding energy per nucleon of uranium is about 7.6 MeV whereas between $A = 90$ and 150 it is about 8.5 MeV. Thus, the energy released in the fission process is about $236(8.5 - 7.6) \approx 200$ MeV, which is many orders of magnitude greater than the energies released in chemical reactions. About 170 MeV is carried away as kinetic energy of the fission fragments; the rest is shared by neutrons emitted by the fragments, by β particles, γ rays, and neutrinos.

The neutrons released in one fission event may be used to induce fissions in other nuclei. Under suitable conditions the process can repeat itself, thereby setting up a **chain reaction.** The energy release is uncontrolled in an atomic bomb and controlled in a nuclear reactor. The first controlled fission reaction took place at the University of Chicago on December 2, 1942, in a reactor designed by Fermi (Fig. 43.14).

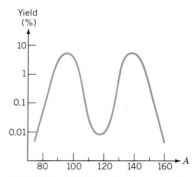

FIGURE 43.13 The distribution of primary fission products from the reaction $n + {}^{235}_{92}U \rightarrow X + Y$. Only a very small percentage of the products have equal mass.

FIGURE 43.14 The first nuclear reactor, designed by E. Fermi, operated for the first time on December 2, 1942.

43.7 FUSION

Figure 43.2 shows that the binding energy of light nuclei increases with atomic number. Therefore, when two light nuclei combine to form a larger nucleus, a process called **fusion,** energy is released. The energy released, per unit mass, is greater in a fusion reaction than in a fission reaction. In order for nuclei to fuse they must overcome the strong potential barrier created by their Coulomb repulsion. Consider two deuterons (^2_1H) separated by twice their radius ($\approx 2 \times 10^{-15}$ m). Their electrostatic potential energy is

$$U = \frac{ke^2}{r} = \frac{(9 \times 10^9 \text{ N} \cdot \text{m}^2/\text{C}^2)(1.6 \times 10^{-19} \text{ C})^2}{(4 \times 10^{-15} \text{ m})}$$

$$\approx 6 \times 10^{-14} \text{ J} \approx 400 \text{ keV}$$

Each deuteron would need a kinetic energy of 200 keV to approach the other closely enough for the nuclear force to bind them. One way to give particles this much energy is to raise a gas to high temperatures. If we set $kT = 200$ keV, we find $T \approx 2 \times 10^9$ K.

Fusion in the interior of the sun is the source of its energy. Actually, the temperature of the interior, about 1.5×10^7 K, means that $kT \approx 1.3$ keV, which is much less than the 200 keV mentioned above. Nonetheless, fusion reactions do occur for two reasons. First, there are always some particles, at the tail of the Maxwell distribution (Fig. 20.8), that have energies far greater than the average. Second, particles can also tunnel through the Coulomb potential barrier. A set of reactions called the **proton–proton** cycle takes place within the sun's interior:

Proton–proton cycle

$$
\begin{array}{lll}
^1\text{H} + {}^1\text{H} \rightarrow {}^2\text{H} + e^+ + \nu & Q = 0.4 \text{ Mev} & \\
^1\text{H} + {}^2\text{H} \rightarrow {}^3\text{He} + \gamma & Q = 5.5 \text{ MeV} & (43.8) \\
^3\text{He} + {}^3\text{He} \rightarrow {}^4\text{He} + {}^1\text{H} + {}^1\text{H} & Q = 12.9 \text{ MeV} &
\end{array}
$$

The first two reactions must occur twice in order for the third to proceed. The total energy released (24.7 MeV) is distributed as kinetic energy among the reaction products. More energy is released when the positron meets an electron: They undergo mutual annihilation with the emission of two γ rays having a total of 1.02 MeV. The probability of the (first) proton–proton reaction is very low, which makes it slow. It takes place in the sun's interior because the particle density there is very high.

The vast amount of energy released in fusion is used in a hydrogen bomb. Such a device, first exploded in 1952, involves an uncontrolled thermonuclear fusion reaction. The necessary thermal energy is provided by the detonation of a fission bomb. The practical use of *controlled* thermonuclear fusion is more elusive. The following reactions using deuterium (D) or tritium (T), are being investigated:

Fusion reactions

$$
\begin{array}{llll}
\text{(D–D)} & ^2\text{H} + {}^2\text{H} \rightarrow {}^3\text{He} + \text{n} & Q = 3.27 \text{ MeV} & \\
\text{(D–D)} & ^2\text{H} + {}^2\text{H} \rightarrow {}^3\text{H} + {}^1\text{H} & Q = 4.03 \text{ MeV} & (43.9) \\
\text{(D–T)} & ^2\text{H} + {}^3\text{H} \rightarrow {}^4\text{He} + \text{n} & Q = 17.6 \text{ MeV} &
\end{array}
$$

In order to produce power from fusion reactions, three criteria must be met.

1. *High temperature.* A temperature greater than 10^8 K is necessary for the particles to get close enough to overcome their electrical repulsion. At such temperatures the atoms are stripped of their electrons, which results in a completely ionized gas called a **plasma.**

2. *High particle density.* This is required to increase the collision rate and thereby the reaction rate.

3. *Long confinement time.* Once the particles have been brought together at a high temperature, they must be kept together long enough for the reaction to occur.

In 1957, J. D. Lawson deduced a necessary, but not sufficient, condition for the net release of energy from fusion in a fusion reactor. If n is the particle density and τ is the confinement time, then the **Lawson criterion** is

$$\text{D–D: } n\tau > 10^{22} \text{ s/m}^3; \qquad \text{D–T: } n\tau > 10^{20} \text{ s/m}^3 \qquad (43.10)$$

If the Lawson criterion is met, the energy obtained from the plasma equals the energy input and losses such as X-ray radiation. Two approaches to harnessing the energy released in fusion are discussed in the special topic on reactors.

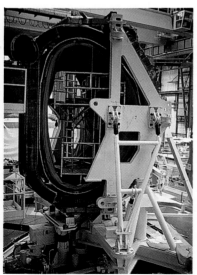

Part of the toroidal chamber used to confine the hot plasma at the Joint European Torus fusion reactor (see p. 907).

SUMMARY

A nucleus is specified by $^A_Z X$, where A is the **mass number** and Z, the **atomic number,** is the number of protons. The mass number $A = N + Z$, where N is the number of neutrons. **Isotopes** have the same Z but different A.

Nuclei are roughly spherical with a radius given by

$$R = 1.2 \, A^{1/3} \text{ fm}$$

where 1 fm $= 10^{-15}$ m.

The **binding energy** (BE) of a nucleus is determined by the difference in mass between the nucleus and the mass of the separated nucleons. It may be found from

$$Q = \Delta m c^2 = (Z m_H + N m_n - m_X)c^2$$

where m_n is the mass of the neutron and m_H and m_X are the masses of the neutral atoms since the electron masses cancel. If the masses are in u, then Q in MeV is given by

$$Q = \Delta m \times 931.5 \text{ MeV/u}$$

There are three types of **radioactive** emission: α decay, β decay, and γ decay. Alpha particles are helium nuclei ($^4_2$He), β particles are either electrons or positrons, and γ rays are high-energy photons.

If there are N_0 radioactive nuclei in a sample at $t = 0$, the number that remain at time t is given by

$$N = N_0 e^{-\lambda t}$$

where λ is called the decay constant. The **decay rate** $R = -dN/dt$ is given by

$$R = R_0 e^{-\lambda t}$$

where $R_0 = \lambda N_0$ is decay rate at $t = 0$. The **half-life,** $T_{1/2}$, is the time it takes for either the number of nuclei or the decay rate to drop to 50% of the value at *any* initial time (not just $t = 0$):

$$T_{1/2} = \frac{0.693}{\lambda}$$

A nuclear reaction in which a target nucleus X is bombarded by a particle a and that results in the production of nucleus Y and a particle b is written as

$$a + X \rightarrow Y + b \qquad \text{or} \qquad X(a, b)Y$$

The **reaction energy** is $Q = \Delta m c^2 = (m_a + m_X - m_Y - m_b)c^2$.

ANSWERS TO IN-CHAPTER EXERCISES

1. BE = $[(6 \times 1.007825 \text{ u}) + (6 \times 1.008665 \text{ u}) - 12 \text{ u})]$ (931.5 MeV/u) = 92.163 MeV. Thus BE/A = 7.68 MeV.

2. The particles have a spread in kinetic energy. As we will see, this was a serious theoretical problem for many years.

3. Q = (226.025406 u − 222.017574 u − 4.002604 u) (931.5 MeV/u) = 4.87 MeV.

4. The decay rate, R, after N half-lives is $R = R_0(0.5)^N$, where R_0 is the initial value. We are given $R/R_0 = 0.15 = (0.5)^N$. Thus, $N = \log(0.15)/\log(0.5) = 2.74$ half-lives. In the case of $^{14}_6$C this would be 2.74 × 5730 y = 15,700 y.

QUESTIONS

1. In what ways are isotopes of a given element (a) similar and (b) different?

2. Do nuclides that lie below the ''line of stability'' in Fig. 43.3 tend to emit electrons or positrons? Why?

3. What is the evidence that electrons emitted in β^- decay come from the nucleus rather than the atomic electrons?

4. The nuclide $^{226}_{88}$Ra has a half-life of only 1640 y, yet it is found in rocks that are billions of years old. How is this possible?

5. (a) Is it possible for a free proton to decay into a neutron? (b) Is it possible for a proton in a nucleus to turn into a neutron? In each case, explain why or why not.

6. Why do heavy radionuclides emit α particles rather than individual neutrons and protons?

7. Could an electron and a proton almost at rest combine to form a neutron?

8. (a) Why are the masses of nuclides numerically close to their mass numbers? (b) Why are some atomic masses in the periodic table not close to integer values?

9. A class of bulbs is rated as having a mean life of 1500 h. In what sense is this ''mean life'' (a) similar to, and (b) different from, the mean life in radioactive decay?

10. Explain why in radioactive decay the energies of α particles are discrete but the energies of β particles span a continuous range.

11. How can one measure the half-life of a radioisotope that has a half-life of over 10^9 years?

12. Why is $^{12}_6$C chosen as the basis for defining the atomic mass unit, u, rather than $^1_1$H?

13. Can one distinguish between isotopes of a given element by examining the optical electronic spectra? Could one use the infrared vibrational spectra of diatomic molecules, if such molecules exist?

14. Nuclei have approximately the same density. From this fact, what can you infer about the nuclear force?

15. What is the effect of heating on the activity of a radioactive sample?

16. Why do fission fragments tend to emit β^- particles rather than β^+ particles?

17. A high temperature is required for fusion reaction. Is the same true for fission?

18. (a) Suggest some ways in which the mass of a proton may be measured. (b) How could the mass of a neutron be measured?

19. Why does the importance of the coulomb force increase relative to the nuclear force as the mass number increases?

20. What is the missing nuclide in the following decays: (a) $^{234}_{94}$Pu → ? + α. (b) $^{64}_{29}$Cu → ? + β+?

21. Why is lead a good shield for γ rays but not useful as a medium in which to thermalize neutrons?

22. Since thermal neutrons have essentially no kinetic energy, where does the energy for fission come from?

23. The temperature of the plasma in a fusion reactor is greater than at the center of the sun. Why has a self-sustaining fusion reaction not yet been achieved?

EXERCISES

43.1 Structure of the Nucleus

1. (I) Use Eq. 43.1 to calculate the radii of the following nuclides: (a) $^{16}_8$O, (b) $^{56}_{26}$Fe, and (c) $^{238}_{92}$U.

2. (I) The radius of the earth is 6400 km and its average density is 5.5 g/cm^3. What is the radius of a sphere of nuclear matter with the same mass but a density of 2.3×10^{17} kg/m^3?

3. (I) What is the density of a neutron star that has a radius of 10 km and a mass equal to that of our sun?

4. (I) The estimated mass of the universe is about 10^{50} kg. What would be the radius of a sphere of the same mass but with the density of nuclear matter (2.3×10^{17} kg/m^3)?

5. (I) By what factor must the mass number A change for the nuclear radius to double?

6. (I) What would be the radius of the nuclide $^{235}_{92}$U if it had the average density of the earth, which is 5.5 g/cm^3?

7. (I) Copper has two stable isotopes, $^{63}_{29}$Cu and $^{65}_{29}$Cu. Their masses are 62.95 u and 64.95 u, respectively. What is the relative abundance of each isotope? The atomic mass of copper in the periodic table is 63.55 u.

8. (I) Neon consists mainly of two isotopes (there are more) with the following relative abundance $^{20}_{10}$Ne (91%) and $^{22}_{10}$Ne (9%). What is the approximate atomic mass of Ne? Check your result with that given in the periodic table.

9. (II) What is the radius of the gold isotope $^{197}_{79}$Au? If the radius of an α particle is 1.8 fm, what initial kinetic energy, in eV, must it have to ''touch'' the surface of the gold nucleus? Assume that the gold stays at rest.

10. (II) Assuming that it is uniformly charged, estimate the charge density of the nuclide $^{56}_{26}$Fe.

43.2 Binding Energy and Nuclear Stability

11. (I) Calculate the average binding energy per nucleon of the following nuclei: (a) $^{40}_{20}$Ca, (b) $^{197}_{79}$Au.

12. (I) Compute the average binding energy per nucleon of the following nuclei: (a) $^6_3$Li, (b) $^{133}_{55}$Cs.

13. (I) In *mirror nuclei* the numbers of neutrons and protons are

interchanged. Find the binding energies of (a) $^{13}_{6}C$, and (b) $^{13}_{7}N$.

14. (I) *Isobars* are nuclei that have the same number of nucleons. What are the binding energies of the isobars (a) $^{15}_{7}N$ and (b) $^{15}_{8}O$?

15. (II) (a) What is the energy required to remove one neutron from $^{7}_{3}Li$? (b) Compare the result of part (a) with the average binding energy per nucleon for this nuclide.

16. (II) (a) What is the energy required to remove one proton from $^{12}_{6}C$? (b) Compare the result of part (a) with the average binding energy per nucleon for this nuclide.

43.3 and 43.4 Radioactivity and the Radioactive Decay Law

17. (I) The radioactive isotope $^{60}_{27}Co$ is used in the treatment of tumors. It undergoes β^{-} decay with a half-life of 5.25 y. What is the initial decay rate of a 0.01 g sample?

18. (I) The nuclide $^{32}_{15}P$ undergoes β^{-} decay with a half-life of 14.3 d. It is used as a tracer isotope in biochemical analysis. What is the initial decay rate of a 1 mg sample?

19. (II) Radon ($^{222}_{86}Rn$) is a radioactive gas with a half-life of 3.82 d. If there are initially 320 decays/s in a sample, how many radon nuclei are left after 1 d?

20. (I) Unlike other light nuclides, $^{8}_{4}Be$ can decay via α particle decay. (a) What is the energy released when $^{8}_{4}Be$ splits into two α particles? (b) Is it possible for $^{12}_{6}C$ to decay into three α particles?

21. (I) If $^{11}_{6}C$ were to emit a positron what would be the remaining nuclide? Is this decay possible?

22. (II) As a result of an accident at a nuclear reactor, the radioactive isotope $^{90}_{38}Sr$ (half-life 29 y) is released into the atmosphere. If the fallout in the vicinity is 1 $\mu g/m^2$, how long would it take for the decay rate to drop to a level of 1 $\mu Ci/m^2$?

23. (I) The decay rate of a freshly prepared sample is 15 μCi. It drops to 9 μCi after 2.5 h. (a) Find the half-life of the nuclide. (b) How many radioactive atoms were initially present?

24. (I) Radioactive $^{239}_{94}Pu$ has a half-life of 24,000 y. What is the initial decay rate in Ci of a 1-g sample?

25. (II) An ancient bone contains 80 g of carbon and has a decay rate of 45 counts/min. How old is it? (Assume that the ratio of isotopes in the atmosphere is $^{14}C/^{12}C = 1.3 \times 10^{-12}$ and has remained constant.)

26. (I) In the early study of radioactivity three radioactive series were discovered. However, before the products were correctly identified, they were given various names. In the following *uranium* series, identify each element:

$$^{238}_{92}U \xrightarrow{\alpha} UX_1 \xrightarrow{\beta^-} UX_2 \xrightarrow{\beta^-} UII \xrightarrow{\alpha} Io \rightarrow \cdots$$

27. (I) In the *actinium* radioactive series that starts with $^{235}_{92}U$, the final (stable) decay product is $^{207}_{82}Pb$. How many α particles and electrons are emitted?

28. (I) The initial decay rate of a sample of $^{131}_{53}I$ (half-life 8.1 d) is 0.2 Ci. (a) What is the initial mass of the sample? (b) What is the number of nuclei present after 10 d?

29. (I) A sample of a radionuclide initially produces 500 counts/min in a geiger counter. Two hours later the rate falls to 320 counts/min. What is the half-life of the nuclide?

30. (I) Tritium ($^{3}_{1}H$), an isotope of hydrogen, has a half-life of 12.3 y. What percentage of a given sample remains after 10 y?

31. (I) Rubidium undergoes the following decay: $^{87}_{37}Rb \rightarrow ^{87}_{38}Sr + \beta^-$ with a half-life of 4.9×10^{10} y. A fossil is discovered in rocks for which the ratio of Sr to Rb is 1.2%. Assuming no Sr existed when the rock was formed, how old is the fossil?

32. (II) What is the decay rate of $^{14}_{6}C$ in an artifact 15,000 y old? State your answer in terms of the number of disintegrations per gram of ^{12}C per minute. Assume that the relative abundance in the atmosphere $^{14}C/^{12}C = 1.3 \times 10^{-12}$ has remained constant.

33. (II) The isotopes of uranium, $^{235}_{92}U$ and $^{238}_{92}U$, are radioactive with half-lives of 7.13×10^8 y and 4.47×10^9 y, respectively. The present ratio of $^{235}_{92}U$ to $^{238}_{92}U$ is about 0.007. What was the ratio 10^9 y ago?

34. (I) The radioactive isotope $^{40}_{19}K$, which decays to $^{40}_{18}Ar$ with a half-life of 2.4×10^8 y, is used in dating rocks. What is the decay rate of a 1-μg sample?

35. (I) A process that competes with β^+ decay is *electron capture* in which a nucleus captures an atomic electron. What is the energy released in the electron capture reaction $^{7}_{4}Be (e^-, \nu) ^{7}_{3}Li$?

36. (I) Identify the daughter nucleus and calculate the energy released when $^{210}_{84}P_0$ undergoes α decay.

37. (I) How many α and β^- particles are involved in the radioactive series that starts with thorium $^{232}_{90}Th$ and ends in the stable product $^{208}_{82}Pb$?

38. (I) What is the energy released in the decay of a free neutron $n \rightarrow p + e^- + \bar{\nu}$? Assume that the antineutrino ($\bar{\nu}$) has no rest mass.

39. (I) What is the energy released in the β^- decay of $^{40}_{19}Kr$?

40. (I) The isotope $^{218}_{84}Po$ can decay either via α particle or β^- particle emission. Find the disintegration energy for each.

43.5 Nuclear Reactions

41. (I) Find the reaction energy for each of the following reactions: (a) Rutherford's discovery of the artificial transmutation of nuclei in 1919:

$$^{14}_{7}N(\alpha, p)^{17}_{8}O$$

(b) The first artificial transmutation of nuclei by accelerated protons accomplished by J. Cockcroft and E. Walton in 1932:

$$^{7}_{3}Li(p, \alpha)^{4}_{2}He$$

42. (I) Calculate the reaction energy in each of the following reactions: (a) The discovery of the neutron by J. Chadwick in 1932:

$$^{9}_{4}Be(\alpha, n)^{12}_{6}C$$

(b) The discovery of artificial radioactivity by F. Joliot and I. Curie in 1934:

$$^{27}_{13}\text{Al}(\alpha, \text{n})^{30}_{15}\text{P}$$

43. (I) Calculate the reaction energy for the following reaction:

$$^{9}_{4}\text{Be}(\text{p}, \alpha)^{6}_{3}\text{Li}$$

44. (I) Identify the missing particle or nuclide in each of the following reactions: (a) $^{10}_{5}\text{B}(\text{n}, ?)^{7}_{3}\text{Li}$; (b) $^{6}_{3}\text{Li}(\text{p}, \alpha)?$; (c) $^{18}_{8}\text{O}(\text{p}, ?)^{18}_{9}\text{F}$; (d) $^{10}_{5}\text{B}(?, \alpha)^{7}_{3}\text{Li}$.

45. (I) Identify the missing particle or nuclide in each of the following reactions: (a) $?(\text{n}, \text{p})^{32}_{15}\text{P}$, (b) $?(\text{p}, \alpha)^{16}_{8}\text{O}$, (c) $^{9}_{4}\text{Be}(\text{n}, \gamma)?$, (d) $^{14}_{7}\text{N}(?, \text{p})^{14}_{6}\text{C}$.

46. (I) The reaction $^{14}\text{N}(\text{n}, \text{p})^{14}\text{C}$ occurs in the upper atmosphere as a result of bombardment of cosmic rays. It serves to replenish the supply of ^{14}C in the atmosphere. What is the reaction energy?

47. (I) Given that the reaction energy $Q = -2.45$ MeV for the reaction $^{18}_{8}\text{O}(\text{p}, \text{n})^{18}_{9}\text{F}$ calculate the mass of $^{18}_{9}\text{F}$. The mass of $^{18}_{8}\text{O}$ is 17.99916 u.

43.6 Fission

48. (I) (a) If the fission of one nucleus of $^{235}_{92}\text{U}$ releases 190 MeV, what would be the total energy released by the fission of 1 kg of this nuclide? (b) How long could this supply the 500 MW requirement of a town if the efficiency of the nuclear reactor is 32%?

49. (II) Assuming all the energy released by the fission of each $^{235}_{92}\text{U}$ nucleus (190 MeV) is absorbed by water, how many $^{235}_{92}\text{U}$ atoms must undergo fission to warm 1 g of water by 1 C°?

50. (I) The isotope $^{235}_{92}\text{U}$ can undergo spontaneous fission with a half-life of about 3×10^{17} y. How many spontaneous fissions would you expect per day in a 1-g sample?

51. (I) Calculate the energy released in the following fission reaction brought about by a thermal neutron:

$$^{1}_{0}\text{n} + {}^{235}_{92}\text{U} \rightarrow {}^{236}_{92}\text{U}^{*} \rightarrow {}^{144}_{56}\text{Ba} + {}^{89}_{36}\text{Kr} + 3^{1}_{0}\text{n}$$

Neglect the kinetic energy of the thermal neutron.

52. (I) Find the change in rest mass that occurs in the explosion of a 20 kiloton fission bomb. One ton of TNT releases 4.2×10^{9} J upon combustion.

53. (II) A "prompt" neutron released in a fission reaction has a kinetic energy of 1 MeV. It passes through a moderator, which reduces its kinetic energy to 0.025 eV. If the neutron loses 50% of its kinetic energy in each collision, how many collisions are needed?

54. (I) What is the de Broglie wavelength of a thermal neutron whose kinetic energy is 0.04 eV?

43.7 Fusion

55. (I) Show that the energy released in the D–D reaction $^{2}_{1}\text{H}(\text{d}, \text{n})^{3}_{2}\text{He}$ is 3.27 MeV.

56. (I) Show that the energy released in the D–T reaction $^{3}_{1}\text{H}(\text{d}, \text{n})^{4}_{2}\text{He}$ is 17.6 MeV.

57. (I) Show that the energy released in the D–D fusion reaction $^{2}_{1}\text{H}(\text{d}, \text{p})^{3}_{1}\text{H}$ is 4.03 MeV.

58. (I) What is the number of fusions needed per second in a fusion reactor to produce an output of 40 MW using the D–T reaction of Eq. 43.9.

59. (II) One D–D fusion reaction releases 4.03 MeV. The ratio of deuterium to hydrogen is 1/6500 in sea water. What is the fusion energy available in 1 kg of sea water?

PROBLEMS

1. (I) The radioisotope $^{90}_{38}\text{Sr}$ has a half-life of 29 y. A sample has an initial decay rate of 24 μCi. How many nuclei decay in the first year?

2. (II) A radioactive nuclide with a decay constant λ has a stable daughter nuclide. There are initially N_{01} parent nuclei and no daughter nuclei. Show that the number N_2 of the daughter nuclide as a function of time is given by

$$N_2 = N_{01}(1 - e^{-\lambda t})$$

3. (I) A sample of a radioactive nuclide has a daughter that is also radioactive. If the decay constants are λ_1 and λ_2, write an expression for rate of increase in the number, N_2, of daughter nuclides. What is the condition for the number of daughter nuclides to stop increasing?

4. (I) The mean life, τ, of a radioactive nuclide is given by

$$\tau = \frac{\int_{N_0}^{0} t \, dN}{\int_{N_0}^{0} dN}$$

where $dN/dt = -\lambda N$. Show that $\tau = 1/\lambda$. (*Hint:* Convert to an integral over time and integrate by parts.)

5. (II) Use conservation of linear momentum to show that in α decay, the kinetic energy of the α particle is given by

$$K_\alpha = \frac{M_\text{D} Q}{(M_\text{D} + m_\alpha)}$$

where M_D is the mass of the daughter nucleus and Q is the disintegration energy. Evaluate this for the decay $^{226}_{88}\text{Ra} \rightarrow {}^{222}_{86}\text{Rn} + \alpha$.

6. (I) (a) Show that in positron decay the disintegration energy is $Q = (m_\text{P} - m_\text{D} - 2m_\text{e})c^2$ where m_P and m_D are the atomic masses of the parent and daughter atoms. Calculate this for the following decay $^{64}_{29}\text{Cu} \rightarrow {}^{64}_{28}\text{Ni} + \beta^+ + \nu$

7. (I) The decay rate of a radioactive sample drops by 40% in 3.5 h. What is the half-life of the nuclide?

8. (II) The electrostatic potential energy of a uniformly charged sphere of radius R and charge Q is $3kQ^2/5R$. How does the total electrostatic potential energy change when the following fission occurs

$$^{236}_{92}U \rightarrow {}^{141}_{56}Ba + {}^{92}_{36}Kr + 3n$$

9. (I) Use momentum conservation to show that the kinetic energy of the α particle in the decay of a nucleus of mass number A initially at rest is

$$K_\alpha \approx \frac{(A - 4)Q}{A}$$

where Q is the disintegration energy. What is the energy of the α particle in the decay $^{236}_{92}U \rightarrow {}^{232}_{90}Th + \alpha$ assuming the $^{236}_{92}U$ nucleus is initially at rest?

10. (II) Use the conservation of energy and momentum to show that the threshold energy for an incoming particle to initiate an endothermic reaction X(a, b)Y, as measured in the laboratory frame, is

$$E_{TH} = \frac{-(M_X + m_a)Q}{M_X}$$

where m_a is the mass of the incoming particle and M_X is the mass of the nucleus, initially at rest. Evaluate this for $^{14}N(\alpha, p)^{17}O$. (*Hint:* Calculate the kinetic energy relative to the center of mass.)

11. (I) Show that the threshold energy of the incoming proton in the reaction $^{13}C(p, n)^{13}N$ is 3.23 MeV. (See the previous problem.)

SPECIAL TOPIC: Fission and Fusion Reactors

We have seen that both the fission and the fusion of nuclei release energy. In this Special Topic we consider some of the principles that underlie the operation of nuclear reactors which are designed to harness this energy.

FISSION REACTORS

A fission reactor is based on the process of fission of heavy nuclei. When a nucleus, such as $^{235}_{92}U$, undergoes fission, it releases neutrons that may be used to initiate fission in other nuclei, thereby creating a chain reaction. In an atomic bomb the chain reaction is uncontrolled; in a fission reactor the chain reaction is controlled.

Moderator

Naturally occurring uranium consists of 0.7% $^{235}_{92}U$ and 99.3% $^{238}_{92}U$. When a $^{238}_{92}U$ nucleus absorbs a neutron, it tends to emit a γ ray rather than undergo fission. In contrast, $^{235}_{92}U$ has a high fission probability for slow neutrons (1 eV or less). The high-energy (≈ 2 MeV) neutrons produced in the fission of $^{235}_{92}U$ must be slowed down before they can induce further fissions. This is accomplished by a material called a **moderator.** In passing through the moderator the average kinetic energy of the neutrons is reduced to the average kinetic energy $\frac{3}{2}kT$ (≈ 0.04 eV at 300 K) characteristic of the temperature of the moderator. Recall from Example 9.7 that in an elastic collision, the maximum transfer of kinetic energy from an incoming particle to a target particle occurs when they have the same mass. Thus protons in water are ideal for this purpose. In passing through water, neutrons are *thermalized* after about 20 collisions within 10^{-3} s. However, protons tend to combine with neutrons to form deuterons: Ordinary "light" water is thereby converted to *heavy water,* D_2O. If the fuel is natural uranium, then heavy water or graphite can be used as moderators. Light water can still be used as a moderator if the uranium is "enriched" by raising the proportion of $^{235}_{92}U$ from 0.7% to about 3% or 4%.

One cannot simply mix the uranium fuel with the moderator because neutrons within the energy range 5 eV to 100 eV have a high probability of being absorbed by $^{238}_{92}U$ nuclei (with later γ emission). They would become unavailable for the fission of the $^{235}_{92}U$ nuclei. Therefore the uranium fuel is packed into zircaloy rods that are arranged in a pattern and immersed in the moderator. Fast neutrons being slowed down find themselves outside the fuel rods as they pass through the 5 eV to 100 eV range. After they have been thermalized they can enter other fuel rods and initiate fission in $^{235}_{92}U$ nuclei.

Critical Size and Control

An important parameter in a chain reaction is the **multiplication factor,** k. This is the ratio of the number of neutrons in one generation of the chain reaction to the number in the previous generation. The production of neutrons is proportional to the volume of the fissile material, whereas the leakage increases with the surface area. When $k = 1$, the number of neutrons produced is equal to the number that are absorbed or leak away. In this condition the system is said to be **critical.**

In an atomic bomb, two subcritical masses of uranium (enriched to 50% $^{235}_{92}U$) are brought together to form a supercritical mass that explodes within 10^{-8} s. Since the enrichment in reactor fuel is much lower (under 4%), a nuclear explosion cannot occur. However, when $k > 1$, the thermal energy generated by the fission events and the radioactivity of the fission fragments can quickly melt the core, which can then melt the concrete below (a possibility sometimes referred to as the China syndrome). In addition, the moderating water would turn to steam and explode, thereby spreading radioactive material.

In order to keep k close to one, **control rods** of cadmium, which has a high absorption cross section for thermal neutrons, are inserted into the core. By carefully raising them, the condition of criticality can be achieved. If $k = 1.01$, the time constant for the increase in the neutron flux is only 0.1 s, which is too fast for human response. The ability to control a reactor depends crucially on a small feature of the fission process. Although almost all neutrons are *prompt*—they are emitted within 10^{-8} s—about 0.7% of the neutrons are *delayed* by between 0.2 s and 55 s. The core of a reactor is designed to be critical only when the contribution of these delayed neutrons is included. Through this approach, the time allowed for control of the reactor becomes greater than human reaction times. In case of an emergency, the control rods are dropped into the core, thereby making it subcritical. However, even after a shutdown, the heat generated by the radioactive decay of the fission fragments continues. In a large reactor the heat production would drop to about 1%, say 20 MW, after a day. But this is still very large.

In the pressurized water reactor shown in Fig. 43.15 the reactor core and the moderating water are contained in the

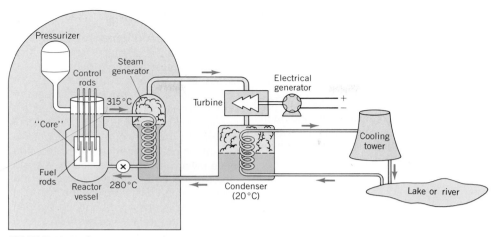

FIGURE 43.15 Some components of a pressurized water reactor.

reactor vessel. The moderating water also serves as the coolant in the primary coolant system. In order to prevent the water ($T = 315$ °C) from boiling, very high pressure (15 MPa or 150 atm) is required. The pipes in the primary coolant system pass through a steam generator where water from the secondary coolant system is converted to high-pressure steam (265 °C, 0.5 MPa) and directed to a turbine, which is connected to an electrical generator. The primary and secondary cooling systems are closed. After the steam passes through the turbine, it is cooled in a condenser with water from a reservoir, such as a river or a lake. The heated water is first cooled by evaporation in towers and then discharged back into the reservoir.

The operation of a nuclear reactor requires the implementation of many safety measures. For example, the reactor vessel and the steam generators are in a steel shell, itself housed in a reinforced concrete building. Nonetheless, the accidents at Three Mile Island (US) and at Chernobyl (USSR) illustrate what happens when proper procedures are not followed. The fission fragments are themselves radioactive. Therefore, even after the uranium fuel has been used, there remains the problem of disposing of these radioactive wastes. Burial in deep salt mines is one possibility. Because the structure within the reactor vessel receives intense neutron bombardment, many elements are "neutron activated," that is, they become radioactive. This limits the useful life of a nuclear reactor to about 30 years.

FUSION REACTORS

In Section 43.7 it was noted that in order to produce useful power from the fusion of nuclei several conditions are required. In particular the particle density and confinement time must satisfy Lawson's criterion, Eq. 43.10. There are two basic approaches to confining a plasma to achieve Lawson's criterion. In the **magnetic confinement** technique, a low particle density is compensated for by a relatively long (1 s) confinement time. In a system based on **inertial confinement,** the particle density is high but only for a short (1 ns) time.

Magnetic Confinement

The **Tokamak** is a magnetic confinement device invented in the USSR. A strong *toroidal field, B_t,* is produced by about 20 coils wrapped around the perimeter of a torus, as shown in Fig. 43.16. A weaker *poloidal field, B_p,* is produced by a large current (10^6 A) that is induced in the plasma by a different, time-varying field generated by coils in the same plane as the torus. The resultant magnetic field lines are helical and serve to confine the plasma. If the plasma were to come into contact with the walls of the containment chamber, the plasma would lose energy and cool down. Furthermore, impurities would be released into the chamber and would severely curtail the operation of the reactor.

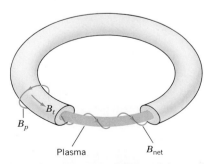

FIGURE 43.16 The plasma in a Tokamak is confined by the combination of magnetic fields. The toriodal field B_t and the poloidal field B_p produce a net field whose lines are helical.

The initial heating of the plasma is accomplished by the induced current mentioned above. Then, beams of high-energy neutral particles (accelerated as ions and then neutralized) are injected into the plasma to deliver about 20 MW, thereby further raising its temperature. Radio frequency coils are also used to heat the plasma.

The 14.1 MeV neutrons from the D–T reaction (Eq. 43.9) are absorbed by a molten lithium "blanket" surrounding the containment chamber. The thermal energy deposited in this blanket can than be used to produce steam for a conventional electrical generator. The tritium produced in the reactions

$$n + {}^7\text{Li} \rightarrow {}^3\text{H} + {}^4\text{He} + n$$
$$n + {}^6\text{Li} \rightarrow {}^3\text{H} + {}^4\text{He}$$

can be extracted and reused.

The **Tokamak fusion test reactor** (TFTR) at Princeton (Fig. 43.17) has operated with a particle density $n = 3 \times 10^{19}$ m^{-3} at a temperature such that $kT = 1.5$ keV, and a confinement time $\tau = 300$ ms. Therefore the product in Lawson's criterion is $n\tau \approx 10^{19}$ s/m^3. In order for such a reactor to produce a 1000-MW electrical output, it would require a plasma temperature such that $kT = 15$ keV and the Lawson product to be $n\tau > 10^{20}$ s/m^3.

In a *driven* fusion reaction, energy is continuously supplied to the plasma. However, in a D–T reaction, 20% of the kinetic energy is carried away by the α particle. These particles may also be used to heat the plasma. If the plasma reaches the **ignition** temperature, it becomes self-sustaining. The ignition mode of operation requires a higher value of $n\tau$, which may be achieved in a few decades.

Inertial Confinement

In the **inertial confinement** approach the fuel is in the form of tiny pellets, of diameter less than 1 mm, that contain a mixture of deuterium and tritium (Fig. 43.18). In the NOVA system at Livermore, California (Fig. 43.19), 0.1 ns pulses from 10 neodymium-doped glass lasers (operating at 1.05 μm) deliver about 200 kJ in 1 ns to each pellet. (This corresponds to a power of 2×10^{14} W, which is greater than the generating capacity of all the stations in the US!) The surface of the pellet vaporizes. As it expands, it sends a shock wave inward, which increases the density of the core by a factor of 10^3 and raises its temperature to over 10^8 K. This occurs within 1.5 ns, before the particles are able to disperse. That is, they are confined by their own inertia. The density of the pellet reaches 10^3 g/cm^3 and its pressure reaches 10^{12} atm (10^{17} Pa)—which is greater than the pressure in the interior of stars. In a sense these are tiny hydrogen bombs. A continuous supply of power would be produced by fusing about 20–50 pellets each second. Charged particles, ions or electrons, may also be used instead of laser beams.

There are several desirable features of fusion power. Deuterium (D) is easily extracted from sea water, where its concentration is 1/6500 of normal hydrogen atoms. Although tritium (T) is scarce and costly ($20 m per kg!), it can be produced by the bombardment of Li by neutrons, as we noted earlier. A "runaway" reaction is not possible because of the low quantity of fuel present at any time. If the magnets or other systems fail, the plasma simply disappears. Radioactive wastes are less of a problem than with fission reactors. Tritium is toxic, but it has a relatively short half-life of 12.3 y. If fusion reactors become viable, we will have tapped the energy source of the stars!

FIGURE 43.17 The Tokamak Test Fusion Reactor at Princeton.

FIGURE 43.18 The tiny pellets used in the inertial confinement approach contain a mixture of deuterium and tritium.

FIGURE 43.19 The NOVA laser fusion reactor at Lawrence Livermore Laboratory, California.

Fuel pellets are bombarded with ions in the Particle Beam Fusion Accelerator II at Sandia National Laboratories. The electrical discharges occur during the firing of a pulse, which delivers 10^{14} W.

CHAPTER 44*

Elementary Particles

The OPAL detector is part of the Large Electron–Positron (LEP) collider at CERN, where these particles are accelerated to 50 GeV in a tunnel 27 km in circumference.

Major Points

1. The classification of elementary particles according to **interaction, spin, strangeness,** and **isospin.**
2. The connection between **symmetry** and **conservation laws.**
3. The **Eightfold Way** and the up, down, and strange **quarks.**
4. **Gauge theory** (Optional).
5. The charmed, top, and bottom quarks (Optional).
6. **Grand unified theory** (Optional).

Ever since Greek times, philosophers and scientists have believed that matter ultimately consists of "elementary" particles with no internal structure. Democritus (ca. 400 B.C.) called these indivisible particles atoms. In 1808, John Dalton proposed that a given element consists of identical atoms and used this model to explain how elements combine to form compounds. A century later, Rutherford showed that atoms do have internal structure: they consist of a tiny nucleus surrounded by electrons. By 1932, it was clear that the nucleus itself is composed of protons and neutrons. As we noted earlier, problems with the conservation laws in β decay were solved by postulating the existence of the neutrino. With the addition of the photon, there were just five elementary particles.

44.1 ANTIMATTER

In 1928, P. A. M. Dirac (Fig. 44.1) incorporated relativity into the framework of quantum mechanics. His theory seemed to imply that free electrons could have negative, as well as positive, energies, where the levels are separated by a gap of $2m_0c^2$ (Fig. 44.2). He suggested that the negative energy states are not normally observed because they are already fully occupied and so the Pauli exclusion principle prevents any downward transitions from the usual positive energy levels. It was noted, however, that a photon of sufficient energy ($> 2m_0c^2 = 1.02$ MeV) should be able to raise a negative-energy electron to a positive level and

FIGURE 44.1 P. A. M. Dirac (1902–1984).

* This chapter has been written as a Special Topic. The optional sections are considerably more challenging but they may give you some idea of recent work in physics.

leave a "hole." The hole would behave like a particle, now called the **positron,** whose mass is equal to that of the electron, but whose charge is $+e$. The positron was detected by C. Anderson in 1932 (Fig. 44.3).

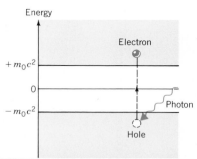

FIGURE 44.2 A photon with sufficient energy can excite an electron with negative energy to a positive energy state.

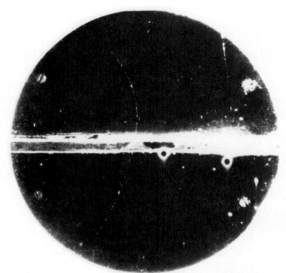

FIGURE 44.3 The photograph taken by C. Anderson that identified the positron. The particle traveled upward and curved to the left in the presence of a magnetic field.

The positron is an example of **antimatter;** it is the antiparticle of the electron. A particle and its antiparticle have certain *intrinsic quantum numbers*, such as charge, of equal magnitude but of opposite signs. We now know that every particle has an antiparticle, although some, like the photon, are their own antiparticles. If sufficient energy is supplied, for example by a photon, a particle–antiparticle pair can be created. Figure 44.4a illustrates this process, called **pair creation.** When a particle meets its antiparticle, they both disappear in a process called **pair annihilation** (Fig. 44.4b). Their mass-energy is used to create photons or other particles. Creation and annihilation are the most dramatic examples of mass-energy equivalence (Fig. 44.5). We have little experience of antimatter since the

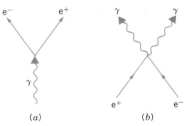

FIGURE 44.4 (a) A photon with sufficient energy can create an electron–positron pair. (b) An electron and positron annihilate each other with the production of two gamma ray photons.

FIGURE 44.5 Electron–positron pairs created by gamma rays (not seen). In the event on the left, the central track is due to an electron knocked out of an atom.

universe seems to consist overwhelmingly of ordinary matter. Any antimatter created naturally or artificially has an extremely short existence because it is annihilated when it comes into contact with ordinary matter.

44.2 EXCHANGE FORCES

The concept of a field was introduced by Faraday and developed by Maxwell. In this view, two charged particles interact via the intermediary of the electromagnetic field: Each particle produces a field, which then acts on the other. Later, **quantum field theory** showed that the energy stored in the field is quantized. Thus two charged particles are said to interact via the exchange of photons emitted and absorbed by the particles. They are called **virtual photons** since the processes occur within 10^{-20} s and are not detected by us. Richard Feynmann (Fig. 44.6) devised a simple way of representing such an interaction between two particles. In Fig. 44.7, called a *Feynmann diagram*, two electrons approach each other, exchange a virtual photon, and change their states.

FIGURE 44.6 Richard Feynmann (1918–1988).

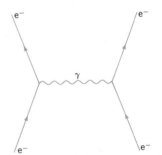

FIGURE 44.7 A Feynmann diagram indicating electron–electron scattering.

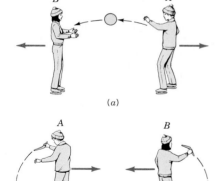

FIGURE 44.8 Two people on a lake. (a) They can produce a repulsive force by throwing a transparent (virtual) ball toward each other. (b) They can produce an effective attractive force by skillfully throwing and catching a transparent (virtual) boomerang.

To see how the exchange of particles gives rise to forces, imagine two skaters A and B at rest on a frozen lake. Skater A throws a transparent ball toward B and recoils backward as a result (Fig. 44.8a). When B catches the ball, she moves in the same direction as the ball, away from A. The ball transfers energy and momentum from A to B and produces an effective repulsive force. We might say that A and B interact via the exchange of invisible (virtual) balls. By skillfully throwing and catching transparent boomerangs, as in Fig. 44.8b, A and B could produce an effective attractive force between them. Of course, these pictures should not be taken literally; the *exchange force* is a purely quantum mechanical effect. A similar (but not identical) type of exchange force arises from the sharing of electrons in the covalent bond—for example, in the hydrogen molecule.

The virtual exchange of particles is made possible by the Heisenberg uncertainty principle, $\Delta E \Delta t > h$. This tells us that the energy of a system can fluctuate by an amount ΔE, provided this fluctuation occurs within an interval $\Delta t < h/\Delta E$. Put differently, a system can "borrow" an energy ΔE beyond that allowed by the conservation of energy, provided it is repaid within Δt. Within this interval, the violation of the conservation of energy will not be detected. The uncertainty in E is at least the mass of the virtual particle, that is, $\Delta E \approx mc^2$. The distance traveled by the particle before it is absorbed is the range, R, of the interaction. If the particle travels at almost the speed of light, the time interval is $\Delta t \approx R/c$. From $\Delta E \Delta t > h$, we see that

$$R \approx \frac{h}{mc}$$

Since the rest mass of the photon is zero, the range of the electromagnetic interaction is infinite.

Field Quanta

The photon is an example of a **field quantum** that *mediates* the *electromagnetic* interaction. The *nuclear* interaction, which binds protons and neutrons in a nucleus, is mediated by the exchange of pions (π^+, π^-, and π^0). For example, a proton can emit a virtual π^+, which is then absorbed by a neutron. In effect, the proton and neutron exchange identities (Fig. 44.9a). Two protons, or two neutrons, exchange neutral pions. The *weak* interaction, which is responsible for β decay, is mediated by the W^+, W^-, and Z^0 particles. This process is shown in Fig. 44.9b where a neutron emits a W^- and is converted into a proton. The W^- subsequently decays into the observed electron and an antineutrino. Finally, the *gravitational* interaction is mediated by the graviton. Field quanta can be made "real" if sufficient energy is supplied—for example, via a collision between particles. These field quanta, with the exception of the graviton, have been detected.

The strength of an interaction may be characterized by a typical time it takes for a reaction or decay to occur. A strong interaction produces fast reactions, whereas a weak interaction is slow. An electromagnetic interaction takes place in approximately 10^{-16} to 10^{-20} s. The time scale for the nuclear interaction is 10^{-23} s, whereas for the weak interaction it is about 10^{-10} s.

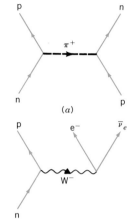

FIGURE 44.9 (*a*) The nuclear interaction: A neutron and a proton interact via the exchange of a positive pion. (*b*) The weak interaction: A neutron emits a W^- particle and is converted to a proton. The W^- subsequently decays into an electron and an antineutrino.

Resonance Particles

After World War II, several new particles were detected in experiments conducted at high altitudes. They were produced as a result of the bombardment of the upper atmosphere by cosmic rays from outer space. At the same time a new breed of **resonance particles,** with very short lifetimes, appeared in particle accelerator experiments. To see how they are detected, consider the scattering of positive pions off protons. At low energies, a pion will be scattered elastically off a proton. As the pion's kinetic energy increases, it may penetrate the coulomb barrier of the proton and interact via the nuclear force. At a certain energy, the pion and proton may combine momentarily to form a new particle, which decays a short time later. The rate at which the new particle is formed is reflected in the rate at which decay products are detected. At still higher energies there is not enough time for the pion and proton to combine, so the reaction rate decreases. The variation in the reaction rate with the available mass-energy has the form of a resonance—which indicates a "preferred" or "natural" energy for the system. For this reason such particles are called resonance particles. Figure 44.10 shows the first resonance particle, which was detected in 1952 by E. Fermi.

Resonance particles have such extremely short lifetimes that they do not leave tracks in detectors. Their existence is inferred from the peak in the energy

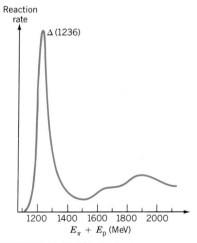

FIGURE 44.10 The first resonance particle was detected by Fermi in 1952 in the scattering of pions by protons. The sharp peak in the reaction rate as a function of the available energy indicates the formation of a short-lived particle.

dependence of the reaction rate. The width of the peak, ΔE, may be used to infer the lifetime of the resonance particle via the Heisenberg uncertainty principle, $\Delta E \Delta t \approx h$. Thus if the measured width of the resonance is $\Delta E = 100$ MeV, the lifetime of the particle is $\Delta t \approx h/\Delta E \approx 10^{-23}$ s.

By the early 1960s, hundreds of resonances had been observed, in addition to several other longer-lived particles. The chaotic situation was reminiscent of chemistry in the 19th century before the work of Mendeleev, when there was no scheme to classify the 60 elements then known. Clearly, the first step in bringing order to the mass of data on elementary particles was to try to find a classification scheme analogous to the periodic table.

44.3 CLASSIFICATION OF PARTICLES

Elementary particles are classified according to several criteria. Table 44.1 is a partial list of relatively stable particles whose lifetimes are greater than 10^{-20} s. It also indicates some decay schemes.

Interactions: Leptons and Hadrons

Particles that take part in the weak and electromagnetic interactions, but not in the nuclear interaction, are called **leptons.** Examples include the electron (e^-), the muon (μ^-), and the neutrino (ν). There are three kinds of neutrino: ν_e, associated with the electron; ν_μ associated with the muon; and ν_τ associated with the τ particle. The lepton family has six members plus their antiparticles.

Reactions involving leptons obey a law of conservation of three **lepton numbers:** $L_e = 1$ for the electron and its associated neutrino, $L_\mu = 1$ for the muon and its neutrino, and $L_\tau = 1$ for the tau particle and its neutrino. A lepton number is an

TABLE 44.1 SOME PARTICLES AND THEIR PROPERTIES

		Symbol	Antiparticle	(MeV) Rest Energy	Mean Life (s)	Typical Decay Modes
Leptons	Electron	e^-	e^+	0.511	Stable	
	Muon	μ^-	μ^+	105.7	2.2×10^{-6}	$\mu^- \rightarrow e^- + \bar{\nu}_e + \nu_\mu$
	Tau	τ^-	τ^+	1784	3×10^{-13} s	$\tau^- \rightarrow e^- + \bar{\nu}_e + \nu_\tau$
	Neutrino	ν_e	$\bar{\nu}_e$	0 (?)	Stable	
		ν_μ	$\bar{\nu}_\mu$	0 (?)	Stable	
		ν_τ	$\bar{\nu}_\tau$	0 (?)	Stable	
Mesons	Pion	π^+	π^-	139.6	2.6×10^{-8}	$\pi^+ \rightarrow \mu^+ + \nu_\mu$
		π^0	Self	135.0	0.83×10^{-16}	$\pi^0 \rightarrow \gamma + \gamma$
	Kaon	K^+	K^-	493.7	1.24×10^{-8}	$K^+ \rightarrow \pi^+ + \pi^0$
		K_S^0	$\bar{K}_S^0$	497.7	0.9×10^{-10}	$K_S^0 \rightarrow \pi^0 + \pi^0$
		K_L^0	$\bar{K}_L^0$	497.7	5.2×10^{-8}	$K_L^0 \rightarrow \pi^0 + \pi^0 + \pi^0$
	Eta	η^0	Self	548.8	7×10^{-19}	$\eta^0 \rightarrow \gamma + \gamma$
Baryons	Proton	p	$\bar{p}$	938.3	Stable	
	Neutron	n	$\bar{n}$	939.6	900	$n \rightarrow p + \bar{e} + \bar{\nu}_e$
	Lambda	Λ^0	$\bar{\Lambda}^0$	1115	2.6×10^{-10}	$\Lambda^0 \rightarrow p^+ + \pi^-$
	Sigma	Σ^+	$\bar{\Sigma}^-$	1189	0.8×10^{-10}	$\Sigma^+ \rightarrow n + \pi^+$
		Σ^0	$\bar{\Sigma}^0$	1192	6×10^{-20}	$\Sigma^0 \rightarrow \Lambda^0 + \gamma$
		Σ^-	$\bar{\Sigma}^+$	1197	1.5×10^{-10}	$\Sigma^- \rightarrow n + \pi^-$
	Xi	Ξ^0	$\bar{\Xi}^0$	1315	2.9×10^{-10}	$\Xi^0 \rightarrow \Lambda^0 + \pi^0$
		Ξ^-	$\bar{\Xi}^+$	1321	1.6×10^{-10}	$\Xi^- \rightarrow \Lambda^0 + \pi^-$
	Omega	Ω^-	Ω^+	1672	0.8×10^{-10}	$\Omega^- \rightarrow \Xi^0 + \pi^-$

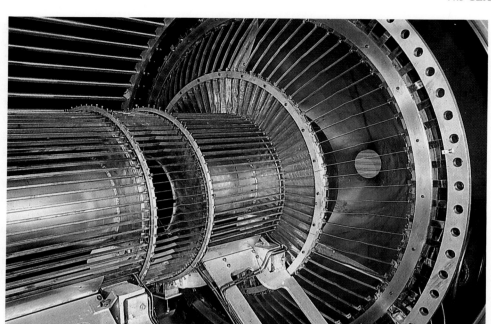

This spectrometer at Los Alamos Laboratories measures electron energies in the b decay of tritium: ${}^3_1H \rightarrow {}^3_2He + e^- + \bar{\nu}_e$. It is designed to determine whether or not the neutrinos have a finite rest mass.

example of an intrinsic quantum number of a particle. The antiparticles are assigned $L = -1$. As an example consider the decay of the muon, which involves two conserved lepton numbers:

$$\mu^- \rightarrow \quad e^- \ + \ \bar{\nu}_e \ + \ \nu_\mu$$
$$L_e: (0) \qquad (+1) + (-1) + (0)$$
$$L_\mu: (1) \qquad (0) \ + \ (0) \ + \ (1)$$

The electron is a stable particle because there is no lighter particle into which it can decay and still conserve electric charge. All the evidence indicates that leptons are truly elementary particles.

Particles that take part in the nuclear interaction, in addition to the weak and electromagnetic interactions, are called **hadrons.** Hadrons that have protons among their final decay products are called **baryons.** Those hadrons that ultimately decay into photons and leptons are called **mesons.** The fact that the proton does not decay into lighter particles is accounted for by the law of **conservation of baryon number,** B. Mesons are assigned $B = 0$, and neutrons and protons have $B = +1$, whereas their antiparticles have $B = -1$. Thus, although a decay such as $p \rightarrow K^+ + \pi^0$ is energetically possible, it is prohibited because the baryon number would decrease by one.

Spin: Fermions and Bosons

Particles with half-integer spin ($\frac{1}{2}\hbar, \frac{3}{2}\hbar, \frac{5}{2}\hbar, \ldots$) are called **fermions** and obey the Pauli exclusion principle (no two identical fermions in a system can have the same set of quantum numbers). Particles with integer spin ($0, 1\hbar, 2\hbar, \ldots$) are called

bosons. Bosons can have identical quantum numbers and tend to congregate in the same level. Leptons and baryons are fermions, whereas mesons and all the field quanta are bosons.

Strangeness

Starting around 1950, a new set of hadrons with unusual properties was discovered. These included K mesons, or kaons, with masses less than that of the proton. The particles with masses greater than that of the proton, such as Λ, Σ, and Ξ, were collectively called "hyperons" (a term no longer used). Consider, for example, the decay of a kaon into two pions: $K^0 \rightarrow \pi^+ + \pi^-$. Since both the kaon and the pion take part in the nuclear interaction, we expect the decay to proceed within about 10^{-23} s. Instead it takes place via the weak interaction in 10^{-10} s, a trillion times longer!

In 1952, A. Pais suggested that during its production via the fast nuclear interaction, a hyperon is always accompanied by a kaon. However, the individual particles can decay only slowly via the weak interaction. Later experiments confirmed this hypothesis of *associated production* of hyperons with kaons. For example:

$$\pi^- + p \rightarrow K^+ + \Sigma^-$$
$$K^- + p \rightarrow K^0 + \Xi^0$$

The phenomenon of associated production and the unusually long lifetimes led to the introduction of a new quantum number. We noted earlier that the stability of the proton follows from the conservation of baryon number. In 1953, M. Gell-Mann (Fig. 44.11) and K. Nishijima independently suggested that both the "strange" stability (the unusually long lifetimes) of the new hadrons and the phenomenon of associated production arise from the conservation of a **strangeness quantum number,** S. The strangeness quantum number is conserved in the nuclear and electromagnetic interactions, but not in the weak interaction. Therefore, a hadron with strangeness ($S \neq 0$) cannot decay into nonstrange particles ($S = 0$) via the strong or electromagnetic interactions; if must do so via the much slower weak interaction. The following reaction shows how strangeness applies to associated production:

$$\pi^- + p \rightarrow K^0 + \Lambda^0$$
$$S: \quad 0 + 0 \quad (+1) + (-1)$$

FIGURE 44.11 Murray Gell-Mann (b.1929).

Even when it is not conserved in a weak interaction, strangeness can change only by one unit at a time ($\Delta S = \pm 1$). For example, Ξ ($S = -2$) does not decay directly to a proton ($S = 0$) but is first converted to a Λ^0 ($S = -1$):

$$\Xi^- \rightarrow \Lambda^0 + \pi^- \quad (\Delta S = +1)$$
$$ \hookrightarrow p + \pi^- \quad (\Delta S = +1)$$

Isospin

It was pointed out in Chapter 43 that the nuclear force is the same for neutrons and protons. Since these particles have nearly the same mass, Heisenberg suggested that if the electromagnetic interaction were "turned off," the neutron and proton could be treated as different states of a single entity, the **nucleon.** By analogy with the intrinsic spin of a particle, he assigned an **isospin,** I, to each particle. A particle with an isospin I has ($2I + 1$) components, I_z, along the z axis in the abstract "isospin space." With $I = \frac{1}{2}$ for the nucleon, there are $2I + 1 = 2$ values for I_z. The proton is assigned the isospin "up" state ($I_z = \frac{1}{2}$) and the neutron is the isospin "down" state ($I_z = -\frac{1}{2}$), as indicated in Fig. 44.12a. The proton and neutron form an isospin doublet.

Gell-Mann used the idea of isospin to identify other families of hadrons with similar properties. In general, a family of hadrons, called a *multiplet*, can be derived from a single "parent" particle with an appropriate value of the isospin I. Each particle in the multiplet is a different state of a single entity. When the isospin vector rotates in the abstract isospin space, it changes the charge on the members of the multiplet—which differ only in the z component of this vector. For example, the three pions (π^+, π^-, π^0) are derived from a single pion whose $I = 1$. The pions form an isospin triplet with $I_z = 0, \pm 1$, as shown in Fig. 44.12b. The existence of several particles was predicted on the basis of isospin classifications.

The introduction and assignment of the new quantum numbers may appear somewhat arbitrary. In fact, the isospin quantum number, the strangeness quantum number, and the baryon number are related to the charge of a particle by a formula proposed by Gell-Mann and Nishijima:

$$Q = I_z + \frac{B + S}{2}$$

We have presented several criteria with which to classify particles. The next step is to find some common ingredient among elementary particles.

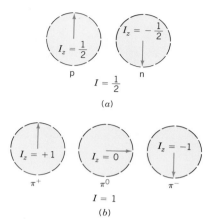

FIGURE 44.12 (a) An isospin doublet, $I = \frac{1}{2}$. The two values of the z component, I_z, represent the proton and the neutron respectively. (b) An isospin triplet, $I = 1$. The three values of I_z represent the three pions.

44.4   SYMMETRY AND CONSERVATION LAWS

Since the dynamical theories of the weak and nuclear interactions are complex and difficult to apply, physicists have turned to other means of extracting information on these interactions. The most powerful has been the investigation of symmetries. We believe that the solution to a problem must reflect the underlying symmetry of a physical system. For example, we used arguments based on spatial symmetry of charge distributions to determine their electric fields from Gauss's law. As we will see, knowledge of the symmetry possessed by a system can lead to profound insights.

The idea of geometrical symmetry is quite familiar. For example if the square in Fig. 44.13 is rotated by 90°, or any multiple thereof, it will appear unchanged. Similarly, reflections in the planes AA' or BB' also leave it unchanged. The square is said to be invariant with respect to a set of rotations and reflections. A circle is more highly symmetric because it is invariant with respect to rotations through any angle. The square has a discrete symmetry, whereas the circle has a continuous symmetry. In general, *a system is said to possess symmetry if it is invariant under a set of operations*.

Symmetry is not confined to physical objects. A mathematical function may retain its form with respect to a set of mathematical operations. For example, if x is replaced by $-x$, the function $y = x^2$ does not change. According to the first postulate of special relativity, all physical laws are invariant with respect to the Lorentz transformation of coordinates. The operations on the square and the Lorentz transformation both involve symmetry operations in physical space and time. Physical theories may contain more abstract symmetries based on different types of mathematical operations.

In 1918, Emmy Noether showed that conservation laws are a consequence of

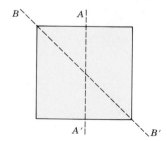

FIGURE 44.13 A square has reflection symmetry about the two lines AA' and BB'—among others.

symmetries possessed by physical laws. For example, the laws of physics are invariant with respect to translation in space. That is, the specific location at which an experiment is conducted does not affect the result—assuming of course that the physical conditions are the same. Noether showed that this translational invariance leads to the conservation of linear momentum. Similarly, angular momentum is conserved because there is no preferred direction in space. That is, conservation of angular momentum is a consequence of the rotational invariance of physical laws. Finally, we believe that the laws of physics are the same today as they were in the distant past, and that they will stay the same into the distant future. This invariance of physical laws with respect to translation in time leads to the conservation of energy.

There is a profound lesson in all this: Whenever physical laws are invariant with respect to a symmetry operation, some quantity is conserved. For example, the nuclear force does not change when neutrons and protons are interchanged. In technical terms, the nuclear force is invariant with respect to rotations of the isospin vector. Consequently, isospin is a conserved quantum number for the nuclear interaction. The knowledge that *every conservation law is associated with some underlying symmetry* has proven to be enormously fruitful in physics. We now go on to see how the investigation of symmetries helped bring order to the chaotic proliferation of "elementary particles."

44.5 THE EIGHTFOLD WAY AND QUARKS

The classification of the mesons and baryons into isospin doublets or triplets showed that each particle in a given multiplet could be generated by rotating the isospin vector. A rotation of the isospin vector is a symmetry operation that changes the charge on the hadrons but leaves the nuclear force invariant.

There is a branch of mathematics called *group theory* that is concerned with sets of symmetry operations that leave a system unchanged. It is useful in dealing with physical systems such as crystals, or with theories that have an underlying symmetry. The operations in a group are performed on vectors (or products of vectors) whose components indicate the possible states of a system. The operations shuffle the components. For example, (a,b,c) can become any one of (b,a,c), (a,c,b), (c,a,b), and so on. Let us see how this is relevant to elementary particles.

The neutron–proton doublet can be represented as the "up" and "down" components of an isospin vector: $(+\frac{1}{2}, -\frac{1}{2})$. The appropriate group for such a two-component vector is called SU(2), which involves three symmetry operations. In the present context, the operations rotate the isospin vector, thereby changing the charge on the nucleon, but leaving the nuclear force unchanged.

In 1961, M. Gell-Mann and Y. Ne'eman independently tried to expand on the idea of isospin multiplets. Consider, for example, the multiplets of the lightest baryons with spin $\frac{1}{2}$, or the mesons with spin 0:

Baryons: $(n; p)$; $(\Xi^+; \Xi^-)$; $(\Sigma^+; \Sigma^-; \Sigma^0)$; Λ^0

Mesons: (K^+, K^0), (K^0, K^-), (π^+, π^-, π^0); η^0

Figure 44.14 shows the baryons plotted with strangeness S on the vertical axis and I_z on the horizontal axis. Could all these eight baryons (or the mesons) be merged into one enlarged family, a "supermultiplet"? The eight baryons would then simply be different manifestations of one fundamental baryon. Given one member of a supermultiplet, a series of symmetry operations should allow one to generate all the others.

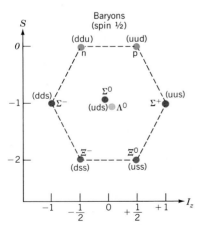

FIGURE 44.14 The octet of spin ½ hadrons. In brackets are the quark assignments.

Gell-Mann and Ne'eman proposed a scheme based on a group called SU(3), for which the fundamental vector has three components. The group SU(3) involves eight symmetry operations that exchange values of charge, Q, baryon number, B, strangeness, S, and isospin, I, within a given supermultiplet. The attractive feature of SU(3) is that it is quite restrictive: It allows only a certain number of multiplets and certain numbers of particles in each multiplet. The product of two basic vectors yields nine combinations (aa, ab, bc, etc.) that split according to their symmetry properties into an octet and a singlet. The product of three vectors yields 27 combinations (abc, cab, abb, etc.) that split into multiplets of size 1, 8, 8, and 10. What is interesting is that the octet can be split further into two doublets, a triplet, and a singlet—which corresponds exactly with the octet structure of the baryons and the mesons listed above. Thus, both the meson and the baryon octet structures arose naturally from the mathematics! This could have been mere coincidence. In any case, the scheme had not yielded any new information up to this point. The real triumph of SU(3) came with the prediction of the Ω^- particle.

A scanner examines particle tracks made in a bubble chamber at Fermilab.

The Ω^- Particle

The decuplet (10 members) from group theory can be split further into a quartet, a triplet, a doublet, and a singlet. By the end of 1963, a set of nine resonance particles with spin $\frac{3}{2}$ appeared to fit into this pattern. This left just one particle ($Q = -1$, $S = -3$) missing (Fig. 44.15). It was named the Ω^- particle. Since the

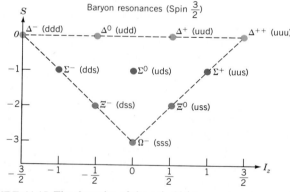

FIGURE 44.15 The decuplet of the spin ½ baryon resonance particles.

average mass of each isospin multiplet differed by about 150 MeV from its neighbors, the mass of the Ω^- was predicted to be about 1680 MeV. The Ω^- was found in February 1964 at Brookhaven, in the scattering of K^- mesons off protons (Fig. 44.16):

$$K^- + p \rightarrow \Omega^- + K^+ + K^0$$
$$\quad\hookrightarrow \Xi^0 + \pi^-$$
$$\qquad\hookrightarrow \Lambda^0 + \qquad\qquad \pi^0$$
$$\qquad\quad\hookrightarrow p + \pi^- \quad \hookrightarrow \gamma \quad + \quad \gamma$$
$$\qquad\qquad\qquad\qquad\quad \hookrightarrow e^+ + e^- \quad \hookrightarrow e^+ + e^-$$

It was astonishing that not only was the Ω^- identified, but even the last two γ rays were nailed down because they both produced electron–positron pairs within the

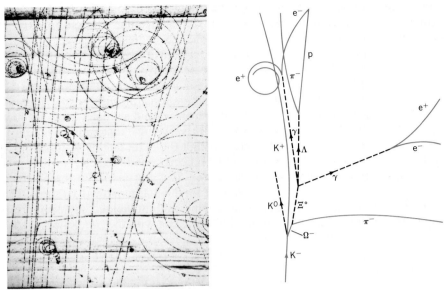

FIGURE 44.16 The bubble chamber photograph in which the Ω^- particle was detected.

detector. The measured mass of the Ω^-, 1675 ± 3 MeV, coincided closely with the predicted value of 1680 MeV.

QUARKS

The obvious question was whether SU(3) is merely a beautiful and useful mathematical scheme or whether it reflects physical reality. It was noted that the particles fit into vectors with 8 or 10 components, but none corresponded to the basic three-component vector of SU(3). In 1964, Gell-Mann and George Zweig independently suggested that the basic vector does indeed represent three particles from which all the others may be constructed: The **up quark** (u), the **down quark** (d), and the **strange quark** (s). Their quantum numbers are shown in Fig. 44.17. The Gell-Mann–Nishijima relation (Eq. 44.1) led to the unexpected conclusion that quarks carry fractional charges: $Q = -e/3$ and $+2e/3$. Mesons are quark–antiquark ($q\bar{q}$) combinations, whereas baryons are qqq combinations, as indicated in Figs. 44.14 and 44.15. Resonance particles are now considered to be excited states of the basic quark combinations.

The prediction and discovery of the Ω^- particle was strong, albeit indirect, evidence for quarks. In general, the reception given the quark model was hostile, especially with regard to the fractional charges. Experimental evidence for quarks first came from the scattering of high-energy electrons off protons at SLAC, in an experiment similar to Rutherford's α particle scattering experiment. The electrons had a de Broglie wavelength of about 5×10^{-17} m, so they could probe deep into the proton. The distribution of the scattered electrons indicated that they were interacting with pointlike centers, which are believed to be the quarks. It was surprising that despite the high energy of the electrons, no quark was ever knocked free. In fact an isolated quark has never been detected.

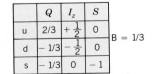

	Q	I_z	S
u	2/3	$+\frac{1}{2}$	0
d	−1/3	$-\frac{1}{2}$	0
s	−1/3	0	−1

B = 1/3

FIGURE 44.17 Some quantum numbers for the u, d, and s quarks. All have $B = \frac{1}{3}$.

Although the three-quark scheme accounted satisfactorily for the known particles and resonances, both experimental and theoretical work suggested that there are more quarks.

44.6 COLOR

Despite the successes of the quark model, there were problems with the assignments of quarks to baryons. The structures of some baryons contained two or three quarks in the same state. For example, the spin $\frac{3}{2}$ baryon Δ^{++} was identified as $u\uparrow u\uparrow u\uparrow$. All three quarks are identical and they appear in the same spin-up state. But quarks have spin $\frac{1}{2}$, which means they are fermions. They should obey the Pauli exclusion principle and thus cannot have an identical set of quantum numbers.

To make the quarks distinguishable particles, an attribute, called **color,** was added by O. W. Greenberg in 1965. Quarks are assigned the primary colors red, blue, or green, while the antiquarks are antired, antiblue, or antigreen. However, these colors cannot be combined arbitrarily. The **color quantum number** represents a new "color charge" on each particle and only "neutral" or colorless (white) particles can be detected. Mesons are composed of color–anticolor combinations, whereas baryons are RBG primary color combinations:

$$\text{Mesons: } q_R\overline{q_R} \text{ or } q_B\overline{q_B} \text{ or } q_G\overline{q_G}; \qquad \text{Baryons: } q_R q_B q_G$$

Neutral mesons are linear superpositions of quark–antiquark configurations in which all three colors are represented. Although the introduction of color charge was based on the Pauli exclusion principle, it turned out to have profound significance for the nuclear interaction.

Wires that make up a drift chamber at SLAC are enclosed in a gas-filled chamber. Electrons released as a result of the ionization caused by a charged particle are detected by the wires. Measurement of the times taken by the electrons to drift to the wires allows the particle's track to be reconstructed.

44.7 GAUGE THEORY (Optional)

One of the most significant advances in recent decades has been the development of what are called gauge theories. In a gauge theory, the ideas of symmetry and invariance are used to extract information on the interactions between particles.

We know that Maxwell's equations are invariant with respect to the Lorentz transformation. Electromagnetic theory possesses other symmetries. For example, if all the positive and negative charges in a system are interchanged, the forces stay the same. As another example, recall that the electric field due to a static distribution of charges may be derived from the potential. One may choose the zero of potential at any convenient value—provided this is done at all points in space—without changing the field. The setting of the zero level is called a choice of gauge. A gauge is a measuring standard, a calibration of the scale by which some quantity is to be measured. (The term is derived from gauge blocks, which serve as length standards in machine shops.)

Local Invariance

The invariance properties of electromagnetic theory cited above are examples of **global symmetry:** *All* the charges had to be interchanged and the zero level of potential had to be the same at *all* points in space. In fact, electromagnetic theory possesses a much more restrictive kind of symmetry in which the setting of the zero of potential can be done arbitrarily at different points in space and time. This is called **local invariance.**

We know that a static electric field can be derived from the electric potential, which depends on the positions of the charges. Similarly, the magnetic field can be derived from a "magnetic" potential that depends on the motion of the charges. Furthermore, a changing magnetic field is associated with an electric field and vice versa. It turns out that if the zero of the electric potential is shifted at a given point, there is an associated change in the magnetic potential that leaves Maxwell's equations unchanged. Therefore the electric and magnetic fields are invariant with respect to the choice of gauge for the electric and magnetic potentials. Electromagnetic theory possesses the simplest kind of local gauge symmetry called U(1). *A theory that displays local invariance is called a **gauge theory.***

The problem can be turned inside out. One might ask what is the theory that satisfies the twin requirements of Lorentz invariance and the simplest, U(1), local gauge symmetry? The results of such a calculation are the equations of electromagnetic theory. Therefore, without knowing anything about the behavior of electric charges, one can derive electromagnetic theory from arguments based solely on symmetry! Thus, instead of searching for details of interactions, it is quite profitable to investigate symmetries and to deduce the interactions and conservation laws from them. For example, it can be shown that the conservation of electric charge is a consequence of the local gauge invariance of electromagnetic theory.

Consider a rubber sheet with the pattern of arrows on a square grid shown in Fig. 44.18a. The pattern will appear the same if the whole sheet rotated by 90° or any multiple thereof. This is a global symmetry: *All* the arrows must be rotated by 90° in the same sense, say clockwise. In the double-arrow pattern of Fig. 44.18b, one could rotate just one double-arrow by 90° and still retain the overall symmetry of the grid. This pattern has local invariance since we may choose different angles (90°, 180°, etc.) at different positions and times. The requirement of local invariance places restrictions on the possible arrow patterns in each square. But something more significant has occurred: The sheet becomes twisted and therefore forces appear between the arrows. In an analogous fashion, the requirement of local invariance gives rise to interactions.

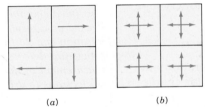

FIGURE 44.18 (a) A pattern with global symmetry. The whole array must be rotated in order for the pattern to appear the same. (b) A pattern with local symmetry. Each pair of arrows may be rotated through 90° or 180° independently of the others.

In quantum mechanics each particle is assigned a wave function. It is a *matter field* that has nothing to do with electric charges. Consider the sheet as the analog of the matter field of a collection of free particles. The operation equivalent to the rotation of the arrows in each square is the choice of phase for the wave function of each particle. The matter field of the free particles has global symmetry: The phase must be shifted by the same amount at all points. The requirement of local gauge invariance on the form of the equation that describes the matter field would mean that the phase of the wave function can be chosen arbitrarily at each point in space and time. The sheet analogy suggests that the price to pay for imposing local invariance is the appearance of an interaction between the particles. Of course, this interaction can also be described as a field. This **gauge field** *compensates for the point to point variation in the phase and maintains the original form of the wave equation.*

The imposition of local invariance on the phase of the electron field requires the introduction of a massless **gauge boson** of spin 1. This is the photon! Thus the electromagnetic field is a gauge field. The quantum field theory is called **quantum electrodynamics** (QED). The general theory of relativity, which deals with gravitation, is also a gauge theory. The corresponding gauge boson is the graviton. The nuclear and weak interactions are also susceptible to similar analysis.

If all the neutrons and protons in a system are interchanged, the nuclear forces stay the same. Put differently, the nuclear interaction has global invariance with respect to rotations of the isospin vector. In 1954, C. N. Yang and R. Mills developed a gauge theory by imposing local invariance [based on SU(2) symmetry] on the rotations of the isospin vector. That is, neutrons or protons were allowed to appear at any point. Their theory required the introduction of three massless gauge bosons of spin 1: one neutral and two charged. The theory was flawed since there is no evidence that massless charged particles exist. Despite this apparent dead end, however, some theorists continued to develop gauge theory for its esthetic beauty.

The weak interaction is responsible for the β decay of the neutron: $n \rightarrow p + e^- + \bar{\nu}_e$ in which the neutron is changed into a proton with the emission of an electron and an antineutrino. If the concept of isospin is extended to the weak interaction, the electron and its neutrino, and the muon and its neutrino, can be treated as two components of a "weak isospin" vector. One can impose local invariance [again SU(2) symmetry] on the rotations of this vector, which means that the electron and its neutrino can be exchanged at any point. In 1958, S. Weinberg, and independently A. Salam and J. Ward, showed that such a gauge theory of the weak interaction involves three massless gauge bosons of spin 1: one neutral and two charged.

44.8 THE ELECTROWEAK INTERACTION

It has long been a dream of physicists to show that the four basic interactions—gravitation, electromagnetism, the weak, and the nuclear—are merely different manifestations of a single basic interaction. S. Weinberg and A. Salam (Fig. 44.19) investigated the possibility of unifying the electromagnetic and weak interactions. They noted that although these interactions differ greatly in strength, range, and other features, all the gauge particles (the photon, W^+, W^-, and Z^0) are spin 1 bosons. Could this be a clue that they belong to a single family?

The electromagnetic interaction has U(1) gauge symmetry, which allows the potential to be chosen at any point. The weak interaction has SU(2) symmetry, which allows the electron and its neutrino to be exchanged at any point. The overall gauge

FIGURE 44.19 S. Weinberg (left) and A. Salam.

symmetry that encompasses both interactions is designated as SU(2) × U(1). With this symmetry, the imposition of local invariance on the exchange of electrons and neutrinos required the introduction of four gauge bosons of spin 1. They were identified as the photon, W^+, W^-, Z^0.

If the electromagnetic and weak interactions are really different aspects of a single interaction, their intrinsic strengths should be the same. However, the weak interaction has an extremely short range (less than 10^{-19} m) and appears to be much weaker than the electromagnetic interaction. The symmetry between these interactions is hidden. One can explain the apparent weakness and extremely short range of the weak interaction if the W and Z particles have large masses. Unfortunately, in a theory of the Yang–Mills type, the gauge bosons are massless. A way had to be found to give the W and Z particles masses without destroying the whole theory.

Broken Symmetry

Consider the ball at the center of the potential energy function in Fig. 44.20. The function has rotational symmetry about the central axis, so all horizontal directions are equivalent. The ball is in unstable equilibrium because a slight disturbance will cause it to roll one way or another. The final state of the system does not display the underlying symmetry of the potential energy function; the symmetry is **hidden.** This is an example of **spontaneous symmetry breaking.**

The concept of hidden symmetry appears in other contexts. For example, the equation that describes the magnetic interaction between atoms in a crystal has no preferred direction. At high temperature this symmetry is manifested by the random orientation of the magnetic moments of the atoms. However, below the Curie temperature the magnetic moments align to form large magnetic domains. A tiny instability forces the system into an equilibrium state with less symmetry than the equations. The symmetric state becomes unstable and the underlying symmetry is hidden. A similar example of broken symmetry occurs in the formation of the superconducting state below the transition temperature.

P. Higgs showed that the introduction of another spin 0 particle, now called the **Higgs boson,** would break the symme-

try between the photon and the other gauge bosons and allow the W^+, W^-, Z^0 to have mass.

The electromagnetic and weak interactions behave differently at low energies and when the distance between particles is large. The symmetry between them is hidden. However, when the available energy is greater than 100 GeV, the photon, W^+, W^-, and Z^0 may be created with equal ease. At higher energies and at distances less than about 10^{-19} m, the electromagnetic and weak interactions have the same strength; their symmetry is restored.

44.9   THE NEW QUARKS

In 1961, S. Glashow (Fig. 44.21) predicted certain decays of K mesons, governed by the weak interaction in which strangeness changes. However, they were not observed. In 1970, S. Glashow, J. Iliopoulos, and L. Maiani developed the Weinberg–Salam theory, keeping in mind this puzzle.

FIGURE 44.21 S. Glashow.

All the evidence indicated that quarks and leptons are truly elementary particles. At that time, there were four leptons and three quarks. The leptons appeared as two doublets (e, ν_e) and (μ, ν_μ), but the quarks appeared as a doublet (u, d) and a singlet (s). Glashow and his co-workers found that certain troublesome anomalies (expressions with infinite values) in the Weinberg–Salam theory cancel only if the sum of the charges on all the fermions is zero. The charges on the electron and the muon sum to $-2e$, but the sum on the up, down, and strange quarks is zero. Their solution involved two innovations.

First, they were guided by the esthetic beauty of a parallel structure of quarks and leptons. They suggested that there is a fourth "c" quark with charge $+2e/3$ that would form an SU(2) doublet with the s quark: (c, s). The absence of the kaon decays predicted earlier by Glashow was accounted for by the conservation of a new **charm quantum number, C.** For the **charmed quark,** $C = +1$, $s = \frac{1}{2}$, $Q = 2e/3$, $S = 0$, and $I = 0$. Like strangeness, the charm quantum number is not conserved in the weak interaction. Second, the problem concerning the anoma-

FIGURE 44.20 A particle in unstable equilibrium in a symmetric potential well. A small instability will cause the particle to roll one way or the other. The final state of the system will not display the underlying symmetry. This is an example of hidden symmetry.

lies would be solved if the four quarks appeared in three varieties—which are just the three colors!

Glashow, Iliopoulos, and Maiani had shown that the extension of gauge theory to the weak interactions of the hadrons *requires* the introduction of both the charm quantum number and color. The theory revealed a deep symmetry between quarks and leptons: Not only do quarks and leptons appear as doublets, but there should be the same number of doublets.

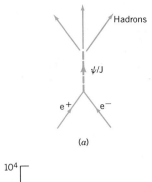

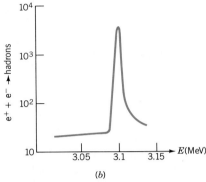

FIGURE 44.23 The sharp increase in the number of hadrons produced in electron–positron collisions observed by B. Richter indicated the formation of the J/Ψ resonance particle.

(a)

(b)

FIGURE 44.22 The J/Ψ particle was discovered independently by (a) B. Richter and (b) S. Ting.

In November 1974, Samuel Ting at Brookhaven and Burton Richter at SLAC (Fig. 44.22) independently discovered a relatively long-lived (10^{-20} s) resonance particle, now called J/Ψ. Richter's data, shown in Figure 44.23, were obtained by observing hadrons produced in electron–electron collisions. It was a spin 1 meson of mass 3.1 GeV, with quantum numbers

$B = 0$, $I = 0$, and $S = 0$, that could not be fitted into the three-quark scheme. It soon became clear that the quark structure of J/Ψ is $c\bar{c}$, which means that $C = 0$ and the new particle had "hidden" charm. Later, other particles with charm ($C \neq 0$), such as the mesons D^0 ($c\bar{u}$), D^- ($\bar{c}d$) and D^+ ($c\bar{d}$), were found. Charmed particles are always created in pairs; one with a c quark, the other with a $\bar{c}$ antiquark.

Three years later, in 1977, L. Lederman at Fermilab found another relatively long-lived resonance particle, now called the Υ (upsilon) particle with a mass of 8.5 GeV. This particle required the introduction of yet another **bottom quark** (b) and was identified as a $b\bar{b}$ combination. The notion of lepton–quark symmetry immediately implied that there should be a **top quark** (t) and two more leptons! The table of quarks shown in Fig. 44.24 was now complete. The heavy τ lepton and its associated

	Q	C	B'	T
c	2/3	1	0	0
b	−1/3	0	1	0
t	2/3	0	0	1

B' : Bottomness
T : Topness
C : Charm

FIGURE 44.24 Quantum numbers for the c, b, and t quarks. All have $S = 0$.

neutrino have been detected, but the top quark has not yet been detected.

The dizzying list of elementary particles (not counting the antiparticles) is now reduced to quarks and leptons. There are six leptons (e, ν_e, μ, ν_μ, τ, ν_τ), whereas quarks come in six **flavors** (u, d, s, c, b, t) and three **colors** (R, G, B).

44.10 QUANTUM CHROMODYNAMICS

Given that hadrons consist of two or three quarks, what holds the quarks together? In 1973, S. Weinberg and others proposed that the interaction between quarks is governed by gauge fields associated with the color charge or quantum number. The "old" SU(3) symmetry interchanged the flavor charges on the u, d, and s quarks. It was a broken symmetry because the strong force was not invariant with respect to all the operations.

The **gauge theory of color** incorporates the "new" and exact SU(3) symmetry of the color force. It involves the exchange of eight massless vector bosons, called **gluons,** between the three colors. Recall that for a baryon to appear colorless, all three colors must be represented at any given time. The imposition of local invariance on the color interaction means that each quark may have any color. Therefore, in order for the baryon as a whole to remain colorless, the gauge gluons must also carry color charges. In fact, gluons carry color–anticolor combinations, but have no flavor charge. Only the first six of the following gluons produce changes in color:

$$g_{B\bar{R}}; \; g_{B\bar{G}}; \; g_{G\bar{R}}; \; g_{G\bar{B}}; \; g_{R\bar{G}}; \; g_{R\bar{B}}; \; g_{01}; \; g_{02}$$

where g_{01} and g_{02} are combinations of $g_{R\bar{R}}$, $g_{G\bar{G}}$ and $g_{B\bar{B}}$. Figure 44.25 shows how the colored quarks u_R and d_B interact via the exchange of a gluon $g_{R\bar{B}}$ moving from left to right or a gluon $g_{B\bar{R}}$ moving from right to left.

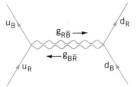

FIGURE 44.25 An interaction between quarks mediated by a gluon.

The strong (color) interaction ignores flavor (it has no taste!) and couples quarks to quarks. The quarks exchange colors, but not flavors, via the gluons. The weak interaction ignores color (it is colorblind!) and, in addition, couples quarks to leptons. It changes the flavor of quarks via the W and Z bosons. The W particle carries the "flavor" charge and interchanges e and ν_e. The W also changes the flavors of the quarks without changing their colors. In the β decay of a neutron: n → p + e + $\bar{\nu}_e$, a neutron changes to a proton with emission of an e and $\bar{\nu}_e$. We can now describe this as a d quark changing to a u quark via the emission of a W particle, which subsequently decays to e + $\bar{\nu}_e$, as in Fig. 44.26. The W of mass 81 GeV was discovered by C. Rubbia in 1982. The Z^0 with a mass of 93 GeV was detected a few months later, in 1983.

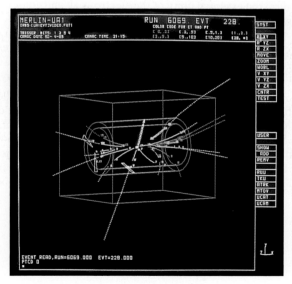

The decay of a W particle into an electron (blue track directed downward) and a neutrino (the thicker blue track directed upward, which was reconstructed after the event). The red and yellow tracks are due to other particles. The white bars indicate where the electron has been registered by the cylindrical electromagnetic calorimeter that measure particle energies. The W was created in a proton–antiproton collision.

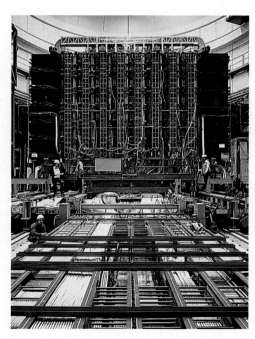

One wall of the giant UA1 detector that was used to detect the W and Z particles.

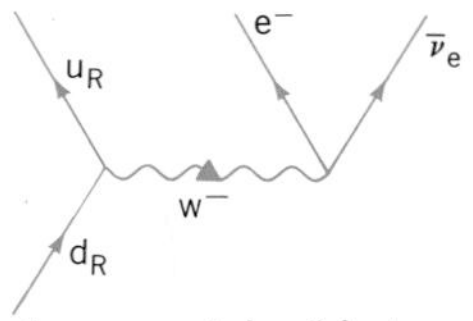

FIGURE 44.26 β^- decay explained in terms of the underlying quark structure of the neutron and proton.

There is a significant difference between the photon and the other gauge bosons. The photon mediates the interaction between electric charges but is itself neutral. On the other hand, the W carries the flavor charge, and the gluons carry the color charge. As a consequence there is interaction between gluons.

What we have been calling the "strong" nuclear force between colorless hadrons is really only a pale remnant of the color force between quarks. A similar situation arises in the electromagnetic interaction. Van der Waals forces between two neutral molecules arise from the induced polarization of the charges. It is only a residual effect, very much weaker than the interaction between net charges as given by Coulomb's law.

Evidence for quarks and gluons comes from reactions such as $e^+ + e^- \rightarrow$ hadrons. At moderate energies, the emerging hadrons are distributed in various directions. At high energy they sometimes spray out in two narrow cones, called *jets*, that emerge in opposite directions. This pattern could be interpreted as being due to the production of a quark–antiquark pair each of which subsequently decays. In 1979, at very high energies, three jets were observed. These might be interpreted as the products of a quark–antiquark pair and a gluon.

44.11 GRAND UNIFIED THEORY

In 1974, H. Georgi and S. Glashow proposed a theory to unify the electroweak and strong interactions. Earlier work had demonstrated the parallel structure of quarks and leptons. Georgi and Glashow took a further step by deciding to place quarks and leptons on an equal footing. This required a more complex symmetry that could encompass both the SU(3) of the color interaction and the SU(2) × U(1) of the electroweak interaction. In the simplest symmetry group, called SU(5), the fundamental vector has five components. Georgi and Glashow chose three quarks and two leptons: d_R, d_B, d_G, e^+, and ν. The gauge theory required the introduction of 24 gauge particles, which split into two groups of 12. Twelve of these are already familiar: four for the electroweak interaction (γ, W^+, W^-, and Z) and eight gluons for the color interaction. The remaining 12, collectively called the Xs, interchange quarks and antiquarks and, most significantly, quarks and leptons. If the SU(5) symmetry were exact, the strengths of the electroweak and color interac-

tions would be identical. Since they are different, the symmetry is hidden.

The masses of the X particles are estimated to be about 10^{15} GeV! The range of the interaction between quarks and leptons is therefore astonishingly small. A virtual X can move only about 10^{-31} m. This energy, or distance, defines the **unification scale**. As the energy increases, the weak force stays roughly constant, and the electromagnetic force increases, but the strong force decreases. The three forces become equally strong at these high energies, or tiny distances, since the difference in mass between the gauge bosons (γ, the Ws, the Xs) would not be important. They could all be created with equal ease. In this **grand unified theory** (GUT) quarks and leptons would be equivalent and could be freely interchanged at the unification scale.

An interesting consequence of GUT is that in order for the exchanges associated with the symmetry operations to be consistent with the conservation of charge, charge must also be quantized in the basic unit $e/3$. It was noted earlier that the gauge invariance of Maxwell's equations leads to the conservation of charge. Now we see that the gauge invariance of GUT explains why it is quantized!

The enormous mass of the Xs makes it impossible for us to detect such a particle. However, there is an indirect test of GUT. Since an X particle changes a quark into a lepton, this means that a proton could decay, and therefore both baryon number and lepton number would not be conserved. The large masses of the Xs implies that the lifetime of a proton due to a decay such as $p \rightarrow e^+ + \gamma$ would be about 10^{32} y. This is an incredibly long time, far greater than the age of the universe. Nonetheless, the present experimental limit is 10^{31} y, which comes pretty close. If the decay of the proton is detected, it will confirm the essential unity of the electroweak and color interactions.

Great strides have been made toward the unification of the forces and in the search for the true elementary particles. We have glimpsed at the astonishing power of gauge theories. If nothing else, the last few years have established that "symmetry rules the world." Yet profound questions remain unanswered. We need only two quarks (u, d) and two leptons (e, ν_e) to construct ordinary matter. To paraphrase a comment made by I. I. Rabi in another context: Who ordered the extra quarks and leptons? What purpose do they serve?

APPENDIX A

SI Units

The *base units* in the Système International are

Meter (m): The meter is the distance traveled by light in vacuum in a time interval of 1/299 792 458 of a second. (1983)

Kilogram (kg): The mass of the international prototype of the kilogram. (1889)

Second (s): The second is the duration of 9 192 631 770 periods of the radiation corresponding to the transition between the two hyperfine levels of the ground state of the cesium-133 atom. (1967)

Ampere (A): The ampere is that constant current which, if maintained in two straight parallel conductors of infinite length, of negligible circular cross section, and placed 1 m apart in vacuum, would produce between these conductors a force equal to 2×10^{-7} newton per meter of length. (1948)

Kelvin (K): The kelvin, the unit of thermodynamics temperature, is the fraction 1/273.16 of the thermodynamics temperature of the triple point of water. (1957)

Candela (Cd): The candela is the luminous intensity, in a given direction, of a source that emits monochromatic radiation of frequency 540×10^{12} hertz and that has a radiant intensity in that direction of 1/683 watt per steradian. (1979)

Mole (mol): The mole is the amount of a substance that contains as many elementary entities as there are atoms in 0.012 kilogram of carbon 12. (1971). The entities may be atoms, molecules, ions, electrons, or other particles.

DERIVED UNITS WITH SPECIAL NAMES

Quantity	Derived Unit	Name
Activity	1 decay/s	Bequerel (Bq)
Capacitance	C/V	farad (F)
Charge	A.s	coulomb (C)
Electric Potential; EMF	J/C	volt (V)
Energy, work	N.m	joule (J)
Force	$kg.m/s^2$	newton (N)
Frequency	l/s	hertz (Hz)
Inductance	V.s/A	henry (H)
Magnetic flux density	Wb/m^2	tesla (T)
Magnetic flux	V.s	weber (Wb)
Power	J/s	watt (W)
Pressure	N/m^2	pascal (Pa)
Resistance	V/A	ohm (Ω)

Mathematics Review

ALGEBRA

Exponents

$$x^m x^n = x^{m+n} \qquad x^{\frac{1}{n}} = \sqrt[n]{x}$$

$$\frac{x^m}{x^n} = x^{m-n} \qquad (x^m)^n = x^{mn}$$

Quadratic Equation

The roots of the quadratic equation

$$ax^2 + bx + c = 0$$

are given by

$$x = \frac{-b \pm \sqrt{b^2 - 4ac}}{2a}$$

If $b^2 < 4ac$, the roots are not real.

Straight Line

The equation of a straight line is

$$y = mx + b$$

where b is the *intercept* on the y axis and m is the *slope* given by

$$m = \frac{y_2 - y_1}{x_2 - x_1} = \frac{\Delta y}{\Delta x}$$

Simultaneous Equations

Suppose we are asked to obtain the solution of the following two equations:

$$3x + 4y = 2 \qquad\qquad\qquad (i)$$

$$5x - 2y = 12 \qquad\qquad\qquad (ii)$$

In one approach, we first solve for x in (i),

$$x = \frac{2 - 4y}{3}$$

and substitute this into (ii):

$$\frac{5(2 - 4y)}{3 - 2y} = 12$$

Thus $y = -1$. Using this in either (i) or (ii) yields $x = 2$.

In the second approach, we eliminate one variable as follows. Multiply (ii) by 2 and add it to (i). This yields $13x = 26$, which means $x = 2$ as before.

Logarithms

If

$$x = a^y$$

then

$$y = \log_a x$$

The quantity y is the logarithm of x to the *base a*. In *common* logarithms $a = 10$ and is written as $\log_{10} x$ or simply $\log x$. In *natural* logarithms $a = e = 2.71828 \ldots$, and is written $\log_e x$ or $\ln x$. (Note that $\ln e = 1$.)

$$\log(AB) = \log A + \log B; \qquad \log(A/B) = \log A - \log B$$

$$\log(A^n) = n \log A$$

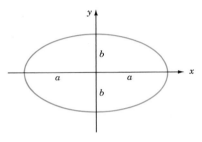

FIGURE A.

GEOMETRY

Triangle: Area $= \frac{1}{2}$ base x height, $\quad A = \frac{1}{2} bh$

Circle: Circumference: $C = 2\pi r$

Area: $A = \pi r^2$

Sphere: Surface area: $A = 4\pi r^2$

Volume: $V = \dfrac{4}{3} \pi r^3$

The equation of a circle of radius R whose center is at the origin is

(Circle) $\qquad\qquad x^2 + y^2 = R^2$

The equation of the ellipse in Fig. A, is

(Ellipse) $\qquad\qquad \dfrac{x^2}{a^2} + \dfrac{y^2}{b^2} = 1$

where $2a$ is the length of the *major* axis and $2b$ is the length of the *minor* axis.

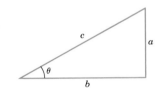

FIGURE B.

TRIGONOMETRY

Using the right-angled triangle in Fig. B, the definitions of the basic trigonometric functions are

$$\sin\theta = \frac{\text{opposite}}{\text{hypotenuse}} = \frac{a}{c}; \qquad \csc\theta = \frac{1}{\sin\theta}$$

$$\cos\theta = \frac{\text{adjacent}}{\text{hypotenuse}} = \frac{b}{c}; \qquad \sec\theta = \frac{1}{\cos\theta}$$

$$\tan\theta = \frac{\text{opposite}}{\text{adjacent}} = \frac{a}{b}; \qquad \cot\theta = \frac{1}{\tan\theta}$$

According to the Pythagorean theorem $c^2 = a^2 + b^2$, thus $\cos^2\theta + \sin^2\theta = 1$.

Using Fig. C, we can state the following two relationships:

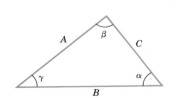

FIGURE C.

Law of cosines $\quad C^2 = A^2 + B^2 - 2AB \cos\gamma$

Law of sines $\quad \dfrac{\sin\alpha}{A} = \dfrac{\sin\beta}{B} = \dfrac{\sin\gamma}{C}$

Some Trigonometric Identities

$$\sin^2\theta + \cos^2\theta = 1 \qquad \sec^2\theta = 1 + \tan^2\theta$$

$$\sin 2\theta = 2 \sin\theta \cos\theta \qquad \cos 2\theta = \cos^2\theta - \sin^2\theta$$

$$= 2 \cos^2\theta - 1$$

$$= 1 - 2 \sin^2\theta$$

$$\tan 2\theta = \frac{2 \tan\theta}{1 - \tan^2\theta}; \qquad \tan\theta = \sqrt{\frac{1 - \cos 2\theta}{1 + \cos 2\theta}}$$

$$\sin(A \pm B) = \sin A \cos B \pm \cos A \sin B$$

$$\cos(A \pm B) = \cos A \cos B \mp \sin A \sin B$$

$$\sin A \pm \sin B = 2 \sin\frac{(A \pm B)}{2} \cos\frac{(A \mp B)}{2}$$

$$\cos A + \cos B = 2 \cos\frac{(A + B)}{2} \cos\frac{(A - B)}{2}$$

$$\cos A - \cos B = 2 \sin\frac{(A + B)}{2} \sin\frac{(B - A)}{2}$$

$$\sin A \cos B = \frac{1}{2}[\sin(A - B) + \sin(A + B)]$$

$$\sin A \cos B = \frac{1}{2}[\cos(A - B) - \cos(A + B)]$$

$$\cos A \cos B = \frac{1}{2}[\cos(A - B) + \cos(A + B)]$$

SERIES EXPANSIONS

$$(a + b)^n = a^n + \frac{n}{1!}a^{n-1}b + \frac{n(n-1)}{2!}a^{n-2}b^2 + \ldots \quad \text{(Binomial)}$$

$$(1 + x)^n = 1 + nx + \frac{n(n-1)}{2!}x^2 + \ldots \quad \text{(Binomial)}$$

$$e^x = 1 + x + \frac{x^2}{2!} + \frac{x^3}{3!} + \cdots$$

$$\ln(1 \pm x) = \pm x - \frac{x^2}{2} \pm \frac{x^3}{3} - \cdots \quad |x| < 1$$

$$\left. \begin{array}{l} \sin x = x - \dfrac{x^3}{3!} + \dfrac{x^5}{5!} - \cdots \\[2mm] \cos x = 1 - \dfrac{x^2}{2!} + \dfrac{x^4}{4!} - \cdots \\[2mm] \tan x = x + \dfrac{x^3}{3} + \dfrac{2x^5}{15} + \cdots \quad |x| < \pi/2 \end{array} \right] \quad x \text{ in radians}$$

APPENDIX C

Calculus Review

DIFFERENTIAL CALCULUS

Product of two functions:

$$\frac{d(uv)}{dx} = u\frac{dv}{dx} + v\frac{du}{dx}$$

Division of two functions:

$$\frac{d}{dx}\left(\frac{u}{v}\right) = \frac{u\frac{dv}{dx} - v\frac{dv}{dx}}{v^2}$$

Chain rule:

Given a function $f(u)$, where u itself is a function of x, then

$$\frac{df}{dx} = \frac{df}{du} \cdot \frac{du}{dx}$$

For example, $\dfrac{d(\sin u)}{dx} = \cos u \cdot \dfrac{du}{dx}$

Some derivatives*

$$\frac{d}{dx}(ax^n) = nax^{n-1}; \qquad \frac{d}{dx}(e^{ax}) = ae^{ax}$$

$$\frac{d}{dx}(\sin ax) = a\cos ax; \qquad \frac{d}{dx}(\cos ax) = -a\sin ax$$

$$\frac{d}{dx}(\tan ax) = a\sec^2 ax; \qquad \frac{d}{dx}(\cot ax) = -a\csc^2 ax$$

$$\frac{d}{dx}(\sec x) = \tan x \sec x; \qquad \frac{d}{dx}(\csc x) = -\cot x \csc x$$

$$\frac{d}{dx}(\ln ax) = \frac{1}{x}$$

INTEGRAL CALCULUS

Integration by parts:

$$\int u\left(\frac{dv}{dx}\right) dx = uv - \int v\left(\frac{du}{dx}\right) dx$$

Some Integrals

(An arbitrary constant may be added to each integral.)*

* For trigonometric functions, x is in radians

$$\int x^n \, dx = \frac{x^{n+1}}{(n+1)} \quad (n \neq -1)$$

$$\int \frac{dx}{x} = \ln x$$

$$\int \frac{dx}{a+bx} = \frac{1}{b} \ln(a+bx)$$

$$\int \frac{dx}{(a+bx)^2} = -\frac{1}{b(a+bx)}$$

$$\int \frac{dx}{a^2+x^2} = \frac{1}{a} \tan^{-1}\left(\frac{x}{a}\right)$$

$$\int \frac{dx}{x^2-a^2} = \frac{1}{2a} \ln\left(\frac{x-a}{x+a}\right) \quad (x^2 > a^2)$$

$$\int \frac{dx}{a^2-x^2} = \frac{1}{2a} \ln\left(\frac{a+x}{a-x}\right) \quad (x^2 < a^2)$$

$$\int \frac{x \, dx}{a^2 \pm x^2} = \pm \tfrac{1}{2} \ln(a^2 \pm x^2)$$

$$\int \frac{dx}{\sqrt{a^2-x^2}} = \sin^{-1}\left(\frac{x}{a}\right)$$

$$= -\cos^{-1}\left(\frac{x}{a}\right) \quad (x^2 < a^2)$$

$$\int \frac{dx}{\sqrt{x^2 \pm a^2}} = \ln[x + \sqrt{x^2 \pm a^2}]$$

$$\int \frac{x \, dx}{\sqrt{a^2-x^2}} = -\sqrt{a^2-x^2}$$

$$\int \frac{x \, dx}{\sqrt{x^2 \pm a^2}} = \sqrt{x^2 \pm a^2}$$

$$\int \frac{dx}{(x^2+a^2)^{3/2}} = \frac{x}{a^2(x^2+a^2)^{1/2}}$$

$$\int \frac{x \, dx}{(x^2+a^2)^{3/2}} = -\frac{1}{(x^2+a^2)^{1/2}}$$

$$\int x\sqrt{x^2 \pm a^2} \, dx = \tfrac{1}{3}(x^2 \pm a^2)^{3/2}$$

$$\int e^{ax} \, dx = \frac{1}{a} e^{ax}$$

$$\int x e^{ax} \, dx = (ax-1)\frac{e^{ax}}{a^2}$$

$$\int x^2 e^{-ax} \, dx = -\frac{1}{a^3}(a^2x^2 + 2ax + 2)e^{-ax}$$

$$\int \ln(ax) \, dx = x \ln(ax) - x$$

$$\int \sin(ax) \, dx = -\frac{1}{a} \cos(ax)$$

$$\int \cos(ax) \, dx = \frac{1}{a} \sin(ax)$$

$$\int \tan(ax) \, dx = \frac{1}{a} \ln[\sec(ax)]$$

$$\int \cot(ax) \, dx = \frac{1}{a} \ln[\sin(ax)]$$

$$\int \sec(ax) \, dx = \frac{1}{a} \ln[\sec(ax) + \tan(ax)]$$

$$\int \csc(ax) \, dx = \frac{1}{a} \ln[\csc(ax) - \cot(ax)]$$

$$\int \sin^2(ax) \, dx = \frac{x}{2} - \frac{\sin(2ax)}{4a}$$

$$\int \cos^2(ax) \, dx = \frac{x}{2} + \frac{\sin(2ax)}{4a}$$

$$\int \frac{1}{\sin^2(ax)} \, dx = -\frac{1}{a} \cot(ax)$$

$$\int \frac{1}{\cos^2(ax)} \, dx = \frac{1}{a} \tan(ax)$$

$$\int \tan^2(ax) \, dx = \frac{1}{a} \tan(ax) - x$$

$$\int \cot^2(ax) \, dx = -\frac{1}{a} \cot(ax) - x$$

$$I_0 = \int_0^\infty e^{-\alpha x^2} \, dx = \frac{1}{2}\sqrt{\frac{\pi}{\alpha}} \quad \text{(Gauss probability integral)}$$

$$I_1 = \int_0^\infty x e^{-\alpha x^2} \, dx = \frac{1}{2\alpha}$$

$$I_2 = \int_0^\infty x^2 e^{-\alpha x^2} \, dx = -\frac{dI_0}{d\alpha} = \tfrac{1}{4}\sqrt{\frac{\pi}{\alpha^3}}$$

$$I_3 = \int_0^\infty x^3 e^{-\alpha x^2} \, dx = -\frac{dI_1}{d\alpha} = \frac{1}{2\alpha^2}$$

$$\vdots$$

$$I_{2n} = (-1)^n \frac{d^n I_0}{d\alpha^n}$$

$$I_{2n+1} = (-1)^n \frac{d^n I_1}{d\alpha^n}$$

APPENDIX D: PERIODIC TABLE OF THE ELEMENTS

Transition elements

Key:

Symbol	C 6
Atomic mass*	12.011
	$2p^2$

Atomic number (6), Electron configuration ($2p^2$)

Main Table

Group I	Group II	Transition → (III–II)										Group III	Group IV	Group V	Group VI	Group VII	Group O
H 1 1.0079 $1s^1$																	**He** 2 4.0026 $1s^2$
Li 3 6.94 $2s^1$	**Be** 4 9.012 $2s^2$											**B** 5 10.81 $2p^1$	**C** 6 12.011 $2p^2$	**N** 7 14.007 $2p^3$	**O** 8 15.999 $2p^4$	**F** 9 18.998 $2p^5$	**Ne** 10 20.18 $2p^6$
Na 11 22.990 $3s^1$	**Mg** 12 24.305 $3s^2$											**Al** 13 26.982 $3p^1$	**Si** 14 28.086 $3p^2$	**P** 15 30.974 $3p^3$	**S** 16 32.06 $3p^4$	**Cl** 17 35.453 $3p^5$	**Ar** 18 39.948 $3p^6$
K 19 39.098 $4s^1$	**Ca** 20 40.08 $4s^2$	**Sc** 21 44.956 $3d^1 4s^2$	**Ti** 22 47.90 $3d^2 4s^2$	**V** 23 50.94 $3d^3 4s^2$	**Cr** 24 51.996 $3d^5 4s^1$	**Mn** 25 54.938 $3d^5 4s^2$	**Fe** 26 55.847 $3d^6 4s^2$	**Co** 27 58.933 $3d^7 4s^2$	**Ni** 28 58.71 $3d^8 4s^2$	**Cu** 29 63.546 $3d^{10} 4s^1$	**Zn** 30 65.38 $3d^{10} 4s^2$	**Ga** 31 69.72 $4p^1$	**Ge** 32 72.59 $4p^2$	**As** 33 74.922 $4p^3$	**Se** 34 78.96 $4p^4$	**Br** 35 79.904 $4p^5$	**Kr** 36 83.80 $4p^6$
Rb 37 85.467 $5s^2$	**Sr** 38 87.62 $5s^2$	**Y** 39 88.906 $4d^1 5s^2$	**Zr** 40 91.22 $4d^2 5s^2$	**Nb** 41 92.906 $4d^4 5s^1$	**Mo** 42 95.94 $4d^5 4s^1$	**Tc** 43 98.9 $4d^5 5s^2$	**Ru** 44 101.07 $4d^7 5s^1$	**Rh** 45 102.906 $4d^8 5s^1$	**Pd** 46 106.4 $4d^{10}$	**Ag** 47 107.868 $4d^{10} 5s^1$	**Cd** 48 112.41 $4d^{10} 5s^2$	**In** 49 114.82 $5p^1$	**Sn** 50 118.69 $5p^2$	**Sb** 51 121.75 $5p^3$	**Te** 52 127.60 $5p^4$	**I** 53 126.90 $5p^5$	**Xe** 54 131.30 $5p^6$
Cs 55 132.905 $6s^1$	**Ba** 56 137.33 $6s^2$	57–71†	**Hf** 72 178.49 $5d^2 6s^2$	**Ta** 73 180.95 $5d^3 6s^2$	**W** 74 183.85 $5d^4 6s^2$	**Re** 75 186.207 $5d^5 6s^2$	**Os** 76 190.2 $5d^6 6s^2$	**Ir** 77 192.22 $5d^7 6s^2$	**Pt** 78 195.09 $5d^9 6s^1$	**Au** 79 196.966 $5d^{10} 6s^1$	**Hg** 80 200.59 $5d^{10} 6s^2$	**Tl** 81 204.37 $6p^1$	**Pb** 82 207.2 $6p^2$	**Bi** 83 208.980 $6p^3$	**Po** 84 (209) $6p^4$	**At** 85 (210) $6p^5$	**Rn** 86 (222) $6p^6$
Fr 87 (223) $7s^1$	**Ra** 88 226.025 $7s^2$	89–103‡	**Rf** 104 (261) $6d^2 7s^2$	**Ha** 105 (260) $6d^3 7s^2$	106 (263)	107 (262)	108 (265)	109 (266)									

Group III (transition column) reference: **Sc** 21 44.956 $3d^1 4s^2$; **Y** 39 88.906 $4d^1 5s^2$; 57–71†; 89–103‡

† **LANTHANIDE SERIES**

La 57	**Ce** 58	**Pr** 59	**Nd** 60	**Pm** 61	**Sm** 62	**Eu** 63	**Gd** 64	**Tb** 65	**Dy** 66	**Ho** 67	**Er** 68	**Tm** 69	**Yb** 70	**Lu** 71
139.906 $5d^1 6s^2$	140.12 $4f^2 6s^2$	140.908 $4f^3 6s^2$	144.24 $4f^4 6s^2$	(145) $4f^5 6s^2$	150.4 $4f^6 6s^2$	151.96 $4f^7 6s^2$	157.25 $5d^1 4f^7 6s^2$	158.925 $4f^9 6s^2$	162.50 $4f^{10} 6s^2$	164.930 $4f^{11} 6s^2$	167.26 $4f^{12} 6s^2$	168.934 $4f^{13} 6s^2$	173.04 $4f^{14} 6s^2$	174.967 $5d^1 4f^{14} 6s^2$

‡ **ACTINIDE SERIES**

Ac 89	**Th** 90	**Pa** 91	**U** 92	**Np** 93	**Pu** 94	**Am** 95	**Cm** 96	**Bk** 97	**Cf** 98	**Es** 99	**Fm** 100	**Md** 101	**No** 102	**Lr** 103
(227) $6d^1 7s^2$	232.038 $6d^2 7s^2$	231.039 $5f^2 6d^1 7s^2$	238.029 $5f^3 6d^1 7s^2$	237.048 $5f^4 6d^1 7s^2$	(244) $5f^6 7s^2$	(243) $5f^7 7s^2$	(247) $5f^7 6d^1 7s^2$	(247) $5f^8 6d^1 7s^2$	(251) $5f^{10} 7s^2$	(253) $5f^{11} 7s^2$	(257) $5f^{12} 7s^2$	(258) $5f^{13} 7s^2$	(259) $5f^{14} 7s^2$	(260) $6d^1 7s^2$

* Average value based on the relative abundance of isotopes on earth. For unstable elements, the mass of the most stable isotope is given in brackets.

APPENDIX E ATOMIC MASSES

Each atomic mass is that of the neutral atom, including its Z electrons.

Atomic Number	Element	Symbol	Mass Number	Atomic Mass (u)
0	(Neutron)	n	1	1.008665
1	Hydrogen	H	1	1.007825
	Deuterium	D	2	2.014102
	Tritium	T	3	3.016050
2	Helium	He	3	3.016029
			4	4.002603
3	Lithium	Li	6	6.015123
			7	7.016005
4	Beryllium	Be	7	7.016930
			8	8.005305
			9	9.012183
5	Boron	B	10	10.012938
			11	11.009305
6	Carbon	C	11	11.011434
			12	12.000000
			13	13.003355
			14	14.003242
7	Nitrogen	N	13	13.005739
			14	14.003074
			15	15.000109
			16	16.006100
			17	17.008450
8	Oxygen	O	15	15.003066
			16	15.994915
			17	16.999130
			18	17.999159
9	Fluorine	F	19	18.998403
10	Neon	Ne	20	19.992439
			22	21.991383
11	Sodium	Na	22	21.994435
			23	22.989768
			24	23.990963
12	Magnesium	Mg	23	22.994127
			24	23.985043
13	Aluminum	Al	27	26.981541
14	Silicon	Si	27	26.986703
			28	27.976928
			30	29.973770
15	Phosphorus	P	30	29.978308
			31	30.973762
			32	31.973906
16	Sulfur	S	32	31.972071
			35	34.969030
17	Chlorine	Cl	35	34.968853
			37	36.965903
18	Argon	Ar	40	39.962384
19	Potassium	K	39	38.963708
			40	39.963996
20	Calcium	Ca	40	39.962592
21	Scandium	Sc	45	44.955910
22	Titanium	Ti	48	47.947947

APPENDIX E ATOMIC MASSES (*continued*)

Atomic Number	Element	Symbol	Mass Number	Atomic Mass (u)
23	Vanadium	V	51	50.943962
24	Chromium	Cr	52	51.940511
25	Manganese	Mn	55	54.938048
26	Iron	Fe	56	55.934940
27	Cobalt	Co	59	58.933198
			60	59.933815
28	Nickel	Ni	58	57.935346
			60	59.930788
			64	63.927968
29	Copper	Cu	63	62.929599
			64	63.929761
			65	64.927792
30	Zinc	Zn	64	63.929144
			66	65.926035
31	Gallium	Ga	69	68.925581
32	Germanium	Ge	72	71.922079
			74	73.921177
33	Arsenic	As	75	74.921594
34	Selenium	Se	80	79.916519
			87	86.9293
35	Bromine	Br	79	78.918336
			86	85.918444
			87	86.920327
36	Krypton	Kr	84	83.911509
			89	88.917555
			92	91.92576
37	Rubidium	Rb	85	84.911794
			87	86.909177
38	Strontium	Sr	86	85.909268
			87	86.909361
			88	87.905618
			90	89.907739
39	Yttrium	Y	89	88.905848
40	Zirconium	Zr	90	89.904702
41	Niobium	Nb	93	92.906376
42	Molybdenum	Mo	98	97.905407
43	Technetium	Tc	98	97.907203
44	Ruthemium	Ru	102	101.904349
45	Rhodium	Rh	103	102.90550
46	Palladium	Pd	106	105.90348
47	Silver	Ag	107	106.905088
			109	108.904755
48	Cadmium	Cd	106	105.90546
			114	113.903358
49	Indium	In	115	114.903876
50	Tin	Sn	120	119.902197
51	Antimony	Sb	121	120.903820
52	Tellurium	Te	130	129.906229
53	Iodine	I	127	126.904475
			131	130.90611
54	Xenon	Xe	132	131.904141
			136	135.90722
			140	139.92143
55	Cesium	Cs	133	132.90544
			140	139.91707
56	Barium	Ba	137	136.90582
			138	137.90524
			140	139.91058

APPENDIX E ATOMIC MASSES (*continued*)

Atomic Number	Element	Symbol	Mass Number	Atomic Mass (u)
			141	140.91413
			144	143.92267
57	Lanthanum	La	139	138.906347
			140	139.90947
58	Cerium	Ce	140	139.905434
59	Praseodymium	Pr	141	140.907648
60	Neodymium	Nd	142	141.907719
61	Promethium	Pm	145	144.91275
62	Samarium	Sm	152	151.919729
63	Europium	Eu	153	152.921226
64	Gadolinium	Gd	158	157.924100
65	Terbium	Tb	159	158.925344
66	Dysprosium	Dy	164	163.929173
67	Holmium	Ho	165	164.930320
68	Erbium	Er	166	165.930292
69	Thulium	Tm	169	168.934213
70	Ytterbium	Yb	174	173.938861
71	Lutecium	Lu	175	174.940771
72	Hafnium	Hf	180	179.946548
73	Tantalum	Ta	181	180.947994
74	Tungsten	W	184	183.950929
75	Rhenium	Re	187	186.955747
76	Osmium	Os	191	190.96101
			192	191.961469
77	Iridium	Ir	191	190.960585
			193	192.962916
78	Platinum	Pt	195	194.964766
79	Gold	Au	197	196.966543
80	Mercury	Hg	202	201.970617
81	Thalium	Tl	205	204.974401
			208	207.98200
82	Lead	Pb	204	203.973020
			206	205.974440
			207	206.975871
			208	207.976627
			209	208.981078
			210	209.984176
			211	210.988736
			212	211.991881
			214	213.999801
83	Bismuth	Bi	209	208.980373
			211	210.98725
84	Polonium	Po	210	209.982848
			218	218.008965
85	Astatine	At	218	218.00868
86	Radon	Rn	220	220.011368
			222	222.017570
87	Francium	Fr	223	223.019736
88	Radium	Ra	224	224.020199
			226	226.025408
			228	228.031072
89	Actinium	Ac	225	225.023218
			227	227.027751
90	Thorium	Th	228	228.02872
			232	232.038054
91	Proactinium	Pa	231	231.035880
92	Uranium	U	232	232.03713
			233	233.039629

APPENDIX E ATOMIC MASSES (*continued*)

Atomic Number	Element	Symbol	Mass Number	Atomic Mass (u)
			235	235.043924
			236	236.045563
			238	238.050785
			239	239.054295
93	Neptunium	Np	237	237.048172
			239	239.052936
94	Plutonium	Pu	239	239.052157
95	Americium	Am	243	243.061374
96	Curium	Cm	245	245.065484
97	Berkelium	Bk	247	247.07030
98	Californium	Cf	249	249.074845
99	Einsteinium	Es	254	254.08802
100	Fermium	Fm	253	253.085173
101	Mendelevium	Md	255	255.09109
102	Nobelium	No	255	255.09323
103	Lawrencium	Lr	257	257.0995
104	Rutherfordium	Rf	261	261.1086
105	Hahnium	Ha	262	262.1138
106			263	263.1182
107			261	261.1217
108			264	264.129
109			266	266.1378

* Data are based on *Nuclides and Isotopes*, 14th Edition, General Electric Co., and *Table of Isotopes*, C. M. Lederer and V. S. Shirley, Eds., 7th Ed., John Wiley & Sons, Inc., 1978.

Chapter 1

Exercises

1. (a) 80.7 ft/s;   (b) 24.6 m/s
3. 10^3 kg/m^3
5. (a) 9.47×10^{12} km;   (b) 7.2 AU/h
7. 0.514 m/s
9. 134 in^3
11. (a) 5;   (b) 3;   (c) 4   (d) 2 to 4
13. (a) 55.4 m^2;   (b) 2.7 m^2;   (c) 52.2 m^3
15. 3.33×10^3
17. (a) 1.495×10^{11} m;   (b) 5.893×10^{-7} m;
    (c) 2×10^{-10} m;   (d) 4×10^{15} m
19. (a) 91.440 m;   (b) 0.40469 hectares
21. (a) 2%;   (b) 4%;   (c) 6%.
23. (a) 5×10^{14} m^2;   (b) 3×10^{20} m^3;   (c) 1×10^6
25. Error of 14.5 min per day
27. 2×10^5 frames
29. (a) 5×10^4 km;   (b) 5×10^4 kg
31. 10^9 m^3
33. 0.1 m^3
35. (a) Correct;   (b) wrong;   (c) correct
37. (a) (2.68 m, 2.25 m);   (b) (−1.15 m, −1.38 m);
    (c) (−1.80 m, 1.26 m);   (d) (1.99 m, −1.67 m)
39. $[\omega] = $ T^{-1}, $[k] = $ MT^{-2}

Problems

1. 2×10^{-10} m, 1 atom
3. $T = C\sqrt{m/k}$, where C is a constant

Chapter 2

Exercises

1. (a) 4.1 m along the +x axis;
   (b) 3.2 m at 70° to the +x axis
3. (a) 3 m at 20° below the +x axis;
   (b) 2 m along the +y axis
7. 60 m at 25° E of N
9. 8.33 m at 24.2° S of E
11. 43.3 cm
13. (a) $\mathbf{A} = -\mathbf{i} - 1.73\mathbf{j}$ m, $\mathbf{B} = 1.53\mathbf{i} + 1.29\mathbf{j}$ m,
    $\mathbf{C} = -1.73\mathbf{i} + \mathbf{j}$ m, $\mathbf{D} = 1.64\mathbf{i} - 1.15\mathbf{j}$ m;
    (b) $0.44\mathbf{i} - 0.59\mathbf{j}$ m;
    (c) 0.74 m at 53.3° below the +x axis.
15. (a) $\mathbf{i} - \mathbf{j}$ m;   (b) 1.41 m;   (c) $0.707\mathbf{i} - 0.707\mathbf{j}$
17. (a) 0;   (b) 180°;   (c) 153°;   (d) 75.5°

19. (a) $-2.47\mathbf{i} - 0.35\mathbf{j}$ km;
    (b) 2.49 km at 8.07° S of W
21. (a) $-6.06\mathbf{i} + 2.63\mathbf{j}$ km;   (b) 3.58 km
25. $-0.920\mathbf{i} + 0.077\mathbf{j} + 0.383\mathbf{k}$
27. (a) $12\mathbf{i} - 4\mathbf{j} + 6\mathbf{k}$;   (b) $\mathbf{A}/7$;   (c) $-4\mathbf{A}/7$
31. (a) $-4.33\mathbf{i} + 2.5\mathbf{j}$ m;   (b) $3.12\mathbf{i} - 1.8\mathbf{j}$ m
33. $-7\mathbf{i} - 3\mathbf{j}$ m
35. (a) $\mathbf{W} = 10\mathbf{i} - 17.3\mathbf{j}$, $\mathbf{T} = -24\mathbf{i} + 18\mathbf{j}$, $\mathbf{F} = 8.66\mathbf{i} + 5\mathbf{j}$;
    (b) $-5.34\mathbf{i} + 5.7\mathbf{j}$
37. (a) $\pm 5\mathbf{k}$;   (b) $\mp 4.16\mathbf{i} \pm 2.77\mathbf{j}$
39. 120°
41. 105°
43. -3.25 m^2
45. $\alpha = 36.7°$, $\beta = 57.7°$, $\gamma = 74.5°$
47. 0.2 m
49. (b) $\mathbf{A} \times \mathbf{B}$ is perpendicular to $\mathbf{A}$
53. $7.36\mathbf{i} - 7.36\mathbf{j} + 4.25\mathbf{k}$

Problems

1. $\pm 4.46\mathbf{i} \mp 2.23\mathbf{j}$ m (use $\mathbf{A} \cdot \mathbf{B} = $ O.)
5. (a) 54.7°;   (b) 60°;   (c) 35.3°
11. $A_x = 4.23$ m, $A_z = 7.66$ m, $A_y = 4.84$ m

Chapter 3

Exercises

1. 10.2 m/s
3. 48 min
5. 13.7 km/h (3.8 m/s)
7. (a) +5 m/s;   (b) +2.5 m/s;   (c) −5 m/s;   (d) 0
9. 4 h 7 min 25 s
11. (a) 6 m/s;   (b) 3.5 m/s;   (c) − 7m/s
13. (a) 18.3 m/s;   (b) 1.67 m/s;   (c) −1.5 m/s^2
15. 1750 m/s^2
17. (a) 74.1 m/s;   (b) 21.1 m/s^2
19. (a) 2.83 m/s^2;   (b) 1.88 m/s^2;   (c) 1.06 m/s^2;
    (d) −6.94 m/s^2
21. (a) 1.04 m to 4.98 m;   (b) 4.25 m/s;
    (c) ≈ 5 m/s;   (d) 5 m/s
23. (a) 13 m/s, 6 m/s^2;   (b) 0.83 s
25. (a) 0, 3 s;   (b) 3 s;   (c) 2 m/s^2;   (d) 4 m/s^2
27. (a) 15 m;   (b) 11.7 m/s
29. (a) 2.5 m/s;   (b) 5.83 m/s
31. (c) 1.67 m/s^2;   (d) 5 m/s^2
35. (b) 14.3 mph/s;   (c) 8.4 s;   (d) 13 s at 105 mph
37. (a) 9.2 mph/s, 4.8 mph/s, 3 mph/s;
    (b) 2000 ft;   (c) 1200 ft
39. 1.28 m/s^2, east
41. (a) 1.87 s, 28.1 m;   (b) 1.87 s, 46.8 m

43. (a) 4 s;   (b) 10 m/s, 12 m/s
45. No, he needs about 1.3 m more
47. (a) -2m/s^2;   (b) 4 m/s
49. (a) -28 m, 16 m/s;
     (b) Average speed = 5 m/s, $v_{av} = -4$ m/s
51. 150 m
53. 6.93 m/s
55. (a) 40.8 m/s^2;   (b) 245 m/s^2
57. (a) 5.05 s;   (b) 44.4 m;   (c) 29.5 m/s
59. (a) 0.6 s, 3.48 s;   (b) 1.02 s, 3.06 s
61. 0.555 m
63. (a) 5.84 s;   (b) -37.4 m/s;   (c) 4.9 s
65. 10 min
67. 5 m, 2.02 s
69. 176 m, 12.0 s

Problems

1. (a) 4.31 s, 53.9 m;   (b) 12.5 m/s, 25 m/s
3. 140 m
5. (a) 210 s;   (b) 10.5 km
7. (a) 3.78 s, 48.9 m;   (b) -32 m/s, -37.4 m/s
9. (a) 32 m/s;   (b) 3.5 s
11. 28 m/s
13. 8 s
15. (a) No;   (b) 26.6 m
17. (a) 3.48 s, 27.7 m;   (b) -39.3 m/s, -9.1 m/s
19. (a) 0.78 s;   (b) 1.84 m
21. 8.6 m
23. (a) 30.6 m;   (b) 28.5 m

31. (a) Yes;   (b) 10.9 m;
     (c) 25.1° below the horizontal
33. 0.1 m less
35. (a) 42.2 m/s;   (b) 1.85 s;   (c) 34.6i + 6.1j m/s
37. (a) 3.37×10^{-2} m/s^2;   (b) 5.9×10^{-3} m/s2;
     (c) 3×10^{-10} m/s^2
39. (a) 1.58g;   (b) 3.55g;   (c) 16.1g;   (d) 0.187g;
     (e) $1.51 \times 10^5 g$
41. 142 s
43. 3.57 m/s2
45. 1260 m/s2
47. 14.2 km
49. 6.71 m/s at 63.4° S of E
51. T_A = 23.1 s; T_B = 26.7 s
53. (a) 232 km/h at 55.2° N of E;   (b) 2.58 h
55. (a) 71.9° N of W;   (b) 21 s
57. 174 km/h at 47.2° W of N
61. (a) Speeding up;   (b) 11.8 m/s
63. (a) 6.32 m/s^2;   (b) 4.9 m/s

Problems

3. (a) $\tan\theta = \sqrt{2v^2/gH}$;   (b) $\sqrt{2v^2H/g}$
5. (a) 76°;   (b) 82.9°
11. 59.4 m, 185.5 m
13. (a) $-3i - 4j$ m/s;   (b) 14.1 km;   (c) 94.5 min
15. 9.78 m/s
19. (a) $-30i + 50j$ km/h;   (b) 3.42 km;
     (c) A: 2.95 km W, B: 1.75 km N

Chapter 4

Exercises

1. (a) 10i − 12j m/s;   (b) 6i − 24j m/s^2;
    (c) 6i − 12j m/s^2
3. (a) 4i + 7j m/s;   (b) 1.2i + 3j m/s^2
5. (a) 0.9i − 0.37j m;   (b) 0.3i − 0.71j m/s;
    (c) −0.71i m/s^2
7. (a) 7 s;   (b) 120 m;   (c) 105 m;
    (d) 15i − 48.6j m/s
9. 4.77 m/s toward thrower
11. (a) 1.82 m too low;   (b) 5.75° (1.84 m)
13. 50 m
15. (a) 9.81 m/s;   (b) 10.9 m/s
17. 7 m/s
19. (a) 8.9 m;   (b) Yes, at 7 m/s
21. (a) 281 m;   (b) 281 m
23. (a) 29.9°;   (b) 1.19 m;   (c) 0.99 s
25. 19.8 m
27. (a) 1.39 cm;   (b) 54.2°
29. (a) 21.0 m/s;   (b) 2.31 s;   (c) 7.6 m

Chapter 5

Exercises

1. T_1 = 34.8 N, T_2 = 53.4 N
3. (a) 14.7 N;   (b) 24.5 N
5. 9i − 8j − 3k N
7. 7.34i + 1.66j m/s
9. (a) 1.37×10^{-15} N;   (b) 4×10^{-9} s
11. (a) 22.5 m/s^2;   (b) 1125 N
13. (a) 3.95×10^5 N;   (b) 3.90×10^5 N
15. 1795 N
17. 5.1 m
19. 840 N
21. (a) 3.13×10^4 N;   (b) 9.38×10^4 N
23. (a) 146 m;   (b) 25.8 s
25. 11.8 km (3.7 km in free−fall)
27. (a) 23.4 N;   (b) 1800 N;   (c) 200 N
29. $m = 2Ma/(a + g)$
31. (a) 4 m/s^2;   (b) 8 N;   (c) 12 N;   (d) 12 N
33. (a) 588 N;   (b) 657 N
35. 4.44 m/s^2, 88.8 N

37. (a) 4.9 N, 2.94 N;   (b) 4.9 N, 2.94 N;
    (c) 5.9 N, 3.54 N;   (d) 3.9 N, 2.34 N;
    (e) 10.2 m/s^2
39. (a) 44.1 N;   (b) 44.1 N;   (c) 46.4 N
41. (a) 118 N;   (b) 4 N;   (c) 47.2 N
43. 0.66 m/s^2 down incline, 12.7 N
45. (a) $\mathbf{v}$ = constant;   (b) 1.8 m/s^2, downward;
    (c) 2.2 m/s^2, upward
47. 955 N at 5.77° to the vertical
49. 226 N

Problems

1. (a) 441 N;   (b) 459 N;   (c) The rope breaks
3. (a) a_2 = 15.2 m/s^2 upward, a_5 = 0.2 m/s^2 upward;
   (b) 50 N
5. $t = 2\sqrt{R/g}$
9. (a) gy/L

Chapter 6

Exercises

1. (a) 0.135 N;   (b) 0.153
3. (a) 0.395;   (b) 7.74 N
5. (a) 27.4 m/s^2;   (b) 627 N
7. (a) No;   (b) 3.71 m/s^2
9. (a) 6.37 m/s^2, downward;
   (b) 9.31 m/s^2 (downward);
   (c) 6.37 m/s^2, downward
11. (a) 83.3 m;   (b) 48.1 m
13. 92.3 m
15. (a) 1.02 m/s^2;   (b) 140 N;   (c) 40 N
17. (a) $\mu mg/(\cos\theta - \mu\sin\theta)$;
    (c) $a = -\mu_k g$
19. (a) Zero;   (b)17.2 N
21. (a) 0.97 m/s^2;   (b) 2.94 m/s^2;   (c) 2.4 kg or 4.8 kg
23. (a) 82.5 N;   (b) 58.9 N;   (c) 0.58 m/s^2 upward
25. 3.31 m/s^2
27. 0.67 m/s^2
29. (a) 14 m/s;   (b) 1470 N
33. 50.9°
35. 791 N at 29.9° to the vertical
37. (a) 1.51 × 10^{-16} s;   (b) 8.3 × 10^{-8} N
39. 84.4 min
41. (a) 14 m/s;   (b) 1180 N
43. (a) $v^2/2\mu g$;   (b) $v^2/\mu g$
45. (a) 7.98 m/s;   (b) 163 N
47. 320
49. 4.74 × 10^{24} kg
51. (a) 6.71 × 10^5 km;   (b) 1.90 × 10^{27} kg
53. (a) 114 min;   (b) 1.52 × 10^4 N
55. 49.5 orbits

57. 2.45×10^{-2} N
59. 1.38 m/s^2
61. 11.4 m/s

Problems

1. $F = \mu mg/(\cos\theta + \mu\sin\theta)$
3. (a) 5.12 kN;   (b) 1.70 m/s^2
7. 7.94 N, 3.04 N
9. 2.04 m/s^2, 1.6 N

Chapter 7

Exercises

1. 24 J
3. 16 kJ
5. 24.1 J
7. (a) 11.8 J;   (b) −11.8 J;   (c) 0
9. (a) 49.7 kJ;   (b) −4 kJ
11. (a) −134 J;   (b) +134 J
13. (a) 30 J;   (b) −10 J
15. (a) +$F_0 A/2$;   (b) −$F_0 A/2$
17. (a) 0.2 J;   (b) 0.6 J
19. (a) $W_1/W_2 = k_1/k_2$;   (b) $W_1/W_2 = k_2/k_1$
21. 6.3 × 10^{23} tons
23. (a) −40 J;   (b) −35.3 J;   (c) No, air resistance
25. (a) 5 × 10^4 J;   (b) 2.25 × 10^5 J;   (c) 6.75 × 10^5 J
27. (a) 3 × 10^7 J;   (b) 7.21 m/s
29. 9.44 kJ
31. 3.2 × 10^4 N
33. (a) 179 J;   (b) −117.6 J;   (c) − 61.4 J;   (d) 0
35. 1.02 m
37. (a) 398 J;   (b) −204 J;   (c) −68.3 J;
    (d) 2.51 m/s
39. (a) 10 J;   (b) 60 J
41. (a) 4.8 J; (b) −0.34 J;   (c) −1.6 J;   (d) 1.38 m/s
43. 784 W
45. 443 W
47. P_E 2.35 × 10^7 W, P_G 4.92 × 10^7 W
49. 43.4 hp
51. 1.5 h
53. 12.1 km
55. 0.74 W
57. 3.3 kW (shot), 2.4 kW (javelin)
59. 23.7%
61. 4.84 min

Problems

1. 1.28 rev
5. (a) −1.6 J;   (b) −0.79 J;   (c) 6.27 J;
   (d) 1.97 m/s;   (e) 1.37 m

7.   392 J
9.   (a) 10.4 kW;   (b) 40.2 kW;   (c) 359 hp

Chapter 8

Exercises

1.   1.83 m/s
3.   1.18 m/s
5.   (a) 2.44 m/s;   (b) 53.6°
7.   (a) 2.53 m/s;   (b) 2.19 m/s;   (c) 28.3 cm
9.   26.6 cm
11.   (a) 1.96 m;   (b) 2.21 m/s
13.   (a) 1.23 m;   (b) 0.95 m/s
15.   (a) Yes;   (b) 1.31 m
17. $4H/7$
19.   (a) 22.3 m/s;   (b) 15.6 m/s
21.   (a) 37.5 m/s;   (b) 60.4 m
23.   (a) 2.45×10^9 J/s;   (b) 2.35×10^7 bulbs
25. 1.35×10^6 hp
27.   23.4 kW
29. 8.74×10^5 J
31.   (a) 1.36 m;   (b) 0.80 m;   (c) 41.6 m
33. -2.0 J
35. $-Cx^4/4$
37. $a/\sqrt{b^2 + x^2}$
39. $+40$ J
43.   (a) -70 J for both. No, it must apply to *any* path;
     (b) -140 J, no
45.   (a) 0.13 nm and 0.33 nm;
     (b) 0.8 nm and 4.1 nm, or 0.47 nm and 0.57 nm;
     (c) 1.6×10^{-20} J
47.   10 km/s
49. 5.16×10^{-4} m/s
53.   (a) 60 MJ;   (b) $\Delta E = mg_E R_E/2 = 31$ MJ
55. 1.22×10^{11} J

Problems

1.   (a) $Cx/(a^2 + x^2)^{3/2}$;   (b) $\pm a/\sqrt{2}$
3.   (a) $r = r_0/2^{1/6}$;   (c) $r = r_0$
9.   (a) $T_2 - T_1$;   (b) $2\pi RN(T_2 - T_1)$
11.   (a) $\cos\theta = 2/3$;   (b) At a lower point, greater θ
13.   (a) Zero;   (b) $L^4/2$;   (c) No
15.   (a) $U = 1/2\ kx^2 - mgx$;   (b) 0.495 m/s;
     (c) 0.1 m;   (d) 0.8 m

Chapter 9

Exercises

1.   (a) 3.5×10^4 m/s;   (b) 0.47 m/s
3.   (a) 3000;   (b) 0.0183

5. $10\mathbf{i} + \mathbf{j}$ m/s
7. $v_1 = 3.1$ m/s, $v_2 = 5.5$ m/s
9.   1.85 m/s
11. $(0.01\mathbf{i} + 1.39\mathbf{j}) \times 10^{-24}$ kg · m/s
13.   (a) 2.25 m/s;   (b) 0.107m/s;   (c) 0.107 m/s
15.   (a) 2/3;   (b) 1/3
17.   (a) 6.86 m/s;   (b) 11.8 m/s
19.   (a) $4.13\mathbf{i} + 3.60\mathbf{j}$ m/s;   (b) 2970 J
21.   (a) 2.56×10^5 m/s;   (b) 1.21×10^{-14} J
23.   (a) 26.6 km/h at 18° W of N;   (b) 1.54×10^9 J
25.   (a) $2.4\mathbf{i}$, 6×10^3 J;   (b) $-0.9\mathbf{i}$; 2.54×10^5 J
27.   (a) 160 m/s;   (b) 99.3%
29.   (a) 7.3 cm;   (b) 750 J
31. 5.97×10^5 m/s
33.   (a) $H/9$;   (b) $H/9$, $4H/9$
35.   (a) 0.89;   (b) 0.28;   (c) 0.019
37. $m_n \approx m_p \approx 1$ u
39.   5620 N
41.   300 N
43.   2000 N
45.   15 N
47.   67.5 N
49.   330 N
51.   (a) $-0.11\mathbf{i} + 1.56\mathbf{j}$ kg · m/s;   (b) $-22\mathbf{i} + 312\mathbf{j}$ N
53.   (a) 435 kg · m/s;   (b) 260 kg · m/s;
     (c) 2.64 m/s;   (d) 0.355 m
55.   (a) 24,000 kg · m/s at 37° S of W;
     (b) 3000 N at 37° S of W
57.   (a) 30°;   (b) $u/\sqrt{3}$ for both;   (c) $K_\alpha = 2K_D/3$
59.   48.1°, 41.8°
61.   2.18 km/s

Problems

9. $h_1 = H/9$, $h_2 = 25H/9$
13.   Two collisons
15. $3m$
17.   (a) 3.17 m/s;   (b) 0.583
19. $3Mgy/L$

Chapter 10

Exercises

1.   (a) 0.126 nm from H;
     (b) 6.76 pm from O along line of symmetry
3.   93.3 cm
5. $x_{CM} = y_{CM} = -0.122R$
7. $x_{CM} = L/2$, $y_{CM} = 0.738L$
9.   4680 km from earth's center
11.   (a) 2.3 m;   (b) 2.3 m from Jack's initial position
13.   (a) $x_{CM} = L/2$; $y_{CM} = 0.29L$;
     (b) $x_{CM} = y_{CM} = 0.354L$

17. (a) 6.67**i** m; 0.67**i** + 2.67**j** m/s;
   (b) $\mathbf{a}_1 = 10\mathbf{j}$ m/s^2; $\mathbf{a}_2 = 4\mathbf{i}$ m/s^2;
   (c) 2.67**i** + 3.33**j** m/s^2;   (e) 13.4**i** + 12**j** m
19. (a) 0.24 m/s;
   (b) 4.44 m from Jack's initial position
21. 500 m
23. (a) 5.36**i** + 6.43**j** m/s;   (b) 16.1**i** + 19.3**j** m
25. (a) −1.8**i**;   (b) $\mathbf{u}_1 = 4.8\mathbf{i}$, $\mathbf{u}_2 = -3.2\mathbf{i}$;
   (c) 18.6 J;   (d) 3.24 J;   (e) 15.36 J
27. (a) 5**i** m/s;   (b) 245 J;   (c) 275 J;
   (d) 245 J;   (e) 245 J;   (f) 0;   (g) 30 J
29. (a) 5**i** m/s; (b) $\mathbf{v}_1' = \mathbf{i}$ m/s, $\mathbf{v}_2' = -2\mathbf{i}$ m/s;
   (c) $\mathbf{v}_1' = -\mathbf{i}$ m/s; $\mathbf{v}_2' = 2\mathbf{i}$ m/s
31. (a) 15 J;   (b) 25 J
33. 40 m/s

Problems

1. $x_{CM} = 2b/3$ from the left, $y_{CM} = h/3$ above the base
3. $x_{CM} = y_{CM} = 0.42R$ relative to the corner
5. $2H/3$
7. (a) $x_{CM} = 0.5$ m from man;   (b) 1.5 m/s;   (c) 3 m
9. (a) 2×10^5 N;   (b) 1.12 km/s
11. $x_{CM} = y_{CM} = z_{CM} = L/3$

Chapter 11

Exercises

1. (a) 1.75 rad/s^2;   (b) 2.22 rev;   (c) 4.6 s;
   (d) 0.46 m/s^2; 0.26 m/s^2;   (e) 1.83 m/s^2; 0
3. (a) 7.27×10^{-5} rad/s;   (b) 1.99×10^{-7} rad/s;
   (c) 29.2 km/s, 30.2 km/s
5. (a) 11 rad/s;   (b) 0.66 m/s;
   (c) 7.26 m/s^2, 0.48 m/s^2
7. 29 rev
9. 2.04×10^5 rpm
11. (a) 6.4 rad/s;   (b) 2.74 rad/s;
   (c) -2.03×10^{-3} rad/s^2
13. (a) −0.105 rad/s^2;   (b) 37.7 m
15. (a) −30.9 rad/s^2;   (b) 31.8 rev
17. −15.9 rad/s^2
19. (a) 24.7 m/s^2;   (b) 2.74 m/s^2
21. 1.67 cm
23. (a) $8Md^2$;   (b) $4M\ell^2$
25. (a) 7 kg · m^2;   (b) 5.71 kg · m^2
27. (a) $6.28\lambda a^3$;   (b) $10.7\lambda a^3$;   (c) $4\lambda a^3$
29. $14.1MR^2$
31. $\pi\sigma R_1^3(2h + R)$
33. $7ML^2/48$
35. $1/2\ MR^2$
37. $\sqrt{3gL}$
39. (a) 0.98 m;   (b) 1.58 m/s

41. (a) One;   (b) 14/15
43. $\sqrt{16gR/3}$
45. (a) 518 kJ;   (b) 276 m
47. 4.98×10^4 N · m
49. (a) 0.01 N · m;   (b) 0.13 N · m
51. (a) 62.8 W;   (b) 1.4 s;   (c) 2.32 rev
53. (a) −20 N · m;   (b) 68 N · m
55. (a) 7.84 rad/s^2;   (b) 2.8 m/s
57. $a_1 = 0.527$ m/s^2; $a_2 = 1.05$ m/s^2, $T_1 = 10.3$ N, $T_2 = 26.2$ N
59. (a) $2g\sin\theta/3$;   (b) $\tan\theta/3$
61. 13.7 mW
63. (a) 29.4 W;   (b) 11 N
65. (a) 49.1 J;   (b) 29.3 s

Problems

1. $M(B^2 + A^2)/2$
3. (a) $\sqrt{3g/L}$;   (b) $v_L = 0$, $v_R = \sqrt{3gL/4}$
5. $2M(A^5 - B^5)/5(A^3 - B^3)$
7. $M(R^4 - a^4 - 2a^2b^2)/2(R^2 - a^2)$
9. $1/2\ Mh^2 \tan^2\alpha$
11. 3.17 kg · m^2
13. $M(a^2 + b^2)/12$
15. (a) $N/20$ rev/s forward;   (b) $N/64$ rev/s backward

Chapter 12

Exercises

1. (a) $\ell_1 = 18\mathbf{k}$ kg · m^2/s, $\ell_2 = -9.70\mathbf{k}$ kg · m^2/s,
   $\ell_3 = -23.5\mathbf{k}$ kg · m^2/s, $\ell_4 = 15\mathbf{k}$ kg ·m^2/s;
   (b) $-0.2\mathbf{k}$ kg · m^2/s
7. (a) $\mathbf{v} = \boldsymbol{\omega} \times \mathbf{r}$;   (b) $\mathbf{a}_r = \boldsymbol{\omega} \times \mathbf{v}$
9. (a) 1.57 N · m;   (b) $0.8v$;   (c) 1.96 m/s^2
11. (a) $MAt^3(2C - Bt)\mathbf{k}$;   (b) $M(6At\mathbf{i} + 2B\mathbf{j})$
13. (a) 1.68 rad/s;   (b) -4×10^{-3} J
15. (a) 3.56 rad/s;   (b) −0.016 J;
   (c) 4.5×10^{-3} N · m
17. (a) 0.909 rad/s;   (b) −341 J
19. (a) 1 rad/s;   (b) 2 rad/s;   (c) +160 J
21. 0.308 rad/s
23. 4.16 rad/s
27. (a) 3.73 cm;   (b) 0.68 cm
29. $T_1 = 417$ N, $T_2 = 298$ N
31. (a) 768 N;   (b) 500 N, 98 N
33. (a) 26.3 N;   (b) 22.8 N; 45.7 N
35. 974 N
37. (a) 1960 N;   (b) 1700 N (horiz.), 2940 N (vert.)
39. (a) 2100 N;   (b) 2060 N
41. 70 N
43. 86.2 cm from the feet

Problems

1. (a) 0.8 m/s;   (b) 2/3 rad/s
3. (a) L_0^2/mr^3;   (b) $(L_0^2/2m)(1/r_2^2 - 1/r_1^2)$;
    (c) $\Delta K = W$;   (d) Yes.
5. (a) $f = Ma$, $fR = I\alpha$;   (c) $(\omega_0 R)^2/18\mu_k g$
7. (b) $24v_0^2/98\mu g$
9. (a) Yes, there is no torque;
    (b) $[C + \sqrt{C^2 + (mbu_0^2)^2}]/mu_0^2$
11. $Mg/4$
13. 5550 N (front), 5480 N (rear)
15. (a) $L/2$;   (b) $L/4$;   (c) $L/6$
17. (a) 0.438 m;   (b) 0.22 m from the central line
19. Wall: 77.3 N; Floor: 588 N (vert.), 77.3 N (horiz.)
21. 53.1°

Chapter 13

Exercises

1. (a) 6.67×10^{-9} m/s²;   (b) 8.17×10^{-5} m/s
3. (a) 2.33×10^{-3} N;   (b) 0.419 N
5. $-1.29 \times 10^{-8}\mathbf{i} + 4.77 \times 10^{-9}\mathbf{j}$ N
7. (a) $(GM/L^2)(-4.84\mathbf{i} + 8.82\mathbf{j})$;
    (b) $(GM/L^2)(-13.1\mathbf{i} - 7.06\mathbf{j})$
9. 6.48 cm/s²
11. (a) 9.796 m/s²;   (b) 4.55 km
13. 4.9 s
15. (a) 0.43 N;   (b) 7.5×10^{17} N;   (c) 2.8×10^{-7} N
17. (a) $4\pi\rho RG/3$;   (b) 4 times larger;
    (c) No change;   (d) 1.26 times larger
21. (a) $v \propto r^{-1/2}$;   (b) $T \propto r^{3/2}$;   (c) $p \propto r^{-1/2}$;
    (d) $L \propto r^{1/2}$
23. 29.3 km/s
25. (a) -2.39×10^9 J;   (b) 96.5 min;   (c) 7.97 km/s
29. (a) 92 min;   (b) -2.25×10^{12} J;
    (c) 7.61 km/s, 7.71 km/s

Problems

1. $T^2 = 3\pi/G\rho$
3. (a) $2GmM/(a^2 + x^2)^{1/2}$;   (b) $2GmMx/(a^2 + x^2)^{3/2}$;
    (c) $x = \pm a/\sqrt{2}$
5. (a) $3.6R$;   (b) $2.6R$
7. (a) $GmM[1/R - 1/(R + h)]$;
    (b) $GmM[1/R - 1/2(R + h)]$;   (c) $h = R/2$
9. $GmMb/(R^2 + b^2)^{3/2}$, GmM/b^2
11. (a) $GmM/2R^2$; (b) $0.44GmM/R^2$;
    (c) $0.32GmM/R^2$
13. $4\pi\rho_o G(r/3 - r^2/8R)$
15. (a) $Gm_1m_2/(r_1 + r_2)2 = m_1\omega^2 r_1 = m_2\omega^2 r_2$

Chapter 14

Exercises

1. (a) 0.860 g/cm³;   (b) 0.851 g/cm³
3. (a) 8.95×10^{14} kg/m³;   (b) 1.17 km
5. 1.75 mm
7. 9.8 mm
9. 3×10^4 N
11. 1.06×10^4 N
13. 1.02×10^5 N
15. 0.21 N
17. 46 cm
19. 1.85×10^4 Pa
21. 1.09 g/cm³
23. 2.46 MPa
25. 104 kPa
27. 1070 kg
29. 6.03 g
31. 857 kg/m³
33. 600 g
35. 750 kg/m³
37. 5600 kg/m³
39. 960 kg/m³
41. 7.5 cm
43. 200.029 g
45. 110 min
47. 70 cm
49. 1.55×10^5 N
51. (a) 3.6 m/s;   (b) 370 kPa
53. 3.97 kPa

Problems

1. 3.53×10^9 N
3. 1078 kg/m³ using 1025 kg/m³ at the surface
7. 0.206 J
11. (b) $H - h$

Chapter 15

Exercises

1. (b) $\cos(\theta + 5\pi/3)$
3. (a) 1.47×10^5 N/m;   (b) 0.656 s
5. 0.284s, 0.866 s, 1.16 s, 1.74 s
7. (a) 0.206 m, -2.33 rad;
    (b) $0.206 \sin(10t - 2.33)$ m;   (c) 0.414 s.
9. (a) $0.08\cos(7.83t)$ m,   (b) 0.414 m/s, 3.68 m/s²
11. (a) ± 0.773 m/s, $+1.87$ m/s;
    (b) ± 0.611 m/s, -3.73 m/s²
13. $x = R\cos(\omega t)$ and $y = R\sin(\omega t)$. Both indicate SHM.
15. 9.8 cm at 34.9 ms, 79.8 ms; -9.8 cm at 150 ms, 194 ms

17. (a) 0.75 kg; (b) 0.24 J; (c) 0.355 s;
    (d) -2.95 m/s^2
19. (a) 0.245 m; (b) 0.351 s
23. (a) 1.64 s ; (b) 1.94 s
25. 20.1 Hz
27. (a) 1.27 s; (b) 0.69 m/s; (c) 11.9 mJ
29. 0.636 s
31. (a) 1.80 s; (b) $(\pi/6)\cos(3.5t)$; (c) 21 mJ;
    (d) 1.27 m/s

Problems

1. $0.25\sin(8t)$ m
5. $T = 2\pi\sqrt{R/g}$
9. (b) $2\pi\sqrt{(M + m/3)/k}$
11. (a) 1.00430 s; (b) 1.01738 s; (c) 1.03963 s;
    (d) 1.07129 s
13. 84.4 min

Chapter 16

Exercises

1. (a) 188 m to 546 m; (b) 2.78 m to 3.41 m
3. (a) 10 Hz; (b) $5\pi/4$ rad; (c) 1/60 s;
    (d) 1.26 m/s
5. 1440 km
7. 0.563 kg
9. (a) $f_2 = f_1/\sqrt{2}$; (b) $v_2 = \sqrt{2}v_1$
11. (b) 1.0 cm/s
13. $(2 \times 10^{-3})/[4 - (x + 12t)^2]$ m
15. (a) $kA\cos(kx - \omega t)$;
    (b) $(\partial y/\partial t)_{max} = v\,(\partial y/\partial x)_{max}$
17. (a) 15.1 cm/s; (b) -15.0 cm/s;
    (c) 379 cm/s^2; (d) -44.5 cm/s^2
19. a, b, d, e
21. (a) 15.7 cm, 0.8 rad; (b) 0.126 s;
    (c) 0.02 cm; (d) -125 cm/s; (e) 0.483 cm/s
23. $0.03\sin[80\pi x + 200\pi t - 0.730]$ m
25. (a) $A\sin[(2\pi/\lambda)(x - vt)]$; (b) $A\sin[2\pi f(x/v - t)]$
    (c) $A\sin[k(x - vt)]$; (d) $A\sin[2\pi(x/\lambda - ft)]$
27. (a) $4\sin(0.4\pi x)\cos(16\pi t)$ cm; (b) 2.5 cm;
    (c) 2.35 cm
29. (a) 144 cm; (b) 17.4 Hz
31. (a) 23.0 cm; (b) 68.9 cm
33. $f_2 = \sqrt{2}f_1$
35. (a) 31.4 cm, 83.3 cm/s; (b) 47.1 cm;
    (c) 1/3, 2/3 points
37. 15 W
39. (a) 6.63 mW; (b) 50.4 N

Problems

1. 160 m/s
3. g/ω^2
15. (b) $\sqrt{m/k}$

Chapter 17

Exercises

1. 0.34 mm
3. 0.375 mm
5. (a) 1.43 km/s; (b) 1.43 m
7. 5.06 km/s
9. 3.04 km/s
13. 19.3 cm, 58.0 cm
15. 78 N
17. 8.5 Hz, 17 Hz, 25.5 Hz
19. (a) 283 Hz; (b) 51.5 cm
21. 15 Hz
23. (a) 78.8 cm; (b) 91.3 cm
25. (a) 227 Hz, 1.50 m; (b) 224 Hz, 1.70 m;
    (c) 225 Hz, 1.6 m
27. (a) 59.6 Hz; (b) 131 Hz
29. Zero, 24.6 Hz
31. 565 Hz
33. (a) 4×10^{-5} W; (b) 4×10^{-17} W
35. 86.2 dB
37. (a) 1.99×10^{-7} W/m2; (b) 1.40×10^{-2} W/m^2
39. 1.41
41. (a) 12.6 cm; (b) 126 m
43. (a) 1.04×10^{-5} m; (b) 0.169 W/m^2
45. 6.35 kHz

Problems

1. (a) 4.97 μW/m^2; (b) 67 dB;
    (c) 6.6×10^{-2} Pa; (c) 9.98×10^{-8} m
3. (b) 414 Hz; (c) 3.5%
5. 22.3 m
9. (a) 404 Hz or 396 Hz; (b) 346 N; (c) 42.1 cm

Chapter 18

Exercises

1. (a) 21.1 °C; (b) 90.6 °C; (c) 37.0 °C
3. -40 °F $= -40$ °C
5. 491 °R
9. (a) 1.25 kg/m^3; (b) 1.42 kg/m^3; (c) 0.089 kg/m^3
11. (a) 24.4 L; (b) 37.2 L
13. (a) 43.8 °C; (b) 4.9 mg
15. 1120 N
17. (a) 16.5 L; (b) 49.5 L

19. (a) 43.9 mm Hg;   (b) 171 K
21. 2.9 °C to 37.1 °C
23. 747 °C
25. 20.6 cm

Problems

1. (a) 1.37 atm;   (b) 1.35 mol;   (c) 0.732 atm
5. 56.5 s loss
7. 0.36 mm
9. (a) 3.53×10^7 N/m^2;   (b) 1.15×10^7 N/m^2
11. 2.61 mm

Chapter 19

Exercises

1. 1055 J
3. 17.7 °C
5. 183 s
7. 107 kJ
9. 1.2×10^4 kg/s
11. (a) 463 kJ;   (b) 61.7 °C
13. (a) 36.6 °C;   (b) No
15. 0.423 C°
17. 8.8 J
19. (a) 2.4 kJ;   (b) −1.3 kJ
21. (a) 229 J;   (b) 39 J
23. (a) 1.22 J;   (b) 3.88 J
25. (a) 150 J;   (b) 130 J;   (c) −100 J;   (d) −130 J
27. (a) 2.96 kJ/kg · K;   (b) 0.713 kJ/kg · K
29. 2.04 °C
31. (a) 5.8 kJ;   (b) 1.66 kJ;   (c) 4.14 kJ
33. (a) −2020 J;   (b) +400 J
35. (a) $W = 3.38$ kJ, $\Delta U = 0$;
     (b) $W = -3910$ J, $\Delta U = +3910$ J
37. −1875 J
39. (a) 10.5 kPa;   (b) 158 K
43. (a) 341 K;   (b) 40.2 kPa, 70.8 kPa;   (c) −2.1 kJ
45. (a) 258 m/s;   (b) 1.31×10^5 N/m^2
47. 30 mW/m^2
49. (a) 75.4 W;   (b) 45 °C
51. 38.5 in

Problems

1. (a) 140 kJ;   (b) 0.132 J;   (c) 140 kJ
5. 565 W

Chapter 20

Exercises

1. (a) 1350 m/s;   (b) 604 m/s;   (c) 181 m/s
3. (a) 4.14×10^{-16} J;   (b) 7.04×10^5 m/s

5. (a) 2740 m/s;   (b) 967 m/s
7. 2.73 km/s
9. (a) 2.07×10^{-14} J;   (b) 3.52×10^6 m/s
11. 273 m/s
13. (a) 517 m/s;   (b) 1.48 MPa
15. 3.34×10^{-9} m
17. 1.17
19. 1.93×10^5 K
21. (a) 2.81 mol;   (b) 9.56 kJ;   (c) 20.8 J/mol.K
23. (a) $Q_V = \Delta U = 1250$ J;   (b) $Q_P = 2080$ J,
     $\Delta U = 1250$ J
25. (a) 316 K;   (b) 310 K
27. 20.1 g/mol, neon
29. 0.304 nm
33. 2.25×10^{19} m

Problems

1. (a) 3.74 m/s;   (b) 3.95 m/s;   (c) 4 m/s
5. (a) 283;   (b) 1780;   (c) 110

Chapter 21

Exercises

1. (a) 800 J;   (b) 600 J
3. (a) 500 W;   (b) 4
5. 40 kWh
7. (a) 136 kW;   (b) 106 kW;   (c) 3.77 gal/h
9. 22 %
11. (a) 5.76%;   (b) 16.4 MW
13. (a) $12.00;   (b) $0.82
15. (a) 1073 J;   (b) 73 J
17. $Q_H = 189$ J, $Q_C = 117$ J
19. (a) 434 J;   (b) 2.30
21. (a) 6.06 kJ/K;   (b) 1.22 kJ/K;   (c) 1.31 kJ/K
23. (a) −61.2 J/K;   (b) 0
25. (a) 20.8 °C;   (b) −3.09 J/K;   (c) +3.4 J/K;
     (d) 0.31 J/K
27. (a) −3 J/K;   (b) 4.8 J/K;   (c) 0;   (d) 1.8 J/K
29. (a) $5nR/2 \ln(T_2/T_1)$;   (b) $3nR/2 \ln(P_2/P_1)$
31. (a) 11.5 J/K;   (b) 11.5 J/K
33. (a) ±1.82 J/K, 0;   (b) 0

Problems

1. (a) $Q_1 = 7.5P_oV_o$, $Q_2 = -6P_oV_o$, $Q_3 = -2.5P_oV_o$,
     $Q_4 = 3P_oV_o$;   (b) $2P_oV_o$;   (c) $W/Q_{in} = 0.19$
3. (a) 431 J;   (b) 0.074
7. (a) 538 J;   (b) 0.4
11. (a) 3.3 J/K;   (b) 1.73 J/K;   (c) 0.9 J/K;
     (d) Let $\Delta T \to 0$
13. 4.92×10^4 J

Chapter 22

Exercises

1. (a) $-191\mathbf{i}$ N   (b) $188\mathbf{i}$ N;
3. (a) $208\mathbf{j}$ N;   (b) $80\mathbf{i} - 277\mathbf{j}$ N
5. (a) 0.75 m;   (b) 1.5 m
7. 1.52×10^{-14} m
9. (a) 4.21×10^{-8} N;   (b) 2.90×10^{-9} N
11. $q_1 = q_2 = -27/80\ \mu C$
13. $F_{elec}/F_{grav} = 2.8 \times 10^{-18}$
15. $\pm 133\ nC\ \pm 267\ nC$
17. 20.5 N
19. 5.76×10^5 N

Problems

1. $q_1 = -3.16\ \mu C,\ q_2 = 1.90\ \mu C,\ q_3 = -5.27\ \mu C$
3. (a) $-2kQqa/(a^2 + x^2)^{3/2}\ \mathbf{j}$;   (b) $x = 0$
5. $q = Q/2$
7. (a) 2.08×10^{13};   (b) 7.57×10^{-12}
9. (a) $v = \sqrt{ke^2/mr}$;
   (c) 5.3×10^{-11} m, 2.12×10^{-10} m, 4.77×10^{-10} m

Chapter 23

Exercises

1. (a) 5.58×10^{-11} N/C;   (b) 1.02×10^{-7} N/C
3. (a) $2500\mathbf{i}$ N/C;   (b) $-1.6 \times 10^{-5}\ \mathbf{i}$ N
5. (a) $-7.64 \times 10^{10}Q/L^2\mathbf{j}$;
   (b) $(-6.44\mathbf{i} - 118\mathbf{j}) \times 10^9\ Q/L^2$
7. 3.06×10^6 N/C
9. (a) $-8050\mathbf{i} + 4020\mathbf{j}$ N/C;   (b) $1920\mathbf{i} - 2880\mathbf{j}$ N/C
11. (a) $-1.08 \times 10^7\mathbf{i} + 6.24 \times 10^6\mathbf{j}$ N/C;
    (b) $32.4\mathbf{i} - 18.7\mathbf{j}$ N   (c) Not at all
17. $(4.66\mathbf{i} + 2.88\mathbf{j}) \times 10^{-3}$ N
19. (a) $(2kq/r^2)[1 - (1 + a^2/r^2)^{-3/2}]$;
    (b) $2kqa^2(a^2 - 3r^2)/r^2(r^2 - a^2)^2$
27. (a) 1.71 ns;   (b) 2.56 cm;   (c) 4.1×10^{-16} J
29. (a) 13.9 cm;   (b) $0.348\ \mu s$
31. (a) 1.78×10^{-12} s;   (b) 8.33×10^{-7} m
33. 3.75×10^6 m/s
35. I and III: $\sigma/2\varepsilon_0$; II: $5\sigma/2\varepsilon_0$; IV: $-\sigma/2\varepsilon_0$
37. (a) 2.26 N;   (b) 12.6 cm from q
39. 2.02×10^4 N/C
41. $(-3\mathbf{i} + 3\mathbf{j}) \times 10^6$ N/C
43. (a) 8×10^{-11} C $\cdot$ m;   (b) 8×10^{-6} J

Problems

3. $-(Cp/x^2)\mathbf{i}$
7. (a) kQ/y^2;   (b) $2kQ/yL$
13. (a) $2kQx/(x^2 + a^2)^{3/2}$;   (b) $2kQ/x^2$;
    (c) $x = \pm a/\sqrt{2}$
17. $25.7°$

Chapter 24

Exercises

1. 10.2 N $\cdot$ m²/C
5. -2.26×10^5 N $\cdot$ m²/C (inward)
3. $\pi R^2 E$
7. (a) 6.78×10^6 N $\cdot$ m²/C;   (b) 1.13×10^6 N $\cdot$ m²/C;
   (c) Yes for (b)
9. (a) 11.3 N/C;   (b) 7.23 N/C
13. (a) Zero;   (b) σ/ε_0
15. (a) σ/ε_0;   (b) Zero
17. $\sigma_2 = -a\sigma_1/b$
19. $\lambda_2 = -\lambda_1$
21. (a) kQ/r^2;   (b) 0
23. $\sigma_a/\sigma_b = -b^2/a^2$

Problems

1. (a) $\rho r/3\varepsilon_0$;   (b) $\rho R^3/3\varepsilon_0 r^2$
3. (a) $-4\pi\sigma R_1^2$;   (b) $-4\pi\sigma(R_2^2 - R_1^2)$;   (c) $-\sigma R_2^2/\varepsilon_0 r^2$
5. (a) $\rho r/2\varepsilon_0$;   (b) $\rho R^2/2\varepsilon_0 r$
7. $ke(1/r^2 - r/R^3)$
9. (a) kQ/r^2;   (b) kQ/r^2
11. (a) $(\rho/2\varepsilon_0)(r - a^2/r)$;   (b) $(\rho/2\varepsilon_0)(R^2/r - a^2/r)$
13. $\rho x/\varepsilon_0$

Chapter 25

Exercises

1. (a) 1.87×10^{28} eV;   (b) 5×10^7 s ≈ 1.58 y
3. 60 V
5. +6 V
7. (a) 3.10×10^{-7} V;   (b) 3.57×10^{-4} V;
   (c) 2.56×10^3 V
9. (a) 2.05×10^6 m/s;   (b) 4.80×10^4 m/s
11. 2×10^{-5} J
13. 150 V
15. (a) 216 V;   (b) 52 kV
17. 1.69 J
19. 6.95×10^{-11} J
21. -2.55 J
23. (a) 12 kV;   (b) 2 km/s
25. (a) 0.8 m, 1.33 m;   (b) -0.8m, $+0.308$ m
27. (a) 3.00 MV;   (b) 2.01 MV;   (c) 1.80 MV
29. Zero
31. (a) $-3\ \mu C$;   (b) $-4.11\ \mu C$
33. (a) 6.59×10^{-11} J;   (b) 6.59×10^{-11} J;
    (c) 5.06×10^{16} per second
35. (a) -9.57 V;   (b) -7.23 V
37. (a) $2kq[1/x - 1/(x^2 + a^2)^{1/2}]$;
    (b) $2kq[1/y - y/(y^2 - a^2)]$
39. (a) 5.26 cm;   (b) 5.88 cm

41. (a) 5.6×10^{-12} J; (b) 5.6×10^{-12} J
43. (a) $2kQy/(y^2 - a^2)$; (b) $2kQ(y^2 + a^2)/(y^2 - a^2)^2$
45. kQr/R^3
47. $2\pi k\sigma[1 - y/(a^2 + y^2)^{1/2}]$
49. (a) 1.44×10^6 V; (b) 27.2 V; (c) No change
51. (a) 30 V; (b) 30 kV; (c) 30 MV

Problems

1. $K_\alpha = 5.51 \times 10^{-12}$ J, $K_{Th} = 9 \times 10^{-14}$ J
3. (a) -30.6 eV; (b) 12.7 eV
5. (a) $k(Q_1/R_1 - Q_2/R_2)$; (b) $k(Q_1 - Q_2)/R_2$;
   (c) $kQ_1 (1/R_1 - 1/R_2)$; (d) $Q_1 = 0$
7. (b) 3.97×10^6 V/m
9. $k\lambda \ln[L + (L^2 + y^2)^{1/2}]/y$
15. (a) 3.38×10^{-11} eV; (b) -3.38×10^{-11} eV;
    (b) -6.76×10^{-11} eV; (d) 6.76×10^{-11} eV

Chapter 26

Exercises

1. (a) 50 pF; (b) 600 pC
3. (a) 54.2 cm²; (b) 167 V; (c) 8.33×10^5 V/m
5. (a) -8.9×10^{-10} C/m²; (b) 91.7 mF;
   (c) 712 μF for the earth
7. $4\varepsilon_0 A/d$
9. 8 nF
11. $Q_1 = 32 \mu$C, $Q_2 = 48 \mu$C, $V_1 = V_2 = 8$ V
13. (a) 4.58 pF; (b) 1.43×10^8
15. 6 μF
17. (a) Two in parallel with two in series;
    (b) Four in series
19. 16 μC, 32 μC; 8 V for both
21. (a) 32.25 μC, 53.75 μC, 10.75 V;
    (b) 5.25 μC, 8.75 μC, 1.75 V
23. 0.22 pF
25. (a) 14.2 pF; (b) 4.08 nJ; (c) 9.6 kV/m;
    (d) 408 μJ/m³
27. 0.192 J/m³
29. $U_1 = 204 \mu$J, $U_2 = 81.6 \mu$J, $U'_1 = 66.7 \mu$J, $U'_2 = 167$ μJ
31. (a) 529 μJ; (b) 6.61 μJ
33. (a) $\varepsilon_0 A/(d - \ell)$; (b) No change
35. (a) Doubles; (b) No change; (c) Doubles
37. 8.96×10^5 J/m³
39. 6.38 V
41. $C_0(\kappa_1 + \kappa_2)/2$
43. (a) 1.72 V; (b) 4.66 pF; (c) No
45. (a) 0.94 cm²; (b) 15 kV

Problems

3. $\frac{1}{2}\varepsilon_0 A\ell V^2/(d - \ell)^2$
5. $0.366C$

7. $Q^2/2\varepsilon_0 A$, attractive
9. $\kappa CV^2/2$

Chapter 27

Exercises

1. (a) 1.19×10^{16} s⁻¹; (b) 2.42×10^3 A/m²
3. (a) 1.07×10^{-5} m/s; (b) 1.49×10^{-3} V/m
5. (a) 2.83×10^6 A/m²; (b) 4.81×10^{-2} V/m
7. 1.06 mA
9. (a) 6.79×10^6 A/m²; (b) 4.24×10^{-4} m/s;
   (c) 0.19 V/m
11. 6.85×10^{-5} A
13. $\rho\ell/\pi(b^2 - a^2)$
15. 148 °C
17. 1.65
19. 0.78
21. 2.83×10^{-8} Ω.m, Al
23. 5.04 Ω
25. (a) 45.6 °C; (b) 5.6 °C
27. 12 V
29. (a) 2.88×10^5 C; (b) 38.4 h
31. 5.23 W, 13.2 W
33. 2.3×10^{19}
35. 720 kW
37. (a) 500 W; (b) 1.25 W
39. 14.5 V
41. 288 Ω; 144 Ω

Problems

1. (a) 41.9 m; (b) 7.00 A
3. (b) 37.4 $\mu\Omega$
5. $\sigma\omega a^2/2$
7. 134 mg
9. (a) 0.77 W, 18.1 W; (b) 1.02×10^{-2} V/m,
   0.240 V/m

Chapter 28

Exercises

1. 11.2 V, 0.706 Ω
3. 12 V, 0.6 Ω
5. (a) 0.525 Ω, (b) 0.215 Ω, 1.295 Ω
7. 0.18 Ω
9. (a) 7.82 Ω; (b) 0.569 V
11. 2 Ω, 3 Ω, 4 Ω, 5 Ω, 6 Ω, 7 Ω, 9 Ω, 6/5 Ω, 4/3 Ω, 12/7 Ω, 12/13 Ω, 26/7 Ω, 13/3 Ω, 26/5 Ω, 20/9 Ω, 7/8 Ω
13. 1 Ω, 2 Ω, 3 Ω and 4 Ω, or 1 Ω, 2 Ω, 4 Ω, 5 Ω
15. 0.5 A, 0.083 A, 8.33 A; 12.5 A
17. (a) 8.94 V; (b) 13.4 V
19. (a) 0.5 A; (b) $P_1 = 0.5$ W, $P_2 = 1$ W;
    (c) 4.5 W, -3 W
21. $I_7 = 0$; $I_3 = 3.33$ A; $I_4 = 2.5$ A

23. $r/2$
25. (a) $I_1 = 3$ A, $V_1 = 6$ V, $I_2 = -1$A, $V_2 = 5$ V, $I_3 = -4$ A; $V = 20$ V;   (b) -20 V
27. $I_1 = 2$ A, $V_1 = 8$ V, $I_2 = 1$ A, $V_2 = 3$ V, $I_3 = 3$ A, $V_3 = 9$ V
29. 18 V, 1 Ω
31. 3 Ω
33. (a) 8.66 V;   (b) $P_1 = 13.5$ W, $P_2 = 3$W
35. 1.44×10^5 Ω
37. 86.8 μF
39. (a) 0.37 mC, 0.37 mA;   (b) 1.71 mJ;   (c) 9.2 mW;   (d) 9.2 mW
41. (a) 1.50 s;   (b) 1 %
43. (a) 0 and 5/3 A;   (b) 21.7 μC
45. 950 Ω, 9 kΩ, 40 kΩ
47. (a) $R_{series} = 200$ kΩ;   (b) $R_{shunt} = 2$ mΩ
49. (a) 9.9 A, 99 V;   (b) 9.9 A, 100 V

Problems

1. $R_1 = 8.2$ mΩ; $R_2 = 32.8$ mΩ; $R_3 = 367$ mΩ
3. $5R/6$
7. 0.409 A
9. (a) 8 V;   (b) 8 V;   (c) 26.7 μs
11. $I_1 = 5$ A, $I_4 = 2$ A, $I_2 = 1.5$ A
13. (a) $I_1 = \xi/R_1$, $I_2 = 0$;   (b) $\xi/(R_1 + R_2)$
17. (a) Series;   (b) Parallel

Chapter 29

Exercises

1. (a) 6.26 m;   (b) 1.31 μs
3. (a) 1.6×10^{-20} kg $\cdot$ m/s;   (b) 4.8×10^5 eV
5. 30° to the $+y$ axis
7. $-0.16\mathbf{i} - 0.32\mathbf{j} - 0.64\mathbf{k}$ N
9. $(-3.13\mathbf{i} - 1.04\mathbf{j}) \times 10^6$ m/s
11. $-0.4\mathbf{j}$ T
13. $\mathbf{F}_1 = IdB_1\mathbf{j}$, $\mathbf{F}_2 = -IdB_1\mathbf{i}$, $\mathbf{F}_3 = IdB_1(\mathbf{i} - \mathbf{j})$
15. $\mathbf{F}_1 = IdB_3\mathbf{k}$; $\mathbf{F}_2 = 0$; $\mathbf{F}_3 = -IdB_3\mathbf{k}$
17. 0.064 N, in the vertical plane 30° below the horizontal
19. (a) $1.85 \times 10^{-2}\mathbf{j}$ T;   (b) $\mathbf{B} = (1.85 \times 10^{-2}\,\mathbf{j} + 3.21 \times 10^{-2}\,\mathbf{k})$ T
21. (a) $\mathbf{F}_1 = -\mathbf{F}_3 = 8\mathbf{k}$ N, $\mathbf{F}_2 = -\mathbf{F}_4 = -40\mathbf{j}$ N;   (b) $\boldsymbol{\mu} = 8\mathbf{i} - 13.8\mathbf{j}$ A $\cdot$ m^2;   (c) $\boldsymbol{\tau} = 6.93\mathbf{k}$ N $\cdot$ m
23. 1.88×10^{-4} N $\cdot$ m
25. (a) $(4.5\mathbf{i} + 3.4\mathbf{j} - 2.25\mathbf{k}) \times 10^{-4}$ N $\cdot$ m;   (b) -1.7×10^{-3} J
27. 3.13°
29. 1.85×10^{-23} A $\cdot$ m^2
31. (a) $r_e/r_p = 5.45 \times 10^{-4}$;   (b) $r_e/r_p = 2.33 \times 10^{-2}$
33. 3.1 mm

35. (a) 11.4 MHz;   (b) 27.6 keV;   (c) 3.84×10^{-21} kg $\cdot$ m/s
37. 5.0 mm
39. (a) $-10^{-4}\mathbf{k}$ T;   (b) 11.4 cm
41. (a) 0.914 T;   (b) 50 kV;   (c) 13.9 MHz
43. (a) 31.3 cm;   (b) 4.1 μs
45. 6.12 cm
47. 4.08 T

Problems

1. (b) $2\pi(I/\mu B)$
5. $\sigma\omega\pi R^4/4$
7. 1.56×10^{-2} T

Chapter 30

Exercises

1. $\mu_0 I_1 I_2 c/2\pi\ [1/a - 1/(a + b)]$ to the right
3. (a) $(6.92\mathbf{i} - 1.54\mathbf{j}) \times 10^{-5}$ T;   (b) $(4.6\mathbf{i} + 20.8\mathbf{j}) \times 10^{-5}$ N
5. 5×10^{-4} T
7. 6.8° W of N
9. $5.14 \times 10^{-7}I/a$, out of page
11. $(\mu_0 I/4)(1/a - 1/b)$, into page
15. (b) $0.766a$
17. (a) $-1.33 \times 10^{-5}\mathbf{j}$ T;   (b) $8.51 \times 10^{-19}\mathbf{k}$ N
19. (a) $\mu_0 Ix/\pi(a^2 + x^2)\mathbf{j}$;   (b) $x = \pm a$
21. $7.14 \times 10^{-7}I/a$
23. (a) 3.14×10^{-5} T;   (b) 0.254 A $\cdot$ m^2
25. 265
29. (a) Parallel to the plate;   (b) $\mu_0 Jt/2$
31. $R/4$ and $4R$

Problems

3. (a) $dI = \sigma\omega r\, dr$;   (b) $dB = \mu_0\sigma\omega\, dr/2$
7. Typical values: 0.034 at -5 cm, 0.498 at 0, 0.960 at $+5$ cm
9. (b) $\mu_0 Id/2\pi R^2$

Chapter 31

Exercises

1. 2.52 mWb
3. 40.2 μV
5. (a) $(3.2t - 2.4t^2) \times 10^{-4}$ Wb;   (b) 1.33 mA
7. $13.7\cos(60\pi t)$ mV
9. (a) 1.96 mWb;   (b) 98.2 mV;   (c) Counterclockwise
11. (a) 1.2 A;   (b) 0.13 N;   (c) 3.9 W;   (d) 3.9 W
13. 71.6 mW
15. (b) $(mg/B\ell)^2R$
17. 3.38 mW

19. 130 rad/s
21. (a) 0.20 V;   (b) 7.15×10^{-4} N.m
23. $eCd/2$
25. (a) 6.03×10^{-5} V/m;   (b) 1.21×10^{-4} V/m
27. (a) 0.81 V;   (b) Zero
29. 7150 rpm

Problems

1. $(\mu_0 Iv/2\pi) \ln[(\ell + d)/d]$
5. 333 μV
7. $\mu_0 \omega I_0 c/2\pi \ln[(a + b)/a] \cos(\omega t)$
9. (b) $L^2 \, dB/dt$

Chapter 32

Exercises

1. (a) 152 μH;   (b) 26.3 A/s
3. (a) 4.8 μWb;   (b) 42 mV
5. 9.38 mV
7. 93.8 mH
9. 8.56 μH
11. (a) 80.4 μH;   (b) 402 μV
13. (a) $\mu_0 h N_1 I_1/2\pi \ln(b/a)$;   (b) $(\mu_0 h N_1 N_2/2\pi) \ln(b/a)$
15. (a) 0.6 mWb;   (b) 0.39 mWb;   (c) 0.14 mWb;
    (d) 80 mV;   (e) 12.6 mV;   (f) 28 mV
17. (a) 0.279 A;   (b) 10.3 V;   (c) 536 ms
19. (a) 6 A/s;   (b) 0.231s;   (c) 0.333 s
21. (a) 13 V;   (b) 11 V
23. (a) 5.5 ms;   (b) 99.3%
25. 0.58 ms
27. (a) 3.98 mJ/m³;   (b) 0.633 A
29. (a) 9.59 W;   (b) 5.59 W;   (c) 15.2 W
31. 5.54 ms
33. 75.2 mA
35. (a) $\mu_0 n^2 I^2/2$;   (b) $\mu_0 n^2 A\ell$
37. (a) 563 Hz;   (b) 0.212 A;   (c) 0.222 ms
39. 1.98 pF to 16.7 pF

Problems

1. (a) $L_1 + L_2$;   (b) $1/L = 1/L_1 + 1/L_2$
5. $(\mu_0 c/2\pi) \ln[(b + a)/a]$
7. (b) 314;   (c) 7.3 μF

Chapter 33

Exercises

1. (a) 7.96 A;   (b) 200 Hz
3. (a) 503 Hz;   (b) 1125 Hz;   (c) 225 Hz
5. (a) 1.7 mC;   (b) 0.64 A
7. (a) 1.3 A;   (b) zero;   (c) ±1.13 A;   (d) 13.2 W
9. (a) 1.40 A, 68.2 W;   (b) 66.3 Hz
11. 5.67 Ω, 44.4 μF

13. 75.6 mH, 724 mH
15. (a) 2.42 A;   (b) 20.6 Ω;   (c) 194 V;   (d) 303 V
17. (a) 300 Ω;   (b) 0.192 H;   (c) 70.5°
19. (a) 49 μF;   (b) 113 kHz
21. 264 Hz
23. (a) 919 Hz;   (b) 621 Hz, 1360 Hz
25. (a) 0.762;   (b) 291 W
27. (a) $X_L = 20$ Ω, $X_C = 50$ Ω;   (b) −63.4°;
    (c) 267 W;   (d) 0.447
29. 150 W
33. 2.74 sin(320t − 1.02) V
35. (a) 16.2 W;   (b) 1.01 W
37. (a) 400;   (b) 2.4 A

Problems

3. (a) 2 V;   (b) $\sqrt{8/3}$ V
5. $4/\sqrt{3}$ V
7. (a) 84.6 V;   (b) 15.7 V;
   (c) 1.59 V. A "low- pass" filter.
9. 37.1 mH
11. (a) $v_{0R} = 7.32$ V, $v_{0C} = 114$ V, $v_{0L} = 14.6$ V;
    (c) 114 V;   (d) 99.7 V

Chapter 34

Exercises

3. 2.9×10^{-7} A
7. 6.28×10^{-13} T
9. (a) 5×10^{-8} T;   (b) 8×10^{-8} T
11. (a) 1.26 cm, 23.9 GHz;
    (b) $E_z = 60\sin(500x + 1.5 \times 10^{11}t)$ V/m
13. (a) 150 V/m;   (b) 0.5 μT
15. (a) 3.33×10^{-6} J/m³;   (b) 4.6×10^{20} J
17. (a) 3.33×10^{-8} T;   (b) 60 W
19. (a) 3.33×10^{-12} N;   (b) 3.33×10^{-6} N
21. 150 kW
23. (a) 5.31×10^{-15} W/m²;   (b) 3.87×10^8 m
25. (a) 3.33×10^{-8} N/m²;   (b) 8×10^{-9} N
27. 19.6 mW

Problems

1. (a) $1/2 \, \mu_0 \varepsilon_0 a(dE/dt)$;   (b) $1/2 \, \varepsilon_0 aE(dE/dt)$
3. (a) 720 W;   (b) 3.82×10^4 W/m²
5. $(S/c)(2 - f)$
7. 2.87×10^{-4} V
9. (a) 0.01 A;   (b) 2 s
11. 8.71×10^{-5} kg.m/s

Chapter 35

Exercises

3. 5
7. 100.1°
9. 2.62 m

Problems

3. (a) 2×10^{-24} kg · m/s; (b) 0.165 nm = $3.2r_B$
5. 0.82
7. $\sqrt{2/L} \cos(n\pi x/L)$, $n = 1, 2, 3$
9. $m\omega/2\hbar$

Chapter 42

Exercises

1. (a) $\sqrt{2}\hbar$; (b) $\sqrt{12}\hbar$
3. $n = 4$, $\ell = 2$, $m_\ell = 0, \pm1, \pm2$, $m_s = \pm1/2$
5. (a) $0, \pm\hbar$; (b) $90°, \pm45°$
7. (a) $\ell = 4$; (b) $n \geq 5$
9. (a) -30.6 eV; (b) $\ell = 0, 1$; $m_\ell = 0, \pm1$
11. $2\hbar$
15. $0.54/r_0 = 1.02 \times 10^9$
17. (a) $0.37/r_0$; (b) $0.54/r_0$; (c) $0.29/r_0$
21. 24.8 kV
23. 5×10^7 Hz$^{1/2}$
25. 5.59 nm
27. $E_1 = -4620$ eV, $E3 = -2650$ eV
29. (n, ℓ, m_ℓ, m_s): $(1, 0, 0, \pm1/2)$, $(2, 0, 0, \pm1/2)$, $(2, 1, 1, \pm1/2)$, $(2, 1, 0, \pm1/2)$
31. 3.21×10^{-23} J/T
33. (a) 2.14 meV; (b) 18.4 T

Problems

5. $\lambda_{\text{Xray}} = 31$ pm; $\lambda_{\text{elec}} = 8.4$ pm

Chapter 43

Exercises

1. (a) 3.0 fm; (b) 4.6 fm; (c) 7.4 fm
3. 4.8×10^{17} kg/m^3

5. 8
7. 70% is ^{63}Cu
9. 7 fm, 26 MeV
11. (a) 8.55 MeV; (b) 7.91 MeV
13. BE(^{13}C) = 97.1 MeV, BE(^{13}N) = 94.1 MeV
15. (a) 6.06 MeV; (b) 5.61 MeV
17. 4.2×10^{11} Bq
19. 1.27×10^8
21. ^{11}B, yes
23. (a) 3.4 h; (b) 9.8×10^9
25. 23,700 y
27. 7 α's, 4 e$^-$'s
29. 3.11 h
31. 8.54×10^8 y
33. 1.58%
35. 6.92 MeV
37. 6 α's, 4 e$^-$'s
39. 1.31 MeV
41. (a) -1.19 MeV; (b) 17.35 MeV
43. 2.13 MeV
45. (a) ^{32}S; (b) ^{19}F; (c) ^{10}Be; (d) n
47. 18.00095 u
49. 1.38×10^{11}
51. 173.6 MeV
53. 26
59. 1.66×10^9 J

Problems

1. 3×10^{13}
3. $\lambda_1 N_1 = \lambda_2 N_2$
5. 4.79 MeV
7. 4.75 h
9. 5.34 MeV

PHOTO CREDITS

Chapter 1

Opener: Photolabs Royal Observatory, Edinburgh. *Page 2*: Michael Holford/Science Museum, London. *Pages 5 and 6*: Courtesy National Bureau of Standards. *Page 14*: (left) Erich Lessing/Culture and Fine Arts Archive, Vienna, Austria; (right) The Granger Collection; (bottom) Courtesy Lowell Observatory. *Page 15*: (left) Foundation Saint-Thomas, Strasbourg, France; (right) Scala/Art Resource.

Chapter 2

Opener: Dave Driscoll/Photo Researchers. *Page 20*: Michel Tcherevkoff/The Image Bank.

Chapter 3

Opener: Aurora French Rail, Inc., NY. *Page 36*: © 1990 Estate of Harold Edgerton; Photo courtesy Palm Press. *Page 45*: Focus on Sports. *Page 46*: Courtesy United States Air Force. *Page 53*: Courtesy NASA.

Chapter 4

Opener: Tom McHugh/Photo Researchers. *Page 56*: Dick Durrance/Woodfin Camp & Associates. *Page 59*: Courtesy United States Air Force. *Page 60*: Courtesy *The Birth of a New Physics* by I. Bernard Cohen. *Page 62*: © Roy Pinney/Picture Library, Inc. *Page 72*: © 1990 Estate of Harold Edgerton; Photo courtesy Palm Press. *Page 74*: Harris Benson. *Page 75*: Bettman Archive. *Page 76*: Ray Stott/The Image Works.

Chapter 5

Opener: UPI. *Page 80*: (bottom) Painting by Sir Godfrey Kneller, National Portrait Gallery, London. *Page 85*: Courtesy NASA. *Page 86*: (top) Courtesy NASA. *Page 87*: Henry Broskinsky, *Time Magazine*.

Chapter 6

Opener: Focus on Sports. *Page 100*: From *Friction and Lubrication of Solids* by F. P. Bowden and D. Tabor, Claredon Press, Oxford, 1950. *Page 104*: Thomas Zimmermann/FPG International. *Page 107*: Diagram from Newton's *Principia*, AIP Niels Bohr Library. *Page 108*: UPI. *Page 113*: Courtesy NASA. *Page 120*: Leo Balterman/FPG International.

Chapter 7

Opener: *Hay Wagon*, Pisa, by Gieli. Florence Museum of Modern Art/Art Resource. *Page 126*: Vandystadt/Photo Researchers. *Page 130*: John P. Kelly/The Image Bank.

Chapter 8

Opener: Jim Strawser/Grant Heilman Photography. *Page 145*: Royal Society, London. *Page 146*: Bernard Giani/Photo Researchers. *Page 151*: © 1990 Estate of Harold Edgerton; Photo courtesy Palm Press.

Chapter 9

Opener: C. Powell, P. Fowler, & D. Perkins/Science Photo Library, Photo Researchers. *Page 168*: Portrait of *René Descartes* by Frans Hals, Louvre, Paris/Art Resource. *Page 173*: (left) Photo by Esther C. Goddard/AIP Niels Bohr Library; (right) Courtesy Clark University Archive. *Page 177*: © 1990 Estate of Harold Edgerton; Photo courtesy Palm Press. *Page 179*: (top) Courtesy Education Development Center; (bottom) Courtesy Trustees of Science Museum, London. *Page 180*: © 1990 Estate of Harold Edgerton; Photo courtesy Palm Press. *Page 181*: (top) Haag Genenr Museum; (bottom) Courtesy New York Public Library Picture Collection.

Chapter 10

Opener: Martha Swope. *Page 194*: © 1990 Estate of Harold Edgerton; Photo courtesy Palm Press. *Page 195*: Courtesy Education Development Center.

Chapter 11

Opener: P. & G. Bowater/The Image Bank. *Page 210*: Harris Benson. *Page 224*: Volvo Car Corporation, Guthenburg, Sweden. *Page 225*: Courtesy NASA. *Page 227*: Courtesy Allied-Signal Aerospace Co, Los Angeles Division. *Page 230*: Courtesy NASA.

Chapter 12

Opener: © NHPA. *Page 241*: S. Fitzgerald/Peter Arnold. *Page 238*: Paul J. Sutton/Duomo. *Page 252*: (top) Courtesy National Hot Rod Association; (center) Courtesy United States Army; (bottom) Courtesy Bell Helicopter.

Chapter 13

Opener: © 1979 R. J. DuFour, Rice University. Published by Planetarium, Salt Lake City, Utah. *Page 267*. Courtesy NASA. *Page 272*: Steve Kaufman/Peter Arnold.

Chapter 14

Opener: Courtesy NASA. *Page 285*: (left) © Barbara Schultz/PAR/NYC, Inc.; (right) Deutsch Museum, Munich. *Page 287*: R. Gaillarde/Gamma-Liaison. *Page 289*: Henri Werle; Courtesy ONERA, Châtillon, France. *Page 294*: (top) Harris Benson; (center) Grant Heilman Photography. *Page 298*: Courtesy NASA.

Chapter 15

Opener: Paul Boyuton/University of Wisconsin. *Page 310*: Harris Benson. *Page 311*: AP/Wide World Photos. *Page 313*: (top) Courtesy NASA.

Chapter 16

Opener: J. M. Loubat/Agence Vandystadt/Photo Researchers. *Page 318*: Harris Benson. *Page 321*: J. R. Berintenstein/Agence Vandystadt/Photo Researchers. *Page 329*: (Top) Courtesy Education Development Center; (bottom) Thomas Rossing.

Chapter 17
Opener: © 1990 Estate of Harold Edgerton; Photo courtesy Palm Press. *Page 329*: Thomas Rossing. *Page 340*: (left) Howard Sochurek/The Stock Market; (center) Courtesy Siemens Corporation; (right) VG Semicon/Science Photo Library/Photo Researchers. *Page 342*: Richard Laird/FPG International.

Chapter 18
Opener: Science Photo Library/Photo Researchers. *Page 358*: Institute e museo di StoRia della Scienza de Firenze, Italy. *Page 363*: (left) Robert Isaacs/Photo Researchers; (right) AP/Wide World Photos. *Page 364*; Courtesy Corning Inc., NY.

Chapter 19
Opener: VANSCAM Thermogram by Daedalus Enterprises, Inc.; Photo courtesy *National Geographic Magazine*. *Page 370*: National Portrait Gallery, London. *Page 373*: The Royal Society, London. *Page 374*: Courtesy Science Museum, London. *Page 384*: (top) Ron Church/Photo Researchers; (center) Lockheed Missiles & Space Co., Inc., CA. *Page 386*: Courtesy FLIR Systems, Portland, Oregon.

Chapter 20
Opener: C. Peterson Merrill/FPG International. *Page 401*: University of Vienna, Niels Bohr Institute.

Chapter 21
Opener: Weldon King/FPG International. *Page 410*: (top) Bettmann Archive; (bottom) The Granger Collection. *Page 411*: "Power from the Sea" By Terry R. Penny and Desikan Bharathan, © 1987 *Scientific American*, Inc. *Page 412*: (top) The Royal Society, London; (bottom) Lande Collection/AIP Niels Bohr Library. *Page 414*: AIP Niels Bohr Library.

Chapter 22
Opener: Michael Holford/Science Museum, London. *Page 433*: Bettmann Archive. *Page 436*: Deutsches Museum, Munich. *Page 439*: (top) E. Scott Barr Collection/AIP Niels Bohr Library; (bottom) Courtesy New York Public Library Picture Collection.

Chapter 23
Opener: Thomas Ives. *Page 450*: Larry Hamill. *Page 451*: Courtesy Dr. Harold Waage, Princeton University. *Page 460*: AIP Niels Bohr Library.

Chapter 24
Opener: Dale E. Boyer/Photo Researchers. *Page 468*: © Archiv/Photo Researchers.

Chapter 25
Opener: Photo by Jon Brenneis for *Fortune Magazine*. *Page 487*: Ira Wyman. *Page 501*: Courtesy Chrysler Corporation. *Page 502*: Courtesy Bethlehem Steel Corporation. *Page 503*: Dr. Erwin Müler, Pennsylvania State University.

Chapter 26
Opener: Dan McCoy/Rainbow. *Page 505*: Courtesy Science Museum, London. *Page 506*: Dr. Harold Waage, Princeton University. *Page 514*: Courtesy MIT High Voltage Research Laboratory.

Chapter 27
Opener: Courtesy CENCO. *Page 523*: (top) The Bettman Archive; (bottom) Lande Collection/AIP Niels Bohr Library. *Page 524*: (top left) Royal Institution, London; (top right) Burndy Library, Norwalk, CT. *Page 525*: Oleg D. Jefimenko, West Virginia University. *Page 530*. E. Scott Barr Collection/AIP Niels Bohr Library. *Page 538*: Painting by Benjamin West, c. 1805, Mraud Mrs. Wharton Sinkler Collection, Philadelphia Museum of Art. *Page 539*: Courtesy NSSL. *Page 540*: J. Zuckerman/Westlight.

Chapter 28
Opener: Lynn Johnson/Black Star. *Page 546*: W. F. Meggers Collection/AIP Niels Bohr Library. *Page 557*: Michael Holford/Science Museum, London.

Chapter 29
Opener: Patrice Loiez/CERN. *Page 569*: (top) From original painting by C. W. Eckersberg; (bottom) Courtesy Education Development Center. *Page 577*: Culver Pictures. *Page 578*: Lawrence Berkeley Laboratory/Science Photo Library/Photo Researchers. *Page 580*: (top) Holiday Film Corp. *Page 581*: From *Mass Spectrometry: Applications in Engineering* by F. A. White and G. M. Wood, John Wiley & Sons, Inc., 1986. *Page 582*: (left) © 1990 Jon Brenneis; (right) Courtesy Lawrence Berkeley Observatory. *Page 583*: (left) David Parker/Science Photo Library Photo Researchers; (right) Fermi National Accelerator Laboratory/Science Photo Library/Photo Researchers. *Page 584*: Photo by Bachrach/AIP Niels Bohr Library. *Page 586*: Cavendish Laboratory, University of Cambridge. *Page 587*: Courtesy Science Museum, London.

Chapter 30
Opener: Stanford Linear Accelerator Center/Science Photo Library/Photo Researchers. *Page 594*: Courtesy Education Development Center. *Page 596*: © John W. Warden. *Page 598*: Courtesy Education Development Center. *Page 599*: Dr. O. Jefimenko. 599 (bottom) courtesy Education Development Center. *Page 600*: AIP Niels Bohr Library. *Page 603*: Lawrence Livermore National Laboratory. *Page 604*: (top left and bottom) Courtesy Smithsonian Institution; (top right) Courtesy New York Public Library Picture Collection. *Page 613* "Magnetic Fields in the Cosmos" by E. N. Parker, *Scientific American*, August 1983. *Page 616*: Courtesy Dr. L. A. Frank.

Chapter 31

Opener: Spectrum Color Library. *Page 617*: Courtesy New York Public Library Picture Collection. *Page 618*: The Royal Institution/AIP Niels Bohr Library. *Page 619*: Michael Holford, Royal Institution Collection, London. *Page 625*: Earl Roberge/Photo Researchers. *Pages 630 and 632*: Michael Holford, Royal Institution Collection, London.

Chapter 32

Opener: Michael S. Weinberg. *Page 640*: Painting by Thomas Leclear, 1877, National Portrait Gallery, Smithsonian Institution. *Page 653*: Courtesy R. W. Dublois. *Page 662*: G. Davis/Sygma.

Chapter 33

Opener: U. S. Dept. of Energy/Science Photo Library/Photo Researchers.

Chapter 34

Opener: Baline Harrington/Stock Market. *Page 679*: AIP Niels Bohr Library. *Page 686*: (top) Photo by Ben Rose, © Miriam Rose; Courtesy Center for Creative Photography; (bottom left) Courtesy Palomar Observatory, California Institute of Technology; (bottom right) Robert McCall. *Page 687*: Deutsches Museum/AIP Niels Bohr Library. *Page 688*: Deutsches Museum, Munich. *Page 689*: Courtesy NASA. *Page 690*: Doug Johnson/Photo Researchers.

Chapter 35

Opener: © Jon Brenneis. *Page 699*: (left) Courtesy NASA; (right) © Michael Freeman/Bendix Aerospace Systems. *Page 701*: Harris Benson. *Page 704*: Courtesy Bausch & Lomb. *Page 705*: (top) Charles S. Allen; (bottom) Foto Forum. *Page 707*: UPI/Bettmann Newsphotos. *Page 710*: Hughes Optical Systems, Danbury, CT. *Page 716*: Courtesy New York Public Library Picture Collection. *Page 717*: (top center) © PAR/NYC; (bottom left) Roy L. Bishop/A.I.P. Niels Bohr Library. *Page 721*: *Time Magazine*.

Chapter 36

Opener: T. Tracy/FPG International. *Page 732*: (top) Science Museum, London; (bottom) Michael Holford; Courtesy Science Museum, London. *Page 733*: Courtesy Science Museum, London. *Page 734*: Courtesy Nikon, Inc. Instrument Group. *Page 735*: Art Resource. *Page 736*: (top) Michael Holford; Courtesy Science Museum, London; (bottom) Bob Gomel/Stock Market. *Page 737*: Lennart Nilsson.

Chapter 37

Opener: Michael Freeman. *Page 748*: Manfred Kage/Peter Arnold. *Page 750*: (top) The Royal Society, London; (bottom) *Atlas of Optical Phenomena*: M. Cagnet, M. Francon, J. Thierr, Springer-Verlag, Berlin. *Page 755*: Courtesy Bausch & Lomb. *Page 756*: Courtesy Bausch &

Lomb. *Page 757*: AIP Niels Bohr Library. *Page 762*: Martin Rogers.

Chapter 38

Opener: Peter Arnold, Inc. *Page 766*: Smithsonian Institution/AIP Niels Bohr Library. *Page 767*: (top and center) *Atlas of Optical Phenomena*: M. Cagnet, M. Francon, J. Thierr, Springer-Verlag, Berlin. *Page 768*: *Atlas of Optical Phenomena*: M. Cagnet, M. Francon, J. Thierr, Springer-Verlag, Berlin. *Page 770*: *Atlas of Optical Phenomena*: M. Cagnet, M. Francon, J. Thierr, Springer-Verlag, Berlin. *Page 771*: (top) The Royal Society/AIP Niels Bohr Library; (center) Courtesy NASA. *Page 773*: © PAR/NYC. *Page 775*: (top left) Courtesy Dr. Eli Brookner, Raytheon Co; (top right) Diego Goldberg/Sygma; (center) Courtesy National Radio Astronomy Observatory. *Page 778*: Photo Researchers. *Page 779*: (top right) From *Seeing the Light*: D. Falk, D. Brill, D. Stork, John Wiley & Sons, Inc., 1986; (bottom left) Joel Arem. *Page 783*: (left) © 1989 David A. Wagner/Stock Market; (right) *Atlas of Optical Phenomena*: M. Cagnet, M. Francon, J. Thierr, Springer-Verlag, Berlin. *Page 784*: Courtesy Cornell University. *Page 790*: (left) *Applications of Holography*: E. S. Barrekette, W. S. Kock, T. Ose, et al., Plenum Press, 1971; (plant) "Acoustics of Violins" by G. M. Hutchins for *Scientific American*, Oct. 1981. *Page 791*: Daniel Quat/Museum of Holography.

Chapter 39

Opener: Science Photo Library/Photo Researchers. *Page 795*: J. L. Charmet/Science Photo Library/Photo Researchers. *Page 810*: Courtesy Los Alamos National Laboratory. *Pages 802 and 803*: © 1989 Hsiung.

Chapter 40

Opener: Larry Keenan/The Image Bank. *Page 821*: Zviki-Eshet/Stock Market. *Page 823*: W. E. Meggers Collection/AIP Niels Bohr Library. *Page 828*: Courtesy Wabash Instrument Corp. *Page 830*: Photograph by Mark Oliphant/AIP Niels Bohr Library. *Page 841*: Theodore Haimar. *Page 843*: (bottom left) Sandia National Laboratories; (bottom right) The Stock Market.

Chapter 41

Opener: *Science Magazine*, February 1989. *Page 844*: AIP Niels Bohr Library. *Page 846*: (top and bottom) Courtesy Film Studio, Education Development Center; (center) C. G. Shull, Massachusetts Institute of Technology. *Page 847*: AIP Niels Bohr Library. *Page 848*: Courtesy Hitachi Ltd. *Page 851*: Courtesy IBM Corp. *Page 852*: Bainbridge Collection/AIP Niels Bohr Library. *Page 860*: Runk/Schoenberger/Grant Heilman Photography. *Page 861*: (top left) Courtesy IBM Research; (top right) © David Scharf/Peter Arnold. *Page 862*: (top) Courtesy Dr. Colin Humphreys, Department of Material Science and Engineering, University of Liverpool; (bottom) Courtesy IBM Corp.

FIGURE CREDITS

Index

CONVERSIONS

Length
1 in. = 2.54 cm (exact)
1 m = 39.37 in. = 3.281 ft
1 mi = 5280 ft = 1.609 km
1 km = 0.6215 mi
1 fermi (fm) = 1×10^{-15} m
1 Angstrom (A) = 1×10^{-10} m
1 nautical mi = 6076 ft = 1.151 mi
1 astronomical unit (AU) = 1.4960×10^{11} m
1 light-year = 9.4607×10^{15} m

Area
1 m^2 = 10^4 cm^2 = 10.76 ft^2
1 ft^2 = 0.0929 m^2
1 in.^{2} = 6.452 cm^2
1 mi^2 = 640 acres
1 hectare = 10^4 m^2 = 2.471 acres
1 acre = 43560 ft^2

Volume
1 m^3 = 10^6 cm^3 = 6.102×10^4 in.^{3}
1 ft^3 = 1728 in.^{3} = 2.832×10^{-2} m^3
1 liter = 10^3 cm^3 = 0.0353 ft^3
$\quad\quad$ = 1.0576 US qt
1 ft^3 = 28.32 liters = 7.481 US gal = 2.832×10^{-2} m^3
1 US gal = 3.786 liters = 231 in.^{3}
1 Imperial gal = 1.201 US gal = 277.42 in.^{3}

Mass
1 u = 1.6605×10^{-27} kg
1 metric ton (tonne) = 10^3 kg
1 slug = 14.59 kg
1 ton (avoirdupois) = 907.2 kg

Time
1 d = 24 h = 1.44×10^3 min = 8.640×10^4 s
1 y = 365.24 d = 3.156×10^7 s

Force
1 N = 10^5 dyne = 0.2248 lb
1 lb = 4.448 N
The weight of 1 kg is 2.205 lb.

Energy
1 J = 10^7 ergs = 0.7376 ft.lb
1 eV = 1.602×10^{-19} J
1 cal = 4.186 J; 1 Cal = 4186 J
1 kWh = 3.600×10^6 J = 3412 Btu
1 Btu = 252.0 cal = 1055 J
1 u is equivalent to 931.5 Mev

Power
1 hp = 550 ft.lb/s = 745.7 W
1 W = 1 J/s = 0.7376 ft.lb/s
1 Btu/h = 0.2931 W

Pressure
1 Pa = 1 N/m^2 = 1.450×10^{-4} lb/in.^{2}
1 atm = 760 mm Hg = 1.013×10^5 N/m^2 = 14.70 lb/in.^{2}
1 bar = 10^5 Pa = 0.9870 atm
1 torr = 1 mm Hg = 133.3 Pa